Lecture Notes in Computer Science 5734

Rastislav Královič Damian Niwiński (Eds.)

Mathematical Foundations of Computer Science 2009

34th International Symposium, MFCS 2009
Novy Smokovec, High Tatras, Slovakia, August 24-28, 2009
Proceedings

Volume Editors

Rastislav Královič
Faculty of Mathematics, Physics and Informatics
Comenius University
Bratislava, Slovakia
E-mail: kralovic@dcs.fmph.uniba.sk

Damian Niwiński
Faculty of Mathematics, Informatics and Mechanics
Warsaw University
Warsaw, Poland
E-mail: niwinski@mimuw.edu.pl

Library of Congress Control Number: 2009932264

CR Subject Classification (1998): F.2, G.1, G.1.2, G.2.1, G.2.2, F.4.3

LNCS Sublibrary: SL 1 – Theoretical Computer Science and General Issues

ISSN 0302-9743
ISBN-10 3-642-03815-8 Springer Berlin Heidelberg New York
ISBN-13 978-3-642-03815-0 Springer Berlin Heidelberg New York

springer.com

Printed in Germany

Typesetting: Camera-ready by author, data conversion by Scientific Publishing Services, Chennai, India
Printed on acid-free paper SPIN: 12740345 06/3180 5 4 3 2 1 0

Preface

The 34th International Symposium on Mathematical Foundations of Computer Science, MFCS 2009, was held in Nový Smokovec, High Tatras (Slovakia) during August 24–28, 2009. This volume contains 7 invited and 56 contributed papers presented at the symposium. The contributed papers were selected by the Program Committee out of a total of 148 submissions.

MFCS 2009 was organized by the Slovak Society for Computer Science and the Faculty of Mathematics, Physics and Informatics of the Comenius University in Bratislava. It was supported by the European Association for Theoretical Computer Science. We acknowledge with gratitude the support of all these institutions.

The series of MFCS symposia has a well-established tradition dating back to 1972. The aim is to encourage high-quality research in all branches of theoretical computer science, and to bring together researchers who do not usually meet at specialized conferences. The symposium is organized on a rotating basis in Poland, Czech Republic, and Slovakia. The previous meetings took place in Jabłonna 1972, Štrbské Pleso 1973, Jadwisin 1974, Mariánske Lázně 1975, Gdańsk 1976, Tatranská Lomnica 1977, Zakopane 1978, Olomouc 1979, Rydzyna 1980, Štrbské Pleso 1981, Prague 1984, Bratislava 1986, Karlovy Vary 1988, Porąbka-Kozubnik 1989, Banská Bystrica 1990, Kazimierz Dolny 1991, Prague 1992, Gdańsk 1993, Košice 1994, Prague 1995, Kraków 1996, Bratislava 1997, Brno 1998, Szklarska Poręba 1999, Bratislava 2000, Mariánske Lázně 2001, Warsaw 2002, Bratislava 2003, Prague 2004, Gdańsk 2005, Stará Lesná 2006, Český Krumlov 2007, Toruń 2008.

The 2009 meeting added a new page to this history, which was possible due to the effort of many people.

We would like to thank the invited speakers Albert Atserias, Didier Caucal, Javier Esparza, Thomas Henzinger, Muthu Muthukrishnan, Pavlos Spirakis, and Peter Widmayer, for presenting their work to the audience of MFCS 2009. The papers provided by the invited speakers appear at the front of this volume. We thank all authors who have submitted their papers for consideration. Many thanks go to the Program Committee, and to all external referees, for their hard work in evaluating the papers. The work of the PC was carried out using the EasyChair system, and we gratefully acknowledge this contribution.

Special thanks are due to the Organizing Committee led by Vanda Hamálková and Dana Pardubská.

June 2009

Rastislav Královič
Damian Niwiński

Preface

Conference Organization

Program Chairs

Rastislav Královič	Comenius U., Bratislava, Slovakia
Damian Niwiński	U. Warsaw, Poland

Program Committee

Marcelo Arenas	U. Santiago, Chile
Luca Becchetti	U. di Roma "La Sapienza", Italy
Mikolaj Bojańczyk	U. Warsaw, Poland
Balder ten Cate	U. Amsterdam, The Netherlands
Thomas Colcombet	U. Paris 7, France
Jurek Czyzowicz	U. du Québec en Outaouais, Canada
Stephane Demri	ENS Cachan, France
Pierre Fraigniaud	U. Paris 7, France
Giuseppe F. Italiano	U. di Roma "Tor Vergata", Italy
Juhani Karhumäki	U. Turku, Finland
Ralf Klasing	LaBRI Bordeaux, France
Michal Koucký	Academy of Sciences, Prague, Czech Republic
Rastislav Královič	Comenius U., Bratislava, Slovakia
Stephan Kreutzer	U. Oxford, UK
Antonín Kučera	U. Brno, Czech Republic
Luděk Kučera	Charles U. Prague, Czech Republic
Salvatore La Torre	U. Salerno, Italy
Jan van Leeuwen	U. Utrecht, The Netherlands
Friedhelm Meyer auf der Heide	U. Paderborn, Germany
Anca Muscholl	U. Bordeaux, France
Damian Niwiński	U. Warsaw, Poland
Alexander Okhotin	U. Turku, Finland
Leszek Pacholski	U. Wrocław, Poland
David Peleg	Weizmann Institute, Rehovot, Israel
Jean-Éric Pin	U. Paris 7, France
Nir Piterman	Imperial College London, UK
Michel Raynal	U. Rennes 1, France
Simona Ronchi Della Rocca	U. Torino, Italy
Jörg Rothe	U. Düsseldorf, Germany

Local Organization

Vanda Hambálková
Jaroslav Janáček
Marke Nagy
Dana Pardubská

External Reviewers

Anil Ada
Jean-Paul Allouche
Omid Amini
Adamatzky Andrew
Eric Angel
Carlos Areces
Yossi Azar
Sebastian Bala
Pablo Barcelo
David Mix Barrington
Dorothea Baumeister
Jason Bell
Dietmar Berwanger
Jean Berstel
Arnab Bhattacharyya
Etienne Birmele
Marek Biskup
Luc Boasson
Hans-Joachim Boeckenhauer
Annalisa De Bonis
Jan Bouda
Vaclav Brozek
Peter Buergisser
Ioannis Caragiannis
Arnaud Carayol
Olivier Carton
Venkat Chakravarthy
Arkadev Chattopadhyay
Kaustuv Chaudhuri
Hana Chockler
Colin Cooper
Win van Dam
Samir Datta
Bastian Degener
Mariangiola Dezani
Mike Domaratzki
Jacques Duparc
Martin Dyer
Edith Elkind
Robert Elsaesser
Gabor Erdelyi
Thomas Erlebach
Lionel Eyraud-Dubois
Piotr Faliszewski
John Fearnley
Alessandro Ferrante
Jiří Fiala
Diego Figueira
Dana Fisman
Vojtech Forejt
Keith Frikken
Pawel Gawrychowski
Giuseppe De Giacomo
Dora Giammarresi
Barbara Di Giampaolo
Emeric Gioan
Fabrizio Grandoni
Serge Grigorieff
Martin Grohe
Stefano Guerrini
Frank Gurski
Kristoffer Arnsfelt Hansen
Tero Harju
Alejandro Hevia
Mika Hirvensalo
John Hitchcock
Petr Hlineny
Christian Hoffmann
Florian Horn
Mathieu Hoyrup
Michael Huth
Tomasz Idziaszek
Radu Iosif
Artur Jez

Jan Jezabek
Stasys Jukna
Tomasz Jurdzinski
Lukasz Kaiser
Panagiotis Kanellopoulos
Jarkko Kari
Shin-ya Katsumata
Tomasz Kazana
Stefan Kiefer
Emanuel Kieronski
Lefteris Kirousis
Christian Knauer
Eryk Kopczynski
Adrian Kosowski
Lukasz Kowalik
Miroslaw Kowaluk
Richard Kralovic
Jan Kratochvil
Matthias Krause
Andrei Krokhin
Michal Kunc
Orna Kupferman
Clemens Kupke
Emmanuelle Lebhar
Sophie Laplante
Slawomir Lasota
Ranko Lazic
Tommi Lehtinen
Leonid Libkin
Nutan Limaye
Claudia Lindner
Maciej Liskiewicz
Sylvain Lombardy
Zvi Lotker
Yoad Lustig
Edita Macajova
Gianluca De Marco
Jean-Yves Marion
Elvira Mayordomo
Frdric Mazoit
Damiano Mazza
Pierre McKenzie
Jakub Michaliszyn
Tobias Moemke
Luca Moscardelli
Filip Murlak
Margherita Napoli
Jelani Nelson
Nicolas Nisse
Nicolas Ollinger
Luca Paolini
Dana Pardubská
Mimmo Parente
Gennaro Parlato
Pawel Parys
Boaz Patt-Shamir
Anthony Perez
Woelfel Philipp
Marcin Pilipczuk
Adolfo Piperno
Wojciech Plandowski
Svetlana Puzynina
Alexander Rabinovich
Frank Radmacher
Tomasz Radzik
Daniel Raible
Christoforos Raptopoulos
Damien Regnault
Christian Reitwiener
Julien Robert
Andrei Romashchenko
Yves Roos
Magnus Roos
Peter Rossmanith
Wojciech Rytter
Robert Samal
Arnaud Sangnier
Mathias Schacht
Bjrn Scheuermann
Sylvain Schmitz
Henning Schnoor
Florian Schoppmann
Celine Scornavacca
Olivier Serre
Jeffrey Shallit
Asaf Shapira
Adi Shraibman
Libor Skarvada
Sagi Snir
Christian Sohler

Holger Spakowski
Wojciech Szpankowski
Siamak Tazari
Pascal Tesson
Michael Thomas
Jacobo Toran
Szymon Torunczyk
Paolo Tranquilli
Hideki Tsuiki
Walter Unger
Paweł Urzyczyn
Carmine Ventre
Nikolay Vereshchagin
Ivan Visconti
Mikhail Volkov
Heribert Vollmer
Jan Vondrak
Rolf Wanka
Rebecca Weber
Frank Wolter
Konrad Zdanowski
Wenhui Zhang
Margherita Zorzi

Table of Contents

Invited Papers

Contributed Papers

Four Subareas of the Theory of Constraints, and Their Links

Albert Atserias

Universitat Politècnica de Catalunya, Barcelona, Spain

Let $V = \{x_1, \ldots, x_n\}$ be a set of variables that range over a set of values $D = \{d_1, \ldots, d_q\}$. A constraint is an expression of the type $R(x_{i_1}, \ldots, x_{i_r})$, where $R \subseteq D^r$ is a relation on the domain set D and $x_{i_1}, \ldots, x_{i_r}$ are variables in V. The space of assignments, or configurations, is the set of all mappings $\sigma : V \to D$. We say that σ satisfies the constraint $R(x_{i_1}, \ldots, x_{i_r})$ if $(\sigma(x_{i_1}), \ldots, \sigma(x_{i_r})) \in R$. Otherwise we say that it falsifies it. On a given *system of constraints* we face a number of important computational problems. Here is a small sample:

1. Is it satisfiable? That is, is there an assignment that satisfies all constraints? If it is satisfiable, can we find a satisfying assignment? If it is unsatisfiable, does it have a small and efficiently checkable proof of unsatisfiability? Can we find it?
2. If it is not satisfiable, what is the maximum number of constraints that can be satisfied simultaneously by some assignment? Knowing that we can satisfy a $1 - \epsilon$ fraction of the constraints simultaneously for some small $\epsilon > 0$, can we find an assignment that satisfies more than, say, a $1 - \sqrt{\epsilon}$ fraction?
3. How many satisfying assignment does the system have? If we can't count it exactly, can we approximate the number of satisfying assignments up to a constant approximation factor? Can we sample a satisfying assignment uniformly or approximately uniformly at random? More generally, if we write $H(\sigma)$ for the number of constraints that are falsified by σ, can we sample an assignment σ with probability proportional to $e^{-\beta H(\sigma)}$ where β is a given *inverse temperature* parameter? Can we compute, exactly or approximately, the so-called partition function of the system, defined as $Z(\beta) = \sum_\sigma e^{-\beta H(\sigma)}$?
4. Knowing that the system has been generated randomly by choosing each R uniformly at random from a fixed set of relations Γ and by choosing each $(x_1, \ldots, x_r)$ uniformly at random in V^r, can we analyze and exploit the typical structure of the assignment space and the constraint system to solve any of the problems above?

These are fundamental problems that can be roughly classified into the following four categories: logic and proof complexity, optimization and approximation, counting and sampling, and analysis of randomly generated instances. Perhaps surprisingly, these four areas of the theory of constraints have been approached through rather different techniques by groups of researchers with small pairwise intersections. However, recent work has shown that these areas might be more related than this state of affairs seems to indicate. We overview these connections with emphasis on the open problems that have the potential of making the connections even tighter.

R. Královič and D. Niwiński (Eds.): MFCS 2009, LNCS 5734, p. 1, 2009.

Synchronization of Regular Automata

Didier Caucal

IGM–CNRS, Université Paris-Est
caucal@univ-mlv.fr

Abstract. Functional graph grammars are finite devices which generate the class of regular automata. We recall the notion of synchronization by grammars, and for any given grammar we consider the class of languages recognized by automata generated by all its synchronized grammars. The synchronization is an automaton-related notion: all grammars generating the same automaton synchronize the same languages. When the synchronizing automaton is unambiguous, the class of its synchronized languages forms an effective boolean algebra lying between the classes of regular languages and unambiguous context-free languages. We additionally provide sufficient conditions for such classes to be closed under concatenation and its iteration.

1 Introduction

An automaton over some alphabet can simply be seen as a finite or countable set of labelled arcs together with two sets of initial and final vertices. Such an automaton recognizes the language of all words labelling an accepting path, i.e. a path leading from an initial to a final vertex. It is well-known that finite automata recognize the regular languages. By applying basic constructions to finite automata, we obtain the nice closure properties of regular languages, namely their closure under boolean operations, concatenation and its iteration. For instance the synchronization product and the determinization of finite automata respectively yield the closure of regular languages under intersection and under complement.

This idea can be extended to more general classes of automata. In this paper, we will be interested in the class of regular automata, which recognize context-free languages and are defined as the (generally infinite) automata generated by functional graph grammars [Ca 07]. Regular automata of finite degree are also precisely those automata which can be finitely decomposed by distance, as well as the regular restrictions of transition graphs of pushdown automata [MS 85], [Ca 07]. Even though the class of context-free languages does not enjoy the same closure properties as regular languages, one can define subclasses of context-free languages which do, using the notion of synchronization.

The notion of synchronization was first defined between grammars [CH 08]. A grammar S is synchronized by a grammar R if for any accepting path μ of (the graph generated by) S, there exists an accepting path λ of R with the same label u such that λ and μ are synchronized: for every prefix v of u, the prefixes

R. Královič and D. Niwiński (Eds.): MFCS 2009, LNCS 5734, pp. 2–23, 2009.

of λ and μ labelled by v lead to vertices of the same level (where the level of a vertex is the minimal number of rewriting steps necessary for the grammar to produce it). A language is synchronized by a grammar R if it is recognized by an automaton generated by a grammar synchronized by R. A fundamental result is that two grammars generating the same automaton yield the same class of synchronized languages [Ca 08]. This way, the notion of synchronization can be transferred to the level of automata: for a regular automaton G, the family $Sync(G)$ is the set of languages synchronized by any grammar generating G.

By extending the above-mentioned constructions from finite automata to grammars, one can establish several closure properties of these families of synchronized languages. The sum of two grammars and the synchronization product of a grammar with a finite automaton respectively entail the closure of $Sync(G)$ under union and under intersection with a regular language for any regular automaton G. The (level preserving) synchronization product of two grammars yields the closure under intersection of $Sync(G)$ when G is unambiguous *i.e.* when any two accepting paths of G have distinct labels. Normalizing of grammar into a grammar only containing arcs and then determinizing it yields, for any unambiguous automaton G, the closure of $Sync(G)$ under complement relative to $L(G)$. This normalization also allows us to express $Sync(G)$ in the case of an infinite degree graph G, by performing the e-closure of $Sync(H)$ for some finite degree automaton H using an extra label e. A final useful normalization only allows the presence of initial and final vertices at level 0. It yields sufficient conditions for the closure of classes of synchronized languages under concatenation and its iteration.

In Section 2, we recall the definition of regular automata. In the next section, we summarize known results on the synchronization of regular automata [Ca 06], [NS 07], [CH 08], [Ca 08]. In the last section, we present a simpler construction for the closure under complement of $Sync(G)$ for unambiguous G [Ca 08] and present new results, especially sufficient conditions for the closure of $Sync(G)$ under concatenation and its iteration.

2 Regular Automata

An automaton is a labelled oriented simple graph with input and output vertices. It recognizes the set of words labelling the paths from an input to an output. Finite automata are automata having a finite number of vertices, they recognize the class of regular languages. Regular automata are the automata generated by functional graph grammars, they recognize the class of context-free languages. A key result, originally due to Muller and Schupp, identifies the regular automata of finite degree with the automata finitely generated by distance.

An automaton over an alphabet (finite set of symbols) T of *terminals* is just a set of arcs labelled over T (a simple labelled oriented graph) with initial and final vertices. We use two symbols ι and o to mark respectively the initial and final vertices. More precisely an *automaton* G is defined by $G \subseteq T{\times}V{\times}V \cup \{\iota,o\}{\times}V$ where V is an arbitrary set such that the set of *vertices*

$$V_G = \{ s \in V \mid \exists\, a \in T\ \exists\, t \in V\ (a,s,t) \in G \ \vee\ (a,t,s) \in G \}$$

of G is finite or countable. Any triple $(a,s,t) \in G$ is an *arc* labelled by a from *source* s to *goal* t; it is identified with the labelled transition $s \xrightarrow[G]{a} t$ or directly $s \xrightarrow{a} t$ if G is understood. Any pair $(c,s) \in G$ is a *coloured vertex* s by $c \in \{\iota, o\}$ also written $c\,s$. A vertex is *initial* (resp. *final*) if it is coloured by ι (resp. o) *i.e.* $\iota\, s \in G$ (resp. $o\, s \in G$). An example of an automaton is given by

$$\begin{aligned} G = \{ n \xrightarrow{a} n+1 \mid n \geq 0 \} &\cup \{ n \xrightarrow{b} x^n \mid n > 0 \} \quad \cup \{ n \xrightarrow{b} y^{2n} \mid n > 0 \} \\ \cup \{ x^{n+1} \xrightarrow{b} x^n \mid n > 0 \} &\cup \{ y^{n+1} \xrightarrow{b} y^n \mid n > 0 \} \\ \cup \{\iota 0, o y\} \qquad\qquad &\cup \{ o x^n \mid n > 0 \} \qquad \cup \{ \iota y^{2n+1} \mid n \geq 0 \} \end{aligned}$$

and is represented (up to isomorphism) below.

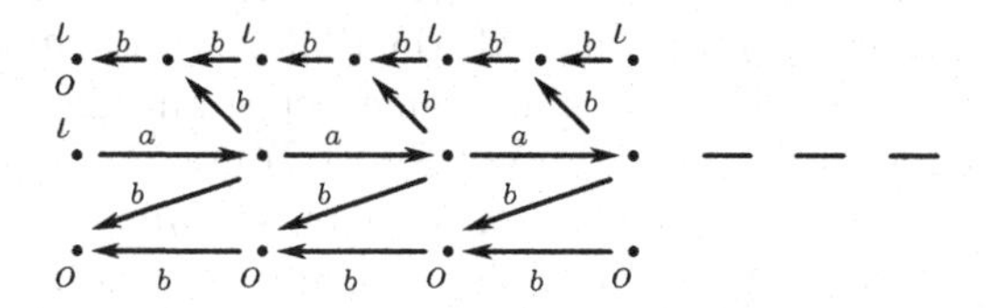

Fig. 1. An automaton

An automaton G is thus a simple vertex- and arc-labelled graph. G has *finite degree* if for any vertex s, the set $\{ t \mid \exists\, a\ (s \xrightarrow{a} t \ \vee\ t \xrightarrow{a} s) \}$ of its adjacent vertices is finite. Recall that $(s_0, a_1, s_1, \ldots, a_n, s_n)$ for $n \geq 0$ and $s_0 \xrightarrow[G]{a_1} s_1 \ldots s_{n-1} \xrightarrow[G]{a_n} s_n$ is a *path* from s_0 to s_n labelled by $u = a_1 \ldots a_n$; we write $s_0 \overset{u}{\underset{G}{\Longrightarrow}} s_n$ or directly $s_0 \overset{u}{\Longrightarrow} s_n$ if G is understood. An *accepting path* is a path from an initial vertex to a final vertex. An automaton is *unambiguous* if two accepting paths have distinct labels. The automaton of Figure 1 is unambiguous. The *language recognized* by an automaton G is the set $L(G)$ of all labels of its accepting paths: $L(G) = \{ u \in T^* \mid \exists\, s,t\ (s \overset{u}{\underset{G}{\Longrightarrow}} t \wedge \iota\, s\,,\, o\, t \in G) \}$.

Note that $\varepsilon \in L(G)$ if there exists a vertex s which is initial and final: $\iota\, s\,,\, o\, s \in G$.

The automaton G of Figure 1 recognizes the language

$$L(G) = \{ a^m b^n \mid 0 < n \leq m \} \cup \{ a^n b^{2n} \mid n > 0 \} \cup \{ b^{2n} \mid n \geq 0 \}.$$

The languages recognized by finite automata are the *regular languages* over T. We generalize finite automata to regular automata using functional graph grammars. To define a graph grammar, we need to extend an arc (resp. a graph) to a hyperarc (resp. a hypergraph). Although such an extension is natural, this may explain why functional graph grammars are not very widespread at the moment. But we will see in the last section that for our purpose, we can restrict to grammars using only arcs.

Let F be a set of symbols ranked by a mapping $\varrho : F \longrightarrow \mathbb{N}$ associating to each $f \in F$ its *arity* $\varrho(f) \geq 0$ such that $F_n = \{\, f \in F \mid \varrho(f) = n \,\}$ is countable for every $n \geq 0$ with $T \subset F_2$ and $\iota, o \in F_1$.

A *hypergraph* G is a subset of $\bigcup_{n\geq 0} F_n \times V^n$ where V is an arbitrary set. Any tuple $(f, s_1, \ldots, s_{\varrho(f)}) \in G$, also written $fs_1 \ldots s_{\varrho(f)}$, is a *hyperarc* of *label* f and of successive *vertices* $s_1, \ldots, s_{\varrho(f)}$. We add the condition that the set of vertices V_G is finite or countable, and the set of labels F_G is finite. An *arc* is a hyperarc fst labelled by $f \in F_2$ and is also denoted by $s \xrightarrow{f} t$. For $n \geq 2$, a hyperarc $fs_1 \ldots s_n$ is depicted as an arrow labelled f and successively linking $s_1, \ldots, s_n$. For $n = 1$ and $n = 0$, it is respectively depicted as a label f (called a *colour*) on vertex s_1 and as an isolated label f called a *constant*. This is illustrated in the next figures. For instance the following hypergraph:

$$G \;=\; \{4 \xrightarrow{b} 1\,,\, 5 \xrightarrow{b} 1\,,\, 2 \xrightarrow{a} 5\,,\, 5 \xrightarrow{b} 3\,,\, 6 \xrightarrow{b} 3\,,\, \iota 4\,,\, o6\,,\, A456\}$$

with $a, b \in F_2$ and $A \in F_3$, is represented below.

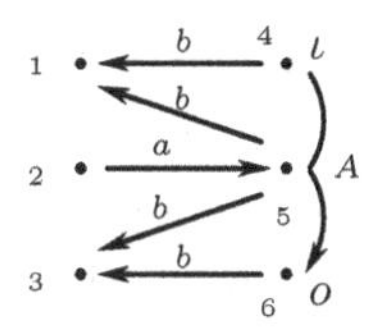

Fig. 2. A finite hypergraph

A (coloured) *graph* G is a hypergraph whose labels are only of arity 1 or 2: $F_G \subset F_1 \cup F_2$. An automaton G over the alphabet T is a graph with a set of labels $F_G \subseteq T \cup \{\iota, o\}$. We can now introduce functional graph grammars to generate regular automata.

A graph *grammar* R is a finite set of rules of the form $fx_1 \ldots x_{\varrho(f)} \longrightarrow H$ where $fx_1 \ldots x_{\varrho(f)}$ is a hyperarc of label f called *non-terminal* joining pairwise distinct vertices $x_1 \neq \ldots \neq x_{\varrho(f)}$ and H is a finite hypergraph.

We denote by N_R the set of non-terminals of R *i.e.* the labels of the left hand sides, by $T_R = \{\, f \in F - N_R \mid \exists\, H \in Im(R),\;\; f \in F_H \,\}$ the *terminals* of R *i.e.* the labels of R which are not non-terminals, and by $F_R = N_R \cup T_R$ the *labels* of R.

We use grammars to generate automata hence in the following, we may assume that $T_R \subseteq T \cup \{\iota, o\}$. Similarly to context-free grammars (on words), a graph grammar has an axiom: an initial finite hypergraph. To indicate this axiom, we assume that any grammar R has a constant non-terminal $Z \in N_R \cap F_0$ which is not a label of any right hand side; the *axiom* of R is the right hand side H of the rule of Z: $Z \longrightarrow H \;\wedge\; Z \notin F_K$ for any $K \in Im(R)$.

Starting from the axiom, we want R to generate a unique automaton up to isomorphism. So we finally assume that any grammar R is *functional* meaning that there is only one rule per non-terminal: if $(X, H)\,,\, (Y, K) \in R$ with $X(1) = Y(1)$ then $(X, H) = (Y, K)$.

For any rule $fx_1 \ldots x_{\varrho(f)} \longrightarrow H$, we say that $x_1, \ldots, x_{\varrho(f)}$ are the *inputs* of f, and $V_{H-[H]}$ is the set of *outputs* of f.

To work with these grammars, it is simpler to assume that any grammar R is *terminal-outside* [Ca 07]: any terminal arc or colour in a right hand side links to at least one non input vertex: $H \cap (T_R{\times}V_X{\times}V_X \cup T_R{\times}V_X) = \emptyset$ for any rule $(X,H) \in R$.

We will use upper-case letters $A, B, C, \ldots$ for non-terminals and lower-case letters $a, b, c \ldots$ for terminals. Here is an example of a (functional graph) grammar R:

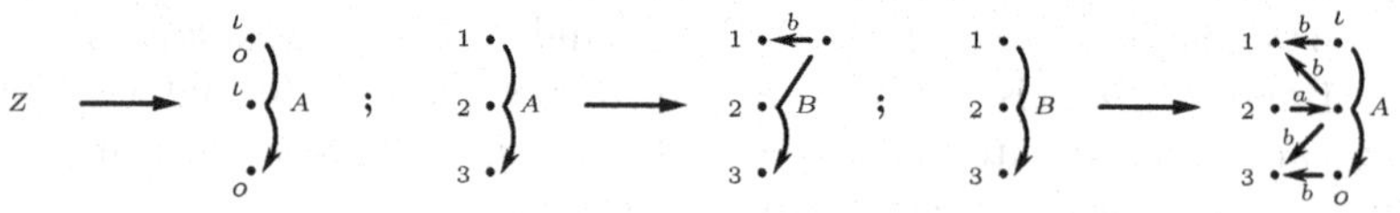

Fig. 3. A (functional graph) grammar

For the previous grammar R, we have $N_R = \{Z, A, B\}$ with Z the axiom and $\varrho(A) = \varrho(B) = 3$, $T_R = \{a, b, \iota, o\}$ and $1, 2, 3$ are the inputs of A and B.

Given a grammar R, the *rewriting* relation $\underset{R}{\longrightarrow}$ is the binary relation between hypergraphs defined as follows: M rewrites into N, written $M\underset{R}{\longrightarrow}N$, if we can choose a non-terminal hyperarc $X = As_1 \ldots s_p$ in M and a rule $Ax_1 \ldots x_p \longrightarrow H$ in R such that N can be obtained by replacing X by H in M: $N = (M-X) \cup h(H)$ for some function h mapping each x_i to s_i, and the other vertices of H injectively to vertices outside of M; this rewriting is denoted by $M\underset{R,\,X}{\longrightarrow}N$. The rewriting $\underset{R,\,X}{\longrightarrow}$ of a hyperarc X is extended in an obvious way to the rewriting $\underset{R,\,E}{\longrightarrow}$ of any set E of non-terminal hyperarcs. The *complete parallel rewriting* $\underset{R}{\Longrightarrow}$ is a simultaneous rewriting according to the set of all non-terminal hyperarcs: $M\underset{R}{\Longrightarrow}N$ if $M\underset{R,\,E}{\longrightarrow}N$ where E is the set of all non-terminal hyperarcs of M. We depict below the first three steps of the parallel derivation of the previous grammar from its constant non-terminal Z:

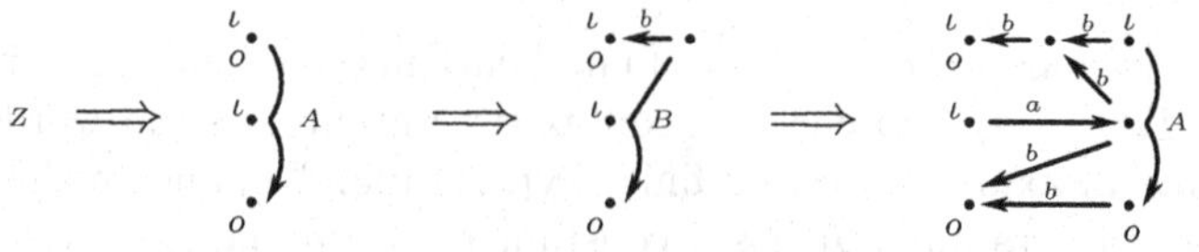

Fig. 4. Parallel derivation for the grammar of Figure 3

Given a grammar R, we restrict any hypergraph H to the automaton $[H]$ of its terminal arcs and coloured vertices:$[H] = H \cap (T{\times}V_H{\times}V_H \cup \{\iota, o\}{\times}V_H)$.

An automaton G is *generated* by R (from its axiom) if G belongs to the following set R^ω of isomorphic automata:

$$R^\omega = \{ \bigcup_{n\geq 0}[H_n] \mid Z \underset{R}{\longrightarrow} H_0 \underset{R}{\Longrightarrow} \ldots H_n \underset{R}{\Longrightarrow} H_{n+1} \ldots \}.$$

Note that in all generality, we need to consider hypergraphs with multiplicities. However using an appropriate normal form, this technicality can be safely omitted [Ca 07].

For instance the automaton of Figure 1 is generated by the grammar of Figure 3. A *regular automaton* is an automaton generated by a (functional graph) grammar. Note that a regular automaton has a finite number of non-isomorphic connected components, and has a finite number of distinct vertex degrees.

Another example is given by the following grammar:

Z ⟶ ι, A, o ; 1, A, 2 ⟶ a, 1, b, A, c, c, A, a, b, 2

which generates the following automaton:

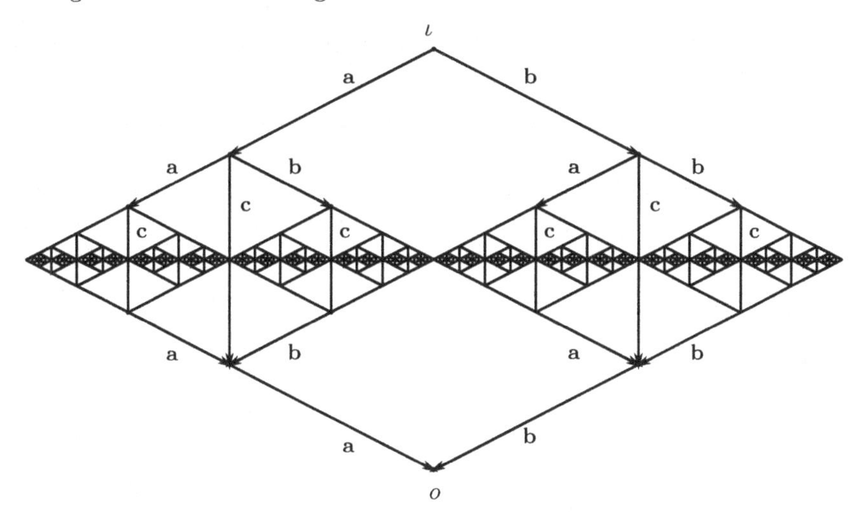

recognizing the language $\{ uc\widetilde{u} \mid u \in \{a,b\}^+ \}$ where $\widetilde{u}$ is the mirror of u.

The *language recognized* by a grammar R is the language $L(R)$ recognized by its generated automaton: $L(R) = L(G)$ for (any) $G \in R^\omega$. This language is well-defined since all automata generated by a given grammar are isomorphic. A grammar R is an *unambiguous grammar* if the automaton it generates is unambiguous.

There is a canonical way to generate the regular automata of finite degree which allows to characterize these automata without the explicit use of grammars. This is the finite decomposition by distance.

The *inverse* G^{-1} of an automaton G is the automaton obtained from G by reversing its arcs and by exchanging initial and final vertices:

$$G^{-1} = \{ t \xrightarrow{a} s \mid s \underset{G}{\xrightarrow{a}} t \} \cup \{ \iota s \mid os \in G \} \cup \{ os \mid \iota s \in G \}.$$

So G^{-1} recognizes the mirror of the words recognized by G. The *restriction* $G_{|I}$

of G to a subset I of vertices is the subgraph of G induced by I:

$$G_{|I} \;=\; G \cap (T{\times}I{\times}I \cup \{\iota,o\}{\times}I).$$

The *distance* $d_I(s)$ of a vertex s to I is the minimal length of the undirected paths between s and I: $d_I(s) = min\{\ |u| \mid \exists\, r \in I,\ r \underset{G \cup G^{-1}}{\overset{u}{\Longrightarrow}} s\ \}$ with $min(\emptyset) = +\infty$.

We take a new colour $\# \in F_1 - \{\iota, o\}$ and define for any integer $n \geq 0$,

$$Dec^{\#}_n(G,I) \;=\; G_{|\{\ s\ \mid\ d_I(s) \geq n\ \}} \cup \{\ \#s \mid d_I(s) = n\ \}$$

In particular $Dec^{\#}_0(G,I) = G \cup \{\ \#s \mid s \in I\ \}$. We say that an automaton G is *finitely decomposable by distance* if for each connected component C of G there exists a finite non empty set I of vertices such that $\bigcup_{n\geq 0} Dec^{\#}_n(C,I)$ has a finite number of non-isomorphic connected components. Such a definition allows the characterization of the class of all automata of finite degree which are regular.

Theorem 1. *An automaton of finite degree is regular if and only if it is finitely decomposable by distance and it has only a finite number of non isomorphic connected components.*

The proof is given in [Ca 07] and is a slight extension of [MS 85] (but without using pushdown automata). Regular automata of finite degree are also the transition graphs of pushdown automata restricted to regular sets of configurations and with regular sets of initial and final configurations. In particular, regular automata of finite degree recognize the same languages as pushdown automata.

Proposition 1. *The (resp. unambiguous) regular automata recognize exactly the (resp. unambiguous) context-free languages.*

This proposition remains true if we restrict to automata of finite degree. We now use grammars to extend the family of regular languages to boolean algebras of unambiguous context-free languages.

3 Synchronization of Regular Automata

We introduce the idea of synchronization between grammars. The class of languages synchronized by a grammar R are the languages recognized by grammars synchronized by R. We show that these families of languages are closed under union by applying the sum of grammars, are closed under intersection with a regular language by defining the synchronization product of a grammar with a finite automaton, and are closed under intersection (in the case of grammars generating unambiguous automata) by performing the synchronization product of grammars. Finally we show that all grammars generating the same automaton synchronize the same languages.

To each vertex s of an automaton $G \in R^{\omega}$ generated by a grammar R, we associate a non negative integer $\ell(s)$ which is the minimal number of rewritings applied from the axiom necessary to reach s. More precisely for $G = \bigcup_{n\geq 0}[H_n]$

with $Z \underset{R}{\longrightarrow} H_0 \underset{R}{\Longrightarrow} \ldots H_n \underset{R}{\Longrightarrow} H_{n+1} \ldots$, the *level* $\ell(s)$ of $s \in V_G$, also written $\ell_G^R(s)$ to specify G and R, is $\ell(s) = min\{\ n \mid s \in V_{H_n}\ \}$.

We depict below the levels of some vertices of the regular automaton of Figure 1 generated by the grammar of Figure 3. This automaton is represented by vertices of increasing level: vertices at a same level are aligned vertically.

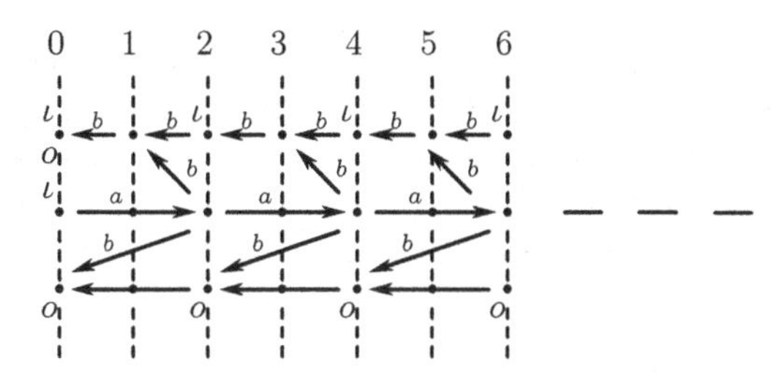

Fig. 5. Vertex levels with the grammar of Figure 3

We say that a grammar S is *synchronized* by a grammar R written $S \lhd R$, or equivalently that R *synchronizes* S written $R \rhd S$, if for any accepting path μ label by u of the automaton generated by S, there is an accepting path λ label by u of the automaton generated by R such that for every prefix v of u, the prefixes of λ and μ labelled by v lead to vertices of the same level: for (any) $G \in R^\omega$ and (any) $H \in S^\omega$ and for any $t_0 \underset{H}{\overset{a_1}{\longrightarrow}} t_1 \ldots \underset{H}{\overset{a_n}{\longrightarrow}} t_n$ with $\iota t_0\,,\, o t_n \in H$, there exists

$$s_0 \underset{G}{\overset{a_1}{\longrightarrow}} s_1 \ldots \underset{G}{\overset{a_n}{\longrightarrow}} s_n \text{ with } \iota s_0\,,\, o s_n \in G \text{ and } \ell_G^R(s_i) = \ell_H^S(t_i) \quad \forall\, i \in [0,n].$$

For instance the grammar of Figure 3 synchronizes the following grammar:

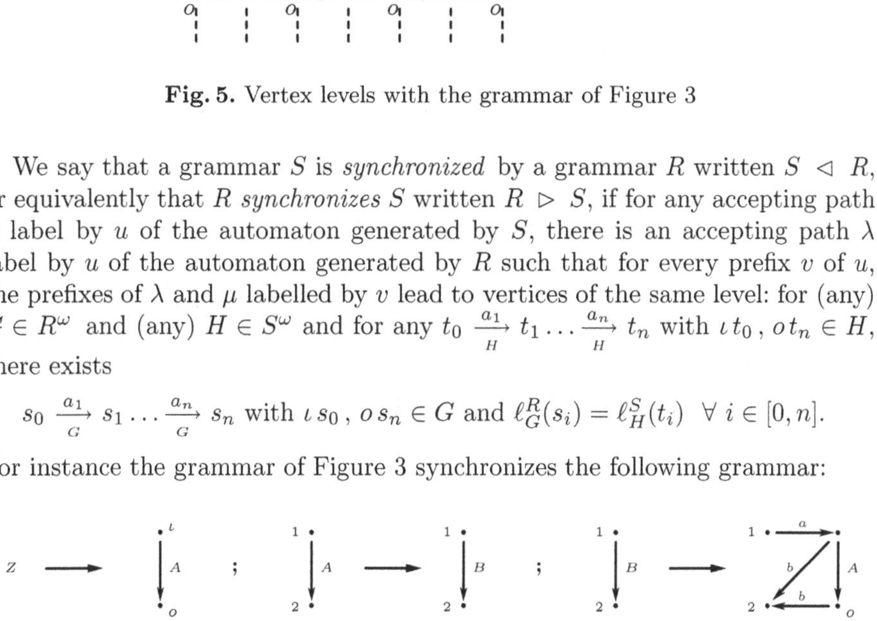

Fig. 6. A grammar synchronized by the grammar of Figure 3

In particular for $S \lhd R$, we have $L(S) \subseteq L(R)$. Note that the *empty grammar* $\{(Z,\emptyset)\}$ is synchronized by any grammar. The synchronization relation $\rhd$ is a reflexive and transitive relation. We denote $\bowtie$ the *bi-synchronization* relation: $R \bowtie S$ if $R \rhd S$ and $S \rhd R$. Note that bi-synchronized grammars $R \bowtie S$ may generate distinct automata: $R^\omega \neq S^\omega$. For any grammar R, the image of R by $\rhd$ is the family $\rhd(R) = \{\ S \mid S \lhd R\ \}$ of grammars synchronized by R and $Sync(R) = \{\ L(S) \mid S \lhd R\ \}$ is the family of languages synchronized by R.

Note that $Sync(R)$ is a family of languages included in $L(R)$ and containing the empty language and $L(R)$. Note also that $Sync(R) = Sync(S)$ for $R \bowtie S$.

Standard operations on finite automata are extended to grammars in order to obtain closure properties of $Sync(R)$. For instance the *synchronization product* of finite automata is extended to arbitrary automata G and H by

$$G{\times}H = \{\ (s,p) \xrightarrow{a} (t,q) \mid s \xrightarrow[G]{a} t \ \wedge\ p \xrightarrow[H]{a} q\ \}$$
$$\cup\ \{\ \iota(s,p) \mid \iota s \in G\ \wedge\ \iota p \in H\ \}\ \cup\ \{\ o(s,p) \mid o\, s \in G\ \wedge\ o\, p \in H\ \}$$

which recognizes $L(G{\times}H) = L(G) \cap L(H)$.

This allows us to define the synchronization product $R{\times}K$ of a grammar R with a finite automaton K [CH 08]. Let $\{q_1,\ldots,q_n\}$ be the vertex set of K. To each $A \in N_R$, we associate a new symbol (A,n) of arity $\varrho(A){\times}n$ except that $(Z,0) = Z$, and to each hyperarc $Ar_1\ldots r_m$ with $m = \varrho(A)$, we associate the hyperarc $(Ar_1\ldots r_m)_K = (A,n)(r_1,q_1)\ldots(r_1,q_n)\ldots(r_m,q_1)\ldots(r_m,q_n)$.

The grammar $R{\times}K$ associates to each rule $(X,H) \in R$ the following rule:

$$X_K \longrightarrow [H]{\times}K\ \cup\ \{\ (BY)_K \mid BY \in H\ \wedge\ B \in N_R\ \}.$$

Example 1. Let us consider the following grammar R:

generating the following (regular) automaton G:

and recognizing the restricted Dyck language $D_1'^*$ over the pair (a,b) [Be 79]: $L(R) = L(G) = D_1'^*$. We consider the following finite automaton K:

recognizing the set of words over $\{a,b\}$ having an even number of a.

So $R{\times}K$ is the following grammar:

generating the automaton $G{\times}K$:

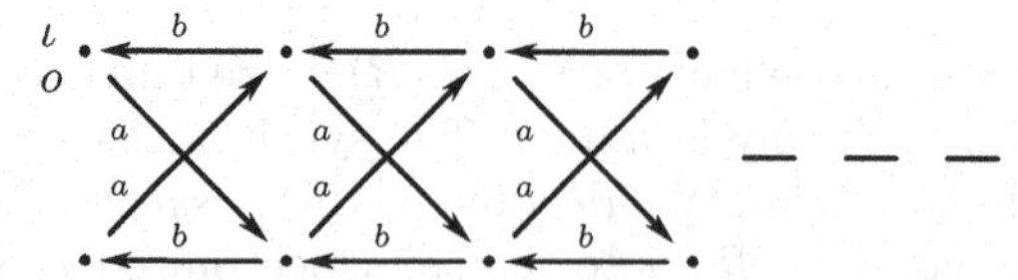

which recognizes $D_1'^*$ restricted to the words with an even number of a. □

The synchronization product of a grammar R with a finite automaton K is synchronized by R *i.e.* $R{\times}K \ \lhd\ R$ and recognizes $L(R{\times}K) = L(R) \cap L(K)$.

Proposition 2. *For any grammar R, the family Sync(R) is closed under intersection with a regular language.*

Propositions 1 and 2 imply the well-known closure property of the family of context-free languages under intersection with a regular language. As $R{\times}K$ is unambiguous for R unambiguous and K deterministic, it also follows Theorem 6.4.1 of [Ha 78] : the family of unambiguous context-free languages is closed under intersection with a regular language.

Another basic operation on finite automata is the disjoint union. This operation is extended to any grammars R_1 and R_2. For any $i \in \{1,2\}$, we denote $R_i' = R_i \times \left(\{\, i \xrightarrow{a} i \mid a \in T\,\} \cup \{\iota i\, ,\, o i\}\right)$ in order to distinguish the vertices of R_1 and R_2. For $(Z,H_1) \in R_1'$ and $(Z,H_2) \in R_2'$, the *sum* of R_1 and R_2 is the grammar

$$R_1 + R_2 \;=\; \{(Z\,,\,H_1 \cup H_2)\} \;\cup\; (R_1' - \{(Z,H_1)\}) \;\cup\; (R_2' - \{(Z,H_2)\})\,.$$

So $(R_1+R_2)^\omega = \{\, G_1 \cup G_2 \mid G_1 \in R_1^\omega \;\wedge\; G_2 \in R_2^\omega \;\wedge\; V_{G_1} \cap V_{G_2} = \emptyset \,\}$ hence $L(R_1+R_2) = L(R_1) \cup L(R_2)$. In particular if $S_1 \;\lhd\; R_1$ and $S_2 \;\lhd\; R_2$ then $S_1 + S_2 \;\lhd\; R_1 + R_2$.

Proposition 3. *For any grammar R, Sync(R) is closed under union.*

The synchronization product of regular automata can be non regular. Furthermore for the regular automaton G:

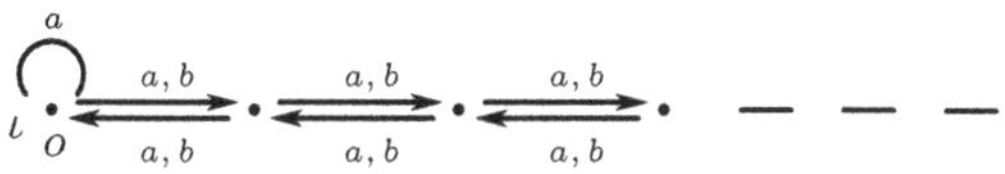

a
a, b a, b a, b
ι o — — —
a, b a, b a, b

the languages $\{\, a^m b^m a^n \mid m,n \geq 0 \,\}$ and $\{\, a^m b^n a^n \mid m,n \geq 0 \,\}$ are in $Sync(G)$ but their intersection $\{\, a^n b^n a^n \mid n \geq 0 \,\}$ is not a context-free language.

The synchronization product of a grammar with a finite automaton is extended for two grammars R and S for generating the *level synchronization product* $G{\times_\ell}H$ of their generated automata $G \in R^\omega$ and $H \in S^\omega$ which is the restriction of $G{\times}H$ to pairs of vertices with same level: $G{\times_\ell}H = (G{\times}H)_{|P}$ for $P = \{\, (s,p) \in V_G{\times}V_H \mid \ell_G^R(s) = \ell_H^S(p) \,\}$. This product can be generated by a grammar $R{\times_\ell}S$ that we define. Let $(A,B) \in N_R{\times}N_S$ be any pair of non-terminals and $E \subseteq [1,\varrho(A)]{\times}[1,\varrho(B)]$ be a binary relation over inputs such that for all $i,j \in [1,\varrho(A)]$, if $E(i) \cap E(j) \neq \emptyset$ then $E(i) = E(j)$, where $E(i) = \{j \mid (i,j) \in E\}$ denotes the *image* of $i \in [1,\varrho(A)]$ by E. Intuitively for a pair $(A,B) \in N_R{\times}N_S$ of non-terminals, a relation $E \subseteq [1,\varrho(A)]{\times}[1,\varrho(B)]$ is used to memorize which entries of A and B are being synchronized.

To any such A, B and E, we associate a new symbol $[A,B,E]$ of arity $|E|$ (where $[Z,Z,\emptyset]$ is assimilated to Z). To each non-terminal hyperarc $Ar_1\ldots r_m$ of R ($A \in N_R$ and $m = \varrho(A)$) and each non-terminal hyperarc $Bs_1\ldots s_n$ of S ($B \in N_S$ and $n = \varrho(B)$), we associate the hyperarc

$$[Ar_1\ldots r_m, Bs_1\ldots s_n, E] = [A,B,E](r_1,s_1)_{_E}\cdots(r_1,s_n)_{_E}\cdots(r_m,s_1)_{_E}\cdots(r_m,s_n)_{_E}$$

with $(r_i,s_j)_{_E} = (r_i,s_j)$ if $(i,j) \in E$, and ε otherwise. The grammar $R{\times_\ell}S$ is then defined by associating to each $(AX,P) \in R$, each $(BY,Q) \in S$, and each $E \subseteq [\varrho(A)]{\times}[\varrho(B)]$, the rule of left hand side $[AX,BY,E]$ and of right hand side

$$\big([P]\times[Q]\big)_{|\overline{E}} \cup \{[CU, DV, E'] \mid CU \in P \wedge C \in N_R \wedge DV \in Q \wedge D \in N_S\}$$

with $\overline{E} = \{\ (X(i), Y(j)) \mid (i, j) \in E\ \} \cup \big(V_P - V_X\big)\times\big(V_Q - V_Y\big)$ and

$E' = \{\ (i, j) \in [\varrho(C)]\times[\varrho(D)] \mid (U(i), V(j)) \in \overline{E}\ \}$.

Note that $R{\times_\ell}S$ is synchronized by R and S, and is bi-synchrnonized with S for $S \lhd R$. Furthermore $R{\times_\ell}S$ generates $G{\times_\ell}H$ for $G \in R^\omega$ and $H \in S^\omega$ hence recognizes a subset of $L(R) \cap L(S)$. However for grammars S and S' synchronized by an unambiguous grammar R, we have $L(S{\times_\ell}S') = L(S) \cap L(S')$.

Proposition 4. *For any unambiguous grammar R, the family $Sync(R)$ is closed under intersection.*

By extending basic operations on finite automata to grammars, it appears that graph grammars are to context-free languages what finite automata are to regular languages. We will continue these extensions in the next section. Let us present a fundamental result concerning grammar synchronization, which states that $Sync(R)$ is independent of the way the automaton R^ω is generated.

Theorem 2. *For any grammars R and S such that $R^\omega = S^\omega$, we have*

$$Sync(R) = Sync(S).$$

Proof sketch

By symmetry of R and S, it is sufficient to show that $Sync(R) \subseteq Sync(S)$.
Let $R' \lhd R$. We want to show that $L(R') \in Sync(S)$.
We have to show the existence of $S' \lhd S$ such that $L(S') = L(R')$.
Note that it is possible that there is no grammar S' synchronized by S and generating the same automaton as R' (*i.e.* $S' \lhd S$ and $S'^\omega = R'^\omega$).
Let $G \in R^\omega = S^\omega$. Any vertex s of G has a level $\ell_G^R(s)$ according to R and a level $\ell_G^S(s)$ according to S.
Let $H \in R'^\omega$ and let $K = (G{\times_\ell}H)_{|P}$ be the automaton obtained by level synchronization product of G with H and restricted to the set P of vertices accessible from ι and co-accessible from o.
The restriction by accessibility from ι and co-accessibility from o can de done by a bi-synchronized grammar [Ca 08]. By definition of $R{\times_\ell}R'$, the automaton K can be generated by a grammar R'' bi-synchronized to R' with

$$\ell_K^{R''}(s,p) = \ell_G^R(s) = \ell_H^{R'}(p) \quad \text{for every} \quad (s,p) \in V_K\,.$$

In particular $L(K) = L(R')$.
Let us show that K is generated by a grammar synchronized by S.
We give the proof for R^ω of finite degree. In that case and for $\|\,\varrho\,\| = \sum_{A \in N_R} \varrho(A)$,

$$|\ell_G^R(s) - \ell_G^R(t)| \ \leq\ \|\,\varrho\,\|.d_G(s,t) \quad \text{for every} \ \ s,t \in V_G\,.$$

Furthermore K is also of finite degree.

We show that K is finitely decomposable not by distance but according to $\ell_K^S(s)$ for the vertices (s,p) of K.

Let $n \geq 0$ and C be a connected component of $K_{|\{\,(s,p)\in V_K \,|\, \ell_G^S(s)\geq n\,\}}$.

So C is fully determined by

$$\text{its } \textit{frontier } : Fr_K(C) = V_C \cap V_{K-C}$$

$$\text{its } \textit{interface} : Int_K(C) = \{\; s \xrightarrow[C]{a} t \mid \{s,t\} \cap Fr_K(C) \neq \emptyset \;\}.$$

Let $(s_0,p_0) \in Fr_K(C)$ and D be the connected component of $G_{\{\,s\,|\,\ell_G^S(s)\geq n\,\}}$ containing s_0. It remains to find a bound b independent of n such that

$$|\ell_K^{R''}(s,p) - \ell_K^{R''}(t,q)| \leq b \quad \text{for every} \quad (s,p)\,,\,(t,q) \in Fr_K(C).$$

For any $(s,p)\,,\,(t,q) \in Fr_K(C)$, we have $s,t \in Fr_G(D)$ hence $d_D(s,t)$ is bounded by the integer

$$c \;=\; max\{\; d_{S^\omega(A)}(i,j) < +\infty \mid A \in N_S \;\wedge\; i,j \in [1,\varrho(A)] \;\}$$

whose $S^\omega(A) \;=\; \{\, \bigcup_{n\geq 0}[H_n] \mid A1\ldots\varrho(A) = H_0 \underset{S}{\Longrightarrow} \ldots\, H_n \underset{S}{\Longrightarrow} H_{n+1} \,\ldots\}$

thus it follows that

$$|\ell_K^{R''}(s,p) - \ell_K^{R''}(t,q)| = |\ell_G^R(s) - \ell_G^R(t)| \leq \|\,\varrho\,\| d_G(s,t) \leq \|\,\varrho\,\| d_D(s,t) \leq \|\,\varrho\,\| c.$$

For G of infinite degree and by Proposition 8, we can express $Sync(G)$ as an ε-closure of $Sync(H)$ for some regular automaton H of finite degree using ε-transitions. □

Theorem 2 allows to transfer the concept of grammar synchronization to the level of regular automata: for any regular automaton G, we can define

$$Sync(G) = Sync(R) \;\text{ for (any) } R \text{ such that } G \in R^\omega.$$

Let us illustrate these ideas by presenting some examples of well-known subfamilies of context-free languages obtained by synchronization.

Example 2. For any finite automaton G, $Sync(G)$ is the family of regular languages included in $L(G)$.

Example 3. For the following regular automaton G:

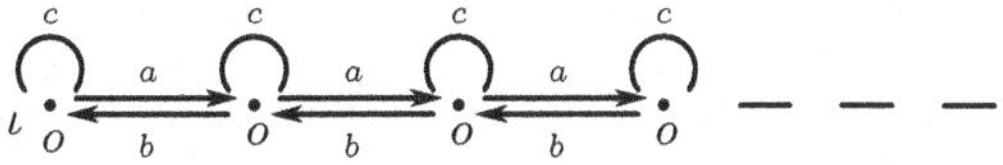

Sync(G) is the family of *input-driven languages* [Me 80] with a pushing, b popping and c internal. As the initial vertex is not source of an arc labelled by b, *Sync*(G) does not contain all the regular languages.

Example 4. We complete the previous automaton by adding an b-loop on the initial vertex to obtain the following automaton G:

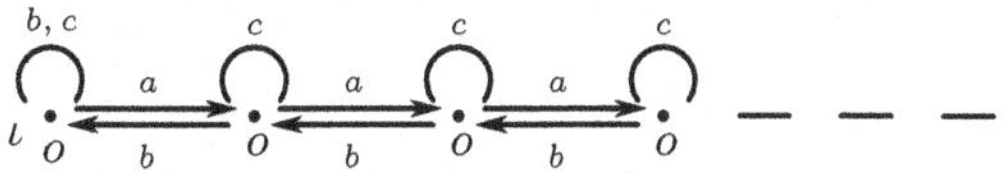

The set *Sync*(G) is the family of *visibly pushdown languages* [AM 04] with a pushing, b popping and c internal.

Example 5. For the following regular automaton G:

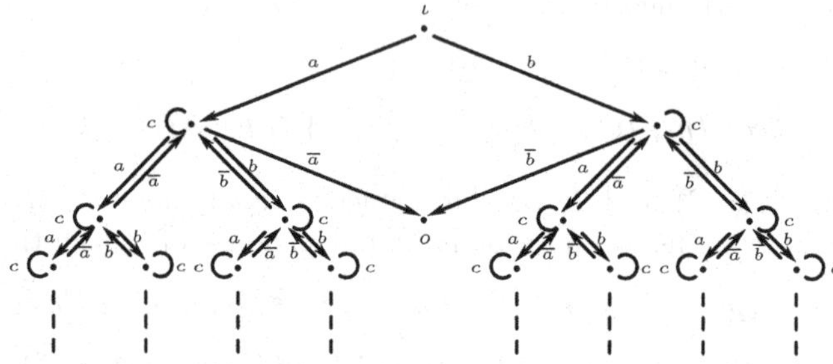

the set *Sync*(G) is the family of *balanced languages* [BB 02] with a, b pushing with their corresponding popping letters $\overline{a}, \overline{b}$, and c is internal.

Example 6. For the grammar R of Figure 6, *Sync*(R) is the family of languages generated by the following linear contex-free grammars:

$$I = P + a^m A(b + \ldots + b^m) \text{ with } m \geq 0 \text{ and } P \subseteq \{\, a^i b^j \mid 1 \leq j \leq i \leq m \,\}$$

$$A = Q + a^n A(b + \ldots + b^n) \;\text{ with } n > 0 \;\text{ and } Q \subseteq \{\, a^i b^j \mid 1 \leq j \leq i \leq n \,\}.$$

For each regular automaton G among the previous examples, *Sync*(G) is a boolean algebra according to $L(G)$ and, except for the last two examples, is also closed under concatenation and its iteration. We now consider new closure properties of synchronized languages for regular automata.

4 Closure Properties

We have seen that the family *Sync*(G) of languages synchronized by a regular automaton G is closed under union and under intersection with a regular language, and under intersection when G is unambiguous. In this section, we consider the closure of *Sync*(G) under complement relative to $L(G)$ and under concatenation and its transitive closure. To obtain these closure properties, we first apply grammar normalizations preserving the synchronized languages. These normalizations also allow us to add ε-arcs to any regular automaton to get a regular automaton of finite degree with the same synchronized languages.

First we put any grammar in an *equivalent* normal form with the same set of synchronized languages. As in the case of finite automata, we transform any automaton G into the *pointed automaton* $G_\perp^\top$ which is language equivalent $L(G_\perp^\top) = L(G)$, with a unique initial vertex $\top \notin V_G$ which is goal of no arc and can be final, and with a unique non initial and final vertex $\perp \notin V_G$ which is source of no arc:

$$\begin{aligned} G_\perp^\top = (G - \{\iota, o\} \times V_G) \,&\cup\, \{\iota\top, o\perp\} \,\cup\, \{\, o\top \mid \exists\, s\,(\iota s, os \in G)\,\} \\ &\cup\, \{\, \top \xrightarrow{a} t \mid \exists\, s\,(s \xrightarrow[G]{a} t \,\wedge\, \iota s \in G)\} \\ &\cup\, \{\, s \xrightarrow{a} \perp \mid \exists\, t\,(s \xrightarrow[G]{a} t \,\wedge\, ot \in G)\} \\ &\cup\, \{\, \top \xrightarrow{a} \perp \mid \exists\, s, t\,(s \xrightarrow[G]{a} t \,\wedge\, \iota s, ot \in G)\}. \end{aligned}$$

For instance, the finite degree regular automaton G of Figure 1 is transformed into the following infinite degree regular automaton $G_\bot^\top$:

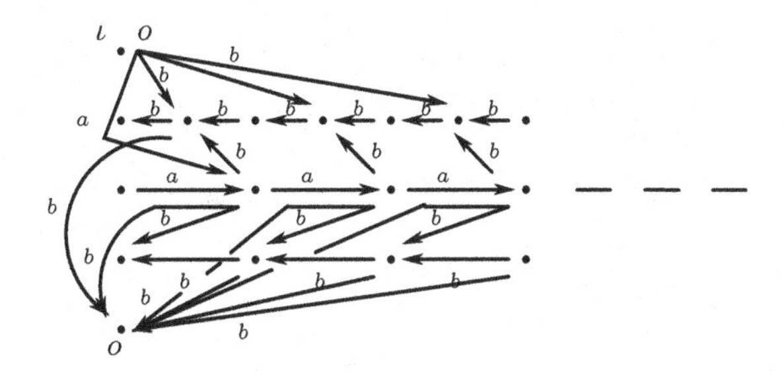

Fig. 7. A pointed regular automaton

Note that if G is unambiguous, $G_\bot^\top$ remains unambiguous. The pointed transformation of a regular automaton remains a regular automaton which can be generated by an 0-*grammar*: only the axiom has initial and final vertices. Let R be any grammar and $\top, \bot$ be two symbols which are not vertices of R. Let $G \in R^\omega$ with $\top, \bot \notin V_G$. We define an 0-grammar $R_\bot^\top$ generating $G_\bot^\top$ and preserving the synchronized languages: $Sync(R_\bot^\top) = Sync(R)$.

First we transform R into a grammar $\widehat{R}$ in which we memorize in the non-terminals the input vertices which are linked to initial or final vertices of the generated automaton. More precisely to any $A \in N_R$ and $I, J \subseteq [1, \varrho(A)]$, we associate a new symbol $A_{I,J}$ of arity $\varrho(A)$ with $Z = Z_{\emptyset,\emptyset}$. We define the grammar $\widehat{R}$ assciating to each $(AX, H) \in R$ and $I, J \subseteq [1, \varrho(A)]$ the following rule:

$$A_{I,J}X \longrightarrow [H] \cup \{\, B_{I',J'}Y \mid BY \in H \ \wedge\ B \in N_R \,\}$$

with $I' = \{\, i \mid Y(i) \in I \vee \iota Y(i) \in H \,\}$ and $J' = \{\, j \mid Y(j) \in J \vee oY(j) \in H \,\}$ and we restrict the rules of $\widehat{R}$ to the non-terminals accessible from Z.

Note that the set $L(R) \cap T$ of letters recognized by R can be determined as

$$\{\, a \mid \exists\, (A_{I,J}X, H) \in \widehat{R} \ \ (\exists\, i \in I\, \exists\, t,\ X(i) \underset{[H]}{\overset{a}{\longrightarrow}} t \ \wedge\ ot \in H)$$

$$\vee\ (\exists\, j \in J\, \exists\, s,\ s \underset{[H]}{\overset{a}{\longrightarrow}} X(j) \ \wedge\ \iota s \in H) \ \vee\ (\exists\, s,t,\ s \underset{[H]}{\overset{a}{\longrightarrow}} t \ \wedge\ \iota s\,, ot \in H) \,\}$$

and $\varepsilon \in L(R) \iff \exists H \in Im(\widehat{R})\ \exists\, s\ (\iota s\,, os \in H)$.

To any $A \in N_R - \{Z\}$ and any $I, J \subseteq [1, \varrho(A)]$, we associate a new symbol $A'_{I,J}$ of arity $\varrho(A) + 2$, and we define the grammar $R_\bot^\top$ containing the axiom rule

$$Z \longrightarrow H_{\emptyset,\emptyset} \cup \{\iota\top\,, o\bot\} \cup \{\, o\top \mid \varepsilon \in L(R) \,\} \cup \{\, \top \overset{a}{\longrightarrow} \bot \mid a \in L(R) \cap T \,\}$$

for $(Z, H) \in \widehat{R}$, and for any $(A_{I,J}X, H) \in \widehat{R}$ with $A \neq Z$, we take in $R_\bot^\top$ the rule $A'_{I,J}\top X\bot \longrightarrow H_{I,J}$ such that $H_{I,J}$ is the following hypergraph:

$$H_{I,J} = ([H] - \{\iota, o\}) \times V_H) \cup \{ B'_{P,Q} \top X \bot \mid B_{P,Q} X \in H \wedge B_{P,Q} \in N_{\widehat{R}} \}$$
$$\cup \{ \top \xrightarrow{a} t \mid \exists\, i \in I\, (X(i) \underset{[H]}{\xrightarrow{a}} t) \vee \exists\, s\, (\iota s \in H \wedge s \underset{[H]}{\xrightarrow{a}} t) \}$$
$$\cup \{ s \xrightarrow{a} \bot \mid \exists\, j \in J\, (s \underset{[H]}{\xrightarrow{a}} X(j)) \vee \exists\, t\, (ot \in H \wedge s \underset{[H]}{\xrightarrow{a}} t) \}$$

and we put $R^{\top}_{\bot}$ into a terminal-outside form [Ca 07].

Example 7. Let us consider the following grammar R:

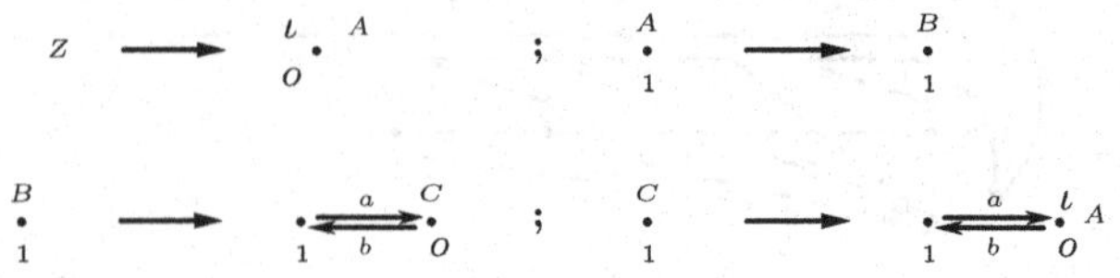

generating the following automaton G (with levels of some vertices):

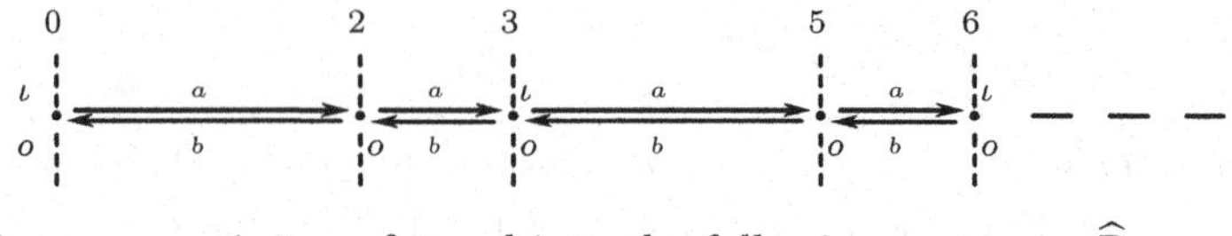

First this grammar is transformed into the following grammar $\widehat{R}$:

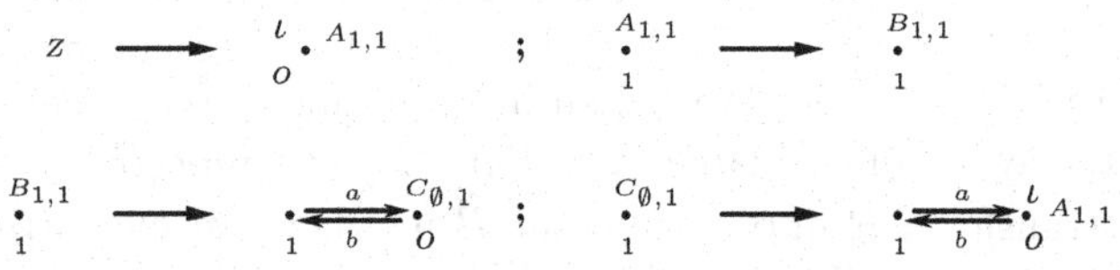

In particular $\varepsilon, a, b \in L(R)$. Then $\widehat{R}$ is transformed into the grammar $R^{\top}_{\bot}$:

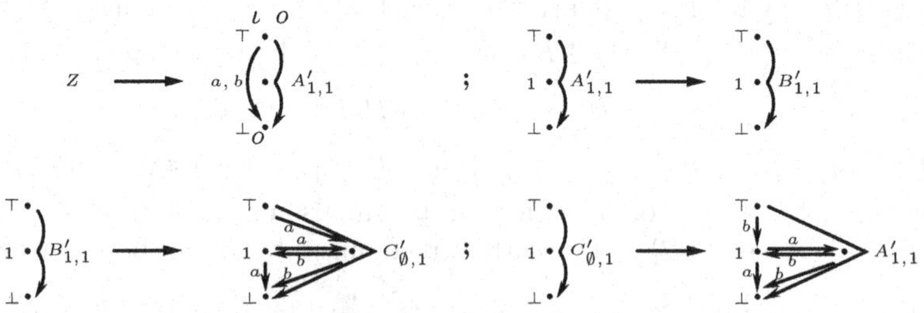

that we put in a terminal-outside form:

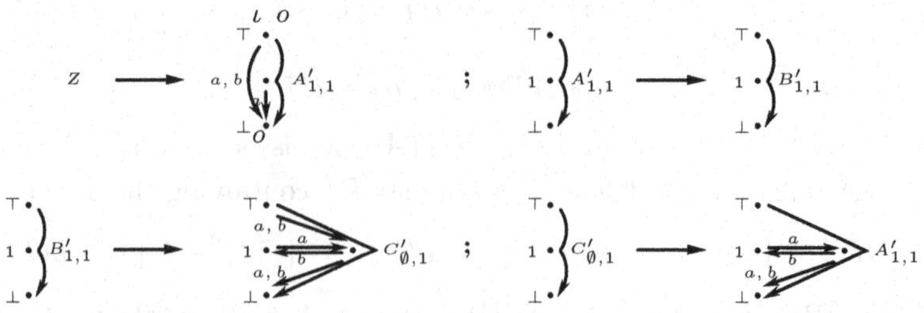

So $R^{\top}_{\bot}$ generates $G^{\top}_{\bot}$:

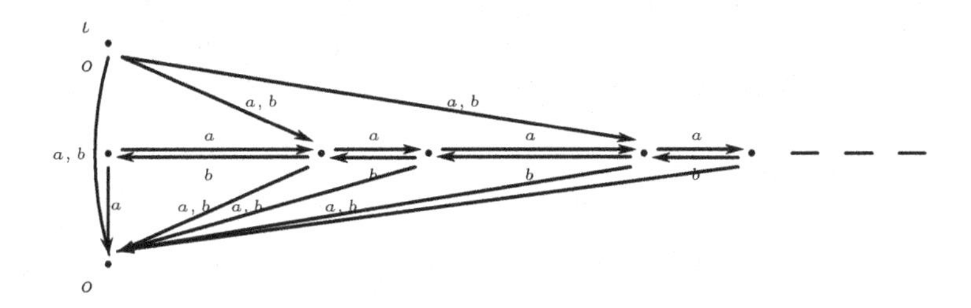

□

The grammars R and $R_{\perp}^{\top}$ synchronize the same languages.

Proposition 5. *For any regular automaton G with $\top, \perp \notin V_G$, the pointed automaton $G_{\perp}^{\top}$ remains regular and $Sync(G_{\perp}^{\top}) = Sync(G)$.*

It follows that, in order to define families of languages by synchronization by a regular automaton G, we can restrict to pointed automata G. A stronger normalization is to transform any grammar R into a grammar S such that $Sync(S) = Sync(R)$ and S is an *arc-grammar* in the following sense: S is an 0-grammar whose any non-terminal $A \in N_S - \{Z\}$ is of arity 2, and for any non axiom rule $Ast \longrightarrow H$, there is no arc in H of goal s or of source t: for any $p \underset{H}{\overset{a}{\longrightarrow}} q$, we have $p \neq t$ and $q \neq s$.

We can transformed any 0-grammar R into a bi-synchronized arc-grammar $\prec R \succ$.

We assume that each rule of R is of the form $A1 \ldots \varrho(A) \longrightarrow H_A$ for any $A \in N_R$. We take a new symbol 0 (not a vertex of R) and a new label $A_{i,j}$ of arity 2 for each $A \in N_R$ and each $i, j \in [1, \varrho(A)]$ in order to generate paths from i to j in $R^\omega(A1 \ldots \varrho(A))$. We define the *splitting* $\prec G \succ$ of any F_R-hypergraph G without vertex 0 as being the graph:

$$\prec G \succ = [G] \cup \{ X(i) \overset{A_{i,j}}{\longrightarrow} X(j) \mid AX \in G \wedge A \in N_R \wedge i, j \in [\varrho(A)] \}$$

and for $p, q \in V_G$ and $P \subseteq V_G$ with $0 \notin V_G$, we define

$$G_{p,P,q} = \big(\{ s \underset{\prec G \succ}{\overset{a}{\longrightarrow}} t \mid t \neq p \wedge s \neq q \wedge s, t \notin P \}\big)_{|I} \quad \text{for } p \neq q$$

$$G_{p,P,p} = \big(\{ s \underset{\prec G \succ}{\overset{a}{\longrightarrow}} t \mid t \neq p \wedge s, t \notin P \} \cup \{ s \overset{a}{\longrightarrow} 0 \mid s \underset{\prec G \succ}{\overset{a}{\longrightarrow}} p \}\big)_{|J}$$

with $I = \{ s \mid p \Longrightarrow s \Longrightarrow q \}$ and $J = \{ s \mid p \Longrightarrow s \Longrightarrow 0 \}$.

This allows to define the *splitting* $\prec R \succ$ of R as being the following arc-grammar:

$$Z \longrightarrow \prec H_Z \succ$$

$$A_{i,j}12 \longrightarrow h_{i,j}\big((H_A)_{i,[\varrho(A)]-\{i,j\},j}\big) \text{ for each } A \in N_R \text{ and } i, j \in [1, \varrho(A)]$$

where $h_{i,j}$ is the vertex renaming defined by

$$h_{i,j}(i) = 1, \; h_{i,j}(j) = 2, \; h_{i,j}(x) = x \text{ otherwise, for } i \neq j$$

$$h_{i,i}(i) = 1, \; h_{i,i}(0) = 2, \; h_{i,i}(x) = x \text{ otherwise.}$$

Thus R and $\prec\!R\!\succ$ are bi-synchronized, and $\prec\!R\!\succ$ is unambiguous when R is unambiguous. Note that we can put $\prec\!R\!\succ$ into a reduced form by removing any non-terminal $A_{i,j}$ such that $\prec\!R\!\succ^{\omega}(A_{i,j})12$ is without path from 1 to 2.

Example 8. The following 0-grammar R:

generates the following automaton G:

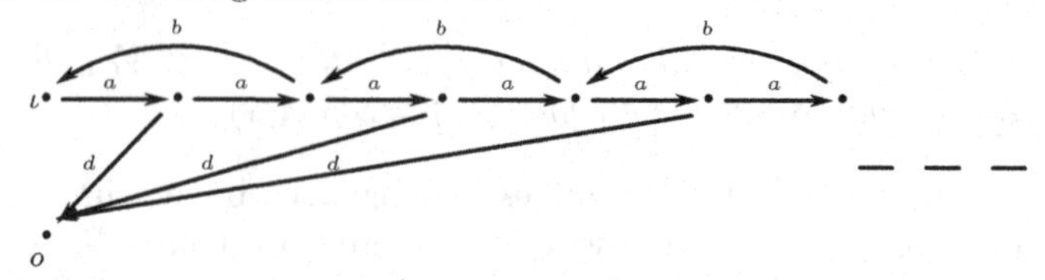

The splitting $\prec\!R\!\succ$ of R is the following grammar:

generating the following automaton:

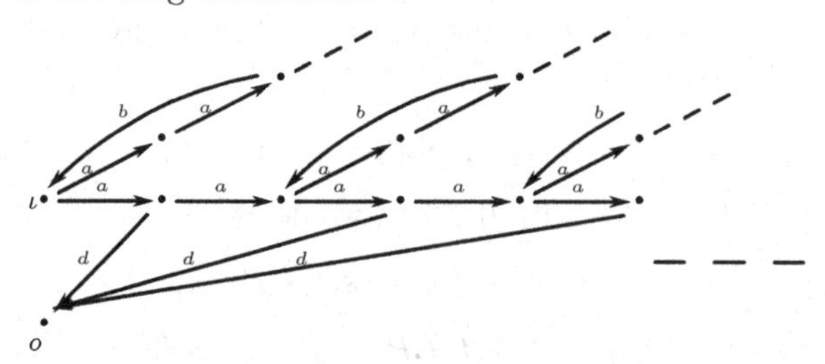

As $R \bowtie \prec\!R\!\succ$, we have $Sync(R) = Sync(\prec\!R\!\succ)$. □

To study closure properties of $Sync(R)$ for any grammar R, we can work with its normal form $\prec\!R^{\top}_{\bot}\!\succ$ which is an arc-grammar generating a pointed automaton. This normalization is really useful to study the closure property of $Sync(R)$ under complement relative to $L(R)$, under concatenation and its iteration.

We have seen that $Sync(R)$ is not closed in general under intersection, hence it is not closed under complement according to $L(R)$ since for any $L, M \subseteq L(R)$, $L \cap M = L(R) - [(L(R) - L) \cup (L(R) - M)]$. For R unambiguous, $Sync(R)$ is closed under intersection, and this remains true under complement according to $L(R)$ [Ca 08]. We give here a simpler construction.

As $\prec R^{\top}_{\perp}\succ$ remains unambiguous, we can assume that R is an arc-grammar. Let $S \lhd R$. We want to show that $L(R) - L(S) \in \mathit{Sync}(R)$. So S is an 0-grammar and S is level-unambiguous as defined in [Ca 08]. Thus $\prec S\succ$ is a level-unambiguous arc-grammar. We take a new colour $c \in F_1 - \{\iota, o\}$ and for any grammar S', we denote S'_c (resp. $S'_{\overline{c}}$) the grammar obtained from S' by replacing the final colour o by c (resp. c by o). So $R + \prec S\succ_c$ is an arc-grammar and $(R + \prec S\succ_c)_{\overline{c}}$ is level-unambiguous. It remains to apply the grammar determinization in [Ca 08] to get the grammar $R/S = \mathit{Det}(R + \prec S\succ_c)$ such that $(R/S)_{\overline{c}}$ is unambiguous and bi-synchronized to $(R + \prec S\succ_c)_{\overline{c}}$. Finally we keep in R/S the final vertices which are not coloured by c to obtain a grammar synchronized by R and recognizing $L(R) - L(S)$.

Theorem 3. *For any unambiguous regular automaton G, the set $\mathit{Sync}(G)$ is an effective boolean algebra according to $L(G)$, containing all the regular languages included in $L(G)$.*

So we can decide the inclusion $L(S) \subseteq L(S')$ for two grammars S and S' synchronized by a common unambiguous grammar. Furthermore for grammars R_1 and R_2 such that $R_1 + R_2$ is level-unambiguous, $\mathit{Sync}(R_1 + R_2) = \{\, L_1 \cup L_2 \mid L_1 \in \mathit{Sync}(R_1) \wedge L_2 \in \mathit{Sync}(R_2) \,\}$ is a boolean algebra included in $L(R_1) \cup L(R_2)$, containing $\mathit{Sync}(R_1)$ and $\mathit{Sync}(R_2)$.

The automata of Examples 2 to 6 are unambiguous hence their families of synchronized languages are boolean algebra. This regular automaton G:

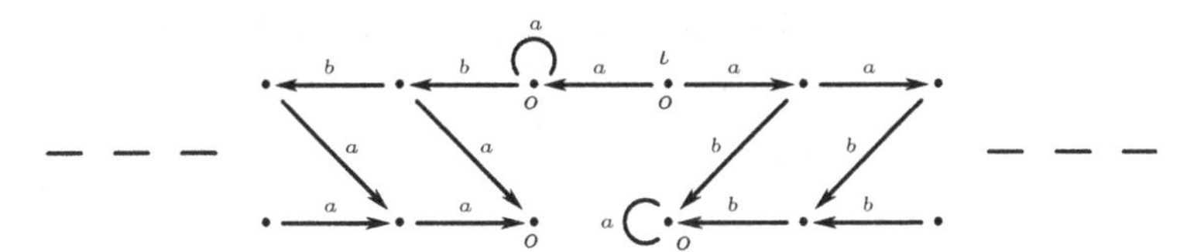

is 2-ambiguous: there are two accepting paths for the words $a^n b^n a^n$ with $n > 0$ and a unique accepting path for the other accepted words. But $\mathit{Sync}(G)$ is not closed under intersection since $\{\, a^m b^m a^n \mid m, n \geq 0 \,\}$ and $\{\, a^m b^n a^n \mid m, n \geq 0 \,\}$ are languages synchronized by G.

For any regular automaton G, the closure of $\mathit{Sync}(G)$ under concatenation $\cdot$ (resp. under its transitive closure $^+$) does not require the unambiguity of G. As $L(G) \in \mathit{Sync}(G)$, a necessary condition is to have $L(G).L(G) \in \mathit{Sync}(G)$ (resp. $L(G)^+ \in \mathit{Sync}(G)$). Note that this necessary condition implies that $L(G)$ is closed under $\cdot$ (resp. $^+$). In particular $\mathit{Sync}(G)$ is not closed under $\cdot$ and $^+$ for the automata of Examples 5 and 6. But this necessary condition is not sufficient since the following regular automaton G:

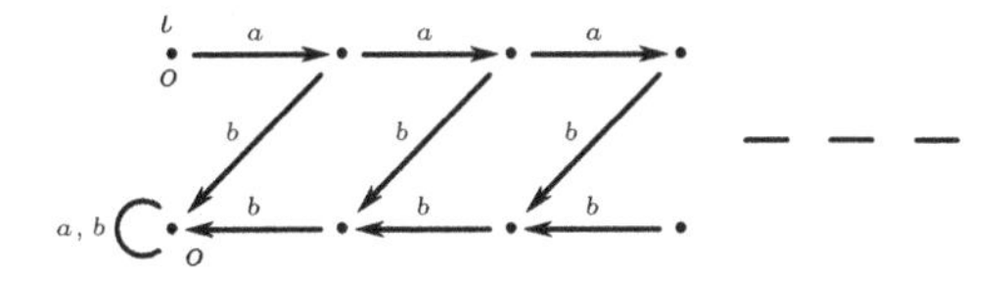

recognizes $L(G) = \varepsilon + M(a+b)^*$ for $M = \{ a^n b^n \mid n > 0 \}$, hence $L(G).L(G) = L(G) = L(G)^+$ but $M \in Sync(G)$ and $M.M, M^+ \notin Sync(G)$.

Let us give a simple and general condition on a grammar R such that $Sync(R)$ is closed under $\cdot$ and $^+$. We say that a grammar is *iterative* if any initial vertex is in the axiom and for (any) $G \in R^\omega$ and any accepting path $s_0 \xrightarrow[G]{a_1} s_1 \ldots \xrightarrow[G]{a_n} s_n$ with ιs_0, $o s_n \in G$ and for any final vertex t *i.e.* $ot \in G$, there exists a path $t \xrightarrow[G]{a_1} t_1 \ldots \xrightarrow[G]{a_n} t_n$ with $o t_n \in G$ such that $\ell(t_i) = \ell(t) + \ell(s_i)$ for all $i \in [1,n]$.

For instance the automaton of Example 3 can be generated by an iterative grammar. And any 0-grammar generating a regular automaton having a unique initial vertex which is the unique final vertex, is iterative. Standard constructions on finite automata for the concatenation and its iteration can be extended to iterative grammars.

Proposition 6. *For any iterative grammar R, the family $Sync(R)$ is closed under concatenation and its transitive closure.*

However the automaton G of Example 4 cannot be generated by an iterated grammar but $Sync(G)$ is closed under $\cdot$ and $^+$. We can also obtain families of synchronized languages which are closed under $\cdot$ and $^+$ by saturating grammars. The *saturation* G^+ of an automaton G is the automaton

$$G^+ \;=\; G \cup \{\, s \xrightarrow{a} r \mid \iota r \in G \;\wedge\; \exists\, t\, (s \xrightarrow[G]{a} t \;\wedge\; ot \in G) \,\}$$

recognizing $L(G^+) = (L(G))^+$.

Note that if G is regular with infinite sets of initial and final vertices, G^+ can be non regular (but is always prefix-recognizable). If G is generated by an 0-grammar R, its saturation G^+ can be generated by a grammar R^+ that we define.

Let (Z,H) be the axiom rule of R and $r_1,\ldots,r_p$ be the initial vertices of H; we can assume that $r_1,\ldots,r_p$ are not vertices of $R - \{(Z,H)\}$. To each $A \in N_R - \{Z\}$ and $I \subseteq [1,\varrho(A)]$, we associate a new symbol A_I of arity $\varrho(A)+p$ and we define R^+ with the following rules:

$$Z \longrightarrow [H]^+ \cup \{\, A_{\{\, i \mid o X(i) \in H \,\}} X r_1 \ldots r_p \mid AX \in H \;\wedge\; A \in N_R \,\}$$

$$A_I X r_1 \ldots r_p \longrightarrow K_I \quad \text{for each } (AX,K) \in R \text{ and } A \neq Z \text{ and } I \subseteq [1,\varrho(A)]$$

whose K_I is the automaton obtained from K as follows:

$$\begin{aligned} K_I = [K] \;&\cup\; \{\, s \xrightarrow{a} r_j \mid j \in [p] \;\wedge\; \exists\, i \in I\ (s \xrightarrow[K]{a} X(i)) \,\} \\ &\cup\; \{\, B_{\{\, j \mid \exists\ i \in I,\ Y(j) = X(i) \,\}} Y r_1 \ldots r_p \mid BY \in K \;\wedge\; B \in N_R \,\}. \end{aligned}$$

So R is synchronized by R^+ and $G^+ \in (R^+)^\omega$ for $G \in R^\omega$.

To characterize $Sync(R^+)$ from $Sync(R)$, we define the *regular closure* $Reg(E)$ of any language family E as being the smallest family of languages containing E and closed under $\cup$, $\cdot$, $+$.

Proposition 7. *For any 0-grammar R,* $Sync(R^+) \;=\; Reg(Sync(R))$.

By Propositions 5, 6 and 7, the following regular automaton G:

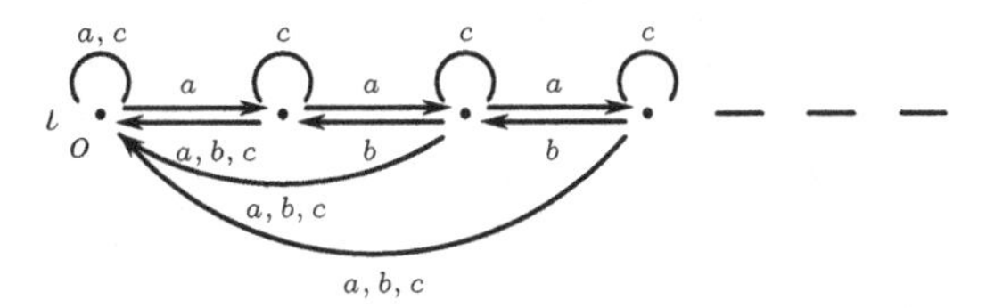

has the same synchronized languages than the automaton of Example 3: $Sync(G)$ is the family of input-driven languages (for a pushing, b popping and c internal). By adding an b-loop on the initial (and final) vertex of G, we obtain an automaton H such that $Sync(H)$ is the family of visibly pushdown languages hence by Proposition 7, is closed under $\cdot$ and $^+$.

Example 9. A natural extension of the visibly pushdown languages is to add reset letters. For a pushing, b popping and c internal, we add a reset letter d to define the following regular automaton G:

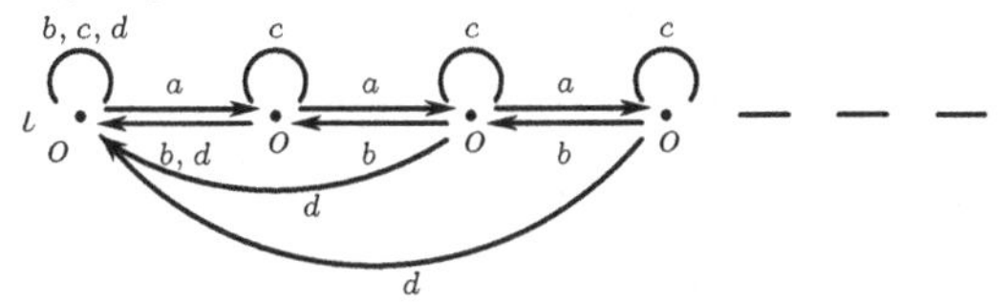

Any language of $Sync(G)$ is a visibly pushdown language taking d as an internal letter, but not the converse: $\{ a^n db^n \mid n \geq 0 \} \notin Sync(G)$. By Theorem 3, $Sync(G)$ is a boolean algebra. Furthermore the following automaton H:

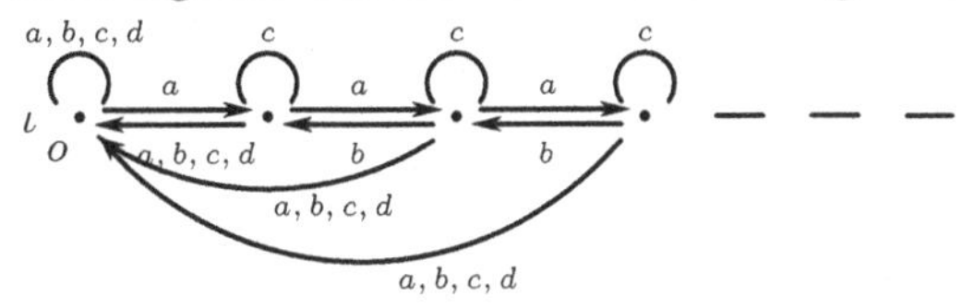

satisfies $Sync(H) = Sync(G)$ and $H^+ = H$ hence by Proposition 7, $Sync(G)$ is also closed under $\cdot$ and $^+$. □

Note that the automata of the previous example have infinite degree. Furthermore for any automaton G of finite degree having an infinite set of initial or final vertices, the pointed automaton $G_{\perp}^{\top}$ is of infinite degree. However any regular automaton of infinite degree (in fact any prefix-recognizable automaton) can be obtained by ϵ-closure from a regular automaton of finite degree using ε-transitions. For instance let us take a new letter $e \notin T$ (instead of the empty word) and let us denote π_e the morphism erasing e in the words over $T \cup \{e\}$: $\pi_e(a) = a$ for any $a \in T$ and $\pi_e(e) = \varepsilon$, that we extend by union to any language $L \subseteq (T \cup \{e\})^* : \pi_e(L) = \{ \pi_e(u) \mid u \in L \}$, and by powerset to any family P of languages: $\pi_e(P) = \{ \pi_e(L) \mid L \in P \}$. The following regular automaton K:

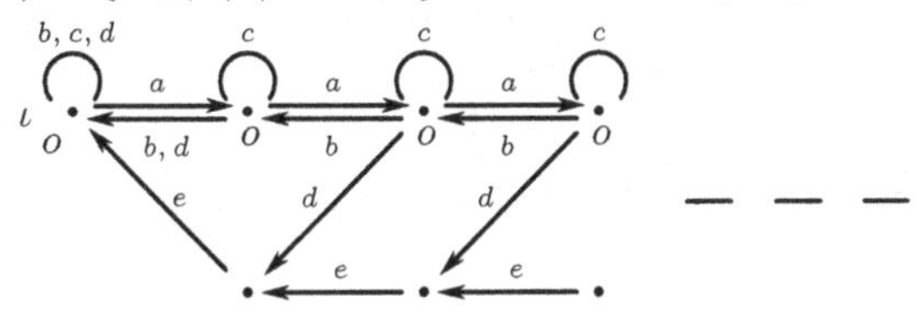

is of finite degree and satisfies $\pi_e(Sync(K)) = Sync(G)$ for the automaton G of Example 9. Let us give a simple transformation of any grammar R to a grammar R_e such that R_e^ω is of finite degree and $\pi_e(Sync(R_e)) = Sync(R)$.

As $Sync(R) = Sync(\prec R_\perp^\top \succ)$, we restrict this transformation to arc-grammars. Let R be an arc-grammar. We define R_e to be an arc-grammar obtained from R by replacing each non axiom rule $Ast \longrightarrow H$ by the rule:

$$Ast \longrightarrow \left([H] \cup \{s \xrightarrow{e} s_e \,,\, t_e \xrightarrow{e} t\} \cup h(H - [H])\right)_{|P}$$

with s_e, t_e be new vertices and h the vertex mapping defined for any $r \in V_H$ by $h(r) = r$ if $r \notin \{s,t\}$, $h(s) = s_e$ and $h(t) = t_e$, and P is the set of vertices accessible from s and co-accessible from t. For instance the arc-grammar R

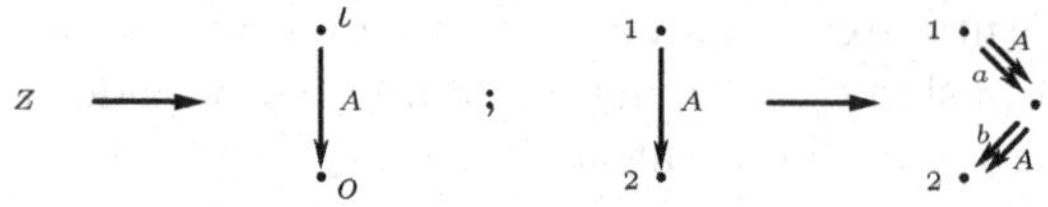

is transformed into the following arc-grammar R_e :

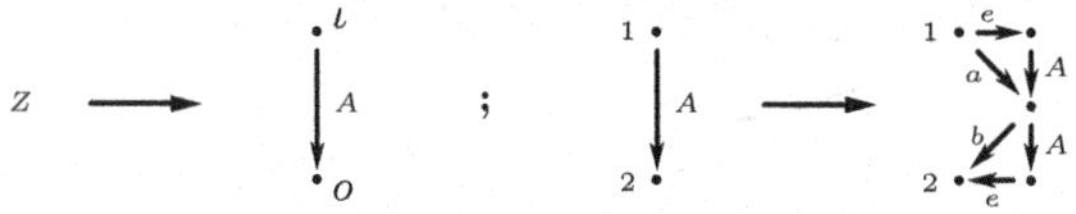

For any rule of R_e, the inputs are separated from the outputs (by e-transitions), hence R_e^ω is of finite degree. Furthermore this transformation preserves the synchronized languages.

Proposition 8. *For any arc-grammar R, $Sync(R) = \pi_e(Sync(R_e))$.*

So for any R, $Sync(R) = \pi_e(Sync(\prec R_\perp^\top \succ_e))$ and $(\prec R_\perp^\top \succ_e)^\omega$ is of finite degree.

All the constructions given in this paper are natural generalizations of usual transformations on finite automata to graph grammars. In this way, basic closure properties could be lifted to sub-families of context-free languages.

5 Conclusion

The synchronization of regular automata is defined through devices generating these automata, namely functional graph grammars. It can also be defined using pushdown automata with ε-transitions [NS 07] because Theorem 2 asserts that the family of languages synchronized by a regular automaton is independent of the way the automaton is generated; it is a graph-related notion. This paper shows that the mechanism of functional graph grammars provides natural constructions on regular automata generalizing usual constructions on finite automata. This paper is also an invitation to extend the notion of synchronization to more general sub-families of automata.

Acknowledgements

Many thanks to Arnaud Carayol and Antoine Meyer for helping me prepare the final version of this paper.

References

[AM 04] Alur, R., Madhusudan, P.: Visibly pushdown languages. In: Babai, L. (ed.) 36th STOC, ACM Proceedings, pp. 202–211 (2004)

[Be 79] Berstel, J.: Transductions and context-free languages. In: Teubner (ed.), pp. 1–278 (1979)

[BB 02] Berstel, J., Boasson, L.: Balanced grammars and their languages. In: Brauer, W., Ehrig, H., Karhumäki, J., Salomaa, A. (eds.) Formal and Natural Computing. LNCS, vol. 2300, pp. 3–25. Springer, Heidelberg (2002)

[Ca 06] Caucal, D.: Synchronization of pushdown automata. In: Ibarra, O.H., Dang, Z. (eds.) DLT 2006. LNCS, vol. 4036, pp. 120–132. Springer, Heidelberg (2006)

[Ca 07] Caucal, D.: Deterministic graph grammars. In: Flum, J., Grädel, E., Wilke, T. (eds.) Texts in Logic and Games 2, pp. 169–250. Amsterdam University Press (2007)

[Ca 08] Caucal, D.: Boolean algebras of unambiguous context-free languages. In: Hariharan, R., Mukund, M., Vinay, V. (eds.) 28th FSTTCS, Dagstuhl Research Online Publication Server (2008)

[CH 08] Caucal, D., Hassen, S.: Synchronization of grammars. In: Hirsch, E.A., Razborov, A.A., Semenov, A., Slissenko, A. (eds.) Computer Science - Theory and Applications. LNCS, vol. 5010, pp. 110–121. Springer, Heidelberg (2008)

[Ha 78] Harrison, M.: Introduction to formal language theory. Addison-Wesley, Reading (1978)

[Me 80] Mehlhorn, K.: Pebbling mountain ranges and its application to DCFL recognition. In: de Bakker, J.W., van Leeuwen, J. (eds.) ICALP 1980. LNCS, vol. 85, pp. 422–432. Springer, Heidelberg (1980)

[MS 85] Muller, D., Schupp, P.: The theory of ends, pushdown automata, and second-order logic. Theoretical Computer Science 37, 51–75 (1985)

[NS 07] Nowotka, D., Srba, J.: Height-deterministic pushdown automata. In: Kučera, L., Kučera, A. (eds.) MFCS 2007. LNCS, vol. 4708, pp. 125–134. Springer, Heidelberg (2007)

Stochastic Process Creation

Javier Esparza

Institut für Informatik, Technische Universität München
Boltzmannstr. 3, 85748 Garching, Germany

In many areas of computer science entities can "reproduce", "replicate", or "create new instances". Paramount examples are threads in multithreaded programs, processes in operating systems, and computer viruses, but many others exist: procedure calls create new incarnations of the callees, web crawlers discover new pages to be explored (and so "create" new tasks), divide-and-conquer procedures split a problem into subproblems, and leaves of tree-based data structures become internal nodes with children. For lack of a better name, I use the generic term *systems with process creation* to refer to all these entities.

In the last months, Tomáš Brázdil, Stefan Kiefer, Michael Luttenberger and myself have started to investigate the behaviour of systems with *stochastic* process creation [3]. We assume that the probability with which an entity will create a new one is known or has been estimated, e.g. as the result of statistical sampling. Using these probabilities we model the reproductive behaviour of the system as a stochastic process, and study the distribution and moments of several random variables. In particular, we are interested in random variables modelling the computational resources needed to completely execute the system, i.e., to execute the initial process and all of its descendants.

Stochastic process creation has been studied by mathematicians for decades under the name *branching (stochastic) processes* [7,1], and so my coauthors and I initially thought that we would find the answers to all basic questions in the literature. Interestingly, this is by no means so. Work on branching processes has been motivated by applications to biology (study of animal populations), physics (study of particle cascades) or chemistry (study of chemical reactions). From a computer scientist's point of view, in these scenarios no process ever waits to be executed, because there is no separation between processes (software) and processor (hardware); for instance, in biology scenarios each individual animal is both a process and the processor executing it, and both are created at the same time. So, in computer science terms, probability theorists have studied systems with an *unbounded* number of processors, in which every new process is immediately assigned an idle processor. The model in which one single processor or, more generally, a fixed number of processors, execute a possibly much larger number of processes, seems to have received little attention. To be more precise, some *urn models* of the literature are equivalent to it [8,11], but they have not been studied through the eyes of a computer scientist, and questions about computational resources have not been addressed.

In this note I rephrase and interpret some results of the theory of branching processes for a computer science audience, and informally present some results of our ongoing work on the single processor case. We have not addressed the

R. Královič and D. Niwiński (Eds.): MFCS 2009, LNCS 5734, pp. 24–33, 2009.

k-processor case so far, because we are only starting to understand it. I avoid formal notations as much as possible, and give no proofs.

1 Some Preliminaries

A system consists initially of exactly one process awaiting execution. We describe systems using a notation similar to that of stochastic grammars. Processes can be of different types. The execution of a process generates new processes with fixed and known probabilities. For instance

$$\begin{array}{lll} X \xhookrightarrow{0.2} \langle X, X \rangle & X \xhookrightarrow{0.3} \langle X, Y \rangle & X \xhookrightarrow{0.5} \emptyset \\ Y \xhookrightarrow{0.7} \langle X \rangle & Y \xhookrightarrow{0.3} \langle Y \rangle & \end{array}$$

denotes a system with two types of processes, X and Y. Processes of type X can generate 2 processes of type X, one process of each type, or zero processes with probabilities 0.2, 0.3, and 0.5, respectively (angular brackets denote multisets). Processes of type Y can generate one process, of type X or Y, with probability 0.7 and 0.3. W.l.o.g. we assume that a process "dies" when it generates its children (systems in which processes continue to exist after spawning children can be simulated by assuming that one of the generated children is the continuation of the parent process). An implicit constraint introduced by our notation is the existence of an upper bound in the possible number of children a process can generate at a point in time, i.e., we assume there is a number k such that for every $n \geq k$ the probability that the process has n children is 0. This is a reasonable assumption for most computer science applications, and simplifies many results.

For the sake of clarity, from this moment on we assume that systems only have one type of processes; while most results can be generalised to the multitype case, their formulation is more complicated and difficult to understand. However, in Section 4.2 we will take this assumption back, since there we present some results which are only meaningful for the multitype case.

In the one-type case a system is completely determined by the probabilities p_d, $d \in \mathbb{N}$, that a process generates d child processes. As running example for the paper we consider the system with $p_0 = 0.6, p_1 = 0.1, p_2 = 0.2, p_3 = 0.1$ and $p_d = 0$ for every $d \geq 4$.

The *probability generating function* (pgf) of a system is the formal power series $f(x) = \sum_{i=0}^{\infty} p_i x^i$. For our running example we get

$$f(x) = 0.6 + 0.1x + 0.2x^2 + 0.1x^3$$

Notice that the pgf is a polynomial with nonnegative coefficients, and so in particular a monotonic function over the nonnegative reals.

The execution history of the initial process can be visualized by means of *family trees*. Instead of a formal definition, we just present a family tree of our running example in Figure 1. In this tree, the initial process creates three "children", which in turn have two, nil, and two "grandchildren", respectively.

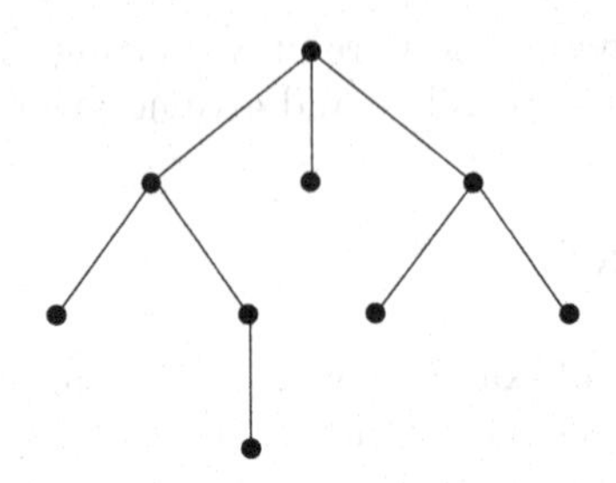

Fig. 1. A family tree

All grandchildren create no further processes with the exception of the second, which has one child; this child terminates without creating any process.

The rest of the note is structured as follows. In Section 2 we briefly discuss the probability that a system terminates. Then, in Sections 3 and 4 we consider the cases of an unbounded number of processors and one processor, respectively.

2 Completion Probability

The *completion probability* of a system is the probability that the initial process will be *completely executed*, i.e., that eventually a point will be reached at which the process itself and all its descendants have terminated. The following theorem was essentially proved by Watson about 150 years ago, and we will sketch a proof a little later:

Theorem 1 ([7]). *The completion probability is the least nonnegative fixed point of the system of pgfs.*

It is not difficult to see that the least nonnegative fixed point of the example above, i.e., the least number a such that $f(a) = a$, is 1. So the initial process is completed with probability 1. But not every system has this property. For instance, the least nonnegative fixed point of the system with pgf $f(x) = 2/3\,x^2 + 1/3$ is only $1/2$.

Critical and subcritical systems. In the rest of the paper we only consider systems with completion probability 1. In particular, this means that the set of all infinite family trees have probability 0, and so infinite family trees can be safely ignored. We can then assign to a system a discrete probability space whose elementary events are the finite family trees; the probability of a tree is the product of the probabilities with which each process in the tree generates its children. For instance, the probability of the family tree of Figure 1 is $(0.6)^5 \cdot 0.1 \cdot (0.2)^2 \cdot 0.1$.

Systems with completion probability 1 can be divided into *critical* and *subcritical*. A system is critical (subcritical) if the average number of children of a process is equal to (strictly smaller than) 1. For our running example the average number of children is $0 \cdot 0.6 + 1 \cdot 0.1 + 2 \cdot 0.2 + 3 \cdot 0.1 = 0.8$, and so the system is subcritical.

3 Unbounded Number of Processors

A family tree specifies which processes will be executed, but not in which order or with which degree of concurrency. This is determined, among other factors, by the number of available processors. In this section we study the case in which this number is unbounded, and every newly created process is immediately assigned a free processor. (Recall that, as discussed in the introduction, these are the branching processes of stochastic theory.) In this model family trees are executed in layers or *generations*: all processes with distance t to the root (the processes of the t-th generation) are executed exactly at time t.

3.1 Completion Time

The completion time is the random variable T that assigns to a family tree τ its number $T(\tau)$ of generations or, in other words, the depth of the tree. The completion time of the tree of Figure 1 is 4.

The main property of the distribution of the completion time T can also be illustrated with our running example. Recall we have $f(x) = 0.6+0.1x+0.2x^2+0.1x^3$. Let P_t denote the probability that a process is completely executed in time at most t, i.e., after at most t generations. If the process generates, say, two children, then it completely executes in at most t time units iff both children completely execute in at most $t-1$ units. Since we assume that children are executed independently, we get that in this case the probability is equal to P_{t-1}^2. So we get $P_t = 0.6 + 0.1P_{t-1} + 0.2P_{t-1}^2 + 0.1P_{t-1}^3$ or, in terms of the pgf, $P_t = f(P_{t-1})$ for every $t \geq 1$. If we denote by $f^t(x)$ the result of iteratively applying t times the function f to x, we obtain the following well-known result [7]:

Theorem 2. *For every* $t \in \mathbb{N}$: $P_t = f^t(0)$.

Since a process becomes completely executed after a possibly large but finite number of generations, the sequence $\{P_t\}_{t\geq 0}$ converges to the completion probability. By Theorem 2, so does the sequence $\{f^t(0)\}_{t\geq 0}$. But the function f is monotonic, and so, by Kleene's theorem, $\{f^t(0)\}_{t\geq 0}$ converges to the least nonnegative fixed point of f. So, in fact, Theorem 1 is an easy corollary of Theorem 2.

The following peculiarity is interesting. Solving fixed point equations is a very common task in computer science, and usually the interest of the sequence $\{f^t(0)\}_{k\geq 0}$ is that, by Kleene's theorem, it provides increasingly accurate approximations to the least nonnegative fixed point. (For this reason, the elements of the sequence are sometimes called the *Kleene approximants.*) In our case it is the other way round: we already know that the least fixed point is 1, and so we are not interested in computing it; it is the approximants themselves we care for! We will come back to this point in Section 4.

3.2 Process Number

The other variable measuring the consumption of resources is the *processor number* N: the minimal number of processors needed to execute the tree if no process must ever wait. Clearly, the processor number $N(\tau)$ of a family tree τ is equal to the size of the largest generation. The process number of the tree of Figure 1 is 4, determined by the size of the third generation.

Determining the distribution of N is harder than for the completion time T, and the problem has been much studied in the literature on stochastic branching processes. We only present here two results for subcritical systems, due to Lindvall and Nerman, that have particular relevance from a computer science point of view. The pgf of a subcritical system has exactly two nonnegative fixed points[1]. The least one gives the completion probability, and is therefore less than or equal to 1. The greatest one is strictly larger than 1, and, surprisingly, it carries important information:

Theorem 3 ([10,12]). *Let $a > 1$ be the greatest nonnegative fixed point of the pgf of a subcritical system. Then*

$$\Pr[N > n] < \frac{a-1}{a^n - 1} \text{ for every } n \geq 1 \quad \text{and} \quad \Pr[N > n] \in \Theta\left(\frac{1}{na^n}\right).$$

In our running example we have $a \approx 1.3722$, and so, for instance, we get $\Pr[N > n] < 0.01$ for $n \geq 12$.

4 Single Processor

In the single processor case, the processor repeatedly selects a process from a pool of processes awaiting execution, executes it, and puts the generated processes (if any) back into the pool. The pool initially contains one process, and the next process to be executed is chosen by a *scheduler*.

We investigate the random variables modelling the time and space needed to completely execute the initial process for different classes of schedulers. Notice that in this case space is a resource, because the local states of the processes in the pool must be stored in memory. For simplicity we assume that storing a process takes a unit of memory. As before, we also assume that executing a process takes one time unit.

The completion time $T(\tau)$ of a family tree τ is given by the total number of executed processes, i.e., by the number of nodes of the tree. Observe that for a fixed tree the completion time is independent of the scheduler: the scheduler determines the order of execution of the processes, but not the number of processes that are executed (this is solely determined by the stochastic choices).

The *completion space* $S(\tau)$ is the maximal size of the pool during the execution of the family tree. The same family tree can be executed in different ways, depending on the order in which the scheduler selects the processes. Consider the family tree of Figure 1. A scheduler that executes all internal nodes before executing any leaves leads to a completion space of 5. Better schedulings can

[1] Strictly speaking, this only holds if $p_d > 0$ for some $d \geq 2$, but the systems that do not satisfy this condition do not exhibit process creation.

be achieved, for instance by completely executing the second child of the initial process, then the first, and then the third; this leads to completion space 3, the pool never contains more than 3 processes.

In the rest of the section we present some results on the distribution and moments of the random variables T and S.

4.1 Completion Time

The expected value and variance of T are easy to compute. We illustrate this with the expected value. In our example, the initial process generates two children with probability 0.2. The expected value conditioned to this first step is $1 + \mathbb{E}[T+T]$ (1 for the step just carried, plus the expected value of executing two children). We get for our example the linear equation

$$\mathbb{E}[T] = 0.6 + 0.1(1+\mathbb{E}[T]) + 0.2(1+2\mathbb{E}[T]) + 0.1(1+3\mathbb{E}[T]) = 1 + 0.8\mathbb{E}[T]$$

and so $\mathbb{E}[T] = 5$.

Theorem 4. *The expected value and variance of T can be computed by solving a linear equation.*

Notice, however, that the expected completion time can be infinite, even if the completion probability is 1. The standard example is the system with pgf $f(x) = 1/2x^2 + 1/2$.

The completion time T has also a natural interpretation in the model with an unbounded number of processors; it corresponds to the total work done by all processors during the execution. Probably for this reason it has also been investigated by branching process theorists. The following theorem about the distribution of T, due to Dwass, reduces the computation of $\Pr[T=j]$ to a rather standard combinatorial problem.

Theorem 5 ([4]). *If $p_0 > 0$ then*

$$\Pr[T=j] = \frac{1}{j} p_{j,j-1}$$

for every $j \geq 0$, where $p_{j,j-1}$ denotes the probability that a generation has $j-1$ processes under the condition that the parent generation has j processes.

Observe that if we are only interested in the value of T then the stochastic process of the system can be embedded in a Markov chain having the different numbers of processes as states. This shows that determining $\Pr[T=j]$ is similar to solving a Gambler's Ruin problem.

4.2 Completion Space

The completion space does not seem to have been studied in the literature. We report on ongoing work [3]. As in the competitive analysis of online algorithms, we have studied the performance of *online* schedulers, and compared it with the performance of an *optimal offline* scheduler. Intuitively, at every point in the computation an online scheduler only knows the part of the family tree containing the processes that have already been executed. On the contrary, the optimal offline scheduler knows in advance which family tree is going to be executed. Consider for instance a scenario in which processes are threads of a deterministic program, but we do not have access to the program code. The only information we have (obtained e.g. through sampling) is the pgf of the threads. While the future of each thread is completely determined by the code and the input values, with the information at our disposal we can only produce a stochastic model of the system, and we can only design online schedulers. Now, assume that we are granted access to the program code, and that by inspecting the code of the initial thread we are able to determine how many threads it and its descendants will generate. Using this information we can now schedule the threads optimally. Of course, in most applications the optimal offline scheduler cannot be implemented (in our scenario, if we allow arbitrary programs then we must solve the halting problem), or has prohibitive computational complexity. This fact does not raise any conceptual problem, because we are interested in the optimal scheduler as a reference against which to compare the performance of other schedulers.

Optimal offline schedulers. Our first results concern the distribution of the completion space S^{op} of an optimal offline scheduler *op* on a fixed but arbitrary system with $f(x)$ as pgf. The optimal offline scheduler is the one that assigns to each family tree the completion space of the space-optimal execution. With this scheduler, the completion space of the family tree of Figure 1 is 3.

Recall how in Theorem 2 we obtained a nice connection between the distribution of the completion time and the Kleene approximants to the least non-negative fixed point of the pgf. We have obtained a very surprising result: the same connection exists between the distribution of the completion space for the optimal scheduler and the *Newton approximants*. The Newton approximants of a pgf are the sequence of values obtained by applying Newton's iterative method for approximating a zero of a differentiable function to the function $f(x) - x$, with 0 as first approximant. Notice that a zero of $f(x) - x$ is always a fixed point of $f(x)$. More precisely, the sequence of Newton approximants is defined as follows

$$\nu^{(0)} = 0 \quad \text{and} \quad \nu^{(k+1)} = \nu^{(k)} + \frac{f(\nu^{(k)}) - \nu^{(k)}}{1 - f'(\nu^{(k)})}$$

where $f'(\nu^{(k)})$ denotes the derivative of f evaluated at $\nu^{(k)}$. We have:

Theorem 6. $\Pr[S^{op} \leq n] = \nu^{(n)}$ *for every* $n \in \mathbb{N}$.

This connection allows to efficiently compute values of the distribution of S^{op}. Moreover, applying recent results on the behaviour and convergence speed of Newton's method [6,9,5], it also leads to the following tail bounds:

Theorem 7. *If a system is subcritical (critical), there are real numbers* $c > 0$ *and* $0 < d < 1$ *such that* $\Pr[S^{op} \geq k] \leq c \cdot d^{2^k}$ *(* $\Pr[S^{op} \geq k] \leq c \cdot d^k$ *) for every* $k \in \mathbb{N}$.

Online schedulers. Recall that online schedulers only know the past of the computation, i.e., they ignore how future stochastic choices will be resolved. We have studied the distribution of S^σ for an arbitrary online scheduler σ. It turns out that the techniques of Theorem 3 can be adapted, and we obtain the following result[2]:

Theorem 8. *Let* $a > 1$ *be the greatest nonnegative fixed point of the pgf of a subcritical system (in a certain normal form). Then*

$$\Pr[S^\sigma \geq n] = \frac{a-1}{a^n - 1}$$

for every online scheduler σ *and for every* $n \geq 1$.

In particular, this theorem proves that all online schedulers have the same distribution. This is intuitively plausible, since online schedulers do not know which process in the pool will have how many descendants. However, this is a particularity of the one-type case. In a multitype system some types can be very prolific in average, while others may quickly disappear from the pool with high probability, and so not all online schedulers are equivalent. Intuitively, a good online scheduler gives priority to process types whose execution "requires less space". This notion, however, is difficult to make mathematically precise, because the space consumed by a process is precisely what we wish to compute! In [3] we generalize Theorem 8 to the multitype case, and the proof suggests a way of assigning weights to process types reflecting how "heavy" or "light" their memory consumption is.

Notice that Theorem 8 also proves a gap between any online and the optimal offline scheduler: $\lim_{n\to\infty}(\Pr[S^\sigma \geq n] / \Pr[S^{op} \geq n]) = \infty$ for every online scheduler σ. The gap also exists in the multitype case.

Depth-first schedulers. So far we have assumed that there are no dependencies between processes requiring a process to be executed before another. We now study a different case, especially important for multithreaded programming scenarios. We assume that when a process creates more than one child, one of the children is in fact the continuation of the parent process, while the other children are spawned processes. We consider the case in which a process can terminate only after all the processes it spawns have terminated. Papers studying space-efficient scheduling of multithreaded programs usually call these computations *strict* (see e.g. [2]). The optimal scheduler of a strict computation in which a thread creates new threads *one at a time* (i.e., a step of the parent process

[2] See [3] for a precise formulation.

can create at most one child process) is easy to determine: it proceeds "depth-first", completely executing the child process before proceeding to execute the continuation of its parent. This results in the familiar stack-based execution.

We have been able to determine the exact asymptotic performance of the depth-first scheduler:

Theorem 9. *For any subcritical, one-at-a-time system and for the depth-first scheduler σ there is $0 < \rho < 1$ such that $\Pr[S^\sigma \leq n] \in \Theta(\rho^n)$ for every $n \geq 1$. Moreover, ρ can be efficiently approximated (see [3] for a precise formulation).*

Notice that in the one-type case all online schedulers are equivalent, and since the depth-first scheduler is online, Theorem 9 is directly implied by Theorem 8. In the multitype case, however, Theorem 8 does not hold, only a weaker version can be shown providing lower and upper bounds for the probability $\Pr[S^\sigma \geq n]$ of an arbitrary online scheduler σ. In this case Theorem 9 does provide additional information, it determines the exact asymptotic performance of the depth-first scheduler.

Expected completion space. We have also investigated the *expected* completion space. Using results on the convergence speed of Newton's method we can show that an optimal offline scheduler *always* has finite expected completion space, even for critical systems. Further, we have proved that, while in a subcritical system every online scheduler has finite expected completion space (an easy result), in a critical system every online scheduler has infinite expected completion space (this is harder). Combining these results we obtain the following interesting dichotomy for critical systems: all optimal offline schedulers have finite expected completion space, but all online schedulers have infinite expected completion space. So a finite expected value can only be obtained by schedulers that, loosely speaking, have access to the code.

5 Conclusions

We have analyzed the behaviour of systems with stochastic process creation. We have shown that the distinction made in computer science between processes and processors introduces new models that have not been studied so far. We have surveyed some basic results about the models that have been studied by mathematicians as part of the theory of stochastic branching processes, and we have reported on our ongoing work on the new models.

The potential applications of our results are in the design of hardware and software systems. Using them we can determine the memory size a system must have in order to accommodate a computation with only a given probability of a memory overflow. However, in order to make the results useful we must consider models in which different process types require different storage and processes are executed by a fixed number of processors. This is where we are currently placing our efforts.

Acknowledgements. As mentioned in the introduction, the new theorems presented in this paper are the result of joint work with Tomáš Brázdil, Stefan Kiefer, and Michael Luttenberger. The proofs can be found in [3], a technical report available online. I am very grateful to Tomáš, Stefan, and Michael for many suggestions and remarks, and for correcting several mistakes. Any remaining errors are my sole responsibility.

References

1. Athreya, K.B., Ney, P.E.: Branching Processes. Springer, Heidelberg (1972)
2. Blumofe, R.D., Leiserson, C.E.: Scheduling multithreaded computations by work stealing. Journal of the ACM 46(5), 720–748 (1999)
3. Brázdil, T., Esparza, J., Kiefer, S., Luttenberger, M.: Space-efficient scheduling of stochastically generated tasks. Technical report, Technische Universität München, Institut für Informatik (April 2009)
4. Dwass, M.: The total progeny in a branching process and a related random walk. Journal of Applied Probability 6, 682–686 (1969)
5. Esparza, J., Kiefer, S., Luttenberger, M.: Convergence thresholds of Newton's method for monotone polynomial equations. In: STACS 2008, pp. 289–300 (2008)
6. Etessami, K., Yannakakis, M.: Recursive markov chains, stochastic grammars, and monotone systems of nonlinear equations. Journal of the ACM 56(1), 1–66 (2009)
7. Harris, T.E.: The Theory of Branching Processes. Springer, Heidelberg (1963)
8. Johnson, N.L., Kotz, S.: Urn Models and Their Application. John Wiley & Sons, Chichester (1977)
9. Kiefer, S., Luttenberger, M., Esparza, J.: On the convergence of Newton's method for monotone systems of polynomial equations. In: STOC 2007, pp. 217–226. ACM, New York (2007)
10. Lindvall, T.: On the maximum of a branching process. Scandinavian Journal of Statistics 3, 209–214 (1976)
11. Mahmoud, H.: Pólya Urn Models. CRC Press, Boca Raton (2008)
12. Nerman, O.: On the maximal generation size of a non-critical galton-watson process. Scandinavian Journal of Statistics 4(3), 131–135 (1977)

Stochastic Games with Finitary Objectives*

Krishnendu Chatterjee[1], Thomas A. Henzinger[2], and Florian Horn[3]

[1] Institute of Science and Technology (IST), Austria
[2] EPFL, Switzerland
[3] CWI, Amsterdam

Abstract. The synthesis of a reactive system with respect to an ω-regular specification requires the solution of a graph game. Such games have been extended in two natural ways. First, a game graph can be equipped with probabilistic choices between alternative transitions, thus allowing the modeling of uncertain behavior. These are called stochastic games. Second, a liveness specification can be strengthened to require satisfaction within an unknown but bounded amount of time. These are called finitary objectives. We study, for the first time, the combination of stochastic games and finitary objectives. We characterize the requirements on optimal strategies and provide algorithms for computing the maximal achievable probability of winning stochastic games with finitary parity or Streett objectives. Most notably, the set of states from which a player can win with probability 1 for a finitary parity objective can be computed in polynomial time, even though no polynomial-time algorithm is known in the nonfinitary case.

1 Introduction

The safety and liveness of reactive systems are usually specified by ω-regular sets of infinite words. Then the *reactive synthesis problem* asks for constructing a winning strategy in a graph game with two players and ω-regular objectives: a player that represents the system and tries to satisfy the specification; and a player that represents the environment and tries to violate the specification. In the presence of uncertain or probabilistic behavior, the graph game is stochastic. Such a *stochastic game* is played on a graph with three kinds of vertices: in player-1 vertices, the first player chooses a successor vertex; in player-2 vertices, the second player chooses a successor vertex; and in probabilistic vertices, a successor vertex is chosen according to a given probability distribution. The result of playing the game ad infinitum is a random walk through the graph. If player 1 has an ω-regular objective ϕ, then she tries to maximize the probability that the infinite path that results from the random walk lies inside the set ϕ. Conversely, player 2 tries to minimize that probability. Since the stochastic games are Borel determined [15], and the ω-regular languages are Borel sets, these games have

* This research was supported in part by the Swiss National Science Foundation under the Indo-Swiss Joint Research Programme, by the European Network of Excellence on Embedded Systems Design (ArtistDesign), and by the European project Combest.

R. Královič and D. Niwiński (Eds.): MFCS 2009, LNCS 5734, pp. 34–54, 2009.

a unique value, i.e., there is a real $v \in [0,1]$ such that player 1 can ensure ϕ with probability arbitrarily close to v, and at the same time, player 2 can ensure $\neg\phi$ with probability arbitrarily close to $1-v$. The computation of v is referred to as the *quantitative value problem* for stochastic games; the decision problem of whether $v = 1$ is referred to as the *qualitative value problem.* In the case of parity objectives, both value problems lie in NP $\cap$ coNP [6], but no polynomial-time solutions are known even if there are no probabilistic vertices. The NP $\cap$ coNP characterization results from the existence of pure (i.e., nonrandomized) positional (i.e., memoryless) optimal strategies for both players. In the case of Streett objectives, optimal player-1 strategies may require memory, and both value problems are coNP-complete [3], which is again the same in the absence of probabilistic vertices.

The specification of liveness for a reactive system by ω-regular sets such as parity or Streett languages has the drawback that, while the synthesized system is guaranteed to be live, we cannot put any bound on its liveness behavior. For example, the liveness objective $\Box(r \rightarrow \Diamond q)$ ensures that every request r issued by the environment is eventually followed by a response q of the synthesized system, but the delay between each request and corresponding response may grow without bound from one request to the next. This is an undesirable behavior, especially in synthesis, where one controls the system to be built and where one would like stronger guarantees. At the same time, it may be impossible to put a fixed bound on the desired response time, because the achievable bound usually is not known. For this reason, the time-scale independent notion of *finitary objectives* was introduced [1]. The finitary version of the liveness objective $\Box(r \rightarrow \Diamond q)$ requires that there exists an unknown bound b such that every request r is followed by a response q within b steps. The synthesized system can have any response time, but its response time will not grow from one request to the next without bound. Finitary versions can be defined for both parity and Streett (strong fairness) objectives [4]. It should be noted that finitary objectives are not ω-regular. While in games with ω-regular objectives, both players have finite-memory strategies, to violate a finitary objective, player 2 may require infinite memory even if there are no probabilistic vertices [4]. Nonetheless, finitary objectives are Borel sets, and thus have well-defined values in stochastic games.

Nonstochastic games with finitary parity and Streett objectives were first studied in [4], and the results of [4] were later significantly improved upon by [12]. This work showed that finitary objectives are not only more desirable for synthesis, but also can be far less costly than their infinitary counterparts. In particular, nonstochastic games with finitary parity objectives can be solved in polynomial time. In the present paper, we study for the first time *stochastic* games with *finitary* objectives. As main results, we show that the qualitative value problem for finitary parity objectives remains polynomial in the stochastic case, and the quantitative value problem can be solved in NP $\cap$ coNP. For stochastic games with finitary Streett objectives, we compute values in exponential time. Yet also here we achieve a significant improvement by solving the qualitative value problem with an exponential term of 2^d (where d is the number of Streett pairs) instead of $n^d \cdot d!$ (where n is the number of vertices), which characterizes the best known algorithm for nonstochastic games with infinitary Streett objectives. Our results follow the pattern of extending properties of stochastic games with infinitary parity and Streett objectives to stochastic games with finitary parity and Streett objectives. However, in the finitary case, the proof techniques are more complicated, because we need to consider infinite-memory strategies.

We now summarize our results in more detail and draw precise comparisons with the two simpler cases of (i) stochastic games with infinitary (rather than finitary) objectives and (ii) nonstochastic (rather than stochastic) games with finitary objectives.

Comparison of finitary and infinitary parity objectives. In case of parity objectives, pure memoryless optimal strategies exist for both players in both nonstochastic (2-player) game graphs [9] and stochastic ($2^1/_2$-player) game graphs [6,17]. For finitary parity objectives on 2-player game graphs, a pure memoryless optimal strategy exists for the player with the finitary parity objective, while the optimal strategy of the other player (with the complementary objective) in general requires infinite memory [5]. We show in this work that the same class of strategies that suffices in 2-player game graphs also suffices for optimality in $2^1/_2$-player game graphs for finitary parity objectives and their complements. The best known complexity bound for 2- and $2^1/_2$-player games with parity objectives is NP $\cap$ coNP [9,6]. In case of $2^1/_2$-player games, the best known complexity bound for the qualitative analysis is also NP $\cap$ coNP. The solution of 2-player game graphs with finitary parity objectives can be achieved in polynomial time (in $O(n^2 \cdot m)$ time [12,5] for game graphs with n states and m edges). In this work we show that the quantitative analysis of $2^1/_2$-player game graphs with finitary parity objectives lies in NP $\cap$ coNP, and the qualitative analysis can be done in $O(n^4 \cdot m)$ time. To obtain a polynomial time solution for the quantitative analysis of $2^1/_2$-player game graphs with finitary parity objectives, one must obtain a polynomial-time solution for the quantitative analysis of $2^1/_2$-player game graphs with Büchi objectives (which is a major open problem).

Comparison of finitary and infinitary Streett objectives. In case of Streett objectives with d pairs, strategies with $d!$ memory is necessary and sufficient for both 2-player game graphs and $2^1/_2$-player game graphs, and for the complementary player pure memoryless optimal strategies exist [8,11,3]. For finitary Streett objectives on 2-player game graphs, an optimal strategy with $d \cdot 2^d$ memory exists for the player with the finitary Streett objective, while the optimal strategy of the other player (with the complementary objective) in general requires infinite memory [5]. We show that the same class of strategies that suffices for 2-player game graphs also suffices for optimality in $2^1/_2$-player game graphs for finitary Streett objectives and their complements. The decision problems for 2- and $2^1/_2$-player games with Streett objectives are coNP-complete. The solution of 2-player game graphs with finitary Streett objectives can be achieved in EXPTIME. In this work we show that both the qualitative and quantitative analysis of $2^1/_2$-player game graphs with finitary Streett objectives can be achieved in EXPTIME. The best known algorithm for 2-player game graphs with Streett objectives is $O(n^d \cdot d!)$ [16], where as in case of $2^1\!/_2$-player game graphs with finitary Streett objectives, we show that the qualitative analysis can be achieved in time $O(n^4 \cdot m \cdot d \cdot 2^d)$. For the quantitative analysis, we present our results for the more general class of *tail* (i.e., prefix-independent) objectives, and obtain the results for finitary parity and Streett objectives as a special case.

2 Definitions

We consider several classes of turn-based games, namely, two-player turn-based probabilistic games ($2^1\!/_2$-player games), two-player turn-based deterministic games (2-player games), and Markov decision processes ($1^1\!/_2$-player games).

Notation. For a finite set A, a *probability distribution* on A is a function $\delta\colon A \to [0,1]$ such that $\sum_{a \in A} \delta(a) = 1$. We denote the set of probability distributions on A by $\mathcal{D}(A)$. Given a distribution $\delta \in \mathcal{D}(A)$, we denote by $\mathrm{Supp}(\delta) = \{x \in A \mid \delta(x) > 0\}$ the *support* of δ.

Game graphs. A *turn-based probabilistic game graph* (*$2^1\!/_2$-player game graph*) $G = ((S,E),(S_1,S_2,S_\bigcirc),\delta)$ consists of a directed graph (S,E), a partition (S_1, S_2, $S_\bigcirc$) of the finite set S of states, and a probabilistic transition function δ: $S_\bigcirc \to \mathcal{D}(S)$, where $\mathcal{D}(S)$ denotes the set of probability distributions over the state space S. The states in S_1 are the *player*-1 states, where player 1 decides the successor state; the states in S_2 are the *player*-2 states, where player 2 decides the successor state; and the states in $S_\bigcirc$ are the *probabilistic* states, where the successor state is chosen according to the probabilistic transition function δ. We assume that for $s \in S_\bigcirc$ and $t \in S$, we have $(s,t) \in E$ iff $\delta(s)(t) > 0$, and we often write $\delta(s,t)$ for $\delta(s)(t)$. For technical convenience we assume that every state in the graph (S,E) has at least one outgoing edge. For a state $s \in S$, we write $E(s)$ to denote the set $\{t \in S \mid (s,t) \in E\}$ of possible successors. The size of a game graph $G = ((S,E),(S_1,S_2,S_\bigcirc),\delta)$ is

$$|G| = |S| + |E| + \sum_{t \in S} \sum_{s \in S_\bigcirc} |\delta(s)(t)|;$$

where $|\delta(s)(t)|$ denotes the space to represent the transition probability $\delta(s)(t)$ in binary.

A set $U \subseteq S$ of states is called *δ-closed* if for every probabilistic state $u \in U \cap S_\bigcirc$, if $(u,t) \in E$, then $t \in U$. The set U is called *δ-live* if for every nonprobabilistic state $s \in U \cap (S_1 \cup S_2)$, there is a state $t \in U$ such that $(s,t) \in E$. A δ-closed and δ-live subset U of S induces a *subgame graph* of G, indicated by $G \upharpoonright U$.

The *turn-based deterministic game graphs* (*2-player game graphs*) are the special case of the $2^1\!/_2$-player game graphs with $S_\bigcirc = \emptyset$. The *Markov decision processes* (*$1^1\!/_2$-player game graphs*) are the special case of the $2^1\!/_2$-player game graphs with $S_1 = \emptyset$ or $S_2 = \emptyset$. We refer to the MDPs with $S_2 = \emptyset$ as *player*-1 *MDPs*, and to the MDPs with $S_1 = \emptyset$ as *player*-2 *MDPs*.

Plays and strategies. An infinite path, or *play*, of the game graph G is an infinite sequence $\omega = \langle s_0, s_1, s_2, \ldots \rangle$ of states such that $(s_k, s_{k+1}) \in E$ for all $k \in \mathbb{N}$. We write Ω for the set of all plays, and for a state $s \in S$, we write $\Omega_s \subseteq \Omega$ for the set of plays that start from the state s.

A *strategy* for player 1 is a function $\sigma\colon S^* \cdot S_1 \to \mathcal{D}(S)$ that assigns a probability distribution to all finite sequences $\boldsymbol{w} \in S^* \cdot S_1$ of states ending in a player-1 state (the sequence represents a prefix of a play). Player 1 follows the strategy σ if in each player-1 move, given that the current history of the game is $\boldsymbol{w} \in S^* \cdot S_1$, she chooses the next state according to the probability distribution $\sigma(\boldsymbol{w})$. A strategy must prescribe only available moves, i.e., for all $\boldsymbol{w} \in S^*$, and $s \in S_1$ we have $\mathrm{Supp}(\sigma(\boldsymbol{w} \cdot s)) \subseteq E(s)$. The strategies for player 2 are defined analogously. We denote by Σ and Π the set of all strategies for player 1 and player 2, respectively.

Once a starting state $s \in S$ and strategies $\sigma \in \Sigma$ and $\pi \in \Pi$ for the two players are fixed, the outcome of the game is a random walk $\omega_s^{\sigma,\pi}$ for which the probabilities of events are uniquely defined, where an *event* $\mathcal{A} \subseteq \Omega$ is a measurable set of paths. Given strategies σ for player 1 and π for player 2, a play $\omega = \langle s_0, s_1, s_2, \ldots \rangle$ is *feasible* if for every $k \in \mathbb{N}$ the following three conditions hold: (1) if $s_k \in S_{\bigcirc}$, then $(s_k, s_{k+1}) \in E$; (2) if $s_k \in S_1$, then $\sigma(s_0, s_1, \ldots, s_k)(s_{k+1}) > 0$; and (3) if $s_k \in S_2$ then $\pi(s_0, s_1, \ldots, s_k)(s_{k+1}) > 0$. Given two strategies $\sigma \in \Sigma$ and $\pi \in \Pi$, and a state $s \in S$, we denote by $\mathrm{Outcome}(s, \sigma, \pi) \subseteq \Omega_s$ the set of feasible plays that start from s given strategies σ and π. For a state $s \in S$ and an event $\mathcal{A} \subseteq \Omega$, we write $\mathrm{Pr}_s^{\sigma,\pi}(\mathcal{A})$ for the probability that a path belongs to $\mathcal{A}$ if the game starts from the state s and the players follow the strategies σ and π, respectively. In the context of player-1 MDPs we often omit the argument π, because Π is a singleton set.

We classify strategies according to their use of randomization and memory. The strategies that do not use randomization are called pure. A player-1 strategy σ is *pure* if for all $\boldsymbol{w} \in S^*$ and $s \in S_1$, there is a state $t \in S$ such that $\sigma(\boldsymbol{w} \cdot s)(t) = 1$. We denote by $\Sigma^P \subseteq \Sigma$ the set of pure strategies for player 1. A strategy that is not necessarily pure is called *randomized*. Let $\mathtt{M}$ be a set called *memory*, that is, $\mathtt{M}$ is a set of memory elements. A player-1 strategy σ can be described as a pair of functions $\sigma = (\sigma_u, \sigma_m)$: a *memory-update* function σ_u: $S \times \mathtt{M} \to \mathtt{M}$ and a *next-move* function σ_m: $S_1 \times \mathtt{M} \to \mathcal{D}(S)$. We can think of strategies with memory as input/output automaton computing the strategies (see [8] for details). A strategy $\sigma = (\sigma_u, \sigma_m)$ is *finite-memory* if the memory $\mathtt{M}$ is finite, and then the size of the strategy σ, denoted as $|\sigma|$, is the size of its memory $\mathtt{M}$, i.e., $|\sigma| = |\mathtt{M}|$. We denote by Σ^F the set of finite-memory strategies for player 1, and by Σ^{PF} the set of *pure finite-memory* strategies; that is, $\Sigma^{PF} = \Sigma^P \cap \Sigma^F$. The strategy (σ_u, σ_m) is *memoryless* if $|\mathtt{M}| = 1$; that is, the next move does not depend on the history of the play but only on the current state. A memoryless player-1 strategy can be represented as a function $\sigma\colon S_1 \to \mathcal{D}(S)$. A *pure memoryless strategy* is a pure strategy that is memoryless. A pure memoryless strategy for player 1 can be represented as a function $\sigma\colon S_1 \to S$. We denote by Σ^M the set of memoryless strategies for player 1, and by Σ^{PM} the set of pure memoryless strategies; that is, $\Sigma^{PM} = \Sigma^P \cap \Sigma^M$. Analogously we define the corresponding strategy families Π^P, Π^F, Π^{PF}, Π^M, and Π^{PM} for player 2.

Counting strategies. We call an infinite memory strategy σ *finite-memory counting* if there is a finite-memory strategy σ' such that for all $j \geq 0$ there exists $k \leq j$ such that the following condition hold: for all $w \in S^*$ such that $|w| = j$ and for all $s \in S_1$ we have $\sigma(w \cdot s) = \sigma'(\text{suffix}(w, k) \cdot s)$, where for $w \in S^*$ of length j and $k \leq j$ we denote by suffix(w, k) the suffix of w of length k. In other words, the strategy σ repeatedly plays the finite-memory strategy σ' in different segments of the play and the switch of the strategy in different segments only depends on the length of the play. We denote by nocount$(|\sigma|)$ the size of the memory of the finite-memory strategy σ' (the memory that is used not for counting), i.e., nocount$(|\sigma|) = |\sigma'|$. We use similar notations for player 2 strategies.

Objectives. An *objective* for a player consists of a Borel set of *winning plays* $\Phi \subseteq \Omega$. In this paper we consider ω-regular objectives, and *finitary parity* and *finitary Streett* objectives (all the objectives we consider in this paper are Borel objectives).

Classical ω-regular objectives. We first present the definitions of various canonical forms of ω-regular objectives and sub-classes of ω-regular objectives. For a play $\omega = \langle s_0, s_1, s_2, \ldots \rangle$, let $\text{Inf}(\omega)$ be the set $\{s \in S \mid s = s_k \text{ for infinitely many } k \geq 0\}$ of states that appear infinitely often in ω.

1. *Reachability and safety objectives.* Given a set $F \subseteq S$ of states, the reachability objective Reach(F) requires that some state in F be visited, and dually, the safety objective Safe(F) requires that only states in F be visited. Formally, the sets of winning plays are $\text{Reach}(F) = \{\langle s_0, s_1, s_2, \ldots \rangle \in \Omega \mid \exists k \geq 0.\ s_k \in F\}$ and $\text{Safe}(F) = \{\langle s_0, s_1, s_2, \ldots \rangle \in \Omega \mid \forall k \geq 0.\ s_k \in F\}$.
2. *Büchi and co-Büchi objectives.* Given a set $F \subseteq S$ of states, the Büchi objective Buchi(F) requires that some state in F be visited infinitely often, and dually, the co-Büchi objective coBuchi(F) requires that only states in F be visited infinitely often. Thus, the sets of winning plays are $\text{Buchi}(F) = \{\omega \in \Omega \mid \text{Inf}(\omega) \cap F \neq \emptyset\}$ and $\text{coBuchi}(F) = \{\omega \in \Omega \mid \text{Inf}(\omega) \subseteq F\}$.
3. *Rabin and Streett objectives.* Given a set $P = \{(E_1, F_1), \ldots, (E_d, F_d)\}$ of pairs of sets of states (i.e, for all $1 \leq j \leq d$, both $E_j \subseteq S$ and $F_j \subseteq S$), the Rabin objective Rabin(P) requires that for some pair $1 \leq j \leq d$, all states in E_j be visited finitely often, and some state in F_j be visited infinitely often. Hence, the winning plays are $\text{Rabin}(P) = \{\omega \in \Omega \mid \exists 1 \leq j \leq d.\ (\text{Inf}(\omega) \cap E_j = \emptyset \text{ and } \text{Inf}(\omega) \cap F_j \neq \emptyset)\}$. Dually, given $P = \{(E_1, F_1), \ldots, (E_d, F_d)\}$, the Streett objective Streett(P) requires that for all pairs $1 \leq j \leq d$, if some state in F_j is visited infinitely often, then some state in E_j be visited infinitely often, i.e., $\text{Streett}(P) = \{\omega \in \Omega \mid \forall 1 \leq j \leq d.\ (\text{Inf}(\omega) \cap E_j \neq \emptyset \text{ or } \text{Inf}(\omega) \cap F_j = \emptyset)\}$.
4. *Parity objectives.* Given a function $p \colon S \to \{0, 1, 2, \ldots, d-1\}$ that maps every state to an integer *priority*, the parity objective Parity(p) requires that of the states that are visited infinitely often, the least priority be even. Formally,

the set of winning plays is $\text{Parity}(p) = \{\omega \in \Omega \mid \min\{p(\text{Inf}(\omega))\} \text{ is even}\}$. The dual, co-parity objective has the set $\text{coParity}(p) = \{\omega \in \Omega \mid \min\{p(\text{Inf}(\omega))\} \text{ is odd}\}$ of winning plays. Parity objectives are closed under complementation: given a function $p : S \to \{0, 1, \ldots, d-1\}$, consider the function $p+1 : S \to \{1, 2, \ldots, d\}$ defined as $p+1(s) = p(s)+1$, for all $s \in S$, and then we have $\text{Parity}(p+1) = \text{coParity}(p)$.

Every parity objective is both a Rabin objective and a Streett objective. The Büchi and co-Büchi objectives are special cases of parity objectives with two priorities, namely, p: $S \to \{0, 1\}$ for Büchi objectives with $F = p^{-1}(0)$, and p: $S \to \{1, 2\}$ for co-Büchi objectives with $F = p^{-1}(2)$. The reachability and safety objectives can be turned into Büchi and co-Büchi objectives, respectively, on slightly modified game graphs.

Finitary objectives. We now define a stronger notion of winning, namely, *finitary winning*, in games with parity and Streett objectives.

Finitary winning for parity objectives. For parity objectives, the finitary winning notion requires that for each visit to an odd priority that is visited infinitely often, the distance to a stronger (i.e., lower) even priority be bounded. To define the winning plays formally, we need the concept of a distance sequence.

Distance sequences for parity objectives. Given a play $\omega = \langle s_0, s_1, s_2, \ldots \rangle$ and a priority function p: $S \to \{0, 1, \ldots, d-1\}$, we define a sequence of distances $dist_k(\omega, p)$, for all $k \geq 0$, as follows:

$$dist_k(\omega, p) = \begin{cases} 0 & \text{if } p(s_k) \text{ is even;} \\ \inf\{k' \geq k \mid p(s_{k'}) \text{ is even and } p(s_{k'}) < p(s_k)\} & \text{if } p(s_k) \text{ is odd.} \end{cases}$$

Intuitively, the distance for a position k in a play with an odd priority at position k, denotes the shortest distance to a stronger even priority in the play. We assume the standard convention that the infimum of the empty set is ∞.

Finitary parity objectives. The finitary parity objective $\text{finParity}(p)$ for a priority function p requires that the sequence of distances for the positions with odd priorities that occur infinitely often be bounded. This is equivalent to requiring that the sequence of all distances be bounded in the limit, and captures the notion that the "good" (even) priorities that appear infinitely often do not appear infinitely rarely. Formally, the sets of winning plays for the finitary parity objective and its complement are $\text{finParity}(p) = \{\omega \in \Omega \mid \limsup_{k \to \infty} dist_k(\omega, p) < \infty\}$ and $\text{cofinParity}(p) = \{\omega \in \Omega \mid \limsup_{k \to \infty} dist_k(\omega, p) = \infty\}$, respectively. Observe that if a play ω is winning for a co-parity objective, then the lim sup of the distance sequence for ω is ∞, that is, $\text{coParity}(p) \subseteq \text{cofinParity}(p)$. However, if a play ω is winning for a (classical) parity objective, then the lim sup of the distance sequence for ω can be ∞ (as shown in Example 1), that is, $\text{finParity}(p) \subsetneq \text{Parity}(p)$.

Example 1. Consider the game shown in Figure 1. The square-shaped states are player 1 states, where player 1 chooses the successor state, and the diamond-shaped states are player 2 states (we will follow this convention throughout this

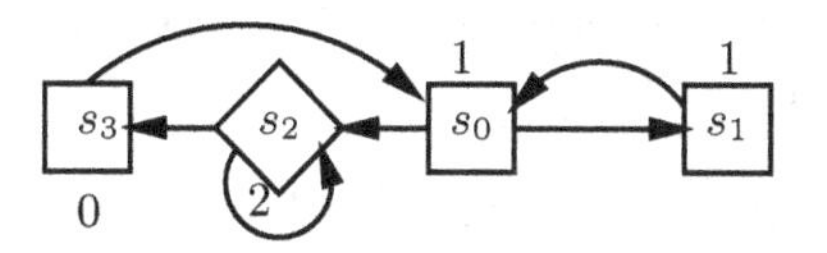

Fig. 1. A simple game graph

paper). The priorities of states are shown next to each state in the figure. If player 1 follows a memoryless strategy σ that chooses the successor s_2 at state s_0, this ensures that against all strategies π for player 2, the minimum priority of the states that are visited infinitely often is even (either state s_3 is visited infinitely often, or both states s_0 and s_1 are visited finitely often). However, consider the strategy π_w for player 2: the strategy π_w is played in rounds, and in round $k \geq 0$, whenever player 1 chooses the successor s_2 at state s_0, player 2 stays in state s_2 for k transitions, and then goes to state s_3 and proceeds to round $k+1$. The strategy π_w ensures that for all strategies σ for player 1, either the minimum priority visited infinitely often is 1 (i.e., both states s_0 and s_1 are visited infinitely often and state s_3 is visited finitely often); or states of priority 1 are visited infinitely often, and the distances between visits to states of priority 1 and subsequent visits to states of priority 0 increase without bound (i.e., the limit of the distances is ∞). Hence it follows that in this game, although player 1 can win for the parity objective, she cannot win for the finitary parity objective. ∎

Finitary winning for Streett objectives. The notion of distance sequence for parity objectives has a natural extension to Streett objectives.

Distance sequences for Streett objectives. Given a play $\omega = \langle s_0, s_1, s_2, \ldots \rangle$ and a set $P = \{(E_1, F_1), \ldots, (E_d, F_d)\}$ of Streett pairs of state sets, the d sequences of distances $dist_k^j(\omega, P)$, for all $k \geq 0$ and $1 \leq j \leq d$, are defined as follows:

$$dist_k^j(\omega, P) = \begin{cases} 0 & \text{if } s_k \notin F_j; \\ \inf\{k' \geq k \mid s_{k'} \in E_j\} & \text{if } s_k \in F_j. \end{cases}$$

Let $dist_k(\omega, P) = \max\{dist_k^j(\omega, P) \mid 1 \leq j \leq d\}$ for all $k \geq 0$.

Finitary Streett objectives. The finitary Streett objective $\text{finStreett}(P)$ for a set P of Streett pairs requires that the distance sequence be bounded in the limit, i.e., the winning plays are $\text{finStreett}(P) = \{\omega \in \Omega \mid \limsup_{k\to\infty} dist_k(\omega, P) < \infty\}$. We use the following notations for the complementary objective: $\text{cofinStreett}(P) = \Omega \setminus \text{finStreett}(P)$.

Tail objectives. An objective Φ is a *tail* objective if the objective is independent of finite prefixes. Formally, an objective Φ is a tail objective if for all $\omega \in \Omega$, we have $\omega \in \Phi$ iff for all ω' obtained by adding or deleting a finite prefix with ω we have $\omega' \in \Phi$ (see [2] for details). The parity, Streett, finitary parity and finitary Streett are independent of finite prefixes and are all tail objectives. Since tail objectives are closed under complementation, it follows that the complementary objectives to finitary parity and Streett are tail objectives as well.

Sure, almost-sure, positive winning, and optimality. Given a player-1 objective Φ, a strategy $\sigma \in \Sigma$ is *sure winning* for player 1 from a state $s \in S$ if for every strategy $\pi \in \Pi$ for player 2, we have $\mathrm{Outcome}(s, \sigma, \pi) \subseteq \Phi$. A strategy σ is *almost-sure winning* for player 1 from the state s for the objective Φ if for every player-2 strategy π, we have $\mathrm{Pr}_s^{\sigma,\pi}(\Phi) = 1$. A strategy σ is *positive winning* for player 1 from the state s for the objective Φ if for every player-2 strategy π, we have $\mathrm{Pr}_s^{\sigma,\pi}(\Phi) > 0$. The sure, almost-sure and positive winning strategies for player 2 are defined analogously. Given an objective Φ, the *sure winning set* $\langle\langle 1 \rangle\rangle_{sure}(\Phi)$ for player 1 is the set of states from which player 1 has a sure winning strategy. Similarly, the *almost-sure winning set* $\langle\langle 1 \rangle\rangle_{almost}(\Phi)$ and the *positive winning set* $\langle\langle 1 \rangle\rangle_{pos}(\Phi)$ for player 1 is the set of states from which player 1 has an almost-sure winning and a positive winning strategy, respectively. The sure winning set $\langle\langle 2 \rangle\rangle_{sure}(\Omega \setminus \Phi)$, the almost-sure winning set $\langle\langle 2 \rangle\rangle_{almost}(\Omega \setminus \Phi)$, and the positive winning set $\langle\langle 2 \rangle\rangle_{pos}(\Omega \setminus \Phi)$ for player 2 are defined analogously. It follows from the definitions that for all $2^1/_2$-player game graphs and all objectives Φ, we have $\langle\langle 1 \rangle\rangle_{sure}(\Phi) \subseteq \langle\langle 1 \rangle\rangle_{almost}(\Phi) \subseteq \langle\langle 1 \rangle\rangle_{pos}(\Phi)$. Computing sure, almost-sure and positive winning sets and strategies is referred to as the *qualitative* analysis of $2^1/_2$-player games [7].

Given objectives $\Phi \subseteq \Omega$ for player 1 and $\Omega \setminus \Phi$ for player 2, we define the *value* functions $\langle\langle 1 \rangle\rangle_{val}$ and $\langle\langle 2 \rangle\rangle_{val}$ for the players 1 and 2, respectively, as the following functions from the state space S to the interval $[0, 1]$ of reals: for all states $s \in S$, let $\langle\langle 1 \rangle\rangle_{val}(\Phi)(s) = \sup_{\sigma \in \Sigma} \inf_{\pi \in \Pi} \mathrm{Pr}_s^{\sigma,\pi}(\Phi)$ and $\langle\langle 2 \rangle\rangle_{val}(\Omega \setminus \Phi)(s) = \sup_{\pi \in \Pi} \inf_{\sigma \in \Sigma} \mathrm{Pr}_s^{\sigma,\pi}(\Omega \setminus \Phi)$. In other words, the value $\langle\langle 1 \rangle\rangle_{val}(\Phi)(s)$ gives the maximal probability with which player 1 can achieve her objective Φ from state s, and analogously for player 2. The strategies that achieve the value are called optimal: a strategy σ for player 1 is *optimal* from the state s for the objective Φ if $\langle\langle 1 \rangle\rangle_{val}(\Phi)(s) = \inf_{\pi \in \Pi} \mathrm{Pr}_s^{\sigma,\pi}(\Phi)$. The optimal strategies for player 2 are defined analogously. Computing values and optimal strategies is referred to as the *quantitative* analysis of $2^1/_2$-player games. The set of states with value 1 is called the *limit-sure winning set* [7]. For $2^1/_2$-player game graphs with ω-regular objectives the almost-sure and limit-sure winning sets coincide [3].

Let $C \in \{P, M, F, PM, PF\}$ and consider the family $\Sigma^C \subseteq \Sigma$ of special strategies for player 1. We say that the family Σ^C *suffices* with respect to a player-1 objective Φ on a class $\mathcal{G}$ of game graphs for *sure winning* if for every game graph $G \in \mathcal{G}$ and state $s \in \langle\langle 1 \rangle\rangle_{sure}(\Phi)$, there is a player-1 strategy $\sigma \in \Sigma^C$ such that for every player-2 strategy $\pi \in \Pi$, we have $\mathrm{Outcome}(s, \sigma, \pi) \subseteq \Phi$. Similarly, the family Σ^C *suffices* with respect to the objective Φ on the class $\mathcal{G}$ of game graphs for (a) *almost-sure winning* if for every game graph $G \in \mathcal{G}$ and state $s \in \langle\langle 1 \rangle\rangle_{almost}(\Phi)$, there is a player-1 strategy $\sigma \in \Sigma^C$ such that for every player-2 strategy $\pi \in \Pi$, we have $\mathrm{Pr}_s^{\sigma,\pi}(\Phi) = 1$; (b) *positive winning* if for every game graph $G \in \mathcal{G}$ and state $s \in \langle\langle 1 \rangle\rangle_{pos}(\Phi)$, there is a player-1 strategy $\sigma \in \Sigma^C$ such that for every player-2 strategy $\pi \in \Pi$, we have $\mathrm{Pr}_s^{\sigma,\pi}(\Phi) > 0$; and (c) *optimality* if for every game graph $G \in \mathcal{G}$ and state $s \in S$, there is a player-1 strategy

$\sigma \in \Sigma^C$ such that $\langle\langle 1\rangle\rangle_{val}(\Phi)(s) = \inf_{\pi\in\Pi} \Pr_s^{\sigma,\pi}(\Phi)$. The notion of sufficiency for size of finite-memory strategies is obtained by referring to the size of the memory M of the strategies. The notions of sufficiency of strategies for player 2 is defined analogously.

Determinacy. For sure winning, the $1^1\!/_2$-player and $2^1\!/_2$-player games coincide with 2-player (deterministic) games where the random player (who chooses the successor at the probabilistic states) is interpreted as an adversary, i.e., as player 2. We present the result formally as a Lemma. We use the following notation: given a $2^1\!/_2$-player game graph $G = ((S,E),(S_1,S_2,S_{\bigcirc}),\delta)$, we denote by $\widehat{G} = Tr_2(G)$ the 2-player game graph defined as follows: $\widehat{G} = ((S,E),(S_1,S_2 \cup S_{\bigcirc}))$.

Lemma 1. *For all $2^1\!/_2$-player game graphs, for all Borel objectives Φ, the sure winning sets for objective Φ for player 1 in the game graphs G and $Tr_2(G)$ coincide.*

Theorem 1 and Theorem 2 state the classical determinacy results for 2-player and $2^1\!/_2$-player game graphs with Borel objectives. It follows from Theorem 2 that for all Borel objectives Φ, for all $\varepsilon > 0$, there exists an ε-optimal strategy σ_ε for player 1 such that for all π and all $s \in S$ we have $\Pr_s^{\sigma,\pi}(\Phi) \geq \langle\langle 1\rangle\rangle_{val}(\Phi)(s) - \varepsilon$.

Theorem 1 (Qualitative determinacy). *The following assertions hold.*

1. *For all 2-player game graphs with state set S, and for all Borel objectives Φ, we have $\langle\langle 1\rangle\rangle_{sure}(\Phi) = S \setminus \langle\langle 2\rangle\rangle_{sure}(\overline{\Phi})$, i.e., the sure winning sets for the two players form a partition of the state space [14].*
2. *The family of pure memoryless strategies suffices for sure winning with respect to Rabin objectives for 2-player game graphs [9]; and the family of pure finite-memory strategies suffices for sure winning with respect to Streett objectives for $2^1\!/_2$-player game graphs [10], and sure winning strategies for Streett objectives in general require memory.*

Theorem 2 (Quantitative determinacy). *The following assertions hold.*

1. *For all $2^1\!/_2$-player game graphs, for all Borel objectives Φ, and for all states s, we have $\langle\langle 1\rangle\rangle_{val}(\Phi)(s) + \langle\langle 2\rangle\rangle_{val}(\overline{\Phi})(s) = 1$ [15].*
2. *The family of pure memoryless strategies suffices for optimality with respect to Rabin objectives for $2^1\!/_2$-player game graphs [3]; and the family of pure finite-memory strategies suffices for optimality with respect to Streett objectives for $2^1\!/_2$-player game graphs [3], and optimal strategies for Streett objectives in general require memory.*

We now present the main results of 2-player games with finitary parity and Streett objectives.

Theorem 3 (Finitary parity games [12,5]). *For all 2-player game graphs with n states and m edges, and all priority functions p the following assertions hold.*

1. *The family of pure memoryless strategies suffices for sure winning with respect to finitary parity objectives. There exist infinite-memory winning strategies* π *for player 2 for the objective* $\text{cofinParity}(p)$ *such that* π *is finite-memory counting with* $\text{nocount}(|\pi|) = 2$*. In general no finite-memory winning strategies exist for player 2 for the objective* $\text{cofinParity}(p)$.
2. *The sure winning sets* $\langle\langle 1\rangle\rangle_{sure}(\text{finParity}(p))$ *and* $\langle\langle 2\rangle\rangle_{sure}(\text{cofinParity}(p))$ *can be computed in* $O(n^2 \cdot m)$ *time.*

Theorem 4 (Finitary Streett games [12,5]). *For all 2-player game graphs with* n *states and* m *edges, and for all sets* $P = \{(E_1, F_1), \ldots, (E_d, F_d)\}$ *with* d *Streett pairs, the following assertions hold.*

1. *There exist finite-memory winning strategies* σ *for player 1 for the objective* $\text{finStreett}(P)$ *such that* $|\sigma| = d \cdot 2^d$*. In general winning strategies for player 1 for the objective* $\text{finStreett}(P)$ *require* $2^{\lfloor \frac{d}{2} \rfloor}$ *memory. There exist infinite-memory winning strategies* π *for player 2 for the objective* $\text{cofinStreett}(P)$ *such that* π *is finite-memory counting with* $\text{nocount}(|\pi|) = d \cdot 2^d$*. In general no finite-memory winning strategies exist for player 2 for the objective* $\text{cofinStreett}(P)$.
2. *The sure winning sets* $\langle\langle 1\rangle\rangle_{sure}(\text{finStreett}(P))$ *and* $\langle\langle 2\rangle\rangle_{sure}(\text{cofinStreett}(P))$ *can be computed in* $O(n^2 \cdot m \cdot d^2 \cdot 4^d)$ *time.*

Remark 1. Recall that Büchi and co-Büchi objectives correspond to parity objectives with two priorities. A finitary Büchi objective is in general a strict subset of the corresponding classical Büchi objective; a finitary co-Büchi objective coincides with the corresponding classical co-Büchi objective. However, it can be shown that for parity objectives with two priorities, the value functions for the classical parity objectives and the finitary parity objectives are the same; that is, for all 2½-player game graphs G and all priority functions p with two priorities, we have $\langle\langle 1\rangle\rangle_{val}(\text{finParity}(p)) = \langle\langle 1\rangle\rangle_{val}(\text{Parity}(p))$ and $\langle\langle 2\rangle\rangle_{val}(\text{cofinParity}(p)) = \langle\langle 2\rangle\rangle_{val}(\text{coParity}(p))$. Note that in Example 1, we have $s_0 \in \langle\langle 1\rangle\rangle_{sure}(\text{Parity}(p))$ and $s_0 \notin \langle\langle 1\rangle\rangle_{sure}(\text{finParity}(p))$. This shows that for priority functions with three or more priorities, the sure winning set for a finitary parity objective can be a strict subset of the sure winning set for the corresponding classical parity objective on 2-player game graphs, that is, $\langle\langle 1\rangle\rangle_{sure}(\text{finParity}(p)) \subsetneq \langle\langle 1\rangle\rangle_{sure}(\text{Parity}(p))$, and in general for 2½-player game graphs we have $\langle\langle 1\rangle\rangle_{val}(\text{finParity}(p)) \leq \langle\langle 1\rangle\rangle_{val}(\text{Parity}(p))$. ∎

3 Qualitative Analysis of Stochastic Finitary Games

In this section we present algorithms for qualitative analysis of 2½-player games with finitary parity and finitary Streett objectives. We first present a few key lemmas that would be useful to prove the correctness of the algorithms.

Lemma 2. *Let* G *be a 2½-player game graph with the set* S *of states, and let* $P = \{(E_1, F_1), (E_2, F_2), \ldots, (E_d, F_d)\}$ *be a set of* d *Streett pairs. If* $\langle\langle 1\rangle\rangle_{sure}(\text{finStreett}(F)) = \emptyset$*, then the following assertions hold:*

1. $\langle\langle 2 \rangle\rangle_{almost}(\text{cofinStreett}(F)) = S$; *and*
2. *there is an almost-sure winning strategy* π *for player 2 with* $\text{nocount}(|\pi|) = d \cdot 2^d$.

Proof. Let $\widehat{G} = Tr_2(G)$ be the 2-player game graph obtained from G. If $\langle\langle 1 \rangle\rangle_{sure}(\text{finStreett}(F)) = \emptyset$ in G, then by Lemma 1 it follows that $\langle\langle 1 \rangle\rangle_{sure}(\text{finStreett}(F)) = \emptyset$ in $\widehat{G}$, and then by Theorem 1 we have $\langle\langle 2 \rangle\rangle_{sure}(\text{cofinStreett}(F)) = S$ for the game graph $\widehat{G}$. If $\langle\langle 2 \rangle\rangle_{sure}(\text{cofinStreett}(F)) = S$ in $\widehat{G}$, then it follows from the results of [5] that there is a pure strategy $\widehat{\pi}$ in $\widehat{G}$ that satisfies the following conditions.

1. For every integer $b \geq 0$, for every strategy $\widehat{\sigma}$ of player 1 in $\widehat{G}$, and from all states s, the play from s given strategies $\widehat{\pi}$ and $\widehat{\sigma}$ satisfies the following condition: there exists position k and $1 \leq j \leq d$, such that the state s_k at the k-th position is in F_j, and for all $k \leq k' < k + b$ the state in k'-th position does not belong to E_j, and $k + b \leq |S| \cdot d \cdot 2^d \cdot (b+1)$.
2. $\text{nocount}(|\widehat{\pi}|) = d \cdot 2^d$.

We obtain an almost-sure winning strategy π^* for player 2 in G as follows: set $b = 1$, the strategy π^* is played in rounds, and in round b the strategy is played according to the following rule:

1. (Step 1). Start play according to $\widehat{\pi}$
   (a) if at any random state the chosen successor is different from $\widehat{\pi}$, then go to the start of step 1 (i.e., start playing like the beginning of round b);
   (b) if for $|S| \cdot d \cdot 2^d \cdot (b+1)$ steps at all random states the chosen successor matches $\widehat{\pi}$, then increment b and proceed to beginning of round $b+1$.

We argue that the strategy π^* is almost-sure winning. Observe that since π^* follows $\widehat{\pi}$ in round b unless there is a deviation at a random state, it follows that if the strategy proceeds from round b to $b+1$, then at round b, there exists a position where the distance is at least b. Hence if the strategy π^* proceeds for infinitely many rounds, then cofinStreett(F) is satisfied. To complete the proof we argue that π^* proceeds through infinitely many rounds with probability 1. For a fixed b, the probability that step 1.(b). succeeds at a given trial is at least $\left(\frac{1}{\delta_{\min}}\right)^{|S| \cdot d \cdot 2^d \cdot (b+1)} > 0$, where $\delta_{\min} = \min\{\delta(s)(t) \mid s \in S_{\bigcirc}, t \in E(s)\} > 0$. Hence it follows that the probability that the strategy gets stuck in step 1.(a). for a fixed b is zero. Since the probability of a countable union of measure zero set is zero, it follows that the probability that the strategy gets stuck in step 1.(a). of any round b is zero. Hence with probability 1 the strategy π^* proceeds through infinitely many rounds, and the desired result follows. ∎

Lemma 2 states that for a finitary Streett objective if the sure winning set for player 1 is empty, then player 2 wins almost-surely everywhere in the game graph. Since parity objectives and finitary parity objectives are a special case of Streett and finitary Streett objectives, respectively, the result of Lemma 2 also holds for finitary parity objectives. This is formalized as the following lemma.

Lemma 3. *Let G be a $2^1/_2$-player game graph with the set S of states, and let p be a priority function. If $\langle\langle 1\rangle\rangle_{sure}(\text{finParity}(p)) = \emptyset$, then the following assertions hold:*

1. *$\langle\langle 2\rangle\rangle_{almost}(\text{cofinParity}(p)) = S$; and*
2. *there is an almost-sure winning strategy π for player 2 with $\text{nocount}(|\pi|) = 2$.*

We now present the notions of attractors in $2^1/_2$-player games and the basic properties of such attractors.

Definition 1 (Attractors). *Given a $2^1/_2$-player game graph G and a set $U \subseteq S$ of states, such that $G \upharpoonright U$ is a subgame, and $T \subseteq S$ we define $Attr_{1,\bigcirc}(T, U)$ as follows:*

$$T_0 = T \cap U; \qquad \text{and for } j \geq 0 \text{ we define } T_{j+1} \text{ from } T_j \text{ as}$$

$$T_{j+1} = T_j \cup \{s \in (S_1 \cup S_\bigcirc) \cap U \mid E(s) \cap T_j \neq \emptyset\} \cup \{s \in S_2 \cap U \mid E(s) \cap U \subseteq T_j\}.$$

and $A = Attr_{1,\bigcirc}(T, U) = \bigcup_{j \geq 0} T_j$. We obtain $Attr_{2,\bigcirc}(T, U)$ by exchanging the roles of player 1 and player 2. A pure memoryless attractor strategy $\sigma^A : (A \setminus T) \cap S_1 \to S$ for player 1 on A to T is as follows: for $i > 0$ and a state $s \in (T_i \setminus T_{i-1}) \cap S_1$, the strategy $\sigma^A(s) \in T_{i-1}$ chooses a successor in T_{i-1} (which exists by definition). ∎

Lemma 4 (Attractor properties). *Let G be a $2^1/_2$-player game graph and $U \subseteq S$ be a set of states such that $G \upharpoonright U$ is a subgame. For a set $T \subseteq S$ of states, let $Z = Attr_{1,\bigcirc}(T, U)$. Then the following assertions hold.*

1. *$G \upharpoonright (U \setminus Z)$ is a subgame.*
2. *Let σ^Z be a pure memoryless attractor strategy for player 1. There exists a constant $c > 0$, such that for all strategies π for player 2 in the subgame $G \upharpoonright U$ and for all states $s \in U$*
   (a) *We have $\Pr_s^{\sigma^Z,\pi}(\text{Reach}(T)) \geq c \cdot \Pr_s^{\sigma^Z,\pi}(\text{Reach}(Z))$; and*
   (b) *if $\Pr_s^{\sigma^Z,\pi}(\text{Buchi}(Z)) > 0$, then $\Pr_s^{\sigma^Z,\pi}(\text{Buchi}(T) \mid \text{Buchi}(Z)) = 1$.*

We now present the second key lemma for the algorithms for the qualitative analysis of $2^1/_2$-player finitary parity and finitary Streett games.

Lemma 5. *Let G be a $2^1/_2$-player game graph with the set S of states, and let Φ be a finitary parity or a finitary Streett objective with d pairs. If $\langle\langle 1\rangle\rangle_{pos}(\Phi) = S$, then the following assertions hold:*

1. *$\langle\langle 1\rangle\rangle_{almost}(\Phi) = S$;*
2. *if Φ is a finitary parity objective, then memoryless almost-sure winning strategies exist; and if Φ is a finitary Streett objective, then an almost-sure winning strategy with memory $d \cdot 2^d$ exists.*

Proof. The proof proceeds by iteratively removing sure winning sets, and the corresponding attractors from the graphs. Let $G^0 = G$, and $S^0 = S$. For $i \geq 0$, let G^i and S^i be the game graph and the set of states at the i-th iteration. Let

$Z_i = \langle\langle 1 \rangle\rangle_{sure}(\Phi)$ in G^i, and $A_i = Attr_{1,\bigcirc}(Z_i, S^i)$. Let $G^{i+1} = G \upharpoonright (S^i \setminus A_i)$, and $X_i = \bigcup_{j \le i} A_i$. We continue this process unless for some k we have $X_k = S$. If for some game graph G^i we have $Z_i = \emptyset$ (i.e., $\langle\langle 1 \rangle\rangle_{sure}(\Phi) = \emptyset$ in G^i), then by Lemma 5 we have that $\langle\langle 2 \rangle\rangle_{almost}(\overline{\Phi}) = S^i$, where $\overline{\Phi}$ is the complementary objective to Φ. This would contradict that $\langle\langle 1 \rangle\rangle_{pos}(\Phi) = S$. It follows that for some k we would have $X_k = S$. The almost-sure winning strategy σ^* for player 1 is defined as follows: in Z_i play a sure winning strategy for Φ in G^i, and in $A_i \setminus Z_i$ play a pure memoryless attractor strategy to reach Z_i. The strategy σ^* ensures the following: (a) from Z_i either the game stays in Z_i and satisfies Φ, or reaches X_{i-1} (this follows since a sure winning strategy is followed in G^i, and player 2 may choose to escape only to X_{i-1}); and (b) if A_i is visited infinitely often, then $X_{i-1} \cup Z_i$ is reached with probability 1 (this follows from the attractor properties, i.e., Lemma 4). It follows from the above two facts that with probability 1 the game settles in some Z_i, i.e., for all strategies π and all states s we have $\Pr_s^{\sigma^*,\pi}(\bigcup_{i \le k} \text{coBuchi}(Z_i)) = 1$. It follows that for all strategies π and all states s we have $\Pr_s^{\sigma^*,\pi}(\Phi) = 1$. By choosing sure winning strategies in Z_i that satisfy the memory requirements (which is possible by Theorem 3 and Theorem 4) we obtain the desired result. ∎

Computation of positive winning set. Given a 2½-player game graph G and a finitary parity or a finitary Streett objective Φ, the set $\langle\langle 1 \rangle\rangle_{pos}(\Phi)$ in G can be computed as follows. Let $G^0 = G$, and $S^0 = S$. For $i \ge 0$, let G^i and S^i be the game graph and the set of states at the i-th iteration. Let $Z_i = \langle\langle 1 \rangle\rangle_{sure}(\Phi)$ in G^i, and $A_i = Attr_{1,\bigcirc}(Z_i, S^i)$. Let $G^{i+1} = G \upharpoonright (S^i \setminus A_i)$, and $X_i = \bigcup_{j \le i} A_i$. If $Z_i = \emptyset$, then $S^i = \langle\langle 2 \rangle\rangle_{almost}(\overline{\Phi})$ and $S \setminus S^i = \langle\langle 1 \rangle\rangle_{pos}(\Phi)$. The correctness follows from Lemma 2.

Computation of almost-sure winning set. Given a 2½-player game graph G and a finitary parity or a finitary Streett objective Φ, the set $\langle\langle 1 \rangle\rangle_{almost}(\Phi)$ in G can be computed as follows. Let $G^0 = G$, and $S^0 = S$. For $i \ge 0$, let G^i and S^i be the game graph and the set of states at the i-th iteration. Let $\overline{Z}_i = \langle\langle 2 \rangle\rangle_{almost}(\overline{\Phi})$ in G^i, and $\overline{A}_i = Attr_{2,\bigcirc}(\overline{Z}_i, S^i)$. Let $G^{i+1} = G \upharpoonright (S^i \setminus \overline{A}_i)$, and $X_i = \bigcup_{j \le i} \overline{A}_i$. In other words, the almost-sure winning set for player 2 and its attractor are iteratively removed from the game graph. If $\overline{Z}_i = \emptyset$, then in G^i we have $\langle\langle 1 \rangle\rangle_{pos}(\Phi) = S^i$, and by Lemma 5 we obtain that $\langle\langle 1 \rangle\rangle_{almost}(\Phi) = S^i$. That is we have $S^i = \langle\langle 1 \rangle\rangle_{almost}(\Phi)$ and $S \setminus S^i = \langle\langle 2 \rangle\rangle_{pos}(\overline{\Phi})$. We have the following theorem summarizing the qualitative complexity of 2½-player games with finitary parity and finitary Streett objectives.

Theorem 5. *Given a 2½-player game graph $G = ((S, E), (S_1, S_2, S_\bigcirc), \delta))$ with n states and m edges, and given a finitary parity or a finitary Streett objective Φ, the following assertions hold.*

1. $\langle\langle 1 \rangle\rangle_{almost}(\Phi) = S \setminus \langle\langle 2 \rangle\rangle_{pos}(\overline{\Phi})$ *and* $\langle\langle 1 \rangle\rangle_{pos}(\Phi) = S \setminus \langle\langle 2 \rangle\rangle_{almost}(\overline{\Phi})$.
2. *The family of pure memoryless strategies suffices for almost-sure and positive winning with respect to finitary parity objectives on 2½-player game graphs. If Φ is a finitary parity objective, then there exist infinite-memory*

almost-sure and positive winning strategies π for player 2 for the complementary infinitary parity objective $\overline{\Phi}$ such that π is finite-memory counting with $\mathrm{nocount}(|\pi|) = 2$. *In general no finite-memory almost-sure and positive winning strategies exist for player 2 for $\overline{\Phi}$.*

3. *If Φ is a finitary Streett objective with d pairs, then there exists a finite-memory almost-sure and positive winning strategy σ for player 1 such that $|\sigma| = d \cdot 2^d$. In general almost-sure and positive winning strategies for player 1 for the objective Φ require $2^{\lfloor \frac{d}{2} \rfloor}$ memory. There exist infinite-memory almost-sure and positive winning strategies π for player 2 for the complementary objective $\overline{\Phi}$ such that π is finite-memory counting with* $\mathrm{nocount}(|\pi|) = d \cdot 2^d$. *In general no finite-memory almost-sure and positive winning strategies exist for player 2 for $\overline{\Phi}$.*
4. *If Φ is a finitary parity objective, then the winning sets $\langle\langle 1 \rangle\rangle_{pos}(\Phi)$ and $\langle\langle 2 \rangle\rangle_{almost}(\overline{\Phi})$ can be computed in time $O(n^3 \cdot m)$, and the sets $\langle\langle 1 \rangle\rangle_{almost}(\Phi)$ and $\langle\langle 2 \rangle\rangle_{pos}(\overline{\Phi})$ can be computed in time $O(n^4 \cdot m)$.*
5. *If Φ is a finitary Streett objective with d pairs, then the winning sets $\langle\langle 1 \rangle\rangle_{pos}(\Phi)$ and $\langle\langle 2 \rangle\rangle_{almost}(\overline{\Phi})$ can be computed in time $O(n^3 \cdot m \cdot d^2 \cdot 4^d)$, and the sets $\langle\langle 1 \rangle\rangle_{almost}(\Phi)$ and $\langle\langle 2 \rangle\rangle_{pos}(\overline{\Phi})$ can be computed in time $O(n^4 \cdot m \cdot d^2 \cdot 4^d)$.*

4 Quantitative Analysis of Stochastic Finitary Games

In this section we consider the quantitative analysis of 2$1/2$-player games with finitary parity and finitary Streett objectives. We start with notion of *value classes.*

Definition 2 (Value classes). *Given a finitary objective Φ, for every real $r \in [0,1]$ the* value class *with value r is* $\mathrm{VC}(\Phi, r) = \{s \in S \mid \langle\langle 1 \rangle\rangle_{val}(\Phi)(s) = r\}$ *is the set of states with value r for player 1. For $r \in [0,1]$ we denote by* $\mathrm{VC}(\Phi, > r) = \bigcup_{q>r} \mathrm{VC}(\Phi, q)$ *the value classes greater than r and by* $\mathrm{VC}(\Phi, < r) = \bigcup_{q<r} \mathrm{VC}(\Phi, q)$ *the value classes smaller than r.* ∎

Definition 3 (Boundary probabilistic states). *Given a set U of states, a state $s \in U \cap S_{\bigcirc}$ is a* boundary probabilistic state *for U if $E(s) \cap (S \setminus U) \neq \emptyset$, i.e., the probabilistic state has an edge out of the set U. We denote by $Bnd(U)$ the set of boundary probabilistic states for U. For a value class* $\mathrm{VC}(\Phi, r)$ *we denote by $Bnd(\Phi, r)$ the set of boundary probabilistic states of value class r.* ∎

Observation. For a state $s \in Bnd(\Phi, r)$ we have $E(s) \cap \mathrm{VC}(\Phi, > r) \neq \emptyset$ and $E(s) \cap \mathrm{VC}(\Phi, < r) \neq \emptyset$, i.e., the boundary probabilistic states have edges to higher and lower value classes.

Reduction of a value class. Given a set U of states, such that U is δ-live, let $Bnd(U)$ be the set boundary probabilistic states for U. We denote by $G_{Bnd(U)}$ the subgame graph $G \upharpoonright U$ where every state in $Bnd(U)$ is converted to an absorbing state (state with a self-loop). Since U is δ-live, we have $G_{Bnd(U)}$ is a subgame graph. We denote by $G_{Bnd(\Phi,r)}$ the subgame graph where every boundary probabilistic state in $Bnd(\Phi, r)$ is converted to an absorbing state. For a tail objective

Φ, we denote by $G_{\Phi,r} = G_{Bnd(\Phi,r)} \upharpoonright \mathrm{VC}(\Phi, r)$: this is a subgame graph since for a tail objective Φ every value class is δ-live, and δ-closed as all states in $Bnd(\Phi, r)$ are converted to absorbing states. We now present a property of tail objectives and we present our results that use the property. Since tail objectives subsume finitary parity and finitary Streett objectives, the desired results would follow for finitary parity and finitary Streett objectives.

Almost-limit property for tail objectives. An objective Φ satisfies the *almost-limit* property if for all $2^1/_2$-player game graphs and for all $F, R \subseteq S$ the following equalities hold:

$$\{s \in S \mid \langle\langle 1\rangle\rangle_{val}(\Phi \cap \mathrm{Safe}(F)) = 1\} = \langle\langle 1\rangle\rangle_{almost}(\Phi \cap \mathrm{Safe}(F));$$

$$\{s \in S \mid \langle\langle 1\rangle\rangle_{val}(\Phi \cup \mathrm{Reach}(R)) = 1\} = \langle\langle 1\rangle\rangle_{almost}(\Phi \cup \mathrm{Reach}(R));$$

$$\{s \in S \mid \langle\langle 2\rangle\rangle_{val}(\overline{\Phi} \cap \mathrm{Safe}(F)) = 1\} = \langle\langle 2\rangle\rangle_{almost}(\overline{\Phi} \cap \mathrm{Safe}(F));$$

$$\{s \in S \mid \langle\langle 2\rangle\rangle_{val}(\overline{\Phi} \cup \mathrm{Reach}(R)) = 1\} = \langle\langle 2\rangle\rangle_{almost}(\overline{\Phi} \cup \mathrm{Reach}(R)).$$

If Φ is a tail objective, then the objective $\Phi \cap \mathrm{Safe}(F)$ can be interpreted as a tail objective $\Phi \cap \mathrm{coBuchi}(F)$ by transforming every state in $S \setminus F$ as a loosing absorbing state. Similarly, if Φ is a tail objective, then the objective $\Phi \cup \mathrm{Reach}(R)$ can be interpreted as a tail objective $\Phi \cup \mathrm{Buchi}(R)$ by transforming every state in R as winning absorbing state. From the results of [13] (Chapter 3) it follows that for all tail objectives Φ we have

$$\{s \in S \mid \langle\langle 1\rangle\rangle_{val}(\Phi)(s) = 1\} = \langle\langle 1\rangle\rangle_{almost}(\Phi);$$

$$\{s \in S \mid \langle\langle 2\rangle\rangle_{val}(\overline{\Phi})(s) = 1\} = \langle\langle 2\rangle\rangle_{almost}(\overline{\Phi}).$$

Hence it follows that all tail objective satisfy the almost-limit property. We now present a lemma, that extends a property of $2^1/_2$-player games with ω-regular objectives to tail objectives (that subsumes finitary parity and Streett objectives).

Lemma 6 (Almost-sure reduction). *Let G be a $2^1/_2$-player game graph and Φ be a tail objective. For $0 < r < 1$, the following assertions hold.*

1. *Player 1 wins almost-surely for objective $\Phi \cup \mathrm{Reach}(Bnd(\Phi, r))$ from all states in $G_{\Phi,r}$, i.e., $\langle\langle 1\rangle\rangle_{almost}(\Phi \cup \mathrm{Reach}(Bnd(\Phi, r))) = \mathrm{VC}(\Phi, r)$ in the subgame graph $G_{\Phi,r}$.*
2. *Player 2 wins almost-surely for objective $\overline{\Phi} \cup \mathrm{Reach}(Bnd(\Phi, r))$ from all states in $G_{\Phi,r}$, i.e., $\langle\langle 2\rangle\rangle_{almost}(\overline{\Phi} \cup \mathrm{Reach}(Bnd(\Phi, r))) = \mathrm{VC}(\Phi, r)$ in the subgame graph $G_{\Phi,r}$.*

Proof. We prove the first part and the second part follows from symmetric arguments. The result is obtained through an argument by contradiction. Let $0 < r < 1$, and let

$$q = \max\{\langle\langle 1\rangle\rangle_{val}(\Phi)(t) \mid t \in E(s) \setminus \mathrm{VC}(\Phi, r), s \in \mathrm{VC}(\Phi, r) \cap S_1\},$$

that is, q is the maximum value a successor state t of a player 1 state $s \in \mathrm{VC}(\Phi, r)$ such that the successor state t is not in $\mathrm{VC}(\Phi, r)$. We must have $q < r$. Hence if player 1 chooses to escape the value class $\mathrm{VC}(\Phi, r)$, then player 1 gets to see a state with value at most $q < r$. We consider the subgame graph $G_{\Phi,r}$. Let $U = \mathrm{VC}(\Phi, r)$ and $Z = Bnd(\Phi, r)$. Assume towards contradiction, there exists a state $s \in U$ such that $s \notin \langle\langle 1 \rangle\rangle_{almost}(\Phi \cup \mathrm{Reach}(Z))$. Then we have $s \in (U \setminus Z)$; and since Φ is a tail objective satisfying the almost-limit property and $s \notin \langle\langle 1 \rangle\rangle_{almost}(\Phi \cup \mathrm{Reach}(Z))$ we have $\langle\langle 2 \rangle\rangle_{val}(\overline{\Phi} \cap \mathrm{Safe}(U \setminus Z))(s) > 0$. Observe that in $G_{\Phi,r}$ we have all states in Z are absorbing states, and hence the objective $\overline{\Phi} \cap \mathrm{Safe}(U \setminus Z)$ is equivalent to the objective $\overline{\Phi} \cap \mathrm{coBuchi}(U \setminus Z)$, which can be considered as a tail objective. Since $\langle\langle 2 \rangle\rangle_{val}(\overline{\Phi} \cap \mathrm{Safe}(U \setminus Z))(s) > 0$, for some state s, it follows from Theorem 1 of [2] that there exists a state $s_1 \in (U \setminus Z)$ such that $\langle\langle 2 \rangle\rangle_{val}(\overline{\Phi} \cap \mathrm{Safe}(U \setminus Z)) = 1$. Then, since Φ is a tail objective satisfying the almost-limit property, it follows that there exists a strategy $\widehat{\pi}$ for player 2 in $G_{\Phi,r}$ such that for all strategies $\widehat{\sigma}$ for player 1 in $G_{\Phi,r}$ we have $\mathrm{Pr}_{s_1}^{\widehat{\sigma},\widehat{\pi}}(\overline{\Phi} \cap \mathrm{Safe}(U \setminus Z)) = 1$. We will now construct a strategy π^* for player 2 as a combination of the strategy $\widehat{\pi}$ and a strategy in the original game G. By Martin's determinacy result (Theorem 2), for all $\varepsilon > 0$, there exists an ε-optimal strategy π_ε for player 2 in G such that for all $s \in S$ and for all strategies σ for player 1 we have

$$\mathrm{Pr}_s^{\sigma,\pi_\varepsilon}(\overline{\Phi}) \geq \langle\langle 2 \rangle\rangle_{val}(\overline{\Phi})(s) - \varepsilon.$$

Let $r - q = \alpha > 0$, and let $\varepsilon = \frac{\alpha}{2}$ and consider an ε-optimal strategy π_ε for player 2 in G. The strategy π^* in G is constructed as follows: for a history w that remains in U, player 2 follows $\widehat{\pi}$; and if the history reaches $(S \setminus U)$, then player 2 follows the strategy π_ε. Formally, for a history $w = \langle s_1, s_2, \ldots, s_k \rangle$ we have

$$\pi^*(w) = \begin{cases} \widehat{\pi}(w) & \text{if for all } 1 \leq j \leq k.\ s_j \in U; \\ \pi_\varepsilon(s_j, s_{j+1}, \ldots, s_k) & \text{where } j = \min\{i \mid s_i \notin U\} \end{cases}$$

We consider the case when the play starts at s_1. The strategy π^* ensures the following: if the game stays in U, then the strategy $\widehat{\pi}$ is followed, and given the play stays in U, the strategy $\widehat{\pi}$ ensures with probability 1 that $\overline{\Phi}$ is satisfied and $Bnd(\Phi, r)$ is not reached. Hence if the game escapes U (i.e., player 1 chooses to escape U), then it reaches a state with value at most q for player 1. We consider an arbitrary strategy σ for player 1 and consider the following cases.

1. If $\mathrm{Pr}_{s_1}^{\sigma,\pi^*}(\mathrm{Safe}(U)) = 1$, then we have $\mathrm{Pr}_{s_1}^{\sigma,\pi^*}(\overline{\Phi} \cap \mathrm{Safe}(U)) = \mathrm{Pr}_{s_1}^{\sigma,\widehat{\pi}}(\overline{\Phi} \cap \mathrm{Safe}(U)) = 1$. Hence we also have $\mathrm{Pr}_{s_1}^{\sigma,\widehat{\pi}}(\overline{\Phi}) = 1$, i.e., we have $\mathrm{Pr}_{s_1}^{\sigma,\pi^*}(\Phi) = 0$.
2. If $\mathrm{Pr}_{s_1}^{\sigma,\pi^*}(\mathrm{Reach}(S \setminus U)) = 1$, then the play reaches a state with value for player 1 at most q and the strategy π_ε ensures that $\mathrm{Pr}_{s_1}^{\sigma,\pi^*}(\Phi) \leq q + \varepsilon$.
3. If $\mathrm{Pr}_{s_1}^{\sigma,\pi^*}(\mathrm{Safe}(U)) > 0$ and $\mathrm{Pr}_{s_1}^{\sigma,\pi^*}(\mathrm{Reach}(S \setminus U)) > 0$, then we condition on both these events and have the following:

$$\begin{aligned}\Pr_{s_1}^{\sigma,\pi^*}(\Phi) &= \Pr_{s_1}^{\sigma,\pi^*}(\Phi \mid \mathrm{Safe}(U)) \cdot \Pr_{s_1}^{\sigma,\pi^*}(\mathrm{Safe}(U)) \\ &\quad + \Pr_{s_1}^{\sigma,\pi^*}(\Phi \mid \mathrm{Reach}(S \setminus U)) \cdot \Pr_{s_1}^{\sigma,\pi^*}(\mathrm{Reach}(S \setminus U)) \\ &\leq 0 + (q+\varepsilon) \cdot \Pr_{s_1}^{\sigma,\pi^*}(\mathrm{Reach}(S \setminus U)) \\ &\leq q + \varepsilon.\end{aligned}$$

The above inequalities are obtained as follows: given the event $\mathrm{Safe}(U)$, the strategy π^* follows $\widehat{\pi}$ and ensures that $\overline{\Phi}$ is satisfied with probability 1 (i.e., Φ is satisfied with probability 0); else the game reaches states where the value for player 1 is at most q, and then the analysis is similar to the previous case.

Hence for all strategies σ we have

$$\Pr_{s_1}^{\sigma,\pi^*}(\Phi) \leq q + \varepsilon = q + \frac{\alpha}{2} = r - \frac{\alpha}{2}.$$

Hence we must have $\langle\langle 1\rangle\rangle_{val}(\Phi)(s_1) \leq r - \frac{\alpha}{2}$. Since $\alpha > 0$ and $s_1 \in \mathrm{VC}(\Phi, r)$ (i.e., $\langle\langle 1\rangle\rangle_{val}(\Phi)(s_1) = r$), we have a contradiction. The desired result follows. ∎

Lemma 7 (Almost-sure to optimality). *Let G be a $2^1/_2$-player game graph and Φ be a tail objective. Let σ be a strategy such that*

- *σ is an almost-sure winning strategy from the almost-sure winning states ($\langle\langle 1\rangle\rangle_{almost}(\Phi)$ in G); and*
- *σ is an almost-sure winning strategy for objective $\Phi \cup \mathrm{Reach}(Bnd(\Phi, r))$ in the game $G_{\Phi,r}$, for all $0 < r < 1$.*

Then σ is an optimal strategy. Analogous result holds for player 2 strategies.

Proof. (Sketch). Consider a strategy σ satisfying the conditions of the lemma, a starting state s, and a counter strategy π. If the play settles in a value-class with $r > 0$, i.e., satisfies $\mathrm{coBuchi}(\mathrm{VC}(\Phi, r))$, for some $r > 0$, then the play satisfies Φ almost-surely. From a value class the play can leave the value class if player 2 chooses to leave to a greater value class, or by reaching the boundary probabilistic states such that average value of the successor states is the value of the value class. Hence it follows that (a) either the event $\bigcup_{r>0} \mathrm{coBuchi}(\mathrm{VC}(\Phi, r))$ holds, and then Φ holds almost-surely; (b) else the event $\mathrm{Reach}(\langle\langle 1\rangle\rangle_{almost}(\Phi) \cup \langle\langle 2\rangle\rangle_{almost}(\overline{\Phi}))$ holds, and by the conditions on leaving the value class it follows that $\Pr_s^{\sigma,\pi}(\mathrm{Reach}(\langle\langle 1\rangle\rangle_{almost}(\Phi)) \mid \mathrm{Reach}(\langle\langle 1\rangle\rangle_{almost}(\Phi) \cup \langle\langle 2\rangle\rangle_{almost}(\overline{\Phi}))) \geq \langle\langle 1\rangle\rangle_{val}(\Phi)(s)$. It follows that for all $s \in S$ and all strategies π we have $\Pr_s^{\sigma,\pi}(\Phi) \geq \langle\langle 1\rangle\rangle_{val}(\Phi)(s)$. The desired result follows. ∎

It follows from Lemma 6 that for tail objectives, strategies satisfying the conditions of Lemma 7 exist. It follows from Lemma 7 that optimal strategies for player 1 for tail objectives (and hence for finitary parity and Streett objectives), and optimal strategies for player 2 for the corresponding complementary objectives is no more complex than the respective almost-sure winning strategies.

Lemma 8. *Let $G = ((S, E), (S_1, S_2, S_\bigcirc), \delta)$ be a $2^1\!/_2$-player game with a tail objective Φ. Let $\mathcal{P} = (V_0, V_1, \ldots, V_k)$ be a partition of the state space S, and let $r_0 > r_1 > r_2 > \ldots > r_k$ be k-real values such that the following conditions hold:*

1. *$V_0 = \langle\langle 1 \rangle\rangle_{almost}(\Phi)$ and $V_k = \langle\langle 2 \rangle\rangle_{almost}(\overline{\Phi})$;*
2. *$r_0 = 1$ and $r_k = 0$;*
3. *for all $1 \le i \le k-1$ we have $Bnd(V_i) \neq \emptyset$ and V_i is δ-live;*
4. *for all $1 \le i \le k-1$ and all $s \in S_2 \cap V_i$ we have $E(s) \subseteq \bigcup_{j \le i} V_j$;*
5. *for all $1 \le i \le k-1$ we have $V_i = \langle\langle 1 \rangle\rangle_{almost}(\Phi \cup \text{Reach}(Bnd(V_i)))$ in $G_{Bnd(V_i)}$;*
6. *let $x_s = r_i$, for $s \in V_i$, and for all $s \in S_\bigcirc$, let x_s satisfy $x_s = \sum_{t \in E(s)} x_t \cdot \delta(s)(t)$.*

Then we have $\langle\langle 1 \rangle\rangle_{val}(\Phi)(s) \ge x_s$ for all $s \in S$. Analogous result holds for player 2.

Proof. (Sketch). We fix a strategy σ such that σ is almost-sure winning from $\langle\langle 1 \rangle\rangle_{almost}(\Phi)$, and in every V_i, for $1 \le i \le k-1$, it is almost-sure winning for the objective $\Phi \cup \text{Reach}(Bnd(V_i))$. Arguments similar to Lemma 7 shows that for $s \in S$ and for all π we have $\Pr_s^{\sigma,\pi}(\Phi)(s) \ge x_s$. ∎

Algorithm for quantitative analysis. We now present an algorithm for quantitative analysis for $2^1\!/_2$-player games with tail objectives. The algorithm is a NP algorithm with an oracle access to the qualitative algorithms. The algorithm is based on Lemma 8. Given a $2^1\!/_2$-player game $G = ((S, E), (S_1, S_2, S_\bigcirc), \delta)$ with a finitary parity or a finitary Streett objective Φ, a state s and a rational number r, the following assertion hold: if $\langle\langle 1 \rangle\rangle_{val}(\Phi)(s) \ge r$, then there exists a partition $\mathcal{P} = (V_0, V_1, V_2, \ldots, V_k)$ of S and rational values $r_0 > r_1 > r_2 > \ldots > r_k$, such that $r_i = \frac{p_i}{q_i}$ with $p_i, q_i \le \delta_u^{4 \cdot |E|}$, where $\delta_u = \max\{q \mid \delta(s)(t) = \frac{p}{q}$ for $p, q \in \mathbb{N}$, $s \in S_\bigcirc$ and $\delta(s)(t) > 0\}$, such that conditions of Lemma 8 are satisfied, and $s \in V_i$ with $r_i \ge r$. The witness $\mathcal{P}$ is the value class partition and the rational values represent the values of the value classes, and the precision of the values can also be proved (we omit details due to lack of space). From the above observation we obtain the algorithm for quantitative analysis as follows: given a $2^1\!/_2$-player game graph $G = ((S, E), (S_1, S_2, S_\bigcirc), \delta)$ with a finitary parity or a finitary Streett objective Φ, a state s and a rational r, to verify that $\langle\langle 1 \rangle\rangle_{val}(\Phi)(s) \ge r$, the algorithm guesses a partition $\mathcal{P} = (V_0, V_1, V_2, \ldots, V_k)$ of S and rational values $r_0 > r_1 > r_2 > \ldots > r_k$, such that $r_i = \frac{p_i}{q_i}$ with $p_i, q_i \le \delta_u^{4 \cdot |E|}$, and then verifies that all the conditions of Lemma 8 are satisfied, and $s \in V_i$ with $r_i \ge r$. Observe that since the guesses of the rational values can be made with $O(|G| \cdot |S| \cdot |E|)$ bits, the guess is polynomial in size of the game. The condition 1 and the condition 5 of Lemma 8 can be verified by any qualitative algorithms, and all the other conditions can be checked in polynomial time. We now summarize the results on quantitative analysis of $2^1\!/_2$-player games with tail objectives, and then present the results for finitary parity and finitary Streett objectives.

Theorem 6. *Given a $2^1\!/_2$-player game graph and a tail objective Φ, the following assertions hold.*

1. *If a family Σ^C of strategies suffices for almost-sure winning for Φ, then the family Σ^C of strategies also suffices for optimality for Φ.*
2. *Given a rational number r and a state s, whether $\langle\langle 1\rangle\rangle_{val}(\Phi)(s) \geq r$ can be decided in $NP^{\mathcal{A}}$, where $\mathcal{A}$ is an oracle for the qualitative analysis of Φ on $2^1/_2$-player game graphs.*

Theorem 7. *Given a $2^1/_2$-player game graph and a finitary parity or a finitary Streett objective Φ, the following assertions hold.*

1. *The family of pure memoryless strategies suffices for optimality with respect to finitary parity objectives on $2^1/_2$-player game graphs. If Φ is a finitary parity objective, then there exist infinite-memory optimal strategies π for player 2 for the complementary infinitary parity objective $\overline{\Phi}$ such that π is finite-memory counting with* $\text{nocount}(|\pi|) = 2$. *In general no finite-memory optimal strategies exist for player 2 for $\overline{\Phi}$.*
2. *If Φ is a finitary Streett objective with d pairs, then there exists a finite-memory optimal strategy σ for player 1 such that $|\sigma| = d \cdot 2^d$. In general optimal strategies for player 1 for the objective Φ require $2^{\lfloor \frac{d}{2} \rfloor}$ memory. There exist infinite-memory optimal strategies π for player 2 for the complementary objective $\overline{\Phi}$ such that π is finite-memory counting with* $\text{nocount}(|\pi|) = d \cdot 2^d$. *In general no finite-memory optimal strategy exists for player 2 for $\overline{\Phi}$.*
3. *If Φ is a finitary parity objective, then given a rational r and a state s, whether $\langle\langle 1\rangle\rangle_{val}(\Phi)(s) \geq r$ can be decided in NP $\cap$ coNP.*
4. *If Φ is a finitary Streett objective, then given a rational r and a state s, whether $\langle\langle 1\rangle\rangle_{val}(\Phi)(s) \geq r$ can be decided in EXPTIME.*

Remark 2. For $2^1/_2$-player games with finitary objectives, the qualitative analysis can be achieved in polynomial time, however, we only prove a NP $\cap$ coNP bound for the quantitative analysis. It may be noted that for $2^1/_2$-player game graphs, the quantitative analysis for finitary and nonfinitary Büchi objectives coincide. The best known bound for quantitative analysis of $2^1/_2$-player games with Büchi objectives is NP $\cap$ coNP, and obtaining a polynomial time algorithm is a major open problem. Hence obtaining a polynomial time algorithm for quantitative analysis of $2^1/_2$-player games with finitary parity objectives would require the solution of a major open problem. ∎

References

1. Alur, R., Henzinger, T.A.: Finitary fairness. In: LICS 1994, pp. 52–61. IEEE, Los Alamitos (1994)
2. Chatterjee, K.: Concurrent games with tail objectives. Theoretical Computer Science 388, 181–198 (2007)
3. Chatterjee, K., de Alfaro, L., Henzinger, T.A.: The complexity of stochastic Rabin and Streett games. In: Caires, L., Italiano, G.F., Monteiro, L., Palamidessi, C., Yung, M. (eds.) ICALP 2005. LNCS, vol. 3580, pp. 878–890. Springer, Heidelberg (2005)

4. Chatterjee, K., Henzinger, T.A.: Finitary winning in ω-regular games. In: Hermanns, H., Palsberg, J. (eds.) TACAS 2006. LNCS, vol. 3920, pp. 257–271. Springer, Heidelberg (2006)
5. Chatterjee, K., Henzinger, T.A., Horn, F.: Finitary winning in ω-regular games. Technical Report: UCB/EECS-2007-120 (2007)
6. Chatterjee, K., Jurdziński, M., Henzinger, T.A.: Quantitative stochastic parity games. In: SODA 2004, pp. 121–130. SIAM, Philadelphia (2004)
7. de Alfaro, L., Henzinger, T.A.: Concurrent omega-regular games. In: LICS 2000, pp. 141–154. IEEE, Los Alamitos (2000)
8. Dziembowski, S., Jurdzinski, M., Walukiewicz, I.: How much memory is needed to win infinite games? In: LICS 1997, pp. 99–110. IEEE, Los Alamitos (1997)
9. Emerson, E.A., Jutla, C.: The complexity of tree automata and logics of programs. In: FOCS 1988, pp. 328–337. IEEE, Los Alamitos (1988)
10. Gurevich, Y., Harrington, L.: Trees, automata, and games. In: STOC 1982, pp. 60–65. ACM Press, New York (1982)
11. Horn, F.: Dicing on the streett. IPL 104, 1–9 (2007)
12. Horn, F.: Faster algorithms for finitary games. In: Grumberg, O., Huth, M. (eds.) TACAS 2007. LNCS, vol. 4424, pp. 472–484. Springer, Heidelberg (2007)
13. Horn, F.: Random Games. PhD thesis, Université Denis-Diderot and RWTH, Aachen (2008)
14. Martin, D.A.: Borel determinacy. Annals of Mathematics 102(2), 363–371 (1975)
15. Martin, D.A.: The determinacy of Blackwell games. The Journal of Symbolic Logic 63(4), 1565–1581 (1998)
16. Piterman, N., Pnueli, A.: Faster solution of Rabin and Streett games. In: LICS 2006, pp. 275–284. IEEE, Los Alamitos (2006)
17. Zielonka, W.: Perfect-information stochastic parity games. In: Walukiewicz, I. (ed.) FOSSACS 2004. LNCS, vol. 2987, pp. 499–513. Springer, Heidelberg (2004)

Stochastic Data Streams

S. Muthukrishnan

Google Inc.

The classical data stream problems are now greatly understood [1]. This talk will focus on problems with stochastic (rather than deterministic) streams where the underlying physical phenomenon generates probabilistic data or data distributions. While basic streaming problems are now getting reexamined with stochastic streams [2,3], we will focus on a few novel problems.

1. Estimating tails. Each item in the data stream random variable $X_i, 1 \leq i \leq n$, $X_i\{0,1\}$, identically distributed such that $E(X_i) = 0$, $E(X_i^2) = \sigma^2 > 0$ but bounded. The query is to estimate $\Pr[\sum X_i \leq c]$. This is a special case of more general probabilistic streams where each X_i may be drawn from a different distribution, and we wish to answer this query using sublinear space. We provide an algorithm using the well known result below (joint work with Krzysztof Onak).

Theorem 1. (Berry-Esseen Theorem*) Let $X_1, X_2, ..., X_n$ be i.i.d. random variables with $E(X_i) = 0$, $E(X_i^2) = \sigma^2 > 0$, and $E(|X_i|^3) = \rho < \infty$. Also, let $Y_n = \sum_i X_i/n$ with F_n the cdf of $Y_n\sqrt{n}/\sigma$, and Φ the cdf of the standard normal distribution. Then there exists a positive constant C such that for all x and n, $|F_n(x) - \Phi(x)| \leq \frac{C\rho}{\sigma^3\sqrt{n}}$.*

2. Max Triggering. The data stream is a sequence of independent random variables $x_1, x_2 \ldots$. The problem is to determine a point τ and trigger an alarm based on x_τ. The goal is to trigger the alarm when x_τ is the maximum possible over a length n. Define $M = E(max_{i=1}^{i=n} x_i)$, the expected value of the maximum. This is target for any streaming algorithm. We will be able to show simply that there exists a streaming algorithm that finds x_τ such that $E(x_\tau) \geq M/2$. Our algorithm will be a suitable implementation of the stopping rule that gives the well known Prophet inequalities [4].

References

1. Muthukrishnan, S.: Data Streams: Algorithms and Applications. In: Foundations and Trends in Theoretical Computer Science. NOW publishers (2005); Also Barbados Lectures (2009), `http://www.cs.mcgill.ca/~denis/notes09.pdf`
2. Jayram, T.S., McGregor, A., Muthukrishnan, S., Vee, E.: Estimating statistical aggregates on probabilistic data streams. ACM Trans. Database Syst. 33(4) (2008)
3. Cormode, G., Garofalakis, M.: Sketching probabilistic data streams. SIGMOD, 281–292 (2007)
4. Samuel-Cahn, E.: Comparisons of optimal stopping values and prophet inequalities for independent non-negative random variables. Ann. Prob. 12, 1213–1216 (1984)

R. Královič and D. Niwiński (Eds.): MFCS 2009, LNCS 5734, p. 55, 2009.

Recent Advances in Population Protocols*

Ioannis Chatzigiannakis[1,2], Othon Michail[1,2], and Paul G. Spirakis[1,2]

[1] Research Academic Computer Technology Institute (RACTI), +302610960200, Patras, Greece
[2] Computer Engineering and Informatics Department (CEID), University of Patras, 26500, Patras, Greece
{ichatz,michailo,spirakis}@cti.gr

Abstract. The *population protocol model* (PP) proposed by Angluin et al. [2] describes sensor networks consisting of passively mobile finite-state agents. The agents sense their environment and communicate in pairs to carry out some computation on the sensed values. The *mediated population protocol model* (MPP) [13] extended the PP model by communication links equipped with a constant size buffer. The MPP model was proved in [13] to be stronger than the PP model. However, its most important contribution is that it provides us with the ability to devise optimizing protocols, approximation protocols and protocols that decide properties of the communication graph on which they run. The latter case, suggests a simplified model, the GDM model, that was formally defined and studied in [11]. GDM is a special case of MPP that captures MPP's ability to decide properties of the communication graph. Here we survey recent advances in the area initiated by the proposal of the PP model and at the same time we provide new protocols, novel ideas and results.

1 Introduction

Most recent advances in microprocessor, wireless communication and sensor/actuator-technologies envision a whole new era of computing, popularly referred to as pervasive computing. Autonomous, ad-hoc networked, wirelessly communicating and *spontaneously interacting* computing devices of *small size* appearing in *great number*, and embedded into environments, appliances and objects of everyday use will deliver services adapted to the person, the time, the place, or the context of their use. The nature and appearance of devices will change to be hidden in the fabric of everyday life, invisibly networked, and will be augmenting everyday environments to form a pervasive computing landscape, in which the physical world becomes merged with a "digital world".

In a seminal work, Angluin et al. [2] (also [3]) considered systems consisting of very small resource limited sensor nodes that are passively mobile. Such nodes, also called agents, have no control over their own movement and interact in pairs, via a local low-power wireless communication mechanism, when they are

* This work has been partially supported by the ICT Programme of the European Union under contract number ICT-2008-215270 (FRONTS).

R. Královič and D. Niwiński (Eds.): MFCS 2009, LNCS 5734, pp. 56–76, 2009.

sufficiently close to each other. The agents form a (usually huge) population that together with the agent's permissible interactions form a communication graph $G = (V, E)$, where V is a population of $|V| = n$ agents and E is the set of permissible (directed) interactions of cardinality denoted by m. In their model, finite-state, and complex behavior of the system as a whole emerges from simple rules governing pairwise interaction of the agents. The most important innovations of the model are inarguably the *constant memory* constraint imposed to the agents and the *nondeterminism* inherent to the interaction pattern. These assumptions provide us with a concrete and realistic model for future systems. Their model is called Population Protocol model and is discussed in Section 2.

The initial goal of the model was to study the *computational limitations* of cooperative systems consisting of many limited devices (agents), imposed to passive (but *fair*) communication by some *scheduler*. Much work showed that there exists an exact characterization of the computable predicates: they are precisely the *semilinear predicates* or equivalently the predicates definable by first-order logical formulas in *Presburger arithmetic* [2,3,5,6,7]. More recent work has concentrated on performance, supported by a random scheduling assumption. [12] proposed a collection of *fair* schedulers and examined the performance of various protocols. [9,10] considered a huge population hypothesis (population going to infinity), and studied the dynamics, stability and computational power of probabilistic population protocols by exploiting the tools of continuous nonlinear dynamics. In [9] it was also proven that there is a strong relation between classical finite population protocols and models given by ordinary differential equations.

There exist a few extensions of the basic model in the relevant literature to more accurately reflect the requirements of practical systems. In [1] they studied what properties of restricted communication graphs are stably computable, gave protocols for some of them, and proposed the model extension with *stabilizing inputs*. The results of [5] show that again the semilinear predicates are all that can be computed by this model. Finally, some works incorporated agent failures [14] and gave to some agents slightly increased computational power [8] (heterogeneous systems). For an excellent introduction to most of the preceding subjects see [7].

In [13] the Population Protocol model was extended in a natural way. Essentially the model was augmented to include a *Mediator*, i.e., a global storage capable of storing very limited information for each communication arc (the state of the arc). When pairs of agents interact, they can read and update the state of the link (arc). Interestingly, although anonymity and uniformity (for the definition of those notions the reader is referred to Section 3) are preserved in the extended model, *the presence of a mediator provides us with significantly more computational power* and gives birth to a new collection of interesting problems in the area of tiny networked and possibly moving artefacts; we can now build systems with the ability of computing subgraphs and solve optimization problems concerning the communication graph. In [13] it was shown that the new model is capable of computing non-semilinear predicates and that any stably computable predicate belongs to $NSPACE(m)$, where m denotes the number of edges of the interaction graph. The extended model is called Mediated Population Protocol model and we present it in Section 3.

One of the most interesting and applicable capabilities of the Mediated Population Protocol model is its ability to decide graph properties. To understand properties of the communication graph is an important step in almost any distributed system. In particular, in [11] the authors temporarily disregarded the input notion of the population and assumed that all agents simply start from a unique initial state (and the same holds for the edges). The obtained model is called GDM. The authors focused on protocols of the GDM model that, when executed fairly on any communication graph G, after a finite number of steps stabilize to a configuration where all agents give 1 as output if G belongs to a graph language L, and 0 otherwise. This is motivated by the idea of having protocols that eventually accept all communication graphs (on which they run) that satisfy a specific property, and eventually reject all remaining communication graphs. The motivation for the proposal of a simplified version of the Mediated Population Protocol model was that it enables us to study what graph properties are stably computable by the mediated model without the need to keep in mind its remaining parameters (which, as a matter of fact, are a lot). The GDM model is discussed in Section 4. Finally, in Section 5 we discuss some future research directions.

2 The Population Protocol Model

2.1 Formal Definition

Definition 1. *A* population protocol *(PP) is a 6-tuple* (X, Y, Q, I, O, δ), *where* X, Y, *and* Q *are all finite sets and*

1. X *is the* input alphabet,
2. Y *is the* output alphabet,
3. Q *is the set of* states,
4. $I : X \to Q$ *is the* input function,
5. $O : Q \to Y$ *is the* output function, *and*
6. $\delta : Q \times Q \to Q \times Q$ *is the* transition function.

If $\delta(a, b) = (a', b')$, *we call* $(a, b) \to (a', b')$ *a* transition *and we define* $\delta_1(a, b) = a'$ *and* $\delta_2(a, b) = b'$.

A population protocol $\mathcal{A} = (X, Y, Q, I, O, \delta)$ runs on a communication graph $G = (V, E)$. Initially, all agents (i.e. the elements of V) receive a global start signal, sense their environment and each one receives an input symbol from X. All agents are initially in a special empty state $\sqcup \notin Q$. When an agent receives an input symbol σ, applies the input function to it and goes to its initial state $I(\sigma) \in Q$. An adversary scheduler selects in each step a directed pair of agents $(u, v) \in E$, where $u, v \in V$ and $u \neq v$, to interact. The interaction happens only if both agents are not in the empty state (they must both have been initialized). Assume that the scheduler selects the pair (u, v), that the current states of u and

v are $a, b \in Q$, respectively, and that $\delta(a,b) = (a', b')$. Agent u plays the role of the *initiator* in the interaction (u, v) and v that of the *responder*. During their interaction u and v apply the transition function to their directed pair of states, and, as a result, u goes to a' and v to b' (both update their states according to δ, and specifically, the initiator applies δ_1 while the responder δ_2).

A *configuration* is a snapshot of the population states. Formally, a configuration is a mapping $C : V \to Q$ specifying the state of each agent in the population. C_0 is the initial configuration (for simplicity we assume that all agents apply the input function at the same time, which is one step before C_0, so in C_0 all empty states have been already replaced, and that' s the reason why we have chosen not to include $\sqcup$ in the model definition) and, for all $u \in V$, $C_0(u) = I(x(u))$, where $x(u)$ is the input symbol sensed by agent u. Let C and C' be configurations, and let u, v be distinct agents. We say that C goes to C' via encounter $e = (u, v)$, denoted $C \xrightarrow{e} C'$, if

$$
\begin{aligned}
C'(u) &= \delta_1(C(u), C(v)), \\
C'(v) &= \delta_2(C(u), C(v)), \text{ and} \\
C'(w) &= C(w) \text{ for all } w \in V - \{u, v\},
\end{aligned}
$$

that is, C' is the result of the interaction of the pair (u, v) under configuration C and is the same as C except for the fact that the states of u, v have been updated according to δ_1 and δ_2, respectively. We say that C can go to C' in one step, denoted $C \to C'$, if $C \xrightarrow{e} C'$ for some encounter $e \in E$. We write $C \xrightarrow{*} C'$ if there is a sequence of configurations $C = C_0, C_1, \ldots, C_t = C'$, such that $C_i \to C_{i+1}$ for all i, $0 \leq i < t$, in which case we say that C' is *reachable* from C.

An *execution* is a finite or infinite sequence of configurations $C_0, C_1, C_2, \ldots$, where C_0 is an initial configuration and $C_i \to C_{i+1}$, for all $i \geq 0$. We have both finite and infinite kinds of executions since the scheduler may stop in a finite number of steps or continue selecting pairs for ever. Moreover, note that, according to the preceding definitions, a scheduler may partition the agents into non-communicating clusters. If that's the case, then it is easy to see that no meaningful computation is possible. To avoid this unpleasant scenario, a strong global *fairness condition* is imposed on the scheduler to ensure the protocol makes progress. An infinite execution is *fair* if for every pair of configurations C and C' such that $C \to C'$, if C occurs infinitely often in the execution, then C' also occurs infinitely often in the execution. A scheduler is fair if it always leads to fair executions. A *computation* is an infinite fair execution.

The following are two critical properties of population protocols:

1. *Uniformity*: Population protocols are uniform. This means that any protocol's description is independent of the population size. Since we assume that the agents have finite storage capacity, and independent of the population size, uniformity enables us to store the protocol code in each agent of the population.
2. *Anonymity*: Population protocols are anonymous. The set of states is finite and does not depend on the size of the population. This impies that there is

no room in the state of an agent to store a unique identifier, and, thus, all agents are treated in the same way by the transition function.

2.2 Stable Computation

Assume a fair scheduler that keeps working forever and a protocol $\mathcal{A}$ that runs on a communication graph $G = (V, E)$. Initially, each agent receives an input symbol from X. An input assignment $x : V \rightarrow X$ is a mapping specifying the input symbol of each agent in the population. Let $\mathcal{X} = X^V$ be the set of all possible input assignments, given the population V and the input alphabet X of $\mathcal{A}$. Population protocols, when controlled by infinitely working schedulers, do not halt. Instead of halting we require any computation of a protocol to *stabilize.* An output assignment $y : V \rightarrow Y$ is a mapping specifying the output symbol of each agent in the population. Any configuration $C \in \mathcal{C} = Q^V$ is associated with an output assignment $y_C = O \circ C$. A configuration C is said to be *output-stable* if for any configuration C' such that $C \xrightarrow{*} C'$ (any configuration reachable from C) $y_{C'} = y_C$. In words, a configuration C is output-stable if all agents maintain the output symbol that have under C in all subsequent steps, no matter how the scheduler proceeds thereafter. A computation $C_0, C_1, C_2, \ldots$ is *stable* if it contains an output-stable configuration C_i, where i is finite.

Definition 2. *A population protocol $\mathcal{A}$ running on a communication graph $G = (V, E)$* stably computes *a predicate $p : \mathcal{X} \rightarrow \{0, 1\}$, if, for any $x \in \mathcal{X}$, every computation of $\mathcal{A}$ on G beginning in $C_0 = I \circ x$ reaches in a finite number of steps an output-stable configuration C_{stable} such that $y_{C_{stable}}(u) = p(x)$ for all $u \in V$. A predicate is* stably computable *if some population protocol stably computes it.*

Assume that a computation of $\mathcal{A}$ on G begins in the configuration corresponding to an input assignment x. Assume, also, that $p(x) = 1$. If $\mathcal{A}$ stably computes p, then we know that after a finite number of steps (if, of course, the scheduler is fair) all agents will give 1 as output, and will continue doing so for ever. This means, that if we wait for a sufficient, but finite, number of steps we can obtain the correct answer of p with input x by querying any agent in the population.

Definition 3. *The* basic population protocol model *(or* standard*) assumes that the communication graph G is always directed and complete.*

Semilinear predicates are predicates whose support is a semilinear set. A *semilinear set* is the finite union of linear sets. A set of vectors in $\mathbb{N}^k$ is *linear* if it is of the form

$$\{\boldsymbol{b} + l_1\boldsymbol{a_1} + l_2\boldsymbol{a_2} + \cdots + l_m\boldsymbol{a_m} \mid l_i \in \mathbb{N}\},$$

where $\boldsymbol{b}$ is a base vector, $\boldsymbol{a_i}$ are basis vectors, and l_i are non-negative integer coefficients. Moreover, semilinear predicates are precisely those predicates that can be defined by first-order logical formulas in *Presburger arithmetic*, as was proven by Ginsburg and Spanier in [15].

In [2] and [3] it was proven that any semilinear predicate is stably computable by the basic population protocol model and in [5] that any stably computable predicate, by the same model, is semilinear, thus together providing an exact characterization of the class of stably computable predicates:

Theorem 1. *A predicate is stably computable by the basic population protocol model iff it is semilinear.*

An immediate observation is that predicates like "the number of c's is the product of the number of a's and the number of b's (in the input assignment)" and "the number of 1's is a power of 2" are not stably computable by the basic model.

A *graph family*, or *graph universe*, is any set of communication graphs. Let $\mathcal{G}$ be a graph family. For any $G \in \mathcal{G}$ and given that X is the input alphabet of some protocol $\mathcal{A}$, there exists a set of all input assignments *appropriate* for G, denoted $\mathcal{X}_G = X^{V(G)}$. Let now $\mathcal{X}_{\mathcal{G}} = \bigcup_{G \in \mathcal{G}}(\mathcal{X}_G \times \{G\})$ or, equivalently, $\mathcal{X}_{\mathcal{G}} = \{(x, G) \mid G \in \mathcal{G}$ and x is an input assignment appropriate for $G\}$. Then we have the following definition:

Definition 4. *A population protocol* $\mathcal{A}$ stably computes a predicate $p : \mathcal{X}_{\mathcal{G}} \rightarrow \{0, 1\}$ *in a family of communication graphs* $\mathcal{G}$*, if, for any* $G \in \mathcal{G}$ *and any* $x \in \mathcal{X}_G$*, every computation of* $\mathcal{A}$ *on* G *beginning in* $C_0 = I \circ x$ *reaches in a finite number of steps an output-stable configuration* C_{stable} *such that* $y_{C_{stable}}(u) = p(x, G)$ *for all* $u \in V(G)$.

Moreover, if p is a mapping from $\mathcal{G}$ to $\{0, 1\}$, that is, a *graph property*, then we say that $\mathcal{A}$ stably computes property p.

Note that we can also consider undirected communication graphs. In the case of an undirected graph we only require that E is symmetric, but we keep the initiator-responder assumption. The latter is important to ensure deterministic transitions, since otherwise we would not be able to now which agent applies/gets the result of δ_1 and which that of δ_2.

2.3 An Example

Let us illustrate what we have seen so far by an example.

Problem 1. (*Undirected Star*) Given a communication graph $G = (V, E)$ from the unrestricted family of undirected graphs (any possible connected and undirected simple graph), find whether the topology is an undirected star.

We devise a protocol, named *UndirectedStar*, that stably computes property Undirected Star, that is, it eventually decides whether the underlying communication graph $G = (V, E)$ taken from the unrestricted family of graphs is an undirected star.

UndirectedStar

- $X = \{0, 1\}$, $Y = \{0, 1\}$,
- $Q = \{(i, j) \mid i \in \{0, 1, 2\}$ and $j \in \{0, 1, 2\}^2\} \cup \{z\}$,

- $I(x) = (0,(0,0))$, for all $x \in X$,
- $O(z) = 0$ and $O(q) = 1$, for all $q \in Q - \{z\}$,
- δ:

$$
\begin{aligned}
((0,(0,0)),(0,(0,0))) &\to ((1,(1,1)),(2,(0,0)))\\
((1,(i,j)),(0,(0,0))) &\to ((1,(0,0)),(2,(i,j+1))),\\
&\qquad \text{if } i \in \{0,1\} \text{ and } j \in \{0,1\} \text{ or } i = 2 \text{ and } j = 0\\
&\to ((1,(0,0)),(2,(i,j))),\ \text{if } i = 1 \text{ and } j = 2\\
&\to (z,z),\ \text{if } i = 2 \text{ and } j = 1\\
((2,(l,k)),(0,(0,0))) &\to ((2,(0,0)),(1,(l+1,k))),\\
&\qquad \text{if } k \in \{0,1\} \text{ and } l \in \{0,1\} \text{ or } k = 2 \text{ and } l = 0\\
&\to ((2,(0,0)),(1,(l,k))),\ \text{if } k = 1 \text{ and } l = 2\\
&\to (z,z),\ \text{if } k = 2 \text{ and } l = 1\\
((1,(i,j)),(2,(l,k))) &\to ((1,(2,j+k)),(2,(0,0))),\ \text{if } j+k < 2 \text{ and } i+l \geq 2\\
&\to ((1,(i+l,2)),(2,(0,0))),\ \text{if } i+l < 2 \text{ and } j+k \geq 2\\
&\to ((1,(i+l,j+k)),(2,(0,0))),\ \text{if } i+l < 2 \text{ and } j+k < 2\\
&\to (z,z),\ \text{if } i+l \geq 2 \text{ and } j+k \geq 2\\
((1,(i,j)),(2,(0,0))) &\to ((1,(0,0)),(2,(i,j)))\\
((1,(0,0)),(2,(l,k))) &\to ((1,(l,k)),(2,(0,0)))\\
((1,(i,j)),(1,(l,k))) &\to (z,z)\\
((2,(i,j)),(2,(l,k))) &\to (z,z)\\
(z,x) &\to (z,z)
\end{aligned}
$$

Note that in the transition $\delta((1,(i,j)),(2,(l,k)))$ we assume that $(i,j) \neq (0,0)$ and $(l,k) \neq (0,0)$.

Definition 5. *An* undirected star *of order n ("n-star") is a tree on n vertices with one vertex having degree $n-1$ and $n-1$ vertices having degree 1.*

Lemma 1. *A connected undirected graph $G = (V,E)$, with $|V| = n \geq 3$, is an undirected star if and only if there is at most one $u \in V$ where $d(u) \geq 2$ (i.e. at most one vertex of degree at least 2).*

Proof. For the only if part, Definition 5 states that an undirected star has only one vertex of degree at least 2. For the other direction, first we note that since G is connected it must have at least $n-1$ edges. Any cycle should contain at least two vertices of degree 2, so G is acyclic. Since G is acyclic and connected it is a tree and therefore it has exactly $n-1$ edges. The latter, together with the fact that each $v \in V$ has $d(v) \geq 1$, but at most one $u \in V$ has $d(u) \geq 2$ implies that $d(u) = n-1$ and $d(v) = 1$, for all $v \in V - \{u\}$, and, according to Definition 5, this completes the proof. □

Corollary 1. *A connected undirected graph $G = (V,E)$, with $|V| = n \geq 3$, is not an undirected star if and only if there are at least two vertices $u, v \in V$ where $d(u) \geq 2$ and $d(v) \geq 2$ (i.e. at least two vertices of degree at least 2).*

So, generally speaking, an algorithm that decides if a connected graph is an undirected star could only check if there is at most one vertex of degree at least 2 (if $n \geq 3$).

Remark 1. Any simple connected graph with only two vertices is an undirected star.

This remark fills the gap of the assumption in Lemma 1 that $n \geq 3$. Note that $n = 1$ is meaningless, since in a graph with a unique vertex no computation can take place (there is not even a single pair of agents to interact).

We can think of states $(i, j) \in Q$ as consisting of two components i and j. We will call i the *basic component* and j the *counter component.* If the interacting pair consists of two agents in the initial state, the protocol assigns basic component 1 to one agent and basic component 2 to the other. Moreover, an agent in basic component 1 gives an agent that is in the initial state basic component 2 and an agent in basic component 2 gives an agent that is in the initial state basic component 1, while the pairs (1,2) and (2,1) do nothing w.r.t. the basic components. Clearly, if the topology is a star and if w.l.o.g. the central vertex (vertex of degree $n - 1$) gets basic component 2, then all peripheral vertices will eventually get basic component 1 and the protocol will output that the topology is a star.

If the topology is not a star, then the protocol must detect that at least two vertices are of degree at least 2. Note, also, that the agents never change their basic component except for the case when they get the reject state z.

Lemma 2. *If G is not a star, then protocol UndirectedStar will eventually create one of the following two situations:*

- *Two neighboring agents with the same basic component, or*
- *at least two agents in basic component 1 and at least two agents in basic component 2.*

Proof. Assume that it won't. Then any neighboring agents will have different basic components and at most one of the basic components 1 and 2 will appear in at least two different agents. So w.l.o.g. $n - 1$ agents will be in component 1 and only one in component 2, since both components always exist in any computation (except before the first interaction and sometime after rejection). But since G is connected it must have at least $n - 1$ edges. In fact, if it has more than $n - 1$ edges it will contain a cycle with at least two vertices (agents) in basic component 1 which will violate the fact that any neighboring vertices will have different basic components and thus G must have exactly $n - 1$ edges. But the latter implies that the $n - 1$ vertices in basic component 1 are directly connected to the unique vertex in basic component 2, which in turn implies that G must be an undirected star, a fact that contradicts the fundamental assumption of the lemma. We could also have proven the statement by contradicting the fact that G not being an undirected star must have at least two vertices $u, v \in V$, where $d(u) \geq 2$ and $d(v) \geq 2$. □

In fact, the protocol always rejects if it finds two neighboring agents in the same basic component and the same does if it finds out that there are at the same time at least two agents in basic component 1 and at least two agents in basic component 2 in the population. The latter is done by the counter component of the states. Thus, according to the preceding lemma, the protocol will eventually reject if G is not a star, because it will provably fall in a situation that leads it to rejection.

Theorem 2. *UndirectedStar stably computes property Undirected Star.*

Proof. The correctness of the statement should be clear after the above discussion. The protocol always reaches an output-stable configuration and at that point any agent outputs the correct answer for the property Undirected Star. □

3 The Mediated Population Protocol Model

In [13] the authors considered the following question: "Is there a way to extend the population protocol model and obtain a stronger model, without violating the uniformity and anonymity properties"? As we shall, in this section, see, the answer to this question is "Yes". Although the idea is simple, it provides us with a model with significantly more computational power and extra capabilities in comparison to the population protocol model. The main modification is to allow the edges of the communication graph to store states from a finite set, whose cardinality is independent of the population size. Two interacting agents read the corresponding edge's state and update it, according to a global transition function, by also taking into account their own states.

3.1 Formal Definition

Definition 6. *A* mediated population protocol *(MPP) is a 12-tuple* $(X, Y, Q, I, O, S, \iota, \omega, r, K, c, \delta)$*, where* X*,* Y*,* Q*,* S*, and* K *are all finite sets and*

1. X *is the* input alphabet,
2. Y *is the* output alphabet,
3. Q *is the set of* agent states,
4. $I : X \to Q$ *is the* agent input function,
5. $O : Q \to Y$ *is the* agent output function,
6. S *is the set of* edge states,
7. $\iota : X \to S$ *is the* edge input function,
8. $\omega : S \to Y$ *is the* edge output function,
9. r *is the* output instruction *(informing the output-viewer how to interpret the output of the protocol),*
10. K *is the totally ordered* cost set,
11. $c : E \to K$ *is the* cost function
12. $\delta : Q \times Q \times K \times S \to Q \times Q \times K \times S$ *is the* transition function.

We assume that the cost remains the same after applying δ *and so we omit specifying an output cost. If* $\delta(q_i, q_j, x, s) = (q'_i, q'_j, s')$ *(which, according to our assumption, is equivalent to* $\delta(q_i, q_j, x, s) = (q'_i, q'_j, x, s')$*), we call* $(q_i, q_j, x, s) \to (q'_i, q'_j, s')$ *a* transition, *and we define* $\delta_1(q_i, q_j, x, s) = q'_i$, $\delta_2(q_i, q_j, x, s) = q'_j$ *and* $\delta_3(q_i, q_j, x, s) = s'$. *We call* δ_1 *the* initiator's acquisition, δ_2 *the* responder's acquisition, *and* δ_3 *the* edge acquisition *(after the corresponding interaction).*

In most cases we assume that $K \subset \mathbb{Z}^+$ and that $c_{max} = \max_{w \in K} \{w\} = \mathcal{O}(1)$. Generally, if $c_{max} = \max_{w \in K} \{|w|\} = \mathcal{O}(1)$ then any agent is capable of storing at most k cumulative costs (at most the value kc_{max}), for some $k = \mathcal{O}(1)$, and we say that the cost function is *useful* (note that a cost range that depends on the population size could make the agents incapable for even a single cost storage and any kind of optimization would be impossible).

A *network configuration* is a mapping $C : V \cup E \to Q \cup S$ specifying the agent state of each agent in the population and the edge state of each edge in the communication graph. Let C and C' be network configurations, and let u, v be distinct agents. We say that C goes to C' via encounter $e = (u, v)$, denoted $C \xrightarrow{e} C'$, if

$$\begin{aligned} C'(u) &= \delta_1(C(u), C(v), x, C(e)) \\ C'(v) &= \delta_2(C(u), C(v), x, C(e)) \\ C'(e) &= \delta_3(C(u), C(v), x, C(e)) \\ C'(z) &= C(z), \text{ for all } z \in (V - \{u, v\}) \cup (E - e). \end{aligned}$$

The definitions of *execution* and *computation* are the same as in the population protocol model but concern network configurations. Note that the mediated population protocol model *preserves both uniformity and anonymity* properties. As a result, any MPP's *code* is of *constant size* and, thus, can be stored in each agent (device) of the population.

A configuration C is called *r-stable* if one of the following conditions holds:

- If the problem concerns a subgraph to be found, then C should fix a subgraph that will not change in any C' reachable from C.
- If the problem concerns a function to be computed by the agents, then an r-stable configuration drops down to an output-stable configuration.

We say that a protocol $\mathcal{A}$ *stably solves* a problem Π, if for every instance I of Π and every computation of $\mathcal{A}$ on I, the network reaches an r-stable configuration C that gives the correct solution for I if interpreted according to the output instruction r. If instead of a problem Π we have a function f to be computed, we say that $\mathcal{A}$ *stably computes* f.

In the special case where Π is an optimization problem, a protocol that stably solves Π is called an *optimizing population protocol* for problem Π.

3.2 An Optimizing Population Protocol

We now give an optimizing population protocol, named *SRLpath*, for the problem of finding the shortest path connecting the root of a directed arborescence to one of its leaves. Formally the problem is the following.

Problem 2. (*Shortest Root-Leaf Path*) Given that the communication graph $G = (V, E)$ is a directed arborescence and a useful cost function $c : E \to K$ on the set of edges, design a protocol that finds the minimum cost path of the (nonempty) set $P = \{p \mid p \text{ is a path from the root to a leaf and } c(p) = O(1)\}$, where $c(p)$ is simply another way to write $\sum_{e \in p} c(e)$.

We assume that the greatest value that any agent is capable of storing is kc_{max}, where both k and $c_{max} = max_{e \in E} c(e)$, are fixed and independent of the size of the population $|V| = n$, given a nonnegative integer-valued cost function $c : E \to K$ (i.e. $K \subset \mathbb{Z}^+$).

If there is at least one path p where $c(p) = \sum_{e \in P} c(e) < kc_{max}$ ($c(p)$ denotes the path length or the total cost of the path), then *SRLpath* will eventually return the shortest path connecting the root to one of the leaves, otherwise it will just return one of the paths with cost at least kc_{max} (one of all root-leaf paths), without guaranteeing that it will be the shortest one, but the output of the root will be 0 indicating that there was no such path.

SRLpath

- $X = \{0, 1\}$,
- $Y = \{0, 1\} \cup Q$,
- $Q = \{q_0, q_s\} \cup \{(i, j) \mid i \in \{q_1, q_2, q_3, q_s\} \text{ and } j \in \{0, 1, 2, \ldots, kc_{max}\}$,
- $I(x) = q_0$, for all $x \in X$,
- $O(q_1, kc_{max}) = 0$, $O(q) = q$, for all $q \in Q - \{(q_1, kc_{max})\}$,
- $S = \{0, 1\}$,
- $\iota(x) = 0$, for all $x \in X$,
- $\omega(s) = s$, for all $s \in S$,
- r: *"If the root outputs 0, fail, else start from the root and follow every edge with output 1, until you reach a leaf"*,
- δ:

$$
\begin{aligned}
(q_0, q_0, c, 0) &\to ((q_1, c), q_0, 1) \\
(q_0, (q_1, c_1), c, 0) &\to ((q_1, kc_{max}), (q_1, c_1), 1), \text{ if } c_1 + c > kc_{max} \\
&\to ((q_1, c_1 + c), (q_1, c_1), 1), \text{ otherwise} \\
((q_1, c_1), (q_1, c_2), c, 1) &\to ((q_1, kc_{max}), (q_1, c_2), 1), \text{ if } c_2 + c > kc_{max} \\
&\to ((q_1, c_2 + c), (q_1, c_2), 1), \text{ otherwise} \\
((q_1, c_1), (q_1, c_2), c, 0) &\to ((q_2, c_2 + c), (q_s, c_2), 1), \text{ if } c_2 + c < c_1 \\
((q_2, c_1), (q_i, c_2), c, 1) &\to ((q_3, c_1), (q_i, c_2), 0), \text{ for } i \in \{1, 2, 3\} \\
((q_3, c_1), (q_s, c_2), c, 1) &\to ((q_1, c_1), (q_1, c_2), 1) \\
((q_1, c_1), q_0, c, 0) &\to ((q_2, c), q_s, 1), \text{ if } c < c_1 \\
((q_2, c_1), q_0, c, 1) &\to ((q_3, c_1), q_0, 0) \\
((q_3, c_1), q_s, c, 1) &\to ((q_1, c_1), q_0, 1)
\end{aligned}
$$

Theorem 3. *If there is at least one root-leaf path p, where $c(p) < kc_{max}$, then SRLpath is an optimizing population protocol for Problem 2. Otherwise, the root outputs 0, indicating that there is no such path.*

Proof. The proof is by induction on the number of nodes of the directed arborescence T. Let $\mathcal{T}_i = \{T \mid T$ is a directed arborescence of i nodes$\}$, i.e. the family of all directed arborescences of i nodes. There is only one directed arborescence in $\mathcal{T}_1$, and only one in $\mathcal{T}_2$, the one that consists of two nodes connected by a directed edge. Obviously, *SRLpath* always finds the shortest root-leaf path for every directed arborescence of at most 2 nodes (finds it trivially). Assume that for every directed arborescence of at most n nodes, *SRLpath* always finds the shortest root-leaf path. Let T_{n+1} be any directed arborescence of $(n+1)$ nodes. By ignoring the root of T_{n+1} and the corresponding edges we get at least one directed arborescence of at most n nodes. On any such subtree we know by the inductive hypothesis that *SRLpath* always finds the shortest root-leaf path. Moreover, it is easy to see that the protocol always keeps in the root of the tree the cost of the selected root-leaf path. Let, u be the removed root, and v_j, where $j = \{1, 2, \ldots, t\}$, its t children. Each child v_j has eventually marked the shortest root-leaf path in the subtree in which v_j is the root and eventually contains the cost of this path in its state, $c(p_j)$. So, eventually u will select the child v_j, for which $\min_{v_j}\{c(p_j) + c(u, v_j)\}$ holds, which will be the shortest root-leaf path of T_{n+1}. Note that if $\min_{v_j}\{c(p_j) + c(u, v_j)\} < kc_{max}$, then at least one such path can be selected by the root (which will never give 0 as output). On the other hand, if there is no such path, then $\min_{v_j}\{c(p_j) + c(u, v_j)\} \geq kc_{max}$ and u can only store kc_{max} which combined with q_1 gives always output 0, indicating that no such path exists. □

3.3 Approximation Protocols

Consider now the following problem:

Problem 3. (*Maximal matching*) Given an undirected communication graph $G = (V, E)$, find a maximal matching, i.e., a set $E' \subseteq E$ such that no two members of E' share a common end point in V and, moreover, there is no $e \in E - E'$ such that e shares no common end point with every member of E'.

A simple protocol to solve this problem is the following: Initially all agents are in state q_0 and all edges in state s_0. When two agents in q_0 interact via an edge in s_0, they both go to q_1 to indicate that they are endpoints of an edge that belongs to the matching formed so far, and the edge goes to s_1 to indicate that it has been put in the matching. All the other transitions have no effect. It is easy to see that eventually the edges in s_1 will form a maximal matching. Moreover, $\omega(s_1) = O(q_1) = 1$ and $\omega(s_0) = O(q_0) = 0$. An appropriate instruction r could be: "*Get each $e \in E$ for which $\omega(s_e) = 1$ (where s_e is the state of e)*", which simply informs the user how to interpret the output of the protocol to get the correct answer (i.e. which edges form the matching).

Definition 7. *Let Π be a minimization problem, and let δ be a positive real number, $\delta \geq 1$. A protocol $\mathcal{A}$ is said to be a δ* factor approximation protocol *for Π if for each instance I of Π, and every computation of $\mathcal{A}$ on I, the network reaches an r-stable configuration C, that, if interpreted according to the output instruction r of $\mathcal{A}$, gives a feasible solution s for I such that*

$$f_{\Pi}(I,s) \leq \delta \cdot OPT(I),$$

where $f_{\Pi}(I,s)$ denotes the objective function value of solution s of instance I, and OPT(I) denotes the objective function value of an optimal solution of instance I.

Now, consider the well-known minimum vertex cover problem defined as follows:

Problem 4. (*Minimum vertex cover*) Given an undirected communication graph $G = (V, E)$, find a minimum cardinality *vertex cover*, i.e., a set $V' \subseteq V$ such that every edge has at least one end point incident at V'.

Let $VertexCover$ be a MPP that agrees on everything to the one already described for Maximal matching, except for the output instruction r, which is now r: "*Get each $v \in V$ for which $O(q_v) = 1$ (where q_v is the state of v)*". Intuitively, we now collect all agents incident to an edge in the maximal mathcing M (for all $e \in M$ we collect the end points of e).

Theorem 4. *$VertexCover$ is a 2 approximation protocol for the minimum vertex cover problem.*

Proof. According to the previous discussion, the edges collected form a maximal matching on G. Moreover, the set C formed by the end points of the edges of M is a minimum vertex cover where $|C| \leq 2 \cdot OPT$, according to the analysis in the introductory chapter of [16]. Thus, $VertexCover$ is a 2 approximation protocol for the minimum vertex cover problem. □

3.4 Computational Power

It is easy to see that the population protocol model is a special case of the mediated population protocol model. In [13] it was proven that there exists a MPP protocol that stably computes the non-semilinear predicate $N_c = N_a \cdot N_b$. In words, it eventually decides whether the number of c's in the input assignment is equal to the product of the number of a's and the number of b's. To do so, the authors stated a composition theorem, that simplifies the proof of existence of MPP protocols. Here we also provide a complete proof of that theorem.

Definition 8. *A MPP $\mathcal{A}$ has* stabilizing states *if in any computation of $\mathcal{A}$, after a finite number of interactions, the states of all agents stop changing.*

Definition 9. *We say that a predicate is* strongly stably computable *by the MPP model, if it is stably computable with the predicate output convention, that is, all agents eventually agree on the correct output value.*

Theorem 5. *Any mediated population protocol $\mathcal{A}$, that stably computes a predicate p with stabilizing states in some family of directed and connected communication graphs $\mathcal{G}$, containing an instruction r that defines a semilinear predicate t on multisets of $\mathcal{A}$'s agent states, can be composed with a provably existing mediated population protocol $\mathcal{B}$, that strongly stably computes t with stabilizing inputs in $\mathcal{G}$, to give a new mediated population protocol $\mathcal{C}$ satisfying the following properties:*

- *$\mathcal{C}$ is formed by the composition of $\mathcal{A}$ and $\mathcal{B}$,*
- *its input is $\mathcal{A}$'s input,*
- *its output is $\mathcal{B}$'s output, and*
- *$\mathcal{C}$ strongly stably computes p (i.e. all agents agree on the correct output) in $\mathcal{G}$.*

Proof. Protocol $\mathcal{A}$ has stabilizing states and an instruction r that defines a semilinear predicate t on multisets of $\mathcal{A}$'s states. Let $X_{\mathcal{A}}$ be the input alphabet of $\mathcal{A}$, $Q_{\mathcal{A}}$ the set of $\mathcal{A}$'s states, $\delta_{\mathcal{A}}$ the transition function of $\mathcal{A}$, and similarly for any other component of $\mathcal{A}$. We will use the indexes $\mathcal{B}$ and $\mathcal{C}$, for the corresponding components of the other two protocols.

Since predicate t is semilinear, according to a result in [1], there is a population protocol $\mathcal{B}'$ that stably computes t with stabilizing inputs in the *unrestricted family of graphs* (denoted by $\mathcal{G}^d_{Unr}$, and consisting of all possible directed and connected communication graphs). Note that $\mathcal{G}$ is a subset of $\mathcal{G}^d_{Unr}$ ($\mathcal{G} \subseteq \mathcal{G}^d_{Unr}$), so any predicate stably computable (both with or without stabilizing inputs) in $\mathcal{G}^d_{Unr}$ is also stably computable in $\mathcal{G}$, since it is stably computable in any possible communication graph. So, $\mathcal{B}'$ stably computes t with stabilizing inputs in $\mathcal{G}$. Moreover, there also exists a mediated population protocol $\mathcal{B}$ (the one that is the same as $\mathcal{B}'$ but simply ignores the additional components of the new model) that strongly stably computes t with stabilizing inputs in $\mathcal{G}$. Note that the input alphabet of $\mathcal{B}$ is $X_{\mathcal{B}} = Q_{\mathcal{A}}$, and its transition function is of the form $\delta_{\mathcal{B}} : (Q_{\mathcal{A}} \times Q_{\mathcal{B}}) \times (Q_{\mathcal{A}} \times Q_{\mathcal{B}}) \to Q_{\mathcal{B}} \times Q_{\mathcal{B}}$, since there is no need to specify edge states (formally we should, but the protocol ignores them). $Q_{\mathcal{A}}$ is the set of $\mathcal{A}$'s agent states and $\mathcal{B}$'s inputs that eventually stabilize.

We define a mediated population protocol $\mathcal{C}$ as follows: $X_{\mathcal{C}} = X_{\mathcal{A}}$, $Y_{\mathcal{C}} = Y_{\mathcal{B}} = \{0,1\}$, $Q_{\mathcal{C}} = Q_{\mathcal{A}} \times Q_{\mathcal{B}}$, $I_{\mathcal{C}} : X_{\mathcal{A}} \to Q_{\mathcal{C}}$ defined as $I_{\mathcal{C}}(x) = (I_{\mathcal{A}}(x), i_{\mathcal{B}})$, for all $x \in Q_{\mathcal{C}}$ where $i_{\mathcal{B}} \in Q_{\mathcal{B}}$ is the initial state of protocol $\mathcal{B}$, $S_{\mathcal{C}} = S_{\mathcal{A}}$, $\iota_{\mathcal{C}} : X_{\mathcal{C}} \to S_{\mathcal{C}}$, that is, $\iota_{\mathcal{C}}(x) = \iota_{\mathcal{A}}(x)$, for all $x \in X_{\mathcal{C}}$, $O_{\mathcal{C}}(a,b) = O_{\mathcal{B}}(b)$, for all $q = (a,b) \in Q_{\mathcal{C}}$, and finally its transition function $\delta_{\mathcal{C}} : Q_{\mathcal{C}} \times Q_{\mathcal{C}} \times S_{\mathcal{C}} \to Q_{\mathcal{C}} \times Q_{\mathcal{C}} \times S_{\mathcal{C}}$ (we omit specifying costs since there is no need for them) is defined as

$$\begin{aligned}\delta_{\mathcal{C}}((a,b),(a',b'),s) = ((&\delta_{\mathcal{A}_1}(a,a',s), \delta_{\mathcal{B}_1}((a,b),(a',b'))),\\ (&\delta_{\mathcal{A}_2}(a,a',s), \delta_{\mathcal{B}_2}((a,b),(a',b'))),\\ &\delta_{\mathcal{A}_3}(a,a',s)),\end{aligned}$$

where for $\delta_{\mathcal{A}}(x,y,z) = (x',y',z')$ (in $\mathcal{A}$'s transition function), we have that $\delta_{\mathcal{A}_1}(x,y,z) = x'$, $\delta_{\mathcal{A}_2}(x,y,z) = y'$, $\delta_{\mathcal{A}_3}(x,y,z) = z'$, and similarly for $\delta_{\mathcal{B}}$.

Intuitively, $\mathcal{C}$ consists of $\mathcal{A}$ and $\mathcal{B}$ running in parallel. The state of each agent is a pair $c = (a, b)$, where $a \in Q_{\mathcal{A}}, b \in Q_{\mathcal{B}}$, and the state of each edge is a member of $S_{\mathcal{A}}$. Initially each agent senses an input x from $X_{\mathcal{A}}$ and this is transformed according to $I_{\mathcal{C}}$ to such a pair, where $a = I_{\mathcal{A}}(x)$ and b is always a special $\mathcal{B}$'s initial state $i_{\mathcal{B}} \in Q_{\mathcal{B}}$. When two agents in states (a, b) and (a', b') interact through an edge in state s, then protocol $\mathcal{A}$ updates the first components of the agent states, i.e. a and a', and the edge state s, as if $\mathcal{B}$ didn't exist. On the other hand, protocol $\mathcal{B}$ updates the second components by taking into account the first components that represent its separate input ports at which the current input symbol of each agent is available at every interaction ($\mathcal{B}$ takes $\mathcal{A}$'s states for agent input symbols that may change arbitrarily between any two computation steps, but the truth is that they change due to $\mathcal{A}$'s computation). Since the first components of $\mathcal{C}$'s agent states eventually stabilize as a result of $\mathcal{A}$'s stabilizing states, protocol $\mathcal{B}$ will eventually obtain stabilizing inputs, consequently will operate correctly, and will strongly stably compute t as if it had began computing on $\mathcal{A}$'s final (output) configuration. But, since t provides the correct answer for p if applied on $\mathcal{A}$'s final configuration, it is obvious that $\mathcal{C}$ must strongly stably compute p in $\mathcal{G}$, and the theorem follows. □

Since MPP strongly stably computes a non-semilinear predicate and PP is a special case of MPP, it follows that the class of stably computable predicates by MPP is a proper superset of the class of stably computable predicates by PP. In other words, *the MPP model is computationally stronger than the PP model.*

In [13] the authors also proved the following result: "Any predicate that is stably computable by the MPP model in any family of communication graphs belongs to the space complexity class $NSPACE(m)$". The idea is simple: By using the MPP that stably computes the predicate we construct a nondeterministic Turing machine that guesses in each step the next selection of the scheduler (thus the next configuration). The machine always replaces the current configuration with a new legal one, and, since any configuration can be represented explicitly with $\mathcal{O}(m)$ space, any branch uses $\mathcal{O}(m)$ space. The machine accepts if some branch reaches a configuration C that satisfies instruction r of the protocol, and if, moreover, no configuration reachable from C violates r (i.e. C must also be r-stable).

4 The GDM Model

In [13] MPP's ability to decide graph languages was implicitly observed. In fact, the authors made a similar observation: "MPPs are able to locate subgraphs". Based on this observation, in [11] the authors considered a special case of the mediated population protocol model, the *graph decision mediated population protocol model*, or simply *GDM*. The purpose of GDM was to simplify the study of decidability capabilities of the MPP model.

4.1 Formal Definition

Definition 10. *A* GDM *is an 8-tuple* $(Y, Q, O, S, r, \delta, q_0, s_0)$, *where* Y, Q, *and* S *are all finite sets and*

1. $Y = \{0,1\}$ *is the* binary output alphabet,
2. Q *is the set of* agent states,
3. $O : Q \to Y$ *is the* agent output function,
4. S *is the set of* edge states,
5. r *is the* output instruction,
6. $\delta : Q \times Q \times S \to Q \times Q \times S$ *is the* transition function,
7. $q_0 \in Q$ *is the* initial agent state, *and*
8. $s_0 \in S$ *is the* initial edge state.

If $\delta(a,b,s) = (a',b',s')$, *we call* $(a,b,s) \to (a',b',s')$ *a* transition *and we define* $\delta_1(a,b,s) = a'$, $\delta_2(a,b,s) = b'$, *and* $\delta_3(a,b,s) = s'$.

Let $\mathcal{U}$ be a graph universe. A *graph language* L is a subset of $\mathcal{U}$ containing communication graphs that share some common property. For example, a common graph universe is the set of all possible directed and weakly connected communication graphs, denoted by $\mathcal{G}$, and $L = \{G \in \mathcal{G} \mid G \text{ has an even number of edges}\}$ is a possible graph language w.r.t. $\mathcal{G}$.

A GDM protocol may run on any graph from a specified graph universe. The graph on which the protocol runs is considered as the *input graph* of the protocol. Note that GDM protocols have no sensed input. Instead, we require each agent in the population to be initially in the initial agent state q_0 and each edge of the communication graph to be initially in the initial edge state s_0. In other words, the initial network configuration, C_0, of any GDM is defined as $C_0(u) = q_0$, for all $u \in V$, and $C_0(e) = s_0$, for all $e \in E$, and any input graph $G = (V, E)$.

We say that a GDM $\mathcal{A}$ *accepts* an input graph G if in any computation of $\mathcal{A}$ on G after finitely many interactions all agents output the value 1 and continue doing so in all subsequent (infinite) computational steps. By replacing 1 with 0 we get the definition of the *reject* case.

Definition 11. *We say that a GDM $\mathcal{A}$* decides *a graph language $L \subseteq \mathcal{U}$ if it accepts any $G \in L$ and rejects any $G \notin L$.*

Definition 12. *A graph language is said to be* decidable *if some GDM decides it.*

4.2 Weakly Connected Graphs

Decidability

The most meaningful graph universe is $\mathcal{G}$ containing all possible directed and weakly connected communication graphs, without self-loops or multiple edges, of any finite number of nodes greater or equal to 2 (we do not allow the empty graph, the graph with a unique node and we neither allow infinite graphs). Here the graph universe is $\mathcal{G}$ and, thus, a graph language can only be a subset of $\mathcal{G}$ (moreover, its elements must share some common property).

In [11] it was proven that the class of decidable graph languages is *closed* under *complement*, *union* and *intersection* operations. Moreover, the authors provided protocols (and proved their correctness) that decide the following graph languages:

1. $N_{even} = \{G \in \mathcal{G} \mid |V(G)| \text{ is even}\}$,
2. $E_{even} = \{G \in \mathcal{G} \mid |E(G)| \text{ is even}\}$,
3. $N_k^{out} = \{G \in \mathcal{G} \mid G$ has some node with at least k outgoing neighbors$\}$ for any $k = \mathcal{O}(1)$,
4. $K_k^{out} = \{G \in \mathcal{G} \mid$ Any node in G has at least k outgoing neighbors$\}$ for any $k = \mathcal{O}(1)$,
5. $M_{out} = \{G \in \mathcal{G} \mid G$ has some node with more outgoing than incoming neighbors$\}$, and
6. $P_k = \{G \in \mathcal{G} \mid G$ has at least one directed path of at least k edges$\}$ for any $k = \mathcal{O}(1)$.

So, all the above languages are decidable by the GDM model, and, by closure under complement, the same holds for their complements. For example, $\overline{N}_k^{out}$ contains all graphs that have no node with at least $k = \mathcal{O}(1)$ outgoing neighbors, and is decidable. In other words, GDM can decide if all nodes have less than k outgoing edges, which is simply the well-known bounded by k out-degree predicate.

To illustrate the formal description of GDM protocols we provide the code of the protocol *DirPath* that was proven in [11] to decide the language $P_k = \{G \in \mathcal{G} \mid G$ has at least one directed path of at least k edges$\}$ for any $k = \mathcal{O}(1)$.

DirPath

- $Q = \{q_0, q_1, 1, \ldots, k\}$, $S = \{0, 1\}$,
- $O(k) = 1$, $O(q) = 0$, for all $q \in Q - \{k\}$,
- r: "*Get any* $u \in V$ *and read its output*",
- δ:

$$\begin{aligned}
(q_0, q_0, 0) &\to (q_1, 1, 1) \\
(q_1, x, 1) &\to (x-1, q_0, 0), \text{ if } x \geq 2 \\
&\to (q_0, q_0, 0), \text{ if } x = 1 \\
(x, q_0, 0) &\to (q_1, x+1, 1), \text{ if } x+1 < k \\
&\to (k, k, 0), \text{ if } x+1 = k \\
(k, \cdot, \cdot) &\to (k, k, \cdot) \\
(\cdot, k, \cdot) &\to (k, k, \cdot)
\end{aligned}$$

Undecidability

If we allow only GDMs with stabilizing states, i.e. GDMs that in any computation after finitely many interactions stop changing their states, then we can prove that a specific graph language w.r.t. $\mathcal{G}$ is *undecidable*. In particular, we can prove that there exists no GDM with stabilizing states to decide the graph language

$$\begin{aligned}
2C = \{G \in \mathcal{G} \mid\ & G \text{ has at least two nodes } u, v \text{ s.t. both } (u, v), (v, u) \\
& \in E(G) \text{ (in other words, } G \text{ has at least one 2-cycle)}\}.
\end{aligned}$$

The proof is based on the following lemma.

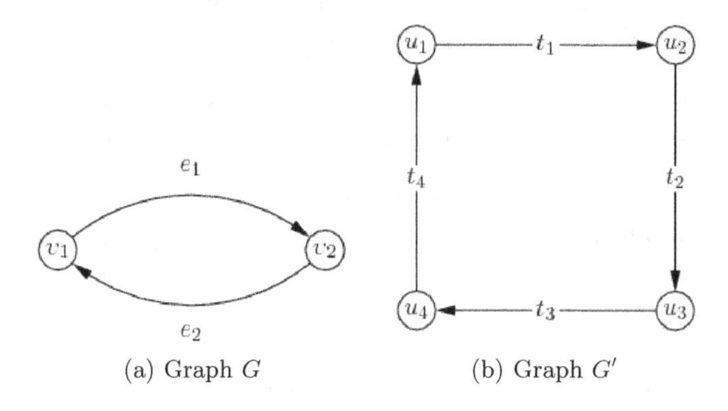

(a) Graph G (b) Graph G'

Fig. 1. $G \in 2C$ and $G' \notin 2C$

Lemma 3. *For any GDM $\mathcal{A}$ and any computation $C_0, C_1, C_2, \ldots$ of $\mathcal{A}$ on G (Figure 1(a)) there exists a computation $C'_0, C'_1, C'_2, \ldots, C'_i, \ldots$ of $\mathcal{A}$ on G' (Figure 1(b)) s.t.*

$$
\begin{aligned}
C_i(v_1) &= C'_{2i}(u_1) = C'_{2i}(u_3) \\
C_i(v_2) &= C'_{2i}(u_2) = C'_{2i}(u_4) \\
C_i(e_1) &= C'_{2i}(t_1) = C'_{2i}(t_3) \\
C_i(e_2) &= C'_{2i}(t_2) = C'_{2i}(t_4)
\end{aligned}
$$

for any finite $i \geq 0$.

Proof. The proof is by induction on i.

Lemma 3 shows that if a GDM $\mathcal{A}$ with stabilizing states could decide $2C$ then there would exist a computation of $\mathcal{A}$ on G' forcing all agents to output incorrectly the value 1 in finitely many steps. But G' does not belong to $2C$, and, since $\mathcal{A}$ decides $2C$, all agents must correct their states to eventually output 0. By taking into account the fact that $\mathcal{A}$ has stabilizing states it is easy to reach a contradiction and prove that no GDM with stabilizing states can decide $2C$. Whether the graph language $2C$ is undecidable by the GDM model in the general case (not only by GDMs with stabilizing states) remains an interesting open problem.

4.3 All Possible Directed Graphs

In [11] it was, also, proven that if we allow the graph universe, $\mathcal{H}$, to contain also disconnected communication graphs, then in this case the GDM model is incapable of deciding even a single nontrivial graph language (we call a graph language L *nontrivial* if $L \neq \emptyset$ and $L \neq \mathcal{H}$). Here we assume the graph universe $\mathcal{H}$ consisting of all possible directed communication graphs, without self-loops or multiple edges of any finite number of nodes greater or equal to 2 (we now also allow graphs that are not even weakly connected). So, now, a graph language can only be a subset of $\mathcal{H}$. We will give the proof idea of that general impossibility result.

First we show that for any nontrivial graph language L, there exists some disconnected graph G in L where at least one component of G does not belong to L or there exists some disconnected graph G' in $\overline{L}$ where at least one component of G' does not belong to $\overline{L}$ (or both). If the statement does not hold then any disconnected graph in L has all its components in L and any disconnected graph in $\overline{L}$ has all its components in $\overline{L}$.

1. All connected graphs belong to L. Then $\overline{L}$ contains at least one disconnected graph (since it is nontrivial) that has all its components in L, which contradicts the fact that the components of any disconnected graph in $\overline{L}$ also belong to $\overline{L}$.
2. All connected graphs belong to $\overline{L}$. The contradiction is symmetric to the previous case.
3. L and $\overline{L}$ contain connected graphs G and G', respectively. Their disjoint union $U = (V \cup V', E \cup E')$ is disconnected, belongs to L or $\overline{L}$ but one of its components belongs to L and the other to $\overline{L}$. The latter contradicts the fact that both components should belong to the same language.

To obtain the impossibility result the reader should use the fact that the class of decidable graph languages is closed under complement and, also, simply notice that GDMs have no way to transmit data between agents of different components when run on disconnected graphs (in fact, it is trivial to see that, when run on disconnected graphs, those protocols essentially run individually on the different components of those graphs).

5 Further Research Directions

Since the Mediated Population Protocol model was proposed very recently, many important questions concerning it remain open and many directions emerging from it are yet unexplored. The first question that comes to one's mind is if there exists some achievable architecture that implements the proposed models. Is there some other notion of fairness (probably weaker) that would be more suitable for real-life applications? Can we give an exact characterization of the stably computable predicates by the Mediated Population Protocol model like the one already given for the Population Protocol model? If we ignore the sensing capabilities of the MPP model and focus on the GDM model, can we give an exact characterization of the class of decidable graph languages? Is there a general method for proving impossibility results that best suits the GDM model? In other words, can we avoid ad-hoc proofs of impossibility results and borrow or modify techniques from classical distributed computing by also taking into account the fact that our systems are uniform, anonymous, have constant protocol descriptions, and have some nondeterminism inherent in the interaction pattern? Can we devise some reliable simulation platform or testbed to extensively test or/and verify our protocols before running them in real application scenarios? Since sensor networks are most of the time used in critical environments (fire detection is a classical example), it is reasonable to wonder whether there exists a unified theoretical framework for fast and reliable protocol verification. Since such protocols are usually small we are hoping for the existence of some fast verification process. Assuming a unique leader in the system has been always helpful in distributed computing (to get an idea of its usefulness in population protocols the reader is referred to [4]). It seems that in the models under consideration assuming a leader in the initial configuration is more helpful than letting the protocol elect one. In particular, it seems that an assumed leader provides us with even more computational power, especially in the case where the goal is the construction of some subgraph or the decision of some specific property of the communication graph. Finally, we believe that it is not so bad to assume that some or all agents have $\mathcal{O}(\log n)$ storage capacity (note that if the population consists of 1 billion agents, i.e. $n = 10^9$, then only about 30 bits of memory are required in each agent!), then it is possible to assume unique identifiers, population protocols are no longer anonymous, and it is of great interest studying this realizable scenario.

References

1. Angluin, D., Aspnes, J., Chan, M., Fischer, M.J., Jiang, H., Peralta, R.: Stably computable properties of network graphs. In: Prasanna, V.K., Iyengar, S.S., Spirakis, P.G., Welsh, M. (eds.) DCOSS 2005. LNCS, vol. 3560, pp. 63–74. Springer, Heidelberg (2005)
2. Angluin, D., Aspnes, J., Diamadi, Z., Fischer, M.J., Peralta, R.: Computation in networks of passively mobile finite-state sensors. In: 23rd Annual ACM Sympsium on Principles of Distributed Computing PODC, pp. 290–299. ACM, New York (2004)
3. Angluin, D., Aspnes, J., Diamadi, Z., Fischer, M.J., Peralta, R.: Computation in networks of passively mobile finite-state sensors. Distributed Computing 18(4), 235–253 (2006)
4. Angluin, D., Aspnes, J., Eisenstat, D.: Fast computation by population protocols with a leader. Distributed Computing 21(3), 183–199 (2008)
5. Angluin, D., Aspnes, J., Eisenstat, D.: Stably computable predicates are semilinear. In: Proc. 25th Annual ACM Symposium on Principles of Distributed Computing, pp. 292–299 (2006)
6. Angluin, D., Aspnes, J., Eisenstat, D., Ruppert, E.: The computational power of population protocols. Distributed Computing 20(4), 279–304 (2007)
7. Aspnes, J., Ruppert, E.: An introduction to population protocols. Bulletin of the European Association for Theoretical Computer Science 93, 98–117 (2007); Mavronicolas, M. (ed.) Columns: Distributed Computing
8. Beauquier, J., Clement, J., Messika, S., Rosaz, L., Rozoy, B.: Self-stabilizing counting in mobile sensor networks. Technical Report 1470, LRI, Université Paris-Sud 11 (2007)
9. Bournez, O., Chassaing, P., Cohen, J., Gerin, L., Koegler, X.: On the convergence of population protocols when population goes to infinity. To appear in Applied Mathematics and Computation (2009)

10. Chatzigiannakis, I., Spirakis, P.G.: The dynamics of probabilistic population protocols. In: Taubenfeld, G. (ed.) DISC 2008. LNCS, vol. 5218, pp. 498–499. Springer, Heidelberg (2008)
11. Chatzigiannakis, I., Michail, O., Spirakis, P.G.: Decidable Graph Languages by Mediated Population Protocols. FRONTS Technical Report FRONTS-TR-2009-16 (May 2009), `http://fronts.cti.gr/aigaion/?TR=80`
12. Chatzigiannakis, I., Michail, O., Spirakis, P.G.: Experimental verification and performance study of extremely large sized population protocols. FRONTS Technical Report FRONTS-TR-2009-3 (January 2009), `http://fronts.cti.gr/aigaion/?TR=61`
13. Chatzigiannakis, I., Michail, O., Spirakis, P.G.: Mediated Population Protocols. FRONTS Technical Report FRONTS-TR-2009-8 (February 2009), `http://fronts.cti.gr/aigaion/?TR=65`, To appear in 36th International Colloquium on Automata, Languages and Programming (ICALP), Rhodes, Greece, July 5-12 (2009)
14. Delporte-Gallet, C., Fauconnier, H., Guerraoui, R., Ruppert, E.: When birds die: Making population protocols fault-tolerant. In: Gibbons, P.B., Abdelzaher, T., Aspnes, J., Rao, R. (eds.) DCOSS 2006. LNCS, vol. 4026, pp. 51–66. Springer, Heidelberg (2006)
15. Ginsburg, S., Spanier, E.H.: Semigroups, Presburger formulas, and languages. Pacific Journal of Mathematics 16, 285–296 (1966)
16. Vazirani, V.: Approximation Algorithms. Springer, Heidelberg (2001)

How to Sort a Train

Peter Widmayer

Institute of Theoretical Computer Science, ETH Zürich, Switzerland
widmayer@inf.ethz.ch

Abstract. This talk attempts to provide an introductory survey of methods for sorting trains in railway yards from an algorithmic perspective, and to highlight open problems in this area.

R. Královič and D. Niwiński (Eds.): MFCS 2009, LNCS 5734, p. 77, 2009.

Arithmetic Circuits, Monomial Algebras and Finite Automata

Vikraman Arvind and Pushkar S. Joglekar

Institute of Mathematical Sciences
C.I.T Campus, Chennai 600 113, India
{arvind,pushkar}@imsc.res.in

Abstract. We study lower bounds for circuit and branching program size over *monomial algebras* both in the noncommutative and commutative setting. Our main tool is automata theory and the main results are:

- An extension of Nisan's noncommutative algebraic branching program size lower bounds [N91] over the free noncommutative ring $\mathbb{F}\langle x_1, x_2, \cdots, x_n\rangle$ to similar lower bounds over the noncommutative monomial algebras $\mathbb{F}\langle x_1, x_2, \cdots, x_n\rangle/I$ for a monomial ideal I generated by subexponential number of monomials.
- An extension of the exponential size lower bounds for monotone commutative circuits [JS82] computing the Permanent in $\mathbb{Q}[x_{11}, x_{12}, \cdots, x_{nn}]$ to an exponential lower bound for monotone commutative circuits computing the Permanent in any monomial algebra $\mathbb{Q}[x_{11}, x_{12}, \cdots, x_{nn}]/I$ such that the monomial ideal I is generated by $o(n/\log n)$ monomials.

1 Introduction

Lower bounds for noncommutative computation were studied in the pioneering paper of Nisan [N91]; he studied noncommutative arithmetic circuits, formulas and algebraic branching programs. Using a rank argument Nisan has shown that the permanent and determinant polynomials in the *free* noncommutative ring $\mathbb{F}\{x_{11}, \cdots, x_{nn}\}$ require exponential size noncommutative formulas (and noncommutative algebraic branching programs). Chien and Sinclair [CS04] explore the same question over other noncommutative algebras. They refine Nisan's rank argument to show exponential size lower bounds for formulas computing the permanent or determinant over the algebra of 2×2 matrices over $\mathbb{F}$, the quaternion algebra, and several other interesting examples.

We study arithmetic circuit lower bounds for noncommutative *monomial algebras.* Recall that an ideal I (more precisely, a 2-sided ideal) of the noncommutative polynomial ring $\mathbb{F}\langle X\rangle$ is a subring that is closed under both left and right multiplication by the ring elements. Our aim is to show arithmetic circuit lower bounds and identity testing in the quotient algebra $\mathbb{F}\langle X\rangle/I$ for different classes of ideals I given by generating sets of polynomials. The circuit size of the polynomial f in the algebra $\mathbb{F}\langle X\rangle/I$ is $C_I(f) = \min_{g\in I} C(f+g)$. Relatedly, we also study polynomial identity testing in monomial algebras $\mathbb{F}\langle X\rangle/I$.

It turns out that the structure of monomial algebras is intimately connected with automata theory. We start with a basic definition. Suppose $X = \{x_1, x_2, \cdots, x_n\}$ is

R. Královič and D. Niwiński (Eds.): MFCS 2009, LNCS 5734, pp. 78–89, 2009.

a set of n noncommuting variables. The free monoid X^* consists of all words over these variables. Let $\mathbb{F}\langle x_1, x_2, \cdots, x_n\rangle$ denote the free noncommutative polynomial ring generated by the variables in X over a field $\mathbb{F}$. The polynomials in this algebra are $\mathbb{F}$-linear combinations of words over X. We denote this noncommutative algebra by $\mathbb{F}\langle X\rangle$. Similarly, if $X = \{x_1, \ldots, x_n\}$ is set of commuting variables then $\mathbb{F}[X] = \mathbb{F}[x_1, \ldots, x_n]$ denotes the commutative ring of polynomials in the variables X over field $\mathbb{F}$. For a polynomial $f \in \mathbb{F}\langle X\rangle$, let $\text{mon}(f) = \{m \in X^* \mid m \text{ is a nonzero monomial in } f\}$ denote the monomial set of f. The monomial set $\text{mon}(f)$ for $f \in \mathbb{F}[X]$ is defined analogously.

Definition 1. *Let $f \in \mathbb{F}\langle X\rangle$ be a polynomial and $\mathcal{A}$ be a finite automaton (deterministic or nondeterministic) accepting a subset of X^*. The* intersection *of $f = \sum c_m m$ by $\mathcal{A}$ is the polynomial $f_{\mathcal{A}} = \sum_{m \in mon(f) \cap L(\mathcal{A})} c_m m$.*

Let $f(mod\ \mathcal{A})$ denote the polynomial $f - f_{\mathcal{A}}$. We refer to $f(mod\ \mathcal{A})$ as the quotient *of f by $\mathcal{A}$. Thus, the automaton $\mathcal{A}$ splits the polynomial f into two parts as $f = f(mod\ \mathcal{A}) + f_{\mathcal{A}}$.*

Given an arithmetic circuit C (or an ABP P) computing a polynomial in $\mathbb{F}\langle X\rangle$ and a deterministic finite automaton $\mathcal{A}$ (a DFA or an NFA) we can talk of the polynomials $C_{\mathcal{A}}$, $C(\text{mod}\mathcal{A})$, $P_{\mathcal{A}}$ and $P(\text{mod}\mathcal{A})$. We show polynomial bounds on the circuit size (respectively, ABP size) of these polynomials in the sizes of C (or P) and $\mathcal{A}$.

If I is a *finitely generated monomial ideal* of $\mathbb{F}\langle X\rangle$, we can design a polynomial-size "pattern matching" DFA $\mathcal{A}$ that accepts precisely the monomials in I. Using this we can reduce the problem of proving lower bounds (and polynomial identity testing) for the monomial algebra $\mathbb{F}\langle X\rangle/I$ to the free noncommutative ring $\mathbb{F}\langle X\rangle$. Applying this idea, we show that the Permanent (and Determinant) in the quotient algebra $\mathbb{F}\langle X\rangle/I$ still requires exponential size ABPs. Hence, we can extend Nisan's lower bound argument to noncommutative monomial algebras. Furthermore, the Raz-Shpilka deterministic identity test for noncommutative ABPs [RS05] also carry over to $\mathbb{F}\langle X\rangle/I$.

In the commutative setting, Jerrum and Snir [JS82] have shown a $2^{\Omega(n)}$ size lower bound for monotone arithmetic circuits computing the $n \times n$ Permanent. We examine the size of monotone arithmetic circuits for any commutative monomial algebra $\mathbb{Q}[x_{11}, x_{12}, \cdots, x_{nn}]/I$ where I is a monomial ideal. Our main result here is a $2^{\Omega(n)}$ lower bound for the $n \times n$ Permanent over $\mathbb{Q}[x_{11}, x_{12}, \cdots, x_{nn}]/I$, where the monomial ideal I is generated by $o(n/\log n)$ monomials.

Next, we study the *Monomial Search Problem.* This is a natural search version of polynomial identity testing: Given a polynomial $f \in \mathbb{F}\langle X\rangle$ (or, in the commutative case $f \in \mathbb{F}[X]$) of total degree d by an arithmetic circuit C or an ABP, the problem is to *find* a nonzero monomial of the polynomial f. Applying our results on intersection of noncommutative ABPs over $\mathbb{F}$ with a DFA, we give a randomized NC^2 algorithm for finding a nonzero monomial and its coefficient. We note that in [AM08, AMS08] there is a similar application of automata theory to isolate a term of the polynomial computed by a given arithmetic circuit. However, the purpose in [AMS08, AM08] is only identity testing and not finding a nonzero monomial. The automata used in [AM08, AMS08] essentially assign random weights to variables to isolate a nonzero monomial. Our notion of intersection and quotienting with automata builds on ideas used in [AM08, AMS08].

We also obtain randomized NC^2 Monomial search algorithm for *commutative* ABPs. For general arithmetic circuits we obtain a randomized NC reduction from Monomial search to identity testing.

We start with some definitions. An *arithmetic circuit* computing a polynomial in the ring $\mathbb{F}[x_1, \cdots, x_n]$ is a directed acyclic graph. Each node if in-degree zero is labelled by a variable x_i or a field element. Each internal node of the circuit has in-degree 2 and is either a $+$ (addition gate) or a $*$ (multiplication gate). The circuit has one special node designated the output gate which computes a polynomial in $\mathbb{F}[x_1, \cdots, x_n]$. A *noncommutative* arithmetic circuit is defined as above except that the inputs to each multiplication gate of the circuit are ordered as left and right (to capture the fact that $*$ is noncommutative). Clearly, such an arithmetic circuit computes a polynomial in the noncommutative ring $\mathbb{F}\langle x_1, \ldots x_n \rangle$.

Definition 2. [N91, RS05] *An* algebraic branching program *(ABP) is a layered directed acyclic graph with one* source *vertex of in-degree zero and one* sink *vertex of out-degree zero. The vertices of the graph are partitioned into layers numbered* $0, 1, \cdots, d$*. Edges may only go from layer* i *to* $i + 1$ *for* $i \in \{0, \cdots, d - 1\}$*. The source is the only vertex at layer* 0 *and the sink is the only vertex at layer* d*. Each edge is labeled with a homogeneous linear form in the input variables. The size of the ABP is the number of vertices.*

We now recall definitions of some complexity classes. Fix a finite input alphabet Σ. A language $L \subseteq \Sigma^*$ is in the class *logspace* (denoted L) if there is a deterministic Turing machine with a read-only input tape and an $O(\log n)$ space-bounded work tape that accepts the language L.

Definition 3. *The complexity class* GapL *is the class of functions* $f : \Sigma^* \rightarrow \mathbb{Z}$*, for which there is a logspace bounded nondeterministic Turing machine* M *such that on any input* $x \in \Sigma^*$*, we have* $f(x) = acc_M(x) - rej_M(x)$*, where* $acc_M(x)$ *and* $rej_M(x)$ *denote the number of accepting and rejecting computation paths of* M *on input* x *respectively.*

A language L is in randomized NC^2 if there is a logspace uniform boolean circuit family $\{C_n\}_{n \geq 1}$ of polynomial size and $\log^2 n$ depth with gates of constant-fanin such that for $x \in \Sigma^n$ we have $\Pr_w[C_n(x, w) = \chi_L(x)] \geq 2/3$, where χ_L is the characteristic function for L.

2 Intersecting and Quotienting by Automata

In this section we focus on the circuit and ABP size complexities of $f(\bmod\ \mathcal{A})$ and $f_{\mathcal{A}}$, in terms of the circuit (resp. ABP) complexity of f and the size of the automaton $\mathcal{A}$, in the case when $\mathcal{A}$ is a deterministic finite automaton. The bounds we obtain are *constructive*: we will efficiently compute a circuit (resp. ABP) for $f(\bmod\ \mathcal{A})$ and $f_{\mathcal{A}}$ from the given circuit (resp. ABP) for f and $\mathcal{A}$.

We first recall the following complexity measures for a polynomial $f \in \mathbb{F}\langle X \rangle$ from Nisan [N91].

Definition 4. [N91] *For $f \in \mathbb{F}\langle X \rangle$, we denote its formula complexity by $F(f)$, its circuit complexity by $C(f)$, and the algebraic branching program complexity by $B(f)$. For $\mathbb{F} = \mathbb{R}$ and a polynomial $f \in \mathbb{F}\langle X \rangle$ with positive coefficients, its* monotone *circuit complexity is denoted by $C^+(f)$,*

We have following theorem relating the complexity of $f(\text{mod } \mathcal{A})$ and $f_\mathcal{A}$ to the complexity of f.

Theorem 1. *Let $f \in \mathbb{F}\langle X \rangle$ and $\mathcal{A}$ be a DFA with s states accepting some subset of X^*. Then, for $g \in \{f(\text{mod } \mathcal{A}), f_\mathcal{A}\}$ we have*

1. $C(g) \leq C(f) \cdot (ns)^{O(1)}$.
2. $C^+(g) \leq C^+(f) \cdot (ns)^{O(1)}$.
3. $B(g) \leq B(f) \cdot (ns)^{O(1)}$.

Furthermore, the circuit (ABP) of the size given above for polynomial g can be computed in deterministic logspace (hence in NC^2*) on input a circuit (resp. ABP) for f and the DFA $\mathcal{A}$.*

Proof. We first describe a circuit construction that proves parts 1 and 2 of the theorem.

Let $\mathcal{A} = (Q, X, \delta, q_0, F)$ be the quintuple describing the given DFA with s states. We can extend the transition function δ to words (i.e. monomials) in X^* as usual: $\delta(a, m) = b$ for states $a, b \in Q$ and a monomial m if the DFA goes from state a to b on the monomial m. In particular, we note that $\delta(a, \epsilon) = a$ for each state a. As in automata theory, this is a useful convention because when we write a polynomial $f \in \mathbb{F}\langle X \rangle$ as $\sum c_m m$, where c_m is the coefficient of the monomial m in f, we can allow for ϵ as the monomial corresponding to the constant term in f.

Let C be the given circuit computing polynomial f. For each gate g of C, let f_g denote the polynomial computed by C at the gate g. In the new circuit C' we will have s^2 gates $\langle g, a, b \rangle, a, b \in Q$ corresponding to each gate g of C. Recall that mon(f) is the set of monomials of f. Let $M_{ab} = \{m \in X^* \mid \delta(a, m) = b\}$ for states $a, b \in Q$. At the gate $\langle g, a, b \rangle$ the circuit C' will compute the polynomial $f_g^{a,b} = \sum_{m \in M_{ab} \cap mon(f_g)} c_m m$, where $f_g = \sum c_m m$.

The input-output connections between the gates of C' are now easy to define. If g is a $+$ gate with input gates h and k so that $f_g = f_h + f_k$, we have $f_g^{a,b} = f_h^{a,b} + f_k^{a,b}$, implying that $\langle h, a, b \rangle$ and $\langle k, a, b \rangle$ are the inputs to the $+$ gate $\langle g, a, b \rangle$. If g is a $\times$ gate, with inputs h and k so that $f_g = f_h \cdot f_k$, we have $f_g^{a,b} = \sum_{c \in Q} f_h^{a,c} \cdot f_k^{c,b}$.

This simple formula can be easily computed by a small subcircuit with $O(s)$ many $+$ gates and $\times$ gates. Finally, let *out* denote the output gate of circuit C, so that $f_{out} = f$. It follows from the definitions that $f(\text{mod } \mathcal{A}) = \sum_{a \notin F} f_{out}^{q_0,a}$ and $f_\mathcal{A} = \sum_{a \in F} f_{out}^{q_0,a}$.

Hence, by introducing a small formula for this computation with suitably designated output gate, we can easily get the circuit C' to compute $f(\text{mod } \mathcal{A})$ or $f_\mathcal{A}$.

The correctness of our construction is immediate. Furthermore, size(C') satisfies the claimed bound. Note that C' will remain a monotone circuit if the given circuit C is monotone. This completes the proof of the first two parts.

Now we prove part 3 of the statement. Let P be an ABP computing polynomial f and $\mathcal{A}$ be a given DFA. The idea for the construction of ABPs that compute $f(\text{mod } \mathcal{A})$

and $f_{\mathcal{A}}$ is quite similar to the construction described for part 1. Consider for instance the ABP P' for $f(\text{mod } \mathcal{A})$. Consider the directed acyclic layered graph underlying the ABP P. In the new ABP P' we will have exactly the same number of layers as for P. However, for each node b in the ith layer of ABP P we will have nodes $\langle b, q\rangle$ for each state q of the DFA $\mathcal{A}$. Now, let f_b denote the polynomial that is computed at the node b by the ABP P. The property that the construction of P' can easily ensure is $f_b = \sum_q f_{\langle b,q\rangle}$, where $f_{\langle b,q\rangle}$ is the polynomial computed at node $\langle b, q\rangle$ by the ABP P'. More precisely, let M_q be the set of all nonzero monomials m of f_b such that on input m the DFA $\mathcal{A}$ goes from start state to state q. Then the polynomial $f_{\langle b,q\rangle}$ will be actually the sum of all the terms of the polynomial f_b corresponding to the monomials in M_q. This construction can now be easily used to obtain ABPs for each of $f(\text{mod } \mathcal{A})$ and $f_{\mathcal{A}}$, and the size of the ABP will satisfy the claimed bound. We omit the easy details of the construction of the ABPs.

A careful inspection of the constructions shows that for a given circuit C and DFA $\mathcal{A}$ we can construct the circuits that compute $C(\text{mod } \mathcal{A})$ and $C(\text{div } \mathcal{A})$ in deterministic logspace (and hence in NC^2). Likewise, the construction of the ABPs for $P(\text{mod } \mathcal{A})$ and $P(\text{div } \mathcal{A})$ for a given ABP P can also be computed in deterministic logspace. ■

The following is an immediate corollary of Theorem 1 by using known results in [RS05],[BW05].

Corollary 1. *1. Given a noncommutative ABP P computing a polynomial $f \in \mathbb{F}\langle X\rangle$ and a deterministic finite automaton $\mathcal{A}$ we can test in deterministic polynomial time whether $f(\text{mod } \mathcal{A})$ is identically zero.*

2. Given a noncommutative poly-degree circuit C computing $f \in \mathbb{F}\langle X\rangle$ and a DFA $\mathcal{A}$ then we can test whether $f(\text{mod } \mathcal{A})$ is identically zero in randomized polynomial time (if C is a monotone circuit then we can test if $f(\text{mod } \mathcal{A})$ is zero in deterministic polynomial time).

In contrast to Corollary 1, in case of NFA's we have:

Theorem 2. *Given a noncommutative formula F computing a polynomial $f \in \mathbb{Q}\langle Z\rangle$ and an NFA A accepting language $L(A) \subseteq Z^*$ then the problem of testing whether the polynomial $f(\text{mod } \mathcal{A})$ is identically zero is* coNP*-hard.*

Proof. We give a reduction from 3CNF-SAT to the complement of the problem. Let $S = C_1 \wedge C_2 \wedge \ldots \wedge C_t$ be a 3CNF formula where $C_i = c_{i1} \vee c_{i2} \vee c_{i3}$ for $1 \le i \le t$, and C_{ij}'s are from $\{w_1, \ldots, w_n\} \cup \{\neg w_1, \ldots, \neg w_n\}$. Let $f = \prod_{i=1}^{t} \sum_{j=1}^{3} z_{ij}$ where $z_{ij} = x_l$ if $c_{ij} = w_l$ and $z_{ij} = y_l$ if $c_{ij} = \neg w_l$ for $1 \le l \le n, 1 \le j \le 3$. Clearly, there is an $O(t)$ size formula F over indeterminates $Z = \{x_1, \ldots, x_n\} \cup \{y_1, \ldots, y_n\}$ for the polynomial f.

Let $L \subseteq Z^*$ be the set of all words of the form $m = ux_ivy_iw$ or $m = uy_ivx_iw$ for some $1 \le i \le n$. Clearly, there is an $O(n)$ size NFA $\mathcal{A}$ such that $L = L(\mathcal{A})$. Notice that the 3CNF formula S is satisfiable if and only if the polynomial $f(\text{mod } \mathcal{A})$ is *not* identically zero. Hence the given problem is coNP-hard. ■

We now prove a similar result for commutative ABPs. However, we need to be careful to consider the right kind of DFAs that capture commutativity so that the constructions of Theorem 1 are meaningful and go through.

Definition 5 (Commutative Automata). *Let $w \in X^d$ be any string of length d over the alphabet $X = \{x_1, \cdots, x_n\}$. Let $C_w \subset X^d$ denote the set of all words w' obtained by shuffling the letters of w.*

A DFA (or NFA) $\mathcal{A}$ over the alphabet $X = \{x_1, \cdots, x_n\}$ is said to be commutative *if for every word $w \in X^*$, w is accepted by $\mathcal{A}$ if and only if every word in C_w is accepted by $\mathcal{A}$.*

The following theorem is the analogue of Theorem 1 for intersecting and quotienting commutative circuits and commutative ABPs by commutative DFAs. We omit the proofs as the constructions are identical to those in the proof of Theorem 1. It can be easily seen from Definition 1 and the proof of Theorem 1 that in the commutative case the constructions are meaningful and work correctly when the DFAs considered are commutative.

Theorem 3. *Let $f \in \mathbb{F}[x_1, x_2, \cdots, x_n]$ and $\mathcal{A}$ be a commutative DFA with s states over alphabet $X = \{x_1, \cdots, x_n\}$. Then, for $g \in \{f(\text{mod } \mathcal{A}), f_{\mathcal{A}}\}$ we have*

1. $C(g) \leq C(f) \cdot (ns)^{O(1)}$.
2. $C^+(g) \leq C^+(f) \cdot (ns)^{O(1)}$.
3. $B(g) \leq B(f) \cdot (ns)^{O(1)}$.

Furthermore, the commutative circuit (ABP) for polynomial g meeting the above size bounds are computable in deterministic logspace (hence in NC^2) on input a circuit (resp. ABP) for f and DFA $\mathcal{A}$.

3 Monomial Algebras and Automata

Definition 6. *A two-sided ideal $I = \langle m_1, m_2, \cdots, m_r \rangle$ of the noncommutative ring $\mathbb{F}\langle X \rangle$ generated by a finite set of monomials $m_1, \cdots, m_r$ is a* finitely generated monomial ideal *of $\mathbb{F}\langle X \rangle$. The quotient algebra $\mathbb{F}\langle X \rangle / I$ is a* finitely generated monomial algebra.

For a polynomial f given by a circuit (or ABP) and a monomial ideal I given by a generating set we are interested in the circuit (resp. ABP) complexity of the polynomial $f(\text{mod } I)$. The corresponding identity testing problem is the *Ideal Membership* problem whether the polynomial $f \in I$. First we consider the noncommutative setting.

Theorem 4. *Let $I = \langle m_1, \cdots, m_r \rangle$ be a monomial ideal in $\mathbb{F}\langle X \rangle$. Let P (resp. C) be a noncommutative ABP (resp. a polynomial degree monotone circuit) computing a polynomial $f \in \mathbb{F}\langle X \rangle$. Then there is a deterministic polynomial-time algorithm to test if the polynomial $f(\text{mod } \mathcal{I})$ is identically zero.*

Proof. Consider monomials m_i as strings over alphabet $\{x_1, \ldots, x_n\}$. Let $d = \max_i\{\text{length}(m_i)\}$. Using the Aho-Corasick pattern matching automaton [AC75] we construct a DFA A with $O(dr)$ states which on input a string $s \in X^*$ accepts s if s contains m_i as a substring for some i. Now using Theorem 1 we obtain an ABP P' (resp. a monotone circuit C') of size $poly(n, d, r)$ which computes the polynomial

$g = f(\text{mod } \mathcal{A})$. Clearly, $f \in \mathcal{I}$ iff $g \equiv 0$. Now we can invoke Corollary 1 to complete the proof. ∎

By constructing a pattern matching automaton as in the proof of Theorem 1, we have an immediate lower bound observation for the Permanent.

Corollary 2. *Let $I = \langle m_1, \cdots, m_r \rangle$ be a monomial ideal and C be a noncommutative ABP (or a polynomial degree noncommutative monotone circuit) over indeterminates $\{x_{ij} \mid 1 \le i, j \le n\}$. If $C = Perm_n(\text{mod } I)$ then either* size(C) *or the number of generating monomials r for I is $2^{\Omega(n)}$.*

3.1 Commutative Monomial Algebras

In this subsection we examine the same problem in the commutative case. Let $I = \langle m_1, \cdots, m_k \rangle$ be a monomial ideal contained in $\mathbb{F}[x_1, \cdots, x_n]$. As before $f(\text{mod } I)$ is a meaningful polynomial in $\mathbb{F}[x_1, \cdots, x_n]$ for $f \in \mathbb{F}[x_1, \cdots, x_n]$.

We consider the problem for monotone circuits. It is useful to understand the connection between monotone noncommutative circuits and context-free grammars. For basics of language theory we refer to [HMU].

Definition 7. *We call a context-free grammar in Chomsky normal form $G = (V, T, P, S)$ an* acyclic CFG *if for any nonterminal $A \in V$ there does not exist any derivation of the form $A \Rightarrow^* uAw$.*

The size $size(G)$ of CFG $G = (V, T, P, S)$ in defined as the total number of symbols (in V, T, S) used in the production rules in P, where V, T, and P are the sets of variables, terminals, and production rules. It is clear that an acyclic CFG generates a finite language.We note the following easy proposition that relates acyclic CFGs to monotone noncommutative circuits over X.

Proposition 1. *For a monotone circuit C of size s computing a polynomial $f \in \mathbb{Q}\langle X \rangle$ let $mon(f)$ denote the set of nonzero monomials of f. Then there is an acyclic CFG G for $mon(f)$ with* size(G) *$= O(s)$. Conversely, if G is an acyclic CFG of size s computing some finite set $L \subset X^*$ of monomials over X, there exists a monotone circuit of size $O(s)$ that computes a polynomial $\sum_{m \in L} a_m m \in \mathbb{Q}\langle X \rangle$, where the positive integer a_m is the number of derivation trees for m in the grammar G.*

Proof. First we prove the forward direction by constructing an acyclic CFG $G = (V, T, P, S)$ for mon(f). Let $V = \{A_g \mid$ g is a gate of circuit $C\}$ be the set of nonterminals of G. We include a production in P for each gate of the circuit C. If g is an input gate with input $x_i, 1 \le i \le n$ include the production $A_g \to x_i$ in P. If the input is a *nonzero* field element then add the production $A_g \to \epsilon$.[1] Let f_g denote the polynomial computed at gate g of C. If g is a $\times$ gate with $f_g = f_h \times f_k$ then include the production $A_g \to A_h A_k$ and if it is $+$ gate with $f_g = f_h + f_k$ include the productions $A_g \to A_h \mid A_k$. Let the start symbol $S = A_g$, where g is the output gate of C. It is easy

[1] If the circuit takes as input 0, we can first propagate it through the circuit and eliminate it.

to see from the above construction that G is acyclic moreover $size(G) = O(s)$ and it generates the finite language $\text{mon}(f)$. The converse direction is similar. ■

We need Lemma 1 to prove monotone circuit lower bound over commutative monomial algebras.

Lemma 1. *Let C be a monotone circuit computing a homogeneous polynomial $f \in \mathbb{Q}[x_1, x_2, \cdots, x_n]$ of degree d and let $\mathcal{A}$ be a commutative NFA of size s computing language $L(\mathcal{A}) \subseteq X^d$. There is a deterministic polynomial (in $size(C), s$) time algorithm to construct a monotone circuit C' which computes polynomial $g \in \mathbb{Q}[x_1, x_2, \cdots, x_n]$ such that $mon(f_{\mathcal{A}}) = mon(g)$.*

Proof. For the given monotone circuit C we can consider it as a noncommutative circuit computing a polynomial f' in $\mathbb{Q}\langle X\rangle$. Notice that f' is *weakly equivalent* to f in the sense of Nisan [N91]: I.e. a monomial m occurs in $\text{mon}(f')$ if and only if some shuffling of m occurs in $\text{mon}(f)$. Applying Proposition 1 we can obtain an acyclic context-free grammar G that generates precisely the finite set $\text{mon}(f')$ of monomials of f'. The context-free grammar G is in Chomsky normal form. We can convert it into a Greibach normal form grammar G' in polynomial time, where the size of G' is $\text{size}(G)^4$ [R67, KB97].

Now, we have a Greibach normal form grammar G' and NFA $\mathcal{A}$. We can apply the standard conversion of a Greibach normal form grammar to a pushdown automaton, to obtain a PDA M that accepts the same set that is generated by G'. The PDA M will encode each symbol of the grammar G' into binary strings of length $O(\log |G'|)$. Hence the PDA M will require an $O(\log |G'|)$ size worktape to simulate the transitions of the PDA apart from an unbounded stack. Now, it is easy to construct a new PDA M' that will simultaneously run the NFA $\mathcal{A}$ on the given input as well as the first PDA M so that M' accepts if and only if both the simulations accept. Clearly, M' will also require an $O(\log |G'|)$ size worktape to carry out this simulation. We can convert the PDA M' back into an acyclic grammar G'' in Chomsky normal form in polynomial time using a standard algorithm. From this acyclic CFG G'' we can obtain a monotone circuit C'', where the gates correspond to nonterminals. By construction it follows that C'' computes a polynomial f'' in $\mathbb{Q}\langle X\rangle$ such that $\text{mon}(f'') = \text{mon}(f') \cap L(\mathcal{A})$. At this point we invoke the fact that $\mathcal{A}$ is a *commutative* NFA. Hence, we can view C'' as a commutative monotone circuit. Let $g \in \mathbb{Q}[x_1, \ldots, x_n]$ is a polynomial computed by C''. So it follows that $\text{mon}(g) = \text{mon}(f_{\mathcal{A}})$ which proves the lemma. ■

Theorem 5. *Let $I = \langle m_1, \ldots, m_k\rangle$ be a commutative monomial ideal in $\mathbb{Q}[x_{11}, \ldots, x_{nn}]$, generated by $k = o(\frac{n}{\lg n})$ many monomials, such that $degree(m_i) \leq n^c$ for a constant c. Suppose C is a monotone circuit computing a polynomial f in $\mathbb{Q}[x_{11}, \ldots, x_{nn}]$ such that the permanent $Perm_n = f(mod\ I)$ then $C^+(f) = 2^{\Omega(n)}$.*

Proof. Let X denote the set of variables $\{x_{11}, \ldots, x_{nn}\}$. For each monomial m_t, $1 \leq t \leq k$ in the generating set for I write $m_t = \prod_{ij} x_{ij}^{e_{ijt}}$, where e_{ijt} are nonnegative integers for $1 \leq i, j \leq n, 1 \leq t \leq k$.

Consider the language $L \subset X^*$ containing all strings m such that for each $t, 1 \leq t \leq k$ there exist $i, j \in [n]$ such that the number of occurrences of x_{ij} in m is strictly less that e_{ijt}. Notice that L is precisely $X^* \setminus I$. Clearly, the language L is *commutative*: if $m \in L$ then so is every reordering of the word m. It is easy to see that there is a *commutative* NFA $\mathcal{A}$ with $n^{O(k)} = 2^{o(n)}$ states such that $L = L(\mathcal{A})$ (the NFA is designed using counters for each t and guessed i, j, note that $e_{ijt} \leq n^c$, for $1 \leq t \leq k, 1 \leq i, j \leq n$). So we have $Perm_n = f_{\mathcal{A}}$.

Suppose the polynomial f can be computed by a monotone circuit C of size $2^{o(n)}$. By Lemma 1 there is a monotone circuit of size $2^{o(n)}$ computing a polynomial g such that $\text{mon}(g) = \text{mon}(f_{\mathcal{A}}) = \text{mon}(Perm_n)$. We observe that the $2^{\Omega(n)}$ size lower bound proof for commutative circuits computing the permanent (specifically, the Jerrum-Snir work [JS82]) also imply the same lower bound for the polynomial g, because the coefficients do not play a role and $\text{mon}(g) = \text{mon}(Perm_n)$. This completes the proof. ■

Corollary 3. *Let C be a commutative monotone arithmetic circuit computing polynomial $f \in \mathbb{Q}[x_1, \ldots, x_n]$ and let $I = \langle m_1, \ldots, m_k \rangle$ be a commutative monomial ideal generated by $k = o(n/\log n)$ monomials, such that for $1 \leq t \leq k$, $degree(m_t) \leq n^c$ for a constant c. Then the problem of testing whether $f \in I$ can be solved in deterministic $2^{o(n)} \cdot poly(size(C))$ time.*

Proof. Let $X = \{x_1, \ldots, x_n\}$. As in the proof of Theorem 5 we can construct an NFA $\mathcal{A}$ of size $2^{o(n)}$ such that $L(\mathcal{A}) = X^* \setminus I$. By Lemma 1 we can construct a monotone commutative circuit C' of size $2^{o(n)} \cdot poly(size(C))$ computing polynomial g such that $mon(g) = mon(f_{\mathcal{A}})$. It is clear that $f \in I$ iff $f_{\mathcal{A}}$ is identically zero iff g is identically zero. We can test if g is identically zero using standard algorithms. ■

Given an ABP (or monotone circuit) of size s computing some polynomial $f \in \mathbb{Q}\langle x_1, \ldots, x_n \rangle$ and a noncommutative monomial ideal $I = \langle m_1, \ldots, m_k \rangle$ we can test if $f \in I$ in deterministic time $2^{o(n)} s^{O(1)}$ even when the number of monomials k generating I is $k = 2^{o(n)}$. On the other hand, in the commutative setting we are able to show a similar result (Corollary 3) only for $k = o(n/\log n)$. Nevertheless, it appears difficult to prove a significantly stronger result. We can show that strengthening Corollary 3 to $k = \frac{n}{2}$ would imply that 3CNF-SAT has a $2^{o(n)}$ time algorithm contradicting the exponential-time hypothesis [IPZ01]. We make this statement precise in the next result (whose proof is similar to that of Theorem 2).

Theorem 6. *Given a commutative monotone circuit C of size s computing a polynomial $f \in \mathbb{Q}[Z]$ in $2n$ variables and a commutative monomial ideal $I = \langle m_1, \ldots, m_k \rangle$, $k \geq n$ then the problem of testing if $f \in I$ is* coNP*-hard. Specifically, for $k = n$ the problem of testing if $f \in I$ does not have a $2^{o(n)} s^{O(1)}$ time algorithm assuming the exponential-time hypothesis.*

In contrast to Theorem 5, we observe that $Perm_n$ can be computed by a small monotone formula modulo a monomial ideal generated by $O(n^3)$ many monomials.

Theorem 7. *There is a monomial ideal $I = \langle m_1, \cdots, m_t \rangle$ of $\mathbb{F}[X]$, where $X = \{x_{ij} \mid 1 \leq i, j \leq n\}$,$t = O(n^3)$ and a polynomial-sized commutative monotone formula $F(x_{11}, \cdots, x_{nn})$ such that $Perm_n = F(mod\, I)$.*

Proof. Let $F = \prod_{i=1}^{n}(x_{i1} + x_{i2} + \cdots + x_{in})$ and I be the monomial ideal generated by the set of monomials $\{x_{ik}x_{jk} \mid 1 \leq i, j, k \leq n\}$. Clearly, $Perm_n = F(\text{mod}\, I)$. ■

4 Monomial Search Problem

We now consider the monomial search problem for ABPs in both commutative and noncommutative setting. The problem is to find a nonzero monomial of the polynomial computed by a given ABP. We apply Theorem 1 to prove these results.

Theorem 8. *Given a noncommutative ABP P computing a polynomial f in $\mathbb{F}\langle X\rangle$ there is a randomized* NC^2 *algorithm that computes a nonzero monomial of f. More precisely, the algorithm is a randomized* FL^{GapL} *algorithm.*

Proof. We can assume wlog that the given ABP P computes a homogeneous degree d polynomial. The proof is by an application of the isolation lemma of [MVV87]. Define the universe $U = \{x_{ij} \mid 1 \leq i \leq n, 1 \leq j \leq d\}$, where the element x_{ij} stands for the occurrence of x_i in the jth position in a monomial. With this encoding every degree d monomial m over X can be encoded as a subset S_m of size d in U, where $S_m = \{x_{ij} \mid x_i \text{ occurs in } j^{th} \text{ position in } m\}$. Following the isolation lemma, we pick a random weight assignment $w : U \longrightarrow [4dn]$. The weight of a monomial m is defined as $w(m) = w(S_m) = \sum_{x_{ij} \in S_m} w(x_{ij})$, and with probability $1/2$ there is a unique minimum weight monomial.

Construction of weight-checking DFA: For any weight value a such that $1 \leq a \leq 4nd^2$, we can easily construct a DFA M_w^a that accepts a monomial $m \in X^*$ iff $m \in X^d$ and $w(m) = a$. This DFA will have $O(4nd^3)$ many states. Furthermore, we can compute this DFA in deterministic logspace. Next, by Theorem 1 we can compute an ABP P_w^a that computes the polynomial $P(\text{div } M_w^a)$ for each of $1 \leq a \leq 4nd^3$. With probability $1/2$ we know that one of $P(\text{div } M_w^a)$ accepts precisely one monomial of the original polynomial f (with the same coefficient).

In order to find each variable occurring in that unique monomial accepted by, say, P_w^a we will design another DFA $\mathcal{A}_{ij}$ which will accept a monomial $m \in X^d$ if and only if x_i occurs in the jth position. Again by Theorem 1 we can compute an ABP $B_{i,j,a,w}$ that accepts precisely $P_w^a(\text{div } \mathcal{A}_{ij})$. Now, the ABP $B_{i,j,a,w}$ either computes the zero polynomial (if x_i does not occur in the jth position of the unique monomial of P_w^a) or it computes that unique monomial of P_w^a. In order to test which is the case, notice that we can *deterministically* assign the values $x_i = 1$ for each variable x_i. Crucially, since P_w^a has a *unique* monomial it will be nonzero even for this deterministic and commutative evaluation. Since the evaluation of an ABP is for commuting values (scalar values), we can carry it out in NC^2 in fact, in FL^{GapL} for any fixed finite field or over $\mathbb{Q}$, (see e.g. [T91], [V91], [MV97]).

Let m be the monomial that is finally constructed. We can construct a DFA $\mathcal{A}_m$ that accepts only m and no other strings. By Theorem 1 we can compute an ABP P' for the polynomial $P(\text{div } \mathcal{A}_m)$ and again check if P' is zero or nonzero by substituting all $x_i = 1$ and evaluating. This will make the algorithm actually a zero-error NC^2 algorithm. The success probability can be boosted by parallel repetition. ■

Next we describe a randomized NC^2 algorithm for the Monomial search problem for commutative ABPs. This is the best we can currently hope for, since deterministic polynomial-time identity testing for commutative ABPs is a major open problem. Our monomial search algorithm is based on a generalized isolation lemma [KS01].

Lemma 2. [KS01, Lemma 4] *Let L be a collection of linear forms over variables $z_1, \ldots, z_n$ with integer coefficients in $\{0, 1, \ldots, K\}$. If each z_i is picked independently and uniformly at random from $\{0, 1, \ldots, 2Kn\}$ then with probability at least $\frac{1}{2}$ there is a unique linear form in L which attains minimum value at $(z_1, z_2, \ldots, z_n)$.*

Theorem 9. *The monomial search problem for commutative algebraic branching programs is in randomized NC^2 (more precisely, it is in randomized $\mathrm{FL}^{\mathrm{GapL}}$).*

Proof. Let P be a commutative algebraic branching program computing a polynomial $f \in \mathbb{F}[x_1, x_2, \ldots, x_n]$. We can assume, without loss of generality, that f is homogeneous of degree d. First, pick a random weight function $w : \{x_1, \cdots, x_n\} \longrightarrow [2dn]$. Next, for each number a such that $0 \le a \le 2d^2n$ we construct a DFA A^a_w which will accept a monomial $m \in X^*$ iff $m \in X^d$ and $w(m) = a$, where $w(m) = \sum_i w(x_i) \cdot \alpha_i$, and x_i occurs exactly α_i times in m. Crucially, notice that A^a_w is a *commutative* DFA. Hence, applying Theorem 3, for each number a we can obtain an ABP P^a_w in deterministic logspace.

By Lemma 2 with probability at least $1/2$ one of the ABPs P^a_w accepts a unique monomial $m = x_1^{\alpha_1} x_2^{\alpha_2} \ldots x_n^{\alpha_n}$ of f. Suppose that value of a is u. Let $c \neq 0$ denote the coefficient of the unique monomial m in f computed by the ABP P^u_w. We need to compute each α_i. We evaluate the ABP P^u_w by setting $x_j = 1$ for all $j \neq i$ to obtain $cx_i^{\alpha_i}$. Evaluating the ABPs P^a_w, for each a, on the inputs $(1, \cdots, 1, x_i, 1, \cdots, 1)$ can be done in NC^2. Indeed, it can be done in $\mathrm{FL}^{\mathrm{GapL}}$, since we only need determinant computation over the field $\mathbb{F}$. This completes the proof sketch. ■

Theorem 10. *There is a deterministic polynomial time algorithm for the monomial search problem for noncommutative algebraic branching programs.*

Proof. W.l.o.g. assume that the input noncommutative ABP P computes a degree d homogeneous polynomial $f = \sum_m a_m m$. The monomial search algorithm is a simple prefix search guided by the Raz-Shpilka deterministic identity test [RS05]. Starting with $w = \epsilon$, we successively compute ABPs $P_\epsilon, P_{w_1}, \cdots, P_{w_d}$, where $|w_k| = k$ and w_k is a prefix of w_{k+1} for each k. Each P_{w_k} is an ABP that computes $f(\text{div } D_{w_k})$ where D_{w_k} is a DFA that accepts all the words with prefix w. The prefix search sets $w_{k+1} = w_k x_i$ for the first indeterminate x_i such that $P_{w_{k+1}}$ computes a nonzero polynomial (to check this we use the Raz-Shpilka identity test on $P_{w_{k+1}}$ [RS05]). Since $f(\text{div } D_{w_k}) \neq 0$ for some indeterminate x_i the polynomial $f(\text{div } D_{w_{k+1}})$ is nonzero. Hence the prefix search will successfully continue. The output of the monomial search will be w_d. ■

Finally, our technique of isolating a monomial using DFAs along with intersecting circuits with DFAs can be applied to get a randomized NC reduction from monomial search for noncommutative (or commutative) circuits to noncommutative (resp. commutative) polynomial identity testing.

Theorem 11. *Monomial search for noncommutative (commutative) circuits is randomized* NC *reducible to noncommutative (resp. commutative) polynomial identity testing.*

References

[AC75] Aho, A.V., Corasick, M.J.: Efficient String Matching: An Aid to Bibliographic Search. Commun. ACM 18(6), 333–340 (1975)

[AM08] Arvind, V., Mukhopadhyay, P.: Derandomizing the isolation lemma and lower bounds for circuit size. In: Goel, A., Jansen, K., Rolim, J.D.P., Rubinfeld, R. (eds.) APPROX and RANDOM 2008. LNCS, vol. 5171, pp. 276–289. Springer, Heidelberg (2008)

[AMS08] Arvind, V., Mukhopadhyay, P., Srinivasan, S.: New results on Noncommutative Polynomial Identity Testing. In: Proc. of Annual IEEE Conference on Computational Complexity, pp. 268–279 (2008)

[BW05] Bogdanov, A., Wee, H.: More on Noncommutative Polynomial Identity Testing. In: Proc. of 20th Annual Conference on Computational Complexity, pp. 92–99 (2005)

[CS04] Chien, S., Sinclair, A.: Algebras with polynomial identities and computing the determinant. In: Proc. Annual IEEE Sym. on Foundations of Computer Science, pp. 352–361 (2004)

[HMU] Hopcroft, J.E., Motawani, R., Ullman, J.D.: Introduction to Automata Theory Languages and Computation, 2nd edn. Pearson Education Publishing Company, London

[IPZ01] Impagliazzo, R., Paturi, R., Zane, F.: Which Problems Have Strongly Exponential Complexity? Journal Computer and System Sciences 63(4), 512–530 (2001)

[JS82] Jerrum, M., Snir, M.: Some Exact Complexity Results for Straight-Line Computations over Semirings. J. ACM 29(3), 874–897 (1982)

[KI03] Kabanets, V., Impagliazzo, R.: Derandomization of polynomial identity test means proving circuit lower bounds. In: Proc. of 35th ACM Sym. on Theory of Computing, pp. 355–364 (2003)

[KB97] Koch, R., Blum, N.: Greibach Normal Form Transformation. In: STACS, pp. 47–54 (1997)

[KS01] Klivans, A., Spielman, D.A.: Randomness efficient identity testing of multivariate polynomials. In: STOC 2001, pp. 216–223 (2001)

[MV97] Mahajan, M., Vinay, V.: A Combinatorial Algorithm for the Determinant. In: SODA 1997, pp. 730–738 (1997)

[MVV87] Mulmuley, K., Vazirani, U.V., Vazirani, V.V.: Matching Is as Easy as Matrix Inversion. In: STOC 1987, pp. 345–354 (1987)

[N91] Nisan, N.: Lower bounds for noncommutative computation. In: Proc. of 23rd ACM Sym. on Theory of Computing, pp. 410–418 (1991)

[R67] Rosenkrantz, D.J.: Matrix equations and normal forms for context-free grammars. J. ACM (14), 501–507 (1967)

[RS05] Raz, R., Shpilka, A.: Deterministic polynomial identity testing in non commutative models. Computational Complexity 14(1), 1–19 (2005)

[T91] Toda, S.: Counting Problems Computationally Equivalent to the Determinant (manuscript)

[V91] Vinay, V.: Counting Auxiliary Pushdown Automata and Semi-unbounded Arithmetic Circuits. In: Proc. 6th Structures in Complexity Theory Conference, pp. 270–284 (1991)

An Improved Approximation Bound for Spanning Star Forest and Color Saving*

Stavros Athanassopoulos, Ioannis Caragiannis, Christos Kaklamanis, and Maria Kyropoulou

Research Academic Computer Technology Institute and
Department of Computer Engineering and Informatics
University of Patras, 26500 Rio, Greece

Abstract. We present a simple algorithm for the maximum spanning star forest problem. We take advantage of the fact that the problem is a special case of complementary set cover and we adapt an algorithm of Duh and Fürer in order to solve it. We prove that this algorithm computes $193/240 \approx 0.804$-approximate spanning star forests; this result improves a previous lower bound of 0.71 by Chen et al. Although the algorithm is purely combinatorial, our analysis defines a linear program that uses a parameter f and which is feasible for values of the parameter f not smaller than the approximation ratio of the algorithm. The analysis is tight and, interestingly, it also applies to complementary versions of set cover such as color saving; it yields the same approximation guarantee of 193/240 that marginally improves the previously known upper bound of Duh and Fürer. We also show that, in general, a natural class of local search algorithms do not provide better than 1/2-approximate spanning star forests.

1 Introduction

We consider the combinatorial problem of decomposing a graph into a forest of stars that span the nodes of the graph. A *star* is a tree of diameter at most two. It consists of a node designated as *center* and (possibly) of *leaves*, i.e., nodes which are connected through edges to the center. The objective of the problem which is known as the *maximum spanning star forest problem* (SSF) is to compute a star decomposition in which the number of leaves (to be considered as the benefit) is maximized. Equivalently, the number of stars has to be minimized. This is equivalent to the *minimum dominating set* problem in which the objective is to compute a minimum-size set of nodes (the dominating set) in the graph so that each node is either part of the dominating set or adjacent to some node in it. The nodes in the dominating set correspond to centers of stars with leaves the nodes outside the dominating set.

* This work is partially supported by the European Union under IST FET Integrated Project FP6-015964 AEOLUS, by the General Secretariat for Research and Technology of the Greek Ministry of Development under programme PENED, and by a "Caratheodory" basic research grant from the University of Patras.

R. Královič and D. Niwiński (Eds.): MFCS 2009, LNCS 5734, pp. 90–101, 2009.

SSF has several applications in the problem of aligning multiple genomic sequences [22], in the comparison of phylogenetic trees [5], and in the diversity problem in the automobile industry [1]. As a combinatorial optimization problem, SSF has been considered in [8,22]. Nguyen et al. [22] prove that the problem is APX-hard by presenting an explicit inapproximability bound of 259/260 and present a combinatorial 0.6-approximation algorithm. Optimal polynomial-time algorithms are presented for special graph classes such as planar graphs and trees. Chen et al. [8] present a better algorithm with approximation ratio 0.71. The main idea is to run both the algorithm of [22], as well as an algorithm that is based on linear programming and randomized rounding and, then, take the best solution. The linear programming relaxation used in [22] has an integrality gap of (at most) 3/4. Interesting generalizations include node-weighted and edge-weighted versions of SSF. [8,22] present approximation algorithms and APX-hardness results for these problems as well. Stronger inapproximability results for these problems recently appeared in [7].

The minimum dominating set problem is a special case of *set cover*. In set cover, we are given a collection $\mathcal{S}$ of sets over a set U of n elements and the objective is to select a minimum-size subcollection $\mathcal{T}$ of sets that contain all elements. The problem is well known to be hard to approximate within a logarithmic factor [13] while the greedy algorithm that iteratively includes in the cover the set that contains the maximum number of uncovered elements achieves a matching upper bound of H_n [19,21]. The same algorithm has an approximation ratio of H_k when the collection is closed under subsets and each set contains at most k elements; this special case of the problem is known as k-set cover. Slightly better results are also known for k-set cover [2,11,14,20]. The most interesting idea that yields these improvements is *semi-local optimization* [11] which applies to 3-set cover. Semi-local optimization is a local search algorithm. The main idea behind it is to start with an empty 3-set covering and augment it by performing local improvements. Note that once the (disjoint) sets of size 3 have been selected, computing the minimum number of sets of size 2 and 1 in order to complete the cover can be done in polynomial time by a maximum matching computation. So, a semi-local (s,t)-improvement step for 3-set cover consists of (i) the deletion of up to t sets of size 3 from the current covering and (ii) the insertion of up to s disjoint sets of size 3 and the minimum number of necessary sets of smaller size that complete the cover, so that the number of sets in the cover decreases; in case of ties, covers with fewer sets of size 1 are preferable. The analysis of [11] shows that the best choice of the parameters (s,t) is $(2,1)$ and that, somehow surprisingly, larger values for these parameters do not yield any further improvements.

As the complement of minimum dominating set, SSF is a special case of *complementary set cover* (in a sense that is explained in detail in Section 2). In complementary set cover, the objective is to maximize the quantity $n-|\mathcal{T}|$. The approximation ratio of a solution for complementary set cover can be thought of as comparing the "distance" of a solution for set cover from the worst possible solution (i.e., the one that uses n sets to cover the elements) to the distance of the best solution of the worst possible one. This yields an alternative performance measure for the analysis of approximation algorithms; such measures have been considered for many combinatorial optimization problems in the context of differential approximation algorithms [10,9] or z-approximations [16] (see also [3,4]). Duh and Fürer [11] also consider the application of their algorithm on instances of complementary set cover in which the collection of sets is not given explicitly. Among them, the most interesting case is when the elements correspond to the nodes of a graph and the sets in the collection are all the independent sets of the graph. In this case, set cover is equivalent to graph coloring (i.e., the problem of coloring the nodes of a graph so that no two adjacent nodes are assigned the same color and the number of colors is minimized). *Complementary graph coloring* is also known as *color saving*.

In this paper, we take advantage of the fact that SSF is a complementary set cover problem in which the collection of sets is closed under subsets and we use the approximation algorithm of Duh and Fürer [11] to solve its instances (Section 2). We obtain an approximation ratio of $193/240 \approx 0.804$ improving the previously best known bound of 0.71 from [8] and beating the integrality gap of the linear programming relaxation used in [8]. Our analysis is tight, it does not exploit the particular structure of SSF, and essentially holds for the more general complementary set cover and color saving problems as well (see Section 3). The result of [11] is a lower bound of $289/360 \approx 0.803$ on the approximation ratio of the same algorithm on general instances of color saving (and complementary set cover). Although our improvement on the approximation ratio is marginal, our proof technique is very interesting and, conceptually, it could be applicable to other contexts. The proof in [11] is based on a detailed accounting of the performance of the algorithm by a case analysis of the different possible ways the elements of the sets in the optimal solution are covered by the algorithm; among the several different cases, only a few are presented in that paper. Our proof is much different in spirit, it is simpler, and does not require any case analysis; it is based on proving feasibility of a linear program. We note that analysis of purely combinatorial algorithms using linear programs whose objective value reveals the approximation factor (in a different way than in the current paper) has also been used for k-set cover [2], wavelength management in optical networks [6], and facility location [18]. The definition of the linear program in this paper follows the terminology of [2] but the resulting linear program is significantly smaller and our analysis is different. In particular, we show that if the algorithm obtains an at most f-approximate solution for some instance, then an appropriately defined linear program that has f as a parameter and its constraints capture the properties of the instance and the way the algorithm is applied on it is feasible. This is stated as a *parameterized LP Lemma*. So, the lower bound on the approximation ratio follows by proving that the corresponding linear program is infeasible for values of f smaller than 193/240.

Furthermore, motivated by well-known set packing heuristics [14,17], we consider a natural family of local search algorithms for SSF. Such an algorithm starts with an initial spanning star forest and repeatedly performs local improvements until this is not possible any more. The resulting solution is a local optimum for the particular algorithm and the question is whether local optima are efficient solutions for SSF as well. We prove that this is not the case for any local search algorithm that belongs to the family and runs in polynomial time. In the proof, we construct almost 1/2-approximate spanning star forests on appropriately defined graphs that are local optima for such algorithms (Section 4).

2 Algorithm Description

In this section, we describe how to exploit the relation of the problem to set cover in order to solve it using a well-known algorithm of Duh and Fürer [11]. We first transform an SSF instance consisting of a graph $G = (V, E)$ into an instance $(V, \mathcal{S})$ of set cover over the nodes. The collection $\mathcal{S}$ consists of *star sets*. A star set is a set of nodes with a common neighbor (assuming that a node neighbors on itself). We distinguish between two types of star sets: A and B. The nodes of a star set s of type A have a node in s as a common neighbor; all other star sets are of type B. Observe that the collection $\mathcal{S}$ that is defined in this way is closed under subsets. We use the term star set cover in order to refer to the particular set cover instance and the term i-star set cover when the collection consists of star sets of size at most i.

We use the algorithm of Duh and Fürer [11] in order to solve the resulting star set cover instance and obtain a disjoint collection of star sets that contain all nodes in V. We refer to star sets of size i as i-star sets. The algorithm first greedily includes disjoint 6-star sets in the solution until no other 6-star set can be included. We use the term maximal to refer to such collections of disjoint sets. After this phase, the nodes that remain uncovered and the star sets that consist only of such nodes form an instance of 5-star set cover. Then, the algorithm executes a restricted phase in order to pick a maximal collection of disjoint 5-star sets and guarantees that the number of 1-star sets in the final solution is not larger than the number of 1-star sets in the optimal solution of the 5-star set cover instance at the beginning of this restricted phase. Then, it applies a similar restricted phase for 4-star sets. After this phase, an instance of 3-star set cover remains to be solved in order to complete the star set covering; the algorithm applies a semi-local optimization phase on it. We use integers $6, 5, 4, 3$ to refer to the phases of the algorithm according to the size of the star sets they consider. In summary, the algorithm can be described as follows.

Phase 6: Choose a maximal collection of disjoint 6-star sets.
Phase 5: Choose a maximal collection of disjoint 5-star sets so that the choice of these star sets does not increase the number of 1-star sets in the final solution.
Phase 4: Choose a maximal collection of disjoint 4-star sets so that the choice of these star sets does not increase the number of 1-star sets in the final solution.
Phase 3: Run the semi-local optimization algorithm on the remaining instance of 3-star set cover.

The output of the algorithm is a disjoint set $\mathcal{T} \subseteq \mathcal{S}$ of star sets. We transform this solution to a spanning star forest as follows. We first consider star sets of $\mathcal{T}$ of type A. For each such set s, we set the common neighbor u of the nodes in s as center and connect the remaining nodes of s to u. Then, we consider the star sets of type B. For each such set s in $\mathcal{T}$, we select a common neighbor u of the nodes in s. If u is already the center of another star, we simply connect the nodes of s that have not been used as centers so far to it. If u is already the leaf of another star, we remove it from that star and connect the nodes of s that have not been used as centers so far to it.

Notice that, once a node has been used as a center, it remains one until the end of the algorithm and each leaf is connected to some center. Since the star sets include all nodes of V, each node is either a center or a leaf of some star and, hence, the solution is indeed a spanning star forest of G. Furthermore, when considering a star set, we increase the number of centers by at most 1. Hence, the total number of leaves is at least $|V| - |\mathcal{T}|$.

3 Analysis

Our main argument for the analysis of the algorithm can be described as follows. We first show that if there is an instance on which the algorithm computes an at most f-approximate solution, then an appropriately defined linear program LP(f) is feasible. Then, we show that LP(f) is infeasible for $f < 193/240$, implying that the approximation ratio of the algorithm is at least $193/240 \approx 0.804$. We consider only non-trivial instances of the problem in which the input graph has at least one edge since any algorithm is optimal for trivial ones.

3.1 The Parameterized LP Lemma

Consider a non-trivial SSF instance I that consists of a graph $G = (V, E)$ on which the algorithm computes an at most f-approximate solution. Let $(V, \mathcal{S})$ be the corresponding star set cover instance. For $i = 5, 4, 3$, denote by $(V_i, \mathcal{S}_i)$ the instance of the i-star set cover problem that has to be solved just before entering phase i. Here, V_i contains the nodes in V that have not been covered in previous phases and $\mathcal{S}_i$ contains the star sets of $\mathcal{S}$ which consist only of nodes in V_i. $\mathcal{S}_i$ contains star sets of the original collection of size at most i. Denote by $\mathcal{O}_i$ an optimal solution of instance $(V_i, \mathcal{S}_i)$; we also denote the optimal solution of $(V, \mathcal{S})$ by $\mathcal{O}$. Since $\mathcal{S}$ is closed under subsets, without loss of generality, we may assume that $\mathcal{O}_i$ contains disjoint sets. Furthermore, it is clear that $|\mathcal{O}_{i-1}| \leq |\mathcal{O}_i|$ for $i = 3, 4, 5$ and $|\mathcal{O}_i| \leq |\mathcal{O}|$.

Set $\theta = \frac{|V|}{|V|-1}$. We denote by T the ratio $|V|/|\mathcal{O}|$. Since the instance is non-trivial, there exists an optimal star cover with at most $|V| - 1$ star sets (i.e., $|\mathcal{O}| \leq |V| - 1$). Hence,

$$T \geq \theta. \tag{1}$$

For the phase i of the algorithm with $i = 3, 4, 5$, we denote by $a_{i,j}$ the ratio of the number of j-star sets in $\mathcal{O}_i$ over the number $|\mathcal{O}|$ of sets in the optimal solution of $(V, \mathcal{S})$. Since $|\mathcal{O}_i| \leq |\mathcal{O}|$ and $|\mathcal{O}_i| = |\mathcal{O}| \sum_{j=1}^{i} a_{i,j}$, we obtain that

$$\sum_{j=1}^{i} a_{i,j} \leq 1. \tag{2}$$

Also, observe that $|V| = T|\mathcal{O}|$ and $|V_5| = |\mathcal{O}| \sum_{j=1}^{5} j a_{5,j}$. Since $V_5 \subseteq V$, we have

$$T \geq \sum_{j=1}^{5} j a_{5,j}. \tag{3}$$

During the phase i with $i = 3, 4, 5$, the algorithm selects a maximal set of i-star sets. This means that any i-star set selected intersects some of the i-star sets of the optimal star set covering of the instance (V_i, S_i) and may intersect with at most i such i-star sets. Since there are $a_{i,i}|\mathcal{O}|$ such star sets in the optimal star set covering of (V_i, S_i), the number $|V_i \setminus V_{i-1}|/i$ of i-star sets selected during phase i is at least $a_{i,i}|\mathcal{O}|/i$. Equivalently, $|V_i \setminus V_{i-1}| \geq a_{i,i}|\mathcal{O}|$. Since $|V_i \setminus V_{i-1}| = \left(\sum_{j=1}^{i} j a_{i,j} - \sum_{j=1}^{i-1} j a_{i-1,j}\right) |\mathcal{O}|$, we obtain that

$$\sum_{j=1}^{i} j a_{i,j} - \sum_{j=1}^{i-1} j a_{i-1,j} \geq a_{i,i}. \tag{4}$$

Phase 3 imposes several extra constraints. Denote by b_3, b_2, b_1 the number of 3-, 2-, and 1-star sets computed by the semi-local optimization phase divided by $|\mathcal{O}|$. We use the main result of the analysis of Duh and Fürer [11] expressed in our notation.

Theorem 1 (Duh and Fürer[11]). *$b_1 \leq \alpha_{3,1}$ and $b_2 + b_1 \leq \alpha_{3,3} + \alpha_{3,2} + \alpha_{3,1}$.*

In addition, the restricted phase i (with $i = 4, 5$) imposes the constraint that the number of the 1-sets in the final solution does not increase compared to the number of 1-sets in the optimal star set covering of (V_i, S_i). Taking into account the first inequality of Theorem 1, we have

$$b_1 \leq a_{i,1}, \text{ for } i = 3, 4, 5. \tag{5}$$

Up to now, we have expressed all the properties of the instance $(V, \mathcal{S})$ as well as the behavior of the algorithm on it, besides the fact that the star set covering obtained implies an at most f-approximate solution for the corresponding SSF instance. We express the benefit of the algorithm in terms of our variables as follows. We denote by t_i the number of star sets computed during the phase i. For phase 6, we have

$$t_6 = \frac{1}{6}|V \setminus V_5| = \frac{1}{6}\left(T - \sum_{j=1}^{5} j a_{5,j}\right) |\mathcal{O}|. \tag{6}$$

For the phase i with $i = 4, 5$, we have

$$t_i = \frac{1}{i}|V_i \setminus V_{i-1}| = \frac{1}{i}\left(\sum_{j=1}^{i} ja_{i,j} - \sum_{j=1}^{i-1} ja_{i-1,j}\right)|\mathcal{O}|. \tag{7}$$

Also, for the semi-local optimization phase, we have

$$\begin{aligned} t_3 = (b_3 + b_2 + b_1)|\mathcal{O}| &= \left(\frac{b_1}{3} + \frac{b_2 + b_1}{3} + \frac{3b_3 + 2b_2 + b_1}{3}\right)|\mathcal{O}| \\ &\leq \left(\frac{b_1}{3} + \frac{a_{3,3} + a_{3,2} + a_{3,1}}{3} + \frac{3a_{3,3} + 2a_{3,2} + a_{3,1}}{3}\right)|\mathcal{O}| \\ &= \left(\frac{1}{3}b_1 + \frac{4}{3}a_{3,3} + a_{3,2} + \frac{2}{3}a_{3,1}\right)|\mathcal{O}|. \end{aligned} \tag{8}$$

The inequality follows by Theorem 1 and since $(3b_3 + 2b_2 + b_1)|\mathcal{O}| = (3a_{3,3} + 2a_{3,2} + a_{3,1})|\mathcal{O}| = |V_3|$.

By the discussion in Section 2, the solution obtained by the algorithm on the SSF instance I has benefit $ALG(I)$ at least $|V| - \sum_{i=3}^{6} t_i$. Consider an optimal spanning star forest of the graph G of instance I and let $OPT(I)$ be its benefit. This naturally corresponds to a star set cover $\mathcal{O}'$ for $(V, \mathcal{S})$ that consists of disjoint star sets of type A. Clearly, $OPT(I) = |V| - |\mathcal{O}'|$ and $|\mathcal{O}| \leq |\mathcal{O}'|$. Hence, $OPT(I) \leq |V| - |\mathcal{O}|$. Since the solution obtained by the algorithm is at most f-approximate (i.e., $ALG(I) \leq f \cdot OPT(I)$), we have $|V| - \sum_{i=3}^{6} t_i \leq f\,(|V| - |\mathcal{O}|)$ and, equivalently,

$$(f-1)|V| + \sum_{i=3}^{6} t_i \geq f|\mathcal{O}|. \tag{9}$$

We use (6), (7), and (8) to upper-bound the left side of inequality (9). We have

$$\begin{aligned} &(f-1)|V| + \sum_{i=3}^{6} t_i \\ &\leq (f-1)T|\mathcal{O}| + \frac{1}{6}\left(T - \sum_{j=1}^{5} ja_{5,j}\right)|\mathcal{O}| + \frac{1}{5}\left(\sum_{j=1}^{5} ja_{5,j} - \sum_{j=1}^{4} ja_{4,j}\right)|\mathcal{O}| \\ &\quad + \frac{1}{4}\left(\sum_{j=1}^{4} ja_{4,j} - \sum_{j=1}^{3} ja_{3,j}\right)|\mathcal{O}| + \left(\frac{1}{3}b_1 + \frac{4}{3}a_{3,3} + a_{3,2} + \frac{2}{3}a_{3,1}\right)|\mathcal{O}| \\ &= \left(\left(f - \frac{5}{6}\right)T + \frac{1}{6}a_{5,5} + \frac{2}{15}a_{5,4} + \frac{1}{10}a_{5,3} + \frac{1}{15}a_{5,2} + \frac{1}{30}a_{5,1} + \frac{1}{5}a_{4,4}\right. \\ &\quad \left. + \frac{3}{20}a_{4,3} + \frac{1}{10}a_{4,2} + \frac{1}{20}a_{4,1} + \frac{7}{12}a_{3,3} + \frac{1}{2}a_{3,2} + \frac{5}{12}a_{3,1} + \frac{1}{3}b_1\right)|\mathcal{O}|. \end{aligned} \tag{10}$$

By (9) and (10), we obtain

$$(f - \frac{5}{6})T + \frac{1}{6}a_{5,5} + \frac{2}{15}a_{5,4} + \frac{1}{10}a_{5,3} + \frac{1}{15}a_{5,2} + \frac{1}{30}a_{5,1} + \frac{1}{5}a_{4,4} + \frac{3}{20}a_{4,3} + \frac{1}{10}a_{4,2} + \frac{1}{20}a_{4,1} + \frac{7}{12}a_{3,3} + \frac{1}{2}a_{3,2} + \frac{5}{12}a_{3,1} + \frac{1}{3}b_1 \geq f. \quad (11)$$

By expressing inequalities (1)-(5) and (11) in standard form, we obtain our parameterized LP lemma.

Lemma 1. *If there exists an instance I of* SSF *for which the algorithm computes a solution of benefit $ALG(I) \leq f \cdot OPT(I)$ for some $f \in [0,1]$, then the following linear program LP(f) has a feasible solution for some $\theta > 1$.*

$$T \geq \theta$$

$$-\sum_{j=1}^{i} a_{i,j} \geq -1, \quad \textit{for } i = 3,4,5$$

$$T - \sum_{j=1}^{5} j a_{5,j} \geq 0$$

$$(i-1)a_{i,i} + \sum_{j=1}^{i-1} j a_{i,j} - \sum_{j=1}^{i-1} j a_{i-1,j} \geq 0, \quad \textit{for } i = 4,5$$

$$a_{i,1} - b_1 \geq 0, \quad \textit{for } i = 3,4,5$$

$$(f - \frac{5}{6})T + \frac{1}{6}a_{5,5} + \frac{2}{15}a_{5,4} + \frac{1}{10}a_{5,3} + \frac{1}{15}a_{5,2} + \frac{1}{30}a_{5,1} + \frac{1}{5}a_{4,4} + \frac{3}{20}a_{4,3} + \frac{1}{10}a_{4,2} + \frac{1}{20}a_{4,1} + \frac{7}{12}a_{3,3} + \frac{1}{2}a_{3,2} + \frac{5}{12}a_{3,1} + \frac{1}{3}b_1 \geq f$$

$$a_{i,j} \geq 0, \quad \textit{for } i = 3,4,5 \textit{ and } j = 1,\ldots,i$$

$$b_1 \geq 0$$

3.2 Proof of the Approximation Bound

The proof of the approximation bound is based on the following lemma.

Lemma 2. *For every $f < 193/240$, LP(f) has no feasible solution.*

Proof. We can assume that LP(f) is a minimization linear program with objective 0. By duality, if it were feasible, then the optimal objective value of the dual maximization linear program should be 0 as well. We show that this is not the case and that the dual has a solution with strictly positive objective value. This implies the lemma.

In the dual LP, we use the eleven variables η, β_3, β_4, β_5, γ_4, γ_5, γ_6, δ_3, δ_4, δ_5 and ζ corresponding to the constraints of LP(f). Variable η corresponds to the first constraint of LP(f), β_i correspond to the second set of constraints, γ_6 corresponds to the third constraint, γ_4 and γ_5 correspond to the fourth set of constraints, δ_i correspond to the fifth set of constraints, and ζ corresponds to the last constraint. So, the dual of LP(f) is

$$\begin{aligned} \text{maximize } & \theta\eta - \beta_3 - \beta_4 - \beta_5 + f\zeta \\ \text{subject to } & \eta + \gamma_6 + (f - 5/6)\zeta \leq 0 \\ & -\delta_3 - \delta_4 - \delta_5 + \zeta/3 \leq 0 \\ & -\beta_3 - \gamma_4 + \delta_3 + 5\zeta/12 \leq 0 \\ & -\beta_3 - 2\gamma_4 + \zeta/2 \leq 0 \\ & -\beta_3 - 3\gamma_4 + 7\zeta/12 \leq 0 \\ & -\beta_4 + \gamma_4 - \gamma_5 + \delta_4 + \zeta/20 \leq 0 \\ & -\beta_4 + 2\gamma_4 - 2\gamma_5 + \zeta/10 \leq 0 \\ & -\beta_4 + 3\gamma_4 - 3\gamma_5 + 3\zeta/20 \leq 0 \\ & -\beta_4 + 3\gamma_4 - 4\gamma_5 + \zeta/5 \leq 0 \\ & -\beta_5 + \gamma_5 - \gamma_6 + \delta_5 + \zeta/30 \leq 0 \\ & -\beta_5 + 2\gamma_5 - 2\gamma_6 + \zeta/15 \leq 0 \\ & -\beta_5 + 3\gamma_5 - 3\gamma_6 + \zeta/10 \leq 0 \\ & -\beta_5 + 4\gamma_5 - 4\gamma_6 + 2\zeta/15 \leq 0 \\ & -\beta_5 + 4\gamma_5 - 5\gamma_6 + \zeta/6 \leq 0 \\ & \beta_i, \delta_i \geq 0, \quad \text{for } i = 3, 4, 5 \\ & \gamma_i \geq 0, \quad \text{for } i = 4, 5, 6 \\ & \zeta, \eta \geq 0 \end{aligned}$$

The solution $\eta = 193/240 - f$, $\beta_3 = 39/72$, $\beta_4 = 11/72$, $\beta_5 = 79/720$, $\gamma_4 = 1/72$, $\gamma_5 = 1/45$, $\gamma_6 = 7/240$, $\delta_3 = 5/36$, $\delta_4 = 1/9$, $\delta_5 = 1/12$, and $\zeta = 1$ satisfies all the constraints. Observe that $\eta - \beta_3 - \beta_4 - \beta_5 + f\zeta = 0$ and, hence, the objective value is $(\theta - 1)\eta = (\theta - 1)(193/240 - f) > 0$. The lemma follows. □

Theorem 2. *The approximation ratio of the algorithm is at least* 193/240.

Proof. By Lemmas 1 and 2, we have that for any $f < 193/240$ and for any instance I of SSF, the algorithm computes a solution of benefit $ALG(I) > f \cdot OPT(I)$. Hence, its approximation ratio is at least 193/240. □

3.3 A Note for Color Saving

We point out that neither the algorithm nor the analysis makes use of the fact that the star sets actually correspond to stars in the graph. Hence, the algorithm applies to the more general complementary set cover problem where the sets are given explicitly. Also, since the algorithm considers sets of constant size, it also applies to color saving. What is required is to replace the term "star set" by the term "independent set". Although computing an independent set of a given size on a graph is a classical NP-hard problem, computing independent sets of size up to 6 as the algorithm requires can be done easily in polynomial time. Our analysis is tight; the lower bound construction is omitted.

4 A Lower Bound for Natural Local Search SSF Algorithms

In this section we consider a natural family of local search algorithms for SSF. We consider solutions of instances of SSF as assignments of values 0 and 1 to the nodes of the input graph where 0 or 1 at a node means that the node is a center or leaf, respectively. Such an assignment corresponds to a feasible solution of SSF when each node that has been assigned value 1 is adjacent to at least one node that has been assigned value 0. The benefit of a feasible assignment is then the number of nodes that have been assigned value 1.

Local search can be used as follows. Starting with any feasible assignment (e.g., with 0 to each node), a local search algorithm repeatedly performs k-changes (for some constant integer k) while this is possible. Performing a k-change means to alter the assignment of $t \leq k$ nodes originally having value 1 to 0 and the assignment of $t+1$ nodes originally assigned value 0 to 1. The algorithm terminates when a local optimum assignment is reached, i.e., one from which no k-change is possible. Note that since k is constant and the benefit increases by 1 at each step, the algorithm terminates in polynomial time. The algorithm is efficient if all local optima have high benefit.

The local search algorithm that performs 0-changes may have very poor approximation ratio. Indeed, consider the instance that consists of two nodes u and v that are assigned the value 1 and $n-2$ nodes u_1, ..., u_{n-2} that are connected to both u and v and are all assigned the value 0. This assignment has a benefit of 2 and is a local optimum since no 0-change is possible, while the solution that assigns 1 to nodes u_1, ..., u_{n-2} and 0 to nodes u and v is feasible and has a benefit of $n-2$. Furthermore, we can show that the local search algorithm that performs 1-changes always computes an 1/2-approximate assignment. This could indicate that better bounds can be obtained by considering k-changes with higher constant values of k. Unfortunately, this is not the case as the following theorem states.

Theorem 3. *For every integer $k \geq 0$ and every $\epsilon \in \left(0, \frac{1}{2(k+2)}\right]$, the local-search algorithm that performs $k-$changes has an $(1/2+\epsilon)$-approximate solution as a local optimum.*

Proof. The proof uses a result of Erdös and Sachs [12] stating that, for every integer $d, g > 0$, there exists a d-regular graph of girth at least g.

Our starting point is a d-regular graph G of girth at least g, where $d = 2\lceil 1/2\epsilon \rceil$ and $g = k+2$. Given G, consider the graph G' that is obtained by replacing each edge (u, v) of G by a path of size 3 (henceforth called a 3-path) $\langle u, Z^1_{uv}, Z^2_{uv}, v \rangle$. So, the graph G has two additional nodes Z^1_{uv} and Z^2_{uv} for each edge (u, v) in G. We refer to these nodes as edge-nodes.

We define an assignment on the nodes of G' which has (asymptotically) half the optimal benefit and which cannot be improved by a k-change. In order to construct this assignment, we make use of the fact that a connected graph has an Euler circuit when all its nodes have even degree. We consider such a directed Euler circuit in G' and assign value 1 to the first node of each 3-path the Euler circuit comes across and value 0 to the other node of the 3-path. Furthermore, we assign value 1 to every non-edge-node. We notice that this is a feasible assignment since every edge-node with value 1 is connected to the other edge-node of the same 3-path having value 0, and every non-edge-node has exactly $d/2 = \lceil 1/2\epsilon \rceil \geq k+2$ neighbors with value 0.

We use proof by contradiction. Suppose the aforementioned bad assignment is not a local optimum and consider another neighboring feasible assignment (that is obtained by applying a $k-$change to the original one). In more detail, assume that the improved assignment is obtained if we assign value 1 to $t+1$ edge-nodes having value 0 originally, as well as change the value of $s \leq t$ edge-nodes and of $t-s$ non-edge-nodes from 1 to 0.

Consider the intermediate assignment, where only the changes on the values of the edge-nodes have been applied. We notice that the constraints regarding the non-edge-nodes are not violated since each such node originally had at least $k+2$ neighbors being assigned value 0 and at most $k+1$ changes from 0 to 1 have been applied. Regarding the edge-nodes, the only ones whose constraint is unsatisfied in the intermediate assignment belong to 3-paths in which both edge-nodes have value 1. The number of such 3-paths is at least $t+1-s$ and at most $k+1$. We consider the edge-induced graph H consisting of the edges of G that correspond to these 3-paths. Since H has at most $k+1$ edges and the girth of G is at least $k+2$, H is a forest. In addition, the nodes of graph H correspond to the $t-s$ non-edge-nodes of G' whose value is 0 at the final assignment. As we have already mentioned, graph H has at least $t+1-s$ edges, which implies that it contains a cycle. Furthermore H has at most $k+1$ edges, therefore its girth is at most $k+1$, which contradicts the original assumption.

We conclude that no $k-$change can be applied to the bad assignment above and it is a local optimum with benefit $n(1+d/2)$ while the benefit of the optimal assignment on G' is dn, which is achieved when only non-edge-nodes are assigned value 0. Hence, the approximation ratio is at most $1/2 + 1/d \leq 1/2 + \epsilon$. $\square$

References

1. Agra, A., Cardoso, D., Cerfeira, O., Rocha, E.: A spanning star forest model for the diversity problem in automobile industry. In: Proceedings of ECCO XVII (2005)
2. Athanassopoulos, S., Caragiannis, I., Kaklamanis, C.: Analysis of approximation algorithms for k-set cover using factor-revealing linear programs. In: Csuhaj-Varjú, E., Ésik, Z. (eds.) FCT 2007. LNCS, vol. 4639, pp. 52–63. Springer, Heidelberg (2007)
3. Ausiello, G., D'Atri, A., Protasi, M.: Structure preserving reductions among convex optimization problems. Journal of Computer and System Sciences 21, 136–153 (1980)
4. Ausiello, G., Marchetti-Spaccamela, A., Protasi, M.: Toward a unified approach for the classification of NP-complete optimation problems. Theoretical Computer Science 12, 83–96 (1980)

5. Berry, V., Guillemot, S., Nicholas, F., Paul, C.: On the approximation of computing evolutionary trees. In: Wang, L. (ed.) COCOON 2005. LNCS, vol. 3595, pp. 115–125. Springer, Heidelberg (2005)
6. Caragiannis, I.: Wavelength management in WDM rings to maximize the number of connections. SIAM Journal on Discrete Mathematics 23(2), 959–978 (2009)
7. Chakrabarty, D., Goel, G.: On the approximability of budgeted allocations and improved lower bounds for submodular welfare maximization and GAP. In: Proceedings of the 49th Annual IEEE Symposium on Foundations of Computer Science (FOCS 2008), pp. 687–696 (2008)
8. Chen, N., Engelberg, R., Nguyen, C.T., Raghavendra, P., Rudra, A., Singh, G.: Improved approximation algorithms for the spanning star forest problem. In: Charikar, M., Jansen, K., Reingold, O., Rolim, J.D.P. (eds.) RANDOM 2007 and APPROX 2007. LNCS, vol. 4627, pp. 44–58. Springer, Heidelberg (2007)
9. Demange, M., Paschos, V.T.: On an approximation measure founded on the links between optimization and polynomial approximation theory. Theoretical Computer Science 158, 117–141 (1996)
10. Demange, M., Grisoni, P., Paschos, V.T.: Differential approximation algorithms for some combinatorial optimization problems. Theoretical Computer Science 209, 107–122 (1998)
11. Duh, R., Fürer, M.: Approximation of k-set cover by semi local optimization. In: Proceedings of the 29th Annual ACM Symposium on Theory of Computing (STOC 1997), pp. 256–264 (1997)
12. Erdös, P., Sachs, H.: Reguläre Graphen gegebener Taillenweite mit minimaler Knotenzahl. Wiss. Z. Univ. Halle, Math.-Nat. 12, 251–258 (1963)
13. Feige, U.: A threshold of $\ln n$ for approximating set cover. Journal of the ACM 45(4), 634–652 (1998)
14. Halldórsson, M.M.: Approximating discrete collections via local improvements. In: Proceedings of the 6th Annual ACM/SIAM Symposium on Discrete Algorithms (SODA 1995), pp. 160–169 (1995)
15. Halldórsson, M.M.: Approximating k-set cover and complementary graph coloring. In: Cunningham, W.H., Queyranne, M., McCormick, S.T. (eds.) IPCO 1996. LNCS, vol. 1084, pp. 118–131. Springer, Heidelberg (1996)
16. Hassin, R., Khuller, S.: z-Approximations. Journal of Algorithms 41(2), 429–442 (2001)
17. Hurkens, C.A.J., Schrijver, A.: On the size of systems of sets every t of which have an SDR, with an application to the worst-case ratio of heuristics for packing problems. SIAM Journal on Discrete Mathematics 2(1), 68–72 (1989)
18. Jain, K., Mahdian, M., Markakis, E., Saberi, A., Vazirani, V.V.: Greedy facility location algorithms analyzed using dual fitting with factor-revealing LP. Journal of the ACM 50(6), 795–824 (2003)
19. Johnson, D.S.: Approximation algorithms for combinatorial problems. Journal of Computer and System Sciences 9, 256–278 (1974)
20. Levin, A.: Approximating the unweighted k-set cover problem: greedy meets local search. SIAM Journal on Discrete Mathematics 23(1), 251–264 (2009)
21. Lovász, L.: On the ratio of optimal integral and fractional covers. Discrete Mathematics 13, 383–390 (1975)
22. Nguyen, C.T., Shen, J., Hou, M., Sheng, L., Miller, W., Zhang, L.: Approximating the spanning star forest problem and its application to genomic sequence alignment. SIAM Journal on Computing 38(3), 946–962 (2008)

Energy-Efficient Communication in Multi-interface Wireless Networks*

Stavros Athanassopoulos, Ioannis Caragiannis, Christos Kaklamanis, and Evi Papaioannou

Research Academic Computer Technology Institute and
Department of Computer Engineering and Informatics
University of Patras, 26500 Rio, Greece

Abstract. We study communication problems in wireless networks supporting multiple interfaces. In such networks, two nodes can communicate if they are close and share a common interface. The activation of each interface has a cost reflecting the energy consumed when a node uses this interface. We distinguish between the symmetric and non-symmetric case, depending on whether all nodes have the same activation cost for each interface or not. For the symmetric case, we present a $(3/2+\epsilon)$-approximation algorithm for the problem of achieving connectivity with minimum activation cost, improving a previous bound of 2. For the non-symmetric case, we show that the connectivity problem is not approximable within a sublogarithmic factor in the number of nodes and present a logarithmic approximation algorithm for a more general problem that models group communication.

1 Introduction

Wireless networks have received significant attention during the recent years. They support a wide range of popular applications and usually constitute parts of larger, global networks, and the Internet. Wireless networks are in general heterogeneous in the sense that they are composed of wireless devices of different characteristics like computational power, energy consumption, radio interfaces, supported communication protocols, etc. Modern wireless devices are equipped with multiple radio interfaces (like most commonly wireless interfaces in use today such as Bluetooth, WiFi and GPRS) and can switch between different communication networks according to connectivity requirements and quality of service constraints (see Fig. 1). Selecting the best radio interfaces for specific connections depends on several factors, like for example, availability of an interface at a particular device, interference constraints, the necessary communication bandwidth, the energy consumed by an active interface and its lifetime, the interfaces available in some neighborhood, topological properties of the network, etc.

* This work is partially supported by the European Union under IST FET Integrated Project FP6-015964 AEOLUS, by the General Secretariat for Research and Technology of the Greek Ministry of Development under programme PENED, and by a "Caratheodory" basic research grant from the University of Patras.

R. Královič and D. Niwiński (Eds.): MFCS 2009, LNCS 5734, pp. 102–111, 2009.

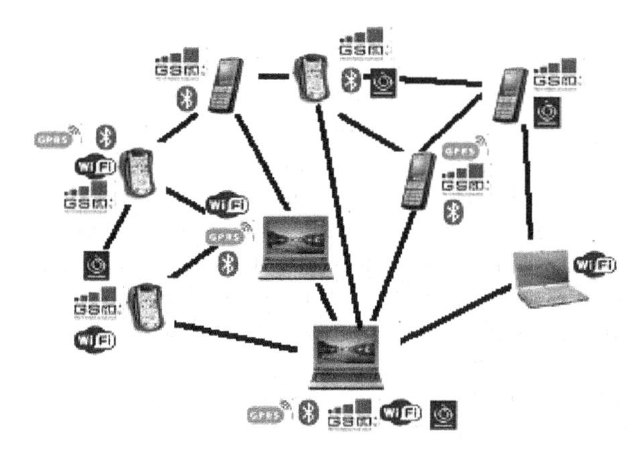

Fig. 1. Modern wireless devices are equipped with multiple radio interfaces and can switch between different communication networks

We study communication problems in wireless networks supporting multiple interfaces. The nodes of such networks are wireless devices equipped with some wireless interfaces. Communication between two such nodes can be established if (i) they are sufficiently close to each other and (ii) both share a common interface. If these requirements are met, then communication is established at a cost equal to the cost of activating a particular interface which is common in both nodes. The activation cost of an interface reflects the energy consumed when a node uses this interface. Our objective is to activate interfaces at the network nodes so that some connectivity property is preserved and the total cost of activated interfaces is minimized. Depending on the required connectivity property, several communication problems in multi-interface wireless networks arise. We consider two such problems: ConMI and GroupMI. In ConMI, we require that the communication is established among all network nodes. In GroupMI, communication must be established among groups of nodes (that do not necessarily include all nodes of the network). ConMI is a special case of GroupMI. We distinguish between two cases. The more general one is when the activation cost for some interface is not the same at all network nodes; this is the non-symmetric case. In the symmetric case of the problem, the cost of activating a particular interface is the same at all network nodes.

Related work. Multi-interface wireless networks have recently attracted research interest since they have emerged as a de facto communication infrastructure and can support a wide range of important and popular applications. In this setting, many basic problems already studied for "traditional" wired and wireless networks have been restated [2], especially those related to network connectivity [5,8] and routing [6] issues. However, the energy efficiency requirements increase the complexity of these problems and pose new challenges.

A combinatorial problem that falls within the general class of communication problems in multi-interface wireless networks has been studied in [11]. In that paper, a graph with desired connections between network nodes is given and the objective is to activate interfaces of minimum total cost at the network nodes so that all the edges in this graph are established. Several variations of the problem are considered depending on the topology of the input graph (e.g., complete graphs, trees, planar graphs, bounded-degree graphs, general graphs) and on whether the number of interfaces is part of the input or a fixed constant. The paper considers both unit-cost interfaces and more general symmetric instances.

ConMI has been introduced in [12] which studies symmetric instances of the problem. ConMI is proved to be APX-hard even when the graph modeling the network has a very special structure and the number of available interfaces is small (e.g., 2). On the positive side, [12] presents a 2-approximation algorithm by exploiting the relation of ConMI on symmetric instances with the minimum spanning tree on an appropriately defined edge-weighted graph. Better approximation bounds are obtained for special cases of ConMI such as the case of unit-cost interfaces.

Our results. We distinguish between the symmetric and non-symmetric case, depending on whether all nodes have the same activation cost for an interface or not. For the symmetric case, we present a $(3/2+\epsilon)$–approximation algorithm for ConMI, improving the previously best known bound of 2 from [12]. The main idea of the algorithm is to use an almost minimum spanning tree (MST) in an appropriately defined hypergraph and transform it to an efficient solution for ConMI. We also consider GroupMI for symmetric instances where we obtain a 4-approximation algorithm; here, we transform instances of the problem to instances of Steiner Forest in a similar way [12] transforms ConMI to MST. For the non-symmetric case, we show that the connectivity problem is not approximable within a sublogarithmic factor in the number of nodes through a reduction from Set Cover, and present a logarithmic approximation algorithm for the more general GroupMI problem. Here, we transform instances of the problem to instance of Node-Weighted Steiner Forest and exploit approximation algorithms of Guha and Khuller [10] (see also [1]) for Node-Weighted Steiner Forest. We remark that techniques for the Node-Weighted Steiner Forest have also been applied either implicitly [3] or explicitly [4] to minimum energy communication problems in ad hoc wireless networks. To the best of our knowledge, neither GroupMI nor non-symmetric instances of multi-interface wireless networks have been studied before.

The rest of the paper is structured as follows. We present some preliminary technical definitions and our notation in Section 2. The upper bound for symmetric instances of ConMI appears in Section 3. The algorithm for GroupMI is presented in Section 4. The results for non-symmetric instances of GroupMI are presented in Section 5.

2 Definitions and Notation

The wireless network is modelled by a graph G in which nodes correspond to network nodes and edges represent potential direct connections between pairs of network nodes. We denote by I the set of available interfaces. Each node u supports a set $I_u \subseteq I$ of interfaces. Two nodes u and v can communicate when they have the same interface activated provided that there exists an edge between them in G. Given a set of activated interfaces $S_u \subseteq I_u$ at each node u, we define the communication graph G_S that has the same set of nodes with G and an edge $e = (u, v)$ of G belongs to G_S if $S_u \cap S_v \neq \emptyset$. Activating interface g at node u has a non-negative cost $c_{u,g}$. Our objective is to activate interfaces at the nodes of G so that the induced communication graph has some connectivity property and the total cost of activated interfaces is minimized. Depending on the required connectivity property, several communication problems in multi-interface wireless networks arise. We consider two such problems: ConMI and GroupMI. In ConMI, we require that the communication graph is connected and spans all nodes of G. In GroupMI, we are additionally given a set of terminal nodes $D \subseteq V$ partitioned into p disjoint subsets $D_1, D_2, ..., D_p$. Here, for $i = 1, ..., p$, we require the communication graph to connect the terminal nodes of D_i. Clearly, ConMI is a special case of GroupMI. We distinguish between two cases. The more general one described above is the non-symmetric case. In the symmetric case of the problem, the cost of activating interface g at each node is the same and equal to c_g.

In the following we usually refer to well-known combinatorial optimization problems such as the Steiner Forest and the Node-Weighted Steiner Forest. In both problems, the input consists of a graph $G = (V, E)$ and a set of terminals $D \subseteq V$ partitioned in p disjoint subsets as in GroupMI, and the objective is to compute a forest of minimum total cost (weight) so that the terminals in the subset D_i belong to the same tree of the forest. In Steiner Forest, each edge e of G has an associated non-negative weight w_e; in Node-Weighted Steiner Forest, the edges are unweighted and each node u has a weight w_u.

3 A ConMI Algorithm for the Symmetric Case

We present a $(3/2+\epsilon)$–approximation algorithm for the symmetric case of ConMI improving the previously known upper bound of 2.

Our algorithm works as follows. Consider an input instance J, a set of available interfaces I and sets I_u of interfaces supported at each node u. First, we transform the graph G into an instance of the problem of computing a minimum spanning tree (MST) in an appropriately defined hypergraph H. Then, we solve almost exactly the MST problem in H using a polynomial-time approximation scheme of Prömel and Steger [13]. We use the resulting tree to determine the interfaces to activate in the nodes of G so that the corresponding communication graph is a connected spanning subgraph of G. This is the output of our algorithm. We show that the algorithm obtains an approximation guarantee of $3/2 + \epsilon$, where ϵ is the approximation guarantee of the MST algorithm in the hypergraph H.

The hypergraph $H = (V, F)$ has the same set of nodes as the graph G. We define the set of edges F of H as follows. We consider all triplets of nodes

v_i, v_j, v_k so that $I_{v_i} \cap I_{v_j} \cap I_{v_k} \neq \emptyset$ which are connected with at least two edges among them in G. We insert the triplet (v_i, v_j, v_k) as a hyperedge f of F. We denote by $s(f)$ the interface in $I_{v_i} \cap I_{v_j} \cap I_{v_k}$ of minimum cost; we call $s(f)$ the interface associated with hyperedge f. We assign to f a weight of $3c_{s(f)}$. This corresponds to the fact that by activating interface $s(f)$ at nodes v_i, v_j, v_k, the edges connecting them are contained in the corresponding communication graph at a cost of $3c_{s(f)}$.

We consider all pairs of nodes v_i, v_j so that $I_{v_i} \cap I_{v_j} \neq \emptyset$ which are connected with an edge in G. We insert the pair (v_i, v_j) as a hyperedge f of F. Again, we denote by $s(f)$ the interface in $I_{v_i} \cap I_{v_j}$ of minimum cost. We assign to f a weight of $2c_{s(f)}$. This corresponds to the fact that by activating $s(f)$ at nodes v_i, v_j, the edge connecting them is contained in the corresponding communication graph at a cost of $2c_{s(f)}$.

Since the edges of the hypergraph H consist of at most 3 nodes, we use the polynomial-time approximation scheme of [13] to obtain a spanning tree T of H. For each edge f of T, we activate interface $s(f)$ at the nodes of G that belong to f. In this way, in the corresponding communication graph the nodes belonging to the same hyperedge of T are connected and since T is connected and spans all nodes of V, the whole communication graph is a spanning subgraph of G, as well.

We denote by $cost(J)$ the total cost of the solution obtained by our algorithm, by $opt(J)$ the cost of the optimal solution, by $mst(H)$ the cost of the minimum spanning tree of H and by $st(H)$ the cost of the spanning tree T. We prove the following two lemmas.

Lemma 1. *$cost(J) \leq st(H)$.*

Proof. Since the set S_u of interfaces activated at each node u consists of interfaces associated with the hyperedges of T that contain u, it holds:

$$cost(J) = \sum_{u \in V} \sum_{g \in S_u} c_g \leq \sum_{u \in V} \sum_{f \in T : u \in f} c_{s(f)} = \sum_{f \in T} w(f) = st(H). \qquad \square$$

Lemma 2. *$mst(H) \leq \frac{3}{2} opt(J)$.*

Proof. Consider an optimal solution to ConMI for instance J that consists of sets of interfaces S_u activated at each node u of G. We denote by S the set of all activated interfaces and by G_S the corresponding communication graph. We decompose G_S into different subgraphs; there is one such subgraph for each interface of S. We denote by G_g the subgraph of G consisting of the set of nodes V_g which have interface g activated in the optimal solution and of the set of edges E_g connecting nodes of V_g in G. Clearly, $opt(J) = \sum_{g \in S} c_g |V_g|$. For each $g \in S$, we compute a minimum spanning tree on each connected component of G_g; the minimum spanning trees on the connected components of G_g form a forest T_g.

We decompose the edges of T_g into special substructures that we call *forks*; a fork is either a set of two edges incident to the same node or a single edge. In each connected component of T_g with m nodes, the procedure that decomposes its edges into forks is the following. If there are two leaves u and v with a common parent, we include the edges incident to u and v in a fork. We remove u, v, and their incident edges from the tree. If no two leaves have a common parent, then some leaf u has a node v of degree 2 as a parent or there is only one remaining edge between two nodes u and v. In the first case, we include the edges incident to u and v in a fork and remove u, v, and their incident edges from the tree. In the second case, we simply include the edge between u and v in a fork and remove it from the tree. We repeat the procedure above until all edges of the tree are included in forks. In each step (possibly besides the last one), 2 among the at most $m-1$ edges of the tree are included in a fork. Hence, the number of forks is at most $m/2$. By repeating this decomposition for each connected component of T_g, we obtain a decomposition of the edges of T_g into at most $|V_g|/2$ forks. The endpoints of each fork of T_g correspond to a hyperedge in H with weight at most $3c_g$. The union of all these hyperedges is a connected spanning subgraph of H (since the union of the T_g's yields G_S). The cost of the minimum spanning tree of H is upper-bounded by the total cost of the hyperedges in this spanning subgraph, i.e.,

$$mst(H) \leq \sum_{g \in S} 3c_g \frac{|V_g|}{2} = \frac{3}{2} opt(J). \qquad \square$$

By Lemmas 1 and 2 and since $st(H) \leq (1+\epsilon)mst(H)$, we obtain the following:

Theorem 1. *For any constant $\epsilon > 0$, there exists a polynomial-time $(3/2+\epsilon)$-approximation algorithm for ConMI.*

4 Group Communication in the Symmetric Case

In this section, we present an algorithm for symmetric instances of GroupMI that has a constant approximation ratio. The main idea of the algorithm is similar to the algorithm of Kosowski et al. [12] for ConMI but instead of using a polynomial-time algorithm for MST, we use the 2-approximation algorithm of Goemans and Williamson [9] for the Steiner Forest problem.

Consider an instance J of GroupMI with a graph $G = (V, E)$ with n nodes, a set of terminal nodes $D \subseteq V$ partitioned into p disjoint subsets $D_1, ..., D_p$ and sets I_u of interfaces supported by each node $u \in V$. We construct an instance J_{SF} of Steiner Forest consisting of a graph $H = (V, A)$ and the set of terminals D partitioned into the same p disjoint subsets of terminals. The set of edges A contains all edges (u, v) of E such that $I_u \cap I_v \neq \emptyset$. Consider an edge $e = (u, v)$ of A and let $s(e)$ be the interface of minimum cost in $I_u \cap I_v$. Then, the weight w_e of e is equal to $2c_{s(e)}$.

We use the algorithm of [9] to solve Steiner Forest for the instance J_{SF} and obtain a forest F which preserves connectivity among the nodes of each terminal set D_i. We obtain the solution to the original instance J as follows. For each interface edge e in F, we activate interface $s(e)$ at the endpoints of e. Clearly, in this way the edges of G that correspond to F are contained in the corresponding communication graph and the required connectivity requirements for instance J are satisfied.

The upper bound on the approximation ratio of the algorithm is given in the following statement. The proof is obtained by extending the arguments used in [12].

Theorem 2. *There exist a 4-approximation algorithm for symmetric instances of GroupMI.*

Proof. Omitted. □

5 Group Communication in the Non-symmetric Case

In this section, we consider the problem GroupMI for non-symmetric instances. In this case, even the simpler problem ConMI does not have constant approximation algorithms as the following statement indicates.

Theorem 3. *Non-symmetric ConMI in networks with n nodes is hard to approximate within $o(\ln n)$.*

Proof. We use a simple reduction from Set Cover. Consider an instance of Set Cover with a ground set U of m elements and a collection $\mathcal{T}$ of subsets of U. The size of the collection $\mathcal{T}$ is polynomial in m. We construct an instance of ConMI as follows. The set of interfaces I has two interfaces 0 and 1. The graph G has a root node r, nodes $u_1, ..., u_{|\mathcal{T}|}$ corresponding to the sets of $\mathcal{T}$, and nodes $v_1, ..., v_m$ corresponding to the elements of U. Node r supports only interface 0 (i.e., $I_r = \{0\}$) with an activation cost 0. Nodes $u_1, ..., u_{|\mathcal{T}|}$ support interfaces 0 and 1 (i.e., $I_{u_i} = \{0, 1\}$) with activation costs 0 for interface 0 and 1 for interface 1. For $i = 1, ..., m$, node v_i supports only interface 1 ($I_{v_i} = \{1\}$) with an activation cost 0. For each set T_i of $\mathcal{T}$, node u_i is connected through an edge with each node v_j so that element j belongs to the set T_i. The root node r has edges to each node u_i, for $i = 1, ..., |\mathcal{T}|$.

We can easily show that any cover of U with C sets from $\mathcal{T}$ yields a solution to ConMI with cost at most C and vice versa. Indeed, consider a solution to the Set Cover instance that consists of a subset of $\mathcal{T}'$ of $\mathcal{T}$. By activating interface 0 at nodes $r, u_1, ..., u_{\mathcal{T}}$ and interface 1 at nodes $v_1, ..., v_m$ and nodes u_i, such that $i \in \mathcal{T}'$, we obtain a solution to the ConMI instance of cost $|\mathcal{T}'|$. Also, given a solution to the ConMI instance, we obtain a cover of U of the same cost by picking the sets of $\mathcal{T}$ that correspond to the nodes u_i which have interface 1 activated. Using well-known inapproximability results for Set Cover [7,14], we obtain an inapproximability bound of $\tau \ln m$. Since the number of nodes n in the instance of ConMI is polynomial in m, we obtain the desired result. □

Next, we present an $O(\ln n)$–approximation algorithm for non-symmetric GroupMI by reducing the problem to instances of Node-Weighted Steiner Forest. The reduction is similar to reductions for minimum energy communication problems in ad hoc wireless networks [4].

Consider an instance J of GroupMI with a graph $G = (V, E)$ with n nodes, a set of terminal nodes $D \subseteq V$ partitioned into p disjoint subsets $D_1, ..., D_p$ and sets I_u of interfaces supported by each node $u \in V$. We construct an instance of Node-Weighted Steiner Forest consisting of a graph $H = (U, A)$ and a set of terminals $D' \subseteq U$ partitioned into p disjoint subsets $D'_1, ..., D'_p$. The graph H is defined as follows. The set of nodes U consists of n disjoint sets of nodes called supernodes. Each supernode corresponds to a node of V. The supernode Z_u corresponding to node $u \in V$ has the following $|I_u| + 1$ nodes: a hub node $Z_{u,0}$ and $|I_u|$ bridge nodes $Z_{u,g}$ for each interface $g \in I_u$. For each pair of nodes $u, v \in V$ and each interface $g \in I_u \cap I_v$, the set A of edges contains an edge between the bridge nodes $Z_{u,g}$ and $Z_{v,g}$. Also, for each node $u \in V$, A contains an edge between the hub node $Z_{u,0}$ and each bridge node $Z_{u,g}$, for $g \in I_u$. Each hub node has weight 0. A bridge node $Z_{u,g}$ corresponding to node $u \in V$ and interface $g \in I_u$ has weight equal to the activation cost $c_{u,g}$ of interface g at node u. The set of terminals D' consists of all the hub nodes. For $i = 1, ..., p$, the set D'_i in the partition of D' consists of the hub nodes $Z_{u,0}$ for each node $u \in D_i$.

We denote by J_{NWSF} the resulting instance of Node-Weighted Steiner Forest. We use a known algorithm for solving Node-Weighted Steiner Forest for the instance J_{NWSF} and obtain a forest F which is a subgraph of H without isolated nodes and which preserves connectivity among the nodes of each terminal set D_i. We obtain the solution S to the original instance J as follows. For each interface g in I_u, we include g in S_u iff $Z_{u,g}$ is a node of F.

The next lemma captures the main property of the reduction.

Lemma 3. *If F is a ρ–approximate solution to J_{NWSF}, then S is a ρ–approximate solution to J.*

Proof. Let u, v be nodes belonging to the same terminal set D_i. Then, the hub nodes $Z_{u,0}$ and $Z_{v,0}$ belong to the set D'_i and there exists a path q from $Z_{u,0}$ to $Z_{v,0}$ in F. The edges of q are either edges connecting a hub node with a bridge node in the same supernode or bridge nodes between different supernodes. Consider the edges of q that connect bridge nodes of different supernodes in the order we visit them by following path q from $Z_{u,0}$ to $Z_{v,0}$. For each such edge $(Z_{x,g}, Z_{y,g})$ the edge (x, y) belongs to G and all of them define a path q' from u to v in G. Since $Z_{x,g}$ and $Z_{y,g}$ belong to F, nodes x and y have interface g activated and hence the edges of q' belong to the induced communication graph.

In the following we show that the total activation cost $cost(J)$ of our solution equals the cost $cost(J_{NWSF})$ of F and that the optimal activation cost $opt(J)$ is lower-bounded by the cost $opt(J_{NWSF})$ of the optimal solution for J_{NWSF}. In this way, we obtain that

$$cost(J) = cost(J_{NWSF}) \leq \rho \cdot opt(J_{NWSF}) \leq \rho \cdot opt(J).$$

Indeed, interface g is activated at node u only if the bridge node $Z_{u,g}$ belongs to F. Since, $w(Z_{u,g}) = c_{u,g}$, we have that the total activation cost of our solution for J is equal to the cost of F.

Now, consider an optimal solution to J consisting of sets S_u of activated interfaces at each node u of G. We construct a subgraph F' of H as follows.

For each edge (u, v) in the communication graph G_S, and for each interface g belonging to $S_u \cap S_v$, we add edge $(Z_{u,g}, Z_{v,g})$ to F'. For each node u and each interface $g \in S_u$, we add edge $(Z_{u,0}, Z_{u,g})$ to F'. Using similar reasoning as above, we obtain that F' maintains the connectivity requirement between nodes of the same terminal set D'_i and the total weight of its nodes equals the total activation cost in the optimal solution of J. Hence, the cost of the optimal solution of J_{NWSF} is not higher than the cost of F'. □

In [10], Guha and Khuller present a $1.61 \ln k$–approximation algorithm for Node-Weighted Steiner Forest, where k is the number of terminals in the graph. We use this algorithm to solve J_{NWSF}, and following the discussion above we obtain a solution of J which is within $1.61 \ln |D|$ of optimal. Thus, we have:

Theorem 4. *There exists a $1.61 \ln |D|$-approximation algorithm for non-symmetric GroupMI, where D is the set of terminals.*

Moreover, if J is an instance of ConMI, then $p = 1$ and the instance J_{NWSF} is actually an instance of Node-Weighted Steiner Tree which can be approximated within $1.35 \ln k$.

Theorem 5. *There exists a $1.35 \ln n$–approximation algorithm for non-symmetric ConMI with n network nodes.*

References

1. Agrawal, A., Klein, P., Ravi, R.: When trees collide: an approximation algorithm for generalized Steiner tree Problem on Networks. In: Proceedings of the 23rd Annual ACM Symposium on Theory of Computing (STOC 1991), pp. 131–144 (1991)
2. Bahl, P., Adya, A., Padhye, J., Walman, A.: Reconsidering wireless systems with multiple radios. SIGCOMM Comput. Commun. Rev. 34(5), 39–46 (2004)
3. Călinescu, G., Kapoor, S., Olshevsky, A., Zelikovsky, A.: Network lifetime and power assignment in ad hoc wireless networks. In: Di Battista, G., Zwick, U. (eds.) ESA 2003. LNCS, vol. 2832, pp. 114–126. Springer, Heidelberg (2003)
4. Caragiannis, I., Kaklamanis, C., Kanellopoulos, P.: Energy-efficient wireless network design. Theory of Computing Systems 39(5), 593–617 (2006)
5. Cavalcanti, D., Gossain, H., Agrawal, D.: Connectivity in multi-radio, multi-channel heterogeneous ad hoc networks. In: Proceedings of the IEEE 16th International Symposium on Personal, Indoor and Mobile Radio Communications (PIMRC 2005), pp. 1322–1326. IEEE, Los Alamitos (2005)
6. Draves, R., Padhye, J., Zill, B.: Routing in multi-radio, multi-hop wireless mesh networks. In: Proceedings of the 10th Annual International Conference on Mobile Computing and Networking (MobiCom 2004), pp. 114–128. ACM, New York (2004)
7. Feige, U.: A Threshold of *lnn* for Approximating Set Cover. Journal of the ACM 45(4), 634–652 (1998)
8. Faragó, A., Basagni, S.: The effect of multi-radio nodes on network connectivity: a graph theoretic analysis. In: Proceedings of the IEEE 19th International Symposium on Personal, Indoor and Mobile Radio Communications (PIMRC 2008). IEEE, Los Alamitos (2008)

9. Goemans, M.X., Williamson, D.P.: A general approximation technique for constrained forest problems. SIAM Journal on Computing 24, 296–317 (1995)
10. Guha, S., Khuller, S.: Improved Methods for Approximating Node Weighted Steiner Trees and Connected Dominating Sets. Information and Computation 150(1), 57–74 (1999)
11. Klasing, R., Kosowski, A., Navarra, A.: Cost minimisation in wireless networks with bounded and unbounded number of interfaces. Networks 53(3), 266–275 (2009)
12. Kosowski, A., Navarra, A., Pinotti, M.C.: Connectivity in multi-interface networks. In: Kaklamanis, C., Nielson, F. (eds.) TGC 2008. LNCS, vol. 5474, pp. 159–173. Springer, Heidelberg (2009)
13. Prömel, H.J., Steger, A.: A new approximation algorithm for the Steiner tree problem with performance ratio 5/3. Journal of Algorithms 36, 89–101 (2000)
14. Raz, R., Safra, S.: A Sub-Constant Error-Probability Low-Degree Test, and a Sub-Constant Error-Probability PCP Characterization of NP. In: Proceedings of the 29th Annual ACM Symposium on the Theory of Computing (STOC 1997), pp. 475–484 (1997)

Private Capacities in Mechanism Design*

Vincenzo Auletta, Paolo Penna, and Giuseppe Persiano

Dipartimento di Informatica ed Applicazioni, Università di Salerno, Italy
{auletta,penna,giuper}@dia.unisa.it

Abstract. Algorithmic mechanism design considers distributed settings where the participants, termed agents, cannot be assumed to follow the protocol but rather their own interests. The protocol can be regarded as an algorithm augmented with a suitable payment rule and the desired condition is termed *truthfulness*, meaning that it is never convenient for an agent to report false information.

Motivated by the applications, we extend the usual one-parameter and multi-parameter settings by considering agents with *private capacities*: each agent can misreport her cost for "executing" a single unit of work *and* the maximum amount of work that each agent can actually execute (i.e., the capacity of the agent). We show that truthfulness in this setting is equivalent to a simple condition on the underlying algorithm. By applying this result to various problems considered in the literature (e.g., makespan minimization on related machines) we show that only some of the existing approaches to the case "without capacities" can be adapted to the case with private capacities. This poses new interesting algorithmic challenges.

1 Introduction

Algorithmic mechanism design considers distributed settings where the participants, termed agents, cannot be assumed to follow the protocol but rather their own interests. The designer must ensure in advance that it is in the agents' interest to behave correctly. The protocol can be regarded as an algorithm augmented with a suitable payment rule and the desired condition is termed *truthfulness*, meaning that it is never convenient for an agent to report false information. We begin with an illustrative example:

Example 1 (scheduling related machines [2]). We have two jobs of size, say, 1 *and* 2 *to be scheduled on two machines. Each allocation specifies the amount of work that is allocated to each machine (the sum of the jobs sizes). Each machine has a* type t_i *which is the time (cost) for processing one unit of work; that is, the type is the inverse of the machine's speed. An allocation, x, assigns an amount of work $w_i(x)$ to machine i and thus its completion time (cost) is equal to*

$$w_i(x) \cdot t_i. \tag{1}$$

* Research funded by the European Union through IST FET Integrated Project AEOLUS (IST-015964).

R. Královič and D. Niwiński (Eds.): MFCS 2009, LNCS 5734, pp. 112–123, 2009.

The goal is to compute an allocation that minimizes the "overall" cost

$$\max\{w_1(x) \cdot t_1, w_2(x) \cdot t_2\} \qquad (2)$$

that is the so called makespan. The type of each machine is only known to its owner (agent) who incurs a cost equal to the completion time of her machine (the quantity in Equation 1). Each agent may find it convenient to misreport her type so to induce the underlying algorithm to assign less work to her machine.

This is a typical *one-parameter* mechanism design problem, meaning that each agent has a *private type* representing the cost for executing one unit of work. The goal is to compute a solution minimizing some "global" cost function which depends on the types of all agents (in the above example, the quantity in Equation 2). The underlying algorithm is then augmented with a suitable payment function so that each agent finds it convenient to report her type truthfully. This important requirement is commonly termed *truthfulness* of the resulting mechanism ("algorithm + payments"). Truthful mechanisms guarantee that the underlying algorithm receives the "correct" input because no agent has a reason to misreport her type.

In this work we introduce and study the natural extension of the one-parameter setting in which agents have *private capacities* and thus can "refuse" allocations that assign them amounts of work above a certain value (see Section 1.1 for the formal model). This setting is naturally motivated by various applications in which agents are not capable or willing to execute arbitrary amounts of work:

- A router can forward packets at a certain rate (per-packet cost), but an amount of traffic exceeding the capacity of the router will "block" the router itself.
- In wireless networks, each node acts as a router and the energy consumption determines the per-packet-cost of the node. The battery capacity of each node determines the maximum amount of packets (work) that can be forwarded.
- Agents can produce identical goods at some cost (per unit) and their capacities represent the maximum amount of goods that each of them can produce.

Truthful mechanisms for the case without capacities need not remain truthful in this new setting. We show that a simple "monotonicity" condition characterizes truthfulness for the case of one-parameter agents with private capacities (Theorem 3). This translates into an algorithmic condition on the underlying algorithm (Corollary 1). We apply this result to the various problems previously considered in the literature to see which of the existing techniques/results extend to our setting. Roughly, all mechanisms for one-parameter settings (without capacities) that are based on "lexicographically optimal" algorithms remain truthful (for the case with capacities) as they satisfy the above monotonicity condition (Section 3.1). This is not true for other mechanisms because the underlying algorithm is no longer monotone when capacities are introduced (Section 3.3).

We then move to *multidimensional* domains which provide a more powerful and general framework. By considering differend "kind" of work and a capacity for each of them one can easily model rather complex problems like, for instance, scheduling with restricted assignment on unrelated machines (i.e., each machine can execute only certain jobs and the execution times change arbitrarily from machine to machine). Here we observe that there is no "simple" monotonicity condition that characterizes truthfulness, even when the problem without capacities has a domain which does have such simple characterization (Section 4).

Connections with existing work. Algorithmic mechanism design questions have been raised in the seminal work by Nisan and Ronen [13]. Mechanism design is a central topic in game theory, with the celebrated Vickrey-Clarke-Groves [17,5,7] mechanisms been probably the most general positive result. These mechanisms work for arbitrary domains, but require the problems' objective to be the so called (weighted) social welfare: essentially, to minimize the (weighted) sum of all agents' costs. Roberts' theorem [14] says that these are the only possible truthful mechanisms for domains that allow for arbitrary valuations. Therefore, most of the research has been focused on specific (restricted) domains and to other global cost functions like, for instance, the makespan in scheduling problems [13,2,1,11,10] or other min-max functions.

Rochet's [15] is able to characterize truthfulness in terms of the so-called "cycle monotonicity" property. In this paper, we refer to the interpretation of cycle monotonicity given by Gui *et al* [8] in terms of graph cycles which gives us a simple way for computing the payments. However, cycle monotonicity is difficult to interpret and to use. To our knowledge, the work by Lavi and Swamy [10] is the first (and only) one to obtain truthful mechanisms for certain two-values scheduling domains directly from Rochet's cycle monotonicity [15].

Bikhchandani *et al* [4] propose the simpler two-cycle monotonicity property (also known as weak-monotonicity) and showed that it characterizes truthfulness for rather general domains. We refer to domains for which two-cycle monotonicity characterizes truthfulness as monotonicity domains. Monotonicity domains turn out to be extremely important because there the construction of the mechanism (essentially) reduces to ensuring that the algorithm obeys relatively simple (two-cycle monotonicity) conditions. Saks and Yu [16] showed that every convex domain is a monotonicity domain. Our main result is that also one-parameter domains with private capacities are monotonicity domains. The resulting characterization generalizes prior results by Myerson [12] and by Archer and Tardos [2] when the set of possible solutions is finite. We remark that domains obtained by adding private capacities are not convex and our result for the two-parameter case implies that the characterization by Saks and Yu [16] cannot be used here. Dobzinski *et al* [6] studied auctions with budget-constrained bidders. Also these domains are different from ours because they put a bound on the payment capability of the agents, while in our problems bounds are put on the assignment of the algorithm. Our positive results on min-max objective functions use ideas by Archer and Tardos [2], Andelman *et al* [1], Mu'alem and Shapira [11], and they extend some results therein to the case with private capacities.

1.1 Agents with Private Capacities

We are given a finite set of feasible solutions. In the *one-parameter* setting, every solution x assigns an amount of *work* $w_i(x)$ to agent i. Agent i has a monetary *cost* equal to $\text{cost}_i(t_i, x) = w_i(x) \cdot t_i$ where $t_i \in \Re^+$ is a private number called the *type* of the agent. This is the cost per unit of work and the value is known only to agent i. We extend the one-parameter setting by introducing *capacities* for the agents. An agent will incur an *infinite* cost whenever she gets an amount of work exceeding her capacity $c_i \in \Re^+$, that is

$$\text{cost}_i(t_i, c_i, x) = \begin{cases} w_i(x) \cdot t_i & \text{if } w_i(x) \leq c_i \\ \infty & \text{otherwise} \end{cases}$$

Each agent makes a *bid* consisting of a type-capacity pair $b_i = (t'_i, c'_i)$, possibly different from the true ones. An *algorithm* A must pick a solution on input the bids $b = (b_1, \ldots, b_n)$ of all agents, and a suitable payment function P assigns a payment $P_i(b)$ to every agent i. Thus, the *utility* of this agent is

$$\text{utility}_i(t_i, c_i, b) := P_i(b) - \text{cost}_i(t_i, c_i, A(b)).$$

Truthtelling is a *dominant strategy* with respect to both types and capacities if the utility of each agent is maximized when she reports truthfully her type and capacity, no matter how we fix the types and the capacities reported by the other agents. Formally, for every i, for every t_i and c_i, and for every b as above

$$\text{utility}_i(t_i, c_i, (t_i, c_i, b_{-i})) \geq \text{utility}_i(t_i, c_i, b)$$

where $(t_i, c_i, b_{-i}) := (b_1, \ldots, b_{i-1}, (t_i, c_i), b_{i+1}, \ldots, b_n)$ is the n-vector obtained by replacing $b_i = (t'_i, c'_i)$ with (t_i, c_i).

Definition 1. *An algorithm A is truthful for one-parameter agents with private capacities if there exists a payment function P such that truthtelling is a dominant strategy with respect to both types and capacities.*

We consider only algorithms that produce an allocation which respects the capacities (no agent gets more work than her reported capacity). A simple (standard) argument reduces truthfulness of A to the truthfulness of the work functions of the single agents (see Section 1.1).

Multidimensional settings. In the *multidimensional* or *k-parameter* setting each type, capacity, and work is a vector of length k. Agent i has a type $t_i = (t_i^1, \ldots, t_i^k)$, a capacity $c_i = (c_i^1, \ldots, c_i^k)$, and she is assigned some amount of work $w_i = (w_i^1, \ldots, w_i^k)$. The resulting cost is

$$w_i \cdot t_i = \sum_j w_i^j \cdot t_i^j$$

provided the j-th amount of work w_i^j does not exceed the corresponding capacity c_i^j for all j ($w_i \leq c_i$ component-wise). The cost is instead ∞ if some capacity is violated ($w_i \not\leq c_i$).

Example 2 (agents with several related machines). Each agent i owns *two* related machines whose processing times are t_i^1 and t_i^2. Every job allocation x assigns an amount of work $w_i^1(x)$ and $w_i^2(x)$ to these machines, respectively. The cost for the agent is the sum of the costs of her machines, that is, $w_i^1(x) \cdot t_i^1 + w_i^2(x) \cdot t_i^2$.

Example 3 (unrelated machines [13]). Each machine corresponds to an agent. The corresponding type is a vector $t_i = (t_i^1, \ldots, t_i^k)$, where t_i^j is the processing time of job j on machine i and k is the number of jobs that need to be scheduled. Each job allocation x specifies a binary vector $w_i(x)$ with $w_i^j(x) = 1$ iff job j is allocated to machine i. The cost for agent i is the completion time of her machine, that is $w_i(x) \cdot t_i = \sum_j w_i^j(x) \cdot t_i^j$. The variant in which machines can execute only certain jobs (restricted assignment) can be modelled with binary capacity vectors c_i: machine i can execute jobs j iff $c_i^j = 1$.

A simple reduction to the single-agent case. In this section we present a simple (standard) reduction that allows us to study truthfulness for the case of a single agent. The truthfulness of an algorithm A can be reduced to a condition on what we call below its single agent work functions: That is, the amount of work that A assigns to a fixed agent i, depending on her reported type and capacity, and having fixed the types and the capacities of all other agents. Each agent receives an amount of work and a payment according to some *work function* f and a suitable *payment function* p. Recall that the algorithm (and thus f) will always assign an amount of work that does not exceed the reported capacity. Hence, infinite costs occur only when the agent misreports her capacity. In the following definition, we consider t' and c' as the type and capacity reported by the agent, and t and c are the true ones.

Definition 2. *A work function f is truthful for a one-parameter agent with private capacity if there exist a payment function p such that, for every types t and t' and capacities c and c',*

$$p(t,c) - f(t,c) \cdot t \geq p(t',c') - \begin{cases} f(t',c') \cdot t & \text{if } f(t',c') \leq c \\ \infty & \text{otherwise} \end{cases}$$

For every i and for every fixed sub-vector $b_{-i} = (b_i, \ldots, b_{i-1}, b_{i+1}, \ldots, b_n)$, agent i receives an amount of work

$$w_i^A(t_i', c_i', b_{-i}) := w_i(A(b_1, \ldots, b_{i-1}, (t_i', c_i'), b_{i+1}, \ldots, b_n)$$

where t_i' and c_i' are the type and the capacity reported by i. Dominant strategies are equivalent to the fact that i's utility is maximized when she is truthtelling, no matter how we have fixed i and b_{-i}. Since the utility of i is given by the work function $f(\cdot) = w_i^A(\cdot, b_{-i})$ and by the payment function $p(\cdot) = P_i(\cdot, b_{-i})$, we have

Fact 1. *An algorithm A is truthful if and only if every single agent work function $f(\cdot) = w_i^A(\cdot, b_{-i})$ is truthful.*

Cycle monotonicity and truthfulness. The material in this section is based on the cycle monotonicity approach by Rochet [15] and its recent interpretation by Gui *et al* [8] in terms of graph cycles. Because of the reduction in the previous subsection, we consider the case of a single agent and its work function f. We shall see that f is truthful if and only if a suitable weighted graph associated to f contains no negative cycles [15,8]. The graph in question is defined as follows. Since the set of feasible solutions is finite, the amount of work that can be allocated to this agent must belong to some finite set $W = \{\dots, w, \dots, w', \dots\}$. We associate to f the following complete directed graph over $|W|$ nodes, one for each possible workload. The length of an edge $w \to w'$ is

$$\delta_{ww'} := \inf\{t \cdot (w' - w)| \quad t \in \Re^+ \text{ and there exists } c \geq w' \text{ such that } f(t, c) = w\}$$

where $\inf \emptyset = \infty$. The length of a cycle in this graph is the sum of the lengths of its edges.

Remark 1. Intuitively speaking, we think of w as the work when reporting the "true" type and capacity (t and c) and w' being the work when reporting some "false" type and capacity (t' and c'). Since every "lie" leading to a work exceeding the true capacity cannot be beneficial, we need to consider only the case $w' \leq c$. Then the condition for being truthful can be rewritten as $p_w + \delta_{ww'} \geq p_{w'}$, where p_w and $p_{w'}$ are the payments received in the two cases, respectively. When there is no "lie" that is potentially beneficial for the agent, we set $\delta_{ww'} = \infty$ which is mathematically equivalent to the fact that we do not add any constraint between these two payments.

The length of a cycle in this graph is the sum of the lengths of its edges.

Definition 3 (monotone [15,8]). *A function f is monotone if its associated graph contains no cycle of negative length.*

Rochet [15] showed that the above condition characterizes truthfulness. In particular, the weaker condition that every two-cycle has nonnegative length is always necessary. We restate the latter (necessary) condition for our setting and Rochet's theorem below.

Definition 4 (two-cycle monotone). *A function f is two-cycle monotone if for every (t, c) and (t', c') it holds that*

$$(t - t') \cdot (w' - w) \geq 0 \quad or \quad c \geq w' \quad or \quad c' \geq w$$

where $w = f(t, c)$ and $w' = f(t', c')$.

Theorem 2 ([15]). *Every truthful function must be two-cycle monotone. Every monotone function is truthful.*

While the above result has been originally stated for finite valuations/costs, it can be easily extended to our setting (where "unfeasible" solutions are modelled by means of infinite costs) using the arguments in [8] (the proof is given in the full version of this work [3]).

Remark 2. The above result applies also to the *multidimensional* case.

2 Characterizations for One-Parameter Agents

We show that two-cycle monotonicity *characterizes* truthfulness for one-parameter agents with private capacities. In particular, this necessary condition is also sufficient:

Theorem 3. *A function is truthful for one-parameter agents with private capacities if and only if it is two-cycle monotone.*

Proof. Since two-cycle monotonicity is a necessary condition (see Theorem 2), we only need to show that it is also sufficient. We prove that every two-cycle monotone function, for one-parameter agents with private capacities, is monotone (and thus truthful because of Theorem 2).

We consider only cycles whose edges have finite length (because otherwise the total length is obviously non-negative). We show that for any cycle with at least three edges, there exists another cycle with fewer edges and whose length is not larger. This fact, combined with the two-cycle monotonicity, implies that there is no cycle of negative length.

Given an arbitrary cycle of three or more edges, we consider the node with maximal work $\hat{w}$ in the cycle. We thus have three consecutive edges in the path, say $w \to \hat{w} \to w'$ with

$$\hat{w} > w \quad \text{and} \quad \hat{w} > w'.$$

If nodes w and w' coincide, then the path $w \to \hat{w} \to w'$ is actually a two-cycle. The two-cycle monotonicity says that $\delta_{w\hat{w}} + \delta_{\hat{w}w} \geq 0$. If we remove these two edges we obtain a cycle with fewer edges and whose length is not larger compared to the original cycle.

Otherwise, we show that a shorter cycle can be obtained by replacing the path $w \to \hat{w} \to w'$ with edge $w \to w'$. Towards this end, we show that

$$\delta_{ww'} \leq \delta_{w\hat{w}} + \delta_{\hat{w}w'}. \tag{3}$$

For every $\epsilon > 0$ and for every $w^{(1)}$ and $w^{(2)}$ such that $\delta_{w^{(1)}w^{(2)}} < \infty$, there exist $t^{(1)}$ and $c^{(1)} \geq w^{(2)}$ such that $w^{(1)} = f(t^{(1)}, c^{(1)})$ and

$$t^{(1)} \cdot (w^{(2)} - w^{(1)}) = \delta_{w^{(1)}w^{(2)}} + \epsilon^*$$

for some ϵ^* satisfying $0 \leq \epsilon^* \leq \epsilon$. In particular, since $\delta_{w\hat{w}}$ and $\delta_{\hat{w}w'}$ are both different from ∞, we can find t, $c \geq \hat{w}$ and $\hat{t}$, $\hat{c} \geq w'$ such that

$$t \cdot (\hat{w} - w) + \hat{t} \cdot (w' - \hat{w}) = \delta_{w\hat{w}} + \delta_{\hat{w}w'} + \epsilon^*$$

where ϵ^* satisfies $0 \leq \epsilon^* \leq \epsilon$. Observe that $\hat{c} \geq \hat{w} > w$ and thus the two-cycle monotonicity

$$(t - \hat{t})(\hat{w} - w) \geq 0$$

implies $\hat{t} \leq t$. This and $\hat{w} > w'$ imply that

$$t \cdot (\hat{w} - w) + \hat{t} \cdot (w' - \hat{w}) \geq t \cdot (w' - w).$$

Since $c \geq \hat{w} > w'$, we have

$$\delta_{ww'} \leq t \cdot (w' - w).$$

By putting things together we obtain

$$\delta_{ww'} \leq \delta_{w\hat{w}} + \delta_{\hat{w}w'} + \epsilon$$

for *every* $\epsilon > 0$. This implies Equation 3. Hence, by replacing the two edges $w \to \hat{w} \to w'$ with edge $w \to w'$ we obtain a cycle with fewer edges and whose length is not larger than the length of the original cycle.

The two-cycle monotonicity condition can be expressed in a more convenient form:

Fact 4. *A function f is two-cycle monotone if and only if for every (t, c) and (t', c') with $t' > t$ it holds that $w' \leq w$ or $w' > c$, where $w = f(t, c)$ and $w' = f(t', c')$.*

We thus obtain a simple algorithmic condition:

Corollary 1. *An algorithm A is truthful for one-parameter agents with private capacities if and only if every work function is two-cycle monotone. That is, for every i and for every b_{-i} the following holds. For any two capacities c_i, and c_i', and for any two types t_i and t_i' with $t_i' > t_i$, it holds that*

$$w_i^A((t_i', c_i'), b_{-i}) \leq w_i^A((t_i, c_i), b_{-i}) \quad or \quad w_i^A((t_i', c_i'), b_{-i}) > c_i.$$

For fixed capacities, this condition boils down to the usual monotonicity of one-parameter agents [2].

3 Applications to Min-max Problems

In this section we apply the characterization result on one-parameter agents with capacities to several optimization problems. We show that exact solutions are possible for min-max objectives (e.g., makespan) and that some (though not all) known techniques for obtaining approximation mechanisms for scheduling can be adapted to the case with private capacities.

3.1 Exact Mechanisms Are Possible

Theorem 5. *Every min-max problem for one-parameter agents (with private capacities) admits an exact truthful mechanism.*

Proof. We show that the optimal lexicographically minimal algorithm is monotone. We prove the theorem for the case of two agents since the proof can be generalized to any number of agents in a straightforward manner. Fix and agent i, and a type $\bar{t}$ and capacity $\bar{c}$ for the other agent. Also let w_{other} and w'_{other} denote the work assigned to the other agent when agent i gets assigned work w and w', respectively (these two values are defined below).

By contradiction, assume that the function associated to this agent is not monotone. By virtue of Theorem 2 and from Fact 4 this means that $t' > t$, $w' > w$, and $c \geq w'$. The latter inequality says that w' is feasible for capacity c and thus the optimality of the algorithm implies

$$\max\{w \cdot t, w_{other} \cdot \bar{t}\} \leq \max\{w' \cdot t, w'_{other} \cdot \bar{t}\}. \quad (4)$$

Similarly, we have $c' \geq w'$ because w' must be feasible for c'. Thus $w' > w$ implies that w is feasible for c' and the optimality of the algorithm yields

$$\max\{w' \cdot t', w'_{other} \cdot \bar{t}\} \leq \max\{w \cdot t', w_{other} \cdot \bar{t}\}. \quad (5)$$

We consider two cases:

1. ($w \cdot t' > w_{other} \cdot \bar{t}$.) Since $w' > w$, we have $\max\{w' \cdot t', w'_{other} \cdot \bar{t}\} \geq w' \cdot t' > w \cdot t' = \max\{w \cdot t', w_{other} \cdot \bar{t}\}$, thus contradicting Inequality (5).
2. ($w \cdot t' \leq w_{other} \cdot \bar{t}$.) Since $t < t'$, we have $w' \cdot t \leq w' \cdot t'$ thus implying that we can chain the inequality in (4) with the one in (5). This and $w \cdot t \leq w \cdot t' \leq w_{other} \cdot \bar{t}$ imply that
$$\max\{w \cdot t, w_{other} \cdot \bar{t}\} = \max\{w \cdot t', w_{other} \cdot \bar{t}\}.$$
Hence, both the inequalities in (4) and (5) hold with '='. This will contradict the fact that the algorithm picks the lexicographically minimal solution. On input t and c assigning work w' is feasible and gives the same cost as assigning work w. Since the algorithm picks w, instead of w', we have that w precedes lexicographically w'. Similarly, on input t' and c', the work w is also feasible and has the same cost as w'. This implies that w' precedes lexicographically w, which is a contradiction.

We conclude that each function associated to some agent must be monotone.

3.2 Makespan on Related Machines in Polynomial Time

Andelman *et al* [1] have obtained a truthful polynomial-time approximation scheme for a constant number of machines. Their idea is that one precomputes, in polynomial-time, a set of allocations and then obtains $(1+\epsilon)$-approximation by picking the best solution out of a precomputed set. We can use the very same idea and pick the solution in a lexicographically minimal fashion as we did to prove Theorem 5 and obtain the following:

Corollary 2. *There exists a polynomial-time $(1+\epsilon)$-approximation truthful mechanism for scheduling selfish machines with private capacities, for any constant number of machines and any $\epsilon > 0$.*

Proof. All we need to show is that we can also compute the payments in polynomial time. Using the characterization by Gui *et al* [8] the payments can be computed as the shortest path in the graph defined in Section 1.1 (for each agent we fix the bids of the others and consider the resulting graph). Notice that the graph has size polynomial because we have precomputed a polynomial number of feasible solutions [1]. The length of each edge corresponds to some breakpoint in which the work assigned to the machine (agent) under consideration reduces from w to some $w' < w$. The breakpoint is the value α for which

$$\max(\alpha \cdot w, M(w)) = \max(\alpha \cdot w', M(w'))$$

where $M(z)$ is the minimum makespan, over all solutions assigning work z to the machine under consideration and ignoring the completion time of this machine (i.e., the makespan with respect to the other machines).

3.3 Limitations of the Greedy Algorithm

We show that the monotone 3-approximation algorithm by Kovacs [9] *cannot* be extended in the "natural" way to the case with private capacities. This algorithm is the greedy LPT algorithm which processes jobs in decreasing order of their sizes; the current job is assigned to the machine resulting in the smallest completion time (ties are broken in a fixed order).

The modified version of the greedy algorithm simply assigns a job under consideration to the "best" machine among those for which adding this job does not exceed the corresponding capacity. It turns out that this modified greedy algorithm is not monotone, for the case with private capacities, even if we restrict to speeds (processing times) that are power of *any* constant $\gamma > 1$. (Kovacs [9] proved the monotonicity for $\gamma = 2$ and obtained a 3-approximation by simply rounding the speeds.)

Theorem 6. *The modified greedy algorithm is not truthful, even for fixed capacities and when restricting to speeds that are power of any $\gamma > 1$.*

Proof. There are three jobs of size 10, 6, and 5, and two machines both having capacity 11. The processing time of the second machine is $\gamma > 1$. We show that the work function corresponding to the first machine is not two-cycle monotone (the theorem then follows from Corollary 1).

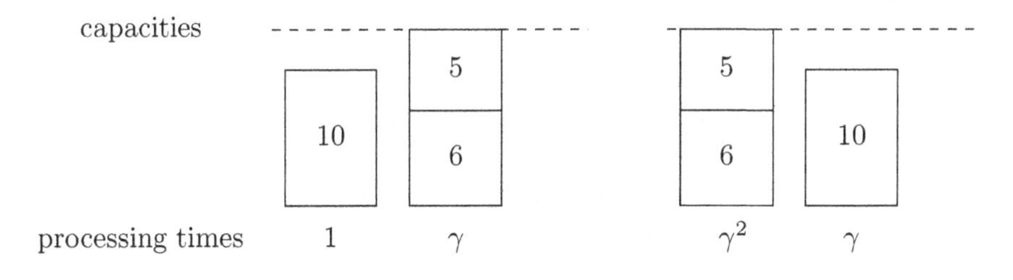

Fig. 1. The proof of Theorem 6

When the first machine has processing time $t_1 = 1$, the algorithm produces the allocation in Figure 1(left) because after the first job is allocated to the fastest machine, the other two jobs must go to the other machine because of the capacity. Now observe that when the first machine has processing time $t'_1 = \gamma^2 > t_1$, the algorithm simply "swaps" the previous allocation and assigns jobs as shown in Figure 1(right). It is easy to see that this violates the (two-cycle monotonicity) condition of Corollary 1 because

$$10 = w_1^A((1, 11), (\gamma, 11)) < w_1^A((\gamma^2, 11), (\gamma, 11)) = 11 = c_1 = c'_1.$$

This concludes the proof.

4 Multidimensional Domains

In this section we show that two-cycle monotonicity does not characterize truthful mechanisms for the *multidimensional* case. We prove the result even for the case of *two-parameter* domains where each agent gets two amounts of different kind of work.

Theorem 7. *Two-cycle monotonicity does not characterize truthfulness for two-parameter agents with private capacities.*

Proof. We show that there exists a function over a domain with three elements such that the associated graph is like in Figure 2 (for the moment ignore the numbers associated to the nodes). Thus the function is two-cycle monotone but not monotone (there exists a cycle with three edges and negative length).

Each node corresponds to some work which is given in output for the type and the capacity shown above this node: for example, $w = (0, 2) = f(t, c)$ where $t = (1, 1)$ and $c = (\infty, \infty)$.

Observe that edge $w' \leftarrow w''$ has length ∞ because work $w'_1 = 1$ exceeds the capacity $c''_1 = 0$. The length of every other edge $w^a \rightarrow w^b$ is given by the formula

$$\delta_{w^a w^b} = t^a \cdot (w^b - w^a) = t_1^a \cdot (w_1^b - w_1^a) + t_2^a \cdot (w_2^b - w_2^a).$$

It is easy to check that the length of each edge is the one shown in Figure 2. This example can be easily extended to a convex domain (details in [3]).

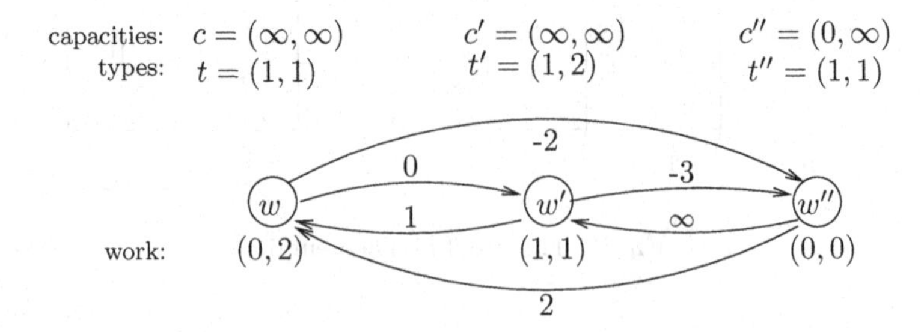

Fig. 2. Proof of Theorem 7

Acknowledgements. We wish to thank Riccardo Silvestri and Carmine Ventre for useful comments on an earlier version of this work. We are also grateful to an anonymous referee for a careful reading of the paper and for several suggestions.

References

1. Andelman, N., Azar, Y., Sorani, M.: Truthful Approximation Mechanisms for Scheduling Selfish Related Machines. Theory of Computing Systems 40(4), 423–436 (2007)
2. Archer, A., Tardos, É.: Truthful mechanisms for one-parameter agents. In: Proc. of the 42nd IEEE Symposium on Foundations of Computer Science (FOCS), pp. 482–491 (2001)
3. Auletta, V., Penna, P., Persiano, G.: Private capacities in mechanism design. Technical report, AEOLUS (2009), `http://www.dia.unisa.it/~penna/papers/capacitated-md-full.pdf`
4. Bikhchandani, S., Chatterji, S., Lavi, R., Mu'alem, A., Nisan, N., Sen, A.: Weak monotonicity characterizes deterministic dominant strategy implementation. Econometrica 74(4), 1109–1132 (2006)
5. Clarke, E.H.: Multipart Pricing of Public Goods. Public Choice, 17–33 (1971)
6. Dobzinski, S., Lavi, R., Nisan, N.: Multi-unit auctions with budget limits. In: FOCS, pp. 260–269. IEEE Computer Society, Los Alamitos (2008)
7. Groves, T.: Incentive in Teams. Econometrica 41, 617–631 (1973)
8. Gui, H., Muller, R., Vohra, R.V.: Dominant strategy mechanisms with multidimensional types. Discussion Papers 1392, Northwestern University, Center for Mathematical Studies in Economics and Management Science (July 2004)
9. Kovács, A.: Fast monotone 3-approximation algorithm for scheduling related machines. In: Brodal, G.S., Leonardi, S. (eds.) ESA 2005. LNCS, vol. 3669, pp. 616–627. Springer, Heidelberg (2005)
10. Lavi, R., Swamy, C.: Truthful mechanism design for multidimensional scheduling via cycle monotonicity. Games and Economic Behavior (2008) (in press), doi:10.1016/j.geb.2008.08.001
11. Mu'alem, A., Schapira, M.: Setting lower bounds on truthfulness. In: Proc. of annual ACM symposium on discrete algorithms (SODA), pp. 1143–1152 (2007)
12. Myerson, R.B.: Optimal auction design. Mathematics of Operations Research 6, 58–73 (1981)
13. Nisan, N., Ronen, A.: Algorithmic Mechanism Design. Games and Economic Behavior 35, 166–196 (2001); Extended abstract in the Proc. of the 31st Annual ACM Symposium on Theory of Computing (STOC), pp. 129–140 (1999)
14. Roberts, K.: The characterization of implementable choice rules. Aggregation and Revelation of Preferences, 321–348 (1979)
15. Rochet, J.-C.: A necessary and sufficient condition for rationalizability in a quasi-linear context. Journal of Mathematical Economics 16(2), 191–200 (1987)
16. Saks, M., Yu, L.: Weak monotonicity suffices for truthfulness on convex domains. In: Proceedings of the 6th ACM conference on Electronic commerce (EC), pp. 286–293. ACM, New York (2005)
17. Vickrey, W.: Counterspeculation, Auctions and Competitive Sealed Tenders. Journal of Finance, 8–37 (1961)

Towards a Dichotomy of Finding Possible Winners in Elections Based on Scoring Rules

Nadja Betzler[1,*] and Britta Dorn[2]

[1] Institut für Informatik, Friedrich-Schiller-Universität Jena
Ernst-Abbe-Platz 2, D-07743 Jena, Germany
nadja.betzler@uni-jena.de
[2] Wilhelm-Schickard-Institut für Informatik, Universität Tübingen
Sand 13, D-72076 Tübingen, Germany
bdorn@informatik.uni-tuebingen.de

Abstract. To make a joint decision, agents (or voters) are often required to provide their preferences as linear orders. To determine a winner, the given linear orders can be aggregated according to a voting protocol. However, in realistic settings, the voters may often only provide partial orders. This directly leads to the POSSIBLE WINNER problem that asks, given a set of partial votes, if a distinguished candidate can still become a winner. In this work, we consider the computational complexity of POSSIBLE WINNER for the broad class of voting protocols defined by scoring rules. A scoring rule provides a score value for every position which a candidate can have in a linear order. Prominent examples include plurality, k-approval, and Borda. Generalizing previous NP-hardness results for some special cases and providing new many-one reductions, we settle the computational complexity for all but one scoring rule. More precisely, for an unbounded number of candidates and unweighted voters, we show that POSSIBLE WINNER is NP-complete for all pure scoring rules except plurality, veto, and the scoring rule defined by the scoring vector $(2, 1, \ldots, 1, 0)$, while it is solvable in polynomial time for plurality and veto.

1 Introduction

Voting scenarios arise whenever the preferences of different parties (*voters*) have to be aggregated to form a joint decision. This is what happens in political elections, group decisions, web site rankings, or multiagent systems. Often, the voting process is executed in the following way: each voter provides his preference as a ranking (linear order) of all the possible alternatives (*candidates*). Given these rankings as an input, a *voting rule* produces a subset of the candidates (*winners*) as an output. However, in realistic settings, the voters may often only provide partial orders instead of linear ones: For example, it might be impossible for the voters to provide a complete preference list because the set of candidates is too large. In addition, not all voters might have given their preferences yet during the aggregation process, or new candidates might be introduced after some voters already have given their rankings. Moreover, one often has to

* Supported by the DFG, Emmy Noether research group PIAF, NI 369/4 and PAWS, NI 369/10.

R. Královič and D. Niwiński (Eds.): MFCS 2009, LNCS 5734, pp. 124–136, 2009.

deal with partial votes due to incomparabilities: for some voters it might not be possible to compare two candidates or certain groups of candidates, be it because of lack of information or due to personal reasons. Hence, the study of partial voting profiles is natural and essential. One question that immediately comes to mind is whether any information on a possible outcome of the voting process can be given in the case of incomplete votes. More specifically, in this paper, we study the POSSIBLE WINNER problem: Given a partial order for each of the voters, can a distinguished candidate c win for at least one extension of the partial orders into linear ones?

Of course, the answer to this question depends on the voting rule that is used. In this work, we will stick to the broad class of voting protocols defined by *scoring rules*. A scoring rule provides a score value for every position that a candidate can take within a linear order. The scores of the candidates are then added up and the candidates with the highest score win. Many well-known voting protocols, including plurality, veto, and Borda, are realized by scoring rules. Other examples are the Formula 1 scoring, which uses the scoring rule defined by the vector $(10, 8, 6, 5, 4, 3, 2, 1, 0, ...)$, or k-approval, which is used in many political elections whenever the voters can express their preference for k candidates within the set of all candidates.

The POSSIBLE WINNER problem was introduced by Konczak and Lang [6] and has been further investigated since then for many types of voting systems [1,7,8,9,10]. Note that the related NECESSARY WINNER problem can be solved in polynomial time for all scoring rules [10]. A prominent special case of POSSIBLE WINNER is MANIPULATION (see e.g. [2,5,11,12]). Here, the given set of partial orders consists of two subsets; one subset contains linearly ordered votes and the other one completely unordered votes. Clearly, all NP-hardness results would carry over from MANIPULATION to POSSIBLE WINNER. However, whereas the case of weighted voters is settled by a full dichotomy [5] for MANIPULATION for scoring rules, we are not aware of any NP-hardness results for scoring rules in the unweighted voter case.

Let us briefly summarize known results for POSSIBLE WINNER for scoring rules. Correcting Konczak and Lang [6] who claimed polynomial-time solvability for all scoring rules, Xia and Conitzer [10] provided NP-completeness results for a class of scoring rules, more specifically, for all scoring rules that have four "equally decreasing score values" followed by another "strictly decreasing score value"; we will provide a more detailed discussion later. Betzler et al. [1] studied the parameterized complexity of POSSIBLE WINNER and, among other results obtained NP-hardness for k-approval in case of two partial orders. However, this NP-hardness result holds only if k is part of the input, and it does not carry over for fixed values of k. Further, due to the restriction to two partial votes, the construction is completely different from the constructions used in this work.

Until now, the computational complexity of POSSIBLE WINNER was still open for a large number of naturally appearing scoring rules. We mention k-approval for small values of k as an example: Assume that one may vote for a board that consists of five members by awarding one point each to five of the candidates (5-approval). Surprisingly, POSSIBLE WINNER turns out to be NP-hard even for 2-approval. Another example is given by voting systems in which each voter is allowed to specify a (small) group of favorites and a (small) group of most disliked candidates.

In this work, we settle the computational complexity of POSSIBLE WINNER for all pure scoring rules except the scoring rule defined by $(2, 1, \ldots, 1, 0)$.[1] For plurality and veto, we provide polynomial-time algorithms. The basic idea to show the NP-completeness for all remaining pure scoring rules can be described as follows. Every scoring vector of unbounded length must either have an unbounded number of positions with different score values or must have an unbounded number of positions with equal score values (or both). Hence, we give many-one reductions covering these two types and then combine them to work for all considered scoring rules. Scoring rules having an unbounded number of positions with different score values are treated in Section 2, where we generalize results from [10]. Scoring rules having an unbounded number of positions with equal score values are investigated in Section 3. Here, we consider two subcases. In one subcase, we consider scoring rules of the form $(\alpha_1, \alpha_2, \ldots, \alpha_2, 0)$, which we deal with in Section 3.2. In the other subcase, we consider all remaining scoring rules with an unbounded number of candidates (Section 3.1). Finally, we combine the obtained results in the main theorem (Section 4).

Preliminaries. Let $C = \{c_1, \ldots, c_m\}$ be the set of *candidates*. A *vote* is a linear order (i.e., a transitive, antisymmetric, and total relation) on C. An n-voter profile P on C consists of n votes $(v_1, \ldots, v_n)$ on C. A *voting rule* r is a function from the set of all profiles on C to the power set of C. *(Positional) scoring rules* are defined by scoring vectors $\vec{\alpha} = (\alpha_1, \alpha_2, \ldots, \alpha_m)$ with integers $\alpha_1 \geq \alpha_2 \geq \cdots \geq \alpha_m$, the *score values*. More specifically, we define that a scoring rule r consists of a sequence of scoring vectors $s_1, s_2, \ldots$ such that for any $i \in \mathbb{N}$ there is a scoring vector for i candidates which can be computed in time polynomial in i.[2] Here, we restrict our results to *pure* scoring rules, that is for every i, the scoring vector for i candidates can be obtained from the scoring vector for $i-1$ candidates by inserting an additional score value at an arbitrary position (respecting the described monotonicity). This definition includes all of the common protocols like Borda or k-approval.[3] We further assume that $\alpha_m = 0$ and that there is no integer that divides all score values. This does not constitute a restriction since for every other voting system there must be an equivalent one that fulfills these constraints [5, Observation 2.2]. Moreover, we only consider *non-trivial* scoring rules, that is, scoring rules with $\alpha_1 \neq 0$ for a scoring vector of unbounded length.

For a vote $v \in P$ and a candidate $c \in C$, let the *score* $s(v, c) := \alpha_j$ where j is the position of c in v. For any profile $P = \{v_1, \ldots, v_n\}$, let $s(P, c) := \sum_{i=1}^{n} s(v_i, c)$. Whenever it is clear from the context which P we refer to, we will just write $s(c)$. The scoring rule will select all candidates c as winners for which $s(P, c)$ is maximized. Famous examples of scoring rules are Borda, that is, $(m-1, m-2, \ldots, 0)$, and k-approval, that is, $(1, \ldots, 1, 0, \ldots, 0)$ starting with k ones. Two relevant special cases of k-approval are plurality, that is $(1, 0, \ldots, 0)$, and veto, that is, $(1, \ldots, 1, 0)$.

[1] The class of pure voting rules still covers all of the common scoring rules. We only constitute some restrictions in the sense that for different numbers of candidates the corresponding scoring vectors can not be chosen completely independently (see Preliminaries).

[2] For scoring rules that are defined for any fixed number of candidates the considered problem can be decided in polynomial time, see [2,9].

[3] Our results can also be extended to broad classes of "non-pure" (hybrid) scoring rules. Due to the lack of space, we defer the related considerations to the full version of this paper.

A *partial order* on C is a reflexive, transitive, and antisymmetric relation on C. We use $>$ to denote the relation given between candidates in a linear order and $\succ$ to denote the relation given between candidates in a partial order. Sometimes, we specify a whole subset of candidates in a partial order, e.g., $e \succ D$. This notation means that $e \succ d$ for all $d \in D$ and there is no specified order among the candidates in D. In contrast, writing $e > D$ in a linear order means that the candidates of D have an arbitrary but fixed order. A linear order v^l *extends* a partial order v^p if $v^p \subseteq v^l$, that is, for any $i, j \leq m$, from $c_i \succ c_j$ in v^p it follows that $c_i > c_j$ in v^l. Given a profile of partial orders $P = (v_1^p, \dots, v_n^p)$ on C, a candidate $c \in C$ is a *possible winner* if there exists an extension $V = (v_1, \dots, v_n)$ such that each v_i extends v_i^p and $c \in r(V)$. The corresponding decision problem is defined as follows.

POSSIBLE WINNER
Given: A set of candidates C, a profile of partial orders $P = (v_1^p, \dots, v_n^p)$ on C, and a distinguished candidate $c \in C$.
Question: Is there an extension profile $V = (v_1, \dots, v_n)$ such that each v_i extends v_i^p and $c \in r(V)$?

This definition allows that multiple candidates obtain the maximal score and we end up with a whole set of winners. If the possible winner c has to be unique, one speaks of a possible *unique winner*, and the corresponding decision problem is defined analogously. As discussed in the following paragraph, all our results hold for both cases.

In all many-one reductions given in this work, one constructs a partial profile P consisting of a set of linear orders V^l and another set of partial votes V^p. Typically, the positions of the distinguished candidate c are already determined in all votes from V^p, that is, $s(P, c)$ is fixed. The "interesting" part of the reductions is formed by the partial orders of V^p in combination with upper bounds for the scores of the non-distinguished candidates. For every candidate $c' \in C \backslash \{c\}$, the *maximum partial score* $s_p^{\max}(c')$ is the maximum number of points c' can make in V^p without beating c in P. The maximum partial scores can be adapted for the unique and for the winner case since beating c in the winner case just means that a candidate makes strictly more points than c and beating c in the unique winner case means that a candidate makes as least as many points as c. Since all reductions only rely on the maximum partial scores, all results hold for both cases. For all reductions given in this work, one can generate an appropriate set of linear votes that implement the required maximum partial scores for each candidate. A general construction scheme of these votes can be found in a long version of this work. For this paper, we will refer to this construction scheme as Construction 1.

Several of our NP-hardness proofs rely on reductions from the NP-complete EXACT 3-COVER (X3C) problem. Given a set of elements $E = \{e_1, \dots, e_q\}$, a family of subsets $\mathcal{S} = \{S_1, \dots, S_t\}$ with $|S_i| = 3$ and $S_i \subseteq E$ for $1 \leq i \leq t$, it asks whether there is a subset $\mathcal{S}' \subseteq \mathcal{S}$ such that for every element $e_j \in E$ there is exactly one $S_i \in \mathcal{S}'$ with $e_j \in S_i$.

Due to the lack of space, several proofs and details had to be deferred to a full version of this paper.

2 An Unbounded Number of Positions with Different Score Values

Xia and Conitzer [10] showed that POSSIBLE WINNER is NP-complete for any scoring rule which contains four consecutive, equally decreasing score values, followed by another strictly decreasing score value. They gave reductions from X3C. Using some non-straightforward gadgetry, we extend their proof to work for scoring rules with an unbounded number of different, not necessarily equally decreasing score values.

We start by describing the basic idea given in [10] (using a slightly modified construction). Given an X3C-instance $(E, \mathcal{S})$, construct a partial profile $P := V^l \cup V^p$ on a set of candidates C where V^l denotes a set of linear orders and V^p a set of t partial orders. To describe the basic idea, we assume that there is an integer $b \geq 1$ such that $\alpha_b > \alpha_{b+1}$ and the difference between the score of the four following score values is equally decreasing, that is, $\alpha_b - \alpha_{b+1} = \alpha_{b+1} - \alpha_{b+2} = \cdots = \alpha_{b+3} - \alpha_{b+4}$, for a scoring vector of appropriate size. Then, $C := \{c, x, w\} \cup E' \cup B$ where $E' := \{e \mid e \in E\}$ and B contains $b-1$ dummy candidates. The distinguished candidate is c. The candidates whose element counterparts belong to the set S_i are denoted by e_{i1}, e_{i2}, e_{i3}. For every $i \in \{1, \dots, t\}$, the partial vote v_i^p is given by $B \succ x \succ e_{i1} \succ e_{i2} \succ e_{i3} \succ C'$, $B \succ w \succ C'$. Note that in v_i^p, the positions of all candidates except $w, x, e_{i1}, e_{i2}, e_{i3}$ are fixed. More precisely, w has to be inserted between positions b and $b+4$ maintaining the partial order $x \succ e_{i1} \succ e_{i2} \succ e_{i3}$. The maximum partial scores are set such that the following three conditions are fulfilled. First, regarding an *element candidate* $e \in E'$, inserting w behind e in two partial orders has the effect that e would beat c, whereas when w is inserted behind e in at most one partial order, c still beats e (Condition 1). Note that e may occur in several votes at different positions, e.g. e might be identical with e_{i1} and e_{j3} for $i \neq j$. However, due to the condition of "equally decreasing" scores, "shifting" e increases its score by the same value in all of the votes. Second, the partial score of x is set such that w can be inserted behind x at most $q/3$ times (Condition 2). Finally, we set $s_p^{\max}(w) = (t - q/3) \cdot \alpha_b + q/3 \cdot \alpha_{b+4}$. This implies that if w is inserted before x in $t - q/3$ votes, then it must be inserted at the last possible position, that is, position $b+4$, in all remaining votes (Condition 3).

Now, having an exact 3-cover for $(E, \mathcal{S})$, it is easy to verify that setting w to position $b+4$ in the partial votes that correspond to the exact 3-cover and to position b in all remaining votes leads to an extension in which c wins. In a yes-instance for (C, P, c), it follows directly from Condition 1 and 2 that w must have the last possible position $b+4$ in exactly $q/3$ votes and position b in all remaining partial votes. Since $|E| = q$ and there are $q/3$ partial votes such that three element candidates are shifted in each of them, due to Condition 1, every element candidate must appear in exactly one of these votes. Hence, c is a possible winner in P if and only if there exists an exact 3-cover of E.

In the remainder of this section, we show how to extend the reduction to scoring rules with strictly, but not equally decreasing scoring values. The problem we encounter is the following: By sending candidate w to the last possible position in the partial vote v_i^p, each of the candidates e_{i1}, e_{i2}, e_{i3} improves by one position and therefore improves its score by the difference given between the corresponding positions. In [10], these differences all had the same value, but now, we have to deal with varying differences. Since the same candidate $e \in E'$ may appear in several votes at different positions, e.g. e might be identical with e_{i1} and e_{j3} for $i \neq j$, it is not clear how to set the maximum partial score of e. In order to cope with this situation, we add two copies $v_i^{p'}$ and $v_i^{p''}$ of every partial vote v_i^p, and permute the positions of candidates e_{i1}, e_{i2}, e_{i3} in these two copies such that each of them takes a different position in $v_i^p, v_i^{p'}, v_i^{p''}$. In this way, if the candidate w is sent to the last possible position in a partial vote and its two copies, each of the candidates e_{i1}, e_{i2}, e_{i3} improves its score by the same value (which is "added" to the maximum partial score). We only have to guarantee that whenever w is sent back in the partial vote v_i^p, then it has to be sent back in the two copies $v_i^{p'}$ and $v_i^{p''}$ as well. We describe how this can be realized using a gadget construction. More precisely, we give a gadget for pairs of partial votes. The case of three votes just uses this scheme for two pairs within the three partial votes.

Given a copy $v_i^{p'}$ for each partial vote v_i^p, we want to force w to take the last possible position in v_i^p if and only if w takes the last possible position in $v_i^{p'}$. We extend the set of candidates C by $2t$ additional candidates $D := \{d_1, \dots, d_t, h_1, \dots, h_t\}$. The set V^p consists of $2t$ partial votes $v_1^p, v_1^{p'}, \dots, v_t^p, v_t^{p'}$ with

$$v_i^p : B \succ x \succ S_i \succ d_1 \succ \dots \succ d_i \succ h_{i+1} \succ \dots \succ h_t \succ C_i', \; B \succ w \succ C_i'$$
$$v_i^{p'} : B \succ x \succ S_i \succ h_1 \succ \dots \succ h_i \succ d_{i+1} \succ \dots \succ d_t \succ C_i'', B \succ w \succ C_i'',$$

for all $1 \leq i \leq t$, with C_i' and C_i'' containing the remaining candidates, respectively. Again, the maximum partial scores are defined such that w can be inserted behind x in at most $2q/3$ votes and must be inserted behind d_t or h_t in at least $2q/3$ votes. Further, for every candidate of D, the maximum partial score is set such that it can be shifted to a better position at most q times in a yes-instance. The candidate set E' is not relevant for the description of the gadget and thus we can assume that each candidate of E' can never beat c. We denote this construction as **Gadget 1**. Using some pigeonhole principle argument, one can show that Gadget 1 works correctly (proof omitted). Combining the construction of [10] with Gadget 1 and by using some further simple padding, one arrives at the following theorem.

Theorem 1. POSSIBLE WINNER *is NP-complete for a scoring rule if, for every positive integer x, there is a number m that is a polynomial function of x and, for the scoring vector of size m, it holds that* $|\{i \mid 1 \leq i \leq m-1 \text{ and } \alpha_i > \alpha_{i+1}\}| \geq x$.

3 An Unbounded Number of Positions with Equal Score Values

In the previous section, we showed NP-hardness for scoring rules with an unbounded number of different score values. In this section, we discuss scoring rules with an unbounded number of positions with equal score value. In the first subsection, we show NP-hardness for POSSIBLE WINNER for scoring rules with an unbounded number of "non-border" positions with the same score. That is, either before or after the group of equal positions, there must be at least two positions with a different score value. In the second subsection, we consider the special type that $\alpha_1 > \alpha_2 = \dots = \alpha_{m-1} > 0$.

3.1 An Unbounded Number of Non-border Positions with Equal Score Values

Here, we discuss scoring rules with non-border equal positions. For example, a scoring rules, such that, for every positive integer x, there is a scoring vector of size m such that there is an i, with $i < m - 2$, and $\alpha_{i-x} = \alpha_i$. This property can be used to construct a basic "logical" tool used in the many-one reductions of this subsection: For two candidates c, c', having $c \succ c'$ in a vote implies that setting c such that it makes less than α_i points implies that also c' makes less than α_i points whereas all candidates placed in the range between $i - x$ and i make exactly α_i points. This can be used to model some implication of the type "$c \Rightarrow c'$" in a vote. For example, for $(m-2)$-approval this condition means that c only has the possibility to make zero points in a vote if also c' makes zero points in this vote whereas other all candidates make one point. Most of the reductions of this subsection are from the NP-complete MULTICOLORED CLIQUE (MC) problem [4]:

Given: An undirected graph $G = (X_1 \cup X_2 \cup \cdots \cup X_k, E)$ with $X_i \cap X_j = \emptyset$ for $1 \leq i < j \leq k$ and the vertices of X_i induce an independent set for $1 \leq i \leq k$.
Question: Is there a clique of size k?

Here, $1, \ldots, k$ are considered as different colors. Then, the problem is equivalent to ask for a *multicolored clique*, that is, a clique that contains one vertex for every color. To ease the presentation, for any $1 \leq i \neq j \leq k$, we interpret the vertices of X_i as red vertices and write $r \in X_i$, and the vertices of X_j as green vertices and write $g \in X_j$.

Reductions from MC are often used to show parameterized hardness results [4]. The general idea is to construct different types of gadgets. Here, the partial votes realize four kinds of gadgets. First, gadgets that choose a vertex of every color (vertex selection). Second, gadgets that choose an edge of every ordered pair of colors, for example, one edge from green to red and one edge from red to green (edge selection). Third, gadgets that check the consistency of two selected ordered edges, e.g. does the chosen red-green candidate refer to the same edge as the choice of the green-red candidate (edge-edge match)? At last, gadgets that check if all edges starting from the same color start from the same vertex (vertex-edge match).

We start by giving a reduction from MC that settles the NP-hardness of POSSIBLE WINNER for $(m-2)$-approval.

Lemma 1. POSSIBLE WINNER *is NP-hard for* $(m-2)$*-approval.*

Proof. Given an MC-instance $G = (X, E)$ with $X = X_1 \cup X_2 \cup \cdots \cup X_k$. Let $E(i, j)$ denote all edges from E between X_i and X_j. W.l.o.g. we can assume that there are integers s and t such that $|X_i| = s$ for $1 \leq i \leq k$, $|E(i, j)| = t$ for all i, j, and that k is odd. We construct a partial profile P on a set C of candidates such that a distinguished candidate $c \in C$ is a possible winner if and only if there is a size-k clique in G. The set of candidates $C := \{c\} \uplus C_X \uplus C_E \uplus D$, where $\uplus$ denotes the disjunctive union, is specified as follows:

- For $i \in \{1, \ldots, k\}$, let $C_X^i := \{r_1, \ldots, r_{k-1} \mid r \in X_i\}$ and $C_X := \bigcup_i C_X^i$.
- For $i, j \in \{1, \ldots, k\}, i \neq j$, let $C_{i,j} := \{rg \mid \{r, g\} \in E(i, j)\}$ and $C'_{i,j} := \{rg' \mid \{r, g\} \in E(i, j)\}$. Then, $C_E := (\bigcup_{i \neq j} C_{i,j}) \uplus (\bigcup_{i \neq j} C'_{i,j})$, i.e., for every edge $\{r, g\} \in E(i, j)$, the set C_E contains the four candidates rg, rg', gr, gr'.

- The set $D := D_X \uplus D_1 \uplus D_2$ is defined as follows. For $i \in \{1, \ldots, k\}$, $D_X^i := \{c_1^r, \ldots, c_{k-2}^r \mid r \in X_i\}$ and $D_X := \bigcup_i D_X^i$. For $i \in \{1, \ldots, k\}$, one has $D_1^i := \{d_1^i, \ldots, d_{k-2}^i\}$ and $D_1 := \bigcup_i D_1^i$. The set D_2 is defined as $D_2 := \{d^i \mid i = 1, \ldots, k\}$.

We refer to the candidates of C_X as *vertex-candidates*, to the candidates of C_E as *edge-candidates*, and to the vertices of D as *dummy-candidates*.

The partial profile P consists of a set of linear votes V^l and a set of partial votes V^p. In each extension of P, the distinguished candidate c gets one point in every partial vote (see definition below). Thus, by using Construction 1, we can set the maximum partial scores as follows. For every candidate $d^i \in D_2$, $s_p^{\max}(d^i) = |V^p| - s + 1$, that is, d^i must get zero points (*take a zero position*) in at least $s - 1$ of the partial votes. For every remaining candidate $c' \in C \backslash (\{c\} \cup D_2)$, $s_p^{\max}(c') = |V^p| - 1$, that is, c' must get zero points in at least one of the partial votes.

In the following, we define $V^p := V_1 \cup V_2 \cup V_3 \cup V_4$. For all our gadgets only the last positions of the votes are relevant. Hence, in the partial votes it is sufficient to explicitly specify the "relevant candidates". More precisely, we define for all partial votes that each candidate that does not appear explicitly in the description of a partial vote is positioned before all candidates that appear in this vote.

The partial votes of V_1 realize the **edge selection gadgets**. Selecting an ordered edge (r, g) with $\{r, g\} \in E$ means to select the corresponding *pair of edge-candidates* rg and rg'. The candidate rg is used for the vertex-edge match check and rg' for the edge-edge match check. For every ordered color pair $(i, j), i \neq j$, V_1 has $t - 1$ copies of the partial vote $\{rg \succ rg' \mid \{r, g\} \in E(i, j)\}$, that is, one of the partial votes has the constraint $rg \succ rg'$ for each $\{r, g\} \in E(i, j)$. The idea of this gadget is as follows. For every ordered color pair we have t edges and $t - 1$ corresponding votes. Within one vote, one pair of edge-candidates can get the two available zero positions. Thus, it is possible to set all but one, namely the selected pair of edge-candidates to zero positions.

The partial votes of V_2 realize the **vertex selection gadgets**. Here, we need $k - 1$ candidates corresponding to a selected vertex to do the vertex-edge match for all edges that are incident in a multicolored clique. To realize this, $V_2 := V_2^a \cup V_2^b$. In V_2^a we select a vertex and in V_2^b, by a cascading effect, we achieve that all $k - 1$ candidates that correspond to this vertex are selected. In V_2^a, for every color i, we have $s - 1$ copies of the partial vote $\{r_1 \succ c_1^r \mid r \in X_i\}$. In V_2^b, for every color i and for every vertex $r \in X_i$, we have the following $k - 2$ votes.

For all *odd* $z \in \{1, \ldots, k-4\}$, $v_z^{r,i} : \{c_z^r \succ c_{z+1}^r, r_{z+1} \succ r_{z+2}\}$.
For all *even* $z \in \{2, \ldots, k-3\}$, $v_z^{r,i} : \{c_z^r \succ c_{z+1}^r, d_{z-1}^i \succ d_z^i\}$.
$v_{k-2}^{r,i} : \{c_{k-2}^r \succ d_{k-2}^i, r_{k-1}^i \succ d^i\}$.

The partial votes of V_3 realize the **vertex-edge match gadgets**. For $i, j \in \{1, \ldots, k\}$, for $j < i$, V_3 contains the vote $\{rg \succ r_j \mid \{r, g\} \in E, r \in X_i, \text{and } g \in X_j\}$ and, for $j > i$, V_3 contains the vote $\{rg \succ r_{j-1} \mid \{r, g\} \in E, r \in X_i, \text{and } g \in X_j\}$.

The partial votes of V_4 realize the **edge-edge match gadgets**. For every unordered color pair $\{i, j\}, i \neq j$ there is the partial vote $\{rg' \succ gr' \mid r \in X_i, g \in X_j\}$.

$$
\begin{array}{lll}
V_1 : & \cdots > rg > rg' & \text{for } i,j \in \{1,\ldots,k\}, i \neq j, r \in X_i \backslash Q, \text{ and } g \in X_j \backslash Q \\
V_2^a : & \cdots > r_1 > c_1^r & \text{for } 1 \leq i \leq k \text{ and } r \in X_i \backslash Q \\
V_2^b : v_z^{r,i} & \cdots > r_{z+1} > r_{z+2} & \text{for } 1 \leq i \leq k, r \in X_i \backslash Q \text{ for all } z \in \{1,3,5,\ldots,k-4\} \\
\quad v_z^{r,i} & \cdots > c_z > c_{z+1} & \text{for } 1 \leq i \leq k, r \in X_i \backslash Q \text{ for all } z \in \{2,4,6,\ldots,k-3\} \\
\quad v_{k-2}^{r,i} & \cdots > r_{k-1} > d^i & \text{for } 1 \leq i \leq k, r \in X_i \backslash Q \\
\quad v_z^{r,i} & \cdots > c_z^r > c_{z+1}^r & \text{for } 1 \leq i \leq k, r \in X_i \cap Q \text{ for all } z \in \{1,3,5,\ldots,k-4\} \\
\quad v_z^{r,i} & \cdots > d_{z-1}^i > d_z^i & \text{for } 1 \leq i \leq k, r \in X_i \cap Q \text{ for all } z \in \{2,4,6,\ldots,k-3\} \\
\quad v_{k-2}^{r,i} & \cdots > c_{k-2}^r > d_{k-2}^i & \text{for } 1 \leq i \leq k, r \in X_i \cap Q \\
V_3 : & \cdots > rg > r_j & \text{for } i,j \in \{1,\ldots,k\}, j < i, r \in X_i \cap Q, \text{ and } g \in X_j \cap Q \\
 & \cdots > rg > r_{j-1} & \text{for } i,j \in \{1,\ldots,k\}, j > i, r \in X_i \cap Q, \text{ and } g \in X_j \cap Q \\
V_4 : & \cdots > rg' > gr' & \text{for } i,j \in \{1,\ldots,k\}, i \neq j, r \in X_i \cap Q, g \in X_j \cap Q
\end{array}
$$

Fig. 1. Extension of the partial votes for the MC-instance. We highlight extensions in which candidates that do not correspond to the solution set Q take the zero positions.

This completes the description of the partial profile. By counting, one can verify a property of the construction that is crucial to see the correctness: In total, the number of zero positions available in the partial votes is exactly equal to the sum of the minimum number of zero position the candidates of $C \backslash \{c\}$ must take such that c is a winner. We denote this property of the construction as *tightness*. It directly follows that if there is a candidate that takes more zero positions than desired, then c cannot win in this extension since then at least one zero position must be "missing" for another candidate.

Claim: The graph G has a clique of size k if and only if c is a possible winner in P.

"$\Rightarrow$" Given a multicolored clique Q of G of size k. Then, extend the partial profile P as given in Figure 1. One can verify that in the given extension every candidate takes the required number of zero positions.

"$\Leftarrow$" Given an extension of P in which c is a winner, we show that the "selected" candidates must correspond to a size-k clique. Recall that the number of zero positions that each candidate must take is "tight" in the sense that if one candidate gets an unnecessary zero position, then for another candidate there are not enough zero positions left.

First (edge selection), for $i, j \in \{1, \ldots, k\}, i \neq j$, we consider the candidates of $C_{i,j}$. The candidates of $C_{i,j}$ can take zero positions in one vote of V_3 and in $t-1$ votes of V_1. Since $|C_{i,j}| = t$ and in the considered votes at most one candidate of $C_{i,j}$ can take a zero position, every candidate of $C_{i,j}$ must take one zero position in one of these votes. We refer to a candidate that takes the zero position in V_3 as solution candidate rg_{sol}. For every non-solution candidate $rg \in C_{i,j} \backslash \{rg_{\text{sol}}\}$, its placement in V_1 also implies that rg' gets a zero position, whereas rg'_{sol} still needs to take one zero position (what is only possible in V_4).

Second, we consider the vertex selection gadgets. Here, analogously to the edge selection, for every color i, we can argue that in V_2^a, out of the set $\{r_1 \mid r \in X_i\}$, we have to set all but one candidate to a zero position. The corresponding *solution vertex* is denoted as r_{sol}. For every vertex $r \in X_i \backslash \{r_{\text{sol}}\}$, this implies that the corresponding dummy-candidate c_1^r also takes a zero position in V_2^a. Now, we show that in V_2^b we have to set all candidates that correspond to non-solution vertices to a zero position whereas all candidates corresponding to r_{sol} must appear only at one-positions. Since for every vertex $r \in X_i \setminus \{r_{\text{sol}}\}$, the vertex c_1^r has already a zero position in V_2^a, it cannot take a zero position within V_2^b anymore without violating the tightness. In contrast, for the selected solution candidate r_{sol}, the corresponding candidates $c_1^{r_{\text{sol}}}$ and r_{sol_1} still need to take one zero position. The only possibility for $c_1^{r_{\text{sol}}}$ to take a zero position is within vote $v_1^{r_{\text{sol}},i}$ by setting $c_1^{r_{\text{sol}}}$ and $c_2^{r_{\text{sol}}}$ to the last two positions. Thus, one cannot set r_{sol_2} and r_{sol_3} to a zero position within V_2. Hence, the only remaining possibility for r_{sol_2} and r_{sol_3} to get zero points remains within the corresponding votes in V_3. This implies for every non-solution vertex r that r_2 and r_3 cannot get zero points in V_3 and, thus, we have to choose to put them on zero positions in the vote $v_1^{r,i}$ from V_2^b. The same principle leads to a cascading effect in the following votes of V_2^b: One cannot choose to set the candidates $c_p^{r_{\text{sol}}}$ for $p \in \{1, \ldots, k-2\}$ to zero positions in votes of V_2^b with even index z and, thus, has to improve upon them in the votes with odd index z. This implies that all vertex-candidates belonging to r_{sol} only appear in one-positions within V_2^b and that all dummy candidates d_p^i for $p \in \{1, \ldots, k-2\}$ are set to one zero position. In contrast, for every non-solution vertex r, one has to set the candidates c_p^r, $p \in \{2, \ldots, k-2\}$, to zero positions in the votes with even index z, and, thus, in the votes with odd index z, one has to set all vertex-candidates belonging to r to zero positions. This further implies that for every non-solution vertex in the last vote of V_2^b one has to set d^i to a zero position and, since there are exactly $s-1$ non-solution vertices, d^i takes the required number of zero positions. Altogether, all vertex-candidates belonging to a solution vertex still need to be placed at a zero position in the remaining votes $V_3 \cup V_4$, whereas all dummy candidates of D and the candidates corresponding to the other vertices have already taken enough zero positions.

Third, consider the vertex-edge match realized in V_3. For $i, j \in \{1, \ldots, k\}, i \neq j$, there is only one remaining vote in which rg_{sol} with $r \in X_i$ and $g \in X_j$ can take a zero position. Hence, rg_{sol} must take this zero-position. This does imply that the corresponding incident vertex x is also set to a zero-position in this vote. If $x \neq r_{sol_i}$, then x has already a zero-position in V_2. Hence, this would contradict the tightness and rg_{sol} and the corresponding vertex must "match". Further, the construction ensures that each of the $k-1$ candidates corresponding to one vertex appears exactly in one vote of V_3 (for each of the $k-1$ candidates, the vote corresponds to edges from different colors). Hence, c can only be possible winner if a selected vertex matches with all selected incident edges.

Finally, we discuss the edge-edge match gadgets. In V_4, for $i, j \in \{1, \ldots, k\}, i \neq j$, one still need to set the solution candidates from $C_{i,j}$ to zero positions. We show that this can only be done if the two "opposite" selected edge-candidates match each other. For two such edges rg_{sol} and gr_{sol}, $r \in X_i, g \in X_j$, there is only one vote in V_4 in which they can get a zero position. If rg_{sol} and gr_{sol} refer to different edges, then in this vote only one of them can get zero points, and, thus, the other one still beats c. Altogether, if c is a possible winner, then the selected vertices and edges correspond to a multicolored clique of size k. □

By generalizing the reduction used for Lemma 1, one can show the following.

Theorem 2. POSSIBLE WINNER *is NP-complete for a scoring rule r if, for every positive integer x, there is a number m that is a polynomial function of x and, for the scoring vector of size m, there is an $i \leq m-1$ such that $\alpha_{i-x} = \cdots = \alpha_{i-1} > \alpha_i$.*

The following theorem is based on further extensions of the MC-reduction used to prove Lemma 1 and some additional reductions from X3C.

Theorem 3. POSSIBLE WINNER *is NP-complete for a scoring rule r if, for every positive integer x, there is a number m that is a polynomial function of x and, for the scoring vector of size m, there is an $i \geq 2$ such that $\alpha_i > \alpha_{i+1} = \cdots = \alpha_{i+x}$.*

3.2 Scoring Rules of the Form $(\alpha_1, \alpha_2, \ldots, \alpha_2, 0)$

The following theorem can be shown by using another type of reductions from X3C.

Theorem 4. POSSIBLE WINNER *is NP-complete for all scoring rules such that there is a constant z and for every $m \geq z$, the scoring vector of size m satisfies the conditions $\alpha_1 > \alpha_2 = \alpha_{m-1} > \alpha_m = 0$ and $\alpha_1 \neq 2 \cdot \alpha_2$.*

4 Main Theorem

To state our main theorem, we still need the results for plurality and veto:

Proposition 1. POSSIBLE WINNER *can be solved in polynomial time for plurality and veto.*

The proof of Proposition 1 can be obtained by using some flow computations very similar to [1, Theorem 6].[4] Finally, we prove our main theorem.

Theorem 5. POSSIBLE WINNER *is NP-complete for all non-trivial pure scoring rules except plurality, veto, and scoring rules with size-m-scoring vector $(2, 1, \ldots, 1, 0)$ for every $m \geq z$ for a constant z. For plurality and veto it is solvable in polynomial time.*

Proof. (Sketch) Plurality and veto are polynomial-time solvable due to Proposition 1. Let r denote a scoring rule as specified in the first part of the theorem. Having any scoring vector different from $(1, 0, \ldots)$, $(1, \ldots, 1, 0)$, and $(2, 1, \ldots, 1, 0)$ for m candidates, it is not possible to obtain a scoring vector of one of these three types for $m' > m$ by inserting scoring values. Hence, since we only consider pure scoring rules, there must be a constant z such that r does not produce a scoring vector of type plurality, veto, or $(2, 1, \ldots, 1, 0)$ for all $m \geq z$. Now, we give a reduction from X3C (restricted to instances of size greater than z) to POSSIBLE WINNER for r that combines the reductions used to show Theorems 1 – 4. Let I with $|I| > z$ denote an X3C-instance. If there is a constant z' such that for all $m \geq z'$ the scoring vector corresponding to r is $(\alpha_1, \alpha_2, \ldots, \alpha_2, 0)$, then we can directly apply the reduction used to show Theorem 4. Otherwise, to make use of the MC-reductions, we apply the following strategy. Since EXACT 3-COVER and MULTICOLORED CLIQUE are NP-complete, there is a polynomial-time reduction from X3C to MC. Hence, let I' denote an MC-instance whose size is polynomial in $|I|$ and that is a yes-instance if and only if I is a yes-instance.

[4] We also refer to Faliszewski [3] for further examples showing the usefulness of employing network flows for voting problems.

Basically, the problem we encounter by showing the NP-hardness for POSSIBLE WINNER for r by using one specific many-one reduction from the previous sections is that such a reduction produces a POSSIBLE WINNER-instance with a certain number m of candidates. Thus, according to the properties of the reductions described in the previous sections, one would need to ensure that the corresponding scoring vector of size m provides a sufficient number of positions with equal/different scores. This seems not to be possible in general. However, for every specific instance I or I', it is not hard to compute a (polynomial) number of positions with equal/different scores that is sufficient for I or I'. For example, for the X3C-reduction used to show Theorem 1, for an instance (E, S), it would be clearly sufficient to have $(|E| + |S|)^2$ positions with equal score since this is a trivial upper bound for the number of candidates used to "encode" (E, S) into an POSSIBLE WINNER-instance. Having computed a sufficient number for all types of reductions either for I or I' (details omitted), we can set x to be the maximum of all these numbers.

Then, we can make use of the following observation (proof omitted). For r, there is a scoring vector whose size is polynomial in x such that either $|\{i \mid \alpha_i > \alpha_{i+1}\}| \geq x$ or such that $\alpha_i = \cdots = \alpha_{i+x}$ for some i. Now, we can distinguish two cases. If $\alpha_i = \cdots = \alpha_{i+x}$ for some i, applying one of the reductions to I' or I given in Theorem 2 or Theorem 3 results in a POSSIBLE WINNER-instance that is a yes-instance (if and only if I' is a yes-instance and, thus, also) if and only if I is a yes-instance. Otherwise, we can apply the reductions given in the proof of Theorem 1 to I resulting in a POSSIBLE WINNER instance that is a yes-instance if and only if I is a yes-instance. Since the NP-membership is obvious, the main theorem follows. □

Acknowledgments. We thank Johannes Uhlmann, Rolf Niedermeier, and the referees of MFCS for constructive feedback that helped to improve this work.

References

1. Betzler, N., Hemmann, S., Niedermeier, R.: A multivariate complexity analysis of determining possible winners given incomplete votes. In: Proc. of 21st IJCAI 2009 (2009)
2. Conitzer, V., Sandholm, T., Lang, J.: When are elections with few candidates hard to manipulate? Journal of the ACM 54(3), 1–33 (2007)
3. Faliszewski, P.: Nonuniform bribery (short paper). In: Proc. 7th AAMAS 2008, pp. 1569–1572 (2008)
4. Fellows, M.R., Hermelin, D., Rosamond, F.A., Vialette, S.: On the parameterized complexity of multiple-interval graph problems. TCS 410(1), 53–61 (2009)
5. Hemaspaandra, E., Hemaspaandra, L.A.: Dichotomy for voting systems. J. Comput. Syst. Sci. 73(1), 73–83 (2007)
6. Konczak, K., Lang, J.: Voting procedures with incomplete preferences. In: Proc. of IJCAI 2005 Multidisciplinary Workshop on Advances in Preference Handling (2005)
7. Lang, J., Pini, M.S., Rossi, F., Venable, K.B., Walsh, T.: Winner determination in sequential majority voting. In: Proc. of 20th IJCAI 2007, pp. 1372–1377 (2007)
8. Pini, M.S., Rossi, F., Venable, K.B., Walsh, T.: Incompleteness and incomparability in preference aggregation. In: Proc. of 20th IJCAI 2007, pp. 1464–1469 (2007)
9. Walsh, T.: Uncertainty in preference elicitation and aggregation. In: Proc. of 22nd AAAI 2007, pp. 3–8. AAAI Press, Menlo Park (2007)

10. Xia, L., Conitzer, V.: Determining possible and necessary winners under common voting rules given partial orders. In: Proc. of 23rd AAAI 2008, pp. 196–201. AAAI Press, Menlo Park (2008)
11. Xia, L., Zuckerman, M., Procaccia, A.D., Conitzer, V., Rosenschein, J.S.: Complexity of unweighted coalitional manipulation under some common voting rules. In: Proc. 21st IJCAI 2009 (2009)
12. Zuckerman, M., Procaccia, A.D., Rosenschein, J.S.: Algorithms for the coalitional manipulation problem. Artificial Intelligence 173(2), 392–412 (2009)

Sampling Edge Covers in 3-Regular Graphs

Ivona Bezáková and William A. Rummler

Rochester Institute of Technology, Rochester, NY, USA
{ib,war5549}@cs.rit.edu

Abstract. An edge cover C of an undirected graph is a set of edges such that every vertex has an adjacent edge in C. We show that a Glauber dynamics Markov chain for edge covers mixes rapidly for graphs with degrees at most three. Glauber dynamics have been studied extensively in the statistical physics community, with special emphasis on lattice graphs. Our results apply, for example, to the hexagonal lattice. Our proof of rapid mixing introduces a new cycle/path decomposition for the canonical flow argument.

1 Introduction

An edge cover of an undirected graph is a subset of its edges such that every vertex has an adjacent edge in the edge cover (i. e., each vertex is "covered"). We initiate the study of sampling (and the related question of counting) of all edge covers of a graph. Our main motivation for this work is to develop new insights that might help with understanding other related combinatorial problems such as matchings and contingency tables.

Glauber dynamics Markov chains on lattice graphs have been studied extensively in both the computer science and statistical physics communities. These single-site-update Markov chains tend to be natural to design but difficult to analyze, even though they are often believed to mix rapidly. Examples of successful theoretical analysis include the works of Randall and Tetali [1] and Vigoda [2]. Randall and Tetali, building on the result of Diaconis and Saloff-Coste [3], proved rapid mixing of Glauber dynamics Markov chains for domino tilings on the grid graph and lozenge tilings on the triangular lattice (these problems can be viewed as perfect matchings on the grid graph and the hexagonal lattice, respectively). Vigoda showed $O(n \log n)$ mixing time of a Glauber dynamics Markov chain for independent sets of graphs with maximum degree 4.

Edge covers are related to matchings in a similar way as vertex covers are related to independent sets. A vertex cover is a set of vertices such that every edge has an end-point in the set. The complement of a vertex cover is an independent set, i. e., a set of vertices that do not share any edges between them. The relationship between edge covers and matchings is not as straightforward but it is still easy to phrase: the size of the maximum matching plus the size of the minimum edge cover equals the number of vertices of the graph. While there is an obvious relationship between the counts of independent sets and vertex covers, no relationship

R. Královič and D. Niwiński (Eds.): MFCS 2009, LNCS 5734, pp. 137–148, 2009.

is known between the respective counts of edge covers and matchings. The problem of sampling and counting of matchings received a lot of attention [4,5,6,1,7], however, the problem remains unsolved for perfect matchings of arbitrary (i. e., non-bipartite) graphs. A perfect matching can be interpreted as an edge cover where every vertex is covered by exactly one edge. Hence, edge covers may be seen as a natural variation on perfect matchings. Jerrum and Sinclair [5] gave an fpras (fully polynomial randomized approximation scheme) for the problem of sampling all matchings (including matchings that are not perfect) of any graph (bipartite or not); however, the problem of sampling all edge covers appears more challenging due to its similarity to the problem of sampling subgraphs with a given degree sequence. The subgraph problem with bipartite input is also known as the binary, or 0/1, contingency tables.

Technical issues arise once vertices are allowed to have degrees larger than 1. In the matching problem, every vertex has degree one (or at most one) and this property helps with applying the canonical path argument for rapid mixing: namely, one needs to define a way of getting from any matching to any other matching using the transitions of the Markov chain (and this needs to be done so that no transition is "overloaded"). The symmetric difference of any two matchings forms a set of alternating cycles and paths (i. e., the edges strictly alternate between the two matchings). This property helps with transforming one matching into the other matching by handling the cycles and paths one by one. With subgraphs that satisfy a given degree sequence, this becomes more challenging but a fairly natural decomposition into cycles and paths is still possible due to the fact that at every vertex, the two subgraphs have the same degree and thus, the edges from the first subgraph can be paired with the edges of the second subgraph. The pairing notion was introduced by Cooper, Dyer, and Greenhill [8] who were building on ideas by Jerrum and Goldberg, and Kannan, Tetali, and Vempala [9].

In contrast, two different edge covers can have different degrees at the same vertex. In this paper we overcome this problem for graphs with degrees up to 3. Our cycle/path decomposition has some interesting characteristics the previous works did not have. We overcome the disproportionateness of the degrees by allowing pairing of edges from the same edge cover. Thus, the components of our decomposition might be non-alternating – they could consist of arbitrary sequences of edges from the two edge covers, including the possibility of all edges coming from the same edge cover. The alternating property is used to keep the intermediate configurations on a canonical path within the state space. Thus, our non-alternating decomposition needs to be done carefully so that the configurations that arise during the process of changing one edge cover into the other edge cover are edge covers as well.

While our restriction on the degrees is rather weak, we hope that the ideas in this work will bring new perspective to other related problems. For example, the results extend to so-called 1-2-edge covers (where every vertex is covered by one or two edges) for certain classes of graphs with arbitrary degrees; these edge covers can be viewed as close relatives of perfect matchings. We defer this extension to the journal version of this paper.

Our main result is:

Theorem 1. *Let G be an undirected graph with n vertices and degrees upper bounded by 3 and let Ω be the set of all its edge covers. Then, there exists a Markov chain with uniform stationary distribution over Ω with polynomial mixing time. More precisely,*

$$\tau(\delta) = O(n^6(n + \log \delta^{-1})),$$

where $\tau(\delta)$ is the mixing time of the chain[1] *and $\delta > 0$ is the sampling error.*

2 Preliminaries

Problem definition. Given is an undirected graph $G = (V_G, E_G)$. An **edge cover** of G is a set of edges $C \subseteq E_G$ such that for every vertex $u \in V_G$, there exists $(u, v) \in C$ for some $v \in V_G$. The **sampling edge covers** problem asks for a randomized polynomial-time algorithm that outputs a uniformly random edge cover of G. The output of the **counting** version of the problem is the number of all edge covers of G.

Markov chain Monte Carlo. In this section we give a very brief overview of the basic Markov chain terminology.

Let (Ω, P) be a Markov chain with state space Ω and transition matrix $P = (p_{i,j})_{|\Omega|\times|\Omega|}$ denoting the probability of going from state $i \in \Omega$ to state $j \in \Omega$. A distribution σ on Ω is called **stationary** if $\omega P = \omega$. A Markov chain is **ergodic** if it is **irreducible** (it is possible to get from every state to every other state with nonzero probability) and **aperiodic** (for every state i, the numbers of steps one can use to get from i to i have greatest common divisor equal to 1). It is well known that for ergodic Markov chains the stationary distribution is unique.

A Markov chain is **reversible** if $\pi(i)p_{i,j} = \pi(j)p_{j,i}$ for every $i, j \in \Omega$ and stationary distribution π. It is not difficult to verify that for an ergodic Markov chain with a symmetric transition matrix P, the stationary distribution is uniform.

The **total variation distance** between two distributions μ, ν on Ω is given by

$$d_{tv}(\mu, \nu) = \frac{1}{2}\sum_{x\in\Omega} |\mu(x) - \nu(x)|.$$

The **mixing time** $\tau_x(\delta)$ of the chain starting at state $x \in \Omega$ is defined as

$$\tau_x(\delta) = \min\{t \geq 0 \mid d_{tv}(P^t(x, \cdot), \pi) \leq \delta\},$$

where $P^t(x, \cdot)$ denotes the distribution after t steps of the chain, starting at the state $x \in \Omega$. Moreover, $\tau(\delta) = \max_{x\in\Omega} \tau_x(\delta)$.

[1] We opted for clarity of presentation over the best possible running time estimate.

One of the techniques for bounding mixing time is called the **canonical path/flow technique.** The canonical path technique requires, for every pair of states $I, F \in \Omega$, to define a path between I and F through other states such that the probability of moving between adjacent states is nonzero, i. e., we define $I = \omega_1, \omega_2, \dots, \omega_k = F$ such that $p_{\omega_i,\omega_{i+1}} > 0$ for every $i \in \{1, \dots, k-1\}$. In the canonical flow technique instead of constructing a single path from I to F, we can define a flow from I to F, i. e., we have the option of defining a set of paths $\mathcal{P}_{I,F}$ from I to F, and for every path $p \in \mathcal{P}_{I,F}$ we set its weight $g(p)$ so that $\sum_{p\in\mathcal{P}_{I,F}} g(p) = 1$. Whenever the volume of paths going through any given transition is not too large, the chain mixes in polynomial time. This notion is formalized as follows, see [10,11] for details. Let us consider all possible paths between any pair of states and let g be the weights (flow values) of a concrete canonical flow instance (i. e., whenever a path is not used by the flow, its weight is set to 0). Moreover, let π be the stationary distribution of the chain. The **congestion** through a transition $T = (M, M')$, $p_{M,M'} > 0$ is defined as

$$\rho_g(T) = \frac{1}{\pi(M)p_{M,M'}} \sum_{p \ni T} \pi(I_p)\pi(F_p)g(p)\ell(g),$$

where the sum ranges through all paths that use T, the path p starts at I_p and ends at F_p, and $\ell(g)$ is the maximum length of any path of nonzero weight in g. Then, the overall congestion is

$$\rho_g = \max_{T=(M,M'):p_{M,M'}>0} \rho_g(T)$$

and the mixing time is bounded by

$$\tau_x(\delta) \le \rho_g(\log \pi(x)^{-1} + \log \delta^{-1}).$$

3 Results

In this section we prove the main theorem of the paper.

3.1 A Markov Chain on Edge Covers

Let $G = (V, E)$ be a graph and let Ω be the set of all of its edge covers. Let $X_i \in \Omega$ be the current state of the Markov chain. The next state is the result of a simple Glauber dynamics-type move:

1. With probability 1/2, let $X_{i+1} = X_i$.
2. Otherwise, choose an edge $e \in E_G$ uniformly at random.
3. If $e \notin X_i$, let $X_{i+1} = X_i \cup \{e\}$.
4. Else, if $X_i \setminus \{e\}$ is an edge cover of G, let $X_{i+1} = X_i \setminus \{e\}$.
5. Else, let $X_{i+1} = X_i$.

In words, the Markov chain chooses a random edge and if it is not in the edge cover, it adds it. If it is already in the edge cover, it removes it if it can (if the result is still a valid edge cover). This Markov chain is symmetric and thus its stationary distribution π is the uniform distribution on Ω. The lazy step (step 1. of the chain) ensures that the chain is aperiodic, and its irreducibility will follow from the definition of the canonical flow. Thus, the chain is ergodic and its stationary distribution is unique.

3.2 Canonical Flow for This Markov Chain

We will define canonical flow between every pair of states $I, F \in \Omega$. Let $I \oplus F$ be the symmetric difference of I and F. We will gradually change the I-edges in $I \oplus F$ into the F-edges (we will leave the edges shared by both I and F, i. e., the edges in $I \cap F$, unchanged).

Cycle/path decomposition. We will first decompose $I \oplus F$ into a set of cycles and paths. Proofs of rapid mixing for matchings and subgraphs with prescribed degree sequence (in the bipartite case known also as binary contingency tables) use decomposition into **alternating** cycles and paths. Our decomposition will **not** necessarily contain only alternating cycles and paths.

Let $v \in V$ be a vertex and let c_I and c_F be the numbers of I-edges and F-edges, respectively, incident to v in $I \oplus F$. For 3-regular graphs we have $c_I, c_F \leq 3$. We will construct a **pairing of the edges in** $I \oplus F$ **at** v as follows. Without loss of generality assume that $c_I \leq c_F$. We will pair all the I-edges with the same number of F-edges. Then, we will pair the remaining F edges with other F edges (hence, the pairing will not be alternating). More precisely, for every I-edge e_I incident to v we select an F-edge e_F incident to v and pair them together. We will be left with $c_F - c_I$ edges (all F-edges) and, if $c_F - c_I = 2$, we will take two of these edges and pair them with each other. If $c_F - c_I = 1$, we leave the remaining F-edge unpaired. (Notice that for 3-regular graphs, $|c_F - c_I| \leq 2$. This follows from the fact that I and F are valid edge covers: suppose that $|c_F - c_I| = 3$, then either F contains all edges incident to v and I contains none, or vice versa. However, either case contradicts the assumption that both I and F are edge covers.) We denote the set of all pairings of the edges in $I \oplus F$ at v by $\Psi_{I,F}(v)$.

A **pairing of** $I \oplus F$ is a set of valid pairings at every vertex $v \in V$. We will denote the set of all (possible) pairings of $I \oplus F$ by $\Psi_{I,F}$.

The **cycle/path decomposition of** $I \oplus F$ **with respect to a pairing from** $\Psi_{I,F}$ is defined as follows. We take an arbitrary edge in $I \oplus F$, let it be (u_0, u_1). We find the edge (u_0, u_1) is paired at u_1 with, let it be (u_1, u_2); similarly, we find the edge (u_{-1}, u_0) that is paired with (u_0, u_1) at u_0 (if these edges exist – it might also be that (u_0, u_1) does not have a paired edge at u_0, or u_1, or both). Next, we find the edge (u_2, u_3) paired with (u_1, u_2) at u_2, etc. Continuing this process, we either find a path or a cycle. More precisely, if we find a path, it will be $u_{-k_1}, u_{-k_1+1}, \ldots, u_{-1}, u_0, u_1, u_2, u_3 \ldots, u_{k_2}$ such that (u_{i-1}, u_i) is paired with (u_i, u_{i+1}) at u_i for every $i \in \{-k_1 + 1, \ldots, k_2 - 1\}$, and there is no edge paired with (u_{-k_1}, u_{-k_1+1}) at u_{-k_1}, and there is no edge paired with (u_{k_2-1}, u_{k_2})

at u_{k_2}. If we find a cycle, it will be $u_0, u_1, u_2, u_3 \ldots, u_k$ such that (u_{i-1}, u_i) is paired with (u_i, u_{i+1}) at u_i for every $i \in \{1, \ldots, k-1\}$, and (u_0, u_1) is paired with (u_{k-1}, u_k) at $u_0 = u_k$. Notice that this path or cycle might repeat vertices[2] We continue finding paths or cycles that use edges that have not yet been used in other paths or cycles.

Definition of the canonical flow. Let $I, F \in \Omega$ be two states and let $\psi \in \Psi_{I,F}$ be a pairing of $I \oplus F$. We first define a **canonical order of cycles/paths** in the cycle/path decomposition of $I \oplus F$ with respect to ψ. Suppose the edges of G are numbered $1, 2, \ldots, |E_G|$, and this numbering is independent of I and F and the pairing ψ. Then we assign a number to every cycle/path identical to its lowest numbered edge, and we order the cycles/paths according to these numbers.

Then, we process each cycle/path following the canonical order, i. e., all the cycles and paths ordered before the current cycle/path have been already processed, and all the cycles and paths ordered after the current cycle/path have not been processed yet. In this context the word "process" means to change the I-edges into the F-edges. Thus, we start at I and after processing every cycle and path, we end up with F.

Processing paths. Let $u_1, u_2, \ldots, u_k$ be a path from the decomposition, and, without loss of generality suppose that (u_1, u_2) is a lower-numbered edge than (u_{k-1}, u_k). We will gradually remove I-edges and add F-edges by dealing with edges (u_i, u_{i+1}) roughly in the order of increasing i. We occasionally need to skip over some i's and come back to them to guarantee that all vertices are always covered (this problem can also be overcome by designing an additional move of the Markov chain corresponding to swapping one edge for another; however, since our goal is to analyze the Glauber dynamics Markov chain, we opted for this small complication in the canonical flow definition). The exact algorithm is described below (and Lemma 1 will show that at any given time all vertices are covered).

1. Let $i = 1$.
2. While $i < k$ do
3. If (u_i, u_{i+1}) is in F, then
4. Use the "add" transition of the Markov chain to add the edge (u_i, u_{i+1}).
5. Else (if the edge is in I)
6. If $i + 1 = k$ or if (u_{i+1}, u_{i+2}) is in I, then
7. Use the "remove" transition of the chain to remove the edge (u_i, u_{i+1}).
8. Else
9. Use the "add" transition to add the edge (u_{i+1}, u_{i+2}).
10. Then use the "remove" transition to remove the edge (u_i, u_{i+1}).
11. Increment i by 1.
12. Increment i by 1.

[2] Technically speaking, it would be more appropriate to use terms "walk" and "tour." We use "path" and "cycle" to be consistent with earlier papers.

Processing cycles. Cycles are processed similarly. Let $u_1, u_2, \ldots, u_{k-1}, u_k = u_1$ be a cycle from the decomposition and let (u_1, u_2) be its lowest numbered F-edge, or let it be the lowest numbered edge if the cycle contains only I-edges. Let the number of (u_2, u_3) be smaller than the number of (u_{k-1}, u_k) (otherwise, we "flip" the cycle, i. e., follow the cycle in the opposite direction, to satisfy this property). We process the cycle as follows.

1. Let $i = 1$ and let $u_{k+1} = u_2$.
2. While $i < k$ do
3. If (u_i, u_{i+1}) is in F, then
4. Use the "add" transition of the Markov chain to add the edge (u_i, u_{i+1}).
5. Else (if the edge is in I)
6. If (u_{i+1}, u_{i+2}) is in I, then
7. Use the "remove" transition of the chain to remove the edge (u_i, u_{i+1}).
8. Else
9. If $i + 1 < k$, then
10. Use the "add" transition to add the edge (u_{i+1}, u_{i+2}).
11. Then use the "remove" transition to remove the edge (u_i, u_{i+1}).
12. Increment i by 1.
13. Increment i by 1.

The following lemma proves that this process always goes only through valid states of the Markov chain.

Lemma 1. *Let $I, F \in \Omega$ and let $\psi \in \Psi_{I,F}$. Moreover, let $I = S_1, S_2, \ldots, S_{\ell-1}, S_\ell = F$ be the sequence of subgraphs of G encountered during the above described process of changing I into F. Then, $S_j \in \Omega$ for every $j \in \{1, 2, \ldots, \ell\}$.*

Proof. The proof goes by induction on j. Clearly, the claim holds for S_1 since we assume that $I \in \Omega$. For the inductive case, let us assume that $S_j \in \Omega$ for some $j \geq 1$. We want to show that $S_{j+1} \in \Omega$, too. The described process obtains S_{j+1} either by adding or by removing an edge. If an edge is added to S_j and all vertices were covered in S_j, all vertices must be covered in S_{j+1} as well. Thus, $S_{j+1} \in \Omega$. It remains to deal with the case when an edge is removed from S_j.

Let (u_i, u_{i+1}) be the removed edge and let $C = u_1, u_2, \ldots, u_k$ be the cycle or path that contains it. Since (u_i, u_{i+1}) was removed, it belongs to I. To prove that $S_{i+1} \in \Omega$, we need to show that after removing the edge, both u_i and u_{i+1} are still covered by another edge in S_{j+1} (and S_j, since S_j and S_{j+1} differ only in the edge (u_i, u_{i+1})).

First we show that u_{i+1} is always covered in S_{j+1}. We have several cases to consider:

- Case 1: $i < k - 1$ and $(u_{i+1}, u_{i+2}) \in I$. Then, after removing (u_i, u_{i+1}), the vertex u_{i+1} is still covered by the edge (u_{i+1}, u_{i+2}) which must be in both S_j and S_{j+1} because the edges are processed in order (with the minor exception of steps 9-11, but this exception does not apply in this case).

- Case 2: $i < k-1$ and $(u_{i+1}, u_{i+2}) \in F$. Then, according to the step 9 in the path processing algorithm, or step 10 in the cycle processing algorithm, S_{j-1} and S_j differ in the addition of edge (u_{i+1}, u_{i+2}), and therefore, the vertex u_{i+1} is covered by that edge in both S_j and S_{j+1}.
- Case 3: $i = k-1$ and C is a path. Since C is a path, the edge $(u_i, u_{i+1}) = (u_{k-1}, u_k)$ is not paired with any other edge at u_k. However, vertex u_k must be covered by at least one edge in I and at least one edge in F. If there is an edge e in $I \cap F$ incident to u_k, then, after removing the edge (u_{k-1}, u_k), vertex u_k will still be covered by e. Otherwise, there is a pair of edges $e_1 \in I$ and $e_2 \in F$ that are paired together at u_k. If e_1, e_2 belong to a different cycle/path than C, then either that cycle/path was processed before C, in which case u_k is covered by e_2 in S_{j+1}, or it will be processed after C, in which case u_k is covered by e_1 in S_{j+1}.

  Finally, let us consider the case when e_1, e_2 belong to C. Since (u_{k-1}, u_k) is the last edge in C and $(u_{k-1}, u_k) \in F$, the edges e_1, e_2 must have been processed earlier. Therefore, $e_1 \notin S_j$ and $e_2 \in S_j$, and, therefore, u_k is covered by e_2 in S_j (and S_{j+1}).
- Case 4: $i = k-1$ and C is a cycle. If $(u_1, u_2) \in F$, then $(u_1, u_2) \in S_j$ since this edge was processed as the first edge of C. Therefore, $u_{i+1} = u_k = u_1$ is covered in S_j (and S_{j+1}). Let us suppose that $(u_1, u_2) \in I$, and, thus, $(u_1, u_2) \notin S_j$. We know that (u_{k-1}, u_k) is paired with (u_1, u_2) at $u_k = u_1$ (forming the cycle C) and both (u_{k-1}, u_k) and (u_1, u_2) are in I. However, the vertex u_k is covered by an edge e in F. Either $e \in I \cap F$, or e must be paired at u_k with an edge $e_2 \in I$ (and e_2 is different from (u_{k-1}, u_k) and (u_1, u_2)) – otherwise we would not have paired together two edges from I. If $e \in I \cap F$, then u_k is always covered by this edge. Otherwise, if e lies on a cycle/path preceding C or on C, then e is in S_j and therefore u_k is covered in S_{j+1}. Alternatively, if e lies on a cycle/path processed after C, then u_k is covered by e_2 in S_{j+1}.

Similar arguments show that u_i is covered in S_{j+1}, we discuss them only briefly:

- Case 1: $i > 1$ and (u_{i-1}, u_i) is in F. Then, u_i is covered by (u_{i-1}, u_i) in S_{j+1}.
- Case 2: $i > 1$ and (u_i, u_{i+1}) is in I, or $i = 1$ and C is a path. In either case, there exists an edge $e \in F$ incident to u_i and it will be either e or the edge e is paired with that will cover u_i in S_{j+1}.
- Case 3: $i = 1$ and C is a cycle. We claim that $(u_{k-1}, u_k) \in I$. Then, the vertex $u_1 = u_k$ is covered by (u_{k-1}, u_k) in S_{j+1}. It remains to show that $(u_{k-1}, u_k) \in I$. By contradiction, suppose this edge is in F. But the edge (u_1, u_2) was chosen as the smallest numbered edge in F, or if there is no edge in F, then the smallest numbered edge in I. Thus, if (u_1, u_2) is in I, then (u_{k-1}, u_k) must be in I as well. □

Splitting the flow. Finally, we are ready to define the canonical flow between I and F. We consider all pairings from $\Psi_{I,F}$. For a pairing $\psi \in \Psi_{I,F}$, we defined a cycle/path decomposition and we described an algorithm for changing I into F by following the canonical order of cycles/paths. We call the sequence of intermediate subgraphs of G during the modification phase the **canonical path from I to F with respect to ψ**. Then, the **canonical flow from I to F** splits equally between all canonical paths from I to F, i. e., for each path p that is a canonical path from I to F, the flow value $g(p) = 1/|\Psi_{I,F}|$.

3.3 Analyzing the Congestion

In this section we bound $\rho(T)$ for every transition T. In words, we are estimating the number of canonical paths (weighted by $1/|\Psi_{I,F}|$) using the transition T. We will use the standard **encoding** technique to bound this number.

Encoding. Let $T = (M, M')$ be a transition from a state M to a state M'. Let I and F be a pair of states and $\psi \in \Psi_{I,F}$ be a pairing of $I \oplus F$ such that the canonical path from I to F associated with ψ uses the transition T. We will encode this canonical path by an "almost state" $E = I \oplus F \oplus M$ and a pairing ψ' of edges in $M \oplus E$ that we will specify later. (Technically, we defined pairings only between a pair of states from Ω, but the same definition easily applies to any subgraphs of G.) The following lemmas show that E has all except possibly up to four vertices covered by edges.

Lemma 2. *Let $I, F \in \Omega$ and $\psi \in \Psi_{I,F}$, let $C_1, C_2, \ldots, C_\ell$ be the components of the cycle/path decomposition of $I \oplus F$ with respect to ψ, and suppose the components are processed in the given order. Let X_i be the state right after processing the i-th component, and let $Y_i = I \oplus F \oplus X_i$. Moreover, let $c_I(v)$, $c_F(v)$, $c_{X_i}(v)$, and $c_{Y_i}(v)$ be the degrees of a vertex v in $I \setminus F$, $F \setminus I$, $X_i \setminus Y_i$, and $Y_i \setminus X_i$, respectively. Then,*

- $\{c_I(v), c_F(v)\} = \{c_{X_i}(v), c_{Y_i}(v)\}$ *for all vertices $v \in V$, and*
- $Y_i \in \Omega$ *for every i.*

Proof. The proof proceeds by induction on i. The base case, $i = 0$, follows from the fact that $X_0 = I$ and $Y_0 = F$. For the inductive case, we assume the statement holds for $i \geq 0$ and we will prove it for $i + 1$.

It will be useful to notice that $0 \leq c_I(v), c_F(v) \leq 2$. This follows from the fact that the graph has maximum degree 3: then, if I and F share an edge adjacent to v, then the degree of v in $I \setminus F$ (and in $F \setminus I$) is at most 2. On the other hand, if there is no common edge adjacent to v, then the degree of v in I must be smaller than 3 because there must exist an edge covering v in F. Thus, $c_I(v) \leq 2$, and, similarly, $c_F \leq 2$.

By the definition of the canonical path with respect to ψ, the state X_i differs from the state X_{i+1} only in the edges that belong to the component C_i. Namely, X_i contains the I-edges of C_i but not the F-edges, and X_{i+1} contains only the F-edges but not the I-edges of C_i. We will show that for every vertex v, we have $\{c_{X_i}(v), c_{Y_i}(v)\} = \{c_{X_{i+1}}(v), c_{Y_{i+1}}(v)\}$, and that v is covered in both X_{i+1} and Y_{i+1}, thus proving the lemma.

Clearly, $c_{X_i}(v) = c_{X_{i+1}}(v)$ and $c_{Y_i}(v) = c_{Y_{i+1}}(v)$ for every vertex v that is not on C_i, thus the claim holds for such vertices. Suppose that v is on C_i. There are several cases:

- Case 1: v is adjacent to two edges on C_i that are paired at v, one edge is in I, and one is in F. Then, we know that $1 \le c_I, c_F \le 2$. Thus, if $c_I = c_F$, we have $c_{X_i} = c_{Y_i} = c_{X_{i+1}} = c_{Y_{i+1}}$ and the claim follows. Otherwise, without loss of generality assume that $c_{X_i} = 1$ and $c_{Y_i} = 2$. Then, $c_{X_{i+1}} = 2$ and $c_{Y_{i+1}} = 1$ and the claim holds.
- Case 2: v is adjacent to two edges on C_i that are paired at v, without loss of generality assume that both are from I. Then, we know that $c_{X_i} = 2$ and $c_{Y_i} = 0$ (otherwise the two I-edges would not have been paired). Thus, $c_{X_{i+1}} = 0$ and $c_{Y_{i+1}} = 2$, and $\{c_{X_i}(v), c_{Y_i}(v)\} = \{c_{X_{i+1}}(v), c_{Y_{i+1}}(v)\}$. It remains to show that v is covered in X_{i+1}. Since v was, by the inductive hypothesis, covered in both X_i and Y_i, the only possibility for this to happen is that X_i and Y_i share an edge adjacent to v. This edge is then in X_{i+1} as well and it covers v.
- Case 3: v is adjacent to a single edge on C_i, without loss of generality suppose it is an edge from I. Then, either $c_{X_i} = 1$ and $c_{Y_i} = 0$, or $c_{X_i} = 2$ and $c_{Y_i} = 1$. In the second case, $c_{X_{i+1}} = 1$ and $c_{Y_{i+1}} = 2$ and the claim holds. For the first case we have $c_{X_{i+1}} = 0$ and $c_{Y_{i+1}} = 1$ and we need to argue that v is covered in X_{i+1}. The argument is analogous to Case 2: there exists an edge shared by both X_i and Y_i (and therefore X_{i+1} and Y_{i+1}) that covers v. □

Lemma 3. *Let $I, F \in \Omega$ and $\psi \in \Psi_{I,F}$. Let $T = (M, M')$ be a transition used on the ψ-defined canonical path from I to F and let $C = u_1, u_2, \ldots, u_k$ be the component of the cycle/path decomposition of $I \oplus F$ that uses T. Let (u_t, u_{t+1}) be the edge in $M \oplus M'$. Then, for every vertex $v \in V \setminus \{u_1, u_{t-1}, u_t, u_{t+1}\}$, we have that $\{c_I(v), c_F(v)\} = \{c_M(v), c_E(v)\}$ and v is covered in $E = I \oplus F \oplus M$.*

The proof of Lemma 3 follows from Lemma 2. We defer the details to the journal version of this paper.

Let us denote the set of "almost states" as Ω', i. e., Ω' is the set of subgraphs of G where at most four vertices are not covered by an edge of the subgraph. The following lemma states a polynomial relationship between the sizes of Ω and Ω' – this relationship will be important for establishing a polynomial bound on the congestion.

Lemma 4. *$|\Omega'| \le (m+1)^4|\Omega|$, where m is the number of edges of G.*

Proof. Let Ω'_k be the subset of Ω' containing graphs with exactly k vertices uncovered. We will give an injection h from Ω'_k to $\Omega \times E_G^k$. Let $H \in \Omega'_k$ and let $v_1, v_2, \ldots, v_k$ be its uncovered vertices. For every $i \in \{1, 2, \ldots, k\}$, take the smallest numbered edge adjacent to v_i, let it be (v_i, v'_i). Add all these edges to H, creating a graph H'. Thus, $h(H) = (H', (v_1, v'_1), (v_2, v'_2), \ldots, (v_k, v'_k))$. Clearly, this map is injective since $H' \setminus \{(v_i, v'_i) \mid i \in \{1, 2, \ldots, k\}\}$ results in H. Therefore, $|\Omega'_k| \le m^k|\Omega|$. Summing over $k \in \{0, 1, 2, 3, 4\}$, we get that $|\Omega'| \le (m^4 + m^3 + m^2 + m + 1)|\Omega| \le (m+1)^4|\Omega|$. □

Specifying the pairing from $\Psi_{E,M}$. By the definition of E we have that $I \oplus F = M \oplus E$. Thus, we need to pair the same set of edges in $M \oplus E$ as in

$I \oplus F$. Recall that the pairing $\psi \in \Psi_{I,F}$ is given. We construct the corresponding pairing ψ' by pairing up the same edges as those paired up in ψ. However, now we need to argue that this pairing is "legal," i. e., that for every vertex, there is at most one unpaired edge (this property follows directly from the fact that it is satisfied for ψ), and if there is a pairing of two edges from E, then no two edges from M are paired together at the same vertex, and vice versa. We allow exceptions at at most four vertices. This statement is summarized in the following lemma and its proof follows from the arguments made in proofs of Lemmas 2 and 3.

Lemma 5. *Let ψ' be constructed as described above and let $C = u_1, u_2, \ldots, u_k$ be the component of the cycle/path decomposition of $I \oplus F$ that uses T. Let (u_s, u_{s+1}) be the edge in $M \oplus M'$ and let $v \in V \setminus \{u_1, u_{s-1}, u_s, u_{s+1}\}$. Finally, recall that $c_M(v)$ is the degree of v in $M \setminus E$ and $c_E(v)$ is its degree in $E \setminus M$. Then, the pairing ψ' pairs $\min\{c_M(v), c_E(v)\}$ edges from M with edges from E at v, plus it pairs the remaining edges so that at most one is left unpaired.*

Notice that if we know E, $T = (M, M')$, and the pairing ψ', we can reconstruct I, F, and ψ. Since ψ' defines a cycle/path decomposition on $E \oplus M = I \oplus F$ and $\psi' = \psi$, we know which components of the decomposition have been already processed, which have not been processed, and which one is partially processed. Therefore, we can uniquely determine which edges of $E \oplus M$ belong to I and which belong to F.

Bounding the congestion. Suppose that $T = (M, M')$ where M and M' are almost identical valid edge covers with the only difference that M does not contain (u, v) and M' does contain this edge, or vice versa.

Notice that $\pi(M) = \pi(I) = \pi(F) = 1/|\Omega|$, and $p_{M,M'} = 1/(2m)$. Also note that the length of any above described canonical path is upper-bounded by m since every edge gets changed (added or removed) at most once. Finally, as we argued above, every path through T can be encoded by a pair of $E \in \Omega'$ and $\psi' \in \Psi_{M,E}$ where the definition of the set of all pairings between two states has been naturally extended to include any two subgraphs of G. Therefore,

$$\rho_g(T) = \frac{1}{\pi(M)p_{M,M'}} \sum_{p \ni T} \pi(I_p)\pi(F_p)g(p)\ell(g) \leq \frac{2m^2}{|\Omega|} \sum_{p \ni T} \frac{1}{|\Psi_{I_p,F_p}|}$$

$$\leq \frac{2m^2}{|\Omega|} \sum_{E \in \Omega'} \sum_{\psi' \in \Psi_{M,E}} \frac{1}{|\Psi_{I_p,F_p}|} \leq \frac{2m^2}{|\Omega|} \sum_{E \in \Omega'} 3^4 \leq 162m^2(m+1)^4 = O(m^6),$$

where the second to last inequality follows from Lemma 5. Namely, we know that the number of pairings of $M \oplus E$ at any given vertex (with a possible exception of up to four vertices) is the same as the number of pairings of $I \oplus F$ at the same vertex. Thus, for every such vertex v, we have $|\Psi_{I_p,F_p}(v)| = |\Psi_{M,E}(v)|$. At the four special vertices, the ratio $|\Psi_{M,E}(v)|/|\Psi_{I_p,F_p}(v)|$ could be arbitrary; however, the largest number of pairings at a vertex v of degree 3 is 3 (if all edges adjacent to v are in M and no edge is in E), the smallest number of pairings is

1. Thus, $|\Psi_{M,E}(v)|/|\Psi_{I_p,F_p}(v)| \leq 3$ and therefore $|\Psi_{M,E}| = \prod_{v \in V_G} |\Psi_{M,E}(v)| \leq 3^4 \prod_{v \in V_G} |\Psi_{I_p,F_p}(v)| = 3^4 |\Psi_{I_p,F_p}|$.

Then, the mixing time is bounded by

$$\tau_x(\delta) \leq \rho_g(\log \pi(x)^{-1} + \log \delta^{-1}) = O(m^6(m + \log \delta^{-1})),$$

where the last equality follows from the fact that the total number of edge covers is upper-bounded by the total number of subgraphs of G. Thus, $\pi(x) = 1/|\Omega|$ and $|\Omega| \leq 2^m$. Also notice that for constant-degree graphs $m = O(n)$. This finishes the proof of Theorem 1.

References

1. Cooper, C., Dyer, M., Greenhill, C.: Sampling regular graphs and a peer- to-peer network. Combinatorics, Probability and Computing 16, 557–593 (2007)
2. Diaconis, P., Saloff-Coste, L.: Comparison Theorems for Reversible Markov Chains. The Annals of Applied Probability 3(3), 696–730 (1993)
3. Diaconis, P., Stroock, D.: Geometric Bounds for Eigenvalues of Markov Chains. The Annals of Applied Probability 1(1), 36–61 (1991)
4. Jerrum, M., Sinclair, A.: Approximating the permanent. SIAM Journal on Computing 18, 1149–1178 (1989)
5. Jerrum, M., Sinclair, A.: Approximate counting, uniform generation and rapidly mixing Markov chains. Information and Computation 82, 93–133 (1989)
6. Jerrum, M., Sinclair, A., Vigoda, E.: A polynomial-time approximation algorithm for the permanent of a matrix with non-negative entries. J. ACM 51(4), 671–697 (2004)
7. Kannan, R., Tetali, P., Vempala, S.: Simple Markov-chain algorithms for generating bipartite graphs and tournaments. Random Structures and Algorithms (RSA) 14(4), 293–308 (1999)
8. Luby, M., Randall, D., Sinclair, A.: Markov chain algorithms for planar lattice structures. SIAM Journal on Computing 31, 167–192 (2001)
9. Randall, D., Tetali, P.: Analyzing Glauber dynamics by comparison of Markov chains. Journal of Mathematical Physics 41, 1598–1615 (2000)
10. Sinclair, A.: Improved Bounds for Mixing Rates of Marked Chains and Multicommodity Flow. Combinatorics, Probability and Computing 1, 351–370 (1992)
11. Valiant, L.G.: The complexity of computing the permanent. Theoretical Computer Science 8, 189–201 (1979)
12. Vigoda, E.: A note on the Glauber dynamics for sampling independent sets. Electronic Journal of Combinatorics 8(1) (2001); Research paper 8

Balanced Paths in Colored Graphs*

Alessandro Bianco, Marco Faella, Fabio Mogavero, and Aniello Murano

Università degli Studi di Napoli "Federico II", Italy
{alessandrobianco,mfaella,mogavero,murano}@na.infn.it

Abstract. We consider finite graphs whose edges are labeled with elements, called *colors*, taken from a fixed finite alphabet. We study the problem of determining whether there is an infinite path where either *(i)* all colors occur with the same asymptotic frequency, or *(ii)* there is a constant which bounds the difference between the occurrences of any two colors for all prefixes of the path. These two notions can be viewed as refinements of the classical notion of fair path, whose simplest form checks whether all colors occur infinitely often. Our notions provide stronger criteria, particularly suitable for scheduling applications based on a coarse-grained model of the jobs involved. We show that both problems are solvable in polynomial time, by reducing them to the feasibility of a linear program.

1 Introduction

In this paper, a colored graph is a finite directed graph whose edges are labeled with tags belonging to a fixed finite set of colors. For an infinite path in a colored graph, we say that the asymptotic frequency of a color is the long-run average number of occurrences of that color. Clearly, a color might have no asymptotic frequency on a certain path, because its long-run average oscillates. We introduce and study the problem of determining whether there is an infinite path in a colored graph where each color occurs with the same asymptotic frequency. We call such a path *balanced*. The existence of such a path in a given colored graph is called the *balance problem*.

Then, we consider the following stronger property: a path has the *bounded difference* property if there is a constant c such that, at all intermediate points, the number of occurrences of any two colors up to that point differ by at most c. The existence of such a path is called the *bounded difference problem* for a given graph. It is easy to prove that each bounded difference path is balanced. Moreover, each path that is both balanced and ultimately periodic (i.e., of the form $\sigma_1 \cdot \sigma_2^{\omega}$, for two finite paths σ_1 and σ_2) is also a bounded difference path. However, there are paths that are balanced but do not have the bounded difference property, as shown in Example 1.

We provide a loop-based characterization for each one of the mentioned decision problems. Both characterizations are based on the notion of *balanced set of loops*. A set of simple loops in the graph is balanced if, using those loops as building blocks, it is possible to build a finite path where all colors occur the same number of times.

We prove that a graph satisfies the balance problem if and only if it contains a balanced set of loops that are mutually reachable (Theorem 1). Similarly, a graph satisfies the bounded difference problem if and only if it contains a balanced set of loops that

* Work partially supported by MIUR PRIN Project 2007-9E5KM8.

R. Královič and D. Niwiński (Eds.): MFCS 2009, LNCS 5734, pp. 149–161, 2009.

are *overlapping*, i.e., each loop has a node in common with another loop in the set (Theorem 2).

Using the above characterizations, for each problem we devise a linear system of equations whose feasibility is equivalent to the solution of the problem. Since the size of these linear systems is polynomial, we obtain that both our problems are decidable in polynomial time. Further, we can compute in polynomial time a finite representation of a path with the required property. We also provide evidence that the problems addressed are non-trivial, by showing that a closely related problem is NP-hard: the problem of checking whether there is a perfectly balanced finite path connecting two given nodes in a graph.

We believe that the two problems that we study and solve in this paper are natural and canonical enough to be of independent theoretical interest. Additionally, they may be regarded as instances of the well established notion of *fairness*.

Balanced paths as fair paths. Colored graphs as studied in this paper routinely occur in the field of computer science that deals with the analysis of concurrent systems [MP91]. In that case, the graph represents the transition relation of a concurrent program and the color of an edge indicates which one of the processes is making progress along that edge. One basic property of interest for those applications is called *fairness* and essentially states that, during an infinite computation, each process is allowed to make progress infinitely often [Fra86]. Starting from this core idea, a rich theory of fairness has been developed, as witnessed by the amount of literature devoted to the subject (see, for instance, [LPS81, Kwi89, dA00]).

Cast in our abstract framework of colored graphs, the above basic version of fairness asks that, along an infinite path in the graph, each color occurs infinitely often. Such requirement does not put any bound on the amount of steps that a process needs to wait before it is allowed to make progress. As a consequence, the asymptotic frequency of some color could be zero even if the path is fair. Accordingly, several authors have proposed stronger versions of fairness. For instance, Alur and Henzinger define *finitary fairness* roughly as the property requiring that there be a fixed bound on the number of steps between two occurrences of any given color [AH98]. A similar proposal, supported by a corresponding temporal logic, was made by Dershowitz et al. [DJP03]. On a finitarily fair path, all colors have positive asymptotic frequency [1].

Our proposed notions of balanced paths and bounded difference paths may be viewed as two further refinements of the notion of fair path. Previous definitions treat the frequencies of the relevant events in isolation and in a strictly qualitative manner. Such definitions only distinguish between zero frequency (not fair), limit-zero frequency (fair, but not finitarily so), and positive frequency (finitarily fair). The current proposal, instead, introduces a quantitative comparison between competing events.

Technically, it is easy to see that bounded difference paths are special cases of finitarily fair paths. On the other hand, finitarily fair paths and balanced paths are incomparable notions.

We believe that the two proposed notions are valuable to some applications, perhaps quite different from the ones in which fairness is usually applied. Both the balance

[1] For the sake of clarity, we are momentarily ignoring those paths that have *no* asymptotic frequency.

property and the bounded difference property are probably too strong for the applications where one step in the graph represents a fine-grained transition of unknown length in a concurrent program. In that case, it may be of little interest to require that all processes make progress with the same (abstract) frequency.

On the other hand, consider a context where each transition corresponds to some complex or otherwise lengthy operation. As an example, consider the model of a concurrent program where all operations have been disregarded, except the access to a peripheral that can only be used in one-hour slots, such as a telescope, which requires some time for re-positioning. Assuming that all jobs have the same priority, it is certainly valuable to find a scheduling policy that assigns the telescope to each job with the same frequency.

As a non-computational example, the graph may represent the rotation of cultures on a crop, with a granularity of 6 months for each transition [Wik09]. In that case, we may very well be interested not just in having each culture eventually planted (fairness) or even planted within a bounded amount of time (finitarily fair), but also occurring with the same frequency as any other culture (balanced or bounded difference).

The rest of the paper is organized as follows. Section 2 introduces the basic definitions. Section 3 establishes connections between the existence of balanced or bounded difference paths in a graph and certain loop-based properties of the graph itself. Section 4 (respectively, Section 5) exploits the properties proved in Section 3 to define a system of linear equations whose feasibility is equivalent to the balance problem (resp., the bounded difference problem).

Due to space limitations, some proofs are omitted and reported in the full version of the current paper.

2 Preliminaries

Let X be a set and i be a positive integer. By X^i we denote the Cartesian product of X with itself i times. By $\mathbb{N}$, $\mathbb{Z}$, $\mathbb{Q}$, and $\mathbb{R}$ we respectively denote the set of non-negative integer, relative integer, rational, and real numbers. Given a positive integer k, let $[k] = \{1,\dots,k\}$ and $[k]_0 = [k] \cup \{0\}$.

A *k-colored graph* (or simply *graph*) is a pair $G = (V,E)$, where V is a set of *nodes* and $E \subseteq V \times [k] \times V$ is a set of *colored edges*. We employ integers as colors for technical convenience. All the results we obtain also hold for an arbitrary set of labels. An edge (u,a,v) is said to be *colored* with a. In the following, we also simply call a k-colored graph a *graph*, when k is clear from the context. For a node $v \in V$ we call ${}_vE = \{(v,a,w) \in E\}$ the set of edges exiting from v, and $E_v = \{(w,a,v) \in E\}$ the set of edges entering in v. For a color $a \in [k]$, we call $E(a) = \{(v,a,w) \in E\}$ the set of edges colored with a. For a node $v \in V$, a *finite v-path* ρ is a finite sequence of edges $(v_i,a_i,v_{i+1})_{i\in\{1,\dots,n\}}$ such that $v_1 = v$. The length of ρ is n and we denote by $\rho(i)$ the i-th edge of ρ. Sometimes, we write the path ρ as $v_1v_2\dots v_n$, when the colors are unimportant. A finite path $\rho = v_1v_2\dots v_n$ is a *loop* if $v_1 = v_n$. A loop $v_1v_2\dots v_n$ is *simple* if $v_i \neq v_j$, for all $i \neq j$, except for $i = 1$ and $j = n$. An *infinite v-path* is defined analogously, i.e., it is an infinite sequence of edges. Let ρ be a finite path and π be a possibly infinite path, we denote by $\rho \cdot \pi$ the *concatenation* of ρ and π. By ρ^ω we denote the infinite path obtained by concatenating ρ with itself infinitely many times. A graph G is *strongly connected* if for each pair (u,v) of nodes there is a finite u-path with last node v and a finite v-path with last node u.

For a finite or infinite path ρ and an integer i, we denote by $\rho^{\leq i}$ the *prefix* of ρ containing i edges. For a color $a \in [k]$, we denote by $|\rho|_a$ the number of edges labeled with a occurring in ρ. For two colors $a,b \in [k]$, we denote the difference between the occurrences of edges labeled with a and b in ρ by $\mathit{diff}_{a,b}(\rho) = |\rho|_a - |\rho|_b$. An infinite path π is *periodic* iff there exists a finite path ρ such that $\pi = \rho^\omega$. A loop σ is *perfectly balanced* iff $\mathit{diff}_{a,b}(\sigma) = 0$ for all $a,b \in [k]$. Finally, we denote by $\mathbf{0}$ and $\mathbf{1}$ the vectors containing only 0's and 1's, respectively. We can now define the following two decision problems.

The balance problem. Let G be a k-colored graph. An infinite path ρ in G is *balanced* if for all $a \in [k]$,

$$\lim_{i\to\infty} \frac{|\rho^{\leq i}|_a}{i} = \frac{1}{k}.$$

The *balance problem* is to determine whether there is a balanced path in G.

The bounded difference problem. Let G be a k-colored graph. An infinite path ρ in G has the *bounded difference property* (or, is a *bounded difference path*) if there exists a number $c \geq 0$, such that, for all $a,b \in [k]$ and $i > 0$,

$$|\mathit{diff}_{a,b}(\rho^{\leq i})| \leq c.$$

The *bounded difference problem* is to determine whether there is a bounded difference path in G.

3 Basic Properties

In this section, we assume that $G = (V,E)$ is a finite k-colored graph, i.e., both V and E are finite. In the following lemma, the proof of item 1 is trivial, while the proof of item 2 can be found in the extended version.

Lemma 1. *The following properties hold:*

1. *if a path has the bounded difference property, then it is balanced;*
2. *a path ρ is balanced if and only if for all $a \in [k-1]$,*

$$\lim_{i\to\infty} \frac{\mathit{diff}_{a,k}(\rho^{\leq i})}{i} = 0.$$

The following example shows that the converse of item 1 of Lemma 1 does not hold.

Example 1. For all $i > 0$, let $\sigma_i = (1\cdot 2)^i \cdot 1 \cdot 3 \cdot (1\cdot 3\cdot 2\cdot 3)^i \cdot 1 \cdot 3 \cdot 3$. Consider the infinite sequence $\sigma = \prod_{i=1}^{\omega} \sigma_i$ obtained by a hypothetic 3-colored graph. On one hand, it is easy to see that for all $i > 0$ it holds $\mathit{diff}_{3,1}(\sigma_i) = 1$. Therefore, $\mathit{diff}_{3,1}(\sigma_1\sigma_2\ldots\sigma_n) = n$, and σ is not a bounded difference path.

On the other hand, since the length of the first n blocks is $\Theta(n^2)$ and the difference between any two colors is $\Theta(n)$, in any prefix $\sigma^{\leq i}$ the difference between any two colors is in $O(\sqrt{i})$. According to item 2 of Lemma 1, σ is balanced. □

Two loops σ, σ' in G are *connected* if there exists a path from a node of σ to a node of σ', and vice-versa. A set $\mathcal{L}$ of loops in G is connected if all pairs of loops in $\mathcal{L}$ are connected. Two loops in G are *overlapping* if they have a node in common. A set $\mathcal{L}$ of loops in G is overlapping if for all pairs of loops $\sigma, \sigma' \in \mathcal{L}$ there exists a sequence $\sigma_1, \dots, \sigma_n$ of loops in $\mathcal{L}$ such that *(i)* $\sigma_1 = \sigma$, *(ii)* $\sigma_n = \sigma'$, and *(iii)* for all $i = 1, \dots, n-1$, (σ_i, σ_{i+1}) are overlapping. Given a set of loops $\mathcal{L}$ in G, the subgraph *induced by* $\mathcal{L}$ is $G' = (V', E')$, where V' and E' are all and only the nodes and the edges, respectively, belonging to a loop in $\mathcal{L}$.

Lemma 2. *Let G be a graph, $\mathcal{L}$ be a set of loops in G, and $G' = (V', E')$ be the subgraph of G induced by $\mathcal{L}$, then the following statements are equivalent:*

1. *$\mathcal{L}$ is* overlapping.
2. *The subgraph G' is strongly connected.*
3. *There exists $u \in V'$ such that for all $v \in V'$ there exists a path in G' from u to v.*

Proof. [$1 \Rightarrow 2$] If $\mathcal{L}$ is overlapping, then, for all pairs of loops σ_1, σ_2, there exists a sequence of loops that links σ_1 with σ_2. Thus, from any node of σ_1, it is possible to reach any node of σ_2. Hence, G' is strongly connected.

[$2 \Rightarrow 3$] Trivial.

[$3 \Rightarrow 2$] Let $u \in V'$ be a witness for (3). Let $v, w \in V'$, we prove that there is a path from v to w. We have that u is connected to both v and w. Since all edges in G' belong to a loop, for all edges $(u', \cdot, v')$ along the path from u to v there is a path from v' to u'. Thus, there is a path from v to u, and, as a consequence, a path from v to w, through u.

[$2 \Rightarrow 1$] If G' is strongly connected, for all $\sigma_1, \sigma_2 \in \mathcal{L}$ there is a path ρ in G' from any node of σ_1 to any node of σ_2. This fact holds since G' is induced by $\mathcal{L}$, so ρ uses only edges of the loops in $\mathcal{L}$. While traversing ρ, every time we move from one loop to the next, these two loops must share a node. Therefore, all pairs of adjacent loops used in ρ are overlapping. Thus $\mathcal{L}$ is overlapping. □

The above lemma implies that if $\mathcal{L}$ is overlapping then it is also connected, since G' is strongly connected.

For all finite paths ρ of G, with a slight abuse of notation let $\mathit{diff}(\rho) = (\mathit{diff}_{1,k}(\rho), \dots, \mathit{diff}_{k-1,k}(\rho))$ be the vector containing the differences between each color and color k, which is taken as a reference. We call $\mathit{diff}(\rho)$ the *difference vector* of ρ[2]. For all finite and infinite paths ρ we call *difference sequence* of ρ the sequence of difference vectors of all prefixes of ρ, i.e., $\{\mathit{diff}(\rho^{\leq n})\}_{n \in \mathbb{N}}$. Given a finite set of loops $\mathcal{L} = \{\sigma_1, \dots, \sigma_l\}$ and a tuple of natural numbers $c_1, \dots, c_l$ not all equal to zero, we call *natural linear combination* (in short, *n.l.c.*) of $\mathcal{L}$ with coefficients $c_1, \dots, c_l$ the vector $x = \sum_{i=1}^{l} c_i \mathit{diff}(\sigma_i)$.

A loop is a *composition* of a finite tuple of simple loops $\mathcal{T}$ if it is obtained by using all and only the edges of $\mathcal{T}$ as many times as they appear in $\mathcal{T}$. Formally, for a loop σ and an edge e, let $n(e, \sigma)$ be the number of occurrences of e in σ. The loop σ is a composition of $(\sigma_1, \dots, \sigma_l)$ if, for all edges e, it holds $n(e, \sigma) = \sum_{i=1}^{l} n(e, \sigma_i)$.

[2] The difference vector is related to the Parikh vector [Par66] of the sequence of colors of the path. Precisely, the difference vector is equal to the first $k-1$ components of the Parikh vector, minus the k-th component.

Lemma 3. *Let σ be a loop of length n containing m distinct nodes. Then, σ is a composition of at least $\lceil \frac{n}{m} \rceil$ simple loops.*

Proof. We use a decomposition algorithm on σ: the algorithm scans the edges of σ from the beginning to the end. As soon as a simple loop is found, i.e., as soon as a node is repeated, such a simple loop is removed from σ and added to the tuple. The tuple given by the removed loops is the decomposition we are looking for. Since a simple loop contains at most m nodes, the tuple contains at least $\lceil \frac{n}{m} \rceil$ loops. □

3.1 The Balance Problem

The following lemma shows that a sequence of integral vectors has sum in $o(n)$ only if there is a finite set of vectors that occur in the sequence which have an n.l.c. with value zero.

Lemma 4. *Let $A \subset \mathbb{Z}^d$ be a finite set of vectors such that there is no subset $A' \subseteq A$ with an n.l.c. of value zero. Let $\{(a_{n,1}, \ldots, a_{n,d})\}_{n \in \mathbb{N}}$ be an infinite sequence of elements of A, and $S_{n,i} = \sum_{j=0}^{n} a_{j,i}$ be the partial sum of the i-th component, for all $n \in \mathbb{N}$ and $i \in [d]$. Then, there exists at least an index h such that $\lim_{n \to \infty} \frac{S_{n,h}}{n} \neq 0$.*

The following result provides a loop-based characterization for the balance problem.

Theorem 1. *A graph G satisfies the balance problem iff there exists a connected set $\mathcal{L}$ of simple loops of G, with zero as an n.l.c.*

Proof. [if] Let $\mathcal{L} = \{\sigma_0, \ldots, \sigma_{l-1}\}$ be a connected set of simple loops having zero as an n.l.c., with coefficients $c_0, \ldots, c_{l-1}$. For all $i = 0, \ldots, l-1$, let v_i be the initial node of σ_i. Since $\mathcal{L}$ is connected, there exists a path ρ_i from v_i to $v_{(i+1) \bmod l}$. For all $j > 0$, define the loop $\pi_j = \sigma_0^{j \cdot c_0} \rho_0 \sigma_1^{j \cdot c_1} \rho_1 \ldots \sigma_{l-1}^{j \cdot c_{l-1}} \rho_{l-1}$. We claim that the infinite path $\pi = \prod_{j>0} \pi_j$ is balanced. Each time a π_j block ends along π, the part of the difference vector produced by the loops of $\mathcal{L}$ is zero. So, when a π_j ends, the difference vector is due only to the paths ρ_i. Since the index of the step $k(j)$ at which π_j ends grows quadratically in j and the difference vector $\mathit{diff}(\pi_1 \ldots \pi_j)$ grows linearly in j, we have that $\lim_{j \to \infty} \mathit{diff}(\pi_1 \ldots \pi_j)/k(j) = \mathbf{0}$. It can be shown that in the steps between $k(j)$ and $k(j+1)$, the i-th component of the difference vector differs from the one of $\mathit{diff}(\pi_1 \ldots \pi_j)$ no more than a function $C_{i,j}$ that grows linearly in j. Specifically, $C_{i,j} = MP_i + jMA_i$, where MP_i is the sum, for all ρ_j, of the maximum modulus of the i-th component of the difference vector along ρ_j, and MA_i is the sum, for all σ_j, of the maximum modulus of the i-th component of the difference vector along σ_j. As a consequence, $\lim_{k \to \infty} \mathit{diff}(\pi^{\leq k})/k = \mathbf{0}$ and π is balanced.

[only if] If there exists an infinite balanced path ρ, since the set of nodes is finite, there is a set V' of nodes occurring infinitely often in ρ. Let ρ' be a suffix of ρ containing only nodes in V'. The path ρ' is balanced and it is composed by an infinite sequence of simple loops on V', plus a remaining simple path (see the proof of Lemma 3 for further details). Let $\mathcal{L}$ be the (finite) set of such simple loops, and let $A \subset \mathbb{Z}^{k-1}$ be the set of difference vectors of the loops in $\mathcal{L}$.

Every time a loop trough V' closes along ρ', the difference vector up to that point is the sum of the difference vectors of the simple loops occurred so far, plus the difference vector of the remaining simple path. Since the remaining simple path cannot have length greater than $|V'|$, the difference vector up to that point differs from a sum of a sequence of elements of A by a constant-bounded term. Let $n(i)$ be the index of the i-th point where a loop is closed along ρ'. Since ρ' is balanced, by statement 2 of Lemma 1, each component of the difference sequence $\{diff(\rho'^{\leq i})\}_{i\in\mathbb{N}}$ is in $o(i)$. Hence, each component of the partial sum of the difference vectors associated to the sequence of loops closed is in $o(n(i))$. By Lemma 4, this is possible only if A has a subset A' with an n.l.c. of value zero. Thus, the set of loops $\mathcal{L}'$ with difference vectors in A' has an n.l.c. with value zero. Moreover, since the loops in $\mathcal{L}'$ are constructed with edges of ρ', they are connected. This concludes the proof. □

3.2 The Bounded Difference Problem

Given a graph, if there exists a perfectly balanced loop σ, it is easy to see that σ^ω is a periodic bounded difference path. Moreover, if ρ is an infinite bounded difference path, then there exists a constant c such that the absolute value of all color differences is smaller than c. Since both the set of nodes and the possible difference vectors along ρ are finite, we can find two indexes $i<j$ such that $\rho(i)=\rho(j)$ and $diff(\rho^{\leq i})=diff(\rho^{\leq j})$. So, $\sigma'=\rho(i)\rho(i+1)\ldots\rho(j)$ is a perfectly balanced loop. Therefore, the following holds.

Lemma 5. *Given a graph G, the following statements are equivalent:*

1. *There exists a bounded difference path.*
2. *There exists a periodic bounded difference path.*
3. *There exists a perfectly balanced loop.*

We now prove the following result.

Lemma 6. *Let G be a graph. There exists a perfectly balanced loop in G iff there exists an overlapping set $\mathcal{L}$ of simple loops of G, with zero as n.l.c.*

Proof. [only if] If there exists a perfectly balanced loop σ, by Lemma 3 the loop is the composition of a tuple $\mathcal{T}$ of simple loops. Let $\mathcal{L}$ be the set of distinct loops occurring in $\mathcal{T}$, and for all $\rho\in\mathcal{L}$, let c_ρ be the number of times ρ occurs in $\mathcal{T}$. Since in the computation of the difference vector of a path it does not matter the order in which the edges are considered, we have $\sum_{\rho\in\mathcal{L}} c_\rho\cdot diff(\rho)=diff(\sigma)=\mathbf{0}$. Finally, since the loops in $\mathcal{L}$ come from the decomposition of a single loop σ, we have that $\mathcal{L}$ is overlapping.

[if] Let $\mathcal{L}=\{\sigma_1,\ldots,\sigma_l\}$ be such that $\sum_{i=1}^{l} c_i\cdot diff(\sigma_i)=\mathbf{0}$. We construct a single loop σ such that $diff(\sigma)=\sum_{i=1}^{l} c_i\cdot diff(\sigma_i)$. The construction proceeds in iterative steps, building a sequence of intermediate paths $\rho_1,\ldots,\rho_l$, such that ρ_l is the wanted perfectly balanced loop. In the first step, we take any loop $\sigma_{i_1}\in\mathcal{L}$ and we traverse it c_{i_1} times, obtaining the first intermediate path $\rho_1=\sigma_{i_1}^{c_{i_1}}$. After the j-th step, since $\mathcal{L}$ is overlapping, there must be a loop $\sigma_{i_{j+1}}\in\mathcal{L}$ that is overlapping with one of the loops in the current intermediate path ρ_j, say in node v. Then, we *reorder* ρ_j in such a way that it

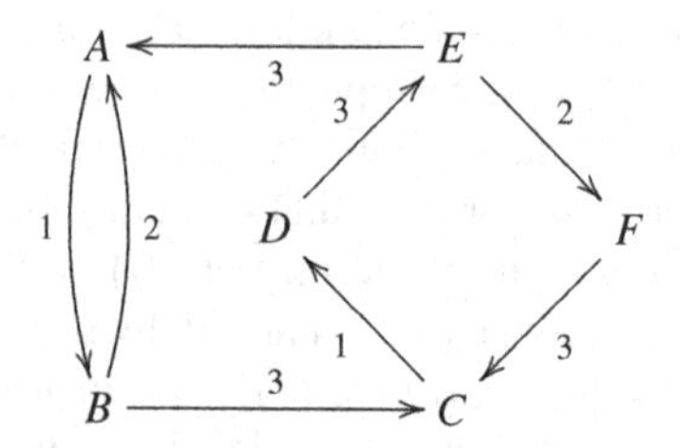

Fig. 1. A 3-colored graph satisfying the balance problem, but not the bounded difference problem

starts and ends in v. Let ρ'_j be such reordering, we set $\rho_{j+1} = \rho'_j \sigma_{i_{j+1}}^{c_{i_{j+1}}}$. One can verify that ρ_l is perfectly balanced. □

The following theorem is a direct consequence of the previous two lemmas.

Theorem 2. *A graph G satisfies the bounded difference problem iff there exists an overlapping set $\mathcal{L}$ of simple loops of G, with zero as n.l.c.*

Example 2. Consider the graph G in Fig. 1. First note that, up to rotation, there are just three simple loops in it: $\sigma_1 = A \cdot B \cdot A$, $\sigma_2 = C \cdot D \cdot E \cdot F \cdot C$, and $\sigma_3 = A \cdot B \cdot C \cdot D \cdot E \cdot A$. It is easy to see that $\mathit{diff}(\sigma_1) = (1,1)$, $\mathit{diff}(\sigma_2) = (-1,-1)$, and $\mathit{diff}(\sigma_3) = (-1,-3)$. On one hand, since the connected set of simple loops $\{\sigma_1, \sigma_2\}$ has zero as n.l.c., we obtain that there is a balanced path in G. Example 1 shows a particular balanced sequence of colors obtained by a non-periodic path of the subgraph G' of G induced by these two loops. On the other hand, for all the three overlapping sets of loops ($\{\sigma_1, \sigma_3\}$, $\{\sigma_2, \sigma_3\}$, and $\{\sigma_1, \sigma_2, \sigma_3\}$) there is no way to obtain a zero n.l.c. with all coefficients different from zero. So, there is no bounded difference path in G. □

3.3 2-Colored Graphs

When the graph G is 2-colored, the difference vector is simply a number. So, if $\mathcal{L}$ is a connected set of simple loops having zero as n.l.c., then there must be either a perfectly balanced simple loop or two loops with difference vectors of opposite sign. Notice that two loops σ, σ' with color differences of opposite sign have the following n.l.c. of value zero: $|\mathit{diff}(\sigma')| \cdot \mathit{diff}(\sigma) + |\mathit{diff}(\sigma)| \cdot \mathit{diff}(\sigma') = 0$. If the two loops are connected but not overlapping, we can construct a sequence of adjacent overlapping simple loops connecting them. In this sequence, we are always able to find a perfectly balanced simple loop or two overlapping simple loops with difference vectors of opposite sign. Therefore, the following holds.

Lemma 7. *Let G be a 2-colored graph. If there exists a* connected *set of simple loops of G with zero as n.l.c., then there exists an* overlapping *set of simple loops of G with zero as n.l.c.*

Due to the above characterization, both decision problems can be solved efficiently, by using a minimum spanning tree algorithm to find two loops of opposite color difference sign, if such exist.

Theorem 3. *A 2-colored graph $G = (V, E)$ satisfies the bounded difference problem iff it satisfies the balance problem. Both problems can be solved in time $O(|V| \cdot |E| \cdot \log |V|)$.*

3.4 A Related NP-Hard Problem

In this section, we introduce an NP-hard problem similar to the bounded difference problem. Given a k-colored graph G and two nodes u and v, the new problem asks whether there exists a perfectly balanced path from u to v. We call this question the *perfectly balanced finite path problem.* To see that this problem is closely related to the bounded difference problem, one can note that it corresponds to the statement of item 3 in Lemma 5, by changing the word *loop* to *finite path.* The following result can be proved using a reduction from 3SAT.

Theorem 4. *The perfectly balanced finite path problem is NP-hard.*

4 Solving the Balance Problem

In this section, we define a system of linear equations whose feasibility is equivalent to the balance problem for a given strongly connected graph.

Definition 1. *Let $G = (V, E)$ be a k-colored graph. We call* balance system *for G the following system of equations on the set of variables $\{x_e \mid e \in E\}$.*

1. for all $v \in V$	$\sum_{e \in E_v} x_e = \sum_{e \in_v E} x_e$
2. for all $a \in [k-1]$	$\sum_{e \in E(a)} x_e = \sum_{e \in E(k)} x_e$
3. for all $e \in E$	$x_e \geq 0$
4.	$\sum_{e \in E} x_e > 0.$

Let $m = |E|$ and $n = |V|$, the balance system has m variables and $m + n + k$ constraints. It helps to think of each variable x_e as a load associated to the edge $e \in E$, and of each constraint as having the following meaning.

1. For each node, the entering load is equal to the exiting load.
2. For each color $a \in [k-1]$, the load on the edges colored by a is equal to the load on the edges colored by k.
3. Every load is non-negative.
4. The total load is positive.

The following lemma justifies the introduction of the balance system.

Lemma 8. *There exists a set $\mathcal{L}$ of simple loops in G with zero as n.l.c. iff the balance system for G is feasible.*

Proof. (Sketch) [only if] If there exists an n.l.c. of $\mathcal{L}$ with value zero, let c_σ be the coefficient associated with a loop $\sigma \in \mathcal{L}$. We can construct a vector $x \in R^m$ that satisfies the balance system. First, define $h(e, \sigma)$ as 1 if the edge e is in σ, and 0 otherwise. Then, we set $x_e = \sum_{\sigma \in \mathcal{L}} c_\sigma h(e, \sigma)$. Considering that, for all $\sigma \in \mathcal{L}$ and $v \in V$, it holds that $\sum_{e\in_v E} h(e,\sigma) = \sum_{e\in E_v} h(e,\sigma)$, it is a matter of algebra to show that x satisfies the balance system.

[if] If the system is feasible, since it has integer coefficients, it has to have a rational solution. Moreover, all constraints are either equalities or inequalities of the type $a^T x \sim 0$, for $\sim \in \{>, \geq\}$. Therefore, if x is a solution then cx is also a solution, for all $c > 0$. Accordingly, if the system has a rational solution, it also has an integer solution $x \in \mathbb{Z}^m$. Due to the constraints (3), such solution must be non-negative. So, in fact $x \in \mathbb{N}^m$.

Then, we consider each component x_e of x as the number of times the edge e is used in a set of loops, and we use x to construct such set with an iterative algorithm. At the first step, we set $x^1 = x$, we take a non-zero component x_e^1 of x^1, we start constructing a loop with the edge e, and then we subtract a unit from x_e^1 to remember that we used it. Next, we look for another non-zero component $x_{e'}^1$ such that e' exits from the node e enters in. It is possible to show that the edge e' can always be found. Then, we add e' to the loop and we subtract a unit from $x_{e'}^1$. We continue looking for edges e' with $x_{e'}^1 > 0$ and exiting from the last node added to the loop, until we close a loop, i.e., until the last edge added enters in the node the first edge e exits from. After constructing a loop, we have a residual vector x^2 for the next step. If such vector is not zero, we construct another loop, and so on until the residual vector is zero. In the end we have a set of (not necessarily simple) loops, and we show that it has zero as n.l.c. Finally, we decompose those loops in simple loops with the algorithm of Lemma 3, and we obtain a set $\mathcal{L}$ of simple loops having zero as a natural linear combination. □

Since in a strongly connected graph all loops are connected, from the previous lemma, we have:

Corollary 1. *If G is strongly connected, there exists a balanced path in G iff the balance system for G is feasible.*

In order to solve the balance problem in G, first we compute the maximal connected components of G using the classical algorithm [CLRS01]. This algorithm is polynomial in n and m. Then, in each component we compute whether the balance system is feasible, by using the polynomial algorithm for feasibility of sets defined by linear constraints [NW88]. This second algorithm is used at most n times and it is polynomial in the number of constraints $(n+m+k)$ and in the logarithm of the maximum modulus of a coefficient in a constraint (in our case, the maximum modulus is 1).

Theorem 5. *The balance problem is in P.*

We remark that the feasibility algorithm can also provide the value of a solution to the system in input. By the proof of Lemma 8, such a solution allows us to compute in polynomial time a set of connected simple loops and the coefficients of an n.l.c. of value zero. As shown in the *if* part of the proof of Theorem 1, this in turn allows us to constructively characterize a balanced path in the graph.

5 Solving the Bounded Difference Problem

In this section, we solve the bounded difference problem using the same approach as in Section 4.

Definition 2. *Let $G = (V,E)$ be a k-colored graph with $m = |E|$, $n = |V|$, and $s_G = \min\{n+k-1,m\}$, and let $u \in V$ be a node. We call* bounded difference system *for (G,u) the following system of equations on the set of variables $\{x_e, y_e \mid e \in E\}$.*

1-4. The same constraints as in the balance system for G	
5. for all $v \in V \setminus \{u\}$	$\sum_{e\in E_v} y_e - \sum_{e\in_v E} y_e = \sum_{e\in_v E} x_e$
6.	$\sum_{e\in_u E} y_e - \sum_{e\in E_u} y_e = \sum_{v\in V\setminus\{u\}} \sum_{e\in_v E} x_e$
7. for all $e \in E$	$y_e \geq 0$
8. for all $e \in E$	$y_e \leq (m \cdot s_G!)x_e$.

The bounded difference system has $2m$ variables and $3m+2n+k$ constraints. It helps to think of the vectors x and y as two loads associated to the edges of G. The constraints 1-4 are the same constraints of the balance problem for G, and they ask that x should represent a set of simple loops of G having zero as a natural linear combination.

The constraints 5-8 are *connection constraints*, asking that y should represent a connection load, from u to every other node of the simple loops defined by x, and carried only on the edges of those loops. Thus, constraints 5-8 ask that the loops represented by x should be overlapping, because of Lemma 2.

5. Each node $v \in V \setminus \{u\}$ absorbs an amount of y-load equal to the amount of x-load traversing it. These constraints ensure that the nodes belonging to the x-solution receive a positive y-load.
6. Node u generates as much y-load as the total x-load on all edges, except the edges exiting u.
7. Every y-load is non-negative.
8. If the x-load on an edge is zero, then the y-load on that edge is also zero. Otherwise, the y-load can be at most $m \cdot s_G!$ times the x-load. More details on the choice of this multiplicative constant follow.

In Lemma 9, we show that if there is a solution x of the balance system, then there is another solution x' whose non-zero components are greater or equal to 1 and less than or equal to $s_G!$, so that $\sum_{e\in E} x'_e \leq m \cdot s_G!$. In this way, the constraints (8) allow each edge that has a positive x-load to carry as its y-load all the y-load exiting from u.

Lemma 9. *Let $G = (V,E)$ be a k-colored graph, with $|V| = n$, $|E| = m$, and $s_G = min\{n+k-1,m\}$. For all solutions x to the balance system for G there exists a solution x' such that, for all $e \in E$, it holds $(x_e = 0 \Rightarrow x'_e = 0)$ and $(x_e > 0 \Rightarrow 1 \leq x'_e \leq s_G!)$. As a consequence, $1 \leq \sum_{e\in E} x'_e \leq m \cdot s_G!$.*

The following lemma states that the bounded difference system can be used to solve the bounded difference problem.

Lemma 10. *There exists an overlapping set of simple loops in G, passing through a node u and having zero as n.l.c. iff the bounded difference system for (G,u) is feasible.*

Proof. [only if] Let $\mathcal{L}$ be an overlapping set of simple loops having an n.l.c. of value zero. Let c_σ be the coefficient associated with the loop $\sigma \in \mathcal{L}$ in such linear combination. We start by constructing a solution $x \in \mathbb{R}^m$ to the balance system as follows. Define

$h(e,\sigma) \in \{0,1\}$ as 1 if the edge e belongs to the loop σ, and 0 otherwise. We set $x_e = \sum_{\sigma \in \mathcal{L}} c_\sigma h(e,\sigma)$. We have that x is a solution to the balance system for G, or equivalently that it satisfies constraints (1)-(4) of the bounded difference system for (G,u).

By Lemma 9, there exists another solution $x' \in \mathbb{R}^m$ to the balance system, such that $x_e = 0 \Rightarrow x'_e = 0$ and $x_e > 0 \Rightarrow 1 \leq x'_e \leq s_G!$. If any loop of the overlapping set $\mathcal{L}$ passes through u, by Lemma 2, there exists a path ρ_v from u to any node v occurring in $\mathcal{L}$. We set $y_e = \sum_{v \in V' - \{u\}} (h(e,\rho_v) \sum_{e \in_v E} x'_e)$. Simple calculations show that (x',y) is a solution to the bounded difference system for (G,u).

[if] If there exists a vector $(x,y) \in \mathbb{R}^{2m}$ satisfying the bounded difference system, then like we did in the second part of Lemma 8, using x, we can construct a set of simple loops $\mathcal{L}$ having zero as n.l.c. Since $\sum_{e \in_u E} y_e - \sum_{e \in E_u} y_e = \sum_{v \in V - \{u\}} \sum_{e \in_v E} x_e$, we have that u belongs to at least one edge used in the construction of $\mathcal{L}$. If we set $G' = (V',E')$ as the subgraph of G induced by $\mathcal{L}$, we are able to show by contradiction that there is a path in G' from u to every other node of V'. Indeed if for some $v \in V' - \{u\}$ there is no path in G' from u to v then there is some load exiting from u that cannot reach its destination using only edges of G'. Since the constraints (8) make it impossible to carry load on edges of G that are not used in $\mathcal{L}$, the connection constraints cannot be satisfied. So, for all $v \in V'$ there is a path in G' from u to v. By Lemma 2, $\mathcal{L}$ is overlapping. □

In order to solve the bounded difference problem in G, for all $u \in V$ we check whether the bounded difference system for (G,u) is feasible, by using a polynomial time algorithm for feasibility of linear systems [NW88]. This algorithm is used at most n times and it is polynomial in the number of constraints $(2n+3m+k)$ and in the logarithm of the maximum modulus M of a coefficient in a constraint. In our case, $M = m \cdot s_G!$. Using Stirling's approximation, we have $\log(m \cdot s_G!) = \log(m) + \Theta(s_G \log(s_G))$. Therefore, we obtain the following.

Theorem 6. *The bounded difference problem is in P.*

Acknowledgements. The second author would like to thank Luca de Alfaro and Krishnendu Chatterjee for the fruitful discussions on a preliminary version of this work.

References

[AH98] Alur, R., Henzinger, T.A.: Finitary fairness. ACM Trans. on Programming Languages and Systems 20(6) (1998)

[CLRS01] Cormen, T.H., Leiserson, C.E., Rivest, R.L., Stein, C.: Introduction to Algorithms, 2nd edn. MIT Press, Cambridge (2001)

[dA00] de Alfaro, L.: From fairness to chance. ENTCS 22 (2000)

[DJP03] Dershowitz, N., Jayasimha, D.N., Park, S.: Bounded fairness. In: Dershowitz, N. (ed.) Verification: Theory and Practice. LNCS, vol. 2772, pp. 304–317. Springer, Heidelberg (2003)

[Fra86] Francez, N.: Fairness. Springer, Heidelberg (1986)

[Kwi89] Kwiatkowska, M.: Survey of fairness notions. Information and Software Technology 31(7), 371–386 (1989)

[LPS81] Lehmann, D., Pnueli, A., Stavi, J.: Impartiality, justice and fairness: The ethics of concurrent termination. In: Even, S., Kariv, O. (eds.) ICALP 1981. LNCS, vol. 115, pp. 264–277. Springer, Heidelberg (1981)

[MP91] Manna, Z., Pnueli, A.: The Temporal Logic of Reactive and Concurrent Systems: Specification. Springer, Heidelberg (1991)

[NW88] Nemhauser, G.L., Wolsey, L.A.: Integer and Combinatorial Optimization. Wiley-Interscience, Hoboken (1988)

[Par66] Parikh, R.J.: On Context-Free Languages. Journal of the ACM 13(4), 570–581 (1966)

[Wik09] Wikipedia. Crop rotation (2009), `http://www.wikipedia.com`

Few Product Gates But Many Zeros

Bernd Borchert[1], Pierre McKenzie[2], and Klaus Reinhardt[1]

[1] WSI, Universität Tübingen, Sand 13, 72076 Tübingen, Germany
{borchert,reinhard}@informatik.uni-tuebingen.de
[2] DIRO, Univ. de Montréal, C.P. 6128 Centre-Ville, Montréal (Qc), H3C 3J7, Canada
mckenzie@iro.umontreal.ca

Abstract. A *d-gem* is a $\{+,-,\times\}$-circuit having very few $\times$-gates and computing from $\{x\} \cup \mathbb{Z}$ a univariate polynomial of degree d having d distinct integer roots. We introduce d-gems because they could help factoring integers and because their existence for infinitely many d would blatantly disprove a variant of the Blum-Cucker-Shub-Smale conjecture. A natural step towards validating the conjecture would thus be to rule out d-gems for large d. Here we construct d-gems for several values of d up to 55. Our 2^n-gems for $n \leq 4$ are skew, that is, each $\{+,-\}$-gate adds an integer. We prove that skew 2^n-gems if they exist require n $\{+,-\}$-gates, and that these for $n \geq 5$ would imply new solutions to the Prouhet-Tarry-Escott problem in number theory. By contrast, skew d-gems over the real numbers are shown to exist for every d.

1 Introduction

Consider the polynomials of degrees 4, 8 and 16 computed by the $\{+,\times\}$-circuits depicted in Figure 1. Each polynomial factors completely over $\mathbb{Z}$ and its roots are distinct. Clearly the number of $\times$-gates used in each case is minimum.

Main question: Do $\{+,-,\times\}$-circuits having n $\times$-gates and computing a polynomial $f(x) \in \mathbb{Z}[x]$ having 2^n distinct integer roots exist for all n?

Crandall [Cr96, Prob. 3.1.13] found a *normalized* circuit for the case $n = 3$ and asked whether such circuits exist for $n > 3$, where a *normalized* circuit starts from x and alternates between a squaring operation and the addition of a constant. Dilcher [Di00] characterized the normalized circuits for $n = 3$. Crandall and Pomerance [CrPo01] constructed an example for $n = 4$ and Bremner [Br08] constructed two infinite families of examples for that case.

Why care? First, polynomials with distinct integer roots may hold the key to factoring integers. For example, knowing [BoMo74] that $\sim n^{1/4}$ operations modulo n suffice to evaluate $x(x-1)(x-2)\cdots(x-n^{1/4}+1)$ at the points $x = n^{1/4}, 2n^{1/4}, 3n^{1/4}, \ldots, n^{1/2}$, Strassen [St76] noted that $\log_2 n$-bit integers can be factored in time $\sim n^{1/4}$. No provably faster deterministic algorithm for factoring is known [GaGe03]. Lipton [Li94] later formulated a hypothesis, on circuits computing polynomials having many distinct integer roots, whose validity would imply that the integer factoring problem is "too easy" to support

R. Královič and D. Niwiński (Eds.): MFCS 2009, LNCS 5734, pp. 162–174, 2009.

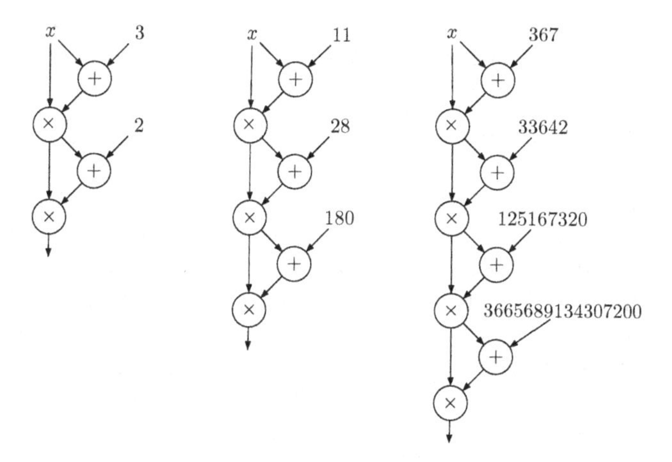

Fig. 1. Circuits computing polynomials having the respective sets of roots $\{0, -1, -2, -3\}$, $\{0, -1, -2, -4, -7, -9, -10, -11\}$ and $\{0, -4, -7, -12, -118, -133, -145, -178, -189, -222, -234, -249, -355, -360, -363, -367\}$

cryptography. A positive answer to our main question (with further size and constructivity assumptions) would validate Lipton's hypothesis. Crandall [Cr96] discusses further connections with factoring.

Second, by Proposition 1, a positive answer would *refute* the L-conjecture of Bürgisser [Bu01]. The L-conjecture states that for some β and any d, any polynomial $f(x) \in \mathbb{Q}[x]$ has at most $(L(f)+d)^\beta$ irreducible factors of degree d or less, where $L(f)$ is the size of a smallest $\{+, -, \times, \div\}$-circuit computing $f(x)$ from $\{x\} \cup \mathbb{Q}$ (see also [Ch04] and [Ro03]). The L-conjecture implies the τ-conjecture [BCSS97], namely that any $f(x) \in \mathbb{Z}[x]$ has at most $(\tau(f)+1)^\beta$ distinct roots in $\mathbb{Z}$, where $\tau(f)$ is the size of a smallest $\{+, -, \times\}$-circuit computing $f(x)$ from $x \cup \{1\}$. The τ-conjecture implies $P_\mathbb{C} \neq NP_\mathbb{C}$ in the Blum-Shub-Smale model of computation over the reals [BCSS97], and Smale named the τ-conjecture the fourth most important millennium mathematical challenge [Sm00]. A compelling micro-step towards validating both the τ-conjecture and the L-conjecture would thus be to provide a negative answer to our main question.

Our main question remains open. Here we partly explain why, by relating the question to a classical number-theoretic problem. We also extend the question to encompass polynomials of any degree $d > 0$ by defining *d-gems*, so named to reflect their "precious and seemingly rare" nature. Let ℓ_d be the length of a shortest addition chain for d, or equivalently, the size of a smallest $\{+\}$-circuit (Section 2 defines circuits formally) computing d from $\{1\}$.

Definition 1. *A $\{+, -, \times\}$-circuit c with inputs from $\{x\} \cup \mathbb{Z}$ is a d-*gem *if c has at most ℓ_d $\times$-gates and if the polynomial $f_c(x) \in \mathbb{Z}[x]$ computed by c has degree d and has precisely d distinct integer roots.*

Hence a 2^n-gem is a $\{+,-,\times\}$-circuit with n product gates that computes a polynomial of maximum degree 2^n and this polynomial factors completely with its 2^n roots integer and distinct. Our contributions are the following:

- we observe, following [PaSt73], that any polynomial of degree d can be computed from $\{x\} \cup \mathbb{Z}$ by a $\{+,-,\times\}$-circuit using $2(\sqrt{d}+1)$ product gates;
- if d-gems exist for infinitely many d, then the L-conjecture fails;
- by theoretical considerations and computer search, we construct *skew* d-gems for $d \leq 22$ and d-gems for $d \leq 31$ and $d = 36, 37, 42, 54, 55$ (see Figure 2), where a circuit is *skew* if each of its $\{+,-\}$-gates merely adds an integer;
- whereas d-gems trivially require ℓ_d product gates when $d = 2^n$, we show that this holds for every $d \leq 71$; we conclude that all the d-gems we are able to construct so far have a minimal number of product gates;
- we observe that a skew 2^n-gems for any $n \geq 5$ would provide new solutions of size 2^{n-1} to the Prouhet-Tarry-Escott problem which has an almost 200-year history in number theory (see for instance [BoIn94]);
- we spell out sufficient conditions implying a 2^n-gem;
- we construct skew *d-gems over the reals*, i.e. with inputs from $\mathbb{R} \cup \{x\}$ and with the requirement of distinct roots in $\mathbb{R}$, for every d;
- we prove that any skew 2^n-*gem over the reals* requires at least n $\{+,-\}$-gates; we conclude that the skew 2^n-gems (over $\mathbb{Z}$) depicted in Figure 1 have a minimal number of $\{+,-\}$-gates among all skew 2^n-gems.

Section 2 defines circuits and proves basic facts. Section 3 relates the existence of d-gems to the Prouhet-Tarry-Escott problem. Section 4 deals with gems over the real numbers. Section 5 describes our d-gem constructions. Section 6 concludes. Proofs left out of this abstract will appear in the full version of the paper.

2 Preliminaries and Basic Facts

By an (*arithmetic*) $\{+,-,\times\}$-*circuit* c we mean a rooted directed acyclic graph with in-degree-2 nodes called *product gates* labeled with $\times$, in-degree-2 nodes called *additive gates* labeled with $+$ or $-$ and in-degree-0 nodes called *input gates* labeled with an integer or the variable x. We write $c_\times$ and c_+ for the numbers of product and additive gates in c respectively. The *size* of c is $c_\times + c_+$. A circuit c *represents* or *computes* a polynomial $f_c(x) \in \mathbb{Z}[x]$. A *zero* or *root* of c is an integer a such that $f_c(a) = 0$. We write $\texttt{izeros}_c$ for the set of zeros of c. For example, if c is the leftmost circuit in Figure 1, then $c_\times = c_+ = 2$ and c represents $f_c(x) = (x(x+3))(x(x+3)+2) = x^4 + 6x^3 + 11x^2 + 6x$ having $\texttt{izeros}_c = \{0,-1,-2,-3\}$.

An *addition chain* for a natural number d is an increasing sequence $d_0 = 1, d_1, \ldots, d_k = d$ of natural numbers such that each d_i for $i > 0$ is the sum of two earlier numbers in the sequence. The polynomial x^d is computable by an optimal $\{\times\}$-circuit having ℓ_d product gates, where ℓ_d is the minimum k for which there is an addition chain such that $d_k = d$ (see [Kn81] for extensive related facts on addition chains, such as $\lceil \log_2 d \rceil \leq \ell_d \leq 2\lfloor \log_2 d \rfloor$ for all d). Recall that ℓ_d

d	$c_\times$	c_+	$f_c(x)$	izeros$_c$
1	0	0	x	$\{0\}$
2	1	1	x^2-1	$\{-1,1\}$
3	2	1	$(x^2-1)x$	$\{-1,0,1\}$
4	2	2	$((x^2-5)^2-16$	$\{-1,1,-3,3\}$
5	3	2	$(((x^2-5)^2-16)x$	$\{0,-2,2,-3,3\}$
5	3	2	$(x^2-1)((x^2-4)x)$	$\{0,-1,1,-2,2\}$
6	3	2	$((x^2-25)^2-24^2)(x^2-25)$	$\{-1,1,-7,7,-5,5\}$
6	3	2	$((x^2-7)x)^2-36$	$\{1,2,-3,-1,-2,3\}$
7	4	2	$(((x^2-7)x)^2-36)x$	$\{0,-1,1,-2,2,-3,3\}$
7	4	2	$((x^2-25)^2-24^2)(x^2-25)x$	$\{0,-1,1,-7,7,-5,5\}$
8	3	3	$(((x^2-65)^2-1696)^2-2073600$	$\{-3,3,-11,11,-7,7,-9,9\}$
9	4	2	$(((x^2-49)x)^2-120^2)((x^2-49)x)$	$\{0,-3,3,-5,5,-8,8,-7,7\}$
10	4	3	$((y^2-236448)^2-123552^2)y$ with $y=(x^2-625)$	$\{\pm 5,35,17,31,25\}$
10	4	3	$(((x^2-250)^2-14436)x)^2-1612802$	$\{\pm 4,8,14,18,20\}$
12	4	3	$(((x^2-91)x)^2-58500)^2-50400^2$	$\{\pm 1,5,6,9,10,11\}$
14	5	3	$((((x^2-7^4)x)^2-...)^2-...)(x^2-7^4)$	$\{\pm 49,16,39,55,21,35,56\}$
15	5	3	$y\times(y^2-34320^2)\times(y^2-41160^2)$ with $y=(x^2-7^4)x$	$\{\pm 0,49,16,39,55,21,35,56\}$
16	4	4	$((((x^2-67405)^2-3525798096)^2-...)^2-...$	$\{\pm 11,367,131,343,77,359,101,353\}$
18	5	5	$f_{c16}\cdot(x^2-1)$	Set$_{16}$ $\cup\{-1,1\}$
18	5	4	$(y^2-2484^2)\times(y^2-4116^2)\times(y^2-5916^2)$ with $y=(x^2-7^2\cdot 13)x$	$\{\pm 4,23,27,7,21,27,12,17,29\}$
20	5	5	$f_{c_{16}}\cdot((x^2-67405)^2-3958423056)$	$\{\pm 67,361,11,367,131,$ $343,77,359,101,353\}$
21	6	4	$y\times(y^2-89760^2)\times(y^2-150480^2)\times$ $(y^2-263640^2)$ with $y=(x^2-7^2\cdot 13^2)x$	$\{\pm 0,91,11,85,96,19,$ $80,99,39,65,104\}$
22	6	6	$f_{c_{20}}\cdot(x^2-1)$	Set$_{20}$ $\cup\{-1,1\}$
23	6		$y\times(y-4838400x^3+208051200x)\times(x^3-16x)$ with $y=z\times(z+45x^3-700x^2-2835x+630)$ and $z=(x^3+x^2-197x+195)\times(x^2-x-42)$	$\{\pm 0,1,2,3,4,6,7,9,10,13,14,15\}$
24	5		$f_{c_4}(y^2)$ with $y=(x^2-7\cdot 13\cdot 19)x$	$\{\pm 3,40,43,8,37,45,$ $15,32,47,23,25,48\}$
24	5	442	$z(z+c_{Prop.3.4})$ with $y=(x^2-11763)^2$ and $z=(y+241x^2+..)(y+195x^2+..)(y+x^2+..)$	$\{\pm 22,61,86,127,140,151,$ $35,47,94,121,146,148\}$
26	6	443	$f_{c24}\cdot(x^2-1)$	Set$_{24}$ $\cup\{-1,1\}$
27	6		$y\times f_{c_4}(y^2)$ with $y=(x^2-7^2\cdot 13^2)x$	Set$_{21}$ $\cup\{\pm 49,56,105\}$
28	6	560	$f_{c24}\cdot(y+117x^2+...)$	Set$_{24}$ $\cup\{-1,1,-153,153\}$
30	6		$f_{c_5}(y^2)$ with $y=(x^2-7^2\cdot 13^2\cdot 19)x$	$\{\pm 13,390,403,35,378,413,70,357,$ $427,103,335,438,117,325,442\}$
36	6		$f_{c_6}(y^2)$ with $y=(x^2-7^2\cdot 13^2\cdot 19)x$	Set$_{30}$ $\cup\{\pm 137,310,447\}$
42	7		$f_{c_7}(y^2)$ with $y=(x^2-7^2\cdot 13^2\cdot 19)x$	Set$_{36}$ $\cup\{\pm 182,273,455\}$
54	7		$f_{c_9}(y^2)$ with $y=(x^2-7^2\cdot 13^2\cdot 19)x$	Set$_{42}$ $\cup\{\pm 202,255,457,225,233,458\}$
55	8		The above $\times x$	The above $\cup\{0\}$

Fig. 2. Some d-gems which are skew for $d \leq 22$. When two examples are given for a given d, these arise from different minimal addition chains for d. The functions f_{c_i} are from Lemma 4 if $i \leq 9$ and from the i-gem in this table otherwise. We omitted the cases $d = 11, 13, 17, 19, 25, 29, 31, 37$ which, like the case $d = 55$ here, are obtained by extending a $(d-1)$-gem; we note that such an extension does not work for 43 which has a shorter addition chain bypassing 42. The constructions not mentioned is this abstract are explained in the full paper.

enters Definition 1 of a d-gem c in the form "$c_\times \leq \ell_d$". We impose "$c_\times \leq \ell_d$" rather than "$c_\times$ is minimal" because of the prospect that Strassen (unpublished work) might have uncovered circuits c with $c_\times < \ell_{\deg(f_c(x))}$ and we do not want declaring the gem status to require a lower bound proof. When $d = 2^n$, then $2^{\ell_{2^n}} = 2^n = d = |\mathtt{izeros}_c| = \deg(f_c(x)) \leq 2^{c_\times} \leq 2^{\ell_d} = 2^{\ell_{2^n}}$ so $c_\times$ is indeed then minimal. We extend this to all d-gems constructed in this paper as follows, since $d < 71$ implies $d > 2^{l_d - 3}$ (see [Kn81, p. 446]):

Lemma 1. *If $d > 2^{l_d-3}$, then any $\{\times, +, -\}$-circuit computing a polynomial $f(x) \in \mathbb{Z}[x]$ of degree d requires at least ℓ_d product gates.*

The L-conjecture clearly fails if skew d-gems exist for infinitely many d, but this also holds for d-gems:

Proposition 1. *If d-gems exist for infinitely many values of d, then the L-conjecture fails.*

To qualify as a d-gem, a circuit c must be extremely efficient in terms of its number of product gates, that is, $c_\times \leq \ell_d \leq 2\log_2 d$. At the opposite end of the spectrum, we have:

Proposition 2. *[PaSt73] Any degree-d polynomial $f(x) \in \mathbb{Z}[x]$ can be computed by a $\{+, -, \times\}$-circuit having at most $2\sqrt{d} + 1$ product gates.*

Proof. Write $f(x) = (\cdots((g_k x^k + g_{k-1})x^k + g_{k-2})x^k + \cdots + g_1)x^k + g_0$ where each $g_i \in \mathbb{Z}[x]$ has degree $k = \lceil \sqrt{d}\, \rceil$. Once $x^2, x^3, \cdots, x^k$ are available, each g_i is computable using additive gates alone. Another k products by x^k suffice. □

3 Gems and the Prouhet-Tarry-Escott Problem

For any $n > 4$, we are unable to rule out the existence of a skew 2^n-gem. In this section we show that any such gem would yield new solutions to a number-theoretic problem having a long history. Then we examine whether solutions to the number-theoretic problem can help constructing gems in return.

Recall from Section 1 that a 2^n-gem is *normalized* if its computation starts from x and iterates the sequence "squaring then adding a constant" at least once. A normalized 2^n-gem thus has n $\times$-gates and n $+$-gates and is entirely described by a sequence $\gamma_1, \ldots, \gamma_n$ of integers with $n \geq 1$. Normalized 8-gems and normalized 16-gems are investigated in [Cr96, Di00, CrPo01, Br08].

Consider an arbitrary skew 2^n-gem c. Consecutive $+$-gates in c can be merged, i.e. $(g + a) + b$ for a gate g and $a, b \in \mathbb{Z}$ can be rewritten $g + (a + b)$. And for c to reach degree 2^n, each $\times$-gate g must reach twice the degree of the nearest $\times$-gate g' having a path to g, i.e. g must perform $(g' + a) \times (g' + b)$ with $a, b \in \mathbb{Z}$. Hence c can be taken to be the circuit $Skew(\alpha_0, \beta_0, \ldots, \alpha_{n-1}, \beta_{n-1}, \alpha_n)$ depicted in Figure 3.

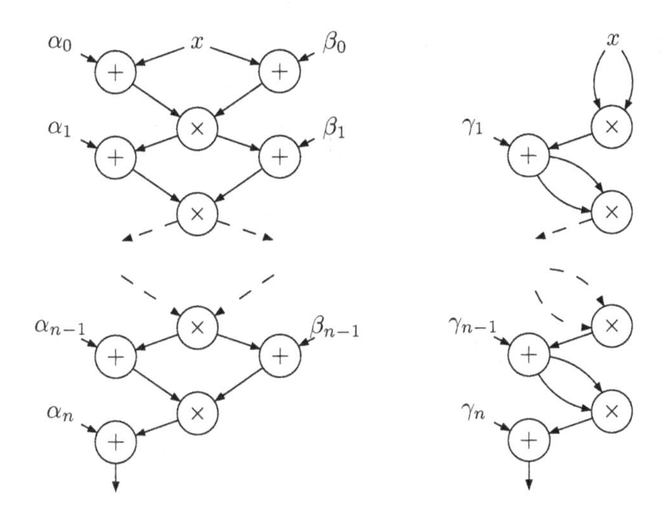

Fig. 3. The skew 2^n-gem $Skew(\alpha_0, \beta_0, \dots, \alpha_{n-1}, \beta_{n-1}, \alpha_n)$ and its normal form

Proposition 3. *Let $\alpha_0, \dots, \alpha_n, \beta_0, \dots, \beta_{n-1}, a_1, \dots, a_{2^n} \in \mathbb{Z}$ be such that $c = Skew(\alpha_0, \beta_0, \dots, \alpha_{n-1}, \beta_{n-1}, \alpha_n)$ computes the polynomial $p(x) = \prod_{i=1}^{2^n}(x - a_i)$. For any $t \in \mathbb{Z} \setminus \{0\}$, $c' = Skew(\alpha_0 t^{2^0}, \beta_0 t^{2^0}, \dots, \alpha_{n-1} t^{2^{n-1}}, \beta_{n-1} t^{2^{n-1}}, \alpha_n t^{2^n})$ computes the polynomial $q(x) = \prod_{i=1}^{2^n}(x - ta_i)$.*

Proof. We claim that c' computes $q(x) = t^{2^n} p(x/t)$. This concludes the proof since then $q(ta_1) = \cdots = q(ta_{2^n}) = 0$. So let $n = 0$. Then c' computes $x + \alpha_0 t^{2^0} = t \cdot (x/t + \alpha_0) = t \cdot p(x/t)$. Now consider $n > 0$ and let q_1, q_2 and p_1, p_2 be computed by the gates input to the lowest $\times$-gate in c' and c respectively. Then $q(x) = q_1(x) \times q_2(x) + \alpha_n t^{2^n} = t^{2^{n-1}} p_1(x/t) \times t^{2^{n-1}} p_2(x/t) + \alpha_n t^{2^n}$ by induction, and the latter equals $t^{2^n}[p_1(x/t) \times p_2(x/t) + \alpha_n] = t^{2^n} p(x/t)$. □

Lemma 2. *(Normal form) Let $n \geq 1$. Given $a_1, \dots, a_{2^n} \in \mathbb{Z}$ and a skew 2^n-gem c such that $f_c(x) = \prod_{i=1}^{2^n}(x - a_i)$, there exist $s \in \mathbb{Z}$ and $t \in \{1, 2\}$ and a normalized 2^n-gem computing the polynomial $\prod_{i=1}^{2^n}(x - ta_i - s)$.*

Proof. Let $c = Skew(\alpha_0, \beta_0, \dots, \alpha_{n-1}, \beta_{n-1}, \alpha_n)$ (see Figure 3). If $\alpha_i + \beta_i$ is odd for some $0 \leq i < n$ then let $t = 2$, else let $t = 1$. By Proposition 3, the skew 2^n-gem $c' = Skew(\alpha_0 t^{2^0}, \beta_0 t^{2^0}, \dots, \alpha_{n-1} t^{2^{n-1}}, \beta_{n-1} t^{2^{n-1}}, \alpha_n t^{2^n})$ computes $q(x) = \prod_{i=1}^{2^n}(x - ta_i)$. We will now normalize c'. First we rewrite each $\times$-gate $(g + a) \times (g + b) = g^2 + (a + b)g + ab$ as $[g + (a + b)/2]^2 + [ab - ((a + b)/2)^2]$, noting that any such $a + b$ occurring in c' is even. Then we merge consecutive $+$-gates since these are skew. The result would be normalized, were it not for an extraneous $x + s$ gate at the input level. We replace $x + s$ with x. This yields a normalized 2^n-gem computing $q(x - s) = \prod_{i=1}^{2^n}(x - ta_i - s)$. □

Definition 2. *(see [BoIn94]) Two sets* $\{a_1, ..., a_m\}, \{b_1, ..., b_m\}$ *solve the* PTE (Prouhet-Tarry-Escott) problem *of degree* k *if* $a_1^i + ... + a_m^i = b_1^i + ... + b_m^i$ *for all* $i \leq k$. *A solution is called* ideal *if* $k = m - 1$. *A solution of the form* $\{a_1, -a_1, \ldots, a_{m/2}, -a_{m/2}\}, \{b_1, -b_1, \ldots, b_{m/2}, -b_{m/2}\}$ *is called* symmetric *and we abbreviate it by* $\{a_1, \ldots, a_{m/2}\}, \{b_1, \ldots, b_{m/2}\}$; *a solution of the form* $\{a_1, \ldots, a_m\}, \{-a_1, \ldots, -a_m\}$ *is also called symmetric.*

Let $p(x) = (x-a_1)(x-a_2)\cdots(x-a_m)$ and $q(x) = (x-b_1)(x-b_2)\cdots(x-b_m)$. Define, for $k = 1, 2, \ldots, m$, $s_k = \sum_{i=1}^{m} a_i^k$ and $t_k = \sum_{i=1}^{m} b_i^k$.

Proposition 4. *([DoBr37], see [BoIn94].) The following are equivalent:*

- $s_1 = t_1$ *and* $s_2 = t_2$ *and* $s_3 = t_3$ *and* $\cdots$ *and* $s_k = t_k$
- *degree*$[p(x) - q(x)] \leq m - (k+1)$.

Corollary 1. *For any* $n \geq 1$ *and any skew* 2^n*-gem* c *such that* $\texttt{izeros}_c = \{a_1, \ldots, a_{2^n}\}$, *there is a partition* $S \uplus T = \{a_1, \ldots, a_{2^n}\}$ *into two equal size sets such that the pair* S, T *is an ideal PTE solution of size* 2^{n-1}.

Proof. We apply Lemma 2 to c and obtain a circuit c' computing a polynomial $(r(x))^2 + \gamma_n = \prod_{i=1}^{2^n}(x - ta_i - s)$ for some $\gamma_n \in \mathbb{Z}$ (Figure 3), $s \in \mathbb{Z}, t \in \{1, 2\}$. Hence $-\gamma_n = e$ for some $e \in \mathbb{N}$. Now $(r(x))^2 - e = p(x)q(x)$, where $p(x) = r(x) + \sqrt{e}$ and $q(x) = r(x) - \sqrt{e}$. Since $\mathbb{Z}[x]$ is a Euclidian ring, $p(x)$ and $q(x)$ must each have 2^{n-1} distinct roots. (So $\sqrt{e} \in \mathbb{N}$.) Now $\deg(p(x) - q(x)) = 0$, so applying Proposition 4 with $k = 2^{n-1} - 1$ shows that $\{a \in \mathbb{Z} : p(a) = 0\}$ and $\{a \in \mathbb{Z} : q(a) = 0\}$ form an ideal PTE solution of size 2^{n-1}. It is well known [BoIn94] that shifting from $ta_i + s$ to ta_i preserves PTE solutions. Further dividing out the ta_i by t to get back to the a_i also preserves PTE solutions. □

A skew 32-gem, if it exists, thus implies an ideal PTE solution of degree 15. Borwein and Ingalls [BoIn94, p. 8] state that "it has been conjectured for a long time that ideal PTE solutions exist for every n", yet the largest such solution known today is due to Shuwen [Shu01] and has degree 11 [We09].

For the converse direction, consider the ideal symmetric PTE solution $\{2, 16, 21, 25\}, \{5, 14, 23, 24\}$ of degree 7 [BoIn94, p. 8]. We could try to unravel the Corollary 1 construction, by expressing $p(x) = (x^2-2^2)(x^2-16^2)(x^2-21^2)(x^2-25^2)$ using 3 products. Here this happens to be possible, by calculating $(x^2-2^2)(x^2-25^2)$ using 2 products, then forming $(x^2-16^2)(x^2-21^2)$ from $(x^2-2^2)(x^2-25^2)$ by repeatedly subtracting x^2 (unavoidable since $2^2+25^2 \neq 16^2+21^2$), and finally obtaining $p(x)$ using one last product. In this special case, we could therefore construct a (non-skew) 16-gem from an ideal symmetric PTE solution.

Figure 2 contains further gems constructed with the help of PTE solutions, such as a 24-gem constructed with the help of a degree-11 solution. But even if PTE solutions of degree 15 were known, these would need to fulfill additional properties in order to yield 32-gems by the strategy described above. We now show that the "sym-perfect" condition below is a sufficient additional condition imposed on ideal symmetric PTE solutions to yield a normalized 2^n gem.

Definition 3. *A pair $S, T \subset \mathbb{Z}$ is called* sym-perfect *if $S = \{a\}$ and $T = \{-a\}$, or if $S = \{a_1, -a_1, \ldots, a_{m/2}, -a_{m/2}\}$, $T = \{b_1, -b_1, \ldots, b_{m/2}, -b_{m/2}\}$ and $\exists s \in \mathbb{Z}$ such that $\{a_1^2 + s, \ldots, a_{m/2}^2 + s\}, \{b_1^2 + s, \ldots, b_{m/2}^2 + s\}$ is sym-perfect.*

If a_1^2 and $a_{m/2}^2$ are respectively smallest and largest in $\{a_1^2, \cdots, a_{m/2}^2\}$, then the shift s occurring in the recursive definition of sym-perfect is necessarily equal to $-(a_1^2 + a_{m/2}^2)/2$ (which is also the average of all the numbers that occur). An example of a sym-perfect pair is $\{-3, 3, -11, 11\}, \{-7, 7, -9, 9\}$, since the pair $\{9 - 65, 121 - 65\}, \{49 - 65, 81 - 65\}$ is sym-perfect by virtue of the pair $\{56^2 - 1696\}, \{16^2 - 1696\}$ in turn being sym-perfect.

Theorem 1. *A set $U \subset \mathbb{Z}$ of size 2^{n+1} can be written as $S \cup T$ for a sym-perfect pair S, T if and only if U is the set of zeros of a normalized 2^{n+1}-gem.*

Proof: The sym-perfect pair $\{a\}, \{-a\}$ corresponds to the 2-gem c with $f_c(x) = x^2 - a^2$ and $\mathtt{izeros}_c = \{a, -a\}$, which forms the induction basis $n = 0$.

Now let $U = \{a_1, -a_1, \ldots, a_{2^n/2}, -a_{2^n/2}\} \cup \{b_1, -b_1, \ldots, b_{2^n/2}, -b_{2^n/2}\}$ where the pair $S' = \{a_1^2 - s, \ldots, a_{2^n/2}^2 - s\}$, $T' = \{b_1^2 - s, \ldots, b_{2^n/2}^2 - s\}$ is sym-perfect. By induction, $S' \cup T'$ is $\mathtt{izeros}_{c'}$ for a 2^n-gem c'. Hence U is $\mathtt{izeros}_c$ for the 2^{n+1}-gem c with $f_c(x) = f_{c'}(x^2 - s)$. Conversely, a normalized 2^{n+1}-gem c with zero set U computes $f_c(x) = g(x)^2 + \gamma_{n+1} = f_{c_S}(x) \times f_{c_T}(x)$ for some normalized 2^n-gems c_S and c_T where γ_i is the same as in c for all $i < n$ (see Figure 3). Then $U = S \cup T$ for their zeros $S = \{a_1, -a_1, \ldots, a_{2^n/2}, -a_{2^n/2}\}$ and $T = \{b_1, -b_1, \ldots, b_{2^n/2}, -b_{2^n/2}\}$. The pair S', T' as above with $s = -\gamma_1$ for γ_1 from the gem c is sym-perfect by induction, which makes S, T sym-perfect. $\square$

Corollary 2. *A sym-perfect pair is an ideal, symmetric PTE solution.*

Proof: We have to show that $a_1^i + (-a_1)^i + \ldots a_{m/2}^i + (-a_{m/2})^i = b_1^i + (-b_1)^i + \ldots b_{m/2}^i + (-b_{m/2})^i$ for all $i < m$. For odd i, neighbours cancel. For $i = 2i'$, this is double the i'-th equation of the PTE $\{a_1^2, \ldots, a_{m/2}^2\}, \{b_1^2, \ldots, b_{m/2}^2\}$ which is just a shift of the sym-perfect PTE from the induction. $\square$

4 The Case of Real Numbers

Exclusively in this section, we extend Definition 1 to the case of real numbers, that is, to $\{+, -, \times\}$-circuits having inputs from $\{x\} \cup \mathbb{R}$ and computing polynomials that factor completely over $\mathbb{R}$. We will bound c_+ from below for skew 2^n-gems c over $\mathbb{R}$ and then construct skew d-gems over $\mathbb{R}$ for every d.

Let A be $\mathbb{Z}$, $\mathbb{Q}$ or $\mathbb{R}$. For short, we will say that a nonzero polynomial $p(x) \in \mathbb{R}[x]$ *crumbles over* A if $\deg(p) = 0$ or if p has $\deg(p)$ distinct roots in A.

Proposition 5. *Let A be $\mathbb{Z}$, $\mathbb{Q}$ or $\mathbb{R}$. Let $p(x) \in \mathbb{R}[x]$ and $q(x) \in \mathbb{R}[x]$. If pq crumbles over A then both p and q crumble over A.*

Fact 2. *(Rolle) Over $\mathbb{R}$, the derivative of a crumbling polynomial crumbles.*

Note that Rolle only applies over $\mathbb{R}$. For example, $(x-1)(x-2)(x-3)$ crumbles over $\mathbb{Z}$ but its derivative $3x^2 - 12x + 11$ crumbles neither over $\mathbb{Z}$ nor over $\mathbb{Q}$. For that reason, the only proof we have of Corollary 3 below is via Theorem 3.

Lemma 3. *Let $e \in \mathbb{R}$ and suppose that a polynomial $p(x) + e \in \mathbb{R}[x]$ of degree 2^n crumbles over $\mathbb{R}$. Then the following holds:*
H1. *Any skew gem over $\mathbb{R}$ for $p(x)$ has at least $n-1$ additive gates.*
H2. *If $e = 0$ then any skew gem over $\mathbb{R}$ for $p(x)$ has at least n additive gates.*

Theorem 3. *A skew 2^n-gem over the reals has at least n additive gates.*

Proof. This follows by applying Lemma 3 with $e = 0$. □

Corollary 3. *Any skew 2^n-gem (over $\mathbb{Z}$) requires n additive gates.*

Proof. Let a skew 2^n-gem c compute $p(x) \in \mathbb{Z}[x]$. If c had fewer than n additive gates, then c as a skew gem over the reals would contradict Theorem 3. □

Rojas [Ro03, p. 4] constructs 2^n-gems over $\mathbb{R}$ for any n. The following variation constructs skew 2^n-gems: $g_1 = x$ and $g_{i+1}(x) := g_i^2(x) - 2$, $1 \leq i < n$, yields $g_n(x)$ having 2^n distinct roots in $[-2, 2]$. We extend this to arbitrary degrees:

Proposition 6. *For all $d > 0$, there exists a skew d-gem over $\mathbb{R}$.*

The construction in the proof of Proposition 6 produces at most ℓ_d additions (exactly ℓ_d when $d = 2^n$). In cases like $d = 3, 7, 9, 27, 81$, it produces $\ell_d/2$ additions. We conjecture that $c_+ \geq \ell_d/2$ for a skew d-gem c over $\mathbb{R}$.

5 Constructing Gems

In this section we construct gems (over $\mathbb{Z}$), at times with the help of a computer. Yet we do not know whether 32-gems or d-gems beyond $d = 55$ exist.

Lemma 4. *For every $d \leq 7$ and $d = 9$, for every d distinct integers $a_1, ..., a_d$, there is a d-gem c_d such that $f_{c_d}(x) = (x - a_1)(x - a_2) \cdots (x - a_d)$.*

Proof. The d-gems are: $f_{c_1}(x) = (x - a_1)$, $f_{c_2}(x) = (x - a_1) \times (x - a_2)$,
$f_{c_3}(x) = f_{c_2}(x) \times (x - a_3)$,
$f_{c_4}(x) = f_{c_2}(x) \times (f_{c_2}(x) + \underbrace{(a_1 + a_2 - a_3 - a_4) \cdot x}_{\text{iterated additions of } x} + (a_3 a_4 - a_1 a_2))$,
$f_{c_5}(x) = f_{c_4}(x) \times (x - a_5)$,
$f_{c_6}(x) = f_{c_3}(x) \times (f_{c_3}(x) + a \cdot f_{c_2}(x) + (a \cdot (a_1 + a_2) - a_1 a_2 - a_1 a_3 - a_2 a_3) \cdot x + (a_4 a_5 a_6 - a_1 a_2 a_3 - a \cdot a_1 a_2))$ with $a = (a_1 + a_2 + a_3 - a_4 - a_5 - a_6)$,
$f_{c_7}(x) = f_{c_6}(x) \times (x - a_7)$,
$f_{c_9}(x) = f_{c_6}(x) \times (f_{c_3}(x) + a \cdot f_{c_2}(x) + (a \cdot (a_1 + a_2) - a_1 a_2 - a_1 a_3 - a_2 a_3) \cdot x + (a_7 a_8 a_9 - a_1 a_2 a_3 - a \cdot a_1 a_2))$ with $a = (a_1 + a_2 + a_3 - a_7 - a_8 - a_9)$. □

The number of additive gates used in proving Lemma 4 can be reduced when the roots satisfy favorable conditions, such as $a_1 = -a_2$ or $a_i = 0$ for some i. Such relationships between the zeros are exploited extensively in Figure 2.

The case $d = 8$ is missing from Lemma 4 because not all polynomials having 8 distinct roots have an 8-gem [Di00]. For any $0 < a < b < c < d$, we can easily construct an 8-gem for $(x+a)(x-a)(x+b)(x-b)(x+c)(x-c)(x+d)(x-d)$ by prepending the 4-gem (available by Lemma 4) for $(y-a^2)(y-b^2)(y-c^2)(y-b^2)$ by "$y \leftarrow x \times x$".

To construct a 16-gem, we seek $0 < a < b < c < d < e < f < g < h$ and an 8-gem for $(y-a^2)(y-b^2)(y-c^2)(y-d^2)(y-e^2)(y-f^2)(y-g^2)(y-h^2)$. We first develop sufficient conditions for the existence of a skew 2^n-gem for any n.

Definition 4. *(Litter conditions.) Let T be the full ordered binary tree with 2^n leaves labeled $a_1, a_2, \ldots, a_{2^n} \in \mathbb{Z}$ in the natural order. Let each internal node in T be labeled with the product of the labels of the leaves subtended by that node. The sequence $a_1, a_2, \ldots, a_{2^n}$* satisfies the litter conditions *if, for $1 < i < n$, each of the 2^i nodes at level i has the same* litter sum, *where the* litter sum *of a node is defined as the sum of the labels of its two children.*

Example 1. The litter conditions are mute for sequences of length 1 or 2. The litter conditions for the sequence a_1, a_2, a_3, a_4 are the equation $a_1 + a_2 = a_3 + a_4$. The litter conditions for the sequence $a^2, b^2, c^2, d^2, e^2, f^2, g^2, h^2$ are

$$a^2 + b^2 = c^2 + d^2 = e^2 + f^2 = g^2 + h^2 \quad \text{and} \quad a^2b^2 + c^2d^2 = e^2f^2 + g^2h^2. \tag{1}$$

Lemma 5. *If the sequence $a_1, a_2, \ldots, a_{2^n} \in \mathbb{Z}$ satisfies the litter conditions, then the following skew 2^n-gem c computes $p(x) = \prod_{1 \le i \le 2^n}(x - a_i)$:*

$$\begin{aligned}
y_0 &\leftarrow x - a_1 \\
y_1 &\leftarrow y_0 \times (y_0 + (a_1 - a_2)) \\
y_2 &\leftarrow y_1 \times (y_1 + (a_3a_4 - a_1a_2)) \\
y_3 &\leftarrow y_2 \times (y_2 + (a_5a_6a_7a_8 - a_1a_2a_3a_4)) \\
&\vdots \qquad\qquad \vdots \\
y_n &\leftarrow y_{n-1} \times (y_{n-1} + (\prod_{i=2^{n-1}+1}^{2^n} a_i - \prod_{i=1}^{2^{n-1}} a_i)).
\end{aligned}$$

□

We return to our quest for a 16-gem. By Lemma 5, any distinct squares $a^2, b^2, c^2, d^2, e^2, f^2, g^2, h^2$ satisfying (1) from Example 1 are the zeros of an 8-gem $p(y)$. In turn, each such 8-gem prepended by "$y \leftarrow x \times x$" is a 16-gem. A small computer in a few hours found several examples, such as:

Proposition 7. *A 16-gem with 4 additive gates exists to compute the polynomial having the 16 roots* $\{\pm 237, \pm 106, \pm 189, \pm 178, \pm 227, \pm 126, \pm 218, \pm 141\}$.

We note that Bremner [Br08] focusses on 16-gems and constructs two infinite families. Turning to 32-gems, we were unable to find a 16-gem having 16 distinct squares as zeros, nor to find a 32-gem by appealing to the litter conditions of a sequence of length 32 directly.

The next lemma is our tool to generate d-gems when $d \neq 2^n$:

Lemma 6. *Let $h(x) \in \mathbb{Z}[x]$ and $m_1, m_2, \ldots, m_d \in \mathbb{Z}$. Suppose that each one of the d polynomials $h(x) - m_i$ is computed by a gem and that no two such polynomials share a root. If $\ell_d + \ell_{\deg(h)} \leq \ell_{d \cdot \deg(h)}$ and for some d-gem c, $f_c(y) = (y - m_1)(y - m_2) \cdots (y - m_d)$, then there is a gem computing $f_c(h(x))$.*

We illustrate the use of Lemma 6 in the following:

Theorem 4. *There exist 36-gems and 54-gems.*

We have filled the rest of Figure 2 largely by trying to combine the mentioned methods along possible shortest addition chains for the degree d. Many cases remain open.

6 Conclusion

The following heuristic for factoring a $2n$-bit integer $N = pq$, for primes p and q of comparable size, is inspired by Lipton [Li94]:

- assume distinct $a_i \in \mathbb{Z}$ and a circuit c computing $f_c(x) = \prod_{i=1}^{2^n} (x - a_i) \in \mathbb{Z}[x]$
- pick $a \in \{0, \ldots, N - 1\}$ at random
- compute $d = f_c(a)$ modulo N by evaluating each gate in c modulo N
- output $\gcd(d, N)$.

This is merely a heuristic because its success probability depends on the distribution of the 2^n integers $(a - a_i)$ modulo N. If this is close to uniform, then indeed $Prob[1 < \gcd(d, N) < N]$ is constant. Of course the heuristic runs in time polynomial in the number of bits required to represent c, and the τ-conjecture [BCSS97] claims that this number is exponential in n.

Here we introduced *gems*. These are circuits that use an almost optimal number of $\times$-gates to compute polynomials that factor completely over $\mathbb{Z}$ with distinct roots. A 2^n-gem could thus serve in the heuristic above if its inputs modulo N can be computed in time polynomial in $n = O(\log N)$, but the L-conjecture [Bu01] claims that even circuits that are much less constrained than d-gems do not exist for large d.

We exhibited d-gems over $\mathbb{R}$ for every d. But the d-gems we care about (over $\mathbb{Z}$) are elusive. In particular, constructing *skew* 32-gems would yield new solutions to the Prouhet-Tarry-Escott problem. These new PTE solutions would even fulfill additional conditions. Yet skew 2^n-gems for any $n \geq 5$ cannot currently be ruled out. This attests to the difficulty of the L-conjecture [Bu01], since skew 2^n-gems would provide the most severe counter-examples imaginable to it.

We constructed d-gems for several d up to $d = 55$. We proved that any skew 2^n-gem requires n additive gates. We showed that for $d \leq 71$, no circuit can compute a degree-d polynomial using fewer than ℓ_d product gates. Hence all the gems constructed so far are $\times$-optimal, and our skew 2^n-gems for $n \leq 4$ are also $\{+,-\}$-optimal among skew gems.

Our main open question is essentially the challenge we started with: do d-gems exist for infinitely many d? A concrete step would be to undertake a search for d-gems of every type for small values of d, extending the systematic approach used by Bremner to study normalized 16-gems [Br08]. This would add entries to Figure 2. Hopefully it could help finding skew 2^n-gems for $n = 5$. As seen above, resolving the existence question for skew 2^n-gems seems like a natural baby step towards resolving the L-conjecture. But even this step would seem to break ground from a number-theoretic perspective.

Acknowledgments. We are grateful to Andrew Granville for insights and for noticing the connection between gems and the Prouhet-Tarry-Escott problem, to Allan Borodin for helpful suggestions and for bringing up Strassen's work, and to Peter Hauck for pointing out the work by Dilcher. We thank Andreas Krebs and Klaus-Jörn Lange, Ken Regan and the referees for their useful comments.

References

[BCSS97] Blum, L., Cucker, F., Shub, M., Smale, S.: Complexity and Real Computation. Springer, Heidelberg (1997)

[BoMo74] Borodin, A., Moenck, B.: Fast modular transforms. Journal of Computer and Systems Science 8(3), 366–386 (1974)

[BoIn94] Borwein, A., Ingalls, C.: The Prouhet-Tarry-Escott Problem Revisited. Enseign. Math. 40, 3–27 (1994)

[Br08] Bremner, A.: When can $(((X^2 - P)^2) - Q)^2 - R)^2 - S^2$ split into linear factors? Experimental Mathematics 17(4), 385–390 (2008)

[Bu01] Bürgisser, P.: On implications between P-NP-hypotheses: Decision versus computation in algebraic complexity. In: Sgall, J., Pultr, A., Kolman, P. (eds.) MFCS 2001. LNCS, vol. 2136, pp. 3–17. Springer, Heidelberg (2001)

[Ch04] Cheng, Q.: Straight Line Programs and Torsion Points on Elliptic Curves. In: Comput. Complex, vol. 12(3-4), pp. 150–161. Birkhauser Verlag, Basel (2004)

[Cr96] Crandall, R.: Topics in advanced scientic computation, TELOS, the Electronic Library of Science. Springer, New York (1996)

[CrPo01] Crandall, R., Pomerance, C.: Primes numbers: a computational perspective. Springer, New York (2001)

[Di00] Dilcher, K.: Nested squares and evaluations of integer products. Experimental Mathematics 9(3), 369–372 (2000)

[DoBr37] Dolwart, H., Brown, O.: The Tarry-Escott problem. Proc. Amer. Math. Soc. 44, 613–626 (1937)

[GaGe03] von zur Gathen, J., Gerhard, J.: Modern Computer Algebra, 2nd edn. Cambridge University Press, Cambridge (2003)

[Kn81] Knuth, D.: The art of computer programming, 2nd edn. Seminumerical algorithms, vol. 2. Addison-Wesley, Reading (1969) (1981)

[Li94] Lipton, R.: Straight-line complexity and integer factorization. In: Huang, M.-D.A., Adleman, L.M. (eds.) ANTS 1994. LNCS, vol. 877, pp. 71–79. Springer, Heidelberg (1994)

[PaSt73] Paterson, M., Stockmeyer, L.: On the number of nonscalar multiplications necessary to evaluate polynomials. SIAM J. Computing 2(1), 60–66 (1973)

[Ro03] Rojas, M.: A Direct Ultrametric Approach to Additive Complexity and the Shub-Smale Tau Conjecture (2003), http://arxiv.org/abs/math/0304100

[Ro93] Rosen, K.: Elementary number theory and its applications, 3rd edn. Addison-Wesley, Reading (1993)

[Shu01] Shuwen, C.: The PTE Problem, http://euler.free.fr/eslp/TarryPrb.htm

[Sm00] Smale, S.: Mathematical problems for the next century. In: Arnold, V., Atiyah, M., Lax, P., Mazur, B. (eds.) Mathematics: Frontiers and Perspectives 2000. AMS, Providence (2000)

[St76] Strassen, V.: Einige Resultate über Berechnungskomplexität. Jahresberichte der DMV 78, 1–8 (1976)

[Usp48] Uspensky, J.V.: Theory of Equations. McGraw-Hill, New York (1948)

[We09] Weisstein, E.: The Prouhet-Tarry-Escott Problem, MathWorld–Wolfram (2009), http://mathworld.wolfram.com/Prouhet-Tarry-EscottProblem.html

[WIMS] http://wims.unice.fr/wims/en_tool~number~twosquares.en.html (1999)

Branching Programs for Tree Evaluation

Mark Braverman[1], Stephen Cook[2], Pierre McKenzie[3], Rahul Santhanam[4], and Dustin Wehr[2]

[1] Microsoft Research
[2] University of Toronto
[3] Université de Montréal
[4] University of Edinburgh

Abstract. The problem $FT_d^h(k)$ consists in computing the value in $[k] = \{1, \ldots, k\}$ taken by the root of a balanced d-ary tree of height h whose internal nodes are labelled with d-ary functions on $[k]$ and whose leaves are labelled with elements of $[k]$. We propose $FT_d^h(k)$ as a good candidate for witnessing $\mathbf{L} \subsetneq \mathbf{LogDCFL}$. We observe that the latter would follow from a proof that k-way branching programs solving $FT_d^h(k)$ require $\Omega(k^{\text{unbounded function}(h)})$ size. We introduce a "state sequence" method that can match the size lower bounds on $FT_d^h(k)$ obtained by the Nečiporuk method and can yield slightly better (yet still subquadratic) bounds for some nonboolean functions. Both methods yield the tight bounds $\Theta(k^3)$ and $\Theta(k^{5/2})$ for deterministic and nondeterministic branching programs solving $FT_2^3(k)$ respectively. We propose as a challenge to break the quadratic barrier inherent in the Nečiporuk method by adapting the state sequence method to handle $FT_d^4(k)$.

1 Introduction

Let T_d^h be the balanced d-ary ordered tree T_d^h of height h, where we take *height* to mean the number of levels in the tree and we number the nodes as suggested by the heap data structure. Thus the root is node 1, and in general the children of node i are (when $d = 2$) nodes $2i, 2i + 1$ (see Figure 1). For every $d, h, k \geq 2$ we define the *Tree Evaluation* problem and its associated decision problem:

Definition 1 ($FT_d^h(k)$ **and** $BT_d^h(k)$)
Given: T_d^h *with each non-leaf node* i *independently labelled with a function* $f_i : [k]^d \to [k]$ *and each leaf node independently labelled with an element from* $[k]$.

Function evaluation problem $FT_d^h(k)$: *Compute the value* $v_1 \in [k]$ *of the root* 1 *of* T_d^h, *where in general* $v_i = a$ *if* i *is a leaf labelled* a, *and* $v_i = f_i(v_{j_1}, \ldots, v_{j_d})$ *if* $j_1, \ldots, j_d$ *are the children of* i.

Boolean evaluation problem $BT_d^h(k)$: *Decide whether* $v_1 = 1$.

In the context of uniform complexity measures such as Turing machine space we rewrite $FT_d^h(k)$ and $BT_d^h(k)$ as $FT_d(h, k)$ and $BT_d(h, k)$ to indicate that d is fixed but h, k are input parameters. It is not hard to show that for each $d \geq 2$ a deterministic logspace-bounded poly-time auxiliary pushdown automaton solves

R. Královič and D. Niwiński (Eds.): MFCS 2009, LNCS 5734, pp. 175–186, 2009.

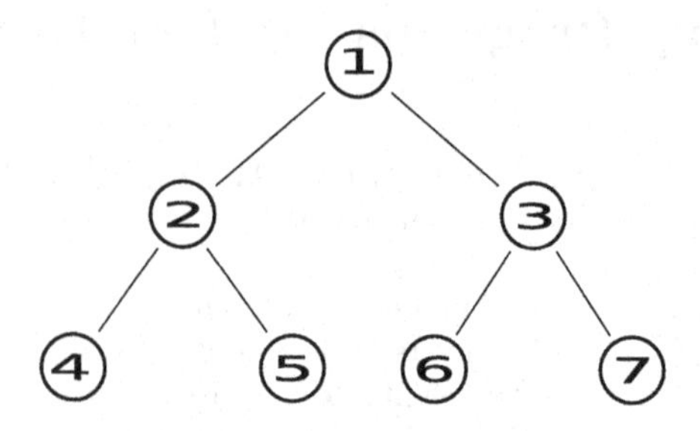

Fig. 1. A height 3 binary tree T_2^3 with nodes numbered heap style

$BT_d(h,k)$, implying by [Sud78] that $BT_d(h,k)$ belongs to the class **LogDCFL** of languages logspace reducible to a deterministic context-free language. We know $\mathbf{L} \subseteq \mathbf{LogDCFL} \subseteq \mathbf{P}$ (see [Mah07] for up to date information on **LogDCFL**). The special case $BT_d(h,2)$ was investigated under a different name in [KRW95] as part of an attempt to separate $\mathbf{NC}^1$ from $\mathbf{NC}^2$. In this paper, we suggest investigating the space complexity of $BT_d(h,k)$ and $FT_d(h,k)$.

We choose to study the Tree Evaluation problem as a particularly interesting candidate for non-membership in **L** or **NL** (deterministic or nondeterministic log space) because pebble games on trees provide natural space bounded algorithms for solving it: Black pebbling provides deterministic algorithms and, though we do not consider these in this paper, black-white pebbling provides nondeterministic algorithms. We choose k-way branching programs (BPs) as our model of Turing machine because the inputs to our problems are tuples of numbers in $[k]$.

For fixed d, h we are interested in how the size (number of states) of BPs solving $FT_d^h(k)$ and $BT_d^h(k)$ grows with k. One of our contributions is an alternative approach to Nečiporuk's lower bound method [Neč66] for this size. Applied to the problem $BT_d^h(k)$, our "state sequence" approach does as well as (but, so far, no better than) Nečiporuk's method. On the other hand, our approach does not suffer in principle from the quadratic limitation inherent in Nečiporuk's method. Hence there is hope that the approach can be extended. The current bottleneck stands at height 4. Proving our conjectured lower bound of $\Omega(k^7/\lg k)$ (writing lg for $\log_2$) for the size of deterministic BPs solving $BT_3^4(k)$ would constitute a breakthrough and would overcome the n^2 Nečiporuk limitation. However we do not yet know how to do this.

The more specific contributions of this paper are the following:

- we observe that for any $d \geq 2$ and unbounded $r(h)$, a lower bound of the form $\Omega(k^{r(h)})$ on the size of BPs solving $FT_d^h(k)$ would prove $BT_d(h,k) \notin \mathbf{L}$;
- we prove tight black pebbling bounds for T_d^h and transfer the upper bounds to size upper bounds of the form $k^{O(h)}$ for deterministic k-way BPs for $FT_d^h(k)$ and $BT_d^h(k)$;
- we prove tight size bounds of $\Theta(k^{2d-1})$ and $\Theta(k^{2d-1}/\lg k)$ for deterministic k-way BPs solving $FT_d^3(k)$ and $BT_d^3(k)$ respectively;
- we prove tight size bounds of $\Theta(k^{3d/2-1/2})$ for nondeterministic k-way BPs solving $BT_d^3(k)$; the argument yields an $\Omega(n^{3/2}/(\lg n)^{3/2})$ bound for the

number of *states* in nondeterministic binary BPs of arbitrary outdegree when n is the input length; a comparable bound of $\Omega(n^{3/2})$ was known but the latter only applied to the number of *edges* [Pud87, Raz91] in such BPs;
- we give examples of functions, such as the restriction $SumMod_2^3(k)$ of $FT_2^3(k)$ in which the root function is fixed to the sum modulo k, and the function $Children_2^4(k)$ which is required to simultaneously compute the root values of two instances of $FT_2^3(k)$, for which the state sequence method yields a better k-way BP size lower bound than a direct application of Nečiporuk's method ($\Omega(k^3)$ versus $\Omega(k^2)$ for $SumMod_2^3(k)$, and $\Omega(k^4)$ versus $\Omega(k^3)$ for $Children_2^4(k)$).

Section 2 defines branching programs and pebbling. Section 3 relates pebbling and branching programs to Turing machine space, and proves the pebbling bounds exploited in Section 4 to prove BP size upper bounds. BP lower bounds obtained using the Nečiporuk method are stated in Subsection 4.1. Our state sequence method is introduced in Subsection 4.2. The proofs left out of this abstract will appear in the full version of the paper.

2 Preliminaries

We assume some familiarity with complexity theory, such as can be found in [Gol08]. We write $[k]$ for $\{1, 2, \ldots, k\}$ and let $k \geq 2$.

Warning: Recall that the *height* of a tree is the number of levels in the tree, as opposed to the distance from root to leaf. Thus T_2^2 has just 3 nodes.

2.1 Branching Programs

Many variants of the branching program model have been studied [Raz91, Weg00]. Our definition below is inspired by Wegener [Weg00, p. 239], by the k-way branching program of Borodin and Cook [BC82] and by its nondeterministic variant [BRS93, GKM08]. We depart from the latter however in two ways: nondeterministic branching program labels are attached to states rather than edges (because we think of branching program states as Turing machine configurations) and cycles in branching programs are allowed (because our lower bounds apply to this more powerful model).

Definition 2 (Branching programs). *A* nondeterministic k-way branching *program B computing a total function $g : [k]^m \to R$, where R is a finite set, is a directed rooted multi-graph whose nodes are called* states. *Every edge has a label from $[k]$. Every state has a label from $[m]$, except $|R|$* final *sink states consecutively labelled with the elements from R. An input $(x_1, \ldots, x_m) \in [k]^m$ activates, for each $1 \leq j \leq m$, every edge labelled x_j out of every state labelled j. A* computation *on input $\boldsymbol{x} = (x_1, \ldots, x_m) \in [k]^m$ is a directed path consisting of edges activated by $\boldsymbol{x}$ which begins with the unique start state (the root), and either it is infinite, or it ends in the final state labelled $g(x_1, \ldots, x_m)$, or it ends in a nonfinal state labelled j with no outedge labelled x_j (in which case we say the* *computation* aborts*). At least one such computation must end in a final state. The* size *of B is its number of states. B is* deterministic k-way *if every non-final state has precisely k outedges labelled* $1, \ldots, k$. *B is* binary *if* $k = 2$.

We say that B solves a decision problem (relation) if it computes the characteristic function of the relation.

A k-way branching program computing the function $FT_d^h(k)$ requires k^d k-ary arguments for each internal node i of T_d^h in order to specify the function f_i, together with one k-ary argument for each leaf. Thus in the notation of Definition 1 $FT_d^h(k)$: $[k]^m \to R$ where $R = [k]$ and $m = \frac{d^{h-1}-1}{d-1} \cdot k^d + d^{h-1}$. Also $BT_d^h(k)$: $[k]^m \to \{0, 1\}$.

We define $\#\mathsf{detFstates}_d^h(k)$ (resp. $\#\mathsf{ndetFstates}_d^h(k)$) to be the minimum number of states required for a deterministic (resp. nondeterministic) k-way branching program to solve $FT_d^h(k)$. Similarly, $\#\mathsf{detBstates}_d^h(k)$ and $\#\mathsf{ndetBstates}_d^h(k)$ denote the number of states for solving $BT_d^h(k)$.

The next lemma is easy to prove and shows that the function problem is not much harder to solve than the Boolean problem.

Lemma 3. $\#\mathsf{detBstates}_d^h(k) \le \#\mathsf{detFstates}_d^h(k) \le k \cdot \#\mathsf{detBstates}_d^h(k)$ *and* $\#\mathsf{ndetBstates}_d^h(k) \le \#\mathsf{ndetFstates}_d^h(k) \le k \cdot \#\mathsf{ndetBstates}_d^h(k)$.

2.2 Pebbling

The pebbling game for dags was defined by Paterson and Hewitt [PH70] and was used as an abstraction for deterministic Turing machine space in [Coo74]. Black-white pebbling was introduced in [CS76] as an abstraction of nondeterministic Turing machine space (see [Nor09] for a recent survey).

We will only make use of a simple 'black pebbling' game in this paper. Here a pebble can be placed on any leaf node, and in general if all children of a node i have pebbles, then one of the pebbles on the children can be moved to i (this is a "sliding" move). The goal is to pebble the root. A *pebbling* of a tree T using p pebbles is any sequence of pebbling moves on nodes of T which starts and ends with no pebbles, and at some point the root is pebbled, and no configuration has more than p pebbles.

We allow "sliding moves" as above (as opposed to placing a new pebble on node i) because we want pebbling algorithms for trees to closely correspond to k-way branching program algorithms for the tree evaluation problem.

We use $\#\mathsf{pebbles}(T)$ to denote the minimum number of pebbles required to pebble T. The following result is proved easily using standard techniques.

Theorem 4. *For every* $d, h \ge 2$, $\#\mathsf{pebbles}(T_d^h) = (d-1)h - d + 2$.

3 Connecting TMs, BPs, and Pebbling

Let $FT_d(h, k)$ be the same as $FT_d^h(k)$ except now the inputs vary with both h and k, and we assume the input to $FT_d(h, k)$ is a binary string X which codes

h and k and codes each node function f_i for the tree T_d^h by a sequence of k^d binary numbers and each leaf value by a binary number in $[k]$, so X has length

$$|X| = \Theta(d^h k^d \lg k) \tag{1}$$

The output is a binary number in $[k]$ giving the value of the root. The problem $BT_d(h,k)$ is the Boolean version of $FT_d(h,k)$: The input is the same, and the instance is true iff the value of the root is 1.

Obviously $BT_d(h,k)$ and $FT_d(h,k)$ can be solved in polynomial time, but we can prove a stronger result.

Theorem 5. *For each $d \geq 2$ the problem $BT_d(h,k)$ is in* **LogDCFL**.

The best known upper bounds on the number of states required by a BP to solve $FT_d^h(k)$ grow as $k^{\Omega(h)}$. The next result shows (Corollary 7) that any provable nontrivial dependency on h, for the power of k expressing the minimum number of such states, would separate **L**, and perhaps **NL** (deterministic and nondeterministic log space), from **LogDCFL**.

Theorem 6. *For each $d \geq 2$, if $BT_d(h,k)$ is in* **L** *(resp.* **NL***) then there is a constant ω_d and a function $c_d(h)$ such that $\#\mathsf{detFstates}_d^h(k) \leq c_d(h)k^{\omega_d}$ (resp. $\#\mathsf{ndetFstates}_d^h(k) \leq c_d(h)k^{\omega_d}$) for all $h, k \geq 2$.*

Proof. By Lemma 3, arguing for $\#\mathsf{detBstates}_d^h(k)$ and $\#\mathsf{ndetBstates}_d^h(k)$ instead of $\#\mathsf{detFstates}_d^h(k)$ and $\#\mathsf{ndetFstates}_d^h(k)$ suffices. In general a Turing machine which can enter at most C different configurations on all inputs of a given length n can be simulated (for inputs of length n) by a binary (and hence k-ary) branching program with C states. Each Turing machine using space $O(\lg n)$ has at most n^c possible configurations on any input of length $n \geq 2$, for some constant c. By (1) the input for $BT_d(h,k)$ has length $n = \Theta(d^h k^d \lg k)$, so there are at most $(d^h k^d \lg k)^{c'}$ possible configurations for a log space Turing machine solving $BT_d(h,k)$, for some constant c'. So we can take $c_d(h) = d^{c'h}$ and $\omega_d = c'(d+1)$. □

Corollary 7. *Fix $d \geq 2$ and any unbounded function $r(h)$. If $\#\mathsf{detFstates}_d^h(k)$ (resp. $\#\mathsf{ndetFstates}_d^h(k)$) $\in \Omega(k^{r(h)})$ then $BT_d(h,k) \notin$* **L** *(resp. $\notin$* **NL***).*

The next result connects pebbling upper bounds with BP upper bounds.

Theorem 8. *If T_d^h can be pebbled with p pebbles, then deterministic branching programs with $O(k^p)$ states can solve $FT_d^h(k)$ and $BT_d^h(k)$.*

Corollary 9. $\#\mathsf{detFstates}_d^h(k) = O(k^{\#\mathsf{pebbles}(T_d^h)})$.

4 Branching Program Bounds

In this section we prove upper bounds for the number of states required for deterministic k-way branching programs to solve both the function problems $FT_d^h(k)$ and the Boolean problems $BT_d^h(k)$. The upper bounds for the function problems come from pebbling via Theorem 8. We prove that these bounds are tight for all trees of heights 2 and 3, and it seems plausible that they might be tight for all trees T_d^h: a proof would separate **L** from **LogDCFL** (see Section 3).

For the Boolean problems the deterministic upper bounds can be improved by a factor of $\lg k$ over the function problems for height $h \geq 3$ (the improvement is tight for height 3).

We also prove tight bounds for nondeterministic BPs solving the Boolean problems for heights 2 and 3.

Theorem 10 (BP Upper Bounds)

$$\#\mathsf{detFstates}_d^h(k) = O(k^{(d-1)h-d+2}) \tag{2}$$
$$\#\mathsf{detBstates}_d^h(k) = O(k^{(d-1)h-d+2}/\lg k), \ \textit{for } h \geq 3 \tag{3}$$
$$\#\mathsf{ndetBstates}_2^3(k) = O(k^{5/2}) \tag{4}$$

We can combine the above upper bounds with the Nečiporuk lower bounds in Subsection 4.1, Figure 2, to obtain the tight bounds in the next theorem for height 3 trees. (The optimal bounds for all cases for height 2 trees are given by the size of the input: $\Theta(k^d)$.)

Corollary 11 (Height 3 trees)

$$\#\mathsf{detFstates}_d^3(k) = \Theta(k^{2d-1})$$
$$\#\mathsf{detBstates}_d^3(k) = \Theta(k^{2d-1}/\lg k)$$
$$\#\mathsf{ndetBstates}_2^3(k) = \Theta(k^{5/2})$$

4.1 The Nečiporuk Method

The Nečiporuk method still yields the strongest explicit binary branching program size lower bounds known today, namely $\Omega(\frac{n^2}{(\lg n)^2})$ for deterministic [Neč66] and $\Omega(\frac{n^{3/2}}{\lg n})$ for nondeterministic (albeit for a weaker nondeterministic model in which states have bounded outdegree [Pud87], see [Raz91]).

By *applying the Nečiporuk method* to a k-way branching program B computing a function $f : [k]^m \to R$, we mean the following well known steps [Neč66]:

1. Upper bound the number $N(s, v)$ of (syntactically) distinct branching programs of type B having s non-final states, each labelled by one of v variables.
2. Pick a partition $\{V_1, \ldots, V_p\}$ of $[m]$.
3. For $1 \leq i \leq p$, lower bound the number $r_{V_i}(f)$ of restrictions $f_{V_i} : [k]^{|V_i|} \to R$ of f obtainable by fixing values of the variables in $[m] \setminus V_i$.
4. Then $\text{size}(B) \geq |R| + \sum_{1 \leq i \leq p} s_i$, where $s_i = \min\{\ s : N(s, |V_i|) \geq r_{V_i}(f)\ \}$.

Theorem 12. *Applying the Nečiporuk method yields Figure 2.*

Model	Lower bound for $FT_d^h(k)$	Lower bound for $BT_d^h(k)$
Deterministic k-way branching program	$\boxed{\frac{d^{h-2}-1}{4(d-1)^2} \cdot k^{2d-1}}$	$\boxed{\frac{d^{h-2}-1}{3(d-1)^2} \cdot \frac{k^{2d-1}}{\lg k}}$
Deterministic binary branching program	$\frac{d^{h-2}-1}{5(d-1)^2} \cdot k^{2d} = \Omega(n^2/(\lg n)^2)$	$\frac{d^{h-2}-1}{4d(d-1)} \cdot \frac{k^{2d}}{\lg k} = \Omega(n^2/(\lg n)^3)$
Nondeterministic k-way BP	$\frac{d^{h-2}-1}{2d-2} \cdot k^{\frac{3d}{2}-\frac{1}{2}}\sqrt{\lg k}$	$\boxed{\frac{d^{h-2}-1}{2d-2} \cdot k^{\frac{3d}{2}-\frac{1}{2}}}$
Nondeterministic binary BP	$\frac{d^{h-2}-1}{2d-2} \cdot k^{\frac{3d}{2}}\sqrt{\lg k} = \Omega(n^{3/2}/\lg n)$	$\frac{d^{h-2}-1}{2d-2} \cdot k^{\frac{3d}{2}} = \Omega(n^{3/2}/(\lg n)^{3/2})$

Fig. 2. Size bounds, expressed in terms of $n = \Theta(k^d \lg k)$ in the binary cases, obtained by applying the Nečiporuk method. Rectangles indicate optimality in k when $h = 3$ (Cor. 11). Improving any entry to $\Omega(k^{\text{unbounded } f(h)})$ would prove $\mathbf{L} \subsetneq \mathbf{P}$ (Cor. 7).

Remark 1. The $\Omega(n^{3/2}/(\lg n)^{3/2})$ binary nondeterministic BP lower bound for the $BT_d^h(k)$ problem and in particular for $BT_2^3(k)$ applies to the number of *states* when these can have arbitrary outdegree. This seems to improve on the best known former bound of $\Omega(n^{3/2}/\lg n)$, slightly larger but obtained for the weaker model in which states have bounded degree, or equivalently, for the switching and rectifier network model in which size is defined as the number of edges [Pud87, Raz91].

Let $Children_d^h(k)$ have the same input as $FT_d^h(k)$ with the exception that the root function is deleted. The output is the tuple $(v_2, v_3, \ldots, v_{d+1})$ of values for the children of the root.

Theorem 13. *For any $d, h \geq 2$, the best k-way deterministic BP size lower bound attainable for $Children_d^h(k)$ by applying the Nečiporuk method is $\Omega(k^{2d-1})$.*

Proof. The function $Children_d^h(k) : [k]^m \to R$ has $m = \Theta(k^d)$. Any partition $\{V_1, \ldots, V_p\}$ of the set of k-ary input variables thus has $p = O(k^d)$. Claim: for each i, the best attainable lower bound on the number of states querying variables from V_i is $O(k^{d-1})$.

Consider such a set V_i, $|V_i| = v \geq 1$. Here $|R| = k^d$, so the number $N_{\text{det}}^{k\text{-way}}(s, v)$ of distinct deterministic BPs having s non-final states querying variables from V_i satisfies

$$N_{\text{det}}^{k\text{-way}}(s, v) \geq 1^s \cdot (s + |R|)^{sk} \geq (1 + k^d)^{sk} \geq k^{dsk}.$$

Hence the estimate used in the Nečiporuk method to upper bound $N_{\text{det}}^{k\text{-way}}(s, v)$ will be at least k^{dsk}. On the other hand, the number of functions $f_{V_i} : [k]^v \to R$ obtained by fixing variables outside of V_i cannot exceed $k^{O(k^d)}$ since the number of variables outside V_i is $\Theta(k^d)$. Hence the best lower bound on the number of states querying variables from V_i obtained by applying the method will be no larger than the smallest s verifying $k^{ck^d} \leq k^{dsk}$ for some c depending on d and k. This proves our claim since then this number is at most $s = O(k^{d-1})$. $\square$

Let $SumMod_d^h(k)$ have the same input as $FT_d^h(k)$ with the exception that the root function is preset to the sum modulo k. In other words the output is $v_2 + v_3 + \cdots + v_{d+1} \mod k$.

Theorem 14. *The best k-way deterministic BP size lower bound attainable for $SumMod_2^3(k)$ by applying the Nečiporuk method is $\Omega(k^2)$.*

4.2 The State Sequence Method

Here we give alternative proofs for some of the lower bounds given in Section 4.1. These proofs are more intricate than the Nečiporuk proofs but they do not suffer a priori from a quadratic limitation. The method also yields stronger lower bounds to $Children_2^4(k)$ and $SumMod_2^3(k)$ than those obtained by applying the Nečiporuk's method as expressed in Subsection 4.1 (see Theorems 13 and 14).

Theorem 15. $\#\mathsf{ndetBstates}_2^3(k) \geq k^{2.5}$ *for sufficiently large k.*

Proof. Consider an input I to $BT_2^3(k)$. We number the nodes in T_2^3 as in Figure 1, and let v_j^I denote the value of node j under input I. We say that a state in a computation on input I *learns* v_j^I if that state queries $f_j^I(v_{2j}^I, v_{2j+1}^I)$ (recall $2j, 2j+1$ are the children of node j).

Definition [Learning Interval]. *Let B be a k-way nondeterministic BP that solves $BT_2^3(k)$. Let $\mathcal{C} = \gamma_0, \gamma_1, \cdots, \gamma_T$ be a computation of B on input I. We say that a state γ_i in the computation is* critical *if one or more of the following holds:*

1. *$i = 0$ or $i = T$.*
2. *γ_i learns v_2^I and there is an earlier state which learns v_3^I with no intervening state that learns v_2^I.*
3. *γ_i learns v_3^I and no earlier state learns v_3^I unless an intervening state learns v_2^I.*

We say that a subsequence $\gamma_i, \gamma_{i+1}, \cdots \gamma_j$ is a learning interval *if γ_i and γ_j are consecutive critical states. The interval is* type 3 *if γ_i learns v_3^I, and otherwise the interval is* type 2.

Thus type 2 learning intervals begin with γ_0 or a state which learns v_2^I, and never learn v_3^I until the last state, and type 3 learning intervals begin with a state which learns v_3^I and never learn v_2^I until the last state.

Now let B be as above, and for $j \in \{2, 3\}$ let Γ_j be the set of all states of B which query the input function f_j. We will prove the theorem by showing that for large k

$$|\Gamma_2| + |\Gamma_3| > k^2\sqrt{k}. \tag{5}$$

For $r, s \in [k]$ let $F_{yes}^{r,s}$ be the set of inputs I to B whose four leaves are labelled r, s, r, s respectively, whose middle node functions f_2^I and f_3^I are identically 0 except $f_2^I(r, s) = v_2^I$ and $f_3^I(r, s) = v_3^I$, and $f_1^I(v_2^I, v_3^I) = 1$ (so $v_1^I = 1$). Thus each such I is a 'YES input', and should be accepted by B.

Note that each member I of $F_{yes}^{r,s}$ is uniquely specified by a triple

$$(v_2^I, v_3^I, f_1^I) \text{ where } f_1^I(v_2^I, v_3^I) = 1 \quad (6)$$

and hence $F_{yes}^{r,s}$ has exactly $k^2(2^{k^2-1})$ members.

For $j \in \{2,3\}$ and $r, s \in [k]$ let $\Gamma_j^{r,s}$ be the subset of Γ_j consisting of those states which query $f_j(r,s)$. Then Γ_j is the disjoint union of $\Gamma_j^{r,s}$ over all pairs (r,s) in $[k] \times [k]$. Hence to prove (5) it suffices to show

$$|\Gamma_2^{r,s}| + |\Gamma_3^{r,s}| > \sqrt{k} \quad (7)$$

for large k and all r, s in $[k]$. We will show this by showing

$$(|\Gamma_2^{r,s}| + 1)(|\Gamma_3^{r,s}| + 1) \geq k/2 \quad (8)$$

for all $k \geq 2$. (Note that given the product, the sum is minimized when the summands are equal.)

For each input I in $F_{yes}^{r,s}$ we associate a fixed accepting computation $\mathcal{C}(I)$ of B on input I.

Now fix $r, s \in [k]$. For $a, b \in [k]$ and $f : [k] \times [k] \to \{0,1\}$ with $f(a,b) = 1$ we use (a,b,f) to denote the input I in $F_{yes}^{r,s}$ it represents as in (6).

To prove (8), the idea is that if it is false, then as I varies through all inputs (a,b,f) in $F_{yes}^{r,s}$ there are too few states learning $v_2^I = a$ and $v_3^I = b$ to verify that $f(a,b) = 1$. Specifically, we can find a, b, f, g such that $f(a,b) = 1$ and $g(a,b) = 0$, and by cutting and pasting the accepting computation $\mathcal{C}(a,b,f)$ with accepting computations of the form $\mathcal{C}(a,b',g)$ and $\mathcal{C}(a',b,g)$ we can construct an accepting computation of the 'NO input' (a,b,g).

We may assume that the branching program B has a unique initial state γ_0 and a unique accepting state δ_{ACC}.

For $j \in \{2,3\}$, $a, b \in [k]$ and $f : [k] \times [k] \to \{0,1\}$ with $f(a,b) = 1$ define $\varphi_j(a,b,f)$ to be the set of all state pairs (γ, δ) such that there is a type j learning interval in $\mathcal{C}(a,b,f)$ which begins with γ and ends with δ. Note that if $j = 2$ then $\gamma \in (\Gamma_2^{r,s} \cup \{\gamma_0\})$ and $\delta \in (\Gamma_3^{r,s} \cup \{\delta_{ACC}\})$, and if $j = 3$ then $\gamma \in \Gamma_3^{r,s}$ and $\delta \in (\Gamma_2^{r,s} \cup \{\delta_{ACC}\})$

To complete the definition, define $\varphi_j(a,b,f) = \varnothing$ if $f(a,b) = 0$.

For $j \in \{2,3\}$ and $f : [k] \times [k] \to \{0,1\}$ we define a function $\varphi_j[f]$ from $[k]$ to sets of state pairs as follows:

$$\varphi_2[f](a) = \bigcup_{b \in [k]} \varphi_2(a,b,f) \ \subseteq S_2$$

$$\varphi_3[f](b) = \bigcup_{a \in [k]} \varphi_3(a,b,f) \ \subseteq S_3$$

where $S_2 = (\Gamma_2^{r,s} \cup \{\gamma_0\}) \times (\Gamma_3^{r,s} \cup \{\delta_{ACC}\})$ and $S_3 = \Gamma_3^{r,s} \times (\Gamma_2^{r,s} \cup \{\delta_{ACC}\})$.

For each f the function $\varphi_j[f]$ can be specified by listing a k-tuple of subsets of S_j, and hence there are at most $2^{k|S_j|}$ distinct such functions as f ranges over the

2^{k^2} Boolean functions on $[k] \times [k]$, and hence there are at most $2^{k(|S_2|+|S_3|)}$ pairs of functions $(\varphi_2[f], \varphi_3[f])$. If we assume that (8) is false, we have $|S_2| + |S_3| < k$. Hence by the pigeonhole principle there must exist distinct Boolean functions f, g such that $\varphi_2[f] = \varphi_2[g]$ and $\varphi_3[f] = \varphi_3[g]$.

Since f and g are distinct we may assume that there exist a, b such that $f(a,b) = 1$ and $g(a,b) = 0$. Since $\varphi_2[f](a) = \varphi_2[g](a)$, if (γ, δ) are the endpoints of a type 2 learning interval in $\mathcal{C}(a,b,f)$ there exists b' such that (γ, δ) are the endpoints of a type 2 learning interval in $\mathcal{C}(a,b',g)$ (and hence $g(a,b') = 1$). Similarly, if (γ, δ) are endpoints of a type 3 learning interval in $\mathcal{C}(a,b,f)$ there exists a' such that (γ, δ) are the endpoints of a type 3 learning interval in $\mathcal{C}(a',b,f)$.

Now we can construct an accepting computation for the 'NO input' (a,b,g) from $\mathcal{C}(a,b,f)$ by replacing each learning interval beginning with some γ and ending with some δ by the corresponding learning interval in $\mathcal{C}(a,b',g)$ or $\mathcal{C}(a',b,g)$. (The new accepting computation has the same sequence of critical states as $\mathcal{C}(a,b,f)$.) This works because a type 2 learning interval never queries v_3 and a type 3 learning interval never queries v_2.

This completes the proof of (8) and the theorem. □

Theorem 16. *Every deterministic branching program that solves $BT_2^3(k)$ has at least $k^3/\lg k$ states for sufficiently large k.*

Proof. We modify the proof of Theorem 15. Let B be a deterministic BP which solves $BT_2^3(k)$, and for $j \in \{2,3\}$ let Γ_j be the set of states in B which query f_j (as before). It suffices to show that for sufficiently large k

$$|\Gamma_2| + |\Gamma_3| \geq k^3/\lg k. \tag{9}$$

For $r, s \in [k]$ we define the set $F^{r,s}$ to be the same as $F^{r,s}_{yes}$ except that we remove the restriction on f_1^I. Hence there are exactly $k^2 2^{k^2}$ inputs in $F^{r,s}$.

As before, for $j \in \{2,3\}$, Γ_j is the disjoint union of $\Gamma^{r,s}$ for $r, s \in [k]$. Thus to prove (9) it suffices to show that for sufficiently large k and all r, s in $[k]$

$$|\Gamma_2^{r,s}| + |\Gamma_3^{r,s}| \geq k/\lg_2 k. \tag{10}$$

We may assume there are unique start, accepting, and rejecting states γ_0, δ_{ACC}, δ_{REJ}. Fix $r, s \in [k]$.

For each root function $f : [k] \times [k] \to \{0,1\}$ we define the functions

$$\psi_2[f] : [k] \times (\Gamma_2^{r,s} \cup \{\gamma_0\}) \to (\Gamma_3^{r,s} \cup \{\delta_{ACC}, \delta_{REJ}\})$$
$$\psi_3[f] : [k] \times \Gamma_3^{r,s} \to (\Gamma_2^{r,s} \cup \{\delta_{ACC}, \delta_{REJ}\})$$

by $\psi_2[f](a,\gamma) = \delta$ if δ is the next critical state after γ in a computation with input (a,b,f) (this is independent of b), or $\delta = \delta_{REJ}$ if there is no such critical state. Similarly $\psi_3[f](b,\delta) = \gamma$ if γ is the next critical state after δ in a computation with input (a,b,f) (this is independent of a), or $\delta = \delta_{REJ}$ if there is no such critical state.

CLAIM: The pair of functions $(\psi_2[f], \psi_3[f])$ is distinct for distinct f.

For suppose otherwise. Then there are f, g such that $\psi_2[f] = \psi_2[g]$ and $\psi_3[f] = \psi_3[g]$ but $f(a,b) \neq g(a,b)$ for some a, b. But then the sequences of critical states in the two computations $C(a,b,f)$ and $C(a,b,g)$ must be the same, and hence the computations either accept both (a,b,f) and (a,b,g) or reject both. So the computations cannot both be correct.

Finally we prove (10) from the CLAIM. Let $s_2 = |\Gamma_2^{r,s}|$ and let $s_3 = |\Gamma_3^{r,s}|$, and let $s = s_2 + s_3$. Then the number of distinct pairs (ψ_2, ψ_3) is at most

$$(s_3 + 2)^{k(s_2+1)}(s_2 + 2)^{ks_3} \leq (s + 2)^{k(s+1)}$$

and since there are 2^{k^2} functions f we have

$$2^{k^2} \leq (s + 2)^{k(s+1)}$$

so taking logs, $k^2 \leq k(s+1)\lg_2(s+2)$ so $k/\lg_2(s+2) \leq s+1$, and (10) follows. □

Recall from Theorem 13 that applying the Nečiporuk method to $Children_2^4(k)$ yields an $\Omega(k^3)$ size lower bound and from Theorem 14 that applying it to $SumMod_2^3(k)$ yields $\Omega(k^2)$. The state sequence method also proves the next two theorems.

Theorem 17. *Any deterministic k-way BP for $Children_2^4(k)$ has at least $k^4/2$ states.*

Theorem 18. *Any deterministic k-way BP for $SumMod_2^3(k)$ requires at least k^3 states.*

5 Conclusion

Our main open question is whether we can adapt the state sequence method to break the $\Omega(n^2)$ barrier for the size of deterministic branching programs. In particular, can the method be extended to handle trees of height 4? Specifically, can we prove a lower bound of $\Omega(k^7/\lg k)$ for $BT_3^4(k)$ (see Theorem 10)?

Another question arises from the $O(k^{5/2})$ upper bound from Theorem 10. Is there a pebbling to justify such a non-integral exponent? As it turns out, the answer is yes. One can introduce fractional black-white pebbling and develop an interesting theory. Our work on that issue will be the subject of another paper.

Acknowledgment. James Cook played a helpful role in the early parts of this research. The second author is grateful to Michael Taitslin for suggesting a version of the tree evaluation problem in which the nodes are labelled by fixed quasi groups (see [Tai05]).

References

[BC82] Borodin, A., Cook, S.: A time-space tradeoff for sorting on a general sequential model of computation. SIAM J. Comput. 11(2), 287–297 (1982)

[BRS93] Borodin, A., Razborov, A., Smolensky, R.: On lower bounds for read-k-times branching programs. Computational Complexity 3, 1–18 (1993)

[Coo74] Cook, S.: An observation on time-storage trade off. J. Comput. Syst. Sci. 9(3), 308–316 (1974)

[CS76] Cook, S., Sethi, R.: Storage requirements for deterministic polynomial time recognizable languages. J. Comput. Syst. Sci. 13(1), 25–37 (1976)

[GKM08] Gál, A., Koucký, M., McKenzie, P.: Incremental branching programs. Theory Comput. Syst. 43(2), 159–184 (2008)

[Gol08] Goldreich, O.: Computational Complexity: A Conceptual Perspective. Cambridge University Press, Cambridge (2008)

[KRW95] Karchmer, M., Raz, R., Wigderson, A.: Super-logarithmic depth lower bounds via direct sum in communication complexity. Computational Complexity 5, 191–204 (1991); An abstract appeared in the 6th Structure in Complexity Theory Conference (1991)

[Mah07] Mahajan, M.: Polynomial size log depth circuits: between $\mathbf{NC}^1$and $\mathbf{AC}^1$. Bulletin of the EATCS 91, 30–42 (2007)

[Neč66] Nečiporuk, È.: On a boolean function. Doklady of the Academy of the USSR 169(4), 765–766 (1966); English translation in Soviet Mathematics Doklady 7(4), 999–1000

[Nor09] Nordström, J.: New wine into old wineskins: A survey of some pebbling classics with supplemental results (2009), `http://people.csail.mit.edu/jakobn/research/`

[PH70] Paterson, M., Hewitt, C.: Comparative schematology. In: Record of Project MAC Conference on Concurrent Systems and Parallel Computations, pp. 119–128. ACM, New Jersey (1970)

[Pud87] Pudlák, P.: The hierarchy of boolean circuits. Computers and artificial intelligence 6(5), 449–468 (1987)

[Raz91] Razborov, A.: Lower bounds for deterministic and nondeterministic branching programs. In: 8th Internat. Symp. on Fundamentals of Computation Theory, pp. 47–60 (1991)

[Sud78] Sudborough, H.: On the tape complexity of deterministic context-free languages. J. ACM 25(3), 405–414 (1978)

[Tai05] Taitslin, M.A.: An example of a problem from PTIME and not in NLogSpace. In: Proceedings of Tver State University (2005); Applied Mathematics, vol. 6(12) (2), pp. 5–22. Tver State University, Tver (2005)

[Weg00] Wegener, I.: Branching Programs and Binary Decision Diagrams. SIAM Monographs on Discrete Mathematics and Applications. Soc. for Industrial and Applied Mathematics, Philadelphia (2000)

A Dichotomy Theorem for Polynomial Evaluation

Irénée Briquel and Pascal Koiran

LIP*, École Normale Supérieure de Lyon, Université de Lyon
{irenee.briquel,pascal.koiran}@ens-lyon.fr

Abstract. A dichotomy theorem for counting problems due to Creignou and Hermann states that or any finite set S of logical relations, the counting problem #SAT(S) is either in FP, or #P-complete. In the present paper we show a dichotomy theorem for polynomial evaluation. That is, we show that for a given set S, either there exists a VNP-complete family of polynomials associated to S, or the associated families of polynomials are all in VP. We give a concise characterization of the sets S that give rise to "easy" and "hard" polynomials. We also prove that several problems which were known to be #P-complete under Turing reductions only are in fact #P-complete under many-one reductions.

1 Introduction

In a seminal paper, Schaefer [13] proved a dichotomy theorem for boolean constraint satisfaction problems: he showed that for any finite set S of logical relations the satisfiability problem SAT(S) for S-formulas is either in P, or NP-complete. Here, an S-formula over a set of n variables is a conjunction of relations of S where the arguments of each relation are freely chosen among the n variables. Schaefer's result was subsequently extended in a number of directions. In particular, dichotomy theorems were obtained for counting problems, optimization problems and the decision problem of quantified boolean formulas. An account of this line of work can be found in the book by Creignou, Khanna and Sudan [6]. In a different direction, constraint satisfaction problems were also studied over non-boolean domains. This turned out to be a surprisingly difficult question, and it took a long time before a dichotomy theorem over domains of size 3 could be obtained [4].

In the present paper we study polynomial evaluation from this dichotomic point of view. Full proofs of our results are presented in a more detailed version of this work [3]. We work within Valiant's algebraic framework: the role of the complexity class NP in Schaefer's dichotomy theorem will be played by the class VNP of "easily definable" polynomial families, and the role of P will be played by the class VP of "easily computable" polynomial families [14,2]. There is a well-known connection between counting problems and polynomial evaluation. For instance, as shown by Valiant the permanent is complete in both settings [15,14]. In the realm of counting problems, a dichotomy theorem was obtained by Creignou and Hermann [5,6].

* UMR 5668 ENS Lyon, CNRS, UCBL associée à l'INRIA.

R. Královič and D. Niwiński (Eds.): MFCS 2009, LNCS 5734, pp. 187–198, 2009.

Theorem 1. *For any finite set S of logical relations, the counting problem* $\#\mathrm{SAT}(S)$ *is either in* FP, *or* #P*-complete.*

In fact, the sets S such that $\#\mathrm{SAT}(S)$ is in FP are exactly the sets containing only affine constraints (a constraint is called affine if it expressible as a system of linear equations over $\mathbb{Z}/2\mathbb{Z}$).

Main Contributions

To a family of boolean formulas (ϕ_n) we associate the multilinear polynomial family

$$P(\phi_n)(\overline{X}) = \sum_{\overline{\varepsilon}} \phi_n(\overline{\varepsilon})\overline{X}^{\overline{\varepsilon}}, \tag{1}$$

where $\overline{X}^{\overline{\varepsilon}}$ is the monomial $X_1^{\varepsilon_1} \cdots X_{k(n)}^{\varepsilon_{k(n)}}$, and $k(n)$ is the number of variables of ϕ_n. Imagine that the ϕ_n are chosen among the S-formulas of a fixed finite set S of logical relations. One would like to understand how the complexity of the polynomials $P(\phi_n)$ depends on S.

Definition 1. *A family (ϕ_n) of S-formulas is called a p-family if ϕ_n is a conjunction of at most $p(n)$ relations from S, for some polynomial p (in particular, ϕ_n depends on polynomially many variables when S is finite).*

Theorem 2 (Main Theorem). *Let S be a finite set of logical relations. If S contains only affine relations of at most two variables, then the families $(P(\phi_n))$ of polynomials associated to p-families of S-formulas (ϕ_n) are in* VP. *Otherwise, there exists a p-family (ϕ_n) of S-formulas such that the corresponding polynomial family $P(\phi_n)$ is* VNP*-complete.*

Note, that the hard cases for counting problems are strictly included in our hard evaluation problems, exactly as the hard decision problems in Schaefer's theorem were strictly included in the hard counting problems.

In our algebraic framework the evaluation of the polynomial associated to a given formula consists in solving a "weighted counting" problem: each assignment $(\varepsilon_1, \ldots, \varepsilon_k)$ of the variables of ϕ comes with a weight $X_1^{\varepsilon_1} \cdots X_k^{\varepsilon_k}$. In particular, when the variables X_i are all set to 1, we obtain the counting problem $\#\mathrm{SAT}(S)$. It is therefore natural that evaluation problems turn out to be harder than counting problems.

The remainder of this paper is mostly devoted to the proof of Theorem 2. Along the way, we obtain several results of independent interest. First, we obtain several new VNP-completeness results. The main ones are about the vertex cover polynomial $\mathrm{VCP}(G)$ and the independent set polynomial $\mathrm{IP}(G)$, associated to a vertex-weighted graph G. Most VNP-completeness results in the literature (and certainly all the results in Chapter 3 of [2]) are about edge-weighted graphs.

Unlike in most VNP-completeness results, we need more general reductions to establish VNP-completeness results than Valiant's p-projection. In Section 4, we use the "c-reductions", introduced by Bürgisser [1,2] in his work on VNP families that are neither p-computable nor VNP-complete. They are akin to the oracle (or Turing) reductions from discrete complexity theory. The c-reduction has not been used widely in VNP-completeness proofs. The only examples that we are aware of are:

(i) A remark in [2] on probability generating functions.
(ii) The VNP-completeness of the weighted Tutte polynomial in [11]. Even there, the power of c-reductions is used in a very restricted way since a single oracle call is performed in each reduction.

By contrast, the power of oracle reductions has been put to good use in #P-completeness theory (mostly as a tool for performing interpolation). Indeed, as pointed out in [9], "interpolation features prominently in a majority of #P-completeness proofs", and "it is not clear whether the phenomenon of #P-completeness would be as ubiquitous if many-one reducibility were to be used in place of Turing." We argue that the importance of Turing reductions in #P-completeness should be revised downwards since, as a byproduct of our VNP-completeness results, we can replace Turing reductions by many-one reductions in several #P-completeness results from the literature. In particular, we obtain a many-one version of Creignou and Hermann's dichotomy theorem[1]. We leave it as an open problem whether the 0/1 partial permanent is #P-complete under many-one reductions (see Section 3 for a definition of the partial permanent, and [8] for a # P-completeness proof under oracle reductions).

Organization of the Paper and Additional Results

Earlier in this section we gave an informal introduction to constraint satisfaction problems. We give more precise definitions at the beginning of Section 2. The remainder of that section is devoted to Valiant's algebraic model of computation. We also deal briefly with the easy cases of Theorem 2 (Remark 1). We then establish the proof of the hard cases of Theorem 2, beginning with the case of non affine constraints. For that case, the high-level structure of the proof is similar to Creignou and Hermann's proof of #P-completeness of the corresponding counting problems in [5]. The singletons $S = \{\mathrm{OR}_2\}$, $S = \{\mathrm{OR}_1\}$ and $S = \{\mathrm{OR}_0\}$ play a special role in the proof. Here OR_2 denotes the negative two-clause $(x, y) \mapsto (\overline{x} \vee \overline{y})$; OR_0 denotes the positive two-clause $(x, y) \mapsto (x \vee y)$; and OR_1 denotes the implicative two-clause $(x, y) \to (\overline{x} \vee y)$. The corresponding VNP-completeness results for $\{\mathrm{OR}_2\}$ and $S = \{\mathrm{OR}_0\}$ are established in section 3; the case of $\{\mathrm{OR}_1\}$ is only treated in the full version [3], since it uses very similar techniques. Together with Creignou and Hermann's results, this suffices to establish the existence of a VNP-complete family for any set S containing

[1] Many-one reductions (Definition 2) are called *many-one counting reductions* in [5,6]. It was already claimed in [5,6] that Theorem 1 holds true for many-one reductions. This was not fully justified since the proof of Theorem 1 is based on many-one reductions from problems which were previously known to be #P-complete under oracle reductions only. The present paper shows that this claim was indeed correct.

non affine clauses (the proof is developed in [3]). Section 4 deals with the affine clauses with at least three variables (Theorem 6). This completes the proof of Theorem 2. In Section 5, we build on our VNP-completeness results to prove #P-completeness under many-one reductions for several problems which were only known to be #P-complete under oracle reductions.

2 Preliminaries

2.1 Constraint Satisfaction Problems

We define a logical relation to be a function from $\{0,1\}^k$ to $\{0,1\}$, for some integer k called the rank of the relation. Let us fix a finite set $S = \{\phi_1, \dots, \phi_n\}$ of logical relations. An S-formula over n variables $(x_1, \dots, x_n)$ is a conjunction of boolean formulas, each of the form $g_i(x_{j_i(1)}, \dots, x_{j_i(k_i)})$ where each g_i belongs to S and k_i is the rank of g_i. In words, each element in the conjunction is obtained by applying a function from S to some variables chosen among the n variables.

An instance of the problem SAT(S) studied by Schaefer [13] is an S-formula ϕ, and one must decide whether ϕ is satisfiable. For instance, consider the 3 boolean relations $\mathrm{OR}_0(x, y) = x \vee y$, $\mathrm{OR}_1(x, y) = \overline{x} \vee y$ and $\mathrm{OR}_2(x, y) = \overline{x} \vee \overline{y}$. The classical problem 2-SAT is SAT(S) where $S = \{\mathrm{OR}_0, \mathrm{OR}_1, \mathrm{OR}_2\}$. The counting problem #SAT(S) was studied by Creignou and Hermann [5]. In this paper we study the complexity of evaluating the polynomials $P(\phi)$ in (1). We establish which sets S give rise to VNP-complete polynomial families, and which one give rise only to easy to compute families. We next define these notions precisely.

2.2 #P-Completeness and VNP-Completeness

Let us introduce the notion of many-one reduction for counting problems :

Definition 2 (Many-one reduction). [17] *Let $f : \{0,1\}^* \to \mathbb{N}$ and $g : \{0,1\}^* \to \mathbb{N}$ be two counting problems. A* many-one reduction *from f to g consists of a pair of polynomial-time computable functions $\sigma : \{0,1\}^* \to \{0,1\}^*$ and $\tau : \mathbb{N} \to \mathbb{N}$ such that for every $x \in \{0,1\}^*$, the equality $f(x) = \tau(g(\sigma(x)))$ holds. When τ is the identity function, this reduction is called* parsimonious.

A counting problem f is #P-hard under many-one reduction if every problem in #P admits a many-one reduction to f.

In Valiant's model one studies the computation of multivariate polynomials. This can be done over any field. In the sequel we fix a field K of characteristic $\neq 2$. All considered polynomials are over K.

A p-family is a sequence $f = (f_n)$ of multivariate polynomials such that the number of variables and the degree are polynomially bounded functions of n. A prominent example of a p-family is the permanent family $\mathrm{PER} = (\mathrm{PER}_n)$, where PER_n is the permanent of an $n \times n$ matrix with independent indeterminate entries.

We define the complexity of a polynomial f to be the minimum number $L(f)$ of nodes of an arithmetic circuit computing f. We recall that the internal nodes of an arithmetic circuit perform additions or multiplications, and each input node is labeled by a constant from K or a variable X_i.

Definition 3 (VP). *A p-family (f_n) is p-computable if $L(f_n)$ is a polynomially bounded function of n. Those families constitute the complexity class* VP.

In Valiant's model, VNP is the analogue of the class NP (or perhaps more accurately, of #P).

Definition 4 (VNP). *A p-family (f_n) is called p-definable if there exists a p-computable family $g = (g_n)$ such that*

$$f_n(X_1, \ldots, X_{p(n)}) = \sum_{\varepsilon \in \{0,1\}^{q(n)}} g_n(X_1, \ldots, X_{p(n)}, \varepsilon_1, \ldots, \varepsilon_{q(n)})$$

The set of p-definable families forms the class VNP.

Clearly, VP is included in VNP. To define VNP-completeness we need a notion of reduction:

Definition 5 (p-projection). *A polynomial f with v arguments is said to be a projection of a polynomial g with u arguments, and we denote it $f \leq g$, if $f(X_1, \ldots, X_v) = g(a_1, \ldots, a_u)$ where each a_i is a variable of f or a constant from K.*

A p-family (f_n) is a p-projection of (g_m) if there exists a polynomially bounded function $t : \mathbb{N} \to \mathbb{N}$ such that: $\exists n_0 \forall n \geq n_0, f_n \leq g_{t(n)}$.

Definition 6 (VNP-completeness). *A p-family $g \in$ VNP is* VNP*-complete if every p-family $f \in$ VNP is a p-projection of g.*

The VNP-completeness of the permanent under p-projections [14,2] is a central result in Valiant's theory.

It seems that p-projections are too weak for some of our completeness results. Instead, we use the more general notion of c-reduction [1,2]. First we recall the notion of oracle computation :

Definition 7. *The* oracle complexity *$L^g(f)$ of a polynomial f with respect to the oracle polynomial g is the minimum number of arithmetic operations $(+, *)$ and evaluations of g over previously computed values that are sufficient to compute f from the indeterminates X_i and constants from K.*

Definition 8 (c-reduction). *Let us consider two p-families $f = (f_n)$ and $g = (g_n)$. We have a polynomial oracle reduction, or c-reduction, from f to g (denoted $f \leq_c g$) if there exists a polynomially bounded function $t : \mathbb{N} \to \mathbb{N}$ such that the map $n \mapsto L^{g_{t(n)}}(f_n)$ is polynomially bounded.*

We can define a more general notion of VNP-completeness based on c-reductions: A p-family f is VNP-hard if $g \leq_c f$ for every p-family $g \in$ VNP. It is VNP-complete if in addition, $f \in$ VNP. The new class of VNP-complete families contains all the classical VNP-complete families since every p-reduction is a c-reduction.

In our completeness proofs we need c-reductions to compute the homogeneous components of a polynomial. This can be achieved thanks to a well-known lemma (see e.g. [2]):

Lemma 1. *Let f be a polynomial in the variables $X_1, \ldots, X_n$. For any δ such that $\delta \leq \deg f$, let denote $f^{(\delta)}$ the homogeneous component of degree δ of f. Then, $L^f(f^{(\delta)})$ is polynomially bounded in the degree of f.*

By Valiant's criterion (Proposition 2.20 in [2]), for any finite set S of logical relations and any p-family (ϕ_n) of S-formulas the polynomials $(P(\phi_n))$ form a VNP family. Furthermore, the only four boolean affine relations with at most two variables are $(x = 0)$, $(x = 1)$, $(x = y)$ and $(x \neq y)$. Since for a conjunction of such relations, the variables are either independent or completely bounded, a polynomial associated to a p-family of such formulas is factorizable. Thus :

Remark 1. For a set S of affine relations with at most two variables, every p-family of polynomials associated to S-formulas is in VP.

All the work in the proof of Theorem 2 therefore goes into the hardness proof.

3 Monotone 2-Clauses

In this section we consider the set $\{\mathrm{OR}_2\} = \{(x, y) \mapsto (\overline{x} \vee \overline{y})\}$ and $\{\mathrm{OR}_0\} = \{(x, y) \mapsto (x \vee y)\}$. For $S = \{\mathrm{OR}_2\}$ and $S = \{\mathrm{OR}_0\}$, we show that there exists a VNP-complete family of polynomials $(P(\phi_n))$ associated to a p-family of S-formulas (ϕ_n).

The partial permanent $\mathrm{PER}^*(A)$ of a matrix $A = (A_{i,j})$ is defined by the formula:

$$\mathrm{PER}^*(A) = \sum_{\pi} \prod_{i \in \mathrm{def}\pi} A_{i\pi(i)}$$

where the sum runs over all injective partial maps from $[1, n]$ to $[1, n]$. It is shown in [2] that the partial permanent is VNP-complete (the proof is attributed to Jerrum). The partial permanent may be written as in (1), where ϕ_n is the boolean formula that recognizes the matrices of partial maps from $[1, n]$ to $[1, n]$. But ϕ_n is a p-family of $\{\mathrm{OR}_2\}$-formulas since

$$\phi_n(\varepsilon) = \bigwedge_{i,j,k : j \neq k} \overline{\varepsilon_{ij}} \vee \overline{\varepsilon_{ik}} \wedge \bigwedge_{i,j,k : i \neq k} \overline{\varepsilon_{ij}} \vee \overline{\varepsilon_{kj}}.$$

Here the first conjunction ensures that the matrix ε has no more than one 1 on each row; the second one ensures that ε has no more than one 1 on each column. We have obtained the following result.

Theorem 3. *The family (ϕ_n) is a p-family of $\{\mathrm{OR}_2\}$-formulas, and the polynomial family $(P(\phi_n))$ is* VNP*-complete under p-projections.*

The remainder of this section is devoted to the set $S = \{\mathrm{OR}_0\} = \{(x, y) \mapsto x \vee y\}$. The role played by the partial permanent in the previous section will be played by vertex cover polynomials. There is more work to do because the corresponding VNP-completeness result is not available from the literature.

Consider a vertex-weighted graph $G = (V, E)$: to each vertex $v_i \in V$ is associated a weight X_i. The vertex cover polynomial of G is

$$\mathrm{VCP}(G) = \sum_{S} \prod_{v_i \in S} X_i \tag{2}$$

where the sum runs over all vertex covers of G (recall that a vertex cover of G is a set $S \subseteq V$ such that for each edge $e \in E$, at least one of the two endpoints of e belongs to S). The univariate vertex cover polynomial defined in [7] is a specialization of ours; it is obtained from $\mathrm{VCP}(G)$ by applying the substitutions $X_i := X$ (for $i = 1, \ldots, n$), where X is a new indeterminate.

Our main result regarding $\{\mathrm{OR}_0\}$-formulas is as follows.

Theorem 4. *There exists a family G_n of polynomial size bipartite graphs such that:*

1. *The family $(\mathrm{VCP}(G_n))$ is VNP-complete.*
2. *$\mathrm{VCP}(G_n) = P(\phi_n)$ where ϕ_n is a p-family of $\{\mathrm{OR}_0\}$-formulas.*

Given a vertex-weighted graph G, let us associate to each $v_i \in V$ a boolean variable ε_i. The interpretation is that v_i is chosen in a vertex cover when ε_i is set to 1. We then have

$$\mathrm{VCP}(G) = \sum_{\varepsilon \in \{0,1\}^{|V|}} \Big[\bigwedge_{(v_i, v_j) \in E} \varepsilon_i \vee \varepsilon_j \Big] \overline{X}^{\overline{\varepsilon}}.$$

The second property in Theorem 4 will therefore hold true for any family (G_n) of polynomial size graphs.

To obtain the first property, we first establish a VNP-completeness result for the independent set polynomial $\mathrm{IP}(G)$. This polynomial is defined like the vertex cover polynomial, except that the sum in (2) now runs over all independent sets S (recall that an independent set is a set $S \subseteq V$ such that there are no edges between any two elements of S).

Theorem 5. *There exists a family (G'_n) of polynomial size graphs such that $\mathrm{IP}(G'_n) = \mathrm{PER}^*_n$ where PER^*_n is the $n \times n$ partial permanent. The family $\mathrm{IP}(G'_n)$ is therefore VNP-complete.*

Proof. The vertices of G'_n are the n^2 edges ij of the complete bipartite graph $K_{n,n}$, and the associated weight is the indeterminate X_{ij}. Two vertices of G'_n are connected by an edge if they share an endpoint in $K_{n,n}$. An independent set in G'_n is nothing but a partial matching in $K_{n,n}$, and the corresponding weights are the same.

Next we obtain a reduction from the independent set polynomial to the vertex cover polynomial. The connection between these two problems is not astonishing since vertex covers are exactly the complements of independent sets. But we deal here with weighted counting problems, so that there is a little more work to do. The connection between independent sets and vertex covers does imply a relation between the polynomials IP(G) and VCP(G). Namely,

$$\mathrm{IP}(G)(X_1,\ldots,X_n) = X_1 \cdots X_n \cdot \mathrm{VCP}(G)(1/X_1,\ldots,1/X_n). \quad (3)$$

Indeed,

$$\mathrm{IP}(G) = \sum_{S \text{ independent}} \frac{X_1 \cdots X_n}{\prod_{v_i \notin S} X_i} = X_1 \cdots X_n \sum_{S' \text{ vertex cover}} \frac{1}{\prod_{v_i \in S'} X_i}.$$

Recall that the incidence graph of a graph $G' = (V', E')$ is a bipartite graph $G = (V, E)$ where $V = V' \cup E'$. In the incidence graph there is an edge between $e' \in E'$ and $u' \in V'$ if u' is one of the two endpoints of e' in G. When G' is vertex weighted, we assign to each V'-vertex of G the same weight as in G and we assign to each E'-vertex of G the constant weight -1.

Lemma 2. *Let G' be a vertex weighted graph and G its vertex weighted incidence graph as defined above. We have the following equalities:*

$$\mathrm{VCP}(G) = (-1)^{e(G')}\mathrm{IP}(G') \quad (4)$$

$$\mathrm{IP}(G) = (-1)^{e(G')}\mathrm{VCP}(G') \quad (5)$$

where $e(G')$ is the number of edges of G'.

Proof. We begin with (4). To each independent set I' of G' we can injectively associate the vertex cover $C = I' \cup E'$. The weight of C is equal to $(-1)^{e(G')}$ times the weight of I'. Moreover, the weights of all other vertex covers of G add up to 0. Indeed, any vertex cover C which is not of this form must contain two vertices $u', v' \in V'$ such that $u'v' \in E'$. The symmetric difference $C\Delta\{u'v'\}$ remains a vertex cover of G, and its weight is opposite to the weight of C since it differs from C only by a vertex $u'v'$ of weight -1.

The equality (5) follow from the combination of (3) and (4).

To complete the proof of Theorem 4 we apply Lemma 2 to the graph $G' = G'_n$ of Theorem 5. The resulting graph $G = G_n$ satisfies $\mathrm{VCP}(G_n) = \mathrm{IP}(G'_n) = \mathrm{PER}^*_n$ since G'_n has an even number of edges: $e(G'_n) = n^2(n-1)$.

4 Affine Relations with at Least Three Variables

Here we consider the case of a set S containing large affine constraints. We first establish the existence of a VNP-complete family of polynomials associated to a p-family of affine formulas, and then show how to reduce this family to each affine constraint with at least three variables. In this section, our VNP-completeness results are in the sense of c-reduction.

Let us consider the $n \times n$ permanent $\mathrm{PER}_n(M)$ of a matrix $M = (M_{i,j})$. It may be expressed as the polynomial associated to the formula accepting the $n \times n$ permutation matrices: $\mathrm{PER}_n(M) = \sum_\varepsilon \phi_n(\varepsilon)\overline{X}^{\overline{\varepsilon}}$.

This formula ϕ_n expresses, that each row and each column of the matrix ε contains exactly one 1. Let us consider the formula φ_n defined by:

$$\varphi_n(\varepsilon) = \bigwedge_{i=1}^{n} \varepsilon_{i1} \oplus \ldots \oplus \varepsilon_{in} = 1 \wedge \bigwedge_{j=1}^{n} \varepsilon_{1j} \oplus \ldots \oplus \varepsilon_{nj} = 1$$

The formula φ_n expresses, that each row and each column of ε contains an odd number of values 1. Thus, φ_n accepts the permutation matrices, and other assignments that contain more values 1. We therefore remark, that the $n \times n$ permanent is exactly the homogeneous component of degree n of $P(\varphi_n)$. But from Lemma 1, this implies a c-reduction from the permanent family to the p-family $(P(\varphi_n))$. Thus:

Lemma 3. *The family $(P(\varphi_n))$ is* VNP*-complete with respect to c-reductions.*

Through c-reductions and p-projections, this suffices to establish the existence of VNP-complete families for affine formulas of at least three variables:

Theorem 6. *1. There exists a* VNP*-complete family of polynomials associated to $\{x \oplus y \oplus z = 0\}$-formulas.*

2. There exists a VNP*-complete family of polynomials associated to $\{x \oplus y \oplus z = 1\}$-formulas.*

3. For every set S containing an affine formula with at least three variables, there exists a VNP*-complete family of polynomials associated to S-formulas.*

Proof. 1. Let us consider the formula φ_n. This formula is a conjunction of affine relations with constant term 1: $x_1 + \ldots + x_k = 1$. Let φ'_n be the formula obtained from φ_n by adding a variable a and replacing such clauses by $x_1 + \ldots + x_k + a = 0$. In the polynomial associated to φ'_n, the term of degree 1 in the variable associated to a is exactly the polynomial $P(\varphi_n)$: when a is assigned to 1, the satisfying assignments of φ'_n are equal to the satisfying assignments of φ_n. Since this term of degree 1 can be recovered by polynomial interpolation of $P(\varphi'_n)$, the family $(P(\varphi_n))$ c-reduces to $(P(\varphi'_n))$. φ'_n is a conjunction of affine relations with constant term 0. The polynomial $P(\varphi'_n)$ is the projection of the polynomial $P(\psi_n)$, where the formula ψ_n is obtained from φ'_n by replacing each affine relation of the type $x_1 \oplus \ldots \oplus x_k = 0$ by the conjunction of relations

$$(x_1 \oplus x_2 \oplus a_1 = 0) \wedge (a_1 \oplus x_3 \oplus a_2 = 0) \wedge \ldots \wedge (a_{k-2} \oplus x_{k-1} \oplus x_k = 0)$$

where the a_i are new variables. In fact, on sees easily, that for a given assignment of the x_i satisfying φ'_n, a single assignment of the a_i gives a satisfying

assignment of ψ_n; and that if the x_i do not satisfy φ'_n, no assignment of the a_i works on. The polynomial $P(\varphi'_n)$ is thus the polynomial obtained by replacing the variables associated to a_i by the value 1 in $P(\psi_n)$; the family $(P(\varphi'_n))$ is a p-projection of $(P(\psi_n))$.
2. The formula ψ_n constructed above is a conjunction of relations of the type $x \oplus y \oplus z = 0$. Let us construct a new formula ψ'_n by introducing two new variables a and b and replacing each of such relations by the conjunction $(x \oplus y \oplus a = 1) \wedge (a \oplus z \oplus b = 1)$. One sees easily, that $P(\psi_n)$ is the projection of $P(\psi'_n)$ obtained by setting the variables associated to a and b to 1 and 0 respectively.
3. Let us suppose, that S contains a relation of the type $x_1 \oplus \ldots \oplus x_k = 0$, with $k \geq 3$. The polynomial $P(\psi_n)$ is the projection of the polynomial associated to the S-formula obtained by replacing each relation $x \oplus y \oplus z = 0$ of ψ_n by a relation $x \oplus y \oplus z \oplus a_1 \oplus \ldots \oplus a_{k-3} = 0$, and setting the variables associated to the a_i to 0. Thus, the family $(P(\psi_n))$ projects on a family of polynomials associated to S-formulas, which is therefore VNP-complete. When S contains a relation with constant term 1, one projects the family $(P(\psi'_n))$ similarly.

5 #P-Completeness Proofs

Up to now, we have studied vertex weighted graphs mostly from the point of view of algebraic complexity theory. Putting weights on edges, or on vertices, can also be useful as an intermediate step in #P-completeness proofs [15,8]. Here we follow this method to obtain new #P-completeness results. Namely, we prove #P-completeness under many-one reductions for several problems which were only known to be #P-complete under oracle reductions.

Theorem 7. *The following problems are* #P*-complete for many-one reductions.*

1. *Vertex Cover: counting the number of vertex covers of a given a graph.*
2. *Independent Set: counting the number of independent sets of a given graph.*
3. *Bipartite Vertex Cover: the restriction of vertex cover to bipartite graphs.*
4. *Bipartite Independent Set: the restriction of independent set to bipartite graphs.*
5. *Antichain: counting the number of antichains of a given poset.*
6. *Ideal: counting the number of ideals of a given poset.*
7. *Implicative* 2-SAT*: counting the number of satisfying assignments of a conjunction of implicative 2-clauses.*
8. *Positive* 2-SAT*: counting the number of satisfying assignments of a conjunction of positive 2-clauses.*
9. *Negative* 2-SAT*: counting the number of satisfying assignments of a conjunction of negative 2-clauses.*

Remark 2. #P-completeness under oracle reductions is established in [12] for the first six problems, in [10] for the 7th problem and in [16] for the last two. In Section 2, the last three problems are denoted #SAT(S) where S is respectively equal to $\{\mathrm{OR}_1\}$, $\{\mathrm{OR}_0\}$ and $\{\mathrm{OR}_2\}$.

Proof. Provan and Ball establish in [12] the equivalence of Problems 1 and 2, 3 and 4, and 5 and 6; they produce many-one reductions from 1 to 8 and from 4 to 5, and Linial gives in [10] a many-one reduction from 6 to 7. Problems 8 and 9 are clearly equivalent. Therefore, to obtain #P-completeness under many-one reductions for all those problems, we just need to show the #P-completeness of Problem 1 and to produce a many-one reduction from Problem 1 to Problem 3 (replacing the oracle reduction from [12]).

In order to prove the #P-completeness of Problem 1, we first establish a many-one reduction from the #P-complete problem of computing the permanent of $\{0,1\}$-matrices (which is known to be #P-complete under many-one reductions [17]) to the problem of computing the vertex cover polynomial of a weighted graph with weights in $\{0,1,-1\}$. In [2], Bürgisser attributes to Jerrum a projection from the permanent to the partial permanent, with the use of the constant -1. Applied to a $\{0,1\}$-matrix, this gives a many-one reduction from the permanent on $\{0,1\}$-matrices to the partial permanent on $\{0,1,-1\}$-matrices. By Theorem 5, the $n \times n$ partial permanent is equal to the independent set polynomial of the graph G'_n; the reduction is obviously polynomial. Moreover, by Lemma 2 this polynomial is the projection of the vertex cover polynomial of G_n, with the use of the constant -1. The partial permanent on entries in $\{0,1,-1\}$ therefore reduces to the vertex cover polynomial on graphs with weights in $\{0,1,-1\}$.

Let G be such a vertex weighted graph, with weights in $\{0,1,-1\}$. A vertex cover of nonzero weight does not contain any vertex v of weight 0, and in order to cover the edges that are incident to v, it must contain all its neighbors. One can therefore remove v, and replace each edge from v to another vertex u by a self-loop (an edge from u to u). Thus, we obtain a graph G' with weights in $\{1,-1\}$ such that $\mathrm{VCP}(G) = \mathrm{VCP}(G')$.

To deal with the weights -1, we use a method similar to [15]. Since $\mathrm{VCP}(G')$ is the value of a partial permanent on a $\{0,1\}$-matrix, it is positive. We will construct an integer N and a graph H such that the number of vertex covers of H modulo N is equal to $\mathrm{VCP}(G')$. This will establish a reduction from the boolean permanent to counting vertex covers.

We choose N larger than the maximum value of the number of vertex covers of G': $N = 2^{v(G')} + 1$ will suit our purposes. Now that we compute the number of vertex covers modulo N, we can replace each -1 weight in G' by the weight $N - 1 = 2^{v(G')}$. But one can simulate such a weight on a vertex by adding to it $v(G')$ leaves.

Finally, we construct a many-one reduction from vertex cover to bipartite vertex cover. By applying two times the transformation of Lemma 4, we have a projection from the vertex cover polynomial of a graph to the vertex cover polynomial of a bipartite graph, with the use of -1 weights. To eliminate these weights, we can follow the method used in our above proof of the #P-completeness of Problem 1. Indeed, since the leaves added to the graph preserve bipartiteness, we obtain a reduction from counting vertex covers in a general graph to counting vertex covers in a bipartite graph.

The proof of Creignou and Hermann's dichotomy theorem [5,6] is based on many-one reductions from the last 3 problems of Theorem 7. We have just shown that these 3 problems are #P-complete under many-one reductions. As a result, we have the following corollary to Theorem 7.

Corollary 1. *Theorem 1 still holds for #P-completeness under many-one reduction.*

Acknowledgements

We thank the anonymous referees for several useful suggestions. In particular, a referee for an earlier version of this paper suggested that affine relations should give rise to polynomial families that are hard to evaluate.

References

1. Bürgisser, P.: On the structure of Valiant's complexity classes. Discrete Mathematics and Theoretical Computer Science 3, 73–94 (1999)
2. Bürgisser, P.: Completeness and Reduction in Algebraic Complexity Theory. Algorithms and Computation in Mathematics, vol. 7. Springer, Heidelberg (2000)
3. Briquel, I., Koiran, P.: A dichotomy theorem for polynomial evaluation, `http://prunel.ccsd.cnrs.fr/ensl-00360974`
4. Bulatov, A.: A dichotomy theorem for constraint satisfaction problems on a 3-element set. Journal of the ACM 53(1), 66–120 (2006)
5. Creignou, N., Hermann, M.: Complexity of generalized satisfiability counting problems. Information and Computation 125, 1–12 (1996)
6. Creignou, N., Khanna, S., Sudan, M.: Complexity classification of boolean constraint satisfaction problems. SIAM monographs on discrete mathematics (2001)
7. Dong, F.M., Hendy, M.D., Teo, K.L., Little, C.H.C.: The vertex-cover polynomial of a graph. Discrete Mathematics 250(1-3), 71–78 (2002)
8. Jerrum, M.: Two-dimensional monomer-dimer systems are computationally intractable. Journal of Statistical Physics 48, 121–134 (1987)
9. Jerrum, M.: Counting, Sampling and Integrating: Algorithms and Complexity. Lectures in Mathematics - ETH Zürich. Birkhäuser, Basel (2003)
10. Linial, N.: Hard enumeration problems in geometry and combinatorics. SIAM Journal of Algebraic and Discrete Methods 7(2), 331–335 (1986)
11. Lotz, M., Makowsky, J.A.: On the algebraic complexity of some families of coloured Tutte polynomials. Advances in Applied Mathematics 32(1), 327–349 (2004)
12. Provan, J.S., Ball, M.O.: The complexity of counting cuts and of computing the probability that a graph is connected. SIAM J. of Comp. 12(4), 777–788 (1983)
13. Schaefer, T.J.: The complexity of satisfiability problems. In: Conference Record of the 10th Symposium on Theory of Computing, pp. 216–226 (1978)
14. Valiant, L.G.: Completeness classes in algebra. In: Proc. 11th ACM Symposium on Theory of Computing, pp. 249–261 (1979)
15. Valiant, L.G.: The complexity of computing the permanent. Theoretical Computer Science 8, 189–201 (1979)
16. Valiant, L.G.: The complexity of enumeration and reliability problems. SIAM Journal of Computing 8(3), 410–421 (1979)
17. Zankó, V.: #P-completeness via many-one reductions. International Journal of Foundations of Computer Science 2(1), 77–82 (1991)

DP-Complete Problems Derived from Extremal NP-Complete Properties

Yi Cao*, Joseph Culberson**, and Lorna Stewart***

Department of Computing Science, University of Alberta
Edmonton, Alberta, T6G 2E8, Canada
{cao1,joe,stewart}@cs.ualberta.ca

Abstract. In contrast to the extremal variants of coNP-complete problems, which are frequently DP-complete, many extremal variants of NP-complete problems are in P. We investigate the extremal variants of two NP-complete problems, the extremal colorability problem with restricted degree and the extremal unfrozen non-implicant problem, and show that both of them are DP-complete.

Keywords: DP-complete, complexity, extremal problem, colorability, implicant, unfrozen.

1 Introduction

Detecting the existence of unsatisfiable subproblems is a significant component of exact solvers for NP-complete problems. In proof complexity, probabilistic limits on the size of minimal unsatisfiable subproblems figures large in establishing exponential lowers for resolution proofs [6]. In logic, determining whether an instance is a minimal unsatisfiable subproblem, also called the critical-UNSAT problem[1], is to determine whether a given Boolean formula is unsatisfiable but deleting any clause makes it satisfiable, and has been shown to be DP-complete [13]. The critical-nonhamilton-path problem [11], the critical-uncolorability problem [3] and many other extremal coNP problems [5] [9] are also known to be DP-complete. Papadimitriou and Yannakakis introduced the class DP, which contains both NP and coNP and is contained in Δ_2^P [12].

The class DP is defined to be all languages that can be expressed as the intersection of a language in NP and a language in coNP. The SAT-UNSAT problem is a typical DP-complete problem [11]: given two Boolean expressions φ, φ', is it true that φ is satisfiable and φ' is not? More such problems are introduced

* The author is supported in part by NSERC, iCORE, Alberta Advanced Education and Technology, Killam Trusts and FGSR of University of Alberta.
** The author is supported by NSERC.
*** The author is supported by NSERC.
[1] The problem is originally called CRITICAL SAT in [13], we changed the name herein for consistency.

R. Královič and D. Niwiński (Eds.): MFCS 2009, LNCS 5734, pp. 199–210, 2009.

in [10]. Many "exact cost" variants of NP-hard optimization problems, for example the exact-TSP problem [13] and the exact-four-colorability problem [15], are also DP-complete. More problems like this can be found in [11] and [14].

Intuitively, extremal coNP problems, such as the critical problems listed above, are in DP because they can be expressed as a coNP language describing the problem instance, while each (complementary) subproblem is in NP. As discussed below, if any extremal coNP-complete problem had a polynomial certificate, then NP=coNP, thus making it likely that such problems are in higher level complexity classes.

Extremal variants of monotonic NP problems are in DP for similar reasons, the instances are in NP, but all monotonic extensions are in coNP. However, in contrast to extremal coNP-complete problems, the extremal instances of many NP-complete problems can be recognized in polynomial time [2].

Some extremal variants of NP-complete problems were previously shown to be in complexity classes that are likely beyond P. The extremal k-complementary subgraph problem is isomorphism complete [2] and the problem of recognizing maximal unfrozen graphs with respect to independent sets is coNP-complete [1]. As far as we know, no extremal variant of an NP-complete problem has been shown to be DP-complete before this paper.

In this paper, we present two new DP-complete problems that are extremal variants of the following NP-complete problems: colorability of a graph with restricted degree, and unfrozen non-implicant. The first illustrates that the addition of what appears to be a trivial condition to a combinatorial problem can vastly change the complexity of recognition. In the second problem, a similar rise in difficulty comes from a requirement for robustness, namely that a non-implicant remain a non-implicant under the addition of any one literal.

Given a set Ω of structures, such as graphs or Boolean formulas, a *property* is a subset $\mathcal{P} \subseteq \Omega$ and the *decision problem* associated with $\mathcal{P}$ is "given an *instance* $I \in \Omega$, is I in $\mathcal{P}$?" A property is said to be *monotone* if whenever an instance I has the property, then any instance obtained from I by adding (deleting) elements to (from) I also has that property. The addition or deletion action is specified according to the property being considered. For example, 3-colorability is a monotone property with respect to the deletion of edges from the graph; i.e. if graph $G = (V, E)$ is 3-colorable, then any graph $G' = (V, E')$ with $E' \subset E$ is also 3-colorable. Hamiltonicity is a monotone property with respect to the addition of edges, and we will discuss monotone properties related to Boolean formulas in Section 3.

Given a monotone property $\mathcal{P}$, an instance I is said to be *unfrozen* with respect to $\mathcal{P}$, written $I \in \mathcal{U}(\mathcal{P})$, if $I \in \mathcal{P}$ and remains in $\mathcal{P}$ after adding (deleting) any element to (from) I. For example a graph G is unfrozen 3-colorable if G is 3-colorable and remains 3-colorable under the addition of any edge. Note that $\mathcal{U}(\mathcal{P})$ is a monotone property, e.g., if a graph is unfrozen 3-colorable then any graph obtained by deleting edges from that graph is also unfrozen 3-colorable.

Note that for a property $\mathcal{P}$, the action (addition or deletion) discussed in the definitions of "unfrozen" and "extremal" is usually the opposite action of the one discussed in the definition of "monotone". Given a non-trivial monotone property $\mathcal{P}$, meaning $\mathcal{P} \neq \emptyset$ and $\mathcal{P} \neq \Omega$, an instance I is *extremal* with respect to $\mathcal{P}$, written $I \in \text{EX}(\mathcal{P})$, if $I \in \mathcal{P}$ but no longer in $\mathcal{P}$ under the addition (deletion) of any element to (from) I. EX($\mathcal{P}$) is a property, but is not monotone.

For many coNP-complete properties $\mathcal{P}$, recognizing instances in EX($\mathcal{P}$) are known to be DP-complete [3] [5] [9] [11] [13]. The following observation partially explains the high complexity of these problems. For any monotone property $\mathcal{P}$, if the instances in EX($\mathcal{P}$) can be verified in polynomial time, then the problem of recognizing the instances of $\mathcal{P}$ is in NP. We can simply guess a subset or superset[2] of the instance and check if it is in EX($\mathcal{P}$). Therefore, recognizing the instances of EX($\mathcal{P}$) for any coNP-complete property $\mathcal{P}$ cannot be solved in polynomial time, unless NP=coNP.

This paper is organized as follows. In Section 2, we will introduce our first example, the extremal 3-colorability problem with maximum degree 8. The DP complexity result can be generalized to the extremal k-colorability problem with maximum degree r, where k is an integer at least 3 and r is a function of k. The extremal k-colorability problem itself is polynomial time solvable; we just need to check if the graph is complete k-partite. It is interesting that adding a polynomial property (the restricted degree) can make a problem much more difficult. In contrast for k-colorability combined with certain NP-complete properties such as Hamiltonicity and independent set, the corresponding extremal problems remain in P [2]. In Section 3, we will give another DP-complete extremal variant of an NP-complete problem, the extremal unfrozen non-implicant problem.

2 Extremal Colorability Problem with Restricted Degree

We first introduce some definitions. Given a graph $G = (V, E)$, a *coloring* of G is a labeling of its vertices, and a *proper* coloring is a labeling such that any pair of adjacent vertices have different labels (*i.e.* colors). Graph G is said to be *k-colorable* if there is a proper coloring of G using k colors, $k \geq 3$. It is well known that the problem to determine whether a graph is k-colorable is NP-complete and remains NP-complete for graphs with restricted maximum degree r, $k \geq 3$ and $r \geq 4$ [8]. A graph G is *extremal k-colorable* if G is k-colorable but after adding any edge (u, v) to G, $u, v \in V$ and $(u, v) \notin E$, the resulting graph is not k-colorable. G is *critical k-uncolorable* if G is not k-colorable, but after deleting any edge (u, v) from G, $(u, v) \in E$, the resulting graph is k-colorable.

Cai and Meyer [4] proved that recognizing critical k-uncolorable graphs is DP-complete. In contrast, the extremal k-colorable graphs can be recognized in polynomial time, because they are just complete k-partite graphs. However, adding a restriction to the maximum degree will make the problem dramatically more difficult.

The *extremal k-colorability problem with maximum degree r* (EkCΔr) is: Given a graph $G = (V, E)$ with maximum degree r and a positive integer k, $r \geq 4$ and $k \geq 3$, is it true that G is k-colorable, but adding any edge (u, v) to G, $u, v \in V$

[2] Assuming that the size of any superset of an instance is polynomially bounded.

and $(u, v) \notin E$, the resulting graph is either not k-colorable or has maximum degree greater than r?

Now, we will show that

Theorem 1. *E3CΔ2t, $t \geq 4$, is DP-complete.*

Proof. We show that the theorem holds for $t = 4$. For t with larger value, the proof is similar; only the device used to control the degrees needs to be modified accordingly. E3CΔ8 is in DP, because there are at most $|V|^2$ possible edges to add and 3-colorability is in NP. To show completeness, we will reduce both the 3-colorable problem with maximum degree 4 (3CΔ4) and the 3-uncolorable problem with maximum degree 4 (3UCΔ4) to E3CΔ8.

Lemma 1. *E3CΔ8 is NP-hard.*

Proof. (*of Lemma 1*) Given a graph G with maximum degree 4, we can simply add $K_{8,8}$'s by breaking an edge and connecting the free ends to vertices of degree less than 8, until every vertex has degree 8. The resulting graph is extremal 3-colorable with maximum degree 8 if and only if G is 3-colorable.

Lemma 2. *E3CΔ8 is coNP-hard.*

Proof. (*of Lemma 2*) Given a graph $G = (V, E)$ with maximum degree 4, we will construct a graph $G' = (V', E')$ such that G' is extremal 3-colorable with maximum degree 8 if and only if G is not 3-colorable. We will need to use the T-device shown in Figure 1(a) throughout the construction. This device has the property that in any proper 3-coloring of a graph containing the T-device as an induced subgraph, if vertices u and v are colored with the same color then x must be colored with that color, but if u and v are colored with different colors then x can be colored with any color. Initially, replace each edge $(u, v) \in E$ by two T-devices in G' as shown in Figure 1(b). Since u and v are not adjacent in G', they can be colored with the same color. In any proper 3-coloring of G', if u and v are colored with the same color, then x and x' must also be colored with that color. We call u and v the *original vertices*, x and x' the *control vertices*, and the rest of the vertices *assistant vertices*. Let $e_1, e_2, \ldots, e_m$ be an arbitrary ordering of the edges of G, and x_i, x'_i be the control vertices in the subgraph of G' corresponding to e_i. We can chain all the control vertices of G' in a cycle with $2m$ additional T-devices as shown in

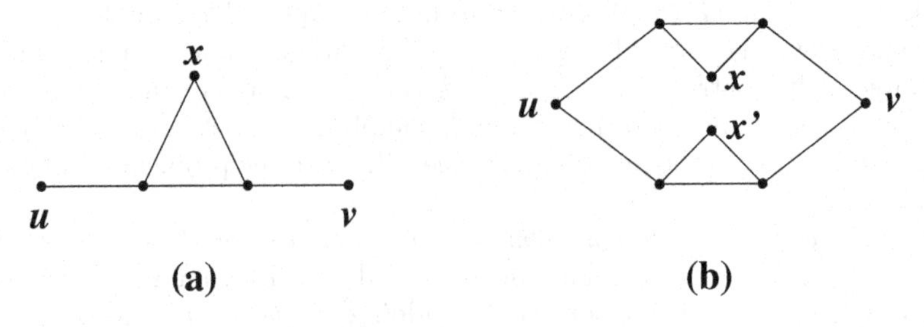

Fig. 1. T-device

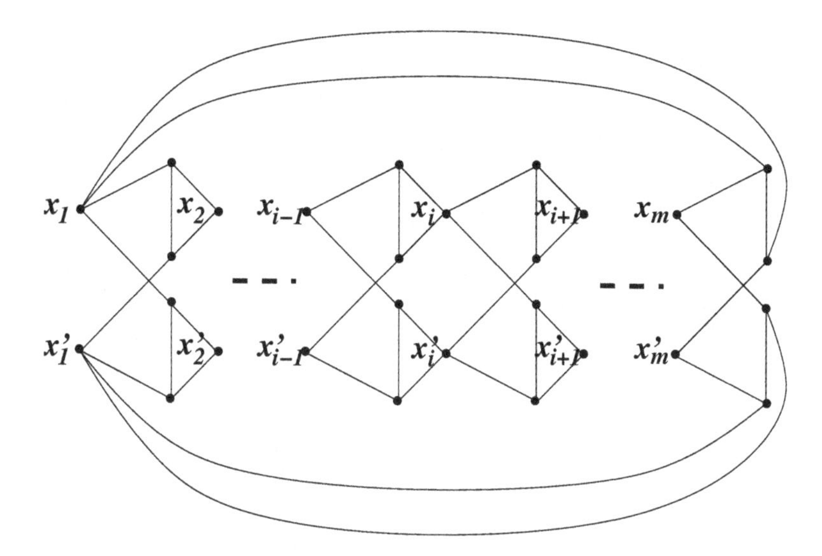

Fig. 2. Chain all the control vertices in a cycle

Figure 2. Note that x_i and x'_i work as the control vertices in the subgraph containing x_{i-1}, x'_{i-1}, x_i and x'_i. Therefore, in any proper 3-coloring of G', if x_{i-1} and x'_{i-1} are colored with the same color, then x_i and x'_i must be also colored with that color. By induction, all the control vertices have to be colored with that color. Please see Figure 3 for an example of the construction.

Graph G' is 3-colorable whether G is 3-colorable or not. We can color all the original and control vertices with one color. Then the graph induced by the assistant vertices is a collection of isolated edges, which can be easily colored by the remaining two colors. At this stage, each control vertex has degree 6, each assistant vertex has degree 3, and each original vertex has degree at most 8 (twice its degree in G). In order to make G' extremal 3-colorable with maximum degree 8 while G is not 3-colorable, we need to add more vertices and edges. Similar to Lemma 1, we can attach $K_{8,8}$'s to the original and assistant vertices to make each of them have degree 8, and leave the control vertices unchanged. This completes the construction of G'.

If G is 3-colorable, then take an arbitrary 3-coloring of G and color the original vertices of G' by this coloring. For each edge of G, in the corresponding induced subgraph of G', the two original vertices are colored with different colors, which means the two control vertices can be colored with any colors. Color the two control vertices with two distinct colors and color the assistant vertices properly. Finally, color the $K_{8,8}$'s according to the colors of their attached vertices. G' is not extremal because we can add an edge between two control vertices with different colors.

If G is not 3-colorable, then for any 3-coloring of G' there is a pair of original vertices u and v such that they have the same color and $(u, v) \in E$, otherwise the coloring of the original vertices of G' is a proper 3-coloring of G. In the induced subgraph corresponding to (u, v), the two control vertices must be colored with

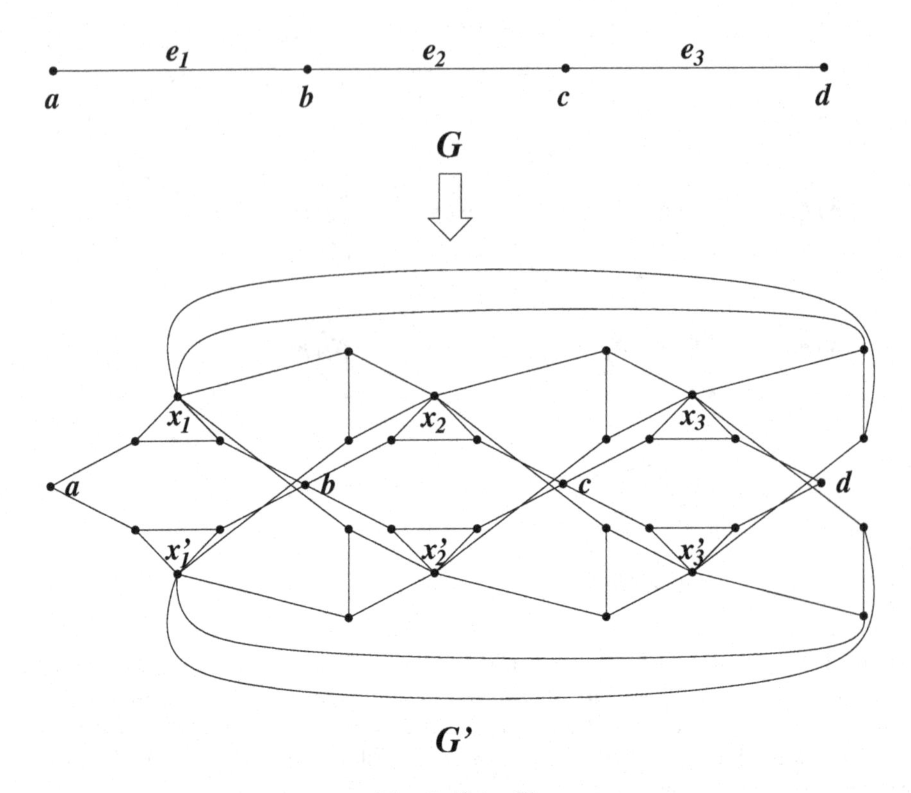

Fig. 3. G to G'

the same color. It follows that all the control vertices are colored with the same color and G' is extremal 3-colorable with maximum degree 8. Therefore, E3CΔ8 is coNP-hard.

To show E3CΔ8 is DP-hard, we can reduce the well-known DP-complete problem SAT-UNSAT to it. Given two formulas φ and φ', we first transform them to two graphs G and G' both with maximum degree 4, such that G is 3-colorable if and only if φ is satisfiable, and G' is not 3-colorable if and only if φ' is not satisfiable [8]. Then by Lemmas 1 and 2, we can construct two graphs H and H' both with maximum degree 8, such that H is extremal 3-colorable with maximum degree 8 if and only if G is 3-colorable, and H' is extremal 3-colorable with maximum degree 8 if and only if G' is not 3-colorable. Let H^* be the graph consisting of two components H and H'. We have that H^* is extremal 3-colorable with maximum degree 8 if and only if φ is satisfiable and φ' is not satisfiable. This completes the proof.

In the remainder of this section, we generalize the complexity results to arbitrary $k \geq 3$. We will show that

Theorem 2. *EkC$\Delta\rho(k)$ is DP-complete, $\rho(k) = k(k + \lceil\sqrt{k}\rceil - 1)$.*

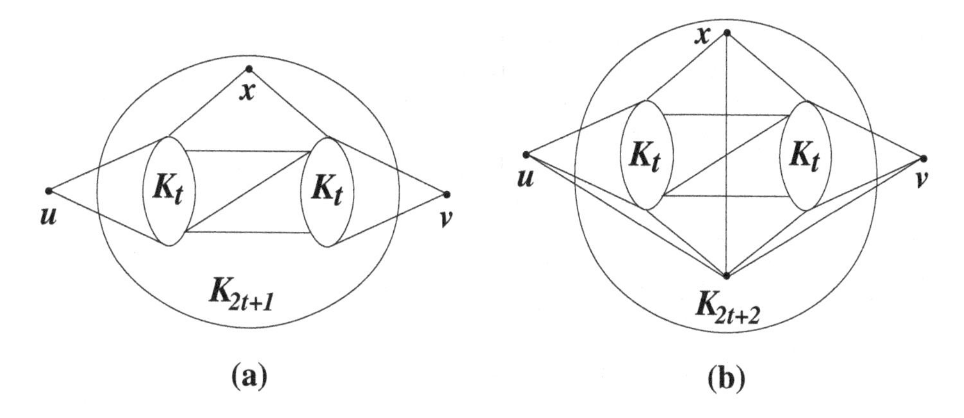

Fig. 4. Generalized T-device

Proof. Emden-Weinert, Hougardy, and Kreuter [7] have shown that determining if a graph G with maximum degree $(k + \lceil\sqrt{k}\rceil - 1 = \rho(k)/k)$ is k-colorable is NP-complete for $k \geq 3$. The same techniques as those used in Theorem 1 will be applied here. We will reduce the $k\mathrm{C}\Delta\rho(k)/k$ and the $k\mathrm{UC}\Delta\rho(k)/k$ to $\mathrm{E}k\mathrm{C}\Delta\rho(k)$, respectively. Again, $\mathrm{E}k\mathrm{C}\Delta\rho(k)$ is in DP because there are at most $|V|^2$ possible edges to add and k-colorability is in NP. To show the NP-hardness, we only need to attach $K_{\rho(k),\rho(k)}$'s to vertices with degree less than $\rho(k)$, so that every vertex in the resulting graph will have degree $\rho(k)$[3]. It follows that the original graph is k-colorable if and only if the resulting graph is extremal k-colorable with maximum degree $\rho(k)$.

Next, we will show the problem is coNP-hard. Given a graph $G = (V, E)$ with maximum degree $\rho(k)/k$, $k \geq 3$, we will construct a graph $G' = (V', E')$ such that G' is extremal k-colorable with maximum degree $\rho(k)$ if and only if G is not k-colorable. The construction of G' is the same as that in Lemma 2 except that the structure of the T-device has to be modified to fit into the new environment. The new T-device consists of two vertices u,v and a k-clique. Each of u and v has $\lfloor \frac{k}{2} \rfloor$ neighbors in the k-clique. There is one vertex x in the k-clique that is adjacent to neither u nor v, and all the other vertices are adjacent to either u or v or both. Vertices u and v have at most one neighbor in common in the k-clique. More specifically, if $k = 2t + 1$, each of u and v has t neighbors in the k-clique, which have no vertex in common (See Figure 4(a)); if $k = 2t + 2$, each of u and v has $t + 1$ neighbors in the k-clique, and there is exactly one vertex that is adjacent to both u and v (See Figure 4(b)). The property of the T-device is maintained in the generalized device, i.e. in any proper k-coloring, if vertices u and v are colored with the same color then x must be colored with that color, but if u and v are colored with different colors then x can be colored with any color.

[3] One vertex may remain at degree $\rho(k) - 1$ if both $\rho(k)$ and n are odd, but adding an edge will still violate the degree restriction.

As in Lemma 2, every edge $(u, v) \in E$ is replaced by two T-devices and all the control vertices are chained together by the T-devices in a way similar to Figure 2. At this stage, each control vertex has degree $2 * (k-1) + 2 * \lfloor \frac{k}{2} \rfloor < 3k$, each assistant vertex has degree k or $k+1$, and each original vertex has degree $2 * \lfloor \frac{k}{2} \rfloor *$ *degree in* $G \leq k * \rho(k)/k = \rho(k)$. By attaching $K_{\rho(k),\rho(k)}$'s, we can make each original and assistant vertex have degree $\rho(k)$ and complete the construction of G'[4]. Graph G is not k-colorable if and only if G' is extremal k-colorable with maximum degree $\rho(k)$.

Now, similar to Theorem 1, we can reduce the SAT-UNSAT problem to EkC$\Delta\rho(k)$ and therefore, EkC$\Delta\rho(k)$ is DP-complete.

3 Extremal Unfrozen Non-implicant Problem

In this section, we will present another example of a DP-complete problem that is the extremal variant of an NP-complete problem. A Boolean formula is an expression written using only Boolean variables, conjunction $\wedge$, disjunction $\vee$, negation $\neg$ and parentheses. A Boolean formula is *satisfiable* if the variables of it can be assigned truth values in such a way as to make the formula evaluate to true. A Boolean formula is *valid* if the formula evaluates to true under any truth assignment of its variables. A *literal* is a variable or a negated variable. A *monomial* is a conjunction of literals, and the empty monomial, denoted λ, is valid. An *implicant* of a formula φ is a monomial C such that $C \rightarrow \varphi$ is valid. A monomial C is a *prime implicant* of φ if C is an implicant of φ but deleting any literal from C will make it no longer an implicant of φ. Given a Boolean formula φ and a monomial C, the problem to decide whether C is an implicant of φ is coNP-complete, because we can reduce the tautology problem to it simply by letting C be an empty monomial. The problem to decide whether C is a prime implicant of φ, which is the extremal variant of a coNP-complete problem, has been shown to be DP-complete [9].

We can define several problems related to the implicant problem. A monomial C is a *non-implicant* of a Boolean formula φ if C is not an implicant of φ, and C is a *critical non-implicant* of φ if C is a non-implicant of φ but adding any literal that is not in C will create an implicant of φ. The non-implicant problem is NP-complete, since it is the complementary problem of the implicant problem. Let us consider the complexity of the critical non-implicant problem. Given a Boolean formula φ and a monomial C, the critical non-implicant problem is to decide whether or not C is a critical non-implicant of φ. We say a variable x is *covered* by a monomial C if either x or $\neg x$ appears in C. Note that if C is a non-implicant of φ and x is not covered in C, $C \wedge x$ and $C \wedge \neg x$ cannot both be implicants of φ. It follows that if C does not cover every variable of φ, then C must not be a critical non-implicant of φ. Another observation about implicant is that if C contains both x and $\neg x$, for any variable x of φ, then C must be an implicant of φ. Therefore, a critical non-implicant of a Boolean formula

[4] One extra "free" vertex may be needed in certain parity situations to complete the construction.

is equivalent to a truth assignment that does not satisfy the formula, and the critical non-implicant problem can be solved in polynomial time like most other extremal NP-complete problems.

We find that the unfrozen non-implicant problem is NP-complete and the extremal unfrozen non-implicant problem is DP-complete. We first introduce the formal definitions of these problems. To define the unfrozen non-implicant and extremal unfrozen non-implicant, we need the concept of non-relevant literal. A literal l is a *non-relevant literal* to a monomial C, if neither l nor $\neg l$ appears in C. A monomial C is an *unfrozen non-implicant* of a Boolean formula φ if C is a non-implicant of φ, and $C \wedge l$ is still a non-implicant of φ for any non-relevant literal l to C. C is an *extremal unfrozen non-implicant* of φ if C is an unfrozen non-implicant of φ, but for any non-relevant literal l to C, $C \wedge l$ is not an unfrozen non-implicant of φ, i.e. there is a literal l' non-relevant to $C \wedge l$ such that $C \wedge l \wedge l'$ is an implicant of φ. Given a Boolean formula φ and a monomial C, the unfrozen non-implicant (resp. extremal unfrozen non-implicant) problem is to determine whether or not C is an unfrozen non-implicant (resp. extremal unfrozen non-implicant) of φ. We now present complexity results for these two problems.

Theorem 3. *The unfrozen non-implicant problem is NP-complete.*

Proof. As we discussed before, the problem to determine whether or not a monomial is a non-implicant of a given formula is NP-complete. Given a Boolean formula φ and a monomial C, the certificate of the unfrozen non-implicant problem is the conjunction of the certificates of the non-implicant problems for $C \wedge l$ for each non-relevant literal l to C. This certificate has size $O(n^2)$ and is polynomial checkable, where n is the number of variables in φ. Therefore, the problem is in NP.

We now reduce the SAT problem to this problem. Given a formula φ, let y be a new variable that does not appear in φ and $\rho = \neg\varphi \wedge y$. The empty monomial λ is an unfrozen non-implicant of ρ if and only if $\varphi \in$SAT. If $\varphi \in$SAT, for any variable x of φ, neither x nor $\neg x$ is an implicant of ρ, because assigning false to y will make ρ unsatisfied; for y, a satisfying truth assignment of φ will make ρ unsatisfied; and $\neg y$ is obviously a non-implicant. Hence, λ is an unfrozen non-implicant of ρ. If $\varphi \notin$SAT, then $\neg\varphi$ is valid and y is an implicant of ρ, from which it follows that λ is not an unfrozen non-implicant of ρ. Therefore, the unfrozen non-implicant problem is NP-complete.

Theorem 4. *The extremal unfrozen non-implicant problem is DP-complete.*

Proof. This problem involves an NP problem to verify an unfrozen non-implicant and a coNP problem to verify at most $2n$ monomials, all of which are not unfrozen non-implicants, where n is the number of variables in the formula. The answer to the extremal unfrozen non-implicant problem is yes if and only if the answers to both of these two problems are yes. Hence, the problem is in DP. To show the DP-completeness, we will reduce both the SAT and the UNSAT problems to the extremal unfrozen non-implicant problem.

Lemma 3. *The extremal unfrozen non-implicant problem is NP-hard.*

Proof. (*of Lemma 3*) Given a formula φ, let $V(\varphi)$ be the set of variables of φ. For a new variable $w \notin V(\varphi)$, let $C_w = (w \wedge a_w) \vee (\neg w \wedge \neg a_w)$, where a_w is another new variable. For each variable $x \in V(\varphi)$, let $C_x = (x \wedge a_x) \vee (\neg x \wedge \neg a_x)$, where a_x is a new variable corresponding to x. Note that C_x has both satisfying assignments and unsatisfying assignments no matter how we assign values to x (a_x). The same property holds for C_w since it has the same structure as C_x. Let $\rho = (\neg\varphi \wedge w) \vee C_w \bigvee_{x \in V(\varphi)} C_x$. We now show that the empty monomial λ is an extremal unfrozen non-implicant of ρ if and only if $\varphi \in$SAT.

If $\varphi \in$SAT, we first prove that λ is an unfrozen non-implicant of ρ by showing that no single literal is an implicant of ρ. For $x \in V(\varphi)$, making $w = false, a_x = false, a_w = true$ and each $C_v = false$, $v \in V(\varphi)$ and $v \neq x$, we will have an unsatisfying truth assignment of ρ. Hence, x is not an implicant of ρ. For w, taking a satisfying truth assignment of φ and making $a_w = false$ and each $C_x = false$, $x \in V(\varphi)$, we will have an unsatisfying truth assignment of ρ. Hence, w is not an implicant of ρ. Similarly, it can be verified that all other literals are non-implicants of ρ. Note that there are many two-literal-monomials in formula ρ and making any of them true will satisfy the entire formula. Also, every literal over variables of ρ is in at least one of these two-literal-monomials, which implies that none of the literals is an unfrozen non-implicant of ρ. Hence, λ is an extremal unfrozen non-implicant of ρ. If $\varphi \notin$SAT, then $\neg\varphi$ is valid and w is an implicant of ρ, which implies that λ is not an unfrozen non-implicant of ρ. Therefore, the extremal unfrozen non-implicant problem is NP-hard.

Lemma 4. *The extremal unfrozen non-implicant problem is coNP-hard.*

Proof. (*of Lemma 4*) Given a formula φ', let $V(\varphi')$ be the set of variables of φ'. For each variable $x \in V(\varphi')$, let $C_x = (x \wedge a_x) \vee (\neg x \wedge \neg a_x)$, where a_x is a new variable corresponding to x. Let $y, z \notin V(\varphi')$ be two new variables, and let $\rho' = (\neg\varphi' \wedge y \wedge z) \vee (\neg y \wedge \neg z) \bigvee_{x \in V(\varphi')} C_x$. The empty monomial λ is an extremal unfrozen non-implicant of ρ' if and only if $\varphi' \in$UNSAT.

If $\varphi' \in$UNSAT, then every literal over variables of ρ' is a non-implicant of ρ'. For a_x, $x \in V(\varphi')$, making $x = false, y = false, z = true$ and each $C_v = false$, $v \in V(\varphi')$ and $v \neq x$, yields an unsatisfying truth assignment of ρ'. For z, making $y = false$ and each $C_x = false$, $x \in V(\varphi')$, yields an unsatisfying truth assignment of ρ'. Other literals can be verified similarly. Also, every literal over the variables of ρ' is in at least one of the two-literal-monomials, which are implicants of ρ'. Hence, λ is an extremal unfrozen non-implicant of ρ'.

If $\varphi' \notin$UNSAT, then y is an unfrozen non-implicant of ρ', which means λ is not an extremal unfrozen non-implicant of ρ'. We can show that for any non-relevant literal l to y, $y \wedge l$ is a non-implicant of ρ'. For example, consider $y \wedge z$, by taking a satisfying truth assignment of φ' and making each $C_x = false$, $x \in V(\varphi')$, we will have an unsatisfying truth assignment of ρ'. Therefore, the extremal unfrozen non-implicant problem is coNP-hard.

To show the extremal unfrozen non-implicant problem is DP-hard, we reduce the SAT-UNSAT problem to it. Given two formulas φ and φ' that have disjoint sets of variables, by Lemmas 3 and 4, we can construct two formulas ρ and ρ' with disjoint variable sets, such that the empty monomial λ is an extremal unfrozen non-implicant of ρ if and only if φ $\in$SAT, and the empty monomial λ is an extremal unfrozen non-implicant of ρ' if and only if φ' $\in$UNSAT. Let $\Pi = \rho \vee \rho'$. Note that any implicant of ρ or ρ' is also an implicant of Π. We now show that λ is an extremal unfrozen non-implicant of Π if and only if (φ, φ') $\in$SAT-UNSAT. If (φ, φ') $\in$SAT-UNSAT, then λ is an extremal unfrozen non-implicant of ρ and ρ'. Because, ρ and ρ' have disjoint sets of variables, λ is also an extremal unfrozen non-implicant of Π. If (φ, φ') $\notin$SAT-UNSAT, there are two cases. If φ $\notin$SAT, then by Lemma 3, w is an implicant of ρ and is also an implicant of Π, which means λ is not an unfrozen non-implicant of Π. If φ $\in$SAT and φ' $\notin$UNSAT, by Lemma 4, y is an unfrozen non-implicant of ρ'. For any literal l over the variables of ρ, because l is a non-implicant of ρ, $y \wedge l$ is a non-implicant of Π. Hence, y is also an unfrozen non-implicant of Π, which means λ is not an extremal unfrozen non-implicant of Π. This completes the proof of Theorem 4 .

References

1. Abbas, N., Culberson, J., Stewart, L.: Recognizing maximal unfrozen graphs with respect to independent sets is coNP-complete. Discrete Mathematics & Theoretical Computer Science 7(1), 141–154 (2005)
2. Beacham, A., Culberson, J.: On the complexity of unfrozen problems. Discrete Applied Mathematics 153(1-3), 3–24 (2005)
3. Cai, J.-Y., Meyer, G.E.: Graph minimal uncolorability is DP-complete. SIAM Journal on Computing 16(2), 259–277 (1987)
4. Cai, J., Meyer, G.E.: On the complexity of graph critical uncolorability. In: Ottmann, T. (ed.) ICALP 1987. LNCS, vol. 267, pp. 394–403. Springer, Heidelberg (1987)
5. Cheng, Y., Ko, K.-I., Wu, W.: On the complexity of non-unique probe selection. Theoretical Computer Science 390(1), 120–125 (2008)
6. Chvátal, V., Szemerédi, E.: Many hard examples for resolution. Journal of the ACM 35, 759–768 (1988)
7. Emden-Weinert, T., Hougardy, S., Kreuter, B.: Uniquely colourable graphs and the hardness of colouring graphs of large girth. Combinatorics, Probability and Computing 7(4), 375–386 (1998)
8. Garey, M.R., Johnson, D.S., Stockmeyer, L.: Some simplified NP-complete graph problems. Theoretical Computer Science 1(3), 237–267 (1976)
9. Goldsmith, J., Hagen, M., Mundhenk, M.: Complexity of DNF and isomorphism of monotone formulas. In: Jedrzejowicz, J., Szepietowski, A. (eds.) MFCS 2005. LNCS, vol. 3618, pp. 410–421. Springer, Heidelberg (2005)
10. Kolaitis, P.G., Martin, D.L., Thakur, M.N.: On the complexity of the containment problem for conjunctive queries with built-in predicates. In: PODS 1998: Proceedings of the 17th ACM SIGACT-SIGMOD-SIGART symposium on Principles of Database Systems, pp. 197–204. ACM, New York (1998)
11. Papadimitriou, C.H.: Computational Complexity. Addison-Wesley, Reading (1994)
12. Papadimitriou, C.H., Yannakakis, M.: The complexity of facets (and some facets of complexity). In: STOC 1982: Proceedings of the 14th annual ACM symposium on Theory of Computing, pp. 255–260. ACM, New York (1982)

13. Papadimitriou, C.H., Wolfe, D.: The complexity of facets resolved. Journal of Computer and System Sciences 37(1), 2–13 (1988)
14. Riege, T., Rothe, J.: Completeness in the boolean hierarchy: Exact-four-colorability, minimal graph uncolorability, and exact domatic number problems - a survey. Journal of Universal Computer Science 12(5), 551–578 (2006)
15. Rothe, J.: Exact complexity of exact-four-colorability. Information Processing Letters 87(1), 7–12 (2003)

The Synchronization Problem for Locally Strongly Transitive Automata*

Arturo Carpi[1] and Flavio D'Alessandro[2]

[1] Dipartimento di Matematica e Informatica, Università degli Studi di Perugia, via Vanvitelli 1, 06123 Perugia, Italy
carpi@dipmat.unipg.it

[2] Dipartimento di Matematica, Università di Roma "La Sapienza" Piazzale Aldo Moro 2, 00185 Roma, Italy
dalessan@mat.uniroma1.it

Abstract. The synchronization problem is investigated for a new class of deterministic automata called locally strongly transitive. An application to synchronizing colorings of aperiodic graphs with a cycle of prime length is also considered.

Keywords: Černý conjecture, road coloring problem, synchronizing automaton, rational series.

1 Introduction

The *synchronization problem* for a deterministic n-state automaton consists in the search of an input-sequence, called a *synchronizing* or *reset word*, such that the state attained by the automaton, when this sequence is read, does not depend on the initial state of the automaton itself. If such a sequence exists, the automaton is called *synchronizing*. If a synchronizing automaton is deterministic and complete, a well-known conjecture by Černý claims that it has a synchronizing word of length not larger than $(n-1)^2$ [7]. This conjecture has been shown to be true for several classes of automata (*cf.* [2,3,4,7,8,9,11,12,15,16,17,18,21]). The interested reader is refered to [13,21] for a historical survey of the Černý conjecture and to [6] for the study of the synchronization problem for unambiguous automata. In [8], the authors have studied the synchronization problem for a new class of automata called *strongly transitive*. An n-state automaton is said to be strongly transitive if it is equipped by a set of n words $\{w_0, \ldots, w_{n-1}\}$, called *independent*, such that, for any pair of states s and t, there exists a word w_i, $0 \leq i \leq n-1$, such that $sw_i = t$. Interesting examples of strongly transitive automata are circular automata and transitive synchronizing automata. The main result of [8] is that any synchronizing strongly transitive n-state automaton has a synchronizing word of length not larger than $(n-2)(n+L-1)+1$,

* This work was partially supported by MIUR project "Aspetti matematici e applicazioni emergenti degli automi e dei linguaggi formali" and by fundings "Facoltà di Scienze MM. FF. NN. 2007" of the University of Rome "La Sapienza".

R. Královič and D. Niwiński (Eds.): MFCS 2009, LNCS 5734, pp. 211–222, 2009.

where L denotes the length of the longest word of an independent set of the automaton. As a straightforward corollary of this result, one can obtain the bound $2(n-2)(n-1)+1$ for the length of the shortest synchronizing word of any n-state synchronizing circular automaton.

In this paper, we consider a generalization of the notion of strong transitivity that we call *local strong transitivity.* An n-state automaton is said to be *locally strongly transitive* if it is equipped by a set of k words $W = \{w_0, \ldots, w_{k-1}\}$ and a set of k distinct states $R = \{q_0, \ldots, q_{k-1}\}$ such that, for all $s \in S$, $\{sw_0, \ldots, sw_{k-1}\} = \{q_0, \ldots, q_{k-1}\}$. The set W is still called *independent* while R is called the *range* of W. A typical example of such kind of automata is that of *one-cluster* automata, recently investigated in [4]. An automaton is called one-cluster if there exists a letter a such that the graph of the automaton has a unique cycle labelled by a power of a. Indeed, denoting by k the length of the cycle, one easily verifies that the words

$$a^{n-1}, a^{n-2}, \ldots, a^{n-k}$$

form an independent set of the automaton whose range is the set of vertices of the cycle. Another more general class of locally strongly transitive automata is that of *word connected* automata. Given a n-state automaton $\mathcal{A} = (S, A, \delta)$ and a word $u \in A^*$, $\mathcal{A}$ is called *u-connected* if there exists a state $q \in S$ such that, for every $s \in S$, there exists $\ell > 0$, with $su^\ell = q$. Define R and W respectively as:

$$R = \{q, qu, \ldots, qu^{k-1}\}, \quad W = \{u^i, u^{i+1}, \ldots, u^{i+k-1}\}, \tag{1}$$

where k is the least positive integer such that $qu^k = q$ and i is the least integer such that, for every $s \in S$, $su^i \in R$. Then one has that W is an independent set of $\mathcal{A}$ with range R.

In this paper, by developing the techniques of [8], we prove that any synchronizing locally strongly transitive n-state automaton has a synchronizing word of length not larger than

$$(k-1)(n+L)+\ell,$$

where k is the cardinality of an independent set W and L and ℓ denote respectively the maximal and the minimal length of the words of W. As a straightforward corollary of this result, we obtain that every n-state synchronizing u-connected automaton has a synchronizing word of length not larger than

$$(k-1)(n+(i+k-1)|u|)+i|u|,$$

where i and k are defined as in (1). In particular, if the automaton is one-cluster, the previous bound becomes $(2k-1)(n-1)$, where k is the length of the unique cycle of the graph of $\mathcal{A}$ labelled by a suitable letter of A.

Another result of this paper is related to the well-known *Road coloring problem.* This problem asks to determine whether any aperiodic and strongly connected graph, with all vertices of the same outdegree, (*AGW graph,* for short), has a synchronizing coloring. The problem was formulated in the context of Symbolic Dynamics by Adler, Goodwyn and Weiss and it is explicitly stated in [1].

In 2007, Trahtman has positively solved it in [19]. Recently Volkov has raised in [20] the problem of evaluating, for any AGW graph G, the minimal length of a reset word for a synchronizing coloring of G. This problem has been called *the Hybrid Černý–Road coloring problem.* It is worth to mention that Ananichev has found, for any $n \geq 2$, a AGW graph of n vertices such that the length of the shortest reset word for any synchronizing coloring of the graph is $(n-1)(n-2)+1$ (see [20]). By applying our main theorem and a result by O' Brien [14], we are able to obtain a partial answer to the Hybrid Černý–Road coloring problem. More precisely, we can prove that, given a AGW graph G of n vertices, without multiple edges, such that G has a simple cycle of prime length $p < n$, there exists a synchronizing coloring of G with a reset word of length not larger than $(2p-1)(n-1)$. Moreover, in the case $p = 2$, that is, if G contains a cycle of length 2, a similar result holds, even in presence of multiple edges. Indeed, for every graph of such kind, we can prove the existence of a synchronizing coloring with a reset word of length not larger than $5(n-1)$.

2 Preliminaries

We assume that the reader is familiar with the theory of automata and rational series. In this section we shortly recall a vocabulary of few terms and we fix the corresponding notation used in the paper.

Let A be a finite alphabet and let A^* be the free monoid of words over the alphabet A. The identity of A^* is called the *empty word* and is denoted by ϵ. The *length* of a word of A^* is the integer $|w|$ inductively defined by $|\epsilon| = 0$, $|wa| = |w| + 1$, $w \in A^*$, $a \in A$. For any finite set W of words of A^*, we denote by L_W and ℓ_W the lengths of the longest word and the shortest word in W respectively.

A finite automaton is a triple $\mathcal{A} = (S, A, \delta)$ where S is a finite set of elements called *states* and δ is a map

$$\delta : S \times A \longrightarrow S.$$

The map δ is called the *transition function* of $\mathcal{A}$. The canonical extension of the map δ to the set $S \times A^*$ is still denoted by δ. For any $u \in A^*$ and $s \in S$, the state $\delta(s, u)$ will be also denoted su. If P is a subset of S and u is a word of A^*, we denote by Pu and Pu^{-1} the sets:

$$Pu = \{su \mid s \in P\}, \quad Pu^{-1} = \{s \in S \mid su \in P\}.$$

If $\{sw : w \in A^*\} = S$, for all $s \in S$, $\mathcal{A}$ is *transitive.* If $n = \mathrm{Card}(S)$, we will say that $\mathcal{A}$ is a n-state automaton. A *synchronizing* or *reset* word of $\mathcal{A}$ is any word $u \in A^*$ such that $\mathrm{Card}(Su) = 1$. The state q such that $Su = \{q\}$ is called *reset state.* A *synchronizing* automaton is an automaton that has a reset word. The following conjecture has been raised in [7].

Černý Conjecture. *Each synchronizing n-state automaton has a reset word of length not larger than $(n-1)^2$.*

We recall that a *formal power series* with rational coefficients and non-commuting variables in A is a mapping of the free monoid A^* into $\mathbb{Q}$. A series $\mathcal{S} : A^* \to \mathbb{Q}$ is *rational* if there exists a triple (α, μ, β) where

- $\alpha \in \mathbb{Q}^{1\times n}$, $\beta \in \mathbb{Q}^{n\times 1}$ are a horizontal and a vertical vector respectively,
- $\mu : A^* \to \mathbb{Q}^{n\times n}$ is a morphism of the free monoid A^* in the multiplicative monoid $\mathbb{Q}^{n\times n}$ of matrices with coefficients in $\mathbb{Q}$,
- for every $u \in A^*$, $\mathcal{S}(u) = \alpha\mu(u)\beta$.

The triple (α, μ, β) is called *a representation* of $\mathcal{S}$ and the integer n is called its *dimension.* With a minor abuse of language, if no ambiguity arises, the number n will be also called the dimension of $\mathcal{S}$. Let $\mathcal{A} = (S, A, \delta)$ be any n-state automaton. One can associate with $\mathcal{A}$ a morphism

$$\varphi_{\mathcal{A}} : A^* \to \mathbb{Q}^{S\times S},$$

of the free monoid A^* in the multiplicative monoid $\mathbb{Q}^{S\times S}$ of matrices over the set of rational numbers, defined as: for any $u \in A^*$ and for any $s, t \in S$,

$$\varphi_{\mathcal{A}}(u)_{st} = \begin{cases} 1 & \text{if } t = su \\ 0 & \text{otherwise.} \end{cases}$$

Let R and K be subsets of S and consider the rational series $\mathcal{S}$ with linear representation $(\alpha, \varphi_{\mathcal{A}}, \beta)$, where, for every $s \in S$,

$$\alpha_s = \begin{cases} 1 & \text{if } s \in R, \\ 0 & \text{otherwise,} \end{cases} \qquad \beta_s = \begin{cases} 1 & \text{if } s \in K, \\ 0 & \text{otherwise.} \end{cases}$$

It is easily seen that, for any $u \in A^*$, one has

$$\mathcal{S}(u) = \mathrm{Card}(Ku^{-1} \cap R). \tag{2}$$

The following well-known result (see [5,10]) extends to rational series a fundamental theorem by Moore and Conway on automata equivalence.

Theorem 1. *Let* $\mathcal{S}_1$, $\mathcal{S}_2 : A^* \to \mathbb{Q}$ *be two rational series with coefficients in* $\mathbb{Q}$ *of dimension* n_1 *and* n_2 *respectively. If, for every* $u \in A^*$ *such that* $|u| \le n_1 + n_2 - 1$, $\mathcal{S}_1(u) = \mathcal{S}_2(u)$, *the series* $\mathcal{S}_1$ *and* $\mathcal{S}_2$ *are equal.*

The following result is a consequence of Theorem 1.

Lemma 1. *Let* $\mathcal{A} = (S, A, \delta)$ *be a synchronizing* n*-state automaton. Assume that* R *and* K *are subsets of* S, *and* t *is an integer such that* $0 < t < \mathrm{Card}(R)$. *Then there exists a word* v *such that*

$$|v| \le n, \quad \mathrm{Card}(Kv^{-1} \cap R) \ne t\,.$$

Proof. Consider the series $\mathcal{S}_1$, $\mathcal{S}_2$ defined respectively by

$$\mathcal{S}_1(v) = \mathrm{Card}(Kv^{-1} \cap R), \quad \mathcal{S}_2(v) = t, \quad v \in A^*\,.$$

In view of (2), $\mathcal{S}_1$ is a rational series of dimension n. Moreover, $\mathcal{S}_2$ is a rational series of dimension 1. We have to prove that $\mathcal{S}_1(v) \neq \mathcal{S}_2(v)$ for some $v \in A^*$ such that $|v| \leq n$. In view of Theorem 1, it is sufficient to show that $\mathcal{S}_1 \neq \mathcal{S}_2$. Let u be a reset word of the automaton $\mathcal{A}$. Then $Ku^{-1} = S$ or $Ku^{-1} = \emptyset$, according to whether the corresponding reset state belongs to K or not. One derives, respectively, $\mathcal{S}_1(u) = \mathrm{Card}(R)$ or $\mathcal{S}_1(u) = 0$ and, in both cases, $\mathcal{S}_1(u) \neq t$. Thus, $\mathcal{S}_1 \neq \mathcal{S}_2$, and the statement follows. □

3 Locally Strongly Transitive Automata

In this section, we study the notion of local strong transitivity. We begin by introducing the following definition.

Definition 1. *Let $\mathcal{A} = (S, A, \delta)$ be an automaton. A set of k words $W = \{w_0, \ldots, w_{k-1}\}$ is called* independent *if there exist k distinct states $q_0, \ldots, q_{k-1}$ of $\mathcal{A}$ such that, for all $s \in S$,*

$$\{sw_0, \ldots, sw_{k-1}\} = \{q_0, \ldots, q_{k-1}\}.$$

The set $R = \{q_0, \ldots, q_{k-1}\}$ will be called the range *of W.*

An automaton is called *locally strongly transitive* if it has an independent set of words. The following example shows that local strong transitivity does not imply transitivity.

Example 1. Consider the 4-state automaton $\mathcal{A}$ over the alphabet $A = \{a, b\}$ defined by the following graph:

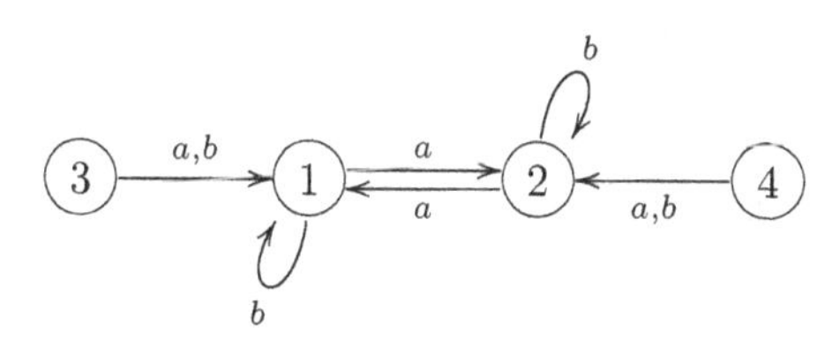

The automaton $\mathcal{A}$ is not transitive. On the other hand, one can easily check that the set $\{a, a^2\}$ is an independent set of $\mathcal{A}$ with range $R = \{1, 2\}$.

The following useful property easily follows from Definition 1.

Lemma 2. *Let $\mathcal{A}$ be an automaton and let W be an independent set of $\mathcal{A}$ with range R. Then, for every $u \in A^*$, the set uW is an independent set of $\mathcal{A}$ with range R.*

Proposition 1. *Let $\mathcal{A} = (S, A, \delta)$ be a n-state automaton and consider an independent set $W = \{w_0, \ldots, w_{k-1}\}$ of $\mathcal{A}$ with range R. Then, for every subset P of R, either*

$$\mathrm{Card}(Pw_i^{-1} \cap R) = \mathrm{Card}(P), \quad \textit{for all } i = 0, \ldots, k-1$$

or there exists j, $0 \leq j \leq k-1$, such that

$$\mathrm{Card}(Pw_j^{-1} \cap R) > \mathrm{Card}(P).$$

Proof. Because of Definition 1, for every $s \in S$ and $r \in R$, there exists exactly one word $w \in W$ such that $s \in \{r\}w^{-1}$. This implies that the sets $\{r\}w_i^{-1}$, $0 \leq i \leq k-1$, give a partition of S. Hence, for any $r \in R$, one has:

$$k = \mathrm{Card}(R) = \sum_{i=0}^{k-1} \mathrm{Card}(R \cap \{r\}w_i^{-1}). \tag{3}$$

Let P be a a subset of R. If P is empty then the statement is trivially true. If $P = \{p_1, \ldots, p_m\}$ is a set of $m \geq 1$ states, then one has:

$$\sum_{i=0}^{k-1} \mathrm{Card}(R \cap Pw_i^{-1}) = \sum_{i=0}^{k-1} \mathrm{Card}\left(\bigcup_{j=1}^{m} R \cap \{p_j\}w_i^{-1} \right).$$

Since $\mathcal{A}$ is deterministic, for any pair p_i, p_j of distinct states of P and for every $u \in A^*$, one has:

$$\{p_i\}u^{-1} \cap \{p_j\}u^{-1} = \emptyset,$$

so that the previous sum can be rewritten as:

$$\sum_{i=0}^{k-1} \sum_{j=1}^{m} \mathrm{Card}(R \cap \{p_j\}w_i^{-1}).$$

The latter equation together with (3) implies that

$$\sum_{i=0}^{k-1} \mathrm{Card}(Pw_i^{-1} \cap R) = k\,\mathrm{Card}(P).$$

The statement follows from the equation above. □

Corollary 1. *Let $\mathcal{A} = (S, A, \delta)$ be a synchronizing n-state automaton and let W be an independent set of $\mathcal{A}$ with range R. Let P be a proper and non empty subset of R. Then there exists a word $w \in A^*$ such that*

$$|w| \leq n + L_W, \quad \mathrm{Card}(Pw^{-1} \cap R) > \mathrm{Card}(P).$$

Proof. Let $W = \{w_0, \ldots, w_{k-1}\}$. We first prove that there exists a word $v \in A^*$ with $|v| \leq n$ such that

$$\mathrm{Card}(P(vw_0)^{-1} \cap R) \neq \mathrm{Card}(P). \tag{4}$$

If $\mathrm{Card}(Pw_0^{-1} \cap R) \neq \mathrm{Card}(P)$, take $v = \epsilon$. Now suppose that

$$\mathrm{Card}(Pw_0^{-1} \cap R) = \mathrm{Card}(P).$$

Since P is a proper and non-empty subset of R and since $\mathcal{A}$ is synchronizing, by applying Lemma 1 with $t = \text{Card}(P)$ and $K = Pw_0^{-1}$, one has that there exists a word $v \in A^*$ such that $|v| \leq n$ and $\text{Card}(P(vw_0)^{-1} \cap R) \neq \text{Card}(P)$. Thus take v that satisfies (4) and let $W' = \{vw_0, \ldots, vw_{k-1}\}$. By Lemma 2, W' is an independent set of $\mathcal{A}$ with range R and $L_{W'} \leq n + L_W$. Therefore, by Proposition 1, taking into account (4),

$$\text{Card}(P(vw)^{-1} \cap R) > \text{Card}(P),$$

for some $w \in W'$. The claim is thus proved. □

As a consequence of Corollary 1, the following theorem holds.

Theorem 2. *Let $\mathcal{A} = (S, A, \delta)$ be a synchronizing n-state automaton and let W be an independent set of $\mathcal{A}$ with range R. Then there exists a reset word for $\mathcal{A}$ of length not larger than*

$$(k-1)(n + L_W) + \ell_W,$$

where $k = \text{Card}(W)$.

Proof. Let $P = \{r\}$ where r is a given state of R. Starting from the set P, by iterated application of Corollary 1, one can find a word v such that

$$|v| \leq (k-1)(n + L_W), \quad Pv^{-1} \cap R = R.$$

Thus we have $Rv = P$. Let u be a word of W of minimal length. Because of Definition 1, we have $Su \subseteq R$ so that $Suv \subseteq Rv = P$. Therefore the word uv is the required word and the statement is proved. □

In the sequel of this section, we will present some results that can be obtained as straightforward corollaries of Theorem 2. We recall that an n-state automaton $\mathcal{A} = (S, A, \delta)$ is *strongly transitive* if there exists an independent set W of n words. Thus, in this case, S is the range of W. The notion of strong transitivity was introduced and studied in [8] where the following result has been proved.

Theorem 3. *Let $\mathcal{A} = (S, A, \delta)$ be a synchronizing strongly transitive n-state automaton with an independent set W. Then there exists a reset word w for $\mathcal{A}$ of length not larger than*

$$1 + (n-2)(n + L_W - 1). \tag{5}$$

Remark 1. Since a strongly transitive automaton $\mathcal{A}$ is also locally strongly transitive, by applying Theorem 2 to $\mathcal{A}$, we obtain the following upper bound on the length of w:

$$(n-1)(n + L_W) + \ell_W,$$

which is larger than that of (5). This gap is the consequence of the following three facts that depend upon the condition $S = R$. The quantity ℓ_W can be obviously deleted in the equation above. The reset word w of $\mathcal{A}$ is factorized as

$w = w_j w_{j-1} \cdots w_0$, where $j \leq n-2$ and, for every $i = 0, \ldots, j$, w_i is obtained by applying Corollary 1. The proof of Corollary 1 is based upon Lemma 1. Under the assumption $S = R$, one can see that the upper bound for the word defined in Lemma 1 is $n - 1$ so that the corresponding upper bound of Corollary 1 can be lowered to $n + L_W - 1$. Finally, since $\mathcal{A}$ is synchronizing, the word w_0 can be chosen as a letter.

Let us now define a remarkable class of locally strongly transitive automata.

Definition 2. *Let $\mathcal{A} = (S, A, \delta)$ be an n-state automaton and let $u \in A^*$. Then $\mathcal{A}$ is called* u-connected *if there exists a state $q \in S$ such that, for every $s \in S$, there exists $k > 0$, such that $su^k = q$.*

Let $\mathcal{A}$ be a u-connected n-state automaton. Define the set R as:

$$R = \{q, qu, \ldots, qu^{k-1}\},$$

where k is the least positive integer such that $qu^k = q$. Let i be the least integer such that, for every $s \in S$, $su^i \in R$. Finally define the set W as:

$$W = \{u^i, u^{i+1}, \ldots, u^{i+k-1}\}.$$

One easily verifies that W is an independent set of $\mathcal{A}$ with range R and, moreover, $\ell_W = i|u|$, $L_W = (i + k - 1)|u|$.

Remark 2. We notice that, by definition of i, there exists a state s such that $s, su, \ldots, su^{i-1} \notin R$ and $su^i \in R$. This implies that the states $s, su, \ldots, su^{i-1}$ are pairwise distinct. Since moreover $\mathrm{Card}(R) = k$, one derives $i + k \leq n$, so that $L_W \leq (n-1)|u|$.

Example 2. Consider the following 6-state automaton $\mathcal{A}$:

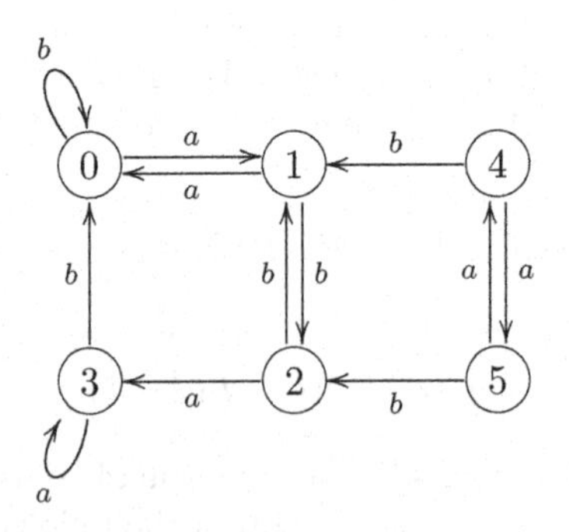

Let $u = ab$ and $q = 0$. One can check that, for all $s \in S$, $su^k = q$, with $k \leq 2$. Since $qu^2 = q$, one has $R = \{0, 2\}$ and one can check that $i = 2$. Thus $W = \{u^2, u^3\}$ is an independent set of $\mathcal{A}$ with range R.

By Remark 2 and by applying Theorem 2, we have

Corollary 2. *Let $\mathcal{A} = (S, A, \delta)$ be a synchronizing n-state automaton. Suppose that $\mathcal{A}$ is u-connected with $u \in A^*$. Let i and k be integers defined as above. Then $\mathcal{A}$ has a reset word of length not larger than*

$$(k-1)(n+(i+k-1)|u|)+i|u|.$$

We say that an automaton $\mathcal{A}$ is *letter-connected* if it is a-connected for some letter $a \in A$. This notion is a natural generalization of that of circular automaton. Indeed, one easily verifies that $\mathcal{A}$ is a-connected if and only if it has a unique cycle labelled by a power of a. By Corollary 2, taking into account that $i+k \leq n$ one derives

Corollary 3. *A synchronizing a-connected n-state automaton, $a \in A$, has a reset word of length not larger than*

$$(2k-1)(n-1),$$

where k is the length of the unique cycle of $\mathcal{A}$ labelled by a power of a.

We remark that the tighter upper bound

$$i+(k-1)(n+k+i-2)$$

for the length of the shortest reset word of a synchronizing a-connected n-state automaton was established in [4].

4 On the Hybrid Černý–Road Coloring Problem

In the sequel, by using the word graph, we will term a finite, directed multigraph with all vertices of outdegree k. A *multiple edge* of a graph is a pair of edges with the same source and the same target. A graph is *aperiodic* if the greatest common divisor of the lengths of all cycles of the graph is 1. A *coloring* of a graph G is a labelling of its edges by letters of a k-letter alphabet that turns G into a complete and deterministic automaton. A coloring of G is *synchronizing* if it transforms G into a synchronizing automaton. The *Road coloring problem* asks the existence of a synchronizing coloring for every aperiodic and strongly connected graph. This problem was formulated in the context of Symbolic Dynamics by Adler, Goodwyn and Weiss and it is explicitly stated in [1]. In 2007, Trahtman has positively solved this problem in [19]. Recently Volkov has raised the following problem [20].

Hybrid Černý–Road coloring problem. *Let G be an aperiodic and strongly connected graph. What is the minimum length of a reset word for a synchronizing coloring of G?*

Now we present a partial answer to the previous problem. For this purpose, we recall the following theorem by O' Brien [14].

Theorem 4. *Let G be an aperiodic and strongly connected graph of n vertices, without multiple edges. Suppose G has a simple cycle C of prime length $p < n$. Then there exists a synchronizing coloring of G such that C is the unique cycle labelled by a power of a given letter a.*

Corollary 4. *Let G be an aperiodic and strongly connected graph of n vertices, without multiple edges. Suppose G has a simple cycle of prime length $p < n$. Then there exists a synchronizing coloring of G with a reset word of length $\leq (2p-1)(n-1)$.*

Proof. The statement follows by applying Theorem 4 and Corollary 3 to G. □

If G contains multiple edges, Theorem 4 cannot be applied so that neither Corollary 4 holds. However, in this case, if G has a cycle of length 2, a result akin to Corollary 4 can be proven. In order to prove this extension, by following [14], we recall some notions and results.

For the sake of simplicity, we assume that all vertices of the graph $G = (S, E)$ have outdegree 2. However, we notice that Proposition 2 stated below remains true also when the common outdegree of the vertices of G is larger.

We suppose that G has a cycle C of length 2 and call s_0, s_1 the vertices of C. A *C-tree* T of G is a subgraph of G that satisfies the following properties:

- the set of vertices of T is S and, for every vertex s of G, exactly one edge outgoing from s is an edge of T;
- C is a subgraph of T;
- for every vertex s of G, there exists a path from s to s_1.

Let T be a given C-tree of G. Define a map

$$C_T : S \longrightarrow \{0,1\}$$

as follows: for every $s \in S$, $C_T(s) = 1$ (resp., $C_T(s) = 0$) if the length of the shortest path in T from s to s_1 is even (resp., odd).

Given a vertex $s \in S$, we say that s is *aperiodic* (with respect to T) if there exists an edge (s,t) of G such that $C_T(t) = C_T(s)$; otherwise the vertex is called *periodic* (with respect to T). One can easily prove that, since G is an aperiodic graph, for every C-tree T of G, there exists an aperiodic vertex.

Let $A = \{a, b\}$ be a binary alphabet and define a coloring of the edges of T as follows: for every edge $e = (s,t)$ of T, label e by the letter a if $C_T(s) = 1$ and by the letter b otherwise. Finally extend, in the obvious way, the latter coloring to the remaining edges of G in order to transform G into an automaton $\mathcal{A}$. We remark that with such a coloring, if $C_T(x) = 1$, then $C_T(xa) = 0$ and, if $C_T(x) = 0$, then $C_T(xb) = 1$. Moreover, if x is aperiodic, then $C_T(xa) = 0$ and $C_T(xb) = 1$. The following lemma can be proved easily.

Lemma 3. *Let x, y be states of $\mathcal{A}$. The following properties hold:*

1. *If $C_T(x) = 1$, then, for every $m \geq \lceil n/2 \rceil - 1$, $x(ab)^m = s_1$;*
2. *If $C_T(x) = 0$, then, for every $m \geq \lceil n/2 \rceil - 1$, $x(ba)^m = s_0$;*

3. *Either* $C_T(x(ab))^m = 0$ *for all* $m \geq 0$ *or* $x(ab)^{n-1} = s_1$.
4. *If* x *is aperiodic, then there exists* $\sigma \in \{a, b\}$ *such that* $C_T(x\sigma) = C_T(y\sigma)$.
5. *There is a word* u *such that* $xu = yu$, *with* $|u| \leq 2n - 2$.

Proof. Conditions 1 and 2 immediately follow from the definition of the coloring of G.

Let us prove Condition 3. Suppose that there exists $m \geq 0$ with $C_T(x(ab)^m) = 1$. Then, by Condition 1, $x(ab)^k = s_1$ for any $k \geq m + \lceil n/2 \rceil - 1$. Let k be the least non-negative integer such that $x(ab)^k = s_1$. The minimality of k implies that the states $x(ab)^i$, $0 \leq i \leq k$ are pairwise distinct. Consequently, $k + 1 \leq n$, so that $x(ab)^{n-1} = s_1(ab)^{n-k-1} = s_1$.

Now let us prove Condition 4. Since x is aperiodic, one has $C_T(xa) = 0$ and $C_T(xb) = 1$. Moreover, either $C_T(ya) = 0$ or $C_T(yb) = 1$, according to the value of $C_T(y)$. The conclusion follows. Let us prove Condition 5. We can find a word v such that $|v| \leq n - 2$ and at least one of the states xv, yv is aperiodic. By Condition 4, $C_T(xv\sigma) = C_T(yv\sigma)$ for some $\sigma = a, b$. Set

$$v' = \begin{cases} (ab)^{\lceil n/2 \rceil - 1} & \text{if } C_T(xv\sigma) = 1\,, \\ (ba)^{\lceil n/2 \rceil - 1} & \text{if } C_T(xv\sigma) = 0\,. \end{cases}$$

According to Conditions 1, 2 one has $xv\sigma v' = yv\sigma v' \in C$ so that the statement is verified with $u = v\sigma v'$. □

Proposition 2. *Let* G *be an aperiodic and strongly connected graph of* n *vertices with outdegree* 2. *Assume that* G *has a cycle of length two. Then there exists a synchronizing coloring of* G *with a reset word of length* $\leq 5(n - 1)$.

Proof. Let C be the cycle of length two of G and let $\mathcal{A}$ be the automaton obtained from G by considering the coloring defined above. By Condition 3, $S(ab)^{n-1} \subseteq \{s_1\} \cup S_0$, where $S_0 = \{x \in S \mid C_T(x) = 0\}$. By Condition 2, $S_0(ba)^{\lceil n/2 \rceil - 1} = \{s_0\}$. Thus, the set $R = S(ab)^{n-1}(ba)^{\lceil n/2 \rceil - 1}$ contains at most 2 states. By Condition 5, there is a word u such that $|u| \leq 2n - 2$ and Ru is reduced to a singleton. We conclude that the word $w = (ab)^{n-1}(ba)^{\lceil n/2 \rceil - 1}u$ is a reset word. Moreover, $|w| \leq 5(n - 1)$. □

References

1. Adler, R.L., Goodwyn, L.W., Weiss, B.: Equivalence of topological Markov shifts. Israel J. Math. 27, 49–63 (1977)
2. Ananichev, D.S., Volkov, M.V.: Synchronizing generalized monotonic automata. Theoret. Comput. Sci. 330(1), 3–13 (2005)
3. Béal, M.-P.: A note on Černý's Conjecture and rational series, technical report, Institut Gaspard Monge, Université de Marne-la-Vallée (2003)
4. Béal, M.-P., Perrin, D.: A quadratic upper bound on the size of a synchronizing word in one-cluster automata. To appear in the proceedings of Developments in Language Theory, DLT 2009, Stuttgart, Germany (2009)

5. Berstel, J., Reutenauer, C.: Rational series and their languages. Springer, Heidelberg (1988)
6. Carpi, A.: On synchronizing unambiguous automata. Theoret. Comput. Sci. 60, 285–296 (1988)
7. Černý, J., Poznámka, k.: Homogénnym eksperimenton s konečnými automatmi. Mat. Fyz. Cas SAV 14, 208–215 (1964)
8. Carpi, A., D'Alessandro, F.: The synchronization problem for strongly transitive automata. In: Ito, M., Toyama, M. (eds.) DLT 2008. LNCS, vol. 5257, pp. 240–251. Springer, Heidelberg (2008)
9. Dubuc, L.: Sur les automates circulaires et la conjecture de Cerny. RAIRO Inform. Théor. Appl. 32, 21–34 (1998)
10. Eilenberg, S.: Automata, Languages and Machines, vol. A. Academic Press, London (1974)
11. Frankl, P.: An extremal problem for two families of sets. Eur. J. Comb. 3, 125–127 (1982)
12. Kari, J.: Synchronizing finite automata on Eulerian digraphs. Theoret. Comput. Sci. 295, 223–232 (2003)
13. Mateescu, A., Salomaa, A.: Many-valued truth functions, Cerny's conjecture and road coloring. EATCS Bull. 68, 134–150 (1999)
14. O' Brien, G.L.: The road coloring problem. Israel J. Math. 39, 145–154 (1981)
15. Pin, J.E.: Le problème de la synchronization et la conjecture de Cerny, Thèse de 3ème cycle, Université de Paris 6 (1978)
16. Pin, J.E.: Sur un cas particulier de la conjecture de Cerny. In: Ausiello, G., Böhm, C. (eds.) ICALP 1978. LNCS, vol. 62, pp. 345–352. Springer, Heidelberg (1978)
17. Rystov, I.: Almost optimal bound of recurrent word length for regular automata. Cybern. Syst. Anal. 31(5), 669–674 (1995)
18. Trahtman, A.N.: The Cerny conjecture for aperiodic automata. Discrete Math. and Theor. Comput. Sci. 9(2), 3–10 (2007)
19. Trahtman, A.N.: The road coloring problem. Israel J. Math. (2008) (to appear)
20. Volkov, M.V.: Communication in: Around the Černý conjecture, International Workshop, University of Wroclaw (Poland) (June 2008)
21. Volkov, M.V.: Synchronizing Automata and the Černý Conjecture. In: Martín-Vide, C., Otto, F., Fernau, H. (eds.) LATA 2008. LNCS, vol. 5196, pp. 11–27. Springer, Heidelberg (2008)

Constructing Brambles

Mathieu Chapelle[1], Frédéric Mazoit[2], and Ioan Todinca[1]

[1] Université d'Orléans, LIFO, BP-6759, F-45067 Orléans Cedex 2, France
{mathieu.chapelle,ioan.todinca}@univ-orleans.fr
[2] Université de Bordeaux, LaBRI, F-33405 Talence Cedex, France
Frederic.Mazoit@labri.fr

Abstract. Given an arbitrary graph G and a number k, it is well-known by a result of Seymour and Thomas [22] that G has treewidth strictly larger than k if and only if it has a bramble of order $k + 2$. Brambles are used in combinatorics as certificates proving that the treewidth of a graph is large. From an algorithmic point of view there are several algorithms computing tree-decompositions of G of width at most k, if such decompositions exist and the running time is polynomial for constant k. Nevertheless, when the treewidth of the input graph is larger than k, to our knowledge there is no algorithm constructing a bramble of order $k+2$. We give here such an algorithm, running in $\mathcal{O}(n^{k+4})$ time. For classes of graphs with polynomial number of minimal separators, we define a notion of *compact brambles* and show how to compute compact brambles of order $k + 2$ in polynomial time, not depending on k.

1 Introduction

Motivation. Treewidth is one of the most extensively studied graph parameters, mainly because many classical $\mathcal{NP}$-hard optimisation problems become polynomial and even linear when restricted to graphs of bounded treewidth. In many applications of treewidth, one needs to compute a tree-decomposition of small width of the input graph. Although determining the treewidth of arbitrary graphs is $\mathcal{NP}$-hard [2], for small values of k one can decide quite efficiently if the treewidth of the input graph is at most k. The first result of this flavour is a rather natural and simple algorithm due to Arnborg, Corneil and Proskurowski [2] working in $\mathcal{O}(n^{k+2})$ time (as usual n denotes the number of vertices of the input graph and m denotes the number of its edges). Using much more sophisticated techniques, Bodlaender [5] solves the same problem by an algorithm running in $\mathcal{O}(f(k)n)$ time, where f is an exponential function. The treewidth problem has been also studied for graph classes, and it was shown [10,11] that for any class of graphs with polynomial number of minimal separators, the treewidth can be computed in polynomial time. In particular the problem is polynomial for several natural graph classes *e.g.* circle graphs, circular arc graphs or weakly chordal graphs. All cited algorithms are able to compute optimal tree-decompositions of the input graph.

One of the main difficulties with treewidth is that we do not have simple certificates for large treewidth. We can easily argue that the treewidth of a graph

R. Královič and D. Niwiński (Eds.): MFCS 2009, LNCS 5734, pp. 223–234, 2009.

G is at most k: it is sufficient to provide a tree-decomposition of width at most k, and one can check in linear time that the decomposition is correct. On the other hand, how to argue that the treewidth of G is (strictly) larger than k? For this purpose, Seymour and Thomas introduced the notion of *brambles*, defined below. The goal of this article is to give algorithms for computing brambles.

Some definitions. A *tree-decomposition* of a graph $G = (V, E)$ is a tree TD such that each node of TD is a bag (vertex subset of G) and satisfies the following properties: (1) each vertex of G appears in at least one bag, (2) for each edge of G there is a bag containing both endpoints of the edge and (3) for any vertex x of G, the bags containing x form a connected subtree of TD. The *width* of the decomposition is the size of its largest bags, minus one. The *treewidth* of G is the minimum width over all tree-decompositions of G.

Without loss of generality, we restrict to tree-decompositions such that there is no bag contained into another. We say that a tree-decomposition TD is finer than another tree-decomposition TD' if each bag of TD is contained in some bag of TD'. Clearly, optimal tree-decompositions are among minimal ones, with respect to the refining relation. A set of vertices of G is a *potential maximal clique* if it is the bag of some minimal tree-decomposition. Potential maximal cliques of a graph are incomparable with respect to inclusion [10]. More important, the number of potential maximal cliques is polynomially bounded in the number of minimal separators of the graph [11], which implies that for several graph classes (circle, circular arc, weakly chordal graphs...) the number of potential maximal cliques is polynomially bounded in the size of the graph.

A *bramble* of order k of $G = (V, E)$ is a function β mapping each vertex subset X of size at most $k - 1$ to a connected component $\beta(X)$ of $G - X$. We require that, for any subsets X, Y of V of size at most $k - 1$, the components $\beta(X)$ and $\beta(Y)$ touch, *i.e.* they have a common vertex or an edge between the two.

Theorem 1 (Treewidth-bramble duality, [22]). *A graph G is of treewidth strictly larger than k if and only if it has a bramble of order $k + 2$.*

Very roughly, if a graph has treewidth larger than k, for each possible bag X of size at most $k + 1$ the bramble points toward a component of $G - X$ which cannot be decomposed by a tree-decomposition of width at most k (restricted to $X \cup \beta(X)$) using X as a bag. Treewidth can also be stated in terms of a cops-and-robber game, and tree-decompositions of width k correspond to winning strategies for $k + 1$ cops, while brambles of order $k + 2$ correspond to escaping strategies for the robber, if the number of cops is at most $k + 1$ (see *e.g.* [6]).

Since optimal tree-decompositions can be obtained using only potential maximal cliques as bags, we introduce now *compact brambles* as follows. Instead of associating to any set X of size less than k a component $\beta(X)$ of $G - X$, we only do this for the sets which are also potential maximal cliques. It is not hard to prove that, in Theorem 1, we can replace brambles by compact brambles, see the full paper [12].

Related work. In the last years, tree-decompositions have been used for practical applications which encouraged the developement of heuristic methods for tree-decompositions of small width [3,13] and, in order to validate the quality of the decompositions, several authors also developed algorithms finding lower bounds for treewidth [9,8,7,13].

A bramble of a graph G can be also defined as a set of connected subsets of $V(G)$, and its order is the minimum cardinality of a hitting set of this family of subsets. Clearly this definition is equivalent to the previous one, but we should point out that, even when the bramble is given, computing its order remains an $\mathcal{NP}$-hard problem. Using the graph minors theory, one can prove that any graph of treewidth larger than k has a bramble (in the sense of this latter definition) of order $k+2$ and of size $f(k)$, for some function f. Nevertheless this function seems extremely huge, even for small value of k. An important result of Grohe and Marx [17] states that there exists a bramble of size polynomial in n *and* k, but of order $\Omega(k^{1/2}/\log^2 k)$. They also emphasize that, if we want a bramble of order $k+2$, its size becomes exponential in k.

Our results. After the seminal paper of Seymour and Thomas proving Theorem 1, there were several results, either with shorter and simpler proofs [4] or for proving other duality theorems of similar flavour, concerning different types of tree-like decompositions, *e.g.* branch-decompositions, path-decompositions or rank-decompositions (see [1] for a survey). Unified versions of these results have been given recently in [1] and [18]. We point out that all these proofs are purely combinatorial: even for a small constant k, it is not clear at all how to use these proofs for constructing brambles on polynomial time (*e.g.* the proofs of [22,4] start by considering the set of all connected subgraphs of the input graph). Nevertheless the recent construction of [18] gives a simpler information on the structure of a bramble (see Theorem 4). Based on their framework which we extend, we build up the following algorithmic result:

Theorem 2. *There is an algorithm that, given a graph G and a number k, computes either a tree-decomposition of G of width at most k, or a bramble of order $k+2$, in $\mathcal{O}(n^{k+4})$ time.*

There is an algorithm that, given a graph G, computes a tree-decomposition of width at most k or a compact bramble of order $k+2$ in $\mathcal{O}(n^4 r^2)$ time, where r is the number of minimal separators of the graph.

These brambles can be used as certificates that allow to check, by a simple algorithm, that a graph has treewidth larger than k. The certificates are of polynomial size for fixed k, and the compact brambles are of polynomial size for graph classes with a polynomial number of minimal separators, like circle, circular-arc or weakly chordal graphs (even for large k). Also, in the area of graph searching (see [16] for a survey), in which it is very common to give optimal strategies for the cops, our construction of brambles provides a simple escape strategy for the robber.

2 Treewidth, Partitioning Trees and the Generalized Duality Theorem

For our purpose, it is more convenient to view tree-decompositions as recursive decompositions of the edge set of the graph, thus we rather use the notion of partitioning trees. The notion is very similar to branch-decompositions [21] except that in our case internal nodes have arbitrary degree; in branch-decompositions, internal nodes are of degree three.

Definition 1 (partitioning trees). *A* partitioning tree *of a graph $G = (V, E)$ is a pair (T, τ) where T is a tree and τ is a one-to-one mapping of the edges of G on the leaves of T.*

Given an internal node i of T, let $\mu(i)$ be the partition of E where each part corresponds to the edges of G mapped on the leaves of a subtree of T obtained by removing node i.

We denote by $\delta(\mu(i))$ the border *of the partition,* i.e. *the set of vertices of G appearing in at least two parts of $\mu(i)$; $\delta(\mu(i))$ is also called the* bag *associated to node i. Let width(T, τ) be the* max $|\delta(\mu(i))| - 1$*, over all internal nodes i of T. The treewidth of G is the minimum width over all partitioning trees of G.*

One can easily transform a partitioning tree (T, τ) into a tree-decomposition such that the bags of the tree-decompositions are exactly the sets $\delta(\mu(i))$. Indeed consider the tree-decomposition with the same tree, associate to each internal node i the bag $\delta(\mu(i))$ and to each leaf j the bag $\{x_j, y_j\}$ where $x_j y_j$ is the edge of G mapped on the leaf j. It is a matter of routine to check that this tree-decomposition satisfies all conditions of tree-decompositions.

Conversely, consider a tree-decomposition TD of G, we describe a partitioning tree (T, τ) obtained from TD without increasing the width. Initially T is a copy TD. For each edge e of G, add a new leaf mapped to this edge, and make it adjacent to one of the nodes of T such that the corresponding bag contains both endpoints of the edge. Eventually, we recursively remove the leaves of T which correspond to nodes of TD (thus there is no edge of G mapped on these leaves). In the end we obtain a partitioning tree of G, and again is not hard to check that (see *e.g.* [19]) for each internal node i of T, we have that $\delta(\mu(i))$ is contained in the bag $X(i)$ corresponding to node i in TD.

Consequently, the treewidth of G is indeed the minimum width over all partitioning trees of G.

Given a graph which is not necessarily of treewidth at most k, we want to capture the "best decompositions" one can obtain with bags of size at most $k+1$. For this purpose we define partial partitioning trees. Roughly speaking, partial partitioning trees of width at most k correspond to tree-decompositions such that all *internal* bags are of size at most $k + 1$ – the leaves are allowed to have arbitrary size.

Definition 2 (partial partitioning trees). *Given a graph $G = (V, E)$, a* partial partitioning tree *is a couple (T, τ), where T is a tree and τ is a one-to-one function from the set of leaves of T to the parts of some partition $(E_1, \ldots, E_p)$*

of E. The bags $\delta(\mu(i))$ of the internal vertices are like in the case of partitioning trees. The bag of the leaf labeled E_i is the set of vertices incident to E_i. The partition $(E_1, \ldots, E_p)$ is called the displayed partition *of (T, τ).*

The width of a partial partitioning tree is $\max |\delta(\mu(i))|$ *over all* internal *nodes of T.*[1]

Partitioning trees are exactly partial partitioning trees such that the corresponding displayed partition of the edge set is a partition into singletons. Given an arbitrary graph, our aim is to characterize displayed partitions corresponding to partial partitioning trees of width at most k. Actually we only consider *connected* partial partitioning trees, and also the more particular case when the labels of the internal nodes are potential maximal cliques, and we will see that this classes contain the optimal decompositions.

The *connected partial partitioning trees* (strongly related to a similar notion on tree-decompositions) are defined as follows. Let X be a set of vertices of G. We say that the set of edges F is a *flap* for X if F is formed by a unique edge with both endpoints in X or if there is a connected component of $G - X$, induced by a vertex subset C of G, and F corresponds exactly to the set of edges of G incident to C (so the edges of F are either between vertices of C or between C and its neighborhood $N_G(C)$). Given a partial partitioning tree (T, τ) and an internal node r which will be considered as the root, let $T(i)$ denote the subtree of T rooted in node i. We denote by $E(i)$ the union of all edge subsets mapped on the leaves of $T(i)$. We say that (T, τ) is a connected partial partitioning tree if and only if, for each internal node j and any son i of j, the edge subset $E(i)$ forms a flap for $\delta(\mu(j))$.

All proofs of this section are given in the full paper [12].

Lemma 1. *Given an arbitrary partial partitioning tree (T, τ) of G, there always exists a connected partial partitioning tree (T', τ') such that each edge subset mapped on a leaf of T' is contained in an edge subset mapped on a leaf of T and each bag of (T', τ') is contained in some bag of (T, τ).*

Let $\mathcal{P}$ be a set of partitions of E. We define a new, larger set of partitions $\mathcal{P}^{\uparrow}$ as follows. Initially, $\mathcal{P}^{\uparrow} = \mathcal{P}$. Then, for any partition $\mu = (E_1, E_2, \ldots, E_p) \in \mathcal{P}^{\uparrow}$ and any partition $\nu = (F_1, F_2, \ldots, F_q, F_{q+1}, \ldots, F_r) \in \mathcal{P}$ such that $F_{q+1} \cup \ldots \cup F_r = E_1$, we add to $\mathcal{P}^{\uparrow}$ the partition $\mu \oplus \nu = (E_2, \ldots, E_p, F_{q+1}, \ldots, F_r)$. The process is iterated until it converges. In terms of partial partitioning trees, each partition $\nu \in \mathcal{P}$ is the displayed partition of a partial partitioning tree with a unique internal node. Initially elements of $\mathcal{P}^{\uparrow}$ correspond to these star-like partial partitioning trees. Then, given any partial partitioning tree T_μ corresponding to an element $\mu = (E_1, E_2, \ldots, E_p) \in \mathcal{P}^{\uparrow}$ and a star-like partial partitioning tree T_ν corresponding to an element $\nu = (F_1, F_2, \ldots, F_q, F_{q+1}, \ldots, F_r) \in \mathcal{P}$, if the leaf E_1 of the first tree is exactly $F_{q+1} \cup \ldots \cup F_r$, then we glue the two trees by identifying the leaf E_1 of T_μ with the internal node of T_ν and then removing the

[1] We emphasize again that a partial partitioning tree might be of small width even if its leaves are mapped on big edge sets.

leaves $F_1, \ldots, F_q$ of T_ν. Thus $\mu \oplus \nu$ is the displayed partition of the new tree. See full paper [12] for a graphical example of this grammar. (A similar but simpler grammar is used in [18].)

The proof of the following statement is an easy consequence of the definitions:

Lemma 2. *Let T be a partial partitioning tree obtained by recursive gluing. The root of T corresponds to the internal node of the first star-like partial partitioning tree used in the recursive gluing. Let x be an internal node of T, and denote by μ_x the partition of $\mathcal{P}$ corresponding to the star-like tree with internal node x – which might be different from the partition $\mu(x)$ introduced in Definition 1. Then for any son y of x, $E(y)$ is a part of μ_x. (Recall that $E(y)$ denotes the union of parts of E mapped on leaves of the subtree $T(y)$ of T rooted in y.) If x is the root, then the sets $E(y)$ for all sons y of x are exactly the parts of μ_x.*

Conversely, let T be a (rooted) partial partitioning tree such that, for each internal node x there is a partition $\mu_x \in \mathcal{P}$ such that for any son y of x, $E(y)$ is a part of μ_x, and, moreover, if x is the root then its sons correspond exactly to the parts of μ_x. Then T is obtained by gluing the partitions μ_x, starting from the root and in a breadth-first search order.

Let $\mathcal{P}_{k-flap}$ be the set of partitions μ of E such that $\delta(\mu)$ is of size at most $k+1$ and the elements of μ are exactly the flaps of $\delta(\mu)$. The set $\mathcal{P}_{k-pmc}$ is the subset of $\mathcal{P}_{k-flap}$ such that for any $\mu \in \mathcal{P}_{k-pmc}$, its border $\delta(\mu)$ is a potential maximal clique. Thus the sets $\mathcal{P}^{\uparrow}_{k-flap}$ and $\mathcal{P}^{\uparrow}_{k-pmc}$ correspond to partial partitioning trees of width at most k.

By Lemma 2, for any graph G, $\mathcal{P}^{\uparrow}_{k-flap}$ is the set of displayed partitions of connected partial partitioning trees of width at most k. Moreover, $\mathcal{P}^{\uparrow}_{k-pmc}$ is the set of displayed partitions of connected partitioning trees of width at most k and such that the bags of all internal vertices are potential maximal cliques. Consequently, we have:

Lemma 3. *G is of treewidth at most k if and only if $\mathcal{P}^{\uparrow}_{k-flap}$ contains the partition into singletons, and if and only if $\mathcal{P}^{\uparrow}_{k-pmc}$ contains the partition into singletons.*

Clearly $\mathcal{P}_{k-flap}$ is of size $\mathcal{O}(n^{k+1})$ and $\mathcal{P}_{k-pmc}$ is of size at most the number of potential maximal cliques of the graph.

We now introduce the notion of orientability, which is useful for the correctness of our algorithm. Given a set of partitions $\mathcal{P}$ and the corresponding set $\mathcal{P}^{\uparrow}$, we say that $\mathcal{P}^{\uparrow}$ is *orientable* if, for any partition $\mu \in \mathcal{P}^{\uparrow}$ and any part F of μ, there is a partition $\nu \in \mathcal{P}^{\uparrow}$ finer than μ (*i.e.* each part of ν is contained in a part of μ) and a partial partitioning tree T_ν displaying ν in which the leaf mapped on F is adjacent to the root.

Lemma 4. *The sets of partitions $\mathcal{P}^{\uparrow}_{k-flap}$ and $\mathcal{P}^{\uparrow}_{k-pmc}$ are orientable.*

Following [18], we define $\mathcal{P}$-brambles, associated to any set $\mathcal{P}$ of partitions of E.

Definition 3 (bramble). *Let $\mathcal{P}$ be an arbitrary set of partitions of E. A $\mathcal{P}$-bramble is a set $\mathcal{B}$ of pairwise intersecting subsets of E, all of them of size at least 2, and such that for any partition $\mu = (E_1, \ldots, E_p) \in \mathcal{P}$, there is a part $E_i \in \mathcal{B}$.*

With this definition, one can see that a bramble of order $k+2$ corresponds exactly to a $\mathcal{P}_{k-flap}$-bramble, and a compact bramble of order $k+2$ corresponds to a $\mathcal{P}_{k-pmc}$-bramble.

We say that a set of partitions $\mathcal{P}^{\uparrow}$ is *refining* if for any two partitions $(A, A_2, \ldots, A_p)$ and $(B, B_2, \ldots, B_q)$ in $\mathcal{P}^{\uparrow}$, with A and B disjoints, there exists a partition $(C_1, \ldots, C_r)$ in $\mathcal{P}^{\uparrow}$ such that each part C_i is contained in some $A_j, 2 \leq j \leq p$, or in some $B_l, 2 \leq l \leq q$.

In [18], the authors show that for the set $\mathcal{P}_k$ defined by the partitions of E having borders of size at most $k+1$ (without any other restriction), $\mathcal{P}_k^{\uparrow}$ is refining. Using Lemma 1, we can easily deduce that $\mathcal{P}_{k-flap}^{\uparrow}$ is refining. On the other hand, much more efforts are required to prove that $\mathcal{P}_{k-pmc}^{\uparrow}$ is also refining (see full paper [12] for details).

Theorem 3. *For any graph G, the sets of partitions $\mathcal{P}_{k-flap}^{\uparrow}$ and $\mathcal{P}_{k-pmc}^{\uparrow}$ are refining.*

The following result implies the "hard part" of Theorem 1. Indeed, by applying Theorem 4 to $\mathcal{P}_{k-flap}^{\uparrow}$, we have that any graph of treewidth greater than k has a bramble of order $k+2$. See [18] or the full paper [12] for a proof of this result.

Theorem 4 ([18]). *Let $\mathcal{P}$ be a set of partitions of E and suppose that $\mathcal{P}$ is refining and does not contain the partition into singletons. Let $\mathcal{B}$ be a set of subsets of E such that:*

1. *Each element of $\mathcal{B}$ is of size at least 2 and it is a part of some $\mu \in \mathcal{P}$;*
2. *For each $\mu = (E_1, \ldots, E_p) \in \mathcal{P}$, there is some part $E_i \in \mathcal{B}$;*
3. *$\mathcal{B}$ is upper closed,* i.e. *for any $F \in \mathcal{B}$, and any superset F' with $F \subset F' \subseteq E$ such that F' is the part of some $\mu' \in \mathcal{P}$, we also have $F' \in \mathcal{B}$;*
4. *$\mathcal{B}$ is inclusion-minimal among all sets satisfying the above conditions.*

Then $\mathcal{B}$ is a $\mathcal{P}$-bramble.

3 The Algorithm

Our goal is to apply Theorem 4 in order to obtain a $\mathcal{P}_{k-flap}^{\uparrow}$-bramble and a $\mathcal{P}_{k-pmc}^{\uparrow}$-bramble. Note that the sets $\mathcal{P}_{k-flap}$ and $\mathcal{P}_{k-pmc}$ are not refining, so we cannot use them directly.

We make an abuse of notation and say that a *flap* of $\mathcal{P}^{\uparrow}$ is a subset of E appearing as the part of some $\mu \in \mathcal{P}^{\uparrow}$. Consequently the flaps of $\mathcal{P}^{\uparrow}$ are exactly the flaps of $\mathcal{P}$. Thus, once we have computed a $\mathcal{P}_{k-flap}^{\uparrow}$-bramble (resp. a $\mathcal{P}_{k-pmc}^{\uparrow}$-bramble), by restricting it to $\mathcal{P}_{k-flap}$ (resp. $\mathcal{P}_{k-pmc}$) we obtain a bramble (resp. a compact bramble) of order $k+2$. The difficulty is that the complexity of our algorithm should be polynomial in the size of $\mathcal{P}_{k-flap}$ (resp. $\mathcal{P}_{k-pmc}$), while the sets $\mathcal{P}^{\uparrow}_{k-flap}$ and $\mathcal{P}^{\uparrow}_{k-pmc}$ may be of exponential size even for small k.

We give now our main algorithmic result. It is stated in a general form, for an arbitrary set of partitions $\mathcal{P}$ such that $\mathcal{P}^{\uparrow}$ is refining and orientable.

Theorem 5 (main theorem). *Let $\mathcal{P}$ be a set of partitions of E. Suppose that $\mathcal{P}^{\uparrow}$ is refining, orientable and does not contain the partition into singletons. Then there is an algorithm constructing a $\mathcal{P}^{\uparrow}$-bramble (and in particular a $\mathcal{P}$-bramble), whose running time is polynomial in the size of E and of $\mathcal{P}$.*

The following algorithm is a straightforward translation of Theorem 4 applied to $\mathcal{P}^{\uparrow}$, so the output $\mathcal{B}_f$ is indeed a $\mathcal{P}^{\uparrow}$-bramble.

```
BRAMBLE(P)
begin
  B ← the set of the flaps of P;
  B_f ← ∅;
  foreach F ∈ B of size one do
    Remove F from B;
  end foreach
  foreach F ∈ B taken in inclusion order do
    if there is a partition μ ∈ P↑ such that F is the unique non-removed
    flap of the partition or ∃F' ∈ B_f : F' ⊆ F then
      Add F to B_f;
    else
      Remove F from B;
    end if
  end foreach
  return B_f;
end
```

Unfortunately the size of $\mathcal{P}^{\uparrow}$ may be exponential in the size of $\mathcal{P}$ and E, and hence the algorithm does not satisfy our complexity requirements because of the test "**if** there is a partition $\mu \in \mathcal{P}^{\uparrow}$ such that F is the unique non-removed flap of the partition", which works on $\mathcal{P}^{\uparrow}$. We would like it to work on $\mathcal{P}$ instead. Thus we replace this test by a marking process working on $\mathcal{P}$ instead of $\mathcal{P}^{\uparrow}$ but giving the same bramble (recall that the flaps of $\mathcal{P}^{\uparrow}$ are exactly the flaps of $\mathcal{P}$).

Let us introduce some definitions for the marking. A flap F is said to be *removed* if it has already been removed from the final $\mathcal{P}$-bramble (instruction "Remove F from $\mathcal{B}$"). Intuitively these *removed* flaps induce some forcing among other flaps: some of the flaps must be added to the final bramble, some others cannot be added to the final bramble. Thus whenever a flap is *removed*, we call the Algorithm UPDATEMARKS. We use two types of markings on the flaps: *forbidden* and *forced*. We prove that a flap F will be marked as *forced* if and only if there is some partition $\mu \in \mathcal{P}^{\uparrow}$ such that all flaps of μ, except F, are *removed*. Thus, in Algorithm BRAMBLE, it suffices to test the mark of the flaps.

UPDATEMARKS
begin
// marking *forbidden* flaps;
while $\exists$ *a flap* F *and a partition* $(F_1, \ldots, F_p, F_{p+1}, \ldots, F_q) \in \mathcal{P}$ **such that** $\left(\cup_i^p F_i\right) = F$ **and** $\forall\ i, 1 \leq i \leq p$, F_i *is* removed **or** F_i *is* forbidden **do**
 Mark F as *forbidden* (if not already marked);
end while
// marking *forced* flaps;
while $\exists$ $(F, F_2, \ldots, F_p) \in \mathcal{P}$ **such that** $\forall\ i, 2 \leq i \leq p\ :\ F_i$ *is* removed **or** F_i *is* forbidden **do**
 Mark F as *forced* (if not already marked);
end while
end

All throughout the algorithm we have the following invariants.

Lemma 5. *A flap* F *is marked as* forbidden *if and only if there exists a subtree* $T(x)$ *of a partial partitioning tree* T *displaying some partition in* $\mathcal{P}^{\uparrow}$ *such that:*

- *Each flap mapped on a leaf of* $T(x)$ *is* removed*;*
- *The union of these flaps is exactly* F*.*

Proof. Suppose first that such a tree exists. We show that the flap F corresponding to the edges mapped on the leaves of $T(x)$ is marked as *forbidden.* Each internal node y of $T(x)$ corresponds to a partition μ_y of $\mathcal{P}$. The edges $E(y)$ of G mapped on the leaves of $T(y)$ form a flap, by construction of T (see also Lemma 2). Consider these internal nodes of $T(x)$, in a bottom up order, we prove by induction that all flaps of such type are marked as *forbidden.* By induction, when node y is considered, for each of its sons $y_1, \ldots, y_p$, either the son is a leaf and hence the corresponding flap is *removed*, or $E(y_i)$ has been previously marked as *forbidden.* Then Algorithm UPDATEMARKS forbids the flap $E(y)$. Consequently, $E(x)$ is marked as *forbidden.*

Conversely, let F be any flap marked as *forbidden*, we construct a partial partitioning tree T with a node x as required. We proceed by induction on the inclusion order over the *forbidden* flaps. When a flap F becomes *forbidden*, it is the union of some flaps $F_1, \ldots, F_p$, each of them being *forbidden* or *removed*, and such that $(F_1, \ldots, F_p, F_{p+1}, \ldots, F_q)$ is an element of $\mathcal{P}$. By induction hypothesis, to each F_i, $i \leq p$, we can associate a tree T_i (which is a subpart of a partial partitioning tree) such that the flaps mapped on the leaves of T_i form a partition of F_i and they are all *removed.* Notice that, if F_i is a *removed* flap, then the tree T_i is only a leaf. Consider now a tree $T(x)$ formed by a root x, corresponding to the partition $(F_1, \ldots, F_p, F_{p+1}, \ldots, F_q) \in \mathcal{P}$, and linked to the roots of $T_1, \ldots, T_p$. Consider any partition $(F, F'_1, \ldots, F'_r) \in \mathcal{P}$ (such a partition exists, since F is a flap). Note that $\cup_j F'_j = \cup_{i \geq p+1} F_i$. The final tree T is obtained by choosing a root z, corresponding to $(F, F'_1, \ldots, F'_r)$, to which root we glue the subtree T_x by adding the edge xz, and for each flap F'_j we add a leaf adjacent to the root

z mapped on F'_j. Thus the final tree T is a gluing between the tree rooted on x and the tree rooted on z. By construction, for each internal node y of this tree, the sets $E(y')$ for the sons y' of y are parts of some partition $\mu_y \in \mathcal{P}$. By Lemma 2, the tree T is a partial partitioning tree displaying an element of $\mathcal{P}^\uparrow$. Clearly all leaves of $T(x)$ are mapped on *removed* flaps. □

Lemma 6. *A flap F is marked as* forced *if and only if there is a partition in $\mathcal{P}^\uparrow$ such that F is the only non-removed flap of the partition.*

Proof. Let us show that F is the unique non-removed flap of some partition in $\mathcal{P}^\uparrow$ if and only if there is a partial partitioning tree T displaying a (possibly another) partition in $\mathcal{P}^\uparrow$ such that all leaves but F correspond to *removed* flaps and, moreover, the leaf mapped on F is adjacent to the root of T. Clearly if such a tree exists, the partition μ displayed by the partial partitioning tree has F as the unique non-removed flap. Suppose now that F is a unique non-removed flap of some partition $\mu' \in \mathcal{P}^\uparrow$. By the fact that $\mathcal{P}^\uparrow$ is orientable, there is a partial partitioning tree T displaying some partition μ, finer than μ', and such that the leaf of T mapped on F is adjacent to the root of the tree. Thus every flap F_i of μ other than F is contained in some flap F'_j of μ'. Since flap F'_j has been *removed* by Algorithm Bramble, flap F_i has also been *removed* (in a previous step, unless the trivial case $F_i = F'_j$): indeed flap F_i has been treated by the algorithm before F'_j, and if F_i is not *removed* it means that it has been added to the bramble $\mathcal{B}_f$, and hence the algorithm will not remove any superset of F_i – contradicting the fact that F'_j is now *removed.* We conclude that all flaps of μ, except F, have been *removed.*

It remains to prove that a flap F is marked as *forced* if and only if there is a partial partitioning tree T displaying a partition in $\mathcal{P}^\uparrow$, such that all leaves but F correspond to *removed* flaps and, moreover, the leaf mapped on F is adjacent to the root of T.

First, if such a tree exists, let z be its root. Then z corresponds to a partition $(F, F_2, \ldots, F_p)$ in $\mathcal{P}$, and each flap F_i corresponds to a subtree T_i of $T - z$. By Lemma 5, every flap F_i, $2 \leq i \leq p$, is *removed* or *forbidden.* Then Algorithm UpdateMarks marks the flap F as forced.

Conversely, suppose that Algorithm UpdateMarks marks the flap F as forced. Let $(F, F_2, \ldots, F_p)$ be the partition in $\mathcal{P}$ which has triggered this mark, so all flaps F_i are *removed* or *forbidden.* Thus to each such flap corresponds a subtree T_i of some partial partitioning tree, such that the leaves of T_i form a partition of F_i and they are all *removed* flaps. The tree T, formed by a root x linked to the roots of each T_i, plus a leaf (mapped on F) adjacent to z satisfies our claim. □

Let us discuss now the time complexity of our algorithm. This complexity is the maximum between the overall complexity of the calls of UpdateMarks, and the complexity of the tests "$\exists F' \in \mathcal{B}_f : F' \subseteq F$" of the Bramble algorithm. Clearly both parts are polynomial in the total number of flaps, the number of elements of $\mathcal{P}$ and the size of the graph. The number of flaps is itself at most $m|\mathcal{P}|$, since each partition has at most m parts. It is easy to see that the overall complexity is quadratic in the size of $\mathcal{P}$, times a small polynomial in the size of the graph. This achieves the proof of Theorem 5.

Our algorithm UPDATEMARKS can be seen as a generalization of the algorithms of [2,10,15]. We can fasten the marking algorithm by using ideas and data structures from [2,15] and especially the fact that the number of good couples (or at least good couples that we really need to use) is moderate. Overall can compute a bramble of maximum order in time $\mathcal{O}(n^{k+4})$ for $\mathcal{P}^{\uparrow}_{k-flap}$. Moreover, for any graph, the number of potential maximal cliques is $O(nr^2)$, where r is the number of minimal separators of the graph [10,11]. For $\mathcal{P}^{\uparrow}_{k-pmc}$, our algorithm can compute a compact bramble of maximum order in time $\mathcal{O}(n^4r^2)$. See the full paper for details and proofs about these complexities [12], which establish Theorem 2.

4 Conclusion

We have presented in this article an algorithm computing brambles of maximum order for arbitrary graphs. The running time for the algorithm is $\mathcal{O}(n^{k+4})$ for computing a bramble of order $k + 2$, and of course we cannot expect drastic improvements since the size of the bramble itself is of order $\Omega(n^{k+1})$.

Treewidth can be defined in terms of graph searching as a game between cops and a robber. As in many games, we can consider the graph of all possible configurations (here it has $\Theta(n^{k+2})$ vertices) and it is possible to compute [14] which are winning configurations for the cops (treewidth at most k) and which are winning for the robber (treewidth larger than k). This can also be considered as a certificate for large treewidth, but clearly more complicated than brambles. As we have pointed out in the introduction, recent results [17] combining linear programming and probabilistic constructions also aim at computing (non optimal) brambles of polynomial size and of large order. To our knowledge, the problem of defining good obstructions to tree-decompositions – in the sense that these obstructions should be of moderate size and easy to manipulate – is largely open.

Another interesting question is whether our algorithm for computing brambles can be used for other tree-like decompositions. As we noted, other parameters (branchwidth, pathwidth, rankwidth...) fit into the framework of [1,18] of partitioning trees and we can define $\mathcal{P}^{\uparrow}_{k-xxx-width}$-brambles in similar ways. The problem is that the size of "basic partitions" (the equivalent of our set $\mathcal{P}_{k-flap}$) may be exponential in n even for small k. Due to results of [19,20], for branchwidth we can also restrict to connected decompositions and thus our algorithm can be used in this case with similar complexity as for treewidth.

Acknowledgement. I. Todinca wishes to thank Fedor Fomin and Karol Suchan for fruitful discussions on this subject.

References

1. Amini, O., Mazoit, F., Thomassé, S., Nisse, N.: Partition Submodular Functions. To appear in Disc. Math. (2008), http://www.lirmm.fr/~thomasse/liste/partsub.pdf

2. Arnborg, S., Corneil, D.G., Proskurowski, A.: Complexity of Finding Embeddings in a k-Tree. SIAM J. on Algebraic and Discrete Methods 8, 277–284 (1987)
3. Bachoore, E.H., Bodlaender, H.L.: New Upper Bound Heuristics for Treewidth. In: Nikoletseas, S.E. (ed.) WEA 2005. LNCS, vol. 3503, pp. 216–227. Springer, Heidelberg (2005)
4. Bellenbaum, P., Diestel, R.: Two Short Proofs Concerning Tree-Decompositions. Combinatorics, Probability & Computing 11(6) (2002)
5. Bodlaender, H.L.: A Linear-Time Algorithm for Finding Tree-Decompositions of Small Treewidth. SIAM J. Comput. 25, 1305–1317 (1996)
6. Bodlaender, H.L.: A Partial k-Arboretum of Graphs with Bounded Treewidth. Theor. Comput. Sci. 209(1-2), 1–45 (1998)
7. Bodlaender, H.L., Grigoriev, A., Koster, A.M.C.A.: Treewidth Lower Bounds with Brambles. Algorithmica 51(1), 81–98 (2008)
8. Bodlaender, H.L., Koster, A.M.C.A.: On the Maximum Cardinality Search Lower Bound for Treewidth. Disc. App. Math. 155(11), 1348–1372 (2007)
9. Bodlaender, H.L., Wolle, T., Koster, A.M.C.A.: Contraction and Treewidth Lower Bounds. J. Graph Algorithms Appl. 10(1), 5–49 (2006)
10. Bouchitté, V., Todinca, I.: Treewidth and Minimum Fill-in: Grouping the Minimal Separators. SIAM J. Comput. 31(1), 212–232 (2001)
11. Bouchitté, V., Todinca, I.: Listing All Potential Maximal Cliques of a Graph. Theor. Comput. Sci. 276(1-2), 17–32 (2002)
12. Chapelle, M., Mazoit, F., Todinca, I.: Constructing Brambles. Technical Report, LIFO, Université d'Orléans (2009)
13. Clautiaux, F., Carlier, J., Moukrim, A., Nègre, S.: New Lower and Upper Bounds for Graph Treewidth. In: Jansen, K., Margraf, M., Mastrolli, M., Rolim, J.D.P. (eds.) WEA 2003. LNCS, vol. 2647, pp. 70–80. Springer, Heidelberg (2003)
14. Fomin, F.V., Fraigniaud, P., Nisse, N.: Nondeterministic Graph Searching: From Pathwidth to Treewidth. Algorithmica 53(3), 358–373 (2009)
15. Fomin, F.V., Kratsch, D., Todinca, I., Villanger, Y.: Exact Algorithms for Treewidth and Minimum Fill-in. SIAM J. Comput. 38(3), 1058–1079 (2008)
16. Fomin, F.V., Thilikos, D.M.: An Annotated Bibliography on Guaranteed Graph Searching. Theor. Comput. Sci. 399(3), 236–245 (2008)
17. Grohe, M., Marx, D.: On Tree Width, Bramble Size, and Expansion. J. Comb. Theory, Ser. B 99(1), 218–228 (2009)
18. Lyaudet, L., Mazoit, F., Thomassé, S.: Partitions Versus Sets: A Case of Duality (submitted 2009), `http://www.lirmm.fr/~thomasse/liste/dualite.pdf`
19. Mazoit, F.: Décompositions Algorithmiques des Graphes. PhD thesis, École Normale Supérieure de Lyon (2004) (in French)
20. Mazoit, F.: The Branch-width of Circular-Arc Graphs. In: Correa, J.R., Hevia, A., Kiwi, M. (eds.) LATIN 2006. LNCS, vol. 3887, pp. 727–736. Springer, Heidelberg (2006)
21. Robertson, N., Seymour, P.D.: Graph Minors X. Obstructions to Tree Decompositions. J. Comb. Theory, Ser. B 52, 153–190 (1991)
22. Seymour, P.D., Thomas, R.: Graph Searching and a Min-Max Theorem for Tree-Width. J. Comb. Theory, Ser. B 58(1), 22–33 (1993)

Self-indexed Text Compression Using Straight-Line Programs

Francisco Claude[1,*] and Gonzalo Navarro[2,**]

[1] David R. Cheriton School of Computer Science, University of Waterloo
fclaude@cs.uwaterloo.ca
[2] Department of Computer Science, University of Chile
gnavarro@dcc.uchile.cl

Abstract. Straight-line programs (SLPs) offer powerful text compression by representing a text $T[1,u]$ in terms of a restricted context-free grammar of n rules, so that T can be recovered in $O(u)$ time. However, the problem of operating the grammar in compressed form has not been studied much. We present a grammar representation whose size is of the same order of that of a plain SLP representation, and can answer other queries apart from expanding nonterminals. This can be of independent interest. We then extend it to achieve the first grammar representation able of extracting text substrings, and of searching the text for patterns, in time $o(n)$. We also give byproducts on representing binary relations.

1 Introduction and Related Work

Grammar-based compression is a well-known technique since at least the seventies, and still a very active area of research. From the different variants of the idea, we focus on the case where a given text $T[1,u]$ is replaced by a context-free grammar (CFG) $\mathcal{G}$ that generates just the string T. Then one can store $\mathcal{G}$ instead of T, and this has shown to provide a universal compression method [18]. Some examples are LZ78 [31], Re-Pair [19] and Sequitur [25], among many others [5].

When a CFG deriving a single string is converted into Chomsky Normal Form, the result is essentially a *Straight-Line Program (SLP)*, that is, a grammar where each nonterminal appears once at the left-hand side of a rule, and can either be converted into a terminal or into the concatenation of two previous nonterminals. SLPs are thus as powerful as CFGs for our purpose, and the grammar-based compression methods above can be straightforwardly translated, with no significant penalty, into SLPs. SLPs are in practice competitive with the best compression methods [11].

There are textual substitution compression methods which are more powerful than those CFG-based [17]. A well-known one is LZ77 [30], which cannot be directly expressed using CFGs. Yet, an LZ77 parsing can be converted into an

* Funded by NSERC of Canada and Go-Bell Scholarships Program.
** Funded in part by Fondecyt Grant 1-080019, and by Millennium Institute for Cell Dynamics and Biotechnology, Grant ICM P05-001-F, Mideplan, Chile.

R. Královič and D. Niwiński (Eds.): MFCS 2009, LNCS 5734, pp. 235–246, 2009.

SLP with an $O(\log u)$ penalty factor in the size of the grammar, which might be preferable as SLPs are much simpler to manipulate [28].

SLPs have received attention because, despite their simplicity, they are able to capture the redundancy of highly repetitive strings. Indeed, an SLP of n rules can represent a text exponentially longer than n. They are also attractive because decompression is easily carried out in linear time. Compression, instead, is more troublesome. Finding the smallest SLP that represents a given text $T[1,u]$ is NP-complete [28,5]. Moreover, some popular grammar-based compressors such as LZ78, Re-Pair and Sequitur, can generate a compressed file much larger than the smallest SLP [5]. Yet, a simple method to achieve an $O(\log u)$-approximation is to parse T using LZ77 and then converting it into an SLP [28], which in addition is *balanced*: the height of the derivation tree for T is $O(\log u)$. (Also, any SLP can be balanced by paying an $O(\log u)$ space penalty factor.)

Compression is regarded nowadays not just as an aid for cheap archival or transmission. Since the last decade, the concept of *compressed text databases* has gained momentum. The idea is to handle a large text collection in compressed form all the time, and decompress just for displaying. Compressed text databases require at least two basic operations over a text $T[1,u]$: *extract* and *find*. Operation *extract* returns any desired portion $T[l,l+m]$ of the text. Operation *find* returns the positions of T where a given search pattern $P[1,m]$ occurs in T. We refer as *occ* to the number of occurrences returned by a *find* operation. *Extract* and *find* should be carried out in $o(u)$ time to be practical for large databases.

There has been some work on random access to grammar-based compressed text, without decompressing all of it [10]. As for finding patterns, there has been much work on *sequential* compressed pattern matching [1], that is, scanning the whole grammar. The most attractive result is that of Kida et al. [17], which can search general SLPs/CFGs in time $O(n+m^2+occ)$. This may be $o(u)$, but still linear in the size of the compressed text. Large compressed text databases require *indexed* searching, where data structures are built on the compressed text to permit searching in $o(n)$ time (at least for small enough m and occ).

Indeed, there has been much work on implementing compressed text databases supporting the operations *extract* and *find* efficiently (usually in $O(m\text{polylog}(n))$ time) [24], but generally based on the Burrows-Wheeler Transform or Compressed Suffix Arrays, not on grammar compression. The only exceptions are based on LZ78-like compression [23,8,27]. These are *self-indexes*, meaning that the compressed text representation itself can support indexed searches. The fact that no (or weak) grammar compression is used makes these self-indexes not sufficiently powerful to cope with highly repetitive text collections, which arise in applications such as computational biology, software repositories, transaction logs, versioned documents, temporal databases, etc. This type of applications require self-indexes based on stronger compression methods, such as general SLPs.

As an example, a recent study modeling a genomics application [29] concluded that none of the existing self-indexes was able to capture the redundancies present in the collection. Even the LZ78-based ones failed, which is not surprising given that LZ78 can output a text exponentially larger than the smallest SLP. The scenario [29] considers a set of r genomes of length n, of individuals of the same species, and can be modeled as r copies of a *base sequence*, where s *edit operations* (substitutions, to simplify) are randomly placed. The most compact self-indexes [24,13,9] occupy essentially nrH_k bits, where H_k is the k-th order entropy of the base sequence, but this is multiplied r times because they are unable of exploiting long-range repetitions. The powerful LZ77, instead, is able to achieve $nH_k + O((r + s) \log n)$ bits, that is, the compressed base sequence plus $O(\log n)$ bits per edit and per sequence. A properly designed SLP can achieve $nH_k + O(r \log n) + O(s \log^2 n)$ bits, which is much better than the current techniques. It is not as good as LZ77, self-indexes based on LZ77 are extremely challenging and do not exist yet.

In this paper we introduce the *first* SLP representation that can support operations *extract* and *find* in $o(n)$ time. More precisely, a plain SLP representation takes $2n \log n$ bits[1], as each new rule expands into two other rules. Our representation takes $O(n \log n) + n \log u$ bits. It can carry out *extract* in time $O((m + h) \log n)$, where h is the height of the derivation tree, and *find* in time $O((m(m + h) + h\,occ) \log n)$ (see the detailed results in Thm. 3). A part of our index is a representation for SLPs which takes $2n \log n(1 + o(1))$ bits and is able of retrieving any rule in time $O(\log n)$, but also of answering other queries on the grammar within the same time, such as finding the rules mentioning a given non-terminal. We also show how to represent a labeled binary relation, which in addition permits a kind of range query.

Our result constitutes a self-index building on much stronger compression methods than the existing ones, and as such, it has the potential of being extremely useful to implement compressed text databases, in particular the very repetitive ones, by combining good compression and efficient indexed searching. Our method is independent on the way the SLP is generated, and as such it can be coupled with different SLP construction algorithms, which might fit different applications.

2 Basic Concepts

2.1 Succinct Data Structures

We make heavy use of succinct data structures for representing sequences with support for *rank/ select* and for range queries. Given a sequence S of length n, drawn from an alphabet Σ of size σ, $rank_S(a, i)$ counts the occurrences of symbol $a \in \Sigma$ in $S[1, i]$, $rank_S(a, 0) = 0$; and $select_S(a, i)$ finds the i-th occurrence of symbol $a \in \Sigma$ in S, $select_S(a, 0) = 0$. We also require that data structures representing S provide operation $access_S(i) = S[i]$.

For the special case $\Sigma = \{0, 1\}$, the problem has been solved using $n + o(n)$ bits of space while answering the three queries in constant time [6]. This was later improved to use $O(m \log \frac{n}{m}) + o(n)$ bits, where m is the number of bits set in the bitmap [26].

[1] In this paper log stands for $\log_2$ unless stated otherwise.

The general case has been proved to be a little harder. Wavelet trees [13] achieve $n \log \sigma + o(n) \log \sigma$ bits of space while answering all the queries in $O(\log \sigma)$ time. Another interesting proposal [12], focused on large alphabets, achieves $n \log \sigma + no(\log \sigma)$ bits of space and answers *rank* and *access* in $O(\log \log \sigma)$ time, while *select* takes $O(1)$ time. Another tradeoff within the same space [12] is $O(1)$ time for *access*, $O(\log \log \sigma)$ time for *select*, and $O(\log \log \sigma \log \log \log \sigma)$ time for *rank*.

Mäkinen and Navarro [20] showed how to use a wavelet tree to represent a permutation π of $[1, n]$ so as to answer *range queries*. We use a trivial variant in this paper. Given a general sequence $S[1, n]$ over alphabet $[1, \sigma]$, we use the wavelet tree of S to find all the symbols of $S[i_1, i_2]$ $(1 \leq i_1 \leq i_2 \leq n)$ which are in the range $[j_1, j_2]$ $(1 \leq j_1 \leq j_2 \leq \sigma)$. The operation takes $O(\log \sigma)$ to count the number of results, and can report each such occurrence in $O(\log \sigma)$ time by tracking each result upwards in the wavelet tree to find its position in S, and downwards to find its symbol in $[1, \sigma]$. The algorithms are almost identical to those for permutations [20].

2.2 Straight-Line Programs

We now define a Straight-Line Program (SLP) and highlight some properties.

Definition 1. *[16] A* Straight-Line Program (SLP) $\mathcal{G} = (X = \{X_1, \ldots, X_n\}, \Sigma)$ *is a grammar that defines a single finite sequence* $T[1, u]$*, drawn from an alphabet* $\Sigma = [1, \sigma]$ *of terminals. It has n rules, which must be of the following types:*

- $X_i \rightarrow \alpha$*, where* $\alpha \in \Sigma$*. It represents string* $\mathcal{F}(X_i) = \alpha$*.*
- $X_i \rightarrow X_l X_r$*, where* $l, r < i$*. It represents string* $\mathcal{F}(X_i) = \mathcal{F}(X_l)\mathcal{F}(X_r)$*.*

We call $\mathcal{F}(X_i)$ *the* phrase generated *by nonterminal* X_i*, and* $T = \mathcal{F}(X_n)$*.*

Definition 2. *[28] The* height *of a symbol* X_i *in the SLP* $\mathcal{G} = (X, \Sigma)$ *is defined as* $height(X_i) = 1$ *if* $X_i \rightarrow \alpha \in \Sigma$*, and* $height(X_i) = 1 + \max(height(X_l), height(X_r))$ *if* $X_i \rightarrow X_l X_r$*. The height of the SLP is* $height(\mathcal{G}) = height(X_n)$*. We will refer to* $height(\mathcal{G})$ *as h when the referred grammar is clear from the context.*

As some of our results will depend on the height of the SLP, it is interesting to recall that an SLP $\mathcal{G}$ of n rules generating $T[1, u]$ can be converted into a $\mathcal{G}'$ of $O(n \log u)$ rules and $height(\mathcal{G}') = O(\log u)$, in $O(n \log u)$ time [28]. Also, as several grammar-compression methods are far from optimal [5], it is interesting that one can find in linear time an $O(\log u)$ approximation to the smallest grammar, which in addition is balanced (height $O(\log u)$) [28].

3 Labeled Binary Relations with Range Queries

In this section we introduce a data structure for labeled binary relations supporting range queries. Consider a binary relation $\mathcal{R} \subseteq A \times B$, where $A =$

$\{1, 2, \ldots, n_1\}$, $B = \{1, 2, \ldots, n_2\}$, a function $\mathcal{L} : A \times B \to L \cup \{\perp\}$, mapping pairs in $\mathcal{R}$ to labels in $L = \{1, 2, \ldots, \ell\}, \ell \geq 1$, and the others to $\perp$. We support the following queries:

- $\mathcal{L}(a, b)$.
- $A(b) = \{a,\ (a, b) \in \mathcal{R}\}$.
- $B(a) = \{b,\ (a, b) \in \mathcal{R}\}$.
- $R(a_1, a_2, b_1, b_2) = \{(a, b) \in \mathcal{R},\ a_1 \leq a \leq a_2,\ b_1 \leq b \leq b_2\}$.
- $\mathcal{L}(l) = \{(a, b) \in \mathcal{R},\ \mathcal{L}(a, b) = l\}$.
- The sizes of the sets: $|A(b)|, |B(a)|$, $|R(a_1, a_2, b_1, b_2)|$, and $|\mathcal{L}(l)|$.

We build on an idea by Barbay et al. [2]. We define, for $a \in A$, $s(a) = b_1 b_2 \ldots b_k$, where $b_i < b_{i+1}$ for $1 \leq i < k$ and $B(a) = \{b_1, b_2, \ldots, b_k\}$. We build a string $S_B = s(1)s(2)\ldots s(n_1)$ and write down the cardinality of each $B(a)$ in unary on a bitmap $X_B = 0^{|B(1)|}10^{|B(2)|}1\ldots0^{|B(n_1)|}1$. Another sequence $S_{\mathcal{L}}$ lists the labels $\mathcal{L}(a, b)$ in the same order they appear in S_B: $S_{\mathcal{L}} = l(1)l(2)\ldots l(n_1)$, $l(a) = \mathcal{L}(a, b_1)\mathcal{L}(a, b_2)\ldots\mathcal{L}(a, b_k)$. We also store a bitmap $X_A = 0^{|A(1)|}10^{|A(2)|}1 \ldots 0^{|A(n_2)|}1$.

We represent S_B using wavelet trees [13], $\mathcal{L}$ with the structure for large alphabets [12], and X_A and X_B in compressed form [26]. Calling $r = |\mathcal{R}|$, S_B requires $r \log n_2 + o(r) \log n_2$ bits, $\mathcal{L}$ requires $r \log \ell + r\, o(\log \ell)$ bits (i.e., zero if $\ell = 1$), and X_A and X_B use $O(n_1 \log \frac{r+n_1}{n_1} + n_2 \log \frac{r+n_2}{n_2}) + o(r + n_1 + n_2) = O(r) + o(n_1 + n_2)$ bits. We answer queries as follows:

- $|A(b)|$: This is just $select_{X_A}(1, b) - select_{X_A}(1, b-1) - 1$.
- $|B(a)|$: It is computed in the same way using X_B.
- $\mathcal{L}(a, b)$: Compute $y \leftarrow select_{X_B}(1, a-1) - a + 1$. Now, if $rank_{S_B}(b, y) = rank_{S_B}(b, y + |B(a)|)$ then a and b are not related and we return $\perp$, otherwise we return $S_{\mathcal{L}}[select_{S_B}(b, rank_{S_B}(b, y + |B(a)|))]$.
- $A(b)$: We first compute $|A(b)|$ and then retrieve the i-th element by doing $y_i \leftarrow select_{S_B}(b, i)$ and returning $1 + select_{X_B}(0, y_i) - y_i$.
- $B(a)$: This is $S_B[select_{X_B}(1, a-1) - a + 2 \ldots select_{X_B}(1, a) - a]$.
- $\mathcal{R}(a_1, a_2, b_1, b_2)$: We first determine which elements in S_B correspond to the range $[a_1, a_2]$. We set $a_1' \leftarrow select_{X_B}(1, a_1 - 1) - a_1 + 2$ and $a_2' \leftarrow select_{X_B}(1, a_2) - a_2$. Then, using range queries in a wavelet tree [20], we retrieve the elements from $S_B[a_1', a_2']$ which are in the range $[b_1, b_2]$.
- $\mathcal{L}(l)$: We retrieve consecutive occurrences of l in $S_{\mathcal{L}}$. For the i-th occurrence we find $y_i \leftarrow select_{S_{\mathcal{L}}}(l, i)$, then we compute $b \leftarrow S_B[y_i]$ and $a \leftarrow 1 + select_{X_B}(0, y_i) - y_i$. Determining $|\mathcal{L}(l)|$ is done via $rank_{S_{\mathcal{L}}}(l, r)$.

We note that, if we do not support queries $\mathcal{R}(a_1, a_2, b_1, b_2)$, we can use also the faster data structure [12] for S_B.

Theorem 1. *Let $\mathcal{R} \subseteq A \times B$ be a binary relation, where $A = \{1, 2, \ldots, n_1\}$, $B = \{1, 2, \ldots, n_2\}$, and a function $\mathcal{L} : A \times B \to L \cup \{\perp\}$, which maps every pair in $\mathcal{R}$ to a label in $L = \{1, 2, \ldots, \ell\}, \ell \geq 1$, and pairs not in $\mathcal{R}$ to $\perp$. Then $\mathcal{R}$ can be indexed using $(r + o(r))(\log n_2 + \log \ell + o(\log \ell) + O(1)) + o(n_1 + n_2)$*

bits of space, where $r = |\mathcal{R}|$. Queries can be answered in the times shown below, where k is the size of the output. One can choose (i) $rnk(x) = acc(x) = \log\log x$ and $sel(x) = 1$, or (ii) $rnk(x) = \log\log x \log\log\log x$, $acc(x) = 1$ and $sel(x) = \log\log x$, independently for $x = \ell$ and for $x = n_2$.

Operation	*Time (with range)*	*Time (without range)*
$\mathcal{L}(a,b)$	$O(\log n_2 + acc(\ell))$	$O(rnk(n_2) + sel(n_2) + acc(\ell))$
$A(b)$	$O(1 + k \log n_2)$	$O(1 + k\, sel(n_2))$
$B(a)$	$O(1 + k \log n_2)$	$O(1 + k\, acc(n_2))$
$\lvert A(b)\rvert, \lvert B(a)\rvert$	$O(1)$	$O(1)$
$R(a_1, a_2, b_1, b_2)$	$O((k+1) \log n_2)$	—
$\lvert R(a_1, a_2, b_1, b_2)\rvert$	$O(\log n_2)$	—
$\mathcal{L}(l)$	$O((k+1) sel(\ell) + k \log n_2)$	$O((k+1) sel(\ell) + k\, acc(n_2))$
$\lvert\mathcal{L}(l)\rvert$	$O(rnk(\ell))$	$O(rnk(\ell))$

We note the asymmetry of the space and time with respect to n_1 and n_2, whereas the functionality is symmetric. This makes it always convenient to arrange that $n_1 \geq n_2$.

4 A Powerful SLP Representation

We provide in this section an SLP representation that permits various queries on the SLP within essentially the same space of a plain representation.

Let us assume for simplicity that all the symbols in Σ are used in the SLP, and thus $\sigma \leq n$ is the effective alphabet size. If this is not the case and $\max(\Sigma) = \sigma' > n$, we can always use a mapping $S[1, \sigma']$ from Σ to the effective alphabet range $[1, \sigma]$, using *rank* and *select* in S. By using Raman et al.'s representation [26], S requires $O(\sigma \log \frac{\sigma'}{\sigma}) = O(n \log \frac{\sigma'}{n})$ bits. Any representation of such an SLP would need to pay for this space.

A plain representation of an SLP with n rules requires at least $2(n-\sigma)\lceil \log n \rceil + \sigma \lceil \log \sigma \rceil \leq 2n \lceil \log n \rceil$ bits. Based on our labeled binary relation data structure of Thm. 1, we give now an alternative SLP representation which requires asymptotically the same space, $2n \log n + o(n \log n)$ bits, and is able to answer a number of interesting queries on the grammar in $O(\log n)$ time. This will be a key part of our indexed SLP representation.

Definition 3. *A* Lexicographic Straight-Line Program (LSLP) $\mathcal{G} = (X, \Sigma, s)$ *is a grammar with nonterminals $X = \{X_1, X_2, \ldots, X_n\}$, terminals Σ, and two types of rules: (i) $X_i \rightarrow \alpha$, where $\alpha \in \Sigma$, (ii) $X_i \rightarrow X_l X_r$, such that:*

1. *The X_is can be renumbered X'_i in order to obtain an SLP.*
2. *$\mathcal{F}(X_i) \preceq \mathcal{F}(X_{i+1}), 1 \leq i < n$, being $\preceq$ the lexicographical order.*
3. *There are no duplicate right hands in the rules.*
4. *X_s is mapped to X'_n, so that $\mathcal{G}$ represents the text $T = \mathcal{F}(X_s)$.*

It is clear that every SLP can be transformed into an LSLP, by removing duplicates and lexicographically sorting the expanded phrases. We will use LSLPs in place of SLPs from now on.

Let us regard a binary relation as a table where the rows represent the elements of set A and the columns the elements of B. In our representation, every row corresponds to a symbol X_l (set A) and every column a symbol X_r (set B). Pairs (l, r) are related, with label i, whenever there exists a rule $X_i \rightarrow X_l X_r$. Since $A = B = L = \{1, 2, \ldots n\}$ and $|\mathcal{R}| = n$, the structure uses $2n \log n + o(n \log n)$ bits. Note that function $\mathcal{L}$ is invertible, thus $|\mathcal{L}(l)| = 1$.

To handle the rules of the form $X_i \rightarrow \alpha$, we set up a bitmap $Y[1, n]$ so that $Y[i] = 1$ if and only if $X_i \rightarrow \alpha$ for some $\alpha \in \Sigma$. Thus we know $X_i \rightarrow \alpha$ in constant time because $Y[i] = 1$ and $\alpha = rank_Y(1, i)$. The total space is $n + o(n) = O(n)$ bits [6]. This works because the rules are lexicographically sorted and all the symbols in Σ are used.

This representation lets us answer the following queries.

- *Access to rules:* Given i, find l and r such that $X_i \rightarrow X_l X_r$, or α such that $X_i \rightarrow \alpha$. If $Y[i] = 1$ we obtain α in constant time as explained. Otherwise, we obtain $\mathcal{L}(i) = \{(l, r)\}$ from the labeled binary relation, in $O(\log n)$ time.
- *Reverse access to rules:* Given l and r, find i such that $X_i \rightarrow X_l X_r$, if any. This is done in $O(\log n)$ time via $\mathcal{L}(l, r)$ (if it returns $\perp$, there is no such X_i). We can also find, given α, the $X_i \rightarrow \alpha$, if any, in $O(1)$ time via $i = select_Y(1, \alpha)$.
- *Rules using a left/right symbol:* Given i, find those j such that $X_j \rightarrow X_i X_r$ (left) or $X_j \rightarrow X_l X_i$ (right) for some X_l, X_r. The first is answered using $\{\mathcal{L}(i, r), r \in B(j)\}$ and the second using $\{\mathcal{L}(l, i), l \in A(j)\}$, in $O(\log n)$ time per each X_i found.
- *Rules using a range of symbols:* Given $l_1 \leq l_2, r_1 \leq r_2$, find those i such that $X_i \rightarrow X_l X_r$ for any $l_1 \leq l \leq l_2$ and $r_1 \leq r \leq r_2$. This is answered, in $O(\log n)$ time per symbol retrieved, using $\{\mathcal{L}(a, b), (a, b) \in \mathcal{R}(l_1, l_2, r_1, r_2)\}$.

Again, if the last operation is not provided, we can choose the faster representation [12] (alternative (i) in Thm. 1), to achieve $O(\log \log n)$ time for all the other queries.

Theorem 2. *An SLP $\mathcal{G} = (X = \{X_1, \ldots, X_n\}, \Sigma)$, $\Sigma = [1, \sigma]$, $\sigma \leq n$, can be represented using $2n \log n + o(n \log n)$ bits, such that all the queries described above (access to rules, reverse access to rules, rules using a symbol, and rules using a range of symbols) can be answered in $O(\log n)$ time per delivered datum. If we do not support the rules using a range of symbols, times drop to $O(\log \log n)$. For arbitrary integer Σ one needs additional $O(n \log \frac{\max(\Sigma)}{n})$ bits.*

5 Indexable Grammar Representations

We now provide an LSLP-based text representation that permits indexed search and random access. We assume our text $T[1, u]$, over alphabet $\Sigma = [1, \sigma]$, is represented with an SLP of n rules.

We will represent an LSLP $\mathcal{G}$ using a variant of Thm. 2. The rows will represent X_l as before, but these will be sorted by *reverse* lexicographic order, as if they represented $\mathcal{F}(X_l)^{rev}$. The columns will represent X_r, ordered lexicographically by $\mathcal{F}(X_r)$. We will also store a permutation π_R, which maps reverse to direct lexicographic ordering. This must be used to translate row positions to nonterminal identifiers. We use Munro et al.'s representation [22] for π_R, with parameter $\epsilon = \frac{1}{\log n}$, so that π_R can be computed in constant time and π_R^{-1} in $O(\log n)$ time, and the structure needs $n \log n + O(n)$ bits of space.

With the LSLP representation and π_R, the space required is $3n \log n + o(n \log n)$ bits. We add other $n \log u$ bits for storing the lengths $|\mathcal{F}(X_i)|$ for all the nonterminals X_i.

5.1 Extraction of Text from an LSLP

To expand a substring $\mathcal{F}(X_i)[j, j']$, we first find position j: We recursively descend in the parse tree rooted at X_i until finding its jth position. Let $X_i \to X_l X_r$, then if $|\mathcal{F}(X_l)| \geq j$ we descend to X_l, otherwise to X_r, in this case looking for position $j - |\mathcal{F}(X_l)|$. This takes $O(height(X_i) \log n)$ time. In our way back from the recursion, if we return from the left child, we fully traverse the right child left to right, until outputting $j' - j + 1$ terminals.

This takes in total $O((height(X_i)+j'-j) \log n)$ time, which is at most $O((h+j'-j) \log n)$. This is because, on one hand, we will follow both children of a rule at most $j' - j$ times. On the other, we will follow only one child at most twice per tree level, as otherwise two of them would share the same parent.

5.2 Searching for a Pattern in an LSLP

Our problem is to find all the occurrences of a pattern $P = p_1 p_2 \ldots p_m$ in the text $T[1, u]$ defined by an LSLP of n rules. As in previous work [15], except for the special case $m = 1$, occurrences can be divided into *primary* and *secondary*. A primary occurrence in $\mathcal{F}(X_i)$, $X_i \to X_l X_r$, is such that it spans a suffix of $\mathcal{F}(X_l)$ and a prefix of $\mathcal{F}(X_r)$, whereas each time X_i is used elsewhere (directly or transitively in other nonterminals that include it) it produces secondary occurrences. In the case $P = \alpha$, we say that the primary occurrence is at $X_i \to \alpha$ and the other occurrences are secondary.

Our strategy is to first locate the primary occurrences, and then track all their secondary occurrences in a recursive fashion. To find primary occurrences of P, we test each of the $m - 1$ possible partitions $P = P_l P_r$, $P_l = p_1 p_2 \ldots p_k$ and $P_r = p_{k+1} \ldots p_m$, $1 \leq k < m$. For each partition $P_l P_r$, we first find all those X_ls such that P_l is a suffix of $\mathcal{F}(X_l)$, and all those X_rs such that P_r is a prefix of $\mathcal{F}(X_r)$. The latter forms a lexicographic range $[r_1, r_2]$ in the $\mathcal{F}(X_r)$s, and the former a lexicographic range $[l_1, l_2]$ in the $\mathcal{F}(X_l)^{rev}$s. Thus, using our LSLP representation, the X_is containing the primary occurrences correspond those labels i found within rows l_1 and l_2, and between columns r_1 and r_2, of the binary relation. Hence a query for *rules using a range of symbols* will retrieve each such X_i in $O(\log n)$ time. If $P = \alpha$, our only primary occurrence is obtained in $O(1)$ time using *reverse access to rules*.

Now, given each primary occurrence at X_i, we must track all the nonterminals that use X_i in their right hand sides. As we track the occurrences, we also maintain the *offset* of the occurrence within the nonterminal. The offset for the primary occurrence at $X_i \to X_l X_r$ is $|\mathcal{F}(X_l)| - k + 1$ (l is obtained with an *access to rule* query for i). Each time we arrive at the initial symbol X_s, the offset gives the position of a new occurrence.

To track the uses of X_i, we first find all those $X_j \to X_i X_r$ for some X_r, using query *rules using a left symbol* for $\pi_R^{-1}(i)$. The offset is unaltered within those new nonterminals. Second, we find all those $X_j \to X_l X_i$ for some X_l, using query *rules using a right symbol* for i. The offset in these new nonterminals is that within X_i plus $|\mathcal{F}(X_l)|$, where again $\pi_R(l)$ is obtained from the result using an *access to rule* query. We proceed recursively with all the nonterminals X_j found, reporting the offsets (and finishing) each time we arrive at X_s.

Note that we are tracking each occurrence individually, so that we can process several times the same nonterminal X_i, yet with different offsets. Each occurrence may require to traverse all the syntax tree up to the root, and we spend $O(\log n)$ time at each step. Moreover, we carry out $m - 1$ range queries for the different pattern partitions. Thus the overall time to find the *occ* occurrences is $O((m + h\, occ) \log n)$.

We remark that we do not need to output all the occurrences of P. If we just want *occ* occurrences, our cost is proportional to this *occ*. Moreover, the *existence problem*, that is, determining whether or not P occurs in T, can be answered just by counting the primary occurrences, and it corresponds to $occ = 0$. The remaining problem is how to find the range of phrases starting/ending with a suffix/prefix of P. This is considered next.

5.3 Prefix and Suffix Searching

We present different time/space tradeoffs, to search for P_l and P_r in the respective sets.

Binary search based approach. We can perform a binary search over the $\mathcal{F}(X_i)$s and over the $\mathcal{F}(X_i)^{rev}$s to determine the ranges where P_r and P_l^{rev}, respectively, belong. We do the first binary search in the nonterminals as they are ordered in the LSLP. In order to do the string comparisons, we extract the first m terminals of $\mathcal{F}(X_i)$, in time $O((m + h) \log n)$ (Sec. 5.1). As the binary search requires $O(\log n)$ comparisons, the total cost is $O((m + h) \log^2 n)$ for the partition $P_l P_r$. The search within the reverse phrases is similar, except that we extract the m rightmost terminals and must use π_R to find the rule from the position in the reverse ordering. This variant needs no extra space.

Compact Patricia Trees. Another option is to build Patricia Trees [21] for the $\mathcal{F}(X_i)$s and for the $\mathcal{F}(X_i)^{rev}$s (adding them a terminator so that each phrase corresponds to a leaf). By using the cardinal tree representation of Benoit et al. [4] for the tree structure and the edge labels, each such tree can be represented using $2n \log \sigma + O(n)$ bits, and traversal (including to a child labeled α) can be carried out in constant time. The ith leaf of the tree for the $\mathcal{F}(X_i)$s corresponds to nonterminal X_i (and the ith of the three for the $\mathcal{F}(X_i)^{rev}$s, to $X_{\pi_R(i)}$). Hence, upon reaching the tree node corresponding to the search string, we obtain the lexicographic range by counting the number of leaves up to the node subtree and past it, which can also be done in constant time [4].

The difficult point is how to store the Patricia tree skips, as in principle they require other $4n \log u$ bits of space. If we do not store the skips at all, we can still compute them at each node by extracting the corresponding substrings for the leftmost and rightmost descendant of the node, and checking for how many more symbols they coincide [6]. This can be obtained in time $O((\ell + h) \log n)$, where ℓ is the skip value (Sec. 5.1). The total search time is thus $O(m \log n + mh \log n) = O(mh \log n)$.

Instead, we can use k bits for the skips, so that skips in $[1, 2^k - 1]$ can be represented, and a skip zero means $\geq 2^k$. Now we need to extract leftmost and rightmost descendants only when the edge length is $\ell \geq 2^k$, and we will work $O((\ell - 2^k + h) \log n)$ time. Although the $\ell - 2^k$ terms still can add up to $O(m)$ (e.g., if all the lengths are $\ell = 2^{k+1}$), the h terms can be paid only $O(1 + m/2^k)$ times. Hence the total search cost is $O((m + h + \frac{mh}{2^k}) \log n)$, at the price of at most $4nk$ extra bits of space. We must also do the final Patricia tree check due to skipped characters, but this adds only $O((m + h) \log n)$ time. For example, using $k = \log h$ we get $O((m + h) \log n)$ time and $4n \log h$ extra bits of space.

As we carry out $m - 1$ searches for prefixes and suffixes of P, as well as $m - 1$ range searches, plus *occ* extraction of occurrences, we have the final result.

Theorem 3. *Let $T[1, u]$ be a text over an effective alphabet $[1, \sigma]$ represented by an SLP of n rules and height h. Then there exists a representation of T using $n(\log u + 3 \log n + O(\log \sigma + \log h) + o(\log n))$ bits, such that any substring $T[l, r]$ can be extracted in time $O((r - l + h) \log n)$, and the positions of occ occurrences of a pattern $P[1, m]$ in T can be found in time $O((m(m + h) + h\,occ) \log n)$. By removing the $O(\log h)$ term in the space, search time raises to $O((m^2 + occ) h \log n)$. By further removing the $O(\log \sigma)$ term in the space, search time raises to $O((m(m + h) \log n + h\,occ) \log n)$. The existence problem is solved within the time corresponding to $occ = 0$.*

Compared with the $2n \log n$ bits of the plain SLP representation, ours requires at least $4n \log n + o(n \log n)$ bits, that is, roughly twice the space. More generally, as long as $u = n^{O(1)}$, our representation uses $O(n \log n)$ bits, of the same order of the SLP size. Otherwise, our representation is superlinear in the size of the SLP (almost quadratic in the extreme case $n = O(\log u)$). Yet, if $u = n^{\omega(1)}$, our representation takes $u^{o(1)}$ bits, which is still much smaller than the *original* text.

We have not discussed construction times for our index (given the SLP). Those are $O(n \log n)$ for the binary relation part, and all the lengths $|\mathcal{F}(X_i)|$ could be easily obtained in $O(n)$ time. Sorting the strings lexicographically, as well as constructing the tries, however, can take as much as $\sum_{i=1}^{n} |\mathcal{F}(X_i)|$, which can be even $\omega(u)$. Yet, as all the phrases are substrings of $T[1, u]$, we can build the suffix array of T in $O(u)$ time [14], record one starting text position of each $\mathcal{F}(X_i)$ (obtained by expanding T from the grammar), and then sorting them in

$O(n \log n)$ time using the inverse suffix array permutation (the ordering when one phrase is a prefix of the other is not relevant for our algorithm). To build the Patricia trees we can build the suffix tree in $O(u)$ time [7], mark the n suffix tree leaves corresponding to phrase beginnings, prune the tree to the ancestors of those leaves (which are $O(n)$ after removing unary paths again), and create new leaves with the corresponding string depths $|\mathcal{F}(X_i)|$. The point to insert the new leaves are found by binary searching the string depths $|\mathcal{F}(X_i)|$ with level ancestor queries [3] from the suffix tree leaves. The process takes $O(u + n \log n)$ time and $O(u \log u)$ bits of space. Reverse phrases are handled identically.

6 Conclusions and Future Work

We have presented the first indexed compressed text representation based on Straight-Line Programs (SLP), which are as powerful as context-free grammars. It achieves space close to that of the bare SLP representation (in many relevant cases, of the same order) and, in addition to just uncompressing, it permits extracting arbitrary substrings of the text, as well as carrying out pattern searches, in time usually sublinear on the grammar size. We also give interesting byproducts related to powerful SLP and binary relation representations.

We regard this as a foundational result on the extremely important problem of achieving self-indexes built on compression methods potentially more powerful than the current ones [24]. As such, there are several possible improvements we plan to work on, such as (1) reducing the $n \log u$ space term; (2) reduce the $O(m^2)$ term in search times; (3) alleviate the $O(h)$ term in search times by restricting the grammar height while retaining good compression; (4) report occurrences faster than one-by-one. We also plan to implement the structure to achieve strong indexes for very repetitive text collections.

References

1. Amir, A., Benson, G.: Efficient two-dimensional compressed matching. In: Proc. 2nd DCC, pp. 279–288 (1992)
2. Barbay, J., Golynski, A., Munro, I., Rao, S.S.: Adaptive searching in succinctly encoded binary relations and tree-structured documents. In: Lewenstein, M., Valiente, G. (eds.) CPM 2006. LNCS, vol. 4009, pp. 24–35. Springer, Heidelberg (2006)
3. Bender, M., Farach-Colton, M.: The level ancestor problem simplified. Theor. Comp. Sci. 321(1), 5–12 (2004)
4. Benoit, D., Demaine, E., Munro, I., Raman, R., Raman, V., Rao, S.S.: Representing trees of higher degree. Algorithmica 43(4), 275–292 (2005)
5. Charikar, M., Lehman, E., Liu, D., Panigrahy, R., Prabhakaran, M., Sahai, A., Shelat, A.: The smallest grammar problem. IEEE TIT 51(7), 2554–2576 (2005)
6. Clark, D.: Compact Pat Trees. PhD thesis, University of Waterloo (1996)
7. Farach-Colton, M., Ferragina, P., Muthukrishnan, S.: On the sorting-complexity of suffix tree construction. J. ACM 47(6), 987–1011 (2000)
8. Ferragina, P., Manzini, G.: Indexing compressed texts. J. ACM 52(4), 552–581 (2005)

9. Ferragina, P., Manzini, G., Mäkinen, V., Navarro, G.: Compressed representations of sequences and full-text indexes. ACM Trans. Alg. 3(2), 20 (2007)
10. Gasieniec, L., Kolpakov, R., Potapov, I., Sant, P.: Real-time traversal in grammar-based compressed files. In: Proc. 15th DCC, p. 458 (2005)
11. Gasieniec, L., Potapov, I.: Time/space efficient compressed pattern matching. Fund. Inf. 56(1-2), 137–154 (2003)
12. Golynski, A., Munro, I., Rao, S.: Rank/select operations on large alphabets: a tool for text indexing. In: Proc. 17th SODA, pp. 368–373 (2006)
13. Grossi, R., Gupta, A., Vitter, J.: High-order entropy-compressed text indexes. In: Proc. 14th SODA, pp. 841–850 (2003)
14. Kärkkäinen, J., Sanders, P.: Simple linear work suffix array construction. In: Baeten, J.C.M., Lenstra, J.K., Parrow, J., Woeginger, G.J. (eds.) ICALP 2003. LNCS, vol. 2719, pp. 943–955. Springer, Heidelberg (2003)
15. Kärkkäinen, J., Ukkonen, E.: Lempel-Ziv parsing and sublinear-size index structures for string matching. In: Proc. 3rd WSP, pp. 141–155. Carleton University Press (1996)
16. Karpinski, M., Rytter, W., Shinohara, A.: An efficient pattern-matching algorithm for strings with short descriptions. Nordic J. Comp. 4(2), 172–186 (1997)
17. Kida, T., Matsumoto, T., Shibata, Y., Takeda, M., Shinohara, A., Arikawa, S.: Collage system: a unifying framework for compressed pattern matching. Theor. Comp. Sci. 298(1), 253–272 (2003)
18. Kieffer, J., Yang, E.-H.: Grammar-based codes: A new class of universal lossless source codes. IEEE TIT 46(3), 737–754 (2000)
19. Larsson, J., Moffat, A.: Off-line dictionary-based compression. Proc. IEEE 88(11), 1722–1732 (2000)
20. Mäkinen, V., Navarro, G.: Rank and select revisited and extended. Theor. Comp. Sci. 387(3), 332–347 (2007)
21. Morrison, D.: PATRICIA – practical algorithm to retrieve information coded in alphanumeric. J. ACM 15(4), 514–534 (1968)
22. Munro, J., Raman, R., Raman, V., Rao, S.S.: Succinct representations of permutations. In: Baeten, J.C.M., Lenstra, J.K., Parrow, J., Woeginger, G.J. (eds.) ICALP 2003. LNCS, vol. 2719, pp. 345–356. Springer, Heidelberg (2003)
23. Navarro, G.: Indexing text using the Ziv-Lempel trie. J. Discr. Alg. 2(1), 87–114 (2004)
24. Navarro, G., Mäkinen, V.: Compressed full-text indexes. ACM Comp. Surv. 39(1), 2 (2007)
25. Nevill-Manning, C., Witten, I., Maulsby, D.: Compression by induction of hierarchical grammars. In: Proc. 4th DCC, pp. 244–253 (1994)
26. Raman, R., Raman, V., Rao, S.: Succinct indexable dictionaries with applications to encoding k-ary trees and multisets. In: Proc. 13th SODA, pp. 233–242 (2002)
27. Russo, L., Oliveira, A.: A compressed self-index using a Ziv-Lempel dictionary. Inf. Retr. 11(4), 359–388 (2008)
28. Rytter, W.: Application of Lempel-Ziv factorization to the approximation of grammar-based compression. Theor. Comp. Sci. 302(1-3), 211–222 (2003)
29. Sirén, J., Välimäki, N., Mäkinen, V., Navarro, G.: Run-length compressed indexes are superior for highly repetitive sequence collections. In: Amir, A., Turpin, A., Moffat, A. (eds.) SPIRE 2008. LNCS, vol. 5280, pp. 164–175. Springer, Heidelberg (2008)
30. Ziv, J., Lempel, A.: A universal algorithm for sequential data compression. IEEE TIT 23(3), 337–343 (1977)
31. Ziv, J., Lempel, A.: Compression of individual sequences via variable length coding. IEEE TIT 24(5), 530–536 (1978)

Security and Tradeoffs of the Akl-Taylor Scheme and Its Variants

Paolo D'Arco[1], Alfredo De Santis[1], Anna Lisa Ferrara[1,2], and Barbara Masucci[1]

[1] Università degli Studi di Salerno, Italy
[2] University of Illinois at Urbana-Champaign, USA

Abstract. In 1983 Akl and Taylor [*Cryptographic Solution to a Problem of Access Control in a Hierarchy*, ACM Transactions on Computer Systems, 1(3), 239–248, 1983] first suggested the use of cryptographic techniques to enforce access control in hierarchical structures. Over time, their scheme has been used in several different contexts, including mobile agents environments and broadcast encryption. However, it has never been fully analyzed from the security point of view.

We provide a rigorous analysis of the Akl-Taylor scheme and prove that it is secure against key recovery. We also show how to obtain different tradeoffs between the amount of public information and the number of steps required to perform key derivation. Moreover, we propose a general construction to set up a key assignment scheme secure w.r.t. key indistinguishability, given *any* key assignment scheme secure against key recovery. Finally, we show how to use our construction, along with our tradeoffs, to obtain a variant of the Akl-Taylor scheme, secure w.r.t key indistinguishability, requiring a constant amount of public information.

1 Introduction

Akl and Taylor [1] designed a hierarchical key assignment scheme where each class is assigned an encryption key that can be used, along with some public parameters, to compute the key assigned to all classes lower down in the hierarchy. Due to its simplicity and versatility, the scheme has been employed to enforce access control in several different domains (e.g., [2,5,6,11,15,16,17]).

Akl and Taylor related the security of their scheme to the infeasibility of extracting r-th roots modulo n, where $r > 1$ is an integer and n is the product of two large unknown primes. However, their analysis gives only an *intuition* for the security of the scheme. Later on, Goldwasser and Micali [9], introduced the use of *security reductions* to provide rigorous security arguments for cryptographic protocols[1]. Despite its use for many years, there has been no attempt to fully analyze the security of the Akl-Taylor scheme according to the Goldwasser-Micali

[1] Security reductions aim at reducing the security of a protocol to the security of a presumed hard computational problem for which no efficient (i.e., probabilistic polynomial time) solving algorithm is known.

R. Královič and D. Niwiński (Eds.): MFCS 2009, LNCS 5734, pp. 247–257, 2009.

paradigm. The issue of providing rigorous security proofs is very important since many key assignment schemes have been shown insecure against collusive attacks (see [4] for references).

We consider two different notions of security for hierarchical key assignment schemes, first proposed by Atallah et al. [3]: security against *key recovery* and with respect to *key indistinguishability*. In the key recovery case, an adversary cannot *compute* a key which cannot be derived by the users he has corrupted; whereas, in the key indistinguishability case, the adversary is not even able to *distinguish* the key from a random string of the same length. Hierarchical key assignment schemes satisfying the above notions of security have been proposed in [3,4,7].

1.1 Our Contribution

In this paper we analyze the Akl-Taylor scheme as well as some of its variants with respect to security and efficiency requirements.

SECURITY. We analyze the Akl-Taylor scheme according to the security definitions in [3]. In particular, we carefully specify how to choose the public parameters in order to get instances of the scheme which are secure against key recovery under the *RSA assumption.*

Motivated by the fact that the Akl-Taylor scheme is not secure w.r.t. key indistinguishability, we propose a general construction to set up a key assignment scheme secure w.r.t. key indistinguishability, given any key assignment scheme secure against key recovery. We use this method to obtain a variant of the Akl-Taylor scheme which achieves security w.r.t. key indistinguishability. Our general method is of independent interest. Indeed, it may be useful for different instantiations.

In the Akl-Taylor scheme the complexity of the key derivation increases with the number of classes in the hierarchy. In order to speed up the key derivation process, MacKinnon et al. [12] and Harn and Lin [10] proposed variants of the Akl-Taylor scheme. Due to lack of space, we will include the security analysis of such schemes in the full version of this paper.

EFFICIENCY. We show that the Akl-Taylor scheme is still secure when only part of the public information as small as a *single prime number* is published. This at the cost of a more expensive key derivation. Thus, we show a tradeoff between the size of the public information and the complexity of the key derivation.

2 Model and Definitions

Consider a set of disjoint classes and a binary relation $\preceq$ that partially orders the set of classes V. The poset $(V, \preceq)$ is called a *partially ordered hierarchy.* For any two classes u and v, the notation $v \preceq u$ is used to indicate that the users in u can access v's data. Clearly, since u can access its own data, it holds that $u \preceq u$, for any $u \in V$. We denote by A_u the set of nodes to whom node u has access to, i.e., $A_u = \{v \in V : v \preceq u\}$, for any $u \in V$. The partially ordered hierarchy

$(V, \preceq)$ can be represented by the directed graph $G^* = (V, E^*)$, where each class corresponds to a vertex in the graph and there is an edge from class u to class v if and only if $v \preceq u$. We denote by $G = (V, E)$ the *minimal representation* of the graph G^*, that is, the directed acyclic graph corresponding to the *transitive and reflexive reduction* of the graph $G^* = (V, E^*)$. Such a graph G has the same transitive and reflexive closure of G^*, i.e., there is a path (of length greater than or equal to zero) from u to v in G if and only if there is the edge (u, v) in E^*.

Definition 1. *Let Γ be a family of graphs corresponding to partially ordered hierarchies. A* hierarchical key assignment scheme *for Γ is a pair (Gen, Der) of algorithms satisfying the following conditions:*

1. *The* information generation algorithm *Gen is probabilistic polynomial-time. It takes as inputs the security parameter 1^τ and a graph $G = (V, E)$ in Γ, and produces as outputs*
   (a) a private information s_u and a key k_u, for any class $u \in V$;
   (b) a public information pub.
   We denote by (s, k, pub) the output of the algorithm Gen, where s and k denote the sequences of private information and of keys, respectively.
2. *The* key derivation algorithm *Der is deterministic polynomial-time. It takes as inputs the security parameter 1^τ, a graph $G = (V, E)$ in Γ, two classes $u \in V$ and $v \in A_u$, the private information s_u assigned to class u and the public information pub, and outputs the key k_v assigned to class v.*
   We require that for each class $u \in V$, each class $v \in A_u$, each private information s_u, each key k_v, each public information pub which can be computed by Gen on inputs 1^τ and G, it holds that $Der(1^\tau, G, u, v, s_u, pub) = k_v$.

In order to evaluate the security of the scheme, we consider a *static adversary* which wants to attack a class $v \in V$ and which is able to corrupt *all* users not allowed to compute the key k_v. We define an algorithm $Corrupt_v$ which, on input the private information s generated by the algorithm Gen, extracts the secret values s_u associated to all classes u in the set of nodes that *do not have access* to node v, i.e., $F_v = \{u \in V : v \notin A_u\}$. We denote by $corr_v$ the sequence output by $Corrupt_v(s)$.

If $A(\cdot, \cdot, \ldots)$ is any probabilistic algorithm, then we denote by $a \leftarrow A(x, y, \ldots)$ the experiment of running A on inputs $x, y, \ldots$ and letting a be the outcome. Similarly, if X is a set, then $x \leftarrow X$ denotes the experiment of selecting an element uniformly from X and assigning x this value. A function $\epsilon : N \rightarrow R$ is *negligible* if, for every constant $c > 0$, there exists an integer τ_c such that $\epsilon(\tau) < \tau^{-c}$, for all $\tau \geq \tau_c$.

We consider two different security goals: against *key recovery* and with respect to *key indistinguishability*. In the key recovery case, the adversary, on input all public information generated by the algorithm Gen, as well as the private information $corr_v$ held by corrupted users, outputs a string k_v^* and succeeds whether $k_v^* = k_v$.

Definition 2. [REC-ST] *Let Γ be a family of graphs corresponding to partially ordered hierarchies, let $G = (V, E) \in \Gamma$ be a graph, and let (Gen, Der) be a* *hierarchical key assignment scheme for Γ. Let* $\mathtt{STAT}_v^{\mathtt{REC}}$ *be a static adversary which attacks a class* v*. Consider the following experiment:*

$$\begin{array}{l}\text{Experiment } \mathbf{Exp}^{\mathtt{REC}}_{\mathtt{STAT}_v}(1^\tau, G)\\ \quad (s,k,pub) \leftarrow Gen(1^\tau, G)\\ \quad corr_v \leftarrow Corrupt_v(s)\\ \quad k_v^* \leftarrow \mathtt{STAT}_v^{\mathtt{REC}}(1^\tau, G, pub, corr_v)\\ \quad \mathbf{return}\ k_v^*\end{array}$$

The advantage of $\mathtt{STAT}_v^{\mathtt{REC}}$ *is defined as* $\mathbf{Adv}^{\mathtt{REC}}_{\mathtt{STAT}_v}(1^\tau, G) = Pr[k_v^* = k_v]$*. The scheme is said to be* secure in the sense of REC-ST *if, for each graph* $G = (V, E)$ *in* Γ *and each class* $v \in V$*, the function* $\mathbf{Adv}^{\mathtt{REC}}_{\mathtt{STAT}_v}(1^\tau, G)$ *is negligible, for each static adversary* $\mathtt{STAT}_v^{\mathtt{REC}}$ *whose time complexity is polynomial in* τ*.*

In the key indistinguishability case, two experiments are considered. In the first one, the adversary is given as a challenge the key k_v, whereas, in the second one, it is given a random string ρ having the same length as k_v. It is the adversary's job to determine whether the received challenge corresponds to k_v or to a random string. We require that the adversary will succeed with probability only negligibly different from 1/2.

Definition 3. [IND-ST] *Let Γ be a family of graphs corresponding to partially ordered hierarchies, let $G = (V, E)$ be a graph in Γ, let (Gen, Der) be a hierarchical key assignment scheme for Γ and let* $\mathtt{STAT}_v^{\mathtt{IND}}$ *be a static adversary which attacks a class* v*. Consider the following two experiments:*

$$\begin{array}{l|l}\text{Experiment } \mathbf{Exp}^{\mathtt{IND-1}}_{\mathtt{STAT}_v}(1^\tau, G) & \text{Experiment } \mathbf{Exp}^{\mathtt{IND-0}}_{\mathtt{STAT}_v}(1^\tau, G)\\ \quad (s,k,pub) \leftarrow Gen(1^\tau, G) & \quad (s,k,pub) \leftarrow Gen(1^\tau, G)\\ \quad corr_v \leftarrow Corrupt_v(s) & \quad corr_v \leftarrow Corrupt_v(s)\\ \quad d \leftarrow \mathtt{STAT}_v^{\mathtt{IND}}(1^\tau, G, pub, corr_v, k_v) & \quad \rho \leftarrow \{0,1\}^{length(k_v)}\\ \quad \mathbf{return}\ d & \quad d \leftarrow \mathtt{STAT}_v^{\mathtt{IND}}(1^\tau, G, pub, corr_v, \rho)\\ & \quad \mathbf{return}\ d\end{array}$$

The advantage of $\mathtt{STAT}_v^{\mathtt{IND}}$ *is defined as*

$$\mathbf{Adv}^{\mathtt{IND}}_{\mathtt{STAT}_v}(1^\tau, G) = |Pr[\mathbf{Exp}^{\mathtt{IND-1}}_{\mathtt{STAT}_v}(1^\tau, G) = 1] - Pr[\mathbf{Exp}^{\mathtt{IND-0}}_{\mathtt{STAT}_v}(1^\tau, G) = 1]|.$$

The scheme is said to be secure in the sense of IND-ST *if, for each graph* $G = (V, E)$ *in* Γ *and each* $v \in V$*, the function* $\mathbf{Adv}^{\mathtt{IND}}_{\mathtt{STAT}_v}(1^\tau, G)$ *is negligible, for each static adversary* $\mathtt{STAT}_v^{\mathtt{IND}}$ *whose time complexity is polynomial in* τ*.*

In Definitions 2 and 3 we have considered a static adversary attacking a class. A different kind of adversary, the *adaptive* one, could also be considered. In [4] it has been proven that security against adaptive adversaries is (polynomially) equivalent to security against static adversaries. Hence, in this paper we will only consider static adversaries.

3 Complexity Assumptions

An *RSA generator* with associated security parameter τ is a randomized algorithm that returns a pair $((n,e),(n,p,q,d))$, where n is the *RSA modulus*, e is the *encryption exponent* and d is the *decryption exponent*, satisfying the following conditions:

- p and q are two distinct large odd primes of τ bits;
- $n = p \cdot q$;
- $e \in Z^*_{\phi(n)}$, where $\phi(n) = (p-1)\cdot(q-1)$;
- $d = e^{-1} \bmod \phi(n)$.

Two strategies to compute the pair $((n,e),(n,p,q,d))$ are used. The former, first chooses the primes p and q, computes n, picks e at random in $Z^*_{\phi(n)}$, and computes d accordingly. Such a strategy yields a *random exponent RSA generator*, denoted $\mathsf{K}^{ran}_{RSA}(1^\tau)$. The latter, fixes the encryption exponent e to be a small odd number, like 3, 17, or $2^{16}+1$, and then generates the other parameters, accordingly[2]. Given a fixed odd number e, such a strategy yields an *RSA generator for exponent* e, denoted $\mathsf{K}^{fix}_{RSA}(1^\tau, e)$.

Let B and $Grsa$ be algorithms where the algorithm $Grsa$ corresponds either to $\mathsf{K}^{ran}_{RSA}(1^\tau)$ or to $\mathsf{K}^{fix}_{RSA}(1^\tau, e)$. Consider the following experiment:

$$
\begin{array}{l}
\textit{Experiment } \mathbf{Exp}_B^{Grsa} \\
\quad ((n,e),(n,p,q,d)) \leftarrow Grsa \\
\quad x \leftarrow Z_n^* \\
\quad y \leftarrow x^e \bmod n \\
\quad x' \leftarrow B(n,e,y) \\
\quad \textbf{if } x' = x \textbf{ then return } 1 \\
\qquad\qquad\quad \textbf{else return } 0
\end{array}
$$

The advantage of B is defined as $\mathbf{Adv}_B^{Grsa} = Pr[\mathbf{Exp}_B^{Grsa} = 1]$.

The RSA generators described above yields the following two assumptions[3].

Random Exponent RSA Assumption. The function $\mathbf{Adv}_B^{\mathsf{K}^{ran}_{RSA}}(1^\tau)$ is negligible, for each probabilistic algorithm B whose time complexity is polynomial in τ.

RSA Assumption for Exponent in a set of odd numbers. Let X be a set of odd numbers. For each $e \in X$, the function $\mathbf{Adv}_B^{\mathsf{K}^{fix}_{RSA}}(1^\tau, e)$ is negligible, for each probabilistic algorithm B with time complexity polynomial in τ.

---

[2] There have been some questions raised about the security of this strategy, since it might be possible that roots of small degree are easier to take than roots of a random degree.

[3] To the best of our knowledge, in the literature there is no analysis of the relationships between the security of these two different strategies.

4 The Akl-Taylor Scheme

In this section we describe the Akl and Taylor scheme [1].

Let Γ be a family of graphs corresponding to partially ordered hierarchies, and let $G = (V, E) \in \Gamma$.

Algorithm $Gen(1^\tau, G)$

1. Randomly choose two distinct large primes p and q having bitlength τ and compute $n = p \cdot q$;
2. For each $v \in V$, choose a distinct prime number p_v and compute the public value t_v as follows:
$$t_v = \begin{cases} 1 & \text{if } A_v = V; \\ \prod_{u \notin A_v} p_u & \text{otherwise.} \end{cases}$$
3. Let *pub* be the sequence of public values computed in the previous step, along with the value n;
4. Randomly choose a secret value k_0, where $1 < k_0 < n$;
5. For each class $v \in V$, compute the private information s_v and the encryption key k_v as follows:
$$s_v = k_v = k_0^{t_v} \bmod n;$$
6. Let s and k be the sequences of private information and keys, respectively, computed in the previous step;
7. Output (s, k, pub).

Algorithm $Der(1^\tau, G, u, v, s_u, pub)$

Extract the values t_v and t_u from *pub* and compute
$$s_u^{t_v/t_u} \bmod n = (k_0^{t_u})^{t_v/t_u} \bmod n = k_v.$$

Akl and Taylor noticed that in order to construct an Akl-Taylor scheme which is resistant to collusive attacks it is needed that, for each $v \in V$ and each $X \subseteq F_v$, $\gcd\{t_u : u \in X\}$ does not divide t_v. Indeed, they showed the following result:

Lemma 1 ([1]). *Let t and $t_1, \ldots, t_m$ be integers, and let $k \in Z_n$, where $n = p \cdot q$ is the product of two large primes. The power $k^t \bmod n$ can be feasibly computed from the set of powers $\{k^{t_1} \bmod n, \ldots, k^{t_m} \bmod n\}$ if and only if $\gcd\{t_1, \ldots, t_m\}$ divides t.*

The proof of Lemma 1 relies on the infeasibility of extracting r-th roots modulo n, where $r > 1$ is an integer and n is the product of two large unknown primes[4]. Lemma 1 gives an *intuition* for the security of the scheme but it has never been shown whether the existence of an efficient adversary breaking the security of the scheme in the sense of REC−ST implies the existence of an efficient procedure which solves a computational hard problem.

[4] In particular, when $\gcd(r, \phi(n)) = 1$, this is the assumption behind the RSA cryptosystem [14]; whereas, if $r = 2$, this assumption is used in the Rabin cryptosystem [13].

4.1 Proving the Security of the Akl-Taylor Scheme

In this section we show that the Akl-Taylor scheme is secure against key recovery provided that the primes associated to the classes are *properly chosen.*

In the following we describe two primes choices. The first choice, denoted *fixed primes choice*, yields instances of the Akl-Taylor scheme secure under the RSA assumption for exponent in a set of odd numbers. The second one, denoted *R-random primes choice*, yields instances secure under the random exponent RSA assumption.

- *Fixed primes choice.* Let $PRIMES_\ell = \{p_1, \ldots, p_\ell\}$ be the set of the first ℓ prime numbers greater than two. Let $u_1, \ldots, u_{|V|}$ be a sorting of V. Associate prime $p_j \in PRIMES_{|V|}$ to class u_j.
- *R-Random primes choice.* Let $\mathtt{R} = \{R_n\}_n$ be a family of sets of integers, where n is an RSA modulus. *Choose uniformly at random*, for each class, a distinct prime number in R_n, for an appropriately chosen family $\mathtt{R}$.

The following result holds:

Theorem 1. *Let $G = (V, E)$ be a partially ordered hierarchy. The Akl-Taylor scheme with the fixed primes choice yields a scheme which is secure in the sense of* REC-ST *under the RSA assumption for exponents in $PRIMES_{|V|}$.*

Let us consider the R-random primes choice. The next result establishes a lower bound for the Euler totient function $\phi(n)$, when n is an RSA modulus obtained as the product of two large distinct primes having bitlength τ. More precisely:

Lemma 2. *Let p and q be two large distinct primes having bitlength τ and let $n = p \cdot q$. The Euler totient function $\phi(n)$ satisfies*

$$\phi(n) > 2^{2\tau-2} - 2^\tau.$$

We fix the family R in the R-random primes choice by setting R_n equal to the set of integers belonging to the interval $[3, w]$, where $w = 2^{2\tau-2} - 2^\tau$. Therefore, to avoid overburdening the notation, in the following we will refer to the R-random primes choice simply as to the *random primes choice.*

The following result holds:

Theorem 2. *Let $G = (V, E)$ be a partially ordered hierarchy. The Akl-Taylor scheme with the random primes choice yields a scheme secure in the sense of* REC-ST *under the random exponent RSA assumption.*

4.2 Reducing Public Information in the Akl-Taylor Scheme

Notice that the primes, which are randomly chosen in $[3, w]$ in the random primes choice, do not need to be independent from each other. Thus, we introduce a *modified random primes choice* for the Akl-Taylor scheme, which works as follows:

1. Let $u_1, \ldots, u_{|V|}$ be a sorting of V;
2. Choose uniformly at random a prime p_1 in $[3, w]$;
3. For $j = 1, \ldots, |V| - 1$, compute $p_{j+1} \in [3, w]$ as the j-th prime greater than p_1;
4. For $j = 1, \ldots, |V|$, assign prime p_j to class u_j.

Notice that if the prime p_1 is too close to w, then $[p_1, w]$ could not contain the $|V| - 1$ primes needed for the other classes. If this event occurs, the generation of primes continues in $[3, p_1]$.

Theorem 3. *The Akl-Taylor scheme with the modified random primes choice yields a scheme which is secure in the sense of* REC-ST *under the random exponent RSA assumption.*

With the modified random primes choice, *only the first prime* needs to be published, along with the modulus n. Indeed, such public information allows each class to compute the other primes and the sequence of integers t_v's needed for key derivation. These computations can be performed efficiently (see [2] where a similar technique was employed). Notice that, in the modified random primes choice, $i \geq 2$ primes of the sequence $p_1, \ldots, p_{|V|}$ might be published. Then, each class v has to compute the $|V| - i$ missing primes accordingly, by performing on average $O((|V| - i) \cdot \tau)$ steps to compute the full sequence. This allows to obtain a tradeoff between the size of the public information and the number of steps required to perform key derivation.

5 Towards Security w.r.t. Key Indistinguishability

In this section we show a general construction for schemes secure with respect to key indistinguishability. The construction uses as a building block, a scheme secure in the sense of REC-ST. Moreover, it makes use of the Goldreich-Levin hard-core bit (GL bit) [8], which is a natural candidate for turning hardness of computation into indistinguishability.

Let τ be the security parameter. Given two strings $x = x_1 \cdots x_\gamma$ and $r = r_1 \cdots r_\gamma$ of length γ polynomially bounded in τ, the GL bit $B_r(x)$ corresponds to the inner product (mod 2) of x and r, i.e., $B_r(x) = \sum_{i=1}^{\gamma} x_i \cdot r_i \bmod 2$.

Goldreich and Levin [8] showed that, for every one-way permutation f, given $f(x)$ and r, it is infeasible to guess $B_r(x)$ with non-negligible advantage.

Let (Gen, Der) be a scheme secure in the sense of REC-ST. The idea of our construction is as follows:

- For each graph $G = (V, E)$, we construct a new graph $G_\gamma = (V_\gamma, E_\gamma)$ by replacing each node u with a chain of $\gamma + 1$ classes denoted $u_0, u_1, \ldots, u_\gamma$. Figure 1 shows an example of the graph $G_\gamma = (V_\gamma, E_\gamma)$ for $G = (V, E)$, where $V = \{a, b, c\}$ and $E = \{(a, b), (a, c)\}$. Notice that, the edge $(u_i, u_{i+1}) \in E_\gamma$, for each $i = 0, \ldots, \gamma - 1$. Moreover, for each $(u, v) \in E$, we place the edge (u_0, v_0) in E_γ.

- Execute the algorithm *Gen* on input $G_\gamma = (V_\gamma, E_\gamma)$.
- For each class $u \in V$, the private information s_u is equal to the the private information s_{u_0} assigned to u_0 by *Gen*, whereas, the key is computed as:

$$k_u = B_r(k_{u_1}) \circ B_r(k_{u_2}) \circ \cdots \circ B_r(k_{u_\gamma}),$$

  where r is a randomly chosen string in $\{0,1\}^\gamma$ and k_{u_i} is the key assigned to u_i by *Gen*, for each $i = 1, \ldots, \gamma$.

It is easy to see that each class $u \in V$, can compute the key k_v assigned to class $v \in A_u$ by using $s_u = s_{u_0}$, the public information output by *Gen*, the string r and the derivation algorithm *Der* of the underlying scheme.

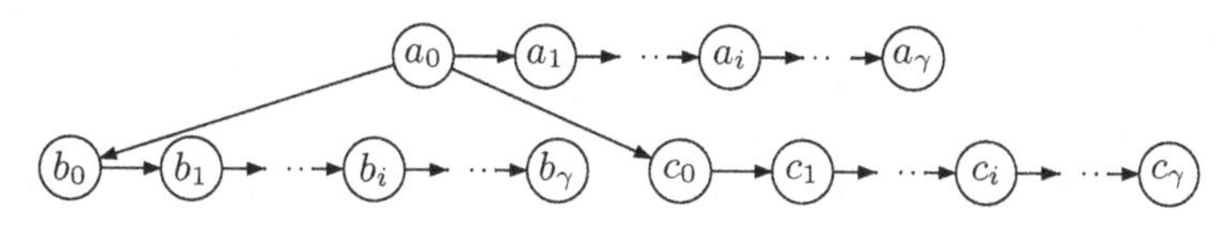

Fig. 1. The graph $G_\gamma = (V_\gamma, E_\gamma)$, where $V = \{a, b, c\}$ and $E = \{(a, b), (a, c)\}$

Intuitively, since, for each $i = 1, \ldots, \gamma$, it is infeasible to guess $B_r(k_{v_i})$ with non-negligible advantage, an adversary for a class $v \in V$ has no information about even a single bit of the key k_v.

Theorem 4. *If (Gen, Der) is a scheme secure in the sense of* REC-ST*, then the scheme resulting by our construction is secure in the sense of* IND-ST.

6 An Akl-Taylor Key Indistinguishable Scheme

In Section 4.1 we have shown that the Akl-Taylor scheme is secure in the sense of REC-ST under either the random exponent or the fixed exponent RSA assumption. Thus, such a scheme can be used as a building block in the construction of Section 5, in order to obtain a key assignment scheme which is secure in the sense of IND-ST under the same assumption.

Notice that the Akl-Taylor scheme is not secure in the sense of IND-ST. Indeed, any adversary which attacks class u knows the key k_v associated to a class v child of class u. In order to check if a value ρ corresponds to the key k_u, the adversary only needs to test whether ρ^{t_v/t_u} is equal to k_v.

In the following we evaluate the parameters of the scheme obtained by using the construction proposed in Section 5 when the underlying scheme is the Akl-Taylor scheme with the random choice of primes. Let τ be the security parameter and let $\gamma = 2\tau$ be the key length. The amount of public information in the resulting scheme corresponds to the $|V|(1+\gamma)$ integers t_v's associated to the classes in V_γ, in addition to the modulus n and the γ-bit string r. On the other hand, the construction requires each class to store one secret value, corresponding to its private information. Finally, a class u, which wants to compute the key held by a class $v \in A_u$ has to perform γ modular exponentiations, to compute the keys $k_{v_1}, \ldots, k_{v_\gamma}$, and γ inner products modulo 2, to compute the GL bits $B_r(k_{v_1}), \ldots, B_r(k_{v_\gamma})$.

According to the results stated in Section 4.2, the amount of public information required by the Akl-Taylor scheme with the modified random prime choice can be reduced to a single prime number, in addition to the modulus n. As explained in Section 4.2, this requires each class to perform on average $O(|V| \cdot \tau^2)$ steps to compute the sequence of primes assigned to the classes in V.

7 Conclusions

We have analyzed the Akl-Taylor scheme with respect to security and efficiency requirements. We have also shown how to obtain different tradeoffs between the amount of public information and the number of steps required to perform key derivation. Motivated by the fact that the Akl-Taylor scheme is not secure w.r.t. key indistinguishability, we have proposed a general construction to setup a key assignment scheme secure w.r.t. key indistinguishability, given any key assignment scheme secure against key recovery. Such a construction is of independent interest and may be useful for different instantiations. As an example, we have shown how to use our construction, along with our tradeoffs, to obtain a variant of the Akl-Taylor scheme, secure w.r.t key indistinguishability, requiring a constant amount of public information.

References

1. Akl, S.G., Taylor, P.D.: Cryptographic Solution to a Problem of Access Control in a Hierarchy. ACM Trans. on Comput. Syst. 1(3), 239–248 (1983)
2. Asano, T.: A Revocation Scheme with Minimal Storage at Receivers. In: Zheng, Y. (ed.) ASIACRYPT 2002. LNCS, vol. 2501, pp. 433–450. Springer, Heidelberg (2002)
3. Atallah, M.J., Blanton, M., Fazio, N., Frikken, K.B.: Dynamic and Efficient Key Management for Access Hierarchies. ACM Trans. Inf. Syst. Secur., Article 18 12(3) (2009); Prelim. version in Proc. of ACM CCS 2005, pp. 190–201 (2005)
4. Ateniese, G., De Santis, A., Ferrara, A.L., Masucci, B.: Provably-Secure Time-Bound Hierarchical Key Assignment Schemes. Rep. 2006/225 at the IACR Cryptology ePrint Archive. Prelim. version in Proc. of ACM CCS 2006, pp. 288–297 (2006)
5. Attrapadung, N., Kobara, K.: Broadcast Encryption with Short Keys and Transmissions. In: Proc. of the 3rd ACM workshop on Digital Rights Management, pp. 55–66 (2003)
6. De Santis, A., Ferrara, A.L., Masucci, B.: Cryptographic Key Assignment Schemes for any Access Control Policy. Inf. Proc. Lett. 92(4), 199–205 (2004)
7. De Santis, A., Ferrara, A.L., Masucci, B.: Efficient Provably-Secure Hierarchical Key Assignment Schemes. In: Kučera, L., Kučera, A. (eds.) MFCS 2007. LNCS, vol. 4708, pp. 371–382. Springer, Heidelberg (2007)
8. Goldreich, O., Levin, L.: A Hard-Core Predicate for All One-Way Functions. In: Proc. of ACM STOC 1989, pp. 25–32 (1989)

9. Goldwasser, S., Micali, S.: Probabilistic Encryption. Journal of Comput. and Syst. Sci. 28, 270–299 (1984)
10. Harn, L., Lin, H.Y.: A Cryptographic Key Generation Scheme for Multilevel Data Security. Comput. and Security 9(6), 539–546 (1990)
11. Lin, I.-C., Oub, H.-H., Hwang, M.-S.: Efficient Access Control and Key Management Schemes for Mobile Agents. Comput. Standards & Interfaces 26(5), 423–433 (2004)
12. MacKinnon, S.J., Taylor, P.D., Meijer, H., Akl, S.G.: An Optimal Algorithm for Assigning Cryptographic Keys to Control Access in a Hierarchy. IEEE Trans. on Computers, C-34(9), 797–802 (1985)
13. Rabin, M.O.: Digitalized Signatures and Public Key Functions as Intractable as Factorization, Tech. Rep. MIT/LCS/TR-212, MIT Lab. for Computer Science (1979)
14. Rivest, R.L., Shamir, A., Adleman, L.: A Method for Obtaining Digital Signatures and Public Key Cryptosystems. Communic. ACM 21, 120–126 (1978)
15. Tzeng, W.-G.: A Secure System for Data Access Based on Anonymous and Time-Dependent Hierarchical Keys. In: Proc. of ACM ASIACCS 2006, pp. 223–230 (2006)
16. Yeh, J.H., Chow, R., Newman, R.: A Key Assignment for Enforcing Access Control Policy Exceptions. In: Proc. of Int. Symposium on Internet Technology, pp. 54–59 (1998)
17. Wang, S.-Y., Laih, C.-S.: Merging: An Efficient Solution for a Time-Bound Hierarchical Key Assignment Scheme. IEEE Trans. on Dependable and Secure Computing 3(1) (2006)

Parameterized Complexity Classes under Logical Reductions

Anuj Dawar[1] and Yuguo He[2]

[1] University of Cambridge Computer Laboratory, Cambridge CB3 0FD, U.K.
anuj.dawar@cl.cam.ac.uk
[2] University of Cambridge Computer Laboratory, Cambridge CB3 0FD, U.K., and
School of Computer Science and Technology, Beijing Institute of Technology, Beijing 100081, China
yuguo.he@cl.cam.ac.uk

Abstract. The parameterized complexity classes of the W-hierarchy are usually defined as the problems reducible to certain natural complete problems by means of fixed-parameter tractable (*fpt*) reductions. We investigate whether the classes can be characterised by means of weaker, logical reductions. We show that each class $W[t]$ has complete problems under slicewise bounded-variable first-order reductions. These are a natural weakening of slicewise bounded-variable LFP reductions which, by a result of Flum and Grohe, are known to be equivalent to *fpt*-reductions. If we relax the restriction on having a bounded number of variables, we obtain reductions that are too strong and, on the other hand, if we consider slicewise quantifier-free first-order reductions, they are considerably weaker. These last two results are established by considering the characterisation of $W[t]$ as the closure of a class of Fagin-definability problems under *fpt*-reductions. We show that replacing these by slicewise first-order reductions yields a hierarchy that collapses, while allowing only quantifier-free first-order reductions yields a hierarchy that is provably strict.

1 Introduction

In the theory of parameterized complexity, the W-hierarchy plays a role similar to NP in classical complexity theory in that many natural parameterized problems are shown intractable by being complete for some level $W[t]$ of the hierarchy. However, one difference between the two, perhaps no more than a historical accident, is that NP was originally defined in terms of resource bounds on a machine model, and the discovery that it has complete problems under polynomial-time reductions (and indeed that many natural combinatorial problems are NP-complete) came as a major advance, which also shows the robustness of the class. On the other hand, the classes $W[t]$ were originally defined as the sets of problems reducible to certain natural complete problems by means of fixed-parameter tractable (*fpt*) reductions [5]. These classes therefore have complete problems by construction. It was only later that a characterisation of these classes in terms of resource-bounded machines was obtained [1]. The robustness of the definition of NP is also demonstrated by the fact that many NP-complete problems are still complete under reductions much weaker than polynomial-time reductions. For instance,

R. Královič and D. Niwiński (Eds.): MFCS 2009, LNCS 5734, pp. 258–269, 2009.

SAT is NP-complete, even under quantifier-free first-order projections, which are reductions even weaker than AC_0 reductions. Thus, the class NP can be characterized as the class of problems reducible to SAT under polynomial-time reductions, or equivalently as the class of problems reducible to SAT under quantifier-free first-order projections. The work we report in this paper is motivated by the question of whether similar robustness results can be shown for the classes $W[t]$. We investigate whether the classes can be characterised by means of weaker reductions, just like NP can.

We concentrate on reductions defined in terms of logical formulas. By a result of Flum and Grohe [7], it is known that *fpt*-reductions can be equivalently characterised, on ordered structures, as slicewise bounded-variable LFP reductions. We consider reductions defined in terms of first-order interpretations and introduce a number of parameterized versions of these. Our main result is that each class $W[t]$ has complete problems under *slicewise bounded-variable first-order* reductions, which are a natural first-order counterpart to slicewise bounded-variable LFP reductions. If we relax the restriction on having a bounded number of variables, we obtain *slicewise first-order* reductions, which are not necessarily *fpt*. Indeed we are able to show that all Fagin definability problems in $W[t]$ are reducible to problems in FPT under such reductions. On the other hand, we show that *slicewise quantifier-free first-order* reductions are considerably weaker in that there are Fagin-definability problems in $W[t+1]$ that cannot reduce to such problems in $W[t]$ under these reductions. This last class of reductions can be seen as the natural parametrization of quantifier-free first-order reductions, for which NP does have complete problems. Thus, our result shows that the definition of $W[t]$ is not quite as robust as that of NP.

We present the necessary background and preliminaries in Section 2. The various kinds of logical reductions are defined in Section 3. Section 4 shows that $W[t]$ contains complete problems under *slicewise bounded-variable first-order* reductions. Section 5 considers the case of the two other kinds of reductions we use. For space reasons, we only give sketches of proofs, omitting details which are often long and tedious coding of reductions as first-order formulas.

2 Preliminaries

We rely on standard definitions and notation from finite model theory (see [6,12]) and the theory of parameterized complexity [9]. We briefly recall some of the definitions we need, but we assume the reader is familiar with this literature.

A relational signature σ consists of a finite collection of relation and constant symbols. A decision problem over σ-structures is an isomorphism-closed class of finite σ-structures. In general, we assume that our structures are ordered. That is to say, that σ contains a distinguished binary relation symbol $\leq$ which is interpreted in every structure as a linear order of the universe. We are often interested in decision problems where the input is naturally described as a structure with additional integer parameters. For instance, the Clique problem requires, given a graph G and an integer k, to decide whether G contains a clique on k vertices. In all such cases that we will be interested in, the value of the integer parameter is bounded by the size of the structure, so it is safe to assume that it is given as an additional constant c in the signature σ, and the position in the linear order $\leq$ of $c^{\mathbb{A}}$ codes the value. However, where it is notationally convenient, we may still write the inputs as pairs $(\mathbb{A}, k)$, where $\mathbb{A}$ is a structure and k is an integer, with the understanding that they are to be understood as such coded structures.

A *parameterized problem* is a pair (Q, κ) where Q is a decision problem over σ-structures and κ a function that maps σ-structures to natural numbers. We say that (Q, κ) is *fixed-parameter tractable* (FPT) if Q is decidable by an algorithm which, given a σ-structure $\mathbb{A}$ of size n runs in time $f(\kappa(\mathbb{A}))n^c$ for some constant c and some computable function f.

Given a pair of parameterized problems, (Q, κ) and (Q', κ') where Q is a decision problem over σ-structures and Q' is a decision problem over σ'-structures, a reduction from (Q, κ) to (Q', κ') is a computable function r from σ-structures to σ'-structures such that:

- for any σ-structure $\mathbb{A}$, $r(\mathbb{A}) \in Q'$ if, and only if, $\mathbb{A} \in Q$; and
- there is a computable function g such that $\kappa'(r(\mathbb{A})) \leq g(\kappa(\mathbb{A}))$.

The reduction r is an *fpt*-reduction if, in addition, r is computable in time $f(\kappa(\mathbb{A}))|\mathbb{A}|^c$ for some constant c and some computable function f. If there is an *fpt*-reduction from (Q, κ) to (Q', κ'), we write $(Q, \kappa) \leq^{\mathtt{fpt}} (Q', \kappa')$ and say that (Q, κ) is *fpt*-reducible to (Q', κ').

FPT is the complexity class of parameterized problems that are regarded as tractable. Above it, there is a hierarchy of complexity classes into which problems that are believed to be intractable are classified. In particular, the W-hierarchy is an increasing (or, at least, non-decreasing) sequence of complexity classes $W[t]$ $(t \geq 1)$ which contain many natural hard problems. These classes were originally defined as the classes of problems *fpt*-reducible to certain weighted satisfiability problems. We use, instead, the equivalent definition from [9] in terms of weighted Fagin-definability, which we give next. For a first-order formula $\varphi(X)$ with a free relational variable X of arity s, we define the weighted Fagin-definability problem for φ as the following parameterized problem.

p-WD$_\varphi$	
Input:	A structure $\mathbb{A}$ and $k \in \mathbb{N}$
Parameter:	k
Problem:	Decide whether there is a relation $S \subseteq A^s$ with $|S| = k$ such that $(\mathbb{A}, S) \models \varphi$.

The complexity class $W[t]$ is then defined as the class of parameterized problems that are *fpt*-reducible to p-WD$_\varphi$ for some Π_t formula φ. Recall that φ is Π_t just in case it is in prenex normal form and its quantifier prefix consists of t alternating blocks, starting with a universal block. These classes are closed under *fpt*-reductions by definition. Indeed, to quote Flum and Grohe [9, p.95]: "for a robust theory, one has to close [...] p-WD-Π_t under *fpt*-reductions". Our aim in this paper is to test this robustness by varying the reductions used in the definition to see whether we still obtain the same classes. We specifically aim to investigate logical reductions and for these, it is convenient to work with descriptive characterisations of the complexity classes. We summarise below such characterisations that have been obtained by Flum and Grohe [7,8].

Recall that LFP is the extension of first-order logic (FO) with an operator **lfp** for least fixed-points of positive operators. We write FO^s for the collection of first-order formulas *with at most s distinct variables* and LFP^s for the collection of formulas of LFP of the form $[\mathbf{lfp}_{X,\mathbf{x}}\varphi](\mathbf{t})$ where $\varphi \in \mathsf{FO}^s$ and $\mathbf{t}$ is a tuple of at most s terms. For any collection Θ of formulas, we say that a parameterized problem (Q, κ) is *slicewise-Θ definable* if, and only if, there is a computable function $\delta : \mathbb{N} \to \Theta$ such that for all $\mathbb{A}$, we have $\mathbb{A} \models \delta(\kappa(\mathbb{A}))$ if, and only if, $\mathbb{A} \in Q$.

Theorem 1 ([7]). *A parameterized problem over* ordered *structures is in FPT if, and only if, for some s it is slicewise-LFP^s definable.*

For a similar characterisation of the classes of the W-hierarchy, we need to introduce some further notation (from [8]). We write $\Sigma_{t,u}$-Bool(LFP^s) for the collection of formulas of LFP of the form

$$\exists x_{11} \cdots \exists x_{1l_1} \forall x_{21} \cdots \forall x_{2l_2} \cdots Qx_{t1} \cdots Qx_{tl_t}\, \chi \tag{1}$$

where χ is a Boolean combination of formulas of LFP^s and for $i \geq 2$, $l_i \leq u$. In other words, the formula consists of a sequence of t alternating blocks of quantifiers, starting with an existential, with the length of all blocks except the first bounded by u, followed by a Boolean combination of LFP^s formulas. Note that all of the variables in the quantifier prefix may appear inside χ though any given formula in the Boolean combination may use at most s of them.

Theorem 2 ([8]). *A parameterized problem over* ordered *structures is in $W[t]$ if, and only if, for some s and u it is slicewise-$\Sigma_{t,u}$-Bool(LFP^s) definable.*

The key to the definition of $\Sigma_{t,u}$-Bool(LFP^s) is the interaction between the unbounded number of variables introduced by the first quantifier block, and the bounded number of variables available inside each LFP^s formula in χ. This is best illustrated with a simple example. The parameterized dominating set problem takes as input a graph G and a parameter k and asks whether G contains a set S of at most k vertices such that every vertex of G is either in S or a neighbour of a vertex in S. For fixed k, this is defined by the following first-order formula.

$$\exists x_1 \cdots \exists x_k \forall y (\bigvee_{1 \leq i \leq k} (y = x_i \vee E(y, x_i)))$$

Here, since each of the formulas $(y = x_i \vee E(y, x_i))$ has only two variables, the whole formula is in $\Sigma_{2,1}$-Bool(LFP^2).

We can somewhat simplify the form of formulas used in Theorem 2. To be precise, we write $\Sigma_{t,u}$-Conj(LFP^s) for those formulas of the form (1) where χ is a conjunction of LFP^s formulas and $\Sigma_{t,u}$-Disj(LFP^s) for those where it is a disjunction. Then, we have the following characterisation.

Theorem 3. *For any* even $t \geq 1$, *a parameterized problem over* ordered *structures is in $W[t]$ if, and only if, for some s and u it is slicewise-$\Sigma_{t,u}$-Disj(LFP^s) definable.*

For any odd $t \geq 1$, *a parameterized problem over* ordered *structures is in $W[t]$ if, and only if, for some s and u it is slicewise-$\Sigma_{t,u}$-Conj(LFP^s) definable.*

Proof. (sketch): Consider the case of odd t, as the other case is dual. The Boolean combination χ in the formula (1) can be written in disjunctive normal form. Now, any formula $\neg\varphi$ where φ is a formula of LFP^s is equivalent to a formula of LFP^{8s}. This follows from Immerman's proof of a normal form for LFP [11]. In particular, one just needs to observe that the increase in the number of variables is bounded by a multiplicative constant. Thus, χ is equivalent to a disjunction of conjunctions of formulas of LFP^{8s}. The idea is now to replace the outermost disjunction with an existential quantifier. For each i, we can write a first-order formula $\varphi_i(x)$ (with just three variables) that asserts that x is the ith element of the linear order $\leq$ (see [4]). We use these to index the m disjuncts in χ. This requires increasing the arity in each fixed-point formula by 1, and the number of variables by at most 3. We thus obtain a formula with one existential quantifier followed by a conjunction of formulas of LFP^{8s+3} that is equivalent to χ on all ordered structures with at least m elements. We can then add a further conjunct to take care of the finitely many small structures. The existential quantifier at the front of the formula is then absorbed into the final block in the prefix, resulting in an increase of the value of u by 1.

For the case of odd t, we begin with a formula of conjunctive normal form and convert the outer conjunction to a universal quantifier. □

3 Logical Reductions

In this section, we introduce reductions that are defined by logical formulas.

Suppose we are given two relational signatures σ and τ and a set of formulas Θ. An m-ary *Θ-interpretation* of τ in σ (with parameters $\mathbf{z}$) is a sequence of formulas of Θ in the signature σ consisting of:

- a formula $\upsilon(\mathbf{x}, \mathbf{z})$;
- a formula $\eta(\mathbf{x}, \mathbf{y}, \mathbf{z})$;
- for each relation symbol R in τ of arity a, a formula $\rho^R(\mathbf{x}_1, \ldots, \mathbf{x}_a, \mathbf{z})$; and
- for each constant symbol c in τ, a formula $\gamma^c(\mathbf{x}, \mathbf{z})$,

where each $\mathbf{x}, \mathbf{y}$ or $\mathbf{x}_i$ is an m-tuple of free variables. We call m the *width* of the interpretation. We say that an interpretation Φ associates a τ-structure $\mathbb{B}$ to a pair $(\mathbb{A}, \mathbf{c})$ where $\mathbb{A}$ is a σ-structure and $\mathbf{c}$ a tuple of elements interpreting the parameters $\mathbf{z}$, if there is a surjective map h from the m-tuples $\{\mathbf{a} \in A^m \mid \mathbb{A} \models \upsilon[\mathbf{a}, \mathbf{c}]\}$ to $\mathbb{B}$ such that:

- $h(\mathbf{a}_1) = h(\mathbf{a}_2)$ if, and only if, $\mathbb{A} \models \eta[\mathbf{a}_1, \mathbf{a}_2, \mathbf{c}]$;
- $R^{\mathbb{B}}(h(\mathbf{a}_1), \ldots, h(\mathbf{a}_a))$ if, and only if, $\mathbb{A} \models \rho^R[\mathbf{a}_1, \ldots, \mathbf{a}_a, \mathbf{c}]$;
- $h(\mathbf{a}) = c^{\mathbb{B}}$ if, and only if, $\mathbb{A} \models \gamma^c[\mathbf{a}, \mathbf{c}]$.

Note that an interpretation Φ associates a τ-structure with $(\mathbb{A}, \mathbf{c})$ only if η defines an equivalence relation on A^m that is a congruence with respect to the relations defined by the formulas ρ^R and γ^c. In such cases however, $\mathbb{B}$ is uniquely determined up to isomorphism and we write $\mathbb{B} = \Phi(\mathbb{A}, \mathbf{c})$. We will only be interested in interpretations that associate a τ-structure to every $(\mathbb{A}, \mathbf{c})$.

We say that a map r from σ-structures to τ-structures is Θ-definable if there is a Θ-interpretation Φ *without parameters* such that for all σ-structures $\mathbb{A}$, $r(\mathbb{A}) = \Phi(\mathbb{A})$.

Thus, we can ask whether a given reduction is LFP-definable or FO^s-definable, for example. It is an easy consequence of the fact that LFP captures P on ordered structures that a reduction is LFP definable *with order* if, and only if, it is a polynomial-time reduction. In the case of the complexity class NP, we know there are complete problems under much weaker reductions such as those defined by quantifier-free formulas in the presence of order, or first-order formulas even without order (see [2,13]).

For reductions between parameterized problems, it is more natural to consider the slicewise definition of interpretations. We say that a reduction r between parameterized problems (Q, κ) and (Q', κ') is *slicewise Θ-definable* if there is an m and a function δ that takes each natural number k to an m-ary Θ-interpretation $\delta(k)$ such that for any σ-structure $\mathbb{A}$ with $r(\mathbb{A}) = \delta(\kappa(\mathbb{A}))(\mathbb{A})$. Note, in particular, that the width m of the interpretation is the same for all k. It is an easy consequence of the proof of Theorem 1 in [7] that a reduction r is an *fpt*-reduction if, and only if, for some s, it is slicewise LFP^s-definable on ordered structures.

The following definition introduces some useful notation for the different classes of reductions we consider.

Definition 1. *For parameterized problems (Q, κ) and (Q', κ'), we write*

1. $(Q, \kappa) \leq^{\texttt{s-fo}} (Q', \kappa')$ *if there is a reduction from (Q, κ) to (Q', κ') that is* slicewise FO-definable*;*
2. $(Q, \kappa) \leq^{\texttt{s-bfo}} (Q', \kappa')$ *if there is a reduction from (Q, κ) to (Q', κ') that is* slicewise FO^s-definable *for some s;*
3. $(Q, \kappa) \leq^{\texttt{s-qf}} (Q', \kappa')$ *if there is a reduction from (Q, κ) to (Q', κ') that is* slicewise Θ-definable*, where Θ is the collection of* quantifier-free *formulas.*

It is clear from the definition that $(Q, \kappa) \leq^{\texttt{s-bfo}} (Q', \kappa')$ implies $(Q, \kappa) \leq^{\texttt{s-fo}} (Q', \kappa')$. Furthermore, since the definition of slicewise reductions requires the interpretations to be of fixed width, and the only variables that occur in a quantifier-free formula are the free variables, it can be easily seen that a $\leq^{\texttt{s-qf}}$ reduction is defined with a bounded number of variables. Thus, $(Q, \kappa) \leq^{\texttt{s-qf}} (Q', \kappa')$ implies $(Q, \kappa) \leq^{\texttt{s-bfo}} (Q', \kappa')$ and the reductions in Definition 1 are increasingly weak as we go down the list. The last two of them are also weaker than *fpt*-reductions, in the sense that, since FO^s formulas are also LFP^s formulas, we have that $(Q, \kappa) \leq^{\texttt{s-bfo}} (Q', \kappa')$ implies $(Q, \kappa) \leq^{\texttt{fpt}} (Q', \kappa')$. As we show in Section 5, it is unlikely that $(Q, \kappa) \leq^{\texttt{s-fo}} (Q', \kappa')$ implies $(Q, \kappa) \leq^{\texttt{fpt}} (Q', \kappa')$ as this would entail the collapse of the W-hierarchy.

4 Bounded-Variable Reductions

In this section, we construct problems that are complete for the class $W[t]$ under $\leq^{\texttt{s-bfo}}$ reductions.

We first consider the decision problem of *alternating reachability*, also known as *game*. We are given a directed graph $G = (V, E)$ along with a bipartition of the vertices $V = V_\exists \uplus V_\forall$ and two distinguished vertices a and b. We are asked to decide whether the pair (a, b) is in the *alternating transitive closure* defined by $(V_\exists, V_\forall, E)$. This is equivalent to asking whether the existential player has a winning strategy in the following two-player token pushing game played on G as follows. The token is initially on a. At each turn, if the token is on an element of $V_\exists$, it is the existential player that moves and, if it is on an element of $V_\forall$, it is the universal player that moves. Each move consists of the player whose turn it is moving the token from a vertex u to a vertex v such that $(u, v) \in E$. If the token reaches b, the existential player has won. In general, we call a directed graph $G = (V, E)$ along with a bipartition $V = V_\exists \uplus V_\forall$ an *alternating graph*; we call the vertices in $V_\exists$ the existential vertices of G and those in $V_\forall$ the universal vertices; and we call a the source vertex and b the target vertex. We can assume without loss of generality that the target vertex has no outgoing edges.

An *alternating path* from a to b is an *acyclic* subgraph $V' \subseteq V, E' \subseteq E$ with $a, b \in V'$ such that for every $u \in V' \cap V_\exists$, $u \neq b$, there is a $v \in V'$ with $(u, v) \in E'$; for every $u \in V' \cap V_\forall$, $u \neq b$ and every $v \in V$ with $(u, v) \in E$ we have that $v \in V'$ and $(u, v) \in E'$; and for every $u \in V'$, there is a path from u to b in (V', E'). It is easily checked that (a, b) is in the alternating transitive closure of $(V_\exists, V_\forall, E)$ if, and only if, there is an alternating path from a to b.

It is known that the alternating reachability problem is complete for P under first-order reductions, in the presence of order (see [10]). Indeed, it is also known that in the absence of order, the problem is still complete for the class of problems that are definable in LFP [3]. Moreover, it is easily shown from the reductions constructed by Dahlhaus in [3] that every problem definable in LFP^s is reducible to alternating reachability by means of a first-order reduction whose width depends only on s, giving us the following lemma.

Lemma 1. *For any s there is an r such that for any formula $\varphi(\mathbf{z})$ of LFP^s in the signature σ, we can find an FO^r-interpretation with parameters $\mathbf{z}$ that takes each $(\mathbb{A}, \mathbf{c})$, where $\mathbb{A}$ is a σ-structure and $\mathbf{c}$ an interpretation of the parameters, to an alternating graph $(V, V_\exists, V_\forall, E, a, b)$ so that $\mathbb{A} \models \varphi[\mathbf{c}]$ if, and only if, there is an alternating path from a to b in $(V_\exists, V_\forall, E)$.*

It is easily checked that alternating reachability is defined by the following formula of LFP^2.

$$[\mathbf{lfp}_{X,x}(x = b \vee (V_\exists(x) \wedge \exists y(E(x,y) \wedge X(y))) \vee \\ (V_\forall(x) \wedge \exists y E(x,y) \wedge \forall y(E(x,y) \rightarrow X(y))))](a)$$

For an alternating graph $G = (V, V_\exists, V_\forall, E)$ and a subset $U \subseteq V$, we say that there is a *U-avoiding alternating path* from a to b if there is an alternating path from a to b which does not include any vertex of U. Note, that this is *not* the same as saying there is an alternating path from a to b in the subgraph of G induced by $V \setminus U$. In particular, a *U-avoiding alternating path* may not include any universal vertex which has an outgoing edge to a vertex in U, though such vertices may appear in an alternating path in the graph $G[V \setminus U]$.

We will now define a series of variants of the alternating reachability problem, which will lead us to the $W[t]$-complete problems we seek to define. In the following definitions, k is a fixed positive integer.

k-conjunctive restricted alternating reachability. Given an alternating graph $G = (V, V_\exists, V_\forall, E)$, along with sets of vertices $C \subseteq U \subseteq V$ and distinguished vertices a and b, where a has at most k outgoing edges, decide whether for every v such that $(a, v) \in E$, there is an $s_v \in C$ and a $(U \setminus \{s_v\})$-avoiding alternating path from v to b. If the answer is yes, we say there is a *k-conjunctive restricted alternating path* from a to b.

In other words, the problem asks whether there is an alternating path from a to b, where $a \in V_\forall$, of a particular restricted kind. The path is not to use the vertices in U apart from C, and these may be used only in a limited way. That is, each outgoing edge from a leads to a path which may use only one vertex of C, though this vertex may be different for the different edges leaving a.

We define a dual to the above for starting vertices a that are existential.

k-disjunctive restricted alternating reachability. Given an alternating graph $G = (V, V_\exists, V_\forall, E)$, along with sets of vertices $C \subseteq U \subseteq V$ and distinguished vertices a and b, where a has at most k outgoing edges, decide whether there is a vertex v with $(a, v) \in E$ and an $s_v \in C$ such that there is a $(U \setminus \{s_v\})$-avoiding alternating path from v to b. If the answer is yes, we say there is a *k-disjunctive restricted alternating path* from a to b.

We next define, by induction on t, the problems of conjunctive and disjunctive k, t-*restricted alternating reachability*, for which the above two problems serve as base cases.

Definition 2 (k, t-restricted alternating reachability). *The* conjunctive $k, 0$-restricted alternating reachability *problem is just the k-conjunctive restricted alternating reachability problem defined above, and similarly, the* disjunctive $k, 0$-restricted alternating reachability *problem is the k-disjunctive restricted alternating reachability problem.*

The conjunctive $k, t+1$-restricted alternating reachability *is the problem of deciding, given an alternating graph $G = (V, V_\exists, V_\forall, E)$, along with sets of vertices $C \subseteq U \subseteq V$ and distinguished vertices a and b, whether for every v such that $(a, v) \in E$, there is a disjunctive k, t-restricted alternating path from v to b.*

Dually, the disjunctive $k, t+1$-restricted alternating reachability *is the problem of deciding whether there is a v such that $(a, v) \in E$ and there is a conjunctive k, t-restricted alternating path from v to b.*

Roughly speaking, the conjunctive k, t-restricted alternating reachability problem asks for an alternating path from a to b, with a a universal node, where we are allowed t alternations before the restrictions on the use of vertices in the sets U and C kick in. The disjunctive version is dual.

We are ready to define the parameterized problems we need. By a *clique* in a directed graph, we just mean a set of vertices such that for each pair of distinct vertices in the set, there are edges in both directions.

Definition 3 (clique-restricted alternating reachability). *For any fixed t, the* parameterized t-clique-restricted alternating reachability problem *is:*

p-t-CLIQUE RESTRICTED ALTERNATING REACHABILITY	
Input:	$G = (V = V_\exists \uplus V_\forall, E)$, $U \subseteq V$, $a, b \in V$ and $k \in \mathbb{N}$.
Parameter:	k
Problem:	Is there a clique $C \subseteq U$ with $\lvert C \rvert = k$ such that $(V, V_\exists, V_\forall, E, U, C)$ admits a conjunctive k, t-restricted alternating path from a to b?

Theorem 4. *For each $t \geq 1$, p-t-clique-restricted alternating reachability is in $W[t]$.*

Proof. This is easily established, using Theorem 2, by writing a formula of $\Sigma_{t,1}$-Bool(LFP^4) that defines the problem for each fixed value of k. This is obtained by taking the prenex normal form of the formula

$$\exists x_1 \cdots x_k (\bigwedge_i U(x_i) \wedge \bigwedge_{i \neq j} E(x_i, x_j) \wedge x_i \neq x_j) \wedge \forall y_1 (E(a, y_1) \rightarrow (\exists y_2 (E(y_1, y_2) \wedge \cdots \Gamma \cdots)$$

where Γ is the formula $\exists y_{t-1}(E(y_{t-2}, y_{t-1}) \wedge \bigwedge_{1 \leq i \leq k}(\bigvee_{1 \leq j \leq k} \theta_i(y_{t-1}, x_j)))$ if t is odd and the formula $\forall y_{t-1}(E(y_{t-2}, y_{t-1}) \rightarrow \bigvee_{1 \leq i,j \leq k} \theta_i(y_{t-1}, x_j))$ if t is even; and $\theta_i(y_{t-1}, x_j)$ is the formula of LFP^4 which states that there is a $U \setminus \{x_j\}$-avoiding alternating path from z to b, where z is the ith (in the linear ordering $\leq$) vertex such that there is an edge from y_{t-1} to z. This formula is obtained as an easy modification of the LFP^2 formula above defining alternating reachability. □

Theorem 5. *For each $t \geq 1$, p-t-clique-restricted alternating reachability problem is $W[t]$-hard.*

Proof. (sketch): Suppose (Q, κ) is in $W[t]$. Assume that t is odd (the case for even t is dual). By Theorem 3, there is a u and an s so that (Q, κ) is slicewise-$\Sigma_{t,u}$-Conj(LFP^s)-definable. Thus, for each k, there is a formula

$$\varphi \equiv \exists x_{11} \cdots \exists x_{1l_1} \forall x_{21} \cdots \forall x_{2l_2} \cdots \exists x_{t1} \cdots \exists x_{tl_t} \bigwedge_{j \in S} \theta_j$$

where each θ_j is in LFP^s, which defines the structures $\mathbb{A}$ such that $\mathbb{A} \in Q$ and $\kappa(\mathbb{A}) = k$. We give an informal description of the reduction that takes $\mathbb{A}$ to an instance $\mathbb{G}$ of t-clique-restricted alternating reachability. The reduction is definable by an FO-interpretation using a number of variables that is a function of s, t and u but independent of k. In what follows, we assume that $\lvert S \rvert \leq \binom{l_1}{s}$. If this is not the case, we can add dummy variables to the first quantifier block without changing the meaning of the formula.

By Lemma 1, we know that each θ_j gives rise to an FO^r-interpretation (for a fixed value of r) that maps $\mathbb{A}$ to an instance of alternating reachability. Note that, as the width of the Since θ_j has (as many as s) free variables, the interpretation will have up to s parameters from among the variables $x_{11}, \ldots, x_{tl_t}$. For notational purposes, we will distinguish between those parameters that are in the variables quantified in the first existential block (i.e. $x_{11} \ldots x_{1l_1}$) and the others. Thus, we write $\mathrm{AR}_j^{\alpha,\beta}$ for the instance of alternating reachability obtained from θ_j, with α the assignment of values to the parameters among $x_{11} \dots x_{1l_1}$ and β the assignment of values to the other parameters. $\mathbb{G}$ will contain the disjoint union of all of these instances (slightly modified as explained below). Note that, since the interpretation taking $\mathbb{A}$ to $\mathrm{AR}_j^{\alpha,\beta}$ has width at most r, the size of $\mathrm{AR}_j^{\alpha,\beta}$ is at most n^r (where n is the size of $\mathbb{A}$). Furthermore, there are at most $|S| \cdot n^s$ such instances. $\mathbb{G}$ also contains a target vertex b with an incoming edge from the target vertex in each $\mathrm{AR}_j^{\alpha,\beta}$.

In addition, for each initial segment $\mathbf{x}$ of the sequence of variables $x_{21} \dots x_{tl_t}$ that ends at a quantifier alternation (i.e. $\mathbf{x} = x_{21} \dots x_{t',l_{t'}}$ for some $t' \leq t$), and each assignment ρ of values from $\mathbb{A}$ to the variables in $\mathbf{x}$, $\mathbb{G}$ contains a new element. Note that the number of such elements is less than $2n^{\sum_{2 \leq i \leq t} l_i}$, which is at most $2n^{u(t-1)}$. For each ρ and ρ', we include an edge from ρ to ρ' just in case ρ' extends the assignment ρ by exactly one quantifier block. If that block is existential, ρ is in $V_\exists$ and if it is universal, ρ is in $V_\forall$. Those ρ which assign a value to every variable in $x_{21} \dots x_{tl_t}$ are in $V_\forall$ and have outgoing edges to a vertex ρ_j, one for each $j \in S$. These vertices are existential and have outgoing edges to the source vertex of each $\mathrm{AR}_j^{\alpha,\beta}$ where the assignment β is consistent with ρ. That is, if a variable x among $x_{21} \dots x_{tl_t}$ occurs among the parameters of θ_j we should have $\beta(x) = \rho(x)$. The unique empty assignment ϵ is the source vertex of $\mathbb{G}$.

Let ψ be the part of φ after the first existential block, i.e. $\varphi \equiv \exists x_{11} \cdots \exists x_{1l_1} \psi$. The structure so far codes the interpretation of ψ with each θ_j replaced by equivalent alternating reachability conditions. If there is an assignment of values to the variables $x_{11} \dots x_{1l_1}$ that makes ψ true, there is an alternating path from source to target in $\mathbb{G}$. However, the converse is not true, as distinct θ_j may share free variables from among $x_{11} \dots x_{1l_1}$ and there is nothing in the alternating path that ensures consistency in the values they assign to these variables. In fact, it can be shown that, in the structure described so far, there is an alternating path from source to target if, and only if, we can assign values to the free variables $x_{11} \dots x_{1l_1}$, *independently for each* θ_j in a way that makes ψ true. We now add a gadget to $\mathbb{G}$ that ensures consistency of the assignment of these values.

$\mathbb{G}$ also contains a set U of vertices, disjoint from those constructed so far. There is one vertex in U for each assignment of values from $\mathbb{A}$ to a subset of the variables $x_{11} \dots x_{1l_1}$ *of size* s. Thus, U contains a total of $\binom{l_1}{s} \cdot n^s$ vertices. All vertices in U are existential (i.e. in $V_\exists$) and for $\alpha, \beta \in U$, there is an edge from α to β if the two assignments agree on all variables they have in common. It is easily checked that the maximal cliques in U are of size $\binom{l_1}{s}$. There is one such clique for each assignment of values to all l_1 variables.

To connect the gadget U with the rest of the construction, we replace every vertex v in each $\mathrm{AR}_j^{\alpha,\beta}$ with two vertices v_{in} and v_{out}. All edges into v now lead to v_{in} and all edges out of v are replaced by edges out of v_{out}. v_{out} is existential if v is existential and universal if v is universal. v_{in} is universal for all v and has exactly two outgoing edges, one to v_{out} and the other to $\alpha \in U$ (i.e. the element of U giving the assignment to the parameters among $x_{11} \dots x_{1l_1}$ that corresponds to the instance $\mathrm{AR}_j^{\alpha,\beta}$). Finally, there is also an edge from $\alpha \in U$ to v_{out} for each vertex v in any instance of the form $\mathrm{AR}_j^{\alpha,\beta}$.

This completes the description of $\mathbb{G}$. It is not difficult to argue that $\mathbb{G}$ contains a clique $C \subseteq U$ of size $\binom{l_1}{s}$ such that $(\mathbb{G}, C)$ admits a conjunctive $\binom{l_1}{s}$, t-restricted alternating path from ϵ to b if, and only if, $\mathbb{A} \models \varphi$. Indeed, if $\mathbb{A} \models \varphi$ and γ is an assignment to the variables $x_{11} \ldots x_{1l_1}$ that witnesses this, we can choose C to consist of all nodes α that are consistent with γ. As we have argued above, this will yield the required alternating path through $\mathbb{G}$. Conversely, any clique $C \subseteq U$ of size $\binom{l_1}{s}$ must correspond to such an assignment γ and thus any alternating path in $\mathbb{G}$ that uses only C will provide a witness that $\mathbb{A} \models \varphi$.

We omit from this sketch the construction of the formulas that show that the interpretation that takes $\mathbb{A}$ to $\mathbb{G}$ can be given by first-order formulas where the number of variables is independent of k (i.e. independent of l_1), but we will make two points in this connection. One is that the total number of vertices in $\mathbb{G}$ is bounded by $\binom{l_1}{s}n^s + |S|n^{s+r} + 2n^{(t-1)u}$. This is a polynomial in n whose degree depends on s, t and u (recall, by Lemma 1, that r is a function of s) but not on l_1. We use this to establish that the width of the interpretation is bounded. One subtle point is that in defining the set U, we need to define not only all s-tuples of elements of $\mathbb{A}$ but to pair them with s-element sets of variables. We do this by identifying the variables with the first l_1 elements of the linear order $\leq$. We then use the fact that any fixed element of a linear order can be identified with a formula of FO^3 (see, for instance, [4]). □

5 Other First-Order Reductions

As we pointed out in Section 2, one standard definition of the class $W[t]$ is as $[p\text{-WD-}\Pi_t]^{\leq^{\text{fpt}}}$, i.e. the class of problems *fpt*-reducible to p-WD$_\varphi$ for some $\varphi \in \Pi_t$. Here we consider the class $[p\text{-WD-}\Pi_t]^{\leq^{\text{s-qf}}}$, i.e. the problems reducible by *slicewise quantifier-free reductions* to p-WD$_\varphi$ for some $\varphi \in \Pi_t$ and show this class is most likely weaker, in the sense that there are problems in $W[t+1]$ that are *provably* not in this class.

It is known that the alternation of quantifiers in a first-order formula yields a strict hierarchy of increasing expressive power, even on ordered structures, in the presence of arithmetic relations [14]. We use this to establish our result.

Say that an alternating graph G is *strictly alternating* if each vertex in $V_\exists$ only has edges to vertices in $V_\forall$ and vice versa. We can write, for each $t \geq 1$, a formula $\varphi_t(X) \in \Pi_t$ with a free set variable X that is satisfied by a strictly alternating graph G with a set S interpreting X if, and only if, the source vertex of G is universal and G contains an alternating path from a to all $b \in S$ (or for some $b \in S$, when t is even) with exactly t alternations. We are able to show, by a reduction from the problems that Sipser [14] uses to establish the strictness of the first-order quantifier alternation hierarchy, that $\varphi_{t+1}(X)$ is not equivalent to any formula of Π_t, even on ordered structures with arithmetic. On the other hand, it is not difficult to show that if p-WD$_{\varphi_{t+1}} \leq^{\text{s-qf}} p$-WD$_\psi$ for a formula $\psi \in \Pi_t$ then, composing ψ with the interpretation, we would obtain a formula of Π_t equivalent to φ_{t+1}. This leads to the following theorem.

Theorem 6. *For each $t \geq 1$, there is a $\varphi \in \Pi_{t+1}$ such that for any $\psi \in \Pi_t$, p-WD$_\varphi \not\leq^{\text{s-qf}} p$-WD$_\psi$.*

On the other hand, for any formula $\varphi(X)$, it is easy to construct a sequence of first-order formulas $\varphi_k (k \in \mathbb{N})$, without the variable X, that define the slices of p-WD$_\varphi$. These can be used to construct a slicewise first-order reduction of p-WD$_\varphi$ to a trivial problem, giving us the following observation.

Theorem 7. *For any $\varphi \in$ FO, there is an FPT problem Q such that p-WD$_\varphi \leq^{s\text{-}fo} Q$.*

6 Concluding Remarks

We have considered varying the notion of reductions used in the definition of the classes of the W-hierarchy. The results of Section 5 show that slicewise quantifier-free reductions are too weak, and slicewise first-order reductions are too strong for the purpose. The intermediate case of slicewise bounded-variable first-order reductions is considered in Section 4 and though these reductions are considerably weaker than *fpt*-reductions, we are able to show the existence of complete problems for the classes of the W-hierarchy. It would be interesting to investigate whether other, natural $W[t]$-complete problems remain complete under these reductions. In particular, is it the case that the closure of p-WD-Π_t under such reductions is all of $W[t]$?

References

1. Chen, Y., Flum, J.: Machine characterization of the classes of the W-hierarchy. In: Baaz, M., Makowsky, J.A. (eds.) CSL 2003. LNCS, vol. 2803, pp. 114–127. Springer, Heidelberg (2003)
2. Dahlhaus, E.: Reduction to NP-complete problems by interpretations. In: Börger, E., Rödding, D., Hasenjaeger, G. (eds.) Rekursive Kombinatorik 1983. LNCS, vol. 171, pp. 357–365. Springer, Heidelberg (1984)
3. Dahlhaus, E.: Skolem normal forms concerning the least fixpoint. In: Börger, E. (ed.) Computation Theory and Logic. LNCS, vol. 270, pp. 101–106. Springer, Heidelberg (1987)
4. Dawar, A.: How many first-order variables are needed on finite ordered structures? In: Artëmov, S.N., Barringer, H., d'Avila Garcez, A.S., Lamb, L.C., Woods, J. (eds.) We Will Show Them! Essays in Honour of Dov Gabbay, vol. 1, pp. 489–520. College Publications (2005)
5. Downey, R.G., Fellows, M.R.: Fixed-parameter tractability and completeness I—basic results. Theoretical Computer Science 141, 109–131 (1995)
6. Ebbinghaus, H.-D., Flum, J.: Finite Model Theory, 2nd edn. Springer, Heidelberg (1999)
7. Flum, J., Grohe, M.: Fixed-parameter tractability, definability, and model checking. SIAM Journal on Computing 31, 113–145 (2001)
8. Flum, J., Grohe, M.: Describing parameterized complexity classes. Information and Computation 187, 291–319 (2003)
9. Flum, J., Grohe, M.: Parameterized complexity theory. Springer, Heidelberg (2006)
10. Immerman, N.: Descriptive Complexity. Springer, Heidelberg (1998)
11. Immerman, N.: Relational queries computable in polynomial time. Inf. Control 68, 86–104 (1986)
12. Libkin, L.: Elements of Finite Model Theory. Springer, Heidelberg (2004)
13. Lovász, L., Gács, P.: Some remarks on generalized spectra. Mathematical Logic Quarterly 23, 547–554 (1977)
14. Sipser, M.: Borel sets and circuit complexity. In: Proceedings of the Fifteenth Annual ACM Symposium on Theory of Computing, pp. 61–69 (1983)

The Communication Complexity of Non-signaling Distributions

Julien Degorre[1], Marc Kaplan[2], Sophie Laplante[2], and Jérémie Roland[3]

[1] CNRS, Laboratoire d'Informatique de Grenoble
[2] LRI, Université Paris-Sud
[3] NEC Laboratories America

Abstract. We study a model of communication complexity that encompasses many well-studied problems, including classical and quantum communication complexity, the complexity of simulating distributions arising from bipartite measurements of shared quantum states, and XOR games. In this model, Alice gets an input x, Bob gets an input y, and their goal is to each produce an output a, b distributed according to some pre-specified joint distribution $p(a, b|x, y)$. Our results apply to any non-signaling distribution, that is, those where Alice's marginal distribution does not depend on Bob's input, and vice versa.

By introducing a simple new technique based on affine combinations of lower-complexity distributions, we give the first general technique to apply to all these settings, with elementary proofs and very intuitive interpretations. The lower bounds we obtain can be expressed as linear programs (or SDPs for quantum communication). We show that the dual formulations have a striking interpretation, since they coincide with maximum violations of Bell and Tsirelson inequalities. The dual expressions are closely related to the winning probability of XOR games. Despite their apparent simplicity, these lower bounds subsume many known communication complexity lower bound methods, most notably the recent lower bounds of Linial and Shraibman for the special case of Boolean functions.

We show that as in the case of Boolean functions, the gap between the quantum and classical lower bounds is at most linear in the size of the support of the distribution, and does not depend on the size of the inputs. This translates into a bound on the gap between maximal Bell and Tsirelson inequality violations, which was previously known only for the case of distributions with Boolean outcomes and uniform marginals. It also allows us to show that for some distributions, information theoretic methods are necessary to prove strong lower bounds.

Finally, we give an exponential upper bound on quantum and classical communication complexity in the simultaneous messages model, for any non-signaling distribution.

1 Introduction

Communication complexity of Boolean functions has a long and rich past, stemming from the paper of Yao in 1979 [1], whose motivation was to study the area of VLSI circuits. In the years that followed, tremendous progress has been made in developing a rich array of lower bound techniques for various models of communication complexity (see e.g. [2]).

R. Královič and D. Niwiński (Eds.): MFCS 2009, LNCS 5734, pp. 270–281, 2009.

From the physics side, the question of studying how much communication is needed to simulate distributions arising from physical phenomena, such as measuring bipartite quantum states, was posed in 1992 by Maudlin, a philosopher of science, who wanted to quantify the non-locality inherent to these systems [3]. Maudlin, and the authors who followed [4,5,6,7,8] (some independently of his work, and of each other) progressively improved upper bounds on simulating correlations of the 2 qubit singlet state. In a recent breakthrough, Regev and Toner [9] proved that two bits of communication suffice to simulate the correlations arising from two-outcome measurements of arbitrary-dimension bipartite quantum states. In the more general case of non-binary outcomes, Shi and Zhu gave a protocol to approximate quantum distributions within constant error, using constant communication [10]. No non-trivial lower bounds are known for this problem.

In this paper, we consider the more general framework of simulating non-signaling distributions. These are distributions of the form $p(a, b|x, y)$, where Alice gets input x and produces an output a, and Bob gets input y and outputs b. The non-signaling condition is a fundamental property of bipartite physical systems, which states that the players gain no information on the other player's input. In particular, distributions arising from quantum measurements on shared bipartite states are non-signaling, and Boolean functions may be reduced to extremal non-signaling distributions with Boolean outcomes and uniform marginals.

Outside of the realm of Boolean functions, a very limited number of tools are available to analyse the communication complexity of distributed tasks, especially for quantum distributions with non-uniform marginals. In such cases, the distributions live in a larger-dimensional space and cannot be cast as communication matrices, so standard techniques do not apply. The structure of non-signaling distributions has been the object of much study in the quantum information community, yet outside the case of distributions with Boolean inputs or outcomes [11,12], or with uniform marginal distributions, much remains to be understood.

Our main contribution is a new method for handling all non-signaling distributions, including the case of non-Boolean outcomes and non-uniform marginals, based on affine combinations of lower-complexity distributions, which we use to obtain both upper and lower bounds on communication. We use the elegant geometric structure of the non-signaling distributions to analyse the communication complexity of Boolean functions, but also non-Boolean or partial functions. Although they are formulated, and proven, in quite a different way, our lower bounds turn out to subsume Linial and Shraibman's factorization norm lower bounds [13], in the restricted case of Boolean functions. Similarly, our upper bounds extend the upper bounds of Shi and Zhu for approximating quantum distributions [10] to all non-signaling distributions (in particular distributions obtained by protocols using entanglement *and* quantum communication).

Our complexity measures can be expressed as linear (or semidefinite) programs, and when we consider the dual of our lower bound expressions, these turn out to correspond precisely to maximal Bell inequality violations in the case of classical communication, and Tsirelson inequality violations for quantum communication. Hence, we have made formal the intuition that large Bell inequalities should lead to large lower bounds on communication complexity.

Many proofs are omitted from this extended abstract, and can be found in the full version of the paper.

2 Preliminaries

2.1 Non-signaling Distributions

Non-signaling, a fundamental postulate of physics, states that any observation on part of a system cannot instantaneously affect a remote part of the system, or similarly, that no signal can travel instantaneously. We consider distributions $p(a,b|x,y)$ where $x \in \mathcal{X}, y \in \mathcal{Y}$ are the inputs of the players, and they are required to each produce an outcome $a \in \mathcal{A}, b \in \mathcal{B}$, distributed according to $p(a,b|x,y)$.

Definition 1 (Non-signaling distributions). *A bipartite, conditional distribution* $\mathbf{p}$ *is non-signaling if* $\forall a,x,y,y', \sum_b p(a,b|x,y) = \sum_b p(a,b|x,y')$*, and* $\forall b,x,x',y$, $\sum_a p(a,b|x,y) = \sum_a p(a,b|x',y)$.

For a non-signaling distribution, the marginal distribution $p(a|x,y) = \sum_b p(a,b|x,y)$ on Alice's output does not depend on y, so we write $p(a|x)$, and similarly $p(b|y)$ for Bob. We denote by $\mathcal{C}$ the set of all non-signaling distributions.

In the case of binary outcomes $\mathcal{A}$=$\mathcal{B}$=$\{\pm 1\}$, a non-signaling distribution is uniquely determined by the (expected) correlations, defined as $C(x,y) = E(a \cdot b|x,y)$, and the (expected) marginals, defined as $M_A(x) = E(a|x), M_B(y) = E(b|y)$. For this reason, we will write $\mathbf{p} = (C, M_A, M_B)$ and use both notations interchangeably when considering distributions over binary outcomes. We denote by $\mathcal{C}_0$ the set of non-signaling distributions with uniform marginals, that is, $\mathbf{p} = (C,0,0)$, and write $C \in \mathcal{C}_0$.

Boolean functions. The communication complexity of Boolean functions is a special case of the problem of simulating non-signaling distributions. As we shall see in Section 2.3, the associated distributions are extremal points of the non-signaling polytope. If the distribution stipulates that the product of the players' outputs equal some function $f : \mathcal{X} \times \mathcal{Y} \to \{\pm 1\}$ then this corresponds to the standard model of communication complexity (up to an additional bit of communication, for Bob to output $f(x,y)$). If we further require that Alice's output be +1 or -1 with equal probability, likewise for Bob, then the distribution is non-signaling and has the following form:

Definition 2. *For a function* $f : \mathcal{X} \times \mathcal{Y} \to \{-1,1\}$*, denote* $\mathbf{p}_f$ *the distribution defined by* $p_f(a,b|x,y) = \frac{1}{2}$ *if* $f(x,y) = a \cdot b$ *and 0 otherwise. Equivalently,* $\mathbf{p}_f = (C_f, 0, 0)$ *where* $C_f(x,y) = f(x,y)$.

In the case of randomized communication complexity, a protocol that simulates a Boolean function with error probability ϵ corresponds to simulating correlations C' scaled down by a factor at most $1-2\epsilon$, that is, $\forall x,y, \mathrm{sgn}(C'(x,y)) = C_f(x,y)$ and $|\, C'(x,y) \,| \geq 1-2\epsilon$. While we will not consider these cases in full detail, non-Boolean functions, partial functions and some relations may be handled in a similar fashion, hence our techniques can be used to show lower bounds in these settings as well.

Quantum distributions. Of particular interest in the study of quantum non-locality are the distributions arising from measuring bipartite quantum states. We will use the following definition (see also [14]):

Definition 3. *A distribution* $\mathbf{p}$ *is* quantum *if there exists a quantum state* $|\psi\rangle$ *in a Hilbert space* $\mathcal{H}$ *and measurement operators* $\{E_a(x) : a \in \mathcal{A}, x \in \mathcal{X}\}$ *and* $\{E_b(y) : b \in \mathcal{B}, y \in \mathcal{Y}\}$*, such that* $p(a, b|x, y) = \langle\psi|E_a(x)E_b(y)|\psi\rangle$*, and*

1. $E_a(x)^\dagger = E_a(x)$ *and* $E_b(y)^\dagger = E_b(y)$*,*
2. $E_a(x) \cdot E_{a'}(x) = \delta_{aa'} E_a(x)$ *and* $E_b(y) \cdot E_{b'}(y) = \delta_{bb'} E_b(y)$*,*
3. $\sum_a E_a(x) = \mathbb{1}$ *and* $\sum_b E_b(x) = \mathbb{1}$*, where* $\mathbb{1}$ *is the identity operators on* $\mathcal{H}$*,*
4. $E_a(x) \cdot E_b(y) = E_b(y) \cdot E_a(x)$*,*

where δ_{ab} *is the Kronecker delta defined by* $\delta_{ab}{=}1$ *if* $a{=}b$ *and 0 otherwise.*

We denote by $\mathcal{Q}$ the set of all quantum distributions. In the restricted case of binary outcomes with uniform marginals, we let $\mathcal{Q}_0$ be the set of all quantum correlations.

2.2 Models of Communication Complexity

We consider the following model of communication complexity of non-signaling distributions $\mathbf{p}$. Alice gets input x, Bob gets input y, and after exchanging bits or qubits, Alice has to output a and Bob b so that the joint distribution is $p(a, b|x, y)$. $R_0(\mathbf{p})$ denotes the communication complexity of simulating $\mathbf{p}$ exactly, using private randomness and classical communication. $Q_0(\mathbf{p})$ denotes the communication complexity of simulating $\mathbf{p}$ exactly, using quantum communication. We use superscripts "pub" and "ent" in the case where the players share random bits or quantum entanglement. For $R_\epsilon(\mathbf{p})$, we are only required to simulate some distribution $\mathbf{p}'$ such that $\delta(\mathbf{p}, \mathbf{p}') \leq \epsilon$, where $\delta(\mathbf{p}, \mathbf{p}') = \max\{|p(\mathcal{E}|x, y) - p'(\mathcal{E}|x, y)| : x, y \in \mathcal{X} \times \mathcal{Y}, \mathcal{E} \subseteq \mathcal{A} \times \mathcal{B}\}$ is the total variation distance (or statistical distance) between two distributions.

For binary outcomes, we write $R_\epsilon(C, M_A, M_B), Q_\epsilon(C, M_A, M_B)$. In the case of Boolean functions, $R_\epsilon(C) = R_\epsilon(C, 0, 0)$ corresponds to the usual notion of computing f with probability at least $1 - \epsilon$, where C is the ± 1 communication matrix of f.

2.3 Non-signaling, Quantum, and Local Distributions

In quantum information, distributions that can be simulated with shared randomness and no communication (also called a local hidden variable model) are called local.

Definition 4. Local deterministic distributions *are of the form* $p(a, b|x, y) = \delta_{a=\lambda_A(x)} \cdot \delta_{b=\lambda_B(y)}$ *where* $\lambda_A : \mathcal{X} \to \mathcal{A}$ *and* $\lambda_B : \mathcal{Y} \to \mathcal{B}$*, and* δ *is the Kronecker delta. A distribution is* local *if it can be written as a convex combination of local deterministic distributions.*

We let Λ be the set of local deterministic distributions $\{\mathbf{p}^\lambda\}_{\lambda \in \Lambda}$ and $\mathcal{L}$ be the set of local distributions. Let $\mathsf{conv}(\Lambda)$ be the convex hull of Λ. In the case of binary outcomes,

Proposition 1. $\mathcal{L} = \mathsf{conv}(\{(u^T v, u, v) : u \in \{\pm 1\}^{\mathcal{X}}, v \in \{\pm 1\}^{\mathcal{Y}}\})$.

We denote by $\mathcal{L}_0$ the set of local correlations over binary outcomes with uniform marginals.

The quantum information literature reveals a great deal of insight into the structure of the classical, quantum, and non-signaling distributions. It is well known that $\mathcal{L}$ and $\mathcal{C}$ are polytopes. While the extremal points of $\mathcal{L}$ are simply the local deterministic distributions, the non-signaling polytope $\mathcal{C}$ has a more complex structure [11,12]. $\mathcal{C}_0$ is the convex hull of the distributions obtained from Boolean functions.

Proposition 2. $\mathcal{C}_0 = \mathsf{conv}(\{(C_f, 0, 0) : C_f \in \{\pm 1\}^{\mathcal{X}\times\mathcal{Y}}\})$.

We show that $\mathcal{C}$ is the affine hull of the local polytope (restricted to the positive orthant since all probabilities $p(a, b|x, y)$ must be positive). This was shown independently of us, first in the quantum logic community [15,16,17,18], and again independently in the physics community [19].

Theorem 1. $\mathcal{C} = \mathsf{aff}^+\{\mathcal{L}\}$, *where* $\mathsf{aff}^+\{\mathcal{L}\}$ *is the restriction to the positive orthant of the affine hull of* $\mathcal{L}$, *and* $\dim \mathcal{C} = \dim \mathcal{L} = |\mathcal{X}| \times |\mathcal{Y}| + |\mathcal{X}| + |\mathcal{Y}|$.

Hence, while local distributions are *convex* combinations of local deterministic distributions, non-signaling distributions are *affine* combinations of these distributions.

Corollary 2 (Affine model). $\mathbf{p} \in \mathcal{C}$ *iff* $\exists q_\lambda \in \mathbb{R}$ *with* $\mathbf{p} = \sum_{\lambda \in \Lambda} q_\lambda \mathbf{p}^\lambda$.

As for the set of quantum distributions $\mathcal{Q}$, it is known to be convex, but not a polytope. Clearly, $\mathcal{L} \subseteq \mathcal{Q} \subseteq \mathcal{C}$. In the case of binary outcomes with uniform marginals, Grothendieck's inequality (see e.g. [20]), together with Tsirelson's theorem [21], implies the following statement.

Proposition 3. $\mathcal{L}_0 \subseteq \mathcal{Q}_0 \subseteq K_G \mathcal{L}_0$, *where* K_G *is Grothendieck's constant.*

3 Lower Bounds for Non-signaling Distributions

Based on the characterization of $\mathcal{C}$ (Theorem 1), we define two new complexity measures for non-signaling distributions, from which we derive a general method for lower bounds in deterministic, randomized, and quantum communication complexity.

Definition 5.
- $\tilde{\nu}(\mathbf{p}) = \min\{\sum_i |q_i| : \exists \mathbf{p}_i \in \mathcal{L}, q_i \in \mathbb{R}, \mathbf{p} = \sum_i q_i \mathbf{p}_i\}$,
- $\tilde{\gamma}_2(\mathbf{p}) = \min\{\sum_i |q_i| : \exists \mathbf{p}_i \in \mathcal{Q}, q_i \in \mathbb{R}, \mathbf{p} = \sum_i q_i \mathbf{p}_i\}$,
- $\tilde{\nu}^\epsilon(\mathbf{p}) = \min\{\tilde{\nu}(\mathbf{p}') : \delta(\mathbf{p}, \mathbf{p}') \leq \epsilon\}$,
- $\tilde{\gamma}_2^\epsilon(\mathbf{p}) = \min\{\tilde{\gamma}_2(\mathbf{p}') : \delta(\mathbf{p}, \mathbf{p}') \leq \epsilon\}$.

These quantities are small when the distribution can be written as a near-convex combination of local (or quantum) distributions, and larger the more negative coefficients are required in such combinations. The set of local distributions $\mathcal{L}$ form the unit sphere of $\tilde{\nu}$, and similarly the set of quantum distributions $\mathcal{Q}$ form the unit sphere of $\tilde{\gamma}_2$.

Lemma 1. $\mathbf{p} \in \mathcal{L} \Longleftrightarrow \tilde{\nu}(\mathbf{p}) = 1$, *and* $\mathbf{p} \in \mathcal{Q} \Longleftrightarrow \tilde{\gamma}_2(\mathbf{p}) = 1$.

3.1 Affine Models for Nonsignaling Distributions

We derive a lower bound by giving an explicit affine model for any non-signaling distribution from a communication protocol.

Let $\mathbf{p}$ be a non-signaling distribution over $\mathcal{A} \times \mathcal{B}$. For $\pi \leq 1$, we define the scaled-down probability distribution $p_\pi(a,b|x,y) = \pi p(a,b|x,y) + (1-\pi)p(a|x)p(b|y)$. For distributions $\mathbf{p}_f$ arising from Boolean functions (Definition 2), this corresponds to scaling down the success probability (more precisely the bias) by a factor of π.

Theorem 3. *1. If $R_0^{\text{pub}}(\mathbf{p}) \leq c$, then $\mathbf{p}_\pi \in \mathcal{L}$ for $\pi = 2^{-c}$.*
2. If $Q_0^{\text{ent}}(\mathbf{p}) \leq q$, then, $\mathbf{p}_\pi \in \mathcal{Q}$ for $\pi = 2^{-2q}$.
3. If $Q_0^{\text{ent}}(C) \leq q$, then $\pi C \in \mathcal{Q}_0$ for $\pi = 2^{-q}$.

The proof idea is as follows. The players can no longer communicate, but they nevertheless wish to follow the protocol. They guess a random transcript, and if both players' messages coincide with the messages in the transcript, they follow the protocol to the end. If a player's messages do not match the transcript, then he outputs independently of the other player, following the marginal distribution corresponding to his input.

Theorem 4. *For any non-signaling distribution* $\mathbf{p}$*,*

1. $R_0^{\text{pub}}(\mathbf{p}) \geq \log(\tilde{\nu}(\mathbf{p})) - 1$, *and* $R_\epsilon^{\text{pub}}(\mathbf{p}) \geq \log(\tilde{\nu}^\epsilon(\mathbf{p})) - 1$.
2. $Q_0^{\text{ent}}(\mathbf{p}) \geq \frac{1}{2}\log(\tilde{\gamma}_2(\mathbf{p})) - 1$, *and* $Q_\epsilon^{\text{ent}}(\mathbf{p}) \geq \frac{1}{2}\log(\tilde{\gamma}_2^\epsilon(\mathbf{p})) - 1$.
3. $Q_0^{\text{ent}}(C) \geq \log(\tilde{\gamma}_2(C))$, *and* $Q_\epsilon^{\text{ent}}(C) \geq \log(\tilde{\gamma}_2^\epsilon(C))$.

Proof. We sketch a proof for the classical case, the quantum case is similar and follows by using teleportation. Let c be the number of bits exchanged. From Theorem 3, we know that $\mathbf{p}_\pi$ is local for $\pi = 2^{-c}$. Let P' be a protocol for $\mathbf{p}_\pi$, using shared randomness and no communication. Notice that $p(a,b|x,y) = \frac{1}{\pi}p_\pi(a,b|x,y) - (\frac{1}{\pi} - 1)p(a|x)p(b|y)$ is an affine model for $p(a,b|x,y)$. Then $\tilde{\nu}(\mathbf{p}) \leq \frac{2}{\pi} - 1$. For the last item we also use the fact that $C = \frac{1}{2}(\frac{1}{\pi}+1)\,\pi C - \frac{1}{2}(\frac{1}{\pi}-1)\,(-\pi C)$.

3.2 Factorization Norm and Related Measures

In the special case of distributions over binary variables with uniform marginals, the quantities $\tilde{\nu}$ and $\tilde{\gamma}_2$ become equivalent to the quantities ν and γ_2 defined in [22,13] (at least for the interesting case of non-local correlations, that is correlations with non-zero communication complexity). When the marginals are uniform we omit them and write $\tilde{\nu}(C)$ and $\tilde{\gamma}_2(C)$. The following are reformulations as Minkowski functionals of the definitions appearing in [22,13].

Definition 6.
- $\nu(C) = \min\{\Lambda > 0 : \frac{1}{\Lambda}C \in \mathcal{L}_0\}$,
- $\gamma_2(C) = \min\{\Lambda > 0 : \frac{1}{\Lambda}C \in \mathcal{Q}_0\}$,
- $\nu^\alpha(C) = \min\{\nu(C') : 1 \leq C(x,y)C'(x,y) \leq \alpha,\ \forall x,y \in \mathcal{X} \times \mathcal{Y}\}$,
- $\gamma_2^\alpha(C) = \min\{\gamma_2(C') : 1 \leq C(x,y)C'(x,y) \leq \alpha,\ \forall x,y \in \mathcal{X} \times \mathcal{Y}\}$.

Lemma 2. *For any correlation* $C : \mathcal{X} \times \mathcal{Y} \to [-1,1]$,

1. $\tilde{\nu}(C) = 1$ *iff* $\nu(C) \leq 1$, *and* $\tilde{\gamma}_2(C) = 1$ *iff* $\gamma_2(C) \leq 1$,

2. $\tilde{\nu}(C) > 1 \Longrightarrow \nu(C) = \tilde{\nu}(C)$,
3. $\tilde{\gamma}_2(C) > 1 \Longrightarrow \gamma_2(C) = \tilde{\gamma}_2(C)$.

The proof crucially uses the fact that the outcomes are Boolean, and that for uniform marginals, local distributions are closed under sign changes. This fails to hold with non-uniform marginals (for instance, quantum measurements on non-maximally-entangled states) since sign changes also change the marginals. In the special case of sign matrices (corresponding to Boolean functions, as shown above), we also have the following correspondence between $\tilde{\nu}^\epsilon$, $\tilde{\gamma}_2^\epsilon$, and $\nu^\alpha, \gamma_2^\alpha$.

Lemma 3. *Let* $0 \leq \epsilon < 1/2$ *and* $\alpha = \frac{1}{1-2\epsilon}$. *For any sign matrix* $C : \mathcal{X} \times \mathcal{Y} \to \{-1,1\}$,

1. $\tilde{\nu}^\epsilon(C) > 1 \Longrightarrow \nu^\alpha(C) = \frac{\tilde{\nu}^\epsilon(C)}{1-2\epsilon}$,
2. $\tilde{\gamma}_2^\epsilon(C) > 1 \Longrightarrow \gamma_2^\alpha(C) = \frac{\tilde{\gamma}_2^\epsilon(C)}{1-2\epsilon}$.

Linial and Shraibman use γ_2^α to derive a lower bound not only on the quantum communication complexity Q_ϵ^{ent}, but also on the classical complexity R_ϵ^{pub}. In the case of binary outcomes with uniform marginals (which includes Boolean functions as a special case), we obtain a similar result by combining our bound for $Q_\epsilon^{\text{ent}}(C)$ with the fact that $Q_\epsilon^{\text{ent}}(C) \leq \lceil \frac{1}{2} R_\epsilon^{\text{pub}}(C) \rceil$, which follows from superdense coding. This implies $R_\epsilon^{\text{pub}}(C) \geq 2\log(\gamma_2^\epsilon(C)) - 1$. In the general case, however, we can only prove that $R_\epsilon^{\text{pub}}(\mathbf{p}) \geq \log(\gamma_2^\epsilon(\mathbf{p})) - 1$. This may be due to the fact that the result holds in the much more general case of arbitrary outcomes and marginals.

Note also that although γ_2 and ν are matrix norms, this fails to be the case for $\tilde{\gamma}_2$ and $\tilde{\nu}$, even in the case of correlations. Nevertheless, it is still possible to formulate dual quantities, which turn out to have sufficient structure, as we show in the next section.

4 Duality, Bell Inequalities, and XOR Games

In their primal formulation, the $\tilde{\gamma}_2$ and $\tilde{\nu}$ methods are difficult to apply since they are formulated as a minimization problem. Transposing to the dual space not only turns the method into a maximization problem; it also has a very natural, well-understood interpretation since it coincides with maximal violations of Bell and Tsirelson inequalities. This is particularly relevant to physics, since it formalizes in very precise terms the intuition that distributions with large Bell inequality violations should require more communication to simulate.

4.1 Bell and Tsirelson Inequalities

Bell inequalities were first introduced by Bell [23], as bounds on correlations achievable by any *local* physical theory. He showed that quantum correlations could violate these inequalities and therefore exhibited non-locality. Tsirelson later proved that quantum correlations should also respect some bound (known as the Tsirelson bound), giving a first example of a "Tsirelson-like" inequality for quantum distributions [24].

Since the set of non-signaling distributions $\mathcal{C}$ lies in an affine space $\mathsf{aff}(\mathcal{C})$, we may consider the isomorphic dual space of linear functionals over this space. The dual quantity $\tilde{\nu}^*$ (technically not a dual norm since $\tilde{\nu}$ itself is not a norm in the general case) is the maximum value of a linear functional in the dual space on local distributions, and $\tilde{\gamma}_2^*$ is the maximum value of a linear functional on quantum distributions. These are exactly what is captured by the Bell and Tsirelson inequalities.

Definition 7 (Bell and Tsirelson inequalities). *Let $B : \mathsf{aff}(\mathcal{C}) \mapsto \mathbb{R}$ be a linear functional on $\mathsf{aff}(\mathcal{C})$, $B(\mathbf{p}) = \sum_{a,b,x,y} B_{abxy} p(a,b|x,y)$. Define $\tilde{\nu}^*(B) = \max_{\mathbf{p}\in\mathcal{L}} B(\mathbf{p})$ and $\tilde{\gamma}_2^*(B) = \max_{\mathbf{p}\in\mathcal{Q}} B(\mathbf{p})$. A Bell inequality is a linear inequality satisfied by any local distribution, $B(\mathbf{p}) \leq \tilde{\nu}^*(B)$ ($\forall\ \mathbf{p} \in \mathcal{L}$), and a Tsirelson inequality is a linear inequality satisfied by any quantum distribution, $B(\mathbf{p}) \leq \tilde{\gamma}_2^*(B)$ ($\forall\ \mathbf{p} \in \mathcal{Q}$).*

By linearity, Bell inequalities are often expressed as linear functionals over the correlations in the case of binary outputs and uniform marginals.

Finally, it follows from (SDP or LP) duality that $\tilde{\gamma}_2$ and $\tilde{\nu}$ amount to finding a maximum violation of a (normalized) Bell or Tsirelson inequality. In the case of $\tilde{\gamma}_2$, since there is no known characterization of $\mathcal{Q}$ by an SDP, we use the SDP hierarchy of [14] to obtain the dual expression.

Theorem 5. *For any distribution $\mathbf{p} \in \mathcal{C}$,*

1. $\tilde{\nu}(\mathbf{p}) = \max\{B(\mathbf{p}) : \forall \mathbf{p}' \in \mathcal{L},\ |B(\mathbf{p}')| \leq 1\}$, *and*
2. $\tilde{\gamma}_2(\mathbf{p}) = \max\{B(\mathbf{p}) : \forall \mathbf{p}' \in \mathcal{Q},\ |B(\mathbf{p}')| \leq 1\}$,

where the maximization is over linear functionals $B : \mathsf{aff}(\mathcal{C}) \mapsto \mathbb{R}$.

4.2 XOR Games

In an XOR game, Alice is given an input x and Bob is given y according to a distribution μ, and they should output $a = \pm 1$ and $b = \pm 1$. They win if $a \cdot b$ equals some ± 1 function $G(x, y)$. Since they are not allowed to communicate, their strategy may be represented as a local correlation matrix $S \in \mathcal{L}_0$. The winning bias given a strategy S with respect to μ is $\epsilon_\mu(G\|S) = \sum_{x,y} \mu(x,y)G(x,y)S(x,y)$, and $\epsilon_\mu^{\text{pub}}(G) = \max_{S\in\mathcal{L}_0} \epsilon_\mu(G\|S)$ is the maximum winning bias of a local (classical) strategy.

Theorem 6. *1. $\nu(C) = \max_{\mu,G} \frac{\epsilon_\mu(G\|C)}{\epsilon_\mu^{\text{pub}}(G)}$ where (μ, G) range over XOR games.*
2. $\nu(C) \geq \frac{1}{\epsilon^{\text{pub}}(C)}$.

5 Comparing $\tilde{\gamma}_2$ and $\tilde{\nu}$

It is known that because of Tsirelson's theorem and Grothendieck's inequality, γ_2 and ν differ by at most a constant. Although neither of these hold beyond the Boolean setting with uniform marginals, we show in this section that this surprisingly also extends to non-signaling distributions.

Theorem 7. *For any non-signaling distribution $\mathbf{p} \in \mathcal{C}$, with inputs in $\mathcal{X} \times \mathcal{Y}$ and outcomes in $\mathcal{A} \times \mathcal{B}$ with $A = |\mathcal{A}|, B = |\mathcal{B}|$,*

1. $\tilde{\nu}(\mathbf{p}) \leq (2K_G + 1)\tilde{\gamma}_2(\mathbf{p})$ *when* $A = B = 2$,
2. $\tilde{\nu}(\mathbf{p}) \leq [2AB(K_G + 1) - 1]\tilde{\gamma}_2(\mathbf{p})$ *for any* A, B.

The proof uses affine combinations to reduce to the case of distributions with binary outcomes and uniform marginals, where Grothendieck's inequality may be applied.

The negative consequence is that one cannot hope to prove separations between classical and quantum communication using this method, except in the case where the number of outcomes is large. For binary outcomes at least, arguments based on analysing the distance to the quantum set, without taking into account the structure of the distribution, will not suffice to prove large separations. For instance, for the promise distribution based on the Deutsch-Jozsa problem for which a linear lower bound is proven in [4], the best lower bound one can achieve using our techniques is at most logarithmic. This is the first example of a problem for which the corruption bound gives an exponentially better lower bound than the Linial and Shraibman family of methods.

On the positive side, this is very surprising and interesting for quantum information, since (by Theorem 5), it tells us that the set of quantum distributions cannot be much larger than the local polytope, for any number of inputs and outcomes.

6 Upper Bounds for Non-signaling Distributions

We have seen that if a distribution can be simulated using t bits of communication, then it may be represented by an affine model with coefficients exponential in t (Theorem 4). In this section, we consider the converse: how much communication is sufficient to simulate a distribution, given an affine model? This approach allows us to show that any (shared randomness or entanglement-assisted) communication protocol can be simulated with simultaneous messages, with an exponential cost to the simulation, which was previously known only in the case of Boolean functions [25,10,26].

Our results imply for example that for any quantum distribution $\mathbf{p} \in \mathcal{Q}$, $Q^{\|}_{\epsilon}(\mathbf{p}) = O(\log(n))$, where n is the input size. This in effect replaces arbitrary entanglement in the state being measured, with logarithmic quantum communication (using no additional resources such as shared randomness).

We use the superscript $\|$ to indicate the simultaneous messages model, where Alice and Bob each send a message to the referee, who without knowing the inputs, outputs the value of the function, or more generally, outputs a, b with the correct probability distribution conditioned on the inputs x, y.

Theorem 8. *For any distribution* $\mathbf{p} \in \mathcal{C}$ *with inputs in* $\mathcal{X} \times \mathcal{Y}$ *with* $|\mathcal{X} \times \mathcal{Y}| \leq 2^n$, *and outcomes in* $\mathcal{A} \times \mathcal{B}$ *with* $A = |\mathcal{A}|, B = |\mathcal{B}|$, *and any* $\epsilon, \delta < 1/2$,

1. $R^{\|,\mathrm{pub}}_{\epsilon+\delta}(\mathbf{p}) \leq 16 \left[\frac{AB\tilde{\nu}^{\epsilon}(\mathbf{p})}{\delta}\right]^2 \ln\left[\frac{4AB}{\delta}\right] \log(AB)$,
2. $Q^{\|}_{\epsilon+\delta}(\mathbf{p}) \leq O\left((AB)^5 \left[\frac{\tilde{\nu}^{\epsilon}(\mathbf{p})}{\delta}\right]^4 \ln\left[\frac{AB}{\delta}\right] \log(n)\right)$.

Proof idea. Let $\mathbf{p} = q_+\mathbf{p}^+ - q_-\mathbf{p}^-$ be an affine model for $\mathbf{p}$. They use shared randomness to generate samples (a, b) from the local distributions $p^+(a, b|x, y)$ and $p^-(a, b|x, y)$, and send these samples to the referee, who then reconstructs the distribution $p(a, b|x, y)$. If the players do not share randomness but are allowed to send quantum messages to the referee (Item 2), they can use quantum fingerprinting.

In the case of Boolean functions, corresponding to correlations $C_f(x,y) \in \{\pm 1\}$ (see Def. 2), the referee's job is made easier by the fact that he only needs to determine the sign of the correlation with probability $1-\delta$, which improves the bounds. Similar improvements can be obtained for other types of promises on the distribution.

Theorem 9. *Let $f : \{0,1\}^n \times \{0,1\}^n \to \{0,1\}$, with sign matrix C_f, and $\epsilon, \delta < 1/2$.*

1. $R_\delta^{\|,\mathrm{pub}}(f) \leq 4\left[\frac{\tilde{\nu}^\epsilon(C_f)}{1-2\epsilon}\right]^2 \ln(\frac{1}{\delta})$,
2. $Q_\delta^{\|}(f) \leq O\left(\log(n)\left[\frac{\tilde{\nu}^\epsilon(C_f)}{1-2\epsilon}\right]^4 \ln(\frac{1}{\delta})\right)$.

From Lemma 3, these bounds may also be expressed in terms of γ_2^α, and the best upper bounds are obtained from $\gamma_2^\infty(C_f) = \frac{1}{\epsilon^{\mathrm{ent}}(C_f)}$. The first item then coincides with the upper bound of [13].

Together with the bound between $\tilde{\nu}$ and $\tilde{\gamma}_2$ from Section 5, and the lower bounds on communication complexity from Section 3, Theorems 8 and 9 immediately imply the following corollaries.

Corollary 10. *Let $f : \{0,1\}^n \times \{0,1\}^n \to \{0,1\}$. If $Q_\epsilon^{\mathrm{ent}}(f) \leq q$, then*

1. $R_\delta^{\|,\mathrm{pub}}(f) \leq K_G^2 \cdot 2^{2q+2} \ln(\frac{1}{\delta}) \frac{1}{(1-2\epsilon)^2}$,
2. $Q_\delta^{\|}(f) \leq O\left(\log(n) 2^{4q} \ln(\frac{1}{\delta}) \frac{1}{(1-2\epsilon)^4}\right)$.

For any distribution $\mathbf{p} \in \mathcal{C}$ with inputs in $\mathcal{X} \times \mathcal{Y}$ with $|\mathcal{X} \times \mathcal{Y}| \leq 2^n$, and outcomes in $\mathcal{A} \times \mathcal{B}$ with $A = |\mathcal{A}|, B = |\mathcal{B}|$, and any $\epsilon, \delta < 1/2$, if $Q_\epsilon^{\mathrm{ent}}(\mathbf{p}) \leq q$,

3. $R_{\epsilon+\delta}^{\|,\mathrm{pub}}(\mathbf{p}) \leq O\left(2^{4q} \frac{(AB)^4}{\delta^2} \ln^2\left[\frac{AB}{\delta}\right]\right)$,
4. $Q_{\epsilon+\delta}^{\|}(\mathbf{p}) \leq O\left(2^{8q} \frac{(AB)^9}{\delta^4} \ln\left[\frac{AB}{\delta}\right] \log(n)\right)$.

The first two items can be compared to results of Yao, Shi and Zhu, and Gavinsky *et al.* [25,10,26], who show how to simulate any (logarithmic) communication protocol for Boolean functions in the simultaneous messages model, with an exponential blowup in communication. The last two items extend these results to arbitrary distributions.

In particular, Item 3 gives in the special case $q = 0$, that is, $\mathbf{p} \in \mathcal{Q}$, a much simpler proof of the constant upper bound on approximating quantum distributions, which Shi and Zhu prove using sophisticated techniques based on diamond norms [10]. Moreover, Item 3 is much more general as it also allows to simulate protocols requiring quantum communication in addition to entanglement. As for Item 4, it also has new interesting consequences. For example, it implies that quantum distributions ($q = 0$) can be approximated with logarithmic quantum communication in the simultaneous messages model, using no additional resources such as shared randomness, and regardless of the amount of entanglement in the bipartite state measured by the two parties.

7 Conclusion and Open Problems

By studying communication complexity in the framework provided by the study of quantum non-locality (and beyond), we have given very natural and intuitive interpretations of the otherwise very abstract lower bounds of Linial and Shraibman. Conversely, it has allowed us to port these very strong and mathematically elegant lower bound methods to the much more general problem of simulating non-signaling distributions.

Since many communication problems may be reduced to the task of simulating a non-signaling distribution, we hope to see applications of this lower bound method to concrete problems for which standard techniques do not apply, in particular for cases that are not Boolean functions, such as non-Boolean functions, partial functions or relations. Let us also note that our method can be generalized to multipartite non-signaling distributions, and will hopefully lead to applications in the number-on-the-forehead model, for which quantum lower bounds seem hard to prove.

In the case of binary distributions with uniform marginals (which includes in particular Boolean functions), Tsirelson's theorem and Grothendieck's inequality (Proposition 3) imply that there is at most a constant gap between ν and γ_2. For this reason, it was known that Linial and Shraibman's factorization norm lower bound technique gives lower bounds of the same of order for classical and quantum communication (note that this is also true for the related discrepancy method). Despite the fact that Tsirelson's theorem and Grothendieck's inequality are not known to extend beyond the case of Boolean outcomes with uniform marginals, we have shown that in the general case of distributions, there is also a constant gap between $\tilde{\nu}$ and $\tilde{\gamma}_2$. While this may be seen as a negative result, this also reveals interesting information about the structure of the sets of local and quantum distributions. In particular, this could have interesting consequences for the study of non-local games.

Acknowledgements

We are grateful to Benjamin Toner for pointing us towards the existing literature on non-signaling distributions as well as very useful discussions of the Linial and Shraibman lower bound on communication complexity. We also thank Peter Høyer, Troy Lee, Oded Regev, Mario Szegedy, and Dieter van Melkebeek with whom we had many stimulating discussions. Part of this work was done while J. Roland was affiliated with FNRS Belgium and U.C. Berkeley. The research was supported by the EU 5th framework program QAP, by the ANR Blanc AlgoQP and ANR Défis QRAC.

References

1. Yao, A.C.C.: Some complexity questions related to distributive computing. In: Proc. 11th STOC, pp. 209–213 (1979)
2. Kushilevitz, E., Nisan, N.: Communication complexity. Cambridge University Press, New York (1997)
3. Maudlin, T.: Bell's inequality, information transmission, and prism models. In: Biennal Meeting of the Philosophy of Science Association, pp. 404–417 (1992)
4. Brassard, G., Cleve, R., Tapp, A.: Cost of Exactly Simulating Quantum Entanglement with Classical Communication. Phys. Rev. Lett. 83, 1874–1877 (1999); quant-ph/9901035
5. Steiner, M.: Towards quantifying non-local information transfer: finite-bit non-locality. Phys. Lett. A 270, 239–244 (2000)
6. Toner, B.F., Bacon, D.: Communication Cost of Simulating Bell Correlations. Phys. Rev. Lett. 91, 187904 (2003)

7. Cerf, N.J., Gisin, N., Massar, S., Popescu, S.: Simulating Maximal Quantum Entanglement without Communication. Phys. Rev. Lett. 94(22), 220403 (2005)
8. Degorre, J., Laplante, S., Roland, J.: Classical simulation of traceless binary observables on any bipartite quantum state. Phys. Rev. A 75(012309) (2007)
9. Regev, O., Toner, B.: Simulating quantum correlations with finite communication. In: Proc. 48th FOCS, pp. 384–394 (2007)
10. Shi, Y., Zhu, Y.: Tensor norms and the classical communication complexity of nonlocal quantum measurement. SIAM J. Comput. 38(3), 753–766 (2008)
11. Jones, N.S., Masanes, L.: Interconversion of nonlocal correlations. Phys. Rev. A 72, 052312 (2005)
12. Barrett, J., Pironio, S.: Popescu-Rohrlich correlations as a unit of nonlocality. Phys. Rev. Lett. 95, 140401 (2005)
13. Linial, N., Shraibman, A.: Lower bounds in communication complexity based on factorization norms. Random Struct. Algorithms 34(3), 368–394 (2009)
14. Navascues, M., Pironio, S., Acin, A.: A convergent hierarchy of semidefinite programs characterizing the set of quantum correlations. New J. Phys. 10(7), 073013 (2008)
15. Randall, C.H., Foulis, D.J.: Operational statistics and tensor products. In: Interpretations and Foundations of Quantum Theory, Volume Interpretations and Foundations of Quantum Theory, pp. 21–28. Wissenschaftsverlag, BibliographischesInstitut (1981)
16. Foulis, D.J., Randall, C.H.: Empirical logic and tensor products. In: Interpretations and Foundations of Quantum Theory, Volume Interpretations and Foundations of Quantum Theory, pp. 1–20. Wissenschaftsverlag, BibliographischesInstitut (1981)
17. Kläy, M., Randall, C.H., Foulis, D.J.: Tensor products and probability weights. Int. J. Theor. Phys. 26(3), 199–219 (1987)
18. Wilce, A.: Tensor products in generalized measure theory. Int. J. Theor. Phys. 31(11), 1915–1928 (1992)
19. Barrett, J.: Information processing in generalized probabilistic theories. Phys. Rev. A 75(3), 032304 (2007)
20. Alon, N., Naor, A.: Approximating the cut-norm via Grothendieck's inequality. SIAM J. Comput. 35(4), 787–803 (2006)
21. Tsirelson, B.S.: Zapiski Math. Inst. Steklov (LOMI) 142, 174–194 (1985); English translation in Quantum analogues of the Bell inequalities. The case of two spatially separated domains. J. Soviet Math. 36, 557–570 (1987)
22. Linial, N., Mendelson, S., Schechtman, G., Shraibman, A.: Complexity measures of sign matrices. Combinatorica 27, 439–463 (2007)
23. Bell, J.S.: On the Einstein Podolsky Rosen paradox. Physics 1, 195 (1964)
24. Tsirelson, B.S.: Quantum generalizations of Bell's inequality. Lett. Math. Phys. 4(2), 93–100 (1980)
25. Yao, A.C.C.: On the power of quantum fingerprinting. In: Proc. 35th STOC, pp. 77–81 (2003)
26. Gavinsky, D., Kempe, J., de Wolf, R.: Strengths and weaknesses of quantum fingerprinting. In: Proc. 21st CCC, pp. 288–295 (2006)

How to Use Spanning Trees to Navigate in Graphs
(Extended Abstract)

Feodor F. Dragan and Yang Xiang

Algorithmic Research Laboratory
Department of Computer Science
Kent State University, Kent, OH 44242, USA
{dragan,yxiang}@cs.kent.edu

Abstract. In this paper, we investigate three strategies of how to use a spanning tree T of a graph G to navigate in G, i.e., to move from a current vertex x towards a destination vertex y via a path that is close to optimal. In each strategy, each vertex v has full knowledge of its neighborhood $N_G[v]$ in G (or, k-neighborhood $D_k(v, G)$, where k is a small integer) and uses a small piece of global information from spanning tree T (e.g., distance or ancestry information in T), available locally at v, to navigate in G. We investigate advantages and limitations of these strategies on particular families of graphs such as graphs with locally connected spanning trees, graphs with bounded length of largest induced cycle, graphs with bounded tree-length, graphs with bounded hyperbolicity. For most of these families of graphs, the ancestry information from a BFS-tree guarantees short enough routing paths. In many cases, the obtained results are optimal up to a constant factor.

1 Introduction

As part of the recent surge of interest in different kind of networks, there has been active research exploring strategies for navigating synthetic and real-world networks (modeled usually as graphs). These strategies specify some rules to be used to advance in a graph from a given vertex towards a target vertex along a path that is close to shortest. Current strategies include (but not limited to): routing with full-tables, interval routing, routing labeling schemes, greedy routing, geographic routing, compass routing, etc. in wired or wireless communication networks and in transportation networks (see [13,15,20,26,31] and papers cited therein); routing through common membership in groups, popularity, and geographic proximity in social networks and e-mail networks (see [9,20,23] and papers cited therein).

Navigation in communication networks is performed using a routing scheme, i.e., a mechanism that can deliver packets of information from any vertex of a network to any other vertex. In most strategies, each vertex v of a graph has full knowledge of its neighborhood and uses a piece of global information

R. Královič and D. Niwiński (Eds.): MFCS 2009, LNCS 5734, pp. 282–294, 2009.

available to it about the graph topology – some "sense of direction" to each destination, stored locally at v. Based only on this information and the address of a destination, vertex v needs to decide whether the packet has reached its destination, and if not, to which neighbor of v to forward the packet.

One of the most popular strategies in wireless (and social) networks is the *geographic routing* (sometimes called also the *greedy geographic routing*), where each vertex forwards the packet to the neighbor geographically closest to the destination (see survey [15] and paper [23]). Each vertex of the network knows its position (e.g., Euclidean coordinates) in the underlying physical space and forwards messages according to the coordinates of the destination and the coordinates of neighbors. Although this greedy method is effective in many cases, packets may get routed to where no neighbor is closer to the destination than the current vertex. Many recovery schemes have been proposed to route around such voids for guaranteed packet delivery as long as a path exists [1,19,22]. These techniques typically exploit planar subgraphs (e.g., Gabriel graph, Relative Neighborhood graph), and packets traverse faces on such graphs using the well-known right-hand rule.

All earlier papers assumed that vertices are aware of their physical location, an assumption which is often violated in practice for various of reasons (see [8,21,27]). In addition, implementations of recovery schemes are either based on non-rigorous heuristics or on complicated planarization procedures. To overcome these shortcomings, recent papers [8,21,27] propose routing algorithms which assign virtual coordinates to vertices in a metric space X and forward messages using geographic routing in X. In [27], the metric space is the Euclidean plane, and virtual coordinates are assigned using a distributed version of Tutte's "rubber band" algorithm for finding convex embeddings of graphs. In [8], the graph is embedded in R^d for some value of d much smaller than the network size, by identifying d beacon vertices and representing each vertex by the vector of distances to those beacons. The distance function on R^d used in [8] is a modification of the ℓ_1 norm. Both [8] and [27] provide substantial experimental support for the efficacy of their proposed embedding techniques – both algorithms are successful in finding a route from the source to the destination more than 95% of the time – but neither of them has a provable guarantee. Unlike embeddings of [8] and [27], the embedding of [21] guarantees that the geographic routing will always be successful in finding a route to the destination, if such a route exists. Algorithm of [21] assigns to each vertex of the network a virtual coordinate in the hyperbolic plane, and performs greedy geographic routing with respect to these virtual coordinates. More precisely, [21] gets virtual coordinates for vertices of a graph G by embedding in the hyperbolic plane a spanning tree of G. The proof that this method guaranties delivery is relied only on the fact that the hyperbolic greedy route is no longer than the spanning tree route between two vertices; even more, it could be much shorter as greedy routes take enough short cuts (edges which are not in the spanning tree) to achieve significant saving in stretch. However, although the experimental results of [21] confirm that the greedy hyperbolic embedding yields routes with low stretch when applied to typical unit-disk graphs, the worst-case stretch is still linear in the network size.

Previous work. Motivated by the work of Robert Kleinberg [21], in paper [6], we initiated exploration of the following strategy in advancing in a graph from a source vertex towards a target vertex. Let $G = (V, E)$ be a (unweighted) graph and T be a spanning tree of G. To route/move in G from a vertex x towards a target vertex y, use the following rule:

TDGR(**T**ree **D**istance **G**reedy **R**outing) strategy: *from a current vertex z (initially $z = x$), unless $z = y$, go to a neighbor of z in G that is closest to y in T.*

In this strategy, each vertex has full knowledge of its neighborhood in G and can use the distances in T to navigate in G. Thus, additionally to standard local information (the neighborhood $N_G(v)$), the only global information that is available to each vertex v is the topology of the spanning tree T. In fact, v can know only a very small piece of information about T and still be able to infer from it the necessary tree-distances. It is known [14,24,25] that the vertices of an n-vertex tree T can be labeled in $O(n \log n)$ total time with labels of up to $O(\log^2 n)$ bits such that given the labels of two vertices v, u of T, it is possible to compute in constant time the distance $d_T(v, u)$, by merely inspecting the labels of u and v. Hence, one may assume that each vertex v of G knows, additionally to its neighborhood in G, only its $O(\log^2 n)$ bit distance label. This distance label can be viewed as a virtual coordinate of v.

For each source vertex x and target vertex y, by this routing strategy, a path, called a *greedy routing path*, is produced (clearly, this routing strategy will always be successful in finding a route to the destination). Denote by $g_{G,T}(x, y)$ the length (i.e., the number of edges) of a longest greedy routing path that can be produced for x and y using this strategy and T. We say that a spanning tree T of a graph G is an *additive r-carcass* for G if $g_{G,T}(x, y) \leq d_G(x, y) + r$ for each ordered pair $x, y \in V$ (in a similar way one can define also a *multiplicative t-carcass* of G, where $g_{G,T}(x, y)/d_G(x, y) \leq t$). Note that this notion differs from the notion of "remote-spanners" introduced recently in [18].

In [6], we investigated the problem, given a graph family $\mathcal{F}$, whether a small integer r exists such that any graph $G \in \mathcal{F}$ admits an additive r-carcass. We showed that rectilinear $p \times q$ grids, hypercubes, distance-hereditary graphs, dually chordal graphs (and, therefore, strongly chordal graphs and interval graphs), all admit additive 0-carcasses. Furthermore, every chordal graph G admits an additive $(\omega(G) + 1)$-carcass (where $\omega(G)$ is the size of a maximum clique of G), each chordal bipartite graph admits an additive 4-carcass. In particular, any k-tree admits an additive $(k + 2)$-carcass. All those carcasses were easy to construct.

This new combinatorial structure, carcass, turned out to be "more attainable" than the well-known structure, tree spanner (a spanning tree T of a graph G is an additive tree r-spanner if for any two vertices x, y of G, $d_T(x, y) \leq d_G(x, y) + r$ holds, and is a multiplicative tree t-spanner if for any two vertices x, y, $d_T(x, y) \leq t\ d_G(x, y)$ holds). It is easy to see that any additive (multiplicative) tree r-spanner is an additive (resp., multiplicative) r-carcass. On the other hand, there is a number of graph families not admitting any tree spanners, yet admitting very good carcasses. For example, any hypercube has an additive 0-carcass (see [6]) but does not have any tree r-spanner (additive or multiplicative) for any constant r. The same holds for 2-trees and chordal bipartite graphs [6].

Results of this paper. All graphs occurring in this paper are connected, finite, undirected, unweighted, loopless and without multiple edges. In a graph $G = (V, E)$ ($n = |V|, m = |E|$) the *length* of a path from a vertex v to a vertex u is the number of edges in the path. The *distance* $d_G(u, v)$ between vertices u and v is the length of a shortest path connecting u and v. The *neighborhood* of a vertex v of G is the set $N_G(v) = \{u \in V : uv \in E\}$ and the *closed neighborhood* of v is $N_G[v] = N_G(v) \cup \{v\}$. The *disk* of radius k centered at v is the set of all vertices at distance at most k to v, i.e., $D_k(v, G) = \{u \in V : d_G(u, v) \leq k\}$.

In this paper we continue investigations of how to use spanning trees to navigate in graphs. Spanning trees are very well understood structures in graphs. There are many results available in literature on how to construct (and maintain) different spanning trees in a number of settings; including in distributed way, in self stabilizing and localized way, etc. (see [10,11,12,17] and papers cited therein).

Additionally to TDGR strategy, we propose to investigate two more strategies. Let $G = (V, E)$ be a graph and T be a spanning tree of G rooted at an arbitrary vertex s. Using T, we associate an interval I_v with each vertex v such that, for any two vertices u and v, $I_u \subseteq I_v$ if and only if u is a descendent of v in T. This can be done in the following way (see [29] and Fig. 1). By depth-first search tour of T, starting at root, assign each vertex u of T a depth-first search number $DFS(u)$. Then, label u by interval $[DFS(u), DFS(w)]$, where w is last descendent of u visited by depth-first search. For two intervals $I_a = [a_L, a_R]$ and $I_b = [b_L, b_R]$, $I_a \subseteq I_b$ if and only if $a_L \geq b_L$ and $a_R \leq b_R$. Let xTy denote the (unique) path of T connecting vertices x and y, and let $N_G[xTy] = \{v \in V : v$ belongs to xTy or is adjacent to a vertex of xTy in $G\}$.

IGR (**I**nterval **G**reedy **R**outing) strategy.
To advance in G from a vertex x towards a target vertex y ($y \neq x$), do:
if there is a neighbor w of x in G such that $y \in I_w$ (i.e., $w \in sTy$),
then go to such a neighbor with smallest (by inclusion) interval;
else (which means $x \notin N_G[sTy]$), go to a neighbor w of x in G
such that $x \in I_w$ and I_w is largest such interval.

IGRF (**I**nterval **G**reedy **R**outing with forwarding to **F**ather) strategy.
To advance in G from a vertex x towards a target vertex y ($y \neq x$), do:
if there is a neighbor w of x in G such that $y \in I_w$ (i.e., $w \in sTy$),
then go to such a neighbor with smallest (by inclusion) interval;
else (which means $x \notin N_G[sTy]$), go to the father of x in T (i.e., a neighbor of x in G interval of which contains x and is smallest by inclusion).

Note that both, IGR and IGRF, strategies are simpler and more compact than the TDGR strategy. In IGR and IGRF, each vertex v, additionally to standard local information (the neighborhood $N_G(v)$), needs to know only $2\lceil \log_2 n \rceil$ bits of global information from the topology of T, namely, its interval I_v. Information

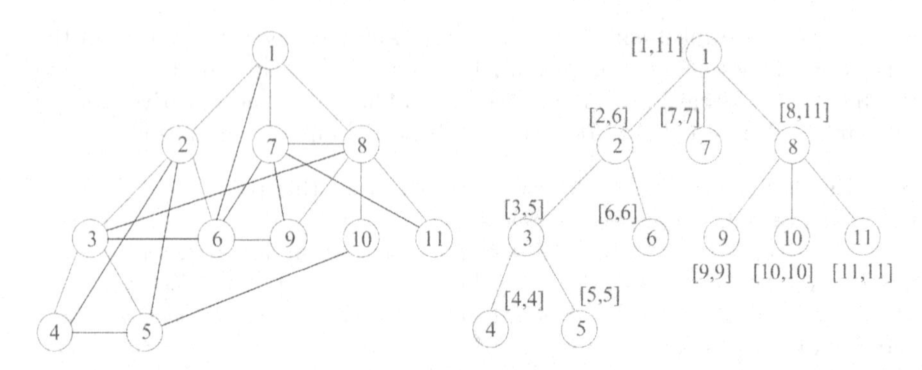

Fig. 1. A graph and its rooted spanning tree with precomputed ancestry intervals. For (ordered) pair of vertices 10 and 4, both IGR and IGRF produce path 10,8,3,4 (TDGR produces 10,5,4). For pair 5 and 8, IGR produces path 5,2,1,8, while IGRF produces path 5,3,8 (TDGR produces 5,10,8). For pair 5 and 7, IGR produces path 5,2,1,7, while IGRF produces path 5,3,2,1,7 (TDGR produces 5,2,1,7).

stored in intervals gives a "sense of direction" in navigation in G (current vertex x either may already know intervals of its neighbors, or it can ask each neighbor w, when needed, whether its interval I_w contains destination y or vertex x itself, and if yes to send I_w to x). On the other hand, as we will show in this paper, routing paths produced by IGR (IGRF) will have, in many cases, almost the same quality as routing paths produced by TDGR. Moreover, in some cases, they will be even shorter than routing paths produced by TDGR.

Let $R_{G,T}(x, y)$ be the routing path produced by IGR strategy (resp., by IGRF strategy) for a source vertex x and a target vertex y in G using T. It will be evident later that this path always exists, i.e., IGR strategy (resp., IGRF strategy) guarantees delivery. Moreover, this path is unique for each ordered pair x, y of vertices (note that, depending on tie breaking rule, TDGR can produce different routing paths for the same ordered pair of vertices). Denote by $g_{G,T}(x, y)$ the length (i.e., the number of edges) of path $R_{G,T}(x, y)$. We say that a spanning tree T of a graph G is an *additive r-frame* (resp., an *additive r-fframe*) for G if the length $g_{G,T}(x, y)$ of the routing path $R_{G,T}(x, y)$ produced by IGR strategy (resp., by IGRF strategy) is at most $d_G(x, y) + r$ for each ordered pair $x, y \in V$. In a similar way one can define also a *multiplicative t-frame* (resp., a *multiplicative t-fframe*) of G, where $g_{G,T}(x, y)/d_G(x, y) \leq t$.

In Sections 2 and 3, we show that each distance-hereditary graph admits an additive 0-frame (0-fframe) and each dually chordal graph (and, hence, each interval graph, each strongly chordal graph) admits an additive 0-frame. In Section 4, we show that each k-chordal graph admits an additive $(k - 1)$-frame ($(k-1)$-fframe), each chordal graph (and, hence, each k-tree) admits an additive 1-frame (1-fframe), each AT-free graph admits an additive 2-frame (2-fframe), each chordal bipartite graph admits an additive 0-frame (0-fframe). Definitions of the graph families will be given in appropriate sections (see also [3] for many equivalent definitions of those families of graphs).

To better understand full potentials and limitations of the proposed routing strategies, in Section 5, we investigate also the following generalizations of them. Let G be a (unweighted) graph and T be a (rooted) spanning tree of G.

k-**localized TDGR** strategy.

To advance in G from a vertex x towards a target vertex y, go, using a shortest path in G, to a vertex $w \in D_k(x,G)$ that is closest to y in T.

In this strategy, each vertex has full knowledge of its disk $D_k(v,G)$ (e.g., all vertices in $D_k(v,G)$ and how to reach each of them via some shortest path of G) and can use the distances in T to navigate in G. Let $g_{G,T}(x,y)$ be the length of a longest path of G that can be produced for x and y using this strategy and T. We say that a spanning tree T of a graph G is a *k-localized additive r-carcass* for G if $g_{G,T}(x,y) \leq d_G(x,y) + r$ for each ordered pair $x, y \in V$ (in a similar way one can define also a *k-localized multiplicative t-carcass* of G.

k-**localized IGR** strategy.

To advance in G from a vertex x towards a target vertex y, do:
if there is a vertex $w \in D_k(x,G)$ such that $y \in I_w$ (i.e., $w \in sTy$),
then go, using a shortest path in G, to such a vertex w
with smallest (by inclusion) interval;
else (which means $d_G(x,sTy) > k$),
go, using a shortest path in G, to a vertex $w \in D_k(x,G)$ such
that $x \in I_w$ and I_w is largest such interval.

k-**localized IGRF** strategy.

To advance in G from a vertex x towards a target vertex y, do:
if there is a vertex $w \in D_k(x,G)$ such that $y \in I_w$ (i.e., $w \in sTy$),
then go, using a shortest path in G, to such a vertex w
with smallest (by inclusion) interval;
else (which means $d_G(x,sTy) > k$), go to the father of x in T.

In these strategies, each vertex has full knowledge of its disk $D_k(v,G)$ (e.g., all vertices in $D_k(v,G)$ and how to reach each of them via some shortest path of G) and can use the DFS intervals I_w to navigate in G. We say that a (rooted) spanning tree T of a graph G is a *k-localized additive r-frame* (resp., a *k-localized additive r-fframe*) for G if the length $g_{G,T}(x,y)$ of the routing path produced by k-localized IGR strategy (resp., k-localized IGRF strategy) is at most $d_G(x,y)+r$ for each ordered pair $x,y \in V$. In a similar way one can define also a *k-localized multiplicative t-frame* (resp., a *k-localized multiplicative t-fframe*) of G.

We show, in Section 5, that any tree-length λ graph admits a λ-localized additive 5λ-fframe (which is also a λ-localized additive 5λ-frame) and any δ-hyperbolic graph admits a 4δ-localized additive 8δ-fframe (which is also a 4δ-localized additive 8δ-frame). Definitions of these graph families will also be given in appropriate sections. Additionally, we show that: for any $\lambda \geq 3$, there exists a tree-length λ graph G with n vertices for which no $(\lambda-2)$-localized additive $\frac{1}{2}\sqrt{\log \frac{n-1}{\lambda}}$-fframe exists; for any $\lambda \geq 4$, there exists a tree-length λ graph G with n vertices for which no $\lfloor 2(\lambda-2)/3 \rfloor$-localized additive $\frac{2}{3}\sqrt{\log \frac{3(n-1)}{4\lambda}}$-frame exists; for any $\lambda \geq 6$, there exists a tree-length λ graph G with n vertices for which no $\lfloor(\lambda-2)/4\rfloor$-localized additive $\frac{3}{4}\sqrt{\log\frac{n-1}{\lambda}}$-carcass exists.

Proofs omitted due to space limitation can be found in the journal version of the paper [7].

2 Preliminaries

Let $G=(V,E)$ be a graph and T be a spanning tree of G rooted at an arbitrary vertex s. We assume that T is given together with the precomputed ancestry intervals. The following facts are immediate from the definitions of IGR and IGRF strategies.

Lemma 1. *Any routing path $R_{G,T}(x,y)$ produced by IGR or IGRF, where x is not an ancestor of y in T, is of the form $x_1 \dots x_k y_l \dots y_1$, where $x_1=x$, $y_1=y$, x_i is a descendent of x_{i+1} in T, and y_i is an ancestor of y_{i-1} in T. In addition, for any $i \in [1,k]$, x_i is not an ancestor of y, and, for any $i \in [1,k-1]$, x_i is not adjacent in G to any vertex of sTy.*
If x is an ancestor of y in T, then $R_{G,T}(x,y)$ has only part $y_l \dots y_1$ with $x=y_l$, $y=y_1$ and y_i being an ancestor of y_{i-1} in T.

In what follows, any routing path produced by IGR (resp., by IGRF, by TDGR) will be called *IGR routing path* (resp., *IGRF routing path, TDGR routing path*).

Corollary 1. *A tale of any IGR routing path (any IGRF routing path) is also an IGR routing path (IGRF routing path, respectively).*

Corollary 2. *Both IGR and IGRF strategies guarantee delivery.*

Corollary 3. *Let T be a BFS-tree (Breadth-First-Search–tree) of a graph G rooted at an arbitrary vertex s, and let x and y be two vertices of G. Then, IGR and IGRF strategies produce the same routing path $R_{G,T}(x,y)$ from x to y.*

Lemma 2. *For any vertices x and y, the IGR routing path (respectively, the IGRF routing path) $R_{G,T}(x,y)$ is unique.*

Lemma 3. *Any IGR routing path $R_{G,T}(x,y)$ is an induced path of G.*

Note that an IGRF routing path $R_{G,T}(x,y)=x_1 \dots x_k y_l \dots y_1$ may not necessarily be induced in the part $x_1 \dots x_k$. In [6], it was shown that routing paths produced by TDGR strategy are also induced paths.

A graph G is called *distance-hereditary* if any induced path of G is a shortest path (see [3] for this and equivalent definitions). By Lemma 3 and Corollary 3, we conclude.

Theorem 1. *Any spanning tree of a distance-hereditary graph G is an additive 0-frame of G, regardless where it is rooted. Any BFS-tree of a distance-hereditary graph G is an additive 0-fframe of G.*

3 Frames for Dually Chordal Graphs

Let G be a graph. We say that a spanning tree T of G is *locally connected* if the closed neighborhood $N_G[v]$ of any vertex v of G induces a subtree in T (i.e., $T \cap N_G[v]$ is a connected subgraph of T). The following result was proven in [6].

Lemma 4. [6] *If T is a locally connected spanning tree of a graph G, then T is an additive 0-carcass of G.*

Here we prove the following lemma.

Lemma 5. *Let G be a graph with a locally connected spanning tree T, and let x and y be two vertices of G. Then, IGR and TDGR strategies produce the same routing path $R_{G,T}(x, y)$ from x to y (regardless where T is rooted).*

Proof. Assume that we want to route from a vertex x towards a vertex y in G, where $x \neq y$. We may assume that $d_G(x, y) \geq 2$, since otherwise both routing strategies will produce path xy. Let x^* (x') be the neighbor of x in G chosen by IGR strategy (resp., by TDGR strategy) to relay the message. We will show that $x' = x^*$ by considering two possible cases. We root the tree T at an arbitrary vertex s.

First assume that $N_G[x] \cap sTy \neq \emptyset$. By IGR strategy, we will choose a neighbor $x^* \in N_G[x]$ such that $y \in I_{x^*}$ and I_{x^*} is the smallest interval by inclusion, i.e., x^* is a vertex from $N_G[x]$ closest in sTy to y. If $d_T(x', y) < d_T(x^*, y)$, then $x' \notin sTy$ and the nearest common ancestor $NCA_T(x', y)$ of x', y in T must be in x^*Ty. Since $T \cap N_G[x]$ is a connected subgraph of T and $x', x^* \in N_G[x]$, we conclude that $NCA_T(x', y)$ must be in $N_G[x]$, too. Thus, we must have $x' = NCA_T(x', y) = x^*$.

Assume now that $N_G[x] \cap sTy = \emptyset$. By IGR strategy, we will choose a neighbor $x^* \in N_G[x]$ such that $x \in I_{x^*}$ and I_{x^*} is the largest interval by inclusion, i.e., x^* is a vertex from $N_G[x]$ closest in sTx to $NCA_T(x, y)$. Consider the nearest common ancestor $NCA_T(x', x^*)$ of x', x^* in T. Since $T \cap N_G[x]$ is a connected subgraph of T and $x', x^* \in N_G[x]$, we conclude that $NCA_T(x', x^*)$ must be in $N_G[x]$, too. Thus, necessarily, we must have $x' = NCA_T(x', x^*) = x^*$.

From these two cases we conclude, by induction, that IGR and TDGR strategies produce the same routing path $R_{G,T}(x, y)$ from x to y. □

Corollary 4. *If T is a locally connected spanning tree of a graph G, then T is an additive 0-frame of G (regardless where T is rooted).*

It has been shown in [2] that the graphs admitting locally connected spanning trees are precisely the dually chordal graphs. Furthermore, [2] showed that the class of dually chordal graphs contains such known families of graphs as strongly chordal graphs, interval graphs and others. Thus, we have the following result.

Theorem 2. *Every dually chordal graph admits an additive 0-frame. In particular, any strongly chordal graph (any interval graph) admits an additive 0-frame.*

Note that, in [2], it was shown that dually chordal graphs can be recognized in linear time, and if a graph G is dually chordal, then a locally connected spanning tree of G can be efficiently constructed.

4 Frames for k-Chordal Graphs and Subclasses

A graph G is called *k-chordal* if it has no induced cycles of size greater than k, and it is called *chordal* if it has no induced cycle of length greater than 3. Chordal graphs are precisely the 3-chordal graphs.

Theorem 3. *Let $G = (V, E)$ be a k-chordal graph. Any BFS-tree T of G is an additive $(k-1)$-fframe (and, hence, an additive $(k-1)$-frame) of G. If G is a chordal graph (i.e., $k = 3$), then any LexBFS-tree T (a special BFS-tree) of G is an additive 1-fframe (and, hence, an additive 1-frame) of G.*

A graph G is called *chordal bipartite* if it is bipartite and has no induced cycles of size greater than 4. Chordal bipartite graphs are precisely the bipartite 4-chordal graphs. A graph is called *AT–free* if it does not have an *asteroidal triple*, i.e. a set of three vertices such that there is a path between any pair of them avoiding the closed neighborhood of the third. It is known that AT-free graphs form a proper subclass of 5-chordal graphs.

Theorem 4. *Every chordal bipartite graph G admits an additive 0-fframe and an additive 0-frame, constructible in $O(n^2)$ time. Any BFS-tree T of an AT-free graph G is an additive 2-fframe (and, hence, an additive 2-frame) of G.*

5 Localized Frames for Tree-Length λ Graphs and δ-Hyperbolic Graphs

In this section, we show that any tree-length λ graph admits a λ-localized additive 5λ-fframe (which is also a λ-localized additive 5λ-frame) and any δ-hyperbolic graph admits a 4δ-localized additive 8δ-fframe (which is also a 4δ-localized additive 8δ-frame). We complement these results with few lower bounds.

Tree-length λ graphs. The *tree-length* of a graph G is the smallest integer λ for which G admits a tree-decomposition into bags of diameter at most λ. It has been introduced and extensively studied in [5]. Chordal graphs are exactly the graphs of tree-length 1, since a graph is chordal if and only if it has a tree-decomposition into cliques (cf. [3]). AT-free graphs and distance-hereditary graphs are of tree-length 2. More generally, k-chordal graphs have tree-length at most $k/2$. However, there are graphs with bounded tree-length and unbounded chordality, like the wheel (here, the *chordality* is the smallest k such that the graph is k-chordal). So, bounded tree-length graphs is a larger class than bounded chordality graphs.

We now recall the definition of *tree-decomposition* introduced by Robertson and Seymour in their work on graph minors [28]. A tree-decomposition of a graph G is a tree T whose vertices, called *bags*, are subsets of $V(G)$ such that: (1) $\cup_{X \in V(T)} X = V(G)$; (2) for all $uv \in E(G)$, there exists $X \in V(T)$ such that $u, v \in X$; and (3) for all $X, Y, Z \in V(T)$, if Y is on the path from X to Z in T then $X \cap Z \subseteq Y$. The *length* of tree-decomposition T of a graph G is $\max_{X \in V(T)} \max_{u,v \in X} d_G(u, v)$, and the *tree-length of* G is the minimum, over all tree-decompositions T of G, of the length of T.

Theorem 5. *If G has the tree-length λ, then any BFS-tree T of G is a λ-localized additive 5λ-fframe (and, hence, a λ-localized additive 5λ-frame) of G.*

Now, we provide some lower bound results.

Lemma 6. *For any $\lambda \geq 3$, there exists a tree-length λ graph without any $(\lambda-2)$-localized additive (λa)-fframe for any constant $a \geq 1$.*

Corollary 5. *For any $\lambda \geq 3$, there exists a tree-length λ graph G with n vertices for which no $(\lambda-2)$-localized additive $\frac{1}{2}\sqrt{\log\frac{n-1}{\lambda}}$-fframe exists.*

Lemma 7. *For any $\lambda \geq 4$, there exists a tree-length λ graph without any $\lfloor 2(\lambda-2)/3 \rfloor$-localized additive (λa)-frame for any constant $a \geq 1$.*

Corollary 6. *For any $\lambda \geq 4$, there exists a tree-length λ graph G with n vertices for which no $\lfloor 2(\lambda-2)/3 \rfloor$-localized additive $\frac{2}{3}\sqrt{\log\frac{3(n-1)}{4\lambda}}$-frame exists.*

Lemma 8. *For any $\lambda \geq 6$, there exists a tree-length λ graph without any $\lfloor (\lambda-2)/4 \rfloor$-localized additive (λa)-carcass for any constant $a \geq 1$.*

Corollary 7. *For any $\lambda \geq 6$, there exists a tree-length λ graph G with n vertices for which no $\lfloor (\lambda-2)/4 \rfloor$-localized additive $\frac{3}{4}\sqrt{\log\frac{n-1}{\lambda}}$-carcass exists.*

δ-hyperbolic graphs. δ-Hyperbolic metric spaces were defined by M. Gromov [16] in 1987 via a simple 4-point condition: for any four points u, v, w, x, the two larger of the distance sums $d(u,v)+d(w,x), d(u,w)+d(v,x), d(u,x)+d(v,w)$ differ by at most 2δ. They play an important role in geometric group theory, geometry of negatively curved spaces, and have recently become of interest in several domains of computer science, including algorithms and networking. For example, (a) it has been shown empirically in [30] that the Internet topology embeds with better accuracy into a hyperbolic space than into an Euclidean space of comparable dimension, (b) every connected finite graph has an embedding in the hyperbolic plane so that the greedy routing based on the virtual coordinates obtained from this embedding is guaranteed to work (see [21]). A connected graph $G=(V,E)$ equipped with standard graph metric d_G is *δ-hyperbolic* if the metric space (V, d_G) is δ-hyperbolic. It is known (see [4]) that all graphs with tree-length λ are λ-hyperbolic, and each δ-hyperbolic graph has the tree–length $O(\delta \log n)$.

Lemma 9. *Let G be a δ-hyperbolic graph. Let s, x, y be arbitrary vertices of G and $P(s,x)$, $P(s,y)$, $P(y,x)$ be arbitrary shortest paths connecting those vertices in G. Then, for vertices $a \in P(s,x)$, $b \in P(s,y)$ with $d_G(s,a) = d_G(s,b) = \lfloor \frac{d_G(s,x)+d_G(s,y)-d_G(x,y)}{2} \rfloor$, the inequality $d_G(a,b) \leq 4\delta$ holds.*

It is clear that δ takes values from $\{0, \frac{1}{2}, 1, \frac{3}{2}, 2, \frac{5}{2}, 3, \ldots\}$, and if $\delta = 0$ then G is a tree. Hence, in what follows. we will assume that $\delta \geq \frac{1}{2}$.

Theorem 6. *If G is a δ-hyperbolic graph, then any BFS-tree T of G is a 4δ-localized additive 8δ-fframe (and, hence, a 4δ-localized additive 8δ-frame) of G.*

Proof. Let T be an arbitrary BFS-tree of G rooted at a vertex s. Let $R_{G,T}(x,y)$ be the routing path from a vertex x to a vertex y produced by 4δ-localized IGRF scheme using tree T. If x is on the T path from y to s, or y is on the T path from x to s, it is easy to see that $R_{G,T}(x,y)$ is a shortest path of G.

Let sTx (resp., sTy) be the path of T from s to x (resp., to y) and $P(y,x)$ be an arbitrary shortest path connecting vertices x and y in G. By Lemma 9, for vertices $a \in sTx$, $b \in sTy$ with $d_G(s,a) = d_G(s,b) = \lfloor \frac{d_G(s,x)+d_G(s,y)-d_G(x,y)}{2} \rfloor$, the inequality $d_G(a,b) \leq 4\delta$ holds. Furthermore, since $d_G(a,x) + d_G(a,s) = d_G(s,x)$ and $d_G(b,y) + d_G(b,s) = d_G(s,y)$, from the choice of a and b, we have $d_G(x,y) \leq d_G(a,x) + d_G(b,y) \leq d_G(x,y) + 1$.

Let x' be a vertex of xTs with $d_G(x', sTy) \leq 4\delta$ closest to x. Clearly, x' belongs to subpath aTx of path sTx. Let y' be a vertex of path yTs with $d_G(x',y') \leq 4\delta$ (i.e., $y' \in D_{4\delta}(x',G)$) closest to y. Then, according to 4δ-localized IGRF scheme, the routing path $R_{G,T}(x,y)$ coincides with $(xTx') \cup$(a shortest path of G from x' to y')$\cup(y'Ty)$. We have $length(R_{G,T}(x,y)) = d_G(x,x') + d_G(x',y') + d_G(y',y)$.

If $y' \in bTy$, then $length(R_{G,T}(x,y)) = d_G(x,x') + d_G(x',y') + d_G(y',y) \leq d_G(x,a) + 4\delta + d_G(b,y) \leq d_G(x,y) + 4\delta + 1$. Assume now that $y' \in bTs$ and $y' \neq b$. Then, we have also $x' \neq a$. Since T is a BFS-tree of G, $d_G(y',b)$ must be at most $d_G(y',x')$ (otherwise, x' is closer than b to s in G, which is impossible). Thus, $d_G(y',b) \leq d_G(y',x') \leq 4\delta$ and, therefore, $length(R_{G,T}(x,y)) = d_G(x,x') + d_G(x',y') + d_G(y',y) \leq d_G(x,a) - 1 + 4\delta + d_G(b,y') + d_G(b,y) \leq d_G(x,y) + 1 - 1 + 8\delta = d_G(x,y) + 8\delta$.

Combining all cases, we conclude that T is a 4δ-localized additive 8δ-fframe (and a 4δ-localized additive 8δ-frame) of G. □

References

1. Bose, P., Morin, P., Stojmenovic, I., Urrutia, J.: Routing with guaranteed delivery in ad hoc wireless networks. In: 3rd International Workshop on Discrete Algorithms and Methods for Mobile Computing and Communic., pp. 48–55. ACM Press, New York (1999)
2. Brandstädt, A., Dragan, F.F., Chepoi, V.D., Voloshin, V.I.: Dually chordal graphs. SIAM J. Discrete Math. 11, 437–455 (1998)
3. Brandstädt, A., Bang Le, V., Spinrad, J.P.: Graph Classes: A Survey, Philadelphia. SIAM Monographs on Discrete Mathematics and Applications (1999)
4. Chepoi, V., Dragan, F.F., Estellon, B., Habib, M., Vaxès, Y.: Diameters, centers, and approximating trees of δ-hyperbolic geodesic spaces and graphs. In: SoCG 2008, pp. 59–68 (2008)
5. Dourisboure, Y., Gavoille, C.: Tree-decompositions with bags of small diameter. Discrete Mathematics 307, 2008–2029 (2007)
6. Dragan, F.F., Matamala, M.: Navigating in a graph by aid of its spanning tree. In: Hong, S.-H., Nagamochi, H., Fukunaga, T. (eds.) ISAAC 2008. LNCS, vol. 5369, pp. 788–799. Springer, Heidelberg (2008)

7. Dragan, F.F., Xiang, Y.: How to use spanning trees to navigate in graphs, full version, http://www.cs.kent.edu/~dragan/MFCS2009-journal.pdf
8. Fonseca, R., Ratnasamy, S., Zhao, J., Ee, C.T., Culler, D., Shenker, S., Stoica, I.: Beacon vector routing: Scalable point-to-point routing in wireless sensornets. In: 2nd USENIX/ACM Symp. on Networked Systems Design and Implementation (2005)
9. Fraigniaud, P.: Small Worlds as Navigable Augmented Networks: Model, Analysis, and Validation. In: Arge, L., Hoffmann, M., Welzl, E. (eds.) ESA 2007. LNCS, vol. 4698, pp. 2–11. Springer, Heidelberg (2007)
10. Fraigniaud, P., Korman, A., Lebhar, E.: Local MST computation with short advice. In: SPAA 2007, 154–160 (2007)
11. Garg, V.K., Agarwal, A.: Distributed maintenance of a spanning tree using labeled tree encoding. In: Cunha, J.C., Medeiros, P.D. (eds.) Euro-Par 2005. LNCS, vol. 3648, pp. 606–616. Springer, Heidelberg (2005)
12. Gartner, F.C.: A Survey of Self-Stabilizing Spanning-Tree Construction Algorithms, Technical Report IC/2003/38, Swiss Federal Institute of Technology (EPFL) (2003)
13. Gavoille, C.: Routing in distributed networks: Overview and open problems. ACM SIGACT News - Distributed Computing Column 32 (2001)
14. Gavoille, C., Peleg, D., Pérennès, S., Raz, R.: Distance labeling in graphs. J. Algorithms 53, 85–112 (2004)
15. Giordano, S., Stojmenovic, I.: Position based routing algorithms for ad hoc networks: A taxonomy. In: Ad Hoc Wireless Networking, pp. 103–136. Kluwer, Dordrecht (2004)
16. Gromov, M.: Hyperbolic Groups. In: Gersten, S.M. (ed.) Essays in group theory. MSRI Series, vol. 8, pp. 75–263 (1987)
17. Holm, J., de Lichtenberg, K., Thorup, M.: Poly-logarithmic deterministic fully-dynamic algorithms for connectivity, minimum spanning tree, 2-edge, and biconnectivity. J. ACM 48(4), 723–760 (2001)
18. Jacquet, P., Viennot, L.: Remote spanners: what to know beyond neighbors. In: IPDPS 2009, pp. 1–15 (2009)
19. Karp, B., Kung, H.T.: GPSR: greedy perimeter stateless routing for wireless networks. In: 6th ACM/IEEE MobiCom., pp. 243–254. ACM Press, New York (2000)
20. Kleinberg, J.M.: The small-world phenomenon: an algorithm perspective. In: STOC 2000, pp. 163–170. ACM, New York (2000)
21. Kleinberg, R.: Geographic routing using hyperbolic space. In: INFOCOM 2007, pp. 1902–1909 (2007)
22. Kuhn, F., Wattenhofer, R., Zhang, Y., Zollinger, A.: Geometric ad-hoc routing: of theory and practice. In: PODC, pp. 63–72. ACM, New York (2003)
23. Liben-Nowell, D., Novak, J., Kumar, R., Raghavan, P., Tomkins, A.: Geographic routing in social networks. PNAS 102, 11623–11628 (2005)
24. Linial, N., London, E., Rabinovich, Y.: The Geometry of Graphs and Some of its Algorithmic Applications. Combinatorica 15, 215–245 (1995)
25. Peleg, D.: Proximity-Preserving Labeling Schemes and Their Applications. J. of Graph Theory 33, 167–176 (2000)
26. Peleg, D.: Distributed Computing: A Locality-Sensitive Approach. SIAM Monographs on Discrete Math. Appl. SIAM, Philadelphia (2000)
27. Rao, A., Papadimitriou, C., Shenker, S., Stoica, I.: Geographical routing without location information. In: Proceedings of MobiCom 2003, pp. 96–108 (2003)
28. Robertson, N., Seymour, P.D.: Graph minors. II. Algorithmic aspects of tree-width. Journal of Algorithms 7, 309–322 (1986)

29. Santoro, N., Khatib, R.: Labelling and Implicit Routing in Networks. The Computer Journal 28(1), 5–8 (1985)
30. Shavitt, Y., Tankel, T.: On internet embedding in hyperbolic spaces for overlay construction and distance estimation. In: INFOCOM 2004 (2004)
31. Thorup, M., Zwick, U.: Compact routing schemes. In: 13th Ann. ACM Symp. on Par. Alg. and Arch., July 2001, pp. 1–10 (2001)

Representing Groups on Graphs

Sagarmoy Dutta and Piyush P. Kurur

Department of Computer Science and Engineering,
Indian Institute of Technology Kanpur,
Kanpur, Uttar Pradesh, India 208016
sagarmoy@cse.iitk.ac.in, ppk@cse.iitk.ac.in

Abstract. In this paper we formulate and study the problem of representing groups on graphs. We show that with respect to polynomial time Turing reducibility, both abelian and solvable group representability are all equivalent to graph isomorphism, even when the group is presented as a permutation group via generators. On the other hand, the representability problem for general groups on trees is equivalent to checking, given a group G and n, whether a nontrivial homomorphism from G to S_n exists. There does not seem to be a polynomial time algorithm for this problem, in spite of the fact that tree isomorphism has polynomial time algorithms.

1 Introduction

Representation theory of groups is a vast and successful branch of mathematics with applications ranging from fundamental physics to computer graphics and coding theory [5]. Recently representation theory has seen quite a few applications in computer science as well. In this article, we study some of the questions related to representation of finite groups on graphs.

A representation of a group G usually means a linear representation, i.e. a homomorphism from the group G to the group $\mathrm{GL}(V)$ of invertible linear transformations on a vector space V. Notice that $\mathrm{GL}(V)$ is the set of *symmetries* or *automorphisms* of the vector space V. In general, by a representation of G on an object X, we mean a homomorphism from G to the automorphism group of X. In this article, we study some computational problems that arise in the representation of *finite groups* on graphs. Our interest is the following group representability problem: Given a group G and a graph X, decide whether G has a nontrivial representation on X. As expected this problem is closely connected to graph isomorphism: We show, for example, that the graph isomorphism problem reduces to representability of abelian groups. In the other direction we show that even for solvable groups the representability on graphs is decidable using a graph isomorphism oracle. The reductions hold true even when the groups are presented as permutation groups. One might be tempted to conjecture that the problem is equivalent to Graph Isomorphism. However we conjecture that this might not be the case. The non-solvable version of this problem seems to be harder than graph isomorphism. For example, we were able to show that

R. Královič and D. Niwiński (Eds.): MFCS 2009, LNCS 5734, pp. 295–306, 2009.

representability of groups on trees, a class of graphs for which isomorphism is decidable in polynomial time, is as hard as checking whether, given an integer n and a group G, the symmetric group S_n has a nontrivial subgroup homomorphic to G, a problem for which no polynomial time algorithm is known.

2 Background

In this section we review the group theory required for the rest of the article. Any standard text book on group theory, for example the one by Hall [4], will contain the required results.

We use the following standard notation: The identity of a group G is denoted by 1. In addition 1 also stands for the singleton group consisting of only the identity. For groups G and H, $H \leq G$ (or $G \geq H$) means that H is a subgroup of G. Similarly by $H \trianglelefteq G$ (or $G \trianglerighteq H$) we mean H is a *normal subgroup* of G.

Let G be any group and let x and y be any two elements. By the *commutator* of x and y, denoted by $[x, y]$, we mean $xyx^{-1}y^{-1}$. The *commutator subgroup* of G is the group generated by the set $\{[x, y] | x, y \in G\}$. We denote the commutator subgroup of G by G'. The following is a well known result in group theory [4, Theorem 9.2.1]

Theorem 1. *The commutator subgroup G' is a normal subgroup of G and G/G' is abelian. Further for any normal subgroup N of G such that G/N is abelian, N contains G' as a subgroup.*

A group is *abelian* if it is commutative, i.e. $gh = hg$ for all group elements g and h. A group G is said to be *solvable* [4, Page 138] if there exists a decreasing chain of groups $G = G_0 \triangleright G_1 \ldots \triangleright G_t = 1$ such that G_{i+1} is the commutator subgroup of G_i for all $0 \leq i < t$.

An important class of groups that play a crucial role in graph isomorphism and related problems are permutation groups. In this paper we follow the notation of Wielandt [11] for permutation groups. Let Ω be a finite set. The *symmetric group* on Ω, denoted by $\mathrm{Sym}(\Omega)$, is the group of all permutations on the set Ω. By a *permutation group* on Ω we mean a subgroup of the symmetric group $\mathrm{Sym}(\Omega)$. For any positive integer n, we will use S_n to denote the symmetric group on $\{1, \ldots, n\}$. Let g be a permutation on Ω and let α be an element of Ω. The image of α under g will be denoted by α^g. For a permutation group G on Ω, the orbit of α is denoted by α^G. Similarly if Δ is a subset of Ω then Δ^g denotes the set $\{\alpha^g | \alpha \in \Delta\}$.

Any permutation group G on n symbols has a generating set of size at most n. Thus for computational tasks involving permutation groups it is assumed that the group is presented to the algorithm via a small generating set. As a result, by efficient algorithms for permutation groups on n symbols we mean algorithms that take time polynomial in the size of the generating set and n.

Let G be a subgroup of S_n and let $G^{(i)}$ denote the subgroup of G that fixes pointwise $j \leq i$, i.e. $G^{(i)} = \{g | j^g = j, 1 \leq j \leq i\}$. Let C_i denote a right *transversal*, i.e. the set of right coset representative, for $G^{(i)}$ in $G^{(i-1)}$. The

$\cup_i C_i$ is a generating set for G and is called the *strong generating set* for G. The corner stone for most polynomial time algorithms for permutation group is the Schreier-Sims [9,10,3] algorithm for computing the *strong generating set* of a permutation group G given an arbitrary generating set. Once the strong generating set is computed, many natural problems for permutation groups can be solved efficiently. We give a list of them in the next theorem.

Theorem 2. *Given a generating set for G there are polynomial time algorithms for the following task.*

1. *Computing the strong generating set.*
2. *Computing the order of G.*

By a graph we mean a *finite undirected graph.* For a graph X, $V(X)$ and $E(X)$ denotes the set of vertices and edges respectively and $\mathrm{Aut}(X)$ denotes the group of all automorphisms of X, i.e. permutations on $V(X)$ that maps edges to edges and non-edges to non-edges.

Definition 1 (Representation). *A representation ρ from a group G to a graph X is a homomorphism from G to the automorphism group* $\mathrm{Aut}(X)$ *of X.*

Alternatively we say that G acts on (the right) of X via the representation ρ. When ρ is understood, we use u^g to denoted $u^{\rho(g)}$.

A representation ρ is *trivial* if all the elements of G are mapped to the identity permutation. A representation ρ is said to be *faithful* if it is an injection as well. Under a faithful action G can be thought of as a subgroup of the automorphism group. We say that G is *representable* on X if there is a nontrivial representation from G to X. We now define the following natural computational problem.

Definition 2 (Group representability problem). *Given a group G and a graph X decide whether G is representable on X nontrivially.*

We will look at various restrictions of the above problem. For example, we study the abelian (solvable) group representability problem where our input groups are abelian (solvable). We also study the group representability problem on trees, by which we mean group representability where the input graph is a tree.

Depending on how the group is presented to the algorithm, the complexity of the problem changes. One possible way to present G is to present it as a permutation group on m symbols via a generating set. In this case the input size is $m + \#V(X)$. On the other hand, we can make the task of the algorithm easier by presenting the group via a multiplication table. In this paper we mostly assume that the group is in fact presented via its multiplication table. Thus polynomial time means polynomial in $\#G$ and $\#V(X)$. However for solvable representability problem, our results extend to the case when G is a permutation group presented via a set of generators.

We now look at the following closely related problem that occurs when we study the representability of groups on trees.

Definition 3 (Permutation representability problem). *Given a group G and an integer n in unary, check whether there is a homomorphism from G to S_n.*

Overview of the Results

Our first result is to show that graph isomorphism reduces to abelian representability problem. In fact we show that graph isomorphism reduces to the representability of prime order cyclic groups on graphs. Next we show that solvable group representability problem reduces to graph isomorphism problem. Thus as far as polynomial time Turing reducibility is concerned abelian group representability and solvable group representability are all equivalent to graph isomorphism. As a corollary we have, solvable group representability on say bounded degree graphs or bounded genus graphs are all in polynomial time.

We then show that group representability on trees is equivalent to permutation representability (Definition 3). This is in contrast to the corresponding isomorphism problem because for trees, isomorphism testing is in polynomial time whereas permutation representability problem does not appear to have a polynomial time algorithm.

3 Abelian Representability

In this section we prove that the graph isomorphism problem reduces to abelian group representability on graphs. Given input graphs X and Y of n vertices each and any prime $p > n$, we construct a graph Z of exactly $p \cdot n$ vertices such that X and Y are isomorphic if and only if the cyclic group of order p is representable on Z. Since for any integer n there is a prime p between n and $2n$ (Bertrand's conjecture), the above constructions gives us a reduction from the graph isomorphism problem to abelian group representability problem.

For the rest of the section, fix the input graphs X and Y. Our task is to decide whether X and Y are isomorphic. Firstly we assume, without loss of generality, that the graphs X and Y are connected, for otherwise we can take their complement graphs X' and Y', which are connected and are isomorphic if and only if X and Y are isomorphic. Let n be the number of vertices in X and Y and let p be any prime greater than n. Consider the graph Z which is the disjoint union of p connected components $Z_1, \ldots, Z_p$ where, for each $1 \leq i < p$, each Z_i is an isomorphic copy of X and Z_p is an isomorphic copy of Y. First we prove the following lemma.

Lemma 1. *If X and Y are isomorphic then $\mathbb{Z}/p\mathbb{Z}$ is representable on Z.*

Proof. Clearly it is sufficient to show that there is an order p automorphism for Z. Let h be an isomorphism from X to Y. For every vertex v in X, let v_i denote its copy in Z_i. Consider the bijection g from $V(Z)$ to itself defined as follows: For all vertices v in $V(X)$ and each $1 \leq i < p-2$, let $v_i^g = v_{i+1}$. Further let g map v_{p-1} to v^h and v^h to v_1. It is easy to verify that g is an automorphism of Z and has order p. □

We now prove the converse

Lemma 2. *If $\mathbb{Z}/p\mathbb{Z}$ can be represented on Z then X and Y are isomorphic.*

Proof. If $\mathbb{Z}/p\mathbb{Z}$ can be represented on Z then there exists a nonidentity automorphism g of Z such that the order of g is p. We consider the action of the cyclic group H, generated by g, on $V(X)$. Since g is nontrivial, there exists at least one H-orbit Δ of $V(X)$ which is of cardinality greater than 1. However by orbit stabiliser formula [11, Theorem 3.2], $\#\Delta$ divides $\#H = p$. Since p is prime, Δ should be of cardinality p.

We prove that no two vertices of Δ belong to the same connected component. Assume the contrary and let α and β be two elements of Δ which also belong to the same connected component of Z. There is some $0 < t < p$ such that $\alpha^{g^t} = \beta$. We assume further, without loss of generality, that $t = 1$, for otherwise we replace g by the automorphism g^t, which is also of order p, and carry out the argument. Therefore $\alpha^g = \beta$ lie in the same component of Z. It follows then that, for each $0 \leq i \leq p-1$, the element $\alpha_i = \alpha^{g^i}$ is in the same component of Z, as automorphisms preserve edges and hence paths. However this means that there is a component of Z that is of cardinality at least p. This is a contradiction as each component of Z has at most $n < p$ vertices as they are copies of either X or Y.

It follows that there is some $1 \leq i < p$, for which g must map at least one vertex of the component Z_i to some vertex of Z_p. As a result the automorphism g maps the entire component Z_i to Z_p. Therefore the components Z_i and Z_p are isomorphic and so are their isomorphic copies X and Y. □

Given two graphs X and Y of n vertices we find a prime p such that $n < p < 2n$, construct the graph Z and construct the multiplication table for $\mathbb{Z}/p\mathbb{Z}$. This requires only logarithmic space in n. Using Lemmas 1 and 2 we have the desired reduction.

Theorem 3. *The graph isomorphism problem logspace many-one reduces to abelian group representability problem.*

4 Solvable Representability Problem

In the previous section we proved that abelian group representability is at least as hard as graph isomorphism. In this section we show that solvable group representability is polynomial time Turing reducible to the graph isomorphism problem. We claim that a solvable group G is representable on X if and only if $\#\mathrm{Aut}\,(X)$ and $\#G/G'$ have a common prime factor, where G' is commutator subgroup of G. We do this in two stages.

Lemma 3. *A solvable group G can be represented on a graph X if $\#G/G'$ and $\#\mathrm{Aut}\,(X)$ have a common prime factor.*

Proof. Firstly notice that it suffices to prove that there is a nontrivial homomorphism, say ρ, from G/G' to $\mathrm{Aut}\,(X)$. A nontrivial representation for G can be obtained by composing the natural quotient homomorphism from G *onto* G/G' with ρ.

Recall that the quotient group G/G' is an abelian group and hence can be represented on X if for some prime p that divides $\#G/G'$, there is an order p automorphism for X. However by the assumption of the theorem, there is a common prime factor, say p, of $\#G/G'$ and $\#\mathrm{Aut}(X)$. Therefore, by Cayley's theorem there is an order p element in $\mathrm{Aut}(X)$. As a result, G/G' and hence G is representable on X. □

To prove the converse, for the rest of the section fix the input, the solvable group G and the graph X. Consider any nontrivial homomorphism ρ from the group G to $\mathrm{Aut}(X)$. Let $H \leq \mathrm{Aut}(X)$ denote the image of the group G under ρ. We will from now on consider ρ as an automorphism from G *onto* H. Since the subgroup H is the homomorphic image of G, H itself is a solvable group.

Lemma 4. *The homomorphism ρ maps the commutator subgroup G' of G onto the commutator subgroup H'.*

Proof. First we prove that $\rho(G') \leq H'$. For this notice that for all x and y in G, since ρ is a homomorphism, $\rho([x, y]) = [\rho(x), \rho(y)]$ is an element of H'. As G' is generated by the set $\{[x, y] | x, y \in G\}$ of all commutators, $\rho(G') \leq H'$. To prove the converse notice that ρ is a surjection on H. Therefore for any element h of H, we have element x_h of G such that $\rho(x_h) = h$. Consider the commutator $[g, h]$ for any two elements g and h of H. We have $\rho([x_g, x_h]) = [g, h]$. This proves that all the commutators of H are in the image of G' and hence $\rho(G') \geq H'$. □

We have the following result about solvable groups that directly follows from the definition of solvable groups [4, Page 138].

Lemma 5. *Let G be any nontrivial solvable group then its commutator subgroup G' is a strict subgroup of G.*

Proof. By the definition of solvable groups, there exist a chain $G = G_0 \rhd G_1 \ldots \rhd G_t = 1$ such that G_{i+1} is the commutator subgroup of G_i for all $0 \leq i < t$. If $G = G' = G_1$ then $G = G_i$ for all $0 \leq i \leq t$ implying $G = 1$ □

We are now ready to prove the converse of Lemma 3.

Lemma 6. *Let G be any solvable group and let X be any graph. The orders $\#G/G'$ and $\#\mathrm{Aut}(X)$ have a common prime factor if G is representable on graph X.*

Proof. Let ρ be any nontrivial homomorphism from G to $\mathrm{Aut}(X)$, and let H be the image of group G under this homomorphism. Since the commutator subgroup G' is strictly contained in the group G (Lemma 5), order of the quotient group $\#G/G' > 1$. Furthermore, the image group H itself is solvable and nontrivial, as it is the image of a solvable group G under a nontrivial homomorphism. Therefore, the commutator subgroup H' is strictly contained in H implying $\#H/\#H' > 1$.

Consider the homomorphism $\tilde{\rho}$ from G *onto* H/H' defined as $\tilde{\rho}(g) = \rho(g)H'$. Since ρ maps G' onto H', we have that G' is in the kernel of $\tilde{\rho}$. Therefore, $\tilde{\rho}$ can be *refined* to a map from G/G' *onto* H/H'. Clearly the prime factors of $\#H/H'$ are all prime factors of $\#G/G'$. However, any prime factor of $\#H/H'$ is a prime factor of $\mathrm{Aut}(X)$, as both H and H' are subgroups of $\mathrm{Aut}(X)$. Therefore, the orders of G/G' and $\mathrm{Aut}(X)$ have a common prime factor. □

The order of the automorphism group of the input graph X can be computed in polynomial time using an oracle to the graph isomorphism problem [7]. Further since the automorphism group is a subgroup of S_n, where n is the cardinality of $V(X)$, all its prime factors are less than n and hence can be determined. Also since G is given as a table, its commutator subgroup G' can be computed in polynomial time and the prime factors of $\#G/G'$ can also be similarly determined. Therefore we can easily check, given the group G via its multiplication table and the graph X, whether the order of the quotient group G/G' has common factors with the order of $\mathrm{Aut}(X)$. We thus have the following theorem.

Theorem 4. *The problem of deciding whether a solvable group can be represented on a given graph Turing reduces to graph isomorphism problem.*

For the reduction in the above theorem to work, it is sufficient to compute the order of G and its commutator subgroup G'. This can be done even when the group G is presented as a permutation group on m symbols via a generating set. To compute $\#G$ we can compute the strong generating set of G and use Theorem 2. Further given a generating set for G, a generating set for its commutator subgroup G' can be compute in polynomial time [3, Theorem 4]. Therefore, the order of G/G' can be computed in polynomial time given the generating set for G. Furthermore, G and G' are subgroups of S_m and hence all their prime factors are less than m and can be determined. We can then check whether $\#G/G'$ has any common prime factors with $\#\mathrm{Aut}(X)$ just as before using the graph isomorphism oracle. Thus we have the following theorem.

Theorem 5. *The solvable group representability problem, where the group is presented as a permutation group via a generating set, reduces to the graph isomorphism problem via polynomial time Turing reduction.*

5 Representation on Tree

In this section we study the representation of groups on trees. It is known that isomorphism of trees can be tested in polynomial time [2]. However we show that the group representability problem over trees is equivalent to permutation representability problem (Definition 3), a problem for which, we believe, there is no polynomial time algorithm.

Firstly, to show that permutation representability problem is reducible to group representability problem on trees, it is sufficient to construct, given and integer n, a tree whose automorphism group is S_n. Clearly a tree with n leaves, all of which is connected to the root, gives such a tree (see Figure 1). Therefore we have the following lemma.

Fig. 1. Tree with automorphism group S_n

Lemma 7. *Permutation representability reduces to representability on tree.*

To prove the converse, we first reduce the group representability problem on an arbitrary tree to the problem of representability on a rooted tree. We then do a divide and conquer on the structure of the rooted tree using the permutation representability oracle. The main idea behind this reduction is Lemma 11 where we show that for any tree T, either there is a vertex which is fixed by all automorphisms, in which case we can choose this vertex as the root, or there are two vertices α and β connected by an edge which together forms an orbit under the action of Aut (T), in which case we can add a dummy root (see Figure 2) to make it a rooted tree without changing the automorphism group.

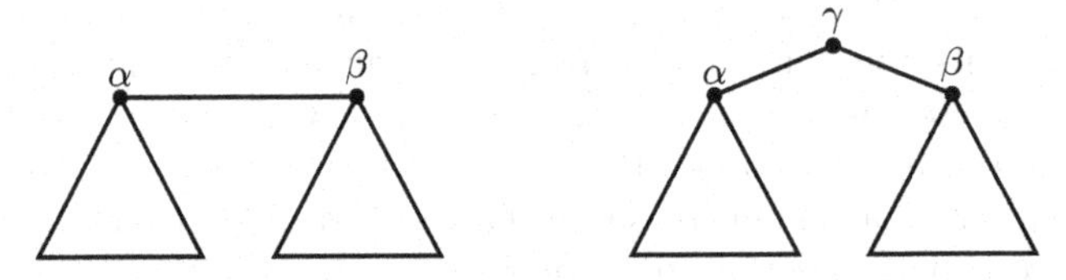

Fig. 2. Minimal orbit has two elements

For the rest of the section fix a tree T. Let Δ be an orbit in the action of Aut (T) on $V(T)$. We define the graph T_Δ as follows: A vertex γ (or edge e) of T belongs to T_Δ if there are two vertices α and β in Δ such that γ (or e) is contained in the path from α to β. It is easy to see that T_Δ contains paths between any two vertices of Δ. Any vertex in T_Δ is connected to some vertex in Δ and all vertices in Δ are connected in T_Δ which implies T_Δ is connected. Furthermore T_Δ has no cycle, as its edge set is a subset of the edge set of T. Therefore T_Δ is a tree.

Lemma 8. *Let g be any automorphism of T and consider any vertex γ (or edge e) of T_Δ. Then the vertex γ^g (or edge e^g) is also in T_Δ.*

Proof. Since γ (or e) is present in T_Δ, there exists α and β in Δ such that γ (or e) is in the path between α and β. Also since automorphisms preserve paths, γ^g (or e^g) is in the path from α^g to β^g. □

Lemma 9. *The orbit Δ is precisely the set of leaves of T_Δ.*

Proof. First we show that all leaf nodes of T_Δ are in orbit Δ. Any node α of T_Δ must lie on a path such that the endpoints are in orbit Δ. If α is a leaf of T_Δ, this can only happen when α itself is in Δ.

We will prove the converse by contradiction. If possible let α be a vertex in the orbit Δ which is not a leaf of T_Δ. Vertex α must lie on the path between two leaves β and γ. Also since β and γ are leaves of T_Δ, they are in the orbit Δ.

Let g be an automorphism of T which maps α to β. Such an automorphism exists because α and β are in the same orbit Δ. The image $\alpha^g = \beta$ must lie on the path between β^g and γ^g and neither β^g or γ^g is β. This is impossible because β is a leaf of T_Δ. □

Lemma 10. *Let γ be a vertex in orbit Σ. If γ is a vertex of the subtree T_Δ then subtree T_Σ is a subtree of T_Δ.*

Proof. Assume that Δ is different from Σ, for otherwise the proof is trivial. First we show that all the vertices of Σ are vertices of T_Δ. The vertex γ lies on a path between two vertices of Δ, say α and β. Take any vertex γ' from the orbit Σ. There is an automorphism g of T which maps γ to γ'. Now $\gamma' = \gamma^g$ lies on the path between α^g and β^g and hence is in the tree T_Δ.

Consider any edge e of T_Σ. There exists γ_1 and γ_2 of Σ such that e is on the path from γ_1 to γ_2. By previous argument, T contains γ_1 and γ_2. Since T is a tree, this path is unique and any subgraph of T, in which γ_1 and γ_2 are connected, must contain this path. Hence T_Δ contains e. □

Lemma 11. *Let T be any tree then either there exists a vertex α that is fixed by all the automorphisms of T or there exists two vertices α and β connected via an edge e such that $\{\alpha, \beta\}$ is an orbit of* $\mathrm{Aut}\,(T)$. *In the latter case every automorphism maps e to itself.*

Proof. Consider the following partial order between orbits of $\mathrm{Aut}\,(T)$: $\Sigma \leq \Delta$ if T_Σ is a subtree of T_Δ. The relation $\leq$ is clearly a partial order because the "subtree" relation is. Since there are finitely many orbits there is always a minimal orbit under the above ordering. From Lemmas 9 and 10 it follows that for an orbit Δ, if Σ is the orbit containing an internal node γ of T_Δ then Σ is strictly less than Δ. Therefore for any minimal orbit Δ, all the nodes are leaves. This is possible if either T_Δ is a singleton vertex α, or consists of exactly two nodes connected via an edge. In the former case all automorphisms of T have to fix α, whereas in the latter case the two nodes may be flipped but the edge connecting them has to be mapped to itself. □

It follows from Lemma 11 that any tree T can be rooted, either at a vertex or at an edge with out changing the automorphism. Given a tree T, since computing the generating set for $\mathrm{Aut}\,(T)$ can be done in polynomial time, we can determine all the orbits of $\mathrm{Aut}\,(T)$ by a simple transitive closure algorithm. Having computed these orbits, we determine whether T has singleton orbit or an orbit of cardinality 2. For trees with an orbit containing a single vertex α, rooting the tree at α does not change the automorphism group. On the other hand if the tree has an orbit with two elements we can add a dummy root as in Figure 2 without changing the automorphism group. Since by Lemma 11 these are the only two possibilities we have the following theorem.

Theorem 6. *There is a polynomial time algorithm that, given as input a tree T, outputs a rooted tree T' such that for any group G, G is representable on T if and only if G is representable on the rooted tree T'.*

For the rest of the section by a tree we mean a rooted tree. We will prove the reduction from representability on rooted trees to permutation representability. First we characterise the automorphism group of a tree in terms of wreath product [Theorem 7] and then show that we can find a nontrivial homomorphism, if there exists one, from the given group G to this automorphism group by querying a permutation representability oracle.

Definition 4 (Semidirect product and wreath product). *Let G and A be any two groups and let φ be any homomorphism from G to $\mathrm{Aut}(A)$, then the semi-direct product $G \ltimes_\varphi A$ is the group whose underlying set is $G \times A$ and the multiplication is defined as $(g, a)(h, b) = (gh, a^{\varphi(h)}b)$.*

We use $W_n(A)$ to denote the wreath product $S_n \wr A$ which is the semidirect product $S_n \ltimes_\varphi A^n$, where A^n is the n-fold direct product of A and $\varphi(h)$, for each h in S_n, permutes $\mathbf{a} \in A^n$ according to the permutation h, i.e. maps $(\ldots, a_i, \ldots) \in A^n$ to $(\ldots, a_j, \ldots)$ where $j^h = i$.

As the wreath product is a semidirect product, we have the following lemma.

Lemma 12. *The wreath product $W_n(A)$ contains (isomorphic copies of) S_n and A^n as subgroups such that A^n is normal and the quotient group $W_n(A)/A^n = S_n$.*

For the rest of the section fix the following: Let T be a tree with root ω with k children. Consider the subtrees of T rooted at each of these k children and partition them such that two subtrees are in the same partition if and only if they are isomorphic. Let t be the number of partitions and let k_i, for $(1 \leq i \leq t)$, be the number of subtrees in the i-th partition. For each i, pick a representative subtree T_i from the i-th partition and let A_i denote the automorphism group of T_i. The following result is well known but a proof is given for completeness.

Theorem 7. *The automorphism group of the tree T is (isomorphic to) the direct product $\prod_{i=1}^{t} W_{k_i}(A_i)$.*

Proof. Let $\omega_1, \ldots, \omega_k$ be the children of the root ω and let X_i denote the subtree rooted at ω_i. We first consider the case when $t = 1$, i.e. all the subtrees X_i are isomorphic. Any automorphism g of T must permute the children ω_i's among themselves and whenever $\omega_i^g = \omega_j$, the entire subtree X_i maps to X_j. As all the subtrees X_i are isomorphic to T_1, the forest $\{X_1, \ldots, X_k\}$ can be thought of as the disjoint union of k copies of the tree T_1 by fixing, for each i, an isomorphism σ_i from T_1 to X_i.

For an automorphism g of T, define the permutation $\tilde{g} \in S_k$ and the automorphisms $a_i(g)$ of T_1 as follows: if $\omega_i^g = \omega_j$ then $i^{\tilde{g}} = j$ and $a_i(g) = \sigma_i g \sigma_j^{-1}$. Consider the map ϕ from $\mathrm{Aut}(T)$ to $W_k(A)$ which maps an automorphism g to the group element $(\tilde{g}, a_1(g), \ldots, a_k(g))$ in $W_k(A)$. It is easy to verify that ϕ is the desired isomorphism.

When the number of partitions t is greater than 1, any automorphism of T fixes the root ω and permutes the subtrees in the i-th partition among themselves. Therefore the automorphism group of T is same as the automorphism group of the collection of forests F_i one for each partition i. Each forest is a disjoint union of k_i copies of T_i and we can argue as before that its automorphism group is (isomorphic to) $W_{k_i}(A)$. Therefore Aut (T) should be the direct product $\prod_{i=1}^{t} W_{k_i}(A_i)$. □

Lemma 13. *If the group G can be represented on the tree T, then there exists $1 \leq i \leq t$ such that there is a nontrivial homomorphism from G to $W_{k_i}(A_i)$.*

Proof. If there is a nontrivial homomorphism from a group G to the direct product of groups $H_1, \ldots, H_t$ then for some i, $1 \leq i \leq t$, there is a nontrivial homomorphism from G to H_i. The lemma then follows from Theorem 7. □

Lemma 14. *If there is a nontrivial homomorphism ρ from a group G to $W_n(A)$ then there is also a nontrivial homomorphism from G either to S_n or to A.*

Proof. Let ρ be a nontrivial homomorphism G to $W_n(A)$. Since A^n is a normal subgroup of $W_n(A)$ and the quotient group $W_n(A)/A^n$ is S_n, there is a homomorphism ρ' from $W_n(A)$ to S_n with kernel A^n. The composition of ρ and ρ' is a homomorphism from G to S_n.

If $\rho' \cdot \rho$ is trivial then ρ' maps all elements of $\rho(G)$ to identity of S_n. Which implies that $\rho(G)$ is a subgroup of the kernel of ρ', that is A^n. So, ρ is a nontrivial homomorphism from G to A^n. Hence there must be a nontrivial homomorphism from G to A. □

Theorem 8. *Given a group G and a rooted tree T with n nodes and an oracle for deciding whether G has a nontrivial homomorphism to S_m for $1 \leq m \leq n$, it can be decided in polynomial time whether G can be represented on T.*

Proof. If the tree has only one vertex then reject. Otherwise let t, $k_1, \ldots, k_t$ and $A_1, \ldots A_t$ be the quantities as defined in Theorem 7. Since there is efficient algorithm to compute tree isomorphism, t and $k_1, \ldots, k_t$ can be computed in polynomial time. If G is representable on T then, by Lemma 13 and Lemma 14, there is a nontrivial homomorphism from G to either S_{k_i} or A_i for some i. Using the oracle, check whether there is a nontrivial homomorphism to any of the symmetric groups. If found then accept, otherwise for all i, decide whether there is a nontrivial homomorphism to A_i by choosing a subtree T_i from the ith partition and recursively asking whether G is representable on T_i. The total number of recursive calls is bounded by the number of vertices of T. Hence the reduction is polynomial time. □

6 Conclusion

In this paper we studied the group representability problem, a computational problem that is closely related to graph isomorphism. The representability problem could be equivalent to graph isomorphism, but the results of Section 5 give some, albeit weak, evidence that this might not be the case. It would be interesting to know what is the exact complexity of this problem vis a vis the graph isomorphism problem. We know from the work of Mathon [7] that the graph isomorphism problem is equivalent to its functional version where, given two graphs X and Y, we have to compute an isomorphism if there exists one. The functional version of group representability, namely give a group G and a graph X compute a nontrivial representation if it exists, does not appear to be equivalent to the decision version. Also it would be interesting to know if the representability problem shares some of lowness of graph isomorphism [8,6,1]. Our hope is that, like the study of group representation in geometry and mathematics, the study of group representability on graphs help us better understand the graph isomorphism problem.

References

1. Arvind, V., Kurur, P.P.: Graph Isomorphism is in SPP. In: 43rd Annual Symposium of Foundations of Computer Science, pp. 743–750. IEEE, Los Alamitos (2002)
2. Babai, L., Luks, E.M.: Canonical labeling of graphs. In: Proceedings of the Fifteenth Annual ACM Symposium on Theory of Computing, pp. 171–183 (1983)
3. Furst, M.L., Hopcroft, J.E., Luks, E.M.: Polynomial-time algorithms for permutation groups. In: IEEE Symposium on Foundations of Computer Science, pp. 36–41 (1980)
4. Hall Jr., M.: The Theory of Groups, 1st edn. The Macmillan Company, New York (1959)
5. Joyner, W.D.: Real world applications of representation theory of non-abelian groups, http://www.usna.edu/Users/math/wdj/repn_thry_appl.htm
6. Köbler, J., Schöning, U., Torán, J.: Graph isomorphism is low for PP. Computational Complexity 2(4), 301–330 (1992)
7. Mathon, R.: A note on graph isomorphism counting problem. Information Processing Letters 8(3), 131–132 (1979)
8. Schöning, U.: Graph isomorphism is in the low hierarchy. In: Symposium on Theoretical Aspects of Computer Science, pp. 114–124 (1987)
9. Sims, C.C.: Computational methods in the study of permutation groups. Computational problems in Abstract Algebra, 169–183 (1970)
10. Sims, C.C.: Some group theoretic algorithms. Topics in Algebra 697, 108–124 (1978)
11. Wielandt, H.: Finite Permutation Groups. Academic Press, New York (1964)

Admissible Strategies in Infinite Games over Graphs*

Marco Faella

Università di Napoli "Federico II", Italy

Abstract. We consider games played on finite graphs, whose objective is to obtain a trace belonging to a given set of accepting traces. We focus on the states from which Player 1 cannot force a win. We compare several criteria for establishing what is the preferable behavior of Player 1 from those states, eventually settling on the notion of *admissible* strategy.

As the main result, we provide a characterization of the goals admitting positional admissible strategies. In addition, we derive a simple algorithm for computing such strategies for various common goals, and we prove the equivalence between the existence of positional winning strategies and the existence of positional subgame perfect strategies.

1 Introduction

Games played on finite graphs have been widely investigated in Computer Science, with applications including controller synthesis [PR89, ALW89, dAFMR05], protocol verification [KR01, BBF07], logic and automata theory [EJ91, Zie98], and compositional verification [dAH01].

These games consist of a finite graph, whose set of states is partitioned into Player-1 and Player-2 states, and a *goal*, which is a set of infinite sequences of states. The game consists in the two players taking turns at picking a successor state, eventually giving rise to an infinite path in the game graph. Player 1 wins the game if she manages to obtain an infinite path belonging to the goal, otherwise Player 2 wins. A (deterministic) *strategy* for a player is a function that, given the current history of the game (a finite sequence of states), chooses the next state. A state s is said to be *winning* if there exists a strategy that guarantees victory to Player 1 regardless of the moves of the adversary, if the game starts in s. A state that is not winning is called *losing*.

The main algorithmic concern of the classical theory of these games is determining the set of winning states. In this paper, we shift the focus to *losing* states, since we believe that many applications would benefit from a theory of best-effort strategies which allowed Player 1 to play in a rational way even from losing states.

For instance, many game models correspond to real-world problems which are not really competitive: the game is just a tool which enables to distinguish internal from external non-determinism. In practice, the behavior of the adversary may turn out to be random, or even cooperative. A strategy of Player 1

* This work was supported by the MIUR PRIN Project 2007-9E5KM8.

R. Královič and D. Niwiński (Eds.): MFCS 2009, LNCS 5734, pp. 307–318, 2009.

which does not "give up", but rather tries its best at winning, may in fact end up winning, even starting from states that are theoretically losing.

In other cases, the game is an over-approximation of reality, giving to Player 2 a wider set of capabilities (i.e., moves in the game) than what most adversaries actually have in practice. Again, a best-effort strategy for Player 1 can thus often lead to victory, even against an adversary which is strictly competitive.

In this paper, we compare several alternative definitions of best-effort strategies, eventually settling on the notion of *admissible* strategy. As a guideline for our investigation, we take the application domain of automated verification and synthesis of open systems. Such a domain is characterized by the fact that, once good strategies for a game have been found, they are intended to be actually implemented in hardware or software.

Best-Effort Strategies. The classical definition of what a "good" strategy is states that a strategy is *winning* if it guarantees victory whenever the game is started in a winning state [Tho95]. This definition does not put any burden on a strategy if the game starts from a losing state. In other words, if the game starts from a losing state, all strategies are considered equivalent.

A first refinement of the classical definition is a slight modification of the game-theoretic notion of *subgame-perfect equilibrium* [OR94]. Cast in our framework, this notion states that a strategy is good if it enforces victory whenever the game history is such that victory can be enforced. We call such strategies *strongly winning*, to avoid confusion with the use of subgame (and subarena) which is common in computer science [Zie98]. It is easy to see that this definition captures the intuitive idea that a good strategy should "enforce victory whenever it can" better than the classical one.

Next, consider games where victory cannot be enforced at any point during the play. Take the Büchi game in Figure 1 [1], whose goal is to visit infinitely often s_0. No matter how many visits to s_0 Player 1 manages to make, he will never reach a point where he can enforce victory. Still, it is intuitively better for him to keep trying (i.e., move to s_1) rather than give up (i.e., move to s_2). To capture this intuition, we resort to the classical game-theoretic notion of *dominance* [OR94]. Given two strategies σ and σ' of Player 1, we say that σ *dominates* σ' if σ is always at least as good as σ', and better than σ' in at least one case. Dominance induces a strict partial order on strategies, whose maximal elements are called *admissible* strategies. In Section 3, we compare the above notions, and we prove that a strategy is admissible if and only if it is simultaneously strongly winning and cooperatively strongly winning (i.e., strongly winning with the help of Player 2).

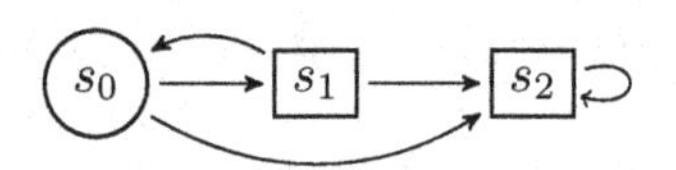

Fig. 1. A game where victory cannot be enforced

To the best of our knowledge, the only paper dealing with admissibility in a context similar to ours is [Ber07], which provides existence results for general

[1] Player-1 states are represented by circles and Player-2 states by squares.

multi-player games of infinite duration, and does not address the issue of the memory requirements of the strategies.

Memory. A useful measure for the complexity of a strategy consists in evaluating how much memory it needs regarding the history of the game. In the simplest case, a strategy requires no memory at all: its decisions are based solely on the current state of the game. Such strategies are called *positional* or *memoryless* [GZ05]. In other cases, a strategy may require the amount of memory that can be provided by a finite automaton (*finite memory*), or more [DJW97].

The memory measure of a strategy is particularly important for the applications that we target in this paper. Since we are interested in actually implementing strategies in hardware or software, the simplest the strategy, the easiest and most efficient it is to implement.

We devote Section 4 to studying the memory requirements for various types of "good" strategies. In particular, we prove that all goals that have positional winning strategies also have positional *strongly* winning strategies. On the other hand, admissible strategies may require an unbounded amount of memory on some of those goals. We then provide necessary and sufficient conditions for a goal to have positional admissible strategies, building on the results of [GZ05].

We also prove that for *prefix-independent* goals, all positional winning strategies are automatically strongly winning. Additionally, prefix-independent goals admitting positional winning strategies also admit positional admissible strategies, as we show by presenting a simple algorithm which computes positional admissible strategies for these goals.

2 Definitions

We treat games that are played by two players on a finite graph, for an infinite number of turns. The aim of the first player is to obtain an infinite trace that belongs to a fixed set of accepting traces. In the literature, such games are termed *two-player*, *turn-based*, and *qualitative*.The following definitions make this framework formal.

A *game* is a tuple $G = (S_1, S_2, \delta, C, F)$ such that: S_1 and S_2 are disjoint finite sets of states; let $S = S_1 \cup S_2$, we have that $\delta \subseteq S \times S$ is the transition relation and $C : S \to \mathbb{N}$ is the coloring function, where $\mathbb{N}$ denotes the set of natural numbers including zero. Finally, $F \subseteq \mathbb{N}^\omega$ is the *goal*, where $\mathbb{N}^\omega$ denotes the set of infinite sequences of natural numbers. We denote by $\neg F$ the complement of F, i.e., $\mathbb{N}^\omega \setminus F$. We assume that games are non-blocking, i.e. each state has at least one successor in δ.

A (finite or infinite) path in G is a (finite or infinite) path in the directed graph (S, δ). With an abuse of notation, we extend the coloring function from states to paths, with the obvious meaning. If a finite path ρ is a prefix of a finite or infinite path ρ', we also say that ρ' *extends* ρ. We denote by $\mathit{first}(\rho)$ the first state of a path ρ and by $\mathit{last}(\rho)$ the last state of a finite path ρ.

Strategies. A *strategy* in G is a function $\sigma : S^* \to S$ such that for all $\rho \in S^*$, $(\mathit{last}(\rho), \sigma(\rho)) \in \delta$. Our strategies are deterministic, or, in game-theoretic terms, *pure*. We denote $Stra_G$ the set of all strategies in G. We do not distinguish *a priori* between strategies of Player 1 and Player 2. However, for sake of clarity, we write σ for a strategy that should intuitively be interpreted as belonging to Player 1, and τ for the (rare) occasions when a strategy of Player 2 is needed.

Consider two strategies σ and τ, and a finite path ρ, and let $n = |\rho|$. We denote by $\mathrm{Outc}_G(\rho, \sigma, \tau)$ the unique infinite path $s_0 s_1 \dots$ such that *(i)* $s_0 s_1 \dots s_{n-1} = \rho$, and *(ii)* for all $i \geq n$, $s_i = \sigma(s_0 \dots s_{i-1})$ if $s_{i-1} \in S_1$ and $s_i = \tau(s_0 \dots s_{i-1})$ otherwise. We set $\mathrm{Outc}_G(\rho, \sigma) = \bigcup_{\tau \in Stra_G} \mathrm{Outc}_G(\rho, \sigma, \tau)$ and $\mathrm{Outc}_G(\rho) = \bigcup_{\sigma \in Stra_G} \mathrm{Outc}_G(\rho, \sigma)$. For all $s \in S$ and $\rho \in \mathrm{Outc}_G(s, \sigma)$, we say that ρ is *consistent* with σ. Similarly, we say that $\mathrm{Outc}_G(s, \sigma, \tau)$ is consistent with σ and τ. We extend the definition of consistent to finite paths in the obvious way.

A strategy σ is *positional* (or *memoryless*) if $\sigma(\rho)$ only depends on the last state of ρ. Formally, for all $\rho, \rho' \in S^*$, if $last(\rho) = last(\rho')$ then $\sigma(\rho) = \sigma(\rho')$.

Dominance. Given two strategies σ and τ, and a state s, we set $val_G(s, \sigma, \tau) = 1$ if $C(\mathrm{Outc}_G(s, \sigma, \tau)) \in F$, and $val_G(s, \sigma, \tau) = 0$ otherwise. Given two strategies σ and σ', we say that σ' *dominates* σ if: *(i)* for all $\tau \in Stra_G$ and all $s \in S$, $val_G(s, \sigma', \tau) \geq val_G(s, \sigma, \tau)$, and *(ii)* there exists $\tau \in Stra_G$ and $s \in S$ such that $val_G(s, \sigma', \tau) > val_G(s, \sigma, \tau)$.

It is easy to check that dominance is an irreflexive, asymmetric and transitive relation. Hence, it is a *strict partial order* on strategies.

Good strategies. In the following, unless stated otherwise, we consider a fixed game $G = (S_1, S_2, \delta, C, F)$ and we omit the G subscript.

For an infinite sequence $x \in \mathbb{N}^\omega$, we say that x is *accepting* if $x \in F$ and *rejecting* otherwise. We reserve the term "winning" to strategies and finite paths, as explained in the following. Let ρ be a finite path in G, we say that a strategy σ is *winning from* ρ if, for all $\rho' \in \mathrm{Outc}(\rho, \sigma)$, we have that $C(\rho')$ is accepting. We say that ρ is *winning* if there is a strategy σ which is winning from ρ. The above definition extends to states, by considering them as length-1 paths. A state that is not winning is called *losing*.

Further, a strategy σ is *cooperatively winning from* ρ if there exists a strategy τ such that $C(\mathrm{Outc}(\rho, \sigma, \tau))$ is accepting. We say that ρ is *cooperatively winning* if there is a strategy σ which is cooperatively winning from ρ. Intuitively, a path is cooperatively winning if the two players together can extend that path into an infinite path that satisfies the goal. Again, the above definitions extend to states, by considering them as length-1 paths.

We can now present the following set of winning criteria. Each of them is a possible definition of what a "good" strategy is.

- A strategy is *winning* if it is winning from all winning states. This criterion intuitively demands that strategies enforce victory whenever the initial state allows it.
- A strategy is *strongly winning* if it is winning from all winning paths that are consistent with it.
- A strategy is *subgame perfect* if it is winning from all winning paths. This criterion states that a strategy should enforce victory whenever the current history of the game allows it.

- A strategy is *cooperatively winning* (in short, *c-winning*) if it is cooperatively winning from all cooperatively winning states. This criterion essentially asks a strategy to be winning with the help of Player 2.
- A strategy is *cooperatively strongly winning* (in short, *cs-winning*) if it is cooperatively winning from all cooperatively winning paths that are consistent with it.
- A strategy is *cooperatively subgame perfect* (in short, *c-perfect*) if it is cooperatively winning from all cooperatively winning paths.
- A strategy is *admissible* if there is no strategy that dominates it. This criterion favors strategies that are maximal w.r.t. the partial order defined by dominance.

The notions of winning and cooperatively winning strategies are customary to computer scientists [Tho95, AHK97]. The notion of subgame perfect strategy comes from classical game theory [OR94]. The introduction of the notion of strongly winning strategy is motivated by the fact that in the target applications game histories that are inconsistent with the strategy of Player 1 cannot occur. Being strongly winning is strictly weaker than being subgame perfect. In particular, there are games for which there is a positional strongly winning strategy, but no positional subgame perfect strategy. The term "strongly winning" seems appropriate since this notion is a natural strengthening of the notion of winning strategy.

We say that a goal F is *positional* if, for all games G with goal F, there is a positional winning strategy in G.

3 Comparing Winning Criteria

In this section, we compare the winning criteria presented in Section 2. Figure 2 summarizes the relationships between the winning criteria under consideration. We start by stating the following basic properties.

Lemma 1. *The following properties hold:*

1. *all strongly winning strategies are winning, but not vice versa;*
2. *all subgame perfect strategies are strongly winning, but not vice versa;*
3. *all cs-winning strategies are c-winning, but not vice versa;*
4. *all c-perfect strategies are cs-winning, but not vice versa;*
5. *all games have a winning (respectively, strongly winning, subgame perfect, c-winning, cs-winning, c-perfect, admissible) strategy.*

Proof. The containments stated in (1) and (2) are obvious by definition. The fact that those containments are strict is proved by simple examples. Similarly for statements (3) and (4).

Regarding statement (5), the existence of a winning (respectively, strongly winning, subgame perfect, c-winning, cs-winning, c-perfect) strategy is obvious by definition. The existence of an admissible strategy can be derived from Theorem 11 from [Ber07]. □

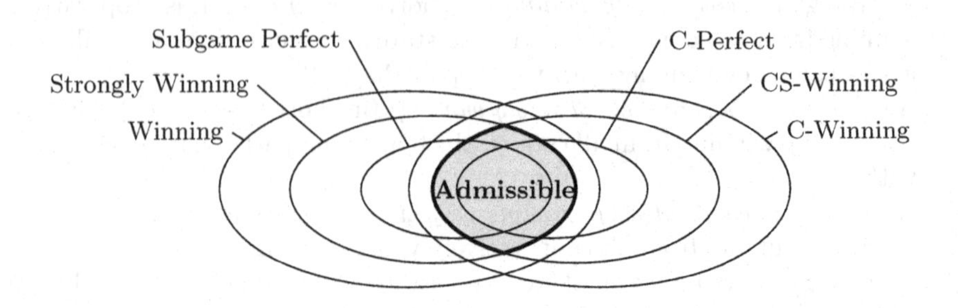

Fig. 2. Comparing winning criteria

The following result provides a characterization of admissibility in terms of the simpler criteria of strongly winning and cooperatively strongly winning. Such characterization will be useful to derive further properties of admissible strategies. The result can be proved as a consequence of Lemma 9 from [Ber07].

Theorem 1. *A strategy is admissible if and only if it is strongly winning and cooperatively strongly winning.*

4 Memory

In this section, we study the amount of memory required by "good" strategies for achieving different kinds of goals. We are particularly interested in identifying those goals which admit positional good strategies, because positional strategies are the easiest to implement.

4.1 Positional Winning Strategies

In this section, we recall the main result of [GZ05], which provides necessary and sufficient conditions for a goal to be positional w.r.t. both players. Such characterization provides the basis for our characterization of the goals admitting positional admissible strategies, in Section 4.3.

We start with some additional notation. For a goal $F \subseteq \mathbb{N}^\omega$, we define its *preference relation* $\preceq_F$ and its strict version $\prec_F$ as follows: for two sequences $x, y \in \mathbb{N}^\omega$,

$$x \prec_F y \overset{\text{def}}{\iff} x \notin F \text{ and } y \in F \qquad\qquad x \preceq_F y \overset{\text{def}}{\iff} \text{if } x \in F \text{ then } y \in F.$$

Next, define the following relations between two languages $X, Y \subseteq \mathbb{N}^\omega$.

$$X \sqsubset_F^b Y \overset{\text{def}}{\iff} \exists y \in Y \,.\, \forall x \in X \,.\, x \prec_F y \quad X \sqsubseteq_F^b Y \overset{\text{def}}{\iff} \forall x \in X \,.\, \exists y \in Y \,.\, x \preceq_F y$$

$$X \sqsubset_F^w Y \overset{\text{def}}{\iff} \exists x \in X \,.\, \forall y \in Y \,.\, x \prec_F y \quad X \sqsubseteq_F^w Y \overset{\text{def}}{\iff} \forall y \in Y \,.\, \exists x \in X \,.\, x \preceq_F y.$$

In the above definitions, the superscripts b and w stand for "best" and "worst", respectively. For instance, $X \sqsubset_F^b Y$ intuitively means that the *best* sequence in Y is strictly better than the best sequence in X. In other words, there is an accepting sequence in Y, while all sequences in X are rejecting. Similarly, $X \sqsubseteq_F^b Y$ means that the best sequence in Y is at least as good as the best sequence in X, i.e., if there is an accepting sequence in X, there is an accepting sequence in Y as well. We omit the subscript "F" when the goal is clear from the context.

A language $M \subseteq \mathbb{N}^*$ is *recognizable* if it is accepted by a finite automaton. We denote by Rec the set of all recognizable languages in $\mathbb{N}^*$. For a language $M \subseteq \mathbb{N}^*$, we denote by $[M]$ the language of all infinite words $x \in \mathbb{N}^\omega$ such that all prefixes of x are prefixes of some word in M.

A goal is *monotone* if for all recognizable sets $M, N \in \text{Rec}$,

$$\exists x \in \mathbb{N}^* . [xM] \sqsubset^b [xN] \implies \forall x \in \mathbb{N}^* . [xM] \sqsubseteq^b [xN].$$

A goal is *selective* iff, for all $x \in \mathbb{N}^*$ and all recognizable sets $M, N, K \in \text{Rec}$,

$$[x(M \cup N)^* K] \sqsubseteq^b [xM^*] \cup [xN^*] \cup [xK].$$

The following result is an adaptation to our setting of Theorem 2 from [GZ05].

Theorem 2 ([GZ05]). *Given a goal F, both players have a positional winning strategy for all games with goal F, if and only if both F and $\neg F$ are monotone and selective.*

4.2 Positional Strongly Winning and Subgame Pefect Strategies

For a game $G = (S_1, S_2, \delta, C, F)$ and a path $\rho = s_0 \dots s_n$ in G, define *detach*(G, ρ) as the game obtained from G by adding a copy of the path ρ to it as a chain of new states ending in the original state s_n. Formally, *detach*$(G, \rho) = (S_1, S_2', \delta', C', F)$, where $S_2' = S_2 \cup \{s_0', s_1', \dots, s_{n-1}'\}$ and $s_0', s_1', \dots, s_{n-1}'$ are new distinct states not belonging to S_2 or to S_1. Then, $(s, t) \in \delta'$ iff either *(i)* $(s, t) \in \delta$, or *(ii)* $s = s_i'$ and $t = s_{i+1}'$, or *(iii)* $s = s_{n-1}'$ and $t = s_n$. Finally, the color labeling is defined by:

$$C'(s) = \begin{cases} C(s_i) & \text{if } s = s_i' \text{ for some } i \in \{0, \dots, n-1\}, \\ C(s) & \text{otherwise.} \end{cases}$$

The key idea of the detach operation consists in converting a path that may need the collaboration of Player 2 to occur, into a path which must occur if the game starts in a certain (new) state. This operation allows us to prove the following result.

Theorem 3. *For a goal F, the following are equivalent:*

1. *F is positional;*
2. *F admits positional strongly winning strategies;*
3. *F admits positional subgame perfect strategies.*

Proof. (Sketch) Since ($3 \implies 2$) and ($2 \implies 1$) are obvious by definition, it remains to prove that $1 \implies 3$. Hence, assume that the goal is positional. Let G be a game and let W be the set of winning paths of G. W may be infinite but it is certainly countable. Consider any ordering of W into $\rho_0, \rho_1, \ldots$. Consider the sequence of games $(G_i)_{i \geq 0}$ defined by $G_0 = G$ and $G_{i+1} = \mathit{detach}(G_i, \rho_i)$. Additionally, consider the sequence of strategies $(\sigma_i)_{i \geq 0}$ defined by: σ_0 is any positional winning strategy in G_0, and

$$\sigma_{i+1} = \begin{cases} \sigma_i & \text{if } \sigma_i \text{ is winning in } G_{i+1}, \\ \text{any positional winning strategy in } G_{i+1} & \text{otherwise.} \end{cases}$$

Due to space constraints, we omit the proof that the sequence $(\sigma_i)_{i \geq 0}$ converges to a subgame-perfect strategy σ^* within a finite number of steps. □

4.3 Positional Admissible Strategies

Since all admissible strategies are winning, admissible strategies require at least as much memory as winning strategies. The following example shows that there are positional goals for which all admissible strategies require an unbounded amount of memory.

Example 1. Consider the goal informally described as follows: an infinite sequence is accepting if and only if either it contains infinitely many 2's, or it contains at least as many 2's as 1's. Such goal is positional, as it is monotone and selective.

Consider the game in Figure 3, with said goal. In the figure, the color of each state appears next to it or above it. The only choice for Player 1 occurs in s_2, where he can choose between s_3 and s_5. One can easily check that all states are losing. However, for all $n > 0$, the initial prefix $s_0^{n+1} s_1^n$ is winning and requires Player 1 to choose s_5 after s_2. On the other hand, the initial prefix $s_0^{n+2} s_1^n$ is cooperatively winning, and requires Player 1 to choose s_3 after s_2.

In conclusion, any admissible strategy must be able to distinguish $s_0^{n+1} s_1^n$ from $s_0^{n+2} s_1^n$, which requires an unbounded amount of memory. □

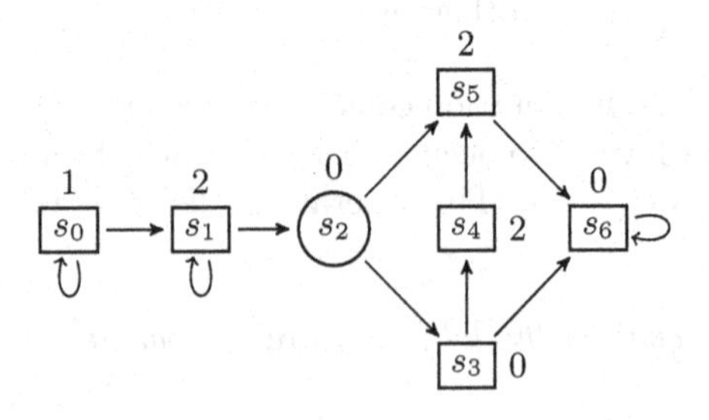

Fig. 3. Admissible strategies may require unbounded memory for a positional goal

As shown in the following, in order to obtain positional admissible strategies, we need the goal to satisfy the following additional property.

Definition 1. *A goal is* strongly monotone *if for all recognizable sets $M, N \in$ Rec,*

$$\exists x \in \mathbb{N}^* . [xM] \sqsubset^b [xN] \implies \forall x \in \mathbb{N}^* . [xM] \sqsubseteq^b [xN] \wedge [xM] \sqsubseteq^w [xN].$$

To gain some intuition, consider the goal of Example 1. We show that it is monotone, but not strongly so. Let x, M, and N be such that $[xM] \sqsubset^b [xN]$. This means that all sequences in $[xM]$ are rejecting, i.e., have a number of 2's smaller than the number of 1's. On the other hand, at least one sequence in $[xN]$ is accepting. Thus, there is a word $z \in [N]$ that has more excess 2's (possibly infinitely many) than any word in $[M]$.

Now, consider any $y \in \mathbb{N}^*$. Assume that $y \cdot y' \in [yM]$ is accepting. Then, $y \cdot z$ must also be accepting. This shows that $[yM] \sqsubseteq^b [yN]$ and the goal is monotone.

Assume instead that there is a rejecting sequence in $[yN]$. This does not imply that there is a rejecting sequence in $[yM]$, as would be required by the definition of strong monotonicity. A concrete counter-example is provided by Example 1. Let $x = 1 \cdot 1 \cdot 1 \cdot 2$, $M = 2 \cdot 0^*$, $N = 0^* + (2 \cdot 2 \cdot 0^*)$. Notice that $x = C(s_0 s_0 s_0 s_1)$, $[M] = C(\text{Outc}(s_5))$, and $[N] = C(\text{Outc}(s_3))$. Since $[xM] = 1{\cdot}1{\cdot}1{\cdot}2{\cdot}2{\cdot}0^\omega$ is rejecting, while $1 \cdot 1 \cdot 1 \cdot 2 \cdot 2 \cdot 2 \cdot 0^\omega \in [xN]$ is accepting, we have that $[xM] \sqsubset^b [xN]$. However, with $y = 1 \cdot 1 \cdot 2$, we have that all paths in $[yM]$ are accepting (there is only one) and there is one rejecting path in $[yN]$ (namely, $1 \cdot 1 \cdot 2 \cdot 0^\omega$). So, $[yM] \not\sqsubseteq^w [yN]$ and the goal is not strongly monotone.

The following is the main result of this section, providing a characterization of the goals admitting positional admissible strategies for both players.

Theorem 4. *Given a goal F, both players have a positional admissible strategy for all games with goal F if and only if both F and $\neg F$ are strongly monotone and selective.*

Due to space constraints, we provide the proof for one direction of Theorem 4, and a proof sketch for the other one.

Lemma 2. *Given a goal F, if Player 1 has a positional admissible strategy for all games with goal F, then F is strongly monotone and selective.*

Proof. By Lemma 5 of [GZ05], F is monotone and selective. It remains to prove that it is strongly monotone. Let $x \in \mathbb{N}^*$, $M, N \in$ Rec such that

$$[xM] \sqsubset^b [xN]. \tag{1}$$

In other words, all paths in $[xM]$ are rejecting and at least one path in $[xN]$ is accepting. Let $y \in \mathbb{N}^*$, we will prove that $[yM] \sqsubseteq^b [yN]$ and $[yM] \sqsubseteq^w [yN]$.

Assume w.l.o.g. that M, N are not empty. Let $\mathcal{A}_x, \mathcal{A}_y, \mathcal{A}_M, \mathcal{A}_N$ be the finite automata recognizing the languages $\{x\}, \{y\}, M, N$, respectively. We can assume w.l.o.g. that these automata are deterministic, and in particular that they have a unique initial state. Let s_x, s_y, s_M, s_N be their respective initial states. Moreover, we can assume that $\mathcal{A}_x$ and $\mathcal{A}_y$ are linearly ordered chains of states, and that there are no edges in $\mathcal{A}_M$ (resp., $\mathcal{A}_N$) that go back to s_M (resp., s_N).

We build a game G as follows. We assign all states of $\mathcal{A}_x, \mathcal{A}_y, \mathcal{A}_M$ to Player 1, and all states of $\mathcal{A}_N$ to Player 2. We remove the two final states of $\mathcal{A}_x$ and $\mathcal{A}_y$ and the two initial states of $\mathcal{A}_M$ and $\mathcal{A}_N$. We connect the four automata as follows. Let t be a brand new state, we connect the penultimate state of $\mathcal{A}_x$ and the penultimate state of $\mathcal{A}_y$ to t. Then, we connect t to all the successors of s_M and s_N. We assign state t to Player 1. A technical difficulty is due to the fact that games are required to be non-blocking, while automata are not. This issue can be overcome using the notion of *essential state*, as proposed in the proof of Lemma 5 of [GZ05].

Let σ^* be a positional admissible strategy for Player 1 in G. By (1), if the game starts in s_x, once in t any cs-winning strategy must choose N. By Theorem 1, we have $\sigma^*(t) = s_N$. Assume that $[yM]$ contains an accepting sequence. Since all states in $\mathcal{A}_M$ have been assigned to Player 1, s_y is a winning state in G. Since σ^* is a winning strategy and $C(\text{Outc}(s_y, \sigma^*)) \subseteq [yN]$, there must be an accepting sequence in $[yN]$ too. Therefore, $[yM] \sqsubseteq^b [yN]$.

Finally, assume that $[yN]$ contains a rejecting sequence. Then, σ^* is not winning from s_y. Since σ^* is a winning strategy, s_y is not a winning state. If we assume that $[yM]$ contains no rejecting sequences, we obtain that $[yM]$, being non-empty, contains at least one accepting sequence. As before, this means that s_y is a winning state, which is a contradiction. Therefore, $[yM] \sqsubseteq^w [yN]$, which concludes the proof. □

Lemma 3. *Given a goal F, if both F and $\neg F$ are strongly monotone and selective, then both players have a positional admissible strategy for all games with goal F.*

Proof. (Sketch) Let $G = (S_1, S_2, \delta, C, F)$, with both F and $\neg F$ strongly monotone and selective. By Theorems 3 and 2, let σ_1 be a positional subgame-perfect strategy for Player 1 in G. Let SW be the set of all states s such that there is a finite path π in G such that: *(i)* $last(\pi) = s$, *(ii)* π is winning, and *(iii)* π can be extended into an infinite rejecting path (i.e., $C(\text{Outc}(\pi)) \setminus F \neq \emptyset$). Let G_1 be the game obtained from G by removing the edges which start in $SW \cap S_1$ and do not belong to σ_1. Let σ^* be a positional cs-winning strategy in G_1. It can be proved that σ^* is subgame-perfect and cs-winning. It follows by Theorem 1 that σ^* is admissible. □

5 Prefix-Independent Goals

A goal F is *prefix-independent* iff for all $x \in \mathbb{N}^\omega$ and all $c \in \mathbb{N}$, $c x \in F$ if and only if $x \in F$. Examples of common prefix-independent goals include Büchi, co-Büchi and parity goals.

Theorem 5. *If a goal F is prefix-independent, then, for all games with goal F, all positional winning strategies are strongly winning, and all positional c-winning strategies are cs-winning.*

The positionality assumption is necessary in the above result. For a prefix-independent goal, it is easy to devise winning strategies that are not positional and not strongly winning. On the other hand, being prefix-independent is not necessary for ensuring that all positional winning strategies are strongly winning. For instance, safety and reachability goals are not prefix-independent, but they ensure said property.

Computing positional admissible strategies. Suppose that we are given a game G with a positional prefix-independent goal F, and that we have an algorithm for computing the set of winning states and a positional winning strategy for all games with goal F. Consider the following procedure, inspired by the proof of Lemma 3.

Procedure 1

1. Compute the set of winning states Win and a positional winning strategy σ for G.
2. Remove from G the edges of Player 1 which start in Win and do not belong to σ.
3. In the resulting game, compute and return a positional cooperatively winning strategy.

The following theorem shows that the strategy returned by Procedure 1 is admissible. As far as the complexity of the procedure is concerned, assuming the usual graph-like adjacency-list representation for games, we obtain the same asymptotical complexity as finding a positional winning strategy for F. In particular, step 3 can easily be performed by attributing all states to Player 1 and then running the algorithm for a positional winning strategy.

Theorem 6. *Assume that F is a positional and prefix-independent goal, and that there is an algorithm for computing the set of winning states and a positional winning strategy for all games G with goal F in time $\mathcal{O}(f(|G|))$. Then, one can compute a positional admissible strategy for all games G with goal F using Procedure 1 in time $\mathcal{O}(f(|G|))$.*

This result allows us to easily compute admissible strategies for several common goals such as Büchi, co-Büchi, and parity. However, prefix-independence is not necessary for Procedure 1 to work. For instance, it is easy to prove that the procedure also returns an admissible strategy for reachability and safety goals.

6 Conclusions

We advanced the claim that computer science applications of game theory, especially in the domain of automatic verification and synthesis of controllers, may benefit from considering various winning criteria, such as admissibility, in addition to the classical one. Given the importance of (the lack of) memory for those applications, and considering that admissible strategies may require unboundedly more memory than plain winning strategies (Example 1), with Theorem 4 we characterize the goals that admit positional admissible strategies.

Further investigation and experimentation is needed to verify our claim in a concrete applicative setting. Moreover, it remains to determine how to compute admissible strategies for goals that are not prefix-independent.

References

[AHK97] Alur, R., Henzinger, T.A., Kupferman, O.: Alternating-time temporal logic. In: Proc. 38th IEEE Symp. Found. of Comp. Sci., pp. 100–109. IEEE Computer Society Press, Los Alamitos (1997)

[ALW89] Abadi, M., Lamport, L., Wolper, P.: Realizable and unrealizable concurrent program specifications. In: Ronchi Della Rocca, S., Ausiello, G., Dezani-Ciancaglini, M. (eds.) ICALP 1989. LNCS, vol. 372, pp. 1–17. Springer, Heidelberg (1989)

[BBF07] Baselice, S., Bonatti, P.A., Faella, M.: On interoperable trust negotiation strategies. In: POLICY 2007: 8th IEEE International Workshop on Policies for Distributed Systems and Networks. IEEE Computer Society, Los Alamitos (2007)

[Ber07] Berwanger, D.: Admissibility in infinite games. In: Thomas, W., Weil, P. (eds.) STACS 2007. LNCS, vol. 4393, pp. 188–199. Springer, Heidelberg (2007)

[dAFMR05] de Alfaro, L., Faella, M., Majumdar, R., Raman, V.: Code aware resource management. In: EMSOFT 2005: 5th Intl. ACM Conference on Embedded Software, pp. 191–202. ACM Press, New York (2005)

[dAH01] de Alfaro, L., Henzinger, T.A.: Interface theories for component-based design. In: Henzinger, T.A., Kirsch, C.M. (eds.) EMSOFT 2001. LNCS, vol. 2211, pp. 148–165. Springer, Heidelberg (2001)

[DJW97] Dziembowski, S., Jurdziński, M., Walukiewicz, I.: How much memory is needed to win infinite games? In: LICS. IEEE Computer Society, Los Alamitos (1997)

[EJ91] Emerson, E.A., Jutla, C.S.: Tree automata, mu-calculus and determinacy (extended abstract). In: Proc. 32nd IEEE Symp. Found. of Comp. Sci., pp. 368–377. IEEE Computer Society Press, Los Alamitos (1991)

[GZ05] Gimbert, H., Zielonka, W.: Games where you can play optimally without any memory. In: Abadi, M., de Alfaro, L. (eds.) CONCUR 2005. LNCS, vol. 3653, pp. 428–442. Springer, Heidelberg (2005)

[KR01] Kremer, S., Raskin, J.-F.: A game-based verification of non-repudiation and fair exchange protocols. In: Larsen, K.G., Nielsen, M. (eds.) CONCUR 2001. LNCS, vol. 2154, pp. 551–565. Springer, Heidelberg (2001)

[OR94] Osborne, M.J., Rubinstein, A.: A Course in Game Theory. MIT Press, Cambridge (1994)

[PR89] Pnueli, A., Rosner, R.: On the synthesis of a reactive module. In: Proceedings of the 16th Annual Symposium on Principles of Programming Languages, pp. 179–190. ACM Press, New York (1989)

[Tho95] Thomas, W.: On the synthesis of strategies in infinite games. In: Mayr, E.W., Puech, C. (eds.) STACS 1995. LNCS, vol. 900, pp. 1–13. Springer, Heidelberg (1995)

[Zie98] Zielonka, W.: Infinite games on finitely coloured graphs with applications to automata on infinite trees. Theoretical Computer Science 200, 135–183 (1998)

A Complexity Dichotomy for Finding Disjoint Solutions of Vertex Deletion Problems

Michael R. Fellows[1,*], Jiong Guo[2,**], Hannes Moser[2,***], and Rolf Niedermeier[2]

[1] PC Research Unit, Office of DVC (Research), University of Newcastle, Callaghan, NSW 2308, Australia
michael.fellows@newcastle.edu.au

[2] Institut für Informatik, Friedrich-Schiller-Universität Jena, Ernst-Abbe-Platz 2, D-07743 Jena, Germany
{jiong.guo,hannes.moser,rolf.niedermeier}@uni-jena.de

Abstract. We investigate the computational complexity of a general "compression task" centrally occurring in the recently developed technique of iterative compression for exactly solving NP-hard minimization problems. The core issue (particularly but not only motivated by iterative compression) is to determine the computational complexity of, given an already inclusion-minimal solution for an underlying (typically NP-hard) vertex deletion problem in graphs, to find a better *disjoint* solution. The complexity of this task is so far lacking a systematic study. We consider a large class of vertex deletion problems on undirected graphs and show that, except for few cases which are polynomial-time solvable, the others are NP-complete. This class includes problems such as VERTEX COVER (here the corresponding compression task is decidable in polynomial time) or UNDIRECTED FEEDBACK VERTEX SET (here the corresponding compression task is NP-complete).

1 Introduction

With the introduction of the iterative compression by Reed et al. [17] in 2004, parameterized complexity analysis has gained a new tool for showing fixed-parameter tractability results for NP-hard minimization problems (cf. [9, 15]). For instance, in 2008, applying iterative compression has led to major breakthroughs concerning the classification of the parameterized complexity of two important problems. First, Chen et al. [4] showed that the NP-complete DIRECTED FEEDBACK VERTEX SET problem is fixed-parameter tractable. Second, Razgon and O'Sullivan [16] proved that the NP-complete ALMOST 2-SAT

* Supported by the Australian Research Council. Work done while staying in Jena as a recipient of the Humboldt Research Award of the Alexander von Humboldt Foundation, Bonn, Germany.

** Partially supported by the DFG, PALG, NI 369/8.

*** Supported by the DFG, project AREG, NI 369/9.

R. Královič and D. Niwiński (Eds.): MFCS 2009, LNCS 5734, pp. 319–330, 2009.

problem is fixed-parameter tractable. Refer to the recent survey [9] for more on iterative compression applied to exactly solving NP-hard minimization problems.

The central idea behind iterative compression is to employ a *compression routine*. This is an algorithm that, given a problem instance and a corresponding solution, either calculates a smaller solution or proves that the given solution is of minimum size. Using a compression routine, one finds an optimal solution to a problem by inductively building up the problem instance and iteratively compressing intermediate solutions. Herein, the essential fact from the viewpoint of parameterized complexity is that if the task performed by the compression routine is fixed-parameter tractable, then so is the problem solved by means of iterative compression. The main strength of iterative compression is that it allows to see the problem from a different angle: The compression routine does not only have the problem instance as input, but also a solution, which carries valuable structural information. The design of a compression routine, therefore, may be simpler than showing that the original problem is fixed-parameter tractable.

While embedding the compression routine into the iteration framework is usually straightforward, finding the compression routine itself is not [9, 15]. For many vertex deletion problems, a common approach to designing a compression routine is to branch on the possible subsets of the uncompressed solution to retain in the compressed solution. This leads to the following generic problem that asks for a *disjoint* compressed solution:

Input: An instance of the underlying NP-hard problem and a solution S.[1]

Question: Is there a solution S' such that $S' \cap S = \emptyset$ and $|S'| < |S|$?

We study the complexity of COMPRESSION TASK depending on what the underlying NP-hard problem is. The computational complexity of COMPRESSION TASK, so far, remains widely unclassified. For instance, the fixed-parameter tractability results (using iterative compression) for VERTEX BIPARTIZATION [17] or UNDIRECTED FEEDBACK VERTEX SET [3, 5, 8] leave open whether the respective COMPRESSION TASK is NP-hard or polynomial-time solvable. By way of contrast, the fixed-parameter tractability result for the NP-complete CLUSTER VERTEX DELETION problem [10] is based on a polynomial-time algorithm for COMPRESSION TASK. Here, extending a framework attributed to Yannakakis [12], we contribute a complete classification of COMPRESSION TASK for a natural class of vertex deletion problems (specified by a graph property Π), including all of the above mentioned problems.

A *graph property* Π is a set of graphs; in the following, we say that a graph G *satisfies* Π if $G \in \Pi$. A graph property Π is *hereditary* if it is closed under vertex deletion, and *non-trivial* if it is satisfied by infinitely many graphs and it is not satisfied by infinitely many graphs.

The classical Π-VERTEX DELETION problem is defined as follows: for a non-trivial hereditary graph property Π testable in polynomial time, given an undirected graph G and a positive integer k, decide whether it is possible to delete

[1] Here, the solution is a set. It is conceivable that the COMPRESSION TASK can be formulated also for other types of solutions.

at most k vertices from the graph such that the resulting graph satisfies Π. For example, UNDIRECTED FEEDBACK VERTEX SET corresponds to the case that Π means "being cycle-free". Yannakakis has shown that Π-VERTEX DELETION is NP-complete for any non-trivial hereditary graph property Π in general graphs [12]. General vertex deletion problems have also been studied in terms of their parameterized complexity [2, 11].

The COMPRESSION TASK restricted to vertex deletion problems with property Π, called DISJOINT Π-VERTEX DELETION, can be formulated as follows:

Input: An undirected graph $G = (V, E)$ and a vertex subset $S \subseteq V$ such that $G[V \setminus S]$ satisfies Π and S is inclusion-minimal under this property, that is, for every proper subset $S' \subset S$ the graph $G[V \setminus S']$ does not satisfy Π.

Question: Is there a vertex subset $S' \subseteq V$ of size at most $|S|$, such that $S \cap S' = \emptyset$ and $G[V \setminus S']$ satisfies Π?

We replace the requirement $|S'| < |S|$ in the definition of COMPRESSION TASK by $|S'| \leq |S|$ without changing the computational complexity, because the corresponding hardness reductions (cf. Lemma 4) work for both cases, and the last case might be of interest if S is already optimal. Moreover, we demand that S is inclusion-minimal; any solution can be made inclusion-minimal in polynomial time if Π can be tested in polynomial time. Thus, this requirement does not change the complexity.

A graph property Π is *determined by the components* if it holds that if every connected component of the graph satisfies Π, then so does the whole graph. The central result of this work can be informally stated as follows:

Main Theorem: *Let Π be any non-trivial hereditary graph property that is determined by the components and that can be tested in polynomial time.* DISJOINT Π-VERTEX DELETION *is NP-complete unless Π is the set of all graphs whose connected components are cliques or Π is the set of all graphs whose connected components are cliques of at most s vertices, $s \geq 1$—in these cases it is polynomial-time solvable.*[2]

The main theorem applies to many natural vertex deletion problems in undirected graphs, including VERTEX COVER, BOUNDED-DEGREE DELETION, UNDIRECTED FEEDBACK VERTEX SET [3, 5, 8], VERTEX BIPARTIZATION [17], CLUSTER VERTEX DELETION [10], CHORDAL DELETION [13], and PLANAR DELETION [14]. Thus, except for VERTEX COVER and CLUSTER VERTEX DELETION, all other problems have an NP-complete COMPRESSION TASK problems.

Our original motivation for this work comes from the desire to better understand the limitations of the iterative compression technique. Beyond this, DISJOINT Π-VERTEX DELETION also seems to be a natural and interesting problem on its own: In combinatorial optimization, one often may be confronted with finding *alternative* good solutions to already found ones. In the setting of DISJOINT Π-VERTEX DELETION, this is put to the extreme in the sense that we ask for solutions that are completely unrelated, that is, disjoint. For instance,

[2] There might exist other polynomial-time solvable cases for *non-hereditary* properties.

this demand also naturally occurs in the context of finding quasicliques [1]. Due to the lack of space, some proofs are deferred to a full version of the paper.

Preliminaries. We only consider undirected graphs $G = (V, E)$ with $n := |V|$ and $m := |E|$. We write $V(G)$ and $E(G)$ to denote, respectively, the vertex and edge set of a graph G. For $v \in V$, let $N_G(v) := \{u \in V \mid \{u, v\} \in E\}$ and let $\deg_G(v) := |N_G(v)|$. For $S \subseteq V$, let $N_G(S) := \bigcup_{v \in S} N(v) \setminus S$. For $S \subseteq V$, let $G[S]$ be the subgraph of G induced by S and $G - S := G[V \setminus S]$. For $v \in V$, let $G - v := G[V \setminus \{v\}]$. For a connected graph G, a *cut-vertex* is a vertex $v \in V$ such that $G - v$ is not connected. A K_3 is a complete graph on three vertices. For $s \geq 1$, the graph $K_{1,s} = (\{u, v_1, \ldots, v_s\}, \{\{u, v_1\}, \ldots, \{u, v_s\}\})$ is a *star*. The vertex u is the *center* of the star and the vertices $v_1, \ldots, v_s$ are the *leaves* of the star.

If a graph H does not satisfy some hereditary property Π, then any supergraph of H does not satisfy Π. We call H a *forbidden subgraph* for Π. For any hereditary property Π there exists a set $\mathcal{H}$ of "minimal" forbidden induced subgraphs, that is, forbidden graphs for which every induced subgraph satisfies Π [7]. For this work, we restrict our attention to non-trivial hereditary properties that are determined by the components. For the corresponding characterization of Π by forbidden induced subgraphs, this means that the set of forbidden subgraphs only contains *connected* graphs.

By simple counting arguments, there exist DISJOINT Π-VERTEX DELETION problems that are not in NP. As [12], we add the stipulation that Π can be tested in polynomial time, hence the corresponding DISJOINT Π-VERTEX DELETION problem is in NP, and our hardness results to come thus will show that it is NP-complete.

A parameterized problem (I, k) is *fixed-parameter tractable* with respect to the parameter k if it can be solved in $f(k) \cdot \text{poly}(|I|)$ time, where I is the input instance and f is some computable function. The corresponding algorithm is called *fixed-parameter algorithm.*

2 Polynomial-Time Solvable Cases

This section covers all cases of DISJOINT Π-VERTEX DELETION that can be decided in polynomial time. The corresponding graph properties are as follows:

Definition 1. *Let Π_s, for $s \geq 1$, be the graph property that contains all graphs whose connected components are cliques of at most s vertices. Furthermore, let Π_∞ be the graph property that contains all graphs whose connected components of G are cliques (of arbitrary size).*

For instance, Π_1, Π_2, and Π_∞ are the properties "being edgeless", "being a graph of maximum degree one", and "every connected component is a clique (of arbitrary size)", respectively. The corresponding sets of forbidden induced subgraphs consist of a single edge (Π_1), a path on three vertices and a clique on three vertices (Π_2), and a path on three vertices (Π_∞). In general, the set of forbidden induced subgraphs of Π_s for $s \geq 2$ contains a path on three vertices and an $(s+1)$-vertex clique. Summarizing, for each property Π_s, $s \geq 1$, and Π_∞, the corresponding set of forbidden induced subgraphs contains a star with at most two leaves, and these are clearly the only properties whose sets of forbidden induced subgraphs contain a star with at most two leaves.

Theorem 1. DISJOINT Π-VERTEX DELETION *is decidable in polynomial time if* $\Pi = \Pi_s$*, for some* $s \geq 1$*, or if* $\Pi = \Pi_\infty$.

Concerning property Π_1, obviously, there can only exist a disjoint solution S', $S' \cap S = \emptyset$, if S forms an independent set in G. Moreover, S' must contain every endpoint of each edge that has one endpoint in S and the other endpoint in $V \setminus S$. Hence, the input is a yes-instance iff S forms an independent set and $|N_G(S)| \leq |S|$. This condition can be tested in polynomial time.

Lemma 1. DISJOINT Π_1-VERTEX DELETION *can be decided in polynomial time.*

DISJOINT Π_∞-VERTEX DELETION is equivalent to the decision version of the compression step for CLUSTER VERTEX DELETION [10].

Lemma 2 ([10]). DISJOINT Π_∞-VERTEX DELETION *can be decided in polynomial time.*

The polynomial-time decidability for the remaining properties Π_s can be proven with similar techniques as in the proof of Lemma 2.

Lemma 3. *For each* $s \geq 2$, DISJOINT Π_s-VERTEX DELETION *can be decided in polynomial time.*

3 NP-Hardness Framework and Simple Proofs

[12] showed that Π-VERTEX DELETION for any non-trivial hereditary property Π is NP-complete. Due to the similarity of Π-VERTEX DELETION to DISJOINT Π-VERTEX DELETION, in some simple cases we can adapt the framework from [12].[3] This section is mainly devoted to this framework and how it is modified to partially use it for DISJOINT Π-VERTEX DELETION.

There are cases, however, where adapting this framework fails; this happens when there is a star with at least three leaves among the family $\mathcal{H}$ of forbidden induced subgraphs, because (as we will see later) a star with at least three leaves does not permit to derive a given solution S for the graph that is constructed by the reduction of the framework. For this case, we have to devise other NP-hardness proofs (if there is a star with at most two leaves, then the problem is polynomial-time decidable). Summarizing, we have to distinguish the following three cases (recall that each graph in $\mathcal{H}$ is connected): (1) $\mathcal{H}$ does not contain a star (NP-hard, this section, Theorem 2), (2) $\mathcal{H}$ contains a star with at least three leaves (NP-hard, Section 4, Theorem 3), and (3) $\mathcal{H}$ contains a star with at most two leaves (polynomial-time decidable, Section 2, Theorem 1).

The main result of this section covers all cases that can be proven by adapting the framework by Yannakakis as described in the remainder of this section.

[3] As made explicit in Lewis and Yannakakis' paper [12], the parts of it we are referring to in our work have been contributed by Yannakakis.

Theorem 2. *Let Π be a non-trivial hereditary property that is determined by the components and let $\mathcal{H}$ be the corresponding set of all forbidden induced subgraphs. If $\mathcal{H}$ contains no star, then* Disjoint Π-Vertex Deletion *is NP-hard.*

The Framework by Yannakakis and its Limitations. In the following, we briefly describe the reduction by Yannakakis [12], which shows that any vertex deletion problem with non-trivial hereditary graph property is NP-hard. Since the hereditary graph properties considered in this paper are assumed to be determined by the components, we present a variant that is restricted to such properties, that is, the forbidden induced subgraphs shall be connected.

Preliminaries. Let $\mathcal{H}$ be the set of forbidden induced subgraphs that correspond to the non-trivial hereditary property Π that is determined by its components. An important concept for the framework is the notion of α-sequences [12].

Definition 2 (α-sequence). *For a connected graph $H \in \mathcal{H}$, if H is 1-connected, then take a cut-vertex c and sort the components of $H - c$ according to their size. If H is not 1-connected, then let c be an arbitrary vertex (in this case, $H - c$ has just one connected component). Sorting the connected components of $H - c$ with respect to their sizes gives a sequence $\alpha = (n_1, \ldots, n_i)$, where $n_1 \geq \ldots \geq n_i$. The sequence depends on the choice of c. The α-sequence of H, $\alpha(H)$, is a sequence which yields a lexicographically smallest such sequence α.*

Let $H \in \mathcal{H}$ be a graph with lexicographically smallest α-sequence among all graphs in $\mathcal{H}$. Note that every induced subgraph of H has a lexicographically smaller α-sequence than H. Since Π is satisfied by all independent sets, the connected graph H must contain at least two vertices, thus a largest component J of $H - c$ contains at least one vertex. Let d be an arbitrary vertex in J, and let H' be the graph resulting by removing all vertices in J from H, and let J' be the graph induced by $V(J) \cup \{c\}$ in H.

Reduction. The reduction by Yannakakis [12] from the NP-complete Vertex Cover[4] problem works as follows. Let G be an instance of Vertex Cover. For every vertex v in G create a copy of H' and identify c and v. Replace every edge $\{u, v\}$ in G by a copy of J', identifying c with u and d with v. Let G' be the resulting graph.

Correctness. The graph G has a size-k vertex cover if and only if G' has a size-k vertex set that obstructs all forbidden induced subgraphs $\mathcal{H}$ in G':

($\Rightarrow$) If A is a vertex cover of G, then $S' := A$ also obstructs all graphs in $\mathcal{H}$: Every connected component of $G' - S'$ is either (1) a copy of $H' - c$ or (2) a copy of H' together with several copies of J', each with either c or d deleted. In the latter case, the copy of H' and the copies of J' intersect exactly in one vertex of $V(G)$. Let C be such a connected component and let v be the

[4] Given a graph $G = (V, E)$ and $k \geq 0$, decide whether there exists a set $S \subseteq V$ of size at most k such that each edge has at least one endpoint in S.

described vertex. In case (1), $\alpha(H' - c)$ is lexicographically smaller than $\alpha(H)$ since $H' - c$ is a subgraph of H. In case (2), v is a cut-vertex and the components of $C - v$ can be divided into a copy of $H' - c$ and several copies of J with one vertex deleted. Since the latter type of components has less than $|V(J)|$ vertices, the cut-vertex v gives an α-sequence for C which is lexicographically smaller than the α-sequence of H. Thus, the connected components in $G' - S'$ have a smaller α-sequence than H, and because H is a forbidden induced subgraph with lexicographically smallest α-sequence, these connected components do not contain forbidden induced subgraphs.

($\Leftarrow$) If S' is a solution for $\mathcal{H}$-Deletion, then one can determine a vertex cover A for G: for each $w \in S'$, if w is in a copy of H' (possibly $w \in V(G)$), then add vertex c of that copy of H' to A, and if w is in a copy of J' (where $w \notin V(G)$), then add vertex c of that copy of J' to A. Obviously, $|A| \leq |S'|$. Suppose that there exists an edge $\{u, v\}$ in $G - A$. Then, S' neither contains any vertex from the two copies of H' corresponding to the vertices u and v nor from the copy of J' that replaced the edge $\{u, v\}$ in the construction of G'. Hence $G' - A$ contains a copy of H, a contradiction. Therefore, A is a vertex cover for G.

Limitations. In some cases, a very similar reduction principle can be applied for Disjoint Π-Vertex Deletion. We simply have to show that there exists an $\mathcal{H}$-obstruction set S in G' with the only restriction that S does not contain any vertex from $V(G)$. Then, in principle, we can use the same arguments as above. However, for some cases this approach fails; for instance, if J' is a clique and some graph of $\mathcal{H}$ is contained in G, then this forbidden induced subgraph, which also exists in G', can only be obstructed by vertices in $V(G)$. For example, this happens when Π is the property "being cycle-free" (Feedback Vertex Set): $\mathcal{H}$ contains all cycles, and the graph H with the smallest α-sequence is K_3. One can deal with this situation by reducing from K_3-free graphs, and using the graph with the smallest α-sequence among all K_3-free graphs in H, as shown in the proof of Lemma 5. The same type of problem, however, also occurs if H is a star. In this case, each connected component of $H - c$ is an isolated vertex. Thus, the vertex d has to be one of these vertices, and G and therefore G' might contain a forbidden induced subgraph with lexicographically higher α-sequence than H. This induced subgraph cannot be obstructed by a set S that is not allowed to contain any vertex from $V(G)$. In this case, the framework by Yannakakis cannot be used and we have to devise other reduction techniques (Section 4).

New Proofs Based on the Reduction Framework by Yannakakis. Recall that we assume here that the set of forbidden induced subgraph corresponding to Π contains no star. We have to distinguish between the cases that (1) all forbidden induced subgraphs contain a K_3 (see Lemma 4), and that (2) not all forbidden induced subgraph contain a K_3 (see Lemma 5).

Lemma 4. *If the set $\mathcal{H}$ of forbidden induced subgraphs corresponding to Π contains only graphs that contain a K_3, then* Disjoint Π-Vertex Deletion *is NP-hard.*

Proof. The proof is by reduction from the NP-complete VERTEX COVER on K_3-free graphs [6]. Let (G, k) be an instance of VERTEX COVER, where G is K_3-free. First, construct a graph G' using the reduction scheme by Yannakakis. Greedily compute a minimal $\mathcal{H}$-obstruction set S_1 for G' such that $S_1 \cap V(G) = \emptyset$. Such a set S_1 always exists, since G is K_3-free and, therefore, does not contain any forbidden induced subgraph.

It remains to take care of the size of the new solution S'; recall that DISJOINT Π-VERTEX DELETION asks for a solution S' such that $|S'| \leq |S|$. First, suppose that $k \leq |S_1|$. Informally speaking, we have to force that only k vertices out of the $|S_1|$ available vertices can be used in G' to obstruct all forbidden induced subgraphs. Let H, c, J, J', and d be defined as in the reduction scheme. We add a padding gadget C constructed as follows to G'. Add a new vertex w and $|S_1|-k+1$ copies of H to G', identify the vertex d of each newly added copy of H with w, and let $S := S_1 \cup \{w\}$. The gadget C is obviously connected and w is a cut-vertex in C. The vertex w obstructs all forbidden induced subgraphs in C, because deleting w (and, thus, d) from each copy of H in C leaves a graph with lexicographically smaller α-sequence (witnessed by c in each copy of H). Hence, S is a minimal $\mathcal{H}$-obstruction set for G'.

An $\mathcal{H}$-obstruction set S' for G' with $S' \cap S = \emptyset$ must contain at least one vertex in each copy of H in C, thus S' must contain at least $|S_1| - k + 1$ vertices of C; putting into S' the vertex c of each copy of H in C obstructs every forbidden induced subgraph in H: every connected component of $C - S'$ either is a copy of $H - c$ or consists of $|S_1| - k + 1$ copies of J that pairwise overlap in vertex w. In the latter case, w is a cut-vertex witnessing that each remaining connected component has size smaller than J, yielding a lexicographically smaller α-sequence. This shows that S', in order to obstruct all forbidden induced subgraphs in C, contains at least $|S_1|-k+1$ vertices. Since $S = S_1 \cup \{w\}$, there remain at most $|S| - |S_1| + k - 1 = k$ vertices to obstruct all forbidden induced subgraphs in $G' - V(C)$.

If $|S_1| < k$, then construct C in the same manner with $k - |S_1| + 1$ copies of H and let S be the union of S_1 and the vertex c of each copy of H. Then, the new solution S' can obstruct all forbidden induced subgraphs in C with the vertex w, and there are $k - |S_1| + 1 + |S_1| - 1 = k$ vertices left to obstruct all forbidden induced subgraphs in $G' - V(C)$.

By these arguments and the reduction scheme, G has a size-k vertex cover if and only if G' has a $\mathcal{H}$-obstruction set S' with $S' \cap S = \emptyset$ and $|S'| \leq |S|$. □

In the following, assume that not all forbidden induced subgraphs contain a K_3, let $\mathcal{H}' \subseteq \mathcal{H}$ be the set of all forbidden induced subgraphs that do not contain a K_3, and let H be a forbidden induced subgraph with lexicographically smallest α-sequence among all graphs in $\mathcal{H}'$.

Lemma 5. *If the set $\mathcal{H}$ of forbidden induced subgraphs corresponding to Π contains no stars, but other graphs that do not contain a K_3, then* DISJOINT Π-VERTEX DELETION *is NP-hard.*

4 Refined Reduction Strategies

Here, we present NP-hardness proofs if H is a star with at least three leaves. The main result of this section is as follows.

Theorem 3. *Let Π be a non-trivial hereditary graph property that is determined by the components and let $\mathcal{H}$ be the corresponding set of all forbidden induced subgraphs. If $\mathcal{H}$ contains a star with at least three leaves, then* DISJOINT Π-VERTEX DELETION *is NP-hard.*

Note that a star has a smaller α-sequence than any other forbidden induced subgraph that is not a star, and there is only one star in $\mathcal{H}$, since the graphs in $\mathcal{H}$ are inclusion-minimal. Therefore, if $\mathcal{H}$ contains a star, then the graph with smallest α-sequence is necessarily the star in $\mathcal{H}$. Let H be the star in $\mathcal{H}$.

The proof of Theorem 3 is based on the following case distinction. (1) H is a star with at least four leaves (Lemma 6), or (2) H is a star with three leaves. In the latter case, we distinguish the following two subcases: (2a) $\mathcal{H}$ contains a P_4 (Lemma 7), and (2b) $\mathcal{H}$ does not contain a P_4 (Lemma 8).

Lemma 6. *If the set $\mathcal{H}$ of forbidden induced subgraphs corresponding to property Π contains a star H, and if H has at least four leaves, then* DISJOINT Π-VERTEX DELETION *is NP-hard.*

Next, we show the NP-hardness of the case that the smallest graph in the set of forbidden induced subgraphs is a star with three leaves. In this case, a reduction from VERTEX COVER seems less promising, since the VERTEX COVER instance we reduce from contains vertices of degree three and therefore copies of the forbidden induced star with three leaves. Hence, we use 3-CNF-SAT. First, we consider the case that the path on four vertices is also forbidden.

Lemma 7. *If the set $\mathcal{H}$ of forbidden induced subgraphs corresponding to property Π contains a star H, and if H has three leaves and $\mathcal{H}$ also contains the path on four vertices, then* DISJOINT Π-VERTEX DELETION *is NP-hard.*

Proof. The proof is by reduction from 3-CNF-SAT. Let $F = c_1 \wedge \cdots \wedge c_m$ be a 3-CNF formula over a variable set $X = \{x_1, \ldots, x_n\}$. We denote the kth literal in clause c_j by l_j^k, for $1 \leq k \leq 3$. An example of the following construction is given in Figure 1. Starting with an empty graph G and $S := \emptyset$, construct an instance (G, S) for DISJOINT Π-VERTEX DELETION as follows. For each variable x_i, introduce a cycle X_i of $12m$ vertices (variable gadget), add every second vertex on X_i to S, and label all the other vertices on the cycle alternately with "+" and "−". For each clause c_j, add a star C_j with three leaves (clause gadget) and add its center vertex to S. Each of the three leaves of C_j corresponds to a literal in c_j, and each leaf is connected to a variable gadget as follows. Suppose that l_j^k is a literal x_i or $\neg x_i$, and let a_k be the leaf of C_j corresponding to l_j^k. Add a star with three leaves (connection gadget), identify one leaf with a_i, identify another leaf with an *unused vertex*[5] on X_i with label "+" if l_j^k is positive

[5] This means that no vertex of an other connection gadget has been identified with this vertex on X_i, that is, it is of degree two.

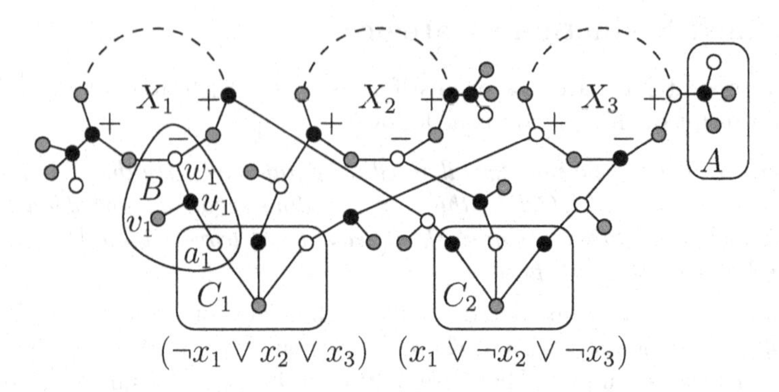

Fig. 1. Example for the reduction in the proof of Lemma 7 for the 3-CNF-SAT formula $(\neg x_1 \vee x_2 \vee x_3) \wedge (x_1 \vee \neg x_2 \vee \neg x_3)$. For illustration, one minimality gadget is labeled with A and one connection gadget is labeled with B. Furthermore, for the connection gadget B the vertices are named according to the definitions of a_k, u_k, v_k and w_k in the proof of Lemma 7 for $k = 1$. The vertices in the given solution S are gray, the vertices in the disjoint solution S', corresponding to the satisfying truth assignment $x_1 =$ true, $x_2 =$ true, $x_3 =$ false, are black.

and with an unused vertex on X_i with label "−" if l_j^k is negative, and add the remaining leaf to S. Finally, for each remaining unused vertex v labeled "+" or "−" in G, add a star with three leaves (minimality gadget), add two of its leaves to S, and add an edge between the center and v. This concludes the construction.

Obviously, $G - S$ only contains paths on three vertices as connected components (cf. Figure 1), that is, $G - S$ satisfies Π. Moreover, S is minimal, that is, for any $v \in S$, $G - (S \setminus \{v\})$ does not satisfy Π. Let q be the number of minimality gadgets. We show that formula F has a satisfying truth assignment if and only if there exists a size-$(q+3nm+3m)$ set S', $S' \cap S = \emptyset$, that obstructs all forbidden induced subgraphs in G. Analogously to the proof of Lemma 6, the construction can be modified (to "correct" the sizes of S and S') by adding a padding gadget based on stars with three leaves. This straightforward modification is omitted.

($\Rightarrow$) We defer the proof of this direction to a full version of the paper.

($\Leftarrow$) Let S', $S' \cap S = \emptyset$, be a size-$(q+3nm+3m)$ vertex set that obstructs every forbidden induced subgraph in G. We may assume that S' does not contain any degree-one vertex in G (since a degree-one vertex in S' could be simply replaced by its neighbor). Recall that the set of forbidden induced subgraphs contains the star with three leaves and the path on four vertices. Each minimality gadget is a star with three leaves, and since we assumed that no degree-one vertex is in S', its center vertex must be in S'. Hence, S' contains exactly q vertices of the minimality gadgets. Since P_4s are forbidden, at least every fourth vertex on the cycle of each variable gadget has to be in S'. However, we will see that S' contains exactly three vertices for each clause (thus, $3m$ vertices for all clauses), and these vertices cannot be vertices on any variable gadget. Therefore, for each variable gadget X_i, the set S' must contain *exactly* every fourth vertex of X_i (in order to obtain a total number of $3mn$ vertices in S' for all n variable gadgets), thus S' either contains all vertices labeled "+" or all vertices labeled "−". If S' contains all vertices labeled "+", then we set $x_i :=$ true, if S' contains all vertices labeled "−", then we set $x_i :=$ false. It remains to show that the assignment defined in this way is a satisfying truth assignment for the formula F.

For a clause gadget C_j, and for each leaf a_k of C_j corresponding to literal l_j^k, let u_k be the center of the corresponding connection gadget, v_k be the degree-one neighbor of u_k, and w_k be the neighbor of u_k on the variable gadget X_i, for some $1 \leq i \leq n$ (cf. Figure 1). There is a P_4 containing the center of C_j, together with a_k, u_k, and v_k. Since the center of C_j is in S, the set S' has to contain at least three vertices to obstruct the three P_4s corresponding to C_j (one for each leaf). Thus, for all clauses, there are at least $3m$ vertices in S' that obstruct these P_4s. In total, S' contains $q + 3nm + 3m$ vertices. Therefore, there are *exactly* $3m$ vertices in S' that obstruct these P_4s. Thus, for a clause gadget C_j, for each leaf a_k, either $a_k \in S'$ or $u_k \in S'$. Which case applies depends on which vertices from X_i are in S': if $w_k \notin S'$, then w_k together with u_k and its two neighbors on X_i induce a star with three leaves, thus $u_k \in S'$. If $w_k \in S'$, then either $a_k \in S'$ or $u_k \in S'$. If $w_k \in S'$ and $u_k \in S'$, however, then one can simply remove u_k from S' and add a_k instead. After that, S' still obstructs all forbidden induced subgraphs. Since S' obstructs all forbidden induced subgraphs, at least one leaf a_k of C_j must be in S', which implies that $w_k \in S'$. Let X_i be the variable gadget that contains w_k. If w_k has label "+", then $x_i =$ true by the definition of the assignment, and by construction $l_j^k = x_i$ is a positive literal, hence c_j is satisfied. If w_k has label "−", then $x_i =$ false, and, by construction, $l_j^k = \neg x_i$ is a negative literal, hence c_j is satisfied. Summarizing, for every clause there is at least one true literal and thus the constructed truth assignment satisfies F. □

Finally, we consider the case that the path on four vertices is not forbidden.

Lemma 8. *If the set $\mathcal{H}$ of forbidden induced subgraphs corresponding to property Π contains a star H, and if H has three leaves and $\mathcal{H}$ does not contain the path on four vertices, then* Disjoint Π-Vertex Deletion *is NP-hard.*

5 Outlook

As indicated in the introductory section, there are important problems amenable to iterative compression that do not fall into the problem class studied here. Among these, in particular, we have Directed Feedback Vertex Set and Almost 2-Sat. Hence, it would be interesting to further generalize our results to other problem classes, among these also being vertex deletion problems on directed graphs or bipartite graphs and edge deletion problems. Our work here has left open the case that a forbidden subgraph may consist of more than one connected component. Finally, one could explore to parameterize Disjoint Π-Vertex Deletion by the number of vertices by which S' should at least differ from S.

References

[1] Abello, J., Resende, M.G.C., Sudarsky, S.: Massive quasi-clique detection. In: Rajsbaum, S. (ed.) LATIN 2002. LNCS, vol. 2286, pp. 598–612. Springer, Heidelberg (2002)
[2] Cai, L.: Fixed-parameter tractability of graph modification problems for hereditary properties. Inf. Process. Lett. 58(4), 171–176 (1996)
[3] Chen, J., Fomin, F.V., Liu, Y., Lu, S., Villanger, Y.: Improved algorithms for feedback vertex set problems. J. Comput. System Sci. 74(7), 1188–1198 (2008)
[4] Chen, J., Liu, Y., Lu, S., O'Sullivan, B., Razgon, I.: A fixed-parameter algorithm for the directed feedback vertex set problem. J. ACM 55(5), Article 21, 19 (2008)
[5] Dehne, F.K.H.A., Fellows, M.R., Langston, M.A., Rosamond, F.A., Stevens, K.: An $O(2^{O(k)}n^3)$ FPT algorithm for the undirected feedback vertex set problem. Theory Comput. Syst. 41(3), 479–492 (2007)
[6] Garey, M.R., Johnson, D.S.: Computers and Intractability: A Guide to the Theory of NP-Completeness. Freeman, New York (1979)
[7] Greenwell, D.L., Hemminger, R.L., Klerlein, J.B.: Forbidden subgraphs. In: Proc. 4th CGTC, pp. 389–394 (1973)
[8] Guo, J., Gramm, J., Hüffner, F., Niedermeier, R., Wernicke, S.: Compression-based fixed-parameter algorithms for feedback vertex set and edge bipartization. J. Comput. System Sci. 72(8), 1386–1396 (2006)
[9] Guo, J., Moser, H., Niedermeier, R.: Iterative compression for exactly solving NP-hard minimization problems. In: Algorithmics of Large and Complex Networks. LNCS, vol. 5515, pp. 65–80. Springer, Heidelberg (2009)
[10] Hüffner, F., Komusiewicz, C., Moser, H., Niedermeier, R.: Fixed-parameter algorithms for cluster vertex deletion. Theory Comput. Syst. (2009); available electronically
[11] Khot, S., Raman, V.: Parameterized complexity of finding subgraphs with hereditary properties. Theor. Comput. Sci. 289(2), 997–1008 (2002)
[12] Lewis, J.M., Yannakakis, M.: The node-deletion problem for hereditary properties is NP-complete. J. Comput. System Sci. 20(2), 219–230 (1980)
[13] Marx, D.: Chordal deletion is fixed-parameter tractable. Algorithmica (2009); available electronically
[14] Marx, D., Schlotter, I.: Obtaining a planar graph by vertex deletion. In: Brandstädt, A., Kratsch, D., Müller, H. (eds.) WG 2007. LNCS, vol. 4769, pp. 292–303. Springer, Heidelberg (2007)
[15] Niedermeier, R.: Invitation to Fixed-Parameter Algorithms. Oxford University Press, Oxford (2006)
[16] Razgon, I., O'Sullivan, B.: Almost 2-SAT is fixed-parameter tractable. J. Comput. System Sci. (2009); available electronically
[17] Reed, B., Smith, K., Vetta, A.: Finding odd cycle transversals. Oper. Res. Lett. 32(4), 299–301 (2004)

Future-Looking Logics on Data Words and Trees

Diego Figueira and Luc Segoufin*

INRIA, LSV, ENS Cachan, France

Abstract. In a data word or a data tree each position carries a label from a finite alphabet and a data value from an infinite domain.

Over data words we consider the logic $\mathsf{LTL}_1^{\downarrow}(\mathsf{F})$, that extends $\mathsf{LTL}(\mathsf{F})$ with one register for storing data values for later comparisons. We show that satisfiability over data words of $\mathsf{LTL}_1^{\downarrow}(\mathsf{F})$ is already non primitive recursive. We also show that the extension of $\mathsf{LTL}_1^{\downarrow}(\mathsf{F})$ with either the backward modality F^{-1} or with one extra register is undecidable. All these lower bounds were already known for $\mathsf{LTL}_1^{\downarrow}(\mathsf{X},\mathsf{F})$ and our results essentially show that the X modality was not necessary.

Moreover we show that over data trees similar lower bounds hold for certain fragments of XPath.

1 Introduction

A data word (data tree) is a word (tree) where each position carries a label from a finite alphabet and a *datum* from some infinite domain. These models have been considsered in the realm of semistructured data [3], timed automata [5] and extended temporal logics [9,8,12]. In this work we consider an infinite domain with no structure where we can only test for equality or inequality between elements.

There have been various logics considered to specify properties over data words and data trees. For example from the standpoint of Temporal Logics (both on data words [8] and trees [12]), of First Order Logics (see [4] for the data words case and [3] for trees), or of logics based on tree patterns [7,2]. The logic $\mathsf{LTL}_1^{\downarrow}(\mathsf{X},\mathsf{F})$ is the extension of $\mathsf{LTL}(\mathsf{X},\mathsf{F})$ with the ability to use one *register* for storing a data value for later comparisons. It has been studied in [9,8] where satisfiability and expressivity issues have been addressed. In [8] it has been established that the satisfiability problem for $\mathsf{LTL}_1^{\downarrow}(\mathsf{X},\mathsf{F})$ is decidable and non primitive recursive on data words. It is also shown in [8] that the *two way* extension $\mathsf{LTL}_1^{\downarrow}(\mathsf{X},\mathsf{F},\mathsf{F}^{-1})$ is undecidable over data words and, similarly, that the extension to 2 registers $\mathsf{LTL}_2^{\downarrow}(\mathsf{X},\mathsf{F})$ is undecidable.

Here we show that even without the X modality, all the aforementioned lower bounds remain valid: the satisfiability problem for $\mathsf{LTL}_1^{\downarrow}(\mathsf{F})$ over data words is non primitive recursive, while for $\mathsf{LTL}_1^{\downarrow}(\mathsf{F},\mathsf{F}^{-1})$ and $\mathsf{LTL}_2^{\downarrow}(\mathsf{F})$ is undecidable.

* The authors acknowledge the financial support of the Future and Emerging Technologies (FET) programme within the Seventh Framework Programme for Research of the European Commission, under the FET-Open grant agreement FOX, number FP7-ICT-233599.

R. Královič and D. Niwiński (Eds.): MFCS 2009, LNCS 5734, pp. 331–343, 2009.

Data trees can be seen as a coding of an XML document [3,12]. Therefore XPath, the node selecting language of W3C [6], can be considered as a logic over data trees. By XPath, we refer to the 1.0 specification without all the domain specific features (arithmetic, string manipulation, etc.) As XPath is at the core of many XML standard languages (like XQuery and XSLT), deciding satisfiability of some of its fragments can be of great help during optimization stages.

A data word can be seen as a special case of an unranked ordered data tree, for instance by adding a root that is the parent of all the positions of the data word. With this consideration the fragment $\mathsf{XPath}(\rightarrow, \rightarrow^+, =)$ of XPath that contains only the axis `next-sibling` ($\rightarrow$) and `following-sibling` ($\rightarrow^+$) can be seen as a fragment of $\mathsf{LTL}_1^\downarrow(\mathsf{X}, \mathsf{F})$. There are nonetheless two important differences between the expressive power of $\mathsf{XPath}(\rightarrow, \rightarrow^+, =)$ and $\mathsf{LTL}_1^\downarrow(\mathsf{X}, \mathsf{F})$. The first one is that the axis $\rightarrow^+$ corresponds to the strict future modality in LTL, denoted $\mathsf{F_s}$ in the sequel. Of course F can be defined using $\mathsf{F_s}$ but the opposite is not true in the absence of X. The second difference lies in the fact that XPath can compare data values in a way strictly more limited than $\mathsf{LTL}_1^\downarrow$. We illustrate this with an example. At a position of a data word, $\mathsf{LTL}_1^\downarrow(\mathsf{F})$ can store the current data value in the register, check for a later symbol a with a data value different from the one in the register, and then further check for a symbol b with a data value matching the one of the register. In this spirit, XPath could only compare the data values at the beginning and the end of the path and could not say anything about the data value of the intermediate a symbol. Hence $\mathsf{XPath}(\rightarrow^+, =)$ should be seen as a fragment of $\mathsf{LTL}_1^\downarrow(\mathsf{F_s})$, incomparable with $\mathsf{LTL}_1^\downarrow(\mathsf{F})$ in terms of expressive power.

Based on these ideas, we exhibit a syntactic fragment $\mathsf{sLTL}_1^\downarrow(\mathsf{F_s})$ of $\mathsf{LTL}_1^\downarrow(\mathsf{F_s})$ that limits the use of register comparisons and has the same expressive power as $\mathsf{XPath}(\rightarrow^+, =)$ on data words. We then show that satisfiability over data words of $\mathsf{sLTL}_1^\downarrow(\mathsf{F_s})$ is non primitive recursive. Similarly the same restriction on $\mathsf{LTL}_1^\downarrow(\mathsf{F_s}, \mathsf{F_s}^{-1})$ yields a logic with the same expressive power as $\mathsf{XPath}(\rightarrow^+, {}^+\!\leftarrow, =)$, which is shown to be undecidable.

Our non primitive recursive results are proved by a reduction from the emptiness problem of *faulty counter automata*. These are counter automata that may have *incrementing errors* in their counters during the run. Non-emptiness for this class of automata was proven to be decidable and not primitive recursive [15]. Our reduction will be centered in a strategy of using data values for coding – with some limitations– a *next step* move of this automaton. We show that the strict future modality together with the limited data comparison capabilities of $\mathsf{sLTL}_1^\downarrow(\mathsf{F_s})$ are sufficient for our coding. With the extra power of $\mathsf{LTL}_1^\downarrow(\mathsf{F})$ for comparing data values, we also show that strictness of the future modality can be avoided. Similar ideas are used for our undecidability results: The extra available expressive power is sufficient to forbid the incrementing errors and thus to code the emptiness problem of a Minsky Counter Automaton.

Related work. There are known complexity results concerning the satisfiability problem of several data-aware fragments of XPath on data trees. When all navigational axes are present, XPath is undecidable [11]. When all vertical axes are present but in the absence of any horizontal axis, the status of the decidability of $\mathsf{XPath}(\downarrow, \downarrow^+, \uparrow, \uparrow^+, =)$ is not yet known [1]. In this paper we show that if decidable, the complexity of this fragment cannot be primitive recursive, even in the absence of the axes $\downarrow$ and $\uparrow$.

In [3] decidability over data trees of the two-variable fragment of first order logic, $\mathsf{FO}_2(+1, \sim)$ is established. As a direct consequence, a fragment of XPath without any `descendant/ancestor` or `following/previous-sibling` axes in the paths of the data test expressions, is shown to be decidable. On the opposite, in the present work we focus on the satisfiability under the absence of the *successor* axis.

In [12] it is shown that the emptiness problem for alternating automata with one register over data trees is decidable. As a direct consequence it is shown that $\mathsf{XPath}(\downarrow, \downarrow^+, \rightarrow, \rightarrow^+, =)$, with some restriction on the expressions with a data value test, is decidable and non primitive recursive. As a consequence of the present work, the hardness result already holds for $\mathsf{XPath}(\downarrow^+, \rightarrow, =)$ and for $\mathsf{XPath}(\rightarrow^+, =)$.

We finally remark that, nevertheless, the satisfiability of $\mathsf{XPath}(\downarrow, \downarrow^+, =)$ is "only" ExpTime-complete [10].

2 Preliminaries

We fix a finite set Σ of *labels* and an infinite set D of *data values*. The models we consider are either data words or data trees. A data word σ over a finite alphabet Σ is a non-empty word of $(\Sigma \times D)^*$. We assume no structure on D and D will be used only to perform tests for (in)equalities. In our examples we will always use data values from $\mathbb{N}$ seen as a *set* of numbers. The data trees we will be using are unranked and ordered. The *domain* of an unranked ordered tree is represented by a prefix-closed set T of elements from $\mathbb{N}^*$ such that whenever $n(i+1) \in T$ then $ni \in T$. The elements of T will be called *nodes* of the tree. A data tree t over Σ, D is a tree domain T together with a labeling function λ assigning an element of $\Sigma \times D$ to any node of t. We use the standard terminology for trees such as descendant, ancestor, sibling, etc.

LTL *with registers.* The most expressive logic for data words we treat here is $\mathsf{LTL}^{\downarrow}_n(\mathsf{F}, \mathsf{X}, \mathsf{F}^{-1}, \mathsf{X}^{-1})$, the Linear Temporal Logic with the freeze quantifier ($\downarrow_i$), test predicate ($\uparrow_i$) and next (X) and future (F) temporal operators together with their inverse modalities ($\mathsf{X}^{-1}, \mathsf{F}^{-1}$). Sentences are defined:

$$\varphi, \psi ::= a \mid \downarrow_i \varphi \mid \uparrow_i \mid \neg\varphi \mid \varphi \wedge \psi \mid \varphi \vee \psi \mid \mathcal{O}\varphi$$

where a is a symbol from a finite alphabet Σ, $i \in \{1, \ldots, n\}$, and $\mathcal{O}$ ranges over $\{\mathsf{F}, \mathsf{X}, \mathsf{F}^{-1}, \mathsf{X}^{-1}\}$. We also use the *strict future* temporal operator $\mathsf{F_s}$ as a shortcut for XF. We will consider fragments of this logics as simple restrictions to certain temporal operators. Intuitively, in the evaluation of $\downarrow_i \varphi$, current data value is 'saved' in the register i, and any appearance of $\uparrow_i$ in φ holds at a position iff its datum is equal to the one stored in register i.

Given a data word σ, we write σ_i for the ith element (couple) of the word, and π_1, π_2 for the projections on Σ and D. A *valuation* is defined as a partial map $v : \{1, \ldots, n\} \to D$. The satisfaction relation $\models$ is inductively defined (we omitted X^{-1} and F^{-1} for succinctness):

$\sigma, i \models_v a$ iff $\pi_1(\sigma_i) = a$ $\qquad$ $\sigma, i \models_v \uparrow_k$ iff $k \in \mathrm{dom}(v)$ and $v(k) = \pi_2(\sigma_i)$

$\sigma, i \models_v \downarrow_k \varphi$ iff $\sigma, i \models_{v[k \mapsto \pi_2(\sigma_i)]} \varphi$ $\qquad$ $\sigma, i \models_v \mathsf{X}\varphi$ iff $i < |\sigma|$ and $\sigma, i+1 \models_v \varphi$

$\sigma, i \models_v \mathsf{F}\varphi$ iff for some $i \leq j \leq |\sigma|$ we have $\sigma, j \models_v \varphi$

where $1 \leq i \leq |\sigma|$. We denote $\sigma, i \models_{v_\emptyset} \varphi$ by $\sigma, i \models \varphi$, with $v_\emptyset$ the empty valuation; and we write $\sigma \models \varphi$ for $\sigma, 1 \models \varphi$. Also, in the case of $\mathsf{LTL}_1^\downarrow$ we use $\downarrow$ and $\uparrow$ instead of $\downarrow_1$ and $\uparrow_1$ for simplicity.

In Section 5 we will lift our results to trees and XPath. In order to do this it is convenient to introduce now a restriction of $\mathsf{LTL}_1^\downarrow$ such that with this restriction $\mathsf{LTL}_1^\downarrow$ corresponds to XPath. A formula of $\mathsf{LTL}_1^\downarrow$ is said to be *simple* if (i) there is at most one occurrence of $\uparrow$ within the scope of each occurrence of $\downarrow$ and, (ii) there is no negation between an occurrence of $\uparrow$ and its matching $\downarrow$, except maybe immediately before $\uparrow$. We denote by $\mathsf{sLTL}_1^\downarrow$ the fragment of $\mathsf{LTL}_1^\downarrow$ containing only simple formulas. The correspondence between $\mathsf{sLTL}_1^\downarrow$ and XPath will be made explicit in Proposition 1 of Section 5.

Faulty counter systems. For proving our lower bounds we will use a reduction from faulty counter automata that we describe here. A *counter automaton* (CA) with zero testing is a tuple $\langle \Sigma, Q, q_0, n, \delta, F \rangle$, where Σ is a finite alphabet, Q is a finite set of states, q_0 is the initial state, $n \in \mathbb{N}$ is the number of counters, $\delta \subset Q \times \Sigma \times L \times Q$ is the transition relation over the instruction set $L = \{\mathtt{inc}, \mathtt{dec}, \mathtt{ifzero}\} \times \{1, \ldots, n\}$, and $F \subset Q$ is the set of accepting states. A *counter valuation* is a function $v : \{1, \ldots, n\} \to \mathbb{N}$. An error-free run over $w \in \Sigma^*$ is a finite sequence $\langle q_0, v_0 \rangle \stackrel{w_0, \ell_0}{\to} \langle q_1, v_1 \rangle \stackrel{w_1, \ell_1}{\to} \cdots$ observing the standard interpretation of the instructions $\ell_0, \ell_1, \cdots$ ($\langle \mathtt{dec}, c \rangle$ can only be performed if c is nonzero), where $v_0, v_1, \ldots$ are counter valuations, v_0 assigns 0 to each counter, and $w = w_0 w_1 \ldots$ such that $w_i \in \Sigma \cup \{\varepsilon\}$ for every i. A run is *accepting* iff it ends with an accepting state.

A *Minsky* CA has error-free runs. For these automata, already with only two counters, finitary emptiness is undecidable [13]. An *Incrementing* CA (from now on ICA) is defined as a Minsky CA except that its runs may contain errors that increase one or more counters non-deterministically. We write that two valuations are in the relation $v \leq v'$ iff, for every counter c, $v(c) \leq v'(c)$. Runs of Incrementing CA are defined by replacing the relation $\stackrel{a,\ell}{\to}$ by $\stackrel{a,\ell}{\rightsquigarrow}$, where $\langle p, u \rangle \stackrel{a,\ell}{\rightsquigarrow} \langle q, v \rangle$ iff there exist valuations u', v' such that $u \leq u'$, $v' \leq v$ and $\langle p, u' \rangle \stackrel{a,\ell}{\to} \langle q, v' \rangle$.

In [8, Theorem 2.9] it is shown that the results of [15] and [14] on Channel Machines can be adapted to prove the following result.

Theorem 1. *Emptiness of* ICA *is decidable and non primitive recursive.*

3 The Case of $\mathsf{sLTL}_1^{\downarrow}$

In this section we show that satisfiability of $\mathsf{sLTL}_1^{\downarrow}(\mathsf{F_s})$ is non primitive recursive over data words. We then prove that satisfiability is undecidable for $\mathsf{sLTL}_1^{\downarrow}(\mathsf{F_s}, \mathsf{F_s}^{-1})$. In the next section we will show that the logic $\mathsf{LTL}_1^{\downarrow}(\mathsf{F})$ is also non primitive recursive and that $\mathsf{LTL}_1^{\downarrow}(\mathsf{F}, \mathsf{F}^{-1})$ is undecidable. In Section 5 we will use the results of this section for proving lower bounds for several fragments of XPath over data trees.

3.1 Lower Bound for $\mathsf{sLTL}_1^{\downarrow}(\mathsf{F_s})$

Theorem 2. *Satisfiability of* $\mathsf{sLTL}_1^{\downarrow}(\mathsf{F_s})$ *on data words is non primitive recursive.*

Proof. We exhibit a PTIME reduction from the non-emptiness of ICA to satisfiability of $\mathsf{sLTL}_1^{\downarrow}(\mathsf{F_s})$. Let $C = \langle \Sigma, Q, q_0, n, \delta, F \rangle$ be an ICA.
Let $L = \{(\mathtt{inc}_i)_{1 \le i \le n}, (\mathtt{dec}_i)_{1 \le i \le n}, (\mathtt{ifzero}_i)_{1 \le i \le n}\}$, and $\hat{\Sigma} = Q \times (\Sigma \cup \{\varepsilon\}) \times L \times Q$. We construct a formula $\varphi_C \in \mathsf{sLTL}_1^{\downarrow}(\mathsf{F_s})$ that is satisfied by a data word iff C accepts the word. We view a run of C of the form:

$$\langle q_0, v_0 \rangle \overset{a,\mathtt{inc}_i}{\rightsquigarrow} \langle q_1, v_1 \rangle \overset{b,\mathtt{dec}_j}{\rightsquigarrow} \langle q_2, v_2 \rangle \overset{b,\mathtt{ifzero}_i}{\rightsquigarrow} \langle q_3, v_3 \rangle \cdots$$

as a string in $\hat{\Sigma}$:

$$\langle q_0, a, \mathtt{inc}_i, q_1 \rangle \langle q_1, b, \mathtt{dec}_j, q_2 \rangle \langle q_2, b, \mathtt{ifzero}_i, q_3 \rangle \cdots$$

The formula φ_C will force that any string that satisfies it codes a run of C. In order to do this, φ_C must ensure that:

- `(begin)` the string starts with q_0,
- `(end)` the string ends with a state of F,
- `(tran)` every symbol of $\hat{\Sigma}$ in the string corresponds to a transition of C,
- `(chain)` the last component of a symbol of $\hat{\Sigma}$ is equal to the first component of the next symbol,
- `(pair)` for each i, every symbol that contains $\mathtt{inc}_i$ occurring in the string to the left of a symbol containing $\mathtt{ifzero}_i$, can be paired with a symbol containing $\mathtt{dec}_i$ and occurring in between the $\mathtt{inc}_i$ and the $\mathtt{ifzero}_i$.

Before continuing let us comment on the `(pair)` condition. If we were coding runs of a perfect Minsky CA (ie, with no incremental errors), to the left of any position containing a $\mathtt{ifzero}_i$, we would require a perfect matching between $\mathtt{inc}_i$ and $\mathtt{dec}_i$ operations in order to make sure that the value of the counter i is indeed zero at the position of the test. But as we are dealing with ICA, we only need to check that each $\mathtt{inc}_i$ has an associated $\mathtt{dec}_i$ to its right and before the test for zero, but we do not enforce the converse, that all $\mathtt{dec}_i$ match an $\mathtt{inc}_i$. This is fortunate because this would require past navigational operators.

The first difficulty comes from the fact that `(pair)` is not a regular relation. The pairing will be obtained using data values. The second difficulty is to enforce `(chain)` without having access to the string successor relation. In order to simulate the successor relation we add extra symbols to the alphabet, with suitable associated data values.

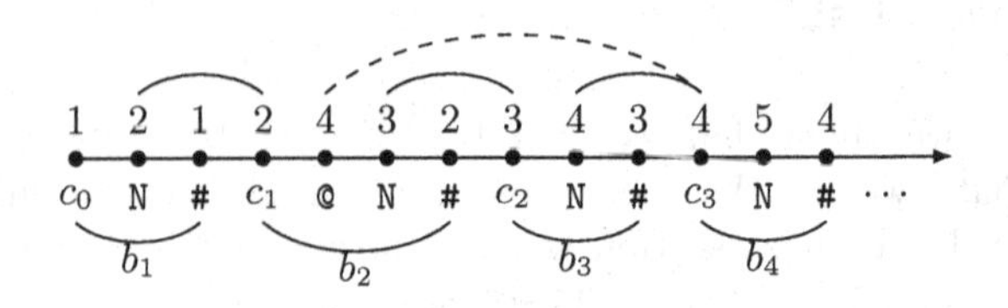

Fig. 1. Coding of an ICA run

Let $\Sigma' = \hat{\Sigma} \cup \{\mathtt{N}, \mathtt{\#}, \mathtt{@}\}$. The coding of a run consists in a succession of *blocks*. Each block is a sequence of 3 or 4 symbols, "c N #" or "c @ N #", with $c \in \hat{\Sigma}$. The data value associated to the c and # symbols of a block is the same and uniquely determines the block: no two blocks may have the same data value. The data value associated with the symbol N of a block is the data value of the *next* block. If a block contains a symbol c that codes a $\mathtt{inc}_i$ operation that is later in the string followed by a $\mathtt{ifzero}_i$, then this block contains a symbol @ whose data value is the that of the block containing $\mathtt{dec}_i$ that is paired with c.

For instance in the example of Fig. 1 one can see four blocks b_1, b_2, b_3, b_4. Each of them starts with a symbol from $\hat{\Sigma}$ coding a transition of the ICA and ends with a # with the same data value marking the end of the block. Inside the block, the data value of N is the same as the data value of the next block. The data value of @ corresponds to that of a future block. In this example c_1 must correspond to a $\mathtt{inc}_i$ while c_3 to $\mathtt{dec}_i$ and there must be a $\mathtt{ifzero}_i$ somewhere to the remaining part of the word (say, b_5). Moreover c_2 can't be a $\mathtt{ifzero}_i$ as otherwise the data value of the symbol @ would refer to a block to the left of c_2.

We now show that the coding depicted above can be enforced in $\mathsf{sLTL}_1^{\downarrow}(\mathsf{F_s})$. By $\pi_1, \pi_2, \pi_3, \pi_4$ we denote the projection of each symbol of $\hat{\Sigma}$ into its corresponding component. To simplify the presentation we use the following abbreviations:

$$\hat{\sigma} \equiv \bigvee_{c \in \hat{\Sigma}} c \qquad \text{INC}(i) \equiv \bigvee_{c \in \hat{\Sigma}, \pi_3(c) = \mathtt{inc}_i} c \qquad \text{INC} \equiv \bigvee_i \text{INC}(i)$$

$$\text{DEC}(i) \equiv \bigvee_{c \in \hat{\Sigma}, \pi_3(c) = \mathtt{dec}_i} c \qquad \text{LAST} \equiv \hat{\sigma} \wedge \neg \mathsf{F_s}\, \hat{\sigma} \qquad \text{IZ}(i) \equiv \bigvee_{c \in \hat{\Sigma}, \pi_3(c) = \mathtt{ifzero}_i} c$$

The formula φ_C that we build is the conjunction of all the folowing formulas.

Forcing the structure

$\mathsf{G}(\text{LAST} \Rightarrow \neg \mathsf{F_s} \top)$: The string ends with the last transition,

$\bigwedge_{c \in \{\mathtt{N},\mathtt{@},\mathtt{\#}\}} \mathsf{G}(c \Rightarrow \neg(\downarrow \mathsf{F_s}(c \wedge \uparrow)))$: the data value associated to each N, # and @ uniquely determines the occurrence of that symbol,

$\mathsf{G}((\hat{\sigma} \wedge \neg\text{LAST}) \Rightarrow (\downarrow \mathsf{F_s}(\mathtt{N} \wedge \mathsf{F_s}(\mathtt{\#} \wedge \uparrow))))$: each occurrence of $\hat{\Sigma}$ (except the last one) is in a block that contains a N and then a #,

$\bigwedge_i \mathsf{G}((\text{INC}(i) \wedge \mathsf{F_s}(\text{IZ}(i))) \Rightarrow \downarrow \mathsf{F_s}(\mathtt{@} \wedge \mathsf{F_s}(\mathtt{N} \wedge \mathsf{F_s}(\mathtt{\#} \wedge \uparrow))))$: every $\mathtt{inc}_i$ block to the left of a $\mathtt{ifzero}_i$ must have a @ before the N,

$\mathsf{G}((\hat{\sigma} \wedge \neg\text{INC}) \Rightarrow \neg \downarrow \mathsf{F_s}(\mathtt{@} \wedge \mathsf{F_s}(\mathtt{\#} \wedge \uparrow)))$: only blocks inc are allowed to have a @,

$\bigwedge_{s\in\{\mathtt{N},\mathtt{@}\}} \mathsf{G}(\hat{\sigma} \Rightarrow \neg(\downarrow \mathsf{F_s}(s \wedge \mathsf{F_s}(s \wedge \mathsf{F_s}(\# \wedge \uparrow)))))$: there is at most one occurrence of N and @ in each block,

$\bigwedge_{s\in\hat{\Sigma}\cup\{\#\}} \mathsf{G}(\hat{\sigma} \Rightarrow \neg \downarrow \mathsf{F_s}(s \wedge \mathsf{F_s}(\# \wedge \uparrow))$: there is exactly one symbol # and one symbol of $\hat{\Sigma}$ per block,

$\mathsf{G}(\mathtt{N} \Rightarrow \downarrow \mathsf{F_s}(\# \wedge \mathsf{F_s}(\hat{\sigma} \wedge \uparrow)))$: each symbol N's datum points to a block to its right,

$\mathsf{G}(\mathtt{N} \Rightarrow \neg \downarrow \mathsf{F_s}(\# \wedge \mathsf{F_s}(\# \wedge \mathsf{F_s}(\hat{\sigma} \wedge \uparrow))))$: in fact N has has to point be the next block.

Once the structure has the expected shape, we can enforce the run as follows. All the formulas below are based on the following trick. In a test of the form $\downarrow \mathsf{F_s}(\mathtt{N} \wedge \mathsf{F_s}(\# \wedge \uparrow))$ which is typically performed at a position of symbol $\hat{\Sigma}$, the last symbol # must have the same data value as the initial position. Hence, because of the structure above, both must be in the same block. Thus the middle symbol N must also be inside that block. From the structure we know that the data value of this N points to the next block. Therefore by replacing the test N by one of the form $\mathtt{N} \wedge (\downarrow \mathsf{F_s}(\uparrow \wedge s))$ we can transfer some finite information from the current block to the next one. This gives the desired successor relation.

Forcing a run

`(begin)`
$$\bigvee_{c\in\hat{\Sigma},\pi_1(c)=q_0} c$$

`(end)`
$$\mathsf{F_s}\Big(\mathrm{LAST} \wedge \bigvee_{c\in\hat{\Sigma},\pi_4(c)\in F} c\Big)$$

`(tran)` All elements used from $\hat{\Sigma}$ correspond to valid transitions. Let $\hat{\Sigma}^C$ be that set of transitions of C,
$$\mathsf{G}\Big(\bigwedge_{c\in\hat{\Sigma}\setminus\hat{\Sigma}^C} \neg c\Big)$$

`(chain)` For every $c \in \hat{\Sigma}$,
$$\mathsf{G}\Big(c \Rightarrow (\downarrow \mathsf{F_s}(\mathtt{N} \wedge \mathsf{F_s}(\# \wedge \uparrow) \wedge \bigvee_{\substack{d\in\hat{\Sigma},\\ \pi_4(c)=\pi_1(d)}} (\downarrow \mathsf{F_s}(d \wedge \uparrow))))\Big)$$

`(pair)` We first make sure that the block of the @ of an $\mathtt{inc}_k$ is matched with a block of a $\mathtt{dec}_k$:

$$\bigwedge_k \mathsf{G}\Big(\mathrm{INC}(k) \Rightarrow \Big(\neg \downarrow \mathsf{F_s}(\mathtt{@} \wedge \mathsf{F_s}(\# \wedge \uparrow)) \vee \downarrow \mathsf{F_s}\big(\mathtt{@} \wedge \downarrow \mathsf{F_s}(\mathrm{DEC}(k) \wedge \uparrow) \wedge \mathsf{F_s}(\# \wedge \uparrow)\big)\Big)\Big)$$

Now, every $\mathtt{inc}_k$ block to the left of a $\mathtt{ifzero}_k$ block:
– 1. The block must contain an @ element:

$$\bigwedge_k \mathsf{G}\big(\mathrm{INC}(k) \Rightarrow (\downarrow \mathsf{F_s}(\mathtt{@} \wedge \mathsf{F_s}(\# \wedge \uparrow)) \vee \neg \mathsf{F_s}(\mathrm{IZ}(k)))\big)$$

– 2. The data value of that @ element must point to a future block *before* any occurrence of $\mathtt{ifzero}_k$:

$$\bigwedge_k \mathsf{G}\big(\mathrm{INC}(k) \Rightarrow \neg(\downarrow \mathsf{F_s}(\mathtt{@} \wedge \downarrow \mathsf{F_s}(\mathrm{IZ}(k) \wedge \mathsf{F_s} \uparrow) \wedge \mathsf{F_s}(\# \wedge \uparrow)))\big)$$

This concludes the construction of φ_C. The correctness proof is standard. □

3.2 Undecidability of $\mathsf{sLTL}_1^{\downarrow}(\mathsf{F_s}, \mathsf{F_s}^{-1})$

We now consider $\mathsf{sLTL}_1^{\downarrow}(\mathsf{F_s}, \mathsf{F_s}^{-1})$. The extra modality can be used to code the run of a (non faulty) Minsky CA.

Theorem 3. *Satisfiability of* $\mathsf{sLTL}_1^{\downarrow}(\mathsf{F_s}, \mathsf{F_s}^{-1})$ *is undecidable.*

Proof. Consider a Minsky CA C. We revisit the proof of Theorem 2. It is easy to see that we can enforce the absence of faulty increments during the run by asking that every $\mathtt{dec}_i$ element is referenced by some previous $\mathtt{inc}_i$ block: $\bigwedge_i \mathsf{G}(\mathrm{DEC}(i) \Rightarrow \downarrow \mathsf{F_s}^{-}(\mathtt{@} \wedge \uparrow))$. We thus make sure that every $\mathtt{dec}$ is related to a corresponding $\mathtt{inc}$. Hence, the coding is that of a *perfect* (non faulty) run. □

4 The Case of $\mathsf{LTL}_1^{\downarrow}$

In this section we lift the lower bounds of the previous section by considering the temporal operator F instead of $\mathsf{F_s}$. We can only do so by removing at the same time the restriction to simple formulas. Hence the results of this section cannot be applied to XPath. Notice that $\mathsf{LTL}_1^{\downarrow}(\mathsf{F})$ and $\mathsf{sLTL}_1^{\downarrow}(\mathsf{F_s})$ are incomparable in terms of expressive power. Indeed, properties like $\downarrow \mathsf{F}(\mathsf{a} \wedge \bar{\uparrow} \wedge \mathsf{F}(\mathsf{b} \wedge \uparrow))$ cannot be expressed in $\mathsf{sLTL}_1^{\downarrow}(\mathsf{F_s})$, while $\mathsf{LTL}_1^{\downarrow}(\mathsf{F})$ cannot express that the model has at least two elements. We do not know whether $\mathsf{sLTL}_1^{\downarrow}(\mathsf{F})$, which is weaker than the two above mentioned logics, is already non primitive recursive. The results of this section improve the results of [8] which show that satisfiability is non primitive recursive for $\mathsf{LTL}_1^{\downarrow}(\mathsf{X}, \mathsf{F})$ and undecidable for $\mathsf{LTL}_1^{\downarrow}(\mathsf{X}, \mathsf{F}, \mathsf{F}^{-1})$.

Theorem 4. *Over data words,*

1. *Satisfiability of* $\mathsf{LTL}_1^{\downarrow}(\mathsf{F})$ *is decidable and non primitive recursive.*
2. *Satisfiability of* $\mathsf{LTL}_1^{\downarrow}(\mathsf{F}, \mathsf{F}^{-1})$ *is undecidable.*

Proof. We only prove Item 1, the proof of Item 2 being similar.
Consider an ICA C and recall the coding of runs of C used in the proof of Theorem 2. In the construction of the formula φ_C, whenever we have "$s \wedge \mathsf{F_s}(s' \wedge \varphi)$" for some φ and $s \neq s'$ two different symbols, $\mathsf{F_s}$ can be equivalently replaced by the F temporal operator. This is the case in all formulas except in three places: (i) The formula saying that $\mathtt{N}$ should point to the next block contains $\mathtt{\#} \wedge \mathsf{F_s}(\mathtt{\#} \wedge \varphi)$. But from the structure that is enforced, this can equivalently be replaced by $\mathtt{\#} \wedge \mathsf{F}(\mathtt{N} \wedge \mathsf{F}(\mathtt{\#} \wedge \varphi))$. (ii) To enforce that each block contains at most one occurrence of symbols in $\hat{\Sigma} \cup \{\mathtt{N}, \mathtt{\#}, \mathtt{@}\}$. (iii) To enforce that no two symbols in $\hat{\Sigma} \cup \{\mathtt{N}, \mathtt{\#}, \mathtt{@}\}$ have the same data value.

In order to cope with (ii) and (iii), we use a slightly different coding for runs of C. This coding is the same as the one for the proof of Theorem 2 except that we allow succession of equal symbols, denoted as *group* in the sequel. Note that in a group two different occurrences of the same symbol in general may have different data values, as we can no longer enforce their distinctness. However, as we will see, we can enforce that the $\hat{\Sigma}$ group of elements of a block have all *the same data value.*

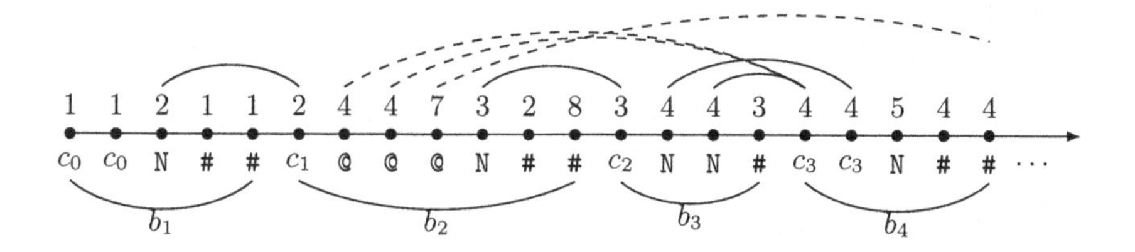

Fig. 2. $\mathsf{LTL}_1^{\downarrow}(\mathsf{F})$ cannot avoid having repeated consecutive symbols

Hence a *block* is now either a group of $c \in \hat{\Sigma}$ followed by a group of `N` followed by a group of `#` or the same with a group of `@` in between. A coding of a run is depicted in Figure 2.

This structure is enforced by modifying the formulas of the proof of Theorem 2 as follows.

(i) The formulas that limit the number of occurrences of symbols in a block are replaced by formulas *limiting the number of groups* in a block.

(ii) The formulas requiring that no two occurrences of a same symbol may have the same data values are replaced by formulas requiring that no two occurrences of a same symbol *in different groups* have the same data value.

(iii) In all other formulas, $\mathsf{F_s}$ is replaced by F.

(iv) Finally, we ensure that although we may have repeated symbols inside a block, all symbols from $\hat{\Sigma}$ have the same data value.

$$\mathsf{G}(\hat{\sigma} \Rightarrow \neg \downarrow \mathsf{F}((\hat{\sigma} \vee \#) \wedge \neg \uparrow \wedge \mathsf{F}(\# \wedge \uparrow)))$$

Note that this implies that each `N` of a group must have the same data values as they all point to the next block. However there could still be `@` symbols with different data values as depicted in Fig. 2.

The new sentences now imply, for instance, that:

1. Every position from a group of $c \in \hat{\Sigma}$ have the same data value which is later matched by an element of a group of `#`.
2. Every position from a group of `N` has the same data value as a position $c \in \hat{\Sigma}$ of the *next* block (and then, it has only *one* possible data value).
3. Every position from a group `@` has the same data value as a position $c \in \hat{\Sigma}$ of a block to its right. Note that the data values of two `@` of the same group may correspond to the data values of symbols in different blocks. This is basically the main conceptual difference with the previous proof.

The proof of correctness of the construction is left to the reader. □

Note that in the previous proof we used the fact that $\mathsf{LTL}_1^{\downarrow}(\mathsf{F})$, although it has only one register, can make (in)equality tests several times throughout a path (as used in the formula of item (iv) in the proof), something that $\mathsf{sLTL}_1^{\downarrow}(\mathsf{F_s})$ and XPath cannot do.

Two registers. When 2 registers are available the previous result can be adapted to code a (non faulty) Minsky CA with a strategy similar to [8, Theorem 5.4].

Theorem 5. *Satisfiability of* $\mathsf{LTL}_2^{\downarrow}(\mathsf{F})$ *is undecidable over data words.*

5 Data Trees and XPath

We now turn to data trees. An XML document can be seen as an unranked tree with *attributes* and data values in its nodes. While a data tree has only *one* data value per node, an XML document element may have *several* attributes each of which with an associated data value. All XPath fragments treated in this paper can force all elements of an XML document to have only one attribute. Therefore all hardness results in the present work hold also for the class of XML documents.

The logic XPath. XPath is a two-sorted language, with *path* expressions $(\alpha, \beta, \ldots)$ and *node* expressions $(\varphi, \psi, \ldots)$. These are defined by mutual recursion:

$$\alpha, \beta ::= \downarrow \mid \downarrow^+ \mid \uparrow \mid \uparrow^+ \mid \rightarrow \mid \rightarrow^+ \mid \leftarrow \mid {}^+\!\!\leftarrow \mid \varepsilon \mid \alpha\beta \mid \alpha \cup \beta \mid \alpha[\varphi]$$
$$\varphi, \psi ::= \sigma \mid \langle\alpha\rangle \mid \neg\varphi \mid \varphi \wedge \psi \mid \alpha = \beta \mid \alpha \neq \beta \qquad (\sigma \in \Sigma)$$

A path expression essentially describes a traversal of the tree by using the axis: `child` ($\downarrow$), `descendant` ($\downarrow^+$), the `next-sibling` ($\rightarrow$), `following-sibling` ($\rightarrow^+$) and their inverses, with the ability to test at any stage for node conditions. Let t be a data tree with domain T and labeling function λ. The semantics of XPath is defined by induction is the usual intuitive way, we only give here some cases:

$$[\![\downarrow]\!]^t = \{(x, xi) \mid xi \in T\} \qquad [\![\alpha\beta]\!]^t = \{(x,z) \mid \exists y.(x,y) \in [\![\alpha]\!]^t \wedge (y,z) \in [\![\beta]\!]^t\}$$
$$[\![\langle\alpha\rangle]\!]^t = \{x \in T \mid \exists y.(x,y) \in [\![\alpha]\!]^t\} \qquad [\![\alpha[\varphi]]\!]^t = \{(x,y) \in [\![\alpha]\!]^t \mid y \in [\![\varphi]\!]^t\}$$
$$[\![\varepsilon]\!]^t = \{(x,x) \mid x \in T\} \qquad [\![\alpha^+]\!]^t = \text{the transitive closure of } [\![\alpha]\!]^t$$
$$[\![\alpha = \beta]\!]^t = \{x \in T \mid \exists y, z.(x,y) \in [\![\alpha]\!]^t, (x,z) \in [\![\beta]\!]^t, \pi_2(\lambda(y)) = \pi_2(\lambda(z))\}$$
$$[\![\alpha \neq \beta]\!]^t = \{x \in T \mid \exists y, z.(x,y) \in [\![\alpha]\!]^t, (x,z) \in [\![\beta]\!]^t, \pi_2(\lambda(y)) \neq \pi_2(\lambda(z))\}$$

A key property for using results of Section 3 is that simple formulas can be translated to XPath and back. The proof of this result is straightforward by induction on the formula.

Proposition 1. *Over data words,* $\mathsf{sLTL}_1^{\downarrow}(\mathsf{F_s})$ *and* $\mathsf{XPath}(\rightarrow^+, =)$ *have the same expressive power. The same holds for* $\mathsf{sLTL}_1^{\downarrow}(\mathsf{F_s}, \mathsf{F_s}^{-1})$ *and* $\mathsf{XPath}({}^+\!\!\leftarrow, \rightarrow^+, =)$. *Moreover, in both cases, the transformation from* $\mathsf{sLTL}_1^{\downarrow}$ *to* XPath *takes polynomial time while it takes exponential time in the other direction.*

The restriction on negations in the definition of $\mathsf{sLTL}_1^{\downarrow}$ corresponds to the fact that XPath path expressions are always positive: any path α is essentially a nesting of operators F with intermediate tests. We remark that there is a big difference between $\mathsf{XPath}(\rightarrow^+, =)$ over data words and $\mathsf{XPath}(\downarrow^+, =)$ over data trees. Indeed $\mathsf{XPath}(\downarrow^+, =)$ is closed under bisimulation and hence it cannot assume that the tree is a vertical path. As the string structure was essential in the proof of Theorem 2, the non primitive recursiveness of $\mathsf{XPath}(\rightarrow^+, =)$ over data words does not lift to $\mathsf{XPath}(\downarrow^+, =)$ over data trees. Actually, satisfiability of $\mathsf{XPath}(\downarrow^+, =)$ is ExpTime-complete [10]. However, if one considers the logic $\mathsf{XPath}(\downarrow^+, \rightarrow, =)$ then the axis $\rightarrow$ can be used to enforce a vertical path $(\neg \downarrow^+ [\rightarrow])$ and therefore it follows from Theorem 2 and Proposition 1 that:

Corollary 1. *Satisfiability of* $\mathsf{XPath}(\downarrow^+, \rightarrow, =)$ *on data trees is at least non primitive recursive.*

Similarly, in $\mathsf{XPath}(\downarrow^+, \uparrow^+, =)$ one can simulate a string by going down to a leaf using $\downarrow^+$ and then use the path from that leaf to the root as a string using $\uparrow^+$.

Corollary 2. *Satisfiability of* $\mathsf{XPath}(\downarrow^+, \uparrow^+, =)$ *on data trees is at least non primitive recursive.*

Note that the decidability of $\mathsf{XPath}(\downarrow^+, \rightarrow, =)$ and of $\mathsf{XPath}(\downarrow^+, \uparrow^+, =)$ is still an open problem.

Open problem: Is $\mathsf{XPath}(\downarrow^+, \uparrow^+, =)$ decidable over data trees?

It would be interesting to know whether the strictness of the axis $\downarrow^+$ is necessary in the above two results. This boils down to know whether $\mathsf{sLTL}_1^{\downarrow}(\mathsf{F})$ is already not primitive recursive over data words. Note that the proof of Theorem 4 uses in an essential way the possibility to make (in)equality tests several times throughout a path. This is exactly what cannot be expressed in $\mathsf{sLTL}_1^{\downarrow}(\mathsf{F})$.

Open problem: Is satisfiability of $\mathsf{sLTL}_1^{\downarrow}(\mathsf{F})$ primitive recursive over data words?

We conclude with some simple consequences of Theorem 3 and Proposition 1:

Corollary 3. *Satisfiability of* $\mathsf{XPath}({}^+\!\!\leftarrow, \rightarrow^+, =)$ *and of* $\mathsf{XPath}(\downarrow^+, \uparrow^+, \rightarrow, =)$ *over data trees is undecidable.*

Corollary 4. *Satisfiability of* $\mathsf{XPath}(\rightarrow^+, \downarrow, \uparrow, =)$ *is undecidable.*

Proof. This is similar to the proof of Theorem 3 with a slight difference. Consider that the coding of the run of the counter machine is done at the first level of the tree (i.e., at distance 1 from the root). Then, the property to ensure that every decrement has a corresponding increment is now: $\bigwedge_i \neg \downarrow [\textsc{dec}(i) \wedge \neg \varepsilon = \uparrow\downarrow [\mathtt{@}]]$.

6 Discussion

By [8] it is known that satisfiability of $\mathsf{LTL}_1^{\downarrow}(\mathsf{X}, \mathsf{F})$ with infinite data words is undecidable. The proof of Theorem 4 can be extended to code runs of ICA over infinite data words, which is known to be undecidable, to show that this result already holds in the absence of X.

Theorem 6. *On infinite data words, the satisfiability problem of* $\mathsf{LTL}_1^{\downarrow}(\mathsf{F})$ *is undecidable.*

Summary of Results

In the table below we summarize the main results and some of the consequences we have mentioned. In this table, $\overline{\mathsf{PR}}$ stands for non primitive recursive.

Logic	Complexity	Details
$\mathsf{LTL}_1^{\downarrow}(\mathsf{F})$	$\overline{\mathsf{PR}}$, decidable	Theorem 4 & [8]
$\mathsf{LTL}_1^{\downarrow}(\mathsf{F}, \mathsf{F}^{-1})$	undecidable	Theorem 4
$\mathsf{LTL}_2^{\downarrow}(\mathsf{F})$	undecidable	Theorem 5
$\mathsf{sLTL}_1^{\downarrow}(\mathsf{F_s})$	$\overline{\mathsf{PR}}$, decidable	Theorem 2 & [8]
$\mathsf{sLTL}_1^{\downarrow}(\mathsf{F_s}, \mathsf{F_s}^{-1})$	undecidable	Theorem 3
$\mathsf{XPath}(\downarrow^+, \rightarrow, =)$	$\overline{\mathsf{PR}}$, decidability unknown	Corollary 1
$\mathsf{XPath}(\downarrow^+, \uparrow^+, =)$	$\overline{\mathsf{PR}}$, decidability unknown	Corollary 2
$\mathsf{XPath}(\downarrow^+, \uparrow^+, \rightarrow, =)$	undecidable	Corollary 3
$\mathsf{XPath}(\rightarrow^+, \downarrow, \uparrow, =)$	undecidable	Corollary 4

References

1. Benedikt, M., Fan, W., Geerts, F.: XPath satisfiability in the presence of DTDs. J. ACM 55(2) (2008)
2. Björklund, H., Martens, W., Schwentick, T.: Optimizing conjunctive queries over trees using schema information. In: Ochmański, E., Tyszkiewicz, J. (eds.) MFCS 2008. LNCS, vol. 5162, pp. 132–143. Springer, Heidelberg (2008)
3. Bojańczyk, M., David, C., Muscholl, A., Schwentick, T., Segoufin, L.: Two-variable logic on data trees and XML reasoning. In: PODS, pp. 10–19 (2006)
4. Bojańczyk, M., Muscholl, A., Schwentick, T., Segoufin, L., David, C.: Two-variable logic on words with data. In: LICS, pp. 7–16 (2006)
5. Bouyer, P., Petit, A., Thérien, D.: An algebraic approach to data languages and timed languages. Inf. Comput. 182(2), 137–162 (2003)
6. Clark, J., DeRose, S.: XML path language (XPath) (1999), W3C Recommendation, `http://www.w3.org/TR/xpath`
7. David, C.: Complexity of data tree patterns over XML documents. In: Ochmański, E., Tyszkiewicz, J. (eds.) MFCS 2008. LNCS, vol. 5162, pp. 278–289. Springer, Heidelberg (2008)
8. Demri, S., Lazić, R.: LTL with the freeze quantifier and register automata. ACM Transactions on Computational Logic (2009)
9. Demri, S., Lazić, R., Nowak, D.: On the freeze quantifier in constraint LTL: Decidability and complexity. In: TIME, pp. 113–121 (2005)
10. Figueira, D.: Satisfiability of downward XPath with data equality tests. In: PODS, Providence, Rhode Island, USA. ACM Press, New York (2009)
11. Geerts, F., Fan, W.: Satisfiability of XPath queries with sibling axes. In: Bierman, G., Koch, C. (eds.) DBPL 2005. LNCS, vol. 3774, pp. 122–137. Springer, Heidelberg (2005)
12. Jurdziński, M., Lazić, R.: Alternating automata on data trees and XPath satisfiability. CoRR, abs/0805.0330 (2008)

13. Minsky, M.L.: Computation: finite and infinite machines. Prentice-Hall, Inc., Englewood Cliffs (1967)
14. Ouaknine, J., Worrell, J.: On Metric temporal logic and faulty Turing machines. In: Aceto, L., Ingólfsdóttir, A. (eds.) FOSSACS 2006. LNCS, vol. 3921, pp. 217–230. Springer, Heidelberg (2006)
15. Schnoebelen, P.: Verifying lossy channel systems has nonprimitive recursive complexity. Information Processing Letters 83(5), 251–261 (2002)

A By-Level Analysis of Multiplicative Exponential Linear Logic

Marco Gaboardi[1], Luca Roversi[1], and Luca Vercelli[2]

[1] Dipartimento di Informatica - Università di Torino
[2] Dipartimento di Matematica - Università di Torino
http://www.di.unito.it/~{gaboardi,rover,vercelli}

Abstract. We study the relations between Multiplicative Exponential Linear Logic (meLL) and Baillot-Mazza Linear Logic by Levels (mL^3). We design a decoration-based translation between propositional meLL and propositional mL^3. The translation preserves the cut elimination. Moreover, we show that there is a proof net Π of second order meLL that cannot have a representative Π' in second order mL^3 under any decoration. This suggests that levels can be an analytical tool in understanding the complexity of second order quantifier.

1 Introduction

The implicit characterization of the polynomial and elementary time computations by means of structural proof theory takes its origins from a predicative analysis of non termination. We recall, indeed, that Girard conceived Elementary Linear Logic (ELL) and Light Linear Logic (LLL) [1] by carefully analyzing the formalization of naïve set theory inside the Multiplicative and Exponential fragment of Linear Logic (meLL). The comprehension scheme could be represented without any paradoxical side effect by forbidding the logical principles *dereliction* $!A \multimap A$ and *digging* $!!A \multimap !A$.

Intuitively, without dereliction and digging the proof nets of both ELL and LLL are *stratified*. Namely, during the cut elimination process, every node of a proof net either disappears or it is always contained in a constant number of *regions*, called boxes. The stratified proof nets of ELL are characterized by a cut elimination cost which is bounded by an elementary function whose parameters are the size of a given net Π and its depth, *i.e.* the maximal number of nested boxes in Π.

Moreover, Girard noted that ruling out the *monoidality* of the functor "!", i.e. $(!A \otimes !B) \multimap !(A \otimes B)$, from ELL yields LLL whose cut elimination cost lowers to a polynomial. The reason is that the logical connective $\otimes$ somewhat allows to count the resources we may need. Commuting $\otimes$ with ! hides the amount of used logical resources because of the *contraction* $!A \multimap (!A \otimes !A)$. So, the absence of monoidality allows to keep counting the needed resource by means of $\otimes$.

In [2], the authors pursue the predicative analysis on meLL by introducing mL^3. This system generalizes ELL by means of explicit indices associated to the edges of the proof nets of meLL. Moreover, further structural restrictions on mL^3 yield a polynomial time sound generalization mL^4 of LLL. The use of indices in meLL analysis is not new and traces back to, at least, [3,4]. The new systems mL^3 and mL^4 still characterize

R. Královič and D. Niwiński (Eds.): MFCS 2009, LNCS 5734, pp. 344–355, 2009.

implicitly the elementary and polynomial computations. Their distinguishing feature lies in a more flexible use of the nodes that change the level, so, somewhat generalizing the notion of box.

Since mL^3 is a restriction of meLL, it is natural that some derivation of the latter cannot be represented inside the former, the reason being we know that the cost of the cut elimination of meLL overwhelms the elementary one of mL^3.

In this paper, we show that indices strongly restrict meLL proof nets in presence of $\exists$ and $\forall$, while they are minor restriction when quantifier do not get used. Indeed, we can show that every proof net Π in the *propositional fragment* of meLL has a representative Π' in mL^3 that preserves the cut elimination. Specifically, Π' is the result of a predicative analysis of Π, based on indices we can use to label every edge of mL^3 proof nets. The proof net Π' of mL^3 is the result of the algorithm @, applied to Π, we introduce in this work. The proof net $@(\Pi)$ is a decoration of both the edges and the formulæ of Π, using the paragraph modality §, whose instances correspond to an index change in the proof net of mL^3 being constructed.

The interest of the translation that @ implements is twofold. Concerning the structural proof theory, @ shows that the modality § internalizes the notion of index at the level of the formulæ. Concerning the implicit characterization of complexity classes, @ offers the possibility of a finer study of normalizations measures of propositional meLL, thanks to the structural aspects that mL^3 supplies.

Finally we answer negatively to the following two natural questions: (i) Is there any extension of @ able to translate every proof net of (full) meLL into mL^3?, and (ii) Is there any translation, alternative to any generalization of @, from meLL to mL^3? The reason of the negative answer lies in the proof net Π of meLL in Figure 6. There is no decoration Π' in mL^3 of Π because to obtain Π' either we should collapse two distinct indices of Π' or we would need a new node able to change indices but not the formulæ. Both solutions would imply a cut elimination cost blow up, unacceptable inside mL^3.

Summing up, the predicative analysis of meLL by means of the indices inside mL^3 identifies as the true source of impredicativity of meLL *the collapse of indices, implicit in the second order quantification of the formulæ of* meLL *itself*. Then, the "side effect" of such a collapse is the huge cut elimination bound of meLL.

2 Second Order meLL

We start by recalling second order Multiplicative Exponential Logic (meLL) in proof nets style. In particular, analogously to [2], we present a meLL version including the paragraph (§) modality.

The formulæ. meLL derives multisets of formulæ that belong to the language generated by the following grammar:

$$F ::= A \mid \flat A \qquad\qquad A ::= \alpha \mid A \otimes A \mid A \parr A \mid \forall\alpha.A \mid \exists\alpha.A \mid\, !A \mid\, ?A \mid \S A \mid A^{\perp}$$

The start symbol F generates both *(standard) formulæ* and *partially discharged formulæ*. Standard formulæ are generated from the start symbol A. Partially discharged formulæ are of kind $\flat A$; the syntax prevents nesting of $\flat$ symbols. We shall use A, B, C,

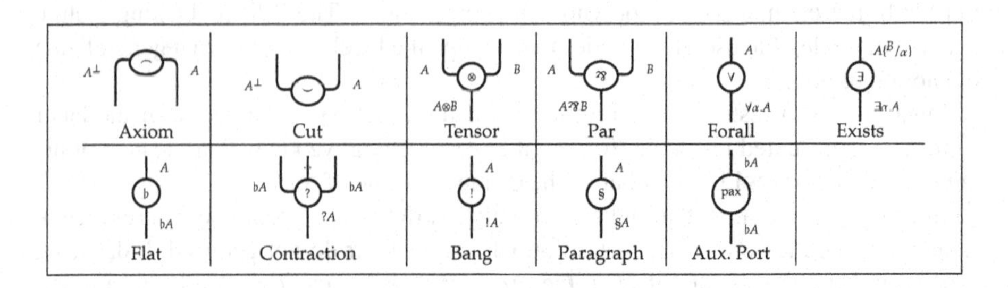

Fig. 1. Nodes for the nets of meLL

possibly with sub or superscripts, to range over standard formulæ, F, G to range over formulæ. Γ, Δ, Ξ range over multisets of formulæ. The standard meLL formulæ are quotiented by the De Morgan rules, where $(A, A^{\perp})$, $(\otimes, \parr)$, $(\forall, \exists)$, $(\S, \S)$ and $(!, ?)$ are the pairs of dual operators. Notice that $\S$ is self dual, namely: $(\S A)^{\perp} = \S(A^{\perp})$.

Proof nets of meLL. Given the nodes in Figure 1, we say that an *Axiom node is a proof net*. Moreover, given two proof nets:

$$\Pi_1 : F_1, \ldots, F_m \qquad \Pi_2 : G_1, \ldots, G_n$$

with $m, n \geq 1$, then all the graphs inductively built from Π_1 and Π_2 by the rule schemas in Figure 2 are proof nets.

Cut elimination in meLL. Every pair of dual *linear* nodes (axiom/cut, $\otimes/\parr$, $\forall/\exists$, $\S/\S$) annihilates in one step of reduction, as usual in literature. The *exponential* pair of dual nodes $!/?$ rewrites by means of the *big-step* in Figure 3.

Basic definitions and properties in meLL. The modality $\S$ is not part of the original version of meLL; it is easy to show that in meLL $\S$ is, essentially, useless, i.e. A and $\S A$ may be proved equivalent in meLL. Nevertheless, $\S$ become useful when handling sublogics of meLL. $\S^k A$ means $\S \ldots \S A$ with k paragraphs.

The original formulation of meLL also contains the **mix** rule and **units**, but for simplicity we omit them.

A **weakening** node is a contraction with 0 premises. We call **axiom-edge**, **weakening-edge**, **cut-edge**, etc. an edge connected to an an axiom node, a weakening node, a cut node, etc..

Fact 1 (About the Structure of the Proof Nets). *Let Π be a proof net of* meLL, *and u one of its cut links or conclusions. Let ρ be a graph-theoretical path along Π from u to an axiom or to a weakening node v, not containing any other axioms. Then ρ does not contain any other cut node.*

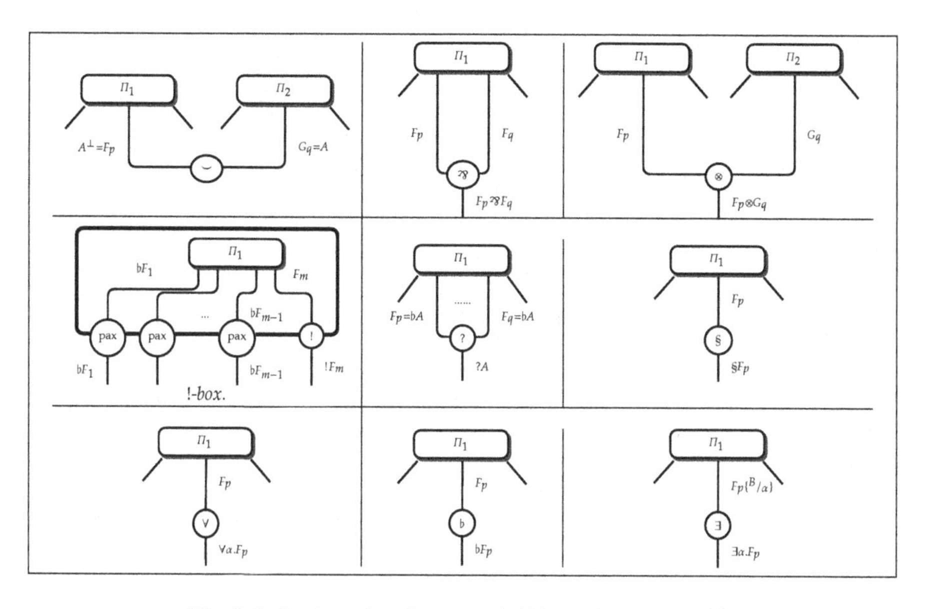

Fig. 2. Inductive rule schemes to build proof nets of meLL

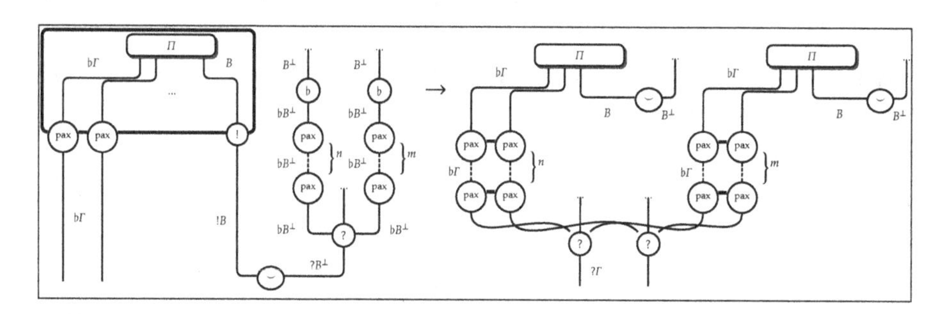

Fig. 3. Big-step reduction. A contraction with k premises in the redex implies k copies of Π in the reduct. For the sake of clarity we do not draw all the boxes in the picture.

Thanks to this Fact, we can state that all the edges of our proof nets are directed downwards, from axioms or weakening nodes towards conclusions or cut nodes, even if we do not draw the corresponding arrows. A **path** inside a meLL proof net Π is a sequence of nodes $\tau = \langle u_0, \ldots, u_k \rangle$ in Π such that (i) each u_i is connected with u_{i+1}, (ii) the direction of such edge is from u_i towards u_{i+1}, and (iii) for every i, $u_i \neq u_{i+1}$. The **size** of a meLL proof net is the number of its nodes.

3 Multiplicative Linear Logic by Levels: mL^3

The system mL^3 is described in [2]. It is the subsystem of all the proof nets of meLL admitting an *indexing*:

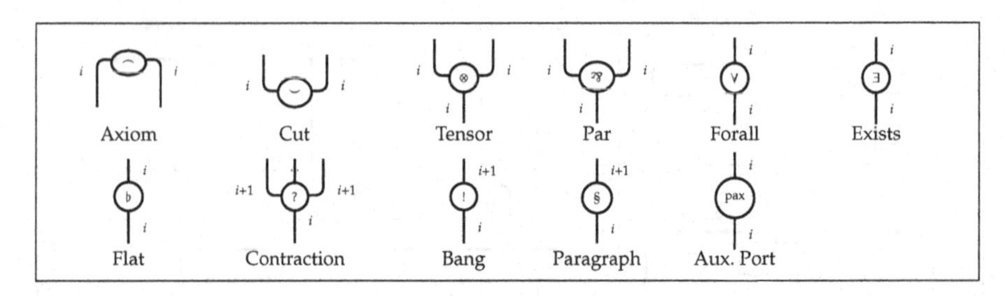

Fig. 4. Costraints for indexing meLL proof nets

Definition 1. *Let Π be a proof net of* meLL. *An* **indexing** *for Π is a function I from the edges of Π to $\mathbb{Z}$ satisfying the constraints in Figure 4 and such that $I(e) = I(e')$ for all the conclusions e, e' of Π.*

Fact 2 (Indexes do not Increase from Axioms to Conclusions). *Let Π be an* mL^3 *proof net, I an indexing for Π, ρ a path from some node u to some node v. Then $I(u) \geq I(v)$.*

It will be convenient to consider a particular kind of indexing.

Definition 2. *Let Π be an* mL^3 *proof net, and I be an indexing for Π. We say that I is* **canonical** *if Π has an edge e such that $I(e) = 0$, and $I(e') \geq 0$ for all edges e' of Π.*

Fact 3 (Existence of Canonical Indexing [2]). *Every proof net of* mL^3 *admits one and only one canonical indexing.*

We can now define a measure on mL^3 proof nets.

Definition 3. *Let Π be an* mL^3 *proof net, and let I_0 be its canonical indexing. The* **level** *of Π is the maximum integer assigned by I_0 to the edges of Π.*

If 2^n_x is the function such that $2^n_0 = 2^n$ and $2^n_m = 2^{2^n_{m-1}}$, then:

Theorem 1 (Elementary bound for mL^3 [2]). *Let Π be an* mL^3 *proof net of size s and level l. Then, the round-by-round cut-elimination procedure reaches a normal form in at most $(l+1)2^s_{2l}$ steps.*

The Theorem above is a result of *weak* polynomial soundness, as it only has been proved for a particular cut-elimination procedure. It is reasonable however that it can be generalized to any reduction strategy, in analogy to what happens in ELL and LLL [5]. The interested reader may find the definition of the round-by-round procedure and a proof of Theorem 1 in [2].

4 Embedding Propositional meLL into mL^3

Definition 4. *Let Π be a proof net of* meLL. *A* **quasi-indexing** *for Π is a function Q from the edges of Π to $\mathbb{Z}$ that respects all the constraints in Figure 4, with the possible exception of the axiom edges, and such that for all conclusion e, e' of Π it holds $Q(e) = Q(e')$.*

Fact 4 (Quasi-Indexing Exists). *Every* meLL *proof net admits a quasi-indexing.*

Proof. Let Π be a proof net of meLL; we want to build some quasi-indexing Q. We call $c_1, \ldots, c_n$ the cut nodes of Π. We arbitrarily choose a value $Q(e) = i$ for all the conclusion edges e of Π, and a value $Q(e_1^j) = Q(e_2^j) = i_j$ for every couple of edges e_1^j, e_2^j incident in c_j. Then, using the rules in Figure 4, we can calculate the value of Q in all the edges of the proof net. The process of calculation terminates when the axiom and weakening nodes are reached. □

For every Π, whose cut nodes are $c_1, \ldots, c_n$, we call $Q(i, i_1, , ...i_n)$ the (unique) quasi-indexing that has value i on the conclusions and value $i_1, \ldots, i_n$ on the cut-edges. This definition is justified looking at the proof of Fact 4.

The coming *level of a formula* is completely unrelated to the levels of Definition 3:

Definition 5. *For every formula A of* meLL *let the* **formula level** $\mathtt{fl}(A)$ *be:*

$$\mathtt{fl}(\alpha) = 0 \qquad \mathtt{fl}(\diamond A) = \mathtt{fl}(A) + 1 \qquad \diamond \in \{!, ?, \S\}$$
$$\mathtt{fl}(\flat A) = \mathtt{fl}(A) \qquad \mathtt{fl}(A \,\square\, B) = \max\{\mathtt{fl}(A), \mathtt{fl}(B)\} \qquad \square \in \{\otimes, \parr\}$$

Definition 6. *Let Π be a proof net of* meLL*, Q a quasi-indexing for it. Let e be an edge in Π, labelled by a formula A. Then, the* **absolute level** *of e in Π is defined as* $\mathtt{al}(e) = Q(e) + \mathtt{fl}(A)$.

Notice that the definition depends on the chosen quasi-indexing.

The following map is crucial in the proof of Proposition 1:

Definition 7. *For every* meLL *formula A let $(A)^*$ be defined as:*

$$(\alpha)^* = \alpha$$
$$(B \,\square\, C)^* = \S^d (B)^* \,\square (C)^* \quad \text{if } d = \mathtt{fl}(C) - \mathtt{fl}(B) \geq 0 \qquad \square \in \{\otimes, \parr\}$$
$$(B \,\square\, C)^* = (B)^* \,\square\, \S^{-d} (C)^* \quad \text{if } d = \mathtt{fl}(C) - \mathtt{fl}(B) \leq 0 \qquad \square \in \{\otimes, \parr\}$$
$$(\diamond A)^* = \diamond((A)^*) \qquad \diamond \in \{!, ?, \flat\}.$$

The algorithm @. The main result of this section concerns the following algorithm @. Let the arguments of @ be a proof net $\Pi : A_1, \ldots, A_n$ of propositional meLL and a quasi-indexing Q for Π. The algorithm returns an mL^3 proof net. We will give a direct proof of this fact. Let the conclusions and the cut edges of Π be $e_1, \ldots, e_n$. Let $K = \max_{1 \leq i \leq n} \{\mathtt{al}(e_i)\}$. For every edge e_i, with $1 \leq i \leq n$, labelled with the formula A_i, we define @ to perform the following steps:

1. Replace A_i by $(A_i)^*$.
2. Add k_i new ($\S$) nodes after the edge e_i where $k_i = K - \mathtt{al}(e_i)$, label the new edges respectively by $\S^1(A_i)^*, \ldots, \S^{k_i}(A_i)^*$ and modify the quasi-indexing accordingly. Note that now $\mathtt{al}\,(e_i) = K$. See Figure 5 a.
3. Apply the subroutine ϑ of @, here below, to the edge e_i.

The subroutine ϑ takes an edge e of (the already modified version of) Π as its argument. ϑ is recursive and is defined by cases on the kind of the edge e:

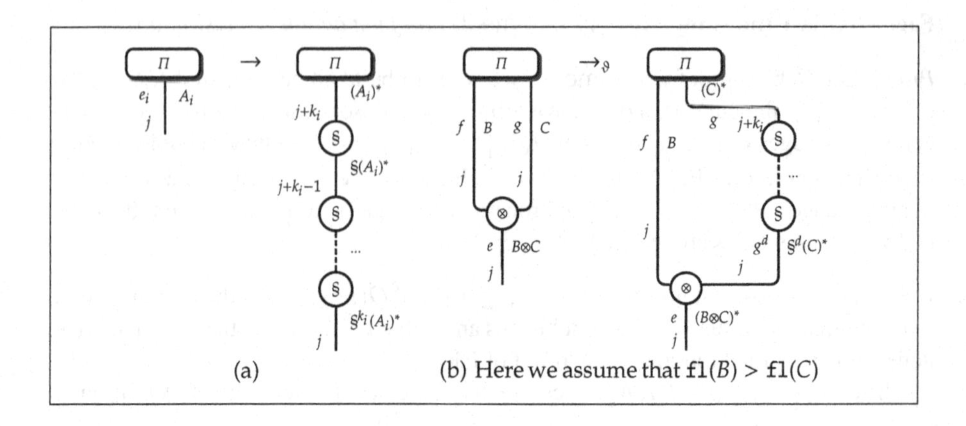

Fig. 5. The main cases of the rewriting steps performed by @ (Proof of Proposition 1)

(a) If e is an axiom edge, then it is done.
(b) If e is the conclusion of a $(\otimes)$ node with premises the edges f and g labelled with formulae B and C respectively, then replace B by $(B)^*$ and C by $(C)^*$ respectively. Let us suppose for clarity that $\texttt{fl}(B) > \texttt{fl}(C)$ (see Figure 5 b). Calling $d = (\texttt{fl}(B) - \texttt{fl}(C))$, we add d new $(\S)$ nodes after the edge g and label the new edges $g^1, \ldots, g^d$ respectively by $\S^1(C_i)^*, \ldots, \S^d(C_i)^*$. Modify Q accordingly, then apply ϑ on f and g.
(c) If e is the conclusion of a $(!)$ node (or $(\flat)$ or (pax)) with premises the edge f labelled with the formula B then replace it by $(B)^*$ and apply ϑ on f.
(d) If e is the conclusion of a $(?)$ node with premises the edges $f_1, \ldots, f_l$ labelled with formulae $B_1, \ldots B_l$, then replace them by $(B_1)^*, \ldots, (B_l)^*$ and apply ϑ on every $f_1, \ldots, f_l$.

Proposition 1 (Embedding Propositional meLL into mL^3). *There is an algorithm $@(\cdot,\cdot)$ that takes every proof net Π of propositional meLL, endowed with a quasi-indexing Q, and returns a proof net $@(Q,\Pi)$ of mL^3. The proof nets Π and $@(Q,\Pi)$ only differ for the possible presence of some new paragraph nodes.*

Proof. $@(\cdot,\cdot)$ is the algorithm already described. $@(\cdot,\cdot)$ transforms a proof net Π of meLL in a new graph $@(Q,\Pi)$, with conclusions labelled by $\S^{k_1}(A_1)^*, \ldots, \S^{k_n}(A_n)^*$, for some $k_1, \ldots, k_n$, to which it is naturally associated a quasi-indexing Q'. The quasi-indexing Q' associates to conclusions and cut edges of $@(Q,\Pi)$ the same indices as Q assigns to conclusions and cut edges of Π. We need to check that $@(Q,\Pi)$ is really a proof net of meLL, and that this proof net is in mL^3.

Let us consider the transformations previously described. The untyped graph is still an untyped proof net of meLL, because we have just added some paragraphs. Moreover, by construction every edge e of Π labelled by A is translated into an edge e' of $@(Q,\Pi)$, labelled by $(A)^*$. So, in particular, axioms, cuts and contractions are labelled correctly. The labelling of the other nodes follows by construction of $@(\cdot,\cdot)$.

At last, we need to show that Q' is an indexing. Let us consider two edges f, g incident into an axiom in $@(Q, \Pi)$, labelled resp. by A and $A^\perp$. Notice that, by construction, for every edge e of $@(Q, \Pi)$ it holds $\mathtt{al}(e) = K$. As a consequence, f and g have the same quasi-index $Q'(e) = \mathtt{al}(e) - \mathtt{fl}(A) = K - \mathtt{fl}(A)$, and so Q' is also an indexing. □

Proposition 2 **($@(\cdot,\cdot)$ preserves the Cut-Elimination).** *For every reduction $\Pi \to^+ \Sigma$ in propositional* meLL*, and for every quasi-indexing Q of Π, there exists a quasi-indexing $\tilde{Q}$ of Σ such that $@(Q, \Pi) \to^+ @(\tilde{Q}, \Sigma)$:*

$$\begin{array}{ccc} \Pi & \to^+ \ \Sigma & \textit{in } \mathsf{meLL} \\ \downarrow & \downarrow & \\ @(Q, \Pi) & \to^+ \ @(\tilde{Q}, \Sigma) & \textit{in } \mathsf{mL}^3 \end{array}$$

Proof. It is enough to prove the result for 1-step reductions $\Pi \to \Sigma$. So, let c be the cut fired during this reduction; c corresponds to a unique cut c' of $@(Q, \Pi)$. By construction of $@(\cdot,\cdot)$, the only difference between Π and $@(Q, \Pi)$ is the possible presence of paragraphs. As many $(\S)$ nodes as T may occur just above c'. If we eliminate all the T $(\S)$ nodes we have that the edges entering c' correspond to the edges entering c. Firing c' yields a proof net Θ of mL^3. We have to show that $\Theta = @(\tilde{Q}, \Sigma)$, for some $\tilde{Q}$. If c was a cut with an axiom, or a cut between a weakening and a closed box, then both c and c' annihilate. Otherwise, we get (at least) one residual c'' of c' inside Θ. We can define $\tilde{Q}$ equal to Q on all the conclusions and cut edges that are not involved in the reduction, and that is defined on the edges entering c'' as follows. We distinguish two cases. If c is not an exponential cut, e is an edge incident to c, and f is an edge incident to c'', then $\tilde{Q}(f) = Q(e) + T$. If c is an exponential cut, $\tilde{Q}(f) = Q(e) + T + 1$. □

Corollary 1 **(Complexity Bound for meLL).** *Let Π be a proof net of* meLL*. Let's call $M = \max\{\mathtt{fl}(A) \mid A$ a formula labelling an edge of $\Pi\}$. Then, the round-by-round cut-elimination procedure of Π terminates in at most $(M+1) \cdot 2_{2M}^{M \cdot |\Pi|}$ steps.*

Proof. Let us fix the quasi indexing $Q = Q(0, 0, \ldots, 0)$, and let us calculate $@(Q, \Pi)$. Notice in particular that (i) the constant $K = \max_{1 \le i \le n}\{\mathtt{al}(e_i)\}$ used defining $@$ in this case is $K = \max_{1 \le i \le n}\{\mathtt{fl}(e_i)\} \le M$; and (ii) the indexing I induced on $@(Q, \Pi)$ is canonical. We want apply Theorem 1 to $@(Q, \Pi)$. The size $|@(Q, \Pi)|$ is bounded by $K \cdot |\Pi|$: indeed, for every node of Π, $@$ adds at most K new $(\S)$ nodes. The level of $@(Q, \Pi)$ is $l = \max\{I(e) \mid e$ is an edge of $@(Q, \Pi)\}$. Every $I(e)$ is bounded by K, so $l \le K$. Thus, $@(Q, \Pi)$ reduces in at most $(K+1) \cdot 2_{2K}^{K \cdot |\Pi|} \le (M+1) \cdot 2_{2M}^{M \cdot |\Pi|}$ steps because of Theorem 1. At last, Proposition 2 tells that Π reduces in at most as many steps as $@(Q, \Pi)$, and the thesis follows. □

5 The Full meLL Case

The Proposition 1 fails for second order meLL proof nets. The counterexample is the proof net Π in Figure 6. The behaviour of Π is analogous to the λ-term $(\lambda x.xx)\overline{2}$. Note

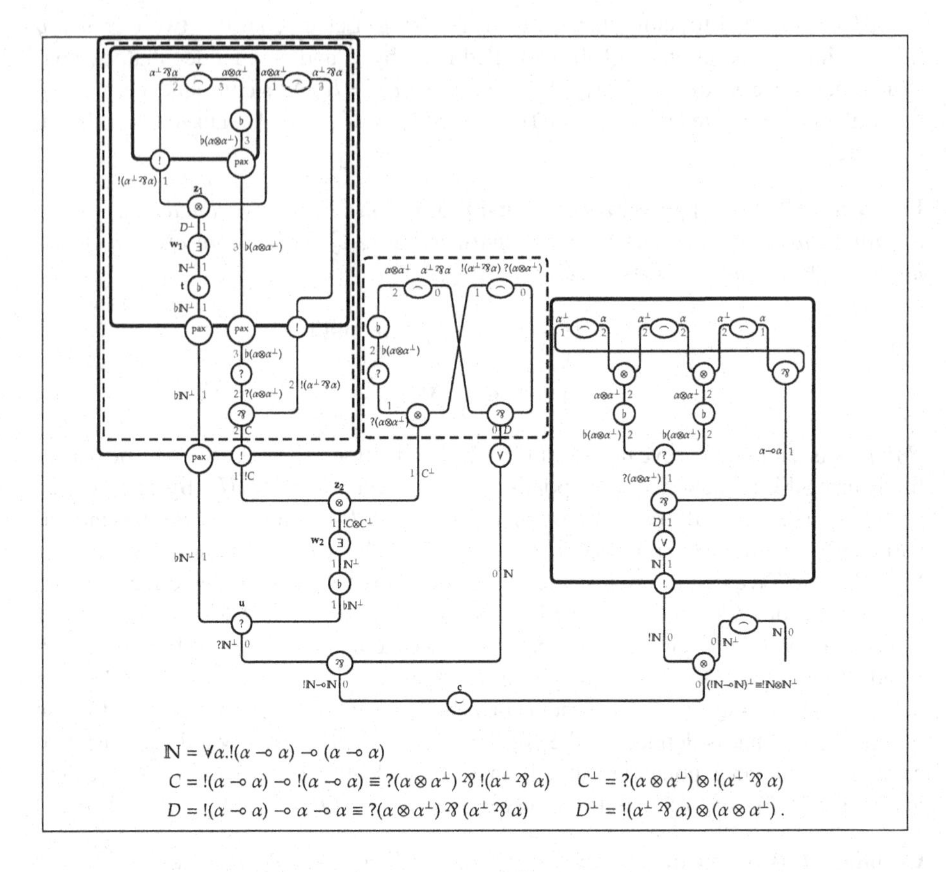

Fig. 6. This proof net represents the λ-term $(\lambda x.xx)\overline{2}$. The two dashed boxes are the proof nets proving $\vdash C^{\perp}, D$ and, essentially, $\vdash ?D^{\perp}, C$.

that the argument $\overline{2}$ of Π is not really necessary, but it is makes evident the dynamic interaction of the two occurrences of x.

We call ρ the path starting from the axiom $\mathbf{v}$ and arriving into the contraction $\mathbf{u}$ passing through the $(\exists)$ node $\mathbf{w_1}$; we call τ the path starting from $\mathbf{v}$ and arriving into $\mathbf{u}$ passing through the $(\exists)$ node $\mathbf{w_2}$.

Firstly, we can imagine to extend the algorithm @ used in the proof of Proposition 1, to a new algorithm $\overline{@}$. It is necessary to extend the definitions of the map $(\cdot)^*$ and of the *formula level*. The most naïve assumption is that $(QA)^* = Q(A)^*$ and $\mathtt{fl}(QA) = \mathtt{fl}(A)$ for each quantifier Q. It will be enough to study the behaviour of $\overline{@}$ along the paths ρ and τ. Starting from the cut node $\mathbf{c}$, $\overline{@}$ would add several new $(\S)$ nodes to Π, in particular over the *right* premise of the $(\otimes)$ nodes $\mathbf{z_1}$ and $\mathbf{z_2}$, but no new nodes over ρ and τ. So, the resulting net would not admit any indexing, because the two edges incident in $\mathbf{v}$ would still have different quasi-indices 2 and 3.

Now, the reader may legitimately think that this problem is due to our particular (and naïve) definition of the algorithm $\overline{@}$. In fact, the problem is more serious. We will show

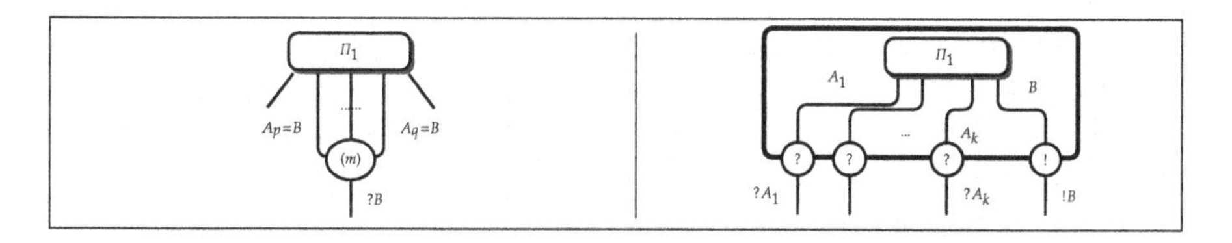

Fig. 7. Exponential inductive rule schemes to build proof nets of SLL

that *there is no way of building an* mL^3 *proof net just adding some* $(\S)$*-nodes to* Π. In order to have the same index on the two sides of **v**, we need to add along the path ρ one $(\S)$ node more than the ones we add along τ. The problem arises as the two formulas labelling the premises of **u** must be equal. Along ρ, a $(\S)$ node can be added only along the four edges connecting **v** to **t**; but whatever edge we choose, if we add a $(\S)$ node along it, we are forced to add another $(\S)$ node along τ to make the premises of **u** agree. And so the resulting proof net cannot be indexed.

6 Concluding Remarks and Further Works

The main contribution of our work is a predicative analysis of meLL by means of the indices inside mL^3. Such an analysis highlights that the source of the huge complexity cost of meLL is due to the use of second order quantifiers that hide and collapse indices. Our analysis is also connected to other problems, that motivate some further developments we outline in the following.

mL^3 as a framework for ICC. We recall that the main reason behind mL^3 is to better understand computations with elementary cost. This work is to support the idea that mL^3 is very useful to characterize other complexity classes. Of course, the simple definition of mL^4 as a subsystem of mL^3, that generalizes a simplified version of LLL, studied in [6,7], already supports such an idea. We strengthen it further by embedding the propositional fragment of SLL [8] in mL^3. We recall that the formulæ of SLL are a subset of the meLL ones. The proof nets of SLL are built using the "linear" nodes of meLL, and the "exponential" nodes in Figure 7. Our embedding of SLL into mL^3 is based on an intermediate embedding of SLL into meLL. Let us call **exponential** every path from a $(\flat)$ node u of a meLL proof net to the first $(?)$ node we may cross, starting from u. SLL can be identified with the subsystem of meLL including all and only the proof nets Π that satisfy the following conditions:

R1: Every exponential path entering a $(?)$ node with one premise crosses at most one (pax) node.
R2: Every exponential path entering a $(?)$ node with more that one premise does not cross any (pax) node.
§N: No $(\S)$ node occurs in Π.

R1 and **R2** simplify analogous conditions in [7]. Basing it on the **R1**, **R2**, and **§N**, we define the following map algorithm from the proof nets of SLL to those ones of meLL. Every (?) node of Π becomes a ($\flat$) node followed by a (pax) node followed by a (?) node. Every multiplexor (m) with k premises becomes a tree composed by k ($\flat$) nodes, followed by a (?) node. Proposition 1 implies that propositional SLL has a corresponding subsystem in mL^3. In particular, it is easy to verify that such a subsystem is the one obtained by considering only the proof nets of propositional mL^3 satisfying exactly **R1** and **R2** since @ preserves them.

Our future work is on the embedding of full SLL into mL^3. This should be possible because the structural constraints that lead from meLL to SLL limit the interaction between second order quantifiers and indices, implicitly hidden by the of-course modality. The proof net in Figure 6, not in SLL, supports this idea, because the second order quantifiers, associated to the duplication-related modality, may require to collapse indices which must be necessarily distinct, as already observed in Section 5.

Complexity bounds for the simply typed λ-calculus. We also aim at a proof theoretical based analysis of the computational complexity of the simply typed λ-calculus, which, under the Curry-Howard analogy, can correspond to intuitionistic propositional meLL. We mean we want to trace back to simply typed λ-calculus the purely structural analysis of the computational complexity that mL^3 supplies for propositional meLL. The point is to avoid any reference to the type of a given simply typed λ-term to infer its normalization cost, as in [9,10]. First steps in this direction are Proposition 1, and a careful inspection of the definition of @. Let Π be a proof net of propositional meLL. Proposition 1 implies that the length of the reduction sequences of $@(\Pi)$ in mL^3 bound those ones of Π. The definition of @ reveals a relation between the structure of Π and the level of $@(\Pi)$. The latter comes from the formulæ levels of formulæ of *only specific axiom nodes* of Π. So, the open points for coming work are at least two: (i) Is there any linear or polynomial function relating the size of Π and the level of $@(\Pi)$?, and (ii) Is there any alternative $@'$ to @ never using the formulæ of the above specific axioms in Π able to yield $@'(\Pi)$ in mL^3?

Acknowledgments. We warmly thank the anonymous referees for the detailed comments and suggestions on the preliminary version of the paper.

References

1. Girard, J.Y.: Light linear logic. Inf. Comput. 143(2), 175–204 (1998)
2. Baillot, P., Mazza, D.: Linear logic by levels and bounded time complexity. Accepted for publication in Theor. Comp. Sci. (2009)
3. Martini, S., Masini, A.: On the fine structure of the exponential rule. In: Girard, J.Y., Lafont, Y., Regnier, L. (eds.) Advances in Linear Logic, pp. 197–210. Cambridge University Press, Cambridge (1995); Proceedings of the Workshop on Linear Logic
4. Martini, S., Masini, A.: A computational interpretation of modal proofs. In: Wansing, H. (ed.) Proof Theory of Modal Logic, Dordrecht, vol. 2, pp. 213–241 (1996)
5. Terui, K.: Light affine lambda calculus and polynomial time strong normalization. Arch. Math. Logic 46(3-4), 253–280 (2007)
6. Mairson, H.G., Terui, K.: On the computational complexity of cut-elimination in linear logic. In: Blundo, C., Laneve, C. (eds.) ICTCS 2003. LNCS, vol. 2841, pp. 23–36. Springer, Heidelberg (2003)

7. Mazza, D.: Linear logic and polynomial time. Mathematical Structures in Computer Science 16(6), 947–988 (2006)
8. Lafont, Y.: Soft linear logic and polynomial time. Theor. Comp. Sci. 318, 163–180 (2004)
9. Schwichtenberg, H.: Complexity of normalization in the pure typed lambda-calculus. In: Troelstra, A.S., van Dalen, D. (eds.) Proc. of Brouwer Centenary Symp. Studies in Logic and the Foundations of Math., vol. 110, pp. 453–457. North-Holland, Amsterdam (1982)
10. Beckmann, A.: Exact bounds for lengths of reductions in typed lambda-calculus. J. Symb. Log. 66(13), 1277–1285 (2001)

Hyper-minimisation Made Efficient

Paweł Gawrychowski and Artur Jeż*

Institute of Computer Science, University of Wrocław, Poland
gawry@cs.uni.wroc.pl, aje@ii.uni.wroc.pl

Abstract. We consider a problem of hyper-minimisation of an automaton [2,3]: given a DFA M we want to compute a smallest automaton N such that the language $L(M)\Delta L(N)$ is finite, where Δ denotes the symmetric difference. We improve the previously known $\mathcal{O}(|\Sigma|n^2)$ solution by giving an expected $\mathcal{O}(|\delta|\log n)$ time algorithm for this problem, where $|\delta|$ is the size of the (potentially partial) transition function. We also give a slightly slower deterministic $\mathcal{O}(|\delta|\log^2 n)$ version of the algorithm.

Then we introduce a similar problem of k-minimisation: for an automaton M and number k we want to find a smallest automaton N such that $L(M)\Delta L(N) \subseteq \Sigma^{<k}$, i.e. the languages they recognize differ only on words of length less than k. We characterise such minimal automata and give algorithm with a similar complexity for this problem.

Keywords: finite automata, minimisation, hyper-minimisation, cover automata.

1 Introduction

DFA is the simplest device recognising languages known in the formal language theory. Studying its properties is motivated by simplicity of the notion, possible applications of the result, connections with various areas in theoretical computer science and the apparent beauty of the results of automata theory.

DFA is defined as a quintuple $\langle Q, \Sigma, \delta, q_0, F\rangle$, where Q is the (finite) state-set, Σ — (finite) alphabet, δ — the transition function, q_0 — starting state, and $F \subseteq Q$ is the set of accepting states. By the usual convention, n denotes $|Q|$.

One of the classical problem in the field is minimisation of a given automaton M: two automatons M, N are *equivalent*, denoted by $M \equiv N$, if $L(M) = L(N)$. An automaton M is *minimal* if for each equivalent N it holds that $|Q(M)| \geq |Q(N)|$). *Minimisation* of an automaton is is a problem of giving a smallest equivalent automaton. A breakthrough was made by Hopcroft [9], who gave an algorithm running in time $\mathcal{O}(n \log n)$. When the alphabet is not fixed, his algorithm runs in time $\mathcal{O}(|\Sigma| n \log n)$, as addressed directly by Gries [7].

A recent development in the area was done by considering a *partial function* δ instead of full transition function: whenever $\delta(q_0, w)$ is not defined, the word

* Supported by MNiSW grants number N206 024 31/3826 2006–2009 and N N206 259035 2008–2010.

R. Královič and D. Niwiński (Eds.): MFCS 2009, LNCS 5734, pp. 356–368, 2009.

w is not accepted. Valmari and Lehtinen [12] gave an $\mathcal{O}(|\delta| \log n)$ algorithm in this case. As $|\delta| = \mathcal{O}(|\Sigma|n)$, this refines the previously existing results.

The question whether any algorithm faster than $\mathcal{O}(|\delta| \log n)$ for automata minimisation is possible remains a challenging open problem. In particular, no argument suggesting that minimisation cannot be done in linear time is known. On the other hand, it is known [5] that there are automata for which all possible executions of the Hopcroft algorithm run in $\Theta(n \log n)$.

A recent notion of f-equivalence [2,3] considers a minimisation of an automaton while allowing the resulting language to differ from the original one on a finite amount of words. Languages L and L' are *f-equivalent*, denoted by $L{\sim}L'$, if $|L \Delta L'| < \infty$, where Δ denotes the symmetric difference of two languages. Similarly, automata M, M' are f-equivalent, denoted by $M{\sim}M'$, if $L(M){\sim}L(M')$. Automaton M is *hyper-minimal*, if for every $M'{\sim}M$ it holds that $|Q(M)| \leq |Q(M')|$. This springs a natural question: how difficult is it to *hyper-minimise* an automaton N, i.e. to construct a hyper-minimal automaton $M{\sim}N$. It is known that such construction can be done in time $\mathcal{O}(n^2)$ [2]. We improve its runtime to (expected) $\mathcal{O}(|\delta| \log n)$, which can be determinized and run in $\mathcal{O}(|\delta| \log^2 n)$. As minimisation reduces to the hyper-minimisation, any substantially faster algorithm would be a major breakthrough in the field.

We then introduce a similar notion of *k-f-equivalence*, denoted by $\sim_k$: $L{\sim_k}L'$ if $\max\{|w| : w \in L \Delta L'\} < k$. An automaton M is *k-minimal* if for all M' such that $L(M){\sim_k}L(M')$ it holds that M has the least number of states. Similarly we study the problem of k-minimisation of a given automaton M.

We introduce relations that allow to better understand the structure of the k-minimal automata and characterise them. Using them we give an algorithm for k-minimisation working in (expected) $\mathcal{O}(|\Sigma| n \log n)$ time. Note that the algorithm reads k as part of the input and the running time does not depend on k. As before it can be determinised and have $\mathcal{O}(|\Sigma| n \log^2 n)$ running time. Since hyper-minimisation of M is equivalent to n-minimisation, one cannot expect that any algorithm faster than $\mathcal{O}(|\delta| \log n)$ can be found.

It should be noted that the $\mathcal{O}(n^2)$ algorithm and our algorithms work in a different way then the Hopcroft's algorithm, which iteratively refined the partition of states. Both Badr's [2] and Badr et al. [3] algorithms calculated the equivalence classes of $\sim$ and then greedily merged the appropriate states. Our algorithm for hyper-minimisation, as well as the one for k-minimisation, works in phases: roughly speaking, in the ℓ-th phase it finds pairs of states q, q' such that $\max\{|w| : w \in L(q) \Delta L(q')\} = \ell$ and merges them.

The notion of k-f-equivalence is somehow complementary to the problem of finding a minimal *cover automata* — for an automaton M such that k is the length of longest word in $L(M)$ we want to find a minimal automaton N such that $L(N) \cap \Sigma^{\leq k} = L(M)$ [4]. It was introduced with practical purpose in mind: if an automaton recognises a finite language then one can store the length of the longest recognised word and a cover automaton. The input is accepted if it is not too long and recognised by the cover automaton. A clever modification of

Hopcroft algorithm can be applied in this setting, yielding a $\mathcal{O}(n \log n)$ algorithm for finding a minimal cover automaton [10].

In parallel, a similar work concerning hyper-minimisation was done by M. Holzer and A. Maletti [8], who independently gave a randomised algorithm for hyper-minimisation running in expected time $\mathcal{O}(|\Sigma| n \log n)$.

2 Preliminaries

For an automaton $M = \langle Q, \Sigma, \delta, q_0, F \rangle$ we define its language $L = L(M)$ in the usual sense. By $M(w)$ we denote $\delta(q_0, w)$. We say that a word w induces a language $L(w) := w^{-1}L$: the language recognised after reading the word w. Let also $L_M(q)$ denote $L(\langle Q, \Sigma, \delta, q, F \rangle)$, where $M = \langle Q, \Sigma, \delta, q_0, F \rangle$, i.e. the language recognized by the automaton M with the starting state set to q.

A standard approach to minimisation is to consider a Myhill-Nerode relation on words, defined for the language $L = L(M)$ as

$$w \equiv_L w' \quad \text{iff} \quad \forall u \in \Sigma^* \; wu \in L \iff wu' \in L.$$

This relation has finitely many equivalence classes for regular languages and each such class corresponds to one state in the minimal DFA recognising the language L. To make this approach more efficient, it should be noted that if $M(w) = M(w')$ then $w \equiv_L w'$, i.e. the relation is in fact defined for states: $q \equiv_L q' \iff L_M(q) = L_M(q')$. The minimisation algorithm starts with partition of states into two classes F and $Q \setminus F$ and iteratively refine the partition until it is left with the set of equivalence classes of $\equiv_L$ [9,12].

It is easy to see that the minimisation reduces to hyper-minimisation: consider an automaton M. Create M' by adding one accepting state *dummy*, one fresh letter \$ to the alphabet and extend the transition function by $\delta_N(q, \$) = \mathit{dummy}$ and for every letter $b \in \Sigma$ $\delta_N(\mathit{dummy}, b) = \mathit{dummy}$. Then for each $q \in Q(M)$ it holds that $L_N(q) = L_M(q) \cup \$(\Sigma \cup \{\$\})^*$, hence it is infinite. Remove the state *dummy* from an automaton hyper-minimal for M'. The obtained automaton is minimal for $L(M)$.

The relation $\sim$ plays a similar role in the problem of hyper-minimisation as $\equiv$ in the problem of minimisation. We refine $\sim$ so that it can be used to the problem of k-minimisation as well.

We employ the classification of states developed in [3]: the *preamble*, denoted by $P(M)$, or simply P, is a set of states q reachable from q_0 by finitely many words, i.e. $L(\langle Q, \Sigma, q_0, \delta, \{q\} \rangle)$ is finite. On the other hand, *kernel* $K(M)$, or simply K, is defined as $Q \setminus P$, i.e. set of states q such that language $L(\langle Q, \Sigma, q_0, \delta, \{q\} \rangle)$ is infinite.

The automaton M' is obtained by *merging state q to state p* in automaton $M = \langle Q, \Sigma, \delta, q_0, F \rangle$ if M' is obtained by changing all transitions ending in q to transitions ending in p and deleting state q. If q was a starting state then p is the new starting state. Formally $M' = \langle Q \setminus \{q\}, \Sigma, \delta', q_0', F \setminus \{q\} \rangle$ where

$$\delta'(r, a) = \begin{cases} \delta(r, a) & \text{if } \delta'(r, a) \neq q \\ p & \text{if } \delta'(r, a) = q \end{cases}, \quad q_0' = \begin{cases} q_0 & \text{if } q_0 \neq q \\ p & \text{if } q_0 = q \end{cases}.$$

For two languages L, L' define their *distance* as

$$d(L, L') = \begin{cases} \max\{|u| : u \in L(w) \Delta L(w')\} + 1 & \text{if } L \neq L' \ , \\ 0 & \text{if } L = L' \ . \end{cases}$$

This definition can be easily extended to states by setting $d(q, q') = d(L(q), L(q'))$. If we fix a language L then the distance between words is defined as $d_L(w, w') = d(L(w), L(w'))$. Usually the language L is clear from the context and so we drop the index L.

Fact 1 (d is a pre-ultrametric). *For all languages L, L', L'' it holds that $d(L, L'') \leq \max(d(L, L'), d(L', L''))$.*

The distance between the words allows relation between words similar to the f-equivalence: $w \sim u$ iff $d(w, u) < \infty$ and $w \not\sim u$ otherwise. This relation is *right invariant*: $u \sim w$ implies $ux \sim wx$ for all words x.

Fact 2. *$\sim$ is a right invariant equivalence relation.*

We extend the f-equivalence to states and automata: $q \sim q'$ iff for every w, w' such that $M(w) = q$ and $M(w') = q'$ it holds that $w \sim w'$; $M \sim M'$ iff $q_0 \sim q'_0$. Not that the definition for automata coincides with the one given in the Introduction: $M \sim M'$ iff $L(M) \Delta L(M')$ is finite.

We characterise the distance between two states in an operational manner. To this end for each $\ell \geq 0$ we introduce relation D^M_ℓ on $Q(M)$ defined by:

- $D^M_0(q, q')$ iff $q = q'$,
- $D^M_\ell(q, q')$ iff for all $a \in \Sigma$ either $D^M_{\ell-1}\big(\delta_M(q, a), \delta_M(q', a)\big)$ or both $\delta_M(q, a)$ and $\delta_M(q', a)$ are not defined.

Denote also $D^M(q, q') \iff \exists_\ell D^M_\ell(q, q')$. By an easy induction it follows that both D^M_ℓ and D^M are equivalence relations. We drop the upper index M whenever the automaton is clear from the context.

Fact 3. *If M is minimal and the transition function is always defined then $D_\ell(q, q')$ iff $d(q, q') \leq \ell$.*

To compute the relation D, one does not need to consider $\ell > n$:

Lemma 1. *Consider an automaton M. Then $D^M = D^M_n$.*

3 Hyper-minimisation

Badr et. al claimed [3, Thm. 3.2] that any (greedy) algorithm that at each step merges p to q such that $p \equiv q \vee (p \sim q \wedge p \in P(M))$ correctly hyper-minimises the automaton. Unfortunately this is not the case, still the argument can be easily corrected — some additional care is needed, when merging two states from the preamble. We say that a linear order $<$ on Q is *valid*, if for every pair of states

$p, q \in P$ and letter $a \in \Sigma$ it holds that $\delta(q, a) = p$ implies $q < p$ and for $q \in P$, $p \in K$ it holds that $q < p$.

Theorem 1. *Any (greedy) algorithm that at each step merges p to q such that* $p \equiv q \vee (p \sim q \wedge p \in P(M) \wedge p < q)$ *correctly hyper-minimises the automaton.*

Such an order can be easily constructed — it is enough to sort topologically the acyclic graph corresponding to the states of the preamble and arbitrarily define it on the kernel.

Our algorithm utilises this approach, similarly to the previously known ones. Its novelty lays in the way pairs of states to be merged are found and the data structures that are employed.

On high level in the ℓ-th phase we calculate for the remaining states the relation D_ℓ on the remaining states, i.e. the distance between them. After that we merge some states and continue to the following phase. This approach result in no more than $|Q|$ phases, by Lemma 1.

3.1 Signatures

Since we deal with partial δ function, it is important to treat differently the states q of the automaton with different sets of defined transitions: for a state q the set $\{a : \delta(q, a) \text{ is defined}\}$ is the *signature* of q, denoted by $\text{sig}(q)$. This allows bounding the running time by $|\delta|$ rather than $n|\Sigma|$.

We would like to think that $\text{sig}(q) \neq \text{sig}(q')$ implies that $q \not\sim q'$ and hence we can minimise states with different signatures separately. This can be achieved, whenever there are no states inducing finite languages.

Fact 4. *Suppose that automaton M has no states q such that L(q) is finite. Then* $\text{sig}(q) \neq \text{sig}(q')$ *implies* $q \not\sim q'$.

The states inducing finite languages are naturally divided into those in the preamble and those in the kernel. The former can be easily removed.

Fact 5. *Let M be an automaton and M′ be obtained from M by removing the set of states* $P(M) \cap \{q : |L(q)| < \infty\}$. *Then for* $q \in Q(M')$ *it holds that* $L_M(q) \sim L_{M'}(q)$. *In particular* $L(M) \sim L(M')$.

Unfortunately, such a straightforward approach cannot be applied to states from the kernel inducing finite languages. On one hand such states cannot be removed, as this causes the language of the automaton to change on infinitely many words. On the other hand, their existence prevents us from using Fact 4. An intermediate approach turns out to work, we can remove the problematic states temporarily, calculate the D classes and then bring back those states.

Lemma 2. *Suppose that automaton M is minimal and has no states* $q \in P(M)$ *such that L(q) is finite. Let M′ be obtained from M by removing the states* $Q' = \{q : q \in K(M) \wedge |L(q)| < \infty\}$. *Then for* $q, q' \in Q \setminus Q'$ *it holds that* $D^{M'}(q, q') \iff L_M(q) \sim L_M(q')$.

3.2 Automata Reduction

Our algorithm has a high time-dependency on $|\Sigma|$. Thus we reduce the input so that we can treat $|\Sigma|$ as a small constant. To this end we transform the input automaton M into other one, denoted by GADGETS(M), using only four-letter alphabet $\Sigma' = \{0, 1, 2, 3\}$. On the other hand, we increase the number of states from n to $\mathcal{O}(|\delta|)$. On a high level, one can imagine that we encode possible letters of Σ as 0–1 sequences. The letters 2, 3 are used to indicate which states simulate the states from the original automaton and which states are technical gadgets and distinguish states of different signatures. For simplicity of the presentation, we add a special non-accepting state *trash* such that each transition not defined explicitly goes to *trash*. Assume that the automaton M is minimised

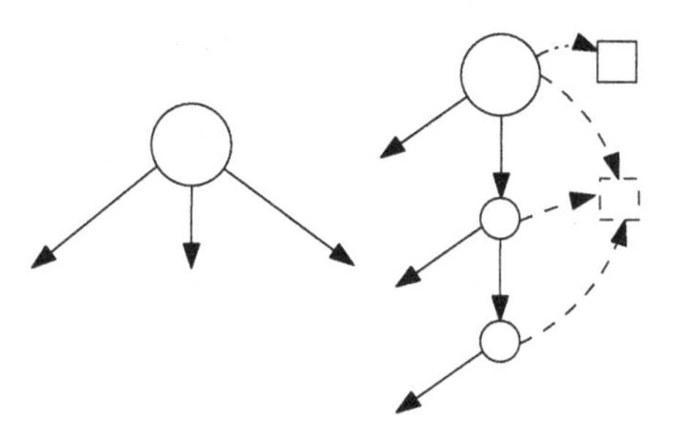

Fig. 1. Example of introducing gadgets on a single state

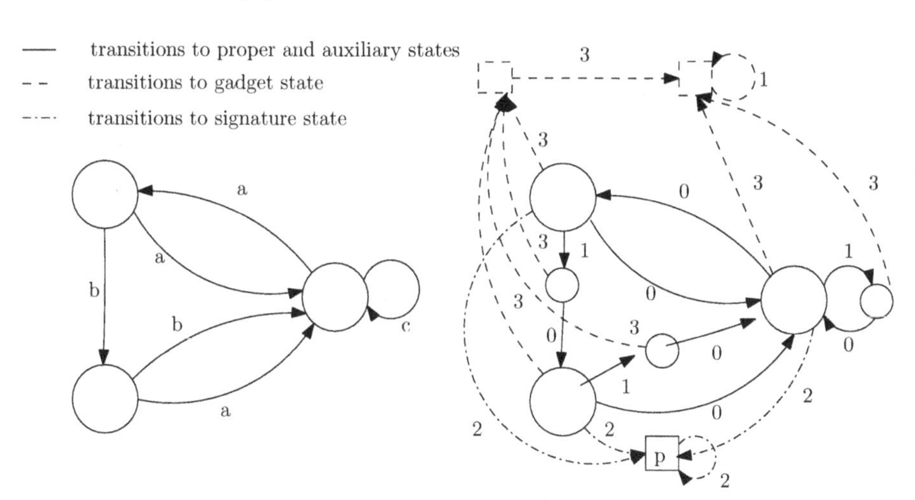

Fig. 2. Automaton M and its M'. There are two signatures.

and does not have states inducing finite language. To shorten the notation, let M' denote GADGETS(M). Please consult Fig. 1 and Fig. 2 for an illustration. Partition the set of states into subsets with the same signature. For each signature sig with ℓ symbols $\{a_0, \ldots, a_{\ell-1}\}$ introduce $\ell - 1$ new *auxiliary states* per each state q of this signature, named $s_{1,q}, s_{2,q}, \ldots s_{\ell-1,q}$. By convention, let $s_{0,q} = q$, we call it a *proper state*. Then define transition function as $\delta_{M'}(s_{i,q}, 0) = \delta_M(q, i)$ for $i = 0, \ldots \ell - 1$; $\delta_{M'}(s_{i,q}, 1) = s_{i+1,q}$ for $i = 0, \ldots \ell - 2$. The newly created states $s_{1,q}, s_{2,q}, \ldots s_{\ell-1,q}$ are assigned the signature $\mathrm{sig}(q)$.

Moreover, we distinguish proper states from auxiliary states by creating a *gadget state* p with $\delta_{M'}(p, 2) = p$ and adding the transition $\delta_{M'}(q, 2) = p$ for each state q. Then we distinguish states corresponding to different signatures by another gadget: let $\{\text{sig}_1, \text{sig}_2, \dots, \text{sig}_k\}$ be the set of all signatures. Introduce *signature states* $\{\text{sig}_1, \text{sig}_2, \dots, \text{sig}_k\}$ and add transitions $\delta_{M'}(\text{sig}_i, 3) = \text{sig}_{i+1}$ for $i = 1, \dots, k-1$, $\delta_{M'}(\text{sig}_k, 1) = \text{sig}_k$ and $\delta_{M'}(s_{j,q}, 3) = \text{sig}(q)$.

The size of the automaton M' is $\Theta(|\delta_M|)$ and M' can be constructed using in similar time bounds. The only nontrivial part is that we need to group states with the same signatures using counting sort. To upper bound the running time by $\mathcal{O}(|\delta| \log n)$ instead of $\mathcal{O}(|\delta| \log |\delta|)$ we show that the $D^{M'}$ classes are of size $\mathcal{O}(n)$.

Lemma 3. *Let $M' = \text{Gadgets}(M)$. Then none of the signature states nor the trash nor the state gadget is $D^{M'}$-equivalent to any other state. If $D^{M'}(s_{i,q}, s_{i',q'})$ then* $\text{sig}(q) = \text{sig}(q')$ *and* $i = i'$.

The automaton M' retains the basic properties of M, meaning that two states in M are f-equivalent iff they are equivalent in M'.

Lemma 4. *Let $q, q' \in Q(M)$ and $M' = \text{Gadgets}(M)$. Then $D^M(q, q')$ iff $D^{M'}(q, q')$.*

3.3 Algorithm and Its Running Time

We now present Algorithm Comp-f-equiv(M) calculating the hyper-minimal automaton f-equivalent to M.

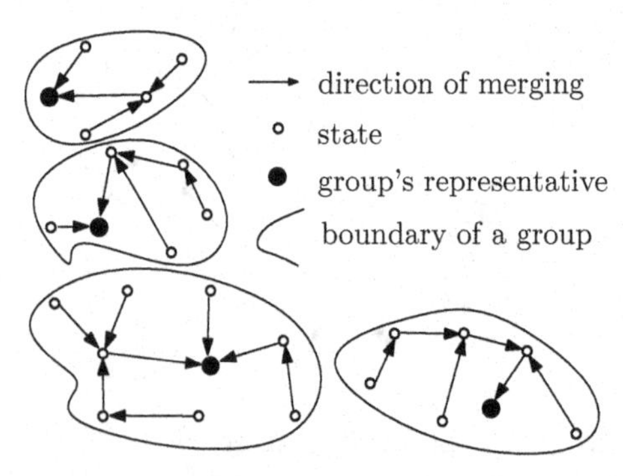

Fig. 3. Example of states in a group

Preprocessing. First of all we calculate $K(M)$ and $P(M)$. This takes just a linear time: we find the strongly connected components of the underlying graph and mark vertices that can be reached from those of them that are nontrivial (i.e. contain either more than two vertices or a loop).

Then we remove the states from the preamble inducing finite languages and minimise the automaton. Let us denote this automaton by M_1. We temporarily remove the states from the kernel of M_1 that induce finite languages and denote the result by M_2. Then we built $M' = \text{Gadgets}(M_2)$.

The algorithm also requires some simple data structures: for each state q a list of its predecessors $l[q]$

$$l[q] = \{q'' : \exists x \in \{0, 1, 2, 3\} \text{ such that } \delta_{M'}(q'', x) = q\}$$

is created. Moreover, we store *rank* of q — the number of its predecessors.

The Theorem 1 requires a pre-computed linear valid order on Q. To fix such an order, sort topologically the states of the preamble and order the states of the kernel arbitrarily, such that they are all greater than the states of the preamble.

For technical reasons, we do not store the states in the dictionary, but rather their *dictionary representatives*, Dic. So a table of representatives is also created. In the beginning, for each state q, $\mathrm{Dic}[q] = q$.

Dictionaries. For the states of M' we built a dictionary mapping a tuple $\langle q_0, q_1, q_2, q_3 \rangle$ into a state q such that $\delta(q, i) = q_i$ for all i. Each state occurs at most once in the structure. To implement this dictionary, we use dynamic hashing with a worst-case constant time lookup and amortized expected constant time for updates (see [6] or a simpler variant with the same performance bounds [11]). We convert each tuple into an integer from $\{1 \mathinner{..} N\}$, where $N = |\delta|^4$ fits in a constant number of machine words, so we can hash it in a constant time. Whenever we insert a tuple already existing in the structure, we have a new pair of states to merge.

The analogous dictionary with $\mathcal{O}(\log n)$ time per operation is very simple for the deterministic case: note that by Lemma 3 if q, q' are merged then they are both either auxiliary states or proper states, they have the same signature and correspond to the same letter. So it is enough to built a separate dictionary for each set of states $Q_{\mathrm{sig}_j, i} = \{s_{i,q} : \mathrm{sig}(q) = \mathrm{sig}_j\}$. As there are only $\mathcal{O}(n)$ elements in such set by Lemma 3, it can be implemented as a balanced binary tree. In order to have a constant time access to the dictionary itself, we create a table of all signatures. For a single signature sig_j it keeps a pointer to the table indexed by $1 \ldots, \ell_j$, where ℓ_j is the size of the alphabet associated with sig_j. For an index i there is a pointer to the dictionary for the set $Q_{\mathrm{sig}_j, i}$. This $\mathcal{O}(\log n)$ bound per operation can be greatly improved if we are allowed to fully use the power of RAM model: plugging in the exponential search trees of [1] gives us a total running time of $\mathcal{O}(|\delta| \log n \frac{\log^2 \log n}{\log \log \log n})$.

Merging states. Suppose two states $q > q'$ are to be merged. Let q_1 be the one of them with higher rank and q_1' the one with lower rank. We remove $\mathrm{Dic}[q_1']$ from the dictionary and keep only $\mathrm{Dic}[q_1]$ and set $\mathrm{Dic}[q]$ to $\mathrm{Dic}[q_1]$. The rank of q is set to be the sum of ranks q and q'.

Then we update the dictionary by reinserting all states q'' such that $\delta_{M'}(q'', x) = q_1'$ for some $x \in \{0, 1, 2, 3\}$. To do this efficiently, we scan $l[q_1']$. After the update $l[q']$ is appended to $l[q]$. We also store the information that q' was merged to q: a *group* consists of the states that were merged to a single state. A *representative* of the group is the unique state that survived the merging.

When all the merging is done, for each q we calculate the representative of q'es group: we inspect a sequence $q = q_0, q_1 \ldots, q_m$ such that q_i was merged to q_{i+1} and q_m was not merged to anything. Then $q_0, q_1 \ldots, q_{m-1}$ were all merged to q_m. All the states that were merged to a single q_m form a $D^{M'}$-class. This information can be used to merge the states from the $P(M_1)$ to their f-equivalent states, i.e. those that are in the same $D^{M'}$-class.

Theorem 2. *COMP-f-EQUIV(M) properly hyper-minimises the automaton M and runs in expected time $\mathcal{O}(|\delta| \log n)$ ($\mathcal{O}(|\delta| \log^2 n)$ worst-case).*

4 k-Minimisation

We now consider the problem of *k-minimising* the automaton. Note that M and N are k-f-equivalent iff $d(L(M), L(N)) \leq k$. The general scheme is the same as previously, this time we are in more difficult situation: there is no notion similar to $\sim$ which works in the case of k-minimisation. Moreover, there is no theoretic characterisation of the k-minimal automaton. We begin with introducing a proper notion and describing the k-minimal automaton from a theoretical point of view. In particular we show that the k-minimal automaton can be obtained by merging some states into the others and the merging can be done in a (somehow) greedy fashion. Then we implement this approach. Unfortunately we were unable to efficiently deal with signatures in this case and the algorithm runs in $\mathcal{O}(|\Sigma| n \log n)$ time. We assume that the transition function is total.

4.1 Relation on States

We start with defining a relation playing the same role as $\sim$:

Definition 1. *We say that $w \sim_k u$ if $d(w, u) = 0$ or $d(w, u) + \min(|w|, |u|) \leq k$ and $w \not\sim_k u$ otherwise.*

The intuition of this relation is similar to the one for $\sim$: consider any regular language L, an automaton M recognising it, and two words w, w'. Let $M(w) = q$, $M(w') = q'$. Suppose q is merged to q'. If $L(q) \neq L(q')$ then $wL(w) \subseteq L(M)$ is changed to $wL(w')$, so it should hold that $|w| + d(L(w), L(w')) \leq k$. On the other hand if we were to merge q' to q then $w'L(w') \subseteq L(M)$ is changed to $w'L(w)$, hence it should hold that $|w'| + d(L(w), L(w')) \leq k$. Choosing the smaller of those terms we obtain $\min(|w|, |w'|) + d(L(w), L(w')) \leq k$, as in the definition of $\sim_k$. On the other hand, if $w \not\sim_k w'$ then it seems that we cannot merge the states q and q'. The second part of this intuition is formalised in Lemma 6. The first one needs some further refinements before it is put to work.

Note that $\sim_k$ is not an equivalence relation for any k. Still, it has some useful properties, which are quite close to being an equivalence relation and are essentially used in proofs of combinatorial properties and in the analysis of the algorithm for calculating the k-hyper-minimal automaton.

Lemma 5. *For all k the relation $\sim_k$ has the following properties*

1. *it is right invariant*
2. *if $w_1 \sim_k w_2$, $w_2 \sim_k w_3$ and $|w_2| \geq \max(|w_1|, |w_3|)$ then $w_1 \sim_k w_3$*
3. *if $w_1 \sim_k w_2$, $w_2 \sim_k w_3$ and $|w_1| \leq \min(|w_2|, |w_3|)$ then $w_1 \sim_k w_3$*

As promised, sets of words $\{w_i\}_{i=1}^{j}$ such that for $w \neq w' \in \{w_i\}_{i=1}^{j}$ we have $w \not\sim_k w'$ can be used to lower-bound the size of the k-minimised automaton:

Lemma 6. *Consider $L' \subseteq L$ such that for each $w \neq w' \in L'$ it holds that $w \not\sim_k w'$. Then for each automaton M such that $L \sim_k L(M)$ it holds that $M(w) \neq M(w')$, in particular M has at least $|L'|$ states.*

The definition of $\sim_k$ can be easily extended to states: $q \sim_k q'$ if for all (w, w') such that $M(w) = q, M(w') = q'$ it holds that $w \sim_k w'$. One can easily see that this is equivalent to $q \sim_k q'$ when

$$d(q, q') = 0 \text{ or } d(q, q') + \min(\max(|w| : M(w) = q), \max(|w| : M(w) = q')) \leq k$$

Instead of considering for a state q all the words w such that $M(w) = q$ it is enough to consider the longest one:

Definition 2. *For every state of $q \in Q$ let its* representative word *(called* word$[q]$*) be a longest word w such that $M(w) = q$ if $q \in P$ or any word with length at least k, if $q \in K$.*

This definition is designed in a way so that the $\sim_k$ equivalence between states could be expressed in terms of representatives of states:

Fact 6. word$[q] \sim_k$ word$[q']$ *iff* $q \sim_k q'$.

4.2 k-Minimal Automaton

We want to define a k-minimal automaton k-f-equivalent to M using relation $\sim_k$ on states of M. Since $\sim_k$ is not an equivalence relation, we need some additional refinement. To this end we construct an equivalence relation $\approx_k$ which refines $\sim_k$ defined on the set of states of M. Its equivalence classes correspond to states in the k-minimal automaton $M' \sim_k M$. The relation has the following properties:

1. $q \approx_k q'$ implies $q \sim_k q'$
2. for each class of abstraction $\{q_i\}_{i \in I}$ of $\approx_k$ we designate its representative word w — the longest of $\{\text{word}[q_i]\}_{i \in I}$. We denote by Rep$[q]$ the representative of q in its class of abstraction and we extend the notion of the representative word to words: Rep$[w]$ = Rep$[M(w)]$.
3. Rep$[q] \neq$ Rep$[q']$ implies Rep$[q] \not\sim_k$ Rep$[q']$

The relation is defined algorithmically. Let us first define Rep$[w_i] = w_i$ for all w_i such that $w_i = \text{word}[q_i]$ for some state q_i. Then consider $\{w_1, \ldots, w_n\} = \{\text{word}[q]\}_{q \in Q}$ in any order. If for considered w_i there exists w_j such that Rep$[w_j] = w_j$, $w_i \sim_k w_j$ and $|w_i| \leq |w_j|$ then for all w such that Rep$[w] = w_i$ set Rep$[w] = w_j$ (choose any such j if there are many possible ones). In the end we extend the notion for states: Rep$[q]$ = Rep[word$[q]$] and set $q \approx_k q'$ if Rep$[q]$ = Rep$[q']$.

Fact 7. *The relation defined above satisfies conditions* (1)–(3).

Consider again minimised automaton M, using relation $\approx_k$ on its states we define an automaton N = EQUIVALENCE-TO-AUTOMATON$(M, \approx_k)$ and later show how to create it efficiently. Construct an automaton N by taking

$$Q_N = \{\langle w \rangle : \text{Rep}[w] = w\} \quad \delta(\langle w \rangle, x) = \langle \text{Rep}[wx] \rangle \quad F_N = \{\langle w \rangle : M(w) \in F\}$$

and set a starting state $\langle \text{Rep}[\epsilon] \rangle$. By property 3 of $\approx_k$ one can easily derive that N is k-minimal.

The natural attempt to show that $L(N) \sim_k L(M)$ is to prove that $N(w) = \langle \text{Rep}[w] \rangle$. Unfortunately this is not the case. We proceed in a slightly more complicated fashion. First we argue that the defined automaton behaves well on short words and then show that $d(L_M(w), L_N(\langle w \rangle))$ can be upper-bounded in terms of $|w|$. Those facts allow proving that $L(N) \sim_k L(M)$.

Theorem 3. *Automaton N is k-minimal and $N \sim_k M$.*

4.3 Algorithm

We show how to efficiently calculate N = EQUIV-TO-AUTO. An algorithm similar to COMP-f-EQUIV is used. As we do not use different signatures for different states, we do not employ gadgets for signatures and additional symbol 3 in the alphabet. On the other hand we have to be a little more subtle now, as we are interested in calculating the classes of relation D_ℓ^M and not only $D^{M'}$. Roughly speaking in the ℓ-th phase we calculate the equivalence classes of D_ℓ^M. To this end we merge states not in some arbitrary fashion but in order representing the inductive definition of D_ℓ.

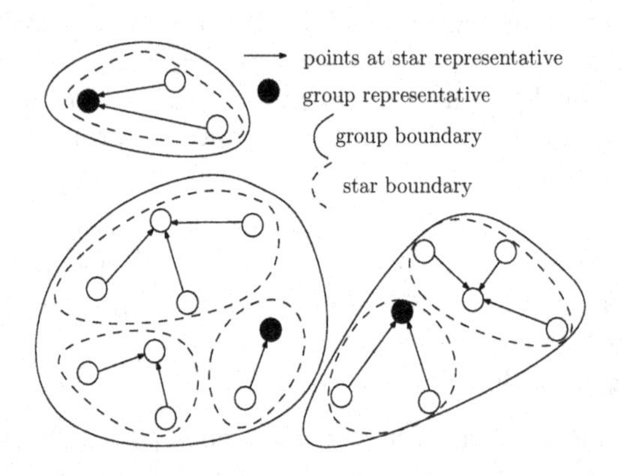

Fig. 4. Example of states in a group and in a star

The algorithm works similar to COMP-EQUIV-k. We list only the important differences. We first minimise the automaton, obtaining M_1 and then introduce gadgets as in Section 3.2, except for the signature gadgets, which are not needed: M' = GADGETS(M_1). The algorithm works in phases. In one phase it first merges all the proper states that could be merged at the beginning of this phase. Then it starts merging the auxiliary states and gadget states. When there are no more auxiliary states to merge, the counter is increased and the next phase begins. Also more information is stored. A group of states that were merged together is subdivided into stars. Each star has its representative (star-representative). In particular group-representative is one of the star-representatives of stars forming this group.

If two groups, represented by q and q', are merged, we check whether $q \sim_k q'$. If so then two stars represented by q and q' are merged and assigned the longer of the two representative words. Note that this representative also becomes the new group representative.

When merging of groups is finished, $N = \textsc{Equiv-To-Auto}(M_1, \text{star})$ is built, treating the states in one star as states in one equivalence class of $\approx_k$.

The running time analysis of COMP-EQUIV-k is the same as the one of COMP-f-EQUIV. All additional operations are done in constant time per operation.

The following lemma formalises the intuition that auxiliary states do not influence the process of merging state and thus phases correspond to calculating the distance between the states of the automaton.

Lemma 7. *Two states $q, q' \in Q(M_1)$ are merged in ℓ-th phase of* COMP-EQUIV-k *iff $d(q, q') \leq \ell$.*

To prove the correctness of the algorithm we show that the following invariants concerning groups and stars are kept: the first two invariants describe properties of stars, the following four the properties of groups.

1. Let $q_1, \ldots, q_i$ be all the states in a star. Then $|\text{word}[q_1]| \geq |\text{word}[q_j]|$ for $j = 2, \ldots, i$,
2. $q_j \sim_k q_{j'}$ for all $j, j' = 1, \ldots, i$.
3. group is a union of stars
4. Let $p_1, \ldots, p_{i'}$ be the star-representatives of star forming a group represented by p_1. Then $|\text{word}[p_1]| \geq |\text{word}[p_j]|$ for $j = 2, \ldots, i'$,
5. $p_j \not\sim_k p_{j'}$ for all $j \neq, j' \in \{1, \ldots, i'\}$,
6. the group consisting of proper states is an equivalence class of $D_\ell^{M_1}$

Theorem 4. COMP-EQUIV-k *correctly k-minimises M.*

Open Problem

Is there a fully deterministic algorithm which hyper-minimize (or maybe even k-minimise) an automaton in time $\mathcal{O}(|\Sigma| n \log n)$?

References

1. Andersson, A., Thorup, M.: Dynamic ordered sets with exponential search trees. J. ACM 54(3), 13 (2007)
2. Badr, A.: Hyper-minimization in $\mathcal{O}(n^2)$. In: Ibarra, O.H., Ravikumar, B. (eds.) CIAA 2008. LNCS, vol. 5148, pp. 223–231. Springer, Heidelberg (2008)
3. Badr, A., Geffert, V., Shipman, I.: Hyper-minimizing minimized deterministic finite state automata. RAIRO - Theoretical Informatics and Applications 43(1), 69–94 (2009)
4. Câmpeanu, C., Santean, N., Yu, S.: Minimal cover-automata for finite languages. Theor. Comput. Sci. 267(1-2), 3–16 (2001)
5. Castiglione, G., Restivo, A., Sciortino, M.: Hopcroft's algorithm and cyclic automata, pp. 172–183 (2008)
6. Dietzfelbinger, M., Karlin, A.R., Mehlhorn, K., auf der Heide, F.M., Rohnert, H., Tarjan, R.E.: Dynamic perfect hashing: Upper and lower bounds. SIAM J. Comput. 23(4), 738–761 (1994)

7. Gries, D.: Describing an algorithm by Hopcroft. Acta Informatica 2, 97–109 (1973)
8. Holzer, M., Maletti, A.: An n log n algorithm for hyper-minimizing states in a (minimized) deterministic automaton. In: CIAA, pp. 4–13 (2009)
9. Hopcroft, J.: An n logn algorithm for minimizing states in a finite automaton. Technical Report, CS-190 (1970)
10. Körner, H.: On minimizing cover automata for finite languages in $\mathcal{O}(n \log n)$ time. In: Champarnaud, J.-M., Maurel, D. (eds.) CIAA 2002. LNCS, vol. 2608, pp. 117–127. Springer, Heidelberg (2003)
11. Pagh, R., Rodler, F.F.: Cuckoo hashing. J. Algorithms 51(2), 122–144 (2004)
12. Valmari, A., Lehtinen, P.: Efficient minimization of DFAs with partial transition. In: Albers, S., Weil, P. (eds.) STACS. Dagstuhl Seminar Proceedings, vol. 08001, pp. 645–656. Internationales Begegnungs- und Forschungszentrum fuer Informatik (IBFI), Schloss Dagstuhl (2008)

Regular Expressions with Counting: Weak versus Strong Determinism

Wouter Gelade[1,*], Marc Gyssens[1], and Wim Martens[2,**]

[1] Hasselt University and Transnational University of Limburg
School for Information Technology
{firstname.lastname}@uhasselt.be
[2] Technical University of Dortmund
{firstname.lastname}@udo.edu

Abstract. We study deterministic regular expressions extended with the counting operator. There exist two notions of determinism, strong and weak determinism, which almost coincide for standard regular expressions. This, however, changes dramatically in the presence of counting. In particular, we show that weakly deterministic expressions with counting are exponentially more succinct and strictly more expressive than strongly deterministic ones, even though they still do not capture all regular languages. In addition, we present a finite automaton model with counters, study its properties and investigate the natural extension of the Glushkov construction translating expressions with counting into such counting automata. This translation yields a deterministic automaton if and only if the expression is strongly deterministic. These results then also allow to derive upper bounds for decision problems for strongly deterministic expressions with counting.

1 Introduction

The use of regular expressions (REs) is quite widespread and includes applications in bioinformatics [17], programming languages [23], model checking [22], XML schema languages [21], etc. In many cases, the standard operators are extended with additional ones to facilitate usability. A popular such operator is the counting operator allowing for expressions of the form "$a^{2,4}$", defining strings containing at least two and at most four a's, which is used for instance in Egrep [9] and Perl [23] patterns and in the XML schema language XML Schema [21].

In addition to expanding the vocabulary of REs, subclasses of REs have been investigated to alleviate, e.g., the matching problem. For instance, in the context of XML and SGML, the strict subclasses of weakly and strongly deterministic regular expressions have been introduced. Weak determinism (also called one-unambiguity [2]) intuitively requires that, when matching a string from left to

* Research Assistant of the Fund for Scientific Research – Flanders (Belgium).
** Supported by the North-Rhine Westphalian Academy of Sciences, Humanities and Arts; and the Stiftung Mercator Essen.

R. Královič and D. Niwiński (Eds.): MFCS 2009, LNCS 5734, pp. 369–381, 2009.

right against an expression, it is always clear against which position in the expression the next symbol must be matched. For example, the expression $(a+b)^*a$ is not weakly deterministic, but the equivalent expression $b^*a(b^*a)^*$ is. Strong determinism intuitively requires additionally that it is also clear *how* to go from one position to the next. For example, $(a^*)^*$ is weakly deterministic, but not strongly deterministic since it is not clear over which star one should iterate when going from one a to the next.

While weak and strong determinism coincide for standard regular expressions [1][1], this situation changes completely when counting is involved. Firstly, the algorithm for deciding whether an expression is weakly deterministic is non-trivial [13]. For instance, $(b?a^{2,3})^{2,2}b$ is weakly deterministic, but the very similar $(b?a^{2,3})^{3,3}b$ is not. So, the amount of non-determinism introduced depends on the concrete values of the counters. Second, as we will show, weakly deterministic expressions with counting are strictly more expressive than strongly deterministic ones. Therefore, the aim of this paper is an in-depth study of the notions of weak and strong determinism in the presence of counting w.r.t. expressiveness, succinctness, and complexity. In particular, our contributions are the following:

- We give a complete overview of the expressive power of the different classes of deterministic expressions with counting. We show that strongly deterministic expressions with counting are equally expressive as standard deterministic expressions. Weakly deterministic expressions with counting, on the other hand, are more expressive than strongly deterministic ones, except for unary languages, on which they coincide. However, not all unary regular languages are definable by weakly deterministic expressions with counting (Section 3).
- We investigate the difference in succinctness between strongly and weakly deterministic expressions with counting, and show that weakly deterministic expressions can be exponentially more succinct than strongly deterministic ones. This result prohibits an efficient algorithm translating a weakly deterministic expression into an equivalent strongly deterministic one, if such an expression exists. This contrasts with the situation of standard expressions where such a linear time algorithm exists [1] (Section 4).
- We present an automaton model extended with counters, counter NFAs (CNFAs), and investigate the complexity of some related problems. For instance, it is shown that boolean operations can be applied efficiently to CDFAs, the deterministic counterpart of CNFAs (Section 5).
- Bruggemann-Klein [1] has shown that the Glushkov construction, translating regular expressions into NFAs, yields a DFA if and only if the original expression is deterministic. We investigate the natural extension of the Glushkov construction to expressions with counters, converting expressions to CNFAs. We show that the resulting automaton is deterministic if and only if the original expression is strongly deterministic (Section 6).

[1] Brüggemann-Klein [1] did not study strong determinism explicitly, although she did study strong unambiguity. However, she gives a procedure to transform expressions into *star normal form* which rewrites weakly deterministic expressions into equivalent strongly deterministic ones in linear time.

- Combining the results of Section 5, concerning CDFAs, with the latter result then also allows to infer better upper bounds on the inclusion and equivalence problem of strongly deterministic expressions with counting. Further, we show that testing whether an expression with counting is strongly deterministic can be done in cubic time, as is the case for weak determinism [13] (Section 7).

The original motivation for this work comes from the XML schema language XML Schema, which uses weakly deterministic expressions with counting. However, it is also noted by Sperberg-McQueen [20], one of its developers, that *"Given the complications which arise from [weakly deterministic expressions], it might be desirable to also require that they be strongly deterministic as well [in XML Schema]."* The design decision for weak determinism is probably inspired by the fact that it is the natural extension of the notion of determinism for standard expressions, and a lack of a detailed analysis of their differences when counting is allowed. A detailed examination of strong and weak determinism of regular expressions with counting intends to fill this gap.

Related work: Apart from the work already mentioned, there are several automata based models for different classes of expressions with counting with as main application XML Schema validation, by Kilpelainen and Tuhkanen [12], Zilio and Lugiez [4], and Sperberg-McQueen [20]. Here, Sperberg-McQueen introduces the extension of the Glushkov construction which we study in Section 6. We introduce a new automata model in Section 5 as none of these models allow to derive all results in Sections 5 and 6. Further, Sperberg-McQueen [20] and Koch and Scherzinger [14] introduce a (slightly different) notion of strongly deterministic expression with and without counting, respectively. We follow the semantic meaning of Sperberg-McQueen's definition, while using the technical approach of Koch and Scherzinger. Finally, Kilpelainen [10] shows that inclusion for weakly deterministic expressions with counting is coNP-hard; and Colazzo, Ghelli, and Sartiani [3] have investigated the inclusion problem involving subclasses of deterministic expressions with counting. Seidl et al. also investigate counting constraints in XML schema languages by adding Presburger constraints to regular languages [18]. Concerning deterministic languages without counting, the seminal paper is by Bruggemann-Klein and Wood [2] where, in particular, it is shown to be decidable whether a language is definable by a deterministic regular expression. Conversely, general regular expressions with counting have also received quite some attention [7,8,11,16].

2 Preliminaries

Let $\mathbb{N}$ denote the natural numbers $\{0, 1, 2, \ldots\}$. For the rest of the paper, Σ always denotes a finite alphabet. The set of regular expressions over Σ, denoted by $\mathrm{RE}(\Sigma)$, is defined as follows: ε and every Σ-symbol is in $\mathrm{RE}(\Sigma)$; and whenever r and s are in $\mathrm{RE}(\Sigma)$, then so are (rs), $(r+s)$, and $(s)^*$. For readability, we usually omit parentheses in examples. The language defined by a regular expression r, denoted by $L(r)$, is defined as usual. By $\mathrm{RE}(\Sigma,\#)$ we denote $\mathrm{RE}(\Sigma)$ extended with *numerical occurrence constraints* or *counting*. That is, when r is an $\mathrm{RE}(\Sigma,\#)$-expression then so is

$r^{k,\ell}$ for $k \in \mathbb{N}$ and $\ell \in \mathbb{N}_0 \cup \{\infty\}$ with $k \leq \ell$. Here, $\mathbb{N}_0$ denotes $\mathbb{N} \setminus \{0\}$. Furthermore, $L(r^{k,\ell}) = \bigcup_{i=k}^{\ell} L(r)^i$. We use $r?$ to abbreviate $(r + \varepsilon)$. Notice that r^* is simply an abbreviation for $r^{0,\infty}$. Therefore, we do not consider the $*$-operator in the context of RE(Σ,#). The *size* of a regular expression r in RE(Σ,#), denoted by $|r|$, is the number of Σ-symbols and operators occurring in r plus the sizes of the binary representations of the integers. An RE(Σ,#) expression r is *nullable* if $\varepsilon \in L(r)$. We say that an RE(Σ,#) r is in *normal form* if for every nullable subexpression $s^{k,l}$ of r we have $k = 0$. Any RE(Σ,#) can easily be normalized in linear time. Therefore, we assume that all expressions used in this paper are in normal form. Sometimes we will use the following observation, which follows directly from the definitions:

Remark 1. A subexpression $r^{k,\ell}$ is nullable if and only if $k = 0$.

Weak determinism. For an RE(Σ,#) r, let Char(r) be the set of Σ-symbols occurring in r. A *marked regular expression with counting over* Σ is a regular expression over $\Sigma \times \mathbb{N}$ in which every $(\Sigma \times \mathbb{N})$-symbol occurs at most once. We denote the set of all these expressions by MRE(Σ,#). Formally, $\overline{r} \in$ MRE(Σ,#) if $\overline{r} \in$ RE($\Sigma \times \mathbb{N}$,#) and, for every subexpression $\overline{s}\,\overline{s}'$ or $\overline{s} + \overline{s}'$ of $\overline{r}$, $\mathrm{Char}(\overline{s}) \cap \mathrm{Char}(\overline{s}') = \emptyset$. A *marked string* is a string over $\Sigma \times \mathbb{N}$ (in which $(\Sigma \times \mathbb{N})$-symbols can occur more than once). When $\overline{r}$ is a marked regular expression, $L(\overline{r})$ is therefore a set of marked strings.

The demarking of a marked expression is obtained by deleting these integers. Formally, the demarking of $\overline{r}$ is $\mathrm{dm}(\overline{r})$, where dm : MRE($\Sigma$,#) $\to$ RE(Σ,#) is defined as $\mathrm{dm}(\varepsilon) := \varepsilon$, $\mathrm{dm}((a,i)) := a$, $\mathrm{dm}(\overline{rs}) := \mathrm{dm}(\overline{r})\mathrm{dm}(\overline{s})$, $\mathrm{dm}(\overline{r+s}) := \mathrm{dm}(\overline{r}) + \mathrm{dm}(\overline{s})$, and $\mathrm{dm}(\overline{r^{k,\ell}}) := \mathrm{dm}(\overline{r})^{k,\ell}$. Any function m : RE($\Sigma$,#) $\to$ MRE(Σ,#) such that for every $r \in$ RE(Σ,#) it holds that $\mathrm{dm}(\mathrm{m}(r)) = r$ is a valid *marking* function. For conciseness and readability, we will from now on write a_i instead of (a,i) in marked regular expressions. For instance, a *marking* of $(a+b)^{1,2}a+bc$ is $(a_1+b_1)^{1,2}a_2+b_2c_1$. The markings and demarkings of strings are defined analogously. For the rest of the paper, we usually leave the actual marking function m implicit and denote by $\overline{r}$ a marking of the expression r. Likewise $\overline{w}$ will denote a marking of a string w. We always use overlined letters to denote marked expressions, symbols, and strings.

Definition 2. An RE(Σ,#) expression r is *weakly deterministic* (also called *one-unambiguous*) if, for all strings $\overline{u}, \overline{v}, \overline{w} \in \mathrm{Char}(\overline{r})^*$ and all symbols $\overline{a}, \overline{b} \in \mathrm{Char}(\overline{r})$, the conditions $\overline{uav}, \overline{ubw} \in L(\overline{r})$ and $\overline{a} \neq \overline{b}$ imply that $a \neq b$.

A regular language is *weakly deterministic with counting* if it is defined by some weakly deterministic RE(Σ,#) expression. The classes of all weakly deterministic languages with counting, respectively, without counting, are denoted by $\mathrm{DET}_W^{\#}(\Sigma)$, respectively, $\mathrm{DET}_W(\Sigma)$.

Intuitively, an expression is weakly deterministic if, when matching a string against the expression from left to right, we always know against which symbol in the expression we must match the next symbol, without looking ahead in the string. For instance, $(a+b)^*a$ and $(a^{2,3}+b)^{3,3}b$ are not weakly deterministic, while $b^*a(b^*a)^*$ and $(a^{2,3}+b)^{2,2}b$ are.

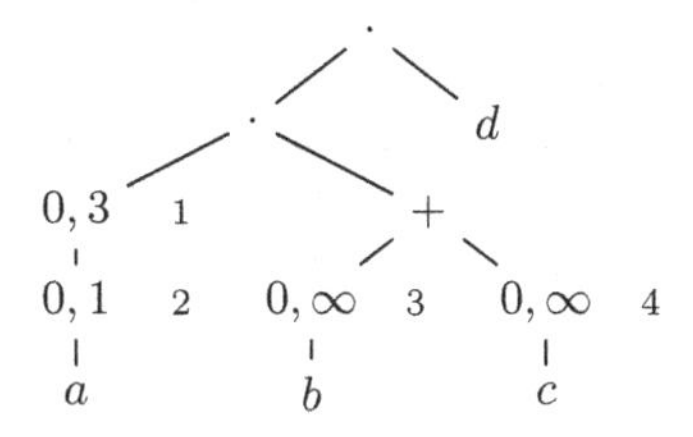

Fig. 1. Parse tree of $(a^{0,1})^{0,3}(b^{0,\infty}+c^{0,\infty})d$. Counter nodes are numbered from 1 to 4.

Strong determinism. Intuitively, an expression is weakly deterministic if, when matching a string from left to right, we always know *where* we are in the expression. For a strongly deterministic expression, we will additionally require that we always know *how* to go from one position to the next. Thereto, we distinguish between going *forward* in an expression and *backward* by *iterating* over a counter. For instance, in the expression $(ab)^{1,2}$ going from a to b implies going forward, whereas going from b to a iterates backward over the counter.

Therefore, an expression such as $((a+\varepsilon)(b+\varepsilon))^{0,2}$ will not be strongly deterministic, although it is weakly deterministic. Indeed, when matching ab, we can go from a to b by either going forward or by iterating over the counter. By the same token, also $(a^{1,2})^{3,4}$ is not strongly deterministic, as we have a choice of counters over which to iterate when reading multiple a's. Conversely, $(a^{2,2})^{3,4}$ is strongly deterministic as it is always clear over which counter we must iterate.

For the definition of strong determinism, we follow the semantic meaning of the definition by Sperberg-McQueen [20], while using the formal approach of Koch and Scherzinger [14] (who called the notion *strong one-unambiguity*)[2]. We denote the *parse tree* of an RE(Σ,#) expression r by pt(r). Figure 1 contains the parse tree of the expression $(a^{0,1})^{0,3}(b^{0,\infty}+c^{0,\infty})d$.

A *bracketing of a regular expression* r is a labeling of the counter nodes of pt(r) by distinct indices. Concretely, we simply number the nodes according to the depth-first left-to-right ordering. The bracketing $\widetilde{r}$ of r is then obtained by replacing each subexpression $s^{k,\ell}$ of r with index i with $(\left[_i s\right]_i)^{k,\ell}$. Therefore, a bracketed regular expression is a regular expression over alphabet $\Sigma \uplus \Gamma$, where $\Gamma := \{\left[_i, \right]_i \mid i \in \mathbb{N}\}$. For example, $(\left[_1(\left[_2 a\right]_2)^{0,1}\right]_1)^{0,3}((\left[_3 b\right]_3)^{0,\infty}+(\left[_4 c\right]_4)^{0,\infty})d$ is a bracketing of $(a^{0,1})^{0,3}(b^{0,\infty}+c^{0,\infty})d$, for which the parse tree is shown in Figure 1. We say that a string w in $\Sigma \uplus \Gamma$ is *correctly bracketed* if w has no substring of the form $\left[_i\right]_i$. That is, we do not allow a derivation of ε in the derivation tree.

Definition 3. A regular expression r is strongly deterministic with counting if r is weakly deterministic and there do not exist strings u, v, w over $\Sigma \cup \Gamma$, strings $\alpha \neq \beta$ over Γ, and a symbol $a \in \Sigma$ such that $u\alpha av$ and $u\beta aw$ are both correctly bracketed and in $L(\widetilde{r})$.

[2] The difference with Koch and Scherzinger is that we allow different derivations of ε while they forbid this. For instance, $a^* + b^*$ is strongly deterministic in our definition, but not in theirs, as ε can be matched by both a^* and b^*.

A standard regular expression (without counting) is strongly deterministic if the expression obtained by replacing each subexpression of the form r^* with $r^{0,\infty}$ is strongly deterministic with counting. The class $\mathrm{DET}^{\#}_S(\Sigma)$, respectively, $\mathrm{DET}_S(\Sigma)$, denotes all languages definable by a strongly deterministic expressions with, respectively, without, counting.

3 Expressive Power

Brüggemann-Klein and Wood [2] proved that for any alphabet Σ $\mathrm{DET}_W(\Sigma)$ forms a strict subclass of the regular languages, denoted $\mathrm{REG}(\Sigma)$. The complete picture of the relative expressive power depends on the size of Σ, as shown by the following theorem.

Theorem 4. *For every alphabet Σ,*

$$DET_S(\Sigma) = DET_W(\Sigma) = DET^{\#}_S(\Sigma) = DET^{\#}_W(\Sigma) \subsetneq REG(\Sigma) \quad (\text{if } |\Sigma| = 1)$$

$$DET_S(\Sigma) = DET_W(\Sigma) = DET^{\#}_S(\Sigma) \subsetneq DET^{\#}_W(\Sigma) \subsetneq REG(\Sigma) \quad (\text{if } |\Sigma| \geq 2)$$

Proof. The equality $\mathrm{DET}_S(\Sigma) = \mathrm{DET}_W(\Sigma)$ is already implicit in the work of Brüggemann-Klein [1]. By this result and by definition, all inclusions from left to right already hold. It therefore suffices to show that (1) $\mathrm{DET}^{\#}_S(\Sigma) \subseteq \mathrm{DET}_S(\Sigma)$ for arbitrary alphabets, (2) $\mathrm{DET}^{\#}_W(\Sigma) \subseteq \mathrm{DET}^{\#}_S(\Sigma)$ for unary alphabets, (3) $\mathrm{DET}^{\#}_S(\Sigma) \subsetneq \mathrm{DET}^{\#}_W(\Sigma)$ for binary alphabets, and (4) $\mathrm{DET}^{\#}_W(\Sigma) \subsetneq \mathrm{REG}(\Sigma)$ for unary alphabets.

(1): We show that each strongly deterministic expression with counting can be transformed into a strongly deterministic expression without counting. This is quite non-trivial, but the crux is to unfold each counting operator in a smart manner, taking special care of nullable expressions.
(2): The crux of this proof lies in Lemma 5. It is well known and easy to see that the minimal DFA for a regular language over a unary alphabet is defined either by a simple chain of states (sometimes also called a *tail* [19]), or a chain followed by a cycle.The languages in $\mathrm{DET}^{\#}_W(\Sigma)$ can be defined in this manner. The following lemma adds to that, that for weakly deterministic regular expressions, only one node in this cycle can be final. The theorem then follows as any such language can be defined by a strongly deterministic expression.

Lemma 5. *Let $\Sigma = \{a\}$, and $L \in REG(\Sigma)$, then $L \in DET^{\#}_W(\Sigma)$ if and only if L is definable by a DFA which is either a chain, or a chain followed by a cycle, for which at most one of the cycle nodes is final.*

(3 and 4): Witnesses for non-inclusion are the languages defined by $(a^{2,3}b?)^*$ and $(aaa)^*(a + aa)$, respectively. Both languages can be shown not to be in $\mathrm{DET}_W(\Sigma)$ [2]. The theorem then follows from the above results.

4 Succinctness

In Section 3 we learned that $\mathrm{DET}^{\#}_{W}(\Sigma)$ strictly contains $\mathrm{DET}^{\#}_{S}(\Sigma)$, prohibiting a translation from weak to strong deterministic expressions with counting. However, one could still hope for an efficient algorithm which, given a weakly deterministic expression known to be equivalent to a strong deterministic one, constructs this expression. However, this is not the case:

Theorem 6. *For every $n \in \mathbb{N}$, there exists an $r \in RE(\Sigma,\#)$ over alphabet $\{a\}$ which is weakly deterministic and of size $\mathcal{O}(n)$ such that every strongly deterministic expression s, with $L(r) = L(s)$, is of size at least 2^n.*

The above theorem holds for the family of languages defined by $(a^{2^n+1,2^{n+1}})^{1,2}$, each of which is weakly deterministic and defines all strings with a's of length from 2^n+1 to 2^{n+2}, except for the string $a^{2^{n+1}+1}$. These expressions, in fact, where introduced by Kilpelainen when studying the inclusion problem for weakly deterministic expressions with counting [10].

5 Counter Automata

Let C be a set of *counter variables* and $\alpha : C \to \mathbb{N}$ be a function assigning a value to each counter variable. We inductively define *guards* over C, denoted $\mathrm{Guard}(C)$, as follows: for every $\mathrm{cv} \in C$ and $k \in \mathbb{N}$, we have that true, false, $\mathrm{cv} = k$, and $\mathrm{cv} < k$ are in $\mathrm{Guard}(C)$. Moreover, when $\phi_1, \phi_2 \in \mathrm{Guard}(C)$, then so are $\phi_1 \wedge \phi_2$, $\phi_1 \vee \phi_2$, and $\neg\phi_1$. For $\phi \in \mathrm{Guard}(C)$, we denote by $\alpha \models \phi$ that α models ϕ, i.e., that applying the value assignment α to the counter variables results in satisfaction of ϕ.

An *update* is a set of statements of the form cv++ and reset(cv) in which every $\mathrm{cv} \in C$ occurs at most once. By $\mathrm{Update}(C)$ we denote the set of all updates.

Definition 7. A non-deterministic *counter automaton* (CNFA) is a 6-tuple $A = (Q, q_0, C, \delta, F, \tau)$ where Q is the finite set of states; $q_0 \in Q$ is the initial state; C is the finite set of counter variables; $\delta : Q \times \Sigma \times \mathrm{Guard}(C) \times \mathrm{Update}(C) \times Q$ is the transition relation; $F : Q \to \mathrm{Guard}(C)$ is the acceptance function; and $\tau : C \to \mathbb{N}$ assigns a maximum value to every counter variable.

Intuitively, A can make a transition (q, a, ϕ, π, q') whenever it is in state q, reads a, and guard ϕ is true under the current values of the counter variables. It then updates the counter variables according to the update π, in a way we explain next, and moves into state q'. To explain the update mechanism formally, we introduce the notion of configuration. Thereto, let $\max(A) = \max\{\tau(c) \mid c \in C\}$. A *configuration* is a pair (q, α) where $q \in Q$ is the current state and $\alpha : C \to \{1, \ldots, \max(A)\}$ is the function mapping counter variables to their current value. Finally, an update π transforms α into $\pi(\alpha)$ by setting $\mathrm{cv} := 1$, when $\mathrm{reset(cv)} \in \pi$, and $\mathrm{cv} := \mathrm{cv} + 1$ when $\mathrm{cv{+}{+}} \in \pi$ and $\alpha(\mathrm{cv}) < \tau(\mathrm{cv})$. Otherwise, the value of cv remains unaltered.

Let α_0 be the function mapping every counter variable to 1. The *initial configuration* γ_0 is (q_0, α_0). A configuration (q, α) is *final* if $\alpha \models F(q)$. A configuration $\gamma' = (q', \alpha')$ *immediately follows* a configuration $\gamma = (q, \alpha)$ by reading $a \in \Sigma$, denoted $\gamma \rightarrow_a \gamma'$, if there exists $(q, a, \phi, \pi, q') \in \delta$ with $\alpha \models \phi$ and $\alpha' = \pi(\alpha)$.

For a string $w = a_1 \cdots a_n$ and two configurations γ and γ', we denote by $\gamma \Rightarrow_w \gamma'$ that $\gamma \rightarrow_{a_1} \cdots \rightarrow_{a_n} \gamma'$. A configuration γ is *reachable* if there exists a string w such that $\gamma_0 \Rightarrow_w \gamma$. A string w is *accepted* by A if $\gamma_0 \Rightarrow_w \gamma_f$ where γ_f is a final configuration. We denote by $L(A)$ the set of strings accepted by A.

A CNFA A is *deterministic* (or a CDFA) if, for every reachable configuration $\gamma = (q, \alpha)$ and for every symbol $a \in \Sigma$, there is at most one transition $(q, a, \phi, \pi, q') \in \delta$ such that $\alpha \models \phi$.

The *size* of a transition θ or acceptance condition $F(q)$ is the number of symbols which occur in it plus the size of the binary representation of each integer occcurring in it. By the same token, the size of A, denoted by $|A|$, is $|Q| + \sum_{q \in Q} \log \tau(q) + |F(q)| + \sum_{\theta \in \delta} |\theta|$.

Theorem 8. *1. Given CNFAs A_1 and A_2, a CNFA A accepting the union or intersection of A_1 and A_2 can be constructed in polynomial time. Moreover, when A_1 and A_2 are deterministic, then so is A.*
2. Given a CDFA A, a CDFA which accepts the complement of A can be constructed in polynomial time.
3. MEMBERSHIP *for word w and CDFA A is in time $\mathcal{O}(|w||A|)$.*
4. MEMBERSHIP *for non-deterministic CNFA is* NP*-complete.*
5. EMPTINESS *for CDFAs and CNFAs is* PSPACE*-complete.*
6. Deciding whether a CNFA A is deterministic is PSPACE*-complete.*

6 From RE(Σ,#) to CNFA

In this section, we show how an RE(Σ,#) expression r can be translated in polynomial time into an equivalent CNFA G_r by applying a natural extension of the well-known Glushkov construction. We emphasize at this point that such an extended Glushkov construction has already been given by Sperberg-McQueen [20]. Therefore, the contribution of this section lies mostly in the characterization given below: G_r is deterministic if and only if r is strongly deterministic. Moreover, as seen in the previous section, CDFAs have desirable properties which by this translation also apply to strongly deterministic RE(Σ,#) expressions. We refer to G_r as the *Glushkov counting automaton of* r.

6.1 Notation and Terminology

We first provide some notation and terminology needed in the construction below. For an RE(Σ,#) expression r, the set first(r) (respectively, last(r)) consists of all symbols which are the first (respectively, last) symbols in some word defined by r. These sets are inductively defined as follows:

- $\text{first}(\varepsilon) = \text{last}(\varepsilon) = \emptyset$ and $\forall a \in \text{Char}(r), \text{first}(a) = \text{last}(a) = \{a\}$;
- $\text{first}(r_1 + r_2) = \text{first}(r_1) \cup \text{first}(r_2)$ and $\text{last}(r_1 + r_2) = \text{last}(r_1) \cup \text{last}(r_2)$;

- If $\varepsilon \in L(r_1)$, $\text{first}(r_1 r_2) = \text{first}(r_1) \cup \text{first}(r_2)$, else $\text{first}(r_1 r_2) = \text{first}(r_1)$;
- If $\varepsilon \in L(r_2)$, $\text{last}(r_1 r_2) = \text{last}(r_1) \cup \text{last}(r_2)$, else $\text{last}(r_1 r_2) = \text{last}(r_2)$;
- $\text{first}(r^{k,\ell}) = \text{first}(r_1)$ and $\text{last}(r^{k,\ell}) = \text{last}(r_1)$.

For a regular expression r, we say that a subexpression of r of the form $s^{k,\ell}$ is an *iterator* or *iterated subexpression of* r. Let $\text{lower}(s^{k,\ell}) := k$, and $\text{upper}(s^{k,\ell}) := \ell$. We say that $s^{k,\ell}$ is *bounded* when $\ell \in \mathbb{N}$, otherwise it is *unbounded*. For instance, an iterator of the form $s^{0,\infty}$ is a nullable, unbounded iterator.

For a marked symbol $\overline{x}$ and an iterator c we denote by $\text{iterators}(\overline{x}, c)$ the list of all iterated subexpressions of c which contain $\overline{x}$, except c itself. For marked symbols $\overline{x}, \overline{y}$, we denote by $\text{iterators}(\overline{x}, \overline{y})$ all iterated subexpressions which contain $\overline{x}$ but not $\overline{y}$. Finally, let $\text{iterators}(\overline{x})$ be the list of all iterated subexpressions which contain $\overline{x}$. Note that all such lists $[c_1, \ldots, c_n]$ contain a sequence of nested subexpressions. Therefore, we will always assume that they are ordered such that $c_1 \prec c_2 \prec \cdots \prec c_n$. Here $c \prec c'$ denotes that c is a subexpression of c'. For example, if $\overline{r} = ((a_1^{1,2} b_1)^{3,4})^{5,6}$, then $\text{iterators}(a_1, \overline{r}) = [a_1^{1,2}, (a_1^{1,2} b_1)^{3,4}]$, $\text{iterators}(a_1, b_1) = [a_1^{1,2}]$, and $\text{iterators}[a_1] = [a_1^{1,2}, (a_1^{1,2} b_1)^{3,4}, ((a_1^{1,2} b_1)^{3,4})^{5,6}]$.

6.2 Construction

We now define the set $\text{follow}(\overline{r})$ for a marked regular expression $\overline{r}$. As in the standard Glushkov construction, this set lies at the basis of the transition relation of G_r. The set $\text{follow}(\overline{r})$ contains triples $(\overline{x}, \overline{y}, c)$, where $\overline{x}$ and $\overline{y}$ are marked symbols and c is either an iterator or null. Intuitively, the states of G_r will be a designated start state plus a state for each symbol in $\text{Char}(\overline{r})$. A triple $(\overline{x}, \overline{y}, c)$ then contains the information we need for G_r to make a transition from state $\overline{x}$ to $\overline{y}$. If $c \neq \text{null}$, this transition iterates over c and all iterators in $\text{iterators}(\overline{x}, c)$ are reset by going to $\overline{y}$. Otherwise, if c equals null, the iterators in $\text{iterators}(\overline{x}, \overline{y})$ are reset. Formally, the set $\text{follow}(\overline{r})$ contains for each subexpression $\overline{s}$ of $\overline{r}$,

- all tuples $(\overline{x}, \overline{y}, \text{null})$ for $\overline{x}$ in $\text{last}(\overline{s_1})$, $\overline{y}$ in $\text{first}(\overline{s_2})$, and $\overline{s} = \overline{s_1}\, \overline{s_2}$; and
- all tuples $(\overline{x}, \overline{y}, \overline{s})$ for $\overline{x}$ in $\text{last}(\overline{s_1})$, $\overline{y}$ in $\text{first}(\overline{s_1})$, and $\overline{s} = \overline{s_1}^{k,\ell}$.

We introduce a counter variable $\text{cv}(c)$ for every iterator c in $\overline{r}$ whose value will always denote which iteration of c we are doing in the current run on the string. We define a number of tests and update commands on these counter variables:

- $\text{value-test}([c_1, \ldots, c_n]) := \bigwedge_{c_i} (\text{lower}(c_i) \leq \text{cv}(c_i)) \wedge (\text{cv}(c_i) \leq \text{upper}(c_i))$. When we leave the iterators $c_1, \ldots, c_n$ we have to check that we have done an admissible number of iterations for each iterator.
- $\text{upperbound-test}(c) := \text{cv}(c) < \text{upper}(c)$ when c is a bounded iterator and $\text{upperbound-test}(c) := \text{true}$ otherwise. When iterating over a bounded iterator, we have to check that we can still do an extra iteration.
- $\text{reset}(c_1, \ldots, c_n) := \{\text{reset}(\text{cv}(c_1)), \ldots, \text{reset}(\text{cv}(c_n))\}$. When leaving some iterators, their values must be reset. The counter variable is reset to 1, because at the time we reenter this iterator, its first iteration is started.
- $\text{update}(c) := \{\text{cv}(c){+}{+}\}$. When iterating over an iterator, we start a new iteration and increment its number of transitions.

We now define the Glushkov counting automaton $G_r = (Q, q_0, C, \delta, F, \tau)$. The set of states Q is the set of symbols in $\overline{r}$ plus an initial state, i.e., $Q := \{q_0\} \uplus \bigcup_{\overline{x} \in \mathrm{Char}(\overline{r})} q_{\overline{x}}$. Let C be the set of iterators occurring in $\overline{r}$. We next define the transition function. For all $\overline{y} \in \mathrm{first}(\overline{r})$, $(q_0, \mathrm{dm}(\overline{y}), \mathit{true}, \emptyset, q_{\overline{y}}) \in \delta$.[3] For every element $(\overline{x}, \overline{y}, c) \in \mathrm{follow}(\overline{r})$, we define a transition $(q_{\overline{x}}, \mathrm{dm}(\overline{y}), \phi, \pi, q_{\overline{y}}) \in \delta$. If $c = \mathrm{null}$, then $\phi := \text{value-test}(\mathrm{iterators}(\overline{x}, \overline{y}))$ and $\pi := \mathrm{reset}(\mathrm{iterators}(\overline{x}, \overline{y}))$. If $c \neq \mathrm{null}$, then $\phi := \text{value-test}(\mathrm{iterators}(\overline{x}, c)) \wedge \text{upperbound-test}(c)$ and $\pi := \mathrm{reset}(\mathrm{iterators}(\overline{x}, c)) \cup \mathrm{update}(c)$. The acceptance criteria of G_r depend on the set $\mathrm{last}(\overline{r})$. For any symbol $\overline{x} \notin \mathrm{last}(\overline{r})$, $F(q_{\overline{x}}) := \mathrm{false}$. For every element $\overline{x} \in \mathrm{last}(\overline{r})$, $F(q_{\overline{x}}) := \text{value-test}(\mathrm{iterators}(\overline{x}))$. Here, we test whether we have done an admissible number of iterations of all iterators in which $\overline{x}$ is located. Finally, $F(q_0) := \mathrm{true}$ if $\varepsilon \in L(r)$. Lastly, for all bounded iterators c, $\tau(\mathrm{cv}(c)) = \mathrm{upper}(c)$ since c never becomes larger than $\mathrm{upper}(c)$, and for all unbounded iterators c, $\tau(\mathrm{cv}(c)) = \mathrm{lower}(c)$ as there are no upper bound tests for $\mathrm{cv}(c)$.

Theorem 9. *For every RE(Σ,#) expression r, $L(G_r) = L(r)$. Moreover, G_r is deterministic iff r is strongly deterministic.*

7 Decidability and Complexity Results

Definition 3, defining strong determinism, is of a semantical nature. Therefore, we provide Algorithm 1 for testing whether a given expression is strongly deterministic, which runs in cubic time. To decide weak determinism, Kilpeläinen and Tuhkanen [13] give a cubic algorithm for RE(Σ,#), while Brüggemann-Klein [1] gives a quadratic algorithm for RE(Σ) by computing its Glushkov automaton and testing whether it is deterministic[4].

Theorem 10. *For any $r \in$ RE(Σ,#), isStrongDeterministic(r) returns true if and only if r is strong deterministic. Moreover, it runs in time $\mathcal{O}(|r|^3)$.*

We next consider the following decision problems, for expressions of class $\mathcal{R}$:
INCLUSION: Given two expressions $r, r' \in \mathcal{R}$, is $L(r) \subseteq L(r')$?
EQUIVALENCE: Given two expressions $r, r' \in \mathcal{R}$, is $L(r) = L(r')$?
INTERSECTION: Given a number of expressions $r_1, \ldots, r_n \in \mathcal{R}$, is $\bigcap_{i=1}^{n} L(r_i) \neq \emptyset$?

Theorem 11. *(1)* INCLUSION *and* EQUIVALENCE *for RE(Σ,#) are* EXPSPACE-*complete [16],* INTERSECTION *for RE(Σ,#) is* PSPACE-*complete [7]. (2)* INCLUSION *and* EQUIVALENCE *for DET$_W$(Σ) are in* PTIME, INTERSECTION *for DET$_W$(Σ) is* PSPACE-*complete [15]. (3)* INCLUSION *for DET$^{\#}_W$(Σ) is* coNP-*hard [11].*

[3] Recall that $\mathrm{dm}(\overline{y})$ denotes the demarking of $\overline{y}$.

[4] There sometimes is some confusion about this result: Computing the Glushkov automaton is quadratic in the expression, while linear in the output automaton (consider, e.g., $(a_1 + \cdots + a_n)(a_1 + \cdots + a_n)$). Only when the alphabet is fixed is the Glushkov automaton of a deterministic expression of size linear in the expression.

Algorithm 1. IsStrongDeterministic. Returns true if r is strong deterministic, false otherwise.

```
    r̄ ← marked version of r
 2: Initialize Follow ← ∅
    Compute first(s̄), last(s̄), for all subexpressions s̄ of r̄
 4: if ∃x̄, ȳ ∈ first(r̄) with x̄ ≠ ȳ and dm(x̄) = dm(ȳ) then return false
    for each subexpression s̄ of r̄, in bottom-up fashion do
 6:     if s̄ = s̄1 s̄2 then
            if last(s̄1) ≠ ∅ and ∃x̄, ȳ ∈ first(s̄1) with x̄ ≠ ȳ and dm(x̄) = dm(ȳ) then
    return false
 8:         F ← {(x̄, dm(ȳ)) | x̄ ∈ last(s̄1), ȳ ∈ first(s̄2)}
        else if s̄ = s̄1^[k,ℓ], with ℓ ≥ 2 then
10:         if ∃x̄, ȳ ∈ first(s̄1) with x̄ ≠ ȳ and dm(x̄) = dm(ȳ) then return false
            F ← {(x̄, dm(ȳ)) | x̄ ∈ last(s̄1), ȳ ∈ first(s̄1)}
12:     if F ∩ Follow ≠ ∅ then return false
        if s̄ = s̄1 s̄2 or s̄ = s̄1^k,ℓ, with ℓ ≥ 2 and k < ℓ then
14:         Follow ← Follow ⊎ F
    return true
```

By combining (1) and (2) of Theorem 11 we get the complexity of INTERSECTION for $\mathrm{DET}^{\#}_{W}(\Sigma)$ and $\mathrm{DET}^{\#}_{S}(\Sigma)$. This is not the case for the INCLUSION and EQUIVALENCE problem, unfortunately. By using the results of the previous sections we can, for $\mathrm{DET}^{\#}_{S}(\Sigma)$, give a PSPACE upperbound for both problems, however.

Theorem 12. *(1)* EQUIVALENCE *and* INCLUSION *for* $DET^{\#}_{S}(\Sigma)$ *are in* PSPACE. *(2)* INTERSECTION *for* $DET^{\#}_{W}(\Sigma)$ *and* $DET^{\#}_{S}(\Sigma)$ *is* PSPACE*-complete.*

8 Conclusion

We investigated and compared the notions of strong and weak determinism in the presence of counting. Weakly deterministic expressions have the advantage of being more expressive and more succinct than strongly deterministic ones. However, strongly deterministic expressions are expressivily equivalent to standard deterministic expressions, a class of languages much better understood than the weakly deterministic languages with counting. Moreover, strongly deterministic expressions are conceptually simpler (as strong determinism does not depend on intricate interplays of the counter values) and correspond naturally to deterministic Glushkov automata. The latter also makes strongly deterministic expressions easier to handle as witnessed by the PSPACE upperbound for inclusion and equivalence, whereas for weakly deterministic expressions only a trivial EXPSPACE upperbound is known. For these reasons, one might wonder if the weak determinism demanded in the current standards for XML Schema should not be replaced by strong determinism. The answer to some of the following open questions can shed more light on this issue: (1) Is it decidable if a language is definable by a weakly deterministic expression with counting? (2) Can the Glushkov construction given in Section 6 be extended such that it translates any weakly deterministic expression with counting into a CDFA? (3) What are the exact complexity bounds for inclusion and equivalence of strongly and weakly deterministic expression with counting?

References

1. Brüggemann-Klein, A.: Regular expressions into finite automata. Theor. Comput. Sci. 120(2), 197–213 (1993)
2. Brüggemann-Klein, A., Wood, D.: One-unambiguous regular languages. Information and Computation 142(2), 182–206 (1998)
3. Colazzo, D., Ghelli, G., Sartiani, C.: Efficient asymmetric inclusion between regular expression types. In: ICDT, pp. 174–182 (2009)
4. Dal-Zilio, S., Lugiez, D.: XML schema, tree logic and sheaves automata. In: Nieuwenhuis, R. (ed.) RTA 2003. LNCS, vol. 2706, pp. 246–263. Springer, Heidelberg (2003)
5. Esparza, J.: Decidability and complexity of Petri net problems – an introduction. In: Petri Nets, pp. 374–428 (1996)
6. Garey, M.R., Johnson, D.S.: Computers and Intractability: A Guide to the Theory of NP-Completeness. Freeman, New York (1979)
7. Gelade, W., Martens, W., Neven, F.: Optimizing schema languages for XML: Numerical constraints and interleaving. In: Schwentick, T., Suciu, D. (eds.) ICDT 2007. LNCS, vol. 4353, pp. 269–283. Springer, Heidelberg (2007)
8. Gelade, W.: Succinctness of regular expressions with interleaving, intersection and counting. In: Ochmański, E., Tyszkiewicz, J. (eds.) MFCS 2008. LNCS, vol. 5162, pp. 363–374. Springer, Heidelberg (2008)
9. Hume, A.: A tale of two greps. Softw. Pract. and Exp. 18(11), 1063–1072 (1988)
10. Kilpeläinen, P.: Inclusion of unambiguous #res is NP-hard (May 2004) (unpublished)
11. Kilpeläinen, P., Tuhkanen, R.: Regular expressions with numerical occurrence indicators — preliminary results. In: SPLST 2003, pp. 163–173 (2003)
12. Kilpeläinen, P., Tuhkanen, R.: Towards efficient implementation of XML schema content models. In: DOCENG 2004, pp. 239–241. ACM, New York (2004)
13. Kilpeläinen, P., Tuhkanen, R.: One-unambiguity of regular expressions with numeric occurrence indicators. Inform. Comput. 205(6), 890–916 (2007)
14. Koch, C., Scherzinger, S.: Attribute grammars for scalable query processing on XML streams. VLDB Journal 16(3), 317–342 (2007)
15. Martens, W., Neven, F., Schwentick, T.: Complexity of decision problems for simple regular expressions. In: Fiala, J., Koubek, V., Kratochvíl, J. (eds.) MFCS 2004. LNCS, vol. 3153, pp. 889–900. Springer, Heidelberg (2004)
16. Meyer, A.R., Stockmeyer, L.J.: The equivalence problem for regular expressions with squaring requires exponential space. In: FOCS, pp. 125–129 (1972)
17. Mount, D.W.: Bioinformatics: Sequence and Genome Analysis. Cold Spring Harbor Laboratory Press (September 2004)
18. Seidl, H., Schwentick, T., Muscholl, A., Habermehl, P.: Counting in trees for free. In: Díaz, J., Karhumäki, J., Lepistö, A., Sannella, D. (eds.) ICALP 2004. LNCS, vol. 3142, pp. 1136–1149. Springer, Heidelberg (2004)
19. Pighizzini, G., Shallit, J.: Unary language operations, state complexity and Jacobsthal's function. Int. J. Found. Comp. Sc. 13(1), 145–159 (2002)

20. Sperberg-McQueen, C.M.: Notes on finite state automata with counters (2004), `http://www.w3.org/XML/2004/05/msm-cfa.html`
21. Sperberg-McQueen, C.M., Thompson, H.: XML Schema (2005), `http://www.w3.org/XML/Schema`
22. Vardi, M.Y.: From monadic logic to PSL. Pillars of Computer Science, 656–681 (2008)
23. Wall, L., Christiansen, T., Orwant, J.: Programming Perl. O'Reilly, Sebastopol (2000)

Choosability of P_5-Free Graphs*

Petr A. Golovach and Pinar Heggernes

Department of Informatics, University of Bergen, PB 7803, 5020 Bergen, Norway
{Peter.Golovach,Pinar.Heggernes}@ii.uib.no

Abstract. A graph is k-choosable if it admits a proper coloring of its vertices for every assignment of k (possibly different) allowed colors to choose from for each vertex. It is NP-hard to decide whether a given graph is k-choosable for $k \geq 3$, and this problem is considered strictly harder than the k-coloring problem. Only few positive results are known on input graphs with a given structure. Here, we prove that the problem is fixed parameter tractable on P_5-free graphs when parameterized by k. This graph class contains the well known and widely studied class of cographs. Our result is surprising since the parameterized complexity of k-coloring is still open on P_5-free graphs. To give a complete picture, we show that the problem remains NP-hard on P_5-free graphs when k is a part of the input.

1 Introduction

Graph coloring is one of the most well known and intensively studied problems in graph theory. The k-COLORING problem asks whether the vertices of an input graph G can be colored with k colors such that no pair of adjacent vertices receive the same color (such coloring is also called a proper coloring). This problem is known to be NP-complete even when $k \geq 3$ is not a part of the input but a fixed constant.

Vizing [19] and Erdős et al. [6] introduced a version of graph coloring called list coloring. In list coloring, a set $L(v)$ of allowed colors is given for each vertex v of the input graph, and we want to decide whether a proper coloring of the graph exists such that each vertex v receives a color from $L(v)$. If G has a list coloring for every assignment of lists of cardinality k to its vertices, then G is said to be k-choosable. Hence the k-CHOOSABILITY problem asks whether an input graph G is k-choosable. List coloring has received increasing attention since the beginning of 90's, and there are very good surveys [1,17] and books [11] on the subject. It is proved to be a very difficult problem; Gutner and Tarsi [9] proved that k-CHOOSABILITY is Π_2^P-complete for bipartite graphs for any fixed $k \geq 3$, whereas 2-CHOOSABILITY can be solved in polynomial time [6]. The 3-CHOOSABILITY and 4-CHOOSABILITY problems remain Π_2^P-complete for planar graphs, whereas any planar graph is 5-choosable [16]. Due to these hardness results, upto the assumption that NP is not equal to co-NP, CHOOSABILITY is strictly harder than COLORING on general graphs [1].

Despite being a difficult problem to deal with, CHOOSABILITY has applications in a large variety of areas, like various kinds of scheduling problems, VLSI design, and frequency assignments [1]. Consequently, any attempt to solve this problem is of interest,

* This work is supported by the Research Council of Norway.

R. Královič and D. Niwiński (Eds.): MFCS 2009, LNCS 5734, pp. 382–391, 2009.

and we attack it using structural information on the input and parameterized algorithms. A problem is fixed parameter tractable (FPT) if its input can be partitioned into a main part (typically the input graph) of size n and a parameter (typically an integer) k so that there is an algorithm that solves the problem in time $O(n^c \cdot f(k))$, where f is a computable function dependent only on k, and c is a fixed constant independent of input [5]. In this case, we say that the problem is FPT when parameterized by k. The field of parameterized algorithms and fixed parameter complexity/tractability has been flourishing during the last decade, with many new results appearing every year in high level conferences and journals, and it has been enriched by several new books [7,14].

In this paper, we show that k-CHOOSABILITY is fixed parameter tractable on P_5-free graphs. These are graphs containing no induced copy of a simple path on 5 vertices, and this graph class contains the class of cographs that has been subject to extensive theoretical study [3]. An interesting point to mention is that the fixed parameter tractability of k-COLORING on P_5-free graphs is still open [10]. As mentioned above, CHOOSABILITY is more difficult than COLORING on general graphs. Our result indicates that the opposite might be true for the class of P_5-free graphs. In last year's MFCS, Hoàng et al. showed that k-COLORING can be solved in polynomial time for any fixed k on P_5-free graphs [10], but in their running time k contributes to the degree of the polynomial. Furthermore, k-COLORING is NP-complete on P_5-free graphs when k is a part of input [12]. To give a complete picture, here we show that k-CHOOSABILITY is NP-hard on P_5-free graphs when k is a part of input. Thus fixed parameter tractability is the best we can expect to achieve for k-CHOOSABILITY on this graph class.

To mention other existing results on the coloring problem on graphs that do not contain long induced paths, 3-COLORING has a polynomial-time solution on P_6-free graphs [15], 5-COLORING is NP-complete for P_8-free graphs, and 4-COLORING is NP-complete for P_{12}-free graphs [20].

2 Definitions and Preliminaries

We consider finite undirected graphs without loops or multiple edges. A graph is denoted by $G = (V, E)$, where $V = V(G)$ is the set of vertices and $E = E(G)$ is the set of edges. For a vertex $v \in V$, the set of vertices that are adjacent to v is called the *neighborhood* of v and denoted by $N_G(v)$ (we may omit index if the graph under consideration is clear from the context). The *degree* of a vertex v is $deg(v) = |N(v)|$. The *average degree* of G is $d(G) = \frac{1}{|V|}\sum_{v \in V} deg(v)$. For a vertex subset $U \subseteq V$ the subgraph of G induced by U is denoted by $G[U]$. A set $U \subseteq V$ is a *clique* if all vertices in U are pairwise adjacent in G. A set of vertices U is a *dominating set* if for each vertex $v \in V$, either $v \in U$ or there is a vertex $u \in U$ such that $v \in N(u)$. We also say that a subgraph H of G is dominating if $V(H)$ is a dominating set. We denote by $G - U$ the graph $G[V \setminus U]$, and by $G - u$ the graph $G[V \setminus \{u\}]$ for $u \in V$.

A *vertex coloring* of a graph $G = (V, E)$ is an assignment $c\colon V \to \mathbb{N}$ of a positive integer (*color*) to each vertex of G. The coloring c is *proper* if adjacent vertices receive distinct colors. Assume that each vertex $v \in V$ is assigned a *color list* $L(v) \subset \mathbb{N}$, which is the set of admissible colors for v. A mapping $c\colon V \to \mathbb{N}$ is a *list coloring* of G if c is a proper vertex coloring and $c(v) \in L(v)$ for every $v \in V$. For a positive integer

k, G is *k-choosable* if G has a list coloring for every assignment of color lists $L(v)$ with $|L(v)| = k$ for all $v \in V$. The *choice number* (also called *list chromatic number*) of G, denoted $ch(G)$, is the minimum integer k such that G is k-choosable. The k-CHOOSABILITY problem asks for a given graph G and a positive integer k, whether G is k-choosable. It is known that dense graphs have large choice number [1], as indicated by the following result.

Proposition 1 ([1]). *Let G be a graph and s be an integer. If*

$$d(G) > 4\binom{s^4}{s}\log(2\binom{s^4}{s})$$

then $ch(G) > s$.

By P_r we denote the graph on vertex set $\{v_1, v_2, \ldots, v_r\}$ and edge set $\{v_1v_2, v_2v_3, \ldots, v_{r-1}v_r\}$. A graph is *$P_r$-free* if it does not contain P_r as an induced subgraph. *Cographs* are the class of P_4-free graphs, and they are contained in the class of P_5-free graphs. These graph classes can be recognized in polynomial time. The following structural property of P_5-free graphs was proved by Bacsó and Tuza [2].

Proposition 2 ([2]). *Every connected P_5-free graph has either a dominating clique or a dominating P_3.*

It follows from the results of [2] that such a clique or path can be constructed in polynomial time.

Finally, we distinguish between the parameterized and the non-parameterized versions of our problem. In the CHOOSABILITY problem, G and k are input. We denote by k-CHOOSABILITY the version of the problem parameterized by k.

3 k-CHOOSABILITY Is FPT on P_5-Free Graphs

In this section we prove that k-CHOOSABILITY is fixed parameter tractable on P_5-free graphs.

Theorem 1. *The k-CHOOSABILITY problem is* FPT *on P_5-free graphs.*

Proof. We give a constructive proof of this theorem by describing a recursive algorithm based on Propositions 1 and 2 that checks whether $ch(G) \leq k$. We assume that $k \geq 3$, since for $k \leq 2$, k-CHOOSABILITY can be solved in polynomial time for general graphs [6]. If G is disconnected, then $ch(G)$ is equal to the maximum choice number of the connected components of G. Thus we also assume that G is connected.

Our algorithm uses as its main tool a procedure called `Color`, given in Algorithm 1. This procedure takes as input a connected P_5-free graph G and a set $W = \{w_1, \ldots, w_r\} \subseteq V(G)$ with a sequence of color lists $\mathcal{L} = (L(w_1), \ldots, L(w_r))$, each of size k. For the notation in this procedure, we let $L = L(w_1) \cup \cdots \cup L(w_r)$, and we denote $l = \max\{\max L(w_1), \ldots, \max L(w_r)\}$. Let also $\mathbb{L} = L(w_1) \times \cdots \times L(w_r)$ and $\mathbb{X} = 2^{\mathbb{L}}$. We say that vertices $w_1, \ldots, w_r$ are colored by $c = (c_1, \ldots, c_r) \in \mathbb{L}$ if

```
Procedure Color(G, W, 𝓛)
  Find a dominating set U = {u_1, ..., u_p} of H = G − W, such that U is a clique or U
  induces a P_3;
  Let 𝒳 = ∅;
  if p > k then Return(NO), Halt;
  if d(G[W ∪ U]) > d then Return(NO), Halt;
  forall Color lists L(u_1), ..., L(u_p) ⊆ {1, ..., l, l+1, ..., l+kp}, s.t. |L(u_i)| = k do
    if U = V(H) then
      Let X = ∅;
      forall List colorings s of H do
        Let X := X ∪ {c ∈ 𝕃: c(w_i) ≠ s(u_j) if w_i u_j ∈ E(G)};
      if X ≠ ∅ then Add(𝒳, X); else Return(NO), Halt;
    if U ≠ V(H) then
      Let H_1, ..., H_q be the connected components of H − U, and let
      F_i = G[W ∪ U ∪ V(H_i)] for i ∈ {1, ..., q};
      Let 𝓛' = (L(u_1), ..., L(u_p)), 𝕃' = 𝕃 × L(u_1) × ··· × L(u_p);
      for i = 1 to q do
        Color(F_i, W ∪ U, 𝓛 ∪ 𝓛');
        if the output is NO then
          Return(NO), Halt;
        else
          Let 𝒳_i be the output;
      Let 𝒴 = 𝒳_1;
      for i = 2 to q do
        Let 𝒵 = ∅;
        forall X ∈ 𝒳_i and Y ∈ 𝒴 do
          if X ∩ Y ≠ ∅ then Add(𝒵, X ∩ X');
          else Return(NO), Halt;
        Let 𝒴 = 𝒵;
      forall Z ∈ 𝒵 do
        Let X = {(c(w_1), ..., c(w_r)): c ∈ Z, c(w_i) ≠ c(u_j) if w_i u_j ∈
        E(G) and c(u_i) ≠ c(u_j) if u_i u_j ∈ E(G)};
        if X ≠ ∅ then Add(𝒳, X);
        else Return(NO), Halt;
  if 𝒳 = ∅ then Return(NO), Halt; else Return(𝒳).
```

Algorithm 1. Pseudo code for the procedure `Color`

each w_i is colored by c_i. Set $H = G - W$. Procedure `Color` produces an output which either contains a list of different sets $\mathcal{X} = (X_1, \ldots, X_s)$, $X_i \in \mathbb{X}$, such that for any assignment of color lists of size k to vertices of H, there is a set X_i with the property that any $c \in X_i$ can be used for coloring of W with respect to adjacencies between vertices in W and vertices in $V(H)$, or the output contains "NO" if there is a list assignment for vertices of H such that no list coloring exists. Denote $d = 4\binom{k^4}{k}\log(2\binom{k^4}{k})$. The subroutine $\texttt{Add}(A, a)$ adds the element a to the set A if $a \notin A$, and the subroutine `Halt` stops the algorithm. Our main algorithm calls Procedure $\texttt{Color}(G, \emptyset, \emptyset)$. To simplify the description of the algorithm it is assumed that for $W = \emptyset$, $\mathbb{L}$ contains unique *zero* coloring (i.e. $\mathbb{L}$ is non empty). If the output is "NO" then G is not k-choosable, and otherwise G is k-choosable.

To prove the correctness of the algorithm, let us analyze one call of Procedure `Color`. Since each induced subgraph of a P_5-free graph is P_5-free, by Proposition 2 it is possible to construct the desired dominating set U in the beginning of the procedure. If $|U| > k \geq 3$ then U is a clique in G and $ch(G) \geq ch(G[U]) > k$. If $d(G[W \cup U]) > d$ then $ch(G) \geq ch(G[W \cup U]) > k$ by Proposition 1. Otherwise we proceed and consider color lists for vertices of U. It should be observed here that it is sufficient to consider only color lists with elements from the set $L \cup \{l+1, \ldots, l+kp\}$, since we have to take into account only intersections of these lists which each other and with lists for vertices of W. If $U = V(H)$ then the output is created by checking all possible list colorings of H. If $U \neq V(H)$ then we proceed with our decomposition of G. Graphs $F_1, \ldots, F_q$ are constructed and Procedure `Color` is called recursively for them. It is possible to consider these graphs independently since vertices of different graphs H_i and H_j are not adjacent. Then outputs for $F_1, \ldots, F_q$ are combined and the output for G is created by checking all possible list colorings of U.

Now we analyze the running time of this algorithm. To estimate the depth of the recursion tree we assume that h sets U are created recursively without halting and denote them by $U_1, \ldots, U_h$. Since $|U_i| \leq k$, $|U_1 \cup \cdots \cup U_h| \leq kh$. Notice that each set U_i is a dominating set for $U_{i+1}, \ldots, U_h$. Hence $\sum_{v \in U_i} \deg_F(v) \geq h - 1$, where $F = G[U_1 \cup \cdots \cup U_h]$, and $\sum_{v \in V(F)} \deg(v) \geq h(h-1)$. This means that $d(F) \geq \frac{h-1}{k}$, and if $h > kd+1 = 4k\binom{k^4}{k}\log(2\binom{k^4}{k})+1$ then Procedure `Color` stops. Therefore the depth of the recursion tree is at most $kd + 1 = 4k\binom{k^4}{k}\log(2\binom{k^4}{k}) + 1$. It can be easily noted that the number of leaves in the recursion tree is at most $n = |V(G)|$, and the number of calls of `Color` is at most $(4k\binom{k^4}{k}\log(2\binom{k^4}{k}) + 1)n = O(k^5 \cdot 2^{k^4} \cdot n)$. Let us analyze the number of operations used for each call of this procedure. The set U can be constructed in polynomial time by the results of [2]. If $|U| > k$ then the algorithm finishes its work. Assume that $|U| \leq k$. Since the depth of the recursion tree is at most $kd+1$, color lists for vertices of U are chosen from the set $\{1, \ldots, (kd+1)k^2\}$, and the number of all such sets is $\binom{(kd+1)k^2}{k}$. So, there are at most $\binom{(kd+1)k^2}{k}^k$ (or $2^{O(k^8 \cdot 2^{k^4})}$) possibilities to assign color lists to vertices of U. The number of all list colorings of vertices of U is at most k^k. Recall that the output of `Color` is either "NO" or a list of different sets $\mathcal{X} = (X_1, \ldots, X_s)$ where $X_i \in \mathbb{X}$. Since the depth of the recursion tree is at most $kd + 1$ and each set U contains at most k elements (if the algorithm does not stop), the size of W is at most $k(kd + 1)$. Hence the output contains at most $2^{k(kd+1)}$ (or $2^{O(k^6 \cdot 2^{k^4})}$) sets. Using these bounds and the observation that $q \leq n$, we can conclude that the number of operations for each call of `Color` is $2^{O(k^8 \cdot 2^{k^4})} \cdot n^c$ for some positive constant c. Taking into account the total number of calls of the procedure we can bound the the running time of our algorithm as $2^{O(k^8 \cdot 2^{k^4})} \cdot n^s$ for some positive constant s.

4 Choosability Is NP-Hard on P_5-Free Graphs

In this section we show that Choosability, with input G and k, remains NP-hard when the input graph is restricted to P_5-free graphs.

Theorem 2. *The* Choosability *problem is* NP*-hard on P_5-free graphs.*

Proof. We reduce the not-all-equal 3-Satisfiability (NAE 3-SAT) problem with only positive literals [8] to Choosability. For a given set of Boolean variables $X = \{x_1, \ldots, x_n\}$, and a set $C = \{C_1, \ldots, C_m\}$ of three-literal clauses over X in which *all literals are positive*, this problem asks whether there is a truth assignment for X such that each clause contains at least one true literal and at least one false literal. NAE 3-SAT is NP-complete [8].

Our reduction has two stages. First we reduce NAE 3-SAT to List Coloring by constructing a graph with color lists for its vertices. Then we build on this graph to complete the reduction from NAE 3-SAT to Choosability.

At the first stage of the reduction we construct a complete bipartite graph ($K_{n,2m}$) H with the vertex set $\{x_1, \ldots, x_n\} \cup \{C_1^{(1)}, \ldots, C_m^{(1)}\} \cup \{C_1^{(2)}, \ldots, C_m^{(2)}\}$, where $\{x_1, \ldots, x_n\}$ and $(\{C_1^{(1)}, \ldots, C_m^{(1)}\} \cup \{C_1^{(2)}, \ldots, C_m^{(2)}\})$ is the bipartition of the vertex set. Hence on the one side of bipartition we have a vertex for each variable, and on the other side we have two vertices for each clause. We define color lists for vertices of H as follows: $L(x_i) = \{2i-1, 2i\}$ for $i \in \{1, \ldots, n\}$, $L(C_j^{(1)}) = \{2p-1, 2q-1, 2r-1\}$ and $L(C_j^{(2)}) = \{2p, 2q, 2r\}$ if the clause C_j contains literals x_p, x_q, x_r for $j \in \{1, \ldots, m\}$.

Lemma 1. *The graph H has a list coloring if and only if there is a truth assignment for the variables in X such that each clause contains at least one true literal and at least one false literal.*

Proof. Assume that H has a list coloring. Set the value of variable $x_i = true$ if vertex x_i is colored by $2i - 1$, and set $x_i = false$ otherwise. For each clause C_j with literals x_p, x_q, x_r, at least one literal has value *true* since at least one color from the list $\{2p, 2q, 2r\}$ is used for coloring vertex $C_j^{(2)}$, and at least one literal has value *false*, since at least one color from the list $\{2p - 1, 2q - 1, 2r - 1\}$ is used for coloring vertex $C_j^{(1)}$.

Suppose now that there is a truth assignment for the variables in X such that each clause contains at least one true literal and at least one false literal. For each variable x_i, we color vertex x_i by the color $2i - 1$ if $x_i = true$, and we color x_i by the color $2i$ otherwise. Then any two vertices $C_j^{(1)}$ and $C_j^{(2)}$, which correspond to the clause C_j with literals x_p, x_q, x_r, can be properly colored, since at least one color from each of lists $\{2p - 1, 2q - 1, 2r - 1\}$ and $\{2p, 2q, 2r\}$ is not used for coloring of vertices $x_1, \ldots, x_n$.

Now we proceed with our reduction and add to H a clique with $k = n + 4nm - 4m$ vertices $u_1, \ldots, u_k$. For each vertex x_i, we add edges $x_i u_\ell$ for $\ell \in \{1, \ldots, k\}$, $\ell \neq 2i - 1, 2i$. For vertices $C_j^{(1)}$ and $C_j^{(2)}$ which correspond to clause C_j with literals

x_p, x_q, x_r, edges $C_j^{(1)}u_\ell$ such that $\ell \neq 2p-1, 2q-1, 2r-1$ and edges $C_j^{(2)}u_\ell$ such that $\ell \neq 2p, 2q, 2r$ are added for $\ell \in \{1, \ldots, k\}$. We denote the obtained graph by G.

We claim that G is k-choosable if and only if there is a truth assignment for the variables in X such that each clause contains at least one true literal and at least one false literal.

For the first direction of the proof of this claim, suppose that for any truth assignment there is a clause all of whose literals have the same value. Then we consider a list coloring for G with same color list $\{1, \ldots, k\}$ for each vertex. Assume without loss of a generality that u_i is colored by color i for $i \in \{1, \ldots, k\}$. Then each vertex x_i can be colored only by colors $2i-1, 2i$, each vertex $C_j^{(1)}$ can be colored only by colors $2p-1, 2q-1, 2r-1$ and each vertex $C_j^{(2)}$ can be colored only by colors $2p, 2q, 2r$ if $C_j^{(1)}, C_j^{(2)}$ correspond to the clause with literals x_p, x_q, x_r. By Lemma 1, it is impossible to extend the coloring of vertices $u_1, \ldots, u_k$ to a list coloring of G.

For the other direction, assume now that there is a truth assignment for the variables in X such that each clause contains at least one true literal and at least one false literal. Assign arbitrarily a color list $L(v)$ of size k to each vertex $v \in V(G)$. We show how to construct a list coloring of G. Denote by U the set of vertices $\{u_{2n+1}, \ldots, u_k\}$. Notice that U is a clique whose vertices are adjacent to all vertices of G. We start coloring the vertices of U and reducing G according to this coloring, using following rules:

1. If there is a non colored vertex $v \in U$ such that $L(v)$ contains a color c which was not used for coloring the vertices of U and there is a vertex $w \in \{x_1, \ldots, x_n\} \cup \{C_1^{(1)}, \ldots, C_m^{(1)}\} \cup \{C_1^{(2)}, \ldots, C_m^{(2)}\}$ such that $c \notin L(w)$, then color v by c. Otherwise choose a non colored vertex $v \in U$ arbitrarily and color it by the first available color.
2. If, after coloring some vertex in U, there is a vertex x_i such that at least $2m-1$ colors that are not included in $L(x_i)$ are used for coloring U, then delete x_i.
3. If, after coloring some vertex in U, there is a vertex $C_j^{(s)}$ with $s \in \{1, 2\}$ such that at least $n-2$ colors that are not included in $L(C_j^{(s)})$ are used for coloring U, then delete $C_j^{(s)}$.

This coloring of U can be constructed due the property that for each $v \in U$, $|L(v)| = k$ and $|U| = k - 2n < k$. Rule 2 is correct since $\deg_G(x_i) = k + 2m - 2$, and therefore if at least $2m-1$ colors that are not included in $L(x_i)$ are used for coloring U, then any extension of the coloring of U to the coloring of $G - x_i$ can be further extended to the coloring of G, since there is at least one color in $L(x_i)$ which is not used for the coloring of neighborhood of this vertex. By same arguments, we can show the correctness of Rule 3 using the fact that $\deg_G(C_j^{(s)}) = k + n - 3$.

If after coloring the vertices of U, all vertices of $\{x_1, \ldots, x_n\} \cup \{C_1^{(1)}, \ldots, C_m^{(1)}\} \cup \{C_1^{(2)}, \ldots, C_m^{(2)}\}$ are deleted then we color remaining vertices $u_1, \ldots, u_{2n}$ greedily, and then we can claim that a list coloring of G exists by the correctness of Rules 2 and 3. Assume that at least one vertex of $\{x_1, \ldots, x_n\} \cup \{C_1^{(1)}, \ldots, C_m^{(1)}\} \cup \{C_1^{(2)}, \ldots, C_m^{(2)}\}$ was not deleted, and denote the set of such remaining vertices by W. Let $v \in U$ be the last colored vertex of U. Since $|U| = k - 2n = n + 4nm - 4m - 2n = n(2m-1) +$

$2m(n-2)$, the color list $L(v)$ contains at least $2n$ colors which are not used for coloring the vertices of U. Furthermore, for each $w \in W$, all these $2n$ colors are included in $L(w)$, due to the way we colored the vertices of U and since w was not deleted by Rules 2 or 3. We denote these unused colors by $1, \ldots, 2n$ and let $L = \{1, \ldots, 2n\}$. We proceed with coloring of G by coloring the vertices $u_1, \ldots, u_{2n}$ by the greedy algorithm using the first available color. Assume without loss of generality that if some vertex u_i is colored by the color from L then it is colored by the color i. Now it remains to color the vertices of W. Notice that $G[W]$ is an induced subgraph of H. For each $w \in W$, denote by $L'(w)$ the colors from $L(w)$ which are not used for coloring vertices from the set $\{u_1, \ldots, u_k\}$ that are adjacent to w. It can be easily seen that for any $x_i \in W$, $2i-1, 2i \in L'(x_1)$, for any $C_j^{(1)} \in W$ which corresponds to clause with literals x_p, x_q, x_r, $2p-1, 2q-1, 2r-1 \in L'(C_j^{(1)})$, and for any $C_j^{(2)} \in W$ which corresponds to clause with literals x_p, x_q, x_r, $2p, 2q, 2r \in L'(C_j^{(2)})$. Since there is a truth assignment for variables X such that each clause contains at least one true literal and at least one false literal, by Lemma 1 we can color the vertices of W.

To conclude the proof of the theorem, it remains to prove that G is P_5-free. Suppose that P is an induced path in G. Since H is a complete bipartite graph, P can contain at most 3 vertices of H and if it contains 3 vertices then these vertices have to be consecutive in P (notice that if P contains vertices only from one set of the bipartition of H, then the number of such vertices is at most 2 since they have to be joined by subpaths of P which go through vertices from the clique $\{u_1, \ldots, u_k\}$). Also P can contain at most 2 vertices from the clique $\{u_1, \ldots, u_k\}$, and if it has 2 vertices then they are consecutive. Hence, P has at most 5 vertices, and if P has 5 vertices then either $P = u_{t_1}u_{t_2}C_{j_1}^{(s_1)}x_iC_{j_2}^{(s_2)}$ or $P = u_{t_1}u_{t_2}x_{i_1}C_j^{(s)}x_{i_2}$. Assume that $P = u_{t_1}u_{t_2}C_{j_1}^{(s_1)}x_iC_{j_2}^{(s_2)}$. Since P is an induced path, vertices u_{t_1}, u_{t_2} are not adjacent to x_i. By the construction of G, it means that $\{t_1, t_2\} = \{2i-1, 2i\}$. But then $C_{j_2}^{(s_2)}$ is adjacent either u_{t_1} or u_{t_2}. Suppose that $P = u_{t_1}u_{t_2}x_{i_1}C_j^{(s)}x_{i_2}$. Again by the construction of G, $\{t_1, t_2\} = \{2i_2-1, 2i_2\}$ and $C_j^{(s)}$ is adjacent to u_{t_1} or u_{t_2}. By these contradictions, P has at most 4 vertices.

5 Conclusion and Open Problems

We proved that the k-CHOOSABILITY problem is FPT for P_5-free graphs when parameterized by k. It can be noted that our algorithm described in the proof of Theorem 1 does not explicitly use the absence of induced paths P_5. It is based on the property that any induced subgraph of a k-choosable P_5-free graph has a dominating set of bounded (by some function of k) size. It would be interesting to construct a more efficient algorithm for k-CHOOSABILITY which actively exploits the fact that the input graph has no induced P_5.

Another interesting question is whether it is possible to extend our result for P_r-free graphs for some $r \geq 6$? Particularly, it is known [18] that any P_6-free graph contains either a dominating biclique or a dominating induced cycle C_6. Is it possible to prove that k-CHOOSABILITY is FPT for P_6-free graphs using this fact?

Also, we proved that k-CHOOSABILITY is NP-hard for P_5-free graphs. Is this problem Π_2^P-complete?

Finally, what can be said about P_4-free graphs or *cographs*? It is possible to construct a more efficient algorithm using same ideas as in the proof of Theorem 1 and the well known fact (see e.g. [3]) that any cographs can be constructed from from isolated vertices by *disjoint union* and *join* operations, and such decomposition of any cograph can be constructed in linear time [4]? Instead of the presence of a dominating clique or a dominating P_3 we can use the property [13] that $ch(K_{r,r^r}) > r$. Unfortunately this algorithm is still double exponential in k. Is it possible to construct a better algorithm?

References

1. Alon, N.: Restricted colorings of graphs. In: Surveys in combinatorics. London Math. Soc. Lecture Note Ser., vol. 187, pp. 1–33. Cambridge Univ. Press, Cambridge (1993)
2. Bacsó, G., Tuza, Z.: Dominating cliques in P_5-free graphs. Period. Math. Hungar. 21, 303–308 (1990)
3. Brandstädt, A., Le, V.B., Spinrad, J.P.: Graph classes: a survey. SIAM Monographs on Discrete Mathematics and Applications. Society for Industrial and Applied Mathematics (SIAM), Philadelphia (1999)
4. Corneil, D.G., Perl, Y., Stewart, L.K.: A linear recognition algorithm for cographs. SIAM J. Comput. 14, 926–934 (1985)
5. Downey, R.G., Fellows, M.R.: Parameterized complexity. Monographs in Computer Science. Springer, Heidelberg (1999)
6. Erdős, P., Rubin, A.L., Taylor, H.: Choosability in graphs. In: Proceedings of the West Coast Conference on Combinatorics, Graph Theory and Computing, Humboldt State Univ., Arcata, Calif, 1980, Utilitas Math., pp. 125–157 (1979)
7. Flum, J., Grohe, M.: Parameterized Complexity Theory. Springer, Heidelberg (2006)
8. Garey, M.R., Johnson, D.S.: Computers and intractability. W. H. Freeman and Co., San Francisco (1979)
9. Gutner, S., Tarsi, M.: Some results on (a:b)-choosability, CoRR, abs/0802.1338 (2008)
10. Hoàng, C.T., Kamiński, M., Lozin, V.V., Sawada, J., Shu, X.: A note on k-colorability of p5-free graphs. In: Ochmański, E., Tyszkiewicz, J. (eds.) MFCS 2008. LNCS, vol. 5162, pp. 387–394. Springer, Heidelberg (2008)
11. Jensen, T.R., Toft, B.: Graph Coloring Problems. Wiley Interscience, Hoboken (1995)
12. Král, D., Kratochvíl, J., Tuza, Z., Woeginger, G.J.: Complexity of coloring graphs without forbidden induced subgraphs. In: Brandstädt, A., Van Le, B. (eds.) WG 2001. LNCS, vol. 2204, pp. 254–262. Springer, Heidelberg (2001)
13. Mahadev, N.V.R., Roberts, F.S., Santhanakrishnan, P.: 3-choosable complete bipartite graphs, Technical Report 49-91, Rutgers University, New Brunswick, NJ (1991)
14. Niedermeier, R.: Invitation to Fixed-Parameter Algorithms. Oxford University Press, Oxford (2006)
15. Randerath, B., Schiermeyer, I.: 3-colorability $\in P$ for P_6-free graphs. Discrete Appl. Math. 136, 299–313 (2004); The 1st Cologne-Twente Workshop on Graphs and Combinatorial Optimization (CTW 2001)
16. Thomassen, C.: Every planar graph is 5-choosable. J. Combin. Theory Ser. B 62, 180–181 (1994)
17. Tuza, Z.: Graph colorings with local constraints—a survey. Discuss. Math. Graph Theory 17, 161–228 (1997)

18. van 't Hof, P., Paulusma, D.: A new characterization of p6-free graphs. In: Hu, X., Wang, J. (eds.) COCOON 2008. LNCS, vol. 5092, pp. 415–424. Springer, Heidelberg (2008)
19. Vizing, V.G.: Coloring the vertices of a graph in prescribed colors. Diskret. Analiz 101, 3–10 (1976)
20. Woeginger, G.J., Sgall, J.: The complexity of coloring graphs without long induced paths. Acta Cybernet. 15, 107–117 (2001)

Time-Bounded Kolmogorov Complexity and Solovay Functions

Rupert Hölzl, Thorsten Kräling, and Wolfgang Merkle

Institut für Informatik, Ruprecht-Karls-Universität, Heidelberg, Germany

Abstract. A Solovay function is a computable upper bound g for prefix-free Kolmogorov complexity K that is nontrivial in the sense that g agrees with K, up to some additive constant, on infinitely many places n. We obtain natural examples of Solovay functions by showing that for some constant c_0 and all computable functions t such that $c_0 n \leq t(n)$, the time-bounded version K^t of K is a Solovay function.

By unifying results of Bienvenu and Downey and of Miller, we show that a right-computable upper bound g of K is a Solovay function if and only if Ω_g is Martin-Löf random. Letting $\Omega_g = \sum 2^{-g(n)}$, we obtain as a corollary that the Martin-Löf randomness of the various variants of Chaitin's Ω extends to the time-bounded case in so far as Ω_{K^t} is Martin-Löf random for any t as above.

As a step in the direction of a characterization of K-triviality in terms of jump-traceability, we demonstrate that a set A is K-trivial if and only if A is $\mathrm{O}(g(n) - \mathrm{K}(n))$-jump traceable for all Solovay functions g, where the equivalence remains true when we restrict attention to functions g of the form K^t, either for a single or all functions t as above.

Finally, we investigate the plain Kolmogorov complexity C and its time-bounded variant C^t of initial segments of computably enumerable sets. Our main theorem here is a dichotomy similar to Kummer's gap theorem and asserts that every high c.e. Turing degree contains a c.e. set B such that for any computable function t there is a constant $c_t > 0$ such that for all m it holds that $\mathrm{C}^t(B \upharpoonright m) \geq c_t \cdot m$, whereas for any nonhigh c.e. set A there is a computable time bound t and a constant c such that for infinitely many m it holds that $\mathrm{C}^t(A \upharpoonright m) \leq \log m + c$. By similar methods it can be shown that any high degree contains a set B such that $\mathrm{C}^t(B \upharpoonright m) \geq^+ m/4$. The constructed sets B have low unbounded but high time-bounded Kolmogorov complexity, and accordingly we obtain an alternative proof of the result due to Juedes, Lathrop, and Lutz [JLL] that every high degree contains a strongly deep set.

1 Introduction and Overview

Prefix-free Kolmogorov complexity K it not computable and in fact does not even allow for computable lower bounds. However, there are computable upper bounds for K and, by a construction that goes back to Solovay [BD, S], there are even computable upper bounds that are nontrivial in the sense that g agrees

R. Královič and D. Niwiński (Eds.): MFCS 2009, LNCS 5734, pp. 392–402, 2009.

with K, up to some additive constant, on infinitely many places n; such upper bounds are called Solovay functions.

For any computable time-bound t, the time-bounded version K^t of K is obviously a computable upper bound for K, and we show that K^t is indeed a Solovay function in case $c_0 n \leq t(n)$ for some appropriate constant c_0. As a corollary, we obtain that the Martin-Löf randomness of the various variants of Chaitin's Ω extends to the time-bounded case in so far as for any t as above, the real number

$$\Omega_{\mathrm{K}^t} = \sum_{n \in \mathbb{N}} \frac{1}{2^{\mathrm{K}^t(n)}}$$

is Martin-Löf random. The corresponding proof exploits the result by Bienvenu and Downey [BD] that a computable function g such that $\Omega_g = \sum 2^{-g(n)}$ converges is a Solovay function if and only if Ω_g is Martin-Löf random. In fact, this equivalence extends by an even simpler proof to the case of functions g that are just right-computable, i.e., effectively approximable from above, and one then obtains as special cases the result of Bienvenu and Downey and a related result of Miller where the role of g is played by the fixed right-computable but noncomputable function K.

An open problem that received some attention recently [BDG, DH, N] is whether the class of K-trivial sets coincides with the class of sets that are $g(n)$-jump-traceable for all computable functions g such that $\sum 2^{-g(n)}$ converges. As a step in the direction of a characterization of K-triviality in terms of jump-traceability, we demonstrate that a set A is K-trivial if and only if A is $\mathrm{O}(g(n) - \mathrm{K}(n))$-jump traceable for all Solovay functions g, where the equivalence remains true when we restrict attention to functions g of the form K^t, either for a single or all functions t as above.

Finally, we consider the time-bounded and unbounded Kolmogorov complexity of the initial segments of sets that are computationally enumerable, or c.e., for short. The initial segments of a c.e. set A have small Kolmogorov complexity, more precisely, by Barzdins' lemma it holds that $\mathrm{C}(A \restriction m) \leq^+ 2 \log m$, where C denotes plain Kolmogorov complexity. Theorem 4, our main result in this section, has a structure similar to Kummer's gap theorem in so far as it asserts a dichotomy in the complexity of initial segments between high and nonhigh c.e. sets. More precisely, every high c.e. Turing degree contains a c.e. set B such that for any computable function t there is a constant $c_t > 0$ such that for all m it holds that $\mathrm{C}^t(B \restriction m) \geq c_t \cdot m$, whereas for any nonhigh c.e. set A there is a computable time bound t and a constant c such that for infinitely many m it holds that $\mathrm{C}^t(A \restriction m) \leq \log m + c$. By similar methods it can be shown that any high degree contains a set B such that $\mathrm{C}^t(B \restriction m) \geq^+ m/4$. The constructed sets B have low unbounded but high time-bounded Kolmogorov complexity, and accordinlgy we obtain an alternative proof of the result due to Juedes, Lathrop, and Lutz [JLL] that every high degree contains a strongly deep set.

Notation. In order to define plain and prefix-free Kolmogorov complexity, we fix additively optimal oracle Turing machines $\mathbb{V}$ and $\mathbb{U}$, where $\mathbb{U}$ has prefix-free domain. We let $\mathrm{C}^A(x)$ denote the Kolmogorov-complexity of x with respect to $\mathbb{V}$ relative to oracle A, let $\mathrm{C}(x) = \mathrm{C}^{\emptyset}(x)$, and similarly define prefix-free Kolmogorov complexity K^A and K with respect to $\mathbb{U}$. In connection with the definition of time-bounded Kolmogorov complexity, we assume that $\mathbb{V}$ and $\mathbb{U}$ both are able to simulate any other Turing machine M running for t steps in $\mathrm{O}(t \cdot \log(t))$ steps for an arbitrary machine M and in $\mathrm{O}(t(n))$ steps in case M has only two work tapes.

For a computable function $t : \mathbb{N} \to \mathbb{N}$ and a machine M, the Kolmogorov complexity relative to M with time bound t is

$$\mathrm{C}^t_M(n) := \min\{|\sigma| : M(\sigma) \downarrow = n \text{ in at most } t(|n|) \text{ steps}\},$$

and we write C^t for $\mathrm{C}^t_{\mathbb{U}}$. The prefix-free Kolmogorov complexity with time bound t denoted by $\mathrm{K}^t_M(n)$ and $\mathrm{K}^t(n) = \mathrm{K}^t_{\mathbb{U}}$ is defined likewise by considering only prefix-free machines and the corresponding universal machine $\mathbb{U}$ in place of $\mathbb{U}$.

We identify strings with natural numbers by the order isomorphism between the length-lexicographical order on strings and the usual order on $\mathbb{N}$, and we write $|m|$ for the length of the string that corresponds to the number m, where then $|m|$ is roughly $\log m$.

2 Solovay Functions and Martin-Löf Randomness

Definition 1 (Li, Vitányi [LV]). *A computable function $f : \mathbb{N} \to \mathbb{N}$ is called a* Solovay function *if* $\mathrm{K}(n) \leq^+ f(n)$ *for all n and* $\mathrm{K}(n) =^+ f(n)$ *for infinitely many n.*

Solovay [S, BD] had already constructed Solovay functions and by slightly variying the standard construction, next we observe that time-bounded prefix-free Kolmogorov complexity indeed provides natural examples of Solovay functions.

Theorem 1. *There is a constant c_0 such that time-bounded prefix-free Kolmogorov complexity K^t is a Solovay function for any computable function $t \colon \mathbb{N} \to \mathbb{N}$ such that $c_0 n \leq t(n)$ holds for almost all n.*

Proof. Fix a standard effective and effectively invertible pairing function $\langle .,. \rangle \colon \mathbb{N}^2 \to \mathbb{N}$ and define a tripling function $[.,.,.] \colon \mathbb{N}^3 \to \mathbb{N}$ by letting

$$[s, \sigma, n] = 1^s 0 \langle \sigma, n \rangle.$$

Let M be a Turing machine with two tapes that on input σ uses its first tape to simulate the universal machine $\mathbb{U}$ on input σ and, in case $\mathbb{U}(\sigma) = n$, to compute $\langle \sigma, n \rangle$, while maintaining on the second tape a unary counter for the number of steps of M required for these computations. In case eventually $\langle \sigma, n \rangle$ had been computed with final counter value s, the output of M is $z = [s, \sigma, n]$, where by construction in this case the total running time of M is in $\mathrm{O}(s)$.

Call z of the form $[s, \sigma, n]$ a Solovay triple in case $M(\sigma) = z$ and σ is an optimal code for n, i.e., $\mathrm{K}(n) = |\sigma|$. For some appropriate constant c_0 and any computable function t that eventually is at least $c_0 n$, for almost all such triples z it then holds that

$$\mathrm{K}(z) =^+ \mathrm{K}^t(z),$$

because given a code for M and σ, by assumption the universal machine $\mathbb{U}$ can simulate the computation of the two-tape machine M with input σ with linear overhead, hence $\mathbb{U}$ uses time $\mathrm{O}(s)$ plus the constant time required for decoding M, i.e., time at most $c_0|z|$. □

Next we derive a unified form of a characterization of Solovay function in terms of Martin-Löf randomness of the corresponding Ω-number due to Bienvenu and Downey [BD] and a result of Miller [M] that asserts that the notions of weakly low and low for Ω coincide. Before, we review some standard notation and facts relating to Ω-numbers.

Definition 2. *For a function $f\colon \mathbb{N} \to \mathbb{N}$, the* Ω-number *of f is*

$$\Omega_f := \sum_{n \in \mathbb{N}} 2^{-f(n)}$$

Definition 3. *A function $f\colon \mathbb{N} \to \mathbb{N}$ is an* information content measure relative to a set *A in case f is right-computable with access to the oracle A and Ω_f converges; furthermore, the function f is an* information content measure *if it is an information content measure relative to the empty set.*

The following remark describes for a given information content measure f an approximation from below to Ω_f that has certain special properties. For the sake of simplicity, in the remark only the oracle-free case is considered and the virtually identical considerations for the general case are omitted.

Remark 1. For a given information content measure f, we fix as follows a non-decreasing computable sequence $a_0, a_1, \ldots$ that converges to Ω_f and call this sequence the canonical approximation of Ω_f.

First, we fix some standard approximation to the given information content measure f from above, i.e., a computable function $(n, s) \mapsto f_s(n)$ such that for all n the sequence $f_0(n), f_1(n), \ldots$ is a nonascending sequence of natural numbers that converges to $f(n)$, where we assume in addition that $f_s(n) - f_{s+1}(n) \in \{0, 1\}$. Then in order to obtain the a_i, let $a_0 = 0$ and given a_i, define a_{i+1} by searching for the next pair of the form $(n, 0)$ or the form $(n, s+1)$ where in addition it holds that $f_s(n) - f_{s+1}(n) = 1$ (with some ordering of pairs understood), let

$$d_i = 2^{-f_0(n)} \qquad \text{or} \qquad d_i = 2^{-f_{s+1}(n)} - 2^{-f_s(n)} = 2^{-f_s(n)},$$

respectively, and let $a_{i+1} = a_i + d_i$. Furthermore, in this situation, say that the increase of d_i from a_i to a_{i+1} occurs due to n.

It is well-known [DH] that among all right-computable functions exactly the information content measures are, up to an additive constant, upper bounds for the prefix-free Kolmogorov complexity K.

Theorem 2 unifies two results by Bienvenu and Downey [BD] and by Miller [M], which are stated below as Corollaries 1 and 2. The proof of the backward direction of the equivalence stated in Theorem 2 is somewhat more direct and uses different methods when compared to the proof of Bienvenu and Downey, and is quite a bit shorter than Miller's proof, though the main trick of delaying the enumeration via the notion of a matched increase is already implicit there [DH, M]. Note in this connection that Bienvenu has independently shown that Miller's result can be obtained as a corollary to the result of Bienvenu and Downey [DH].

Theorem 2. *Let f be an information content measure relative to a set A. Then f has the Solovay property with respect to K^A, i.e.,*

$$\lim_{n\to\infty} (f(n) - \mathrm{K}^A(n)) \neq +\infty \tag{1}$$

if and only if Ω_f is Martin-Löf random relative to A.

Proof. We first show the backwards direction of the equivalence asserted in the theorem, where the construction and its verification bear some similarities to Kučera and Slaman's [KS] proof that left-computable sets that are not Solovay complete cannot be Martin-Löf random. We assume that (1) is false and construct a sequence $U_0, U_1, \ldots$ of sets that is a Martin-Löf test relative to A and covers Ω_f. In order to obtain the component U_c, let $a_0, a_1, \ldots$ be the canonical approximation to Ω_f where in particular $a_{i+1} = a_i + d_i$ for increases d_i that occur due to some n. Let b_i be the sum of all increases d_j such that $j \leq i$ and where d_j and d_i are due to the same n. With the index c understood, say an increase d_i due to n is matched if it holds that

$$2^{c+1} b_i \leq 2^{-\mathrm{K}^A(n)}.$$

For every d_i for which it could be verified that d_i is matched, add an interval of size $2d_i$ to U_c where this interval either starts at a_i or at the maximum place that is already covered by U_c, whichever is larger. By construction the sum of all matched d_i is at most $\Omega^A/2^{c+1} \leq 2^{-(c+1)}$ and the sets U_c are uniformly c.e. relative to A, hence $U_0, U_1, \ldots$ is a Martin-Löf test relative to A. Furthermore, this test covers Ω_f because by the assumption that (1) is false, for any c almost all increases are matched.

For ease of reference, we review the proof of the forward direction of the equivalence asserted in the theorem, which follows by the same line of standard argument that has already been used by Bienvenu and Downey and by Miller. For a proof by contraposition, assume that Ω_f is not Martin-Löf random relative to A, i.e., for every constant c there is a prefix σ_c of Ω_f such that $\mathrm{K}^A(\sigma_c) \leq |\sigma_c| - 2c$. Again consider the canonical approximation $a_0, a_1, \ldots$ to Ω_f where $f(n, 0), f(n, 1), \ldots$ is the corresponding effective approximation from above to $f(n)$ as in Remark 1. Moreover, for σ_c as above we let s_c be the least index s such that a_s exceeds σ_c (assuming that the expansion of Ω_f is not eventually constant and leaving the similar considerations for this case to the reader). Then the sum over all values $2^{-f(n)}$ such that none of the increases d_0 through d_s

was due to n is at most $2^{-|\sigma_c|}$, hence all pairs of the form $(f(n,s) - |\sigma_c| + 1, n)$ for such n and s where either $s = 0$ or $f(n,s)$ differs from $f(n, s-1)$ form a sequence of Kraft-Chaitin axioms, which is uniformly effective in c and σ_c relative to oracle A. Observe that by construction for each n, there is an axiom of the form $(f(n) - |\sigma_c| + 1, n)$ and the sum of all terms 2^{-k} over all axioms of the form (k, n) is less than $2^{-f(n)-|\sigma_c|}$.

Now consider a prefix-free Turing machine M with oracle A that given codes for c and σ_c and some other word p as input, first computes c and σ_c, then searches for s_c, and finally outputs the word that is coded by p according to the Kraft-Chaitin axioms for c, if such a word exists. If we let d be the coding constant for M, we have for all sufficiently large c and x that $\mathrm{K}^A(n) \leq 2\log c + \mathrm{K}^A(\sigma_c) + f(n) - |\sigma_c| + 1 + d \leq f(n) - c$. □

As special cases of Theorem 2 we obtain the following results by Bienvenu and Downey [BD] and by Miller [M], where the former one is immediate and for the latter one it suffices to observe that the definition of the notion low for Ω in terms of Chaitin's Ω number

$$\Omega := \sum_{\{x\colon \mathbb{U}(x)\downarrow\}} 2^{-|x|}.$$

is equivalent to a definition in terms of Ω_{K}.

Corollary 1 (Bienvenu and Downey). *A computable information content measure f is a Solovay function if and only if Ω_f is Martin-Löf random.*

Corollary 2 (Miller). *A set A is weakly low if and only if A is low for Ω.*

Proof. In order to see the latter result, it suffices to let $f = \mathrm{K}$ and to recall that for this choice of f the properties of A that occur in the two equivalent assertions in the conclusion of Theorem 2 coincide with the concepts weakly low and low for Ω_K. But the latter property is equivalent to being low for Ω, since for any set A, it is equivalent to require that some or that all left-computable Martin-Löf random set are Martin-Löf random relative to A [N, Proposition 8.8.1]. □

By Corollary 1 and Theorem 1 it is immediate that the known Martin-Löf randomness of Ω_{K} extends to the time-bounded case.

Corollary 3. *There is a constant c_0 such that $\Omega_{\mathrm{K}^t} := \sum_{x\in\mathbb{N}} 2^{-\mathrm{K}^t(x)}$is Martin-Löf random for any computable function twhere $c_0 n \leq t(n)$ for almost all n.*

3 Solovay Functions and Jump-Traceability

In an attempt to define K-triviality without resorting to effective randomness or measure, Barmpalias, Downey and Greenberg [BDG] searched for characterizations of K-triviality via jump-traceability. They demonstrated that K-triviality is not implied by being h-jump-traceable for all computable functions h such that $\sum_n 1/h(n)$ converges. Subsequently, the following question received some attention: Can K-triviality be characterized by being g-jump traceable for all computable functions g such that $\sum 2^{-g(n)}$ converges, that is, for all computable functions g that, up to an additive constant term, are upper bounds for K?

We will now argue that Solovay functions can be used for a characterization of K-triviality in terms of jump traceability. However, we will not be able to completely avoid the notion of Kolmogorov complexity.

Definition 4. *A set A is* K-trivial *if* $\mathrm{K}(A \upharpoonright n) \leq^+ \mathrm{K}(n)$ *for all* n.

Definition 5. *Let $h : \mathbb{N} \to \mathbb{N}$ be a computable function. A set A is* $\mathrm{O}(h(n))$-jump-traceable *if for every function Φ partially computable in A there is a function $h \in \mathrm{O}(h(n))$ and a sequence $(T_n)_{n \in \mathbb{N}}$ of uniformly c.e. finite sets, which is called a trace, such that for all n*

$$|T_n| \leq h(n), \quad \text{and} \quad \Phi(n) \in T_n$$

for all n such that $\Phi(n)$ is defined.

Theorem 3. *There is a constant c_0 such that the following assertions are equivalent for any set A.*

(i) A is K*-trivial.*
(ii) A is $\mathrm{O}(g(n) - \mathrm{K}(n))$-jump-traceable for every Solovay function g.
(iii) A is $\mathrm{O}(\mathrm{K}^t(n) - \mathrm{K}(n))$-jump-traceable for all computable functions t where $c_0 n \leq t(n)$ for almost all n.
(iv) A is $\mathrm{O}(\mathrm{K}^t(n) - \mathrm{K}(n))$-jump-traceable for some computable function t where $c_0 n \leq t(n)$ for almost all n.

Proof. The implication (ii)⇒(iii) is immediate by Theorem 1, and the implication (iii)⇒(iv) is trivially true. So it suffices to show the implications (i)⇒(ii) and (iv)⇒(i), where due to space considerations we only sketch the corresponding proofs.

First, let A be K-trivial and let Φ^A be any partially A-computable function. Let $\langle .,. \rangle$ be some standard effective pairing function. Since A is K-trivial and hence low for K, we have

$$\mathrm{K}(\langle n, \Phi^A(n) \rangle =^+ \mathrm{K}^A(\langle n, \Phi^A(n) \rangle =^+ \mathrm{K}^A(n) =^+ \mathrm{K}(n) ,$$

whenever $\Phi^A(n)$ is defined. Observe that the constant that is implicit in the relation $=^+$ depends only on A in the case of the first and last relation symbol, but depends also on Φ in case of the middle one.

By the coding theorem there can be at most constantly many pairs of the form (n, y) such that $\mathrm{K}(n, y)$ and $\mathrm{K}(n)$ differ at most by a constant, and given n, $\mathrm{K}(n)$ and the constant, we can enumerate all such pairs. But then given a Solovay function g, for given n we can enumerate at most $g(n) - \mathrm{K}(n) + 1$ possible values for $\mathrm{K}(n)$ and for each such value at most constantly many pairs (n, y) such that some y is equal to $\Phi^A(n)$, hence A is $\mathrm{O}(g(n) - \mathrm{K}(n))$-jump-traceable.

Next, let c_0 be the constant from Theorem 1 and let t be a computable time bound such that (iv) is true for this value of c_0. Then K^t is a Solovay function by choice of c_0.

Recall the tripling function $[.,.,.]$ and the concept of a Solovay triple $[s, \sigma, n]$ from the proof of Theorem 1, and define a partial A-computable function Φ that maps any Solovay triple $[s, \sigma, n]$ to $A \upharpoonright n$. Then given an optimal code σ for n, one can compute the corresponding Solovay triple $z = [s, \sigma, n]$, where then $\mathrm{K}^t(z)$ and $\mathrm{K}(z)$ differ only by a constant, hence the trace of Φ^A at z has constant size and contains the value $A \upharpoonright n$, i.e., we have $\mathrm{K}(A \upharpoonright n) \leq^+ |\sigma| = \mathrm{K}(n)$, hence A is K-trivial. □

4 Time Bounded Kolmogorov Complexity and Strong Depth

The initial segments of a c.e. set A have small Kolmogorov complexity, by Barzdins' lemma [DH] it holds for all m that

$$\mathrm{C}(A \upharpoonright m \mid m) \leq^+ \log m \quad \text{and} \quad \mathrm{C}(A \upharpoonright m) \leq^+ 2 \log m.$$

Furthermore, there are infinitely many initial segments that have considerably smaller complexity. The corresponding observation in the following remark is extremely easy, but apparently went unnoticed so far and, in particular, improves on corresponding statements in the literature [DH, Lemma attributed to Solovay in Chapter 14].

Remark 2. Let A be a c.e. set. Then there is a constant c such that for infinitely many m it holds that

$$\mathrm{C}(A \upharpoonright m \mid m) \leq c, \quad \mathrm{C}(A \upharpoonright m) \leq^+ \mathrm{C}(m) + c, \quad \text{and } \mathrm{C}(A \upharpoonright m) \leq \log m + c.$$

For a proof, it suffices to fix an effective enumeration of A and to observe that there are infinitely many $m \in A$ such that m is enumerated after all numbers $n \leq m$ that are in A, i.e., when knowing m one can simulate the enumeration until m appears, at which point one then knows $A \upharpoonright m$.

Barzdins [Ba] states that there are c.e. sets with high time-bounded Kolmogorov complexity, and the following lemma generalizes this in so far as such sets can be found in every high Turing degree.

Lemma 1. *For any high set A there is a set B where $A =_{\mathrm{T}} B$ such that for every computable time bound t there is a constant $c_t > 0$ where*

$$\mathrm{C}^t(B \upharpoonright m) \geq^+ c_t \cdot m \quad \text{and} \quad \mathrm{C}(B \upharpoonright m) \leq^+ 2 \log m.$$

Moreover, if A is c.e., B can be chosen to be c.e. as well.

Proof. Let A be any high set. We will construct a Turing-equivalent set B as required. Recall the following equivalent characterization of a set A being high:

there is a function g computable in A that majorizes any computable function f, i.e., $f(n) \leq g(n)$ for almost all n. Fix such a function g, and observe that in case A is c.e., we can assume that g can be effectively approximated from below. Otherwise we may replace g with the function g' defined as follows. Let M_g be an oracle Turing machine that computes g if supplied with oracle A. For all n, let

$$\tilde{g}(n,s) := \max\{M_g^{A_i}(n) \mid i \leq s\},$$

where A_i is the approximation to A after i steps of enumeration, and let $g'(n) := \lim_{s\to\infty} \tilde{g}(n,s)$. We have $g(n) \leq g'(n)$ for all n and by construction, g' can be effectively approximated from below.

Partition $\mathbb{N}$ into consecutive intervals $I_0, I_1, \ldots$ where interval I_j has length 2^j and let $m_j = \max I_j$. By abuse of notation, let $t_0, t_1, \ldots$ be an effective enumeration of all partial computable functions. Observe that it is sufficient to ensure that the assertion in the theorem is true for all $t = t_i$ such that t_i is computable, nondecreasing and unbounded. Assign the time bounds to the intervals $I_0, I_1, \ldots$ such that t_0 will be assigned to every second interval including the first one, t_1 to every second interval including the first one of the *remaining* intervals, and so on for $t_2, t_3, \ldots$, and note that this way t_i will be assigned to every 2^{i+1}-th interval.

We construct a set B as required. In order to code A into B, for all j let $B(m_j) = A(j)$, while the remaining bits of B are specified as follows. Fix any interval I_j and assume that this interval is assigned to $t = t_i$. Let B have empty intersection with $I_j \setminus \{m_j\}$ in case the computation of $t(m_j)$ requires more than $g(j)$ steps. Otherwise, run all codes of length at most $|I_j| - 2$ on the universal machine $\mathbb{V}$ for $2t(m_j)$ steps each, and let w_j be the least word of lenght $|I_j| - 1$ that is not output by any of these computations, hence $\mathrm{C}^{2t}(w_j) \geq |w_j| - 1$. Let the restriction of B to the first $|w_j|$ places in I_j be equal to w_j.

Now let v_j be the initial segment of B of length $m_j + 1$, i.e., up to and including I_j. In case $t = t_i$ is computable, nondecreasing and unbounded, for almost all intervals I_j assigned to t, we have $\mathrm{C}^t(v_j) > |v_j|/3$, because otherwise, for some appropriate constant c, the corresponding codes would yield for almost all j that $\mathrm{C}^{2t}(w_j) \leq |v_j|/3 + c \leq |I_j| - 2$. Furthermore, by construction for every such t there is a constant $c_t > 0$ such that for almost all m, there is some interval I_j assigned to t such that $m_j \leq m$ and $c_t m \leq m_j/4$, hence for almost all m the initial segment of B up to m cannot have Kolmogorov complexity of less than $c_t m$.

We omit the routine proof that A and B are Turing-equivalent and that if A was c.e., then B is c.e. as well, where for the latter fact we need the assumption that g can be effectively approximated from below.

Finally, to see that $\mathrm{C}(B \upharpoonright m) \leq^+ 2 \log m$, notice that in order to determine $B \upharpoonright m$ without time bounds it is enough to know on which of the intervals I_j the assigned time bounds t_i terminate before their computation is canceled by g, which requires one bit per interval, plus another one describing the bit of A coded into B at the end of each interval. □

Lemma 2. *Every high degree contains for every computable, nondecreasing and unbounded function h a set B such that for every computable time bound t and almost all m,*

$$\mathrm{C}^t(B \upharpoonright m) \geq^+ \frac{1}{4}m \quad \text{and} \quad \mathrm{C}(B \upharpoonright m) \leq h(m) \cdot \log m.$$

Proof. The argument is similar to the proof of Lemma 1, but now, when considering interval I_j, we diagonalize against the largest running time among $t_0(m_j), \ldots, t_{h(j)-2}(m_j)$ such that the computation of this value requires not more than $g(j)$ steps. This way we ensure – for any computable time bound t – that at the end of almost all intervals I_j compression by a factor of at most $1/2$ is possible, and that within interval I_j, we have compressibility by a factor of at most $1/4$, up to a constant additive term, because interval I_{j-1} was compressible by a factor of at most $1/2$. □

Kummer's gap theorem asserts that any array noncomputable c.e. Turing degree contains a c.e. set A such that there are infinitely many m such that $\mathrm{C}(A \upharpoonright m) \geq 2\log m$, whereas all c.e. sets in an array computable Turing degree satisfy $\mathrm{C}(A \upharpoonright m) \leq (1+\varepsilon)\log m$ for all $\varepsilon > 0$ and almost all m. Similarly, Theorem 4, the main result of this section, asserts a dichotomy for the time-bounded complexity of intial segments between high and nonhigh sets.

Theorem 4. *Let A be any c.e. set.*

(i) If A is high, then there is a c.e. set B with $B =_{\mathrm{T}} A$ such that for every computable time bound t there is a constant $c_t > 0$ such that for all m, it holds that $\mathrm{C}^t(B \upharpoonright m) \geq c_t \cdot m$.
(ii) If A is not high, then there is a computable time bound t such that $\mathrm{C}^t(A \upharpoonright m) \leq^+ \log m$.

Proof. The first assertion is immediate from Lemma 1. In order to demonstrate the second assertion, it suffices to observe that for the modulus of convergence s of the c.e. set A there is a computable function f such that $s(m) \leq f(m)$ for infinitely many m. □

As another easy consequence of Lemma 1, we get an alternative proof of the result due to Juedes, Lathrop and Lutz [JLL] that every high degree contains a strongly deep set. We omit details due to lack of space.

References

[BDG] Barmpalias, G., Downey, R., Greenberg, N.: K-trivial degrees and the jump-traceability hierarchy. Proc. Amer. Math. Soc.; posted on January 22, 2009, PII S 0002-9939(09)09761-5 (to appear in print)
[Ba] Barzdins, J.M.: Complexity of programs to determine whether natural numbers not greater than n belong to a computably enumerable set. Soviet Math. Dokl. 9, 1251–1254 (1968)

[BD] Bienvenu, L., Downey, R.: Kolmogorov Complexity and Solovay Functions. In: Symposium on Theoretical Aspects of Computer Science , STACS 2009, pp. 147–158 (2009), `http://drops.dagstuhl.de/opus/volltexte/2009/1810`

[DH] Downey, R., Hirschfeldt, D.: Algorithmic Randomness (manuscript, 2009)

[JLL] Juedes, D., Lathrop, J., Lutz, J.: Computational depth and reducibility. Theor. Comput. Sci. 132(2), 37–70 (1994)

[K] Kummer, M.: Kolmogorov complexity and instance complexity of recursively enumerable sets. SIAM Journal on Computing 25, 1123–1143 (1996)

[KS] Kučera, A., Slaman, T.: Randomness and recursive enumerability. SIAM Journal on Computing 31, 199–211 (2001)

[LV] Li, M., Vitányi, P.: An Introduction to Kolmogorov Complexity and Its Applications, 3rd edn. Springer, Heidelberg (2008)

[M] Miller, J.S.: The K-degrees, low for K-degrees and weakly low for K-degrees. Notre Dame Journal of Formal Logic (to appear)

[N] Nies, A.: Computability and Randomness. Oxford University Press, Oxford (2008)

[S] Solovay, R.: Draft of a paper (or series of papers) on Chaitin's work. Unpublished notes, 215 pages (1975)

The Longest Path Problem Is Polynomial on Interval Graphs

Kyriaki Ioannidou[1,⋆], George B. Mertzios[2,⋆⋆], and Stavros D. Nikolopoulos[1,⋆]

[1] Department of Computer Science, University of Ioannina, Greece
{kioannid,stavros}@cs.uoi.gr
[2] Department of Computer Science, RWTH Aachen University, Germany
mertzios@cs.rwth-aachen.de

Abstract. The longest path problem is the problem of finding a path of maximum length in a graph. Polynomial solutions for this problem are known only for small classes of graphs, while it is NP-hard on general graphs, as it is a generalization of the Hamiltonian path problem. Motivated by the work of Uehara and Uno in [20], where they left the longest path problem open for the class of interval graphs, in this paper we show that the problem can be solved in polynomial time on interval graphs. The proposed algorithm runs in $O(n^4)$ time, where n is the number of vertices of the input graph, and bases on a dynamic programming approach.

Keywords: Longest path problem, interval graphs, polynomial algorithm, complexity, dynamic programming.

1 Introduction

A well studied problem in graph theory with numerous applications is the Hamiltonian path problem, i.e., the problem of determining whether a graph is Hamiltonian; a graph is said to be Hamiltonian if it contains a Hamiltonian path, that is, a simple path in which every vertex of the graph appears exactly once. Even if a graph is not Hamiltonian, it makes sense in several applications to search for a longest path, or equivalently, to find a maximum induced subgraph of the graph which is Hamiltonian. However, finding a longest path seems to be more difficult than deciding whether or not a graph admits a Hamiltonian path. Indeed, it has been proved that even if a graph has a Hamiltonian path, the problem of finding a path of length $n - n^{\varepsilon}$ for any $\varepsilon < 1$ is NP-hard, where n is the number of vertices of the graph [15]. Moreover, there is no polynomial-time constant-factor approximation algorithm for the longest path problem unless P=NP [15]. For related results see also [7, 8, 9, 22, 23].

It is clear that the longest path problem is NP-hard on every class of graphs on which the Hamiltonian path problem is NP-complete. The Hamiltonian path

⋆ This research is co-financed by E.U.-European Social Fund (80%) and the Greek Ministry of Development-GSRT (20%).
⋆⋆ This research is partially supported by Empirikion Foundation, Greece.

R. Královič and D. Niwiński (Eds.): MFCS 2009, LNCS 5734, pp. 403–414, 2009.

problem is known to be NP-complete in general graphs [10, 11], and remains NP-complete even when restricted to some small classes of graphs such as split graphs [13], chordal bipartite graphs, split strongly chordal graphs [17], circle graphs [5], planar graphs [11], and grid graphs [14]. However, it makes sense to investigate the tractability of the longest path problem on the classes of graphs for which the Hamiltonian path problem admits polynomial time solutions. Such classes include interval graphs [16], circular-arc graphs [6], convex bipartite graphs [17], and co-comparability graphs [4]. Note that the problem of finding a longest path on proper interval graphs is easy, since all connected proper interval graphs have a Hamiltonian path which can be computed in linear time [2]. On the contrary, not all interval graphs are Hamiltonian; in the case where an interval graph has a Hamiltonian path, it can be computed in linear time [16]. However, in the case where an interval graph is not Hamiltonian, there is no known algorithm for finding a longest path on it.

In contrast to the Hamiltonian path problem, there are few known polynomial time solutions for the longest path problem, and these restrict to trees and some small graph classes. Specifically, a linear time algorithm for finding a longest path in a tree was proposed by Dijkstra around 1960, a formal proof of which can be found in [3]. Later, through a generalization of Dijkstra's algorithm for trees, Uehara and Uno [20] solved the longest path problem for weighted trees and block graphs in linear time and space, and for cacti in $O(n^2)$ time and space, where n and m denote the number of vertices and edges of the input graph, respectively. More recently, polynomial algorithms have been proposed that solve the longest path problem on bipartite permutation graphs in $O(n)$ time and space [21], and on ptolemaic graphs in $O(n^5)$ time and $O(n^2)$ space [19].

Furthermore, Uehara and Uno in [20] introduced a subclass of interval graphs, namely interval biconvex graphs, which is a superclass of proper interval and threshold graphs, and solved the longest path problem on this class in $O(n^3(m + n \log n))$ time. As a corollary, they showed that a longest path of a threshold graph can be found in $O(n + m)$ time and space. They left open the complexity of the longest path problem on interval graphs.

In this paper, we resolve the open problem posed in [20] by showing that the longest path problem admits a polynomial time solution on interval graphs. Interval graphs form an important and well-known class of perfect graphs [13]; a graph G is an interval graph if its vertices can be put in a one-to-one correspondence with a family of intervals on the real line, such that two vertices are adjacent in G if and only if their corresponding intervals intersect. In particular, we propose an algorithm for solving the longest path problem on interval graphs which runs in $O(n^4)$ time using a dynamic programming approach. Thus, not only we answer the question left open by Uehara and Uno in [20], but also improve the known time complexity of the problem on interval biconvex graphs, a subclass of interval graphs [20].

Interval graphs form a well-studied class of perfect graphs, have important properties, and admit polynomial time solutions for several problems that are NP-complete on general graphs (see e.g. [1, 13, 16]). Moreover, interval graphs have received a lot of attention due to their applicability to DNA physical mapping problems [12], and find many applications in several fields and disciplines such as genetics, molecular biology, scheduling, VLSI circuit design, archaeology and psychology [13].

2 Theoretical Framework

We consider finite undirected graphs with no loops or multiple edges. For a graph G, we denote its vertex and edge set by $V(G)$ and $E(G)$, respectively. An undirected edge is a pair of distinct vertices $u, v \in V(G)$, and is denoted by uv. We say that the vertex u is adjacent to the vertex v or, equivalently, the vertex u sees the vertex v, if there is an edge uv in G. Let S be a set of vertices of a graph G. Then, the cardinality of the set S is denoted by $|S|$ and the subgraph of G induced by S is denoted by $G[S]$. The set $N(v) = \{u \in V(G) : uv \in E(G)\}$ is called the *neighborhood* of the vertex $v \in V(G)$ in G, sometimes denoted by $N_G(v)$ for clarity reasons. The set $N[v] = N(v) \cup \{v\}$ is called the *closed neighborhood* of the vertex $v \in V(G)$.

A *simple path* of a graph G is a sequence of distinct vertices $v_1, v_2, \ldots, v_k$ such that $v_i v_{i+1} \in E(G)$, for each i, $1 \leq i \leq k-1$, and is denoted by $(v_1, v_2, \ldots, v_k)$; throughout the paper all paths considered are simple. We denote by $V(P)$ the set of vertices in the path P, and define the *length* of the path P to be the number of vertices in P, i.e., $|P| = |V(P)|$. We call *right endpoint* of a path $P = (v_1, v_2, \ldots, v_k)$ the last vertex v_k of P. Moreover, let $P = (v_1, v_2, \ldots, v_{i-1}, v_i, v_{i+1}, \ldots, v_j, v_{j+1}, v_{j+2}, \ldots, v_k)$ and $P_0 = (v_i, v_{i+1}, \ldots, v_j)$ be two paths of a graph. Sometimes, we shall denote the path P by $P = (v_1, v_2, \ldots, v_{i-1}, P_0, v_{j+1}, v_{j+2}, \ldots, v_k)$.

2.1 Structural Properties of Interval Graphs

A graph G is an *interval graph* if its vertices can be put in a one-to-one correspondence with a family F of intervals on the real line such that two vertices are adjacent in G if and only if the corresponding intervals intersect; F is called an *intersection model* for G [1]. The class of interval graphs is *hereditary*, that is, every induced subgraph of an interval graph G is also an interval graph. Ramalingam and Rangan [18] proposed a numbering of the vertices of an interval graph; they stated the following lemma.

Lemma 1. *(Ramalingam and Rangan [18]): The vertices of any interval graph G can be numbered with integers $1, 2, \ldots, |V(G)|$ such that if $i < j < k$ and $ik \in E(G)$, then $jk \in E(G)$.*

As shown in [18], the proposed numbering, which results after sorting the intervals of the intersection model of a graph G on their right ends [1], can be obtained in $O(|V(G)| + |E(G)|)$ time. An ordering of the vertices according to this numbering is found to be quite useful in solving some graph-theoretic problems on interval graphs [1,18]. Throughout the paper, such an ordering is called a *right-end ordering* of G. Let u and v be two vertices of G; if π is a right-end ordering of G, denote $u <_\pi v$ if u appears before v in π. In particular, if $\pi = (u_1, u_2, \ldots, u_{|V(G)|})$ is a right-end ordering of G, then $u_i <_\pi u_j$ if and only if $i < j$.

Lemma 2. *Let G be an interval graph, and let π be a right-end ordering of G. Let $P = (v_1, v_2, \ldots, v_k)$ be a path of G, and let $v_\ell \notin V(P)$ be a vertex of G such that $v_1 <_\pi v_\ell <_\pi v_k$ and $v_\ell v_k \notin E(G)$. Then, there exist two consecutive vertices v_{i-1} and v_i in P, $2 \leq i \leq k$, such that $v_{i-1}v_\ell \in E(G)$ and $v_\ell <_\pi v_i$.*

2.2 Normal Paths

Our algorithm for constructing a longest path of an interval graph G uses a specific type of paths, namely normal paths.

Definition 1. *Let G be an interval graph, and let π be a right-end ordering of G. The path $P = (v_1, v_2, \ldots, v_k)$ of G is called a* normal path, *if v_1 is the leftmost vertex of $V(P)$ in π, and for every i, $2 \leq i \leq k$, the vertex v_i is the leftmost vertex of $N(v_{i-1}) \cap \{v_i, v_{i+1}, \ldots, v_k\}$ in π.*

The notion of a normal path of an interval graph G is a generalization of the notion of a typical path of G; the path $P = (v_1, v_2, \ldots, v_k)$ of an interval graph G is called a *typical* path, if v_1 is the leftmost vertex of $V(P)$ in π. The notion of a typical path was introduced by Arikati and Rangan [1], in order to solve the path cover problem on interval graphs; they proved the following result.

Lemma 3. *(Arikati and Rangan [1]): Let P be a path of an interval graph G. Then, there exists a typical path P' in G such that $V(P') = V(P)$.*

The following lemma is the basis of our algorithm for solving the longest path problem on interval graphs.

Lemma 4. *Let P be a path of an interval graph G. Then, there exists a normal path P' of G, such that $V(P') = V(P)$.*

3 Interval Graphs and the Longest Path Problem

In this section we present our algorithm, which we call Algorithm LP_Interval, for solving the longest path problem on interval graphs; it consists of three phases and works as follows:

- Phase 1: it takes an interval graph G and constructs the auxiliary interval graph H;
- Phase 2: it computes a longest path P on H using Algorithm LP_on_H;
- Phase 3: it computes a longest path $\widehat{P}$ on G from the path P;

The proposed algorithm computes a longest path P of the graph H using dynamic programming techniques and, then, computes a longest path $\widehat{P}$ of G from the path P. We next describe in detail the three phases of our algorithm and prove properties of the constructed graph H which will be used for proving the correctness of the algorithm.

3.1 The Interval Graph H

In this section we present Phase 1 of the algorithm: given an interval graph G and a right-end ordering π of G, we construct the interval graph H and a right-end ordering σ of H.

▶ **Construction of H and σ:** Let G be an interval graph and let $\pi = (v_1, v_2, \ldots, v_{|V(G)|})$ be a right-end ordering of G. Initially, set $V(H) = V(G)$, $\sigma = \pi$, and $A = \emptyset$. Traverse the vertices of π from left to right and do the following: for every vertex v_i add two vertices $a_{i,1}$ and $a_{i,2}$ to $V(H)$ and make both these vertices to be adjacent to every vertex in $N_G[v_i] \cap \{v_i, v_{i+1}, \ldots, v_{|V(G)|}\}$; add $a_{i,1}$ and $a_{i,2}$ to A. Update σ such that $a_{1,1} <_\sigma a_{1,2} <_\sigma v_1$, and $v_{i-1} <_\sigma a_{i,1} <_\sigma a_{i,2} <_\sigma v_i$ for every i, $2 \leq i \leq |V(G)|$.

We call the constructed graph H the *stable-connection graph* of the graph G. Hereafter, we will denote by n the number $|V(H)|$ of vertices of the graph H and by $\sigma = (u_1, u_2, \ldots, u_n)$ the constructed ordering of H. By construction, the vertex set of the graph H consists of the vertices of the set $C = V(G)$ and the vertices of the set A. We will refer to C as the set of the *connector vertices* c of the graph H and to A as the set of *stable vertices* a of the graph H; we denote these sets by $C(H)$ and $A(H)$, respectively. Note that $|A(H)| = 2|V(G)|$.

By the construction of the stable-connection graph H, all neighbors of a stable vertex $a \in A(H)$ are connector vertices $c \in C(H)$, such that $a <_\sigma c$. Moreover, observe that all neighbors of a stable vertex form a clique in G and, thus, also in H. For every connector vertex $u_i \in C(H)$, we denote by $u_{f(u_i)}$ and $u_{h(u_i)}$ the leftmost and rightmost neighbor of u_i in σ, respectively, which appear before u_i in σ, i.e., $u_{f(u_i)} <_\sigma u_{h(u_i)} <_\sigma u_i$. Note that $u_{f(u_i)}$ and $u_{h(u_i)}$ are distinct stable vertices, for every connector vertex u_i.

Lemma 5. *Let G be an interval graph. The stable-connection graph H of G is an interval graph, and the vertex ordering σ is a right-end ordering of H.*

Definition 2. *Let H be the stable-connection graph of an interval graph G, and let $\sigma = (u_1, u_2, \ldots, u_n)$ be the right-end ordering of H. For every pair of indices i, j, $1 \leq i \leq j \leq n$, we define the graph $H(i,j)$ to be the subgraph $H[S]$ of H, induced by the the set $S = \{u_i, u_{i+1}, \ldots, u_j\} \setminus \{u_k \in C(H) : u_{f(u_k)} <_\sigma u_i\}$.*

The following properties hold for every induced subgraph $H(i,j)$, $1 \leq i \leq j \leq n$, and they are used for proving the correctness of Algorithm LP_on_H.

Observation 1. *Let u_k be a connector vertex of $H(i,j)$, i.e., $u_k \in C(H(i,j))$. Then, for every vertex $u_\ell \in V(H(i,j))$, such that $u_k <_\sigma u_\ell$ and $u_k u_\ell \in E(H(i,j))$, u_ℓ is also a connector vertex of $H(i,j)$.*

Observation 2. *No two stable vertices of $H(i,j)$ are adjacent.*

Lemma 6. *Let $P = (v_1, v_2, \ldots, v_k)$ be a normal path of $H(i,j)$. Then:*

Algorithm LP_on_H

Input: a stable-connection graph H, a right-end ordering $\sigma = (u_1, u_2, \ldots, u_n)$ of H.
Output: a longest binormal path of H.

```
for j = 1 to n
    for i = j downto 1
        if i = j and u_i ∈ A(H) then
            ℓ(u_i; i, i) ← 1;  P(u_i; i, i) = (u_i);
        if i ≠ j then
            for every stable vertex u_k ∈ A(H), i ≤ k ≤ j − 1
                ℓ(u_k; i, j) ← ℓ(u_k; i, j−1);  P(u_k; i, j) = P(u_k; i, j−1);  {initialization}
            if u_j is a stable vertex of H(i, j), i.e., u_j ∈ A(H) then
                ℓ(u_j; i, j) ← 1;  P(u_j; i, j) = (u_j);
            if u_j is a connector vertex of H(i, j), i.e., u_j ∈ C(H) and i ≤ f(u_j) then
                execute process(H(i, j));
compute the max{ℓ(u_k; 1, n) : u_k ∈ A(H)} and the corresponding path P(u_k; 1, n);
```

where the procedure `process()` is as follows:

```
process(H(i, j))

for y = f(u_j) + 1 to j − 1
    for x = f(u_j) to y − 1        {u_x and u_y are adjacent to u_j}
        if u_x, u_y ∈ A(H) then
            w_1 ← ℓ(u_x; i, j − 1);  P'_1 = P(u_x; i, j − 1);
            w_2 ← ℓ(u_y; x + 1, j − 1);  P'_2 = P(u_y; x + 1, j − 1);
            if w_1 + w_2 + 1 > ℓ(u_y; i, j) then
                ℓ(u_y; i, j) ← w_1 + w_2 + 1;  P(u_y; i, j) = (P'_1, u_j, P'_2);
return the value ℓ(u_k; i, j) and the path P(u_k; i, j), ∀ u_k ∈ A(H(f(u_j) + 1, j − 1));
```

Fig. 1. The algorithm for finding a longest binormal path of H

(a) For any two stable vertices v_r and v_ℓ in P, v_r appears before v_ℓ in P if and only if $v_r <_\sigma v_\ell$.
(b) For any two connector vertices v_r and v_ℓ in P, if v_ℓ appears before v_r in P and $v_r <_\sigma v_\ell$, then v_r does not see the previous vertex $v_{\ell-1}$ of v_ℓ in P.

3.2 Finding a Longest Path on H

In this section we present Phase 2 of Algorithm LP_Interval. Let G be an interval graph and let H be the stable-connection graph of G constructed in Phase 1. We next present Algorithm LP_on_H, which computes a longest path of the graph H. Let us first give some definitions and notations necessary for the description of the algorithm.

Definition 3. *Let H be a stable-connection graph, and let P be a path of $H(i, j)$, $1 \le i \le j \le n$. The path P is called* binormal *if P is a normal path of $H(i, j)$,*

ALGORITHM LP_INTERVAL

Input: an interval graph G and a right-end ordering π of G.
Output: a longest path $\widehat{P}$ of G.

1. Construct the stable-connection graph H of G and the right-end ordering σ of H; let $V(H) = C \cup A$, where $C = V(G)$ and A are the sets of the connector and stable vertices of H, respectively;
2. Compute a longest binormal path P of H, using Algorithm LP_on_H; let $P = (v_1, v_2, \ldots, v_{2k}, v_{2k+1})$, where $v_{2i} \in C$, $1 \leq i \leq k$, and $v_{2i+1} \in A$, $0 \leq i \leq k$;
3. Compute a longest path $\widehat{P} = (v_2, v_4, \ldots, v_{2k})$ of G, by deleting all stable vertices $\{v_1, v_3, \ldots, v_{2k+1}\}$ from the longest binormal path P of H;

Fig. 2. The algorithm for solving the longest path problem on an interval graph G

both endpoints of P are stable vertices, and no two connector vertices are consecutive in P.

Notation 1. *Let H be a stable-connection graph, and let $\sigma = (u_1, u_2, \ldots, u_n)$ be the right-end ordering of H. For every stable vertex $u_k \in A(H(i,j))$, we denote by $P(u_k; i, j)$ a longest binormal path of $H(i,j)$ with u_k as its right endpoint, and by $\ell(u_k; i, j)$ the length of $P(u_k; i, j)$.*

Since any binormal path is a normal path, Lemma 6 also holds for binormal paths. Moreover, since $P(u_k; i, j)$ is a binormal path, it follows that its right endpoint u_k is also the rightmost stable vertex of P in σ, due to Lemma 6(a).

Algorithm LP_on_H, which is presented in Figure 1, computes for every induced subgraph $H(i,j)$ and for every stable vertex $u_k \in A(H(i,j))$, the length $\ell(u_k; i, j)$ and the corresponding path $P(u_k; i, j)$. Since $H(1,n) = H$, it follows that the maximum among the values $\ell(u_k; 1, n)$, where $u_k \in A(H)$, is the length of a longest binormal path $P(u_k; 1, n)$ of H. In Section 4.2 we prove that the length of a longest path of H equals to the length of a longest binormal path of H. Thus, the binormal path $P(u_k; 1, n)$ computed by Algorithm LP_on_H is also a longest path of H.

3.3 Finding a Longest Path on G

During Phase 3 of our Algorithm LP_Interval, we compute a path $\widehat{P}$ from the longest binormal path P of H, computed by Algorithm LP_on_H, by simply deleting all the stable vertices of P. In Section 4.2 we prove that the resulting path $\widehat{P}$ is a longest path of the interval graph G.

In Figure 2, we present our Algorithm LP_Interval for solving the longest path problem on an interval graph G; note that Steps 1, 2, and 3 of the algorithm correspond to the presented Phases 1, 2, and 3, respectively.

4 Correctness and Time Complexity

In this section we prove the correctness of our algorithm and compute its time complexity. More specifically, in Section 4.1 we show that Algorithm LP_on_H computes a longest binormal path P of the graph H (in Lemma 13 we prove that this path is also a longest path of H), while in Section 4.2 we show that the length of a longest binormal path P of H is equal to $2k+1$, where k is the length of a longest path of G. Finally, we show that the path $\widehat{P}$ constructed at Step 3 of Algorithm LP_Interval is a longest path of G.

4.1 Correctness of Algorithm LP_on_H

We next prove that Algorithm LP_on_H correctly computes a longest binormal path of the graph H. The following lemmas appear useful in the proof of the algorithm's correctness.

Lemma 7. *Let H be a stable-connection graph, and let $\sigma = (u_1, u_2, \ldots, u_n)$ be the right-end ordering of H. Let P be a longest binormal path of $H(i,j)$ with u_y as its right endpoint, let u_k be the rightmost connector vertex of $H(i,j)$ in σ, and let $u_{f(u_k)+1} \leq_\sigma u_y \leq_\sigma u_{h(u_k)}$. Then, there exists a longest binormal path P' of $H(i,j)$ with u_y as its right endpoint, which contains the connector vertex u_k.*

Lemma 8. *Let H be a stable-connection graph, and let σ be the right-end ordering of H. Let $P = (P_1, v_\ell, P_2)$ be a binormal path of $H(i,j)$, and let v_ℓ be a connector vertex of $H(i,j)$. Then, P_1 and P_2 are binormal paths of $H(i,j)$.*

Lemma 9. *Let H be a stable-connection graph, and let $\sigma = (u_1, u_2, \ldots, u_n)$ be the right-end ordering of H. Let P_1 be a binormal path of $H(i, j-1)$ with u_x as its right endpoint, and let P_2 be a binormal path of $H(x+1, j-1)$ with u_y as its right endpoint, such that $V(P_1) \cap V(P_2) = \emptyset$. Suppose that u_j is a connector vertex of H and that $u_i \leq_\sigma u_{f(u_j)} \leq_\sigma u_x$. Then, $P = (P_1, u_j, P_2)$ is a binormal path of $H(i,j)$ with u_y as its right endpoint.*

Lemma 10. *Let H be a stable-connection graph, and let σ be the right-end ordering of H. For every induced subgraph $H(i,j)$ of H, $1 \leq i \leq j \leq n$, and for every stable vertex $u_y \in A(H(i,j))$, Algorithm LP_on_H computes the length $\ell(u_y; i, j)$ of a longest binormal path of $H(i,j)$ which has u_y as its right endpoint and, also, the corresponding path $P(u_y; i, j)$.*

Proof (sketch). Let P be a longest binormal path of the stable-connection graph $H(i,j)$, which has a vertex $u_y \in A(H(i,j))$ as its right endpoint. Consider first the case where $C(H(i,j)) = \emptyset$; the graph $H(i,j)$ is consisted of a set of stable vertices $A(H(i,j))$, which is an independent set, due to Observation 2. Therefore, in this case Algorithm LP_on_H sets $\ell(u_y; i, j) = 1$ for every vertex $u_y \in A(H(i,j))$, which is indeed the length of the longest binormal path $P(u_y; i, j) = (u_y)$ of $H(i,j)$ which has u_y as its right endpoint. Therefore, the lemma holds for every induced subgraph $H(i,j)$, for which $C(H(i,j)) = \emptyset$.

We examine next the case where $C(H(i,j)) \neq \emptyset$. Let $C(H) = \{c_1, c_2, \ldots, c_k, \ldots, c_t\}$ be the set of connector vertices of H, where $c_1 <_\sigma c_2 <_\sigma \ldots <_\sigma c_k <_\sigma \ldots <_\sigma c_t$. Let $\sigma = (u_1, u_2, \ldots, u_n)$ be the vertex ordering of H constructed in Phase 1. Recall that, by the construction of H, $n = 3t$, and $A(H) = V(H) \setminus C(H)$ is the set of stable vertices of H.

Let $H(i,j)$ be an induced subgraph of H, and let c_k be the rightmost connector vertex of $H(i,j)$ in σ. The proof of the lemma is done by induction on the index k of the rightmost connector vertex c_k of $H(i,j)$. More specifically, given a connector vertex c_k of H, we prove that the lemma holds for every induced subgraph $H(i,j)$ of H, which has c_k as its rightmost connector vertex in σ. To this end, in both the induction basis and the induction step, we distinguish three cases on the position of the stable vertex u_y in the ordering σ: $u_i \leq_\sigma u_y \leq_\sigma u_{f(c_k)}$, $u_{h(c_k)} <_\sigma u_y \leq_\sigma u_j$, and $u_{f(c_k)+1} \leq_\sigma u_y \leq_\sigma u_{h(c_k)}$. In each of these three cases, we examine first the length of a longest binormal path of $H(i,j)$ with u_y as its right endpoint and, then, we compare this value to the length of the path computed by Algorithm LP_on_H. Moreover, we prove that the path computed by Algorithm LP_on_H is indeed a binormal path with u_y as its right endpoint. □

Due to Lemma 10, and since the output of Algorithm LP_on_H is the maximum among the lengths $\ell(u_y; 1, n)$, $u_y \in A(H(1,n))$, along with the corresponding path, it follows that Algorithm LP_on_H computes a longest binormal path of $H(1,n)$ with right endpoint a vertex $u_y \in A(H(1,n))$. Thus, since $H(1,n) = H$, we obtain the following result.

Lemma 11. *Let G be an interval graph. Algorithm LP_on_H computes a longest binormal path of the stable-connection graph H of the graph G.*

4.2 Correctness of Algorithm LP_Interval

We next show that Algorithm LP_Interval correctly computes a longest path of an interval graph G. The correctness proof is based on the following property: for any longest path P of G there exists a longest binormal path P' of H, such that $|P'| = 2|P| + 1$ and vice versa (this property is proved in Lemma 12). Therefore, we obtain that the length of a longest binormal path P of H computed by Algorithm LP_on_H is equal to $2k + 1$, where k is the length of a longest path $\widehat{P}$ of G. Next, we show that the length of a longest binormal path of H equals to the length of a longest path of H. Finally, we show that the path $\widehat{P}$ computed at Step 3 of Algorithm LP_Interval is indeed a longest path of G.

Lemma 12. *Let H be the stable-connection graph of an interval graph G. Then, for any longest path P of G there exists a longest binormal path P' of H, such that $|P'| = 2|P| + 1$ and vice versa.*

Proof. Let σ be the right-end ordering of the graph H constructed in Phase 1. ($\Longrightarrow$) Let $P = (v_1, v_2, \ldots, v_k)$ be a longest path of G, i.e., $|P| = k$. We will show that there exists a binormal path P' of H such that $|P'| = 2k + 1$. Since G is an induced subgraph of H, the path P of G is a path of H as well. We construct a path

$\widehat{P}$ of H from P, by adding to P the appropriate stable vertices, using the following procedure. Initially, set $\widehat{P} = P$ and for every subpath (v_i, v_{i+1}) of the path $\widehat{P}$, $1 \leq i \leq k-1$, do the following: consider first the case where $v_i <_\sigma v_{i+1}$; then, by the construction of H, v_{i+1} is adjacent to both stable vertices $a_{i,1}$ and $a_{i,2}$ associated with the connector vertex v_i. If $a_{i,1}$ has not already been added to $\widehat{P}$, then replace the subpath (v_i, v_{i+1}) by the path $(v_i, a_{i,1}, v_{i+1})$; otherwise, replace the subpath (v_i, v_{i+1}) by the path $(v_i, a_{i,2}, v_{i+1})$. Similarly, in the case where $v_{i+1} <_\sigma v_i$, replace the subpath (v_i, v_{i+1}) by the path $(v_i, a_{i+1,1}, v_{i+1})$ or $(v_i, a_{i+1,2}, v_{i+1})$, respectively. Finally, consider the endpoint v_1 (resp. v_k) of $\widehat{P}$. If $a_{1,1}$ (resp. $a_{k,1}$) has not already been added to $\widehat{P}$, then add $a_{1,1}$ (resp. $a_{k,1}$) as the first (resp. last) vertex of $\widehat{P}$; otherwise, add $a_{1,2}$ (resp. $a_{k,2}$) as the first (resp. last) vertex of $\widehat{P}$.

By the construction of $\widehat{P}$ it is easy to see that for every connector vertex v of P we add two stable vertices as neighbors of v in $\widehat{P}$, and since in H there are exactly two stable vertices associated with every connector vertex v, it follows that every stable vertex of H appears at most once in $\widehat{P}$. Furthermore, since we add in total $k+1$ stable vertices to P, where $|P| = k$, it follows that $|\widehat{P}| = 2k+1$. Denote now by P' a normal path of H such that $V(P') = V(\widehat{P})$. Such a path exists, due to Lemma 4. Due to the above construction, the path $\widehat{P}$ is consisted of $k+1$ stable vertices and k connector vertices. Thus, since no two stable vertices are adjacent in H due to Observation 2, and since P' is a normal path of H, it follows that P' is a binormal path of H. Thus, for any longest path P of G there exists a binormal path P' of H, such that $|P'| = 2|P| + 1$.

($\Longleftarrow$) Consider now a longest binormal path $P' = (v_1, v_2, \ldots, v_\ell)$ of H. Since P' is binormal, it follows that $\ell = 2k+1$, and that P' has k connector vertices and $k+1$ stable vertices, for some $k \geq 1$. We construct a path P by deleting all stable vertices from the path P' of H. By the construction of H, all neighbors of a stable vertex a are connector vertices and form a clique in G; thus, for every subpath (v, a, v') of P', v is adjacent to v' in G. It follows that P is a path of G. Since we removed all the $k+1$ stable vertices of P', it follows that $|P| = k$, i.e., $|P'| = 2|P| + 1$.

Summarizing, we have constructed a binormal path P' of H from a longest path P of G such that $|P'| = 2|P|+1$, and a path P of G from a longest binormal path P' of H such that $|P'| = 2|P| + 1$. This completes the proof. □

Lemma 13. *For any longest path P and any longest binormal path P' of H, it holds $|P'| = |P|$.*

Let P be the longest binormal path of H computed in Step 2 of Algorithm LP_Interval, using Algorithm LP_on_H. Then, in Step 3 Algorithm LP_Interval computes the path $\widehat{P}$ by deleting all stable vertices from P. By the construction of H, all neighbors of a stable vertex a are connector vertices and form a clique in G; thus, for every subpath (v, a, v') of P, v is adjacent to v' in G. It follows that $\widehat{P}$ is a path of G. Moreover, since P is binormal, it has k connector vertices and $k+1$ stable vertices, i.e., $|P| = 2k + 1$, where $k \geq 1$. Thus, since we have removed all $k+1$ stable vertices of P, it follows that $|\widehat{P}| = k$ and, thus, $\widehat{P}$ is a longest path of G due to Lemma 12. Thus, we have proved the following result.

Theorem 1. *Algorithm LP_Interval computes a longest path of an interval graph G.*

4.3 Time Complexity

Let G be an interval graph on $|V(G)| = n$ vertices and $|E(G)| = m$ edges. It has been shown that we can obtain the right-end ordering π of G, which results from numbering the intervals after sorting them on their right ends, in $O(n+m)$ time [1,18].

First, we show that Step 1 of Algorithm LP_Interval, which constructs the stable-connection graph H of the graph G, takes $O(n^2)$ time. Indeed, for every connector vertex u_i, $1 \leq i \leq n$, we can add two stable vertices in $V(H)$ in $O(1)$ time and we can compute the specific neighborhood of u_i in $O(n)$ time.

Step 2 of Algorithm LP_Interval includes the execution of Algorithm LP_on_H. The subroutine `process()` takes $O(n^2)$ time, due to the $O(n^2)$ pairs of the neighbors u_x and u_y of the connector vertex u_j in the graph $H(i,j)$. Additionally, the subroutine `process()` is executed at most once for each subgraph $H(i,j)$ of H, $1 \leq i \leq j \leq n$, i.e., it is executed $O(n^2)$ times. Thus, Algorithm LP_on_H takes $O(n^4)$ time.

Step 3 of Algorithm LP_Interval can be executed in $O(n)$ time since we simply traverse the vertices of the path P, constructed by Algorithm LP_on_H, and delete every stable vertex.

Theorem 2. *A longest path of an interval graph can be computed in $O(n^4)$ time.*

5 Concluding Remarks

In this paper we presented a polynomial-time algorithm for solving the longest path problem on interval graphs, which runs in $O(n^4)$ time and, thus, provided a solution to the open problem stated by Uehara and Uno in [20] asking for the complexity status of the longest path problem on interval graphs. It would be interesting to see whether the ideas presented in this paper can be applied to find a polynomial solution to the longest path problem on convex and biconvex graphs, the complexities of which still remain open [20].

References

1. Arikati, S.R., Pandu Rangan, C.: Linear algorithm for optimal path cover problem on interval graphs. Inform. Proc. Lett. 35, 149–153 (1990)
2. Bertossi, A.A.: Finding Hamiltonian circuits in proper interval graphs. Inform. Proc. Lett. 17, 97–101 (1983)
3. Bulterman, R., van der Sommen, F., Zwaan, G., Verhoeff, T., van Gasteren, A., Feijen, W.: On computing a longest path in a tree. Inform. Proc. Lett. 81, 93–96 (2002)
4. Damaschke, P., Deogun, J.S., Kratsch, D., Steiner, G.: Finding Hamiltonian paths in cocomparability graphs using the bump number algorithm. Order 8, 383–391 (1992)

5. Damaschke, P.: The Hamiltonian circuit problem for circle graphs is NP-complete. Inform. Proc. Lett. 32, 1–2 (1989)
6. Damaschke, P.: Paths in interval graphs and circular arc graphs. Discrete Math. 112, 49–64 (1993)
7. Feder, T., Motwani, R.: Finding large cycles in Hamiltonian graphs. In: Proc. 16th annual ACM-SIAM Symp. on Discrete Algorithms (SODA), pp. 166–175. ACM, New York (2005)
8. Gabow, H.N.: Finding paths and cycles of superpolylogarithmic length. In: Proc. 36th annual ACM Symp. on Theory of Computing (STOC), pp. 407–416. ACM, New York (2004)
9. Gabow, H.N., Nie, S.: Finding long paths, cycles and circuits. In: Hong, S.-H., Nagamochi, H., Fukunaga, T. (eds.) ISAAC 2008. LNCS, vol. 5369, pp. 752–763. Springer, Heidelberg (2008)
10. Garey, M.R., Johnson, D.S.: Computers and Intractability: A Guide to the Theory of NP-completeness. W.H. Freeman, San Francisco (1979)
11. Garey, M.R., Johnson, D.S., Tarjan, R.E.: The planar Hamiltonian circuit problem is NP-complete. SIAM J. Computing 5, 704–714 (1976)
12. Goldberg, P.W., Golumbic, M.C., Kaplan, H., Shamir, R.: Four strikes against physical mapping of DNA. Journal of Computational Biology 2, 139–152 (1995)
13. Golumbic, M.C.: Algorithmic Graph Theory and Perfect Graphs. Annals of Discrete Mathematics, vol. 57. North-Holland Publishing Co., Amsterdam (2004)
14. Itai, A., Papadimitriou, C.H., Szwarcfiter, J.L.: Hamiltonian paths in grid graphs. SIAM J. Computing 11, 676–686 (1982)
15. Karger, D., Motwani, R., Ramkumar, G.D.S.: On approximating the longest path in a graph. Algorithmica 18, 82–98 (1997)
16. Keil, J.M.: Finding Hamiltonian circuits in interval graphs. Inform. Proc. Lett. 20, 201–206 (1985)
17. Müller, H.: Hamiltonian circuits in chordal bipartite graphs. Discrete Math. 156, 291–298 (1996)
18. Ramalingam, G., Pandu Rangan, C.: A unified approach to domination problems on interval graphs. Inform. Proc. Lett. 27, 271–274 (1988)
19. Takahara, Y., Teramoto, S., Uehara, R.: Longest path problems on ptolemaic graphs. IEICE Trans. Inf. and Syst. 91-D, 170–177 (2008)
20. Uehara, R., Uno, Y.: Efficient algorithms for the longest path problem. In: Fleischer, R., Trippen, G. (eds.) ISAAC 2004. LNCS, vol. 3341, pp. 871–883. Springer, Heidelberg (2004)
21. Uehara, R., Valiente, G.: Linear structure of bipartite permutation graphs and the longest path problem. Inform. Proc. Lett. 103, 71–77 (2007)
22. Vishwanathan, S.: An approximation algorithm for finding a long path in Hamiltonian graphs. In: Proc. 11th annual ACM-SIAM Symp. on Discrete Algorithms (SODA), pp. 680–685. ACM, New York (2000)
23. Zhang, Z., Li, H.: Algorithms for long paths in graphs. Theoret. Comput. Sci. 377, 25–34 (2007)

Synthesis for Structure Rewriting Systems*

Łukasz Kaiser

Mathematische Grundlagen der Informatik, RWTH Aachen, Germany
kaiser@logic.rwth-aachen.de

Abstract. The description of a single state of a modelled system is often complex in practice, but few procedures for synthesis address this problem in depth. We study systems in which a state is described by an arbitrary finite structure, and changes of the state are represented by structure rewriting rules, a generalisation of term and graph rewriting. Both the environment and the controller are allowed to change the structure in this way, and the question we ask is how a strategy for the controller that ensures a given property can be synthesised.

We focus on one particular class of structure rewriting rules, namely on separated structure rewriting, a limited syntactic class of rules. To counter this restrictiveness, we allow the property to be ensured by the controller to be specified in a very expressive logic: a combination of monadic second-order logic evaluated on states and the modal μ-calculus for the temporal evolution of the whole system. We show that for the considered class of rules and this logic, it can be decided whether the controller has a strategy ensuring a given property, and in such case a finite-memory strategy can be synthesised. Additionally, we prove that the same holds if the property is given by a monadic second-order formula to be evaluated on the limit of the evolution of the system.

1 Introduction

Structure rewriting is a generalisation of graph rewriting and graph grammars, which have been widely studied in computer science [1] and even used as a basis for software development environments.[1] Since unrestricted graph rewriting constitutes a programming language, most questions about unrestricted rewriting systems are necessarily undecidable. While the Graph Minor Theorem has recently allowed a basic analysis of large classes of single-pushout graph transformation systems [3], to define structure rewriting for which the synthesis problem remains decidable, it is necessary to strongly limit the allowed rewriting rules.

Choosing a restricted class of structure rewriting rules, we want to preserve the original motivation of practical applicability. In the context of software verification, this means that we want to allow at least basic manipulations on graphs that often appear as memory structures on the heap. One candidate for such a

* This work was partially supported by the DFG Graduiertenkolleg 1298 AlgoSyn.

[1] It is interesting to note that the idea to rewrite relational structures was introduced in 1973 by Rajlich [2] and it preceded most of the work on graph grammars.

R. Královič and D. Niwiński (Eds.): MFCS 2009, LNCS 5734, pp. 415–426, 2009.

class, used for example in the verification of Pointer Assertion Logic programs [4], are graphs of bounded clique-width. The class of rewriting rules that corresponds to such graphs, separated handle hypergraph rewriting rules, was identified in [5] in the context of hypergraph grammars.

We study the synthesis problem, which we view as a two-player zero-sum game, in the course of which a structure is manipulated using similar separated structure rewriting rules. We consider two possibilities to define the property to be ensured, i.e. the winning condition in such games. One possibility is to use a μ-calculus formula evaluated on the sequence of structures that constitutes a play. In such a formula, instead of the usual predicates assigned to states, we allow arbitrary monadic second-order formulas to be evaluated on the "current" structure. The other possibility is to give a single monadic second-order formula and evaluate it on the limit of all structures that appear during the play.

As our main result, we show that for the games described above it is decidable which player has a winning strategy and that a winning strategy can be constructed. In fact, we prove that conditions expressed by monadic second-order formulas over the structures can be reduced to ω-regular conditions over the game arena. Thus, we identify a class of structure rewriting games that have the same nice properties as ω-regular games.

2 Preliminaries

For any set A we denote by A^* and A^ω the set of finite, and respectively infinite, sequences of elements of A. Given a (finite or infinite) sequence $\alpha = a_0 a_1 \dots$ we write $\alpha[i]$ to denote the $(i+1)$st element of α, i.e. $\alpha[i] = a_i$.

A (relational) *structure* over a finite signature $\tau = \{R_1, \dots, R_n\}$ (with R_i having arity r_i) is a tuple $\mathfrak{A} = (A, R_1^{\mathfrak{A}}, \dots, R_n^{\mathfrak{A}})$ where A is the universe of $\mathfrak{A}$ and each relation $R_i^{\mathfrak{A}} \subseteq A^{r_i}$. We often write $a \in \mathfrak{A}$ when $a \in A$ is meant.

Given two structures $\mathfrak{A}, \mathfrak{B}$ over the same signature τ we say that a function $f : \mathfrak{A} \hookrightarrow \mathfrak{B}$ is an *embedding* if f is injective and for each $R_i \in \tau$ it holds that $(a_1, \dots, a_{r_i}) \in R_i^{\mathfrak{A}} \iff (f(a_1), \dots, f(a_{r_i})) \in R_i^{\mathfrak{B}}$.

Given a sequence of structures $\mathfrak{A}_0 \mathfrak{A}_1 \dots$ we define the limit of this sequence $\mathfrak{A}_\infty = \lim \mathfrak{A}_i$. The universe of $\mathfrak{A}_\infty$ consists of all elements that remain in all $\mathfrak{A}_i$ from some n on, $A_\infty = \bigcup_{n \in \mathbb{N}} \bigcap_{i \geq n} A_i$. The relations are defined similarly, i.e. a tuple is in $R_k^{\mathfrak{A}_\infty}$ if for some n it is in all $R_k^{\mathfrak{A}_i}$ for $i > n$, so $R_k^{\mathfrak{A}_\infty} = \bigcup_{n \in \mathbb{N}} \bigcap_{i \geq n} R_k^{\mathfrak{A}_i}$.

There are many ways to describe properties of structures, and the most general way that we use is monadic second-order logic, MSO. We omit the standard formal definition of the semantics of MSO here, let us only mention the syntax. Atomic formulas are built using first-order variables $x_0, x_1, \dots$ and second-order variables $X_0, X_1, \dots$ in the expressions $R_i(x_1, \dots, x_{r_i})$ or $x \in X$. Formulas can be negated, connected by disjunction and conjunction, and both first and second-order quantification is allowed. So if φ and ψ are MSO formulas, then $\varphi \wedge \psi$, $\varphi \vee \psi$, $\neg\varphi$, $\exists x \varphi$, $\forall x \varphi$ and $\exists X \varphi$, $\forall X \varphi$ are MSO-formulas as well. We write $\mathfrak{A} \models \varphi$ if the formula φ is satisfied by the structure $\mathfrak{A}$.

A play of a structure rewriting game corresponds to a sequence of structures, so to express properties of plays we not only need to express properties of structures, but also the ways they change through the sequence. One of the most expressive logics used for such temporal properties, which subsumes for example the linear time logic, is the modal μ-calculus, L_μ. We use an extension of L_μ where arbitrary MSO formulas are allowed instead of predicates. The syntax of $L_\mu[\text{MSO}]$ is given by

$$\varphi \;=\; \psi_{\text{MSO}} \mid Y \mid \varphi \wedge \varphi \mid \varphi \vee \varphi \mid \Diamond\varphi \mid \mu Y\,\varphi \mid \nu Y\,\varphi,$$

where ψ_{MSO} is any MSO sentence. The semantics of $L_\mu[\text{MSO}]$, i.e. the notion that a sequence of structures $\mathfrak{A}_0\mathfrak{A}_1\ldots$ satisfies an $L_\mu[\text{MSO}]$ formula φ, is defined analogously to the standard semantics of L_μ (cf. Chapter 10 of [6]), with the only change that instead of using predicates we say that ψ_{MSO} holds at position i if and only if it is satisfied by the structure $\mathfrak{A}_i$. We do not repeat the formal semantics of $L_\mu[\text{MSO}]$ here, let us only mention the intuition, namely that $\Diamond\varphi$ holds in a sequence $\mathfrak{A}_0\mathfrak{A}_1\ldots$ if φ holds from the *next* step on, i.e. in $\mathfrak{A}_1\mathfrak{A}_2\ldots$, and that $\mu Y\,\varphi$ denotes the least fixed-point and $\nu Y\,\varphi$ the greatest fixed-point.

In addition to MSO and $L_\mu[\text{MSO}]$, we use the standard definitions of alternating parity ω-word automata and non-deterministic parity ω-tree automata, with the slight modification that our word automata have priorities on transitions and not on states, and tree automata have final positions for the case that some branch of the tree is finite.

3 Structure Rewriting Games

We will define two-player games in the course of which a structure is manipulated by the players using separated structure rewriting rules, similar to the ones presented in [5]. In this section we introduce such rules, the rewriting process and the corresponding games.

3.1 Structure Rewriting Rules

Let us fix the signature τ and partition it into two disjoint subsets: the set τ_{t} of *terminal* relation symbols and the set τ_{n} of *non-terminal* symbols. We say that a structure $\mathfrak{A}$ is *separated* if no element appears in two non-terminal relations, i.e. if for all $(a_1,\ldots,a_{r_k}) \in R_k^{\mathfrak{A}}$, $(b_1,\ldots,b_{r_l}) \in R_l^{\mathfrak{A}}$ with $R_k, R_l \in \tau_{\text{n}}$ it holds that $a_i \neq b_j$ for all $i \leq r_k, j \leq r_l$ (except if $k = l$ and $\overline{a} = \overline{b}$ of course).

A *structure rewriting rule* $\mathfrak{L} \to \mathfrak{R}$ consists of a finite structure $\mathfrak{L}$ over the signature τ and a finite structure $\mathfrak{R}$ over the extended signature $\tau \cup \{P_l\}_{l\in\mathfrak{L}}$, where each P_l is a unary predicate, i.e. $P_l^{\mathfrak{R}} \subseteq \mathfrak{R}$, and we assume that $P_l^{\mathfrak{R}}$ are pairwise disjoint (this assumption can be omitted, as we explain in Section 5).

A *match* of the rule $\mathfrak{L} \to \mathfrak{R}$ in another structure $\mathfrak{A}$ is an embedding $\sigma : \mathfrak{L} \hookrightarrow \mathfrak{A}$, which induces the following mapping relation M_σ on $\mathfrak{R} \times \mathfrak{A}$:

$$(r,a) \in M_\sigma \iff a = \sigma(l) \text{ and } r \in P_l^{\mathfrak{R}} \text{ for some } l \in \mathfrak{L}.$$

Intuitively, we consider the elements of $\mathfrak{R}$ that belong to $P_l^{\mathfrak{R}}$ as replacements for l, and thus M_σ contains the pairs (r,a) for which a is to be replaced by r.

We define the result of an application of $\mathfrak{L} \to \mathfrak{R}$ to $\mathfrak{A}$ on the match σ as a structure $\mathfrak{B} = \mathfrak{A}[\mathfrak{L} \to \mathfrak{R}/\sigma]$, such that the universe of $\mathfrak{B}$ is given by $(A \setminus \sigma(L)) \dot{\cup} R$, and the relations as follows (writing bM_σ^I for $\{a \mid (b,a) \in M_\sigma^I\}$):

$$(b_1, \ldots, b_{r_i}) \in R_i^{\mathfrak{B}} \iff (b_1 M_\sigma^I \times \ldots \times b_{r_i} M_\sigma^I) \cap R_i^{\mathfrak{A}} \neq \emptyset,$$

where $M_\sigma^I = M_\sigma \cup \{(a,a) \mid a \in \mathfrak{A}\}$. Intuitively, we remove $\mathfrak{L}$, insert $\mathfrak{R}$, and connect all $r \in P_l^{\mathfrak{R}}$ in places where l was before in $\mathfrak{A}$, as given by σ.

A *separated structure rewriting rule* is a rewriting rule $\mathfrak{L} \to \mathfrak{R}$ where $\mathfrak{R}$ is separated and $\mathfrak{L}$ consists of only one tuple of elements in a non-terminal relation, i.e. there exists an $R_i \in \tau_{\mathrm{n}}$ such that $\mathfrak{L} = (\{l_1, \ldots, l_{r_i}\}, R_1^{\mathfrak{L}}, \ldots, R_n^{\mathfrak{L}})$ with $R_i^{\mathfrak{L}} = \{(l_1, \ldots, l_{r_i})\}$ and $R_j^{\mathfrak{L}} = \emptyset$ for $j \neq i$. (Note that $l_i = l_j$ is possible.) We denote the set of all separated rules over $\tau = \tau_{\mathrm{n}} \cup \tau_{\mathrm{t}}$ by $\mathbb{S}(\tau)$.

An example of a separated rewriting of a structure with one binary terminal relation R_0 (depicted as unlabelled edges) and one non-terminal binary relation R_1 is given in Figure 1.

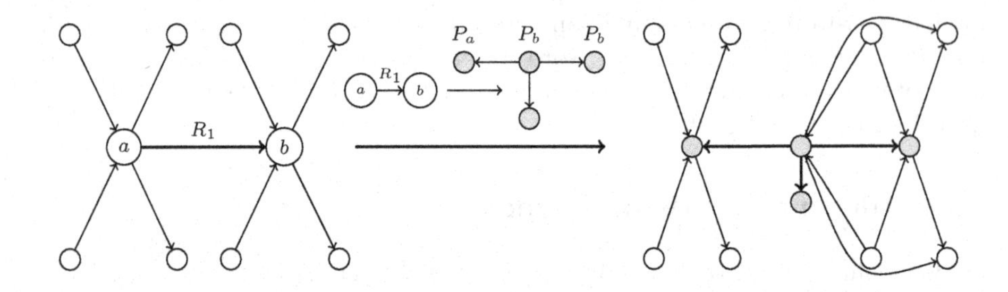

Fig. 1. Rewriting the tuple $(a,b) \in R_1$ in a structure

Applications of a single separated rewriting rule to a separated structure are confluent and yield again a separated structure (cf. [5]). Thus, if $\mathfrak{A}$ is a finite separated structure and $\mathfrak{L} \to \mathfrak{R}$ is a separated rule, we can define $\mathfrak{A}[\mathfrak{L} \to \mathfrak{R}]$, the separated structure resulting from applying the rule to *all* tuples in $R_i^{\mathfrak{A}}$ in any order (where R_i is the single non-empty relation in $\mathfrak{L}$). Note that if $R_i^{\mathfrak{A}} = \emptyset$, then $\mathfrak{A}[\mathfrak{L} \to \mathfrak{R}] = \mathfrak{A}$.

3.2 Games Played with Structures

Let $\alpha = r_0 r_1 \ldots$ be a sequence of separated rewriting rules, all belonging to a finite set. For a non-terminal symbol $R_k \in \tau_{\mathrm{n}}$ we define the R_k-starting structure $\mathfrak{S}_k$ as one with only one R_k-tuple, $\mathfrak{S}_k = (\{a_1, \ldots, a_{r_k}\}, R_1^{\mathfrak{S}_k}, \ldots, R_n^{\mathfrak{S}_k})$ with $R_k^{\mathfrak{S}_k} = \{(a_1, \ldots, a_{r_i})\}$ and $R_j^{\mathfrak{S}_k} = \emptyset$ for $j \neq k$. Having a starting relation symbol (and structure) and a sequence of rules, we can define the corresponding sequence of finite separated structures $\mathrm{st}_k(\alpha) = \mathfrak{A}_0 \mathfrak{A}_1 \ldots$ such that $\mathfrak{A}_0 = \mathfrak{S}_k$ and for each $i \in \mathbb{N}$ we set $\mathfrak{A}_{i+1} = \mathfrak{A}_i[r_i]$. We will be interested either in properties of the whole sequence of structures expressed in $\mathrm{L}_\mu[\mathrm{MSO}]$ or in a property of the limit structure $\lim \mathfrak{A}_i$ expressed in MSO.

Definition 1. *A* separated structure rewriting game $\mathcal{G} = (V_0, V_1, E, \varphi)$ *consists of disjoint sets* V_0 *and* V_1 *of positions of Player* 0 *and Player* 1*, respectively, a set of moves* $E \subseteq V \times S \times V$*, where* $V = V_0 \cup V_1$ *and* $S \subseteq \mathbb{S}(\tau)$ *is a* finite *set of separated rewriting rules, and either an* $\mathrm{L}_\mu[\mathrm{MSO}]$ *or an* MSO *formula* φ *that describes the winning condition of the game.*

A *play* of $\mathcal{G} = (V_0, V_1, E, \varphi)$ starts in a distinguished position $v_0 \in V$ and with a starting structure $\mathfrak{A}_0 = \mathfrak{S}_k$ for some non-terminal $R_k \in \tau_\mathrm{n}$. If the play is in a position $v \in V_i$ with structure $\mathfrak{A}$, player i must choose a move $(v, r, w) \in E$ (for simplicity we assume that such a move always exists). The play continues from the position w with the structure $\mathfrak{A}[r]$. Formally, a play $\pi = v_0 e_0 v_1 e_1 \dots$ of $\mathcal{G}$ is an infinite sequence of positions and moves, $\pi \in (VE)^\omega$, such that $e_i = (v_i, r_i, v_{i+1})$, i.e. the ith move goes from the ith position to the $(i+1)$st position. A play π as above induces a sequence of rules $r_0 r_1 \dots$ seen during the play, which we denote by $\mathrm{rules}(\pi)$.

A *strategy* of player i is a function that assigns to each history of a play ending in a position of player i, i.e. to each $h = v_0 \dots v_n \in (VE)^* V_i$, the next move $(v_n, r, w) \in E$. Note that the structure corresponding to each position v_n is a function of h and the starting symbol R_k, so we can omit the constructed structures in the definition of a strategy. We say that a play $\pi = v_0 e_0 v_1 \dots$ is consistent with a strategy σ_i of player i if for each prefix $h = v_0 e_0 \dots v_k$ of π with $v_k \in V_i$ it holds that $e_k = (v_k, r, v_{k+1}) = \sigma_i(h)$. When the starting position v_0, the non-terminal symbol R_k, and the strategies of both players σ_0 and σ_1 are fixed, there exists a unique play $\pi = v_0 e_0 \dots$ that starts in v_0 and is consistent with both these strategies. This play induces a unique sequence of structures $\mathrm{st}_k(\mathrm{rules}(\pi))$, which we will denote by $\pi_k(\sigma_0, \sigma_1, v_0)$. We say that Player 0 wins the play π if either $\mathrm{st}_k(\mathrm{rules}(\pi)) \models \varphi$, in case the winning condition is given by an $\mathrm{L}_\mu[\mathrm{MSO}]$ formula φ, or if $\lim \mathrm{st}_k(\mathrm{rules}(\pi)) \models \varphi$, if φ is an MSO formula to be evaluated on the limit structure.

We say that Player 0 wins the game $\mathcal{G}$ from v_0 and R_k if she has a strategy σ_0 such that for all strategies σ_1 of her opponent, $\pi_k(\sigma_0, \sigma_1, v_0) \models \varphi$ (or $\lim \pi_k(\sigma_0, \sigma_1, v_0) \models \varphi$). If Player 1 has a strategy σ_1 such that for all strategies σ_0 of Player 0, $\pi_k(\sigma_0, \sigma_1, v_0) \not\models \varphi$, then we say that Player 1 wins the game $\mathcal{G}$. We will prove the following main result about separated graph rewriting games.

Theorem 1. *Let* $\mathcal{G}$ *be a finite separated structure rewriting game,* v_0 *a position in* $\mathcal{G}$ *and* $R_k \in \tau_\mathrm{n}$ *a non-terminal symbol. Then either Player* 0 *wins* $\mathcal{G}$ *starting from* v_0 *and* R_k *or Player* 1 *does, it is decidable which player is the winner and a winning strategy for this player can be constructed.*

The theorem above is a consequence of the following stronger theorem, which allows us to reduce questions about separated structure rewriting games to questions about ω-regular games.

Theorem 2. *Let* S *be a finite set of separated structure rewriting rules over a signature* $\tau = \tau_\mathrm{n} \cup \tau_\mathrm{t}$ *and let* $R_k \in \tau_\mathrm{n}$*. For any* MSO *formula* φ *the set of finite sequences of rules which end in a structure satisfying* φ*, i.e. the set*

$$\{r_0 \dots r_i \mid \mathrm{st}_k(r_0 \dots r_i)[i] \models \varphi\}$$

is a regular subset of S^*. *Moreover, the set* $\{\pi \subseteq S^\omega \mid \lim \mathrm{st}_k(\pi) \models \varphi\}$ *is an* ω*-regular subset of* S^ω. *Both these statements are effective, i.e. the automata can be algorithmically constructed from* S, R_k *and* φ.

By Theorem 2, if φ is a formula of $\mathrm{L}_\mu[\mathrm{MSO}]$, then, for each MSO-sentence ψ_{MSO} occurring in φ, there is a corresponding regular language $\mathcal{L}(\psi_{\mathrm{MSO}}) \subseteq S^*$. By the standard correspondence of regular languages and MSO (and L_μ as well) on words, this implies that the set $\{\pi \subseteq S^\omega \mid \mathrm{st}_k(\pi) \models \varphi\}$ is ω-regular as well. Thus, both in the case of an $\mathrm{L}_\mu[\mathrm{MSO}]$ formula evaluated on the whole sequence, and in the case of an MSO formula evaluated on the limit structure, the set of winning sequences of rules is ω-regular, and the automaton recognising it can be effectively constructed.

Therefore, for any separated structure rewriting game $\mathcal{G}$, we get an equivalent ω-regular winning condition over the same game arena. Since ω-regular games are determined, establishing the winner in such games is decidable and finite-memory strategies are sufficient to win [7], Theorem 1 follows. Note that any result on ω-regular games can be transferred to separated structure games in the same way: for example, players could be allowed to take moves concurrently or one could consider multi-player games and ask for admissible strategies [8].

Let us remark[2] that defining $\mathfrak{A}[\mathfrak{L} \to \mathfrak{R}]$ as the structure with *all* occurrences of $\mathfrak{L}$ rewritten to $\mathfrak{R}$ is crucial for Theorem 2 and its consequences. Note that this is in *contrast* to the case of graph grammars [5], where any rule can be applied at any position. If we allowed the players to pick both a position to rewrite and a rewriting rule, it would be possible to simulate active context-free games, which were proven undecidable in [9]. To simulate these games, one would represent a word as a directed line with unary non-terminal predicates representing letters, e.g. the word *aba* would be represented as $(a) \rightarrow (b) \rightarrow (a)$.

4 Proving Regularity: From Structures to Words

In this section, we prove Theorem 2 in a few steps. First, we reduce structure rewriting to tree rewriting in a way reminding of the tight connection between separated handle rewriting of graphs and the vertex replacement algebra [5]. In addition to the standard vertex replacement methods, we also preserve the exact sequence of rewriting steps. Next, we are concerned with checking an MSO property on a tree constructed by a sequence of applications of simple tree rewriting rules. To do this, we take a tree automaton that checks this property and construct an alternating word automaton running on the sequence of tree rewriting rules that simulates the tree automaton.

4.1 From Structures to Trees

We represent the structures that the players manipulate by binary trees with labelled nodes. The leaves of the tree represent the elements of the structure,

[2] Thanks to Anca Muscholl for pointing out this remark and reduction.

and the labels describe which tuples of elements are in which relations. Note that this is a standard representation for graphs of bounded clique-width.

For our purposes, a labelled binary tree $\mathcal{T} = (T, \preceq, \lambda)$ consists of a prefix-closed set $T \subseteq \{0,1\}^*$, the prefix relation $\preceq$ and the labelling function $\lambda : T \to \Sigma_S$. The set of labels Σ_S depends on a number k_S that we will later compute from the considered set of separated rewrite rules S, and contains the following types of labels (with an intuition on how they will be used later).

- The symbols 'n' for all $n \leq k_S$ (used to label leaves of $\mathcal{T}$).
- The symbol '$\oplus$' (denoting the disjoint sum).
- The symbols '$i \leftarrow j$' for all $i, j \leq k_S$ (used to re-label i to j).
- The symbols '$R_k(i_1, \ldots, i_{r_k})$' for each $R_k \in \tau$ and each number $i_j \leq k_S$ (for adding to the relation R_k *all* tuples $(a_1, \ldots, a_{r_k})$ if a_j is labelled by i_j).

We consider only labellings that obey a few simple structural properties.

(1) The label $\lambda(v)$ is a number if and only if v is a leaf of $\mathcal{T}$.
(2) A node $v \in T$ has both successors $v0, v1 \in T$ if and only if $\lambda(v) = \oplus$.

Moreover, for separated structures, one additional property holds.

(3) For each $R_k \in \tau_{\rm n}$ the subtree below every node v with $\lambda(v) = R_k(i_1, \ldots, i_{r_k})$ has exactly $|\{i_1, \ldots, i_{r_k}\}|$ leaves and all its other nodes are labelled by $\oplus$.

With each tree $\mathcal{T} = (T, \preceq, \lambda)$ that fulfils the properties (1) and (2) we now associate a structure $\mathfrak{A} = S(\mathcal{T})$, and if (3) is fulfilled, then $\mathfrak{A}$ is separated. As said before, the universe of $\mathfrak{A}$ consists of the leaves of $\mathcal{T}$. For each $R_k \in \tau$, a tuple $(v_1, \ldots, v_{r_k})$ belongs to $R_k^{\mathfrak{A}}$ iff there exists a node $v \in T$ such that:

- $v \preceq v_j$ for all $j \in \{1, \ldots, r_k\}$,
- $\lambda(v) = R_k(i_1, \ldots, i_{r_k})$ for some tuple $i_1, \ldots, i_{r_k} \leq k_S$, such that
- for $j \in \{1, \ldots, r_k\}$, each v_j is re-labelled to i_j on the path from v_j to v in $\mathcal{T}$.

To define the label l_n of a leaf w to which it is re-labelled on a path $w = w_0 \ldots w_n$, we start with $l_0 = \lambda(w)$ and set

$$l_{k+1} = \begin{cases} j & \text{if } \lambda(w_k) = j \leftarrow i \text{ and } l_k = i, \\ l_k & \text{otherwise.} \end{cases}$$

Note that the condition that no $R_k(\bar{i})$-labels appear beneath any $R_l(\bar{i})$-label for all $R_k, R_l \in \tau_{\rm n}$ guarantees, that no elements in $R_k^{\mathfrak{A}}$ will appear in any other non-terminal relation, so the structure $\mathfrak{A}$ is separated in such a case.

For finite structures, the converse of the above remark also holds. The following lemma is obtained by constructing the tree $\mathcal{T}$ bottom-up, starting with the separated non-terminal relations of $\mathfrak{A}$, as shown for an example structure in Figure 2 (where there is one terminal binary relation R_0 drawn as unlabelled edges, and one explicitly marked non-terminal binary relation R_1; some re-labelling nodes in the tree are not strictly necessary).

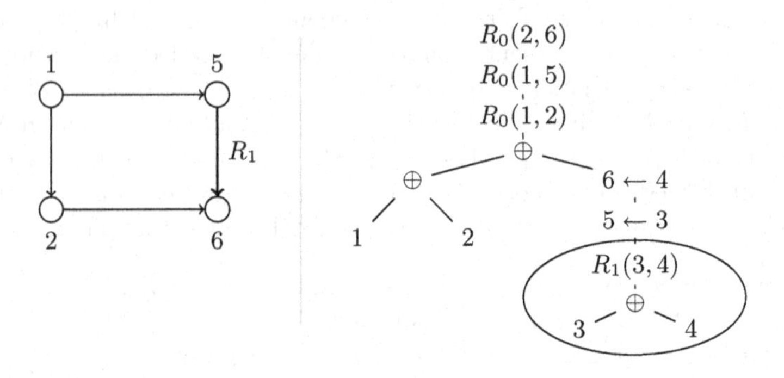

Fig. 2. Representing a separated structure, with marked subtree for the relation R_1

Lemma 1. *For every finite* separated *structure* $\mathfrak{A}$ *there exists a tree* $\mathcal{T}$ *such that* $\mathfrak{A} = S(\mathcal{T})$, *the properties* (1)–(3) *above are satisfied, and each element of the structure is re-labelled to a unique label at the root of* $\mathcal{T}$.

For a finite set of separated rewriting rules $S = \{\mathfrak{L}_1 \to \mathfrak{R}_1, \dots, \mathfrak{L}_m \to \mathfrak{R}_m\}$, let $\mathcal{T}_k$ be a tree that represents $\mathfrak{R}_k$ (without P_l), as constructed above. Let m_k be the maximal number that appears in the labels of $\mathcal{T}_k$ and set $k_S = \max_{k=1\dots m} m_k + 1$. A crucial observation is that replacing all $R_k(\bar{\imath})$-labelled subtrees in a representation of a separated structure by an extended tree $\mathcal{T}_k$ (see the added re-labellings in Figure 2) exactly corresponds to structure rewriting.

To extend $\mathcal{T}_k$, we construct, for a label $R_j(i_1, \dots, i_{r_j})$, the replacement tree $\mathcal{T}_k[R_j(i_1, \dots, i_{r_j})]$ as follows. Let $l(r)$ be the unique label that every element r of $\mathfrak{R}_k$ gets at the root of $\mathcal{T}_k$ (guaranteed by Lemma 1). We create a sequence of nodes that, for each $r \in \mathfrak{R}_k$, contains exactly one node with label '$n \leftarrow l(r)$'. The number n is equal to k_S if r is in no set $P_l^{\mathfrak{R}_k}$, and $n = i_m$ if $r \in P_l^{\mathfrak{R}_k}$ and the element l corresponds to the i_m-leaf in the representation of $R_j(i_1, \dots, i_{r_j})$.

The relationship between structure rewriting and tree rewriting is formalised in the following lemma which is a consequence of the definitions of structure rewriting and interpretation of a structure in a tree.

Lemma 2. *Let* $\mathfrak{A} = S(\mathcal{T})$ *be a separated structure represented by a tree* $\mathcal{T}$ *satisfying properties* (1)–(3) *and such that the maximal label number* k_S *does not appear in* $\bar{\imath}$ *in any label* $R_k(\bar{\imath})$ *in* $\mathcal{T}$. *Then, for each rule* $\mathfrak{L}_k \to \mathfrak{R}_k$ *from* S *with* $R_l \in \tau_{\mathrm{n}}$ *being the non-empty relation in* $\mathfrak{L}$, *the tree* T' *obtained from* $\mathcal{T}$ *by replacing each* $R_l(\bar{\imath})$ *subtree by* $\mathcal{T}_k[R_l(\bar{\imath})]$, *represents the structure* $\mathfrak{A}[\mathfrak{L}_k \to \mathfrak{R}_k]$.

It also follows from the construction, that if $\mathfrak{A}_0\mathfrak{A}_1 \dots$ is a sequence of rewritten structures and $\mathcal{T}_0\mathcal{T}_1 \dots$ the corresponding sequence of trees representing them, then $\lim \mathfrak{A}_n$ is represented by $\lim \mathcal{T}_n$.

Thus, to complete the transition from structure rewriting to tree rewriting, we only need to translate the MSO properties of structures to MSO properties of trees that represent them. This is done in an analogous way to interpreting bounded clique-width graphs in the tree. By the definition of $S(\mathcal{T})$ given above, elements of $S(\mathcal{T})$ are leaves of $\mathcal{T}$ and a tuple $\overline{v}$ belongs to $R_l^{S(\mathcal{T})}$ if an inductively defined condition is fulfilled. The property of being a leaf is easy to express in MSO, and the inductive definition for $\overline{v} \in R_l^{S(\mathcal{T})}$ can be expressed as well, because MSO is strong enough to allow fixed-point definitions and there are only finitely many labels in use. Thus, we can state the following lemma.

Lemma 3. *Fix a signature τ and k_S. For every* MSO *formula φ over τ there exists a (computable)* MSO *formula ψ over the signature $\{\preceq, P_x \mid x \in \Sigma_S\}$ of the Σ_S-labelled trees such that for each such tree $\mathcal{T}$, $S(\mathcal{T}) \models \varphi \iff \mathcal{T} \models \psi$.*

4.2 Simplifying Tree Rewriting

Above, we translated separated structure rewriting to rewriting trees, where only specific subtrees are replaced. This restriction is important as alternating reachability is undecidable on arbitrary ground tree rewriting systems [10].

Before we proceed to words, let us reduce the problem to a simplified version of tree rewriting: one where only leaves are rewritten. Previously, we have been rewriting a subtree of an $R_l(\overline{i})$-labelled node to some other tree. But property (3), fulfilled by all trees representing separated structures, guarantees that there are only finitely many isomorphic subtrees $\mathcal{S}$ rooted at $R_l(\overline{i})$-labelled nodes. Thus, we replace such nodes and the whole subtree by new leaves labelled by $R_l^{\mathcal{S}}(\overline{i})$, and from now on we operate on such reduced trees.

To rewrite trees, we use the same notation as for structure rewriting. Thus, if $\mathcal{T}$ is a tree, then $\mathcal{T}[c \to \mathcal{T}']$ denotes $\mathcal{T}$ with all c-labelled leaves replaced by $\mathcal{T}'$. Note that this is a special case of separated structure rewriting if leaves are labelled by non-terminal predicates. (To preserve the partial order on the tree, the whole tree $\mathcal{T}'$ must be included in the new predicate P_c.)

By the classical result of Rabin, for each MSO formula ψ over a labelled binary tree, there exists a non-deterministic tree automaton $\mathcal{A}_\psi$ that accepts a labelled tree $\mathcal{T}$ if and only if $\mathcal{T} \models \psi$.

Given a sequence of separated rewriting rules $\pi = r_0 r_1 \ldots$ that generates a sequence of structures $\mathrm{st}_k(\pi) = \mathfrak{A}_0\mathfrak{A}_1 \ldots$, we have shown above how to reduce the question whether $\mathfrak{A}_n \models \varphi$ (or $\lim \mathfrak{A}_n \models \varphi$) to the question whether $\mathcal{T}_n \models \psi$ (or $\lim \mathcal{T}_n \models \psi$), where $\mathcal{T}_n$ are the corresponding trees.

If $\mathcal{A}_\psi$ is the tree automaton corresponding to ψ, we construct an automaton $\mathcal{A}'_\psi$ that accepts the reduced tree (with all $R_k(\overline{i})$-labelled nodes with subtrees $\mathcal{S}$ replaced by $R_i^{\mathcal{S}}(\overline{i})$-labelled leaves) if and only if $\mathcal{A}_\psi$ accepts the original one. This is done by letting $\mathcal{A}'_\psi$, in an $R_i^{\mathcal{S}}(\overline{i})$-leaf, simulate any run of $\mathcal{A}_\psi$ on the subtree $\mathcal{S}$ (which is possible, as $\mathcal{S}$ has bounded size).

4.3 From Trees to Words

For the sequence of structures and rules considered above, let $\mathcal{T}_0$ be a tree such that $\mathfrak{A}_0 = S(\mathcal{T}_0)$. We replace each rule $r_i = \mathfrak{L}_k \to \mathfrak{R}_k$, where R_l is the only non-empty relation in $\mathfrak{L}_k$, by $s_i = R_l^{\mathcal{S}}(\bar{i}) \to \mathcal{T}_k[R_l(\bar{i})]^3$. Rewriting the tree $\mathcal{T}_0$ using the rules s_i generates a sequence of reduced trees $\mathcal{T}_0\mathcal{T}_1\dots$ and, as shown previously, $\mathfrak{A}_n \models \varphi$ (or $\lim \mathfrak{A}_n \models \varphi$) if and only if $\mathcal{A}'_\psi$ accepts $\mathcal{T}_n$ (or $\lim \mathcal{T}_n$).

We will show how to simulate the run of $\mathcal{A}'_\psi$ on the tree $\mathcal{T}_n$ (or $\lim \mathcal{T}_n$) by a run of an alternating word automaton $\mathcal{B}$ on the sequence $s_0s_1\dots s_n$ (or $s_0s_1\dots$) of rules. By the correspondence between s_i and r_i, the same automaton $\mathcal{B}$ (with swapped alphabet) accepts the corresponding sequences $r_0r_1\dots$ of separated structure rewriting rules as required in Theorem 2.

The construction of $\mathcal{B}$ from $\mathcal{A}'_\psi$ and the starting tree $\mathcal{T}_0$ proceeds in two steps. We first reduce the problem to tree rewriting rules where the right-hand side is either a constant or has height one, and the starting tree has one vertex. After this easy reduction, we construct the alternating automaton $\mathcal{B}$ in Lemma 4.

For the first step, observe that rewriting with a rule $s = c \to \mathcal{T}$ can be represented as a sequence of rewritings with rules having a smaller right-hand side, building the tree $\mathcal{T}$ step by step. For this, we need to add new labels corresponding to every proper subtree of $\mathcal{T}$. In this way, a single rule s is replaced by a sequence of rules $s'_1 \dots s'_m$ with simpler right-hand sides and using more labels, such that applying $s'_1 \dots s'_m$ in sequence gives the same result as applying s once. Since the number m of smaller rules needed to replace a given rule s is constant, this operation preserves regularity, i.e. for a regular set $\mathcal{L}$ of sequences of the simpler rules s', the set of sequences of full rules such that their expansion is in $\mathcal{L}$ is regular as well. Thus, it is enough to show that the set of sequences of simple rules resulting in a tree accepted by $\mathcal{A}'_\psi$ is regular.

Let R be a finite set of tree rewriting rules of the simple form $c \to c'$, $c \to g(c')$ or $c \to f(c_1, c_2)$. For such tree rewriting rules, we make the second step, in which an alternating word automaton is constructed that simulates the automaton running on the tree. The existence of such an automaton, stated in the following lemma, is another instance of the classical relationship between tree automata and games.

Lemma 4. *Let $\mathcal{A}$ be a non-deterministic parity tree automaton and s a label. There exists an alternating parity word automaton $\mathcal{B}$ over the alphabet R such that $\mathcal{B}$ accepts $s_0s_1\dots \in R^\omega$ (or $s_0\dots s_n \in R^*$) if and only if $\mathcal{A}$ accepts the limit tree* $\lim \mathcal{T}_i$ *(or $\mathcal{T}_n$), where $\mathcal{T}_0$ consists of one node labelled a_0 and $\mathcal{T}_{i+1} = \mathcal{T}_i[s_i]$.*

Since alternating parity word automata accept exactly the ω-regular languages, the above lemma completes the proof of Theorem 2.

5 Consequences

In this section we state a few consequences of Theorem 2 that illustrate the usefulness of structure rewriting games. To start with, let us remark that many

[3] Note that a priori s_i is not a single rule because there can be different trees $\mathcal{S}$ and sequences $\bar{i}$. Thus, formally, we should replace r_i by a sequence of all possible s_i-rules, with added checks that not too much is rewritten. But the differences in $\mathcal{S}$ and $\bar{i}$ are in fact irrelevant: one could as well pick any single option and use it everywhere consistently. Therefore we take the liberty and consider s_i as a single rule.

operations on structures of bounded size can be represented by separated structure rewriting in a more natural way than by listing all possible states.

To move away from finite-state systems, let us show how decidability of MSO over pushdown graphs is a direct consequence of Theorem 1. It follows from the fact that every pushdown graph can be constructed as limit graph in a simple game with two positions and two kinds of moves, one with rules that construct the configuration of the pushdown system when the stack is empty and another one used to construct the next configurations with more symbols on the stack.

The intimate connection of ω-regularity and MSO, together with Theorem 2, allows us to make direct use of the above construction to generalise Theorem 1 to games played on pushdown arenas. Indeed, ω-regular winning conditions can be expressed in MSO, so if MSO is decidable on a class of graphs, so is establishing the winner in games with ω-regular winning conditions.

In addition to the synthesis problem we considered, one might ask whether an $L_\mu[\mathrm{MSO}]$ formula (of the full μ-calculus, allowing the $\Box$ operator as well) holds on the whole abstract reduction graph generated by a separated structure rewriting system. This verification problem can also be solved with our methods, using the standard translation between μ-calculus and parity games.

Let us finally explain why the assumption that the predicates $P_l^{\mathfrak{R}}$ are pairwise disjoint, made in the definition of separated rewrite rules, is not necessary. In the proofs, the only use of this disjointness was when a node in a tree representing an element $r \in \mathfrak{R}$ was re-labelled. The newly assigned label guaranteed that r will appear in the correct tuples in all relations. If r belonged to various different predicates $P_l^{\mathfrak{R}}$, it would be necessary to assign to it a *set* of labels at the same time, instead of a single one. Technically, this is a change as the trees representing structures would have to be labelled by sets of numbers, and the MSO formulas interpreting the structure in such tree would have to account for that. Substantially, it is exactly analogous to the case we presented. Similarly, one could extend rewriting rules to include special predicates $P_l^{R_k,i}$, which would add the marked elements only to the relation R_k and only at ith position.

6 Perspectives

We proved that in the special case of separated rewriting rules, the synthesis problem remains decidable even if the expressive logic $L_\mu[\mathrm{MSO}]$ is used to specify the winning condition. It is natural to ask about other, less restricted classes of structure rewrite rules and logics, for which this problem is decidable.

One interesting logic to consider is the extension of first-order logic by simple reachability. It was shown in [11] that this logic is decidable on a class of graphs that can be represented by trees similar to the ones considered in this paper, but with an additional node label for asynchronous product. We ask whether there is a syntactic class of structure rewrite rules that corresponds to such graphs.

Another example is the class of structure rewrite rules $\mathfrak{L} \to \mathfrak{R}$ where both in $\mathfrak{L}$ and in $\mathfrak{R}$ the only non-empty relations are unary (but now $\mathfrak{L}$ can contain more than one element, so the rule is not necessarily separated). If the starting structure contains only unary relations as well, then only unary relations appear in all rewritten structures. In this case, it is enough to count the number of elements in each combination of the unary predicates, and thus the occurring structures are just another representation of Petri nets, and rewriting represents changes of the marking of the net. Thus, for this special case, it is known precisely which problems are and which are not decidable. But if one allows only unary predicates on the left-hand side and any separated structure on the right-hand side, then the question which problems remain decidable is open.

In addition to other classes of rules and logics, it is interesting to ask how restricted structure rewriting rules can be used to approximate systems where more complex rewriting takes place. One example of this kind is the use of hyperedge replacement grammars for abstraction of data in pointer-manipulating programs [12]. Graphs obtained by hyperedge replacement are a subclass of structures generated by separated rewriting which we considered. This justifies our view of the presented results as a first step towards algorithmic synthesis for general structure rewriting systems.

References

1. Nagl, M.: A tutorial and bibliographical survey on graph grammars. In: Ng, E.W., Ehrig, H., Rozenberg, G. (eds.) Graph Grammars 1978. LNCS, vol. 73, pp. 70–126. Springer, Heidelberg (1979)
2. Rajlich, V.: Relational structures and dynamics of certain discrete systems. In: Proc. of MFCS 1973, High Tatras, September 3-8, pp. 285–292 (1973)
3. Joshi, S., König, B.: Applying the graph minor theorem to the verification of graph transformation systems. In: Gupta, A., Malik, S. (eds.) CAV 2008. LNCS, vol. 5123, pp. 214–226. Springer, Heidelberg (2008)
4. Møller, A., Schwartzbach, M.I.: The pointer assertion logic engine. In: Proc. ACM Conf. on Programming Language Design and Implementation (2001)
5. Courcelle, B., Engelfriet, J., Rozenberg, G.: Context-free handle-rewriting hypergraph grammars. In: Ehrig, H., Kreowski, H.-J., Rozenberg, G. (eds.) Graph Grammars 1990. LNCS, vol. 532, pp. 253–268. Springer, Heidelberg (1991)
6. Grädel, E., Thomas, W., Wilke, T. (eds.): Automata, Logics, and Infinite Games. LNCS, vol. 2500. Springer, Heidelberg (2002)
7. McNaughton, R.: Infinite games played on finite graphs. Annals of Pure and Applies Logic 65(2), 149–184 (1993)
8. Berwanger, D.: Admissibility in infinite games. In: Thomas, W., Weil, P. (eds.) STACS 2007. LNCS, vol. 4393, pp. 188–199. Springer, Heidelberg (2007)
9. Muscholl, A., Schwentick, T., Segoufin, L.: Active context-free games. In: Diekert, V., Habib, M. (eds.) STACS 2004. LNCS, vol. 2996, pp. 452–464. Springer, Heidelberg (2004)
10. Löding, C.: Infinite Graphs Generated by Tree Rewriting. PhD thesis (2003)
11. Colcombet, T.: On families of graphs having a decidable first order theory with reachability. In: Widmayer, P., Triguero, F., Morales, R., Hennessy, M., Eidenbenz, S., Conejo, R. (eds.) ICALP 2002. LNCS, vol. 2380, pp. 98–109. Springer, Heidelberg (2002)
12. Rieger, S., Noll, T.: Abstracting complex data structures by hyperedge replacement. In: Ehrig, H., Heckel, R., Rozenberg, G., Taentzer, G. (eds.) ICGT 2008. LNCS, vol. 5214, pp. 69–83. Springer, Heidelberg (2008)

On the Hybrid Extension of CTL and CTL^+

Ahmet Kara[1], Volker Weber[1], Martin Lange[2], and Thomas Schwentick[1]

[1] Technische Universität Dortmund
[2] Ludwig-Maximilians-Universität München

Abstract. The paper studies the expressivity, relative succinctness and complexity of satisfiability for hybrid extensions of the branching-time logics CTL and CTL^+ by variables. Previous complexity results show that only fragments with *one variable* do have elementary complexity. It is shown that H^1CTL^+ and H^1CTL, the hybrid extensions with one variable of CTL^+ and CTL, respectively, are expressively equivalent but H^1CTL^+ is exponentially more succinct than H^1CTL. On the other hand, $HCTL^+$, the hybrid extension of CTL with arbitrarily many variables does not capture CTL^*, as it even cannot express the simple CTL^* property $EGFp$. The satisfiability problem for H^1CTL^+ is complete for triply exponential time, this remains true for quite weak fragments and quite strong extensions of the logic.

1 Introduction

Reasoning about trees is at the heart of many fields in computer science . A wealth of sometimes quite different frameworks has been proposed for this purpose, according to the needs of the respective application. For reasoning about computation trees as they occur in verification, branching-time logics like CTL and tree automata are two such frameworks. In some settings, the ability to mark a node in a tree and to refer to this node turned out to be useful. As neither classical branching-time logics nor tree automata provide this feature, many different variations have been considered, including tree automata with pebbles [8,22,25], memoryful CTL^* [15], branching-time logics with forgettable past [17,18], and logics with the "freeze" operator [12]. It is an obvious question how this feature can be incorporated into branching-time logics *without losing their desirable properties* which made them prevailing in verification [23].

This question leads into the field of hybrid logics, where such extensions of temporal logics are studied [3]. In particular, a hybrid extension of CTL has been introduced in [25]. As usual for branching-time logics, formulas of their hybrid extensions are evaluated at nodes of a computation tree, but it is possible to bind a variable to the current node, to evaluate formulas relative to the root and to check whether the current node is bound to a variable. As an example, the HCTL-formula $\downarrow x @_{\text{root}} \text{EF}(p \wedge \text{EF}x)$ intuitively says "I can place x at the current node, jump back to the root, go to a node where p holds and follow some (downward) path to reach x again. Or, equivalently: "there was a node fulfilling p in the past of the current node".

R. Královič and D. Niwiński (Eds.): MFCS 2009, LNCS 5734, pp. 427–438, 2009.

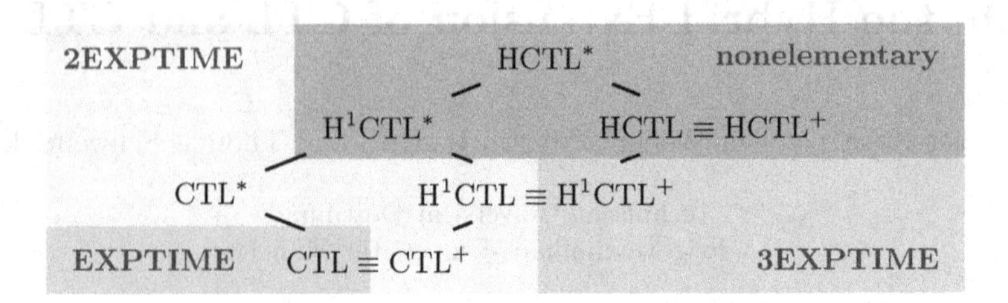

Fig. 1. Expressivity and complexity of satisfiability for hybrid branching-time logics. The lines indicate strict inclusion, unrelated logics are incomparable.

In this paper we continue the investigation of hybrid extensions of classical branching-time logics started in [25]. The main questions considered are (1) expressivity, (2) complexity of the satisfiability problem, and (3) succinctness. Figure 1 shows our results in their context.

Classical branching-time logics are CTL (with polynomial time model checking and exponential time satisfiability) and CTL* (with polynomial space model checking and doubly exponential time satisfiability test). As CTL is sometimes not expressive enough[1] and CTL* is considered too expensive for some applications, there has been an intense investigation of intermediate logics. We take up two of them here: CTL$^+$, where a path formula is a Boolean combination of basic path formulas[2] and ECTL, where fairness properties can be stated explicitly.

Whereas (even simpler) hybrid logics are undecidable over arbitrary transition systems [1], their restriction to trees is decidable via a simple translation to Monadic Second Order logic. However, the complexity of the satisfiability problem is high even for simple hybrid temporal logics over the frame of natural numbers: nonelementary [9] , even if only two variables are allowed [21,25]. The one variable extension of CTL, H^{1}CTL, behaves considerably better, its satisfiability problem can be solved in **2EXPTIME** [25]. This is the reason why this paper concentrates on natural extensions of this complexity-wise relatively modest logic. Even H^{1}CTL can express properties that are not bisimulation-invariant (e.g., that a certain configuration can be reached along two distinct computation paths) and is thus not captured by CTL*. In fact, [25] shows that H^{1}CTL captures and is strictly stronger than CTL with past, another extension of CTL studied in previous work [14]. One of our main results is that H^{1}CTL (and actually even HCTL$^+$) does not capture ECTL (and therefore not CTL*) as it cannot express simple fairness properties like EGFp. To this end, we introduce a simple Ehrenfeucht-style game (in the spirit of [2]). We show that existence of a winning strategy for the second player in the game for a property P implies that P cannot be expressed in HCTL$^+$.

In [25] it is also shown that the satisfiability problem for H^{1}CTL* has nonelementary complexity. We show here that the huge complexity gap between

[1] Some things cannot be expressed at all, some only in a very verbose way.

[2] Precise definitions can be found in Section 2.

H^1CTL and H^1CTL^* does not yet occur between H^1CTL and H^1CTL^+: we prove that there is only an exponential complexity gap between H^1CTL and H^1CTL^+, even when H^1CTL^+ is extended by past modalities and fairness operators. We pinpoint the exact complexity by proving the problem complete for **3EXPTIME**.

The exponential gap between the complexities for satisfiability of H^1CTL and H^1CTL^+ already suggests that H^1CTL^+ might be exponentially more succinct than H^1CTL. In fact, we show an exponential succinctness gap between the two logics by a proof based on the height of finite models. It should be noted that an $\mathcal{O}(n)!$-succinctness gap between CTL and H^1CTL was established in [25]. We mention that there are other papers on hybrid logics and hybrid tree logics that do not study expressiveness or complexity issue, e.g., [10,20].

The paper is organized as follows. Definitions of the logics we use are in Section 2. Expressivity results are presented in Section 3. The complexity results can be found in Section 4, the succinctness results in Section 5. Proofs omitted due to space constraints can be found in the full version of this paper [13].

Note. We mourn the loss of Volker Weber, who died suddenly and unexpectedly on the 7th of April 2009. He was 30 years old. Volker contributed a lot to the present paper which we prepared and submitted after his death.

2 Definitions

Tree logics. We first review the definition of CTL and CTL* [5]. Formulas of CTL* are composed from *state formulas* φ and *path formulas* ψ. They have the following abstract syntax.

$$\varphi ::= p \mid \neg\varphi \mid \varphi \vee \varphi \mid \varphi \wedge \varphi \mid \mathrm{E}\psi \mid \mathrm{A}\psi$$
$$\psi ::= \varphi \mid \neg\psi \mid \psi \vee \psi \mid \psi \wedge \psi \mid \mathrm{X}\psi \mid \psi\mathrm{U}\psi$$

We use the customary abbreviations $\mathrm{F}\psi$ for $\top\mathrm{U}\psi$ and $\mathrm{G}\psi$ for $\neg\mathrm{F}\neg\psi$. The semantics of formulas is defined inductively. The semantics of path formulas is defined relative to a tree[3] $\mathcal{T}$, a path π of $\mathcal{T}$ and a position $i \geq 0$ of this path. E.g., $\mathcal{T},\pi,i \models \psi_1\mathrm{U}\psi_2$ if there is some $j \geq i$ such that $\mathcal{T},\pi,j \models \psi_2$ and, for each $l, i \leq l < j$, $\mathcal{T},\pi,l \models \psi_1$. The semantics of state formulas is defined relative to a tree $\mathcal{T}$ and a node v of $\mathcal{T}$. E.g., $\mathcal{T},v \models \mathrm{E}\psi$ if there is a path π in $\mathcal{T}$, starting from v such that $\mathcal{T},\pi,0 \models \psi$. A state formula φ holds in a tree $\mathcal{T}$ if it holds in its root. Thus, sets of trees can be defined by CTL* state formulas.

CTL is a strict sub-logic of CTL*. It allows only path formulas of the forms $\mathrm{X}\varphi$ and $\varphi_1\mathrm{U}\varphi_2$ where $\varphi,\varphi_1,\varphi_2$ are state formulas. CTL$^+$ is the sub-logic of CTL* where path formulas are Boolean combinations of formulas of the forms $\mathrm{X}\varphi$ and $\varphi_1\mathrm{U}\varphi_2$ and $\varphi,\varphi_1,\varphi_2$ are state formulas.

[3] In general, we consider finite and infinite trees and, correspondingly, finite and infinite paths in trees. It should always be clear from the context whether we restrict attention to finite or infinite trees.

Hybrid logics. In hybrid logics, a limited use of variables is allowed. For a general introduction to hybrid logics we refer to [3]. As mentioned in the introduction, we concentrate in this paper on hybrid logic formulas with *one* variable x. However, as we also discuss logics with more variables, we define hybrid logics $\mathrm{H^kCTL^*}$ with k variables. For each $k \geq 1$, the syntax of $\mathrm{H^kCTL^*}$ is defined by extending $\mathrm{CTL^*}$ with the following rules for state formulas.

$$\varphi ::= \downarrow x_i \, \varphi \mid x_i \mid @_{x_i} \varphi \mid \mathrm{root} \mid @_{\mathrm{root}} \varphi$$

where $i \in \{1, \ldots, k\}$. The semantics is now relative to a vector $\boldsymbol{u} = (u_1, \ldots, u_k)$ of nodes of $\mathcal{T}$ representing an assignment $x_i \mapsto u_i$. For a node v and $i \leq k$ we write $\boldsymbol{u}[i/v]$ to denote $(u_1, \ldots, u_{i-1}, v, u_{i+1}, \ldots, u_k)$. For a tree $\mathcal{T}$ a node v and a vector $\boldsymbol{u}$, the semantics of the new state formulas is defined as follows.

$$\begin{array}{ll}
\mathcal{T}, v, \boldsymbol{u} \models \downarrow x_i \, \varphi & \text{if } \ \mathcal{T}, v, \boldsymbol{u}[i/v] \models \varphi \\
\mathcal{T}, v, \boldsymbol{u} \models x_i & \text{if } \ v = u_i \\
\mathcal{T}, v, \boldsymbol{u} \models @_{x_i} \varphi & \text{if } \ \mathcal{T}, u_i, \boldsymbol{u} \models \varphi \\
\mathcal{T}, v, \boldsymbol{u} \models \mathrm{root} & \text{if } \ v \text{ is the root of } \mathcal{T} \\
\mathcal{T}, v, \boldsymbol{u} \models @_{\mathrm{root}} \varphi & \text{if } \ \mathcal{T}, r, \boldsymbol{u} \models \varphi, \text{ where } r \text{ is the root of } \mathcal{T}
\end{array}$$

Similarly, the semantics of path formulas is defined relative to a tree $\mathcal{T}$, a path π of $\mathcal{T}$, a position $i \geq 0$ of π and a vector $\boldsymbol{u}$. Intuitively, to evaluate a formula $\downarrow x_i \, \varphi$ one puts a pebble x_i on the current node v and evaluates φ. During the evaluation, x_i refers to v (unless it is bound again by another $\downarrow x_i$-quantifier).

The hybrid logics $\mathrm{H^kCTL^+}$ and $\mathrm{H^kCTL}$ are obtained by restricting $\mathrm{H^kCTL^*}$ in the same fashion as for $\mathrm{CTL^+}$ and CTL, respectively. The logic HCTL is the union of all logics $\mathrm{H^kCTL}$, likewise $\mathrm{HCTL^+}$ and $\mathrm{HCTL^*}$.

(Finite) satisfiability of formulas, the notion of a model and equivalence of two (path and state) formulas ψ and ψ' (denoted $\psi \equiv \psi'$) are defined in the obvious way. We say that a logic $\mathcal{L}'$ is at least as expressive as $\mathcal{L}$ (denoted as $\mathcal{L} \leq \mathcal{L}'$) if for every $\varphi \in \mathcal{L}$ there is a $\varphi' \in \mathcal{L}'$ such that $\varphi \equiv \varphi'$. $\mathcal{L}$ and $\mathcal{L}'$ have the *same expressive power* if $\mathcal{L} \leq \mathcal{L}'$ and $\mathcal{L}' \leq \mathcal{L}$. $\mathcal{L}'$ is *strict more expressive* than $\mathcal{L}$ if $\mathcal{L} \leq \mathcal{L}'$ but not $\mathcal{L}' \leq \mathcal{L}$.

Size, depth and succinctness. For each formula φ, we define its *size* $|\varphi|$ as usual and its *depth* $d(\varphi)$ as the nesting depth with respect to path quantifiers.

The formal notion of *succinctness* is a bit delicate. We follow the approach of [11] and refer to the discussion there. We say that a logic $\mathcal{L}$ is *h-succinct in* a logic $\mathcal{L}'$, for a function $h : \mathbb{N} \to \mathbb{R}$, if for every formula φ in $\mathcal{L}$ there is an equivalent formula φ' in $\mathcal{L}'$ such that $|\varphi'| \leq h(|\varphi|)$. $\mathcal{L}$ is *$\mathcal{F}$-succinct in $\mathcal{L}'$* if $\mathcal{L}$ is h-succinct in $\mathcal{L}'$, for some h in function class $\mathcal{F}$. We say that $\mathcal{L}$ is *exponentially more succinct* than $\mathcal{L}'$ if $\mathcal{L}$ is *not* h-succinct in $\mathcal{L}'$, for any function $h \in 2^{o(n)}$.

Normal forms. We say that a $\mathrm{H^kCTL}$ formula is in E-*normal form*, if it does not use the path quantifier A at all. A formula is in U-*normal form* if it only uses the combinations EX, EU and AU (but not, e.g., EG and AX).

Proposition 1. *Let $k \geq 1$. For each* $\mathrm{H^kCTL}$ *formula φ there is an equivalent* $\mathrm{H^kCTL}$*-formula of linear size in* U*-normal form and an equivalent* $\mathrm{H^kCTL}$*-formula in* E*-normal form.*

3 Expressivity of HCTL and HCTL$^+$

3.1 The Expressive Power of HCTL$^+$ Compared to HCTL

Syntactically CTL$^+$ extends CTL by allowing Boolean combinations of path formulas in the scope of a path quantifier A or E. Semantically this gives CTL$^+$ the ability to fix a path and test its properties by *several* path formulas. However in [6] it is shown that every CTL$^+$-formula can be translated to an equivalent CTL-formula. The techniques used there are applicable to the hybrid versions of these logics.

Theorem 2. *For every* $k \geq 1$, H^{k}CTL *has the same expressive power as* H^{k}CTL$^+$.

Proof (Sketch). For a given $k \geq 1$ it is clear that every H^{k}CTL-formula is also a H^{k}CTL$^+$-formula. It remains to show that every H^{k}CTL$^+$-formula can be transformed into an equivalent H^{k}CTL-formula. In [6] , rules for the transformation of a CTL$^+$ formula into an equivalent CTL formula are given. Here, we have to consider the additional case in which a subformula in the scope of the $\downarrow x$-operator is transformed. However, it is not hard to see that the transformation extends to this case as any assignment to a variable x can be viewed as a proposition that only holds in one node. It should be noted that for a H^{k}CTL$^+$-formula φ the whole transformation constructs a H^{k}CTL-formula of size $2^{\mathcal{O}(|\varphi| \log |\varphi|)}$. □

The transformation algorithm in Theorem 2 also yields an upper bound for the succinctness between H^{1}CTL$^+$ and H^{1}CTL.

Corollary 3. H^{1}CTL$^+$ *is* $2^{\mathcal{O}(n \log n)}$*-succinct in* H^{1}CTL.

3.2 Fairness Is Not Expressible in HCTL$^+$

In this subsection, we show the following result.

Theorem 4. *There is no formula in* HCTL$^+$ *which is logically equivalent to* $\mathrm{E}\overset{\infty}{\mathrm{F}}p$.

Here, $\mathcal{T}, v, \boldsymbol{u} \models \mathrm{E}\overset{\infty}{\mathrm{F}}\varphi$ if there is a path π starting from v that has infinitely many nodes v' with $\mathcal{T}, v', \boldsymbol{u} \models \varphi$. As an immediate consequence of this theorem, HCTL$^+$ does not capture CTL*.

In order to prove Theorem 4, we define an Ehrenfeucht-style game that corresponds to the expressive power of HCTL. A game for a different hybrid logic was studied in [2]. We show that if a set L of trees can be characterized by a HCTL-formula, the spoiler has a winning strategy in the game for L. We expect the converse to be true as well but do not attempt to prove it as it is not needed for our purposes here.

Let L be a set of (finite or infinite) trees. The HCTL-*game* for L is played by two players, the *spoiler* and the *duplicator*. First, the spoiler picks a number k which will be the number of rounds in the core game. Afterwards, the duplicator chooses two trees, $\mathcal{T} \in L$ and $\mathcal{T}' \notin L$. The goal of the spoiler is to make use of the difference between $\mathcal{T}$ and $\mathcal{T}'$ in the core game.

The *core game* consists of k rounds of moves, where in each round i a node from $\mathcal{T}$ and a node from $\mathcal{T}'$ are selected according to the following rules. The spoiler can choose whether she starts her move in $\mathcal{T}$ or in $\mathcal{T}'$ and whether she plays a node move or a path move.

In a *node move* she simply picks a node from $\mathcal{T}$ (or $\mathcal{T}'$) and the duplicator picks a node in the other tree. We refer to these two nodes by a_i (in $\mathcal{T}$) and a_i' (in $\mathcal{T}'$), respectively, where i is the number of the round.

In a *path move*, the spoiler first chooses one of the trees. Let us assume she chooses $\mathcal{T}$, the case of $\mathcal{T}'$ is completely analogous. She picks an already selected node a_j of $\mathcal{T}$, for some $j < i$ and a path π starting in a_j. However, a node a_j can only be selected if there is no other node a_l, $l < i$ below a_j. The duplicator answers by selecting a path π' from a_j'. Then, the spoiler selects some node a_i' from π' and the duplicator selects a node a_i from π.

The duplicator wins the game if at the end the following conditions hold, for every $i, j \leq k$:

- a_i is the root iff a_i' is the root;
- $a_i = a_j$ iff $a_i' = a_j'$;
- for every proposition p, p holds in a_i iff it holds in a_i';
- there is a (downward) path from a_i to a_j iff there is a path from a_i' to a_j';
- a_j is a child of a_i iff a_j' is a child of a_i'.

Theorem 5. *If a set L of (finite and infinite) trees can be characterized by a* HCTL*-formula, the spoiler has a winning strategy on the* HCTL*-game for L.*

The proof of Thm. 5 is by induction on the structure of the HCTL-formula [13].

Now we turn to the proof of Thm. 4. It makes use of the following lemma which is easy to prove using standard techniques (see, e.g., [19]). The lemma will be used to show that the duplicator has certain move options on paths starting from the root. The parameter S_k given by the lemma will be used below for the construction of the structures $\mathcal{B}_k$.

For a string $s \in \Sigma^*$ and a symbol $a \in \Sigma$ let $|s|$ denote the length of s and $|s|_a$ the number of occurrences of a in s.

Lemma 6. *For each $k \geq 0$ there is a number $S_k \geq 0$ such that, for each $s \in \{0,1\}^*$ there is an $s' \in \{0,1\}^*$ such that $|s'| \leq S_k$ and $s \equiv_k s'$.*

Here, $\equiv_k$ is equivalence with respect to the k-round Ehrenfeucht game on strings (or equivalently with respect to first-order sentences of quantifier depth k). It should be noted that, if $k \geq 3$ and $s \equiv_k s'$, then the following conditions hold.

- $s \in \{0\}^*$ implies $s' \in \{0\}^*$.
- If the first symbol of s is 1 the same holds for s'.
- If s does not have consecutive 1's, s' does not either.

We fix some S_k, for each k.

The proof of Thm. 4 uses the HCTL-game defined above. Remember that the spoiler opens the game with the choice of a $k \in \mathbb{N}$ and the duplicator responds with two trees $\mathcal{T} \in L$ and $\mathcal{T}' \notin L$. We want to show that the duplicator has a winning strategy so we need to construct such trees, and then need to show that the duplicator has a winning strategy for the k-round core game on $\mathcal{T}$ and $\mathcal{T}'$.

We will use transition systems in order to finitely represent infinite trees. A transition system is a $\mathcal{K} = (V, E, v_0, \ell)$ where (V, E) is a directed graph, $v_0 \in V$, and ℓ labels each state $v \in V$ with a finite set of propositions. The *unraveling* $T(\mathcal{K})$ is a tree with node set V^+ and root v_0. A node $v_0 \dots v_{n-1}v_n$ is a child of $v_0 \dots v_{n-1}$ iff $(v_{n-1}, v_n) \in E$. Finally, the label of a node $v_0 \dots v_n$ is $\ell(v_n)$.

Inspired by [7] we define transition systems $\mathcal{A}_i$, for each $i \geq 0$, as depicted in Fig. 2 (a). Nodes in which p holds are depicted black, the others are white (and we subsequently refer to them as black and white nodes, respectively).

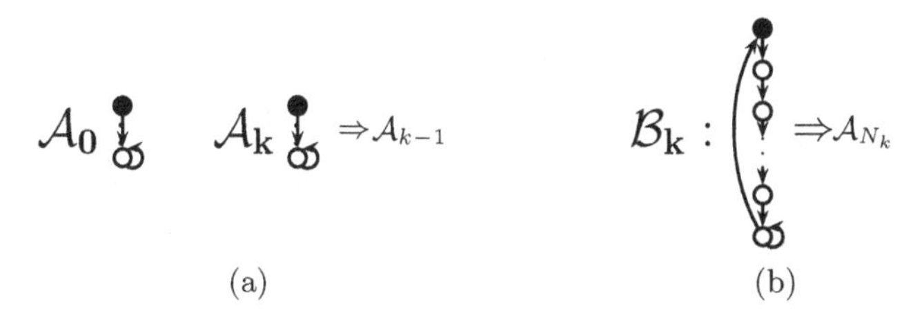

Fig. 2. Illustration of the definition of (a) $\mathcal{A}_k$ and (b) $\mathcal{B}_k$. The path of white nodes in $\mathcal{B}_k$ consists of S_k nodes. The double arrow $\Rightarrow$ indicates that every white node on the left is connected to every black node on the right.

Thus, $\mathcal{A}_0$ has a black (root) node and a white node with a cycle. $\mathcal{A}_i$ has a black (root) node, a white node with a cycle and a copy of $\mathcal{A}_{i-1}$. Furthermore, there is an edge from the white node below the root of $\mathcal{A}_i$ to each black node in the copy of $\mathcal{A}_{i-1}$ (as indicated by $\Rightarrow$). Let $\mathcal{T}_i := T(\mathcal{A}_i)$. We first introduce some notation and state some simple observations concerning the tree $\mathcal{T}_i$.

(1) For a node v in $\mathcal{T}_i$ we denote the maximum number of black nodes on a path starting in v (and not counting v itself) the *height* $h(v)$ of v. Then the root of $\mathcal{T}_i$ has height i.
(2) If u and v are black nodes of some $\mathcal{T}_i$ with $h(u) = h(v)$ then the subtrees $T(u)$ and $T(v)$ induced by u and v are isomorphic.
(3) The height of a tree is defined as the height of its root.
(4) A white node v of height i has one white child (of height i) and i black children of heights $0, \dots, i-1$. A black node has exactly one white son.
(5) Each finite path π of $\mathcal{T}_i$ induces a string $s(\pi) \in \{0,1\}^*$ in a natural way: $s(\pi)$ has one position, for each node of π, carrying a 1 iff the corresponding node is black.
(6) The root of $\mathcal{T}_i$ has only one child. We call the subtree induced by this (white!) child $\mathcal{U}_i$. If v is a white node of height i then $T(v)$ is isomorphic to $\mathcal{U}_i$.

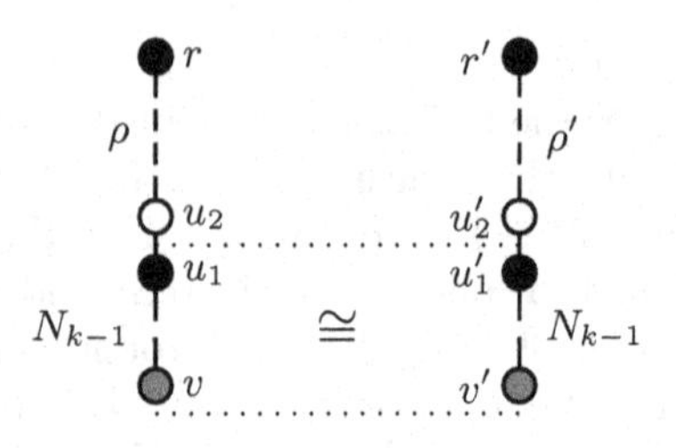

Fig. 3. Illustration of the case where $h(v) \leq N_{k-1}$. The colors of v and v' are not known a priori.

Next we define numbers N_k inductively as follows: $N_0 := 0$ and $N_k := N_{k-1} + \max(S_3, S_k) + 1$.

The following lemma shows that the duplicator has a winning strategy in two structures of the same kind, provided they both have sufficient depth.

Lemma 7. *Let i, j, k be numbers such that $i, j \geq N_k$. Then the duplicator has a winning strategy in the k-round core game on (a) $\mathcal{T}_i$ and $\mathcal{T}_j$, and (b) $\mathcal{U}_i$ and $\mathcal{U}_j$.*

Proof (Sketch). In both cases, the proof is by induction on k, the case $k = 0$ being trivial. We consider (a) first. Let $k > 0$ and let us assume that the spoiler chooses $v \in \mathcal{T}_i$ in her first node move. We distinguish two cases based on the height of v.

$h(v) > N_{k-1}$: Let π denote the path from r to v. By Lemma 6 there is a string s' with $|s'| \leq S_l$ such that $s(\pi) \equiv_l s'$, where $l = \max(k, 3)$. Here, $l \geq 3$ guarantees in particular that s' does not have consecutive 1's. As $j \geq N_k = N_{k-1} + S_l + 1$, there is a node v' of height $\geq N_{k-1}$ in $\mathcal{T}_j$ such that the path π' from r' to v' satifies $s(\pi') = s'$. The duplicator chooses v' as her answer in this round. By a compositional argument, involving the induction hypothesis, it can be shown that the duplicator has a winning strategy for the remaining $k - 1$ rounds.

$h(v) \leq N_{k-1}$: Let π be the path from r to v, and u_1 be the highest black node on π with $h(u_1) \leq N_{k-1}$. Then we must have $h(u_1) = N_{k-1}$ because π contains black nodes of height up to $i \geq N_k$. Hence, u_1 has a white parent u_2 s.t. $h(u_2) > N_{k-1}$. We determine a node u_2' in T' in the same way we picked v' for v in the first case. In particular, $h(u_2') \geq N_{k-1}$ and for the paths ρ leading from r to u_2 and ρ' leading from r' to u_2' we have $s(\rho) \equiv_k s(\rho')$.

Let u_1' be the black child of u_2' of height $h(u_1)$. As $h(u_1) = h(u_1')$ there is an isomorphism σ between $T(u_1)$ and $T(u_2)$ and we choose $v' := \sigma(v)$. An illustration is given in Figure 3.

The winning strategy of the duplicator for the remaining $k - 1$ rounds follows σ on $T(u_1)$ and $T(u_2)$ and is analogous to the first case in the rest of the trees. The case of path moves is very similar, see [13].

□

We are now prepared to prove Thm. 4.

Proof (of Thm. 4). By Thm. 2 it is sufficient to show that no formula equivalent to $\mathrm{E}\overset{\infty}{\mathrm{F}}p$ exists in HCTL. To this end, we prove that the duplicator has a winning strategy in the HCTL-game for the set of trees fulfilling $\mathrm{E}\overset{\infty}{\mathrm{F}}p$.

We define transition systems $\mathcal{B}_k$, for $k \geq 0$. As illustrated in Figure 2 (b), $\mathcal{B}_k$ has a black root from which a path of length S_k of white nodes starts. The last of these white nodes has a self-loop and an edge back to the root. Furthermore, $\mathcal{B}_k$ has a copy of $\mathcal{A}_{N_k}$ and there is an edge from each white node of the initial path to each black node of the copy of $\mathcal{A}_{N_k}$. Clearly, for each k, $T(\mathcal{B}_k) \models \mathrm{E}\overset{\infty}{\mathrm{F}}p$ and $T(\mathcal{A}_k) \not\models \mathrm{E}\overset{\infty}{\mathrm{F}}p$.

It can be shown that, for each k, the duplicator has a winning strategy in the k-round core game on $\mathcal{T} = T(\mathcal{B}_k)$ and $\mathcal{T}' = T(\mathcal{A}_{N_k})$ [13]. □

4 Satisfiability of H^{1}CTL$^+$

Theorem 8. *Satisfiability of* H^{1}CTL$^+$ *is hard for* **3EXPTIME**.

Proof. The proof is by reduction from a tiling game (with **3EXPTIME** complexity) to the satisfiability problem of H^{1}CTL$^+$. Actually we show that the lower bound even holds for the fragment of H^{1}CTL$^+$ without the U-operator (but with the F-operator instead).

An instance $I = (T, H, V, F, L, n)$ of the *2EXP-corridor tiling game* consists of a finite set T of *tile types*, two relations $H, V \subseteq T \times T$ which constitute the *horizontal and vertical constraints*, respectively, two sets $F, L \subseteq T$ which describe the starting and end conditions, respectively, and a number n given in unary. The game is played by two players, E and A, on a board consisting of 2^{2^n} columns and (potentially) infinitely many rows. Starting with player E and following the constraints H, V and F the players put tiles to the board consecutively from left to right and row by row. The constraints prescribe the following conditions:

- A tile t' can only be placed immediately to the right of a tile t if $(t, t') \in H$.
- A tile t' can only be placed immediately above a tile t if $(t, t') \in V$.
- The types of all tiles in the first row belong to the set F.

Player E wins the game if a row is completed containing only tiles from L or if A makes a move that violates the constraints. On the other hand, player A wins if E makes a forbidden move or the game goes on ad infinitum.

A winning strategy for E has to yield a countermove for all possible moves of A in all possible reachable situations. Furthermore, the starting condition and the horizontal and vertical constraints have to be respected. Finally, the winning strategy must guarantee that either player A comes into a situation where he can no longer make an allowed move or a row with tiles from L is completed.

The problem to decide for an instance I whether player E has a winning strategy on I is complete for **3EXPTIME**. This follows by a straightforward extension of [4].

□

We can obtain, by simple instantiation, a consequence of this lower complexity bound which will be useful later on in proving the exponential succinctness of $\mathrm{H}^1\mathrm{CTL}^+$ in $\mathrm{H}^1\mathrm{CTL}$.

Corollary 9. *There are finitely satisfiable* $\mathrm{H}^1\mathrm{CTL}^+$ *formulas* φ_n, $n \in \mathbb{N}$, *of size* $\mathcal{O}(n)$ *s.t. every tree model* $\mathcal{T}_n$ *of* φ_n *has height at least* $2^{2^{2^n}}$.

Proof. It is not difficult to construct instances I_n, $n \in \mathbb{N}$, of the 2EXP-tiling game with $|I| = \mathcal{O}(n)$ over a set T of tiles with $|T| = \mathcal{O}(1)$ such that player E has a winning strategy and any successful tiling of the 2^{2^n}-corridor requires $2^{2^{2^n}}$ rows. In order to achieve this, one encodes bits using tiles and forms the constraints in a way that enforces the first row to encode the number 0 in binary of length 2^{2^n}, and each other row to encode the successor in the natural number of the preceding row, while winning requires the number $2^{2^{2^n}}$ to be reached. The construction in the proof of Thm. 8 then maps each such I_n to a formula φ_n of size $\mathcal{O}(n)$ that is finitely satisfiable such that every finite model $\mathcal{T}_n$ of φ_n encodes a winning strategy for player E in the 2^{2^n}-tiling game. Such a strategy will yield a successful tiling of the 2^{2^n}-corridor for any counterstrategy of player A, and any such tiling is encoded on a path of $\mathcal{T}_n$ which contains each row of length 2^{2^n} as a segment of which there are $2^{2^{2^n}}$ many. Thus, $\mathcal{T}_n$ has to have height at least $2^{2^n} \cdot 2^{2^{2^n}}$. □

Using the ideas of the transformation mentioned in Theorem 2 we can show that the lower bound for $\mathrm{H}^1\mathrm{CTL}^+$ is optimal. Even for strictly more expressive logics than $\mathrm{H}^1\mathrm{CTL}^+$ the satisfiability problem remains in **3EXPTIME**.

Theorem 10. *The satisfiability problem for* $\mathrm{H}^1\mathrm{CTL}^+$ *is* **3EXPTIME**-*complete.*

Proof. The lower bound follows from Thm. 8. The upper bound of **3EXPTIME** also holds when $\mathrm{H}^1\mathrm{CTL}^+$ is extended by the fairness operators $\overset{\infty}{\mathrm{F}}$ and $\overset{\infty}{\mathrm{G}}$ and the operators Y (previous) and S (since) [14] which are the past counterparts of X and U. The proof is by an exponential reduction to the satisfiability problem of $\mathrm{H}^1\mathrm{CTL}$ extended by Y and S which is **2EXPTIME**-complete [24]. It should be noted that because of Thm. 4 the extension of $\mathrm{H}^1\mathrm{CTL}^+$ by $\overset{\infty}{\mathrm{F}}$ yields a strictly more expressive logic. □

5 The Succinctness of $\mathrm{H}^1\mathrm{CTL}^+$ w.r.t. $\mathrm{H}^1\mathrm{CTL}$

In Corollary 3 an upper bound of $2^{\mathcal{O}(n \log n)}$ for the succinctness of $\mathrm{H}^1\mathrm{CTL}^+$ in $\mathrm{H}^1\mathrm{CTL}$ is given. In this section we establish the lower bound for the succinctness between the two logics. Actually we show that $\mathrm{H}^1\mathrm{CTL}^+$ is exponentially more succinct than $\mathrm{H}^1\mathrm{CTL}$. The model-theoretic approach we use in the proof is inspired by [16]. We first establish a kind of small model property for $\mathrm{H}^1\mathrm{CTL}$.

Theorem 11. *Every finitely satisfiable* $\mathrm{H}^1\mathrm{CTL}$-*formula* φ *with* $|\varphi| = n$ *has a model of depth* $2^{2^{\mathcal{O}(n)}}$.

Proof. In [24] it was shown that for every H^1CTL-formula φ, an equivalent non-deterministic Büchi tree automaton A_φ with $2^{2^{\mathcal{O}(|\varphi|)}}$ states can be constructed. It is easy to see by a pumping argument that if A_φ accepts some finite tree at all, it accepts one of depth $2^{2^{\mathcal{O}(|\varphi|)}}$. It should be noted that the construction in [24] only constructs an automaton that is equivalent to φ with respect to satisfiability. However, the only non-equivalent transformation step is from φ to a formula φ' without nested occurrences of the $\downarrow$-operator (Lemma 4.3 in [24]). It is easy to see that this step only affects the propositions of models but not their shape let alone depth. □

Corollary 9 and Theorem 11 together immediately yield the following.

Corollary 12. H^1CTL^+ *is exponentially more succinct than* H^1CTL.

6 Conclusion

The aim of this paper is to contribute to the understanding of one-variable hybrid logics on trees, one of the extensions of temporal logics with reasonable complexity. We showed that H^1CTL^+ has no additional power over H^1CTL but is exponentially more succinct, we settled the complexity of H^1CTL^+ and showed that hybrid variables do not help in expressing fairness (as $HCTL^+$ cannot express EGFp).

However, we leave a couple of issues for further study, including the following.

- We conjecture that the succinctness gap between H^1CTL^+ and H^1CTL is actually $\theta(n)!$.
- We expect the HCTL-game to capture exactly the expressive power of HCTL. Remember that here we needed and showed only one part of this equivalence.
- The complexity of Model Checking for HCTL has to be explored thoroughly, on trees and on arbitrary transition systems. In this context, two possible semantics should be explored: the one, where variables are bound to nodes of the computation tree and the one which binds nodes to states of the transition system (the latter semantics makes the satisfiability problem undecidable on arbitrary transition systems [2])

References

1. Areces, C., Blackburn, P., Marx, M.: The computational complexity of hybrid temporal logics. Logic Journal of the IGPL 8(5), 653–679 (2000)
2. Areces, C., Blackburn, P., Marx, M.: Hybrid logics: Characterization, interpolation and complexity. J. of Symbolic Logic 66(3), 977–1010 (2001)
3. Areces, C., ten Cate, B.: Hybrid logics. In: Handbook of Modal Logic. Studies in Logic, vol. 3, pp. 821–868. Elsevier, Amsterdam (2007)
4. Chlebus, B.S.: Domino-tiling games. J. Comput. Syst. Sci. 32(3), 374–392 (1986)
5. Clarke, E.M., Emerson, E.A.: Design and synthesis of synchronization skeletons using branching-time temporal logic. In: Kozen, D. (ed.) Logic of Programs 1981. LNCS, vol. 131, pp. 52–71. Springer, Heidelberg (1982)

6. Emerson, E.A., Halpern, J.Y.: Decision procedures and expressiveness in the temporal logic of branching time. In: STOC 1982, pp. 169–180. ACM, New York (1982)
7. Emerson, E.A., Halpern, J.Y.: "Sometimes" and "not never" revisited: on branching versus linear time temporal logic. J. ACM 33(1), 151–178 (1986)
8. Engelfriet, J., Hoogeboom, H.J.: Tree-walking pebble automata. In: Jewels are Forever, pp. 72–83. Springer, Heidelberg (1999)
9. Franceschet, M., de Rijke, M., Schlingloff, B.-H.: Hybrid logics on linear structures: Expressivity and complexity. In: TIME-ICTL 2003, pp. 192–202. IEEE, Los Alamitos (2003)
10. Goranko, V.: Temporal logics with reference pointers and computation tree logics. Journal of Applied Non-Classical Logics 10(3-4) (2000)
11. Grohe, M., Schweikardt, N.: The succinctness of first-order logic on linear orders. Logical Methods in Computer Science 1(1) (2005)
12. Jurdziński, M., Lazić, R.: Alternation-free mu-calculus for data trees. In: Proc. of the 22th LICS 2007. IEEE, Los Alamitos (2007)
13. Kara, A., Weber, V., Lange, M., Schwentick, T.: On the hybrid extension of CTL and CTL^+, arXiv:0906.2541 [cs.LO]
14. Kupferman, O., Pnueli, A.: Once and for all. In: Proc. of the 10th LICS 1995, pp. 25–35. IEEE, Los Alamitos (1995)
15. Kupferman, O., Vardi, M.Y.: Memoryful branching-time logic. In: Proc. of the 21st LICS 2006, pp. 265–274. IEEE, Los Alamitos (2006)
16. Lange, M.: A purely model-theoretic proof of the exponential succinctness gap between CTL^+ and CTL. Information Processing Letters 108, 308–312 (2008)
17. Laroussinie, F., Schnoebelen, P.: A hierarchy of temporal logics with past. Theor. Comput. Sci. 148(2), 303–324 (1995)
18. Laroussinie, F., Schnoebelen, P.: Specification in CTL+Past for verification in CTL. Inf. Comput. 156(1-2), 236–263 (2000)
19. Libkin, L.: Elements of Finite Model Theory. Springer, Heidelberg (2004)
20. Sattler, U., Vardi, M.Y.: The hybrid μ-calculus. In: Goré, R.P., Leitsch, A., Nipkow, T. (eds.) IJCAR 2001. LNCS (LNAI), vol. 2083, pp. 76–91. Springer, Heidelberg (2001)
21. Schwentick, T., Weber, V.: Bounded-variable fragments of hybrid logics. In: Thomas, W., Weil, P. (eds.) STACS 2007. LNCS, vol. 4393, pp. 561–572. Springer, Heidelberg (2007)
22. ten Cate, B., Segoufin, L.: XPath, transitive closure logic, and nested tree walking automata. In: Proc. of the 27th PODS 2008, pp. 251–260. ACM, New York (2008)
23. Vardi, M.Y.: From church and prior to psl. In: Grumberg, O., Veith, H. (eds.) 25 Years of Model Checking. LNCS, vol. 5000, pp. 150–171. Springer, Heidelberg (2008)
24. Weber, V.: Hybrid branching-time logics. CoRR, abs/0708.1723 (2007)
25. Weber, V.: Branching-time logics repeatedly referring to states. Accepted to JoLLI (2009); An extended abstract appeared in the proceedings of HyLo 2007

Bounds on Non-surjective Cellular Automata

Jarkko Kari[1], Pascal Vanier[2], and Thomas Zeume[3]

[1] University of Turku
jkari@utu.fi
[2] Laboratoire d'Informatique Fondamentale de Marseille
pascal.vanier@ens-lyon.fr
[3] Gottfried Wilhelm Leibniz Universität Hannover
thomas-zeume@web.de

Abstract. Cellular automata (CA) are discrete, homogeneous dynamical systems. Non-surjective one-dimensional CA have finite words with no preimage (called *orphans*), pairs of different words starting and ending identically and having the same image (*diamonds*) and words with more/fewer preimages than the average number (*unbalanced* words). Using a linear algebra approach, we obtain new upper bounds on the lengths of the shortest such objects. In the case of an n-state, non-surjective CA with neighborhood range 2 our bounds are of the orders $O(n^2)$, $O(n^{3/2})$ and $O(n)$ for the shortest orphan, diamond and unbalanced word, respectively.

1 Introduction

Non-surjective cellular automata (CA) have *Garden of Eden*-configurations: configurations without a preimage. By compactness, there exist also finite patterns that do not appear in any image configuration [2]. These we call *orphans*. The *Garden of Eden* -theorem by *E. F. Moore* [5] and *J. Myhill* [7] proves the existence of *mutually erasable* patterns in non surjective CA. These are two finite words such that two configurations, having the same prefix and suffix and only differing on those words, have the same image. Mutually erasable words are also called *diamonds*. The existence of diamonds is equivalent to being non-surjective. Non-surjectivity is also equivalent to the existence of finite words of the same size with a different number of preimages [2]. Patterns that do not have the average number of preimages are called *unbalanced*.

One will naturally think about the size of these objects. The objective of this paper is to bound the size of the smallest orphan, diamond and unbalanced word of any non-surjective one-dimensional CA, in terms of the number of states in the automaton. For dimensions greater than one, it is known that it is not decidable whether a given CA is surjective or not [3]. Therefore only bounds for one-dimensional CA will be studied: in higher dimensions no recursive bounds exist.

Using a standard blocking technique, one can convert any one-dimensional CA into a CA with the range-2 neighborhood, so we will only study the bounds in this range-2 case. The previously known bounds were an exponential bound

R. Královič and D. Niwiński (Eds.): MFCS 2009, LNCS 5734, pp. 439–450, 2009.

for shortest orphans, a quadratic bound for shortest diamonds and an $O(n^2 \ln n)$ bound for the shortest unbalanced words, where n is the number of states. These can be found in, or easily deduced from [9] and [6].

We will use tools based on linear algebra to obtain a linear upper bound for the size of the shortest unbalanced words, a quadratic bound for the size of the shortest orphans and an $2n^{\frac{3}{2}}$ bound for the length of the shortest diamonds. By using some combinatorial arguments, we are able to reduce the bounds for orphans and diamonds slightly. For all problems we also tried to bound the best possible upper bounds from below. However, the only non-constant lower bounds we found are for the case of shortest orphans.

The paper is organized as follows: In Section 2 we give the definitions of the main concepts, and in Section 3 we explain our basic linear algebra tools that we use to obtain bounds. In Section 4 we provide the bounds that the linear algebra approach immediately provides for the shortest orphans, diamonds and unbalanced words. All bounds are better than any previously known bounds. In the case of orphans the improvement is even from exponential to polynomial. In Section 5 we fine-tune the bounds by looking in detail at the first steps of the dimension reductions. In Section 6 we consider the algorithmic aspects and show that an orphan, a diamond and an unbalanced word can be found in polynomial time. Section 7 reports the existence for every $n \geq 2$ of an n-state range-2 CA whose shortest orphan is of length $2n-1$. In contrast, for diamonds and unbalanced words we have no non-constant examples. Finally, in Section 8 we formulate open problems and discuss some related questions.

2 Definitions

A *configuration* is a function c that assigns a state from a finite *state set* S to every point in $\mathbb{Z}$. We denote the state of a point $p \in \mathbb{Z}$ of a configuration c by c_p. The set of all configurations is $S^{\mathbb{Z}}$. The *range-r neighborhood* is the tuple $N = (0, \ldots, r-1)$. A *local rule* is a function $f : S^{|N|} \to S$.

Formally, a *cellular automaton* G is a 3-tuple (S, N, f). At each time step a new configuration $G(c)$ is computed from c by updating the states with f at each point p:

$$G(c)_p = f(c_p, c_{p+1} \ldots, c_{p+r-1}) \text{ for all } p \in \mathbb{Z}.$$

The hereby defined function G is called the *global function* of the CA.

We will only consider range-2 CA G with a non-surjective global function throughout the whole paper. The generalization of our results to general range-r are quite straightforward, and will only be briefly discussed in Section 8.

Applications of the range-2 local rule $f : S \times S \longrightarrow S$ to finite words $w \in S^*$ will be denoted by the same symbol G as the global function. In this case

$$G(s_1 s_2 \ldots s_n) = t_1 t_2 \ldots t_{n-1}$$

where $t_i = f(s_i, s_{i+1})$ for all $i = 1, 2, \ldots, n-1$.

Configurations without a preimage are called *Garden of Eden*-configurations. We will be interested in *finite* words without a preimage, that is to say words u

for which there exists no word w such that $G(w) = u$. Such u exist if and only if the CA is non-surjective [2] and are called *orphans.*

By the *Garden of Eden*-theorem of *E. F. Moore* and *J. Myhill*, see [5] and [7], we know that non-surjective CA have configurations which differ only in a finite number of cells and which have the same image under G. We consider the finite differing part of such configurations: Two words $w = p\, c_1 \ldots c_l\, s$ and $w' = p\, c'_1 \ldots c'_l\, s \in S^{l+2}$ with $G(w) = G(w')$ are said to form a *diamond* if $(c_1, \ldots, c_l) \neq (c'_1, \ldots, c'_l)$. The length of the diamond is l. A cellular automaton has a diamond iff it is non-surjective.

There are $|S|^k$ different words of length k and $|S|^{k+1}$ different preimages for words of length k. We call G *k-balanced*, if all words of length k have the same number of predecessor words, i.e. $|G^{-1}(u)| = |S|$ for all $u \in S^k$. Words u with $|G^{-1}(u)| \neq |S|$ are called *unbalanced.* A CA has unbalanced words, iff it is non-surjective [2].

3 Linear Algebra Tools

3.1 Vectorial Interpretation of Sets

The proofs are mainly based on the vectorial interpretation of sets of states, introduced in [1] for the CA context: let us denote by $S = \{1, \ldots, n\}$ the set of all states. A subset $X \subseteq S$ is interpreted as the 0-1 vector x in $\mathbb{R}^n$ whose i-th coordinate is 1 if $i \in X$ and 0 otherwise. The vectors corresponding to single element sets are the unit coordinate vectors e_i and they form a basis of the vector space $\mathbb{R}^n$.

We define $f_a : \mathbb{R}^n \longrightarrow \mathbb{R}^n$ as the linear transformation such that

$$f_a(e_i) = \sum_{f(i,j)=a} e_j,$$

where f is the local transition function of the automaton and a a state. We define $\phi : \mathbb{R}^n \longrightarrow \mathbb{R}$ as the linear form defined by

$$\phi(x) = x \cdot (1, \ldots, 1).$$

We call $\phi(x)$ the *weight* of a vector x since it is the sum of its coordinates. If $w = a_1 \ldots a_k$ is a word[1] on the alphabet S, let f_w denote the composition $f_{a_k} \circ \cdots \circ f_{a_1}$. The analogous notation for compositions is also used with any other family of functions indexed by letters of an alphabet.

Lemma 1. ***Balance*** *The following equality holds for all* $x \in \mathbb{R}^n$ *:*

$$\sum_{a \in S} \phi(f_a(x)) = \phi(x)|S|$$

[1] The word could be empty, in which case $f_w(x) = x$.

Proof. For each e_i we have

$$\sum_{a \in S} f_a(e_i) = (1, 1, \dots, 1)$$

because for each $j \in S$ there exists a unique $a \in S$ such that $f(i,j) = a$. By the linearity of ϕ we have

$$\sum_{a \in S} \phi(f_a(e_i)) = \phi(\sum_{a \in S} f_a(e_i)) = |S| = \phi(e_i)|S|.$$

Now, due to the linearity of f_a and ϕ this extends to any vector x of $\mathbb{R}^n$ in place of e_i. □

Note, that if x is a vector corresponding to $X \subseteq S$, then $\phi(f_w(x))$ is the number of preimages of w that start in a state of X. This can be proved by an easy induction on the length of w.

The main application of the balance lemma is the following: If $\phi(f_w(x)) \neq \phi(x)$ for some word w of length k then $\phi(f_u(x)) < \phi(x)$ and $\phi(f_{u'}(x)) > \phi(x)$ for some words u and u' of that same length k.

3.2 A Very Useful Lemma

For any $x \in \mathbb{R}^n$ with $\phi(x) > 0$ we want to establish an upper bound for the length of shortest words u, u' such that $\phi(f_u(x)) < \phi(x)$ and $\phi(f_{u'}(x)) > \phi(x)$. Therefore, the following lemma will be crucial. Strongly inspired by [8] and [4], it can be considered as the basis for all our results.

The lemma concerns affine subspaces of $\mathbb{R}^n$ and the minimal number of applications of given linear transformations that take a given point outside the subspace. Recall that an *affine subspace* of $\mathbb{R}^n$ of dimension d is a set $x + V$ where $x \in \mathbb{R}^n$ and $V \subseteq \mathbb{R}^n$ is a linear subspace of dimension d. In particular, singleton sets $\{x\}$ are affine subspaces of dimension 0. Affine subspaces relevant in our setup are the sets $\{x \in \mathbb{R}^n \mid \phi(x) = c\}$ of vectors having a fixed weight $c \in \mathbb{R}$. Their dimension is $n - 1$.

An *affine combination* of vectors is a linear combination where the coefficients sum up to one. Affine subspaces are closed under affine combinations of their elements, and conversely, any set closed under affine combinations is affine. The affine subspace *generated* by a set $X \subseteq \mathbb{R}^n$ of vectors consists of all affine combinations of elements of X.

Lemma 2. *Let A be an affine subspace of $\mathbb{R}^n$, and let $x \in A$. Let Σ be an alphabet and for every $a \in \Sigma$, let $\psi_a : \mathbb{R}^n \to \mathbb{R}^n$ be a linear transformation. Then, if there is a word w such that $\psi_w(x) \notin A$ then there exists such a word of length at most* $\dim A + 1$.

Proof. Consider the affine spaces $A_0 \subseteq A_1 \subseteq \dots$ defined by $A_0 = \{x\}$, and A_{i+1} is generated by $A_i \cup \bigcup_{a \in S} \psi_a(A_i)$. Equivalently, A_i is generated by $\{\psi_w(x) \mid |w| \leq i\}$.

By definition, $A_{i+1} = A_i$ means that $\psi_a(A_i) \subseteq A_i$ for all $a \in \Sigma$. Hence if $A_{i+1} = A_i$ for some i, then $A_j = A_i$ for every $j \geq i$.

Let i be the smallest number such that $\psi_w(x) \notin A$ for some w of length i, that is, the smallest i such that $A_i \not\subseteq A$. This means that in $A_0 \subset A_1 \subset \cdots \subset A_i$ all inclusions are proper. In terms of dimensions of affine spaces we have that

$$0 = \dim A_0 < \dim A_1 < \cdots < \dim A_i.$$

This means that $\dim A_{i-1} \geq i-1$. But $A_{i-1} \subseteq A$ so that we also have $\dim A_{i-1} \leq \dim A$. We conclude that $i \leq \dim A + 1$. □

For our case we get the following corollary.

Corollary 1. *Let G be a non-surjective CA and $x \in \mathbb{R}^n$ a vector such that $\phi(x) > 0$. Then there exist words u, u' of length at most $n = |S|$ such that $\phi(f_u(x)) < \phi(x)$ and $\phi(f_{u'}(x)) > \phi(x)$.*

Proof. This is a direct application of Lemma 2. Let $\Sigma = S$, $\psi_a = f_a$ and $A = x + \ker \phi$. Note that A is an affine subspace of $\mathbb{R}^n$ of dimension $n-1$, and $\phi(f_w(x)) \neq \phi(x)$ iff $f_w(x) \notin A$. Since G is non-surjective, there exists an orphan v. As $f_v(e_i) = 0$ for all unit coordinate vectors e_i, we have $f_v(x) = 0$. In particular, $f_v(x) \notin A$. Thus, by Lemma 2, there exists a word w of length at most n such that $\phi(f_w(x)) \neq \phi(x)$. Now, Lemma 1 guarantees the existence of words u, u' of the same length $|w|$ with $\phi(f_u(x)) < \phi(x)$ and $\phi(f_{u'}(x)) > \phi(x)$. □

4 Basic Bounds

We will prove first a quadratic bound for the size of orphans using the previously defined linear algebraic tools.

Theorem 1. *If a range-2 non-surjective CA has n states, then it has an orphan of length at most n^2.*

Proof. We start with $x = (1, \ldots, 1)$. By applying Corollary 1, we know that there exists a word w of length at most n such that $\phi(f_w(x)) < \phi(x) = n$. As $f_w(x)$ is a vector of non-negative integers, $0 \leq \phi(f_w(x)) \leq \phi(x) - 1$. Repeating this argument on $f_w(x)$ in place of x, and continuing likewise, we successively decrease the weight of the vectors. After at most n iterations we obtain a vector of weight 0. Concatenating all words w gives an orphan of length at most n^2. □

We are not aware of any previously published polynomial bound for the length of the shortest orphan. An exponential bound 2^n can be easily seen by combinatorial arguments.

In [6], *Moothathu* proves an $O(n^2 \ln n)$ upper bound for the shortest unbalanced words and asks whether this can be improved. Our linear algebra tools lead straightforwardly to a better bound:

Theorem 2. *Let G be a non-surjective range-2 CA with n states. Then its shortest unbalanced words have at most length n.*

Proof. Let $x = (1, \ldots, 1)$. Because G is non-surjective, by Corollary 1 there is a word w of length n such that $\phi(f_w(x)) \neq \phi(x) = n$. Since $\phi(f_w(x))$ is the number of predecessors of w starting with an arbitrary state of S, we have that w is unbalanced. □

For diamonds also, we obtain a better bound than the previously existing one.

Theorem 3. *Let $G = (S, N, f)$ be a non-surjective range-2 CA with n states. Then $2(\lfloor\sqrt{n}\rfloor - 1)n + 2$ is an upper bound for shortest diamonds of G.*

Proof. Let $k = \lfloor\sqrt{n}\rfloor + 1$ and let $\diamond = p\ c_1 \ldots c_m\ s, p\ d_1 \ldots d_m\ s$ be a diamond of G where $c_1 \neq d_1$ and $c_m \neq d_m$. Let us denote $a = f(p, c_1) = f(p, d_1)$ and $b = f(c_m, s) = f(d_m, s)$. Let x be the 0/1 -vector corresponding to $\{c_1, d_1\}$, so $\phi(x) = 2$.

By Corollary 1, we can successively increase the weight of x by reading words of length n. Thus there is a word w of length at most $(k-2)n$ such that $\phi(f_w(x)) \geq k$, i.e. there are more than k preimages of aw that start in p. If two of them end with the same letter, we already found a diamond. Therefore without loss of generality, we can assume that all of them have different last letters. Denote the set of these last letters by L. Symmetrically we can find a word $\widetilde{w}$ of length $(k-2)n$ with at least k preimages, where the preimages end either in c_m or d_m and have different first letters. We denote the set of first letters of those preimages by R.

As $|L \times R| = k^2 > n$, there are two distinct pairs $(l, r), (l', r') \in L \times R$ with $f(l, r) = f(l', r')$. Let $w_l, w_{l'}$ be words of length $|w|$ such that $G(p\ w_l\ l) = G(p\ w_{l'}\ l') = aw$. Analogously, let $w_r, w_{r'}$ be words of length $|\widetilde{w}|$ such that $G(r\ w_r\ s) = G(r'\ w_{r'}\ s) = \widetilde{w}b$. Then $p\ w_l\ l\ r\ w_r\ s$ and $p\ w_{l'}\ l'\ r'\ w_{r'}\ s$ form a diamond, and the length of the diamond is at most

$$|w_l\ l\ r\ w_r| = |w| + |\widetilde{w}| + 2 \leq 2(k-2)n + 2.$$ □

5 Improved Bounds

5.1 Improving the Algebraic Tools

We can improve Lemma 2 under conditions that apply for diamonds and orphans. The idea is to prove a lower bound for the dimension of the first affine subspace A_1. Let us call a 0/1 -square matrix *k-regular* if every row and every column contains exactly k ones.

Lemma 3. *Let M be a k-regular 0/1 matrix of size $n \times n$, where $1 \leq k \leq n-1$. Then* $\operatorname{rank} M \geq \max\{\lceil \frac{n}{k} \rceil, \lceil \frac{n}{n-k} \rceil\}$.

Proof. It follows from the assumptions that every column contains k non-zero elements. To any collection of i columns one can then add another linearly independent column, provided $i < \frac{n}{k}$. This follows from the fact that some row r contains a zero in each of the i columns, so any column with a non-zero element in row r is linearly independent of the i columns. Hence the rank of the matrix is at least $\frac{n}{k}$.

Analogously, the rank of the $(n-k)$-regular matrix $M' = \mathbf{1} - M$ is at least $\frac{n}{n-k}$, where $\mathbf{1}$ is the matrix that contains only ones. The ranks of M and M' are easily seen equal, so the result follows. □

Now we state an improvement of Corollary 1, that can be applied to diamonds and orphans of range-2 CA.

Corollary 2. *Let $G = (S, N, f)$ be a range-2 non-surjective CA with $S = \{1, \ldots, n\}$, and let x be a 0/1 -vector. Let $k = \phi(x)$ and assume that $1 \le k \le n-1$. If for all $s \in S$, $f_s(x)$ is a 0/1 -vector then there are words u, u' of length at most $n - \max\{\lceil \frac{n}{k} \rceil, \lceil \frac{n}{n-k} \rceil\} + 2$ with $\phi(f_u(x)) < \phi(x)$ and $\phi(f_{u'}(x)) > \phi(x)$.*

Proof. We follow the proof of Lemma 2 and use the same notation. Here, A is the affine space $x + \ker \phi$. Since G is not surjective, there is a w with $f_w(x) \notin A$. We give a lower bound for the dimension of A_1, the affine space generated by vectors x and $f_s(x)$ over all $s \in S$.

Define the 0/1 -matrix $M = (f_1(x), \ldots, f_n(x))$. Without loss of generality, we can assume that $\phi(f_s(x)) = k$ for all $s \in S$, as otherwise $f_s(x) \notin A$. Thus, every column of M contains exactly k ones and $n-k$ zeros. Further, as in the proof of Lemma 1 we see that the sum of the columns is $(k, k, \ldots, k)^T$ so the matrix M is k-regular.

Lemma 3 gives $r = \max\{\lceil \frac{n}{k} \rceil, \lceil \frac{n}{n-k} \rceil\}$ as a lower bound for the rank of M. Thus the affine subspace

$$A_1 = x + \langle \{f_s(x) - x \mid s \in S\} \rangle$$

of Lemma 2 has at least dimension $r - 1$. Hence the chain of dimensions in Lemma 2 becomes

$$r - 1 \le \dim A_1 < \ldots < \dim A_{i-1} \le \dim A = n - 1$$

Hence $r + i - 3 \le \dim A_{i-1} \le n - 1$ and therefore there is an u with $|u| \le n - r + 2$ and $\phi(f_u(x)) \ne \phi(x)$. The claim follows from Lemma 1. □

To obtain the bounds for the lengths of the shortest orphan and diamond, we apply the corollary above at each weight reduction step. This will lead to expressions of the form $\overline{H}_{k,n} = \sum_{i=1}^{k} \lceil \frac{n}{i} \rceil$. We immediately see that $\overline{H}_{k,n} > n \ln k$, since $\sum_{i=1}^{k} \frac{1}{i} > \ln k$. Further, we note that

$$\sum_{i=1}^{n-1} \max\left\{ \left\lceil \frac{n}{i} \right\rceil, \left\lceil \frac{n}{n-i} \right\rceil \right\} = 2\overline{H}_{\lfloor \frac{n-1}{2} \rfloor, n} + b$$

where $b = 2$ if n is even otherwise $b = 0$.

5.2 Improved Results

Combinatorial methods as well as Corollary 2 allow us to lower the bounds by looking at each weight reduction step. Let $F_a : 2^S \to 2^S$ be the set mapping

$$F_a(X) = \{y \in S \mid \exists x \in X : f(x,y) = a\}$$

for all $a \in S$ and $X \subseteq S$. Function F_a corresponds to the linear mapping f_a, with the difference that F_a ignores possible multiplicities.

The next lemma lowers the bound for the first step of weight reduction.

Lemma 4. *If G is non-surjective then there exists a letter a such that $|F_a(S)| < |S|$.*

Proof. If $F_a(S) = S$ for all $a \in S$ then $F_w(S) = S$ for all $w \in S^*$, so the CA would be surjective. □

For the second step and the last step of the weight reduction, Corollary 2 provides a reducing word of length at most 2. More generally, we have the following:

Lemma 5. *If $|X| = k$ and $1 \le k \le n-1$ then there exists a word u of length at most $n - \max\{\lceil \frac{n}{k} \rceil, \lceil \frac{n}{n-k} \rceil\} + 2$ such that $|F_u(X)| < |X|$.*

Proof. Let $x \in \mathbb{R}^n$ be the 0/1 -vector corresponding to set X. If $f_s(x)$ is not a 0/1 -vector for some $s \in S$ then either $\phi(f_s(x)) > k$ or $|F_s(X)| < k$. In both cases $|F_a(X)| < |X|$ for some $a \in S$.

If all $f_s(x)$ are 0/1 -vectors then the conditions of Corollary 2 are satisfied, so the bound follows from the Corollary. □

Now we easily get the following upper bounds for shortest orphans and diamonds.

Theorem 4. *Let $G = (N, S, f)$ be a non-surjective range-2 CA with $|S| = n$. Then $n^2 - 2\overline{H}_{\lfloor \frac{n-1}{2} \rfloor, n} + n - b - 1$ is an upper bound for the length of a shortest orphan, where $b = 2$ if n is even otherwise $b = 0$. Especially, a smallest orphan has at most length $n^2 - 2n \ln\lfloor \frac{n-1}{2} \rfloor + n$.*

Proof. We modify the proof of Theorem 1. We use Lemma 4 to improve the first reduction step and Lemma 5 on the remaining steps. We get the following upper bound for the length of a shortest orphan w:

$$\begin{aligned} |w| &\le 1 + \textstyle\sum_{i=1}^{n-1} (n - \max\{\lceil \frac{n}{i} \rceil, \lceil \frac{n}{n-i} \rceil\} + 2) \\ &= 1 + (n-1)(n+2) - \textstyle\sum_{i=1}^{n-1} \max\{\lceil \frac{n}{i} \rceil, \lceil \frac{n}{n-i} \rceil\} \\ &= n^2 + n - 2\overline{H}_{\lfloor \frac{n-1}{2} \rfloor, n} - b - 1 \end{aligned}$$

where $b = 2$ if n is even otherwise $b = 0$. □

Theorem 5. *Let $G = (S, N, f)$ be a non-surjective range-2 CA with $|S| = n$, its shortest diamond has at most length $2(\lfloor \sqrt{n} \rfloor n - n \ln(\lfloor \sqrt{n} \rfloor) + 2\lfloor \sqrt{n} \rfloor - 2)$.*

Proof. We modify the proof of Theorem 3. Note, that the conditions of Corollary 2 are fulfilled for every increment of weight: If at any stage $f_s(x)$ is not a $0/1$ -vector then a diamond is found.

Thus the step increasing the weight from i takes a word of length at most $n - \lceil \frac{n}{i} \rceil + 2$. Therefore we obtain the upper bound

$$\sum_{i=2}^{k-1} \left(n - \left\lceil \frac{n}{i} \right\rceil + 2\right) = (k-2)(n+2) + n - \overline{H}_{k-1,n}$$

for the lengths of the words w and $\widetilde{w}$ in the proof of Theorem 3. Recall that $k = \lfloor \sqrt{n} \rfloor + 1$. We now obtain the result from this, the fact that $|w| + |\widetilde{w}| + 2$ is an upper bound for the length of the shortest diamond, and the fact that $\overline{H}_{k-1,n} > n \ln(k-1)$. □

6 Algorithms

The proof of Lemma 2 gives us algorithms for finding orphans, diamonds and unbalanced words whose running times are polynomial in the number of states.

Here, only one step will be described in Algorithm 1. That is to say, the algorithm is for obtaining from a vector x a word $w = w_1 \ldots w_k$ such that $\phi(f_w(x)) < \phi(x)$. Note that we can easily have the converse, by only changing the inequality in the algorithm. The full algorithms can be easily deduced from this by looking at the proofs of Theorems 1, 2 and 3.

Algorithm 1. How to decrease the weight of a vector

Data: A non-surjective CA $G = (S, (0,1), f)$ and $x \in \{0,1\}^n \setminus 0^n$
Result: A word w with $\phi(f_w(x)) < \phi(x)$.
$F := \{(x, \epsilon)\}$ `% Set of independent vectors reached.`
while $\exists a \in S, \exists (x,w) \in F$ `such that` $f_a(x)$ `is not in the affine space generated by` F **do**
  if $\phi(f_a(x)) < \phi(x)$ **then**
    return wa ;
  end
  if $\phi(f_a(x)) = \phi(x)$ **then**
    $F := F \cup \{(f_a(x), wa)\}$;
  end
end

It is important to note that the algorithms based on this method will not necessarily find the shortest orphans, diamonds or unbalanced words. De Bruijn-graph based algorithms of [9] can be used to find a shortest diamond in polynomial time, using a breadth-first search on the product of the de Bruijn automaton with itself. Our algorithm above finds a shortest unbalanced word if the search in the **while** -loop is done in the breadth-first order, starting with vector $x = (1, 1, \ldots, 1)$. We are not aware of a polynomial time method to find a shortest orphan.

7 Tightness

It is an interesting question to determine the best possible upper bounds for the lengths of shortest orphans, diamonds and unbalanced words. To see how tight our bounds are, we tried to discover families of CA with long shortest orphans, diamonds and unbalanced words.

For the shortest orphans we were able to construct, for every $n \geq 1$, an n-state, range-2 CA $\mathcal{A}_n$ whose shortest orphan has length $2n-1$. This example is still well below our quadratic upper bound, but we conjecture that $2n-1$ is the proper bound for the length of the shortest orphan. Through exhaustive search we verified that no 2, 3 or 4 state automaton exceeds this bound.

The transition table of $\mathcal{A}_n$ is given in figure 1. The state set is $S = \{0, 1, 2, \dots, n-1\}$ and the local rule f is obtained from the function $f'(a, b) = b - a \pmod{n}$ by changing the entry $(0, 0)$ from 0 to 1.

f	0	1	$\dots$	$n-2$	$n-1$
0	$\underline{1}$	1	$\dots$	$n-2$	$n-1$
1	$n-1$	0	$\dots$	$n-3$	$n-2$
$\vdots$	$\vdots$	$\vdots$		$\vdots$	$\vdots$
$n-2$	2	3	$\dots$	0	1
$n-1$	1	2	$\dots$	$n-1$	0

Fig. 1. Local transition function of the automaton $\mathcal{A}_n$ which has a shortest orphan of length $2n-1$

Theorem 6. *For any number of states $n \geq 2$, the shortest orphan of automaton $\mathcal{A}_n$ has length $2n-1$.*

Proof. Word $w = 0(10)^{n-1}$ is an orphan: the only preimages of 0 are aa for $a \neq 0$, so a preimage of w cannot contain 0's and should be of the form

$$a\ a\ a+1\ a+1\ \dots\ a-1\ a-1.$$

But such a word necessarily contains letter 0, a contradiction.

Let us prove that every word u of length $2n-2$ has a preimage. The related local rule $f'(a, b) = a - b \pmod{n}$ makes a CA surjective, so u has a preimage v under the local rule f'. Moreover, adding any constant $c \pmod{n}$ to all the letters of v provides another preimage of u under f'. Since the length of v is $2n-1$, some letter $s \in S$ appears at most once in v. Hence the preimage v' obtained by subtracting s from every letter of v contains at most one 0. But on such words f' and f are identical, so v' is a pre-image to u under f. □

Concerning the shortest diamond, a computer search found an 8-state CA whose shortest diamond is of length 6. No automata with longer shortest diamond has been found. For every $n \leq 6$ a CA with n states and a shortest diamond of length $n-1$ exists, and we conjecture this to be the optimal bound.

8 Conclusion and Open Problems

For all problems—size of a smallest unbalanced word, orphan and diamond—the linear algebra approach leads to better bounds than the ones obtained with combinatorial methods. For some of the problems, we decreased the upper bound further by looking closer into the dimensional argument of our approach. The results can be extended straightforward to arbitrary neighborhood ranges. For unbalanced words and orphans one can use exactly the same approach. For diamonds a standard blocking technique can be used.

The following table gives an overview of the results, for range-2 and the general range.

	$r = 2$	general r
Non-balanced word	n	n^{r-1}
Diamond	$2n^{\frac{3}{2}} - n \ln n + 2\sqrt{n} - 2$	$(r-1)(n^{\frac{3(r-1)}{2}} - n^{r-1}\frac{r-1}{2} \ln n)$
Orphan	$n^2 - 2n \ln \frac{n}{2} + n$	$n^{2(r-1)}$

We reported in Section 7 a family of n-state CA with shortest orphans of length $2n - 1$, and also results of preliminary computer experiments. These lead us to formulate the following conjectures:

Conjecture 1. The tight upper bound of the length for the shortest orphan of an n -state non-surjective range-2 CA is $2n - 1$.

Conjecture 2. The tight upper bound of the length for the shortest diamond of an n -state non-surjective range-2 CA is $n - 1$.

An interesting related problem is the size of words for which many orphans occur. It can be seen rather easily that more than half of the words in S^l are orphans when l is exponential in the size of shortest orphans. Can a polynomial bound be obtained? This is relevant for the complexity of converting a given non-surjective CA to a lattice gas automaton, see [10].

References

1. Czeizler, E., Kari, J.: A tight linear bound on the neighborhood of inverse cellular automata. In: Caires, L., Italiano, G.F., Monteiro, L., Palamidessi, C., Yung, M. (eds.) ICALP 2005. LNCS, vol. 3580, pp. 410–420. Springer, Heidelberg (2005)
2. Hedlund, G.: Endomorphisms and automorphisms of shift dynamical systems. In: Mathematical Systems Theory, vol. 3, pp. 320–375. Springer, Heidelberg (1969)
3. Kari, J.: Reversibility and surjectivity problems of cellular automata. J. Comput. Syst. Sci. 48, 149–182 (1994)
4. Kari, J.: Synchronizing finite automata on eulerian digraphs. Theor. Comput. Sci. 295(1-3), 223–232 (2003)

5. Moore, E.F.: Machine models of self reproduction. In: Mathematical Society Proceedings of Symposia in Applied Mathematics, vol. 14, pp. 17–33 (1962)
6. Subrahmonian Moothathu, T.K.: Studies in Topological Dynamics with Emphasis on Cellular Automata. PhD thesis, Department of Mathematics and Statistics, School of MCIS, University of Hyderabad (2006)
7. Myhill, J.: The converse of Moore's garden-of-eden theorem. Proc. Amer. Math. Soc. 14, 685–686 (1963)
8. Pin, J.-E.: Utilisation de l'algèbre linéaire en théorie des automates. In: Actes du 1er Colloque AFCET-SMF de Mathématiques Appliquées, pp. 85–92. AFCET (1978)
9. Sutner, K.: Linear cellular automata and de bruijn automata. In: Mathematics and Its Applications 4, vol. 460, pp. 303–320. Kluwer, Dordrecht (1999)
10. Toffoli, T., Capobianco, S., Mentrasti, P.: When–and how–can a cellular automaton be rewritten as a lattice gas? Theor. Comput. Sci. 403, 71–88 (2008)

FO Model Checking on Nested Pushdown Trees

Alexander Kartzow

TU Darmstadt, Fachbereich Mathematik, Schlossgartenstr. 7, 64289 Darmstadt

Abstract. Nested Pushdown Trees are unfoldings of pushdown graphs with an additional jump-relation. These graphs are closely related to collapsible pushdown graphs. They enjoy decidable μ-calculus model checking while monadic second-order logic is undecidable on this class. We show that nested pushdown trees are tree-automatic structures, whence first-order model checking is decidable. Furthermore, we prove that it is in 2-EXPSPACE using pumping arguments on runs of pushdown systems. For these arguments we also develop a Gaifman style argument for graphs of small diameter.

1 Introduction

Nested pushdown trees were introduced in [1] as an expansion of trees generated by pushdown systems with nested jump-edges. They were proposed for software verification as jump-edges may be used to reason about matching pairs of calls and returns in a program. Another approach to software verification checks pushdown trees (without jump-edges) against specifications given by automata or μ-calculus formulas. But these methods even lack the ability to express that every call has a matching return. Alur et al. showed that nested pushdown trees are tame structures with respect to the μ-calculus, in the sense that μ-calculus model checking on nested pushdown trees is decidable. On the other hand they proved the undecidability of monadic second-order logic on nested pushdown trees. These results make nested pushdown trees an interesting class from a model theoretic point of view because there are few natural classes that separate μ-calculus and monadic second-order logic with respect to model checking. In fact, the author knows of only one similar result, namely, for the class of collapsible pushdown graphs [7]. The hierarchy of collapsible pushdown graphs forms an extension of the hierarchy of higher-order pushdown graphs by using a new operation called collapse. There is a close relation between nested pushdown trees and collapsible pushdown graphs: the former are first-order interpretable in collapsible pushdown graphs of order two. [1] In this sense, jump-edges form a very weak form of collapse-edges. For both classes nothing is known so far

[1] As the proof of this claim is unpublished, we give an idea: A node in a nested pushdown tree is a run, i.e., a list of pairs of states and stacks. Push the state onto the stack. This list of stacks can be seen as a level 2 stack and every edge in the nested tree can then be simulated by up to four operations of the collapsible pushdown system.

R. Královič and D. Niwiński (Eds.): MFCS 2009, LNCS 5734, pp. 451–463, 2009.

about the decidability of first-order model checking. In the following we are going to settle the problem for nested pushdown trees with the positive answer that first-order model checking for nested pushdown trees is in 2-EXPSPACE. Furthermore, we show that nested pushdown trees are tree-automatic. The notion of tree-automatic structures was developed in [2] and generalises the concept of automatic structures to the tree case. These are (usually) infinite structures that allow a finite representation by tree-automata. Due to the good algorithmic behaviour of tree-automata the class of tree-automatic structures has nice properties, e.g., first-order model checking is decidable. But in general even automatic structures, and hence also tree-automatic structures, have non-elementary lower bounds for FO model checking [4]. Nevertheless we can show that model checking on nested pushdown trees is elementary by using pumping techniques for pushdown systems.

Here is an outline of the paper. In Section 2 we present an Ehrenfeucht-Fraïssé-game argument for the equivalence of certain structures with parameters for first-order logic up to a fixed quantifier rank. This argument is a form of locality argument on structures of small diameter, despite the fact that small diameters normally prohibit the use of locality arguments. We use local isomorphisms on subgraphs which are nicely embedded into the full graph. Later, this is a main tool in our pumping arguments. Section 3.1 provides the definition of nested pushdown trees and Section 3.2 contains the proof that these structures are tree-automatic. In order to show that first-order model checking on nested pushdown trees is in 2-EXPSPACE (Section 3.4), we develop pumping arguments on nested pushdown trees in 3.3.

2 A Gaifman Style Lemma on Graphs of Small Diameter

In this section we present a game argument showing that certain tuples of a given graph have the same $\simeq_\rho$-type, where $\simeq_\rho$ is equivalence for first-order formulas up to quantifier rank ρ. This argument forms the back-bone of the transformations we are going to use on tuples in a nested pushdown tree. It is a kind of Gaifman-locality argument for certain graphs with possibly small diameter. The crucial property of these graphs is that there are some generic edges that make the diameter small in the sense that a lot of vertices are connected to the same vertex, but when these edges are removed the diameter becomes large. Therefore, on the graph with these generic edges removed we can apply Gaifman-like arguments in order to establish partial isomorphisms and $\simeq_\rho$-equivalence. As disjoint but isomorphic neighbourhoods in such a graph have generic edges to the same vertices (in the full graph) moving a tuple from one neighbourhood to the other does not change the $\simeq_\rho$-type of the tuple.

We use the following definitions and notation. By FO we denote *first-order logic* and we write FO_ρ for the restriction of FO to formulas of quantifier rank up to ρ. We write $\bar{a} = a_1, a_2 \dots, a_n \in A$ for a tuple of elements from a set A. For structures $\mathfrak{A}$ and $\mathfrak{B}$ with n parameters $\bar{a} \in A^n$ and $\bar{b} \in B^n$ we write $\mathfrak{A}, \bar{a} \simeq_\rho \mathfrak{B}, \bar{b}$ for the fact that $\mathfrak{A}, \models \varphi(\bar{a})$ if and only if $\mathfrak{B} \models \varphi(\bar{b})$ for all $\varphi \in \mathrm{FO}_\rho$. For some structure $G = (V, E_1, E_2, \dots, E_n)$ with binary relations $E_1, E_2, \dots, E_n$ and sets $A, B \subseteq V$ we say that A and B *touch* if $A \cap B \neq \emptyset$ or there are $a \in A$, $b \in B$ such that $(a,b) \in E_i$ or $(b,a) \in E_i$ for some $i \leq n$. For a tuple $\bar{a} \in A$ we define inductively the *l-neighbourhood* of $\bar{a}$ with respect to A setting $A_0(\bar{a}) := \{a_i \in \bar{a}\}$, and

$$A_{l+1}(\bar{a}) := A_l(\bar{a}) \cup \{b \in A : \text{there are } i \leq n \text{ and } c \in A_l(\bar{a}) \text{ s.t. } (b,c) \in E_i \text{ or } (c,b) \in E_i\} \ .$$

We write $N_l(\bar{a})$ for the l-neighbourhood with respect to the whole universe V.

We say that A and B are isomorphic over $C \subseteq V$ and write $A \simeq_C B$ if there is some isomorphism $\varphi : G{\restriction}A \simeq G{\restriction}B$ such that for all $a \in A$ and $c \in C$

$$(a,c) \in E_i \text{ iff } (\varphi(a), c) \in E_i \qquad \text{and} \qquad (c,a) \in E_i \text{ iff } (c, \varphi(a)) \in E_i \ .$$

Lemma 1. *Let $G = (V, E_1, E_2, \dots, E_n)$ be some structure, $A, B \subseteq V$ not touching and let $\varphi : A \simeq B$ be an isomorphism of the induced subgraphs. Let $\bar{a} \in A$ and $\bar{c} \in C := G \setminus \big(A_{2^\rho}(\bar{a}) \cup B_{2^\rho}(\varphi(\bar{a}))\big)$.*

$$\varphi{\restriction}A_{2^\rho-1}(\bar{a}) : A_{2^\rho-1}(\bar{a}) \simeq_C B_{2^\rho-1}(\varphi(\bar{a})) \quad \textit{implies} \quad G, \bar{a}, \varphi(\bar{a}), \bar{c} \simeq_\rho G, \varphi(\bar{a}), \bar{a}, \bar{c}.$$

Proof. We prove the claim by induction on ρ using Ehrenfeucht-Fraïssé-Game-terminology. By symmetry, we may assume that Spoiler extends the left-hand side, i.e., extending $\bar{a}, \varphi(\bar{a}), \bar{c}$ by some $d \in V$. The general idea is that Spoiler either chooses an element in $A \cup B$ that is close to $\bar{a}$ or $\varphi(\bar{a})$ and Duplicator responds with applying the isomorphism φ. Otherwise, Duplicator just responds choosing the same element as Spoiler.
Local case: if $d \in A_{2^{\rho-1}}(\bar{a})$ set $a' := d$ and if $d \in \varphi(A_{2^{\rho-1}}(\bar{a}))$ set $a' := \varphi^{-1}(d)$. Then we set $\bar{a}' := \bar{a}, a'$.

As $A_{2^{\rho-1}}(\bar{a}') \subseteq A_{2^\rho}(\bar{a})$, we have $\bar{c} \in C' := G \setminus \big(A_{2^{\rho-1}}(\bar{a}') \cup \varphi(A_{2^{\rho-1}}(\bar{a}'))\big)$. Since A and B do not touch and $C' = C \cup D$ for $D \subseteq \big(A \setminus A_{2^{\rho-1}}(\bar{a}')\big) \cup \big(B \setminus B_{2^{\rho-1}}(\varphi(\bar{a}'))\big)$ we get $A_{2^{\rho-1}-1}(\bar{a}') \simeq_{C'} \varphi(A_{2^{\rho-1}-1}(\bar{a}'))$. Hence, we obtain by induction hypothesis

$$G, \bar{a}', \varphi(\bar{a}'), \bar{c} \simeq_{\rho-1} G, \varphi(\bar{a}'), \bar{a}', \bar{c}$$

Nonlocal case: otherwise, $d \in C' := G \setminus \big(A_{2^{\rho-1}}(\bar{a}) \cup \varphi(A_{2^{\rho-1}}(\bar{a}))\big)$ and we set $\bar{c}' := \bar{c}_\rho, d$.
Note that $A_{2^{\rho-1}-1}(\bar{a}) \simeq_{C'} \varphi(A_{2^{\rho-1}-1}(\bar{a}))$ as A and B are not touching and the distance of elements in $A_{2^{\rho-1}-1}(\bar{a})$ and elements in $C' \cap A$ is at least 2. Hence, by induction hypothesis

$$G, \bar{a}, \varphi(\bar{a}), \bar{c}' \simeq_{\rho-1} G, \varphi(\bar{a}), \bar{a}, \bar{c}'. \qquad \square$$

3 Nested Pushdown Trees

Nested pushdown trees are generated by pushdown systems in the following way. We unfold the configuration graph of a pushdown system and we add a *jump relation* that connects every push- with the corresponding pop-operations.

After formally introducing nested pushdown trees, we show that this class of structures is tree-automatic. This already implies that FO model checking for nested pushdown trees is decidable. But it does not yield an elementary bound for the complexity since the model checking for tree-automatic structures is in general non-elementary [4].

We give a separate argument that yields an elementary bound. This argument is based on pumping techniques. In Section 3.3 we present these techniques which shorten long runs but preserve their $\simeq_\rho$-type in the nested pushdown tree. Due to this result, we only have to inspect finitely many short runs in order to find witnesses for existential quantifications. Section 3.4 shows that this search may be done in 2-EXPSPACE.

3.1 Definition

Definition 1 (Pushdown System). *A tuple $P = (Q, \Sigma, \Delta, (q_0, \bot))$ with a finite set of states Q, a finite set of stack symbols Σ, an initial configuration $(q_0, \bot) \in Q \times \Sigma$ and a transition relation $\Delta \subseteq Q \times \Sigma \times Q \times \{\mathrm{pop}, \mathrm{id}, \mathrm{push}_\sigma$ for each $\sigma \in \Sigma\}$ is called a* pushdown system.

Definition 2. *A* run *r of P is a function $r : \{0, 1, 2, \dots, n\} \to Q \times \Sigma^*$ such that for all $i < n$ there is some $(q, \sigma, p, op) \in \Delta$ and some $w_i \in \Sigma^*$ such that $r(i) = (q, w_i\sigma)$ and $r(i+1) = (p, op(w_i\sigma))$, where $\mathrm{pop}(w_i\sigma) = w_i, \mathrm{id}(w_i\sigma) = w_i\sigma$, and $\mathrm{push}_\tau(w_i\sigma) = w_i\sigma\tau$. We call r a run from $r(0)$ to $r(n)$. We say that the* length *of r is $\mathrm{length}(r) := n$.*

For runs r and r' of length n and m, respectively, such that $r(n) = r'(0)$ we call

$$s : \{0, 1, \dots, n+m\} \to Q \times \Sigma^* \qquad s(i) := \begin{cases} r(i) & \text{if } i \leq n \\ r'(i-n) & \text{otherwise} \end{cases}$$

the composition of r and r'. *We also say that s* decomposes into r and r'.

Note that a run does not necessarily start in the initial configuration $(q_0, \bot)$ of the pushdown system P. The next definition summarises some useful notation about runs.

Definition 3. *Let r be a run of a pushdown system $P = (Q, \Sigma, \Delta, (q_0, \bot))$ and let w, v be words over Σ.*

- *If r has length n, then $\mathrm{last}(r) := r(n)$.*
- *By $w \leq v$ we mean that w is a prefix of v.*
- *For $r(i) = (q, v)$, we set $\mathrm{Stck}(r(i)) := v$. We write $|r(i)|$ for $|v|$ and $w \leq r(i)$ if $w \leq v$.*
- *We say that r is w-*prefixed *if $w \leq r(i)$ for all $i \in \mathrm{dom}(r)$.*
- *We set $\max(r) := \max\{|r(i)| : i \in \mathrm{dom}(r)\}$.*

Remark 1 (Prefix Replacement). Let r be a w-prefixed run of some pushdown system P for some word $w \in \Sigma^*$. If $w' \in \Sigma^*$ ends with the same letter as w then the function

$$r[w/w'] : \mathrm{dom}(r) \to Q \times \Sigma^*$$
$$r[w/w'](i) := (q_i, w'w_i) \quad \text{if } r(i) = (q_i, ww_i)$$

is a run of P, where ww_i denotes the usual concatenation of the words w and w_i.

Definition 4 (Nested Pushdown Tree (NPT)). *Let $P = (Q, \Sigma, \Delta, (q_0, \bot))$ be a pushdown system. Then the* nested pushdown tree *generated by P is* NPT $(P) := (R, \to, \hookrightarrow)$ *where $(R, \to)$ is the unfolding of the configuration graph of P, i.e., R is the set of all runs of P starting at the configuration $q_0, \bot$. For two runs $r_1, r_2 \in R$, we have $r_1 \to r_2$ if r_2 extends r_1 by exactly one configuration. The binary relation $\hookrightarrow$ is called* jump relation *and is defined as follows: let $r_1, r_2 \in R$ and* $\mathrm{last}(r_1) = (q, w) \in Q \times \Sigma^*$. *Then $r_1 \hookrightarrow r_2$ if r_1 is an initial segment of r_2,* $\mathrm{last}(r_2) = (q', w)$ *for some $q' \in Q$ and w is a proper prefix of all stacks between* $\mathrm{last}(r_1)$ *and* $\mathrm{last}(r_2)$, *i.e., $w < r_2(i)$ for all* $\mathrm{length}(r_1) < i < \mathrm{length}(r_2)$.

3.2 NPT Are Tree-Automatic

We start with the notion of a tree-automatic structure which was introduced in [2]. A *tree* is a finite, prefix closed subset of $\{0,1\}^*$, where ε represents the root and we assume the successors at each vertex to be ordered. For a finite set Σ, a Σ-labelled tree is a map $c : T \to \Sigma$ for some tree T. The *convolution* of two Σ-labelled trees c_1 and c_2 is defined as $c_1 \otimes c_2 : \mathrm{dom}(t_1) \cup \mathrm{dom}(t_2) \to (\Sigma \cup \{\Box\})^2$, where $\Box$ represents undefined elements, and

$$(c_1 \otimes c_2)(t) = \begin{cases} (c_1(t), c_2(t)) & \text{if } t \in \mathrm{dom}(c_1) \cap \mathrm{dom}(c_2) \ , \\ (c_1(t), \Box) & \text{if } t \in \mathrm{dom}(c_1) \setminus \mathrm{dom}(c_2) \ , \\ (\Box, c_2(t)) & \text{if } t \in \mathrm{dom}(c_2) \setminus \mathrm{dom}(c_1) \ . \end{cases}$$

A *tree-automaton* is a tuple $A = (Q, \Sigma, \Delta, F)$ where Q is a finite set of states, $\Sigma \subseteq Q$ a finite set of labels, $\Delta \subseteq Q^2 \times \Sigma \times Q$ the transition relation, and $F \subseteq Q$ the set of final states. A run of A on a Σ-labelled tree $c : T \to \Sigma$ is a function $r : T \to Q$ such that for each leaf $l \in T$ we have $r(l) = c(l)$ and for inner nodes $n \in T$ we have $r(n) = q$ if there is some $(q_0, q_1, \sigma, q) \in \Delta$ such that $r(ni) = q_i$ for $i \in \{0,1\}$ and $c(n) = \sigma$. A run r is accepting if $r(0) \in F$. Note that we require $\Sigma \subseteq Q$ as A has no special initial state but starts at every leaf of the tree initialised with the label of this leaf.

A structure $\mathfrak{B} = (B, E_1, E_2, \ldots, E_n)$ with binary relations E_i is *tree-automatic* if there are automata $A_B, A_{E_1}, A_{E_2}, \ldots, A_{E_n}$ such that

1. A_B accepts a set C of Σ-labelled trees.
2. There is a bijection $f : C \to B$.

3. for $c_1, c_2 \in C$, the automaton A_{E_i} accepts $c_1 \otimes c_2$ if and only if $(f(c_1), f(c_2)) \in E_i$.

Theorem 1. *Nested pushdown trees are tree-automatic.*[2]

By the decidability of the FO model checking for arbitrary tree-automatic structures [2] we obtain that FO model checking on nested pushdown trees is decidable.

For the proof of the theorem, we use the fact that, for every context-free grammar G, there is a tree-automaton A which accepts exactly the derivation trees of G. In order to prove tree-automaticity of a NPT, it therefore suffices to give a context free grammar of runs of a pushdown system P (starting at the initial configuration) and to provide grammars generating all pairs of derivations of runs that are connected by $\rightarrow$ or $\hookrightarrow$, respectively.

Let $P = (Q, \Sigma, \Delta, (q_0, \bot))$ be a pushdown system. The following context-free grammar generates all runs of P which start at the initial configuration $q_0, \bot$. The terminal symbols are the transitions of P, i.e., $T := \Delta$. We use the non-terminal symbols $N := \{X_{(q,\sigma)} : q \in Q, \sigma \in \Sigma\} \cup \{C^p_{(q,\sigma)} : q, p \in Q, \sigma \in \Sigma\}$. The idea of the coding is the following. A non-terminal $X_{(q,\sigma)}$ generates a subrun starting from (q, σ) and a $C^p_{(q,\sigma)}$ generates a subrun starting at (q, σ), ending at (p, σ) and in between this element σ is never removed from the stack. Note that such a subrun may be extended by prefixing some push$_\sigma$- and postfixing some pop-operation that deletes this symbol σ again. For $q, p, r, s \in Q$ and $\sigma, \tau \in \Sigma$, the productions are

$$\begin{aligned}
X_{(q,\sigma)} \rightarrow &(q,\sigma,p,\mathrm{id}) \mid (q,\sigma,p,\mathrm{id})X_{(p,\sigma)} \mid (q,\sigma,p,\mathrm{push}_\tau) \mid (q,\sigma,p,\mathrm{push}_\tau)X_{(p,\tau)} \\
&\mid (q,\sigma,p,\mathrm{push}_\tau)C^r_{(p,\tau)}(r,\tau,s,\mathrm{pop}) \mid (q,\sigma,p,\mathrm{push}_\tau)C^r_{(p,\tau)}(r,\tau,s,\mathrm{pop})X_{(s,\sigma)} \\
&\mid (q,\sigma,p,\mathrm{push}_\tau)(p,\tau,r,\mathrm{pop}) \mid (q,\sigma,p,\mathrm{push}_\tau)(p,\tau,r,\mathrm{pop})X_{(r,\sigma)} \\
\text{and } C^p_{(q,\sigma)} \rightarrow &(q,\sigma,r,\mathrm{id})C^p_{(r,\sigma)} \mid (q,\sigma,p,\mathrm{id}) \mid (q,\sigma,r,\mathrm{push}_\tau)C^s_{(r,\tau)}(s,\tau,u,\mathrm{pop})C^p_{(u,\sigma)} \\
&\mid (q,\sigma,s,\mathrm{push}_\tau)(s,\tau,u,\mathrm{pop})C^p_{(u,\sigma)} \mid (q,\sigma,r,\mathrm{push}_\tau)C^s_{(r,\tau)}(s,\tau,p,\mathrm{pop}) \\
&\mid (q,\sigma,r,\mathrm{push}_\tau)(r,\tau,p,\mathrm{pop})
\end{aligned}$$

Note that for every run r of P starting in $(q_0, \bot)$ there is a unique derivation tree starting from $X_{(q_0,\bot)}$ and the leaves of this derivation tree – read from left to right – are the transitions of r. Vice versa, every derivation tree codes a valid run.

As a next step we show that the set of convolutions of the derivation trees of runs r_1, r_2 such that r_2 extends r_1 by exactly one transition may also be defined via some context free grammar. Note that if a run r_2 extends another run r_1 by a push$_\sigma$- or id-transition, the derivation trees only differ in the subtree that starts at the end of the unique longest path that is labelled by non-terminals

[2] I thank Dietrich Kuske for proposing a useful coding of runs in trees.

$X_{(q,\sigma)}$ (where q and σ may vary along the path). The coding of r_2 contains an isomorphic copy of this subtree in the coding of r_1, and extends this subtree by a new rightmost successor with label $X_{(q,\sigma)}$ for some $q \in Q$ and $\sigma \in \Sigma$ and this new rightmost successor has a successor itself which is labelled by the last transition of r_2. The case of a pop-transition is a bit more involved as the subrun between this pop-operation and the corresponding push$_\sigma$-operation is derived from a $C^p_{(q,\sigma)}$ symbol in the derivation of r_2, while it is derived from $X_{(q,\sigma)}$ in the derivation of r_1. But in fact, for both derivations, the form of this subtree is the same and the terminal symbols coincide. The only difference is that the non-terminals of the form $X_{(r,\tau)}$ in the derivation of r_1 are replaced by $C^s_{(r,\tau)}$ for some $s \in Q$ in the derivation of r_2.

We use the following notation. For some terminal or non-terminal a we write a^2 as an abbreviation for the pair (a, a) and we write $Z^p_{(q,\sigma)}$ for the pair $(X_{(q,\sigma)}, C^p_{(q,\sigma)})$. The productions are

$$
\begin{aligned}
(X_{(q,\sigma)})^2 \to& a_1^2 a_2^2 \dots a_n^2 (X_{(p,\tau)})^2 \quad \text{for every } X_{(q,\sigma)} \to a_1 a_2 \dots a_n X_{(p,\tau)} \\
(C^p_{(q,\sigma)})^2 \to& a_1^2 a_2^2 \dots a_n^2 \quad \text{for every } C^p_{(q,\sigma)} \to a_1 a_2 \dots a_n \\
(X_{(q,\sigma)})^2 \to& (q,\sigma,p,\mathrm{id})^2 \big(\square, X_{(p,\sigma)}\big) \mid (q,\sigma,p,\mathrm{push}_\tau)^2 \big(\square, X_{(p,\tau)}\big) \\
&\mid (q,\sigma,p,\mathrm{push}_\tau)^2 (C^r_{(p,\tau)})^2 (r,\tau,s,\mathrm{pop})^2 \big(\square, X_{(s,\sigma)}\big) \\
&\mid (q,\sigma,p,\mathrm{push}_\tau)^2 Z^r_{(p,\tau)} \big(\square, (r,\tau,s,\mathrm{pop})\big) \mid (q,\sigma,p,\mathrm{push}_\tau)^2 \big(\square, (p,\tau,s,\mathrm{pop})\big) \\
Z^p_{(q,\sigma)} \to& (q,\sigma,r,\mathrm{id})^2 Z^p_{(r,\sigma)} \mid (q,\sigma,p,\mathrm{id})^2 \mid (q,\sigma,r,\mathrm{push}_\tau)^2 (C^s_{(r,\tau)})^2 (s,\tau,u,\mathrm{pop})^2 Z^p_{(u,\sigma)} \\
&\mid (q,\sigma,s,\mathrm{push}_\tau)^2 (s,\tau,u,\mathrm{pop})^2 Z^p_{(u,\sigma)} \mid (q,\sigma,r,\mathrm{push}_\tau)^2 (C^s_{(r,\tau)})^2 (s,\tau,p,\mathrm{pop})^2 \\
&\mid (q,\sigma,r,\mathrm{push}_\tau)^2 (r,\tau,p,\mathrm{pop})^2 \\
(\square, X_{(q,\sigma)}) \to& \big(\square, (q,\sigma,p,\mathrm{id})\big) \mid \big(\square, (q,\sigma,p,\mathrm{push}_\tau)\big)
\end{aligned}
$$

Analogously to the $\to$-case, we can give a grammar for runs r_1, r_2 such that $r_1 \hookrightarrow r_2$. If $r_1 \hookrightarrow r_2$, then r_1 is an initial segment of r_2. Thus, the derivation of r_2 contains that of r_1. It extends the derivation of r_1 by a derivation of the form $(q, \sigma, p, \mathrm{push}_\tau) C^r_{(p,\tau)} (r, \tau, s, \mathrm{pop})$. The following productions describe this:

$$
\begin{aligned}
(X_{(q,\sigma)})^2 \to& a_1^2 a_2^2 \dots a_n^2 (X_{(p,\tau)})^2 \quad \text{for every } X_{(q,\sigma)} \to a_1 a_2 \dots a_n X_{(p,\tau)} \\
(C^p_{(q,\sigma)})^2 \to& a_1^2 a_2^2 \dots a_n^2 \quad \text{for every } C^p_{(q,\sigma)} \to a_1 a_2 \dots a_n \\
(X_{(q,\sigma)})^2 \to& (q,\sigma,p,\mathrm{id})^2 (\square, X_{(p,\sigma)}) \mid (q,\sigma,p,\mathrm{push}_\tau)^2 (\square, X_{(p,\tau)}) \\
&\mid (q,\sigma,p,\mathrm{push}_\tau)^2 (C^r_{(p,\tau)})^2 (r,\tau,s,\mathrm{pop})^2 (\square, X_{(s,\sigma)}) \\
(\square, X_{(p,\sigma)}) \to& (\square, (q,\sigma,p,\mathrm{push}_\tau)) (\square, (p,\tau,r,\mathrm{pop})) \\
&\mid (\square, (q,\sigma,p,\mathrm{push}_\tau)) (\square, C^s_{(r,\tau)}) (\square, (p,\tau,r,\mathrm{pop}))
\end{aligned}
$$

The productions of $(\square, C^p_{(q,\sigma)})$ are exactly as for $C^p_{(q,\sigma)}$ in the second component with first component always marked $\square$, i.e., the first run is already finished and the second run extends the first one by some "closed" subrun, i.e., a subrun that starts and ends with the same stack content.

3.3 $\simeq_\rho$-Pumping on NPT

In this section we present several pumping lemmas on runs of a pushdown system P. The aim is to show that for every run of a pushdown system there is another one of bounded length which represents a node with the same $\simeq_\rho$-type in the NPT generated by P. We use these lemmas later to prove an elementary bound for the complexity of FO model checking on nested pushdown trees. As every $\simeq_\rho$-type has a witness of bounded length, a model checking algorithm for an FO_ρ-formula only has to check runs of bounded length in order to find a witness for an existential quantification.

We bound the length of a run in three steps. The first one reduces the size of the last stack of a run, the second one reduces the size of the maximal stack passed along the run and the last one gives us basically a bound on the number of occurrences of every given stack along the run. We will see that these conditions are sufficient for bounding its length.

We start with a general observation about the structure of runs that are related by some edge. We will use this lemma in several of our pumping lemmas.

Lemma 2. *Let $r = r_1 \circ r_2 \circ r_3$ be a run of a pushdown system P, $w \in \Sigma^*$ and $\sigma \in \Sigma$ such that r_2 is w-prefixed, r_3 is $(w\sigma)$-prefixed and* $\mathrm{Stck}(\mathrm{last}(r_1)) = w$. *If $r * s$ for $* \in \{\hookrightarrow, \hookleftarrow, \rightarrow, \leftarrow\}$ then $s = r_1 \circ r_2'$ for some w-prefixed run r_2'.*

Proof. As $w\sigma \leq \mathrm{last}(r)$ we have $w \leq \mathrm{last}(s)$. Hence, the only non-trivial case is $s \hookrightarrow r$. By definition of $\hookrightarrow$, we have $w\sigma \leq r(i)$ for all $i \in \mathrm{dom}(r) \setminus \mathrm{dom}(s)$ and s is an initial segment of r. Thus, r_1 is an initial segment of s. □

Now we can state our first pumping lemma, that reduces the size of the last configuration of a given run, while preserving its $\simeq_\rho$-type.

Lemma 3 (First $\simeq_\rho$-Pumping Lemma). *Let $\bar{r} = r_1, r_2, \ldots, r_m \in \mathrm{NPT}(P)$ and*
$r \in \mathrm{NPT}(P)$ such that

$$|\mathrm{last}(r)| > |\mathrm{last}(r_i)| + (2 + 2^{\rho+1})|Q| \cdot |\Sigma| + 2^\rho + 1 \quad \textit{for all } i \leq m .$$

There is an $s \in \mathrm{NPT}(P)$ such that $|\mathrm{last}(s)| < |\mathrm{last}(r)|$ and $\mathrm{NPT}(P), \bar{r}, r \simeq_\rho \mathrm{NPT}(P), \bar{r}, s$.

Proof. Because of the length of $v := \mathrm{Stck}(\mathrm{last}(r))$ there are $w_1 < w_2 \leq v$ and decompositions of r as $r = r_{w_1} \circ s_{w_1} = r_{w_2} \circ s_{w_2}$ such that

1. $s_{w_i}(0) = (w_i, q)$ for some $q \in Q$ and all $i \in \{1, 2\}$;
2. s_{w_i} is w_i-prefixed;
3. $|w_1| > |\mathrm{last}(r_i)|$ and $|\mathrm{last}(r)| > |w_2| + 2^\rho$;
4. w_1 and w_2 end with the same letter $\sigma \in \Sigma$;
5. $|w_2| - |w_1| > 1 + 2^{\rho+1}$.

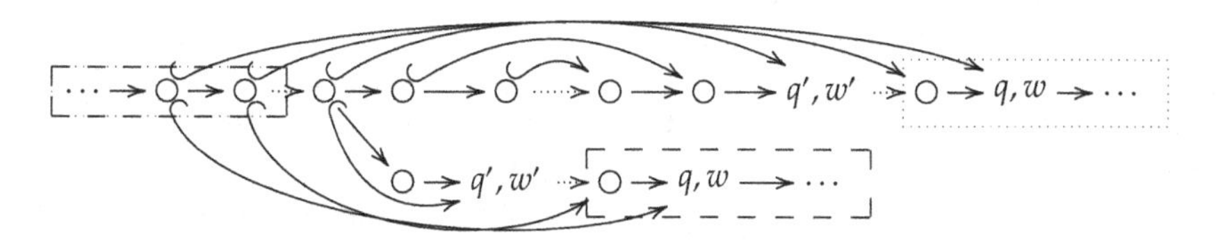

Fig. 1. Second pumping lemma: we replace the upper q, w by the lower one. The dotted / dashed boxes mark the neighbourhood of the upper / lower q, w

Then $s := r_{w_1} \circ s_{w_2}[w_2/w_1]$ is well defined by Remark 1. Note that $N_{2^\rho}(r)$ and $N_{2^\rho}(s)$ do not touch because $|\text{last}(r)| - |\text{last}(s)| = |w_2| - |w_1| > 1 + 2^{\rho+1}$ and for runs connected by a path of length $2 \cdot 2^\rho + 1$ the height of their last stacks does not differ by more than $2 \cdot 2^\rho + 1$. Furthermore, due to 3., Lemma 2, and Remark 1, it follows that for all $r' \in N_{2^\rho}(r)$ we have $r' = r_{w_2} \circ r'_{w_2}$ for some w_2-prefixed r'_{w_2} and the function $\varphi : r' \mapsto r_{w_1} \circ r'_{w_2}[w_2/w_1]$ is an embedding of $N_{2^\rho}(r)$ into $N_{2^\rho}(s)$. For the same reasons, $\varphi^{-1} : r_{w_1} \circ r'_{w_1} \mapsto r_{w_2} \circ r'_{w_1}[w_1/w_2]$ for a w_1-prefixed run r'_{w_1} forms an embedding of $N_{2^\rho}(s)$ into $N_{2^\rho}(r)$. Finally, as $|\text{last}(r)| > |\text{last}(s)| \geq |w_1| > |\text{last}(r_i)| + 2^\rho$, again by Lemma 2, r_i cannot be in the 2^ρ-neighbourhood of r and s. Hence, we may apply Lemma 1. □

Now we are going to prove a second $\simeq_\rho$-type preserving pumping lemma that preserves the last configuration of a run r, but reduces $\max(r)$.

Lemma 4 (Second $\simeq_\rho$-Pumping Lemma). *Let* $\bar{r} = r_1, r_2, \ldots, r_m \in \text{NPT}(P)$ *and* $r \in \text{NPT}(P)$ *such that* $\max(r) > \max(r_i) + |Q|^2|\Sigma| + 1$ *for all* $1 \leq i \leq m$, *and such that* $\max(r) > |\text{last}(r)| + |Q|^2|\Sigma| + 2^\rho + 1$. *Then there is some* $s \in \text{NPT}(P)$ *such that* $\text{last}(s) = \text{last}(r)$, $\max(s) < \max(r)$, *and* $\bar{r}, r \simeq_\rho \bar{r}, s$.

Proof. We eliminate the last occurrence of a stack of length $\max(r)$ in r. For this purpose, let $i \in \text{dom}(r)$ be maximal with $\text{Stck}(r(i)) = w$ for $w \in \Sigma^*$ with $|w| > |Q^2||\Sigma| + 2^\rho + 1 + |\text{last}(r)|$. Then for all $\text{last}(r) \leq v \leq w$, the run r decomposes as $r = r_v \circ s_v \circ t_v$ such that $i \in \text{dom}(r_v \circ s_v)$, s_v is v-prefixed, $s_v(0) = (q_v, v)$, $\text{last}(s_v) = (p_v, v)$ for some $q_v, p_v \in Q$, and $|t(i)| < |v|$ for all $1 \leq i \leq \text{length}(t)$. Then there are $v_1 < v_2 \leq w$ with

1. $\max(r_i) < |v_1|$
2. $|v_2| > |v_1| > |\text{last}(r)| + 2^\rho$
3. $q_{v_1} = q_{v_2}, p_{v_1} = p_{v_2}$
4. the last letter of v_1 and v_2 is the same $\sigma \in \Sigma$.

Then we set $s'_{v_2} := s_{v_2}[v_2/v_1]$. Note that $s := r_{v_1} \circ s'_{v_2} \circ t_{v_1}$ is a well defined run. We use Lemma 1 to show that $\bar{r}, r \simeq_\rho \bar{r}, s$. We set

$$A := \{t \in N_{2^\rho}(r) : t = r_{v_1} \circ s_{v_1} \circ t', t' \text{ run}\} \quad B := \{t \in N_{2^\rho}(s) : t = r_{v_1} \circ s'_{v_2} \circ t', t' \text{ run}\} .$$

Note that $r_i \notin A \cup B$ as for all $t \in A \cup B$, we have $\max(t) \geq |v_1| > \max(r_i)$. From Lemma 2 it follows that for each run t' such that $r_{v_1} \circ s_{v_1} \circ t' \in N_{2^\rho}(r)$ or $r_{v_1} \circ s'_{v_2} \circ t' \in N_{2^\rho}(s)$ we have $|t'(i)| < |v_1|$ for $1 \leq i \leq \text{length}(t')$. Hence, for $j := \text{length}(r_{v_1} \circ s_{v_1})$ and every $t \in A$ we have $\text{Stck}(t(j)) = v_1$, while for all $t \in B$ we have $|t(j)| < |v_1|$ as $\text{length}(s_{v_1}) > \text{length}(s'_{v_2})$. Thus, for all $a \in A$ and $b \in B$ the runs a and b disagree on a proper prefix of both elements, whence A and B cannot touch.

Now we claim that there is an isomorphism of the induced subgraphs $\varphi : A_{2^\rho}(r) \simeq B_{2^\rho}(s)$, given by $r_{v_1} \circ s_{v_1} \circ t' \mapsto r_{v_1} \circ s'_{v_2} \circ t'$. For this note that for any two runs t', t'' and for $* \in \{\rightarrow, \leftarrow, \hookrightarrow, \hookleftarrow\}$ we have

$$(r_{v_1} \circ s_{v_1} \circ t') * (r_{v_1} \circ s_{v_1} \circ t'') \quad \text{iff} \quad t' * t'' \quad \text{iff} \quad (r_{v_1} \circ s'_{v_2} \circ t') * (r_{v_1} \circ s'_{v_2} \circ t'').$$

In order to apply the game argument, we finally have to show that edges between $A_{2^\rho-1}(r)$ and $\text{NPT}(P) \setminus A_{2^\rho}(r)$ are preserved under φ. Assume that $a \in A_{2^\rho-1}(r)$ and $c \in \text{NPT}(P) \setminus \big(A_{2^\rho}(r) \cup B_{2^\rho}(s)\big)$. Note that $a \rightarrow c$ or $a \hookrightarrow c$ implies that a is a subrun of c and thus $c \in A_{2^\rho}(r)$ by definition of A. Assume that $c \rightarrow a$. then $|\text{last}(c)| \leq |\text{last}(r) + 2^\rho| < |v_1|$. Hence $c \neq r_{v_1} \circ s_{v_1}$. But as $r_{v_1} \circ s_{v_1}$ is a proper initial segment of a, this results in $c \in A_{2^\rho}(r)$. Thus, if $c \in \text{NPT}(P) \setminus A_{2^\rho}(r)$ is connected to a then $c \hookrightarrow a$ and c is an initial segment of $r_{v_1} \circ s_{v_1}$. But as the last stack of a and c agree and $|\text{last}(a)| < |v_1|$ then c is an initial segment of r_{v_1}. Thus, $c \hookrightarrow \varphi(a)$ as s_{v_1} and s'_{v_2} are both v_1-prefixed and $\text{last}(a) = \text{last}(\varphi(a)) < v_1$.

Now we found an $\simeq_\rho$-equivalent run s that is shorter than r. Iterating this process leads eventually to some run s with the desired properties □

Now we state our last pumping lemma, which decreases the number of occurrences of a given stack in a run r without affecting its $\simeq_\rho$-type. In order to do this we have to define what it means for a given stack w to occur often in a run r. We are going to count the occurrences of w as a stack in a w-prefixed subrun of r. Afterwards, we will see that bounding this number and $\max(r)$ is a sufficient condition to bound the total number of occurrences of a stack w in r.

Definition 5. *Let r be a run of the pushdown system $P = (Q, \Sigma, \Delta, (q_0, \bot))$ of length n. The number of occurrences of w in r is denoted $|r|_w := \big|\{i \in \mathbb{N} : \text{Stck}(r(i)) = w\}\big|$. We set $\Xi(r) := \max\big\{|s|_w : w \in \Sigma^*$ and s is a w-prefixed subrun of $r\big\}$.*

Lemma 5 (Third $\simeq_\rho$-Pumping Lemma). *Let $\bar{r} \in \text{NPT}(P)$ such that $\Xi(r_i) \leq B$ for all $r_i \in \bar{r}$ and some $B \in \mathbb{N}$. For $r \in \text{NPT}(P)$, there is some $s \in \text{NPT}(P)$ such that $\max(s) \leq \max(r), \text{last}(s) = \text{last}(r), \Xi(s) \leq B + (2^{\rho+1} + 2)|Q| + 2^\rho + 1$, and $\bar{r}, r \simeq_\rho \bar{r}, s$.*

Figure 2 gives an idea of the proof which is similar to that of the Lemma 4.

3.4 FO Model Checking on NPT Is in 2-EXPSPACE

Using the three pumping lemmas we can now establish a "dynamic small witness property" for NPT: given the length of the runs representing parameters in a

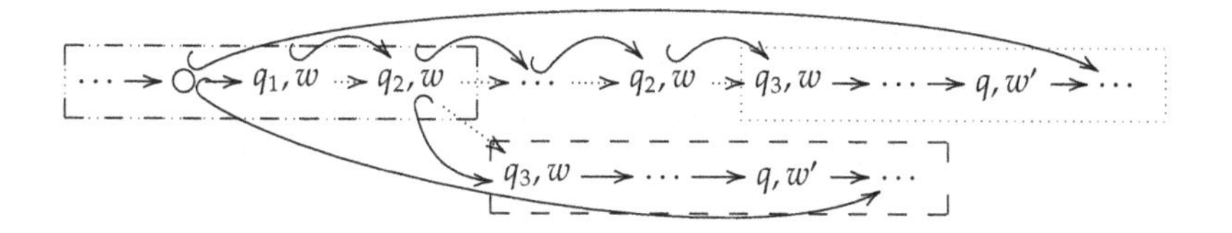

Fig. 2. Third pumping lemma: we replace the upper q, w' by the lower one. The dotted / dashed boxes mark the neighbourhood of the upper / lower q, w'.

formula of quantifier rank ρ, we can bound the length of the run representing a witness for the first existential quantification occurring in the formula, if there is some witness for this quantification at all. The crucial point is that a bound on $\max(r)$ and a bound on $\Xi(r)$ yield a bound on the length of r:

Lemma 6. *Let $P = (Q, \Sigma, \Delta, (q_0, \bot))$ be a pushdown system and r a run of P such that $\max(r) = h$ and $\Xi(r) = b$, then $\mathrm{length}(r) \leq \frac{b^{h+2}-b}{b-1}$.*

Proof. Let $m_h := b$. For every $w \in \Sigma^h$ and some w-prefixed subrun s of r we have $\mathrm{length}(s) \leq m_h$ as the height of all stacks in s is h, whence all elements in s have stack w.

Now assume that every subrun t of r which is w-prefixed for some $w \in \Sigma^{n+1}$ has $\mathrm{length}(t) \leq m_{n+1}$. Let $w \in \Sigma^n$ be an arbitrary word and let s be a maximal w-prefixed subrun of r. Then there are $0 = e_1 < e_2 < \ldots < e_f < e_{f+1} = \mathrm{length}(s) + 1$ such that for $0 \leq i \leq f$ we have $\mathrm{Stck}(s(e_i)) = w$ and s restricted to (e_i, e_{i+1}) is w_i-prefixed for some $w_i \in \Sigma^{n+1}$. We have $f \leq b$ due to $\Xi(s) \leq \Xi(r) \leq b$. By assumption we get $\mathrm{length}(s) \leq (1 + m_{n+1})b$. Note that r is ε-prefixed, hence

$$\mathrm{length}(r) \leq m_0 = b + bm_1 = b + b^2 + b^2 m_2 = \ldots = m_h \sum_{i=0}^{h} b^i = \frac{b^{h+2} - b}{b - 1}. \qquad \square$$

In the following we define our notion of a small run. Let $P = (Q, \Sigma, \Delta, (q_0, \bot))$ be a pushdown system. For $j \leq k \in \mathbb{N}$ we say that some $r \in \mathrm{NPT}(P)$ is *(j, k)-small* if

$$|\mathrm{last}(r)| \leq 6|P|^2 j 2^k, \qquad \max(r) \leq 8|P|^3 j 2^k, \qquad \text{and } \Xi(r) \leq 6|P| j 2^k \ .$$

Lemma 7. *Let $P = (Q, \Sigma, \Delta, (q_0, \bot))$ be a pushdown system, $\bar{r} = r_1, r_2, \ldots, r_i \in \mathrm{NPT}(P)$ and $i \leq k \in \mathbb{N}$. Then there are $\bar{r}' = r'_1, r'_2, \ldots, r'_i \in \mathrm{NPT}(P)$ such that every r'_j is (j, k)-small and $\bar{r} \simeq_{k-i} \bar{r}'$.*

The proof is by induction on i using the pumping lemmas.

With the bounds on the length of runs we can do FO model checking by brute force inspection of short runs. In order to check for an existential witness we only have to test all runs of bounded length. The bound depends on the number of parameters chosen before and on the size of the formula which we check. This

Algorithm: `ModelCheck`(P, α, φ)
Input: pushdown system P, $\varphi \in \mathrm{FO}_\rho$, an assignment $\alpha : \mathrm{free}(\varphi) \to \mathrm{NPT}(P)$ such that $n = |\mathrm{dom}(\alpha)|$ and $\alpha(x_j)$ is $(j, \rho + n)$-small for each $j \leq n$
if φ *is an atom or negated atom* **then**
  if $\mathrm{NPT}(P) \models \varphi[\alpha]$ **then** accept **else** reject;
if $\varphi = \varphi_1 \vee \varphi_2$ **then** **guess** $i \in \{1, 2\}$, and `ModelCheck`(P, α, φ_i);
if $\varphi = \varphi_1 \wedge \varphi_2$ **then** **universally choose** $i \in \{1, 2\}$, and `ModelCheck`(P, α, φ_i);
if $\varphi = \exists x_i \varphi_1$ **then**
  guess an $(i, k + n)$-small a of $\mathrm{NPT}(P)$ and `ModelCheck`$(P, \alpha[x_i \mapsto a], \varphi_1)$;
if $\varphi = \forall x_i \varphi_1$ **then**
  universally choose an $(i, k + n)$-small a of $\mathrm{NPT}(P)$ and `ModelCheck`$(P, \alpha[x_i \mapsto a], \varphi_1)$;

Algorithm 1. ModelCheck used in the proof of Theorem 2

means for a fixed quantifier in some formula φ we only have to check a finite initial part of the nested pushdown tree under consideration. Thus, we can give an alternating algorithm for FO model checking on NPT that works similar to the FO model checking algorithm on finite structures explained in [6].

Theorem 2. *The structure complexity of* FO *model checking on* NPT *is in* EXPSPACE, *while its expression and combined complexity are in* 2-EXPSPACE.

Proof. We assume that the i-th quantifier with respect to quantifier depth binds x_i. The algorithm ModelCheck (see next page), decides $\mathrm{NPT}(P) \models \varphi$. Due to Lemma 7, a straightforward induction shows that ModelCheck is correct. We analyse the space that this algorithm uses. Due to Lemma 6 an (i, k)-small run r has bounded length and we can store it as a list of $\exp(O(i|P|^4 k \exp(k)))$ many transitions. Thus, we need $\exp(O(i|P|^4 k \exp(k))) \log(P)$ space for storing one run. Additionally, we need space for checking whether such a list of transitions forms a valid run and for checking the atomic type of the runs. We can do this by simulation of P. The size of the stack is bounded by the size of the runs. Thus, the alternating algorithm ModelCheck is in

$$\mathrm{ASPACE}\Big(|\varphi| \log(|P|) \exp(O(|P|^4 |\varphi|^2 \exp(|\varphi|)))\Big) \subseteq \mathrm{ASPACE}\Big(\exp(O(|P|^4 \exp(2|\varphi|)))\Big) .$$

As the number of alternations is bounded by $|\varphi|$, we see by [5](Theorem 4.2) that FO model checking for NPT is in $\mathrm{DSPACE}\big(\exp(O(|P|^4 \exp(2|\varphi|)))\big)$.

4 Conclusions

By tree-automaticity as well as pumping techniques we showed decidability of the FO model checking on NPT. Both approaches are transferable to some extent to the case of collapsible pushdown graphs. The tree-automaticity argument applies at least to level 2 of the hierarchy of collapsible pushdown automata.[3]

[3] We obtained this result recently and hope to publish it soon.

But for arguments in the spirit of generation growth [4] combined with a result about counting abilities of higher-order pushdown systems[3], one obtains level 5 collapsible pushdown systems that are not tree-automatic. This raises the question of a characterisation of all tree-automatic collapsible pushdown graphs, especially for levels 3 and 4. Another open problem is effective FO model checking on collapsible pushdown graphs and whether pumping techniques lead to effective model checking algorithms on these graphs.

References

1. Alur, R., Chaudhuri, S., Madhusudan, P.: Languages of nested trees. In: Ball, T., Jones, R.B. (eds.) CAV 2006. LNCS, vol. 4144, pp. 329–342. Springer, Heidelberg (2006)
2. Blumensath, A.: Automatic structures. Diploma thesis, RWTH Aachen (1999)
3. Blumensath, A.: On the structure of graphs in the caucal hierarchy. Theor. Comput. Sci. 400(1-3), 19–45 (2008)
4. Blumensath, A., Grädel, E.: Automatic structures. In: Proc. 15th IEEE Symp. on Logic in Computer Science, pp. 51–62. IEEE Computer Society Press, Los Alamitos (2000)
5. Chandra, A.K., Kozen, D.C., Stockmeyer, L.J.: Alternation. J. ACM 28(1), 114–133 (1981)
6. Grädel, E.: Finite model theory and descriptive complexity. In: Finite Model Theory and Its Applications, pp. 125–230. Springer, Heidelberg (2007)
7. Hague, M., Murawski, A.S., Ong, C.-H.L., Serre, O.: Collapsible pushdown automata and recursion schemes. In: LICS 2008: Proceedings of the 2008 23rd Annual IEEE Symposium on Logic in Computer Science, pp. 452–461 (2008)

The Prismoid of Resources

Delia Kesner and Fabien Renaud

PPS, CNRS and Université Paris Diderot

Abstract. We define a framework called the *prismoid of resources* where each vertex refines the λ-calculus by using a different choice to make explicit or implicit (meta-level) the definition of the contraction, weakening, and substitution operations. For all the calculi in the prismoid we show simulation of β-reduction, confluence, preservation of β-strong normalisation and strong normalisation for typed terms. Full composition also holds for all the calculi of the prismoid handling explicit substitutions. The whole development of the prismoid is done by making the set of resources a parameter, so that the properties for each vertex are obtained as a particular case of the general abstract proofs.

1 Introduction

Linear Logic [5] gives a logical framework to formalise the notion of control of resources by means of weakening, contraction and linear substitution. A succinct representation of Linear Logic proofs is given by Proof-Nets [5] which are often used as a semantical support to define λ-calculi with explicit control operators [19,18,9,7].

In this paper we develop an homogeneous framework of λ-calculi called the *prismoid of resources*. Each vertex is a specialised λ-calculus parametrised by a set of *sorts* wich are of two kinds : resources $\mathtt{w}$ (weakening) and $\mathtt{c}$ (contraction), and cut-elimination operation $\mathtt{s}$ (substitution). If a sort in $\{\mathtt{c}, \mathtt{s}, \mathtt{w}\}$ belongs to a given calculus, then management of the corresponding operations is explicit in this calculus. Explicit resources will allow more refined cut-elimination procedures. Each edge is an operation to simulate and/or project one vertex into the other one. The eight calculi of the prismoid correspond to 2^3 different ways to combine sorts by means of explicit or implicit (meta-level) operations.

The asymmetry between different sorts will be reflected in the prismoid by means of its two bases. The base $\mathfrak{B}_I$ contains all the calculi without explicit substitutions and the base $\mathfrak{B}_E$ only contains those with explicit substitutions. The bases are of different nature as they will not enjoy exactly the same properties.

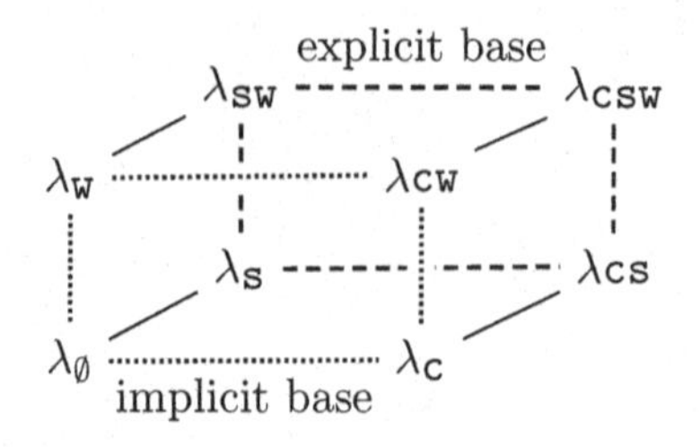

R. Královič and D. Niwiński (Eds.): MFCS 2009, LNCS 5734, pp. 464–476, 2009.

Thus for example, the $\lambda_{\mathtt{cs}}$-calculus has only explicit control of contraction and substitution, the λ-calculus has no explicit control at all, and the $\lambda_{\mathtt{csw}}$-calculus – a slight variation of $\lambda\mathtt{lxr}$ [9] – has explicit control of everything.

For all calculi of the prismoid we show simulation of β-reduction, confluence, preservation of β-strong normalisation (PSN) and strong normalisation (SN) for simply typed terms. Thus in particular, none of the calculi suffers from Mellies' counter-example [14]. Full composition, stating that explicit substitution is able to implement the underlying notion of higher-order substitution, is also shown for all calculi with sort $\mathtt{s}$, ie. those included in the explicit substitution base. Each property is stated and proved by making the set of sorts a parameter, so that the properties for each vertex of the prismoid turn out to be a particular case of some general abstract proof, which may hold for the whole prismoid or just for only one base.

While both implicit and explicit substitutions are usually [1,6,13] defined by means of the propagation of an operator through the structure of terms, the behaviour of calculi of the prismoid can be understood as a mechanism to decrease the multiplicity of variables that are affected by substitutions. This notion is close in spirit to MELL Proof-Nets, and shares common ideas with calculi by Milner [16] and Accattoli and Guerrini [2]. However their formalisms only handle the substitution operation as explicit, leaving weakening and contraction as implicit functions.

Road Map: Section 2 introduces syntax and operational semantics of the prismoid. Section 3 explores how to enrich the λ-calculus by adding more explicit control of resources, while Section 4 deals with the dual operation which forgets rich information given by explicit weakening and contraction. Section 5 is devoted to PSN and confluence on untyped terms. Finally, typed terms are introduced in Section 6 together with a SN proof for them. We conclude and give future directions of work in Section 7.

2 Terms and Rules of the Prismoid

We assume a denumerable set of variable symbols $x, y, z, \ldots$. Lists and sets of variables are denoted by capital Greek letters $\Gamma, \Delta, \Pi, \ldots$. We write $\Gamma; y$ for $\Gamma \cup \{y\}$ when $y \notin \Gamma$. We use $\Gamma \setminus \Delta$ for **set difference** and $\Gamma \setminus\!\!\setminus \Delta$ for **obligation set difference** which is only defined if $\Delta \subseteq \Gamma$.

Terms are given by the grammar $t, u ::= x \mid \lambda x.t \mid tu \mid t[x/u] \mid \mathcal{W}_x(t) \mid \mathcal{C}_x^{y|z}(t)$. The terms x, $\lambda x.t$, tu, $t[x/u]$, $\mathcal{W}_x(t)$ and $\mathcal{C}_x^{y|z}(t)$ are respectively called **term variable**, **abstraction**, **application**, **closure**, **weakening** and **contraction**. **Free** and **bound** variables of t, respectively written $\mathtt{fv}(t)$ and $\mathtt{bv}(t)$, are defined as usual: $\lambda x.u$ and $u[x/v]$ bind x in u, $\mathcal{C}_x^{y|z}(u)$ binds y and z in u, x is free in $\mathcal{C}_x^{y|z}(u)$ and in $\mathcal{W}_x(t)$.

Given three lists of variables $\Gamma = x_1, \ldots, x_n$, $\Delta = y_1, \ldots, y_n$ and $\Pi = z_1, \ldots, z_n$ of the same length, the notations $\mathcal{W}_\Gamma(t)$ and $\mathcal{C}_\Gamma^{\Delta|\Pi}(t)$ mean, respectively, $\mathcal{W}_{x_1}(\ldots \mathcal{W}_{x_n}(t))$ and $\mathcal{C}_{x_1}^{y_1|z_1}(\ldots \mathcal{C}_{x_n}^{y_n|z_n}(t))$. These notations will extend naturally to sets of variables of same size thanks to the equivalence relation in Figure 1. The particular cases $\mathcal{C}_{\emptyset}^{\emptyset|\emptyset}(t)$ and $\mathcal{W}_{\emptyset}(t)$ mean simply t.

Given lists $\Gamma = x_1, \ldots, x_n$ and $\Delta = y_1, \ldots, y_n$, the **renaming** of Γ by Δ in t, written $R_{\Delta}^{\Gamma}(t)$, is the capture-avoiding simultaneous substitution of y_i for every free occurrence of x_i in t. For example $R_{y_1 y_2}^{x_1 x_2}(\mathcal{C}_{x_1}^{y|z}(x_2 y z)) = \mathcal{C}_{y_1}^{y|z}(y_2 y z)$.

Alpha-conversion is the (standard) congruence generated by *renaming* of bound variables. For example, $\lambda x_1.x_1\mathcal{C}_x^{y_1|z_1}(y_1 z_1) \equiv_\alpha \lambda x_2.x_2\mathcal{C}_x^{y_2|z_2}(y_2 z_2)$. All the operations defined along the paper are considered modulo alpha-conversion so that in particular capture of variables is not possible.

The set of **positive free variables** of a term t, written $\texttt{fv}^+(t)$, denotes the free variables of t which represent a term variable at the end of some (possibly empty) contraction chain. Formally,

$$\begin{array}{ll}
\texttt{fv}^+(y) & := \{y\} \\
\texttt{fv}^+(\lambda y.u) & := \texttt{fv}^+(u) \setminus \{y\} \\
\texttt{fv}^+(uv) & := \texttt{fv}^+(u) \cup \texttt{fv}^+(v) \\
\texttt{fv}^+(\mathcal{W}_y(u)) & := \texttt{fv}^+(u) \\
\texttt{fv}^+(u[y/v]) & := (\texttt{fv}^+(u) \setminus \{y\}) \cup \texttt{fv}^+(v) \\
\texttt{fv}^+(\mathcal{C}_y^{z|w}(u)) & := (\texttt{fv}^+(u) \setminus \{z,w\}) \cup \{y\} \text{ if } z \in \texttt{fv}^+(u) \text{ or } w \in \texttt{fv}^+(u) \\
\texttt{fv}^+(\mathcal{C}_y^{z|w}(u)) & := \texttt{fv}^+(u) \setminus \{z,w\} \text{ otherwise}
\end{array}$$

The **number of occurrences** of the positive free variable x in the term t is written $|\texttt{fv}^+(t)|_x$. We extend this definition to sets by $|\texttt{fv}^+(t)|_\Gamma = \Sigma_{x\in\Gamma}|\texttt{fv}^+(t)|_x$. Thus for example, given $t = \mathcal{W}_{x_1}(xx)\ \mathcal{W}_x(y)\ \mathcal{C}_z^{z_1|z_2}(z_2)$, we have $x, y, z \in \texttt{fv}^+(t)$ with $|\texttt{fv}^+(t)|_x = 2$, $|\texttt{fv}^+(t)|_y = |\texttt{fv}^+(t)|_z = 1$ but $x_1 \notin \texttt{fv}^+(t)$.

We write $t_{[y]_x}$ for the **non-deterministic replacement** of *one positive* occurrence of x in t by a *fresh* variable y. Thus for example, $(\mathcal{W}_x(t)\ x\ x)_{[y]_x}$ may denote either $\mathcal{W}_x(t)\ y\ x$ or $\mathcal{W}_x(t)\ x\ y$, but neither $\mathcal{W}_y(t)\ x\ x$ nor $\mathcal{W}_x(t)\ y\ y$.

The **deletion** function removes non positive free variables in Γ from t:

$$\begin{array}{lll}
\texttt{del}_\Gamma(y) & := y & \\
\texttt{del}_\Gamma(u\ v) & := \texttt{del}_\Gamma(u)\ \texttt{del}_\Gamma(v) & \\
\texttt{del}_\Gamma(\lambda y.u) & := \lambda y.\texttt{del}_\Gamma(u) & \text{if } y \notin \Gamma \\
\texttt{del}_\Gamma(u[y/v]) & := \texttt{del}_\Gamma(u)[y/\texttt{del}_\Gamma(v)] & \text{if } y \notin \Gamma \\
\texttt{del}_\Gamma(\mathcal{W}_x(u)) & := \begin{cases} u \\ \mathcal{W}_x(\texttt{del}_\Gamma(u)) \end{cases} & \begin{array}{l} \text{if } x \in \Gamma \\ \text{if } x \notin \Gamma \end{array} \\
\texttt{del}_\Gamma(\mathcal{C}_x^{y|z}(u)) & := \begin{cases} \texttt{del}_{\Gamma\setminus x\cup\{y,z\}}(u) \\ \mathcal{C}_x^{y|z}(\texttt{del}_\Gamma(u)) \end{cases} & \begin{array}{l} \text{if } x \in \Gamma\ \&\ y,z \notin \Gamma\ \&\ x \notin \texttt{fv}^+(\mathcal{C}_x^{y|z}(u)) \\ \text{otherwise} \end{array}
\end{array}$$

This operation does not increase the size of the term. Moreover, if $x \in \texttt{fv}(t) \setminus \texttt{fv}^+(t)$, then $\texttt{size}(\texttt{del}_x(t)) < \texttt{size}(t)$.

Now, let us consider a set of **resources** $\mathcal{R} = \{\texttt{c}, \texttt{w}\}$ and a set of **sorts** $\mathcal{S} = \mathcal{R} \cup \{\texttt{s}\}$. For every subset $\mathcal{B} \subseteq \mathcal{S}$, we define a calculus $\lambda_\mathcal{B}$ in the **prismoid of resources** which is equipped with a set $\mathcal{T}_\mathcal{B}$ of **well-formed** terms, called $\mathcal{B}$-terms, together with a reduction relation $\rightarrow_\mathcal{B}$ given by a *subset* of the reduction system described in Figure 1. Each calculus belongs to a **base** : the explicit substitution base $\mathfrak{B}_E$ which contains all the calculi having at least sort $\mathtt{s}$ and the implicit substitution base $\mathfrak{B}_I$ containing all the other calculi. A term t is in $\mathcal{T}_{\mathcal{B}}$ iff $\exists\, \Gamma$ s.t. $\Gamma \Vdash_{\mathcal{B}} t$ is derivable in the following system :

$$\frac{}{x \Vdash_{\mathcal{B}} x} \qquad \frac{\Gamma \Vdash_{\mathcal{B}} u \quad \Delta \Vdash_{\mathcal{B}} v}{\Gamma \uplus_{\mathcal{B}} \Delta \Vdash_{\mathcal{B}} uv} \qquad \frac{\Gamma \Vdash_{\mathcal{B}} u}{\Gamma \setminus\!\!\setminus_{\mathcal{B}} x \Vdash_{\mathcal{B}} \lambda x.u} \qquad \frac{\Gamma \Vdash_{\mathcal{B}} u}{\Gamma; x \Vdash_{\mathcal{B}} \mathcal{W}_x(u)}\ (\mathtt{w} \in \mathcal{B})$$

$$\frac{\Gamma \Vdash_{\mathcal{B}} v \quad \Delta \Vdash_{\mathcal{B}} u}{\Gamma \uplus_{\mathcal{B}} (\Delta \setminus\!\!\setminus_{\mathcal{B}} x) \Vdash_{\mathcal{B}} u[x/v]}\ (\mathtt{s} \in \mathcal{B}) \qquad \frac{\Gamma \Vdash_{\mathcal{B}} u}{x; (\Gamma \setminus\!\!\setminus_{\mathcal{B}} \{y, z\}) \Vdash_{\mathcal{B}} \mathcal{C}_x^{y|z}(u)}\ (\mathtt{c} \in \mathcal{B})$$

In the previous rules, $\uplus_{\mathcal{B}}$ means standard union if $\mathtt{c} \notin \mathcal{B}$ and disjoint union if $\mathtt{c} \in \mathcal{B}$. Similarly, $\Gamma \setminus\!\!\setminus_{\mathcal{B}} \Delta$ is used for $\Gamma \setminus \Delta$ if $\mathtt{w} \notin \mathcal{B}$ and for $\Gamma \setminus\!\!\setminus \Delta$ if $\mathtt{w} \in \mathcal{B}$.

Notice that variables, applications and abstractions belong to all calculi of the prismoid while weakening, contraction and substitutions only appear in calculi having the corresponding sort. If t is a $\mathcal{B}$-term, then $\mathtt{w} \in \mathcal{B}$ implies that bound variables of t cannot be useless, and $\mathtt{c} \in \mathcal{B}$ implies that no free variable of t has more than one free occurrence. Thus for example the term $\lambda z.xy$ belongs to the calculus $\lambda_{\mathcal{B}}$ only if $\mathtt{w} \notin \mathcal{B}$ (thus it belongs to $\lambda_\emptyset$, $\lambda_{\mathtt{c}}$, $\lambda_{\mathtt{s}}$, $\lambda_{\mathtt{cs}}$), and $(xz)[z/yx]$ belongs to $\lambda_{\mathcal{B}}$ only if $\mathtt{s} \in \mathcal{B}$ and $\mathtt{c} \notin \mathcal{B}$ (thus it belongs to $\lambda_{\mathtt{s}}$ and $\lambda_{\mathtt{sw}}$). A useful property is that $\Gamma \Vdash_{\mathcal{B}} t$ implies $\Gamma = \mathtt{fv}(t)$.

In order to introduce the reduction rules of the prismoid we need a meta-level notion of substitution defined on alpha-equivalence classes; it is the one implemented by the explicit control of resources. A $\mathcal{B}$-**substitution** is a pair of the form $\{x/v\}$ with $v \in \mathcal{T}_{\mathcal{B}}$. The **application of a** $\mathcal{B}$**-substitution** $\{x/u\}$ to a $\mathcal{B}$-term t is defined as follows: if $|\mathtt{fv}^+(t)|_x = 0$ we have to check if x occurs negatively. If $|\mathtt{fv}(t)|_x = 0$ or $\mathtt{w} \notin \mathcal{B}$ then $t\{x/u\} := \mathtt{del}_x(t)$. Otherwise, $t\{x/u\} := \mathcal{W}_{\mathtt{fv}(u)\setminus\mathtt{fv}(t)}(\mathtt{del}_x(t))$. If $|\mathtt{fv}^+(t)|_x = n+1 \geq 2$, then $t\{x/u\} := t_{[y]_x}\{y/u\}\{x/u\}$. If $|\mathtt{fv}^+(t)|_x = 1$, $t\{x/u\} := \mathtt{del}_x(t)\{\!\{x/u\}\!\}$ where $\{\!\{x/u\}\!\}$ is defined as follows :

$$\begin{array}{lll}
x\{\!\{x/u\}\!\} & := u & \\
y\{\!\{x/u\}\!\} & := y & x \neq y \\
(s\ v)\{\!\{x/u\}\!\} & := s\{\!\{x/u\}\!\}\ v\{\!\{x/u\}\!\} & \\
(\lambda y.v)\{\!\{x/u\}\!\} & := \lambda y.v\{\!\{x/u\}\!\} & x \neq y\ \&\ y \notin \mathtt{fv}(u) \\
s[y/v]\{\!\{x/u\}\!\} & := s\{\!\{x/u\}\!\}[y/v\{\!\{x/u\}\!\}] & x \neq y\ \&\ y \notin \mathtt{fv}(u) \\
\mathcal{W}_y(v)\{\!\{x/u\}\!\} & := \mathcal{W}_{y\setminus\mathtt{fv}(u)}(v\{\!\{x/u\}\!\}) & x \neq y \\
\mathcal{C}_y^{z|w}(v)\{\!\{x/u\}\!\} & := \begin{cases} \mathcal{C}_\Gamma^{\Delta|\Pi}(v\{z/R_\Delta^\Gamma(u)\}\{w/R_\Pi^\Gamma(u)\}) \\ \mathcal{C}_y^{z|w}(v\{\!\{x/u\}\!\}) \end{cases} & \begin{cases} x = y\ \&\ \Gamma = \mathtt{fv}(u) \\ \Delta, \Pi \text{ are fresh} \end{cases} \\
 & & x \neq y\ \&\ z, w \notin \mathtt{fv}(u)
\end{array}$$

This definition looks complex, this is because it is covering all the calculi of the prismoid by a unique homogeneous specification. The restriction of this operation to particular subsets of resources results in simplified notions of substitutions. As a typical example, the previous definition can be shown to be equivalent to the well-known notion of higher-order substitution on $\mathtt{s}$-terms [8].

We now introduce the reduction system of the prismoid. In the last column of Figure 1 we use the notation $\mathcal{A}^+$ (resp. $\mathcal{A}^-$) to specify that the equation/rule belongs to the calculus $\lambda_{\mathcal{B}}$ iff $\mathcal{A} \subseteq \mathcal{B}$ (resp. $\mathcal{A} \cap \mathcal{B} = \emptyset$). Thus, each calculus $\lambda_{\mathcal{B}}$ contains only a strict subset of the reduction rules and equations in Figure 1.

All the equations and rules can be understood by means of MELL Proof-Nets reduction (see for example [9]). The reduction rules can be split into four groups: the first one fires implicit/explicit substitution, the second one implements substitution by decrementing multiplicity of variables and/or performing propagation, the third one pulls weakening operators as close to the top as possible and the fourth one pushes contractions as deep as possible. Alpha-conversion

Equations :

$(\mathtt{CC}_{\mathcal{A}})$	$\mathcal{C}_w^{x\mid z}(\mathcal{C}_x^{y\mid p}(t))$	$\equiv \mathcal{C}_w^{x\mid y}(\mathcal{C}_x^{z\mid p}(t))$		$\mathtt{c}^+$
$(\mathtt{C}_{\mathcal{C}})$	$\mathcal{C}_x^{y\mid z}(t)$	$\equiv \mathcal{C}_x^{z\mid y}(t)$		$\mathtt{c}^+$
$(\mathtt{CC}_{\mathcal{C}})$	$\mathcal{C}_a^{b\mid c}(\mathcal{C}_x^{y\mid z}(t))$	$\equiv \mathcal{C}_x^{y\mid z}(\mathcal{C}_a^{b\mid c}(t))$	$x \neq b, c$ & $a \neq y, z$	$\mathtt{c}^+$
$(\mathtt{WW}_{\mathcal{C}})$	$\mathcal{W}_x(\mathcal{W}_y(t))$	$\equiv \mathcal{W}_y(\mathcal{W}_x(t))$		$\mathtt{w}^+$
$(\mathtt{SS}_{\mathcal{C}})$	$t[x/u][y/v]$	$\equiv t[y/v][x/u]$	$y \notin \mathtt{fv}(u)$ & $x \notin \mathtt{fv}(v)$	$\mathtt{s}^+$

Rules :

(β)	$(\lambda x.t)\ u$	$\to t\{x/u\}$		$\mathtt{s}^-$
$(\mathtt{B})$	$(\lambda x.t)\ u$	$\to t[x/u]$		$\mathtt{s}^+$
$(\mathtt{V})$	$x[x/u]$	$\to u$		$\mathtt{s}^+$
$(\mathtt{SGc})$	$t[x/u]$	$\to t$	$x \notin \mathtt{fv}(t)$	$\mathtt{s}^+$ & $\mathtt{w}^-$
$(\mathtt{SDup})$	$t[x/u]$	$\to t_{[y]_x}[x/u][y/u]$	$\lvert\mathtt{fv}^+(t)\rvert_x > 1$ & y fresh	$\mathtt{s}^+$ & $\mathtt{c}^-$
$(\mathtt{SL})$	$(\lambda y.t)[x/u]$	$\to \lambda y.t[x/u]$		$\mathtt{s}^+$
$(\mathtt{SA_L})$	$(t\ v)[x/u]$	$\to t[x/u]\ v$	$x \notin \mathtt{fv}(v)$	$\mathtt{s}^+$
$(\mathtt{SA_R})$	$(t\ v)[x/u]$	$\to t\ v[x/u]$	$x \notin \mathtt{fv}(t)$	$\mathtt{s}^+$
$(\mathtt{SS})$	$t[x/u][y/v]$	$\to t[x/u[y/v]]$	$y \in \mathtt{fv}^+(u) \setminus \mathtt{fv}(t)$	$\mathtt{s}^+$
$(\mathtt{SW_1})$	$\mathcal{W}_x(t)[x/u]$	$\to \mathcal{W}_{\mathtt{fv}(u)\setminus\mathtt{fv}(t)}(t)$		$(\mathtt{sw})^+$
$(\mathtt{SW_2})$	$\mathcal{W}_y(t)[x/u]$	$\to \mathcal{W}_{y\setminus\mathtt{fv}(u)}(t[x/u])$	$x \neq y$	$(\mathtt{sw})^+$
$(\mathtt{LW})$	$\lambda x.\mathcal{W}_y(t)$	$\to \mathcal{W}_y(\lambda x.t)$	$x \neq y$	$\mathtt{w}^+$
$(\mathtt{AW_l})$	$\mathcal{W}_y(u)\ v$	$\to \mathcal{W}_{y\setminus\mathtt{fv}(v)}(u\ v)$		$\mathtt{w}^+$
$(\mathtt{AW_r})$	$u\ \mathcal{W}_y(v)$	$\to \mathcal{W}_{y\setminus\mathtt{fv}(u)}(u\ v)$		$\mathtt{w}^+$
$(\mathtt{SW})$	$t[x/\mathcal{W}_y(u)]$	$\to \mathcal{W}_{y\setminus\mathtt{fv}(t)}(t[x/u])$		$(\mathtt{sw})^+$
$(\mathtt{SCa})$	$\mathcal{C}_x^{y\mid z}(t)[x/u]$	$\to \mathcal{C}_\Gamma^{\Delta\mid\Pi}(t[y/R_\Delta^\Gamma(u)][z/R_\Pi^\Gamma(u)])$	$y, z \in \mathtt{fv}^+(t)$; $\Gamma = \mathtt{fv}(u)$; Δ and Π are fresh	$(\mathtt{cs})^+$
$(\mathtt{CL})$	$\mathcal{C}_w^{y\mid z}(\lambda x.t)$	$\to \lambda x.\mathcal{C}_w^{y\mid z}(t)$		$\mathtt{c}^+$
$(\mathtt{CA_L})$	$\mathcal{C}_w^{y\mid z}(t\ u)$	$\to \mathcal{C}_w^{y\mid z}(t)\ u$	$y, z \notin \mathtt{fv}(u)$	$\mathtt{c}^+$
$(\mathtt{CA_R})$	$\mathcal{C}_w^{y\mid z}(t\ u)$	$\to t\ \mathcal{C}_w^{y\mid z}(u)$	$y, z \notin \mathtt{fv}(t)$	$\mathtt{c}^+$
$(\mathtt{CS})$	$\mathcal{C}_w^{y\mid z}(t[x/u])$	$\to t[x/\mathcal{C}_w^{y\mid z}(u)]$	$y, z \in \mathtt{fv}^+(u)$	$(\mathtt{cs})^+$
$(\mathtt{SCb})$	$\mathcal{C}_w^{y\mid z}(t)[x/u]$	$\to \mathcal{C}_w^{y\mid z}(t[x/u])$	$x \neq w$ & $y, z \notin \mathtt{fv}(u)$	$(\mathtt{cs})^+$
$(\mathtt{CW_1})$	$\mathcal{C}_w^{y\mid z}(\mathcal{W}_y(t))$	$\to R_w^z(t)$		$(\mathtt{cw})^+$
$(\mathtt{CW_2})$	$\mathcal{C}_w^{y\mid z}(\mathcal{W}_x(t))$	$\to \mathcal{W}_x(\mathcal{C}_w^{y\mid z}(t))$	$x \neq y, z$	$(\mathtt{cw})^+$
$(\mathtt{CGc})$	$\mathcal{C}_w^{y\mid z}(t)$	$\to R_w^z(t)$	$y \notin \mathtt{fv}(t)$	$\mathtt{c}^+$ & $\mathtt{w}^-$

Fig. 1. The reduction rules and equations of the prismoid

guarantees that no capture of variables occurs during reduction. The use of positive conditions (conditions on positive free variables) in some of the rules will become clear when discussing projection at the end of Section 4.

The notations $\Rightarrow_{\mathcal{R}}$, $\equiv_{\mathcal{E}}$ and $\to_{\mathcal{R}\cup\mathcal{E}}$, mean, respectively, the rewriting (resp. equivalence and rewriting modulo) relation generated by the rules $\mathcal{R}$ (resp. equations $\mathcal{E}$ and rules $\mathcal{R}$ modulo equations $\mathcal{E}$). Similarly, $\Rightarrow_{\mathcal{B}}$, $\equiv_{\mathcal{B}}$ and $\to_{\mathcal{B}}$ mean, respectively, the rewriting (resp. equivalence and rewriting modulo) relation generated by the rules (resp. the equations and rules modulo equations) of the calculus $\lambda_{\mathcal{B}}$. Thus for example the reduction relation $\to_{\emptyset}$ is only generated by the β-rule exactly as in λ-calculus. Another example is $\to_{\mathtt{c}}$ which can be written $\to_{\{\beta,\mathtt{CL},\mathtt{CA_L},\mathtt{CA_R},\mathtt{CGc}\}\cup\{\mathtt{CC}_{\mathcal{A}},\mathtt{C}_{\mathcal{C}},\mathtt{CC}_{\mathcal{C}}\}}$. Sometimes we mix both notations to denote particular subrelations, thus for example $\to_{\mathtt{c}\setminus\beta}$ means $\to_{\{\mathtt{CL},\mathtt{CA_L},\mathtt{CA_R},\mathtt{CGc}\}\cup\{\mathtt{CC}_{\mathcal{A}},\mathtt{C}_{\mathcal{C}},\mathtt{CC}_{\mathcal{C}}\}}$.

Among the eight calculi of the prismoid we can distinguish the $\lambda_{\emptyset}$-calculus, known as λ-calculus, which is defined by means of the $\to_{\emptyset}$-reduction relation on $\emptyset$-terms. Another language of the prismoid is the $\lambda_{\mathtt{csw}}$-calculus, a variation of $\lambda\mathtt{lxr}$ [9], defined by means of the $\to_{\{\mathtt{c},\mathtt{s},\mathtt{w}\}}$-reduction relation on $\{\mathtt{c},\mathtt{s},\mathtt{w}\}$-terms. A last example is the $\lambda_{\mathtt{w}}$-calculus given by means of $\to_{\mathtt{w}}$-reduction, that is, $\to_{\{\beta,\mathtt{LW},\mathtt{AW_l},\mathtt{AW_r}\}\cup\{\mathtt{WW}_{\mathcal{C}}\}}$.

A $\mathcal{B}$-term t is in **$\mathcal{B}$-normal form** is there is no u s.t. $t \to_{\mathcal{B}} u$. A $\mathcal{B}$-term t is said to be **$\mathcal{B}$-strongly normalising**, written $t \in \mathcal{SN}_{\mathcal{B}}$, iff there is no infinite $\mathcal{B}$-reduction sequence starting at t.

The system enjoys the following properties.

Lemma 1 (Preservation of Well-Formed Terms by Reduction). *If $\Gamma \Vdash_{\mathcal{B}} t$ and $t \to_{\mathcal{B}} u$, then $\exists\ \Delta \subseteq \Gamma$ s.t. $\Delta \Vdash_{\mathcal{B}} u$. Moreover $\mathtt{w} \in \mathcal{B}$ implies $\Delta = \Gamma$.*

Lemma 2 (Full Composition). *Let $t[y/v] \in \mathcal{T}_{\mathcal{B}}$. Then $t[y/v] \to^*_{\mathcal{B}} t\{y/v\}$.*

3 Adding Resources

This section is devoted to the simulation of the $\lambda_{\emptyset}$-calculus into richer calculi having more resources. The operation is only defined in the calculi of the base $\mathfrak{B}_I$. We consider the function $\mathtt{AR}_{\mathcal{A}}(\_) : \mathcal{T}_{\emptyset} \mapsto \mathcal{T}_{\mathcal{A}}$ for $\mathcal{A} \subseteq \mathcal{R}$ which enriches a $\lambda_{\emptyset}$-term in order to fulfill the constraints needed to be an $\mathcal{A}$-term. Adding is done not only on a static level (the terms) but also on a dynamic level (reduction).

$$
\begin{array}{lll}
\mathtt{AR}_{\mathcal{A}}(x) & := x & \\
\mathtt{AR}_{\mathcal{A}}(\lambda x.t) & := \begin{cases} \lambda x.\mathcal{W}_x(\mathtt{AR}_{\mathcal{A}}(t)) \\ \lambda x.\mathtt{AR}_{\mathcal{A}}(t) \end{cases} & \begin{array}{l} \mathtt{w} \in \mathcal{A}\ \&\ x \notin \mathtt{fv}(t) \\ \text{otherwise} \end{array} \\
\mathtt{AR}_{\mathcal{A}}(t\ u) & := \begin{cases} \mathcal{C}_{\Gamma}^{\Delta|\Pi}(R_{\Delta}^{\Gamma}(\mathtt{AR}_{\mathcal{A}}(t))R_{\Pi}^{\Gamma}(\mathtt{AR}_{\mathcal{A}}(u))) \\ \mathtt{AR}_{\mathcal{A}}(t)\ \mathtt{AR}_{\mathcal{A}}(u) \end{cases} & \begin{array}{l} \begin{cases} \mathtt{c} \in \mathcal{A}\ \&\ \Gamma = \mathtt{fv}(t) \cap \mathtt{fv}(u) \\ \Delta \text{ and } \Pi \text{ are fresh} \end{cases} \\ \text{otherwise} \end{array}
\end{array}
$$

For example, adding resource $\mathtt{c}$ (resp. $\mathtt{w}$) to $t = \lambda x.yy$ gives $\lambda x.\mathcal{C}_y^{y_1|y_2}(y_1y_2)$ (resp. $\lambda x.\mathcal{W}_x(yy)$), while adding both of them gives $\lambda x.\mathcal{W}_x(\mathcal{C}_y^{y_1|y_2}(y_1y_2))$.

We now establish the relation between $\mathtt{AR}_{\mathcal{A}}()$ and implicit substitution, which is the technical key lemma of the paper.

Lemma 3. *Let $t, u \in \mathcal{T}_{\emptyset}$. Then*

- *If $\mathtt{c} \notin \mathcal{A}$, then $\mathtt{AR}_{\mathcal{A}}(t)\{x/\mathtt{AR}_{\mathcal{A}}(u)\} = \mathtt{AR}_{\mathcal{A}}(t\{x/u\})$.*
- *If $\mathtt{c} \in \mathcal{A}$, then $\mathcal{C}_{\Gamma}^{\Delta|\Pi}(R_{\Delta}^{\Gamma}(\mathtt{AR}_{\mathcal{A}}(t))\{x/R_{\Pi}^{\Gamma}(\mathtt{AR}_{\mathcal{A}}(u))\}) \to_{\mathcal{A}}^{*} \mathtt{AR}_{\mathcal{A}}(t\{x/u\})$, where $\Gamma = (\mathtt{fv}(t) \setminus x) \cap \mathtt{fv}(u)$ and Δ, Π are fresh sets of variables.*

Proof. By induction on t, using the usual lambda-calculus substitution definition [3] for x, as the source calculus is the lambda-calculus.

Theorem 1 (Simulation). *Let $t \in \mathcal{T}_{\emptyset}$ such that $t \to_{\emptyset} t'$.*

- *If $\mathtt{w} \in \mathcal{A}$, then $\mathtt{AR}_{\mathcal{A}}(t) \to_{\mathcal{A}}^{+} \mathcal{W}_{\mathtt{fv}(t)\setminus\mathtt{fv}(t')}(\mathtt{AR}_{\mathcal{A}}(t'))$.*
- *If $\mathtt{w} \notin \mathcal{A}$, then $\mathtt{AR}_{\mathcal{A}}(t) \to_{\mathcal{A}}^{+} \mathtt{AR}_{\mathcal{A}}(t')$.*

Proof. By induction on the reduction relation $\to_{\beta}$ using Lemma 3.

While Theorem 1 states that adding resources to the $\lambda_{\emptyset}$-calculus is well behaved, this does not necessarily hold for *any* arbitrary calculus of the prismoid. Thus for example, what happens when the $\lambda_{\mathtt{S}}$-calculus is enriched with resource $\mathtt{w}$? Is it possible to simulate each $\mathtt{s}$-reduction step by a sequence of $\mathtt{sw}$-reduction steps? Unfortunately the answer is no: we have $t_1 = (x\ y)[z/v] \to_{\mathtt{S}} x\ y[z/v] = t_2$ but $\mathtt{AR}_{\mathtt{W}}(t_1) = \mathcal{W}_z(x\ y)[z/v] \not\to_{\mathtt{SW}} x\ \mathcal{W}_z(y)[z/v] = \mathtt{AR}_{\mathtt{W}}(t_2)$.

4 Removing Resources

In this section we give a mechanism to remove resources, that is, to change the status of weakening and/or contraction from explicit to implicit. This is dual to the operation allowing to add resources to terms presented in Section 3. Whereas adding is only defined within the implicit base, removing is defined in both bases. As adding, removing is not only done on a static level, but also on a dynamic one. Thus for example, removing translates any $\mathtt{csw}$-reduction sequence into a $\mathcal{B}$-reduction sequence, for any $\mathcal{B} \in \{\mathtt{s}, \mathtt{cs}, \mathtt{sw}\}$.

Given two lists of variables $\Gamma = y_1 \ldots y_n$ (with all y_i distinct) and $\Delta = z_1 \ldots z_n$, then $(\Gamma \mapsto \Delta)(y)$ is y if $y \notin \Gamma$, or z_i if $y = y_i$ for some i. The **collapsing** function of a term without contractions is then defined modulo α-conversion as follows:

$$
\begin{array}{lll}
\mathtt{S}_{\Delta}^{\Gamma}(y) & := (\Gamma \mapsto \Delta)(y) & \\
\mathtt{S}_{\Delta}^{\Gamma}(uv) & := \mathtt{S}_{\Delta}^{\Gamma}(u)\mathtt{S}_{\Delta}^{\Gamma}(v) & \\
\mathtt{S}_{\Delta}^{\Gamma}(\lambda y.u) & := \lambda y.\mathtt{S}_{\Delta}^{\Gamma}(u) & y \notin \Gamma \\
\mathtt{S}_{\Delta}^{\Gamma}(u[y/v]) & := \mathtt{S}_{\Delta}^{\Gamma}(u)[y/\mathtt{S}_{\Delta}^{\Gamma}(v)] & y \notin \Gamma \\
\mathtt{S}_{\Delta}^{\Gamma}(\mathcal{W}_y(v)) & := \begin{cases} \mathtt{S}_{\Delta}^{\Gamma}(v) \\ \mathcal{W}_y(\mathtt{S}_{\Delta}^{\Gamma}(v)) \end{cases} & \begin{array}{l} (\Gamma \mapsto \Delta)(y) \in \mathtt{fv}(\mathtt{S}_{\Delta}^{\Gamma}(v)) \\ (\Gamma \mapsto \Delta)(y) \notin \mathtt{fv}(\mathtt{S}_{\Delta}^{\Gamma}(v)) \end{array}
\end{array}
$$

This function renames the variables of a term in such a way that every occurrence of $\mathcal{W}_x(t)$ in the term implies $x \notin \mathtt{fv}(t)$. For example $\mathtt{S}_{x,x}^{y,z}(\mathcal{W}_y(\mathcal{W}_z(x))) = x$.

The function $\mathtt{RR}_{\mathcal{A}}(\_) : \mathcal{T}_{\mathcal{B}} \mapsto \mathcal{T}_{\mathcal{B}\setminus\mathcal{A}}$ removes $\mathcal{A} \subseteq \mathcal{R}$ from a $\mathcal{B}$-term .

$$\begin{array}{llll}
\mathtt{RR}_{\mathcal{A}}(x) := x & \mathtt{RR}_{\mathcal{A}}(t[x/u]) := \mathtt{RR}_{\mathcal{A}}(t)[x/\mathtt{RR}_{\mathcal{A}}(u)] & & \\
\mathtt{RR}_{\mathcal{A}}(\lambda x.t) := \lambda x.\mathtt{RR}_{\mathcal{A}}(t) & \mathtt{RR}_{\mathcal{A}}(\mathcal{W}_x(t)) := \begin{cases} \mathtt{RR}_{\mathcal{A}}(t) & \text{if } \mathtt{w} \in \mathcal{A} \\ \mathcal{W}_x(\mathtt{RR}_{\mathcal{A}}(t)) & \text{if } \mathtt{w} \notin \mathcal{A} \end{cases} & & \\
\mathtt{RR}_{\mathcal{A}}(t\ u) := \mathtt{RR}_{\mathcal{A}}(t)\ \mathtt{RR}_{\mathcal{A}}(u) & \mathtt{RR}_{\mathcal{A}}(\mathcal{C}_x^{y|z}(t)) := \begin{cases} \mathtt{S}_{x,x}^{y,z}(\mathtt{RR}_{\mathcal{A}}(t)) & \text{if } \mathtt{c} \in \mathcal{A} \\ \mathcal{C}_x^{y|z}(\mathtt{RR}_{\mathcal{A}}(t)) & \text{if } \mathtt{c} \notin \mathcal{A} \end{cases} & &
\end{array}$$

Lemma 4. *Let $t, u \in \mathcal{T}_{\mathcal{A}}$ and $\mathtt{b} \in \mathcal{R}$. Then $\mathtt{RR}_{\mathtt{b}}(t\{x/u\}) = \mathtt{RR}_{\mathtt{b}}(t)\{x/\mathtt{RR}_{\mathtt{b}}(u)\}$.*

Calculi of the prismoid include rules/equations to handle substitution but also other rules/equations to handle resources $\{\mathtt{c}, \mathtt{w}\}$. Moreover, implicit (resp. explicit) substitution is managed by the β-rule (resp. the whole system $\mathtt{s}$). We can then split the reduction relation $\to_{\mathcal{B}}$ in two different parts: one for (implicit or explicit) substitution, which can be strictly projected into itself, and another one for weakening and contraction, which can be projected into a more subtle way given by the following statement.

Theorem 2. *Let $\mathcal{A} \subseteq \mathcal{R}$ such that $\mathcal{A} \subseteq \mathcal{B} \subseteq \mathcal{S}$ and let $t \in \mathcal{T}_{\mathcal{B}}$. If $t \equiv_{\mathcal{B}} u$, then $\mathtt{RR}_{\mathcal{A}}(t) \equiv_{\mathcal{B}\setminus\mathcal{A}} \mathtt{RR}_{\mathcal{A}}(u)$. Otherwise, we sum up in the following array :*

$\mathtt{s} \notin \mathcal{B}$	$t \Rightarrow_{\beta} u$	$\mathtt{RR}_{\mathcal{A}}(t) \to_{\beta}^{+} \mathtt{RR}_{\mathcal{A}}(u)$	$\mathtt{s} \in \mathcal{B}$	$t \Rightarrow_{\mathtt{s}} u$	$\mathtt{RR}_{\mathcal{A}}(t) \to_{\mathtt{s}}^{+} \mathtt{RR}_{\mathcal{A}}(u)$
	$t \Rightarrow_{\mathcal{B}\setminus\beta} u$	$\mathtt{RR}_{\mathcal{A}}(t) \to_{\mathcal{B}\setminus\beta\setminus\mathcal{A}}^{*} \mathtt{RR}_{\mathcal{A}}(u)$		$t \Rightarrow_{\mathcal{B}\setminus\mathtt{s}} u$	$\mathtt{RR}_{\mathcal{A}}(t) \to_{\mathcal{B}\setminus\mathtt{s}\setminus\mathcal{A}}^{*} \mathtt{RR}_{\mathcal{A}}(u)$
		$\mathtt{RR}_{\mathcal{B}}(t) = \mathtt{RR}_{\mathcal{B}}(u)$			$\mathtt{RR}_{\mathcal{B}}(t) = \mathtt{RR}_{\mathcal{B}}(u)$

Proof. By induction on the reduction relation using Lemma 4. For the points involving $\mathtt{RR}_{\mathcal{A}}(\_)$, one can first consider the case where $\mathcal{A}$ is a singleton. Then the general result follows from two successive applications of the simpler property.

It is now time to discuss the need of positive conditions (conditions involving positive free variables) in the specification of the reduction rules of the prismoid. For that, let us consider a relaxed form of $\mathtt{SS}_1$ rule $t[x/u][y/v] \to t[x/u[y/v]]$ if $y \in \mathtt{fv}(u) \setminus \mathtt{fv}(t)$ (instead of $y \in \mathtt{fv}^{+}(u) \setminus \mathtt{fv}(t)$).

The need of the condition $y \in \mathtt{fv}(u)$ is well-known [4], otherwise PSN does not hold. The need of the condition $y \notin \mathtt{fv}(t)$ is also natural if one wants to preserve well-formed terms. Now, the reduction step $t_1 = x[x/\mathcal{W}_y(z)][y/y'] \to_{\mathtt{SS}_1} x[x/\mathcal{W}_y(z)[y/y']] = t_2$ in the calculus with sorts $\{\mathtt{s}, \mathtt{w}\}$ cannot be projected into $\mathtt{RR}_{\mathtt{w}}(t_1) = x[x/z][y/y'] \to_{\mathtt{SS}_1} x[x/z[y/y']] = \mathtt{RR}_{\mathtt{w}}(t_2)$ since $y \notin \mathtt{fv}(z)$. Similar examples can be given to justify positive conditions in rules $\mathtt{SDup}, \mathtt{SCa}$ and $\mathtt{CS}$.

Lemma 5. *Let $t \in \mathcal{T}_{\emptyset}$ and let $\mathcal{A} \subseteq \mathcal{R}$. Then $\mathtt{RR}_{\mathcal{A}}(\mathtt{AR}_{\mathcal{A}}(t)) = t$.*

The following property states that administration of weakening and/or contraction is terminating in any calculus. The proof can be done by interpreting reduction steps by a strictly decreasing arithmetical measure.

Lemma 6. *If* $\mathtt{s} \notin \mathcal{B}$*, then the reduction relation* $\rightarrow_{\mathcal{B}\setminus\beta}$ *is terminating. If* $\mathtt{s} \in \mathcal{B}$*, then the reduction relation* $\rightarrow_{\mathcal{B}\setminus\mathtt{s}}$ *is terminating.*

We conclude this section by relating adding and removing resources :

Corollary 1. *Let* $\emptyset \neq \mathcal{A} \subseteq \mathcal{R}$*. Then, the unique* $\mathcal{A}$*-normal form of* $t \in \mathcal{T}_{\mathcal{A}}$ *is* $\mathtt{AR}_{\mathcal{A}}(\mathtt{RR}_{\mathcal{A}}(t))$ *if* $\mathtt{w} \notin \mathcal{A}$*, and* $\mathcal{W}_{\mathtt{fv}(t)\setminus\mathtt{fv}(\mathtt{RR}_{\mathcal{A}}(t))}(\mathtt{AR}_{\mathcal{A}}(\mathtt{RR}_{\mathcal{A}}(t)))$ *if* $\mathtt{w} \in \mathcal{A}$*.*

Proof. Suppose $\mathtt{w} \in \mathcal{A}$. Termination of $\rightarrow_{\mathcal{A}}$ (Lemma 6) implies that there is t' in $\mathcal{A}$-normal form such that $t \rightarrow^*_{\mathcal{A}} t'$. By Lemma 1, $\mathtt{fv}(t) = \mathtt{fv}(t')$ and by Theorem 2, $\mathtt{RR}_{\mathcal{A}}(t) = \mathtt{RR}_{\mathcal{A}}(t')$. Since t' is in $\mathcal{A}$-normal form, then we get $t' \equiv_{\mathcal{A}} \mathcal{W}_{\mathtt{fv}(t')\setminus\mathtt{fv}(\mathtt{RR}_{\mathcal{A}}(t'))}(\mathtt{AR}_{\mathcal{A}}(\mathtt{RR}_{\mathcal{A}}(t')))$ by a simple induction. Hence, $t' \equiv_{\mathcal{A}} \mathcal{W}_{\mathtt{fv}(t)\setminus\mathtt{fv}(\mathtt{RR}_{\mathcal{A}}(t))}(\mathtt{AR}_{\mathcal{A}}(\mathtt{RR}_{\mathcal{A}}(t)))$. To show uniqueness, let us consider two $\mathcal{A}$-normal forms t'_1 and t'_2 of t. By the previous remark, both t'_1 and t'_2 are congruent to the term $\mathcal{W}_{\mathtt{fv}(t)\setminus\mathtt{fv}(\mathtt{RR}_{\mathcal{A}}(t))}(\mathtt{AR}_{\mathcal{A}}(\mathtt{RR}_{\mathcal{A}}(t)))$ which concludes the case. The case $\mathtt{w} \notin \mathcal{A}$ is similar.

5 Untyped Properties

We first show PSN for all the calculi of the prismoid. The proof will be split in two different subcases, one for each base. This dissociation comes from the fact that redexes are erased by β-reduction in base $\mathfrak{B}_I$ while they are erased by $\mathtt{SGc}$ and/or $\mathtt{SW}_1$-reduction in base $\mathfrak{B}_E$.

Theorem 3 (PSN for the prismoid). *Let* $\mathcal{B} \subseteq \mathcal{S}$ *and* $\mathcal{A} = \mathcal{B} \setminus \{\mathtt{s}\}$*. If* $t \in \mathcal{T}_{\emptyset}$ & $t \in \mathcal{SN}_{\emptyset}$*, then* $\mathtt{AR}_{\mathcal{A}}(t) \in \mathcal{SN}_{\mathcal{B}}$*.*

Proof. There are three cases, one for $\mathfrak{B}_I$ and two subcases for $\mathfrak{B}_E$.

- Suppose $\mathtt{s} \notin \mathcal{B}$. We first show that $u \in \mathcal{T}_{\mathcal{B}}$ & $\mathtt{RR}_{\mathcal{B}}(u) \in \mathcal{SN}_{\emptyset}$ imply $u \in \mathcal{SN}_{\mathcal{B}}$. For that we apply Theorem 6 in the appendix with $\mathtt{A}_1 = \rightarrow_{\beta}$, $\mathtt{A}_2 = \rightarrow_{\mathcal{B}\setminus\beta}$, $\mathtt{A} = \rightarrow_{\beta}$ and $\mathcal{R} = \mathtt{RR}_{\mathcal{B}}(\_)$, using Theorem 2 and Lemma 6. Take $u = \mathtt{AR}_{\mathcal{B}}(t)$. Then $\mathtt{RR}_{\mathcal{B}}(\mathtt{AR}_{\mathcal{B}}(t)) =_{L.5} t \in \mathcal{SN}_{\emptyset}$ by hypothesis. Thus, $\mathtt{AR}_{\mathcal{B}}(t) \in \mathcal{SN}_{\mathcal{B}}$.
- Suppose $\mathcal{B} = \{\mathtt{s}\}$. The proof of $\mathtt{AR}_{\mathtt{s}}(t) = t \in \mathcal{SN}_{\mathtt{s}}$ follows a modular proof technique to show PSN of calculi with full composition which is completely developed in [8]. Details concerning the $\mathtt{s}$-calculus can be found in [17].
- Suppose $\mathtt{s} \in \mathcal{B}$. Then $\mathcal{B} = \{\mathtt{s}\} \cup \mathcal{A}$. We show that $u \in \mathcal{T}_{\mathcal{B}}$ & $\mathtt{RR}_{\mathcal{A}}(u) \in \mathcal{SN}_{\mathtt{s}}$ imply $u \in \mathcal{SN}_{\mathcal{B}}$. For that we apply Theorem 6 in the appendix with $\mathtt{A}_1 = \rightarrow_{\mathtt{s}}$, $\mathtt{A}_2 = \rightarrow_{\mathcal{B}\setminus\mathtt{s}}$, $\mathtt{A} = \rightarrow_{\mathtt{s}}$ and $\mathcal{R} = \mathtt{RR}_{\mathcal{A}}(\_)$, using Theorem 2 and Lemma 6. Now, take $u = \mathtt{AR}_{\mathcal{A}}(t)$. We have $\mathtt{RR}_{\mathcal{A}}(\mathtt{AR}_{\mathcal{A}}(t)) = \mathtt{RR}_{\mathcal{A}}(\mathtt{AR}_{\mathcal{A}}(t)) =_{L.5} t \in \mathcal{SN}_{\emptyset}$ by hypothesis and $t \in \mathcal{SN}_{\mathtt{s}}$ by the previous point. Thus, $\mathtt{AR}_{\mathcal{A}}(t) \in \mathcal{SN}_{\mathcal{B}}$.

Confluence of each calculus of the prismoid is based on that of the $\lambda_{\emptyset}$-calculus [3]. For any $\mathcal{A} \subseteq \mathcal{R}$, consider $\mathtt{xc} : \mathcal{T}_{\{\mathtt{s}\}\cup\mathcal{A}} \mapsto \mathcal{T}_{\mathcal{A}}$ which replaces explicit by implicit substitution.

$$
\begin{array}{llll}
\mathtt{xc}(y) & := y & \mathtt{xc}(\mathcal{W}_y(t)) & := \mathcal{W}_y(\mathtt{xc}(t)) \\
\mathtt{xc}(t\ u) & := \mathtt{xc}(t)\ \mathtt{xc}(u) & \mathtt{xc}(\mathcal{C}_y^{y_1|y_2}(t)) & := \mathcal{C}_y^{y_1|y_2}(\mathtt{xc}(t)) \\
\mathtt{xc}(\lambda y.t) & := \lambda y.\mathtt{xc}(t) & \mathtt{xc}(t[y/u]) & := \mathtt{xc}(t)\{y/\mathtt{xc}(u)\}
\end{array}
$$

Lemma 7. *Let* $t \in \mathcal{T}_{\mathcal{B}}$. *Then 1)* $t \to^*_{\mathcal{B}} \mathtt{xc}(t)$, *2)* $\mathtt{RR}_{\mathcal{B}\setminus\mathtt{s}}(\mathtt{xc}(t)) = \mathtt{xc}(\mathtt{RR}_{\mathcal{B}\setminus\mathtt{s}}(t))$. *3) if* $t \to_{\mathtt{s}} u$, *then* $\mathtt{xc}(t) \to^*_{\beta} \mathtt{xc}(u)$.

Proof. The first and the second property are shown by induction on t using respectively Lemmas 2 and 4. The third property is shown by induction on $t \to_{\mathtt{s}} u$.

Theorem 4. *All the languages of the prismoid are confluent.*

Proof. Let $t \to_{\mathcal{B}} t_1$ and $t \to_{\mathcal{B}} t_2$. We remark that $\mathcal{B} = \mathcal{A}$ or $\mathcal{B} = \{\mathtt{s}\} \cup \mathcal{A}$, with $\mathcal{A} \subseteq \mathcal{R}$. We have $\mathtt{RR}_{\mathcal{A}}(t) \to^*_{\mathcal{B}\setminus\mathcal{A}} \mathtt{RR}_{\mathcal{A}}(t_i)$ (i=1,2) by Theorem 2; $\mathtt{xc}(\mathtt{RR}_{\mathcal{A}}(t)) \to^*_{\beta} \mathtt{xc}(\mathtt{RR}_{\mathcal{A}}(t_i))$ (i=1,2) by Lemma 7; and $\mathtt{xc}(\mathtt{RR}_{\mathcal{A}}(t_i)) \to^*_{\beta} t_3$ (i=1,2) for some $t_3 \in \mathcal{T}_{\emptyset}$ by confluence of the λ-calculus [3]. Also, $\mathtt{AR}_{\mathcal{A}}(\mathtt{RR}_{\mathcal{A}}(\mathtt{xc}(t_i))) =_{L.\,7} \mathtt{AR}_{\mathcal{A}}(\mathtt{xc}(\mathtt{RR}_{\mathcal{A}}(t_i))) \to^*_{\mathcal{A}} \mathcal{W}_{\Delta_i}(\mathtt{AR}_{\mathcal{A}}(t_3))$ for some Δ_i (i=1,2) by Theorem 1.

But $t_i \to^*_{\mathcal{B}}$ $(L.\ 7)$ $\mathtt{xc}(t_i) \to^*_{\mathcal{A}}$ $(C.\ 1)$ $\mathcal{W}_{\Gamma_i}(\mathtt{AR}_{\mathcal{A}}(\mathtt{RR}_{\mathcal{A}}(\mathtt{xc}(t_i))))$ for some Γ_i (i=1,2). Then $\mathcal{W}_{\Gamma_i}(\mathtt{AR}_{\mathcal{A}}(\mathtt{RR}_{\mathcal{A}}(\mathtt{xc}(t_i)))) \to^*_{\mathcal{A}} \mathcal{W}_{\Gamma_i \cup \Delta_i}(\mathtt{AR}_{\mathcal{A}}(t_3))$ (i=1,2). Now, $\to^*_{\mathcal{A}} \subseteq \to^*_{\mathcal{B}}$ so in order to close the diagram we reason as follows.

If $\mathtt{w} \notin \mathcal{B}$, then $\Gamma_1 \cup \Delta_1 = \Gamma_2 \cup \Delta_2 = \emptyset$ and we are done. If $\mathtt{w} \in \mathcal{B}$, then $\to_{\mathcal{B}}$ preserves free variables by Lemma 1 so that $\mathtt{fv}(t) = \mathtt{fv}(t_i) = \mathtt{fv}(\mathcal{W}_{\Gamma_i \cup \Delta_i}(\mathtt{AR}_{\mathcal{A}}(t_3)))$ (i=1,2) which gives $\Gamma_1 \cup \Delta_1 = \Gamma_2 \cup \Delta_2$

6 Typing

We now introduce simply typed terms for all the calculi of the prismoid, and show that they all enjoy strong normalisation. **Types** are built over a countable set of atomic symbols and the type constructor $\to$.

An **environment** is a finite set of pairs of the form $x : T$. If $\Gamma = \{x_1 : T_1, ..., x_n : T_n\}$ is an environment then the domain of Γ is $\mathtt{dom}(\Gamma) = \{x_1, ..., x_n\}$. Two environments Γ and Δ are said to be **compatible** if $x : T \in \Gamma$ and $x : U \in \Delta$ imply $T = U$. Two environments Γ and Δ are said to be **disjoint** if there is no common variable between them. Compatible union (resp. disjoint union) is defined to be the union of compatible (resp. disjoint) environments.

Typing judgements have the form $\Gamma \vdash t : T$ for t a term, T a type and Γ an environment. **Typing rules** extend the inductive rules for well-formed terms (Section 2) with type annotations. Thus, typed terms are necessarily well-formed and each set of sorts $\mathcal{B}$ has its own set of typing rules.

$$\frac{}{x : T \vdash_{\mathcal{B}} x : T} \qquad \frac{\Gamma \vdash_{\mathcal{B}} t : T}{x : U; (\Gamma \setminus\!\!\setminus_{\mathcal{B}} \{y : U, z : U\}) \vdash_{\mathcal{B}} \mathcal{C}^{y|z}_x(t) : T}\ (\mathtt{c} \in \mathcal{B})$$

$$\frac{\Gamma \vdash_{\mathcal{B}} t : T}{\Gamma; x : U \vdash_{\mathcal{B}} \mathcal{W}_x(t) : T}\ (\mathtt{w} \in \mathcal{B}) \qquad \frac{\Gamma \vdash_{\mathcal{B}} u : U \quad \Delta \vdash_{\mathcal{B}} t : T}{\Gamma \uplus_{\mathcal{B}} (\Delta \setminus\!\!\setminus_{\mathcal{B}} x : U) \vdash_{\mathcal{B}} t[x/u] : T}\ (\mathtt{s} \in \mathcal{B})$$

$$\frac{\Gamma \vdash_{\mathcal{B}} t : U}{\Gamma \setminus\!\!\setminus_{\mathcal{B}} x : T \vdash_{\mathcal{B}} \lambda x.t : T \to U} \qquad \frac{\Gamma \vdash_{\mathcal{B}} t : T \to U \quad \Delta \vdash_{\mathcal{B}} u : T}{\Gamma \uplus_{\mathcal{B}} \Delta \vdash_{\mathcal{B}} tu : U}$$

A term $t \in \mathcal{T}_{\mathcal{B}}$ is said to **have type** T (written $t \in \mathcal{T}_{\mathcal{B}}^{T}$) iff there is Γ s.t. $\Gamma \vdash_{\mathcal{B}} t : T$. A term $t \in \mathcal{T}_{\mathcal{B}}$ is said to be **well-typed** iff there is T s.t. $t \in \mathcal{T}_{\mathcal{B}}^{T}$. Remark that every well-typed $\mathcal{B}$-term has a unique type.

Lemma 8. *If $\Gamma \vdash_{\mathcal{B}} t : T$, then 1) $\mathtt{fv}(t) = \mathtt{dom}(\Gamma)$, 2) $\Gamma \setminus \Pi; \Delta \vdash_{\mathcal{B}} R_{\Delta}^{\Pi}(t) : T$, for every $\Pi \subseteq \Gamma$ and fresh Δ, 3) $\mathtt{RR}_{\mathcal{A}}(t) \in \mathcal{T}_{\mathcal{B}\setminus\mathcal{A}}^{T}$, for every $\mathcal{A} \subseteq \mathcal{R}$.*

Proof. By induction on $\Gamma \vdash_{\mathcal{B}} t : T$.

Theorem 5 (Subject Reduction). *If $t \in \mathcal{T}_{\mathcal{B}}^{T}$ & $t \rightarrow_{\mathcal{B}} u$, then $u \in \mathcal{T}_{\mathcal{B}}^{T}$.*

Proof. By induction on the reduction relation using Lemma 8.

Corollary 2 (Strong normalization). *Let $t \in \mathcal{T}_{\mathcal{B}}^{T}$, then $t \in \mathcal{SN}_{\mathcal{B}}$.*

Proof. Let $\mathcal{A} \subseteq \mathcal{R}$ so that $\mathcal{B} = \mathcal{A}$ or $\mathcal{B} = \mathcal{A} \cup \{\mathtt{s}\}$. It is well-known that (simply) typed $\lambda_{\emptyset}$-calculus is strongly normalising (see for example [3]). It is also straightforward to show that PSN for the $\lambda_{\mathtt{s}}$-calculus implies strong normalisation for well-typed $\mathtt{s}$-terms (see for example [7]). By Theorem 2 any infinite $\mathcal{B}$-reduction sequence starting at t can be projected into an infinite $(\mathcal{B}\setminus\mathcal{A})$-reduction sequence starting at $\mathtt{RR}_{\mathcal{A}}(t)$. By Lemma 8 $\mathtt{RR}_{\mathcal{A}}(t)$ is a well-typed $(\mathcal{B} \setminus \mathcal{A})$-term, that is, a well-typed term in $\lambda_{\emptyset}$ or $\lambda_{\mathtt{s}}$. This leads to a contradiction.

7 Conclusion and Future Work

The prismoid of resources is proposed as an homogeneous framework to define λ-calculi being able to control weakening, contraction and linear substitution. The formalism is based on MELL Proof-Nets so that the computational behaviour of substitution is not only based on the propagation of substitution through terms but also on the decreasingness of the multiplicity of variables that are affected by substitutions. All calculi of the prismoid enjoy sanity properties such as simulation of β-reduction, confluence, preservation of β-strong normalisation and strong normalisation for typed terms.

The technology used in the prismoid could also be applied to implement higher-order rewriting systems. Indeed, it seems possible to extend these ideas to different frameworks such as CRSs [11], ERSs [10] or HRSs [15].

Another open problem concerns meta-confluence, that is, confluence for terms with meta-variables. This could be useful in the framework of Proof Assistants.

Finally, a more technical question is related to the operational semantics of the calculi of the prismoid. It seems possible to extend the ideas in [2] to our framework in order to identify those reduction rules of the prismoid that could be transformed into equations. Equivalence classes will be bigger, but reduction rules will coincide exactly with those of Nets [2]. While the operational semantics proposed in this paper is more adapted to implementation issues, the opposite direction would give a more abstract and flexible framework to study denotational properties.

References

1. Abadi, M., Cardelli, L., Curien, P.L., Lévy, J.-J.: Explicit substitutions. Journal of Functional Programming 4(1), 375–416 (1991)
2. Accattoli, B., Guerrini, S.: Jumping Boxes. Representing lambda-calculus boxes by jumps. To appear in Proceedings of CSL (2009)
3. Barendregt, H.: The Lambda Calculus: Its Syntax and Semantics. Studies in Logic and the Foundations of Mathematics, vol. 103. North-Holland, Amsterdam (1984)
4. Bloo, R.: Preservation of Termination for Explicit Substitution. PhD thesis, Eindhoven University of Technology (1997)
5. Girard, J.-Y.: Linear logic. Theoretical Computer Science 50(1), 1–101 (1987)
6. Kamareddine, F., Ríos, A.: A λ-calculus à la de Bruijn with explicit substitutions. In: Swierstra, S.D. (ed.) PLILP 1995. LNCS, vol. 982. Springer, Heidelberg (1995)
7. Kesner, D.: The theory of calculi with explicit substitutions revisited. In: Duparc, J., Henzinger, T.A. (eds.) CSL 2007. LNCS, vol. 4646, pp. 238–252. Springer, Heidelberg (2007)
8. Kesner, D.: Perpetuality for full and safe composition (in a constructive setting). In: Aceto, L., Damgård, I., Goldberg, L.A., Halldórsson, M.M., Ingólfsdóttir, A., Walukiewicz, I. (eds.) ICALP 2008, Part II. LNCS, vol. 5126, pp. 311–322. Springer, Heidelberg (2008)
9. Kesner, D., Lengrand, S.: Resource operators for lambda-calculus. Information and Computation 205(4), 419–473 (2007)
10. Khasidashvili, Z.: Expression reduction systems. In: Proceedings of IN Vekua Institute of Applied Mathematics, vol. 36, Tbilisi (1990)
11. Klop, J.-W.: Combinatory Reduction Systems. Mathematical Centre Tracts, vol. 127. PhD Thesis. Mathematisch Centrum, Amsterdam (1980)
12. Lengrand, S.: Normalisation and Equivalence in Proof Theory and Type Theory. PhD thesis, University Paris 7 and University of St Andrews (November 2006)
13. Lescanne, P., Rouyer-Degli, J.: The calculus of explicit substitutions $\lambda\nu$. Technical report, INRIA, Lorraine (1994)
14. Melliès, P.-A.: Typed λ-calculi with explicit substitutions may not terminate. In: Dezani-Ciancaglini, M., Plotkin, G. (eds.) TLCA 1995. LNCS, vol. 902. Springer, Heidelberg (1995)
15. Nipkow, T.: Higher-order critical pairs. In: 6th Annual IEEE Symposium on Logic in Computer Science (LICS) (1991)
16. Milner, R.: Local bigraphs and confluence: two conjectures. In: EXPRESS. ENTCS, vol. 175 (2006)
17. Renaud, F.: Preservation of strong normalisation for lambda-s (2008), `http://www.pps.jussieu.fr/~renaud`
18. Sinot, F.-R., Fernández, M., Mackie, I.: Efficient reductions with director strings. In: Baaz, M., Makowsky, J.A. (eds.) CSL 2003. LNCS, vol. 2803. Springer, Heidelberg (2003)
19. van Oostrom, V.: Net-calculus. Course Notes (2001), `http://www.phil.uu.nl/~oostrom/cmitt/00-01/net.ps`

A Appendix

Theorem 6 ([12]). *Let* $\mathtt{A}_1$ *and* $\mathtt{A}_2$ *be two reduction relations on the set* $\mathtt{k}$ *and let* $\mathtt{A}$ *be a reduction relation on the set* $\mathtt{K}$*. Let* $\mathcal{R} \subseteq \mathtt{k} \times \mathtt{K}$*. Suppose*

- *For every* u, v, U $(u\ \mathcal{R}\ U\ \&\ u\ \mathtt{A}_1\ v$ *imply* $\exists V$ *s.t.* $v\ \mathcal{R}\ V$ *and* $U\ \mathtt{A}^+\ V)$*.*
- *For every* u, v, U $(u\ \mathcal{R}\ U\ \&\ u\ \mathtt{A}_2\ v$ *imply* $\exists V$ *s.t.* $v\ \mathcal{R}\ V$ *and* $U\ \mathtt{A}^*\ V)$*.*
- *The relation* $\mathtt{A}_2$ *is well-founded.*

Then, $t\ \mathcal{R}\ T\ \&\ T \in \mathcal{SN}_{\mathtt{A}}$ *imply* $t \in \mathcal{SN}_{\mathtt{A}_1 \cup \mathtt{A}_2}$*.*

A Dynamic Algorithm for Reachability Games Played on Trees

Bakhadyr Khoussainov, Jiamou Liu, and Imran Khaliq

Department of Computer Science, University of Auckland, New Zealand
bmk@cs.auckland.ac.nz
{jliu036,ikha020}@aucklanduni.ac.nz

Abstract. Our goal is to start the investigation of dynamic algorithms for solving games that are played on finite graphs. The dynamic game determinacy problem calls for finding efficient algorithms that decide the winner of the game when the underlying graph undergoes repeated modifications. In this paper, we focus on turn-based reachability games. We provide an algorithm that solves the dynamic reachability game problem on trees. The amortized time complexity of our algorithm is $O(\log n)$, where n is the number of nodes in the current graph.

1 Introduction

We start to investigate dynamic algorithms for solving games that are played on finite graphs. Games played on graphs, with reachability, Büchi, Muller, Streett, parity and similar type of winning conditions, have recently attracted a great attention due to connections with model checking and verification problems, automata and logic [6][11][13][17]. Given one of these games, *to solve the game* means to design an (efficient) algorithm that tells us from which nodes a given player wins the game. Polynomial time algorithms exist to solve some of these games, while efficient algorithms for other games remain unknown. For example, on a graph with n nodes and m edges, the reachability game problem is in $O(n+m)$ and is PTIME-complete [8], and Büchi games are in $O(n \cdot m)$ [1]. Parity games are known to be in NP$\cap$ Co-NP but not known to be in P.

An algorithm for solving the games is *static* if the games remain unchanged over time. We pose the *dynamic game determinacy problem*:

> We would like to maintain the graph of the game that undergoes a sequence of update and query operations in such a way that facilitates an efficient solution of the current game.

Contrary to the static case, the dynamic determinacy problem takes as input a game $\mathcal{G}$ and a (finite or infinite) sequence $\alpha_1, \alpha_2, \alpha_3, \ldots$ of *update* or *query* operations. The goal is to optimize the average running time per operation over a worst-case sequence of operations. This is known as the *amortized running time* of the operations.

R. Královič and D. Niwiński (Eds.): MFCS 2009, LNCS 5734, pp. 477–488, 2009.

There has recently been increasing interest in dynamic graph algorithms (See, for example, [4][5]). The dynamic reachability problem on graphs have been investigated in a series of papers by King [9], Demetrescu and Italiano [3], Roditty [14] and Roditty and Zwick [15][16]. In [16], it is shown that for directed graphs with m edges and n nodes, there is a dynamic algorithm for the reachability problem which has an amortized update time of $O(m+n\log n)$ and a worst-case query time of $O(n)$. This paper extends this line of research to dynamic reachability game algorithms. In the setting of games, for a given directed graph G and a player σ, a set of nodes T is *reachable* from a node u in G means that there is a strategy for player σ such that starting from u, all paths produced by player σ following that strategy reach T, regardless of the actions of the opponent. Hence, the dynamic reachability game problem can be viewed as a generalization of the dynamic reachability problem for graphs.

We now describe two-person *reachability games* played on directed finite graphs. The two players are *Player 0* and *Player 1*. The *arena* $\mathcal{A}$ of the game is a directed graph (V_0, V_1, E), where V_0 is a finite set of *0-nodes*, V_1 is a finite set of *1-nodes* disjoint from V_0, and $E \subseteq V_0 \times V_1 \cup V_1 \times V_0$ is the edge relation. We use V to denote $V_0 \cup V_1$. A *reachability game* $\mathcal{G}$ is a pair $(\mathcal{A}, T)$ where $\mathcal{A}$ is the arena and $T \subseteq V$ is the set of *target nodes* for Player 0.

The players start by placing a token on some *initial node* $v \in V$ and then move the token in rounds. At each round, the token is moved along an edge by respecting the direction of the edge. If the token is placed at $u \in V_\sigma$, where $\sigma \in \{0,1\}$, then Player σ moves the token from u to a v such that $(u,v) \in E$. The play stops when the token reaches a node with no out-going edge or a target node. Otherwise, the play continues forever. Formally, a play is a (finite or infinite) sequence $\pi = v_0\ v_1\ v_2\ \ldots$ such that $(v_i, v_{i+1}) \in E$ for all i. Player 0 *wins the play* π if π is finite and the last node in π is in T. Otherwise, Player 1 wins the play.

In this paper we provide an algorithm that solves the dynamic reachability game played on trees. We investigate the *amortized time complexity* of the algorithm. We concentrate on trees because: (1) Trees are simple data structures, and the study of dynamic algorithms on trees is the first step towards the dynamic game determinacy problem. (2) Even in the case of trees the techniques one needs to employ is non-trivial. (3) The amortized time analysis for the dynamic reachability game problem on graphs, in general case, is an interesting hard problem. (4) Finally, we give a satisfactory solution to the problem on trees. We show that the amortized time complexity of our algorithm is of order $O(\log n)$, where n is the number of nodes on the tree. The space complexity of our algorithm is $O(n)$.

2 A Static Algorithm for Reachability Games

Let $\mathcal{G} = (\mathcal{A}, T)$ be a reachability game. A *(memoryless) strategy* for Player σ is a partial function $f_\sigma : V_\sigma \to V_{1-\sigma}$. A play $\pi = v_0\, v_1 \ldots$ *conforms* f_σ if $v_{i+1} = f_\sigma(v_i)$ whenever $v_i \in V_\sigma$ and f_σ is defined on v_i for all i. All strategies in this paper are memoryless. A *winning strategy for Player σ from v* is a strategy f_σ such that Player σ wins all plays starting from v that conform f_σ. A node u is a *winning position* for Player σ, if Player σ has a winning strategy from u. The *σ-winning region*, denoted W_σ, is the set of all winning positions for Player σ. Note that the winning regions are defined for memoryless strategies. A game *enjoys memoryless determinacy* if the regions W_0 and W_1 partition V.

Theorem 1 (Reachability game determinacy[7]). *Reachability games enjoy memoryless determinacy. Moreover, there is an algorithm that computes W_0 and W_1 in time $O(n+m)$.*

Proof. For $Y \subseteq V$, set $\text{Pre}(Y) = \{v \in V_0 \mid \exists u[(v,u) \in E \wedge u \in Y]\} \cup \{v \in V_1 \mid \forall u[(v,u) \in E \rightarrow u \in Y]\}$. Define a sequence $T_0, T_1, ...$ such that $T_0 = T$, and for $i > 0$, $T_i = \text{Pre}(T_{i-1}) \cup T_{i-1}$. There is an s such that $T_s = T_{s+1}$. We say a node u has *rank* r, $r \geq 0$, if $u \in T_r - T_{r-1}$. A node u has infinite rank if $u \notin T_s$. Once checks that a node $u \in W_0$ if and only if u has a finite rank. Computing W_0 takes $O(n+m)$ time. □

3 Dynamic Reachability Game Problem: A Set-Up

As mentioned above, the dynamic game determinacy problem takes as input a reachability game $\mathcal{G} = (\mathcal{A}, T)$ and a sequence $\alpha_1, \alpha_2, \ldots$ of *update* and *query* operations. The operations produce the sequence of games $\mathcal{G}_0, \mathcal{G}_1, \ldots$ such that $\mathcal{G}_i$ is obtained from $\mathcal{G}_{i-1}$ by applying the operation α_i. A dynamic algorithm should solve the game $\mathcal{G}_i$ for each i. We use the notation

$$\mathcal{A}_i = (V_{0,i} \cup V_{1,i}, E_i),\ V_i,\ T_i,\ W_{\sigma,i}$$

to denote the arena, the set of nodes, the target set and σ-winning region for $\mathcal{G}_i$. We define the following six *update operations* and one *query* operation:

1. *InsertNode*(u,i,j) operation, where $i,j \in \{0,1\}$, creates a new node u in V_i. Set u as a target if $j=1$ and not a target if $j=0$.
2. *DeleteNode*(u) deletes $u \in V$ where u is *isolated*, i.e., with no incoming or outgoing edge.
3. *InsertEdge*(u,v) operation inserts an edge from u to v.
4. *DeleteEdge*(u,v) operation deletes the edge from u to v.
5. *SetTarget*(u) operation sets node u as a target.
6. *UnsetTarget*(u) operation sets node u not as a target.
7. *Query(u)* operation returns **true** if $u \in W_0$ and **false** if $u \in W_1$

By Theorem 1, each node of V_i belongs to either $W_{0,i}$ or $W_{1,i}$.

Definition 1. *A node u is in* state σ *if $u \in W_\sigma$. The node u* changes its state *at stage $i+1$, if u is moved either from $W_{0,i}$ to $W_{1,i+1}$ or from $W_{1,i}$ to $W_{0,i+1}$.*

Using the *static algorithm* from Theorem 1, one produces two *lazy* dynamic algorithms for reachability games. The first algorithm runs the static algorithm after each update and therefore query takes constant time at each stage. The second algorithm modifies the game graph after each update operation *without* re-computing the winning positions, but the algorithm runs the static algorithm for the *Query(u)* operation. In this way, the update operations take constant time, but *Query(u)* takes the same time as the static algorithm. The amortized time complexity in both algorithms is the same as the static algorithm.

4 Reachability Game Played on Trees

This section is the main technical contribution of the paper. All trees are directed, and we assume that the reader is familiar with basic terminology for trees. A *forest* consists of pairwise disjoint trees. Since the underlying tree of the game undergos changes, the game will in fact be played on forests. We however still say that a *reachability game* $\mathcal{G}$ *is played on trees* if its arena is a forest F. We describe a fully dynamic algorithm for reachability games played on trees.

A forest F is implemented as a doubly linked list List(F) of nodes. A node u is represented by the tuple $(p(u)$, pos(u), tar$(u))$ where $p(u)$ is a pointer to the parent of u ($p(u)$ = null if u is a root), pos$(u) = \sigma$ if $u \in V_\sigma$ and a boolean variable tar(u) = **true** iff u is a target.

Our algorithm supports all the operations listed in Section 3. At stage s, the algorithm maintains a forest F_s obtained from F_{s-1} after performing an operation. We briefly discuss the operations and their implementations:

- Inputs of the update and query operation are given as pointers to their representatives in the linked list List(F_s).
- The operations *InsertNode* and *DeleteNode* are performed in constant time. The *InsertNode*(u, i, j) operation links the last node in List(F_s) to a new node u. The *DeleteNode*(u) operation deletes u from List(F_s).
- All other update operations and the query operation have amortized time $O(\log n)$, where n is the number of nodes in V.
- The *InsertEdge*(u, v) operation is performed only when v is the root of a tree not containing u. *InsertEdge*(u, v) links the trees containing u and v. *DeleteEdge*(u, v) does the opposite by splitting the tree containing u and v into two trees. One contains u and the other has v as its root.

4.1 Splay Trees

We describe *splay trees* (see [10] for details) which we will use in our algorithm. The *splay trees* form a dynamic data structure for maintaining elements drawn from a totally ordered domain D. Elements in D are arranged in a collection P_D of splay trees, each of which is identified by the root element.

- *Splay*(A, u): Reorganize the splay tree A so that u is at the root if $u \in A$.
- *Join*(A, B): Join two splay trees $A, B \in P_D$, where each element in A is less than each element in B, into one tree.

- *Split*(A, u): Split the splay tree $A \in P_D$ into two new splay trees

$$\text{Above}(u) = \{x \in A \mid x > u\} \text{ and } \text{Below}(u) = \{x \in A \mid x \leq u\}.$$

- *Max*(A)/*Min*(A): Returns the Max/Min element in $A \in P_D$.

Theorem 2 (splay trees[12]). *For the splay trees on P_D, the amortized time of the operations above is $O(\log n)$, where n is the cardinality of D.* □

4.2 Dynamic Path Partition

Recall that a node u *is in state* σ at stage s if $u \in W_{\sigma,s}$. Denote the state of u at stage s by $\text{State}_s(u)$. The update operations may change the state of u. This change may trigger a series of state changes on the ancestors of u. The state of u *does not change* if no update operation is applied to a descendant of u. We need to have a *ChangeState*(u) algorithm which carries out the necessary updates when the state of u is changed. The *ChangeState(u)* algorithm will not be executed alone, but will rather be a subroutine of other update operations.

One may propose a naive *ChangeState*(u) algorithm as follows. We hold for each node v its current state. When *ChangeState*(u) is called, it first changes the state of u, then checks if the parent $p(u)$ of u needs to change its state. If so, we changes the state of $p(u)$, and checks if the parent of $p(u)$ needs to change its state, etc. This algorithm takes $O(n)$ amortized time. Our goal is to improve this time bound. For this, we introduce *dynamic path partition* explained below.

Let $x <_F y$ denote the fact that x is an ancestor of y in forest F. Set $Path(x, y) = \{z \mid x \leq_F z \leq_F y\}$.

Definition 2. *A* path partition *of a forest F is a collection P_F of sets such that P_F partitions the set of nodes in F and each set in P_F is of the form $Path(x, y)$ for some $x, y \in F$.*

Nodes in $Path(x, y)$ are linearly ordered by $<_F$. Call the element of P_F that contains u the *block* of u. A block is *homogeneous* if all its elements have the same state. A path partition P_F is *homogeneous* if each block in P_F is homogeneous. For any u in F_s, set $\nu_s(u) = |\{v \mid (u, v) \in E_s \wedge \text{State}_s(u) = \text{State}_s(v)\}|$.

Definition 3. *A node u in F_s is* stable *at stage s if $u \in T_s$ or for some $\sigma \in \{0, 1\}$, $u \in V_{\sigma,s} \cap W_{\sigma,s}$ and $\nu_s(u) \geq 2$. We use Z_s to denote the set of stable nodes at stage s.*

The following lemma shows how state changes influence non-stable nodes in homogeneous fragments of blocks. The proof follows from the definitions.

Lemma 1. *Suppose x is stable, $x \leq_{F_s} y$, $Path(x, y)$ is homogeneous and there is no stable node in $\{z \mid x <_{F_s} z \leq_{F_s} y\}$. If y changes state at stage $s+1$ then the nodes in the set $\{z \mid x <_{F_s} z \leq_{F_s} y\}$ are precisely those nodes that need to change their states.* □

Definition 4. *A path partition of F is* stable *if whenever a node u is stable, it is the $\leq_F$-maximum element in its own block.*

From this definition, for a stable path partition, if u is stable, all elements $x \neq u$ in the block of u are *not* stable. An example of a stable and homogeneous partition is the trivial partition consisting of singletons.

At stage s, the algorithm maintains two data structures, one is the linked list List(F_s) as described above, and the other is the path partition P_{F_s}. Each node u in List(F_s) has an extra pointer to its representative in P_{F_s}. The path partition is maintained to be homogeneous and stable. Denote the block of u at stage s by $B_s(u)$. To obtain logarithmic amortized time, each $B_s(u)$ is represented by a splay tree.

4.3 *ChangeState*(*u*) Algorithm

The *ChangeState*(u) algorithm carries out two tasks. One is that it changes the states of all the necessary nodes of the underlying forest once u changes its state. The second is that it changes the path partition by preserving its homogeneity and stability properties. For the ease of notations, in this subsection we do not use the subscripts s for the forest F_s and the target set T_s. We write F for F_s and T for T_s.

We explain our *ChangeState(u)* algorithm informally. Suppose u changes its state at stage $s+1$. The algorithm defines a sequence of nodes $x_1 >_F x_2 >_F \ldots$ where $x_1 = u$. For $i \geq 1$, the algorithm splits $B_s(x_i)$ into two disjoint sets Above(x_i) and Below(x_i) and temporarily sets $B_{s+1}(x_i) = \text{Below}(x_i)$. By homogeneity, all nodes in Below(x_i) have the same state at stage s. Change the state of all nodes in $B_{s+1}(x_i)$ (this can be done by Lemma 1) and join the current two blocks containing u and $B_{s+1}(x_i)$ into one. If $\min\{B_{s+1}(x_i)\}$ is the root, stop the process. Otherwise, consider the parent of $\min\{B_{s+1}(x_i)\}$ which is $w_i = p(\min\{B_{s+1}(x_i)\})$. If $\text{State}_s(w_i) \neq \text{State}_s(x_i)$ or $w_i \in Z_s$, stop the loop, do not define x_{i+1} and $B_{s+1}(u)$ is now determined. Otherwise, set $x_{i+1} = w_i$ and repeat the above process for x_{i+1}. Consider the last w_i in the process described above that did not go into the block of u. If $w_i \notin Z_s$ and it becomes stable after the change, split $B_s(w_i)$ into Above(w_i) and Below(w_i) and declare that $w_i \in Z_{s+1}$. These all determine the new partition at stage $s+1$.

Algorithm 1 implements the *ChangeState(u)* procedure. The current block of v is denoted by $B(v)$. Elements of $B(v)$ are stored in a splay tree with order $\leq_F$. With the root of each splay tree $B(v)$ the variable $q(B(v)) \in \{0,1\}$ is associated to denote the current state of nodes in $B(v)$. The **while** loop computes the sequence $x_1, x_2, \ldots$ and $w_1, w_2, \ldots$ described above using the variables x and w. The boolean variable ChangeNext decides if the **while** loop is active. The boolean variable Stable(v) indicates whether v is stable. With each stable node v, the variable $n(v)$ equals $\nu(v)$ at the given stage.

The next two lemmas imply that the path partition obtained after the execution of *ChangeState(u)* remains homogeneous and stable.

Lemma 2. $B_{s+1}(u) = \{z \mid \mathit{State}_s(z) \neq \mathit{State}_{s+1}(z)\}$.

Algorithm 1. ChangeState(u)

1: ChangeNext $\leftarrow$ **true**; $x \leftarrow u$.
2: **while** ChangeNext **do**
3: Split($B(x), x$); $q(B(x)) \leftarrow 1 - q(B(x))$; Join($B(u), B(x)$).
4: **if** $p(\min\{B(x)\}) =$ null **then** Stop the process. **end if**
5: $w \leftarrow p(\min\{B(x)\})$.
6: **if** Stable(w) $\vee$ $q(B(w)) = q(B(u))$ **then** ChangeNext $\leftarrow$ **false**. **end if**
7: $x \leftarrow w$.
8: **end while**
9: Run **UpdateStable**(x, u)

Algorithm 2. UpdateStable(x, u)

1: **if** Stable(x) **then**
2: $n(x) \leftarrow [q(B(u)) = q(B(x))?\ n(x) + 1 : n(x) - 1]$.
3: **if** $n(x) < 2 \wedge \neg$ tar(x) **then** Stable(x) $\leftarrow$ **false**. **end if**
4: **else**
5: Stable(x) $\leftarrow$ **true**; Split($B(x), x$); $n(x) \leftarrow 2$.
6: **end if**

Proof. If $z = u$ then $\text{State}_s(z) \neq \text{State}_{s+1}(z)$. Assume $z \neq u$ and $z \in B_{s+1}(u)$. Let x be the $\leq_F$-maximum node in $B_s(z)$. By definition of $B_{s+1}(u)$, $x \notin Z_s$ and $\text{State}_s(x) = \text{State}_s(u)$. Therefore by Lemma 1 and the construction of the algorithm, $B_{s+1}(u) \subseteq \{z \mid \text{State}_s(z) \neq \text{State}_{s+1}(z)\}$.

Conversely, if $\text{State}_s(z) \neq \text{State}_{s+1}(z)$, z must be an ancestor of u. Let x be the $\leq_F$-minimum node in $B_{s+1}(u)$. If x is the root, $\{z \mid \text{State}_s(z) \neq \text{State}_{s+1}(z)\} \subseteq B_{s+1}(u)$. If x is not the root, either $p(x) \in Z_s$ or $\text{State}_s(p(x)) = \text{State}_{s+1}(u)$. Again by Lemma 1 and description of the algorithm, $\text{State}_s(x) = \text{State}_{s+1}(x)$ and thus $\{z \mid \text{State}_s(z) \neq \text{State}_{s+1}(z)\} \subseteq B_{s+1}(u)$. □

Lemma 3. *Suppose $v \notin Z_s$. The node $v \in Z_{s+1}$ if and only if v is the parent of the $\leq_F$-least node that changes its state at stage $s + 1$.*

Proof. Suppose for simplicity that $v \in V_{0,s}$. The case when $v \in V_{1,s}$ can be proved in a similar way. Suppose one of v's children, say v_0, is the $\leq_F$-least node that changes its state at stage $s + 1$. If $\text{State}_s(v) = 1$, then all of its children are in $W_{1,s}$. Thus $\text{State}_{s+1}(v_0)$=0, and v should also changes to state 0. This contradicts with the $\leq_F$-minimality of v_1. Therefore $v \in W_{0,s}$. Since $v \notin Z_s$, exactly one of its child, say v_1 is in $W_{0,s}$. If $v_1 = v_0$, then none of v's children is in $W_{0,s+1}$ and $\text{State}_s(v) \neq \text{State}_{s+1}(v)$. Therefore $v_1 \neq v_0$. Thus at stage $s + 1$, v has exactly two children in $W_{0,s+1}$ and $v \in Z_{s+1}$.

On the other hand, suppose $v \in Z_{s+1}$. This means that, $v \in W_{0,s+1}$ and v has two children v_0, v_1 in $W_{0,s+1}$. Note that for any x, at most one child of x may change state at any given stage. Therefore, at most one of v_0 and v_1, say v_1, is in $W_{0,s}$. Hence $v \in W_{0,s}$, and $\text{State}_s(v)$ =$\text{State}_{s+1}(v)$. Therefore v_0 is the $\leq_F$-least node that changes state at stage $s + 1$. □

4.4 Update and Query Operations

We describe the update and query operations. The query operation takes a parameter u and returns $q(B(u))$, which is the state variable associated with the root of the splay tree representing $B(u)$. We use an extra variable $c(u)$ to denote the current number of children of u.

In principle, the algorithms for operations *InsertEdge*(u, v), *DeleteEdge*(u, v), *SetTarget*(u) and *UnsetTarget*(u) perform the following three tasks. (1) Firstly, it carries out the update operation on u and v. (2) Secondly, it calls *ChangeState*(u) in the case when u needs to change state. (3) Lastly, it updates $c(z), n(z)$ and *Stable*(z) for each z. Task (1) is straightforward by changing the values of $p(v)$ and tar(u). Task (2) (3) can be done by using a fixed number of **if** statements, each with a fixed boolean condition involving comparisons on the variables. We illustrate this using the *InsertEdge*(u, v) operation as follows.

Algorithm 3. InsertEdge(u, v)

1: $p(v) \leftarrow u$; $c(u) \leftarrow c(u) + 1$.
2: **if** $\neg$ tar$(u) \wedge q(B(u)) \neq q(B(v)) \wedge (c(u) = 1$ or $q(B(u)) \neq$ Pos$(u))$ **then**
3:     Run **ChangeState**(u).
4: **else if** Stable$(u) \wedge q(B(u)) = q(B(v))$ **then**
5:     $n(u) \leftarrow n(u) + 1$.
6: **else if** $\neg$Stable$(u) \wedge c(u) > 1 \wedge q(B(u)) = q(B(v)) =Pos(u)$ **then**
7:     Stable$(u) \leftarrow$ **true**; Split$(B(u), u)$; $n(u) \leftarrow 2$.
8: **end if**

The *InsertEdge*(u, v) operation is described in Algorithm 3. Suppose the path partition on F_s is homogeneous and stable, and

1. $q(B(z)) = \text{State}_s(z)$.
2. $n(z) = |\{z' \mid (z, z') \in E_s \wedge \text{State}_s(z) = \text{State}_s(z')\}|$.
3. $c(z) = |\{z' \mid (z, z') \in E_s\}|$.
4. Stable(z) is *true* if and only if $z \in Z_s$.

Suppose the edge (u, v) is inserted at stage $s + 1$. We prove the following two lemmas which imply the correctness of the algorithm.

Lemma 4. ChangeState(u) *is called in the* InsertEdge(u, v) *algorithm if and only if State*$_s(u) \neq$ *State*$_{s+1}(u)$.

Proof. Note that *ChangeState*(u) is called if and only if the condition for the **if** statement at Line 2 of Algorithm 3 holds. We prove one direction of the lemma, the other direction is straightforward.

Suppose $\text{State}_s(u) \neq \text{State}_{s+1}(u)$. It must be that $u \notin T_s$ and $\text{State}_s(u) \neq \text{State}_s(v)$. Suppose further that u is not a leaf at stage s. If $u \in V_{0,s} \cap W_{0,s}$, there is a child w of u which is also in $W_{0,s}$. This means that $\text{State}_s(u) = \text{State}_{s+1}(u)$. Similarly, one may conclude that u does not change state if $u \in V_{1,s} \cap W_{1,s}$. Therefore $u \in (V_{0,s} \cap W_{1,s}) \cup (V_{1,s} \cap W_{0,s})$. Therefore the condition for the **if** statement at Algorithm 3 Line 2 holds. □

Lemma 5. *Stable*(u) *is* **true** *at stage* $s+1$ *if and only if* $u \in Z_{s+1}$.

Proof. It is easy to see that if $u \in Z_s$ then $u \in Z_{s+1}$ and Stable(u) is set to **true** at stage $s+1$. Suppose $u \notin Z_s$. Note that Stable(u) is set to **true** at stage $s+1$ if and only if Line 6 in Algorithm 3 is reached and the condition for the **if** statement at this line holds, if and only if u is not a leaf at stage s, $\text{State}_s(u) = \text{State}_{s+1}(u) = \text{State}_s(v)$ and $u \in V_{\sigma,s} \cap W_{\sigma,s}$ for some $\sigma \in \{0,1\}$.

If u is not a leaf, $\text{State}_s(u) = \text{State}_{s+1}(u) = \text{State}_s(v)$ and $u \in V_{\sigma,s} \cap W_{\sigma,s}$ for some $\sigma \in \{0,1\}$, then there is a child w of u in $W_{\sigma,s}$ and thus $u \in Z_{s+1}$. On the other hand, suppose $u \in Z_{s+1}$. If u changes state, then $u \in V_{\sigma,s} \cap W_{1-\sigma,s}$ for some $\sigma \in \{0,1\}$. This means all children of u are in $W_{1-\sigma,s}$ and $u \notin Z_{s+1}$. Therefore it must be that u did not change state. By definition of a stable node, u is not a leaf, $\text{State}_s(u) = \text{State}_s(v)$ and $u \in V_{\sigma,s} \cap W_{\sigma,s}$. □

Algorithm 4. DeleteEdge(u, v)

1: $p(v) \leftarrow$ null; $c(u) \leftarrow c(u) - 1$.
2: **if** $B(u) = B(v)$ **then** Split($B(u), u$). **end if**
3: **if** $\neg$Stable(u) $\wedge$ $q(B(u)) = q(B(v)) \wedge [(c(u) = 0 \wedge q(B(u)) = 0) \vee (q(B(u)) =$ Pos$(u))]$ **then**
4: Run **ChangeState**(u).
5: **end if**
6: **if** Stable(u) $\wedge$ $q(B(u)) = q(B(v))$ **then**
7: $n(u) \leftarrow n(u) - 1$.
8: **if** $n(u) < 2$ **then** Stable(u) $\leftarrow$ **false**. **end if**
9: **end if**

The *DeleteEdge*(u, v) operation is described in Algorithm 4. Suppose that the edge (u, v) is deleted at stage $s+1$. We also have the following two lemmas which imply the correctness of the algorithm. The proofs are similar in spirit to the proofs of Lemma 4 and Lemma 5.

Lemma 6. ChangeState(u) *is called in the* DeleteEdge(u, v) *algorithm if and only if* $\text{State}_s(u) \neq \text{State}_{s+1}(u)$. □

Lemma 7. *Stable*(u) *is* **true** *at stage* $s+1$ *if and only if* $u \in Z_{s+1}$. □

The *SetTarget*(u) and *UnsetTarget*(u) operations are described in Algorithm 5 and Algorithm 6, respectively.

4.5 Correctness

The next lemma implies the correctness of the algorithms 1-6.

Lemma 8. *At each stage* s, $\text{State}_s(z) = q(B(z))$ *for all node* $z \in F_s$.

Algorithm 5. SetTarget(u)

1: **if** $c(u) = 0$ **then** $n(u) = 0$.
2: **else if** Pos(u) $\neq q(B(u))$ **then** $n(u) \leftarrow$ [Pos(u) = 0? 0 : $c(u)$].
3: **else if** Stable(u) **then** $n(u) \leftarrow$ [Pos(u) = 0? $n(u)$: $c(u) - n(u)$].
4: **else** $n(u) \leftarrow$ [Pos(u) = 0? 1 : $c(n) - 1$]. **end if**
5: tar(u) $\leftarrow$ **true**; Stable(u) $\leftarrow$**true**; Split($B(u), u$).
6: **if** $q(B(u)) = 1$ **then** Run **ChangeState**(u). **end if**

Algorithm 6. UnsetTarget(u)

1: **if** (Pos(u) = 0 $\wedge$ $n(u) = 0$) $\vee$ (Pos(u) = 1 $\wedge$ $n(u) < c(u)$) **then**
2: Run **ChangeState**(u); $n(u) \leftarrow c(u) - n(u)$.
3: **end if**
4: Stable(u) $\leftarrow$ [($n(u) > 1 \wedge$ Pos(u) = $q(B(u))$)? **true**: **false**].
5: tar(u) $\leftarrow$ **false**.

Proof. The proof proceeds by induction on s. For simplicity, we assume that the initial forest F_0 contains only isolated nodes. We set for each node z, $p(z) =$ null, $c(z) = n(z) = 0$, $B(z) = \{z\}$, Stable(z) = **true** if and only if $q(B(z)) = 0$ if and only if tar(z) = **true**. For the case when F_0 is an arbitrary forest, we may assume that the variables $c(z), n(z), B(z)$, Stable(z) and $q(B(z))$ has been pre-set to their respect values as described in the previous subsection. We use $\gamma_s(u)$ to denote the number of children of u in stage s and recall that $\nu_s(u)$ is the number of u's children in the same state as u. The lemma follows from the fact that the following six inductive assumptions are preserved at each stage s.

(1) The set $\{B(v) \mid v \in V_{0,s} \cup V_{1,s}\}$ forms a homogeneous path partition of F_s.
(2) For each z, $c(z) = \gamma_s(z)$.
(3) For each z, $n(z) = \nu_s(z)$ whenever z is stable.
(4) For each z, Stable(z) is set to True if and only if z is a stable node.
(5) The path partition $\{B(v) \mid v \in V\}$ is stable.
(6) For each z, State$_s(z) = q(B(z))$. □

5 Complexity

We analyze the amortized complexity of our algorithm. Each *InsertEdge*(u, v), *DeleteEdge*(u, v), *SetTarget*(u), and *UnsetTarget*(u) algorithm runs at most once the *ChangeState*(u) algorithm, a fixed number of splay tree operations, and a fixed number of other low-level operations such as pointer manipulations and comparisons. By Theorem 2, each splay tree operation takes amortized time $O(\log n)$ where n is the number of nodes in the forest. Therefore, the amortized time complexity for these operations is $O(\log n)$ plus the amortized time taken by the *ChangeState*(u) algorithm.

We now focus on the amortized time complexity of the *ChangeState*(u) algorithm. The algorithm runs in iterations. At each iteration, it processes the current block of nodes $B(x)$ and examines the parent w of the $\leq_F$-least node in $B(x)$. If w does not exist or does not need to change state, the algorithm stops and calls *UpdateStable*(w, x); otherwise, the algorithm sets x to w and starts another iteration. Each iteration in the algorithm and the *UpdateStable*(w, x) algorithm both run a fixed number of splay tree operations and therefore takes amortized time $O(\log n)$. Therefore the algorithm takes time $O(\tau \log n + \log n)$ to execute a sequence of k update operations, where τ is the number of iterations of *ChangeState*(u). We prove the following lemma which implies the $O(\log n)$ amortized time complexity of the above update operations.

Lemma 9. *For any sequence of k update operations, the total number of iterations ran by the* ChangeState(u) *algorithm is $O(k)$.*

Proof. Given a forest $F = (V, E)$ and path partition P_F. For a node $u \in V$, let $B(u)$ denote the block of u. Define $E_F^P \subseteq P_F^2$ such that for all $u, v \in V$

$$(B(u), B(v)) \in E_F^P \text{ if and only if } (\exists w \in B(u))(\exists w' \in B(v))(w, w') \in E$$

The pair (P_F, E_F^P) also forms a forest, which we call the *partition forest* of P_F.

We prove the lemma using the *accounting method* (see [2]). At each stage s, we define the *credit function* $\rho_s : P_{F_s} \to \mathbb{N}$. For block $B \in P_{F_0}$, let $\rho_0(B)$ be the number of children of B in P_{F_0}. At each stage $s+1$, the credit function ρ_{s+1} is obtained from ρ_s with the following requirements:

- We create for each execution of *InsertEdge*(u, v) and *ChangeState*(u) an *amortized cost* of 1. This cost contributes towards the *credit* $\rho_s(B(u))$. Note that we can create amortized cost at most 2 at each stage.
- For each iteration of *ChangeState*(u), we take out 1 from $\rho_s(B)$ of some $B \in P_{F_s}$.

Let t_s be the total number of iterations ran at stage s. Our goal is to define ρ_s in such a way that for any $s > 0$,

$$\sum_{B \in P_{F_s}} \rho_s(B) \leq \sum_{B \in P_{F_{s-1}}} \rho_{s-1}(B) + 2 - t_s$$

and $\rho_s(B) \geq 0$ for any $B \in P_{F_s}$. Note that the existence of such a credit function ρ_s implies for any k,

$$\sum_{s=1}^{k} t_s \leq 2k \in O(k).$$

We define our desired credit function ρ_s by preserving, for every stage s, the following invariant:

$$(\forall B \in P_{F_s}) \;\; \rho_s(B) = |\; \{B' \in P_{F_s} \mid (B, B') \in E_{F_s}^P\} \;|.$$

The detailed procedure for defining ρ_s is omitted due to space restriction. $\square$

Theorem 3. *There exists a fully dynamic algorithm to solve reachability games played on trees which supports* InsertNode(u, i, j) *and* DeleteNode(u) *in constant time, and* InsertEdge(u, v), DeleteEdge(u, v), SetTarget(u), UnsetTarget(u) *and* Query(u) *in amortized* $O(\log n)$*-time where* n *is the number of nodes in the forest. The space complexity of the algorithm is* $O(n)$.

Proof. By Lemma 9, the update operations have amortized time complexity $O(\log n)$. The splay tree data structures take space $O(n)$ at each stage. □

References

1. Chatterjee, K., Henzinger, T., Piterman, N.: Algorithms for büchi games. In: Proceedings of the 3rd Workshop of Games in Design and Verification (GDV 2006) (2006)
2. Cormen, T., Leiserson, C., Rivest, R., Stein, C.: Introduction to algorithms, 2nd edn. MIT Press and McGraw-Hill (2001)
3. Demetrescu, C., Italiano, G.: Fully dynamic transitive closure: Breaking through the $O(n^2)$ barrier. In: Proceedings of FOCS 2000, pp. 381–389 (2000)
4. Demetrescu, C., Italiano, G.: A new approach to dynamic all pairs shortest paths. In: Proceedings of STOC 2003, pp. 159–166 (2003)
5. Eppstein, D., Galil, Z., Italiano, G.: Dynamic graph algorithms. In: Atallah, M. (ed.) Algorithms and Theoretical Computing Handbook. CRC Press, Boca Raton (1999)
6. Grädel, E.: Model checking games. In: Proceedings of WOLLIC 2002. Electronic Notes in Theoretical Computer Science, vol. 67. Elsevier, Amsterdam (2002)
7. Grädel, E., Thomas, W., Wilke, T. (eds.): Automata, Logics, and Infinite Games. LNCS, vol. 2500. Springer, Heidelberg (2002)
8. Immerman, N.: Number of quantifiers is better than number of tape cells. Journal of Computer and System Sciences 22, 384–406 (1981)
9. King, V.: Fully dynamic algorithms for maintaining all-pairs shortest paths and transitive closure in digraphs. In: Proceedings of FOCS 1999, pp. 81–91 (1999)
10. Kozen, D.: The design and analysis of algorithms. Monographs in computer science. Springer, New York (1992)
11. Murawski, A.: Reachability games and game semantics: Comparing nondeterministic programs. In: Proceedings of LICS 2008, pp. 353–363 (2008)
12. Tarjan, R.: Data structures and network algorithms. Regional Conference Series in Applied Mathematics, vol. 44. SIAM, Philadelphia (1983)
13. Radmacher, F., Thomas, W.: A game theoretic approach to the analysis of dynamic networks. In: Proceedings of the 1st Workshop on Verification of Adaptive Systems, VerAS 2007. Electronic Notes in Theoretical Computer Science, vol. 200(2), pp. 21–37. Kaiserslautern, Germany (2007)
14. Rodity, L.: A faster and simpler fully dynamic transitive closure. In: Proceedings of SODA 2003, pp. 401–412 (2003)
15. Roditty, L., Zwick, U.: Improved dynamic reachability algorithm for directed graphs. In: Proceedings of FOCS 2002, pp. 679–689 (2002)
16. Roditty, L., Zwick, U.: A fully dynamic reachability algorithm for directed graphs with an almost linear update time. In: Proceedings of 36th ACM Symposium on Theory of Computing (STOC 2004), pp. 184–191 (2004)
17. Thomas, W.: Infinite games and verification (Extended abstract of a tutorial). In: Brinksma, E., Larsen, K.G. (eds.) CAV 2002. LNCS, vol. 2404, pp. 58–64. Springer, Heidelberg (2002)

An Algebraic Characterization of Semirings for Which the Support of Every Recognizable Series Is Recognizable

Daniel Kirsten

University Leipzig, Institute for Computer Science,
Postfach 10 09 20, 04009 Leipzig, Germany
www.informatik.uni-leipzig.de/~kirsten/

Abstract. We show that for a semiring $\mathbb{K}$, the support of all recognizable series over $\mathbb{K}$ are recognizable if and only if in every finitely generated subsemiring of $\mathbb{K}$ there exists a finite congruence such that $0_{\mathbb{K}}$ is a singleton congruence class.

1 Introduction

One stream in the rich theory of formal power series deals with connections to formal languages. To each formal power series, one associates a certain language, called the support, which consists of all words which are not mapped to zero.

It is well known that recognizable series do not necessarily have recognizable supports. In the recent decades, one investigated sufficient conditions for the recognizability of the support of recognizable series, see [3, 9] for recent surveys.

Let us call some semiring $\mathbb{K}$ an *SR-semiring* (support-recognizable), if the support of every recognizable series over $\mathbb{K}$ is a recognizable language. It is well-known that all positive semirings (semirings which are both zero-divisor free and zero-sum free) and all locally finite semirings are SR-semirings [2, 3, 9].

WANG showed that commutative, quasi-positive semirings (that is, for every $k \in \mathbb{K} \setminus \{0_{\mathbb{K}}\}$, $\ell \in \mathbb{K}$, $n \in \mathbb{N}$, it holds $k^n + \ell \neq 0_{\mathbb{K}}$) are SR-semirings [13]. Very recently, the author showed that zero-sum free, commutative semirings are SR-semirings [5].

The standard examples for semirings which are not SR-semirings are $\mathbb{Q}$ and $\mathbb{Z}$ with usual addition and multiplication [2, 3, 9].

Despite the research in the recent decades, a characterization of SR-semirings was never achieved. One does not even know a necessary condition for a semiring to be an SR-semiring. The proof techniques for the above subclasses of SR-semirings are entirely different. Another confusing fact is all finite fields as well as $\mathbb{Q}^+$ are SR-semirings, but $\mathbb{Q}$ is not an SR-semiring.

In the present paper, we show an algebraic characterization of SR-semirings: some semiring $\mathbb{K}$ is an SR-semiring iff in every finitely generated subsemiring of $\mathbb{K}$ there exists a finite congruence such that $0_{\mathbb{K}}$ is a singleton congruence class. Equivalently, $\mathbb{K}$ is an SR-semiring iff in every finitely generated subsemiring of $\mathbb{K}$, the set $\{0_{\mathbb{K}}\}$ is recognizable according to MEZEI and WRIGHT [7]. Roughly

R. Královič and D. Niwiński (Eds.): MFCS 2009, LNCS 5734, pp. 489–500, 2009.

spoken, we prove this characterization by assuming some finite subset $C \subseteq \mathbb{K}$, defining a certain recognizable series S_C, and using the syntactic monoid of the support of S_C to show that the syntactic congruence of the set $\{0_\mathbb{K}\}$ in the subsemiring generated by C is finite.

From this characterization, we obtain some closure properties: the class of SR-semirings is closed under direct product, and the semiring of $(n \times n)$-matrices over some SR-semiring is an SR-semiring. In particular, the $(n \times n)$-matrices over $\mathbb{Q}^+$ are an SR-semiring which was not known in the literature.

The paper is organized as follows: In Section 2, we deal with some preliminaries. In Section 3.1, we present known subclasses of SR-semirings. In Section 3.2, we present Proposition 1 which characterizes the recognizability of R-supports and includes the key idea of the paper. In Section 3.3, we consider finitely generated semirings. In Section 3.4, we develop our characterization of SR-semirings as a particular case of the results from Section 3.2. In Section 3.5, we discuss some open questions.

2 Preliminaries

2.1 Notations

Let $\mathbb{N} = \{0, 1, \dots\}$. Let M be some set.

We call some mapping $f : \mathbb{N} \to M$ *ultimately constant* if there is some $k \in \mathbb{N}$ such that for every $k' \geq k$, $f(k') = f(k)$.

2.2 Monoids and Semirings

A *monoid* $(\mathbb{M}, \cdot, 1_\mathbb{M})$ consists of a set $\mathbb{M}$, a binary associative operation $\cdot$, and some $1_\mathbb{M} \in \mathbb{M}$ which is an identity for $\cdot$. If no confusion arises, we denote a monoid by $\mathbb{M}$.

Given some alphabet Σ, we denote by Σ^* the *free monoid* over Σ.

Let $\mathbb{M}$ be some monoid and $C \subseteq \mathbb{M}$. We denote by $[\![\,]\!] : C^* \to \mathbb{M}$ the unique homomorphism which arises from the identity on letters. We denote by $\langle C, \cdot \rangle$ the least subset of $\mathbb{M}$ which includes C and $1_\mathbb{M}$, and is closed under $\cdot$. Clearly, $\langle C, \cdot \rangle$ together with $\cdot$ and $1_\mathbb{M}$ is itself a monoid called the *submonoid generated by* C. We have $\langle C, \cdot \rangle = \{[\![w]\!] \mid w \in C^*\}$.

A *semiring* $(\mathbb{K}, +, \cdot, 0_\mathbb{K}, 1_\mathbb{K})$ consists of a set $\mathbb{K}$ together with two binary operations $+$ and $\cdot$ such that $(\mathbb{K}, +, 0_\mathbb{K})$ is a commutative monoid, $(\mathbb{K}, \cdot, 1_\mathbb{K})$ is a monoid with zero $0_\mathbb{K}$, and $(\mathbb{K}, \cdot, 1_\mathbb{K})$ distributes over $(\mathbb{K}, +, 0_\mathbb{K})$. As above, we denote a semiring by $\mathbb{K}$ if no confusion arises.

We call some semiring $\mathbb{K}$ *commutative* if $(\mathbb{K}, \cdot, 1_\mathbb{K})$ is a commutative monoid.

We call $\mathbb{K}$ *zero-divisor free* if for every $k, \ell \in \mathbb{K} \setminus \{0_\mathbb{K}\}$, we have $k\ell \neq 0_\mathbb{K}$. We call $\mathbb{K}$ *zero-sum free* if for every $k, \ell \in \mathbb{K} \setminus \{0_\mathbb{K}\}$, we have $k + \ell \neq 0_\mathbb{K}$. We call semirings which are zero-divisor free and zero-sum free *positive* semirings.

Let $C \subseteq \mathbb{K}$. We denote by $\langle C, \cdot \rangle$ the submonoid of $(\mathbb{K}, \cdot, 1_\mathbb{K})$ generated by C. We denote by $\langle C, +, \cdot \rangle$ the least subset of $\mathbb{K}$ which contains C, $1_\mathbb{K}$, and $0_\mathbb{K}$, and is closed under $+$ and $\cdot$. We call $\langle C, +, \cdot \rangle$ the *subsemiring of* $\mathbb{K}$ *generated by* C.

Let $k \in \mathbb{K}$. A *decomposition of k over C* consists of some $n \in \mathbb{N}$ and $w_1, \ldots, w_n \in C^*$ such that

$$k = \sum_{i \in \{1,\ldots,n\}} [\![w_i]\!].$$

There exists a decomposition of k over C iff $k \in \langle C, +, \cdot \rangle$.

If $\mathbb{K} = \langle C, +, \cdot \rangle$ for some finite subset $C \subseteq \mathbb{K}$, then $\mathbb{K}$ is called *finitely generated*. If for every finite $C \subseteq \mathbb{K}$, the subsemiring $\langle C, +, \cdot \rangle$ is finite, then $\mathbb{K}$ is called locally finite.

2.3 Monoids and Congruences

Let $(\mathbb{M}, \cdot, 1_\mathbb{M})$ be some monoid. We call an equivalence relation $\sim$ on $\mathbb{M}$ a *congruence* iff for every $k, \ell, m \in \mathbb{M}$, $k \sim \ell$ implies $km \sim \ell m$ and $mk \sim m\ell$. Equivalently, we can say that an equivalence relation $\sim$ on $\mathbb{M}$ is a *congruence* iff for every $k, k', \ell, \ell' \in \mathbb{M}$ satisfying $k \sim k'$ and $\ell \sim \ell'$, we have $k\ell \sim k'\ell'$.

A congruence is called *finite*, if it has finitely many congruence classes.

A congruence $\sim$ is called *coarser* than a congruence $\sim'$, if for every $k, \ell \in \mathbb{M}$, $k \sim' \ell$ implies $k \sim \ell$.

Let $L \subseteq \mathbb{M}$. For $k, \ell \in \mathbb{M}$, let $k \sim_L \ell$ if for every $m, m' \in \mathbb{M}$, we have $mkm' \in L$ iff $m\ell m' \in L$. The relation $\sim_L$ is a congruence and called the *syntactic congruence of* L. The quotient of $\mathbb{M}$ under $\sim_L$ is denoted by $\mathbb{M}_L = \mathbb{M}/_{\sim_L}$ and called the *syntactic monoid* of L. The canonical homomorphism $\eta_L : \mathbb{M} \to \mathbb{M}_L$ is called the *syntactic homomorphism* of L.

A congruence $\sim$ *saturates* L if L is a union of congruence classes of $\sim$. It is well-known that $\sim_L$ is the coarsest congruence which saturates L. Consequently, there exists some finite congruence which saturates L if and only if $\sim_L$ is finite. In this case, L is called a *recognizable* set over $\mathbb{M}$. If $\mathbb{M}$ is a free monoid Σ^*, then this notion of recognizability coincides with recognizability by finite automata.

2.4 Semirings and Congruences

Let $(\mathbb{K}, +, \cdot, 0_\mathbb{K}, 1_\mathbb{K})$ be some semiring.

An equivalence relation $\sim$ on $\mathbb{K}$ is called a *congruence*, if $\sim$ is a congruence for both the monoids $(\mathbb{K}, +, 0_\mathbb{K})$ and $(\mathbb{K}, \cdot, 1_\mathbb{K})$.

Let $R \subseteq \mathbb{K}$. For $k, \ell \in \mathbb{K}$, let $k \sim_R \ell$ if for every $p, q, m \in \mathbb{K}$, we have $pkq + m \in R$ iff $p\ell q + m \in R$.

It is easy to show that $\sim_R$ is an equivalence on $\mathbb{K}$. Moreover, we can show that $\sim_R$ is a congruence. For this, let $k, \ell \in \mathbb{K}$ satisfy $k \sim_R \ell$. Let $x \in \mathbb{K}$ be arbitrary. At first, we want to show $kx \sim_R \ell x$ and $xk \sim_R x\ell$. Let $p, q, m \in \mathbb{K}$ be arbitrary. Since $k \sim_R \ell$, we have $p(kx)q+m = pk(xq)+m \in R$ iff $p(\ell x)q+m = p\ell(xq)+m \in R$, and hence, $kx \sim_R \ell x$. By a symmetric argument, we get $xk \sim_R x\ell$. Thus, $\sim_R$ is a congruence for $(\mathbb{K}, \cdot, 1_\mathbb{K})$.

Next, we want to show $k + x \sim_R \ell + x$. We have $p(k+x)q+m = pkq+(pxq+m) \in R$ iff $p(\ell+x)q+m = p\ell q+(pxq+m) \in R$. Hence, $k+x \sim_R \ell+x$, and by the commutativity of $+$, it follows $x+k \sim_R x+\ell$.

Thus, $\sim_R$ is a congruence. We call $\sim_R$ the *syntactic congruence of* R.

Given $k, \ell \in \mathbb{K}$ satisfy $k \sim_R \ell$, we have $1_\mathbb{K} k 1_\mathbb{K} + 0_\mathbb{K} = k \in R$ iff $1_\mathbb{K} \ell 1_\mathbb{K} + 0_\mathbb{K} = \ell \in R$, i.e., $k \in R$ iff $\ell \in R$. Thus, $\sim_R$ saturates R.

Lemma 1. *The relation* $\sim_R$ *is the coarsest congruence which saturates* R.

Proof. Let $\sim$ be some congruence on $\mathbb{K}$ which saturates R. Let $k, \ell \in \mathbb{K}$ satisfy $k \sim \ell$. Let $p, q, m \in \mathbb{K}$. Since $\sim$ is a congruence, we have $pkq + m \sim p\ell q + m$. Since $\sim$ saturates R, we have $pkq + m \in R$ iff $p\ell q + m \in R$. Hence, $k \sim_R \ell$.

According to Mezei and Wright [7], some subset $M \subseteq \mathbb{K}$ is called *recognizable over* $(\mathbb{K}, +, \cdot, 0_\mathbb{K}, 1_\mathbb{K})$ if there is a finite semiring K', a subset $F \subseteq \mathbb{K}'$ and a homomorphism $h : \mathbb{K} \to \mathbb{K}'$ such that $M = h^{-1}(F)$.

We have then the following classic coincidence:

Theorem 1. *Let* $R \subseteq \mathbb{K}$. *The following assertions are equivalent:*

1. *The set* R *is recognizable.*
2. *The set* R *is saturated by a finite congruence* $\sim$ *on* $\mathbb{K}$.
3. *The syntactic congruence of* R *is finite.*

Proof (sketch). (3) $\Rightarrow$ (2) is obvious. (2) $\Rightarrow$ (3) follows from Lemma 1. To show (1) $\Rightarrow$ (2), one sets $k \sim \ell$ iff $h(k) = h(\ell)$. To show (2) $\Rightarrow$ (1), one uses the quotient $\mathbb{K}/_\sim$, the canonical homomorphism, and the image of R as $\mathbb{K}'$, h, and F, respectively.

2.5 Weighted Finite Automata

We recall some notions on (weighted automata) and recommend [1, 2, 3, 4, 6, 8, 10] for overviews.

We denote by Σ some finite alphabet. Let $(\mathbb{K}, +, \cdot, 0, 1)$ be a semiring. Mappings from Σ^* to $\mathbb{K}$ are often called *series*. We denote the class of all series from Σ^* to $\mathbb{K}$ by $\mathbb{K}\langle\langle\Sigma^*\rangle\rangle$. A *weighted finite automaton over* Σ *and* $\mathbb{K}$ (for short *WFA*) is a tuple $[Q, \mu, \lambda, \varrho]$, where

- Q is a non-empty, finite set of *states*,
- $\mu : \Sigma^* \to \mathbb{K}^{Q \times Q}$ is a homomorphism whereas $\mathbb{K}^{Q \times Q}$ are the $Q \times Q$ matrices over $\mathbb{K}$ with usual matrix multiplication, and
- $\lambda, \varrho \in \mathbb{K}^Q$.

Let $\mathcal{A} = [Q, \mu, \lambda, \varrho]$ be a WFA. It defines a series $|\mathcal{A}|$ by $|\mathcal{A}|(w) = \lambda\mu(w)\varrho$ for $w \in \Sigma^*$. As usual, $\lambda\mu(w)\varrho$ is understood as a product of the row λ, the matrix $\mu(w)$, and the column ϱ.

A series S is called *recognizable over* $\mathbb{K}$ if $S = |\mathcal{A}|$ for some WFA $\mathcal{A}$. The class of all recognizable series is denoted by $\mathbb{K}^{\text{rec}}\langle\langle\Sigma^*\rangle\rangle$.

Let $C \subseteq \mathbb{K}$. If for every $p, q \in Q$, we have $\lambda[p], \mu[p, q], \varrho[q] \in C \cup \{0_\mathbb{K}\}$, then we call $\mathcal{A}$ a WFA *with weights in* C.

3 Overview and Main Results

3.1 Historical Background

Let $\mathbb{K}$ be a semiring and $S \in \mathbb{K}\langle\langle \Sigma^* \rangle\rangle$. The *support of* S is defined by

$$\mathsf{supp}(S) = \{w \in \Sigma^* \mid S(w) \neq 0_\mathbb{K}\}.$$

It is well-known that if S is recognizable, $\mathsf{supp}(S)$ is not necessarily recognizable. The standard example is the series S over the semiring of the rational numbers $(\mathbb{Q}, +, \cdot, 0, 1)$ defined by $S(w) = 2^{|w|_a} 0.5^{|w|_b} - 0.5^{|w|_a} 2^{|w|_b}$.[1] For $w \in \Sigma^*$, we have $S(w) = 0$ iff $|w|_a = |w|_b$. Hence, $\mathsf{supp}(S) = \{w \in \Sigma^* \mid |w|_a \neq |w|_b\}$ which is not a recognizable language. However, S is well known to be recognizable [2, 3, 9].

We could also define $S(w) = 2^{|w|_a} 3^{|w|_b} - 3^{|w|_a} 2^{|w|_b}$ which gives a counter example in the semiring $(\mathbb{Z}, +, \cdot, 0, 1)$.

From the beginnings of the research on formal power series in the sixties it was studied under which conditions the support of a recognizable series is recognizable [10, 2, 3, 9]. One is particularly interested in classes of semirings for which the support of any recognizable series is recognizable. This problem was already considered in [10].

Let us call a semiring $\mathbb{K}$ an *SR-semiring* (support-recognizable), if for every finite alphabet[2] Σ and every $S \in \mathbb{K}^{\mathrm{rec}}\langle\langle \Sigma^* \rangle\rangle$, $\mathsf{supp}(S)$ is a recognizable language.

Today, one knows three classes of SR-semirings.

1. The most obvious statement was already mentioned in [10] (see also [2, 9]): Every positive semiring is an SR-semiring. The proof is straightforward, starting from a WFA $\mathcal{A} = [Q, \mu, \lambda, \varrho]$ one "removes the weights from $\mathcal{A}$" and obtains a non-deterministic automaton recognizing $\mathsf{supp}(|\mathcal{A}|)$.
   Let $\mathbb{K}_1$ be the semiring of finite languages over a finite alphabet Δ, equipped with union and concatenation, $(\mathcal{P}_\mathrm{f}(\Delta^*), \cup, \cdot, \emptyset, \{\varepsilon\})$. The semiring $\mathbb{K}_1$ is a typical example of a positive semiring.
2. Every locally finite semiring is an SR-semiring [2, 3, 9]. Given some WFA $\mathcal{A} = [Q, \mu, \lambda, \varrho]$ over a locally finite semiring, the range of μ is finite. Hence, we can define a finite congruence $\sim$ on Σ^* by setting $u \sim v$ iff $\mu(u) = \mu(v)$. This congruence saturates $\mathsf{supp}(|\mathcal{A}|)$, and hence, $\mathsf{supp}(|\mathcal{A}|)$ is recognizable. This class includes all finite semirings.
   Let $\mathbb{K}_2$ be the set of all ultimately constant[3] mappings $f : \mathbb{N} \to \mathbb{Z}/4\mathbb{Z}$. With componentwise $+$ and $\cdot$, $(\mathbb{K}_2, +, \cdot, 0, 1)$ is a locally finite semiring.
3. There is another, less known class of SR-semirings: In 1998 [13], Wang introduced *quasi-positive* semirings $\mathbb{K}$ (that is, for every $k \in \mathbb{K} \setminus \{0_\mathbb{K}\}$, $\ell \in \mathbb{K}$, $n \in \mathbb{N}$, it holds $k^n + \ell \neq 0_\mathbb{K}$), and showed that all commutative, quasi-positive semirings are SR-semirings. Very recently, the author generalized Wang's result by showing that every commutative, zero-sum free semiring is an SR- semiring [5]. The proof relies on counting factors and Dickson's lemma.

[1] For $w \in \Sigma^*, x \in \Sigma$, $|w|_x$ denotes the number of occurrences of x in w.

[2] We could equivalently use some fixed Σ with at least two letters (see Section 3.4).

[3] The restriction to ultimately constant mappings is just because the author prefers countable examples.

Let $\mathbb{Q}^+ = \{q \in \mathbb{Q} \mid q \geq 0\}$ and $\mathbb{K}_3 = \big(\mathbb{Q}^+ \times \mathbb{Q}^+, +, \cdot, (0,0), (1,1)\big)$ whereas $+$ and $\cdot$ are defined componentwise. The semiring $\mathbb{K}_3$ is a typical commutative, zero-sum free semiring which has zero-divisors.

These three classes are incomparable: The semiring $\mathbb{K}_1$ belongs exclusively to the first class, because it is not locally finite, and it is not commutative for $|\Delta| > 1$. The semiring $\mathbb{K}_2$ belongs exclusively to the second class, because it has zero-sums. The semiring $\mathbb{K}_3$ belongs exclusively to the third class, because it has zero-divisors and it is not locally finite.

To the authors knowledge, there is no known SR-semiring in the literature beyond these three classes.

Let us mention that there is an important related result: in 1961 and 1975 M. P. Schützenberger and E. D. Sontag showed that if $\mathbb{K}$ is a ring, then for every recognizable series S with a finite image and every $R \subseteq \mathbb{K}$, the set $\mathsf{supp}_R(S) = \{w \in \Sigma^* \mid S(w) \in R\}$ is recognizable [11, 12, 2, 9].

The SR-semirings under (1) and (2) above, and the result by Schützenberger/Sontag are the fundamental cornerstones in this area for 25 years.

Up to now, it was not possible to characterize the SR-semirings. To the authors knowledge, one does not even know a necessary condition for a semiring to be an SR-semiring. In the authors opinion, there are two confusing observations:

The first confusing observation arises around the rational numbers: On the one hand, the field $(\mathbb{Q}, +, \cdot, 0, 1)$ is the standard example of a non-SR-semiring. On the other hand, all finite fields are SR-semirings. Moreover, $(\mathbb{Q}^+, +, \cdot, 0, 1)$ is a positive semiring and thus an SR-semiring.

The other confusing observation is that the proof ideas for these three classes are quite different on a first glance, and it seems to be hopeless to give one coherent proof for all the three classes.

3.2 On R-Supports

Let $\mathbb{K}$ be a semiring and $R \subseteq \mathbb{K}$. The *R-support* of some $S \in \mathbb{K}\langle\langle \Sigma^* \rangle\rangle$ is defined as

$$\mathsf{supp}_R(S) = \{w \in \Sigma^* \mid S(w) \in R\}.$$

In Proposition 1, we show a crucial characterization for the recognizability of R-supports. All the other results in the paper are more or less conclusions from Proposition 1. In particular, we will achieve results on the support by considering the particular case $R = \mathbb{K} \setminus \{0_\mathbb{K}\}$ of Proposition 1.

Proposition 1. *Let C be a finite subset of $\mathbb{K}$ and assume $1_\mathbb{K} \in C$. Let $R \subseteq \mathbb{K}$. The following assertions are equivalent:*

1. *For every WFA $\mathcal{A}$ over some finite alphabet Σ with weights in C, the set $\mathsf{supp}_R(|\mathcal{A}|)$ is recognizable.*
2. *For every WFA $\mathcal{A}$ over $\{a, b\}$ with weights in C, the set $\mathsf{supp}_R(|\mathcal{A}|)$ is recognizable.*
3. *The set $R \cap \langle C, +, \cdot \rangle$ is recognizable over the semiring $\langle C, +, \cdot \rangle$.*

Proof. (1) $\Rightarrow$ (2) is obvious.

(2) $\Rightarrow$ (1) (sketch) Let Σ be an arbitrary, finite alphabet. Let $\mathcal{A} = [Q, \mu, \lambda, \varrho]$ be a WFA over Σ with weights in C. Let $h : \Sigma^* \to \{a, b\}^*$ be some injective homomorphism. We can construct a WFA $\mathcal{A}'$ over $\{a, b\}$ with weights in C such that for every $w \in \{a, b\}^*$

$$|\mathcal{A}'|(w) = \begin{cases} |\mathcal{A}|\big(h^{-1}(w)\big) & \text{if } h^{-1}(w) \text{ is defined} \\ 0_{\mathbb{K}} & \text{if } h^{-1}(w) \text{ is not defined.} \end{cases}$$

To construct $\mathcal{A}'$, we use the states Q (from $\mathcal{A}$) and some additional states. The key idea is: if $\mathcal{A}$ can read some letter $c \in \Sigma$ from some $p \in Q$ to some $q \in Q$ with a weight $\mu(c)[p, q]$, then $\mathcal{A}'$ can read $h(c)$ from p to q with the same weight, i.e., $\mu'\big(h(c)\big)[p, q] = \mu(c)[p, q]$. To proceed the construction of $\mathcal{A}'$ and prove its correctness formally, we need the path semantics of WFA which is not introduced here for the reasons of space.

Clearly, $\mathsf{supp}_R(|\mathcal{A}|) = h^{-1}\big(\mathsf{supp}_R(|\mathcal{A}'|)\big)$. By (2), $\mathsf{supp}_R(|\mathcal{A}'|)$ is recognizable. Since recognizable are closed under inverse homomorphisms, $\mathsf{supp}_R(|\mathcal{A}|)$ is recognizable.

(3) $\Rightarrow$ (1) Let $\mathcal{A} = [Q, \mu, \lambda, \varrho]$ be a WFA with weights in C. Clearly, μ is a homomorphism $\mu : \Sigma^* \to \langle C, +, \cdot \rangle^{Q \times Q}$. By (3), there are a homomorphism $h : \langle C, +, \cdot \rangle \to \mathbb{K}_\mathsf{f}$ into a finite semiring $\mathbb{K}_\mathsf{f}$ and a subset $F \subseteq \mathbb{K}_\mathsf{f}$ such that $R \cap \langle C, +, \cdot \rangle = h^{-1}(F)$. The homomorphism h extends to a homomorphism $h : \langle C, +, \cdot \rangle^{Q \times Q} \to \mathbb{K}_\mathsf{f}^{Q \times Q}$.

We define a relation $\sim$ on Σ^* by setting $u \sim v$ if $h(\mu(u)) = h(\mu(v))$. Since $h \circ \mu : \Sigma^* \to \mathbb{K}_\mathsf{f}^{Q \times Q}$ is a homomorphism, $\sim$ is a congruence. Since $h \circ \mu$ maps into the finite monoid $\mathbb{K}_\mathsf{f}^{Q \times Q}$, $\sim$ is a finite congruence.

We show that $\sim$ saturates $\mathsf{supp}_R(|\mathcal{A}|)$. Let $u, v \in \Sigma^*$ satisfy $u \sim v$, i.e., $h(\mu(u)) = h(\mu(v))$. Consequently, $h(\lambda) \cdot h(\mu(u)) \cdot h(\varrho) = h(\lambda) \cdot h(\mu(v)) \cdot h(\varrho)$, $h(\lambda\mu(u)\varrho) = h(\lambda\mu(v)\varrho)$, i.e., $h\big(|\mathcal{A}|(u)\big) = h\big(|\mathcal{A}|(v)\big)$. By the choice of h, we have $|\mathcal{A}|(u) \in R \cap \langle C, +, \cdot \rangle$ iff $|\mathcal{A}|(v) \in R \cap \langle C, +, \cdot \rangle$. Since $|\mathcal{A}|(u), |\mathcal{A}|(v) \in \langle C, +, \cdot \rangle$, it follows $|\mathcal{A}|(u) \in R$ iff $|\mathcal{A}|(v) \in R$, i.e., $u \in \mathsf{supp}_R(|\mathcal{A}|)$ iff $v \in \mathsf{supp}_R(|\mathcal{A}|)$.

(1) $\Rightarrow$ (3) We define a suitable series S_C over the alphabet $\Sigma = C \uplus \{\#\}$. From the recognizability of $\mathsf{supp}_R(S_C)$, we conclude that the syntactic congruence of $R \cap \langle C, +, \cdot \rangle$ in the semiring $\langle C, +, \cdot \rangle$ is finite.

Let $n \geq 0$ and $w_1, \ldots, w_n \in C^*$. We define

$$S_C(\#w_1\#w_2 \ldots \#w_n) = \sum_{i=1}^{n} [\![w_i]\!].$$

(Recall that $[\![\,]\!] : C^* \to \langle C, \cdot \rangle \subseteq \langle C, +, \cdot \rangle$ is the canonical homomorphism.)

For $w \in C\Sigma^*$, we define $S_C(w) = 0_{\mathbb{K}}$.

Clearly, $\{S_C(w) \mid w \in \Sigma^*\} = \langle C, +, \cdot \rangle$.

The series S_C is recognized by the following WFA with weights in C. The label $C|C$ at state 3 means that for every $x \in C$, there is a transition from 3 to 3 which is labeled with the letter x and weighted with x.

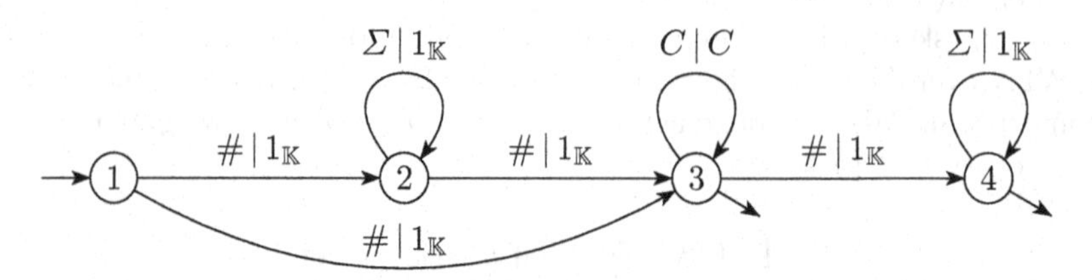

Let $w = \#w_1\#w_2 \dots \#w_n$. The key idea of the WFA is, that for every $1 \le i \le n$, there is a path which reads w_i from state 3 to state 3 and is weighted with $[\![w_i]\!]$.

For the construction of this WFA, we need the assumption $1_{\mathbb{K}} \in C$.

For abbreviation, we denote $R' = R \cap \langle C, +, \cdot \rangle$.

Let $\mathbb{M}$ and $\eta : \Sigma^* \to \mathbb{M}$ be the syntactic monoid and the syntactic homomorphism of $\mathsf{supp}_R(S_C)$. By (1), $\mathbb{M}$ is finite. By a counting argument, there is some $n \le |\mathbb{M}|!$ such that for every $k \in \mathbb{M}$, we have $k^n = k^{2n}$.

We show two claims (a) and (b). For $k \in \mathbb{K}$ and $m \in \mathbb{N}$, let

$$m \otimes k \;=\; \underbrace{k + \cdots + k}_{m\text{-times}}.$$

(a) For every $w, w' \in C^*$ satisfying $\eta(w) = \eta(w')$, we have $[\![w]\!] \sim_{R'} [\![w']\!]$.
(b) For every $w \in C^*$, we have $2n \otimes [\![w]\!] \;\sim_{R'}\; n \otimes [\![w]\!]$.

We show (a). Let $p, q, m \in \langle C, +, \cdot \rangle$. We have to show $p[\![w]\!]q + m \in R'$ iff $p[\![w']\!]q + m \in R'$.

Let $n_p \ge 0$ and $p_1, \dots, p_{n_p} \in C^*$ be a decomposition of p. Assume similar decompositions for q and for m. If $n_p = 0$ or $n_q = 0$, then $p = 0_{\mathbb{K}}$ or $q = 0_{\mathbb{K}}$, and we are done. Assume $n_p \ge 1$ and $n_q \ge 1$. Let $x \in \{w, w'\}$, and consider the word:[4]

$$\left(\bigodot_{i=1}^{n_p} \bigodot_{j=1}^{n_q} \#p_i x q_j \right) \left(\bigodot_{i=1}^{n_m} \#m_i \right). \tag{1}$$

If we set $x = w$ (resp. $x = w'$), then expression (1) defines a word which we call y (resp. y'). We have $S_C(y) =$

$$\sum_{i=1}^{n_p} \sum_{j=1}^{n_q} [\![p_i w q_j]\!] + \sum_{i=1}^{n_m} [\![m_i]\!] = \left(\sum_{i=1}^{n_p} [\![p_i]\!] \right) [\![w]\!] \left(\sum_{j=1}^{n_q} [\![q_j]\!] \right) + \sum_{i=1}^{n_m} [\![m_i]\!] = p[\![w]\!]q + m,$$

and similarly, $S_C(y') = p[\![w']\!]q + m$. From $\eta(w) = \eta(w')$, it follows $\eta(y) = \eta(y')$. Hence, $y \in \mathsf{supp}_R(S_C)$ iff $y' \in \mathsf{supp}_R(S_C)$. Thus, $S_C(y) \in R$ iff $S_C(y') \in R$. Since $S_C(y), S_C(y') \in \langle C, +, \cdot \rangle$, it follows, $S_C(y) \in R'$ iff $S_C(y') \in R'$, i.e., $p[\![w]\!]q + m \in R'$ iff $p[\![w']\!]q + m \in R'$.

We show (b). Let $p, q, m \in \langle C, +, \cdot \rangle$ and assume decompositions as above. Let $x \in \{n, 2n\}$ and consider the word

$$\left(\underbrace{\bigodot_{i=1}^{n_p} \bigodot_{j=1}^{n_q} \bigodot_{\ell=1}^{x} \#p_i w q_j} \right) \left(\bigodot_{i=1}^{n_m} \#m_i \right). \tag{2}$$

[4] The large symbol $\odot$ in Expressions (1) and (2) denotes the concatenation of words.

If we set $x = 2n$ (resp. $x = n$), then expression (2) defines a word which we call y (resp. y'). We have $S_C(y) = p(2n \otimes [\![w]\!])q + m$, and $S_C(y') = p(n \otimes [\![w]\!])q + m$. We have $\eta(y) = \eta(y')$. (Note that for fixed i and j, the η-image of the underbraced part in (2) is the same for $x = n$ and for $x = 2n$ by the choice of n.) Now, we can argue as for (a).

Now, we have shown (a) and (b), and we can show that $\sim_{R'}$ is finite.

Let $X \subseteq C^*$ such that for every $w \in C^*$ there is exactly one $\sim_{R'}$-equivalent[5] word in X. We have $|X| \leq |\mathbb{M}|$ by (a).

Let $p \in \langle C, +, \cdot \rangle$. Let $n_p \geq 0$ and $p_1, \ldots, p_{n_p} \in C^*$ be a decomposition of p. By (a) we can replace in $p_1, \ldots, p_{n_p}$ every word by a $\sim_{R'}$-equivalent word in X. If there are at least $2n$ equal words in the list, we can utilize (b) and erase n of the $2n$ equal words. In this way, we achieve for every $x \in X$ some $n_x < 2n$ such that

$$p \sim_{R'} \sum_{x \in X} n_x \otimes [\![x]\!].$$

Consequently, $\sim_{R'}$ has at most $(2n)^{|X|} \leq \big(2(|\mathbb{M}|!)\big)^{|\mathbb{M}|}$ classes. □

Theorem 2. *Let $R \subseteq \mathbb{K}$. The following assertions are equivalent:*

1. *For every WFA $\mathcal{A}$ over some finite Σ, the set $\mathsf{supp}_R(|\mathcal{A}|)$ is recognizable.*
2. *For every WFA $\mathcal{A}$ over $\{a, b\}$, the set $\mathsf{supp}_R(|\mathcal{A}|)$ is recognizable.*
3. *For every finitely generated subsemiring $\mathbb{K}' \subseteq \mathbb{K}$, the set $R \cap \mathbb{K}'$ is recognizable over the semiring $\mathbb{K}'$.*

Proof. (1) $\Leftrightarrow$ (2) and (3) $\Rightarrow$ (1) follow immediately from Proposition 1(1) $\Leftrightarrow$ (2) and Proposition 1(3) $\Rightarrow$ (1), respectively.

(1) $\Rightarrow$ (3) Just let $C \subseteq \mathbb{K}'$ be a finite set of generators of $\mathbb{K}'$, and apply Proposition 1(1) $\Rightarrow$ (3) for the set $C = C' \cup \{1_{\mathbb{K}}\}$.

3.3 On Finitely Generated Semirings

For finitely generated semirings, Theorem 2 simplifies as follows:

Corollary 1. *Let $\mathbb{K}$ be a finitely generated semiring and let $R \subseteq \mathbb{K}$. The following assertions are equivalent:*

1. *For every WFA $\mathcal{A}$ over $\mathbb{K}$ and some Σ, the set $\mathsf{supp}_R(|\mathcal{A}|)$ is recognizable.*
2. *The set R is recognizable over $\mathbb{K}$.*

Proof. (1)$\Rightarrow$(2) follows immediately from Theorem 2(1)$\Rightarrow$(3).

Assume (2). Clearly, for every subsemiring $\mathbb{K}' \subseteq \mathbb{K}$, the set $R \cap \mathbb{K}'$ is recognizable over $\mathbb{K}'$. (To obtain a finite congruence on $\mathbb{K}'$ which saturates $R \cap \mathbb{K}'$, we can simply restrict a finite congruence on $\mathbb{K}$ which saturates R to $\mathbb{K}'$.) Now, (1) follows from Theorem 2(3)$\Rightarrow$(1). □

[5] Let us call $u, v \in C^*$ $\sim_{R'}$-equivalent if $[\![u]\!] \sim_{R'} [\![v]\!]$.

Corollary 1(1)⇒(2) does not hold for arbitrary semirings. Just consider $\mathbb{K}_2$ from Section 3.1 and let $R = \{0_{\mathbb{K}_2}\}$. Every finitely generated subsemiring $\mathbb{K}'$ of $\mathbb{K}_2$ is finite, and hence, $R = R \cap \mathbb{K}'$ is recognizable over $\mathbb{K}'$. By Theorem 2, $\mathbb{K}_2$ and R satisfy Corollary 1(1).

However, $\mathbb{K}_2$ and R do not satisfy Corollary 1(2): just let $f \neq g \in \mathbb{K}_2$. We have $(1_{\mathbb{K}_2} \cdot f \cdot 1_{\mathbb{K}_2}) + (-f) = 0_{\mathbb{K}_2}$ but $(1_{\mathbb{K}_2} \cdot g \cdot 1_{\mathbb{K}_2}) + (-f) = g - f \neq 0_{\mathbb{K}_2}$. Hence, $f \not\sim_R g$, i.e., $\sim_R$ is infinite. Thus, $R = \{0_{\mathbb{K}_2}\}$ is not recognizable over $\mathbb{K}_2$.

In the particular case of the semiring $(\mathbb{Z}, +, \cdot, 0, 1)$, it was observed in [2, 3] that for every $a, b \in \mathbb{Z}$, $R = a + b\mathbb{Z}$, and every $S \in \mathbb{Z}^{\text{rec}}\langle\langle \Sigma^* \rangle\rangle$, $\mathsf{supp}_R(S)$ is recognizable. We can extend this observation.

Corollary 2. *Let $R \subseteq \mathbb{Z}$. The following assertions are equivalent:*

1. *For every WFA $\mathcal{A}$ over $\mathbb{Z}$ and some Σ, the set $\mathsf{supp}_R(|\mathcal{A}|)$ is recognizable.*
2. *There is some $b \geq 1$ such that for every $x \in R$, we have $x - b \in R$ and $x + b \in R$.*

Proof. By Corollary 1, it remains to show that (2) characterizes the recognizable sets of $\mathbb{Z}$. Indeed, if R satisfies (2), then R is saturated by the finite congruence $\sim$ defined by $x \sim y$ iff $x - y$ is a multiple of b.

Conversely, if R is recognizable, then R is saturated by a finite congruence $\sim$. There are $y \in \mathbb{Z}$, $b \geq 1$ such that $y \sim y + b$. It follows $0 \sim b$, and (2).

Clearly, we could also fix $\Sigma = \{a, b\}$ in Corollaries 1 and 2.

3.4 On SR-Semirings

We present our characterization of SR-semirings:

Theorem 3. *Let $\mathbb{K}$ be a semiring. The following assertions are equivalent:*

1. *The semiring $\mathbb{K}$ is an SR-semiring.*
2. *For every finitely generated subsemiring $\mathbb{K}' \subseteq \mathbb{K}$, $\{0_{\mathbb{K}}\}$ is a recognizable set over $\mathbb{K}'$.*
3. *In every finitely generated subsemiring $\mathbb{K}' \subseteq \mathbb{K}$, there is a finite congruence such that $\{0_{\mathbb{K}}\}$ is a singleton congruence class.*

Proof. (2)⇔(3) follows from Theorem 1(1)⇔(2).

Let $R = \mathbb{K} \setminus \{0_{\mathbb{K}}\}$. We show (1)⇒(2). Let $\mathbb{K}' \subseteq \mathbb{K}$ be a finitely generated subsemiring. From Theorem 2(2)⇒(3), it follows that the set $R \cap \mathbb{K}' = \mathbb{K}' \setminus \{0_{\mathbb{K}}\}$ as well as its complement $\{0_{\mathbb{K}}\}$ are recognizable.

We show (2)⇒(1). From (2), it follows that for every finitely generated subsemiring $\mathbb{K}' \subseteq \mathbb{K}$, the set $\mathbb{K}' \setminus \{0_{\mathbb{K}}\} = R \cap \mathbb{K}'$ is recognizable over $\mathbb{K}'$. Thus, (1) follows from Theorem 2(3)⇒(1). □

The reader should be aware that in the proof of (1)⇒(2) in Theorem 3, we just require that for every recognizable series S over the alphabet $\{a, b\}$ over $\mathbb{K}$, $\mathsf{supp}(S)$ is recognizable. In the proof of (2)⇒(1) in Theorem 3, we prove that for every recognizable series S over some finite Σ over $\mathbb{K}$, $\mathsf{supp}(S)$ is recognizable. Consequently, the size of the alphabet Σ is not relevant in the definition of an SR-semiring in Section 3.1.

Let us reconsider the known cases of SR-semirings from Section 3.1. If $\mathbb{K}$ is a positive semiring, then every subsemiring $\mathbb{K}'$ is a positive semiring. The definition of a positive semiring just says that there is a congruence on $\mathbb{K}'$ which consists of the congruence classes $\{0_\mathbb{K}\}$ and $\mathbb{K}' \setminus \{0_\mathbb{K}\}$. This congruence is somehow an extremal case of a congruence in Theorem 3(3).

If $\mathbb{K}$ is a locally finite semiring, then every finitely generated subsemiring $\mathbb{K}'$ is finite. Hence, we can show that $\mathbb{K}$ is an SR-semiring by using the identity relation on $\mathbb{K}'$ as a congruence in Theorem 3(3). This congruence is somehow the other extremal case of a congruence in Theorem 3(3).

Consequently, the known cases of SR-semirings, the positive semirings and the locally finite semirings, are two contrary extremal cases of Theorem 3.

Theorem 4. *For SR-semirings $\mathbb{K}, \mathbb{K}_1, \mathbb{K}_2$, the semirings $\mathbb{K}_1 \times \mathbb{K}_2$ and $\mathbb{K}^{n \times n}$ for $n \geq 1$ are SR-semirings.*

Proof. We show the claim for $\mathbb{K}_1 \times \mathbb{K}_2$. It suffices to show that $\mathbb{K}_1 \times \mathbb{K}_2$ satisfies Theorem 3(3). So let $\mathbb{K}' \subseteq \mathbb{K}_1 \times \mathbb{K}_2$ be finitely generated. Let $C \subseteq \mathbb{K}'$ be a finite set of generators of $\mathbb{K}'$. Let $i \in \{1, 2\}$. Let $C_i \subseteq \mathbb{K}_i$ be the i-th components of the members of C, i.e., $C \subseteq C_1 \times C_2$. Since $\mathbb{K}_i$ is an SR-semiring, there is by Theorem 3(1)$\Rightarrow$(3) a finite congruence $\sim_i$ on $\langle C_i, +, \cdot \rangle$ such that $\{0_{\mathbb{K}_i}\}$ is a singleton congruence class. We define a relation $\sim$ on $\mathbb{K}'$ by setting $(k_1, k_2) \sim (\ell_1, \ell_2)$ iff $k_1 \sim_1 \ell_1$ and $k_2 \sim_2 \ell_2$ for every $(k_1, k_2), (\ell_1, \ell_2) \in \mathbb{K}'$. It is easy to verify that $\sim$ is a congruence as in Theorem 3(3).

We show the claim for $\mathbb{K}^{n \times n}$. Again, we show that $\mathbb{K}^{n \times n}$ satisfies Theorem 3(3). Let $\mathbb{K}' \subseteq \mathbb{K}^{n \times n}$ be a finitely generated subsemiring. Let C' be a finite set of generators of $\mathbb{K}'$. Let

$$C = \{A[i,j] \mid A \in C', 1 \leq i, j \leq n\}.$$

Clearly, C is finite and $\mathbb{K}' \subseteq \langle C, +, \cdot \rangle^{n \times n}$. Since $\mathbb{K}$ is an SR-semiring, there is by Theorem 3(1)$\Rightarrow$(3) a congruence $\sim$ on $\langle C, +, \cdot \rangle$ such that $\{0_\mathbb{K}\}$ is a singleton congruence class. We can extend $\sim$ to $\mathbb{K}'$: for $A, B \in \mathbb{K}'$, we set $A \sim B$ if $A[i,j] \sim B[i,j]$ for every $1 \leq i, j \leq n$. Clearly, $\sim$ on $\mathbb{K}'$ is a finite congruence and the null-matrix is a singleton congruence class. $\square$

From Theorem 4, we can construct SR-semirings which were not known previously. For example, the product $\mathcal{P}_\mathsf{f}(\Delta^*) \times \mathbb{Z}/4\mathbb{Z}$ (with componentwise operation) and the semiring of $n \times n$ matrices over $\mathbb{Q}^+$ for $n \geq 1$ are SR-semirings by Theorem 4, but they do not fall in any class in Section 3.1.

3.5 Open Questions

It is not clear whether or how one can characterize the semirings for which the support (or R-support) of every recognizable series with finite image is recognizable. As mentioned in Section 3.1, this class includes all rings (in particular $\mathbb{Q}$) [11, 12, 2, 9], and hence, this class strictly includes all SR-semirings.

Given some language $L \subseteq \Sigma^*$, the *characteristic series of* L is defined by $1_L(w) = 1_{\mathbb{K}}$ if $w \in L$ and $1_L(w) = 0_{\mathbb{K}}$ if $w \notin L$. It is well-known that for every semiring, the characteristic series of a recognizable language is recognizable [2, 9]. It is a wide open problem to characterize the semirings for which the converse holds, i.e., to characterize the semirings for which the support of every recognizable series with image $\{0_{\mathbb{K}}, 1_{\mathbb{K}}\}$ is recognizable. See [13] for recent results.

Moreover, one could consider the case of a single letter alphabet. If for some semiring $\mathbb{K}$ and a one-letter alphabet Σ the support of every series in $\mathbb{K}^{\text{rec}}\langle\langle\Sigma^*\rangle\rangle$ is recognizable, is $\mathbb{K}$ necessarily an SR-semiring?

Finally, it is not clear whether Proposition 1 holds for finite sets $C \subseteq \mathbb{K}$ which do not include $1_{\mathbb{K}}$.

Acknowledgments

The author thanks anonymous referees for their useful remarks.

References

[1] Berstel, J.: Transductions and Context-Free Languages. B. G. Teubner, Stuttgart (1979)

[2] Berstel, J., Reutenauer, C.: Rational Series and Their Languages. EATCS Monographs on Theoretical Computer Science, vol. 12. Springer, Heidelberg (1984)

[3] Berstel, J., Reutenauer, C.: Noncommutative rational series with applications (prelimary electronic version) (2009), http://www.igm.univ-mlv.fr/~berstel/

[4] Droste, M., Kuich, W., Vogler, H. (eds.): Handbook of Weighted Automata. Monographs in Theoretical Computer Science. An EATCS Series. Springer, Heidelberg (2009)

[5] Kirsten, D.: The support of a recognizable series over a zero-sum free, commutative semiring is recognizable. In: Diekert, V., Nowotka, D. (eds.) DLT 2009. LNCS, vol. 5583. Springer, Heidelberg (2009)

[6] Kuich, W.: Semirings and formal power series. In: Rozenberg, G., Salomaa, A. (eds.) Handbook of Formal Languages. Word, Language, Grammar, vol. 1, pp. 609–677. Springer, Berlin (1997)

[7] Mezei, J., Wright, J.B.: Algebraic automata and context-free sets. Information and Control 11, 3–29 (1967)

[8] Reutenauer, C.: A survey on noncommutative rational series. DIMACS Series in Discrete Mathematics and Theoretical Computer Science 24, 159–169 (1996)

[9] Sakarovitch, J.: Rational and recognisable power series. In: [4], ch. 4 (2009)

[10] Salomaa, A., Soittola, M.: Automata-Theoretic Aspects of Formal Power Series. Texts and Monographs on Computer Science. Springer, Heidelberg (1978)

[11] Schützenberger, M.-P.: On the definition of a family of automata. Information and Control 4, 245–270 (1961)

[12] Sontag, E.D.: On some questions of rationality and decidability. Journal of Computer and System Sciences 11(3), 375–382 (1975)

[13] Wang, H.: On rational series and rational languages. Theoretical Computer Science 205(1-2), 329–336 (1998)

Graph Decomposition for Improving Memoryless Periodic Exploration*

Adrian Kosowski[1,2] and Alfredo Navarra[3]

[1] LaBRI - Université Bordeaux 1, 351 cours de la Libération, 33405 Talence, France
[2] Department of Algorithms and System Modeling, Gdańsk University of Technology, Narutowicza 11/12, 80233 Gdańsk, Poland
adrian@kaims.pl
[3] Dipartimento di Matematica e Informatica, Università degli Studi di Perugia, Via Vanvitelli 1, 06123 Perugia, Italy
navarra@dmi.unipg.it

Abstract. We consider a general framework in which a memoryless robot periodically explores all the nodes of a connected anonymous graph by following local information available at each vertex. For each vertex v, the endpoints of all edges adjacent to v are assigned unique labels from the range 1 to $\deg(v)$ (the degree of v). The generic exploration strategy is implemented using a right-hand-rule transition function: after entering vertex v via the edge labeled i, the robot proceeds with its exploration, leaving via the edge having label $[i \bmod \deg(v)] + 1$ at v.

A lot of attention has been given to the problem of labeling the graph so as to achieve a periodic exploration having the minimum possible length π. It has recently been proved [Czyzowicz et al., *Proc. SIROCCO'09* [1]] that $\pi \leq 4\frac{1}{3}n$ holds for all graphs of n vertices. Herein, we provide a new labeling scheme which leads to shorter exploration cycles, improving the general bound to $\pi \leq 4n - 2$. This main result is shown to be tight with respect to the class of labelings admitting certain connectivity properties. The labeling scheme is based on a new graph decomposition which may be of independent interest.

1 Introduction

The problem of finding Eulerian cycles and Hamiltonian cycles is at the heart of graph theory. In fact, many applications arise in different contexts. One of these can be found in the field of graph exploration by means of a mobile entity. The entity might be represented, for instance, by a software agent or a robot. The task of periodic visiting of all vertices of a network is particularly useful in network maintenance, where the status of every vertex has to be checked regularly. According to the imposed model as well as to the robot capabilities, the problem may vary in its complexity.

* The research was partially funded by the State Committee for Scientific Research (Poland) Grant 4 T11C 047 25, by the ANR-project "ALADDIN" (France), and by the project "CEPAGE" of INRIA (France).

R. Královič and D. Niwiński (Eds.): MFCS 2009, LNCS 5734, pp. 501–512, 2009.

Model assumptions. We assume that the explored graph $G = (V, E)$ is simple, undirected, connected, and anonymous, i.e., the vertices in the graph are neither labeled nor colored. However, while visiting a vertex the robot can distinguish between its adjacent edges. This is achieved by using a predefined local ordering of edges known as a *local orientation*, which can be given in one of two ways:

- An implicit *cyclic ordering*, given at each node $v \in V$ as a local cyclic ordering of adjacent edges. This is encoded through a *NextPort* function which naturally defines a transition operation known as the *right-hand-rule* (or *basic walk* [2]): a robot entering vertex v by an adjacent edge e leaves this vertex by the next edge adjacent to v in the specified cyclic order, $NextPort(e)$. The local ordering of edges is cyclic in the sense that successive applications of *NextPort* iterate through all of the edges adjacent to v.
- An explicit *port labeling*, in which, for each vertex $v \in V$, there exist consecutive integer labels (starting from 1), also called *port numbers*, preassigned to all the edges adjacent to v, next to the endpoint v of each edge. Thus, each edge of the graph is assigned two port numbers, one for each endpoint. The labels at node v are always distinct and form the discrete interval $[1, \deg(v)]$, where $\deg(v)$ is the degree of v in G. In such a setting, the natural definition of *NextPort* for port number l at vertex v is $NextPort(l) := [l \bmod \deg(v)] + 1$.

A robot, initially located at a generic vertex v, starts the exploration of G by traversing the edge having label 1 at endpoint v. Once it has reached the other endpoint of this edge, say u, it reads the associated label, say l, and enters to the neighboring vertex u of v. In order to keep on with the exploration, the robot now follows the right-hand rule, leaving vertex u by the edge labeled $NextPort(u)$. In doing so, eventually the robot will re-enter port 1 at vertex v, and the traversal will proceed periodically from then on. We will say that the robot *explores* the graph if its route goes through each vertex of the graph at least once; from now on, we will only consider port labelings leading to valid explorations. It is known that all graphs admit a port labeling leading to an exploration [3].

Studied parameter and its motivation. For a given port labeling, the *exploration period* π is defined as the total number of steps made by the robot before returning to the initial port (or equivalently, as the total number of arcs of the form (u, v), for $\{u, v\} \in E$, used during the exploration). In this paper, we focus on finding labelings which lead to valid explorations of minimum possible period. This immediately leads to the natural definition of the graph parameter $\pi(G)$ known as the *minimum exploration period* of the graph.

The parameter $\pi(G)$ obviously characterizes the best-case behavior of the basic walk on G, but its studies have in fact some much stronger motivation: $\pi(G)$ is the minimum possible exploration period for any oblivious robot (i.e., a robot which is not equipped with any state information which survives when traversing an edge), even if the labeling of the graph is given using explicit port numbers, and the choice of the next edge is not necessarily governed by the

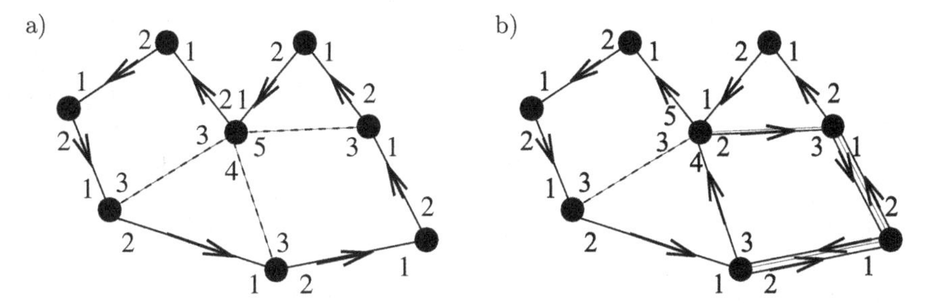

Fig. 1. Exploration cycles obtained for different labelings: (a) a labeling leading to a Hamiltonian cycle, (b) another exemplary labeling

right-hand-rule [1]. In other words, in the context of these studies, the basic walk is fully representative of all oblivious exploration strategies.

Moreover, the value of $\pi(G)$, expressed in relation to the number of nodes n, exposes certain interesting structural properties of the graph. For example, we have that $\pi(G) = n$ if and only if G is Hamiltonian, and for a Hamiltonian graph an appropriate labeling can be defined so as to direct ports 1 and 2 of all nodes along the edges of the Hamiltonian cycle (Fig. 1a). It is also known that $\pi(G) < 2n$ for all graphs admitting a spanning tree T such that $G \setminus T$ has no isolated vertices [1].

Related work. Periodic graph exploration problems have been studied in a wide variety of contexts; we confine ourselves to a brief survey of directly related results in the model of anonymous undirected graphs with local port labels, and robots are deterministic and equipped with no memory (or a very small number of bits of memory).

When no assumptions are made about the port labeling, it is a well-established fact [4] that no oblivious robot can explore all graphs. In [5], the impossibility result was extended to a finite team of robots, showing that they cannot explore all planar cubic graphs. This result is improved in [6], where the authors introduce a powerful tool, called the Jumping Automaton for Graphs (JAG). A JAG is a finite team of finite automata that permanently cooperate and that can use *teleportation* to move from their current location to the location of any other automaton. However, even JAGs cannot explore all graphs. The proof that a robot requires at least n states (and thus $\Omega(\log n)$ bits of state memory) to explore all graphs of order n can be found in [7]. On the other hand, by a seminal result of Reingold [8], a robot equipped with $\Theta(\log n)$ memory can perform a deterministic exploration of any graph, and the resulting exploration period is thus polynomial with respect to n. In [9], the authors investigate the graph exploration capability with respect to the memory size provided to a robot. Another way of assisting exploration is described in [10], where it is shown that an appropriate pre-coloring of the graph using 3 colors always allows a robot with constant memory to successfully complete the exploration.

A natural problem consists in manually setting up the local orientation of the port numbers in order to allow an oblivious robot to efficiently explore the input graph. This line of study, which is also pursued herein, was introduced in [3]. That paper provided the first constructions of port labelings leading to short exploration periods for an oblivious robot, showing that for any graph on n nodes, we have $\pi(G) \leq 10n$. Recently, by applying a clever graph decomposition technique in order to build an appropriate exploration cycle, [1] have improved this bound to $\pi(G) \leq 4\frac{1}{3}n$. They have also shown a strong worst-case lower bound: for arbitrarily large values of n, there exist n-node graphs G_n such that $\pi(G_n) \geq 2.8n - O(1)$.

An interesting variation to this problem was proposed in [11], where the robot is equipped with few extra memory bits; we will denote the exploration periods in such a model by π_c. In [11] it is shown how to obtain an exploration period $\pi_c(G) \leq 4n - 2$, regardless of the starting vertex of the robot. The obtained bound has since been improved in [12] to $\pi_c(G) < 3.75n - 2$ by exploiting some particular graph properties, still allowing only constant memory. The constant memory model was also addressed in [1] and the bound was further improved to $\pi_c(G) < 3.5n - 2$ by using a combination of the properties from [12] and the new decomposition technique also used in [12] for the oblivious case. Interestingly enough, apart from the relation $\pi_c(G_n) \geq 2n - 2$ which clearly holds whenever G_n is a tree on n nodes, there are to date no known non-trivial lower bounds on the worst case value of parameter π_c.

Our results. The central result of this paper is to establish an improved bound on the minimum exploration period of any graph G on n nodes, namely, $\pi(G) \leq 4n - 2$. The proof is constructive, and we also show that a port labeling which guarantees an exploration period of at most $4n - 2$ for the basic walk can be computed in polynomial time in a centralized setting. The result is obtained by applying a new graph decomposition technique which may be of interest in its own right, and a completely new analysis. Our labeling preserves a structural property first introduced in [1]: the set of edges which are traversed twice (in opposite directions) during one exploration period is a connected spanning subgraph of G. We present an example showing that the worst-case analysis of our approach is tight, and that our approach is the best possible when restricted to the class of explorations having the stated structural property.

Outline. Section 2 describes the structural properties defined in [1] which we use in order to construct a suitable graph decomposition. The main proofs are given in Section 3. Subsection 3.1 provides lemmata which characterize the decomposition that we introduce, while Subsection 3.2 formally describes the corresponding algorithmic procedure. Further properties of the labelings are discussed in Section 4. Subsection 4.1 describes a modification of our construction, leading to a polynomial time algorithm for computing an appropriate labeling. Subsection 4.2 shows that for some graphs, the proposed construction is the best possible with respect to the applied graph decomposition. Final remarks are given in Section 5.

Preliminaries and notation. For a graph or multigraph H, we will denote by $V(H)$ its vertex set, by $E(H)$ its edge multiset, and by $|E(H)|$ the number of its edges (including multiple edges). The number of edges adjacent in H to a vertex

$v \in V(H)$ is denoted by $\deg_H(v)$. The notation $2e$ denotes 2 copies of an edge e; the notation $2H$ denotes a multigraph with vertex set $V(H)$ and each edge $e \in E(H)$ replaced by $2e$. An edge $e \in H$ is called *double* if $2e$ belongs to H, and *single* otherwise. Throughout the paper we will never consider multigraphs with more than two parallel edges.

Let $G = (V, E)$ be a connected simple graph, with $|V| = n$. Any labeling scheme for G uniquely determines an exploration cycle, understood as a sequence of directed edges traversed in a single period of the exploration, i.e., a sequence in which the directed edge (u, v) corresponds to a transition of the robot from some node u to another node v, where $\{u, v\} \in E$. The corresponding *exploration multigraph* H is defined as the undirected submultigraph of $2G$ given by the edges of G traversed by the robot during one exploration cycle (each edge is included as it is traversed, possibly twice if it is traversed in both directions). Let H_2 be the spanning subgraph of H consisting of its double edges only, and let $H_1 = H \setminus H_2$. A vertex $v \in V$ is called *saturated* in H with respect to G if $\deg_H(v) = \deg_{2G}(v)$.

In the next section, we show how to build the exploration multigraph H by keeping H_2 as small as possible, so as not to allow the robot to traverse too many edges twice.

2 The Graph Decomposition

In order to provide the new construction of the exploration multigraph H, we need to briefly discuss some of the structural properties of exploration multigraphs. The following simple observation holds (cf. e.g. [1] for a more detailed discussion).

Proposition 1 ([1]). *Any exploration multigraph $H \subseteq 2G$ has the following properties:*

A. For each vertex $v \in V$, $\deg_H(v)$ is even.
B. Each vertex $v \in V$ having $\deg_{H_1}(v) = 0$ is saturated in H with respect to G.

The converse of the above proposition does not hold in general, but one more additional property can also be formulated.

C. H_2 is connected.

Then, the following structural theorem has recently been shown.

Theorem 1 ([1]). *Any multigraph $H \subseteq 2G$ fulfilling properties A, B, and C is a valid exploration multigraph, i.e. induces an exploration cycle on G of length at most $|E(H)|$.*

We can thus concentrate on defining a multigraph which satisfies properties A, B, and C. To achieve this, in graph G we select an arbitrary spanning tree T_0. Let $G' = G \setminus T_0$. Then, in multigraph $2G'$ we find a spanning (not necessarily connected) submultigraph H' satisfying properties analogous to A and B:

A'. For each vertex $v \in V$, $\deg_{H'}(v)$ is even.
B'. Each vertex $v \in V$ having $\deg_{H'_1}(v) = 0$ is saturated in H' with respect to G'.

The final multigraph H is given as $H = H' \cup 2T_0$, thus $2T_0 \subseteq H_2$. It is clear that H satisfies properties A, B, C, and that $|E(H)| = |E(H')| + 2(n-1)$. Note that the construction of H' can be performed independently for each connected component of G'; throughout the rest of the discussion, w.l.o.g. we assume that G' is connected. Hence, in order to obtain an exploration cycle with period $\pi(G) \leq 4n - 2$, we confine ourselves to constructing an appropriate submultigraph $H' \subseteq 2G'$ with $|E(H')| \leq 2n$. So, it remains to show the following theorem, which constitutes the main result of our paper.

Theorem 2. *For any connected graph G' with vertex set V, $|V| = n$, there exists a multigraph $H' \subseteq 2G'$ such that $|E(H')| \leq 2n$, and H' satisfies properties A' and B'.*

3 Proof of Theorem 2

Let T be a rooted spanning tree in G' with root r. We will call a vertex $v \in V$ *tree-saturated* in T if $\deg_T(v) = \deg_{G'}(v)$. For tree T, let $s(T)$ denote the number of tree-saturated vertices in T, and let $s_h(T)$, for $0 \leq h < n$, be the number of tree-saturated vertices in T at height (i.e. distance in tree T from root r to the vertex) not greater than h. The vertex adjacent to v on the path in T leading from v to root r will be called the *parent* $p(v)$, while the edge connecting these two vertices will be called the *parent edge* $pe(v) = \{p(v), v\}$. For convenience of notation, we will occasionally augment the edge set of tree T by the fictional parent edge of the root $pe(r)$, and then denote $E^+(T) = E(T) \cup pe(r)$.

Consider the following partial order on rooted spanning trees in graph G'. We will say that $T_a < T_b$ if one of the following conditions is fulfilled.

1. $s(T_a) < s(T_b)$,
2. $s(T_a) = s(T_b)$, and for some h, $0 \leq h < n$, we have $\forall_{0 \leq l < h}\ s_l(T_a) = s_l(T_b)$ and $s_h(T_a) > s_h(T_b)$.

Now, multigraph H' can be determined by the following algorithm:

Algorithm 1: Computing multigraph H'

1. Let T be a minimal spanning tree in G' with respect to order $(<)$.
2. Let S be a subgraph in $G' \setminus T$, whose connected components are stars, such that for each $v \in V$ either v is tree-saturated in T or $\deg_S(v) > 0$.
3. Find a submultigraph $H' \subseteq S \cup 2T$ fulfilling properties A' and B', such that $|E(H')| \leq 2n$, and return it as output.

We have to show that all the steps of the above algorithm are well defined. Step (1) requires no comment. For step (2), notice that graph S is well defined because any graph admits a subgraph which is a set of stars, touching all non-isolated vertices; for graph $G' \setminus T$, the only isolated vertices are those which were tree-saturated in T. Before we proceed to describe the procedure to be used in Step (3), we first need to show some structural lemmata.

3.1 Properties of Graph $\mathcal{S}$

Let S be an arbitrary star which is a connected component of $\mathcal{S}$ with vertex set $\{v_1, \ldots, v_k\}$. Whenever $k \geq 3$ (i.e. when S is not a single edge) we assume that v_1 is the center of the star. Note that none of the vertices v_i can be tree-saturated in T.

Lemma 1. *Assume that $k \geq 3$. Then for any v_i, $1 < i \leq k$, if $p(v_i)$ is tree-saturated in T, then v_i lies on the path in T from v_1 to root r.*

Proof. Suppose that, to the contrary, $p(v_i)$ is tree-saturated in T and, that v_i does not lie on the path from v_1 to root r. We have two cases.

1. Center v_1 lies on the path from v_i to root r (Fig. 2a). Then, consider the tree $T^* = T \cup \{v_1, v_i\} \setminus pe(v_i)$. T^* is a valid spanning tree in G'; moreover, $s(T^*) < s(T)$ (since vertex $p(v_i)$ is no longer tree-saturated in T^* and no new vertices are tree-saturated), a contradiction with the minimality of T.
2. Vertices v_1 and v_i lie on paths to root r, neither of which is contained in the other (Fig. 2b). Then we obtain a contradiction by defining tree T^* in the same way as in the previously considered case. □

Lemma 2. *Assume that $k \geq 3$. Then there can exist at most one v_i, $1 \leq i \leq k$, such that $p(v_i)$ is tree-saturated in T.*

Proof. Suppose that, to the contrary, there are two vertices v_i and v_j, $i \neq j$, such that $p(v_i)$ and $p(v_j)$ are tree-saturated (note that we do not necessarily assume that $p(v_i) \neq p(v_j)$). Once again, we need to consider two cases.

1. The parent $p(v_1)$ of the center of the star is tree-saturated (i.e. we can put $i = 1$; Fig. 3a). Then by Lemma 1, v_j lies on the path in T leading from v_1 to root r. Then, consider the tree $T^* = T \cup \{v_1, v_j\} \setminus pe(v_1)$. T^* is a valid spanning tree in G'. We now have $s(T^*) \leq s(T)$, since vertex $p(v_1)$ is no longer tree-saturated in T^*, vertex v_j may have become tree-saturated, while the tree-saturation of the other vertices does not change. However, if $s(T^*) = s(T)$, then v_j must be tree-saturated in T^*, so denoting by h

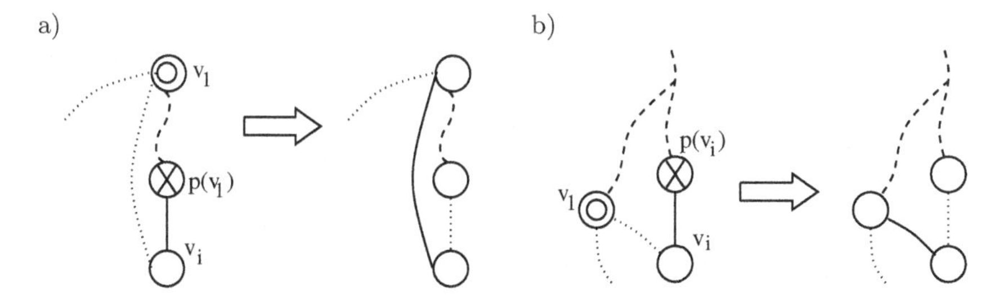

Fig. 2. Arrangements of a star center with respect to saturated vertices (solid edges belong to the tree, dashed edges belong to a star; a cross denotes a saturated vertex, a double circle denotes the star center)

the height of v_j in T (equivalently T^*) we have $\forall_{0 \le l < h}\ s_l(T^*) = s_l(T)$ and $s_h(T^*) > s_h(T)$. Thus, we obtain that $T^* < T$, a contradiction with the minimality of T.

2. The parents $p(v_i)$ and $p(v_j)$, for some $i, j > 1$, are tree-saturated (Fig. 3b). Then by Lemma 1, both v_i and v_j lie on the path in T leading from v_1 to root r; without loss of generality we can assume that v_j is closer than v_i to root r. Then, consider the tree $T^* = T \cup \{v_1, v_j\} \setminus pe(v_i)$. T^* is a valid spanning tree in G', and using the same arguments as in the previous case we obtain that $T^* < T$, a contradiction with the minimality of T. □

Lemma 3. *Let S be a two-vertex star component consisting of vertices v_1, v_2, at heights h_1 and h_2 in tree T, respectively. If $p(v_1)$ and $p(v_2)$ are both tree-saturated, then $|h_1 - h_2| \le 1$.*

Proof. Without loss of generality let $h_1 \le h_2$. Suppose that $p(v_1)$ and $p(v_2)$ are both tree-saturated, and that $h_1 < h_2 - 1$. Similarly to the proof of Lemma 1, let $T^* = T \cup \{v_1, v_2\} \setminus pe(v_2)$. It is clear that in T^* vertex $p(v_2)$ (at level $h_2 - 1$) is no longer tree-saturated, vertex v_1 (at level $h_1 < h_2 - 1$) may possibly become tree-saturated, while the tree-saturation of the remaining vertices does not change. Hence $T^* < T$, a contradiction with the minimality of T. □

3.2 Construction of Multigraph H'

We now describe the routine used in Step (3) of Algorithm 1 to construct submultigraph $H' \subseteq \mathcal{S} \cup 2T$ by iteratively adding edges. Taking into account Lemma 3, we can write $E(\mathcal{S}) = E(\mathcal{S})_+ \cup \bigcup_{h=1}^{n} E(\mathcal{S})_{h,h} \cup \bigcup_{h=1}^{n} E(\mathcal{S})_{h,h-1}$, where $E(\mathcal{S})_+$ denotes the set of edges belonging to stars of more than 2 vertices,

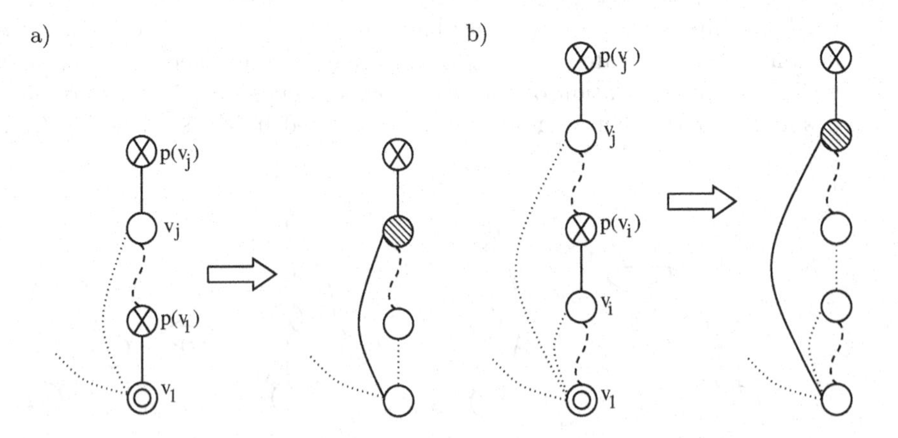

Fig. 3. Arrangements of two saturated parents for a star (solid edges belong to the tree, dashed edges belong to a star; a cross denotes a saturated vertex, a shaded circle denotes a possible saturated vertex, while a double circle denotes the star center)

$E(\mathcal{S})_{h,h}$ is the set of edges of two-vertex stars with both vertices at height h in tree T, and finally $E(\mathcal{S})_{h,h-1}$ is the set of edges of two-vertex stars with one vertex at height h and the other at height $h-1$ in tree T. Likewise, since T is a tree, we can write $E(2T) = \bigcup_{h=1}^{n} E(2T)_{h,h-1}$, where $E(2T)_{h,h-1}$ contains edges connecting a vertex at height h with a vertex at height $h-1$ in tree T.

We start by putting $E(H') = E(\mathcal{S})_{+}$. Then, for all levels $h \in (n, n-1, \ldots, 1)$, considered in decreasing order, we choose which edges from $E(\mathcal{S})_{h,h}$, $E(\mathcal{S})_{h,h-1}$ and $E(2T)_{h,h-1}$ to add to $E(H')$:

(a) Add all edges from $E(\mathcal{S})_{h,h} \cup E(\mathcal{S})_{h,h-1}$ to $E(H')$.
(b) For each vertex v at level h, add to $E(H')$ a subset of the two copies of edge $pe(v)$ from $E(2T)_{h,h-1}$ so that the degree of v is even in H'. When an even number of edges is required, 2 edges should be used if v is tree-saturated or $p(v)$ is tree-saturated, and 0 edges should be used otherwise.
(c) Successively consider all edges $\{v_1, v_2\} \in E(\mathcal{S})_{h,h} \cup E(\mathcal{S})_{h,h-1}$ which were added in step (a). If both vertex v_1 and vertex v_2 are currently of even degree in H', and both $p(v_1)$ and $p(v_2)$ are tree-saturated, then remove from $E(H')$ edge $\{v_1, v_2\}$.
(d) For each vertex v at level h affected by changes in step (c), remove from $E(H')$ one of the two copies of edge $pe(v)$ so that the degree of v is even in H'.

To show that graph H' fulfills properties A' and B' it suffices to prove the following lemma.

Lemma 4. *For each vertex $v \in V$, $\deg_{H'}(v)$ is even. Moreover, if $\deg_{H_1'}(v) = 0$, then v is tree-saturated and $\deg_{H_2'}(v) = \deg_{2T}(v)$.*

Proof. For each height h, the degree of all vertices at height h in H' is always even after completion of steps (b) and (d) for height h, and is never changed afterwards. (For the root r at height $h = 0$, the degree must be even by the Handshaking lemma). Consequently, the degree of all vertices in H' is even.

If $\deg_{H_1'}(v) = 0$, then since all edges of stars with more than two vertices appear in $E(H_1')$, v can only be adjacent to a 2-vertex star or tree-saturated in T. The former case is impossible, since by definition of steps (c) and (d) either the edge of the 2-vertex star, or $pe(v)$ belongs to $E(H_1')$. In the latter case, the definition of steps (a)-(d) is such that each edge of tree T adjacent to a tree-saturated vertex appears either in 1 or 2 copies in $E(H') = E(H_1') \cup E(H_2')$, hence if $\deg_{H_1'}(v) = 0$, then clearly $\deg_{H_2'}(v) = \deg_{2T}(v)$. □

The relation $|E(H')| \leq 2n$ will now be shown using a local cost-based argument. First, we will assign costs c to vertices and edges of T as follows.

1. For each edge $e \in E(T)$, cost $c(e) \in \{0, 1, 2\}$ is set as the number of times edge e appears in multigraph H'.
2. For each vertex $v \in V$, we set cost $c(v) \in \{0, 1\}$ by considering the following cases:
   (a) if v is tree-saturated in T, then $c(v) = 0$,

(b) if v belongs to a 2-vertex star from $\mathcal{S}$, then $c(v) = 1$ if $pe(v)$ appears in H' at most once, and $c(v) = 0$ otherwise,
(c) if v belongs to a star from $\mathcal{S}$ having at least 3 vertices, then $c(v) = 0$ if $p(v)$ is tree-saturated, and $c(v) = 1$ otherwise.

Lemma 5. *For the cost assignment c we have $\sum_{x \in E(T) \cup V} c(x) \geq |E(H')| = |E(H' \cap 2T)| + |E(H' \cap \mathcal{S})|$.*

Proof. Indeed, we have $\sum_{x \in E(T)} c(x) = |E(H' \cap 2T)|$ by definition of the costs of edges. Next, notice that $|E(H' \cap \mathcal{S})|$ can be redistributed over the costs of vertices as follows (note that the stars are vertex-disjoint). For a k-vertex star, with $k > 2$, we pay for $k-1$ edges by assigning a cost of 1 to all vertices of the star, except perhaps one vertex v for which $p(v)$ is tree-saturated (by Lemma 2 there can be at most one such vertex for each star). For an edge of a 2-vertex star which appears in H', we pay for the edge using the cost assigned to at least one of its end-vertices having cost 1; an end-vertex with such cost must exist by the construction of steps (c)-(d) of the algorithm. □

Lemma 6. *For each vertex $v \in V$, $c(v) + c(pe(v)) \leq 2$.*

Proof. Since $c(v) \leq 1$, the claim is clearly true if $pe(v) < 2$. If $c(pe(v)) = 2$, then edge $pe(v)$ appears twice in H', hence by the construction of steps (a)-(d) of the algorithm it is clear that vertex v or vertex $p(v)$ is tree-saturated. In either case, we have $c(v) = 0$ by the definition of costs, which completes the proof. □

Combining Lemmata 5 and 6, summing over all vertices, immediately gives the sought bound, $|E(H')| \leq 2n$, which completes the proof of Theorem 2.

4 Construction of the Labeling

Recalling from Section 2 that $|E(H)| = |E(H')| + 2(n-1)$, we can rephrase Theorem 2 in the original language of graph exploration.

Theorem 3. *For any graph G of size n there exists a port labeling leading to an exploration period $\pi \leq 4n - 2$.*

It is natural to ask about the runtime of the procedure required to obtain a labeling with such an exploration period, and about the tightness of the obtained bound; we address these questions in the following subsections.

4.1 Runtime of the Labeling Procedure

Whereas the construction of the appropriate cycle can always be performed using Algorithm 1 (in finite time), this does not necessarily mean that a solution can be found in polynomial time. The problem consists in computing an appropriate spanning tree, minimal in the sense of order ($<$), in step (1). In general, finding a spanning tree having a minimum number of saturated vertices is already *NP*-hard. (The proof of this observation proceeds by reduction from the problem of finding a Hamiltonian path in a 3-regular graph: a 3-regular graph has a spanning tree without saturated vertices if and only if it admits a Hamiltonian path.)

In order to obtain polynomial time complexity, we apply a slight modification of step (1): instead of requiring tree T to be minimal, we only require that Lemmata 1, 2 and 3 hold for this tree. If there appears a contradiction in the proof of either lemmata, we replace tree T by tree T^*, using the modifications shown in Fig. 2 for Lemma 1. For Lemma 2, we use the replacement from Fig. 3a, while instead of the replacement from Fig. 3b, define tree T^* by putting $T^* = T \cup \{v_1, v_j\} \cup \{v_1, v_i\} \setminus (pe(v_1) \cup pe(v_i))$. For Lemma 3, we simply perform the operations described in the proof.

We now observe that the modification for Lemma 1 decreases the value of parameter $s(T)$, whereas the modifications for Lemmas 2 and 3 either decrease the value of parameter $s(T)$, or decrease the value of parameter $\sum_{0 \leq l < n}(l \cdot s_l(T))$. The initial value of parameter $s(T)$ is at most n; clearly, the modification process stops when $s(T) = 0$. Next, since we have $\sum_{0 \leq l < n}(l \cdot s_l(T)) < n^2$, then this parameter may only be decreased less than n^2 times for a given value of $s(T)$. Overall, the total number of tree modifications is less than n^3. This guarantees polynomial runtime of the whole algorithm.

Theorem 4. *There exists a polynomial time algorithm which, given a graph G, determines a port labeling leading to an exploration period $\pi \leq 4n - 2$.*

4.2 Tightness of the Bound

In this section, we show that there actually exist instances for which the length of the computed exploration cycle by means of our approach is $4n - O(1)$. Consider the following property of multigraph H_2, obtained by a slight modification of property C, which is also satisfied by our construction.

D. H_2 spans vertex set V, i.e. for all $v \in V$, $\deg_{H_2}(v) > 0$.

We make the following claim, which is partially complementary to Theorem 1.

Theorem 5. *For all values of $n \geq 3$, there exists a graph G of order n, such that any exploration multigraph $H \subseteq 2G$ fulfilling properties A, B, and D, has $|E(H)| \geq 4n - 8$ edges.*

Proof. Consider the complete bipartite graph $G = K_{2,l}$ on $n = l + 2$ vertices, for any $l \geq 1$. By property D, for each of the l vertices $v \in V$ such that $\deg_G(v) = 2$, we have $\deg_{H_2}(v) > 0$, hence $\deg_{H_1}(v) \leq 1$. Taking into account property A we obtain $\deg_{H_1}(v) = 0$, and so by property B, v is saturated in H_2. This means that $H = 2G$, and so $|E(H)| = 2|E(G)| = 4n - 8$. □

5 Conclusions

We have shown that by applying a special construction of a spanning tree, dedicated to the considered exploration problem, it is possible to find in polynomial time an exploration cycle of length at most $4n - 2$. Moreover, we have shown that the obtained bound on the exploration period is sometimes tight up to an additive factor. The best known existential lower bound on the length of an exploration cycle is within an additive factor of $2.8n$ [1]. However, obtaining cycles significantly shorter than $4n$ would require some completely new insight; in particular, the construction would need to avoid the condition imposed on the double-edge subgraph of multigraph H (property D in Theorem 5).

Acknowledgement. The authors are most grateful to Jurek Czyzowicz, Leszek Gąsieniec, David Ilcinkas, and Ralf Klasing for helpful comments and discussions.

References

1. Czyzowicz, J., Dobrev, S., Gąsieniec, L., Ilcinkas, D., Jansson, J., Klasing, R., Lignos, Y., Martin, R., Sadakane, K., Sung, W.: More efficient periodic traversal in anonymous undirected graphs. In: Proc. 16th Colloquium on Structural Information and Communication Complexity (SIROCCO). LNCS. Springer, Heidelberg (to appear, 2009)
2. Gąsieniec, L., Radzik, T.: Memory efficient anonymous graph exploration. In: Broersma, H., Erlebach, T., Friedetzky, T., Paulusma, D. (eds.) WG 2008. LNCS, vol. 5344, pp. 14–29. Springer, Heidelberg (2008)
3. Dobrev, S., Jansson, J., Sadakane, K., Sung, W.-K.: Finding short right-hand-on-the-wall walks in graphs. In: Pelc, A., Raynal, M. (eds.) SIROCCO 2005. LNCS, vol. 3499, pp. 127–139. Springer, Heidelberg (2005)
4. Budach, L.: Automata and labyrinths. Mathematische Nachrichten 86, 195–282 (1978)
5. Rollik, H.: Automaten in planaren graphen. Acta Informatica 13, 287–298 (1980)
6. Cook, S., Rackoff, C.: Space lower bounds for maze threadability on restricted machines. SIAM Journal on Computing 9(3), 636–652 (1980)
7. Fraigniaud, P., Ilcinkas, D., Peer, G., Pelc, A., Peleg, D.: Graph exploration by a finite automaton. Theoretical Computer Science 345(2-3), 331–344 (2005)
8. Reingold, O.: Undirected st-connectivity in log-space. In: Proc. 37th Annual ACM Symposium on Theory of Computing (STOC), pp. 376–385 (2005)
9. Fraigniaud, P., Ilcinkas, D., Pelc, A.: Impact of memory size on graph exploration capability. Discrete Applied Mathematics 156(12), 2310–2319 (2008)
10. Cohen, R., Fraigniaud, P., Ilcinkas, D., Korman, A., Peleg, D.: Label-guided graph exploration by a finite automaton. ACM Transactions on Algorithms 4(4), 1–18 (2008)
11. Ilcinkas, D.: Setting port numbers for fast graph exploration. Theoretical Computer Science 401(1-3), 236–242 (2008)
12. Gąsieniec, L., Klasing, R., Martin, R., Navarra, A., Zhang, X.: Fast periodic graph exploration with constant memory. Journal of Computer and System Sciences 74(5), 802–822 (2008)

On FO^2 Quantifier Alternation over Words*

Manfred Kufleitner[1] and Pascal Weil[2,3]

[1] Institut für Formale Methoden der Informatik, Universität Stuttgart, Germany
[2] LaBRI, Université de Bordeaux and CNRS, France
[3] Department of Computer Science and Engineering, IIT Delhi, India
manfred.kufleitner@fmi.uni-stuttgart.de
pascal.weil@labri.fr

Abstract. We show that each level of the quantifier alternation hierarchy within $\mathsf{FO}^2[<]$ on words is a variety of languages. We use the notion of condensed rankers, a refinement of the rankers defined by Weis and Immerman, to produce a decidable hierarchy of varieties which is interwoven with the quantifier alternation hierarchy – and conjecturally equal to it. It follows that the latter hierarchy is decidable within one unit, a much more precise result than what is known about the quantifier alternation hierarchy within $\mathsf{FO}[<]$, where no decidability result is known beyond the very first levels.

First-order logic is an important object of study in connection with computer science and language theory, not least because many important and natural problems are first-order definable: our understanding of the expressive power of this logic and the efficiency of the solution of related algorithmic problems are of direct interest in such fields as verification. Here, by first-order logic, we mean the first-order logic of the linear order, $\mathsf{FO}[<]$, interpreted on finite words.

In this context, there has been continued interest in fragments of first-order logic, defined by the limitation of certain resources, e.g. the quantifier alternation hierarchy (which is closely related with the dot-depth hierarchy of star-free languages). It is still an open problem whether each level of this hierarchy is decidable.[1] Another natural restriction concerns the number of variables used (and re-used!) in a formula. It is interesting, notably because the trade-off between formula size and number of variables is known to be related with the trade-off between parallel time and number of processes, see [18,5,1,4].

In this paper, we concentrate on $\mathsf{FO}^2[<]$, the 2-variable fragment of $\mathsf{FO}[<]$. It is well-known that every $\mathsf{FO}[<]$-formula is logically equivalent with a formula using only 3 variables, but that $\mathsf{FO}^2[<]$ is properly less expressive than $\mathsf{FO}[<]$. The expressive power of $\mathsf{FO}^2[<]$ was characterized in many interesting fashions (see [12,14,15,3]), and in particular, we know how to decide whether an $\mathsf{FO}[<]$-formula is equivalent to one in $\mathsf{FO}^2[<]$.

* Both authors acknowledge support from the ANR project DOTS, the ESF program AUTOMATHA and the Indo-French P2R project MODISTE-COVER.

[1] On the other hand, the quantifier alternation hierarchy collapses at level 2 for the first-order logic of the successor $\mathsf{FO}[S]$ [16,9].

R. Královič and D. Niwiński (Eds.): MFCS 2009, LNCS 5734, pp. 513–524, 2009.

A recent result of Weis and Immerman refined a result of Schwentick, Thérien and Vollmer [12] to give a combinatorial description of the $\mathsf{FO}^2_m[<]$-definable languages (those that can be defined by an $\mathsf{FO}^2[<]$-formula with quantifier alternation bounded above by m), using the notion of rankers. Rankers are finite sequences of instructions of the form *go to the next a-position to the right* (resp. *left*) *of the current position.*

Our first set of results shows that $\mathcal{FO}^2_m$ (the $\mathsf{FO}^2_m[<]$-definable languages), and the classes of languages defined by rankers having m alternations of directions (right *vs.* left), are varieties of languages. This means that membership of a language L in these classes depends only on the syntactic monoid of L, which justifies an algebraic approach of decidability.

Our investigation shows that rankers are actually better suited to characterize a natural hierarchy within unary temporal logic, and we introduce the new notion of a condensed ranker, that is more adapted to discuss the quantifier alternation hierarchy within $\mathsf{FO}^2[<]$. There again, the alternation of directions in rankers defines hierarchies of varieties of languages $\mathcal{R}_m$ and $\mathcal{L}_m$, with particularly interesting properties. Indeed, we show that these varieties are decidable, that they admit a neat characterization in terms of closure under deterministic and co-deterministic products, and that $\mathcal{R}_m \cup \mathcal{L}_m \subsetneq \mathcal{FO}^2_m \subsetneq \mathcal{R}_{m+1} \cap \mathcal{L}_{m+1}$. The latter containments show that we can effectively compute, given a language $L \in \mathcal{FO}^2$, an integer m such that L is in $\mathcal{FO}^2_{m+1}$, possibly in $\mathcal{FO}^2_m$, but not in $\mathcal{FO}^2_{m-1}$. This is much more precise than the current level of knowledge on the general quantifier alternation hierarchy in $\mathsf{FO}[<]$.[2]

1 An Algebraic Approach to Study $\mathbf{FO}^2_m$

If $u \in A^+$ is a non-empty word, we denote by $u[i]$ the letter of u in position i ($1 \le i \le |u|$), and by $u[i,j]$ the factor $u[i] \cdots u[j]$ of u ($1 \le i \le j \le |u|$). Then we identify the word u with the logical structure $(\{1,\ldots,|u|\}, (\mathbf{a})_{a \in A})$, where $\mathbf{a}$ denotes the set of integers i such that $u[i] = a$.

Let $\mathsf{FO}[<]$ (resp. $\mathsf{FO}^k[<]$, $k \ge 0$) denote the set of first-order formulas using the unary predicates $\mathbf{a}$ ($a \in A$) and the binary predicate $<$ (resp. and at most k variable symbols). It is well-known that $\mathsf{FO}^3[<]$ is as expressive as $\mathsf{FO}[<]$ and that $\mathsf{FO}^2[<]$ is properly less expressive.

In the sequel, we omit the predicate $<$ and we write simply FO or FO^k. The classes of FO- and FO^2-definable languages have well-known beautiful characterizations [12,14,15,3]. Two are of particular interest in this paper.

- The algebraic characterization in terms of recognizing monoids: a language is FO-definable if and only if it is recognized by a finite aperiodic monoid, i.e., one in which $x^n = x^{n+1}$ for each element x and for all n large enough (Schützenberger and McNaughton-Ladner, see [13]); and a language is FO^2-definable if and only if it is recognized by a finite monoid in **DA** (see [14,3]), a class of monoids with many interesting characterizations, which will be discussed later. These algebraic

[2] Unfortunately, it cannot help directly with the general problem since a language L is $\mathsf{FO}^2[<]$-definable if and only if L and its complement are Σ_2-definable [11].

characterizations prove the decidability of the corresponding classes of languages: L is FO (resp. FO^2)-definable if and only if the (effectively computable) syntactic monoid of L is in the (decidable) class of aperiodic monoids (resp. in $\mathbf{DA}$).

- The language-theoretic characterization: a language is FO-definable if and only if it is star-free, i.e., it can be obtained from singletons using Boolean operations and concatenation products (Schützenberger, see [8]); a language is FO^2-definable if and only if it can be written as the disjoint union of unambiguous products of the form $B_0^*a_1B_1^* \cdots a_kB_k^*$, where $k \geq 0$, the a_i are letters and the B_i are subsets of the alphabet. Such a product is called *unambiguous* if each word $u \in B_0^*a_1B_1^* \cdots a_kB_k^*$ admits a unique factorization in the form $u = u_0a_1u_1 \cdots a_ku_k$ such that $u_i \in B_i^*$ for each i.

We now concentrate on FO^2-formulas and we define two important parameters concerning such formulas. To simplify matters, we consider only formulas where negation is used only on atomic formulas so that, in particular, no quantifier is negated. This is naturally possible up to logical equivalence. Now, with each formula $\varphi \in \mathsf{FO}^2$, we associate in the natural way a parse tree: each occurrence of a quantification, $\exists x$ or $\forall x$, yields a unary node, each occurrence of $\vee$ or $\wedge$ yields a binary node, and the leaves are labeled with atomic or negated atomic formulas. Each path from root to leaf in this parse tree has a *quantifier label*, which is the sequence of quantifier node labels ($\exists$ or $\forall$) encountered along this path. A *block* in this quantifier label is a maximal factor consisting only of $\exists$ or only of $\forall$. The *quantifier depth* of φ is the maximum length of the quantifier label of a path in the parse tree of φ, and the *number of blocks* of φ is the maximum number of blocks in the quantifier label of a path in its parse tree.

We let $\mathsf{FO}^2_{m,n}$ denote the set of first-order formulas with quantifier depth at most n and with at most m blocks and let FO^2_m denote the union of the $\mathsf{FO}^2_{m,n}$ for all n. We also denote by $\mathcal{FO}^2$ ($\mathcal{FO}^2_m$) the class of FO^2 (FO^2_m)-definable languages. Weis and Immerman's characterization of the expressive power of $\mathsf{FO}^2_{m,n}[<]$ in terms of rankers [18], see Thm. 1 below, forms the basis of our own results.

1.1 Rankers and Logic

A *ranker* [18] is a non-empty word on the alphabet $\{\mathsf{X}_a, \mathsf{Y}_a \mid a \in A\}$.[3] Rankers may define positions in words: given a word $u \in A^+$ and a letter $a \in A$, we denote by $\mathsf{X}_a(u)$ (resp. $\mathsf{Y}_a(u)$) the least (resp. greatest) integer $1 \leq i \leq |u|$ such that $u[i] = a$. If a does not occur in u, we say that $\mathsf{Y}_a(u)$ and $\mathsf{X}_a(u)$ are not defined. If in addition q is an integer such that $1 \leq q \leq |u|$, we let

$$\mathsf{X}_a(u,q) = \mathsf{X}_a(u[q+1,|u|])$$
$$\mathsf{Y}_a(u,q) = \mathsf{Y}_a(u[1,q-1]).$$

These definitions are extended to all rankers: if r' is a ranker, $\mathsf{Z} \in \{\mathsf{X}_a, \mathsf{Y}_a \mid a \in A\}$ and $r = r'\mathsf{Z}$, we let $r(u,q) = \mathsf{Z}(u, r'(u,q))$ if $r'(u,q)$ and $\mathsf{Z}(u, r'(u,q))$ are defined, and we say that $r(u,q)$ is undefined otherwise.

[3] Weis and Immerman write $\rhd_a$ and $\lhd_a$ instead of X_a and Y_a. We rather follow the notation in [3], where X and Y refer to the future and past operators of LTL.

Finally, if r starts with an X- (resp. Y-) letter, we say that r defines the position $r(u) = r(u, 0)$ (resp. $r(u) = r(u, |u| + 1)$), or that it is undefined on u if this position does not exist. Then $L(r)$ is the language of all words on which r is defined. We say that the words u and v *agree on a class* R of rankers if exactly the same rankers from R are defined on u and v.

Rankers can be seen as sequences of directional instructions: *go to the next a to the right*, resp. *to the left*. We classify them by the number of instructions and by the number of changes of direction. The *depth* of a ranker r is defined to be its length (as a word). A *block* in r is a maximal factor in $\{\mathsf{X}_a \mid a \in A\}^+$ (an X-block) or in $\{\mathsf{Y}_a \mid a \in A\}^+$ (a Y-block). If $n \geq m$, we denote by $R^{\mathsf{X}}_{m,n}$ (resp. $R^{\mathsf{Y}}_{m,n}$) the set of m-block, depth n rankers, starting with an X -(resp. Y-) block, and we let $R_{m,n} = R^{\mathsf{X}}_{m,n} \cup R^{\mathsf{Y}}_{m,n}$ and $\underline{R}^{\mathsf{X}}_{m,n} = \bigcup_{n' \leq n} R^{\mathsf{X}}_{m,n'} \cup \bigcup_{m' < m, n' < n} R_{m',n'}$. We define $\underline{R}^{\mathsf{Y}}_{m,n}$ dually and we let $\underline{R}^{\mathsf{X}}_{m} = \bigcup_{n \geq m} \underline{R}^{\mathsf{X}}_{m,n}$, $\underline{R}^{\mathsf{Y}}_{m} = \bigcup_{n \geq m} \underline{R}^{\mathsf{Y}}_{m,n}$ and $\underline{R}_m = \underline{R}^{\mathsf{X}}_m \cup \underline{R}^{\mathsf{Y}}_m$.[4]

Rankers and temporal logic. Let us depart for a moment from the consideration of FO^2-formulas, to observe that rankers are naturally suited to describe the different levels of a natural class of temporal logic. The symbols X_a and Y_a $(a \in A)$ can be seen as modal (temporal) operators, with the *future* and *past* semantics respectively. We denote the resulting temporal logic (known as *unary temporal logic*) by TL: its only atomic formula is $\top$, the other formulas are built using Boolean connectives and modal operators. Let $u \in A^+$ and let $0 \leq i \leq |u| + 1$. We say that $\top$ holds at every position i, $(u, i) \models \top$; Boolean connectives are interpreted as usual; and $(u, i) \models \mathsf{X}_a\varphi$ (resp. $\mathsf{Y}_a\varphi$) if and only if $(u, j) \models \varphi$, where j is the least a-position such that $i < j$ (resp. the greatest a-position such that $j < i$). We also say that $u \models \mathsf{X}_a\varphi$ (resp. $\mathsf{Y}_a\varphi$) if $(u, 0) \models \mathsf{X}_a\varphi$ (resp. $(u, 1 + |u|) \models \mathsf{Y}_a\varphi$).

TL is a fragment of *propositional temporal logic* PTL; the latter is expressively equivalent to FO and TL is expressively equivalent to FO^2, see [14].

As in the case of FO^2-formulas, one may consider the parse tree of a TL-formula and define inductively its depth and number of alternations (between past and future operators). If $n \geq m$, the fragment $\mathsf{TL}^{\mathsf{X}}_{m,n}$ (resp. $\mathsf{TL}^{\mathsf{Y}}_{m,n}$) consists of the TL-formulas with depth n and with m alternations, in which every branch (of the parse tree) with exactly m alternations starts with future (resp. past) operators. The fragments $\mathsf{TL}_{m,n}$, $\underline{\mathsf{TL}}^{\mathsf{X}}_{m,n}$, $\underline{\mathsf{TL}}^{\mathsf{Y}}_{m,n}$, $\underline{\mathsf{TL}}^{\mathsf{X}}_{m}$, $\underline{\mathsf{TL}}^{\mathsf{Y}}_{m}$ and $\underline{\mathsf{TL}}_m$ are defined according to the same pattern as in the definition of $R_{m,n}$, $\underline{R}^{\mathsf{X}}_{m,n}$, $\underline{R}^{\mathsf{Y}}_{m,n}$, $\underline{R}^{\mathsf{X}}_{m}$, $\underline{R}^{\mathsf{Y}}_{m}$ and $\underline{R}_m$. We also denote by $\mathcal{TL}^{\mathsf{X}}_{m,n}$ ($\mathcal{TL}^{\mathsf{X}}_{m}$, $\underline{\mathcal{TL}}_m$, etc) the class of $\mathsf{TL}^{\mathsf{X}}_{m,n}$ ($\mathsf{TL}^{\mathsf{X}}_{m}$, $\underline{\mathcal{TL}}^{\mathsf{X}}_{m}$, etc)-definable languages. The following result is elementary.

Proposition 1. *Let* $1 \leq m \leq n$. *Two words satisfy the same* $\mathsf{TL}^{\mathsf{X}}_{m,n}$ *formulas if and only if they agree on rankers from* $R^{\mathsf{X}}_{m,n}$. *A language is in* $\mathcal{TL}^{\mathsf{X}}_{m,n}$ *if and only if it is a Boolean combination of languages of the form* $L(r)$, $r \in R^{\mathsf{X}}_{m,n}$.

---

[4] Readers familiar with [18] may notice a small difference between our $\underline{R}^{\mathsf{X}}_{m,n}$ and their analogous $R^{\star}_{m\triangleright,n}$; introduced for technical reasons, it creates no difference between $\underline{R}_{m,n}$ and $R^{\star}_{m,n}$, the classes which intervene in crucial Thm. 1 below.

Similar statements hold for $\mathsf{TL}^{\mathsf{Y}}_{m,n}$, $\mathsf{TL}_{m,n}$, $\underline{\mathsf{TL}}^{\mathsf{X}}_{m,n}$, $\underline{\mathsf{TL}}^{\mathsf{Y}}_{m,n}$, $\underline{\mathsf{TL}}^{\mathsf{X}}_{m}$, $\underline{\mathsf{TL}}^{\mathsf{Y}}_{m}$ *and* $\underline{\mathsf{TL}}_{m}$, *relative to the corresponding classes of rankers.*

Rankers and FO^2. The connection established by Weis and Immerman [18] between rankers and formulas in $\mathsf{FO}^2_{m,n}$, Thm. 1 below, is deeper. If x, y are integers, we let $\mathsf{ord}(x, y)$, the *order type* of x and y, be one of the symbols $<$, $>$ or $=$, depending on whether $x < y$, $x > y$ or $x = y$.

Theorem 1. *Let* $u, v \in A^*$ *and let* $1 \leq m \leq n$. *Then* u *and* v *satisfy the same formulas in* $\mathsf{FO}^2_{m,n}$ *if and only if*

(WI 1) u *and* v *agree on rankers from* $R_{m,n}$,
(WI 2) *if the rankers* $r \in \underline{R}_{m,n}$ *and* $r' \in \underline{R}_{m-1,n-1}$ *are defined on* u *and* v, *then* $\mathsf{ord}(r(u), r'(u)) = \mathsf{ord}(r(v), r'(v))$.
(WI 3) *if* $r \in \underline{R}_{m,n}$ *and* $r' \in \underline{R}_{m,n-1}$ *are defined on* u *and* v *and end with different direction letters, then* $\mathsf{ord}(r(u), r'(u)) = \mathsf{ord}(r(v), r'(v))$.

Corollary 1. *For each* $n \geq m \geq 1$, $\underline{\mathcal{TL}}_{m,n} \subseteq \mathcal{FO}^2_{m,n}$ *and* $\underline{\mathcal{TL}}_{m} \subseteq \mathcal{FO}^2_{m}$.

FO^2_m and $\underline{\mathsf{TL}}_m$-definable languages form varieties. Our first result is the following. We refer the reader to [8] and to Section 1.2 below for background and discussion on varieties of languages.

Proposition 2. *For each* $n \geq m \geq 1$, *the classes* $\underline{\mathcal{TL}}^{\mathsf{X}}_{m,n}$ $\underline{\mathcal{TL}}^{\mathsf{Y}}_{m,n}$, $\underline{\mathcal{TL}}^{\mathsf{Y}}_{m}$, $\underline{\mathcal{TL}}^{\mathsf{Y}}_{m}$, $\underline{\mathcal{TL}}_{m,n}$, $\underline{\mathcal{TL}}_{m}$, $\mathcal{FO}^2_{m,n}$ *and* $\mathcal{FO}^2_{m}$ *are varieties of languages.*

Sketch of proof. Let $\rho_{m,n}$ be the relation for two words to agree on $\underline{\mathsf{TL}}^{\mathsf{X}}_{m,n}$-formulas. Using Prop. 1, one verifies that $\rho_{m,n}$ is a finite index congruence. Then a language is $\underline{\mathsf{TL}}^{\mathsf{X}}_{m,n}$-definable if and only if it is a union of $\rho_{m,n}$-classes, if and only if it is recognized by the finite monoid $A^*/\rho_{m,n}$. It follows that these languages are exactly those accepted by the monoids in the pseudovariety generated by the $A^*/\rho_{m,n}$, for all finite alphabets A, and hence they form a variety of languages. $\underline{\mathcal{TL}}^{\mathsf{X}}_{m}$, being the increasing union of the $\underline{\mathcal{TL}}^{\mathsf{X}}_{m,n}$ $(n \geq m)$, is a variety as well.

The proof for the other fragments of TL is similar. For the fragments of FO^2, we use Thm. 1 instead of Prop. 1. □

This result shows that, for a given regular language L, $\underline{\mathsf{TL}}^{\mathsf{X}}_{m}$- (resp. $\underline{\mathsf{TL}}_{m}$-, FO^2_{m}-, etc) definability is characterized algebraically, that is, it depends only on the syntactic monoid of L. This justifies using the algebraic path to tackle decidability of these definability problems. Eilenberg's theory of varieties provides the mathematical framework.

1.2 A Short Survey on Varieties and Pseudovarieties

We summarize in this section the information on monoid and variety theory that will be relevant for our purpose, see [8,2,14,15] for more details.

A language $L \subseteq A^*$ is *recognized* by a monoid M if there exists a morphism $\varphi\colon A^* \to M$ such that $L = \varphi^{-1}(\varphi(L))$. For instance, if $u \in A^*$ and $B \subseteq A$, let $\mathsf{alph}(u) = \{a \in A \mid u = vaw \text{ for some } v, w \in A^*\}$ and $[B] = \{u \in A^* \mid \mathsf{alph}(u) = B\}$. Then $[B]$ is recognized by the direct product of $|B|$ copies of the 2-element monoid $\{0, 1\}$ (multiplicative).

A *pseudovariety* of monoids is a class of finite monoids closed under taking direct products, homomorphic images and submonoids. Pseudovarieties of subsemigroups are defined similarly. A *class of languages* $\mathcal{V}$ is a collection $\mathcal{V} = (\mathcal{V}(A))_A$, indexed by all finite alphabets A, such that $\mathcal{V}(A)$ is a set of languages in A^*. If $\mathbf{V}$ is a pseudovariety of monoids, we let $\mathcal{V}(A)$ be the set of languages of A^* recognized by a monoid in $\mathbf{V}$. The class $\mathcal{V}$ is closed under Boolean operations, residuals and inverse homomorphic images. Classes of recognizable languages with these properties are called *varieties* of languages, and Eilenberg's theorem (see [8]) states that the correspondence $\mathbf{V} \mapsto \mathcal{V}$, from pseudovarieties of monoids to varieties of languages, is one-to-one and onto. Moreover, the decidability of membership in the pseudovariety $\mathbf{V}$, implies the decidability of the variety $\mathcal{V}$: indeed, a language is in $\mathcal{V}$ if and only if its (effectively computable) syntactic monoid is in $\mathbf{V}$.

For every finite semigroup S and $s \in S$, we denote by s^ω the unique power of s which is idempotent. Another important monoid-theoretic concept is the following: if S is a monoid and $s, t \in S$, we say that $s \leq_{\mathcal{J}} t$ (resp. $s \leq_{\mathcal{R}} t$, $s \leq_{\mathcal{L}} t$) if $s = utv$ (resp. $s = tv$, $s = ut$) for some $u, v \in S \cup \{1\}$. We also say that $s \mathrel{\mathcal{J}} t$ is $s \leq_{\mathcal{J}} t$ and $t \leq_{\mathcal{J}} s$. The relations $\mathcal{R}$ and $\mathcal{L}$ are defined similarly.

Pseudovarieties that will be important in this paper are the following.

- $\mathbf{J}_1$, the pseudovariety of idempotent and commutative monoids; the corresponding variety is the Boolean algebra generated by the languages $[B]$, $B \subseteq A$.
- $\mathbf{R}$, $\mathbf{L}$ and $\mathbf{J}$, the pseudovarieties of $\mathcal{R}$-, $\mathcal{L}$- and $\mathcal{J}$-trivial monoids; a monoid is, say, $\mathcal{R}$-trivial if each of its $\mathcal{R}$-classes is a singleton.
- $\mathbf{DA}$, the pseudovariety of all finite monoids in which $(xy)^\omega x (xy)^\omega = (xy)^\omega$ for all x, y; $\mathbf{DA}$ has a great many characterizations in combinatorial, algebraic and logical terms [2,3,11,12,14,15].
- $\mathbf{K}$ (resp. $\mathbf{D}$, $\mathbf{LI}$) is the pseudovariety of semigroups in which $x^\omega y = x^\omega$ (resp. $y x^\omega = x^\omega$, $x^\omega y x^\omega = x^\omega$) for all x, y.

Finally, if $\mathbf{V}$ is a pseudovariety of semigroups and $\mathbf{W}$ is a pseudovariety of monoids, we say that a finite monoid M lies in the *Mal'cev product* $\mathbf{W} \circledm \mathbf{V}$ if there exists a finite monoid T and onto morphisms $\alpha\colon T \to M$ and $\beta\colon T \to N$ such that $N \in \mathbf{V}$ and $\beta^{-1}(e) \in \mathbf{W}$ for each idempotent e of N. Then $\mathbf{W} \circledm \mathbf{V}$ is a pseudovariety of monoids and we have in particular [8,2,10]:

$$\mathbf{K} \circledm \mathbf{J}_1 = \mathbf{K} \circledm \mathbf{J} = \mathbf{R},\ \ \mathbf{D} \circledm \mathbf{J}_1 = \mathbf{D} \circledm \mathbf{J} = \mathbf{L},\ \ \mathbf{LI} \circledm \mathbf{J}_1 = \mathbf{LI} \circledm \mathbf{J} = \mathbf{DA}.$$

We denote by $\underline{\mathbf{TL}}^{\mathsf{X}}_{m,n}$, $\underline{\mathbf{TL}}^{\mathsf{Y}}_{m,n}$, $\underline{\mathbf{TL}}^{\mathsf{Y}}_{m}$, $\underline{\mathbf{TL}}^{\mathsf{Y}}_{m}$, $\underline{\mathbf{TL}}_{m,n}$, $\underline{\mathbf{TL}}_{m}$, $\mathbf{FO}^2_{m,n}$ and $\mathbf{FO}^2_m$ the pseudovarieties corresponding to the language varieties discovered in Prop. 2.

2 Main Results

Our main tool to approach the decidability of FO^2_m-definability lies in a variant of rankers, which we borrow from a proof in Weis and Immerman's paper [18]. As in the turtle language of [12], a ranker can be seen as a sequence of instructions: go to the next a to the right, go to the next b to the left, etc. We say that a ranker r is *condensed on* u if it is defined on u, and if the sequence of positions visited *zooms in* on $r(u)$, never crossing over a position already visited. Formally, $r = \mathsf{Z}_1 \cdots \mathsf{Z}_n$ is condensed on u if there exists a chain of open intervals

$$(0, |u|+1) = (i_0, j_0) \supset (i_1, j_1) \supset \cdots \supset (i_{n-1}, j_{n-1}) \ni r(u)$$

such that for all $1 \leq \ell \leq n-1$ the following properties are satisfied:

- If $\mathsf{Z}_\ell \mathsf{Z}_{\ell+1} = \mathsf{X}_a \mathsf{X}_b$ then $(i_\ell, j_\ell) = (\mathsf{X}_a(u, i_{\ell-1}), j_{\ell-1})$.
- If $\mathsf{Z}_\ell \mathsf{Z}_{\ell+1} = \mathsf{Y}_a \mathsf{Y}_b$ then $(i_\ell, j_\ell) = (i_{\ell-1}, \mathsf{Y}_a(u, j_{\ell-1}))$.
- If $\mathsf{Z}_\ell \mathsf{Z}_{\ell+1} = \mathsf{X}_a \mathsf{Y}_b$ then $(i_\ell, j_\ell) = (i_{\ell-1}, \mathsf{X}_a(u, i_{\ell-1}))$.
- If $\mathsf{Z}_\ell \mathsf{Z}_{\ell+1} = \mathsf{Y}_a \mathsf{X}_b$ then $(i_\ell, j_\ell) = (\mathsf{Y}_a(u, j_{\ell-1}), j_{\ell-1})$.

For instance, the ranker $\mathsf{X}_a \mathsf{Y}_b \mathsf{X}_c$ is defined on the words bac and bca, but it is condensed only on bca. Rankers in $\underline{R}_1$, or of the form $\mathsf{X}_a \mathsf{Y}_{b_1} \cdots \mathsf{Y}_{b_k}$ or $\mathsf{Y}_a \mathsf{X}_{b_1} \cdots \mathsf{X}_{b_k}$, are condensed on all words on which they are defined. We denote by $L_c(r)$ the set of all words on which r is condensed.

Condensed rankers form a natural notion, equally well-suited to the description of FO^2_m-definability (see Thm. 2 below). With respect to TL, for which Prop. 1 shows a perfect match with the notion of rankers, they can be interpreted as adding a strong notion of unambiguity, see Section 3 below and the work of Lodaya, Pandya and Shah [7] on unambiguous interval temporal logic.

2.1 Condensed Rankers Determine a Hierarchy of Pseudovarieties

Let us say that two words u and v *agree on condensed rankers from a set* R of rankers, if the same rankers are condensed on u and v. We write $u \triangleright_{m,n} v$ (resp. $u \triangleleft_{m,n} v$) if u and v agree on condensed rankers in $\underline{R}^{\mathsf{X}}_{m,n}$ (resp. $\underline{R}^{\mathsf{Y}}_{m,n}$).

These relations turn out to have a very nice recursive characterization. For each word $u \in A^*$ and letter a occurring in u, the *a-left* (resp. *a-right*) *factorization* of u is the factorization that isolates the leftmost (resp. rightmost) occurrence of a in u; that is, the factorization $u = u_- a u_+$ such that a does not occur in u_- (resp. u_+). We say that the word $a_1 \cdots a_r$ is a *subword* of u if u can be factored as $u = u_0 a_1 u_1 \cdots a_r u_r$, with the $u_i \in A^*$.

Proposition 3. *The relations $\triangleright_{m,n}$ and $\triangleleft_{m,n}$ ($n \geq m \geq 1$) are uniquely determined by the following properties.*

- $u \triangleright_{1,n} v$ if and only if $u \triangleleft_{1,n} v$, if and only if u and v have the same subwords of length at most n.

- If $m \geq 2$, then $u \triangleright_{m,n} v$ if and only if $\mathsf{alph}(u) = \mathsf{alph}(v)$, $u \triangleleft_{m-1,n-1} v$ and for each letter $a \in \mathsf{alph}(u)$, the a-left factorizations $u = u_- a u_+$ and $v = v_- a v_+$ satisfy $u_- \triangleleft_{m-1,n-1} v_-$ and $u_+ \triangleright_{m,n-1} v_+$ (if $n > m$).

- *If $m \geq 2$, then $u \lhd_{m,n} v$ if and only if $\mathsf{alph}(u) = \mathsf{alph}(v)$, $u \rhd_{m-1,n-1} v$ and for each letter $a \in \mathsf{alph}(u)$, the a-right factorizations $u = u_- a u_+$ and $v = v_- a v_+$ satisfy $u_+ \rhd_{m-1,n-1} v_+$ and $u_- \lhd_{m,n-1} v_-$ (if $n > m$).*

Corollary 2. *The relations $\rhd_{m,n}$ and $\lhd_{m,n}$ are finite-index congruences.*

For each $m \geq 1$, let us denote by $\mathbf{R}_m$ (resp. $\mathbf{L}_m$) the pseudovariety generated by the quotients $A^*/\rhd_{m,n}$ (resp. $A^*/\lhd_{m,n}$), where $n \geq m$ and A is a finite alphabet. Corollary 2 shows that a language L is in the corresponding variety $\mathcal{R}_m$ (resp. $\mathcal{L}_m$) if and only if L is a Boolean combination of languages of the form $L_c(r)$, with $r \in \underline{R}^{\mathsf{X}}_m$ (resp. $\underline{R}^{\mathsf{Y}}_m$).

By definition, $\mathbf{R}_m$ and $\mathbf{L}_m$ are contained in both $\mathbf{R}_{m+1}$ and $\mathbf{L}_{m+1}$ for all m. Moreover, Prop. 3 shows that $\rhd_{1,n} = \lhd_{1,n}$ is the congruence defining the piecewise n-testable languages studied by Simon in the early 1970s, and that, in consequence, $\mathbf{R}_1 = \mathbf{L}_1 = \mathbf{J}$, the pseudovariety of $\mathcal{J}$-trivial monoids [8].

In addition, one can show that if a position in a word u is defined by a ranker $r \in \underline{R}^{\mathsf{X}}_{m,n}$ (resp. $\underline{R}^{\mathsf{Y}}_{m,n}$), then the same position is defined by a ranker $s \in \underline{R}^{\mathsf{X}}_{m,n}$ (resp. $\underline{R}^{\mathsf{Y}}_{m,n}$) which is condensed on u. This leads to the following result[5].

Proposition 4. *Let $n \geq m \geq 1$. If the words u and v agree on condensed rankers in $\underline{R}^{\mathsf{X}}_{m,n}$ (resp. $\underline{R}^{\mathsf{Y}}_{m,n}$), then they agree on rankers from the same class. In particular, $\underline{\mathbf{TL}}^{\mathsf{X}}_m \subseteq \mathbf{R}_m$ and $\underline{\mathbf{TL}}^{\mathsf{Y}}_m \subseteq \mathbf{L}_m$.*

As indicated above, condensed rankers allow for a description of FO^2_m-definability, as neat as with ordinary rankers: we show that the statement of Weis and Immerman's theorem can be modified to use condensed rankers instead.

Theorem 2. *Let $u, v \in A^*$ and let $1 \leq m \leq n$. Then u and v satisfy the same formulas in $\mathsf{FO}^2_{m,n}$ if and only if*

(WI 1c) *u and v agree on condensed rankers from $R_{m,n}$,*
(WI 2c) *if the rankers $r \in \underline{R}_{m,n}$ and $r' \in \underline{R}_{m-1,n-1}$ are condensed on u and v, then $\mathsf{ord}(r(u), r'(u)) = \mathsf{ord}(r(v), r'(v))$.*
(WI 3c) *if $r \in \underline{R}_{m,n}$ and $r' \in \underline{R}_{m,n-1}$ are condensed on u and v and end with different direction letters, then $\mathsf{ord}(r(u), r'(u)) = \mathsf{ord}(r(v), r'(v))$.*

Thus there is a connection between $\mathcal{FO}^2_m$ and the varieties $\mathcal{R}_m$ and $\mathcal{L}_m$. But much more can be said about the latter varieties.

2.2 Language Hierarchies

Prop. 3 also leads to a description of the language varieties $\mathcal{R}_m$ and $\mathcal{L}_m$ in terms of deterministic and co-deterministic products. Recall that a product of languages $L = L_0 a_1 L_1 \cdots a_k L_k$ ($k \geq 1$, $a_i \in A$, $L_i \subseteq A^*$) is *deterministic* if, for $0 \leq i \leq k$, each word $u \in L$ has a unique prefix in $L_0 a_1 L_1 \cdots L_{i-1} a_i$. If for

[5] Whose converse does not hold, see Ex. 2 below.

each i, the letter a_i does not occur in L_{i-1}, the product $L_0a_1L_1\cdots a_kL_k$ is called *visibly deterministic*: this is obviously also a deterministic product.

The definition of a *co-deterministic* or *visibly co-deterministic* product is dual, in terms of suffixes instead of prefixes. If $\mathcal{V}$ is a class of languages and A is a finite alphabet, let $\mathcal{V}^{det}(A)$ (resp. $\mathcal{V}^{vdet}(A)$, $\mathcal{V}^{codet}(A)$, $\mathcal{V}^{vcodet}(A)$) be the set of all Boolean combinations of languages of $\mathcal{V}(A)$ and of deterministic (resp. visibly deterministic, co-deterministic, visibly co-deterministic) products of languages of $\mathcal{V}(A)$. Schützenberger gave algebraic characterizations of the closure operations $\mathcal{V} \longmapsto \mathcal{V}^{det}$ and $\mathcal{V} \longmapsto \mathcal{V}^{codet}$, see [8]: if $\mathcal{V}$ is a variety of languages and if $\mathbf{V}$ is the corresponding pseudovariety of monoids, then $\mathcal{V}^{det}$ and $\mathcal{V}^{codet}$ are varieties of languages and the corresponding pseudovarieties are, respectively, $\mathbf{K} ⓜ \mathbf{V}$ and $\mathbf{D} ⓜ \mathbf{V}$. Then we show the following.

Proposition 5. *For each $m \geq 1$, we have $\mathcal{R}_{m+1} = \mathcal{L}_m^{vdet} = \mathcal{L}_m^{det}$, $\mathbf{R}_{m+1} = \mathbf{K} ⓜ \mathbf{L}_m$, $\mathcal{L}_{m+1} = \mathcal{R}_m^{vcodet} = \mathcal{R}_m^{codet}$ and $\mathbf{L}_{m+1} = \mathbf{D} ⓜ \mathbf{R}_m$. In particular, $\mathbf{R}_2 = \mathbf{R}$ and $\mathbf{L}_2 = \mathbf{L}$.*

Sketch of proof. Prop. 3 shows that $\mathcal{R}_{m+1} \subseteq \mathcal{L}_m^{vdet}$, which is trivially contained in $\mathcal{L}_m^{det}$. The last containment is proved algebraically, by showing that if $\gamma\colon A^* \to M$ is an onto morphism, and $M \in \mathbf{K} ⓜ \mathbf{L}_m$, then for some large enough n, $u \rhd_{m+1,n} v$ implies $\gamma(u) = \gamma(v)$: thus M is a quotient of $A^*/\rhd_{m+1,n}$ and hence, $M \in \mathbf{R}_{m+1}$. This proof relies on a technical property of semigroups in $\mathbf{DA}$: if $a \in A$ occurs in $\mathsf{alph}(v)$ and $\gamma(u) \mathrel{\mathcal{R}} \gamma(uv)$, then $\gamma(uva) \mathrel{\mathcal{R}} \gamma(u)$. □

It turns out that the $\mathbf{R}_m$ and the $\mathbf{L}_m$ were studied in the semigroup-theoretic literature (Kufleitner, Trotter and Weil, [17,6]). In [6], it is defined as the hierarchy of pseudovarieties obtained from $\mathbf{J}$ by repeated applications of the operations $\mathbf{X} \mapsto \mathbf{K} ⓜ \mathbf{X}$ and $\mathbf{X} \mapsto \mathbf{D} ⓜ \mathbf{X}$. Prop. 5 shows that it is the same hierarchy as that considered in this paper[6]. The following results are proved in [6, Section 4].

Proposition 6. *The hierarchies $(\mathbf{R}_m)_m$ and $(\mathbf{L}_m)_m$ are infinite chains of decidable pseudovarieties, and their unions are equal to $\mathbf{DA}$. Moreover, every m-generated monoid in $\mathbf{DA}$ lies in $\mathbf{R}_{m+1} \cap \mathbf{L}_{m+1}$.*

The decidability statement in Prop. 6 is in fact a consequence of a more precise statement (see [17,6]) which gives defining pseudoidentities for the $\mathbf{R}_m$ and $\mathbf{L}_m$. Let $x_1, x_2, \ldots$ be a sequence of variables. If u is a word on that alphabet, we let $\bar{u}$ be the mirror image of u, that is, the word obtained from reading u from right to left. We let

$$\begin{aligned} & G_2 = x_2x_1, \qquad I_2 = x_2x_1x_2, \\ \text{for } n > 2, \quad & G_n = x_n\overline{G_{n-1}}, \qquad I_n = G_nx_n\overline{I_{n-1}}, \\ & \varphi(x_1) = (x_1^\omega x_2^\omega x_1^\omega)^\omega, \quad \varphi(x_2) = x_2^\omega, \\ \text{and, for } n > 2, \quad & \varphi(x_n) = (x_n^\omega \varphi(\overline{G_{n-1}}G_{n-1})^\omega x_n^\omega)^\omega. \end{aligned}$$

[6] More precisely, the pseudovarieties $\mathbf{R}_m$ and $\mathbf{L}_m$ in [6] are pseudovarieties of semigroups, and the $\mathbf{R}_m$ and $\mathbf{L}_m$ considered in this paper are the classes of monoids in these pseudovarieties.

Proposition 7 ([6]). *For each* $m \geq 2$, $\mathbf{R}_m = \mathbf{DA} \cap [\![\varphi(G_m) = \varphi(I_m)]\!]$ *and* $\mathbf{L}_m = \mathbf{DA} \cap [\![\varphi(\overline{G_m}) = \varphi(\overline{I_m})]\!]$.

Example 1. For $\mathbf{R}_2$, this yields $x_2^\omega(x_1^\omega x_2^\omega x_1^\omega)^\omega = x_2^\omega(x_1^\omega x_2^\omega x_1^\omega)^\omega x_2^\omega$. One can verify that, together with the pseudo-identity defining $\mathbf{DA}$, this is equivalent to the usual pseudo-identity describing $\mathbf{R} = \mathbf{R}_2$, namely $(st)^\omega s = (st)^\omega$.

For $\mathbf{R}_3 = \mathbf{K} ⓜ \mathbf{L}$, no pseudo-identity was known in the literature. We get

$$\begin{aligned}\varphi(G_3) &= (x_3^\omega((x_1^\omega x_2^\omega x_1^\omega)^\omega x_2^\omega(x_1^\omega x_2^\omega x_1^\omega)^\omega)^\omega x_3^\omega)^\omega \\ \varphi(I_3) &= (x_3^\omega((x_1^\omega x_2^\omega x_1^\omega)^\omega x_2^\omega(x_1^\omega x_2^\omega x_1^\omega)^\omega)^\omega x_3^\omega)^\omega \\ &\quad (x_3^\omega((x_1^\omega x_2^\omega x_1^\omega)^\omega x_2^\omega(x_1^\omega x_2^\omega x_1^\omega)^\omega)^\omega x_3^\omega)^\omega x_2^\omega(x_1^\omega x_2^\omega x_1^\omega)^\omega x_2^\omega.\end{aligned}$$

2.3 Connection with the $\mathbf{TL}_m$ and the $\mathbf{FO}^2_m$ Hierarchies

Prop. 4 established a containment between the $\mathbf{R}_m$ (resp. $\mathbf{L}_m$) and the $\underline{\mathbf{TL}}_m$ hierarchies. A technical analysis allows us to prove a containment in the other direction, but one that is not very tight – showing the difference between the consideration of condensed rankers and that of ordinary rankers.

Proposition 8. $\mathbf{R}_2 = \underline{\mathbf{TL}}_2^{\mathsf{X}}$ *and* $\mathbf{L}_2 = \underline{\mathbf{TL}}_2^{\mathsf{Y}}$. *If* $m \geq 3$ *and if two words agree on rankers in* $\underline{R}^{\mathsf{X}}_{\lfloor 3m/2 \rfloor}$ *(resp.* $\underline{R}^{\mathsf{Y}}_{\lfloor 3m/2 \rfloor}$*), then they agree on condensed rankers in* $\underline{R}^{\mathsf{X}}_m$ *(resp.* $\underline{R}^{\mathsf{Y}}_m$*). In particular* $\mathbf{R}_m \subseteq \underline{\mathbf{TL}}^{\mathsf{X}}_{\lfloor 3m/2 \rfloor}$ *and* $\mathbf{L}_m \subseteq \underline{\mathbf{TL}}^{\mathsf{Y}}_{\lfloor 3m/2 \rfloor}$.

Example 2. The language $L_c(\mathsf{X}_a\mathsf{Y}_b\mathsf{X}_c)$ is in $\mathcal{R}_3$ and not in $\underline{\mathcal{TL}}^{\mathsf{X}}_3$.

The connection between the $\mathbf{R}_m$, $\mathbf{L}_m$ and $\mathbf{FO}^2_m$ hierarchies is tighter.

Theorem 3. *Let* $m \geq 1$. *Every language in* $\mathcal{R}_m$ *or* $\mathcal{L}_m$ *is* FO^2_m*-definable, and every* FO^2_m*-definable language is in* $\mathcal{R}_{m+1} \cap \mathcal{L}_{m+1}$. *Equivalently, we have*

$$\mathbf{R}_m \vee \mathbf{L}_m \subseteq \mathbf{FO}^2_m \subseteq \mathbf{R}_{m+1} \cap \mathbf{L}_{m+1},$$

where $\mathbf{V} \vee \mathbf{W}$ *denotes the least pseudovariety containing* $\mathbf{V}$ *and* $\mathbf{W}$.

Sketch of proof. The containment $\mathbf{R}_m \vee \mathbf{L}_m \subseteq \mathbf{FO}^2_m$ follows directly from Property (**WI 1c**) in Thm. 2. The proof of the converse containment also relies on that theorem. We show that if $u \rhd_{m+1,2n}$ or $u \lhd_{m+1,2n}$, then Properties (**WI 1c**), (**WI 2c**) and (**WI 3c**) hold for m, n. This is done by a complex and quite technical induction. □

If $m = 1$, we know that $\mathbf{R}_2 \cap \mathbf{L}_2 = \mathbf{R} \cap \mathbf{L} = \mathbf{J} = \mathbf{R}_1 \vee \mathbf{L}_1$: this reflects the elementary observation that FO^2_1-definable languages, like FO_1-definable languages, are the piecewise testable languages. For $m \geq 2$, we conjecture that $\mathbf{R}_m \vee \mathbf{L}_m$ is properly contained in $\mathbf{R}_{m+1} \cap \mathbf{L}_{m+1}$. The following shows it holds for $m = 2$.

Example 3. It is elementary to find an FO^2_2-formula defining the language $L = \{b, c\}^*ca\{a, b\}^*$ (every c is before every a, every b has an a in its past or a c in its future, and there is at least an a and a c). The words $u_n = (bc)^n(ab)^n$ are in L, while the words $v_n = (bc)^n b(ca)^n$ are not. Almeida and Azevedo showed that $\mathbf{R}_2 \vee \mathbf{L}_2$ is defined by the pseudo-identity $(bc)^\omega(ab)^\omega = (bc)^\omega b(ab)^\omega$ [2, Thm. 9.2.13 and Exerc. 9.2.15]). In particular, for each language K recognized by a monoid in $\mathbf{R}_2 \vee \mathbf{L}_2$, the words u_n and v_n are eventually all in K, or all in the complement of K. Therefore L is not recognized by such a monoid, which proves that $\mathbf{R}_2 \vee \mathbf{L}_2$ is strictly contained in $\mathbf{FO}^2_2$, and hence also in $\mathbf{R}_3 \cap \mathbf{L}_3$. It also shows that $\underline{\mathcal{TL}}_2$ is properly contained in $\mathcal{FO}^2_2$.

Finally, we formulate the following conjecture.

Conjecture 1. For each $m \geq 1$, $\mathbf{FO}^2_m = \mathbf{R}_{m+1} \cap \mathbf{L}_{m+1}$.

3 Consequences

The main consequence we draw of Thm. 3 and of the decidability of the pseudovarieties $\mathbf{R}_m$ and $\mathbf{L}_m$ is summarized in the next statement.

Theorem 4. *Given an FO^2-definable language L, one can compute an integer m such that L is FO^2_{m+1}-definable but not FO^2_{m-1}-definable. That is: we can decide the quantifier alternation level of L within one unit.*

Sketch of proof. If $M \in \mathbf{DA}$, we can compute the largest m such that $M \notin \mathbf{R}_m \cap \mathbf{L}_m$ (Prop. 6). Then $M \in \mathbf{FO}^2_{m+1} \setminus \mathbf{FO}^2_{m-1}$ by Thm. 3. □

The fact that the $\mathbf{R}_m$ and $\mathbf{L}_m$ form strict hierarchies (Prop. 6), together with Thm. 3, proves that the $\mathcal{FO}^2_m$ hierarchy is infinite. Weis and Immerman had already proved this result by combinatorial means [18], whereas our proof is algebraic. From that result on the $\mathcal{FO}^2_m$, it is also possible to recover the strict hierarchy result on the $\mathbf{R}_m$ and $\mathbf{L}_m$ and the fact that their union is equal to $\mathbf{DA}$. By the same token, Prop. 4 and 8 show that the $\underline{\mathcal{TL}}_m$ (resp. $\underline{\mathbf{TL}}_m$) hierarchy is infinite and that its union is all of $\mathcal{FO}^2$ (resp. $\mathbf{DA}$).

Similarly, the fact that an m-generated element of $\mathbf{DA}$ lies in $\mathbf{R}_{m+1} \cap \mathbf{L}_{m+1}$ (Prop. 6), shows that an FO^2-definable language in A^* lies in $\mathcal{R}_{|A|+1} \cap \mathcal{L}_{|A|+1}$, and hence in $\mathcal{FO}^2_{m+1}$ – a fact that was already established by combinatorial means by Weis and Immerman [18, Thm. 4.6]. It also shows that such a language is in $\underline{\mathcal{TL}}_{\frac{3}{2}(|A|+1)}$ by Prop. 8.

Finally we note the following result. It was mentioned in the introduction that the languages in $\mathcal{FO}^2$ are disjoint unions of unambiguous products of the form $B_0^* a_1 B_1^* \cdots a_k B_k^*$, where each B_i is a subset of A. Props. 5 and 6 imply the following statement.[7]

[7] The weaker statement with the word *visibly* deleted was proved by the authors in [6], as well as by Lodaya, Pandya and Shah [7].

Proposition 9. *The least variety of languages containing the languages of the form B^* ($B \subseteq A$) and closed under visibly deterministic and visibly co-deterministic products, is $\mathcal{FO}^2$.*

Every unambiguous product of languages of the form $B_0^* a_1 B_1^* \cdots a_k B_k^*$ (with each $B_i \subseteq A$), can be expressed in terms of the B_i^* and the a_i using only Boolean operations and at most $|A| + 1$ applications of visibly deterministic and visibly co-deterministic products, starting with a visibly deterministic (resp. co-deterministic) product.

References

1. Adler, M., Immerman, N.: An $n!$ lower bound on formula size. ACM Trans. Computational Logic 4, 296–314 (2003)
2. Almeida, J.: Finite Semigroups and Universal Algebra. World Scientific, Singapore (1994)
3. Diekert, V., Gastin, P., Kufleitner, M.: A survey on small fragments of first-order logic over finite words. Internat. J. Found. Comput. Sci. 19(3), 513–548 (2008)
4. Grohe, M., Schweikardt, N.: The succinctness of first-order logic on linear orders. Logical Methods in Computer Science 1 (2005)
5. Immerman, N.: Descriptive Complexity. Springer, Heidelberg (1999)
6. Kufleitner, M., Weil, P.: On the lattice of sub-pseudovarieties of DA (to appear)
7. Lodaya, K., Pandya, P.K., Shah, S.S.: Marking the chops: an unambiguous temporal logic. In: IFIP TCS 2008, pp. 461–476 (2008)
8. Pin, J.-É.: Varieties of Formal Languages. North Oxford Academic (1986)
9. Pin, J.-É.: Expressive power of existential first-order sentences of Büchi's sequential calculus. Discrete Maths 291, 155–174 (2005)
10. Pin, J.-É., Weil, P.: Profinite semigroups, Mal'cev products and identities. J. Algebra 182, 604–626 (1996)
11. Pin, J.-É., Weil, P.: Polynomial closure and unambiguous product. Theory Comput. Systems 30, 383–422 (1997)
12. Schwentick, T., Thérien, D., Vollmer, H.: Partially-ordered two-way automata: A new characterization of DA. In: Kuich, W., Rozenberg, G., Salomaa, A. (eds.) DLT 2001. LNCS, vol. 2295, pp. 239–250. Springer, Heidelberg (2002)
13. Straubing, H.: Finite Automata, Formal Logic, and Circuit Complexity. Birkhäuser, Basel (1994)
14. Tesson, P., Thérien, D.: Diamonds are forever: The variety DA. In: Gomes, G., Ventura, P., Pin, J.-É. (eds.) Semigroups, Algorithms, Automata and Languages, pp. 475–500. World Scientific, Singapore (2002)
15. Tesson, P., Thérien, D.: Logic meets algebra: the case of regular languages. Logical Methods in Computer Science 3, 1–37 (2007)
16. Thomas, W.: Classifying regular events in symbolic logic. J. Comput. Systems and Science 25, 360–376 (1982)
17. Trotter, P., Weil, P.: The lattice of pseudovarieties of idempotent semigroups and a non-regular analogue. Algebra Universalis 37, 491–526 (1997)
18. Weis, P., Immerman, N.: Structure theorem and strict alternation hierarchy for FO^2 on words. In: Duparc, J., Henzinger, T.A. (eds.) CSL 2007. LNCS, vol. 4646, pp. 343–357. Springer, Heidelberg (2007)

On the Recognizability of Self-generating Sets

Tomi Kärki*, Anne Lacroix, and Michel Rigo

University of Liège, Institute of Mathematics, Grand Traverse 12 (B 37),
B-4000 Liège, Belgium
{T.Karki,A.Lacroix,M.Rigo}@ulg.ac.be

Abstract. Let I be a finite set of integers and F be a finite set of maps of the form $n \mapsto k_i\, n + \ell_i$ with integer coefficients. For an integer base $k \geq 2$, we study the k-recognizability of the minimal set X of integers containing I and satisfying $\varphi(X) \subseteq X$ for all $\varphi \in F$. In particular, solving a conjecture of Allouche, Shallit and Skordev, we show under some technical conditions that if two of the constants k_i are multiplicatively independent, then X is not k-recognizable for any $k \geq 2$.

1 Introduction

In the general framework of numeration systems, the so-called recognizable sets of integers have been extensively studied. Let $k \geq 2$ be an integer. The function $\mathrm{rep}_k \colon \mathbb{N} \to \{0, \ldots, k-1\}^*$ maps a non-negative integer onto its k-ary representation (without leading zeros). A set $X \subseteq \mathbb{N}$ is *k-recognizable* if the language $\mathrm{rep}_k(X) = \{\mathrm{rep}_k(n) \mid n \in X\}$ is regular; see, for instance, [3]. A similar definition can be given for the k-recognizable subsets of $\mathbb{Z}$ using convenient conventions to represent negative numbers, like adding a symbol "$-$" to the alphabet or considering the positive and the negative elements separately. Since the seminal work of Cobham [4], it is well-known that the recognizability of a set depends on the choice of the base k — except for the ultimately periodic sets, i.e., the union of a finite set and a finite number of infinite arithmetic progressions, which are easily seen to be k-recognizable for all $k \geq 2$. The celebrated theorem of Cobham can be stated as follows. Let $k, \ell \geq 2$ be two multiplicatively independent bases, i.e., $\log k / \log \ell$ is irrational. If a set $X \subseteq \mathbb{N}$ is both k-recognizable and ℓ-recognizable, then it is ultimately periodic.

Kimberling introduced the so-called *self-generating* sets of integers [10]. They can be defined as follows. Let $r \geq 1$ and $G = \{\varphi_1, \varphi_2, \ldots, \varphi_r\}$ be a set of affine maps where $\varphi_i : n \mapsto k_i\, n + \ell_i$ with $k_i, \ell_i \in \mathbb{Z}$ and $2 \leq k_1 \leq k_2 \leq \cdots \leq k_r$. The set generated by G and a finite set of integers I is the minimal subset X of $\mathbb{Z}$ containing I and such that $\varphi_i(X) \subseteq X$ for all $i = 1, \ldots, r$. For any subset $S \subseteq \mathbb{Z}$, we set $G(S) := \{\varphi(s) \mid s \in S, \varphi \in G\}$, $G^0(S) := S$ and $G^{m+1}(S) := G(G^m(S))$ for all $m \geq 0$. Otherwise stated $X = \bigcup_{m \geq 0} G^m(I)$ is the set of all integers n such that there exist $m \geq 0$, $a \in I$ and a finite sequence $(\varphi_{i_1}, \varphi_{i_2}, \ldots, \varphi_{i_m})$ of maps in G such that

$$n = \varphi_{i_m} \circ \varphi_{i_{m-1}} \circ \cdots \circ \varphi_{i_1}(a) = \varphi_{i_m}(\varphi_{i_{m-1}}(\cdots \varphi_{i_1}(a) \cdots)). \tag{1}$$

* Supported by Osk. Huttunen Foundation.

R. Královič and D. Niwiński (Eds.): MFCS 2009, LNCS 5734, pp. 525–536, 2009.

Example 1. In [10], for $G = \{n \mapsto 2n, n \mapsto 4n - 1\}$ and $I = \{1\}$, it is shown that the corresponding self-generating set

$$\mathcal{K}_1 = \{1, 2, 3, 4, 6, 7, 8, 11, 12, 14, 15, 16, \ldots\}$$

is closely related to the Fibonacci word. Notice that for $I = \{0\}$, we get a subset containing negative integers: $\mathcal{K}_0 = \{0, -1, -2, -4, -5, -8, -9, \ldots\}$. In particular, for $I = \{0, 1\}$, the corresponding self-generating set is $\mathcal{K}_0 \cup \mathcal{K}_1$.

These self-generating sets are also called *affinely recursive* in [11] where the correspondence between words $i_1 i_2 \cdots i_m$ over the alphabet $\{1, 2, \ldots, r\}$ and integers $\varphi_{i_m}(\varphi_{i_{m-1}}(\cdots \varphi_{i_1}(1) \cdots))$ is studied. For example, conditions under which this correspondence is one-to-one are given, which in turn implies that the natural ordering of the integers induces an ordering on the set of non-empty words over $\{1, 2, \ldots, r\}$ providing a kind of abstract numeration system [12].

In [2] a general framework for self-generating sets is considered. The k-ary representations of the elements in a self-generating set are related to words over $\Sigma_k = \{0, 1, \ldots, k-1\}$ where some fixed block of digits is missing. As an illustration, one can notice that the set $\mathcal{K}_1 - 1 = \{0, 1, 2, 3, 5, 6, 7, 10, \ldots\}$ introduced in Example 1 consists of all integers whose binary expansion does not contain "00" as factor. Recall that the *characteristic sequence* $(\mathbf{c}_X(n))_{n \geq 0}$ of a set $X \subseteq \mathbb{N}$ is defined by $\mathbf{c}_X(n) = 1$, if $n \in X$ and $\mathbf{c}_X(n) = 0$, otherwise. In particular, X is k-recognizable (resp., ultimately periodic) if and only if $(\mathbf{c}_X(n))_{n \geq 0}$ is k-automatic (resp., an ultimately periodic infinite word). These self-generating sets are consequently studied from the point of view of automatic and morphic sequences as well as in relation to non-standard numeration systems; for the definitions and further information, see [1,13]. Moreover, Allouche, Shallit and Skordev ask the following question: *Under what conditions is the characteristic sequence of a self-generating set k-automatic* ? They also present the following conjecture.

Conjecture 1. With "mixed base" rules, such as $G = \{n \mapsto 2n + 1, n \mapsto 3n\}$, the set generated from $I = \{1\}$ is not k-recognizable for any integer base $k \geq 2$.

Let us fix notation once and for all.

Definition 1. *In this paper, instead of considering a set G of maps as described above, we will moreover consider the extended set of $r + 1 \geq 2$ maps*

$$F = G \cup \{\varphi_0\} = \{\varphi_0, \varphi_1, \ldots, \varphi_r\} \text{ where } \varphi_0 : n \mapsto n$$

and $\varphi_i : n \mapsto k_i n + \ell_i$ with $k_i, \ell_i \in \mathbb{Z}$ and $2 \leq k_1 \leq k_2 \leq \cdots \leq k_r$. Having identity function at our disposal, for any set $S \subseteq \mathbb{Z}$, we have $F^m(S) \subseteq F^{m+1}(S)$. Therefore, for any finite set I of integers, the set

$$F^\omega(I) := \lim_{m \to \infty} F^m(I)$$

is exactly the self-generating set *with respect to G and I.*

The content of the paper is the following.

1. If we add to F an extra map $\psi : n \mapsto n+\ell$ with $\ell \neq 0$, then the corresponding self-generating set $F^\omega(I)$ is ultimately periodic and therefore k-recognizable for all $k \geq 2$.
2. If all the multiplicative constants k_i are pairwise multiplicatively dependent, then we give a general method to build a finite automaton recognizing $\text{rep}_k(F^\omega(I))$ for any k that is multiplicatively dependent on every k_i. Let us note that the case where the constants k_i are powers of a fixed base is considered in [8].
3. If there exist i, j such that k_i and k_j are multiplicatively independent and if $\sum_{i=1}^{r} k_i^{-1} < 1$, then $F^\omega(I)$ is not k-recognizable for any $k \geq 2$. In particular, this condition always holds for sets F where $r = 2$ and $k_1 < k_2$ are multiplicatively independent, answering Conjecture 1 in the affirmative.

The techniques rely on a classical gap theorem; see Theorem 3. We study differences and ratios of consecutive elements in the considered self-generating set.

2 Ultimately Periodic Self-generating Sets

Theorem 1. *If we add to F in Definition 1 an extra map $\psi : n \mapsto n + \ell$ with $\ell \neq 0$, then the corresponding self-generating set $F^\omega(I)$ is ultimately periodic of period ℓ.*

Proof. Denote by $F^j(I) \mod \ell$ the set $\{n \mod \ell \mid n \in F^j(I)\}$. Recall that the identity function φ_0 belongs to F. Since there are finitely many congruence classes modulo ℓ and $F^j(I) \mod \ell \subseteq F^{j+1}(I) \mod \ell$, there must exist an integer J such that $F^{J+1}(I) \mod \ell = F^J(I) \mod \ell$. Moreover, this means that $F^j(I) \mod \ell = F^J(I) \mod \ell$ for every $j \geq J$, and, consequently,

$$F^\omega(I) \mod \ell = F^J(I) \mod \ell. \tag{2}$$

On the other hand, if $n \in F^\omega(I)$, then $\psi^t(n) = n + t\,\ell \in F^\omega(I)$. Since $n + t\,\ell \equiv n \mod \ell$, we conclude by (2), for any $n \geq \max F^J(I)$, that

$$\mathbf{c}_{F^\omega(I)}(n) = \begin{cases} 1, & \text{if } n \mod \ell \in F^J(I) \mod \ell; \\ 0, & \text{otherwise.} \end{cases}$$

Hence, the characteristic sequence of $F^\omega(I)$ is ultimately periodic with preperiod $\max F^J(I)$ and period ℓ.

Remark 1. In Definition 1 and in what follows, we always assume that all the multiplicative constants k_i of the affine maps $\varphi_1, \ldots, \varphi_r$ in F are at least 2. This condition does not guarantee that the corresponding self-generating set is not ultimately periodic. For example, if $\varphi_i(x) = r\,x + i$ for $i = 1, \ldots, r$, then we easily see that $F^\omega(\{0\}) = \mathbb{N}$.

Let $y \geq 0$. Recall (for instance, see [3]) that a set $Y \subseteq \mathbb{N}$ is k-recognizable if and only if $Y + y$ is k-recognizable. As explained by the following lemma, from the point of view of recognizability of subsets of $\mathbb{N}$, one can assume that all the additive constants ℓ_i are non-negative.

Lemma 1. *Let $X = F^\omega(I)$ be a self-generating set as given in Definition 1. There exist a non-negative integer y and a self-generating set $Y = \widehat{F}^\omega(I - y)$, where $\widehat{F} = \{\varphi_0, \widehat{\varphi}_1, \ldots, \widehat{\varphi}_r\}$, such that $X = Y + y$ and $\widehat{\varphi}_i : n \mapsto k_i n + \widehat{\ell}_i$ for every $i = 1, 2, \ldots, r$ with some non-negative constants $\widehat{\ell}_i$ completely determined by F.*

Proof. Assume that at least for some function $\varphi_i \in F$ the constant ℓ_i is negative. Otherwise, the claim is trivial. Let $y = \max\{|\ell_i| \mid \ell_i < 0\}$ and set, for $i = 1, 2, \ldots, r$,

$$\widehat{\ell}_i := \ell_i + (k_i - 1)y.$$

Since $k_i \geq 2$, the constants $\widehat{\ell}_i$ are non-negative for every i.

We show by induction on the number of applied maps that x belongs to $F^\omega(I)$ if and only if $x - y$ belongs to $\widehat{F}^\omega(I - y)$. First, for any $x \in I$, it is obvious that $x - y$ belongs to $I - y$. Recall that $F = G \cup \{\varphi_0\}$ and assume now that $x \in G^m(I)$ for some $m \geq 1$. Otherwise stated, x is obtained by applying m maps in $\{\varphi_1, \ldots, \varphi_r\}$. Therefore there exist $z \in G^{m-1}(I)$ and $i \in \{1, \ldots, r\}$ such that $x = \varphi_i(z)$. By induction hypothesis, $z - y$ belongs to $\widehat{G}^{m-1}(I - y)$ where $\widehat{G} = \widehat{F} \setminus \{\varphi_0\}$. Then we have $\varphi_i(z) = k_i z + \ell_i$ and $\widehat{\varphi}_i(z - y) = k_i(z - y) + \ell_i + (k_i - 1)y = \varphi_i(z) - y$. This proves that $x - y$ belongs to $\widehat{G}^m(I - y)$. Assume now that $x - y \in \widehat{G}^m(I - y)$ for some $m \geq 1$. There exist $z \in \widehat{G}^{m-1}(I - y)$ and $i \in \{1, \ldots, r\}$ such that $x - y = \widehat{\varphi}_i(z)$. Then $x = k_i(z + y) + \ell_i = \varphi_i(z + y)$ and by induction hypothesis $z + y$ belongs to $G^{m-1}(I)$. This concludes the proof.

Example 2. Consider the set $X = \mathcal{K}_1$ given in Example 1 and generated from $\{1\}$ by the maps $n \mapsto 2n$ and $n \mapsto 4n - 1$. Applying the constructions given in the previous proof, set $y = 1$ and consider the maps $2n + 1$ and $4n + 2$. These two maps generate from $\{1\} - 1 = \{0\}$, the set $\{0, 1, 2, 3, 5, 6, 7, 10, \ldots\}$ which is equal to $X - 1$.

3 Multiplicatively Dependent Case

In this section, we assume that the multiplicative coefficients k_i appearing in Definition 1 are all pairwise multiplicatively dependent, i.e., for every pair (i, j), there exist positive integers e_i and e_j such that $k_i^{e_i} = k_j^{e_j}$. Note that k_i and k_j are multiplicatively dependent if and only if there exist an integer $n \geq 2$ and two integers $d_i, d_j \geq 1$ such that $k_i = n^{d_i}$ and $k_j = n^{d_j}$. By this characterization, it is easy to see that if the coefficients k_i are pairwise multiplicatively dependent, then there exists an integer k such that every k_i is a power of k. Our aim is to build a finite automaton showing that the set $F^\omega(I)$ is k-recognizable.

Recall that $\Sigma_k = \{0, 1, \ldots, k-1\}$ and that $\mathrm{rep}_k \colon \mathbb{N} \to \Sigma_k^*$ maps an integer n to its k-ary representation without leading zeros. For any finite alphabet $A \subseteq \mathbb{Z}$, the function $\mathrm{val}_{A,k} \colon A^* \to \mathbb{Z}$ maps a word $w = w_n w_{n-1} \cdots w_0$ over A to the corresponding numerical value

$$\mathrm{val}_{A,k}(w) = \sum_{i=0}^{n} w_i\, k^i .$$

The function defined over the set of words $w \in A^*$ such that $\mathrm{val}_{A,k}(w) \geq 0$ and which maps w to $\mathrm{rep}_k(\mathrm{val}_{A,k}(w))$ is called *normalization* over A. In the special case $A = \Sigma_k$, we simply write val_k instead of $\mathrm{val}_{\Sigma_k,k}$.

Theorem 2. *Let F given in Definition 1 be such that the multiplicative coefficients $k_1, \ldots, k_r$ are all pairwise multiplicatively dependent. The corresponding self-generating set $X = F^\omega(I)$ is k-recognizable if k_i is a power of k for every $i = 1, 2, \ldots, r$.*

We first sketch a proof relying on Frougny's normalization theorem.

Proof. Assume that $X \subseteq \mathbb{N}$ and that the maps are of the kind $\varphi_i : n \mapsto k^{e_i}\, n + \ell_i$ with $e_1 \geq 1$ for all $i \in \{1, \ldots, r\}$. Let $n = \varphi_{i_m}(\varphi_{i_{m-1}}(\cdots \varphi_{i_1}(a) \cdots))$ for some $a \in I$. With that integer, we associate the word

$$w = a\, 0^{e_{i_1}-1} \ell_{i_1} \cdots 0^{e_{i_m}-1} \ell_{i_m}$$

over the finite alphabet $I \cup \{0, \ell_1, \ldots, \ell_r\} \subset \mathbb{Z}$. One can notice that $\mathrm{val}_k(w) = n$. Apply Proposition 7.1.4 in [13] (see also [7]) and Theorem 4.3.6 in [1] to the language $I\{0^{e_1-1}\ell_1, \ldots, 0^{e_r-1}\ell_r\}^*$ to get the regular language $\mathrm{rep}_k(F^\omega(I))$.

We give below another proof which is independent of Frougny's normalization theorem. It describes a way to build an automaton recognizing the k-ary representations of $F^\omega(I)$. We denote by $\mathbb{Z}_{\geq 0}$ (resp., $\mathbb{Z}_{\leq 0}$) the set of non-negative (resp., non-positive) integers.

Remark 2. The set of non-negative elements (resp., the set of absolute values of non-positive elements) in $F^\omega(I)$ can be obtained from a finite set of non-negative elements. Let $m_\ell = \max\{|\ell_i| \mid i = 1, 2, \ldots, r\}$ and denote by M_ℓ the interval of integers $[\![-m_\ell, m_\ell]\!]$. Define $I_j := F^j(I) \cap M_\ell$ for $j \geq 0$. Since $k_i \geq 2$ for all $i \in \{1, 2, \ldots, r\}$, it follows that if n does not belong to M_ℓ, then $\varphi_i(n) \notin M_\ell$ for all $i \in \{0, 1, \ldots, r\}$. By this property and since $F^j(I) \subseteq F^{j+1}(I)$, there must exist an integer J such that $I_j = I_J$ for all $j \geq J$. Hence, the integers of $F^\omega(I)$ falling into the interval M_ℓ are exactly the ones in I_J. We set $I^+ := (I_J \cup I) \cap \mathbb{Z}_{\geq 0}$ and $I^- := (I_J \cup I) \cap \mathbb{Z}_{\leq 0}$. By the property above, we conclude that

$$F^\omega(I) = (F^\omega(I^+) \cap \mathbb{Z}_{\geq 0}) \cup (F^\omega(I^-) \cap \mathbb{Z}_{\leq 0}).$$

Hence, $F^\omega(I^+) \cap \mathbb{Z}_{\geq 0}$ is obtained from I^+ by considering only non-negative images of the maps φ_i. Let $\overline{F} = \{\varphi_0, \overline{\varphi}_1, \overline{\varphi}_2, \ldots, \overline{\varphi}_r\}$, where $\overline{\varphi}_i : n \mapsto k_i\, n - \ell_i$ for $i = 1, 2, \ldots, r$. Then we have $F^\omega(I^-) \cap \mathbb{Z}_{\leq 0} = -(\overline{F}(-I^-) \cap \mathbb{Z}_{\geq 0})$. Otherwise stated, the negation of the elements in $F^\omega(I^-) \cap \mathbb{Z}_{\leq 0}$ are obtained from $-I^-$ by considering only non-negative images of the maps $\overline{\varphi_i}$.

Proof. From the previous remark, we may assume without loss of generality that I and X are subsets of $\mathbb{N}$. Let k be an integer such that all the coefficients k_i are powers of k. Note that $\text{val}_k^{-1}(n)$ contains all the representations of the integer n in base k over Σ_k^*, including those with leading zeros. We define a non-deterministic finite automaton $\mathcal{A} = (Q, \{q_0\}, \Sigma_k, \Delta, T)$ accepting the reversal of the elements in $\text{val}_k^{-1}(F^\omega(I))$, so we may allow leading zeros in front of the most significant digit. The transition relation Δ is a finite subset of $Q \times \Sigma_k^* \times Q$. If (p, w, q) belongs to Δ, we write $p \xrightarrow{w} q$. An input $x \in \Sigma_k^*$ is accepted if and only if there is a sequence of states $q_0, q_1, \ldots, q_i$ such that $q_i \in T$, x can be factored as $u_1 \cdots u_i$ and $(q_0, u_1, q_1), (q_1, u_2, q_2), \ldots (q_{i-1}, u_i, q_i) \in \Delta$.

Let M be the maximal element in $I \cup \{k_1, \ldots, k_r, |\ell_1|, \ldots, |\ell_r|\}$ and $m = |\text{rep}_k(M)|$. Define $Q = \{q_0\} \cup (\{-1, 0, +1\} \times \Sigma_k^{m+1})$. A state $q = (c, x) \in Q \setminus \{q_0\}$ is final if and only if $c = 0$ and $x \in \text{val}_k^{-1}(I)$. From the initial state q_0, we have all the transitions

$$q_0 \xrightarrow{w} (0, \widetilde{w})$$

where $w \in \Sigma_k^{m+1}$ and $\widetilde{w}$ is the reversal of w. Recall that entries are read in $\mathcal{A}$ the least significant digit first, that is from right to left. This explains why we consider the reversals in the encoding. From each state $Q \setminus \{q_0\}$ there are transitions corresponding to the maps φ_i, $i = 1, 2, \ldots, r$. The idea is to guess the sequence of maps $(\varphi_{i_1}, \varphi_{i_2}, \ldots, \varphi_{i_m})$ that was used to obtain the integer corresponding to the input belonging to $\text{val}_k^{-1}(F^\omega(I))$ and apply the inverses of these maps in reversed order to get back the representation of one of the initial values in I. The first component of a state $q = (c, x_m x_{m-1} \cdots x_0)$ corresponds to a carry bit and the second component represents the last $m+1$ digits of a number, x_0 being the least significant one. We show how to simulate the multiplications and additions in the successive applications of the affine functions φ_i using only the carry bit c and the digits $x_m x_{m-1} \cdots x_0$.

Consider first a state $p = (0, x_m x_{m-1} \cdots x_0)$ and φ_i-transitions, where $\varphi_i : n \mapsto k_i\, n + \ell_i$ and $\ell_i \geq 0$. For the inverse of φ_i, we want to subtract ℓ_i and then divide by $k_i = k^t$ for some positive integer t. We do this using the classical paper-and-pencil method as illustrated in Figure 1(a), where $\text{val}_k(y_m y_{m-1} \cdots y_0) = \ell_i$ and $x = 1$, if $\text{val}_k(x_m x_{m-1} \cdots x_0) < \ell_i$ and $x = 0$, otherwise. Note that, by the definition of m, we have $y_m = 0$. Hence, a "carry" bit x might be needed only if $x_m = 0$. Multiplying an integer n by k^t corresponds to adding t zeros at the end of the k-ary representation of n. Hence, if φ_i is the correct guess, $z_m z_{m-1} \cdots z_0$ should have at least t zeros as suffix. If this is not the case, we choose to have

$$\begin{array}{r} x x_m x_{m-1} \cdots x_1 x_0 \\ -\ y_m\, y_{m-1} \cdots y_1\, y_0 \\ \hline z_m\, z_{m-1} \cdots z_1\, z_0 \end{array} \qquad\qquad \begin{array}{r} x_m x_{m-1} \cdots x_1 x_0 \\ +\ y_m\, y_{m-1} \cdots y_1\, y_0 \\ \hline z z_m\, z_{m-1} \cdots z_1\, z_0 \end{array}$$

(a) subtraction (b) addition

Fig. 1. The paper-and-pencil subtraction and addition

no φ_i-transitions starting from p. If $x = 0$, then φ_i-transitions are of the form

$$p \xrightarrow{w} (0, \widetilde{w} z_m \cdots z_t), \tag{3}$$

where w is any word over Σ_k of length t. If $x = 1$, then we have two cases depending on the form of $w \in \Sigma_k^t$:

1. If $w = 0^t$, then the transition is

$$p \xrightarrow{w} \left(-1, (k-1)^t z_m \cdots z_t\right), \tag{4}$$

   where the first component -1 indicates that a carry was needed in a "previous" subtraction and it must be borrowed from the first non-zero digit of the input that will be read in the future.
2. Otherwise $\widetilde{w} = vu0^s$, where $s < t$, $u \in \{1, 2, \ldots, k-1\}$ and $v \in \Sigma_k^{t-s-1}$, then the transition is

$$p \xrightarrow{w} \left(0, v(u-1)(k-1)^s z_m \cdots z_t\right). \tag{5}$$

   Here the carry $x = 1$ was borrowed from u and no carry is postponed to future calculations.

Consider next a state $p = (0, x_m x_{m-1} \cdots x_0)$ and φ_i-transitions, where $\varphi_i : n \mapsto k_i\, n + \ell_i$ and $\ell_i < 0$. Instead of subtraction, we consider now addition by the paper-and-pencil method where $\mathrm{val}_k(y_m y_{m-1} \cdots y_0) = |\ell_i|$. This is illustrated in Figure 1(b). Note that since $y_m = 0$ by the definition of m, a carry $z = 1$ can occur only if $x_m = k - 1$. As above, the φ_i-transitions exist only if the last t digits of $z_m z_{m-1} \cdots z_0$ are zeros. This holds also for any transition considered in the sequel. If $z = 0$, then we have the transitions of the form (3). If $z = 1$ we have again two cases depending on the digits of $w \in \Sigma_k^t$:

1. If $w = (k-1)^t$, then the carry is shifted to future calculations.

$$p \xrightarrow{w} \left(+1, 0^t z_m \cdots z_t\right). \tag{6}$$

2. If $\widetilde{w} = vu(k-1)^s$, where $s < t$, $u \in \{0, 1, \ldots, k-2\}$ and $v \in \Sigma_k^{t-s-1}$, then

$$p \xrightarrow{w} \left(0, v(u+1)0^s z_m \cdots z_t\right). \tag{7}$$

   Here the carry is added to the digit u and no carry is postponed to future calculations.

Secondly, consider a state of the form $p = (-1, x_m x_{m-1} \cdots x_0)$ and assume that $\ell_i \geq 0$. The carry component -1 means that we have borrowed a carry in a subtraction and after the subtraction we have read only zeros, which have been turned into digits $k-1$. Otherwise, if non-zero digits were read, there would be no longer a carry -1. Hence, we can be sure that $x_m = k-1$ in Figure 1(a), and consequently, we have $x = 0$, since $y_m = 0$. This is important, since it means that no "new" carry is borrowed. Again, assume that $z_{t-1} \cdots z_0 = 0^t$. If $w = 0^t$, then the transition is of type (4). Otherwise, the transitions are of type (5).

If $\ell_i < 0$, then we perform addition as in Figure 1(b) and assume $z_{t-1} \cdots z_0 = 0^t$. If $z = 0$ and $w = 0^t$, no new carries occur and again the transition is of type (4). If $z = 0$ and $w \neq 0^t$, then the transitions are of type (5). If $z = 1$, then this positive carry and the negative carry borrowed in a previous calculation annihilate each other. Hence, the transition is of type (3).

Finally, consider the states of the form $p = (+1, x_m x_{m-1} \cdots x_0)$. The carry component $+1$ means that we obtained a carry in an addition and after the addition we have read only digits $k-1$, which have been turned into zeros. Otherwise, if a digit $u \neq k-1$ were read, it would have been turned into $u+1$ and there would be no longer a carry $+1$. Hence, we conclude that $x_m = 0$ in Figures 1(a) and 1(b). If $\ell_i \geq 0$, then consider Figure 1(a) and assume $z_{t-1} \cdots z_0 = 0^t$. If $x = 0$ and $w = (k-1)^t$, then the transition is of type (6). If $x = 0$ and $w \neq (k-1)^t$, then the transitions are of type (7). If $x = 1$, then the negative and positive carry annihilate each other and the transitions are of type (3). Assume now that $\ell_i < 0$ and $z_{t-1} \cdots z_0 = 0^t$. In Figure 1(b) no new carry $z = 1$ can occur, since both $x_m = 0$ and $y_m = 0$. Hence, we have only two cases. If $w = (k-1)^t$, then the transition is of type (6). Otherwise, it is of the type (7).

If $n = \varphi_{i_m}(\varphi_{i_{m-1}}(\cdots \varphi_{i_1}(a) \cdots))$ for some $a \in I$, then using the above transitions and k-ary representations, we are able to correctly simulate the calculation $n \mapsto \varphi_{i_m}^{-1}(n) \mapsto \varphi_{i_{m-1}}^{-1}(\varphi_{i_m}^{-1}(n)) \mapsto \cdots \mapsto a$ as long as the k-ary representation of n given as input contains enough leading zeros. However, we may fix this by replacing the set of final states T by an enlarged set T'. A state $q' \in Q$ belongs to T' if there exists a path with label 0^t, $t \geq 0$, from q' to some state $q \in T$. Hence, with the modified final states the automaton $\mathcal{A}$ accepts all the reversals of the words in $\mathrm{val}_k^{-1}(F^\omega(I))$. On the other hand, it cannot accept any other word. Namely, for any word w accepted by $\mathcal{A}$ there is a sequence $(\varphi_{i_1}, \varphi_{i_2}, \ldots, \varphi_{i_m})$ such that (1) holds for $n = \mathrm{val}_k(w)$. It is well-known that any non-deterministic finite automaton can be turned into a DFA, e.g., by the subset construction. Hence, $F^\omega(I)$ is k-recognizable.

Remark 3. The set $F^\omega(I)$ considered in the above theorem is k-recognizable and therefore k^n-recognizable for all $n \geq 1$; again, see [3] for details. But usually this set is not ultimately periodic and therefore, by Cobham's Theorem, not ℓ-recognizable for any $\ell \geq 2$ such that k and ℓ are multiplicatively independent. Indeed, if Theorem 4 described below can be applied, then $F^\omega(I)$ contains arbitrarily large gaps.

4 Multiplicatively Independent Case

In this section, our aim is to show that $F^\omega(I) \subseteq \mathbb{N}$ given in Definition 1 is not recognizable in any base $k \geq 2$ provided that $\sum_{i=1}^r k_i^{-1} < 1$ and that there are at least two multiplicatively independent coefficients k_i. For the proof, we introduce the following notation. Let $X = \{x_0 < x_1 < x_2 < \cdots\}$ be an infinite ordered subset of $\mathbb{N}$. Then we denote

$$R_X = \limsup_{i \to \infty} \frac{x_{i+1}}{x_i} \text{ and } D_X = \limsup_{i \to \infty} (x_{i+1} - x_i).$$

In order to prove that a set is not k-recognizable for any base $k \geq 2$, we use the following result from [5], see also Eilenberg's book [6, Chapter V, Theorem 5.4].

Theorem 3 (Gap Theorem). *Let $k \geq 2$. If X is a k-recognizable infinite subset of $\mathbb{N}$, then either $R_X > 1$ or $D_X < \infty$.*

Note that $D_X < \infty$ means that X is *syndetic*, i.e., there exists a constant C such that the gap $x_{i+1} - x_i$ between any two consecutive elements x_i, x_{i+1} in X is bounded by C. Let us first show that if $\sum_{i=1}^{r} k_i^{-1} < 1$, then the set $F^\omega(I)$ given in Definition 1 contains arbitrarily large gaps.

Theorem 4. *Let $X = F^\omega(I)$ be a self-generating subset of $\mathbb{N}$ given in Definition 1. If $\sum_{i=1}^{r} k_i^{-1} < 1$, then X is not syndetic.*

Proof. Let $n \geq 1$ and $K = k_1 k_2 \cdots k_r$. Let $g = g_1 \circ g_2 \circ \cdots \circ g_n$ be a composite function, where g_j belongs to $\{\varphi_1, \varphi_2, \ldots, \varphi_r\}$ for every $j = 1, 2, \ldots, n$ and $g_j = \varphi_i$ for exactly n_i integers $j \in \{1, \ldots, n\}$. Note that $n_1 + n_2 \cdots + n_r = n$. By definition, we have $g(x) = k_1^{n_1} k_2^{n_2} \cdots k_r^{n_r} x + c_g$, where c_g is some constant depending only on g. Since $k_1^{n_1} k_2^{n_2} \cdots k_r^{n_r}$ divides K^n, we get

$$\#\{g(x) \mod K^n \mid x \in \mathbb{Z}\} = k_1^{n-n_1} k_2^{n-n_2} \cdots k_r^{n-n_r}.$$

The set $F^n(I)$ contains exactly the integers obtained by at most n applications of maps in F. For any interval of integers $[\![N, N + K^n - 1]\!]$ where $N > \max F^n(I)$, the elements in X belonging to this interval have been obtained by applying at least $n+1$ maps. Hence, in the interval $[\![N, N + K^n - 1]\!]$ there can be at most $k_1^{n-n_1} k_2^{n-n_2} \cdots k_r^{n-n_r}$ integers $x \in X$ such that the last n maps which produce x correspond to the composite function g, i.e., such that there exists $y \in X$ satisfying $g(y) = x$. For fixed numbers n_i, $i = 1, 2, \ldots, r$, there are $n!/(n_1! n_2! \cdots n_r!)$ functions g of the type described above. Thus, the number of integers in $X \cap [\![N, N + K^n - 1]\!]$ for any large enough N is at most

$$\sum_{n_1, n_2, \ldots, n_r} \left(\frac{n!}{n_1! n_2! \cdots n_r!} \right) k_1^{n-n_1} k_2^{n-n_2} \cdots k_r^{n-n_r} = K^n \left(\frac{1}{k_1} + \frac{1}{k_2} + \cdots + \frac{1}{k_r} \right)^n$$

where the sum is over $n_1, n_2, \ldots, n_r \geq 0$ satisfying $n_1 + n_2 + \cdots + n_r = n$. Hence, the biggest gap $x_{i+1} - x_i$ between two consecutive elements $x_i, x_{i+1} \in X$ in the interval $[\![N, N + K^n - 1]\!]$ is at least

$$d(n) = \frac{K^n}{K^n \left(\frac{1}{k_1} + \frac{1}{k_2} + \cdots + \frac{1}{k_r} \right)^n} = \left(\frac{1}{k_1} + \frac{1}{k_2} + \cdots + \frac{1}{k_r} \right)^{-n}.$$

Since $\sum_{i=1}^{r} k_i^{-1} < 1$, the function $d(n)$ tends to infinity as n tends to infinity. This means that there are arbitrarily large gaps in X. In other words, the self-generating set X is not syndetic.

Before showing that $R_X = 1$ let us first recall the density property of multiplicatively independent integers. A set S is dense in an interval I if every subinterval of I contains an element of S.

Theorem 5. *If $k, \ell \geq 2$ are multiplicatively independent, $\{k^p/\ell^q \mid p, q \geq 0\}$ is dense in $[0, \infty)$.*

This is a consequence of Kronecker's theorem, which states that for any irrational number θ the sequence $(\{n\theta\})_{n\geq 0}$ is dense in the interval $[0, 1)$. Here $\{x\}$ denotes the fractional part of the real number x. The proof of Kronecker's theorem as well as the proof of Theorem 5 can be found in [1, Section 2.5] or [9]. As an easy consequence of the previous theorem, we obtain the following result.

Corollary 1. *Let $\alpha > 0$ and β be two real numbers. If k and ℓ are multiplicatively independent, then the set $\{(\alpha k^p + \beta)/\ell^q \mid p, q \geq 0\}$ is dense in $[0, \infty)$.*

Proof. We show how to get arbitrarily close to any positive real number x. Let $\epsilon > 0$. By Theorem 5, there exists integers p and q such that

$$\left|\frac{x}{\alpha} - \frac{k^p}{\ell^q}\right| < \frac{\epsilon}{2\alpha} \quad \text{and} \quad \left|\frac{\beta}{\ell^q}\right| < \frac{\epsilon}{2}.$$

Hence, it follows that

$$\left|x - \frac{\alpha k^p + \beta}{\ell^q}\right| \leq \left|x - \frac{\alpha k^p}{\ell^q}\right| + \left|\frac{\beta}{\ell^q}\right| < \frac{\epsilon}{2\alpha}\alpha + \frac{\epsilon}{2} = \epsilon.$$

Let us next consider the ratio R_X of a self-generating set X.

Theorem 6. *For any self-generating set $X = F^\omega(I)$ given in Definition 1 where k_i and k_j are multiplicatively independent for some i and j, we have $R_X = 1$.*

Proof. Without loss of generality, we may assume that $F = \{\varphi_0, \varphi_1, \varphi_2\}$, where $\varphi_1 : n \mapsto k_1 n + \ell_1$, $\varphi_2 : n \mapsto k_2 n + \ell_2$, and k_1 and k_2 are multiplicatively independent. Namely, for $F \subseteq F'$, it is obvious that $F^\omega(I) \subseteq F'^\omega(I)$ and consequently, $R_{F^\omega(I)} = 1$ implies $R_{F'^\omega(I)} = 1$. By Lemma 1, we may also assume that ℓ_1 and ℓ_2 are non-negative. Moreover, with Remark 2, we may consider that both I and X are subsets of $\mathbb{N}$.

Let $a \in X$ be a positive integer and set $X_n := X \cap [\varphi_1^{n-1}(a), \varphi_1^n(a)]$ for all $n > 0$. Note that $\cup_{n\in\mathbb{N}} X_n = X \cap [a, \infty)$. Recall that $X = \{x_0 < x_1 < x_2 < \cdots\}$ and define

$$r_n := \max\left\{\frac{x_{i+1} - x_i}{x_i} \middle| x_{i+1}, x_i \in X_n\right\}.$$

Note that, for all x and for $j = 1, 2$, if we set $b_j := \ell_j/(k_j - 1)$, then we have

$$\varphi_j^n(x) = k_j^n x + \ell_j \sum_{i=0}^{n-1} k_j^i = (x + b_j)\, k_j^n - b_j. \tag{8}$$

Let $m \geq 0$ and x_i, x_{i+1} be two consecutive elements belonging to the set X_m. By Corollary 1, there exist infinitely many positive integers p and q such that

$$\frac{\varphi_2^p(a)}{k_1^q} = \frac{(a + b_2)k_2^p - b_2}{k_1^q} \in \left[x_{i+1} + b_1 - \frac{3}{4}(x_{i+1} - x_i), x_i + b_1 + \frac{3}{4}(x_{i+1} - x_i)\right].$$

Therefore $\varphi_2^p(a)$ is an element of X belonging to the interval

$$[c,d] := \left[k_1^q(x_{i+1}+b_1) - \frac{3}{4}k_1^q(x_{i+1}-x_i), k_1^q(x_i+b_1) + \frac{3}{4}k_1^q(x_{i+1}-x_i)\right],$$

which is a sub-interval[1] of the interval $[\varphi_1^q(x_i), \varphi_1^q(x_{i+1})]$. In other words, we have

$$\varphi_1^q(x_i) < c < \varphi_2^p(a) < d < \varphi_1^q(x_{i+1})$$

Hence, for all $t > q$, the difference $x_{j+1} - x_j$ of any two consecutive elements x_j, x_{j+1} of X in the interval $[\varphi_1^t(x_i), \varphi_1^t(x_{i+1})]$ is at most

$$\begin{aligned}\max\{&\varphi_1^{t-q}(\varphi_1^q(x_{i+1})) - \varphi_1^{t-q}(\varphi_2^p(a)), \varphi_1^{t-q}(\varphi_2^p(a)) - \varphi_1^{t-q}(\varphi_1^q(x_i))\} \\ &\leq \max\{\varphi_1^t(x_{i+1}) - \varphi_1^{t-q}(c), \varphi_1^{t-q}(d) - \varphi_1^t(x_i)\} = \frac{3}{4}k_1^t(x_{i+1}-x_i) + b_1 k_1^{t-q}.\end{aligned}$$

Thus, the ratio $(x_{j+1}-x_j)/x_j$ is at most

$$\frac{3\,k_1^t(x_{i+1}-x_i)}{4\,\varphi_1^t(x_i)} + \frac{b_1 k_1^{t-q}}{\varphi_1^t(x_i)} = \frac{3\,k_1^t(x_{i+1}-x_i)}{4\,\varphi_1^t(x_i)} + \frac{1}{k_1^q}\frac{b_1 k_1^t}{(x_i+b_1)k_1^t - b_1}. \tag{9}$$

The latter term in this sum can be taken as small as possible for q and t large enough ($1/k_1^q$ tends to 0 and the other factor tends to a constant $b_1/(x_i+b_1)$). In particular, for q and t large enough, we have

$$\frac{b_1 k_1^{t-q}}{\varphi_1^t(x_i)} < \frac{x_{i+1}-x_i}{12x_i}.$$

Moreover, we have

$$\frac{3\,k_1^t(x_{i+1}-x_i)}{4\,\varphi_1^t(x_i)} = \frac{3\,(x_{i+1}-x_i)}{4\,(x_i+b_1-b_1/k^t)} < \frac{3\,(x_{i+1}-x_i)}{4\,x_i} < \frac{10\,(x_{i+1}-x_i)}{12\,x_i}.$$

Thus, by (9), we obtain

$$\frac{x_{j+1}-x_j}{x_j} < \frac{11\,(x_{i+1}-x_i)}{12\,x_i}. \tag{10}$$

Since the above holds for any consecutive elements x_i and x_{i+1} in X_m and there are only finitely many such pairs, we conclude that there exists an integer N_1 such that (10) holds for any consecutive elements $x_j, x_{j+1} \in X_n$ where $n \geq N_1$. Hence, we obtain $r_n < \frac{11}{12} r_m$ for every $n \geq N_1$. Moreover, by repeating this procedure, we conclude that there exists an integer N_k such that

$$r_n < \left(\frac{11}{12}\right)^k r_m$$

for every $n \geq N_k$. This implies that $\limsup_{n\to\infty} r_n = 0$ and, consequently,

$$R_X = 1 + \limsup_{n\to\infty} r_n = 1.$$

[1] $c - \varphi_1^q(x_i) = \frac{1}{4}k_1^q(x_{i+1}-x_i) + b_1$ and $\varphi_1^q(x_{i+1}) - d = \frac{1}{4}k_1^q(x_{i+1}-x_i) - b_1$ which is positive for large enough q.

Our main result is a straightforward consequence of the previous theorems.

Theorem 7. *Let* $X = F^\omega(I)$ *be given in Definition 1. If* $\sum_{t=1}^{r} k_t^{-1} < 1$ *and there exist* i, j *such that* k_i *and* k_j *are multiplicatively independent, then* $F^\omega(I)$ *is not* k*-recognizable for any integer base* $k \geq 2$.

Proof. Let $X = F^\omega(I)$ satisfy the assumptions of the theorem. By Theorem 4, we have $D_X = \infty$ and, by Theorem 6, we have $R_X = 1$. Thus, Theorem 3 implies that X is not k-recognizable for any k.

As a corollary, we have solved the conjecture presented in [2].

Corollary 2. *Let* $F = \{\varphi_0, n \mapsto k_1 n + \ell_1, n \mapsto k_2 n + \ell_2\}$, *where* k_1 *and* k_2 *are multiplicatively independent. Then any infinite self-generating set* $F^\omega(I)$ *given in Definition 1 is not* k*-recognizable for any* $k \geq 2$.

Proof. This follows directly from Theorem 7. Namely, if k_1 and k_2 are multiplicatively independent, then $k_1 \geq 2$ and $k_2 \geq 3$ and $k_1^{-1} + k_2^{-1} \leq 1/2 + 1/3 = 5/6 < 1$.

References

1. Allouche, J.-P., Shallit, J.: Automatic Sequences: Theory, Applications, Generalizations. Cambridge University Press, Cambridge (2003)
2. Allouche, J.-P., Shallit, J., Skordev, G.: Self-generating sets, integers with missing blocks, and substitutions. Discrete Math. 292, 1–15 (2005)
3. Bruyère, V., Hansel, G., Michaux, C., Villemaire, R.: Logic and p-recognizable sets of integers. Bull. Belg. Math. Soc. 1, 191–238 (1994)
4. Cobham, A.: On the base-dependence of sets of numbers recognizable by finite automata. Math. Systems Theory 3, 186–192 (1969)
5. Cobham, A.: Uniform tag sequences. Math. Systems Theory 6, 186–192 (1972)
6. Eilenberg, S.: Automata, languages, and machines. Pure and Applied Mathematics, vol. A, 58. Academic Press, New York (1974)
7. Frougny, C.: Representations of numbers and finite automata. Math. Systems Theory 25, 37–60 (1992)
8. Garth, D., Gouge, A.: Affinely self-generating sets and morphisms. J. Integer Seq. 10, Article 07.1.5 (2007)
9. Hardy, G.H., Wright, E.M.: Introduction to the Theory of Numbers. Oxford Univ. Press, Oxford (1985)
10. Kimberling, C.: A self-generating set and the golden mean. J. Integer Seq. 3, Article 00.2.8 (2000)
11. Kimberling, C.: Affinely recursive sets and orderings of languages. Discrete Math. 274, 147–159 (2004)
12. Lecomte, P.B.A., Rigo, M.: Numeration systems on a regular language. Theory Comput. Syst. 34, 27–44 (2001)
13. Lothaire, M.: Algebraic Combinatorics on Words. In: Encyclopedia of Mathematics and its Applications, vol. 90. Cambridge University Press, Cambridge (2002)

The Isomorphism Problem for k-Trees Is Complete for Logspace

Johannes Köbler and Sebastian Kuhnert

Institut für Informatik, Humboldt Universität zu Berlin, Germany
{koebler,kuhnert}@informatik.hu-berlin.de

Abstract. We show that k-tree isomorphism can be decided in logarithmic space by giving a logspace canonical labeling algorithm. This improves over the previous StUL upper bound and matches the lower bound. As a consequence, the isomorphism, the automorphism, as well as the canonization problem for k-trees are all complete for deterministic logspace. We also show that even simple structural properties of k-trees are complete for logspace.

Keywords: graph isomorphism, graph canonization, k-trees, space complexity, logspace completeness.

1 Introduction

Two graphs G and H are called *isomorphic* if there is a bijective mapping ϕ between the vertices of G and the vertices of H that preserves the adjacency relation, i.e., ϕ relates edges to edges and non-edges to non-edges. *Graph Isomorphism* (GI) is the problem of deciding whether two given graphs are isomorphic. The problem has received considerable attention since it is one of the few natural problems in NP that are neither known to be NP-complete nor known to be solvable in polynomial time.

It is known that GI is contained in coAM [GS86, Sch88] and in SPP [AK06] providing strong evidence that GI is not NP-complete. On the other hand, the strongest known hardness result due to Torán [Tor04] says that GI is hard for the class DET (cf. [Coo85]). DET is a subclass of NC^2 (even of TC^1) and contains NL as well as all logspace counting classes [AJ93, BDH$^+$92].

For some restricted graph classes the known upper and lower complexity bounds for the isomorphism problem match. For example, a linear time algorithm for tree isomorphism was already known in 1974 to Aho, Hopcroft and Ullman [AHU74]. In 1991, an NC algorithm was developed by Miller and Reif [MR91], and one year later, Lindell [Lin92] obtained an L upper bound. On the other hand, in [JKM$^+$03] it is shown that tree isomorphism is L-hard (provided that the trees are given in pointer notation). In [ADK08], Lindell's log-space upper bound has been extended to the class of partial 2-trees, a class of planar graphs also known as generalized series-parallel graphs. Very recently, it has been shown that even the isomorphism problem for all planar graphs is

R. Královič and D. Niwiński (Eds.): MFCS 2009, LNCS 5734, pp. 537–548, 2009.

in logspace [DLN+08]. Much of the recent progress on logspace algorithms for graphs has only become possible through Reingold's result that connectivity in undirected graphs can be decided in deterministic logspace [Rei05]. Our result does not depend on this, yielding a comparatively simple algorithm.

In this paper we show that the isomorphism problem for k-trees is in logspace for each fixed $k \in \mathbb{N}^+$. This improves the previously known upper bound of StUL [ADK07] and matches the lower bound. In fact, we prove the formally stronger result that a canonical labeling for a given k-tree is computable in logspace. Recall that the *canonization problem* for graphs is to produce a *canonical form* $canon(G)$ for a given graph G such that $canon(G)$ is isomorphic to G and $canon(G_1) = canon(G_2)$ for any pair of isomorphic graphs G_1 and G_2. Clearly, graph isomorphism reduces to graph canonization. A *canonical labeling* for G is any isomorphism between G and $canon(G)$. It is not hard to see that even the search version of GI (i.e., computing an isomorphism between two given graphs in case it exists) as well as the automorphism group problem (i.e., computing a generating set of the automorphism group of a given graph) are both logspace reducible to the canonical labeling problem.

The parallel complexity of k-tree isomorphism has been previously investigated by Del Greco, Sekharan, and Sridhar [GSS02] who introduced the concept of the *kernel* of a k-tree in order to restrict the search for an isomorphism between two given k-trees. We show that the kernel of a k-tree can be computed in logspace and exploit this fact to restrict the search for a canonical labeling of a given k-tree G. To be more precise, we first transform G into an undirected tree $T(G)$ whose nodes are formed by the k-cliques and $(k+1)$-cliques of G. Then we compute the center node of $T(G)$ which coincides with the kernel of G and try all labelings of the vertices in $\ker(G)$. In order to extend a labeling of the kernel vertices of G to the other vertices of G in a canonical way, we color the nodes of the tree $T(G)$ to encode additional structural information about G. Finally, we apply a variant of Lindell's algorithm to compute canonical labelings for the colored versions of $T(G)$ and derive from them a canonical labeling for the k-tree G.

Our tree representation $T(G)$ is similar to the construction used in [ADK07]. The main advantage of our construction lies in the fact that the tree $T(G)$ can be directly constructed in logspace from G, whereas the tree representation of [ADK07] is obtained as a reachable subgraph of a mangrove[1] based on G and hence can only be derived from G with the help of an StUL oracle.

2 Preliminaries

As usual, **L** is the class of all languages decidable by Turing machines with read-only input tape and an $\mathcal{O}(\log n)$ bound on the space used on the working tapes. **FL** is the class of all functions computable by Turing machines that additionally have a write-only output tape.

[1] A mangrove is a digraph with at most one directed path between each pair of nodes.

Given a graph G, we use $\boldsymbol{V(G)}$ and $\boldsymbol{E(G)}$ to denote its vertex and edge sets, respectively. We define the following notations for subgraphs of G. For $M \subseteq V(G)$, $\boldsymbol{G[M]}$ denotes the subgraph of G induced by M and we use $\boldsymbol{G - M}$ as a shorthand for $G[V(G) \setminus M]$.

Given a graph G and two vertices $u, v \in V(G)$, the distance $\boldsymbol{d_G(u, v)}$ is the length of the shortest path from u to v. The **eccentricity** of a vertex $v \in V(G)$ is the longest distance to another vertex, i.e., $\boldsymbol{ecc_G(v)} = \max\{d_G(u, v) | u \in V(G)\}$. The **center** of G consists of all vertices with minimal eccentricity.

Given two graphs G and H, an **isomorphism** from G to H is a bijection $\phi\colon V(G) \to V(H)$ with $\{u, v\} \in E(G) \Leftrightarrow \{\phi(u), \phi(v)\} \in E(H)$. On colored graphs, an isomorphism must additionally preserve colors. G and H are called **isomorphic**, in symbols $G \cong H$, if there is an isomorphism from G to H. Given a graph class $\mathcal{G}$, a function f defined on $\mathcal{G}$ computes an **invariant** for $\mathcal{G}$ if

$$\forall G, H \in \mathcal{G} : G \cong H \Rightarrow f(G) = f(H) .$$

If the reverse implication also holds, f is a **complete invariant** for $\mathcal{G}$. If additionally $f(G) \cong G$ for all $G \in \mathcal{G}$, f computes **canonical forms** for $\mathcal{G}$. Given a function f that computes canonical forms, an isomorphism ψ_G from G to its canonical form $f(G)$ is called a **canonical labeling**.

The isomorphisms from a graph G to itself are called **automorphisms** and they form a group, which we denote by $\mathbf{Aut}(\boldsymbol{G})$. An automorphism is called **non-trivial** if it is not the identity. The **graph automorphism problem (GA)** is to decide if a graph has non-trivial automorphisms. A graph without non-trivial automorphisms is called **rigid**.

In the next section, we present an FL algorithm that, given a k-tree G, computes a canonical labeling ψ_G.

3 Canonizing k-Trees

Fix any $k \in \mathbb{N}^+$. The class of **k-trees** is inductively defined as follows. Any k-clique is a k-tree. Further, given a k-tree G and a k-clique C in G, one can construct another k-tree by adding a new vertex v and connecting v to every vertex in C. The initial k-clique is called **base** of G, and the k-clique C the new vertex v is connected to is called **support** of v. Note that each k-clique of a k-tree G can be used as base for constructing G – but once the base is fixed, the support of each vertex is uniquely determined.

An interesting special case of k-trees are **k-paths**, where the support C_i of any new vertex v_i (except the first vertex added to G) must either contain the vertex v_{i-1} added in the previous step or be equal to the support C_{i-1} of the latter. Fig. 1 shows a 2-tree that is a 2-path as well.

We note that k-trees can be recognized in logspace [ADK07], so we can safely assume that the input is indeed a k-tree.

We first define a tree representation $T(G)$ for k-trees G.

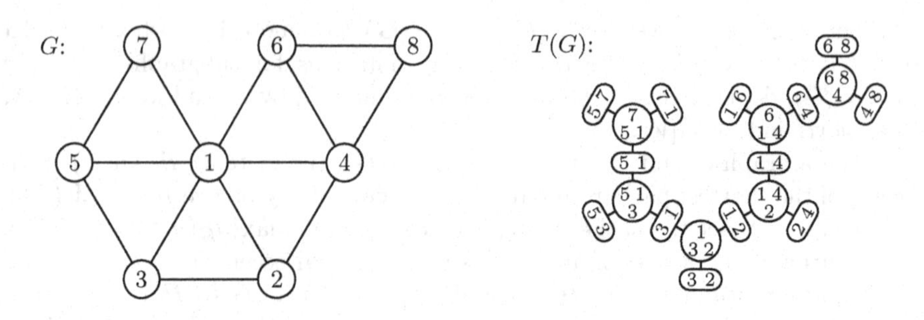

Fig. 1. A 2-tree G and its tree representation $T(G)$

Definition 1. *For a k-tree G, its **tree representation $T(G)$** is defined by*

$$V(T(G)) = \{M \subseteq V(G) \mid M \text{ is a } k\text{-clique or a } (k+1)\text{-clique}\}$$
$$E(T(G)) = \{\{M_1, M_2\} \subseteq V(T(G)) \mid M_1 \subsetneq M_2\} .$$

Note that $T(G)$ reflects the iterative construction of G: The base of G is a k-clique and thus a node in $T(G)$. Each time a new vertex u is added to G, it is connected to all vertices of its support P_u (a k-clique), forming a new $(k+1)$-clique C_u that is a superset of P_u. In $T(G)$, the addition of u results in a new node C_u being added and connected to P_u. Additionally, the k many k-cliques in C_u that contain the new vertex u are added as new nodes to $T(G)$ and connected to C_u. From these observations it is clear that $T(G)$ is indeed a tree.

We continue by proving some basic properties of our tree representation $T(G)$.

Lemma 2. *For any k-tree G and any vertex $v \in V(G)$, the nodes of $T(G)$ that contain v form a subtree of $T(G)$.*

Proof. We prove by induction over the construction of G that any node M added to $T(G)$ with $v \in M$ either is the unique node first introducing v or is hooked up to a previously added node that contains v. If M is a k-clique in G this is immediately clear as it is either the base node in $T(G)$ or it is a subset of a $(k+1)$-clique node and hence does not introduce any new vertices. So assume that M is a $(k+1)$-clique C_u, which was added to $T(G)$ upon the addition of some vertex u to G. If $u = v$ then M is the single node of $T(G)$ introducing v. If $u \neq v$ we have $v \in C_u \setminus \{u\} = P_u$ and thus there is an edge to a previously added k-clique node P_u that contains v. □

Lemma 3. *For any k-tree G, the center of $T(G)$ is a single node.*

Proof. Suppose not. Then the center consists of two adjacent nodes, one a k-clique and one a $(k+1)$-clique. This leads to a contradiction because k-clique nodes have even eccentricity while that of $(k+1)$-clique nodes is odd: All leaves are k-clique nodes, and k-cliques and $(k+1)$-cliques alternate on every path. □

Definition 4. *The clique corresponding to the center node of $T(G)$ is called* ***kernel*** *of G and denoted* $\mathbf{ker}(G)$.

Note that $\ker(G)$ can be either a k-clique or a $(k+1)$-clique, depending on the structure of G. The concept of the kernel of a k-tree was introduced in [GSS02]. The definition there is slightly different but the equivalence can be easily verified.

We continue by recalling some basic facts concerning undirected trees.

Fact 5. *Given an undirected tree T and two nodes $u, v \in V(T)$, the distance $d_T(u, v)$ can be computed in* FL.

Proof. Think of T as rooted in u and all edges directed away from u. The direction of an edge e can be determined in logspace by computing the lexicographically-first Euler tour starting at u that visits each edge once per direction (cf. [AM04]). Then the unique path from v to u can be found by always choosing the unique incoming edge as next step. Only the current node and the number of steps have to be remembered. Upon reaching u, output the number of steps taken. □

Fact 6. *The center of an undirected tree T can be computed in* FL.

Proof. We first show that the eccentricity $ecc_T(u)$ of each node $u \in V(T)$ is computable in logspace. This can be done by iterating over all $v \in V(T)$, each time calculating $d_T(u, v)$ (this is possible in logspace by Fact 5). Only the maximum distance to u has to be remembered, the result being $ecc_T(u)$.

Observe now that also the maximum eccentricity $ecc_{\max}$ of all nodes $u \in V(T)$ is computable in logspace by iterating over all $u \in V(T)$. Then compute again the eccentricity of all nodes u, this time outputting u if $ecc_T(u) = ecc_{\max}$. □

Our goal is to canonize G by using Lindell's algorithm [Lin92] to canonize $T(G)$. To achieve this, we declare the kernel K of G as the root of $T(G)$. As a consequence, we can identify each $(k+1)$-clique $M \in V(T(G)) \setminus \{K\}$ with the unique vertex $v \in M$ that is not present in the k-clique M' that lies next to M on the path from K to M in $T(G)$. For later use, we denote this vertex by $v(M)$ and for each $v \in V(G) \setminus K$, we use M_v to denote the unique $(k+1)$-clique $M \in V(T(G)) \setminus \{K\}$ with $v(M) = v$.

It is clear that $T(G)$ does not provide complete structural information about G, since the vertices in the kernel K are indistinguishable in $T(G)$ and further, only one out of the k edges between each added vertex u and its support can be recovered from $T(G)$. To add the missing information, we give individual colors to the kernel vertices and color the nodes of $T(G)$ as well. Since the kernel K of a given k-tree G can be determined in logspace, we can simplify the notation by assuming that K consists of the vertices $1, \ldots, k'$, where $k' = \|K\| \in \{k, k+1\}$ equals the size of K.

Definition 7. *Let G be a k-tree with vertex set $V(G) = \{1, \ldots, n\}$ and kernel $K = \{1, \ldots, k'\}$. For each vertex v of G, we denote by*

$$\boldsymbol{l_G(v)} = \min\left\{d_{T(G)}(K, M) \mid M \in V(T(G)), v \in M\right\}$$

*the **level** of v in G. Further, for any permutation $\pi \in S_{k'}$, let $\boldsymbol{T(G,\pi)}$ denote the directed colored tree obtained from $T(G)$ by choosing K as the root and coloring each node $M \in V(T(G))$ by the set $\mathbf{c(M)} = \{c(v) \mid v \in M\}$, where*

$$\mathbf{c}(\boldsymbol{v}) = \begin{cases} \pi(v) & \text{if } v \in \ker(G), \\ l_G(v) + k' & \text{otherwise.} \end{cases}$$

The definition of $T(G,\pi)$ is similar to the construction of the colored tree $T(G,B,\theta)$ in [ADK07]. The main advantage of our construction lies in the fact that $T(G,\pi)$ can be directly constructed from G in logspace, whereas the tree representation used in [ADK07] (which in turn is related to the decomposition defined in [KCP82]) is defined as the reachable subgraph of a mangrove derived from G. This allows us to decide st-reachability in the tree $T(G,\pi)$ in logspace, an essential step to achieve our upper bound.

Another advantage of our construction comes with the usage of the kernel K as a canonical base. This makes it superfluous to cycle through all k-cliques of G (as in [ADK07]), leaving only the permutations of the vertices within K to enumerate.

Lemma 8. *For a k-tree G and a permutation π on the kernel K of G, $T(G,\pi)$ can be computed in* FL.

Proof. It is clear that the nodes and edges of $T(G)$ can be determined in logspace: First iterate over all subsets M of $V(G)$ of size k (this requires space $k \log n$) and output M as a node if M is a k-clique in G. Likewise, find and output all $(k+1)$-cliques M, each time adding edges to all $k+1$ many k-cliques contained in M. The (intermediate) result $T(G)$ cannot be stored due to space limitations, but it is possible to recompute it as needed (as long as only a constant number of operations is chained).

Next determine the kernel K of G (Fact 6) and think of all edges in $T(G)$ directed away from K. As described in Fact 5, the direction can be determined in logspace. It remains to compute the color $c(M)$ of each node $M \in V(T(G))$.

For each $v \in M$ calculate $c(v)$ by examining the unique path from M to K in $T(G)$ (the path can be found by following the unique incoming edge at each node). Store the length ℓ of the path and the position p_v where v was last found (this can be done in parallel for all $v \in M$). If $p_v = \ell$ (i. e. $v \in K$), then add the number $c(v) = \pi(v)$ to the color $c(M)$ of M. If $p_v < \ell$, add the number $c(v) = \ell - p_v + \|K\|$ to $c(M)$. The latter is correct, because by Lemma 2 the nodes containing v form a subtree of $T(G)$ and thus the node that is closest to K and contains v is on the path from K to M. □

We will need to compute a canonical labeling of $T(G,\pi)$. We observe the following generalization of the logspace tree canonization algorithm.

Lemma 9. *Lindell's algorithm [Lin92] can be extended to colored trees and to output not only a canonical form, but also a canonical labeling. This modification preserves the logarithmic space bound.*

Proof sketch. Colors can be handled by extending the *tree isomorphism order* defined in [Lin92] by using $color(s) < color(t)$ as additional condition (where s and t are the roots of the trees to compare). The canonical labeling can be computed by using a counter i initialized to 0: Instead of printing (the first letter of) the canon of a node v, increment i and print "$v \mapsto i$". □

Next we show that the colored tree representations of isomorphic k-trees are also isomorphic, provided that the kernels are labeled accordingly.

Lemma 10. *Let $\phi \in S_n$ be an isomorphism between two k-trees G and H with $V(G) = V(H) = \{1, \ldots, n\}$ and $\ker(G) = \ker(H) = K$. Then ϕ (viewed as a mapping from $V(T(G))$ to $V(T(H))$) is an isomorphism between $T(G, \pi_1)$ and $T(H, \pi_2)$, provided that $\pi_1(u) = \pi_2(\phi(u))$ for all $u \in K$.*

Proof. It can be easily checked that any isomorphism between G and H is also an isomorphism between $T(G)$ and $T(H)$ that maps the kernel K of G to the kernel of H, which equals K by assumption. In order to show that the color of a node $M \in V(T(G, \pi_1))$ coincides with the color of $\phi(M) \in V(T(H, \pi_2))$, we prove the stronger claim that $c(v) = c(\phi(v))$ for all $v \in V(G)$. For $v \in K$, we have $c(\phi(v)) = \pi_2(\phi(v)) = \pi_1(v) = c(v)$ by assumption. Since ϕ must preserve the level of the vertices, it follows further for $v \in V(G) \setminus K$ that

$$c(\phi(v)) = l_H(\phi(v)) + \|K\| = l_G(v) + \|K\| = c(v) .$$

This completes the proof of the lemma. □

Conversely, the next lemma shows that from any isomorphic copy T of $T(G, \pi_1)$ we can easily derive an isomorphic copy G' of G. Moreover, any isomorphism ϕ between $T(G, \pi)$ and T can be efficiently converted into an isomorphism between G and G'.

Lemma 11. *Let G be a k-tree and let π be a permutation on the kernel K of G. Then from any colored tree T that is isomorphic to $T(G, \pi)$, an isomorphic copy G' of G can be computed in logspace. Further, it is possible to compute in logspace an isomorphism between G and G' from any given isomorphism between $T(G, \pi)$ and T.*

Proof sketch. Construct G' as follows. Let $V(G') = \{1, \ldots, n\}$, where n is k plus the number of $(k+1)$-clique nodes in T (we call $m \in V(T)$ an ***l*-clique node**, if $l = \|c(m)\|$). This is correct due to the one-to-one correspondence between the vertices $v \in V(G) \setminus K$ and the $(k+1)$-clique nodes $M_v \in T(G, \pi) \setminus \{K\}$. Next determine the center node z of T (see Lemma 6) and make $\{1, \ldots, k'\}$ a clique in G', where $k' = \|c(z)\|$. Further, for any non-center $(k+1)$-clique node $m \in V(T) \setminus \{z\}$, let $v(m)$ denote the corresponding vertex in $V(G')$ (to make this mapping unique, let $v(m)$ preserve the order of $(k+1)$-clique nodes in $V(T)$). Based on the color $c(m) = \{c_1, \ldots, c_{k+1}\}$ of m add the following edges to $E(G')$: For each $c_i \le k'$ add an edge $\{c_i, v(m)\}$ and for each $c_i > k'$ with $c_i < c_{\max} = \max\{c_i \mid c_i \in c(m)\}$ add an edge $\{v(m), v(m')\}$, where m' is the

$(c_i - k')$-th node on the path from z to m. This completes the construction of G'.

Now let ϕ be an isomorphism from $T(G, \pi)$ to T. Construct an isomorphism ϕ' from G to G' as follows. For $v \in K$, let $\phi'(v) = \pi(v)$, and for $v \notin K$, let $\phi'(v) = v(\phi(M_v))$. By induction on the level of v in G, it can be proven that this is indeed an isomorphism. Both constructions can easily be seen to be in logspace. □

Now we are ready to prove our main result.

Theorem 12. *Given a k-tree G with vertex set $V(G) = \{1, \ldots, n\}$ and kernel $K = \{1, \ldots, k'\}$, a canonical labeling $\psi_G \in S_n$ can be computed in* FL.

Proof. In order to compute ψ_G we iterate over all permutations $\pi \in S_{k'}$ and compute a canonical labeling $\psi_{T(G,\pi)}$ for the colored tree $T(G, \pi)$ using the algorithm from Lemma 9. Let π_1 be one of the permutations that give rise to the lexicographically smallest colored tree $\psi_{T(G,\pi_1)}(T(G, \pi_1))$. By applying Lemma 11, we can reconstruct from this tree an isomorphic copy $canon(G)$ of G together with an isomorphism ψ_G between G and $canon(G)$. By Lemmas 8 and 9, it is clear that ψ_G is computable in logspace.

It remains to show that the canonical labelings of any two isomorphic k-trees G and H map these graphs to the same canon $\psi_G(G) = \psi_H(H)$. To see this, let $\pi_1, \pi_2 \in S_{k'}$ be two permutations that give rise to the lexicographically smallest trees $\psi_{T(G,\pi_1)}(T(G, \pi_1))$ and $\psi_{T(H,\pi_2)}(T(H, \pi_2))$, respectively. Since by Lemma 10 for any tree $T(G, \pi_1)$ that can be derived from G via some permutation π_1 there is an isomorphic tree $T(H, \pi_2)$ that can be derived from H via some permutation π_2 (and vice versa), it follows that $\psi_{T(G,\pi_1)}(T(G, \pi_1))$ and $\psi_{T(H,\pi_2)}(T(H, \pi_2))$ are equal, implying that $canon(G) = canon(H)$. □

We note that the above construction can be extended to colored k-trees as follows. Let $\zeta\colon V(G) \to C$ be a vertex coloring of G. Modify the coloring of $T(G, \pi)$ (cf. Definition 7) by replacing $c(v)$ with the pair $c'(v) = (c(v), \zeta(v))$.

Theorem 12 immediately yields the following corollaries.

Corollary 13. *For any fixed k, k-tree canonization is in* FL.

Corollary 14. *For any fixed k, k-tree isomorphism is* L*-complete.*

The L-hardness can be seen by a reduction from the isomorphism problem for trees in pointer notation, which is known to be L-hard [JKM+03]. The reduction transforms a tree T into a k-tree $\boldsymbol{E_k(T)}$ by adding a $(k-1)$-clique C and connecting C to all nodes in $V(T)$ (cf. Fig. 2). It can easily be seen that $E_k(T)$ is a k-tree and that $T_1 \cong T_2 \Leftrightarrow E_k(T_1) \cong E_k(T_2)$.

We note that fixing k is essential, as the isomorphism problem for the class of all k-trees, $k \in \mathbb{N}^+$, is isomorphism complete [KCP82] and thereby unlikely to be decidable in polynomial time.

Furthermore, there is a standard Turing reduction of the automorphism group problem (i.e., computing a generating set of the automorphism group of a given graph) to the search version of GI for colored graphs (cf. [Hof82, KST93]). It is not hard to see that this reduction can be performed in logspace.

Corollary 15. *For any fixed k, computing a generating set of the automorphism group of a given k-tree, and hence computing a canonical labeling coset for a given k-tree is in* FL.

Corollary 16. *For any fixed k, the k-tree automorphism problem (i. e., deciding whether a given k-tree has a non-trivial automorphism) is* L*-complete.*

Observe that the mapping $T \mapsto E_k(T)$ does not provide a correct reduction of tree automorphism (which is L-complete [JKM$^+$03]) to k-tree automorphism, as the vertices within the newly added clique can always be permuted without changing the graph. To sidestep this difficulty, we use a transformation E'_k that preserves rigidity. Let T be a rooted tree with $n = \|V(T)\|$ and root $r \in V(T)$. Then $\boldsymbol{E'_k(T, r)}$ is defined as follows (cf. Fig. 2):

$$
\begin{aligned}
V(E'_k(T,r)) &= V(T) \cup \{u_i \mid 1 \le i \le k+n\} \\
E(E'_k(T,r)) &= E(T) \cup \{\{v,u_i\} \mid v \in V(T), 1 \le i \le k-1\} \\
&\quad \cup \{\{r,u_k\}\} \cup \{\{u_i,u_j\} \mid 1 \le i < j \le k+n, j-i \le k\}
\end{aligned}
$$

It is easy to see that $E'_k(T,r)$ is a k-tree and that any non-trivial automorphism of (T,r) induces a non-trivial automorphism of $E'_k(T,r)$. To see that $E'_k(T,r)$ is rigid whenever (T,r) is rigid, assume $n > k$ (all smaller trees can be hard-coded in the reduction). Any automorphism of $E'_k(T,r)$ must fix all newly added vertices u_i: Each of the vertices u_i, $1 \le i \le k-1$, is uniquely determined by its degree $n+k-2+i$ (unless r is connected to all vertices of T, but then T is a star and not rigid anyway). The vertices u_i, $k+1 \le i \le k+n$, are the only ones not adjacent to u_1 and uniquely identified by the structure of $E'_k(T,r)$ (examine the tree representation $T(E'_k(T,r))$ to see this). Finally, the vertex u_k is unique among the remaining ones by the shortest distance to u_{k+n}.

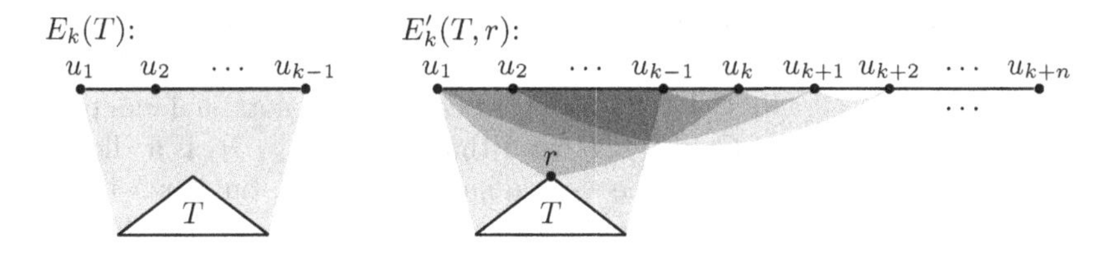

Fig. 2. The transformations $E_k(T)$ and $E'_k(T,r)$

4 Complete Problems for Logspace

In this section we prove some additional completeness results for logspace that are related to our main result. The hardness is under $\mathsf{DLOGTIME}$-uniform AC^0-reductions. We first recall that ORD is L-complete, where **ORD** is the problem of deciding for a directed line graph P and two vertices $s, t \in V(P)$ if there is a path from s to t [Ete97].

In Lemma 6 we have seen that the center of an undirected tree can be computed in FL. We now show that the decision variant is hard for L even when restricted to paths.

Theorem 17. *Given an undirected path P and a vertex $c \in V(P)$, it is* L*-hard to decide if c belongs to the center of P.*

This implies the L-hardness of the following problem: Given a k-tree (or k-path) G and a vertex $c \in V(G)$, decide whether c belongs to the kernel of G. The reduction for this is $(P, c) \mapsto (E_k(P), c)$, where E_k is as defined above.

Proof. We reduce from ORD using $(P, s, t) \mapsto (P', n)$ as reduction, where

$$\begin{aligned} V(P') &= V(P) \cup \{i' \mid i \in V(P)\} \cup \{s''\} \\ E(P') &= \{\{i, j\} \mid (i, j) \in E(P) \wedge j \neq t\} \cup \{\{n, n'\}\} \\ &\cup \{\{i', j'\} \mid (i, j) \in E(P) \wedge j \notin \{s, t\}\} \\ &\cup \{\{i', s''\} \mid (i, s) \in E(P)\} \cup \{\{s'', s'\}\} \end{aligned}$$

and n is the vertex without successor in P. P' is the undirected path that consists of two copies of P that are twisted before t, connected at their ends and have the second copy of s duplicated (cf. Fig. 3). If s precedes t in P (left side), then n is the center of P', but if t precedes s then n' is the center of P' (right side). □

Finally, we examine two problems related to the structure of k-trees. Let G be a graph. A vertex $v \in V(G)$ is called **simplicial** in G, if its neighborhood induces a clique. A bijective mapping $\sigma\colon \{1, \ldots, \|V(G)\|\} \to V(G)$ of the vertices of G is called **perfect elimination order (PEO)**, if for all i, $\sigma(i)$ is simplicial in $G - \bigcup_{j<i}\{\sigma(j)\}$. Note that a graph can have several perfect elimination orders, so finding a PEO is not a functional but a search problem. It is well-known that a graph has a PEO if and only if it is chordal. As k-trees are a subclass of chordal graphs, each k-tree has a PEO.

A related problem is the **fast reordering problem (FRP)** which is defined in [GSS02] as a preprocessing step for parallel algorithms. It consists of finding a sequence of sets $R_0, \ldots, R_k \subseteq V(G)$, such that each R_i is a maximal independent set of simplicial vertices of $G - \bigcup_{j<i} R_j$ and that $G - \bigcup_{0 \le j \le k} R_j$ is a clique. For general chordal graphs there can be several such sequences, but for k-trees this sequence is unique and the remaining clique is the kernel. In [GSS02] it was shown that if the input graphs are restricted to k-trees, the FRP can be solved in NC. We improve this and show logspace completeness for both problems:

Theorem 18. *For k-trees (k fixed), it is logspace complete to find a perfect elimination order and to solve the fast reordering problem.*

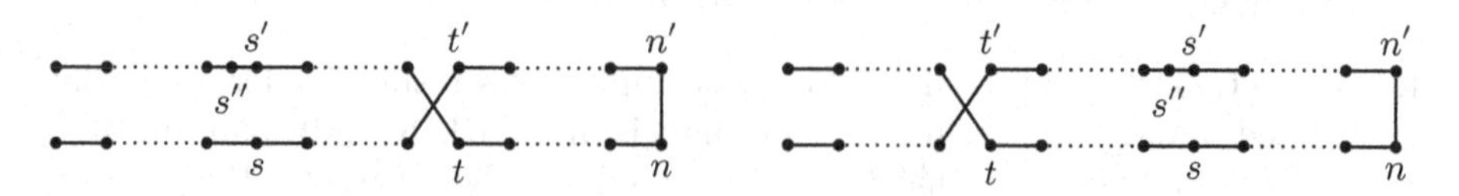

Fig. 3. The reduction of ORD to verifying the center

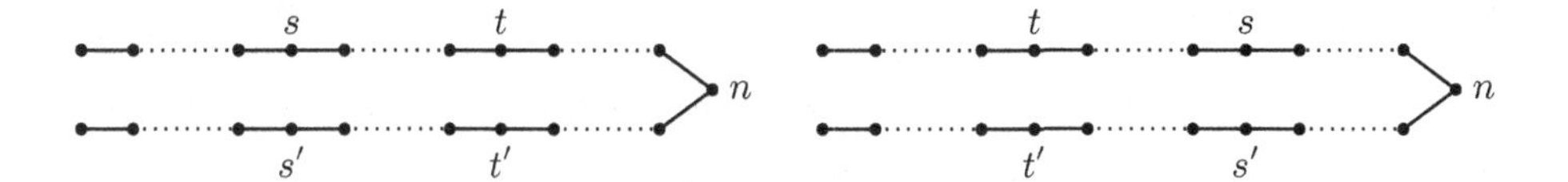

Fig. 4. The reduction of ORD to finding a PEO

Proof. We first show FRP $\in$ FL: Let G be a k-tree. We compute the level $l_G(v)$ for each $v \in V(G)$ (cf. Definition 7). As observed in Lemma 8, this is possible in logspace. Let $l_{\max} = \max\{l_G(v) \mid v \in V(G)\}$. Output $R_i := \{v \in V(G) \mid l_G(v) = l_{\max} - 2i\}$ for $i = 0, \ldots, \lceil l_{\max}/2 \rceil - 1$. The correctness follows from the structure of $T(G)$.

Next, we note that a perfect elimination order can be efficiently computed when a solution to the FRP is known (i. e. finding a PEO reduces to solving the FRP): Take the members of the R_i in ascending order (first those from R_0, then those from R_1 and so on up to R_k) and finally those from $\ker(G) = V(G) \setminus \bigcup_i R_i$. No matter which order is chosen within the R_i and the kernel, the result is a PEO, as each R_i is independent and $\ker(G)$ is a clique.

Finally we show that finding a perfect elimination order is hard for logspace even for paths. The result for k-trees (and k-paths) can again be obtained using the construction of E_k given above. We solve an ORD instance (P, s, t) in DLOGTIME-uniform AC^0 with a single oracle gate for computing a PEO of the path P' given by

$$\begin{aligned} V(P') &= V(P) \cup \{i' \mid i \in V(P) \setminus \{n\}\} \\ E(P') &= \{\{i,j\} \mid (i,j) \in E(P)\} \\ &\quad \cup \{\{i',j'\} \mid (i,j) \in E(P), j \neq n\} \cup \{\{i',n\} \mid (i,n) \in E(P)\} \end{aligned}$$

where n is the vertex in P without successor. We claim that for any PEO σ of P' (where p_i is a shorthand for the position $\sigma^{-1}(i)$ of a vertex in σ):

$$(P, s, t) \in \text{ORD} \Leftrightarrow p_s \leq p_t \leq p_n \vee p_{s'} \leq p_{t'} \leq p_n$$

If $(P, s, t) \notin$ ORD, then s is between t and n, and s' is between t' and n in P' (right side in Fig. 4). Thus σ cannot satisfy both $p_s \leq p_t$ and $p_{s'} \leq p_{t'}$. If $(P, s, t) \in$ ORD (left side of Fig. 4), n does not become simplicial until at least one copy of P is completely removed. Similarly, if the first copy is completely removed before n, t does not become simplicial before s is removed; and if the second copy is completely removed before n, t' does not become simplicial before s' is removed. □

References

[ADK07] Arvind, V., Das, B., Köbler, J.: The space complexity of k-tree isomorphism. In: Tokuyama, T. (ed.) ISAAC 2007. LNCS, vol. 4835, pp. 822–833. Springer, Heidelberg (2007)

[ADK08] Arvind, V., Das, B., Köbler, J.: A logspace algorithm for partial 2-tree canonization. In: Hirsch, E.A., Razborov, A.A., Semenov, A., Slissenko, A. (eds.) Computer Science – Theory and Applications. LNCS, vol. 5010, pp. 40–51. Springer, Heidelberg (2008)

[AHU74] Aho, A., Hopcroft, J., Ullman, J.: The design and analysis of computer algorithms. Addison-Wesley, Reading (1974)

[AJ93] Àlvarez, C., Jenner, B.: A very hard log-space counting class. Theoretical Computer Science 107(1), 3–30 (1993)

[AK06] Arvind, V., Kurur, P.P.: Graph isomorphism is in SPP. Information and Computation 204(5), 835–852 (2006)

[AM04] Allender, E., Mahajan, M.: The complexity of planarity testing. Information and Computation 139(1) (February 2004)

[BDH+92] Buntrock, G., Damm, C., Hertrampf, U., Meinel, C.: Structure and importance of logspace-MOD classes. Mathematical Systems Theory 25, 223–237 (1992)

[Coo85] Cook, S.A.: A taxonomy of problems with fast parallel algorithms. Information and Control 64, 2–22 (1985)

[DLN+08] Datta, S., Limaye, N., Nimbhorkar, P., Thierauf, T., Wagner, F.: A log-space algorithm for canonization of planar graphs. CoRR (2008), `http://arxiv.org/abs/0809.2319`

[Ete97] Etessami, K.: Counting quantifiers, successor relations, and logarithmic space. Journal of Computer and System Sciences 54(3), 400–411 (1997)

[GS86] Goldwasser, S., Sipser, M.: Private coins versus public coins in interactive proof systems. In: Randomness and Computation. Advances in Computing Research, vol. 5, pp. 73–90. JAI Press (1989)

[GSS02] Del Greco, J.G., Sekharan, C.N., Sridhar, R.: Fast parallel reordering and isomorphism testing of k-trees. Algorithmica 32(1), 61–72 (2002)

[Hof82] Hoffmann, C.M.: Group-Theoretic Algorithms and Graph Isomorphism. LNCS, vol. 136. Springer, Heidelberg (1982)

[JKM+03] Jenner, B., Köbler, J., McKenzie, P., Torán, J.: Completeness results for graph isomorphism. Journal of Computer and System Sciences 66, 549–566 (2003)

[KCP82] Klawe, M.M., Corneil, D.G., Proskurowski, A.: Isomorphism testing in hookup classes. SIAM Journal on Algebraic and Discrete Methods 3(2), 260–274 (1982)

[KST93] Köbler, J., Schöning, U., Torán, J.: The Graph Isomorphism Problem: Its Structural Complexity. Progress in Theoretical Computer Science. Birkhäuser, Boston (1993)

[Lin92] Lindell, S.: A logspace algorithm for tree canonization. extended abstract. In: Proceedings of the 24th STOC, pp. 400–404. ACM, New York (1992)

[MR91] Miller, G., Reif, J.: Parallel tree contraction part 2: further applications. SIAM Journal on Computing 20, 1128–1147 (1991)

[Rei05] Reingold, O.: Undirected st-connectivity in log-space. In: Proceedings of the 37th STOC, pp. 376–385. ACM, New York (2005)

[Sch88] Schöning, U.: Graph isomorphism is in the low hierarchy. Journal of Computer and System Sciences 37, 312–323 (1988)

[Tor04] Torán, J.: On the hardness of graph isomorphism. SIAM Journal on Computing 33(5), 1093–1108 (2004)

Snake-Deterministic Tiling Systems*

Violetta Lonati[1] and Matteo Pradella[2]

[1] Dipartimento di Scienze dell'Informazione, Università degli Studi di Milano
Via Comelico 39/41, 20135 Milano, Italy
lonati@dsi.unimi.it
[2] IEIIT, Consiglio Nazionale delle Ricerche
Via Golgi 40, 20133 Milano, Italy
matteo.pradella@polimi.it

Abstract. The concept of determinism, while clear and well assessed for string languages, is still matter of research as far as picture languages are concerned. We introduce here a new kind of determinism, called *snake*, based on the *boustrophedonic* scanning strategy, that is a natural scanning strategy used by many algorithms on 2D arrays and pictures. We consider a snake-deterministic variant of tiling systems, which defines the so-called *Snake-DREC* class of languages. Snake-DREC properly extends the more traditional approach of diagonal-based determinism, used e.g. by deterministic tiling systems, and by online tessellation automata. Our main result is showing that the concept of snake-determinism of tiles coincides with row (or column) unambiguity.

Keywords: picture language, 2D language, tiling systems, online tessellation automata, determinism, unambiguity.

1 Introduction

Picture languages are a generalization of string languages to two dimensions: a picture is a two-dimensional array of elements from a finite alphabet. Several classes of picture languages have been considered in the literature [8,10,6,12]. In particular, here we refer to class REC introduced in [8] with the aim to generalize to 2D the class of regular string languages. REC is a robust class that has various characterizations; in particular, it is the class of picture languages that can be generated by *tiling systems*, a model introduced in [7], where pictures are specified as alphabetic projection of a local 2D language defined by a set of tiles.

For string regular languages, two central notions are those of *determinism* and *unambiguity*. Going towards 2D, the concept of unambiguity is straightforward and yields to class UREC [7]. UREC defines unambiguously tiling recognizable languages, whose pictures are the projection of a unique element in the corresponding local language. In an effort to go towards determinism, the authors of [1] introduced an intermediate

* This work has been supported by the MIUR PRIN project "Mathematical aspects and emerging applications of automata and formal languages", and CNR RSTL 760 "2D grammars for defining pictures".

R. Královič and D. Niwiński (Eds.): MFCS 2009, LNCS 5734, pp. 549–560, 2009.

notion of "line" unambiguity, embodied in classes Row-UREC and Col-UREC, and based on backtracking at most linear in one dimension of the picture.

The concept of determinism for picture languages is far from being well understood. The most relevant difficulty is that in 2D any notion of determinism seems to require some pre-established "scanning strategy" for reading the picture. Tiling systems are implicitly nondeterministic: REC is not closed under complement, and the membership problem is NP-complete [11]. Clearly, this latter fact severely hinders the potential applicability of the notation. The identification of a reasonably "rich" deterministic subset of REC would spur its application, since it would allow linear parsing w.r.t. the number of pixels of the input picture.

In past and more recent years, several different deterministic subclasses of REC have been studied, e.g. the classes defined by deterministic 4-way automata [10] or deterministic online tessellation automata [9]. This latter model inspired the notion of determinism of [1], that relies on four diagonal-based scanning strategies, each starting from one of the four corners of the picture. To mark this aspect, in this paper we will call the corresponding deterministic class Diag-DREC[1].

In a effort to generalize their approach, the same authors in [2] suggest other kinds of strategies. Inspired by their work, we introduce here a new kind of determinism for tiles, based on a boustrophedonic scanning strategy, that is a natural scanning strategy used by many algorithms on pictures and 2D arrays (such as shearsort) [4,2,5]. This leads to a class called Snake-DREC, which can be defined equivalently in terms of tiling systems or online tessellation acceptors.

Snake-DREC properly extends Diag-DREC while keeping some important closure properties. For instance, it is still closed under complement, rotation and symmetries. However, like Diag-DREC, it is not closed under intersection. When pictures of only one row (or column) are considered, this model reduces to deterministic finite state automata. Quite surprisingly, we found that our notion of determinism coincides with line unambiguity of Row-UREC (or Col-UREC): our main result is showing that the languages of this class can actually be recognized deterministically by following a boustrophedonic scanning strategy.

The paper is organized as follows. In Section 2 we recall some basic definitions and properties on two-dimensional languages and tiling systems. In Section 3 we introduce snake-deterministic tiling systems. In Section 4 we present our main result. In the last section we define and characterize class Snake-DREC.

2 Preliminaries

2.1 Tiling Recognizable Picture Languages

The following definitions are taken and adapted from [8].

Let Σ be a finite alphabet. A two-dimensional array of elements of Σ is a *picture* over Σ. The set of all pictures over Σ is Σ^{++}. A picture language is a subset of Σ^{++}. If C denotes some kind of picture-accepting device, then $\mathcal{L}(C)$ denotes the class of picture languages recognized by such devices.

[1] The original name is DREC.

For $h, k \geq 1$, $\Sigma^{h,k}$ denotes the set of pictures of size (h, k); $\# \notin \Sigma$ is used when needed as a *boundary symbol*; $\hat{p}$ refers to the bordered version of picture p. That is, for $p \in \Sigma^{h,k}$, it is

$$p = \begin{array}{|c|c|c|} \hline p(1,1) & \dots & p(1,k) \\ \hline \vdots & \ddots & \vdots \\ \hline p(h,1) & \dots & p(h,k) \\ \hline \end{array} \qquad \hat{p} = \begin{array}{|c|c|c|c|c|} \hline \# & \# & \dots & \# & \# \\ \hline \# & p(1,1) & \dots & p(1,k) & \# \\ \hline \vdots & \vdots & \ddots & \vdots & \vdots \\ \hline \# & p(h,1) & \dots & p(h,k) & \# \\ \hline \# & \# & \dots & \# & \# \\ \hline \end{array}.$$

A *pixel* is an element $p(i, j)$ of p. We call (i, j) the *position* in p of the pixel. We will sometimes use position (i, j) with i or j equal to 0, or $h + 1$, or $k + 1$ for referring to borders.

We will sometimes consider the 90° clockwise *rotation*, the *horizontal mirror*, and the *vertical mirror* of a picture p. E.g. if $p = \begin{array}{|c|c|} \hline a & b \\ \hline c & d \\ \hline \end{array}$, then $\begin{array}{|c|c|} \hline c & a \\ \hline d & b \\ \hline \end{array}$, $\begin{array}{|c|c|} \hline c & d \\ \hline a & b \\ \hline \end{array}$, and $\begin{array}{|c|c|} \hline b & a \\ \hline d & c \\ \hline \end{array}$ are its rotation, horizontal mirror and vertical mirror, respectively. Naturally, the same operations can be applied to languages, and classes of languages, too.

We call *tile* a square picture of size (2,2). We denote by $T(p)$ the set of all tiles contained in a picture p.

Let Σ be a finite alphabet. A (two-dimensional) language $L \subseteq \Sigma^{++}$ is *local* if there exists a finite set Θ of tiles over the alphabet $\Sigma \cup \{\#\}$ such that $L = \{p \in \Sigma^{++} \mid T(\hat{p}) \subseteq \Theta\}$. We will refer to such language as $L(\Theta)$.

Let $\pi : \Gamma \to \Sigma$ be a mapping between two alphabets. Given a picture $p \in \Gamma^{++}$, the *projection* of p by π is the picture $\pi(p) \in \Sigma^{++}$ such that $\pi(p)\ (i, j) = \pi(p(i, j))$ for every position (i, j). Analogously, the projection of a language $L \subseteq \Gamma^{++}$ by π is the set $\pi(L) = \{\pi(p) \mid p \in \Gamma^{++}\} \subseteq \Sigma^{++}$.

A *tiling system* (TS) is a 4-tuple $\tau = \langle \Sigma, \Gamma, \Theta, \pi \rangle$ where Σ and Γ are two finite alphabets, Θ is a finite set of tiles over the alphabet $\Gamma \cup \{\#\}$ and $\pi : \Gamma \to \Sigma$ is a projection. A picture language $L \subseteq \Sigma^{++}$ is *tiling recognizable* if there exists a tiling system $\langle \Sigma, \Gamma, \Theta, \pi \rangle$ such that $L = \pi(L(\Theta))$. We say that τ generates L and denote by REC the class of picture languages that are tiling recognizable, i.e, REC $= \mathcal{L}$(TS). Notice in particular that any local language is tiling recognizable.

Example 1. The language L_{center} of square pictures over $\{0, 1\}$ with odd size, greater than 2, and having 1 only in the center is generated by the tiling system $\langle \Sigma, \Gamma, \Theta, \pi \rangle$, where: $\Gamma = \{1, \diagdown, \diagup, \cdot\}$; $\pi(1) = 1$, $\pi(x) = 0$ for $x \neq 1$, and the set of tiles is $\Theta = T(\hat{p})$,

$$p = \begin{array}{|c|c|c|c|c|c|c|} \hline \diagdown & \cdot & \cdot & \cdot & \cdot & \cdot & \diagup \\ \hline \cdot & \diagdown & \cdot & \cdot & \cdot & \diagup & \cdot \\ \hline \cdot & \cdot & \diagdown & \cdot & \diagup & \cdot & \cdot \\ \hline \cdot & \cdot & \cdot & 1 & \cdot & \cdot & \cdot \\ \hline \cdot & \cdot & \diagup & \cdot & \diagdown & \cdot & \cdot \\ \hline \cdot & \diagup & \cdot & \cdot & \cdot & \diagdown & \cdot \\ \hline \diagup & \cdot & \cdot & \cdot & \cdot & \cdot & \diagdown \\ \hline \end{array}.$$

Notice that it is straightforward to extend the previous tiling system to define the language L'_{center} of square pictures with odd size, and having 1 not only in the center, but possibly elsewhere. E.g., we may set $\Gamma = (\{0,1\} \times \{\diagdown, \diagup, \cdot\}) \cup \{(1,1)\}$ and π: $\pi(0,y) = 0, \pi(1,x) = 1$.

REC coincides with the class of languages recognized by online tessellation acceptors (OTA), that are special acceptors related to cellular automata [9]. Informally, an online tessellation acceptor can be described as an infinite two-dimensional array of identical finite-state automata, where the computation proceeds by counter-diagonals starting from top-left towards bottom-right corner of the input picture. A run of a OTA on a picture consists in associating a state to each position of the picture. At the beginning, an initial state is assigned to all top and left border positions. The state at position (i, j) is given by the transition function and depends both on the symbol of the picture at that position, and on the states already associated with positions $(i, j-1), (i-1, j-1)$ and $(i-1, j)$. The picture is accepted if the state associated with the bottom-right corner is final.

A natural subclass of REC, already introduced in [7], is UREC consisting of the tiling recognizable languages whose pictures are the projection of a unique element in the corresponding local language. Formally, a tiling system $\langle\Sigma, \Gamma, \Theta, \pi\rangle$ is called *unambiguous* if, for every $q, q' \in L(\Theta)$, $\pi(q) = \pi(q')$ implies $q = q'$. UREC is the class of all unambiguous languages. It is known that UREC $\subset$ REC and that it is undecidable whether a tiling system is unambiguous [3].

2.2 Diagonal-Deterministic Languages

Here we present the notion of determinism proposed in [1]. This is inspired by the deterministic version of online tessellation acceptors [9], which are directed according to a corner-to-corner direction (namely, from top-left to bottom-right, or *tl2br*).

Consider a scanning strategy that respects the tl2br direction: any position (x, y) is read only if all the positions that are above and to the left of (x, y) have already been read. Roughly speaking, tl2br determinism means that, given a picture $p \in \Sigma^{++}$, its preimage $p' \in L(\Theta) \subseteq \Gamma^{++}$ can be build deterministically when scanning p with any such strategy. Formally, a tiling system $\tau = \langle\Sigma, \Gamma, \Theta, \pi\rangle$ is called *tl2br-deterministic* if for any $X, Y, Z \in \Gamma \cup \{\#\}$ and $a \in \Sigma$, there exists at most one tile

$$\begin{array}{|c|c|}\hline X & Y \\ \hline Z & A \\ \hline\end{array} \in \Theta \qquad \text{with } \pi(A) = a.$$

By rotation, one can define *d-deterministic* tiling systems (*d*-DTS) for any corner-to-corner direction d in {tl2br, tr2bl, bl2tr, br2tl}, where t, b, l, and r stand for top, bottom, left, and right, respectively.

Example 2. The language $L_{fr=fc}$ of square pictures where the first row equals the first column is in $\mathcal{L}$(tl2br-DTS) $\cap$ $\mathcal{L}$(tr2bl-DTS) $\cap$ $\mathcal{L}$(bl2tr-DTS), but does not belong to $\mathcal{L}$(br2tl-DTS) [1].

We use Diag-DREC to denote the family of languages recognized by some d-DTS (for all corner-to-corner directions d). Diag-DREC is equal to the closure by rotation of the class of languages recognized by deterministic OTAs (denoted as DOTAs).

Example 3. The language $L_{\exists r=lr}$ of square pictures where there is one row that equals the last one cannot be recognized neither by any tl2br-DTS [9, Theorem 3.1], nor (symmetrically) by any tr2bl-DTS. However, $L_{\exists r=lr} \in$ Diag-DREC since one can prove that it is recognized both by bl2tr-DTS and br2tl-DTS.

Diag-DREC is properly included in UREC, as the following example testifies.

Example 4. Let $L_{\text{frames}} = L_{fr=fc} \cap L_{lr=lc} \cap L_{2fr=\overline{2lc}} \cap L_{2lr=\overline{2fc}}$ be the language of square pictures such that: the first row equals the first column, the last row equals the last column, the second row equals the reverse of the second last column, the second last row equals the reverse of the second column. Then L_{frames} is in UREC but not in Diag-DREC [1].

2.3 Row and Column Unambiguity

In [1] a hierarchy of classes between determinism and unambiguity is also exhibited. Here we adapt the basic definitions, results, and examples from [1].

Consider the side-to-side direction t2b (from top to bottom). A tiling system $\langle \Sigma, \Gamma, \Theta, \pi \rangle$ is called t2b-*unambiguous* if, for any rows $\mathbf{X} = (X_1, X_2, \cdots, X_m) \in \Gamma^{1,m} \cup \{\#\}^{1,m}$ and $\mathbf{a} = (a_1, a_2, \cdots, a_m) \in \Sigma^{1,m}$, there exists at most one row $\mathbf{A} = (A_1, A_2, \cdots, A_m) \in \Gamma^{1,m}$ such that

$$\pi(\mathbf{A}) = \mathbf{a} \quad \text{and} \quad T\left(\begin{array}{|c|c|c|c|c|c|}\hline \# & X_1 & X_2 & \ldots & X_m & \# \\ \hline \# & A_1 & A_2 & \ldots & A_m & \# \\ \hline \end{array}\right) \subseteq \Theta\,. \tag{1}$$

Example 5. Let $L = L_{\exists c=fc} \cap L_{\exists c=lc}$ of square pictures where there are one column that equals the first one and one column that equals the last one. L is not in Diag-DREC, but it can be recognized by a t2b-unambiguous tiling system.

Similar properties define d-unambiguous tiling systems (d-UTS) for any side-to-side direction $d \in \{t2b, b2t, l2r, r2l\}$. Row-UREC (resp. Col-UREC) denotes the class of *row-unambiguous* (resp. *column-unambiguous*) languages, i.e., the languages generated by t2b-UTS or b2t-UTS (resp. l2r-UTS or r2l-UTS). The following relation holds, with all strict inclusions:

$$\text{Diag-DREC} \subset (\text{Col-UREC} \cap \text{Row-UREC}) \subset (\text{Col-UREC} \cup \text{Row-UREC}) \subset \text{UREC}. \tag{2}$$

In the rest of the paper we will use the informal term *line unambiguity* for referring both to row and to column unambiguity.

3 Snake-Deterministic Tiling Systems

Given a tiling system $\tau = \langle \Sigma, \Gamma, \Theta, \pi \rangle$ and a picture $p \in \Sigma^{++}$, imagine to build one preimage $p' \in L(\Theta)$, $\pi(p') = p$, by scanning p with a boustrophedonic strategy. More precisely, start from the top-left corner, scan the first row of p rightwards, then scan the second row leftwards, and so on, like in the following picture, where the number in each pixel denotes its scanning order:

1	2	3	4
8	7	6	5
9	10	11	12

.

This means that we scan odd rows rightwards and even row leftwards, assigning a symbol in Γ to each position. The choice of the symbol clearly depends on the symbols in the neighbourhood, and it is determined by a tile in Θ. In general, the choice is not unique and hence the procedure may not be deterministic. We introduce the following definition to guarantee such condition.

Definition 1. *A tiling system* $\tau = \langle \Sigma, \Gamma, \Theta, \pi \rangle$ *is* snake-deterministic *if* Γ *and* Θ *can be partitioned as* $\Gamma = \Gamma_1 \cup \Gamma_2$, $\Theta = \Theta_1 \cup \Theta_2$, *where*

– $\langle \Sigma, \Gamma, \Theta_1, \pi \rangle$ *is tl2br-deterministic and* Θ_1 *contains only tiles like*

$$\begin{array}{|c|c|}\hline a_2 & b_2 \\ \hline a_1 & b_1 \\ \hline \end{array}, \quad \text{with } a_i, b_i \in \Gamma_i \cup \{\#\} \text{ for } i = 1,2;$$

– $\langle \Sigma, \Gamma, \Theta_2, \pi \rangle$ *is tr2bl-deterministic and* Θ_2 *contains only tiles like*

$$\begin{array}{|c|c|}\hline a_1 & b_1 \\ \hline a_2 & b_2 \\ \hline \end{array}, \quad \text{with } a_i, b_i \in \Gamma_i \cup \{\#\} \text{ for } i = 1,2, \text{ and } (a_1, b_1) \neq (\#, \#).$$

Snake-deterministic tiling systems are abbreviated as snake-DTS.

In the following we shall always use this notation: symbols on light gray background belong to $\Gamma_1 \cup \{\#\}$, symbols on dark gray background belong to $\Gamma_2 \cup \{\#\}$. Hence tiles in Θ_1 or Θ_2 will appear as $\begin{array}{|c|c|}\hline a & b \\ \hline c & d \\ \hline \end{array}$ or $\begin{array}{|c|c|}\hline c & d \\ \hline a & b \\ \hline \end{array}$, respectively.

In [1], it is proved that tl2br-deterministic tiling systems are equivalent to deterministic online tessellation acceptors (DOTA). Analogously, here we introduce a similar model of acceptor equivalent to snake-deterministic tiling systems.

Definition 2. *A* deterministic snake online tessellation acceptor *(ZOTA) is a 7-tuple* $\langle \Sigma, Q_1, Q_2, q_{01}, q_{02}, F, \delta \rangle$ *where:*

- Σ *is the input alphabet;*
- Q_1 *and* Q_2 *are two disjoint set of states;*
- $q_{0i} \in Q_i$ *are the initial states;*
- $F \subset Q_1 \cup Q_2 = Q$;
- $\delta : Q \times Q \times Q \times \Sigma \mapsto Q$ *is the transition function satisfying* $\delta(p_1, q_1, p_2, a) \in Q_2$, $\delta(p_2, q_2, p_1, a) \in Q_1$, *for every* $p_i, q_i \in Q_i$ *and* $a \in \Sigma$.

A run of any ZOTA on a picture consists in scanning the picture following the snake-like strategy, associating, at each step, a state with the current position in the picture. At step 0, the initial state q_{01} is assigned to all the border positions $(0, j)$ and $(i, 0)$ with i even, whereas the initial state q_{02} is assigned to all positions $(i, 0)$ with i odd. The state at position (i, j) is given by the transition function and depends on the input symbol at that position and on the states already associated with some of the neighbouring positions: for odd i, the positions considered are $(i-1, j)$, $(i-1, j-1)$, and $(i, j-1)$; for even i, $(i-1, j)$, $(i-1, j+1)$, and $(i, j+1)$. The picture is accepted if the state associated with the last position (i.e. the bottom-rightmost for pictures with an odd number of rows, the bottom-leftmost otherwise) is in F.

Reasoning as in [8, Theorem 8.1] one can easily prove that deterministic snake tessellation automata are equivalent to snake-deterministic tiling systems.

Proposition 1. *$\mathcal{L}$(ZOTA) = $\mathcal{L}$(snake-DTS).*

Proposition 2. *$\mathcal{L}$(snake-DTS) is a boolean algebra.*

Proof (sketch). If L is recognized by a ZOTA, then so is its complement (it is sufficient to exchange final states with non-final ones). Hence, by Proposition 1, $\mathcal{L}$(snake-DTS) is closed under complement.

Moreover, given two snake-DTSs recognizing two languages L_1 and L_2 respectively, one can follow the construction defined in [8] to build a new TS recognizing the intersection $L_1 \cap L_2$. Such construction preserves snake-determinism, hence we get the closure under intersection. □

The following example shows how a snake-DTS, or, equivalently, a ZOTA, can "propagate signals" both in tl2br and tr2bl corner-to-corner directions. This property does not hold in general for tl2br-DTSs or tr2bl-DTSs. Indeed, we will show in Section 4 that they are strictly less powerful than snake-DTSs.

Example 6. The language L_{center} described in Example 1 is recognized by the following snake-DTS. $\Gamma = \Gamma_1 \cup \Gamma_2$ where

$$\Gamma_0 = \{\diagdown, \diagup, \searrow, \swarrow, \rightarrow, \leftarrow, -, \cdot, 1\}, \qquad \Gamma_1 = \{1\} \times \Gamma_0, \qquad \Gamma_2 = \{2\} \times \Gamma_0,$$

and π is such that $\pi(x, 1) = 1$, $\pi(x, y) = 0$, for $y \neq 1$. $\Theta = T(\hat{p}) \cup T(\hat{q})$ where p and q are the following pictures. For better readability, the first component of symbols in Γ_1 is depicted as a light gray background, instead of the symbol 1; analogously, the first component of symbols in Γ_2 is depicted as a dark gray background.

$p =$ 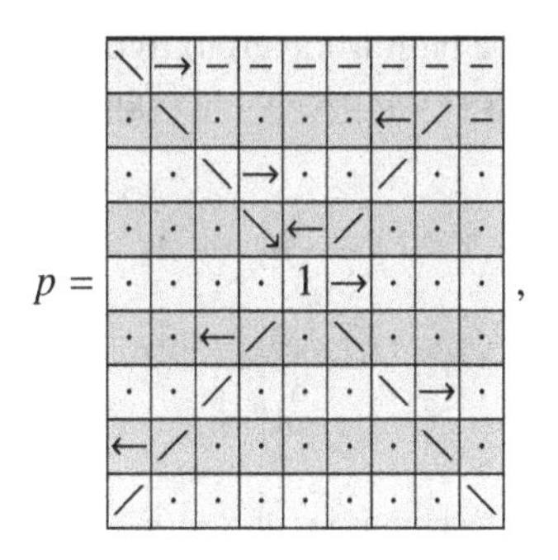 , $q =$ 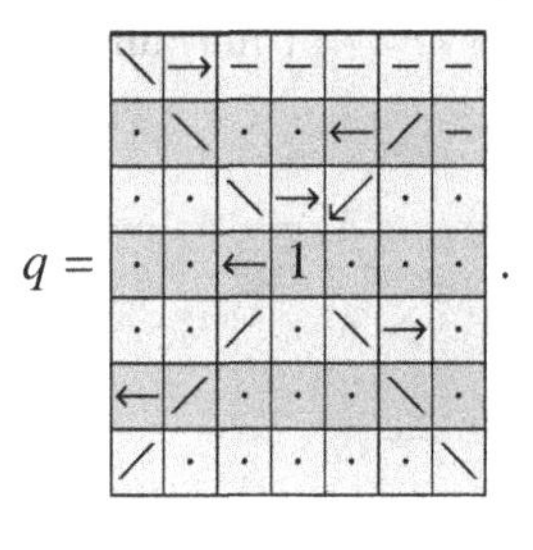 .

The basic mechanism of this tiling system is the same as the one of Example 1: the two diagonals are used to identify the center. To make the tiles snake-deterministic, we have first to distinguish odd and even rows, by using in Γ a first component 1, and 2, respectively. First, notice that we have to use two prototypal pictures p and q to define the tile-set: p represents pictures in which the center symbol is found during a left-to-right scan, while q represents the other direction. Symbols $\rightarrow$ and $\leftarrow$ mark the fact that there is a diagonal at the position immediately below. This information is needed for the right-to-left component of the boustrophedonic movement, to mark the top-left to bottom-right diagonal (symbol $\rightarrow$), and analogously for the other direction and diagonal (symbol $\leftarrow$). Symbol $-$ is used to identify the start of the top-right to bottom-left diagonal. Symbols $\searrow$ and $\swarrow$ are used both to mark diagonals, and to state that at the following row the center will be found. Notice that the language generated by this tiling system does not contain pictures having side less than 7 - it is clearly straightforward to extend it to cover those cases as well.

A simple extension of the same structure can be used to define a snake-deterministic tiling system for the language L'_{center}, mentioned at the end of Example 1. □

4 Snake Determinism Is Equivalent to Line Unambiguity

In this section we prove our main result, showing that snake-deterministic tiling systems are equivalent to t2b-unambiguous tiling systems.

Theorem 1. *$\mathcal{L}$(snake-DTS) = $\mathcal{L}$(t2b-UTS).*

In one direction, the result is easy (any snake-deterministic tiling system is also t2b-unambiguous). The converse is less intuitive; in order to prove it, from now on let $\tau = \langle \Sigma, \Gamma, \Theta \pi \rangle$ be a t2b-UTS.

First of all, let Γ_1 (resp. Γ_2) be the set of symbols in Γ that may appear only in odd (resp. even) rows, and w.l.o.g assume that Γ_1 and Γ_2 are disjoint. (Otherwise we can mark with subscript i all elements that may appear in Γ_i, possibly duplicating symbols and tiles.) Consequently, as in the definition of snake-DTS, split the set of tiles into two sets Θ_1 and Θ_2. If the resulting tiling system is not snake-deterministic, then we build a snake-deterministic tiling system $\tilde{\tau} = \langle \Sigma, \tilde{\Gamma}, \tilde{\Theta}, \tilde{\pi} \rangle$ that simulates τ. Before formally defining $\tilde{\tau}$, let us first point out some important remarks.

Given any $\mathbf{X} = (X_1, X_2, \ldots, X_m) \in \Gamma^{1,m} \cup \{\#\}^{1,m}$ and $\mathbf{a} = (a_1, a_2, \ldots, a_m) \in \Sigma^{1,m}$, there exists at most one preimage $\mathbf{A} = (A_1, A_2, \ldots, A_m) \in \Gamma^{1,m}$ satisfying relation (1). However, we have no guarantees that $\mathbf{A}$ can be built from left to right deterministically. For instance, for $m = 4$, τ may allow the choices represented in Figure 1 (left).

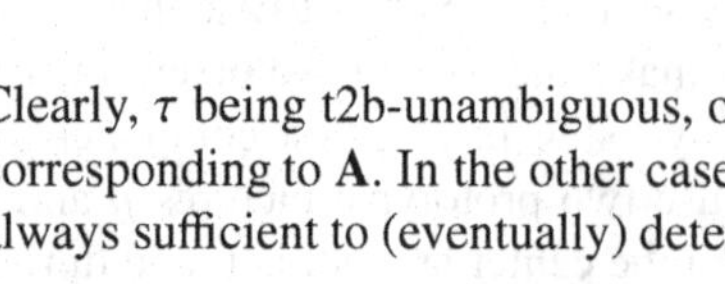

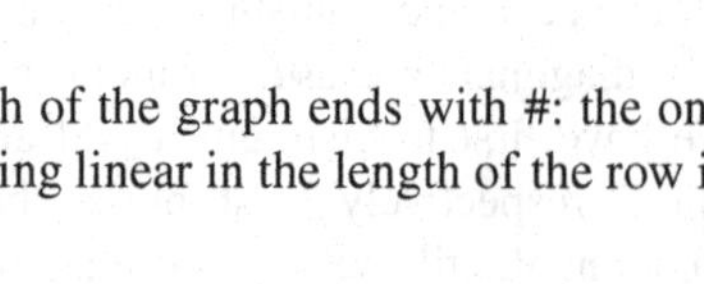

$$\# \to A_1 \to A_2 \to A_3 \to A_4 \to \# \qquad\qquad \# \to A_1 \to A_2 \to A_3 \to A_4 \to \#$$
$$A'_1 \quad A'_2 \to A'_3 \quad A'_4 \qquad\qquad A'_1 \quad A'_2 \to A'_3 \quad A'_4$$
$$A''_1 \to A''_2 \to A''_3 \qquad\qquad A''_1$$

Fig. 1. Graph (left) and tree (right) of the preimages of row (a_1, a_2, a_3, a_4)

Clearly, τ being t2b-unambiguous, only one branch of the graph ends with #: the one corresponding to $\mathbf{A}$. In the other cases, a backtracking linear in the length of the row is always sufficient to (eventually) determine $\mathbf{A}$.

Remark 1. Since we are building a preimage of a fixed row $\mathbf{a}$, at each position we can choose among symbols that all have the same image through π. E.g., $\pi(A_1) = \pi(A'_1) = \pi(A''_1) = a_1$.

Remark 2. Since τ is t2b-unambiguous, the branch corresponding to $\mathbf{A}$ cannot contain symbols with in-degree greater that one (otherwise there would exist two different preimages of row **a** satisfying relation (1), a contradiction). In other words, if two or more branches "collapse", the successive symbols may be ignored. Then we can assume that the graph of the preimages of **a** is actually a tree, where each symbol has exactly one predecessor. We call it the *tree of the preimages of* $\boldsymbol{a}$ *in* Γ. For the previous example, the tree is depicted in Figure 1 (right).

Similar remarks can be done if we try to build the preimage of row **a** from right to left.

To simulate τ deterministically on a picture p, we proceed as follows. Let p' be the unique preimage of p in $L(\Theta)$. When scanning rightwards the first row of p, we compute and keep trace of the tree of its preimages in Γ; at the end of the row, we determine which branch is successful (i.e, the one that corresponds to the first row of p'). When scanning the second row backwards, we use such information (together with the traces we left in the previous scan) to reconstruct backwards the successful branch and, at the same time, we compute and keep trace of the tree of the preimages in Γ of the current row. This procedure continues till the last row has been scanned.

To represent locally the tree of preimages of the current row, we store at each position the set of symbols of Γ that may appear at the corresponding position of p', together with their predecessors in the tree. To represent the correspondence between a symbol and its predecessors, which is unique by Remark 2, we use partial functions. For instance, the tree of the previous example is represented by the sequence of partial functions $(\alpha_1, \alpha_2, \alpha_3, \alpha_4, \alpha_5)$, where $\alpha_1(A_1) = \alpha_1(A'_1) = \alpha_1(A''_1) = \#$, $\alpha_2(A_2) = \alpha_2(A'_2) = A_1$, $\alpha_3(A_3) = A_2, \alpha_3(A'_3) = A'_2$, $\alpha_4(A_4) = \alpha_4(A'_4) = A_3$, and $\alpha_5(\#) = A_4$.

We shall need some notation. Given a partial function $f : X \to Y$, we set $I_f = f(\Delta_f)$; moreover, we write $f(x) = \perp$ if $f(x)$ is not defined, set $\Delta_f = \{x \in X \mid f(x) \neq \perp\}$, and say that f is non-empty if $\Delta_f \neq \emptyset$. For $i = 1, 2$, let $\hat{\Gamma}_i = \Gamma_i \cup \{\#\}$ and call Φ_i the set of non-empty partial functions $\varphi : \hat{\Gamma}_i \to \hat{\Gamma}_i$ such that $|\pi(\Delta_\varphi)| = |\pi(I_\varphi)| = 1$ (this last condition is the formalization of Remark 1). In particular, for every $A \in \Gamma_i$, let $\sharp_A$ be the function in Φ_i with domain $\{\#\}$ such that $\sharp_A(\#) = A$; moreover, let $\sharp$ be the function with domain $\{\#\}$ such that $\sharp(\#) = \#$. Finally, we abbreviate $\pi(\Delta_\varphi)$ by $\pi(\varphi)$.

Recall that during the simulation we perform two operations at the same time: we reconstruct the successful branch in the tree of preimages of the previous row, and compute the tree of preimages of the current row. Hence the local alphabet of $\tilde{\tau}$ must contain both pieces of information. This leads to the following definition:

$$\tilde{\Gamma} = \tilde{\Gamma}_1 \cup \tilde{\Gamma}_2 \text{ where } \tilde{\Gamma}_1 = \hat{\Gamma}_2 \times \Phi_1, \text{ and } \tilde{\Gamma}_2 = \hat{\Gamma}_1 \times \Phi_2, \tag{3}$$

$$\forall (A, \varphi) \in \tilde{\Gamma} : \tilde{\pi}(A, \varphi) = \pi(\varphi). \tag{4}$$

The role of symbol (A, φ) is the following: A is the correct symbol that one should have chosen when scanning the above position (i.e., the symbol appearing at that position in p'), whereas φ keeps trace of all possible symbols that may appear in the current position, together with their predecessors in the computation. Notice that w.l.o.g we define more than one border symbol in $\tilde{\Gamma}$, i.e., all pairs (A, φ) with $\pi(\varphi) = \#$.

In order to define the set of tiles, we need some other notations. For any $b \in \Sigma \cup \{\#\}$, we introduce the partial function r-next$_b : \hat{\Gamma}_2 \times \hat{\Gamma}_2 \times \Phi_1 \to \Phi_1$ by setting r-next$_b(X, Y, \alpha) =$

β, where, for every $B \in \hat{\Gamma}_1$:

$$\beta(B) = \begin{cases} A & \text{if } \pi(B) = b \text{ and } A \text{ is the unique element in } \Delta_\alpha \text{ s.t. } \begin{array}{|c|c|}\hline X & Y \\ \hline A & B \\ \hline\end{array} \in \Theta_1, \\ \perp & \text{otherwise.} \end{cases}$$

Informally, r-next$_b(X, Y, \alpha)$ represents all possible symbols that can appear in next position, when going rightwards, reading symbol b, and given previous neighbours like $\begin{array}{|c|c|}\hline X & Y \\ \hline A & \\ \hline\end{array}$, with $A \in \Delta_\alpha$.

Symmetrically, for any $d \in \Sigma$ let l-next$_d : \hat{\Gamma}_1 \times \hat{\Gamma}_1 \times \Phi_2 \to \Phi_2$ be the partial function defined by l-next$_d(A, B, \gamma) = \delta$, where, for every $D \in \hat{\Gamma}_2$:

$$\delta(D) = \begin{cases} C & \text{if } \pi(D) = d \text{ and } C \text{ is the unique element in } \Delta_\gamma \text{ s.t. } \begin{array}{|c|c|}\hline A & B \\ \hline D & C \\ \hline\end{array} \in \Theta_2, \\ \perp & \text{otherwise.} \end{cases}$$

Lemma 1. *Let $\tau = \langle \Sigma, \Gamma, \Theta, \pi \rangle$ be t2b-unambiguous and let $\boldsymbol{X} = (X_1, X_2, \ldots, X_m)$, $\boldsymbol{A} = (A_1, A_2, \ldots, A_m)$ and $\boldsymbol{a} = (a_1, a_2, \ldots, a_m)$ satisfying relation (1), with $\boldsymbol{X}_i \in \Gamma_2 \cup \{\#\}$ for every i. Moreover set*

$$\alpha_1 = r\text{-}next_{a_1}(\#, X_1, \sharp), \qquad \forall j = 2, \ldots, m: \ \alpha_j = r\text{-}next_{a_j}(X_{j-1}, X_j, \alpha_{j-1})$$

Then, for every $j = 1, 2, \ldots, m$, $A_j \in \Delta_{\alpha_i}$, $\alpha_1(A_1) = \#$, $\alpha_j(A_j) = A_{j-1}$ for $j \neq 1$. Similar results hold for the symmetric direction.

Proof. We reason by induction on $j = 1, 2, \ldots, m$. For sake of brevity, we use Δ_i to denote Δ_{α_i}. Clearly, $A_1 \in \Delta_1$ with $\alpha_1(A_1) = \#$. Now, assuming that the statement holds for $k \leq j$, we prove it for $j + 1$. We have $\begin{array}{|c|c|}\hline X_j & X_{j+1} \\ \hline A_j & A_{j+1} \\ \hline\end{array} \in \Theta$. If $\begin{array}{|c|c|}\hline X_j & X_{j+1} \\ \hline A'_j & A_{j+1} \\ \hline\end{array} \in \Theta$ for some other $A'_j \in \Delta_j$ then, setting $A'_{k-1} = \alpha_k(A'_k)$ for every $k = j, \ldots, 2, 1$, we obtain that $\begin{array}{|c|c|c|c|c|c|c|c|}\hline \# & X_1 & \ldots & X_j & X_{j+1} & \ldots & X_m & \# \\ \hline \# & A'_1 & \ldots & A'_j & A_{j+1} & \ldots & A_m & \# \\ \hline\end{array} \in \tilde{\Theta}_1$. This yields a contradiction, since also relation (1) holds but τ is 2tb-unambiguous. Thus, A_j is unique and hence $A_{j+1} \in \Delta_{j+1}$ with $\alpha_{j+1}(A_{j+1}) = A_j$. □

We are ready to prove Theorem 1, as a straightforward consequence of the following proposition.

Proposition 3. *Given a t2b-unambiguous tiling system $\tau = \langle \Sigma, \Gamma, \Theta, \pi \rangle$, let $\tilde{\tau}$ be the tiling system $= \langle \Sigma, \tilde{\Gamma}, \tilde{\Theta}, \tilde{\pi} \rangle$ where $\tilde{\Gamma}$ and $\tilde{\pi}$ are defined as in (3) and (4), while $\tilde{\Theta} = \tilde{\Theta}_1 \cup \tilde{\Theta}_2$, where*

$$\tilde{\Theta}_1 = \left\{ \begin{array}{|c|c|}\hline (A, \delta) & (B, \gamma) \\ \hline (D, \lambda) & (\delta(D), \mu) \\ \hline\end{array} \;\middle|\; \begin{array}{c} (\pi(\delta), \pi(\gamma)) = (\#, \#) \Rightarrow (A, B) = (\#, \#), \\ D \in \Delta_\delta, \ \delta(D) \in \Delta_\gamma, \ \pi(\lambda) = \# \Rightarrow \lambda = \sharp, \\ \mu = r\text{-}next_{\tilde{\pi}(\mu)}(D, \delta(D), \lambda) \end{array} \right\}$$

$$\tilde{\Theta}_2 = \left\{ \begin{array}{|c|c|}\hline (X, \alpha) & (Y, \beta) \\ \hline (\beta(B), \delta) & (B, \gamma) \\ \hline\end{array} \;\middle|\; \begin{array}{c} (\pi(\alpha), \pi(\beta)) \neq (\#, \#), \\ B \in \Delta_\beta, \ \beta(B) \in \Delta_\alpha, \ \pi(\gamma) = \# \Rightarrow \gamma = \sharp, \\ \delta = l\text{-}next_{\tilde{\pi}(\delta)}(\beta(B), B, \gamma) \end{array} \right\}$$

Then, $\tilde{\tau}$ is a snake-DTS equivalent to τ.

$(\#, \sharp)$	$(\#, \sharp)$	$(\#, \sharp)$	$\cdots$	$(\#, \sharp)$	$(\#, \sharp)$
$(\#, \sharp)$	$(\#, \alpha_{1,1})$	$(\#, \alpha_{1,2})$	$\cdots$	$(\#, \alpha_{1,m})$	$(\#, \sharp_{A_{1,m}})$
$(\#, \sharp_{A_{2,1}})$	$(A_{1,1}, \alpha_{2,1})$	$(A_{1,2}, \alpha_{2,2})$	$\cdots$	$(A_{1,m}, \alpha_{2,m})$	$(\#, \sharp)$
$(\#, \sharp)$	$(A_{2,1}, \alpha_{3,1})$	$(A_{2,2}, \alpha_{3,2})$	$\cdots$	$(A_{2,m}, \alpha_{3,m})$	$(\#, \sharp_{A_{3,m}})$
$\cdots$	$\cdots$	$\cdots$	$\cdots$	$\cdots$	$\cdots$
$(\#, \sharp)$	$(A_{n-1,1}, \alpha_{n,1})$	$(A_{n-1,2}, \alpha_{n,2})$	$\cdots$	$(A_{n-1,m}, \alpha_{n,m})$	$(\#, \sharp_{A_{n,m}})$
$(\#, \sharp)$	$(A_{n,1}, \sharp)$	$(A_{n,2}, \sharp)$	$\cdots$	$(A_{n,m}, \sharp)$	$(\#, \sharp)$

#	#	#	$\cdots$	#	#	#
#	$A_{1,1}$	$A_{1,2}$	$\cdots$	$A_{1,m-1}$	$A_{1,m}$	#
#	$A_{2,1}$	$A_{2,2}$	$\cdots$	$A_{2,m-1}$	$A_{2,m}$	#
#	$A_{3,1}$	$A_{3,2}$	$\cdots$	$A_{3,m-1}$	$A_{3,m}$	#
$\cdots$	$\cdots$	$\cdots$	$\cdots$	$\cdots$	$\cdots$	$\cdots$
#	$A_{n,1}$	$A_{n,2}$	$\cdots$	$A_{n,m-1}$	$A_{n,m}$	#
#	#	#	$\cdots$	#	#	#

Fig. 2. Examples of bordered pictures in $L(\tilde{\Theta})$ (left), and $L(\Theta)$ (right)

Proof. The TS $\tilde{\tau}$ is snake-deterministic by definition. We prove that $\tilde{\pi}(L(\tilde{\Theta}) = \pi(L(\Theta))$. First let $\tilde{p} \in L(\tilde{\Theta})$. W.l.o.g assume that the number of rows of $\tilde{p}$ is odd; then $\widehat{\tilde{p}}$ is as in Figure 2 (left). By the definition of $\tilde{\Theta}$, this implies that the picture p in Figure 2 (right) belongs to $L(\Theta)$. Moreover, one can easily see that $\pi(p) = \tilde{\pi}(\tilde{p})$. Hence, $\tilde{\pi}(L(\tilde{\Theta})) \subseteq \pi(L(\Theta))$.

On the other hand, consider a picture p as in Figure 2 (right). Then, let $\tilde{p}$ be a picture as in Figure 2 (left), where symbols $A_{i,j}$ are from p, whereas the partial functions $\alpha_{i,j}$ are defined inductively according to the boustrophedonic order of positions (i, j):

$$
\begin{array}{lll}
\alpha_{1,1} = \text{r-next}_{\pi(A_{1,1})}(\#, \#, \sharp), & \alpha_{1,j} = \text{r-next}_{\pi(A_{1,j})}(\#, \#, \alpha_{1,j-1}) & j = 2, \ldots, m, \\
\alpha_{2,m} = \text{l-next}_{\pi(A_{2,m})}(A_{1,m}, \#, \sharp), & \alpha_{2,j} = \text{l-next}_{\pi(A_{2,j})}(A_{1,j}, A_{a,j+1}, \alpha_{2,j+1}) & j = m-1, \ldots, 2, \\
\alpha_{3,1} = \text{r-next}_{\pi(A_{3,1})}(\#, A_{2,1}, \sharp), & \alpha_{3,j} = \text{r-next}_{\pi(A_{3,j})}(A_{2,j-1}, A_{2,j}, \alpha_{1,j-1}) & j = 2, \ldots, m, \\
\ldots & &
\end{array}
$$

One can verify that each α_{ij} is well defined. Indeed, using Lemma 1 one can prove that, for every $i = 1, 2, \ldots, n$ and $j = 1, 2, \ldots, m$, $A_{i,j} \in \Delta_{\alpha_{i,j}}$ and $\alpha_{i,j}(A_{i,j})$ is $A_{i,j-1}$ if i is odd, or $A_{i,j+1}$ if i is even. By the definition of $\tilde{\Theta}$, this implies that $\tilde{p} \in L(\tilde{\Theta})$. Since obviously $\tilde{\pi}(\tilde{p}) = \pi(p)$, we get $\pi(L(\Theta) \subseteq \tilde{\pi}(L(\tilde{\Theta}))$ and this concludes the proof. □

5 Class Snake-DREC

Theorem 1 implies that snake-DTS can simulate both tl2br-DTS and tl2br-DTS. Actually, this extension is proper as shown in next proposition.

Proposition 4. *$\mathcal{L}$(snake-DTS) properly extends $\mathcal{L}$(tl2br-DTS) ∪ $\mathcal{L}$(tr2bl-DTS).*

Proof. Since both tl2br-DTS and tl2br-DTS are t2b-unambiguous, the inclusion is a consequence of Theorem 1. The inclusion is proper as testified by the language $L = L_{\exists c=fc} \cap L_{\exists c=lc}$ described in Example 5. □

Notice that $\mathcal{L}$(snake-DTS) does not extend the whole class Diag-DREC. For instance the language $L_{\exists r=lr}$ described in Example 3 is in Diag-DREC but, reasoning as in [1], one can prove that it does not belong to $\mathcal{L}$(snake-DTS). On the contrary, by Proposition 4 we have that the closure under horizontal mirror of $\mathcal{L}$(snake-DTS) properly includes Diag-DREC. However, it is not closed by rotation: for instance

$L_{\exists c=lc} \in \mathcal{L}$(snake-DTS) since it is in $\mathcal{L}$(tr2bl-DTS), but its rotation is not (see again Example 3). This leads to the following definition.

Definition 3. *Snake-DREC is the closure under rotation of $\mathcal{L}$(snake-DTS). The languages in Snake-DREC are called* snake-deterministic.

We conclude characterizing Snake-DREC and summarizing its properties in the following theorem.

Theorem 2. *Snake-DREC = Row-UREC ∪ Col-UREC. Snake-DREC is properly included between Diag-DREC and UREC. Snake-DREC is closed under complement, rotation and mirrors, but not under intersection.*

Proof. The first identity follows by Theorem 1, by applying rotations. Then, the inclusions are a straightforward consequence of relation (2). Proposition 2 implies the closure under complement; the closure under rotation is obvious by definition; the closure under mirrors follows by the closure under mirrors of both Row-UREC and Col-UREC. $L_{\exists r=fr}$ is in Snake-DREC, but its intersection with all its rotations is not [1]. □

Acknowledgments. We thank Alberto Bertoni and Massimiliano Goldwurm for their useful comments.

References

1. Anselmo, M., Giammarresi, D., Madonia, M.: From determinism to non-determinism in recognizable two-dimensional languages. In: Harju, T., Karhumäki, J., Lepistö, A. (eds.) DLT 2007. LNCS, vol. 4588, pp. 36–47. Springer, Heidelberg (2007)
2. Anselmo, M., Giammarresi, D., Madonia, M.: A computational model for recognizable two-dimensional languages. Theoretical Computer Science (to appear, 2009)
3. Anselmo, M., Giammarresi, D., Madonia, M., Restivo, A.: Unambiguous recognizable two-dimensional languages. Theoretical Informatics and Applications 40(2), 277–293 (2006)
4. Behrooz, P.: Introduction to Parallel Processing: Algorithms and Architectures. Kluwer Academic Publishers, Norwell (1999)
5. Bertoni, A., Goldwurm, M., Lonati, V.: On the complexity of unary tiling-recognizable picture languages. Fundamenta Informaticae 91(2), 231–249 (2009)
6. Cherubini, A., Crespi Reghizzi, S., Pradella, M.: Regional languages and tiling: A unifying approach to picture grammars. In: Ochmański, E., Tyszkiewicz, J. (eds.) MFCS 2008. LNCS, vol. 5162, pp. 253–264. Springer, Heidelberg (2008)
7. Giammarresi, D., Restivo, A.: Recognizable picture languages. International Journal Pattern Recognition and Artificial Intelligence 6(2-3), 241–256 (1992); Special Issue on Parallel Image Processing
8. Giammarresi, D., Restivo, A.: Two-dimensional languages. In: Salomaa, A., Rozenberg, G. (eds.) Handbook of Formal Languages. Beyond Words, vol. 3, pp. 215–267. Springer, Heidelberg (1997)
9. Inoue, K., Nakamura, A.: Some properties of two-dimensional on-line tessellation acceptors. Information Sciences 13, 95–121 (1977)
10. Inoue, K., Takanami, I.: A survey of two-dimensional automata theory. Information Sciences 55(1-3), 99–121 (1991)
11. Lindgren, K., Moore, C., Nordahl, M.: Complexity of two-dimensional patterns. Journal of Statistical Physics 91(5-6), 909–951 (1998)
12. Matz, O.: On piecewise testable, starfree, and recognizable picture languages. In: Nivat, M. (ed.) FOSSACS 1998. LNCS, vol. 1378, pp. 203–210. Springer, Heidelberg (1998)

Query Automata for Nested Words

P. Madhusudan and Mahesh Viswanathan

University of Illinois, Urbana-Champaign
{madhu,vmahesh}@cs.uiuc.edu

Abstract. We study visibly pushdown automata (VPA) models for expressing and evaluating queries on words with a nesting structure. We define a query VPA model, which is a 2-way deterministic VPA that can mark in one run *all* positions in a document that satisfy a query, and show that it is equi-expressive as unary monadic queries. This surprising result parallels a classic result by Hopcroft and Ullman for queries on regular word languages. We also compare our model to query models on unranked trees, and show that our result is fundamentally different from those known for automata on trees.

1 Introduction

A nested word is a word endowed with a nesting structure that captures hierarchically structured segments of the word. Applications of nested words abound in computer science— terms and expressions are naturally nested (the bracketing capturing the nesting), XML/HTML/SGML documents are nested words capturing hierarchically structured data elements (the open and close tags capture the nesting), and even runs of recursive sequential programs can be seen as nested words capturing the nested calling structure (the call to and return from procedures capture the nesting).

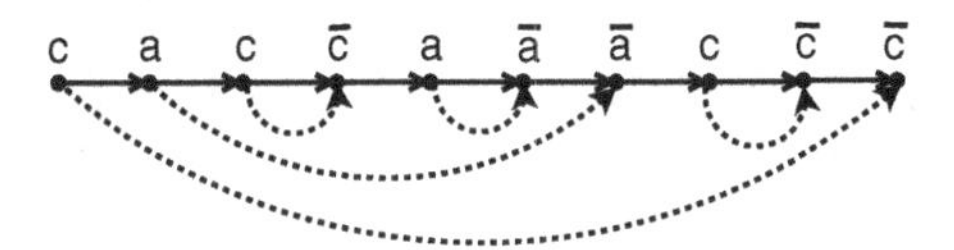

Fig. 1. A nested word

Trees have been the traditional approach to model nested structures. The rich results in the automata theory of trees is a robust theory that captures tractable representations of nested structures. Nested words are an alternative way to describe nested structures, where the linear arrangement of data is emphasized (as is common in a document representing this data, like an XML document). Automata on nested words process a document (word) along this linear order, but also exploit the nesting edges.

The systematic study of nested word structures and finite automata working on them was first done using *visibly pushdown automata* (where the automaton

R. Královič and D. Niwiński (Eds.): MFCS 2009, LNCS 5734, pp. 561–573, 2009.

processes the word left to right, and uses a stack to relay information flow along the nesting edges) [1]. An alternative and mathematically equivalent model is that of nested word automata [2], which are *finite state automata* (no stack) and process a nested word linearly, but where the automata are additionally allowed to refer to nested-edge-predecessors in order to update the state.

Visibly pushdown automata and nested word automata were first introduced in the context of formal verification (since runs of recursive programs are nested words). Since its introduction in 2004, this model has become quite popular, and a rich theory of visibly pushdown languages has been developed, ranging from applications to model-checking, monitoring, temporal logics, programming languages, security, XML, and complexity theory[1].

In this paper, we study the power of visibly pushdown automata in expressing and answering *queries* on nested words. We introduce an automaton model for defining queries, called query visibly pushdown automata (query VPA). A query VPA is a *deterministic* two-way (can move left and right) visibly pushdown automaton that can mark positions of a word. When moving right, a query VPA pushes onto the stack when reading open-tags and pops from the stack reading close-tags, like a visibly pushdown automaton. However, when moving left it does the opposite: it pushes onto the stack reading close-tags and pops from the stack reading open-tags. Hence it preserves the invariant that the height of the stack when at a particular position in the word is precisely the number of unmatched calls before that position (which is the number of unmatched returns after that position).

A query VPA runs (deterministically) on a word and *marks* a set of positions; these positions are to be seen as the set of all answers to a query. Note that a query VPA, when given a word, gives all answers to the query in a *single* run.

Our main result is that query VPAs have the right expressive power for defining queries: we show that a query is expressible as a unary formula in monadic second-order logic (MSO) if and only if it is implemented by a query VPA. Both directions are non-trivial, and the direction of implementing any MSO query using query VPAs relies on a beautiful observation by Hopcroft and Ullman [4].

We find it remarkable that the simple definition of two-way VPAs exactly captures all monadic queries. This result actually parallels a classic result in the theory of automata on finite *non-nested* words, which equates the power of unary monadic queries on (non-nested) words to that of two-way automata on words [4,10].

A query VPA, when viewed as an algorithm working on a word, traverses the word back and forth, and outputs *all* positions that answer the query, using only space $O(d)$, where d is the depth of the word (the depth of nesting in the word). Our result hence implies that any unary MSO query can be answered on a word of length n using only $O(d)$ space (with *no* dependence on n) and in time $O(n^2)$. As far as we know (and also based on discussions [11]), there is no such parallel result using the theory of tree automata (it is however well-known that given a

[1] See the VPL/Nested word languages page at `http://www.cs.uiuc.edu/~madhu/vpa` for a comprehensive list of papers and references.

particular position, checking whether it is an answer to a query can be done in $O(d)$-space; see also the related work below).

Related Work. Theoretical query models for XML have been studied using tree models. While most tree automata models work by passing states along the edges of the tree, there are tree automaton models that work in the linear order corresponding to its word representation (see [3] and the more recent [6]).

The most closely related work to our result on query VPAs is that of the query automaton model on unranked trees [10]. Query automata (more precisely, S2DTA^u) on unranked trees select positions in a tree and is exactly as powerful as unary monadic queries on trees [10] (which is the same as that on nested words). This model works like a two-way tree automaton on unranked trees, has parallel copies of automata that go up and down the tree, with automata processing children of a node synchronizing when going up to their parent. However, they are complex due to a special class of transitions (called *stay transitions*) that rewrite, using regular relations, the states labeling the children of a node. Further, there is a *semantic* restriction that requires that that the children of any node be processed by a stay transition at most once.

We believe that query VPA are significantly simpler and a different model than S2DTA^u, and it is not easy to convert between the two models (in either direction). Note that we also do not have any semantic restriction of the kind imposed on stay transitions in our setting, which is an important difference between the two models.

In [5], *selecting tree-automata* are defined, which are simpler than query automata and can return positions that satisfy any MSO query. However, these automata are *nondeterministic* in nature, and thus fundamentally different from our model.

Another line of work that is related is the work of Neumann and Seidl: in [9] (see also [8]), it was shown that a single-pass from left-to-right is sufficient to answer all queries that pick an element by referring only to properties of the document that occur before the element; these queries do not handle future predicates and hence can work in one pass using only $O(d)$ space.

Yet another work that is relevant is that reported in Neumann's thesis ([8], Chapter 7), where it is shown that for any query, there is a pushdown automaton that does two passes on a document, the first pass left-to-right, and the second pass right-to-left, such that any node being an answer to a query is determined solely by the states of the automaton on the two passes. Note that this is quite different from ours; an algorithm implementing this scheme would have to store states of the automaton at *all positions* in the first pass, and hence will require $O(n)$-space to output all answers to a query.

2 Preliminaries

For simplicity of exposition, we will assume that every letter in the word is the source or target of a nested edge; extending our results to the general class of nested words is straightforward. Nested words will be modeled as words over an alphabet where a single letter of the alphabet encodes an open/close tag.

Let Σ be a fixed finite alphabet of "open tags", and let $\overline{\Sigma} = \{\overline{c} \mid c \in \Sigma\}$ be the corresponding alphabet of "close tags". Let $\widehat{\Sigma} = \Sigma \cup \overline{\Sigma}$. A *well-matched* word is any word generated by the grammar: $W \to cW\overline{c}$, $W \to WW$, $W \to \epsilon$, where we have a rule $W \to cW\overline{c}$ for every $c \in \Sigma$. The set of all well-matched words over $\widehat{\Sigma}$ will be denoted by $WM(\widehat{\Sigma})$.

Nested Words, Monadic Second-Order Logic

A well-matched word $w \in WM(\widehat{\Sigma})$ can be seen as a nested structure: a linear labeled structure with nesting edges. For example, the structure corresponding to the word $cac\overline{c}a\overline{a}\overline{a}c\overline{c}\overline{c}$ is shown in Figure 1. The linear skeleton (denoted by solid edges) encodes the word and the nesting edges (denoted by dotted edges) relate open-tags with their matching close-tags. We skip the formal definition, but denote the nested structure associated with a word w as $nw(w) = (\{1, \ldots, |w|\}, \{Q_a\}_{a \in \widehat{\Sigma}}, \leq, \nu)$, where the universe is the set of positions in w, each Q_a is a unary predicate that is true at the positions labeled a, the $\leq$ relation encodes the linear order of the word, and ν is a binary relation encoding the nesting edges.

Monadic second-order logic (MSO_ν) over nested structures is defined in the standard manner, with interpreted relations $\leq$ and ν: Formally, fix a countable set of first-order variables FV and a countable set of monadic second-order (set) variables SV. Then the syntax of MSO formulas over $\widehat{\Sigma}$ labeled nested structures is defined as:

$$\begin{aligned} \varphi ::- & \; x \in X \mid Q_i(x) \mid x \leq y \mid \nu(x,y) \mid \varphi \vee \varphi \mid \neg\varphi \mid \\ & \exists x(\varphi) \mid \exists X(\varphi) \\ & \textit{where } x, y \in FV, X \in SV. \end{aligned}$$

Automata on Nested Words

There are two definitions of automata on nested words which are roughly equivalent: *nested word automata* and *visibly pushdown automata*. In this paper, we prefer the latter formalism. Intuitively, a visibly pushdown automaton is a pushdown automaton that reads a nested word left to right, and pushes a symbol onto the stack when reading open-tags and pops a symbol from the stack when reading a closed tag. Note that a symbol pushed at an open tag is popped at the matching closed tag. Formally,

Definition 1 (VPA). *A visibly pushdown automaton (*VPA*) over $(\Sigma, \overline{\Sigma})$ is a tuple $A = (Q, q_0, \Gamma, \delta, F)$, where Q is a finite set of states, $q_0 \in Q$ is the initial state, $F \subseteq Q$ is the set of final states, Γ is a finite stack alphabet, and $\delta = \langle \delta^{open}, \delta^{close} \rangle$ is the transition relation, where:*

– $\delta^{open} \subseteq ((Q \times \Sigma) \times (Q \times \Gamma))$;

– $\delta^{close} \subseteq ((Q \times \overline{\Sigma} \times \Gamma) \times Q)$.

A transition $(q, c, q', \gamma) \in \delta^{open}$ (denoted $q \xrightarrow{c/\gamma} q'$) is a push-transition, where the automaton reading c changes state from q to q', pushing γ onto the stack. Similarly, a transition $(q, \overline{c}, \gamma, q')$ (denoted $q \xrightarrow{\overline{c}/\gamma} q'$) is a pop-transition, allowing the automaton, when in state q reading $\overline{c}$ with γ on the top of the stack, to pop γ off the stack and change state to q'. A *configuration* of a VPA A is a pair $(q, s) \in Q \times \Gamma^*$. If $a \in \widehat{\Sigma}$, we say that $(q_1, s_1) \xrightarrow{a}_A (q_2, s_2)$ if one of the following conditions are true:

– $a = c \in \Sigma$, $s_2 = \gamma.s_1$ and $(q_1, c, q_2, \gamma) \in \delta^{open}$, or
– $a = \overline{c} \in \overline{\Sigma}$, $s_1 = \gamma.s_2$ and $(q_1, \overline{c}, \gamma, q_2) \in \delta^{close}$.

Note that the height of the stack after reading a prefix u of a well-matched word w is precisely the number of unmatched calls in u. We extend the definition of $\xrightarrow{a}_A$ to words over $\widehat{\Sigma}^*$ in the natural manner. The language $L(A)$ accepted by VPA A is the set of words $w \in \widehat{\Sigma}^*$ such that $(q_0, \epsilon) \xrightarrow{w}_A (q, \epsilon)$ for some $q \in Q^F$. One important observation about VPAs, made in [1], is that deterministic VPAs are as expressive as non-deterministic VPAs (defined above). Finally, a language L of well-matched words is called a *visibly pushdown language* (VPL) if there some VPA A such that $L = L(A)$.

Monadic Queries and Automata

A (unary) *query* is a function $f : WM(\widehat{\Sigma}) \to 2^{\mathbb{N}}$ such that for every $w \in WM(\widehat{\Sigma})$, $f(w) \subseteq [|w|]$. In other words, a query is a function that maps any well-matched word to a set of positions in the word.

A *unary monadic query* is a formula $\varphi(x_0)$ in MSO_ν that has precisely one free variable, the first-order variable x_0. Such a formula defines a query f_φ: for any word w, $f_\varphi(w)$ is the set of positions i such that the nested structure corresponding to w satisfies $\varphi(x_0)$ when x_0 is interpreted to be the i'th position. We will say query f is *expressible in* MSO_ν if there is a unary monadic query $\varphi(x_0)$ such that $f = f_\varphi$. We will consider unary monadic queries as the standard way to specify queries on nested words in this paper.

Any query f over $\widehat{\Sigma}$-labeled nested structures can be *encoded* as a language of well-matched words over a modified alphabet. If $\widehat{\Sigma} = (\Sigma, \overline{\Sigma})$, then let $\widehat{\Sigma}' = (\Sigma', \overline{\Sigma'})$ where $\Sigma' = \Sigma \cup (\Sigma \times \{*\})$. A *starred-word* is a well-matched word of the form $u(a, *)v$ where $u, v \in \widehat{\Sigma}^*$, i.e. it is a well matched over $\widehat{\Sigma}$ where precisely one letter has been annotated with a $*$.

A query f then corresponds to a set of starred-words:
$L_*(f) = \{a_1 a_2 \ldots a_{i-1}(a_i, *)a_{i+1} \ldots a_n \mid i \in f(a_1 \ldots a_{i-1} a_i a_{i+1} \ldots a_n)$ *and each* $a_j \in \widehat{\Sigma}\}$. Intuitively, $L_*(f)$ contains the set of all words w where a single position of w is marked with a $*$, and this position is an answer to the query f on w. We refer to $L_*(f)$ as the *starred-language* of f. It is easy to see that the above is a 1-1 correspondence between unary queries and starred-languages.

From results on visibly pushdown automata, in particular the equivalence of MSO_ν formulas and visibly pushdown automata [1], the following lemma follows:

Theorem 1. *A query f is expressible in MSO_ν iff $L_*(f)$ is a visibly pushdown language.*

Hence we can view unary monadic queries as simply visibly pushdown starred-languages, which will help in many proofs in this paper.

The main result of this paper is as follows. We define the automaton model of *query VPA* over nested words, which is a two-way visibly pushdown automaton that answers unary queries by marking positions in an input word. We show that a unary query is expressible in MSO_ν iff it is computed by some query VPA. Notice that this result is very different from Theorem 1; the query VPA is a machine that marks *all* positions that are answers to a query, as opposed to a VPA that can check if a *single* marked position is an answer to a query.

3 Query VPA

The goal of this section is to define an automaton model for nested words called a *query VPA*. A query VPA is a pushdown automaton that can move both left and right over the input string and store information in a stack. The crucial property that ensures tractability of this model is that the stack height of such a machine is pre-determined at any position in the word. More precisely, any query VPA P working over a well-matched input w has the property that, for any partition of w into two strings u and v (i.e., $w = uv$), the stack height of P at the interface of u and v is the same as the number of unmatched open-tags in u (which is the same as the number of unmatched close-tag in v). In order to ensure this invariant, we define the two-way VPA as one that pushes on open-tags and pops on close-tags while moving right, *but pushes on close-tags and pops on open-tags while moving left.* Finally, in addition to the ability to move both left and right over the input, the query VPA can *mark* some positions by entering special *marking states*; intuitively, the positions in a word that are marked will be answers to the unary query that the automaton computes.

We will now define query VPA formally. We will assume that there is a left-end-marker $\triangleright$ and a right-end-marker $\triangleleft$ for the input to ensure that the automaton doesn't fall off its ends; clearly, $\triangleright, \triangleleft \notin \widehat{\Sigma}$.

Definition 2 (Query VPA). *A query VPA (QVPA) over $(\Sigma, \overline{\Sigma})$ is a tuple $P = (Q, q_0, \Gamma, \delta, Q_*, S, C)$, where Q is a finite set of states, $q_0 \in Q$ is the initial state, $Q_* \subseteq Q$ is a set of marking states, Γ is a finite stack alphabet, (S, C) is a partition of $Q \times \widehat{\Sigma}$, and $\delta = \langle \delta^{open}, \delta^{close}, \delta^{chng} \rangle$ is the transition relation, where:*

- $\delta^{open}_{right} : S \cap (Q \times \Sigma) \to (Q \times \Gamma)$
- $\delta^{open}_{left} : (S \cap (Q \times \Sigma)) \times \Gamma \to Q$

- $\delta^{close}_{right} : (S \cap (Q \times \overline{\Sigma})) \times \Gamma) \rightarrow Q$
- $\delta^{close}_{left} : S \cap (Q \times \overline{\Sigma}) \rightarrow (Q \times \Gamma)$
- $\delta^{chng} : C \cup (Q \times \{\triangleright, \triangleleft\}) \rightarrow Q$

The δ^{open}_{right} and δ^{close}_{left} functions encode push-transitions of the automaton reading an open-tag and moving right, and reading a close-tag and moving left, respectively. The δ^{open}_{left} and δ^{close}_{right} functions encode pop-transitions when the automaton reads an open-tag and moves left, and reads a close-tag and moves right, respectively. On the other hand, the δ^{chng} function encodes transitions where the automaton changes the direction of its head movement. Observe that we force the automaton to change direction whenever it reads either $\triangleright$ or $\triangleleft$. Note that the query VPA has no final states. Finally, the definition above describes a deterministic model, which we will focus on in this paper. The non-deterministic version of the above automaton can also be defined, but they do not increase the expressive power.

We will now define the execution of a query VPA on a word $x = \triangleright w \triangleleft$. A *configuration* of the query VPA is a tuple $\langle p, d, q, s \rangle$, where $p \in [|x|]$ is the position in the input currently being scanned, $d \in \{\mathit{left}, \mathit{right}\}$ is the direction in which the head moving currently, $q \in Q$ is the current state, and $s \in \Gamma^*$ is the current stack contents. The *initial configuration* is $\langle 1, \mathit{right}, q_0, \epsilon \rangle$, i.e., initially the automaton is reading the leftmost symbol of w (not $\triangleright$), it is moving right, in the initial state, with an empty stack. A *run* is a sequence $c_0, c_1, \ldots c_n$, where c_0 is the initial configuration, and for each i, if $c_i = \langle p_i, d_i, q_i, s_i \rangle$ and $c_{i+1} = \langle p_{i+1}, d_{i+1}, q_{i+1}, s_{i+1} \rangle$, then one of the following holds

- If $(q_i, x[p_i]) \in S \cap (Q \times \Sigma)$, and $d_i = \mathit{right}$ then $d_{i+1} = d_i$, $p_{i+1} = p_i + 1$, $q_{i+1} = q$ and $s_{i+1} = \gamma s_i$, where $\delta^{open}_{right}(q_i, x[p_i]) = (q, \gamma)$
- If $(q_i, x[p_i]) \in S \cap (Q \times \overline{\Sigma})$, and $d_i = \mathit{right}$ then $d_{i+1} = d_i$, $p_{i+1} = p_i + 1$, $q_{i+1} = q$ and $s_i = \gamma s_{i+1}$, where $\delta^{close}_{right}(q_i, x[p_i], \gamma) = q$
- If $(q_i, x[p_i]) \in S \cap (Q \times \Sigma)$, and $d_i = \mathit{left}$ then $d_{i+1} = d_i$, $p_{i+1} = p_i - 1$, $q_{i+1} = q$ and $s_i = \gamma s_{i+1}$, where $\delta^{open}_{left}(q_i, x[p_i], \gamma) = q$
- If $(q_i, x[p_i]) \in S \cap (Q \times \overline{\Sigma})$, and $d_i = \mathit{left}$ then $d_{i+1} = d_i$, $p_{i+1} = p_i - 1$, $q_{i+1} = q$ and $s_{i+1} = \gamma s_i$, where $\delta^{close}_{left}(q_i, x[p_i]) = (q, \gamma)$
- If $(q_i, x[p_i]) \in C$ then $s_i = s_{i+1}$, and $q_{i+1} = \delta^{chng}(q_i, x[p_i])$. To define the new position and direction, there are two cases to consider. If $d_i = \mathit{right}$ then $d_{i+1} = \mathit{left}$ and $p_{i+1} = p_i - 1$. On the other hand, if $d_i = \mathit{left}$ then $d_{i+1} = \mathit{right}$ and $p_{i+1} = p_i + 1$.

Observe that the way the run is defined, the stack height in any configuration is determined by the word w. More precisely, in any configuration $c = \langle p, d, q, s \rangle$ of the run with $p \in \{1, \ldots |w|\}$, if $d = \mathit{right}$ then $|s|$ is equal to the number of unmatched open-tags in the word $x[1] \cdots x[p-1]$ (which is the same as the number of unmatched close-tags in $x[p] \cdots x[|w|]$). On the other hand, if $d = \mathit{left}$ then $|s|$ is equal to the number of unmatched open-tags in the word $x[1] \cdots x[p]$. When scanning the left-end-marker ($p = 0$) or the right-end-marker ($p = |x|+1$), the stack height is always 0.

Finally, the query VPA P is said to *mark* a position j in a well-matched word w, where $1 \leq j \leq |w|$, if the unique run $c_0 \ldots c_n$ of P on the input $\rhd w \lhd$ is such that for some i, $c_i = (j, d, q, s)$ where $q \in Q_*$. The query *implemented by* the query VPA P is the function f_P, where $f_P(w)$ is the set of all positions marked by P when executed on $\rhd w \lhd$.

We now state the main result of this paper: the set of queries implemented by query VPA is precisely the set of unary monadic queries.

Theorem 2. *A query f is expressible in MSO_ν if and only if there is query VPA P such that $f_P = f$.*

The proof of Theorem 2 follows from Lemmas 1 and 2 that are proved in the next two sections.

3.1 Implementing Monadic Queries on Query VPA

In this section we prove one direction of Theorem 2, namely, that every monadic query can be implemented on a query VPA.

Lemma 1. *For any monadic query f, there is a query VPA P such that $f = f_P$.*

Proof. Let f be a monadic unary query. From Theorem 1, we know that there is a deterministic VPA A such that $L_*(f) = L(A)$. This suggests a very simple algorithm that will mark all the answers to query f by repeatedly simulating A on the word w. First the algorithm will simulate the VPA A on the word w, assuming that the starred position is the rightmost symbol of w; the algorithm marks position $|w|$ of $\rhd w \lhd$ only if A accepts. Then the algorithm simulates A assuming that the starred position is $|w| - 2$, and so on, each time marking a position if the run of A on the appropriate starred word is accepting. A naïve implementation of this algorithm will require maintaining the starred position, as well as the current position in the word that the simulation of A is reading, and the ability to update these things. It is unclear how this additional information can be maintained by a query VPA that is constrained to update its stack according to whether it is reading an open-tag or a close-tag. The crux of the proof of this direction is demonstrating that this can indeed be accomplished. While we draw on ideas used in a similar proof for queries on regular word languages (see [10] for a recent exposition), the construction is more involved due to the presence of a stack.

Before giving more details about the construction, we will give two technical constructions involving VPAs. First given any VPA $B = (Q, q_0, \Gamma, \delta, F)$ there is a VPA B' with a canonical stack alphabet that recognizes the same language; the VPA $B' = (Q, q_0, Q \times \Sigma, \delta', F)$ which pushes (q, c) whenever it reads an open-tag c in state q. Details of this construction can be found in [1]. Next, given any VPA $B = (Q, q_0, \Gamma, \delta, F)$, there is a VPA B^{pop} recognizing the same language, which remembers *the symbol last popped since the last unmatched open-tag* in its control state. We can construct this: $B^{pop} = (Q \times (\Gamma \cup \{\bot\}), (q_0, \bot), \Gamma, \delta', F \times (\Gamma \cup \{\bot\}))$, where the new transitions are as follows. If $q \xrightarrow{c/\gamma_1}_B q'$ then $(q,\gamma) \xrightarrow{c/\gamma_1}_{B^{pop}} (q',\bot)$. If $q \xrightarrow{\overline{c}/\gamma_1}_B q'$ then $(q,\gamma) \xrightarrow{\overline{c}/\gamma_1} (q',\gamma_1)$.

For the rest of this proof, let us fix A to be the deterministic VPA with a canonical stack alphabet recognizing $L_*(f)$, and A^{pop} to be the (deterministic) version of A that remembers the last popped symbol in the control state. We will now describe the query VPA P for f. Let us fix the input to be $\triangleright w \triangleleft$, where $w = a_1 a_2 \cdots a_n$.

P will proceed by checking for each i, i starting from n and decremented in each phase till it becomes 1, whether position i is an answer to the query. To do this, it must check if A accepts the starred word where the star is in position i. P will achieve this by maintaining an invariant, which we describe below.

The Invariant. Let $w = a_1 \dots a_n$, and consider a position i in w. Let $w = u a_i v$ where $u = a_1 \dots a_{i-1}$ and $v = a_{i+1} \dots a_n$.

Recall that the suffix from position $i+1$, v, can be uniquely written as $w_k \overline{c_k} w_{k-1} \overline{c_{k-1}} \cdots w_1 \overline{c_1} w_0$, where for each j, w_j is a well-matched word, and $\overline{c_k}, \dots \overline{c_1}$ are the unmatched close-tags in v.

In phase i, the query VPA will navigate to position i with stack σ such that (a) its control has information of a pair (q,γ) such that this state with stack σ is the configuration reached by A^{pop} on reading the prefix u, and (b) its control has the set B of states of A that is the precise set of states q' such that A accepts v from the configuration (q',σ). Hence the automaton has a summary of the way A would have processed the unstarred prefix up to (but not including) position i and a summary of how the unstarred suffix from position $i+1$ would be processed. Under these circumstances, the query VPA can very easily decide whether position i is an answer to the query — if A on reading $(a_i,*)$ can go from state q to some state in B, then position i must be marked.

Technically, in order to ensure that this invariant can be maintained, the query VPA needs to maintain more information: a set of pairs of states of A, S, that summarizes how A would behave on the first well-matched word in v (i.e. w_k), a stack symbol *StackSym* that accounts for the difference in stack heights, and several components of these in the stack to maintain the computation. We skip these technical details here.

Marking position i. If $a_i \in \Sigma$ then the query VPA P will mark position i iff $q \xrightarrow{(a_i,*)/(q,(a_i,*))}_A q'$ where $q' \in B$. Similarly, if $a_i \in \overline{\Sigma}$ then P marks i iff $q \xrightarrow{(a_i,*)/\gamma'}_A q'$ with $q' \in B$.

Maintaining the Invariant. Initially the query VPA P will simulate the VPA A^{pop} on the word w from left to right. Doing this will allow it to obtain the invariant for position n. So what is left is to describe how the invariant for position $i-1$ can be obtained from the invariant for position i. Determining the components of the invariant, except for the new control state of A^{pop}, are easy and follow from the definitions; the details are deferred to the appendix. Computing the new state of A^{pop} at position $i-1$ is interesting, and we describe this below.

Determining the state of A^{pop}. Recall that we know the state of A^{pop} after reading $a_1 \cdots a_{i-1}$, which is (q, γ). We need to compute the state (q', γ') of A^{pop} after reading $a_1 \cdots a_{i-2}$. The general idea is as follows. The query VPA P will simulate A^{pop} backwards on symbol a_{i-1}, i.e., it will compute $Prev = \{p \mid p \xrightarrow{a_{i-1}}_{A^{pop}} (q, \gamma)$. If $|Prev| = 1$ then we are done. On the other hand, suppose $|Prev| = k > 1$. In this case, P will continue simulating A^{pop} backwards on symbols a_{i-2}, a_{i-3} and so on, while maintaining for each state $p \in Prev$ the set of states of A^{pop} that reach p. If at some position j the sets associated with all states $p \in Prev$ become empty except one, or $j = 1$ (we reach the beginning of the word), then we know which state $p \in Prev$ is state of A^{pop} after reading $a_1 \cdots a_{i-2}$ — it is either the unique state whose set is non-empty or it is state whose set includes the initial state. However P now needs to get back to position $i - 1$. This is done by observing the following. We know at position $j + 1$ at least two different threads of the backward simulation are still alive. Position $i - 1$ is the unique position where these two threads converge if we now simulate A^{pop} forwards. The idea just outlined works for queries on regular word languages, but does not work for query VPA due to one problem. If we go too far back in the backwards simulation (like the beginning of the word), we will lose all the information stored in the stack. Therefore, we must ensure that the backward simulation does not result in the stack height being lower than what it is supposed after reading $a_1 \cdots a_{i-2}$. To do this we use the special properties of the VPA A^{pop}. Observe that if we go backwards on an unmatched open-tag (in $a_1 \cdots a_{i-1}$), the state of the VPA A at the time the open-tag was read is on the stack. Thus, the state of A^{pop} *after* reading the open-tag is uniquely known. Next if $a_{i-1} \in \overline{\Sigma}$ is a matched close-tag, then since we keep track of the last symbol popped after reading a_{i-1}, we know the symbol that was popped when a_{i-1} was read, which allows us to know the state of A, when it read the open-tag that matches a_{i-1}. These two observations ensure that we never pop symbols out of the stack. The details are as follows.

$a_{i-1} \in \Sigma$: Simulate backwards until (in the worst case) the rightmost unmatched open-tag symbol in the word $a_1 \ldots a_{i-2}$, and then simulate forward to determine the state (q', γ') as described above.

$a_{i-1} \in \overline{\Sigma}$: γ is symbol that is popped by A^{pop} when it reads a_{i-1}. So γ encodes the state of A when the matching open-tag a_j to a_{i-1} was read. So simulate backwards until a_j is encountered and then simulate forwards.

This completes the description of the query VPA.

3.2 Translating Query VPA to Monadic Queries

We now complete the proof of Theorem 2, by showing that any query implemented on a query VPA can be described as a unary monadic query.

Lemma 2. *For any query VPA A, there is an MSO_ν formula $\varphi(x)$ such that $f_\varphi = f_A$.*

Proof. Let A be a query VPA. The query defined by A will be translated into an MSO_ν formula through several intermediate stages.

Let f be the query defined by the query VPA A. We first construct a two-way (non-marking) VPA B that accepts the starred-language of f. B accepts a word w with a $*$ in position i if and only if $i \in f(w)$. Constructing B is easy. B simulates A on a word w with a $*$ in position i and accepts the word if A reaches position i in a marking state. B also ensures in a first run over the word that the word has a unique position that is marked with a $*$. The language accepted by B is $L_*(f)$, the starred-language of f.

Any nested word w can be represented as a tree called a *stack tree*. A stack tree is a $\hat{\Sigma}$ binary tree that has one node for every position in w, and the node corresponding to position i is labeled by $w[i]$. The stack tree corresponding to a word w is defined inductively as follows: (a) if $w = \epsilon$, then the stack tree of w is the empty tree, and (b) if $w = cw_1\bar{c}w_2$, then the stack tree corresponding to w has its root labeled c, has the stack-tree corresponding to w_1 rooted at its left child, the right child is labeled $\bar{c}$ which has no left child, but has a right child which has the stack-tree corresponding to w_2 rooted at it.

We now show that the set of stack-trees corresponding the starred words accepted by B can be accepted using a *pushdown tree-walking automaton* [7]. A pushdown tree-walking automaton works on a tree by starting at the root and walking up and down the tree, and has access to a stack onto which it always pushes a symbol when going down the tree and pops the stack when coming up an edge. Note that the height of the stack when at a node of the tree is hence always the depth of the node from the root. From B, we can build a tree-walking automaton C that reads the tree corresponding to a starred word, and simulates B on it. C can navigate the tree and effectively simulate moving left or right on the word. When B moves right reading a call symbol, C moves to the left child of the call and pushes the symbol B pushes onto its stack. When B moves right to read the corresponding return, C would go up from the left subtree to this call and pop the symbol from the stack and use it to simulate the move on the return. The backward moves on the word that B makes can also be simulated: for example, when B reads a return and moves left, C would go to the corresponding node on the left-subtree of the call node corresponding to the return, and when doing so push the appropriate symbol that B pushed onto the stack. When C moves down from an internal or return node, or from a call node to the right, it pushes in a dummy symbol onto the stack.In summary, whenever B is in a position i with stack $\gamma_1 \dots \gamma_k$, C would be reading the node corresponding to i in the tree, and the stack would have $\gamma_1 \dots \gamma_k$ when restricted to non-dummy symbols.

It is known that pushdown tree-walking automata precisely accept regular tree languages. Hence we can construct an MSO formula on trees that precisely is true on all trees that correspond to starred words accepted by B. This MSO formula can be translated to MSO_ν ψ on nested words, which is true on precisely the set of starred nested words that B accepts. Assuming x is not a variable in ψ, we replace every atomic formula of the form $Q_{(a,*)}(y)$ (the atomic formula checking whether position y is labeled a and is starred) by the formula $x = y \wedge Q_a(y)$, to get a formula $\varphi(x)$, with a free variable x. Intuitively, we replace every check the formula does for a starred label by a check as to whether that position is x. It is easy to see then that the formula $\varphi(x)$ is an MSO_ν formula on $\widehat{\Sigma}$-labeled (unstarred) nested words, which precisely defines the query defined by B, and hence the original query VPA A. This concludes the proof.

4 Conclusions

The query automaton model we have defined on nested words is an elegant model that answers queries using the least space possible. While our result is theoretical in nature, it may have implications on applications: our model shows that unary queries on XML (like logical XPath queries) can be answered using only $O(d)$-space; we also believe that our model could have applications in verification where, given a run of a sequential program, we can build efficient algorithms that answer queries such as "which positions x of the run satisfy a temporal formula $\varphi(x)$?", with applications to debugging error traces.

Finally, our query automaton model can be adapted to an analogous MSO-complete unary query automata on unranked trees as well: we can define a 2-way pushdown automaton tree-walking automaton that processes it by traversing it according to the linear order (determined by its serialization as a word), pushing onto the stack when going down a tree and popping the stack when coming up; this will essentially be an encoding of the query VPA on the tree.

References

1. Alur, R., Madhusudan, P.: Visibly pushdown languages. In: STOC, pp. 202–211. ACM Press, New York (2004)
2. Alur, R., Madhusudan, P.: Adding nesting structure to words. In: Ibarra, O.H., Dang, Z. (eds.) DLT 2006. LNCS, vol. 4036, pp. 1–13. Springer, Heidelberg (2006)
3. Carme, J., Niehren, J., Tommasi, M.: Querying unranked trees with stepwise tree automata. In: van Oostrom, V. (ed.) RTA 2004. LNCS, vol. 3091, pp. 105–118. Springer, Heidelberg (2004)
4. Engelfriet, J., Hoogeboom, H.J.: Mso definable string transductions and two-way finite-state transducers. ACM Trans. Comput. Logic 2(2), 216–254 (2001)
5. Frick, M., Grohe, M., Koch, C.: Query evaluation on compressed trees (extended abstract). In: LICS, p. 188. IEEE Computer Society, Los Alamitos (2003)
6. Gauwin, O., Niehren, J., Roos, Y.: Streaming tree automata. Inf. Process. Lett. 109(1), 13–17 (2008)
7. Kamimura, T., Slutzki, G.: Parallel and two-way automata on directed ordered acyclic graphs. Information and Control 49(1), 10–51 (1981)
8. Neumann, A.: Parsing and querying xml documents in sml. Universität Trier, Trier, Germany (1999)

9. Neumann, A., Seidl, H.: Locating matches of tree patterns in forests. In: Arvind, V., Sarukkai, S. (eds.) FST TCS 1998. LNCS, vol. 1530, pp. 134–146. Springer, Heidelberg (1998)
10. Neven, F., Schwentick, T.: Query Automata over Finite Trees. Theoretical Computer Science 275(1-2), 633–674 (2002)
11. Schwentick, T.: Personal communication (2008)

A General Class of Models of $\mathcal{H}^*$

Giulio Manzonetto*

INRIA
giulio.manzonetto@inria.fr

Abstract. We provide sufficient conditions for categorical models living in arbitrary cpo-enriched cartesian closed categories to have the maximal consistent sensible λ-theory as their equational theory. Finally, we prove that a model of pure λ-calculus we have recently introduced in a cartesian closed category of sets and (multi-)relations fulfils these conditions.

Keywords: λ-calculus, λ-theories, non-well-pointed categories, cpo-enriched categories, relational semantics, Approximation Theorem.

Introduction

The first model of λ-calculus, namely $\mathscr{D}_\infty$, was postulated by Scott in 1969 in the category of complete lattices and continuous functions. After Scott's $\mathscr{D}_\infty$, a large number of models have been introduced in various categories of domains. For example, the *continuous semantics* [14] is given in the cartesian closed category (ccc, for short) whose objects are complete partial orders and morphisms are Scott continuous functions. The *stable semantics* [3] and the *strongly stable semantics* [5] are refinements of the continuous semantics which have been introduced to capture the notion of 'sequential' continuous function.

Although these semantics are very rich (in each of them it is possible to build up $2^{\aleph_0}$ models having pairwise distinct λ-theories) they are also hugely *incomplete*: there is a continuum of λ-theories that cannot be presented as equational theories of continuous, stable, or strongly stable models (see [13]). For this reason, researchers are today shifting their attention towards less canonical structures and categories [7,11,12]. This is also due to a widespread growing interest in two branches of computer science which are strongly related to the semantics of λ-calculus: *game semantics* and *linear logic*. The categories naturally arising in these fields are often *non-standard* since they can have morphisms which are not functions and/or they can be non-well-pointed.

At the moment, there is a lack of general methods for a uniform treatment of models living in non-standard semantics. For instance, the classic method for turning a categorical model into a λ-model asked for well-pointed categories [2, Sec. 5.5], whilst, in collaboration with Bucciarelli and Ehrhard, we have recently shown that such a requirement was unnecessary [6]. In the same paper we have also built an extensional model $\mathscr{D}$ of λ-calculus living in a (highly) non-well-pointed ccc of sets and (multi-)relations, which has been previously studied as a

* Work supported by ANR ParSec grant ANR-06-SETI-010-02 and CONCERTO.

R. Královič and D. Niwiński (Eds.): MFCS 2009, LNCS 5734, pp. 574–586, 2009.

semantic framework for linear logic [8,4]. We conjectured that $\mathscr{D}$ can be seen as a "relational version" of Scott's $\mathscr{D}_\infty$ and, hence, that its equational theory is the maximal consistent sensible λ-theory $\mathcal{H}^*$ (just like for $\mathscr{D}_\infty$). Unfortunately, the classic methods to characterize the equational theory of a model are not directly applicable to our model $\mathscr{D}$, since it lives in a non-well-pointed category.

In the present paper, we provide sufficient conditions for categorical models living in possibly non-well-pointed, but cpo-enriched, ccc's to have $\mathcal{H}^*$ as their equational theory. The idea of the proof is that we want to find a class of models (as large as possible) satisfying an Approximation Theorem. More precisely, we want to be able to characterize the interpretation of a λ-term M as the least upper bound of the interpretations of its approximants. These approximants are particular terms of an auxiliary calculus, due to Wadsworth [15], and called here the *labelled* $\lambda\bot$*-calculus*, which is strongly normalizable and Church-Rosser.

Then we define the "well stratifiable $\bot$-models", and we show that they model Wadsworth's calculus and satisfy the Approximation Theorem. As a consequence, we get that every well stratifiable $\bot$-model $\mathscr{U}$ equates all λ-terms having the same Böhm tree; in particular, $\mathscr{U}$ is sensible, i.e., it equates all unsolvable λ-terms. Finally we prove, under the additional hypothesis that $\mathscr{U}$ is extensional, that the theory of $\mathscr{U}$ is $\mathcal{H}^*$.

At the end of the paper, we show that our relational model $\mathscr{D}$ of [6] fulfils these conditions and we conclude that its equational theory is $\mathcal{H}^*$.

1 Preliminaries

To keep this article self-contained, we summarize some definitions and results. With regard to the λ-calculus we follow the notation and terminology of [2]. Our main reference for category theory is [1].

Cartesian Closed Categories. Let $\mathbf{C}$ be a *cartesian closed category* (ccc, for short). We denote by $A \times B$ the *product* of A and B, by $[A \Rightarrow B]$ the *exponential object* and by $ev \in \mathbf{C}([A \Rightarrow B] \times A, B)$ the *evaluation morphism*. For any C and $f \in \mathbf{C}(C \times A, B)$, $\Lambda(f) \in \mathbf{C}(C, [A \Rightarrow B])$ stands for the (unique) morphism such that $ev \circ (\Lambda(f) \times \mathrm{Id}) = f$. Finally, $\mathbb{1}$ denotes the terminal object and $!_A$ the only morphism in $\mathbf{C}(A, \mathbb{1})$. We recall that in a ccc the following equalities hold:

$$\text{(pair)} \quad \langle f, g\rangle \circ h = \langle f \circ h, g \circ h\rangle \qquad \Lambda(f) \circ g = \Lambda(f \circ (g \times \mathrm{Id})) \quad \text{(Curry)}$$
$$\text{(beta)} \quad ev \circ \langle \Lambda(f), g\rangle = f \circ \langle \mathrm{Id}, g\rangle \qquad \Lambda(ev) = \mathrm{Id} \quad \text{(Id-Curry)}$$

We say that $\mathbf{C}$ is *well-pointed* if, for all $f, g \in \mathbf{C}(A, B)$, whenever $f \neq g$, there exists a morphism $h \in \mathbf{C}(\mathbb{1}, A)$ such that $f \circ h \neq g \circ h$.

The ccc $\mathbf{C}$ is *cpo-enriched* if every homset is a cpo $(\mathbf{C}(A, B), \sqsubseteq_{(A,B)}, \bot_{(A,B)})$, composition is continuous, pairing and currying are monotonic, and the following strictness conditions hold: (l-strict) $\bot \circ f = \bot$, (ev-strict) $ev \circ \langle \bot, f\rangle = \bot$.

The λ-Calculus. Let Var be a countably infinite set of variables. The set Λ of λ*-terms* is inductively defined as usual: $x \in \Lambda$, for each $x \in \mathrm{Var}$; if $M, N \in \Lambda$ then $MN \in \Lambda$; if $M \in \Lambda$ then $\lambda x.M \in \Lambda$, for each $x \in \mathrm{Var}$.

Concerning specific λ-terms, we set $\mathbf{I} \equiv \lambda x.x$ and $\Omega \equiv (\lambda x.xx)(\lambda x.xx)$.

Given a reduction rule $\rightarrow_R$ we write $\twoheadrightarrow_R$ ($=_R$) for its transitive and reflexive (and symmetric) closure. A λ-term M is *solvable* if $M \twoheadrightarrow_\beta \lambda x_1 \ldots x_n.yN_1 \cdots N_k$ for some $x_1, \ldots, x_n \in \mathrm{Var}$, $N_1, \ldots, N_k \in \Lambda$ ($n, k \geq 0$); otherwise M is *unsolvable.*

A λ-*theory* is any congruence on Λ, containing $=_\beta$. A λ-theory is: *consistent* if it does not equate all λ-terms; *extensional* if it contains $=_\eta$; *sensible* if it equates all unsolvable λ-terms. The set of all λ-theories, ordered by inclusion, forms a complete lattice. We denote by $\mathcal{H}^*$ the greatest consistent sensible λ-theory.

The *Böhm tree* $\mathrm{BT}(M)$ of a λ-term M is defined as follows: if M is unsolvable, then $\mathrm{BT}(M) = \bot$, that is, $\mathrm{BT}(M)$ is a tree with a unique node labelled by $\bot$; if M is solvable, then $M \twoheadrightarrow_\beta \lambda x_1 \ldots x_n.yN_1 \cdots N_k$ (with $n, k \geq 0$) and:

$$\mathrm{BT}(M) = \begin{array}{ccccc} & & \lambda x_1 \ldots x_n.y & & \\ & \swarrow & & \searrow & \\ \mathrm{BT}(N_1) & & \cdots & & \mathrm{BT}(N_k) \end{array}$$

We call $\mathcal{B}$ the minimum λ-theory equating all λ-terms having the same Böhm tree. Given two Böhm trees t, t' we define $t \subseteq_{BT} t'$ if, and only if, t results from t' by replacing some subtrees with $\bot$. The relation $\subseteq_{BT}$ is transferred on λ-terms by setting $M \sqsubseteq_{BT} N$ if, and only if, $\mathrm{BT}(M) \subseteq_{BT} \mathrm{BT}(N)$.

We write $M \sqsubseteq_{\eta,\infty} N$ if $\mathrm{BT}(N)$ is a (possibly infinite) η-expansion of $\mathrm{BT}(M)$ (see [2, Def. 10.2.10(iii)]). For example, let us consider $J \equiv \Theta(\lambda jxy.x(jy))$, where Θ is Turing's fixpoint combinator [2, Def. 6.1.4]. Then $x \sqsubseteq_{\eta,\infty} Jx$, since

$$\begin{aligned} Jx &=_\beta \lambda z_0.x(Jz_0) =_\beta \lambda z_0.x(\lambda z_1.z_0(Jz_1)) \\ &=_\beta \lambda z_0.x(\lambda z_1.z_0(\lambda z_2.z_1(Jz_2))) =_\beta \ldots \end{aligned}$$

Using $\sqsubseteq_{\eta,\infty}$, we can define another relation on λ-terms which will be useful in Subsec. 2.6. For all $M, N \in \Lambda$ we set $M \precsim_\eta N$ if there exist M', N' such that $M \sqsubseteq_{\eta,\infty} M' \sqsubseteq_{BT} N' \sqsupseteq_{\eta,\infty} N$. Let us provide an example of this situation:

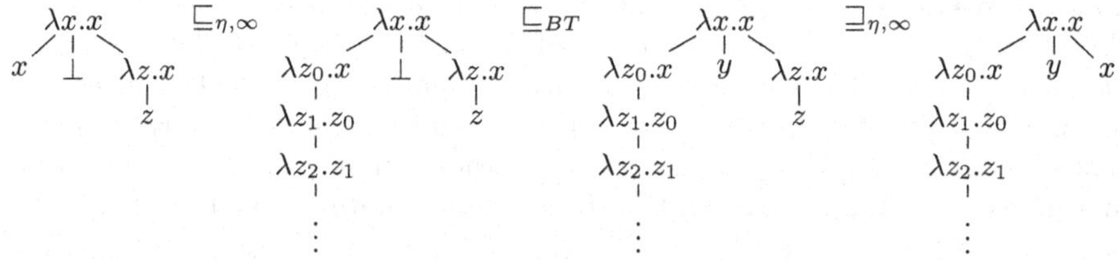

$$\begin{array}{ccccccc}
\begin{array}{ccc} & \lambda x.x & \\ x & \bot & \lambda z.x \\ & & | \\ & & z \end{array}
& \sqsubseteq_{\eta,\infty} &
\begin{array}{ccc} & \lambda x.x & \\ \lambda z_0.x & \bot & \lambda z.x \\ | & & | \\ \lambda z_1.z_0 & & z \\ | & & \\ \lambda z_2.z_1 & & \\ | & & \\ \vdots & & \end{array}
& \sqsubseteq_{BT} &
\begin{array}{ccc} & \lambda x.x & \\ \lambda z_0.x & y & \lambda z.x \\ | & & | \\ \lambda z_1.z_0 & & z \\ | & & \\ \lambda z_2.z_1 & & \\ | & & \\ \vdots & & \end{array}
& \sqsupseteq_{\eta,\infty} &
\begin{array}{ccc} & \lambda x.x & \\ \lambda z_0.x & y & x \\ | & & \\ \lambda z_1.z_0 & & \\ | & & \\ \lambda z_2.z_1 & & \\ | & & \\ \vdots & & \end{array}
\end{array}$$

Finally, we write $M \simeq_\eta N$ for $M \precsim_\eta N \precsim_\eta M$.

2 Well Stratifiable Categorical Models

The λ-theory $\mathcal{H}^*$ was first introduced by Hyland [10] and Wadsworth [15], who proved (independently) that the theory of $\mathscr{D}_\infty$ is $\mathcal{H}^*$. This proof has been extended by Gouy in [9] with the aim of showing that also the stable analogue of $\mathscr{D}_\infty$ had $\mathcal{H}^*$ as equational theory. Actually, his result is more powerful and covers many suitably stratifiable models living in "regular" ccc's (in the sense of [9]). However, all regular ccc's have (particular) cpo's as objects and (particular) continuous functions as morphisms, hence only concrete categories can be regular. Concerning models in non-well-pointed categories, Di Gianantonio et al. [7] provided a similar proof, but it only works for non-concrete *categories of games*.

In this section we provide sufficient conditions for models living in (possibly non-well-pointed) cpo-enriched ccc's to have $\mathcal{H}^*$ as equational theory.

2.1 A Uniform Interpretation of λ-Terms

A *model of λ-calculus* $\mathscr{U}$ is a reflexive object in a ccc $\mathbf{C}$, i.e., a triple $(U, \mathrm{Ap}, \lambda)$ such that U is an object of $\mathbf{C}$, and $\lambda \in \mathbf{C}([U \Rightarrow U], U)$ and $\mathrm{Ap} \in \mathbf{C}(U, [U \Rightarrow U])$ satisfy $\mathrm{Ap} \circ \lambda = \mathrm{Id}_{[U \Rightarrow U]}$. $\mathscr{U}$ is called *extensional* when moreover $\lambda \circ \mathrm{Ap} = \mathrm{Id}_U$.

A λ-term M is usually interpreted as a morphism $|M|_I \in \mathbf{C}(U^I, U)$ for some finite subset $I \subset \mathrm{Var}$ containing the free variables of M. The arbitrary choice of I is tedious to treat when dealing with the equalities induced by a model. Fortunately, when the underlying category has countable products, we are able to interpret all λ-terms in the homset $\mathbf{C}(U^{\mathrm{Var}}, U)$ just slightly modifying the usual definition of interpretation (see [2, Def. 5.5.3(vii)]). Indeed, given $M \in \Lambda$, we can define $|M|_{\mathrm{Var}} \in \mathbf{C}(U^{\mathrm{Var}}, U)$ by structural induction on M, as follows:

- $|x|_{\mathrm{Var}} = \pi_x^{\mathrm{Var}}$,
- $|NP|_{\mathrm{Var}} = |N|_{\mathrm{Var}} \bullet |P|_{\mathrm{Var}}$, where $a \bullet b = ev \circ \langle \mathrm{Ap} \circ a, b\rangle$,
- $|\lambda x.N|_{\mathrm{Var}} = \lambda \circ \Lambda(|N|_{\mathrm{Var}} \circ \eta_x)$, where $\eta_x = \Pi^{\mathrm{Var}}_{\mathrm{Var}-\{x\}} \times \mathrm{Id} \in \mathbf{C}(U^{\mathrm{Var}} \times U, U^{\mathrm{Var}})$.

It is easy to check that $|M|_{\mathrm{Var}} = |M|_I \circ \Pi_I^{\mathrm{Var}}$ for all finite sets I containing the free variables of M. In other words, $|M|_{\mathrm{Var}}$ is the morphism $|M|_I \in \mathbf{C}(U^I, U)$ 'seen' as an element of $\mathbf{C}(U^{\mathrm{Var}}, U)$. Moreover, $|M|_{\mathrm{Var}} = |N|_{\mathrm{Var}}$ iff $|M|_I = |N|_I$.

Hence, for the sake of simplicity, we will work in ccc's having countable products and we will use the interpretation $|-|_{\mathrm{Var}}$. We insist on the fact that this is just a simplification: all the work done in this section could be adapted to cover also categorical models living in ccc's without countable products but the statements and the proofs would be significantly more technical.

We set $\mathrm{Th}(\mathscr{U}) = \{(M, N) : |M|_{\mathrm{Var}} = |N|_{\mathrm{Var}}\}$. $\mathrm{Th}(\mathscr{U})$ is called the *λ-theory induced by* $\mathscr{U}$, or just *the (equational) theory of* $\mathscr{U}$. It is easy to check that if $\mathscr{U}$ is an extensional model then $\mathrm{Th}(\mathscr{U})$ is an extensional λ-theory.

2.2 Stratifiable Models in Cpo-Enriched Ccc's

The classic methods for proving that the theory of a categorical model is $\mathcal{H}^*$ require that the λ-terms are interpreted as elements of a cpo and that the morphisms involved in the definition of the interpretation are continuous functions. Thus, working *possibly outside* well-pointed categories, it becomes natural to consider categorical models living in cpo-enriched ccc's.

From now on, and until the end of the section, we consider a fixed non-trivial categorical model $\mathscr{U} = (U, \mathrm{Ap}, \lambda)$ living in a cpo-enriched ccc $\mathbf{C}$ having countable products.

Since in a cpo-enriched ccc pairing and currying are monotonic we get the following lemma.

Lemma 1. *The operations* $\bullet$ *and* $\lambda \circ \Lambda(- \circ \eta_x)$ *are continuous.*

To lighten the notation we write $\sqsubseteq$ and $\bot$ respectively for $\sqsubseteq_{(U^{\mathrm{Var}},U)}$ and $\bot_{(U^{\mathrm{Var}},U)}$.

Definition 1. *The model* $\mathcal{U}$ *is a* $\bot$-model *if the following two conditions hold:*
(i) $\bot \bullet a = \bot$ *for all* $a \in \mathbf{C}(U^{\mathrm{Var}}, U)$,
(ii) $\lambda \circ \Lambda(\bot_{(U^{\mathrm{Var}} \times U, U)}) = \bot$.

Stratifications of models are done by using special morphisms, acting at the level of $\mathbf{C}(U, U)$ and called 'projections'.

Definition 2. *Given an object* U *of a cpo-enriched ccc* $\mathbf{C}$, *a morphism* $p \in \mathbf{C}(U, U)$ *is a* projection from U into U *if* $p \sqsubseteq_{(U,U)} \mathrm{Id}_U$ *and* $p \circ p = p$.

From now on, we also fix a family $(p_k)_{k \in \mathbb{N}}$ of projections from U into U such that $(p_k)_{k \in \mathbb{N}}$ is increasing with respect to $\sqsubseteq_{(U,U)}$ and $\sqcup_{k \in \mathbb{N}} p_k = \mathrm{Id}_U$.

Notation 1. *Given a morphism* $a \in \mathbf{C}(U^{\mathrm{Var}}, U)$ *we write* a_k *for* $p_k \circ a$.

Remark 1. Since the p_k's are increasing, $\sqcup_{k \in \mathbb{N}} p_k = \mathrm{Id}_U$, and composition is continuous, we have for every morphism $a \in \mathbf{C}(U^{\mathrm{Var}}, U)$:
(i) $a_k \sqsubseteq a$,
(ii) $a = \sqcup_{k \in \mathbb{N}} a_k$.

Definition 3. *The model* $\mathcal{U}$ *is called:*
(i) stratified *(by* $(p_k)_{k \in \mathbb{N}}$*) if* $a_{k+1} \bullet b = (a \bullet b_k)_k$*;*
(ii) well stratified *(by* $(p_k)_{k \in \mathbb{N}}$*) if, moreover,* $a_0 \bullet b = (a \bullet \bot)_0$.

Of course, the fact that $\mathcal{U}$ is a (well) stratified model depends on the family $(p_k)_{k \in \mathbb{N}}$ we are considering. Hence, it is natural and convenient to introduce the notion of (well) stratifiable model.

Definition 4. *The model* $\mathcal{U}$ *is* stratifiable *(*well stratifiable*) if there exists a family* $(p_k)_{k \in \mathbb{N}}$ *making* $\mathcal{U}$ *stratified (well stratified).*

The aim of this section is in fact to prove that every extensional well stratifiable $\bot$-model has $\mathcal{H}^*$ as equational theory.

2.3 Modelling the Labelled $\lambda\bot$-Calculus in $\mathcal{U}$

We recall now the definition of the *labelled* $\lambda\bot$*-calculus* (see [15] or [2, Sec. 14.1]). We consider a set $C = \{c_k : k \in \mathbb{N}\}$ of constants called *labels*, together with a constant $\bot$ to indicate lack of information.

The set $\Lambda_\bot^{lab}$ of *labelled* $\lambda\bot$*-terms* is inductively defined as follows: $\bot \in \Lambda_\bot^{lab}$; $x \in \Lambda_\bot^{lab}$, for every $x \in \mathrm{Var}$; if $M, N \in \Lambda_\bot^{lab}$ then $MN \in \Lambda_\bot^{lab}$; if $M \in \Lambda_\bot^{lab}$ then $\lambda x.M \in \Lambda_\bot^{lab}$, for every $x \in \mathrm{Var}$; if $M \in \Lambda_\bot^{lab}$ then $c_k M \in \Lambda_\bot^{lab}$, for every $c_k \in C$.

We will denote by $\Lambda_\bot$ the subset of $\Lambda_\bot^{lab}$ consisting of those terms that do not contain any label; note that $\Lambda \subsetneq \Lambda_\bot \subsetneq \Lambda_\bot^{lab}$.

The labelled $\lambda\bot$-terms can be interpreted in $\mathcal{U}$; the intuitive meaning of $c_k M$ is the k-th projection applied to the meaning of M. Hence, we define the interpretation function as the unique extension of the interpretation function of λ-terms such that:

- $|\bot|_{\mathrm{Var}} = \bot$,
- $|c_k M|_{\mathrm{Var}} = p_k \circ |M|_{\mathrm{Var}} = (|M|_{\mathrm{Var}})_k$, for all $M \in \Lambda_\bot^{lab}$ and $k \in \mathbb{N}$.

Since the ccc $\mathbf{C}$ is cpo-enriched, all labelled $\lambda\bot$-terms are interpreted in the cpo $(\mathbf{C}(U^{\mathrm{Var}}, U), \sqsubseteq, \bot)$. Hence we can transfer this ordering, and the corresponding equality, on $\Lambda_\bot^{lab}$ as follows.

Definition 5. *For all $M, N \in \Lambda_\bot^{lab}$ we set $M \sqsubseteq_{\mathscr{U}} N$ iff $|M|_{\mathrm{Var}} \sqsubseteq |N|_{\mathrm{Var}}$. Moreover, we write $M =_{\mathscr{U}} N$ iff $M \sqsubseteq_{\mathscr{U}} N$ and $N \sqsubseteq_{\mathscr{U}} M$.*

It is straightforward to check that both $\sqsubseteq_{\mathscr{U}}$ and $=_{\mathscr{U}}$ are contextual.

The notion of substitution can be extended to $\Lambda_\bot^{lab}$ by setting: $\bot[M/x] = \bot$ and $(c_k M)[N/x] = c_k(M[N/x])$ for all $M, N \in \Lambda_\bot^{lab}$. We now show that $\mathscr{U}$ is sound for the β-conversion extended to $\Lambda_\bot^{lab}$.

Lemma 2. *For all $M, N \in \Lambda_\bot^{lab}$ we have $(\lambda x.M)N =_{\mathscr{U}} M[N/x]$.*

Proof. By [2, Prop. 5.5.5] we know that $(\lambda x.M)N =_{\mathscr{U}} M[N/x]$ still holds for λ-calculi extended with constants c, if $|c|_{\mathrm{Var}} = u \circ !_{U^{\mathrm{Var}}}$ for some $u \subset \mathbf{C}(\mathbb{1}, U)$. Hence, this lemma holds since the interpretation defined above is equal to that obtained by setting: $|c_k|_{\mathrm{Var}} = \lambda \circ \Lambda(p_k) \circ !_{U^{\mathrm{Var}}}$ and $|\bot|_{\mathrm{Var}} = \bot_{(\mathbb{1},U)} \circ !_{U^{\mathrm{Var}}}$.

We now introduce the reduction rules on labelled $\lambda\bot$-terms which generate the labelled $\lambda\bot$-calculus.

Definition 6

The ω-reduction *is defined by:*

$\bot M \to_\omega \bot$

$\lambda x.\bot \to_\omega \bot$

The γ-reduction *is defined by:*

$c_0(\lambda x.M)N \to_\gamma c_0(M[\bot/x])$

$c_{k+1}(\lambda x.M)N \to_\gamma c_k(M[c_k N/x])$.

The ϵ-reduction *is defined by:*

$c_k \bot \to_\epsilon \bot$,

$c_k(c_n M) \to_\epsilon c_{\min(k,m)} M$.

The calculus on $\Lambda_\bot^{lab}$ generated by the ω-, γ-, ϵ-reductions is called *labelled $\lambda\bot$-calculus*. Note that the β-reduction is not considered here.

The main properties of this calculus are summarized in the next theorem.

Theorem 1. *[2, Thm. 14.1.12 and 14.2.3] The labelled $\lambda\bot$-calculus is strongly normalizable and Church Rosser.*

We now show that the interpretation of a labelled $\lambda\bot$-term, in a well stratified $\bot$-model, is invariant along its ω-, ϵ-, γ-reduction paths.

Proposition 1. *If $\mathscr{U}$ is a well stratified $\bot$-model, then for all $M, N \in \Lambda_\bot^{lab}$:*

(i) $\bot M =_{\mathscr{U}} \bot$,

(ii) $\lambda x.\bot =_{\mathscr{U}} \bot$,

(iii) $c_k \bot =_{\mathscr{U}} \bot$,

(iv) $c_n(c_m M) =_{\mathscr{U}} c_{\min(n,m)} M$,

(v) $(c_0 \lambda x.M)N =_{\mathscr{U}} c_0(M[\bot/x])$,

(vi) $(c_{k+1} \lambda x.M)N =_{\mathscr{U}} c_k(M[c_k N/x])$.

Proof. (*i*) $|\underline{\perp} M|_{\mathrm{Var}} = |\underline{\perp}|_{\mathrm{Var}} \bullet |M|_{\mathrm{Var}} = \perp \bullet |M|_{\mathrm{Var}}$, which is $\perp$ by Def. 1(*i*).
(*ii*) $|\lambda x.\underline{\perp}|_{\mathrm{Var}} = \lambda \circ \Lambda(|\underline{\perp}|_{\mathrm{Var}} \circ \eta_x) = \lambda \circ \Lambda(\perp \circ \eta_x)$. Using (l-strict) this is equal to $\lambda \circ \Lambda(\perp_{(U^{\mathrm{Var}} \times U, U)})$, which is $\perp$ by Def. 1(*ii*). On the other side, $|\underline{\perp}|_{\mathrm{Var}} = \perp$.
(*iii*) $|c_k \underline{\perp}|_{\mathrm{Var}} = \perp_k$, hence by Rem. 1 we obtain $\perp_k \sqsubseteq \sqcup_{k\in\mathbb{N}} \perp_k = \perp$. The other inequality is clear.
(*iv*) $|c_n(c_m M)|_{\mathrm{Var}} = p_n \circ p_m \circ |M|_{\mathrm{Var}}$. By continuity of $\circ$, and since the sequence $(p_k)_{k\in\mathbb{N}}$ is increasing and every $p_k \sqsubseteq_{(U,U)} \mathrm{Id}_U$ we obtain $p_n \circ p_m = p_{\min(n,m)}$.
(*v*)
$$\begin{aligned} |(c_0 \lambda x.M)N|_{\mathrm{Var}} &= (|\lambda x.M|_{\mathrm{Var}})_0 \bullet |N|_{\mathrm{Var}} && \text{by def. of } | - |_{\mathrm{Var}} \\ &= (|\lambda x.M|_{\mathrm{Var}} \bullet \perp)_0 && \text{by Def. 3}(ii) \\ &= |c_0((\lambda x.M)\underline{\perp})|_{\mathrm{Var}} && \text{by def. of } | - |_{\mathrm{Var}} \\ &= |c_0(M[\underline{\perp}/x])|_{\mathrm{Var}} && \text{by Lemma 2.} \end{aligned}$$
(*vi*)
$$\begin{aligned} |(c_{k+1} \lambda x.M)N|_{\mathrm{Var}} &= (|\lambda x.M|_{\mathrm{Var}})_{k+1} \bullet |N|_{\mathrm{Var}} && \text{by def. of } | - |_{\mathrm{Var}} \\ &= (|\lambda x.M|_{\mathrm{Var}} \bullet (|N|_{\mathrm{Var}})_k)_k && \text{by Def. 3}(i) \\ &= |c_k((\lambda x.M)(c_k N))|_{\mathrm{Var}} && \text{by def. of } | - |_{\mathrm{Var}} \\ &= |c_k(M[c_k N/x])|_{\mathrm{Var}} && \text{by Lemma 2.} \end{aligned}$$

Corollary 1. *If $\mathscr{U}$ is a well stratified $\perp$-model, then for all $M, N \in \Lambda_{\underline{\perp}}^{lab}$, $M =_{\omega\gamma\epsilon} N$ implies $M =_{\mathscr{U}} N$.*

Proof. The result follows from Prop. 1, since the relation $=_{\mathscr{U}}$ is contextual.

Thus, every well stratifiable $\perp$-model is a model of the labelled $\lambda\perp$-calculus.

2.4 Completely Labelled $\lambda\perp$-Terms

We now study the properties of those labelled $\lambda\perp$-terms M which are completely labelled. This means that every subterm of M "has" a label.

Definition 7. *The set of* completely labelled $\lambda\perp$*-terms is defined by induction: $c_k \underline{\perp}$ is a completely labelled $\lambda\perp$-term, for every k; $c_k x$ is a completely labelled $\lambda\perp$-term, for every x and k; if $M, N \in \Lambda_{\underline{\perp}}^{lab}$ are completely labelled then also $c_k(MN)$ and $c_k(\lambda x.M)$ are completely labelled for every x and k.*

Note that every completely labelled $\lambda\perp$-term is β-normal, since every lambda abstraction is "blocked" by a c_k.

Definition 8. *A* complete labelling L *of a term $M \in \Lambda_{\underline{\perp}}$ is a map which assigns to each subterm of M a natural number. We will write $\mathcal{L}_M$ for the set of all complete labellings of M.*

Notation 2. *Given a term $M \in \Lambda_{\underline{\perp}}$ and a complete labelling L of M, we denote by M^L the resulting completely labelled $\lambda\perp$-term.*

It is easy to check that $\mathcal{L}_M$ is directed w.r.t. the following partial ordering: $L_1 \sqsubseteq_{lab} L_2$ iff for each subterm N of M we have $L_1(N) \leq L_2(N)$. By structural induction on the subterms of M one proves that $L_1 \sqsubseteq_{lab} L_2$ implies $M^{L_1} \sqsubseteq_{\mathscr{U}} M^{L_2}$. Thus, the set of the M^L 's such that $L \in \mathcal{L}_M$, is also directed w.r.t. $\sqsubseteq_{\mathscr{U}}$.

Lemma 3. *If $\mathscr{U}$ is a well stratified $\perp$-model, then for all $M \in \Lambda_{\underline{\perp}}$ we have $|M|_{\mathrm{Var}} = \sqcup_{L\in\mathcal{L}_M} |M^L|_{\mathrm{Var}}$.*

Proof. By straightforward induction on M, using $a = \sqcup_{k\in\mathbb{N}} a_k$ and Lemma 1.

2.5 The Approximation Theorem and Applications

Approximation theorems are an important tool in the analysis of the λ-theories induced by the models of λ-calculus. In this section we provide an Approximation Theorem for the class of well stratified $\perp$-models: we show that the interpretation of a λ-term in a well stratified $\perp$-model $\mathscr{U}$ is the least upper bound of the interpretations of its direct approximants. From this it follows first that $\mathrm{Th}(\mathscr{U})$ is sensible, and second that $\mathcal{B} \subseteq \mathrm{Th}(\mathscr{U})$.

Definition 9. *Let $M, N \in \Lambda_\perp$, then:* (1) *N is an* approximant *of M if there is a context $C[-_1, \ldots, -_k]$ over $\Lambda_\perp$, with $k \geq 0$, and $M_1, \ldots, M_k \in \Lambda_\perp$ such that $N \equiv C[\perp, \ldots, \perp]$ and $M \equiv C[M_1, \ldots, M_k]$;* (2) *$N$ is an* approximate normal form *(*app-nf*, for short) of M if, furthermore, it is $\beta\omega$-normal.*

Given $M \in \Lambda$, we define the set $\mathcal{A}(M)$ of *all direct approximants* of M as follows: $\mathcal{A}(M) = \{W \in \Lambda_\perp : \exists N, (M \twoheadrightarrow_\beta N) \text{ and } W \text{ is an app-nf of } N\}$.

It is easy to check that if M is unsolvable then $\mathcal{A}(M) = \{\perp\}$.

The proof of the following lemma is straightforward once recalled that, if $N \in \mathcal{A}(M)$, then M results (up to β-conversion) from N by replacing some $\perp$ in N by other terms.

Lemma 4. *If $\mathscr{U}$ is a well stratifiable $\perp$-model and $M \in \Lambda$, then for all $N \in \mathcal{A}(M)$ we have $N \sqsubseteq_{\mathscr{U}} M$.*

Given $M \in \Lambda_\perp^{lab}$ we will denote by $\overline{M} \in \Lambda_\perp$ the term obtained from M by erasing all labels.

Lemma 5. *For all $M \in \Lambda_\perp^{lab}$, we have that $M \sqsubseteq_{\mathscr{U}} \overline{M}$.*

Proof. By Rem. 1(i) we have $(|M|_{\mathrm{Var}})_k \sqsubseteq |M|_{\mathrm{Var}}$, and this implies $c_k M \sqsubseteq_{\mathscr{U}} M$. We conclude the proof since $\sqsubseteq_{\mathscr{U}}$ is contextual.

The following syntactic property is a consequence of the results in [2, Sec. 14.3].

Proposition 2. *[9, Prop. 1.9] Let $M \in \Lambda$ and L be a complete labelling of M. If $\mathrm{nf}(M^L)$ is the $\omega\gamma\epsilon$-normal form of M^L, then $\overline{\mathrm{nf}(M^L)} \in \mathcal{A}(M)$.*

Theorem 2. *(Approximation Theorem) If $\mathscr{U}$ is a well stratified $\perp$-model, then for all $M \in \Lambda$:*

$$|M|_{\mathrm{Var}} = \bigsqcup \mathcal{A}(M),$$

where $\bigsqcup \mathcal{A}(M) = \bigsqcup \{|W|_{\mathrm{Var}} : W \in \mathcal{A}(M)\}$.

Proof. Let L be a complete labelling for M. From Thm. 1 there is a unique $\omega\epsilon\gamma$-normal form of M^L. We denote this normal form by $\mathrm{nf}(M^L)$. Since $M^L \twoheadrightarrow_{\epsilon\gamma\omega} \mathrm{nf}(M^L)$, and $\mathscr{U}$ is a model of the labelled $\lambda\perp$-calculus (Cor. 1), we have $M^L =_{\mathscr{U}} \mathrm{nf}(M^L)$. Moreover, Prop. 2 implies that $\overline{\mathrm{nf}(M^L)} \in \mathcal{A}(M)$ and hence $\mathrm{nf}(M^L) \sqsubseteq_{\mathscr{U}} \overline{\mathrm{nf}(M^L)}$ by Lemma 5. This implies that $|\mathrm{nf}(M^L)|_{\mathrm{Var}} \sqsubseteq \bigsqcup \mathcal{A}(M)$. Since L is an arbitrary complete labelling for M, we have: $|M|_{\mathrm{Var}} = \sqcup_{L \in \mathcal{L}_M} |M^L|_{\mathrm{Var}}$, by Lemma 3 this is equal to $\sqcup_{L \in \mathcal{L}_M} |\mathrm{nf}(M^L)|_{\mathrm{Var}} \sqsubseteq \bigsqcup \mathcal{A}(M)$. The opposite inequality is clear.

Corollary 2. *Let $\mathscr{U}$ be a non-trivial well stratifiable $\bot$-model. A λ-term M is unsolvable if, and only if, $M =_{\mathscr{U}} \bot$.*

Proof. ($\Rightarrow$) If M is unsolvable, then $\mathcal{A}(M) = \{\bot\}$. Hence, $M =_{\mathscr{U}} \bot$ by Thm. 2. ($\Leftarrow$) If M is solvable, then by [2, Thm. 8.3.14] there exist $N_1, \dots, N_k \in \Lambda$, with $k \geq 0$, such that $MN_1 \cdots N_k =_{\mathscr{U}} \mathbf{I}$. Since $\mathscr{U}$ is a $\bot$-model, $M =_{\mathscr{U}} \bot$ would imply $\mathbf{I} =_{\mathscr{U}} \bot$ (by Def. 1(i)) and $\mathscr{U}$ would be trivial. Contradiction.

Corollary 3. *If $\mathscr{U}$ is a well stratifiable $\bot$-model, then* $\mathrm{Th}(\mathscr{U})$ *is sensible.*

We show that the notion of Böhm tree can be also generalized to terms in $\Lambda_\bot$.

Definition 10. *For all $M \in \Lambda_\bot$ we write* $\mathrm{BT}(M)$ *for the Böhm tree of the λ-term obtained by substituting Ω for all occurrences of $\bot$ in M.* Vice versa, *for all $M \in \Lambda$ we denote by $M^{[k]} \in \Lambda_\bot$ the (unique) $\beta\omega$-normal form such that* $\mathrm{BT}(M^{[k]}) = \mathrm{BT}^k(M)$ *(where* $\mathrm{BT}^k(M)$ *is the Böhm tree of M pruned at level k).*

It is straightforward to check that, for every λ-term M, $M^{[k]} \in \mathcal{A}(M)$. *Vice versa,* the following proposition is a consequence of the Approximation Theorem.

Proposition 3. *If $\mathscr{U}$ is a well stratifiable $\bot$-model then, for all $M \in \Lambda$, $|M|_{\mathrm{Var}} = \sqcup_{k\in\mathbb{N}} |M^{[k]}|_{\mathrm{Var}}$.*

Proof. For all $W \in \mathcal{A}(M)$, there exists a $k \in \mathbb{N}$ such that all the nodes in $\mathrm{BT}(W)$ have depth less than k. Thus $W \sqsubseteq_{BT} M^{[k]}$ and $W \sqsubseteq_{\mathscr{U}} M^{[k]}$ by Thm. 2.

Corollary 4. *If $N \sqsubseteq_{BT} M$ then $N \sqsubseteq_{\mathscr{U}} M$.*

Proof. If $N \sqsubseteq_{BT} M$ then for all $k \in \mathbb{N}$ we have $N^{[k]} \sqsubseteq_{BT} M$. By Lemma 4 $N^{[k]} \sqsubseteq_{\mathscr{U}} M$. Thus $|N|_{\mathrm{Var}} = \sqcup_{k\in\mathbb{N}} |N^{[k]}|_{\mathrm{Var}} \sqsubseteq |M|_{\mathrm{Var}}$ by Prop. 3.

As a direct consequence we get the following result.

Theorem 3. *If $\mathscr{U}$ is a well stratifiable $\bot$-model, then $\mathcal{B} \subseteq$* $\mathrm{Th}(\mathscr{U})$.

2.6 A General Class of Models of $\mathcal{H}^*$

The definition of $\simeq_\eta$ has been recalled in Sec. 1, together with those of $\sqsubseteq_{BT}$, $\sqsubseteq_{\eta,\infty}$, $\precsim_\eta$. However, for proving that $\mathrm{Th}(\mathscr{U}) = \mathcal{H}^*$, the following alternative characterization of $\sqsubseteq_{\eta,\infty}$ will be useful.

Theorem 4. *[2, Lemma 10.2.26] The following conditions are equivalent:*
- *$M \sqsubseteq_{\eta,\infty} N$,*
- *for all $k \in \mathbb{N}$ there exists $P_k \in \Lambda$ such that $P_k \twoheadrightarrow_\eta M$, and $P_k^{[k]} = N^{[k]}$.*

Lemma 6. *If $\mathscr{U}$ is an extensional well stratified $\bot$-model then, for all $M \in \Lambda_\bot$ and $x \in$* Var, *$x \sqsubseteq_{\eta,\infty} M$ implies $c_n x \sqsubseteq_{\mathscr{U}} M$ for all $n \in \mathbb{N}$.*

Proof. From [2, Def. 10.2.10], we can assume that $M \equiv \lambda y_1 \dots y_m.xM_1 \cdots M_m$ with $y_i \sqsubseteq_{\eta,\infty} M_i$. The proof is done by induction on n. If $n = 0$, then:

$$\begin{aligned}
c_0 x &=_{\mathscr{U}} \lambda y_1 \dots y_m.c_0 x y_1 \cdots y_m && \text{since } \mathscr{U} \text{ is extensional,}\\
&=_{\mathscr{U}} \lambda y_1 \dots y_m.c_0(x\bot) y_2 \cdots y_m && \text{since } \mathscr{U} \text{ is well stratified (Def. 3}(ii)),\\
&\vdots && \vdots\\
&=_{\mathscr{U}} \lambda y_1 \dots y_m.c_0(x\bot \cdots \bot) && \text{since } \mathscr{U} \text{ is well stratified (Def. 3}(ii)),\\
&\sqsubseteq_{\mathscr{U}} \lambda y_1 \dots y_m.x\bot \cdots \bot && \text{by Lemma 5,}\\
&\sqsubseteq_{\mathscr{U}} \lambda y_1 \dots y_m.xM_1 \cdots M_m && \text{by } \bot \sqsubseteq_{\mathscr{U}} M_i.
\end{aligned}$$

If $n > 0$, then:

$$\begin{aligned}
c_n x &=_{\mathscr{U}} \lambda y_1 \dots y_m.c_n x y_1 \cdots y_m && \text{since } \mathscr{U} \text{ is extensional,}\\
&=_{\mathscr{U}} \lambda y_1 \dots y_m.c_{n-1}(x(c_{n-1}y_1)) y_2 \dots y_m && \text{since } \mathscr{U} \text{ is stratified (Def. 3}(i)),\\
&\vdots && \vdots\\
&=_{\mathscr{U}} \lambda y_1 \dots y_m.c_{n-m}(x(c_{n-1}y_1) \cdots (c_{n-m}y_m)) && \text{since } \mathscr{U} \text{ is stratified (Def. 3}(i)).
\end{aligned}$$

Recalling that $y_i \sqsubseteq_{\eta,\infty} M_i$, we have:

$$\begin{aligned}
&\lambda y_1 \dots y_m.c_{n-m}(x(c_{n-1}y_1) \cdots (c_{n-m}y_m))\\
\sqsubseteq_{\mathscr{U}}\ &\lambda y_1 \dots y_m.c_{n-m}(xM_1 \cdots M_m) && \text{since } c_{n-i}y_i \sqsubseteq_{\mathscr{U}} M_i \text{ by I.H.,}\\
\sqsubseteq_{\mathscr{U}}\ &\lambda y_1 \dots y_m.xM_1 \cdots M_m && \text{by Lemma 5.}
\end{aligned}$$

Lemma 7. *Let $\mathscr{U}$ be an extensional well stratified $\bot$-model and $M, N, W \in \Lambda_\bot$. If W is a $\beta\omega$-normal form such that $W \sqsubseteq_{BT} M$ and $M \sqsubseteq_{\eta,\infty} N$, then $W \sqsubseteq_{\mathscr{U}} N$.*

Proof. The proof is done by induction on the structure of W.
If $W \equiv \bot$, then it is trivial.
If $W \equiv x$ then $M \equiv x$ and we conclude by Lemma 6 since $|x|_{\mathrm{Var}} = \sqcup_{n\in\mathbb{N}}(|x|_{\mathrm{Var}})_n$.
If $W \equiv \lambda x_1 \dots x_m.yW_1 \cdots W_r$, then $M =_\beta \lambda x_1 \dots x_m.yM_1 \cdots M_r$ and every W_i is a $\beta\omega$-normal form such that $W_i \sqsubseteq_{BT} M_i$ (for $i \leq r$). By $M \sqsubseteq_{\eta,\infty} N$, we can assume that $N =_{\beta\eta} \lambda x_1 \dots x_{m+s}.yN_1 \cdots N_{r+s}$, with $x_{m+k} \sqsubseteq_{\eta,\infty} N_{r+k}$ (for $1 \leq k \leq s$) and $M_i \sqsubseteq_{\eta,\infty} N_i$ (for $i \leq r$). From $x_{m+k} \sqsubseteq_{\eta,\infty} N_{r+k}$ we obtain, using the previous lemma, that $x_{m+k} \sqsubseteq_{\mathscr{U}} N_{r+k}$. Moreover, since $W_i \sqsubseteq_{BT} M_i \sqsubseteq_{\eta,\infty} N_i$, the induction hypothesis implies $W_i \sqsubseteq_{\mathscr{U}} N_i$. Hence, $W \sqsubseteq_{\mathscr{U}} N$.

Lemma 8. *If $\mathscr{U}$ is an extensional well stratifiable $\bot$-model then for all $M, N \in \Lambda$:*
(i) $M \sqsubseteq_{\eta,\infty} N$ *implies* $M =_{\mathscr{U}} N$,
(ii) $M \precsim_\eta N$ *implies* $M \sqsubseteq_{\mathscr{U}} N$.

Proof. (i) Suppose that $M \sqsubseteq_{\eta,\infty} N$. Since all $W \in \mathcal{A}(M)$ are $\beta\omega$-normal forms such that $W \sqsubseteq_{BT} M$, the Approximation Theorem and Lemma 7 imply that $M \sqsubseteq_{\mathscr{U}} N$. We prove now that also $N \sqsubseteq_{\mathscr{U}} M$ holds. By the characterization of $\sqsubseteq_{\eta,\infty}$ given in Thm. 4 we know that for all $k \in \mathbb{N}$ there exists a λ-term P_k such that $P_k \twoheadrightarrow_\eta M$ and $P_k^{[k]} = N^{[k]}$. Since every $P_k^{[k]} \in \mathcal{A}(P_k)$, we have $P_k^{[k]} \sqsubseteq_{\mathscr{U}} P_k$; also, from the extensionality of $\mathscr{U}$, $P_k =_{\mathscr{U}} M$. Thus, by Prop. 3, we have $|N|_{\mathrm{Var}} = \sqcup_{k\in\mathbb{N}}|N^{[k]}|_{\mathrm{Var}} = \sqcup_{k\in\mathbb{N}}|P_k^{[k]}|_{\mathrm{Var}} \sqsubseteq |M|_{\mathrm{Var}}$. This implies $N \sqsubseteq_{\mathscr{U}} M$.

(ii) Suppose now that $M \precsim_\eta N$. By definition, there exist two λ-terms M' and N' such that $M \sqsubseteq_{\eta,\infty} M' \sqsubseteq_{BT} N' \sqsupseteq_{\eta,\infty} N$. We conclude as follows: $M =_{\mathscr{U}} M'$ by (i), $M' \sqsubseteq_{\mathscr{U}} N'$ by Thm. 3, and $N' =_{\mathscr{U}} N$, again by (i).

We recall that the λ-theory $\mathcal{H}^*$ can be defined in terms of Böhm trees as follows: $M =_{\mathcal{H}^*} N$ if, and only if, $M \simeq_\eta N$ (see [2, Thm. 16.2.7]).

Theorem 5. *If $\mathscr{U}$ is a non-trivial well stratifiable extensional $\bot$-model living in a cpo-enriched ccc (having countable products), then* $\mathrm{Th}(\mathscr{U}) = \mathcal{H}^*$.

Proof. By Lemma 8(*ii*) we have that $M \simeq_\eta N$ implies $M =_{\mathscr{U}} N$. Thus, $\mathcal{H}^* \subseteq \mathrm{Th}(\mathscr{U})$. We conclude since $\mathcal{H}^*$ is the maximal consistent sensible λ-theory.

3 An Extensional Relational Model of λ-Calculus

In this section we recall the definition of our model $\mathscr{D}$ of [6], which is extensional by construction. Finally, we prove that $\mathrm{Th}(\mathscr{D}) = \mathcal{H}^*$ by applying Thm. 5

Multisets and Sequences. Let S be a set. A *multiset* m over S can be defined as an unordered list $m = [a_1, a_2, \dots]$ with repetitions such that $a_i \in S$ for all i. A multiset m is called *finite* if it is a finite list, we denote by $[]$ the empty multiset. We will write $\mathcal{M}_f(S)$ for the set of all finite multisets over S. Given two multisets $m_1 = [a_1, a_2, \dots]$ and $m_2 = [b_1, b_2, \dots]$ the *multiset union* of m_1, m_2 is defined by $m_1 \uplus m_2 = [a_1, b_1, a_2, b_2, \dots]$.

A $\mathbb{N}$-indexed sequence $\sigma = (m_1, m_2, \dots)$ of multisets is *quasi-finite* if $m_i = []$ holds for all but a finite number of indices i. If S is a set, we denote by $\mathcal{M}_f(S)^{(\omega)}$ the set of all quasi-finite $\mathbb{N}$-indexed sequences of multisets over S.

MRel: a Relational Semantics. We shortly present the category **MRel**. The objects of **MRel** are all the sets. A morphism from S to T is a relation from $\mathcal{M}_f(S)$ to T, in other words, $\mathbf{MRel}(S,T) = \mathscr{P}(\mathcal{M}_f(S) \times T)$. The identity of S is the relation $\mathrm{Id}_S = \{([a], a) : a \in S\} \in \mathbf{MRel}(S,S)$. The composition of $s \in \mathbf{MRel}(S,T)$ and $t \in \mathbf{MRel}(T,U)$ is defined by:

$$t \circ s = \{(m,c) : \exists (m_1,b_1),\dots,(m_k,b_k) \in s \text{ such that } m = m_1 \uplus \dots \uplus m_k \text{ and } ([b_1,\dots,b_k],c) \in t\}.$$

The categorical product $S \times T$ of two sets S and T is their disjoint union. The terminal object $\mathbb{1}$ is the empty set, and $!_S$ is the empty relation.

MRel is cartesian closed, non-well-pointed and has countable products [6, Sec. 4].

A Relational Analogue of $\mathscr{D}_\infty$. We build a family of sets $(D_n)_{n\in\mathbb{N}}$ as follows: $D_0 = \emptyset$, $D_{n+1} = \mathcal{M}_f(D_n)^{(\omega)}$. Since the operation $S \mapsto \mathcal{M}_f(S)^{(\omega)}$ is monotonic and $D_0 \subseteq D_1$, we have $D_n \subseteq D_{n+1}$ for all $n \in \mathbb{N}$. Finally, we set $D = \bigcup_{n\in\mathbb{N}} D_n$.

To define an isomorphism in **MRel** between D and $[D \Rightarrow D](= \mathcal{M}_f(D) \times D)$ just remark that every element $\sigma = (\sigma_0, \sigma_1, \sigma_2, \dots) \in D$ stands for the pair $(\sigma_0, (\sigma_1, \sigma_2, \dots))$ and *vice versa.* Given $\sigma \in D$ and $m \in \mathcal{M}_f(D)$, we write $m \cdot \sigma$ for the element $\tau \in D$ such that $\tau_1 = m$ and $\tau_{i+1} = \sigma_i$. This defines a bijection between $\mathcal{M}_f(D) \times D$ and D, and hence an isomorphism in **MRel** as follows:

Proposition 4. *The triple $\mathcal{D} = (D, \mathrm{Ap}, \lambda)$ where:*

- $\lambda = \{([(m,\sigma)], m \cdot \sigma) : m \in \mathcal{M}_f(D), \sigma \in D\} \in \mathbf{MRel}([D \Rightarrow D], D)$,

– $\mathrm{Ap} = \{([m \cdot \sigma], (m, \sigma)) : m \in \mathcal{M}_f(D), \sigma \in D\} \in \mathbf{MRel}(D, [D \Rightarrow D])$,
is a (non-trivial) extensional categorical model of λ*-calculus.*

We now prove that $\mathrm{Th}(\mathscr{D}) = \mathcal{H}^*$. From Thm. 5 it is enough to check that **MRel** is cpo-enriched and $\mathscr{D}$ is a well stratifiable $\perp$-model.

Theorem 6. *The ccc* **MRel** *is cpo-enriched.*

Proof. It is clear that, for all sets S, T, the homset $(\mathbf{MRel}(S, T), \subseteq, \emptyset)$ is a cpo, that composition is continuous, and pairing and currying are monotonic. Finally, it is easy to check that the strictness conditions hold.

Theorem 7. $\mathscr{D}$ *is an extensional well stratifiable* $\perp$*-model, thus* $\mathrm{Th}(\mathscr{D}) = \mathcal{H}^*$.

Proof. $\mathscr{D}$ is an extensional model by Prop. 4. By definition of Ap and λ it is straigthforward to check that $\emptyset \bullet a = \emptyset$, for all $a \in \mathbf{MRel}(D^{\mathrm{Var}}, D)$, and that $\lambda \circ \Lambda(\emptyset) = \emptyset$, hence $\mathscr{D}$ is a $\perp$-model. Let now $p_n = \{([\sigma], \sigma) : \sigma \in D_n\}$, where $(D_n)_{n \in \mathbb{N}}$ is the family of sets which has been used to build D. Since $(D_n)_{n \in \mathbb{N}}$ is increasing also $(p_n)_{n \in \mathbb{N}}$ is, and furthermore $\bigsqcup_{n \in \mathbb{N}} p_n = \{([\sigma], \sigma) : \sigma \in D\} = \mathrm{Id}_D$. Then, easy calculations show that $\mathscr{D}$ enjoys conditions (i) and (ii) of Def. 3.

Acknowledgements. Thanks to C. Berline, J.-J. Lévy and A. Salibra for invaluable discussions. Thanks to the anonymous reviewers for careful reading and suggestions.

References

1. Amadio, R., Curien, P.-L.: Domains and lambda-calculi. Cambridge Tracts in Theor. Comp. Sci., vol. 46. Cambridge University Press, New York (1998)
2. Barendregt, H.P.: The lambda calculus: Its syntax and semantics. North-Holland Publishing Co., Amsterdam (1984)
3. Berry, G.: Stable models of typed lambda-calculi. In: Ausiello, G., Böhm, C. (eds.) ICALP 1978. LNCS, vol. 62. Springer, Heidelberg (1978)
4. Bucciarelli, A., Ehrhard, T.: On phase semantics and denotational semantics: the exponentials. Ann. Pure Appl. Logic 109(3), 205–241 (2001)
5. Bucciarelli, A., Ehrhard, T.: Sequentiality and strong stability. In: LICS 1991, pp. 138–145 (1991)
6. Bucciarelli, A., Ehrhard, T., Manzonetto, G.: Not enough points is enough. In: Duparc, J., Henzinger, T.A. (eds.) CSL 2007. LNCS, vol. 4646, pp. 298–312. Springer, Heidelberg (2007)
7. Di Gianantonio, P., Franco, G., Honsell, F.: Game semantics for untyped $\lambda\beta\eta$-calculus. In: Girard, J.-Y. (ed.) TLCA 1999. LNCS, vol. 1581, pp. 114–128. Springer, Heidelberg (1999)
8. Girard, J.-Y.: Normal functors, power series and the λ-calculus. Annals of pure and applied logic 37, 129–177 (1988)
9. Gouy, X.: Etude des théories équationnelles et des propriétés algébriques des modèles stables du λ-calcul. PhD Thesis, University of Paris 7 (1995)

10. Hyland, J.M.E.: A syntactic characterization of the equality in some models for the lambda calculus. J. London Math. Soc. (2) 12(3), 361–370 (1975)
11. Hyland, M., Nagayama, M., Power, J., Rosolini, G.: A category theoretic formulation for Engeler-style models of the untyped λ-calculus. ENTCS 161, 43–57 (2006)
12. Ker, A.D., Nickau, H., Ong, C.-H.L.: A universal innocent game model for the Böhm tree lambda theory. In: Flum, J., Rodríguez-Artalejo, M. (eds.) CSL 1999. LNCS, vol. 1683, pp. 405–419. Springer, Heidelberg (1999)
13. Salibra, A.: A continuum of theories of lambda calculus without semantics. In: Proc. LICS 2001, pp. 334–343 (2001)
14. Scott, D.S.: Continuous lattices. Toposes, algebraic geometry and logic, Berlin (1972)
15. Wadsworth, C.P.: The relation between computational and denotational properties for Scott's D_∞-models of the lambda-calculus. SIAM J. Comp. 5(3), 488–521 (1976)

The Complexity of Satisfiability for Fragments of Hybrid Logic—Part I*

Arne Meier[1], Martin Mundhenk[2], Thomas Schneider[3], Michael Thomas[1], Volker Weber[4], and Felix Weiss[2]

[1] Theoretical Computer Science, University of Hannover, Germany
{meier,thomas}@thi.uni-hannover.de
[2] Institut für Informatik, Universität Jena, Germany
{martin.mundhenk,felix.weiss}@uni-jena.de
[3] Computer Science, University of Manchester, UK
schneider@cs.man.ac.uk
[4] Fakultät für Informatik, Technische Universität Dortmund, Germany

Abstract. The satisfiability problem of hybrid logics with the downarrow binder is known to be undecidable. This initiated a research program on decidable and tractable fragments.

In this paper, we investigate the effect of restricting the propositional part of the language on decidability and on the complexity of the satisfiability problem over arbitrary, transitive, total frames, and frames based on equivalence relations. We also consider different sets of modal and hybrid operators. We trace the border of decidability and give the precise complexity of most fragments, in particular for all fragments including negation. For the monotone fragments, we are able to distinguish the easy from the hard cases, depending on the allowed set of operators.

Keywords: hybrid logic, satisfiability, decidability, complexity, Post's lattice.

1 Introduction

Hybrid logics are well-behaved extensions of modal logic. However, their expressive power often has adverse effects on their computational properties: for instance, the satisfiability problem for basic modal logic extended with the $\downarrow$ binder is undecidable [9,15,1], as opposed to PSPACE-complete for basic modal logic [17] and modal logic extended with nominals and the satisfaction operator @ [1].

In order to regain decidability, many restrictions of the hybrid binder language have been considered. On the syntax side, it has been shown in [30] that restricting the interactions between $\downarrow$ and universal operators (such as $\wedge$, $\Box$) makes satisfiability decidable again. On the semantics side, the satisfiability problem for the $\downarrow$ language has been investigated over different frame classes. It becomes

* Supported in part by the grants DFG VO 630/6-1, DFG SCHW 678/4-1, BC-ARC 1323, DAAD-ARC D/08/08881.

R. Královič and D. Niwiński (Eds.): MFCS 2009, LNCS 5734, pp. 587–599, 2009.

decidable over frames with bounded width [30], over transitive and complete frames [21], and over frames with an equivalence relation [20]. In the latter case, decidability is not lost if @ or the global modality is added to the language [20], which is not the case over transitive frames [21]. Furthermore, over linear frames and transitive trees, where $\downarrow$ on its own is useless, extensions of the $\downarrow$ language have been shown to be decidable, albeit nonelementarily, in [14,21]. But elementarily decidable fragments over these frame classes have been obtained by bounding the number of state variables [29,32,11]. An overview of complexity results for hybrid logics can be found in [27].

Our aim is to obtain a more fine-grained distinction between decidable and undecidable hybrid logics by restricting the set of Boolean operators allowed in formulae. This is interesting in its own right because it will outline sources of "bad" behaviour (i.e., undecidability) more precisely. Furthermore, it is interesting in view of the relation between modal and description logic (DL). Concept satisfiability, the DL-counterpart of modal satisfiability, plays an important role because other useful decision problems for DLs are reducible to it. For a number of DLs without full Boolean expressivity, notably the $\mathcal{EL}$ and DL-Lite families, this problem is tractable [4,5,12], and other relevant decision problems have lower complexity than for the standard DL $\mathcal{ALC}$, the counterpart of the modal logic K. In the case of these restricted DLs, there are also fine-grained analyses of additional features that increase complexity and those which do not [5,3]. Our study can be seen as a general framework which accommodates restrictions of different types—on Boolean operators systematically, and also on modal operators and frame classes. As one possible application of the obtained results, we will gain insights into the complexity of extensions of modal and description logics with hybrid operators—among them the above mentioned restricted DLs.

For the sake of generality, we will systematically replace the usual $\wedge$, $\neg$ with arbitrary, not necessarily complete, sets of Boolean operators. All such possible sets are captured in Post's lattice [24,10], which consists of all clones, i. e., all closed sets of Boolean functions. Each clone corresponds to a set of Boolean operators closed under nesting, and vice versa. The lattice allows for transferring upper and lower complexity bounds between clones. It will thus be possible to prove finitely many results that will be valid for an infinite number of sets of operators—and hence for infinitely many satisfiability problems. This technique has been used for analysing the complexity of satisfiability for propositional logic [18] and modal logic [6], satisfiability and model checking for linear temporal logic [8,7], and satisfiability of constraint satisfaction problems [26,28].

Using Post's lattice, we will investigate the complexity of the satisfiability problem for hybrid logics containing the modal operators $\Diamond$, $\Box$ and the following hybrid features: nominals, the satisfaction operator @ and the hybrid binder $\downarrow$. We will consider subsets of these operators, as well as the above described systematic restrictions to the Boolean operators allowed. We will carry out this analysis over four different frame classes: all frames, transitive frames, total frames (where every state has at least one successor), and frames with equivalence relations (ER frames). The work presented here is part of ongoing work that also includes acyclic frame classes such as transitive trees and linear structures.

While our analysis is complete with respect to the sets of Boolean operators covered, it is far from complete for sets of modal and hybrid operators, as well as for frame classes. This is because the latter "dimensions" of expressivity are much more difficult to systematise. Therefore, we are currently restricting ourselves to the most prominent sets of modal/hybrid operators and frame classes. It should also be noted that a fourth dimension is possible, namely allowing for multiple accessibility relations in models, i.e., multiple modalities of each kind. We have omitted this consideration from the present paper mostly for the sake of

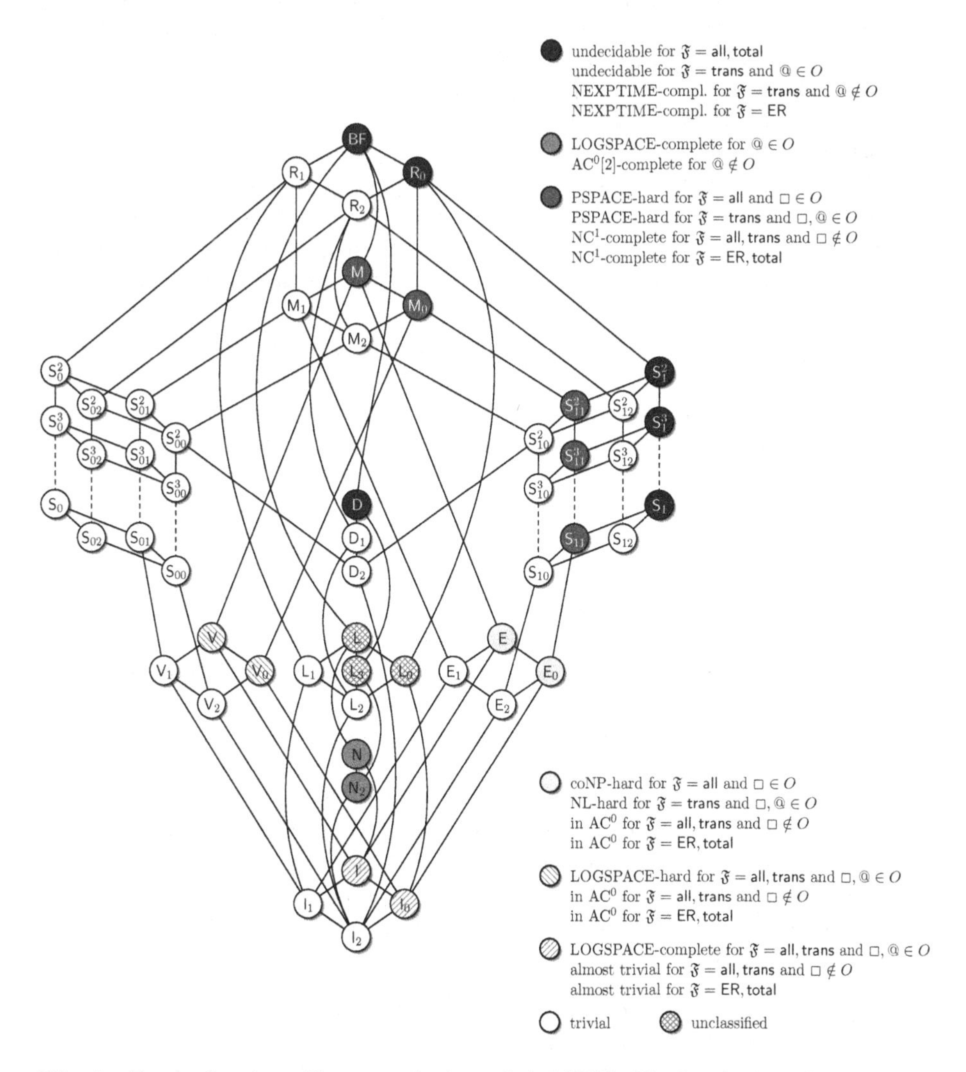

Fig. 1. Post's Lattice. The complexity of $\mathfrak{F}$-$\mathsf{SAT}(O, B)$ for frame classes $\mathfrak{F} \in \{\mathsf{all}, \mathsf{trans}, \mathsf{total}, \mathsf{ER}\}$ and sets O of Boolean operators with $\{\Diamond, \downarrow\} \subseteq O \subseteq \{\Diamond, \Box, \downarrow, @\}$.

a clearer presentation. However, we believe that many of the upper bounds can be straightforwardly extended to the multi-modal case—and will therefore indeed be helpful to gain insights into the behaviour of more expressive description logics.

This paper contains the most complete subset of our results obtained so far (see Figure 1), namely the following. We will show that, over each of the four above frame classes, satisfiability is as hard as in the full Boolean case whenever the negation of the implication or self-dual Boolean operators are allowed. (A Boolean function is self-dual if negating all of its arguments will always negate its value.) This means that, in these cases, satisfiability remains undecidable over arbitrary frames, total frames and, if the @-operator is present, over transitive frames; and NEXPTIME-complete over transitive frames without @ and over ER frames. These results can be found in Section 3.3.

In Section 3.3, we also completely classify the complexity of fragments including only negation and the Boolean constants. We obtain completeness for LOGSPACE if the @-operator is included and for $\mathrm{AC}^0[2]$ otherwise.

For all monotone fragments including the Boolean constant $\mathsf{0}$, we obtain a duality between easy cases, which are all included in NC^1, and hard cases, for which we obtain lower bounds ranging from LOGSPACE to PSPACE (Section 3.2). Satisfiability for fragments not including $\mathsf{0}$, but possibly all $\mathsf{1}$-reproducing functions, turns out to be trivial as shown in Section 3.1.

For the fragments that are based on the binary xor operator, the complexity is open. This case has turned out to be difficult to handle in [6,8,7]. A list of still open questions can be found in Section 4. Due to the page limit, this paper does not contain proofs. Our proofs can be found in [19].

2 Preliminaries

Boolean Functions and Clones. We can identify an n-ary propositional operator (connector) c with the n-ary *Boolean function* $f_c\colon \{\mathsf{0},\mathsf{1}\}^n \to \{\mathsf{0},\mathsf{1}\}$ defined by $f_c(a_1,\ldots,a_n) = \mathsf{1}$ if and only if $c(x_1,\ldots,x_n)$ becomes true when assigning a_i to x_i for all $1 \leq i \leq n$. The Boolean values *false* and *true* correspond to constants, i. e., nullary functions, and will be denoted by $\mathsf{0}$ and $\mathsf{1}$.

A set of Boolean functions is called a *clone* if it contains all projections and is closed under arbitrary composition [23, Chapter 1]. The set of all Boolean clones forms a lattice, which has been completely classified by Post [24]. For a set B of Boolean functions, we denote by $[B]$ the smallest clone containing B and call B a *base* for $[B]$. Whenever we use B for a set, we assume that B is finite.

The clones relevant to this paper are listed in Table 1. We use the following notions for n-ary Boolean functions f:

- f is *t-reproducing* if $f(t,\ldots,t) = t$, $t \in \{\mathsf{0},\mathsf{1}\}$.
- f is *monotone* if $a_1 \leq b_1,\ldots,a_n \leq b_n$ implies $f(a_1,\ldots,a_n) \leq f(b_1,\ldots,b_n)$.
- f is *t-separating* if there exists an $i \in \{1,\ldots,n\}$ such that $f(a_1,\ldots,a_n) = t$ implies $a_i = t$, $t \in \{\mathsf{0},\mathsf{1}\}$.

– f is *self-dual* if $f \equiv \text{dual}(f)$, where $\text{dual}(f)(x_1, \ldots, x_n) = \neg f(\neg x_1, \ldots, \neg x_n)$.

The definition of all Boolean clones can be found, e. g., in [10]. Notice that $[B \cup \{1\}] = \mathsf{BF}$ if and only if $[B] \supseteq \mathsf{S}_1$ or $[B] \supseteq \mathsf{D}$.

Table 1. Boolean clones relevant to this paper, with definitions and bases

Name	Definition	Base
BF	All Boolean functions	$\{\wedge, \neg\}$
R_1	1-reproducing functions	$\{\vee, \rightarrow\}$
M	monotone functions	$\{\vee, \wedge, 0, 1\}$
S_1	1-separating functions	$\{x \wedge \overline{y}\}$
S_{11}	$\mathsf{S}_1 \cap \mathsf{M}$	$\{x \wedge (y \vee z), 0\}$
D	self-dual functions	$\{(x \wedge \overline{y}) \vee (x \wedge \overline{z}) \vee (\overline{y} \wedge \overline{z})\}$
V	constant or n-ary OR functions	$\{\vee, 0, 1\}$
E	constant or n-ary AND functions	$\{\wedge, 0, 1\}$
E_0		$\{\wedge, 0\}$
N	functions depending on at most one variable	$\{\neg, 0, 1\}$
N_2		$\{\neg\}$
I	constant or identity functions	$\{\text{id}, 0, 1\}$
I_0		$\{\text{id}, 0\}$
I_1		$\{\text{id}, 1\}$
I_2		$\{\text{id}\}$

Hybrid Logic. In the following, we will introduce the notions and definitions of hybrid logic. The terminology is largely taken from [2].

Let PROP be a countable set of *atomic propositions*, NOM be a countable set of *nominals*, SVAR be a countable set of *variables* and $\mathsf{ATOM} = \mathsf{PROP} \cup \mathsf{NOM} \cup \mathsf{SVAR}$. We will stick with the common practice to denote atomic propositions by $p, q, \ldots$, nominals by $i, j, \ldots$, and variables by $x, y, \ldots$. We define the language of *hybrid (modal) logic* $\mathcal{HL}$ as the set of well-formed formulae of the form

$$\varphi ::= a \mid c(\varphi, \ldots, \varphi) \mid \Diamond\varphi \mid \Box\varphi \mid \downarrow x.\varphi \mid @_t\varphi$$

where $a \in \mathsf{ATOM}$, c is a Boolean operator, $x \in \mathsf{SVAR}$ and $t \in \mathsf{NOM} \cup \mathsf{SVAR}$. Note that the usual cases $\top$ and $\bot$ are covered by the Boolean constants 1 and 0.

Formulae of $\mathcal{HL}$ are interpreted on *(hybrid) Kripke structures* $K = (W, R, \eta)$, consisting of a set of *states* W, a *transition relation* $R\colon W \times W$, and a *labeling function* $\eta\colon \mathsf{PROP} \cup \mathsf{NOM} \rightarrow \wp(W)$ that maps PROP and NOM to subsets of W with $|\eta(i)| = 1$ for all $i \in \mathsf{NOM}$. In order to evaluate $\downarrow$-formulae, an assignment $g\colon \mathsf{SVAR} \rightarrow W$ is necessary. Given an assignment g, a state variable x and a state w, an *x-variant g^x_w of g* is defined by $g^x_w(x) = w$ and $g^x_w(x') = g(x')$ for all $x \neq x'$. For any $a \in \mathsf{ATOM}$, let $[\eta, g](a) = \{g(a)\}$ if $a \in \mathsf{SVAR}$ and $[\eta, g](a) = \eta(a)$, otherwise. The satisfaction relation of hybrid formulae is defined by

$K, g, w \models a$ iff $w \in [\eta, g](a)$, $a \in \mathsf{ATOM}$,
$K, g, w \models c(\varphi_1, \ldots, \varphi_n)$ iff $f_c(t_1, \ldots, t_n) = 1$, where t_i is the truth value of $K, g, w \models \varphi_i$, $1 \leq i \leq n$,

$$\begin{array}{ll}K,g,w \models \Diamond\varphi & \text{iff } K,g,w' \models \varphi \text{ for some } w' \in W \text{ with } wRw',\\ K,g,w \models \Box\varphi & \text{iff } K,g,w' \models \varphi \text{ for all } w' \in W \text{ with } wRw',\\ K,g,w \models @_t\varphi & \text{iff } K,g,w \models \varphi \text{ for } w \in W \text{ such that } w \in \eta(t),\\ K,g,w \models {\downarrow}x.\varphi & \text{iff } K,g^x_w,w \models \varphi.\end{array}$$

A hybrid formula φ is said to be *satisfiable* if there exists a Kripke structure $K = (W, R, \eta)$, a $w \in W$ and an assignment $g\colon \mathsf{SVAR} \to W$ such that $K, g, w \models \varphi$.

The *at* operator $@_t$ shifts evaluation to the state named by $t \in \mathsf{NOM} \cup \mathsf{SVAR}$. The *downarrow binder* $\downarrow x.$ binds the state variable x to the current state. The symbols $@_x$, $\downarrow x.$ are called *hybrid operators* whereas the symbols $\Diamond$ and $\Box$ are called *modal operators.*

For considering fragments of hybrid logics, we define subsets of the language $\mathcal{HL}$ as follows. Let B be a finite set of Boolean functions and O a set of hybrid and modal operators. We define $\mathcal{HL}(O, B)$ to denote the set of well-formed hybrid formulae using the operators in O and the Boolean connectives in B only.

Properties of Frames. A *frame* F is a pair (W, R), where W is a set of states and $R \subseteq W \times W$ a transition relation. We will refer to a frame as being *transitive, total* or *ER* whenever its transition relation R is transitive ($uRv \wedge vRw \to uRw$), total ($\forall u \exists v(uRv)$), or an equivalence relation, i. e., reflexive (uRu), transitive and symmetric ($uRv \to vRu$). In this paper we will consider the class all of all frames, the class trans of all transitive frames, the class total of all total frames, and the class ER of all ER frames.

The Satisfiability Problem. Let $K = (W, R, \eta)$ be a Kripke structure. Say that K *is based on a frame* F iff F is the frame underlying K, i. e., $F = (W, R)$. We define the *satisfiability problems* for the fragments of $\mathcal{HL}$ over frame classes defined above as follows.

Problem: $\mathfrak{F}$-$\mathsf{SAT}(O, B)$
Input: an $\mathcal{HL}(O, B)$-formula φ
Output: is there a Kripke structure $K = (W, R, \eta)$ based on a frame from $\mathfrak{F}$, an assignment $g\colon \mathsf{SVAR} \to W$ and a $w \in W$ such that $K, g, w \models \varphi$?

In case $\mathfrak{F} = \mathsf{all}$, we will omit the prefix and simply write $\mathsf{SAT}(O, B)$.

Complexity Theory. We assume familiarity with the standard notions of complexity theory as, e. g., defined in [22]. In particular, we will make use of the classes LOGSPACE, NL, P, coNP, PSPACE, NEXPTIME, and coRE.

We will now introduce the notions of circuit complexity required for this paper, for more information on circuit complexity the reader is referred to [31]. The class NC^1 is defined as the set of languages recognizable by a logtime-uniform Boolean circuits of logarithmic depth and polynomial size over $\{\wedge, \vee, \neg\}$, where the fan-in of $\wedge$ and $\vee$ gates is fixed to 2.

The class AC^0 is defined as the set of languages recognizable by a logtime-uniform Boolean circuits of constant depth and polynomial size over $\{\wedge, \vee, \neg\}$, where the fan-in of gates of the first two types is not bounded. If, in addition, modulo-2 gates are allowed, then the corresponding class is $AC^0[2]$. Both AC^0 and $AC^0[2]$ are strictly contained in NC^1. Altogether, the following inclusions are known: $AC^0 \subseteq AC^0[2] \subset NC^1 \subseteq$ LOGSPACE $\subseteq$ NL $\subseteq$ P $\subseteq$ coNP $\subseteq$ PSPACE $\subset$ NEXPTIME $\subset$ coRE.

A language A is *constant-depth reducible* to D, $A \leq_{cd} D$, if there is a logtime-uniform AC^0-circuit family with oracle gates for D that decides membership in A. Unless otherwise stated, all reductions in this paper are $\leq_{cd}$-reductions.

Known results. The following theorem summarizes results for hybrid binder languages with Boolean operators $\wedge, \vee, \neg$ that are known from the literature.

Theorem 1 **([1,21,20])**

(1) $\mathsf{SAT}(\{\Diamond, \downarrow\}, \{\wedge, \vee, \neg\})$ *and* $\mathsf{SAT}(\{\Diamond, \Box, \downarrow, @\}, \{\wedge, \vee, \neg\})$ *are* coRE*-complete.*
(2) $\mathsf{trans\text{-}SAT}(\{\Diamond, \downarrow\}, \{\wedge, \vee, \neg\})$ *is* NEXPTIME*-complete.*
(3) $\mathsf{trans\text{-}SAT}(\{\Diamond, \downarrow, @\}, \{\wedge, \vee, \neg\})$ *is* coRE*-complete.*
(4) $\mathsf{ER\text{-}SAT}(\{\Diamond, \downarrow\}, \{\wedge, \vee, \neg\})$ *is* NEXPTIME*-complete.*
(5) $\mathsf{ER\text{-}SAT}(\{\Diamond, \downarrow, @\}, \{\wedge, \vee, \neg\})$ *is* NEXPTIME*-complete.*

3 Results

In this section, we present our results ordered by clones. Section 3.1 considers clones containing only 1-reproducing functions. Clones containing the Boolean constant 0 but not negation are considered in Section 3.2. Finally, in Section 3.3, we study satisfiability problems based on clones with negation.

This arrangement is motivated by the observation that the availability of the Boolean constant 0 and/or negation has a very strong impact on our results. Although we obtain different complexities for the clones including 0 but not negation (namely, I, V, E, and M), the results for these clones follow a certain pattern. But if we add negation, this picture changes completely.

Please note that opposed to the importance of the presence of 0, which makes the difference between trivial and nontrivial problems, hybrid languages can always express the constant 1 as $\downarrow x.x$ or $@_x x$. Therefore, we only have to consider clones including 1.

3.1 Why We Cannot Say Anything without Saying "False"

We start our investigation at the clone I_2, which contains only the identity function.[1] Obviously, every hybrid I_2-formula is satisfied by the model consisting of a singleton reflexive state to which all propositions, nominals, and state variables are labeled.

But this observation takes us much further, as we can add conjunction, disjunction, and implication for example, and still satisfy every formula by the same model. In fact, we can add every 1-reproducing function, i. e., every function that produces 1 if all parameters are 1, obtaining the following result.

[1] Since $1 \equiv \downarrow x.x$, there is no difference between the satisfiability problems for I_2 and I_1.

Theorem 2. *$\mathfrak{F}$-SAT$(\{\Diamond, \Box, \downarrow, @\}, B)$ for $[B] \subseteq \mathsf{R}_1$ and all considered frame classes $\mathfrak{F} \in \{\mathsf{all}, \mathsf{trans}, \mathsf{total}, \mathsf{ER}\}$ is trivial.*

It is interesting to note which Boolean operations are not contained in R_1. The most basic ones are the Boolean constant $\mathsf{0}$ and negation, as every clone in Post's lattice that is not below R_1 contains one of these.

As hybrid languages can always express the Boolean constant $\mathsf{1}$, the presence of negation implies the availability of $\mathsf{0}$. Therefore, there are two kinds of clones remaining: those containing $\mathsf{0}$ but not negation, and those containing negation. In the following subsection, we will consider the first kind, i. e., the monotone clones below M. Clones with negation will be considered in Section 3.3.

3.2 Everything But Negation – The Monotone Clones

In this section, we consider the clones below M that contain the Boolean constant $\mathsf{0}$; satisfiability for the clones without $\mathsf{0}$ is trivial by Theorem 2. Roughly speaking, we consider the clones I, V, E, and M. We start with I and then jump to M. Clones containing either disjunction or conjunction are considered last, as some results will easily follow from the preceding cases.

The clone I. The clone I is of particular interest, as it allows us to study the effect of having the Boolean constant $\mathsf{0}$ at our disposal, yielding the following two observations. First, the Boolean constant $\mathsf{0}$ distinguishes trivial from non-trivial satisfiability problems. While all satisfiability problems for clones without $\mathsf{0}$ are trivial (Theorem 2), all problems for clones with $\mathsf{0}$ are not. The precise complexity of the latter problems will vary from almost trivial cases (Theorem 3) to LOGSPACE-completeness (Theorem 4), depending on the modal and hybrid operators allowed. Higher complexities and even undecidability occur if we add further Boolean functions as discussed in the following sections.

Second, Theorems 3 and 4 demonstrate a duality between easy and hard cases, which we will see in all results for clones below M. For the full set of modal and hybrid operators, satisfiability problems over the class of all frames and the class of transitive frames will be considerably harder than those over total frames and equivalence relations. Furthermore, if we drop the $\Box$-operator when considering arbitrary or transitive frames, complexity will drop to where it is for total frames and equivalence relations.

Intuitively speaking, we might say that the complexity gap we observe in the results for monotone clones is due to the ability to express that a state has no successor by $\Box\mathsf{0}$. On the one hand, if we cannot express this property because of the absence of $\Box$ or if there are no such states because we only consider frames with a total accessibility relation, satisfiability for the clone I is *almost trivial*, i. e., we only need to look at one symbol of a formula to determine its satisfiability.

Theorem 3. *The following satisfiability problems are almost trivial.*[2]

1. total-SAT$(\{\Diamond, \Box, \downarrow, @\}, B)$ *and* ER-SAT$(\{\Diamond, \Box, \downarrow, @\}, B)$ *for* $[B] \subseteq \mathsf{I}$.

[2] More precisely, they are in $\Delta_0^{\mathcal{R}}$, a class strictly below DLOGTIME [25].

2. $\mathsf{SAT}(\{\Diamond,\downarrow,@\},B)$ *and* $\mathsf{trans\text{-}SAT}(\{\Diamond,\downarrow,@\},B)$ *for* $[B]\subseteq\mathsf{I}$.

On the other hand, the proof of the following theorem shows how to use $\Box 0$ to obtain LOGSPACE-hardness, without using any further Boolean connectives. A matching upper bound will be presented in Theorem 15.

Theorem 4. $\mathsf{SAT}(\{\Diamond,\Box,\downarrow,@\},B)$ *and* $\mathsf{trans\text{-}SAT}(\{\Diamond,\Box,\downarrow,@\},B)$ *for* $[B]\supseteq\mathsf{I}_0$ *are* LOGSPACE-*hard.*

The clone M. Let us now consider the clone M of all monotone functions. Here, more precisely for all clones between S_{11} and M, we obtain the same duality as in the previous section, only at a higher level of complexity. For the "hard cases", i.e., those satisfiability problems where we consider non-total frame classes and all modal and hybrid operators, we obtain PSPACE-hardness. For the class of all frames, this follows immediately from the corresponding result for modal logic.

Theorem 5 ([6,16]). $\mathsf{SAT}(\{\Diamond,\Box\},B)$ *is* PSPACE-*hard for* $[B]\supseteq\mathsf{S}_{11}$.

Unfortunately, the proof of this result does not generalize to transitive frames.

Theorem 6. $\mathsf{trans\text{-}SAT}(\{\Diamond,\Box,\downarrow,@\},B)$ *is* PSPACE-*hard for* $[B]\supseteq\mathsf{S}_{11}$.

The proof of Theorem 6 crucially depends on the existence of states without successor, and the ability to express this property: the truth values 0 (resp. 1) are encoded as states having no (resp. at least one) successor. If there are no such states (Theorem 7) or if we cannot express this property (Corollary 8), complexity drops to NC^1.

Theorem 7. $\mathsf{total\text{-}SAT}(O,B)$ *and* $\mathsf{ER\text{-}SAT}(O,B)$ *are* NC^1-*complete under* $\leq_{\mathrm{cd}}$-*reductions for* $O\subseteq\{\Diamond,\Box,\downarrow,@\}$ *and* $\mathsf{S}_{11}\subseteq[B]\subseteq\mathsf{M}$.

The proof of Theorem 7, shows that deciding $\varphi\in\mathsf{total\text{-}SAT}(\{\Diamond,\Box,\downarrow,@\},\mathsf{M})$ is equivalent to deciding whether $K_1,g_1,w_1\models\varphi$, for the singleton reflexive model K_1 mapping all atomic propositions into state w_1. In order to decide the latter, all hybrid and modal operators of φ can be ignored, as K_1 is a singleton model. There, only the treatment of the $\Box$-operator depends on the transition relation being total or an equivalence relation. If this operator is not allowed, the same argumentation goes through for our other frame classes, too.

Corollary 8. $\mathsf{SAT}(O,\ B)$ *and* $\mathsf{trans\text{-}SAT}(O,\ B)$ *are* NC^1 *-complete for* $O\subseteq\{\Diamond,\downarrow,@\}$ *and* $\mathsf{S}_{11}\subseteq[B]\subseteq\mathsf{M}$.

The clones V and E. If we consider conjunction or disjunction only separately, the complexity of formula evaluation decreases from NC^1-complete to below AC^0. As the complexity for the "easy cases" for M was determined by this complexity (Theorem 7 and Corollary 8), the following results are not too surprising.

Theorem 9. $\mathsf{total\text{-}SAT}(\{\Diamond,\Box,\downarrow,@\},B)$ *and* $\mathsf{ER\text{-}SAT}(\{\Diamond,\Box,\downarrow,@\},B)$ *are in* AC^0 *for* $[B]\subseteq\mathsf{E}$ *or* $[B]\subseteq\mathsf{V}$.

Corollary 10. $\mathsf{SAT}(O, B)$ *and* $\mathsf{trans\text{-}SAT}(O, B)$ *are in* AC^0 *for* $O \subseteq \{\Diamond, \downarrow, @\}$ *and* $[B] \subseteq \mathsf{E}$ *or* $[B] \subseteq \mathsf{V}$.

This result is optimal in the sense that including all modal and hybrid operators we immediately get LOGSPACE lower bounds from Theorem 4.

For the case of conjunctions, this result can be improved. Considering arbitrary frames, a coNP lower bound is already known for the modal satisfiability problem.

Theorem 11 ([6,13]). $\mathsf{SAT}(\{\Diamond, \Box\}, B)$ *is* coNP*-hard for* $[B] \supseteq \mathsf{E}_0$.

For transitive frames, we are able to show NL-hardness.

Theorem 12. $\mathsf{trans\text{-}SAT}(\{\Diamond, \Box, \downarrow, @\}, B)$ *is* NL*-hard for* $[B] \supseteq \mathsf{E}_0$.

We conjecture that all lower bounds provided in this section (except, perhaps, the last one) are optimal. Nevertheless, matching upper bounds are missing.

3.3 Clones Including Negation

Negation immediately limits the number of relevant satisfiability problems in two ways. First, as $\Box\varphi \equiv \neg\Diamond\neg\varphi$, we cannot exclude the $\Box$-operator and keep $\Diamond$ as we did for monotone clones. Therefore, we have to consider only two hybrid languages: with and without @. Second, as $\mathsf{1}$ and $\mathsf{0}$ are always expressible by $\downarrow x.x$ and $\neg\downarrow x.x$, we only need to consider clones with both constants. These are N (only negation), L (exclusive or), and BF (all Boolean functions).

While we will completely classify all satisfiability problems based on N and BF, we will not provide any specific results for L.

Negation only. The results for the satisfiability problems based on N stick out from our other results, as N is the only clone (besides those for which satisfiability is trivial) where all complexity results are the same for all frame classes we consider. We show that satisfiability for the hybrid language including @ is LOGSPACE-complete, while it is $\mathrm{AC}^0[2]$-complete for the language without @.

Theorem 13. $\mathfrak{F}\text{-}\mathsf{SAT}(\{\Diamond, \Box, \downarrow\}, B)$ *is* $\mathrm{AC}^0[2]$*-complete for* $[B] \subseteq \mathsf{N}$ *and all considered frame classes* $\mathfrak{F} \in \{\mathsf{all}, \mathsf{trans}, \mathsf{total}, \mathsf{ER}\}$.

The key to the proof of this theorem is that a given formula can be transformed into negation normal form by an $\mathrm{AC}^0[2]$-circuit. Subsequently determining satisfiability is easy. We now turn to the hybrid language including the @-operator.

Theorem 14. $\mathfrak{F}\text{-}\mathsf{SAT}(\{\Diamond, \downarrow, @\}, B)$ *is* LOGSPACE*-hard for* $[B] \supseteq \mathsf{N}_2$ *and all considered frame classes* $\mathfrak{F} \in \{\mathsf{all}, \mathsf{trans}, \mathsf{total}, \mathsf{ER}\}$.

We again provide a matching upper bound for all considered frame classes, which yields LOGSPACE-completeness of the respective satisfiability problems.

Theorem 15. $\mathfrak{F}\text{-}\mathsf{SAT}(\{\Diamond, \Box, \downarrow, @\}, B)$ *is in* LOGSPACE *for* $[B] \subseteq \mathsf{N}$ *and all considered frame classes* $\mathfrak{F} \in \{\mathsf{all}, \mathsf{trans}, \mathsf{total}, \mathsf{ER}\}$.

All Boolean functions. Finally, let us consider the clones between D, S_1 and BF, the clone of all Boolean functions. For the classes of all frames, transitive frames, and equivalence relations, we can transfer results obtained for the set $\{\wedge, \vee, \neg\}$ of Boolean functions to these clones using a technical Lemma (see Appendix). Additionally, we show that we can reduce the satisfiability problem over the class of all frames to the one over the class of total frames, establishing undecidability for all hybrid languages in this case.

Theorem 16. *Let* $[B \cup \{1\}] = \mathsf{BF}$. *Then:*

(1) $\mathsf{SAT}(O, B)$ *and* $\mathsf{total\text{-}SAT}(O, B)$ *are* coRE*-complete, for any* $O \supseteq \{\Diamond, \downarrow\}$.
(2) $\mathsf{trans\text{-}SAT}(\{\Diamond, \downarrow, @\}, B)$ *is* coRE*-complete.*
(3) $\mathsf{trans\text{-}SAT}(\{\Diamond, \downarrow\}, B)$ *is* NEXPTIME*-complete.*
(4) $\mathsf{ER\text{-}SAT}(\{\Diamond, \downarrow\}, B)$ *and* $\mathsf{ER\text{-}SAT}(\{\Diamond, \downarrow, @\}, B)$ *are* NEXPTIME*-complete.*

4 Conclusions

We have almost completely classified the complexity of hybrid binder logics over four frame classes with respect to all possible combinations of Boolean operators, see Figure 1. The main open question is for tight upper bounds for the monotone fragments including the $\Box$-operator over the classes of all and of transitive frames.

Another open questions concerns the hybrid languages with $\Box$ but without @ over the class of transitive frames. The complexity for the respective satisfiability problems based on V, E, and M is open; in the case of V even for the class of all frames. For I, containment in AC^0 follows from an analysis of the proof of Theorem 13. Finally, we could not obtain any bounds on the complexity for problems based on L, besides LOGSPACE-hardness inherited from Theorem 4.

We are currently investigating the same problems over frame classes important for representing modal properties, such as transitive trees, linear frames and the natural numbers. Here, satisfiability for $\downarrow$, @, and arbitrary Boolean operators is already decidable, but with a nonelementary lower bound; hence, a complexity analysis is worthwile as well. Because each such frame is acyclic, the fact that certain formulae are always satisfied in the singleton reflexive frame is not helpful any longer. This makes obtaining upper bounds more difficult. On the other hand, we can also express the constant 0 by $\downarrow x.\Diamond x$, which reduces the sets of Boolean operators to consider. We plan to publish these results in "Part II".

References

1. Areces, C., Blackburn, P., Marx, M.: A road-map on complexity for hybrid logics. In: Flum, J., Rodríguez-Artalejo, M. (eds.) CSL 1999. LNCS, vol. 1683, pp. 307–321. Springer, Heidelberg (1999)
2. Areces, C., Blackburn, P., Marx, M.: The computational complexity of hybrid temporal logics. Logic Journal of the IGPL 8(5), 653–679 (2000)

3. Artale, A., Calvanese, D., Kontchakov, R., Zakharyaschev, M.: DL-Lite in the light of first-order logic. In: Proc. AAAI 2007, pp. 361–366 (2007)
4. Baader, F.: Terminological cycles in a description logic with existential restrictions. In: Proc. IJCAI 2003, pp. 325–330 (2003)
5. Baader, F., Brandt, S., Lutz, C.: Pushing the $\mathcal{EL}$ envelope. In: Proc. IJCAI 2005, pp. 364–369 (2005)
6. Bauland, M., Hemaspaandra, E., Schnoor, H., Schnoor, I.: Generalized modal satisfiability. In: Durand, B., Thomas, W. (eds.) STACS 2006. LNCS, vol. 3884, pp. 500–511. Springer, Heidelberg (2006)
7. Bauland, M., Mundhenk, M., Schneider, T., Schnoor, H., Schnoor, I., Vollmer, H.: The tractability of model checking for LTL: the good, the bad, and the ugly fragments. In: Proc. M4M-5 (2007). ENTCS, vol. 231, pp. 277–292 (2009)
8. Bauland, M., Schneider, T., Schnoor, H., Schnoor, I., Vollmer, H.: The complexity of generalized satisfiability for linear temporal logic. In: Seidl, H. (ed.) FOSSACS 2007. LNCS, vol. 4423, pp. 48–62. Springer, Heidelberg (2007)
9. Blackburn, P., Seligman, J.: Hybrid languages. JoLLI 4, 41–62 (1995)
10. Böhler, E., Creignou, N., Reith, S., Vollmer, H.: Playing with Boolean blocks, part I: Post's lattice with applications to complexity theory. ACM-SIGACT Newsletter 34(4), 38–52 (2003)
11. Bozzelli, L., Lanotte, R.: Complexity and succinctness issues for linear-time hybrid logics. In: Hölldobler, S., Lutz, C., Wansing, H. (eds.) JELIA 2008. LNCS, vol. 5293, pp. 48–61. Springer, Heidelberg (2008)
12. Calvanese, D., De Giacomo, G., Lembo, D., Lenzerini, M., Rosati, R.: DL-Lite: Tractable description logics for ontologies. In: Proc. AAAI 2005, pp. 602–607 (2005)
13. Donini, F., Hollunder, B., Lenzerini, M., Nardi, D., Nutt, W., Spaccamela, A.: The complexity of existential quantification in concept languages. Artificial Intelligence 53(2-3), 309–327 (1992)
14. Franceschet, M., de Rijke, M., Schlingloff, B.: Hybrid logics on linear structures: Expressivity and complexity. In: Proc. 10th TIME, pp. 166–173 (2003)
15. Goranko, V.: Hierarchies of modal and temporal logics with reference pointers. Journal of Logic, Language and Information 5(1), 1–24 (1996)
16. Hemaspaandra, E.: The complexity of poor man's logic. Journal of Logic and Computation 11(4), 609–622 (2001); Corrected version available at arXiv (2005)
17. Ladner, R.: The computational complexity of provability in systems of modal propositional logic. SIAM Journal on Computing 6(3), 467–480 (1977)
18. Lewis, H.: Satisfiability problems for propositional calculi. Math. Sys. Theory 13, 45–53 (1979)
19. Meier, A., Mundhenk, M., Schneider, T., Thomas, M., Weber, V., Weiss, F.: The complexity of satisfiability for fragments of hybrid logic — Part I (2009), `http://arxiv.org/abs/0906.1489`
20. Mundhenk, M., Schneider, T.: The complexity of hybrid logics over equivalence relations. In: Proc. HyLo, pp. 81–90 (2007); To appear in J. of Logic, Language and Information
21. Mundhenk, M., Schneider, T., Schwentick, T., Weber, V.: Complexity of hybrid logics over transitive frames. In: Proc. M4M-4 (2005), `http://arxiv.org/abs/0806.4130`
22. Papadimitriou, C.H.: Computational Complexity. Addison-Wesley, Reading (1994)
23. Pippenger, N.: Theories of Computability. Cambridge University Press, Cambridge (1997)
24. Post, E.: The two-valued iterative systems of mathematical logic. Annals of Mathematical Studies 5, 1–122 (1941)

25. Regan, K., Vollmer, H.: Gap-languages and log-time complexity classes. Theoretical Computer Science 188, 101–116 (1997)
26. Schaefer, T.J.: The complexity of satisfiability problems. In: Proc. STOC, pp. 216–226. ACM Press, New York (1978)
27. Schneider, T.: The Complexity of Hybrid Logics over Restricted Classes of Frames. PhD thesis, Univ. of Jena (2007)
28. Schnoor, H.: Algebraic Techniques for Satisfiability Problems. PhD thesis, Univ. of Hannover (2007)
29. Schwentick, T., Weber, V.: Bounded-variable fragments of hybrid logics. In: Thomas, W., Weil, P. (eds.) STACS 2007. LNCS, vol. 4393, pp. 561–572. Springer, Heidelberg (2007)
30. ten Cate, B., Franceschet, M.: On the complexity of hybrid logics with binders. In: Ong, L. (ed.) CSL 2005. LNCS, vol. 3634, pp. 339–354. Springer, Heidelberg (2005)
31. Vollmer, H.: Introduction to Circuit Complexity. Springer, Heidelberg (1999)
32. Weber, V.: Hybrid branching-time logics. In: Proc. of HyLo, pp. 51–60 (2007); Accepted for a special issue of the J. of Logic, Language and Information

Colouring Non-sparse Random Intersection Graphs*

Sotiris Nikoletseas[1,2], Christoforos Raptopoulos[3], and Paul G. Spirakis[1,2]

[1] Computer Technology Institute, P.O. Box 1122, 26110 Patras, Greece
[2] University of Patras, 26500 Patras, Greece
[3] Heinz Nixdorf Institute, Computer Science Departement, University of Paderborn, 33095 Paderborn, Germany
nikole@cti.gr, raptopox@hni.uni-paderborn.de, spirakis@cti.gr

Abstract. An intersection graph of n vertices assumes that each vertex is equipped with a subset of a global label set. Two vertices share an edge when their label sets intersect. Random Intersection Graphs (RIGs) (as defined in [18,32]) consider label sets formed by the following experiment: each vertex, independently and uniformly, examines all the labels (m in total) one by one. Each examination is independent and the vertex succeeds to put the label in her set with probability p. Such graphs nicely capture interactions in networks due to sharing of resources among nodes. We study here the problem of efficiently coloring (and of finding upper bounds to the chromatic number) of RIGs. We concentrate in a range of parameters not examined in the literature, namely: (a) $m = n^{\alpha}$ for α less than 1 (in this range, RIGs differ substantially from the Erdös-Renyi random graphs) and (b) the selection probability p is quite high (e.g. at least $\frac{\ln^2 n}{m}$ in our algorithm) and disallows direct greedy colouring methods.

We manage to get the following results:

- For the case $mp \leq \beta \ln n$, for any constant $\beta < 1 - \alpha$, we prove that np colours are enough to colour most of the vertices of the graph with high probability (whp). This means that even for quite dense graphs, using the same number of colours as those needed to properly colour the clique induced by any label suffices to colour almost all of the vertices of the graph. Note also that this range of values of m, p is quite wider than the one studied in [4].
- We propose and analyze an algorithm CliqueColour for finding a proper colouring of a random instance of $\mathcal{G}_{n,m,p}$, for any $mp \geq \ln^2 n$. The algorithm uses information of the label sets assigned to the vertices of $G_{n,m,p}$ and runs in $O\left(\frac{n^2 mp^2}{\ln n}\right)$ time, which is polynomial in n and m. We also show by a reduction to the uniform random intersection graphs model that the number of colours required by the algorithm are of the correct order of magnitude with the actual chromatic number of $G_{n,m,p}$.

* This work was partially supported by the ICT Programme of the European Union under contract number ICT-2008-215270 (FRONTS). Also supported by Research Training Group GK-693 of the Paderborn Institute for Scientific Computation (PaSCo).

R. Královič and D. Niwiński (Eds.): MFCS 2009, LNCS 5734, pp. 600–611, 2009.

- We finally compare the problem of finding a proper colouring for $G_{n,m,p}$ to that of colouring hypergraphs so that no edge is monochromatic. We show how one can find in polynomial time a k-colouring of the vertices of $G_{n,m,p}$, for any integer k, such that no clique induced by only one label in $G_{n,m,p}$ is monochromatic.

Our techniques are novel and try to exploit as much as possible the hidden structure of random intersection graphs in this interesting range.

1 Introduction

We study random intersection graphs (RIGs), a relatively recent combinatorial model, that nicely captures interactions between nodes in distributed networks. Such interactions may occur for example when nodes blindly select resources (such as frequencies) from a limited globally available domain. For this graph model, we investigate the important combinatorial problem of vertex colouring, namely assigning integers (colours) to the vertices of the graph such that no adjacent vertices get the same colour. Colouring of sparse random intersection graphs was studied in [4]. The range of values that we consider here is different and gives quite denser graphs. Furthermore, our techniques are different than those used by the authors in [4]. Colouring properties provide useful insight to algorithmic design for important problems (like frequency assignment and concurrency control) in distributed networks characterized by dense interactions and resource limitations, such as wireless mobile and sensory networks.

1.1 Importance and Motivation

Random intersection graphs may be used to model several real-life applications characterized by dense, blind, possibly local interactions quite accurately (compared to the well known $G_{n,\hat{p}}$ model where edges appear independently with probability $\hat{p}$). In particular, the $G_{n,\hat{p}}$ model seems inappropriate for describing some real world networks (like mobile, sensor and social networks) because it lacks certain features of those networks, such as a scale free degree distribution and the emergence of local clusters. One of the underlying reasons for this mismatch is the independence between edges, in other words the missing transitivity that characterizes such networks: if vertices x and y exhibit a relationship of some kind in a real world network and so do vertices y and z, then this suggests a connection between vertices x and z, too.

For example, we consider the following scenario concerning efficient and secure communication in sensor networks: The vertices in our model correspond to sensor devices that blindly choose a limited number of resources among a globally available set of shared resources (such as communication channels, encryption keys etc). Whenever two sensors select at least one resource in common (e.g. a common communication channel, a common encryption key), a communication link is implicitly established (represented by a graph edge); this gives rise to communication graphs that look like random intersection graphs. Particularly for security purposes, the random selection of elements in our graphs can be seen as a way to establish local common keys on-line, without any global scheme for predistribution of keys. In such a case, the set of labels can be a global set of large primes (known to all) but each node selects uniformly at random only a few. Two nodes that have selected a common prime can communicate securely. Notice that no other node can know what numbers a different node has selected. Thus, the local communication is guaranteed to be secure. In the case when the shared resource is the wireless spectrum, then nodes choosing the same label (frequency) may interfere, and the corresponding link in the intersection graph abstracts a conflict, while a colour class (e.g. the vertices with the same colour) corresponds to wireless devices that can simultaneously access the wireless medium.

Random intersection graphs are also relevant to and capture quite nicely social networking. Indeed, a social network is a structure made of nodes (individuals or organizations) tied by one or more specific types of interdependency, such as values, visions, financial exchange, friends, conflicts, web links etc. Social network analysis views social relationships in terms of nodes and ties. Nodes are the individual actors within the networks and ties are the relationships between the actors. In particular, as [8] explicitly suggests proposing RIGs with a power law degree distribution, people with common friends may become friends as well, since they probably share common attributes and attributes can be obtained quite randomly (such as social preference, hobbies etc). So when a person A connects to B and B connects to C, the probability of a connection between A and C is higher because of more probable attributes in common. Thus RIGs can abstract such tendency for triangle clusterings.

Other applications may include oblivious resource sharing in a (general) distributed setting, interactions of mobile agents traversing the web etc. Even epidemiological phenomena (like spread of disease) tend to be more accurately captured by these "interaction-sensitive" random graph models.

1.2 Related Work

Random intersection graphs, denoted by $G_{n,m,p}$, were first defined in [18,32]. In this model, to each of the n vertices of the graph, a random subset of a universal set of m elements is assigned, by independently choosing elements with the same probability p. Two vertices u, v are then adjacent in the $G_{n,m,p}$ graph if and only if their assigned sets of elements have at least one element in common. Various properties of $G_{n,m,p}$ such as connectivity, degree distribution, independent sets, Hamilton cycles and its relation to the well known Bernoulli random graph model were investigated in [18,33,26,12,14].

In [4] the authors propose algorithms that whp probability colour sparse instances of $G_{n,m,p}$. In particular, for $m = n^\alpha, \alpha > 0$ and $p = o\left(\sqrt{\frac{1}{nm}}\right)$ they show that $G_{n,m,p}$ can be coloured optimally. Also, in the case where $m = n^\alpha, \alpha < 1$ and $p = o\left(\frac{1}{m \ln n}\right)$ they show that $\chi(G_{n,m,p}) \sim np$ whp. To do this, they prove that $G_{n,m,p}$ is chordal whp (or equivalently, the label graph does not contain cycles) and so a perfect elimination scheme can be used to find a colouring in polynomial time. The range of values we consider here is different than the one needed for the algorithms in [4] to work. In particular, we study colouring $G_{n,m,p}$ for the wider range $mp \leq (1-\alpha)\ln n$, as well as the denser range $mp \geq \ln^2 n$. We have to note also, that the proof techniques used in [4] cannot be used in the range we consider, since the properties that they examine do not hold in our case. Hence a completely new approach is needed.

The book [22] contains several techniques for upper bounding the chromatic number $\chi(G)$ of arbitrary graphs G. For general graphs it seems that one cannot easily beat the bound $\chi(G) \leq \Delta$, where Δ is the maximum degree of G (or $\chi(G) \leq \Delta+1$ if G is a clique or an odd cycle). However, assuming that the graph G has some additional structure, many interesting and advanced techniques for bounding the chromatic number exist for proving bounds for the chromatic number. These techniques are in fact algorithms that have (a) an iterative part that can generally be implemented in polynomial time which is followed almost always by (b) an application of the Local Lemma, which does not always lead to an algorithm that runs in polynomial time. Using a technique like that, Johansson (see chapter 13 of [22]) proved that for triangle free graphs we have $\chi(G) \leq \frac{160\Delta}{\ln \Delta}$, which is the strongest result known so far. Using a modification of Johansson's technique, Frieze and Mubayi [15] proved a quite strong bound on the chromatic number of simple Hypergraphs.

Concerning part (b) of the above technique, when some additional assumptions are true (that can in general be thought of as a stricter form of the Local Lemma assumptions), then Beck's technique [3] (see also [24] where the authors put Beck's technique in a more general framework) can be used to actually convert the Local Lemma existential proof into a polynomial running time algorithm! We have to say here that there are not many polynomial time algorithms that colour graphs with a relatively small number of colours. The best known approximation algorithm gives an approximation of $O\left(n\frac{(\log\log n)^2}{(\log n)^3}\right)$. Also, Molloy and Reed [23] used Beck's technique to find a polynomial algorithm to colour optimally graphs whose chromatic number is close to their maximum degree Δ. Another notable algorithm is the one proposed by Alon and Kahale [2] that uses the second to last eigenvalue of a special case of random graphs that are 3-colourable (more specifically they are constructed starting with 3 sets of n vertices each and then drawing edges between any pair of vertices that lie on different sets independently with probability p; of course these graphs are 3-colourable by definition and it is assumed that one is given the graph as it is, but without any information concerning the 3 sets).

Colouring Bernoulli random graphs was considered in [5] and also [21]. As it seems to be implied by these two works, randomness sometimes allows for smaller chromatic number than maximum degree whp. For $G_{n,\hat{p}}$, it is shown that whp $\chi(G_{n,\hat{p}}) \sim \frac{d}{\log d}$, where d is the mean degree. We have to point out here that both [5] and [21] prove that there exists a colouring of $G_{n,\hat{p}}$ using around $\frac{d}{\log d}$, but their proof does not lead to polynomial time algorithms. In fact, to the best of our knowledge, the problem of constructing a colouring of $G_{n,\hat{p}}$ using $\Theta\left(\frac{d}{\log d}\right)$ colours remains open for non-trivial values of $\hat{p}$.

Distributed Computing Related Work. From a distributed computing perspective, our work is related to collision avoidance and message inhibition methods ([20]) as well as range assignment problems in directional antennas' optimization ([7]). The (distant-2) chromatic number of random proximity and random geometric graphs has been studied in [9]. Furthermore, our colouring results can be applied in coordinating MAC access in sensor networks (see [6]). Our results also relate to distributed colouring and channel utilization ([19,31]). Finally, the RIG modeling can be useful in the efficient blind selection of few encryption keys for secure communications over radio channels ([10]), as well as in k-Secret sharing between swarm mobile devices (see [11]).

1.3 Our Contribution

In this paper we study the problem of colouring a random instance of the random intersection graphs model $\mathcal{G}_{n,m,p}$, mainly for the interesting range $m = n^{\alpha}, \alpha < 1$, where the model seems to differ the most from Bernoulli random graphs (see [14] and [29]). In particular

- For the case $mp \leq \beta \ln n$, for any constant $\beta < 1-\alpha$, we prove that np colours are enough to colour most of the vertices of the graph with high probability (whp). This means that even for quite dense graphs, using the same number of colours as those needed to properly colour the clique induced by any label suffices to colour almost all of the vertices of the graph. Note also that this range of values of m, p is quite wider than the one studied in [4].
- We propose and analyze an algorithm CliqueColour for finding a proper colouring of a random instance of $\mathcal{G}_{n,m,p}$, for any $mp \geq \ln^2 n$. The algorithm uses information of the label sets assigned to the vertices of $G_{n,m,p}$ and runs in $O\left(\frac{n^2mp^2}{\ln n}\right)$ time, which is polynomial in n and m. We also show by a reduction to the uniform random intersection graphs model that the number of colours required by the algorithm are of the correct order of magnitude with the actual chromatic number of $G_{n,m,p}$.
- We finally compare the problem of finding a proper colouring for $G_{n,m,p}$ to that of colouring hypergraphs so that no edge is monochromatic. We show how one can find in polynomial time a k-colouring of the vertices of $G_{n,m,p}$, for any integer k, such that no clique induced by only one label in $G_{n,m,p}$ is monochromatic.

Our proof techniques try to take advantage of the special randomness of $G_{n,m,p}$ and the way that edges appear in it as part of cliques. Especially in the design of algorithm CliqueColor, by carefully colouring a few vertices we were able to reduce the complex problem of colouring the whole graph, to the problem of colouring a simpler one.

2 Definition of the Model

We now formally define the model of random intersection graphs.

Definition 1 (Random Intersection Graph - $G_{n,m,p}$ [18,32]). *Consider a universe $\mathcal{M} = \{1, 2, \dots, m\}$ of elements and a set of vertices $V(G) = \{v_1, v_2, \dots, v_n\}$. If we assign independently to each vertex v_j, $j = 1, 2, \dots, n$, a subset S_{v_j} of $\mathcal{M}$ choosing each element $i \in \mathcal{M}$ independently with probability p and put an edge between two vertices v_{j_1}, v_{j_2} if and only if $S_{v_{j_1}} \cap S_{v_{j_2}} \neq \emptyset$, then the resulting graph is an instance of the random intersection graph $G_{n,m,p}$.*

Consider now the bipartite graph with vertex set $V(G) \cup \mathcal{M}$ and edge set $\{(v_j, i) : i \in S_{v_j}\}$. We will refer to this graph as the bipartite random graph $B_{n,m,p}$ associated to $G_{n,m,p}$.

In this model we also denote by L_l the set of vertices that have chosen label $l \in M$. The degree of $v \in V(G)$ will be denoted by $d_G(v)$.

By the above definition, one can realize that the edges in a random intersection graph appear as parts of cliques. In particular, the sets $L_l, l \in \mathcal{M}$ are in fact a (not necessarily minimal) clique cover of $G_{n,m,p}$. The size of each such clique is a binomial random variable with parameters n, p. Similarly, the number of cliques that a vertex v belongs to (i.e. $|S_v|$) is a binomial random variable with parameters m, p. In general, one can imagine that the smaller p is, the smaller the intersections between different (label) cliques will be in the clique cover implied by the sets $L_l, l \in \mathcal{M}$. In the extreme case where the cliques $L_l, l \in \mathcal{M}$ are disjoint, one could colour $G_{n,m,p}$ optimally by using the sets L_l. The authors in [4] show that one can also optimally colour $G_{n,m,p}$ even in the case where the sets $L_l, l \in \mathcal{M}$ intersect, provided that there is no induced cycle of size more than 3 in the instance graph. In this paper we consider a different range of values for the parameters of $\mathcal{G}_{n,m,p}$ that give whp intersection graphs in which the intersection between the sets $L_l, l \in \mathcal{M}$ is much higher and the clique structure of the graph is more complex. Consequently, the techniques used in [4] cannot be used here.

A closely related model to $\mathcal{G}_{n,m,p}$ is the *Uniform Random Intersection Graphs Model*, denote by $\mathcal{G}_{n,m,\lambda}$, where λ is a positive integer, which was first defined in [17]. In this model, every vertex chooses independently, uniformly at random a set of exactly λ labels and then we connect vertices that have at least one label in common. It is worth mentioning here that, apart from the case where the number of labels chosen by a vertex in $\mathcal{G}_{n,m,p}$ is concentrated around its mean value, the probabilistic behavior of $\mathcal{G}_{n,m,\lambda}$ seems a lot different than the one of $\mathcal{G}_{n,m,p}$.

3 Colouring Almost All Vertices

We are going to consider the case where $m = n^{\alpha}$, for $\alpha \in (0, 1)$ some fixed constant. The area $mp = o\left(\frac{1}{\ln n}\right)$ gives almost surely instances in which the label graph (i.e. the dual graph where the labels in $\mathcal{M}$ play the role of vertices and the vertices in V play the role of labels) is quite sparse and can be coloured optimally using $\max_{l \in \mathcal{M}} |L_l|$ colours (see [4]). We will here consider the denser area $mp = \Omega\left(\frac{1}{\ln n}\right)$. In this range of values, it is easy to see that the values of $|L_l|$ are concentrated around np. We were able to prove that even for values of the parameters m, p that give quite denser graphs, we can still use np colours to properly colour most of the graph.[1]. Our proof technique is inspired by analogous ideas of Frieze in [16] (see also [21]). Before presenting the main result, we state an auxiliary lemma that was proved in [26,27] and will be useful in the proof.

Lemma 1 ([26,27]). *Let $G_{n,m,p}$ be a random instance of the random intersection graphs model. Then the conditional probability that a set of k_0 vertices is an independent set, given that k_i of them are already an independent set is equal to*

$$\left((1-p)^{k_0-k_i} + (k_0-k_i)p(1-p)^{k_0-k_i-1}\left(1-\frac{k_i p}{1+(k_i-1)p}\right)\right)^m.$$

Proof. See [26,27]. □

We are now ready to present our theorem.

Theorem 1. *When $m = n^{\alpha}, \alpha < 1$ and $mp \leq \beta \ln n$, for any constant $\beta < 1-\alpha$. Then a random instance of the random intersection graphs model $G_{n,m,p}$ contains a subset of at least $n - o(n)$ vertices that can be coloured using np colours, with probability at least $1 - e^{-n^{0.99}}$.*

Proof. Due to lack of space we refer the interested reader to the full version of the paper [25]. □

It is worth noting here that the proof of Theorem 1 can also be used similarly to prove that $\Theta(np)$ colours are enough to colour $n - o(n)$ vertices even in the case where $mp = \beta \ln n$, for any constant $\beta > 0$. However, finding the exact constant multiplying np is technically more difficult.

4 A Polynomial Time Algorithm for the Case $mp \geq \ln^2 n$

In the following algorithm every vertex chooses i.u.a.r (independently, uniformly at random) a preference in colours, denoted by $shade(\cdot)$ and every label l chooses a preference in the colours of the vertices in L_l, denoted by $c_l(\cdot)$.

Algorithm CliqueColour
Input: An instance $G_{n,m,p}$ of $\mathcal{G}_{n,m,p}$ and its associated bipartite $B_{n,m,p}$.
Output: A proper colouring $G_{n,m,p}$.

1. for every $v \in V$ choose a colour denoted by $shade(v)$ independently, uniformly at random among those in $\mathcal{C}$;
2. for every $l \in \mathcal{M}$ choose a colouring $c_l(\cdot)$ of the vertices in L_l such that for every colour in $\{c \in \mathcal{C} : \exists v \in L_l \text{ with } shade(v) = c\}$ there is exactly one vertex in the set $\{u \in L_l : shade(u) = c\}$ having $c_l(u) = c$ while the rest remain uncoloured;

[1] Note however, that this does not mean that the chromatic number is close to np, since the part that is not coloured could be a clique in the worst case.

3. set $U = \emptyset$ and $C = \emptyset$;
4. **for** $l = 1$ **to** m **do** {
5. colour every vertex in $L_l \backslash \{U \cup C\}$ according to $c_l(\cdot)$ iff there is no collision with the colour of a vertex in $L_l \cap C$;
6. include every vertex in L_l coloured that way in C and the rest in U; }
7. let $\mathcal{H}$ denote the (intersection) subgraph of $G_{n,m,p}$ induced by the vertices in U;
8. give a proper colouring of $\mathcal{H}$ using a new set of colours $\mathcal{C}'$;
9. **output** a colouring of $G_{n,m,p}$ using $|\mathcal{C} \cup \mathcal{C}'|$ colours;

It is easy to see now that the above algorithm provides a proper colouring of its input graph. The number of colours that it needs (i.e. the cardinality of the sets $\mathcal{C}$ and $\mathcal{C}'$) and the time needed to colour $\mathcal{H}$ in step 8 are considered in Theorem 3.

Theorem 2 (Correctness). *Given an instance $G_{n,m,p}$ of the random intersection graphs model, algorithm CliqueColour always finds a proper colouring.*

Proof. Due to lack of space we refer the interested reader to the full version of the paper [25]. □

The following theorem concerns the efficiency of algorithm CliqueColour, provided that additionally $mp \geq \ln^2 n$ and $p = o\left(\frac{1}{\sqrt{m}}\right)$. Notice that for p larger than $\frac{1}{\sqrt{m}}$, every instance of the random intersection graphs model $\mathcal{G}_{n,m,p}$, with $m = n^\alpha, \alpha < 1$, is complete whp.

Theorem 3 (Efficiency). *Algorithm CliqueColour succeeds in finding a proper $\Theta\left(\frac{nmp^2}{\ln n}\right)$-colouring using of $G_{n,m,p}$ in polynomial time, provided that $mp \geq \ln^2 n, p = o\left(\frac{1}{\sqrt{m}}\right)$ and $m = n^\alpha, \alpha < 1$.*

Proof. For $s \in \mathcal{C}$, let Z_c denote the number of vertices $v \in V$ such that $shade(v) = c$. Z_c is a binomial random variable, so by Chernoff bounds we can see that, for any positive constant β_1 that can be arbitrarily small

$$\Pr\left(\left|Z_c - \frac{n}{|\mathcal{C}|}\right| \geq \frac{\beta_1 n}{|\mathcal{C}|}\right) \leq 2e^{-\frac{\beta^2 n}{3|\mathcal{C}|}}.$$

For $|\mathcal{C}| = \Theta\left(\frac{mnp^2}{\ln n}\right)$ and $p = o\left(\frac{1}{\sqrt{m}}\right)$, we can then use Boole's inequality to see that there is no $c \in \mathcal{C}$ such that $\left|Z_c - \frac{n}{|\mathcal{C}|}\right| \geq \frac{\beta_1 n}{|\mathcal{C}|}$, with probability $1 - o(1)$, i.e. almost surely.

Using the same type of arguments, we can also verify that for arbitrarily small positive constants β_2 and β_3, we have that $\Pr(\exists v \in V : ||S_v| - mp| \geq \beta_2 mp) = o(1)$ and $\Pr(\exists l \in \mathcal{M} : ||L_l| - np| \geq \beta_3 np) = o(1)$, for all $mp = \omega(\ln n)$ and $m = n^\alpha, \alpha < 1$.

We will now prove that the maximum degree of the graph $\mathcal{H}$ is small enough to allow a proper colouring of $\mathcal{H}$ using $\mathcal{C}' = \Theta\left(\frac{nmp^2}{\ln n}\right)$ colours. For a label

$l \in \mathcal{M}$ let Y_l denote the number of vertices $v \in L_l$ such that $c_l(v) \neq shade(v)$. In order for a label l not to be able to assign colour $shade(v)$ to $v \in L_l$, it should be the case that it has assigned colour $shade(v)$ to another vertex $u \in L_l$ with $shade(u) = shade(v)$. Hence, the only way to have a collision is when two or more vertices with the same shade have all chosen label l. Notice also that in order to have $Y_l \geq k$, the number of different shades appearing among the vertices that have chosen label l should be at most $|L_l| - k$. This means that $\Pr(Y_l \geq k) \leq \binom{|L_l|}{k}\left(\frac{|L_l|-k}{|\mathcal{C}|}\right)^k$. Given the concentration bound for $|L_l|$, we have that $\Pr(\exists l : Y_l \geq k)$ is at most

$$m\binom{(1+\beta_3)np}{k}\left(\frac{(1+\beta_3)np-k}{|\mathcal{C}|}\right)^k + o(1) \leq m\left(\frac{3np}{k}\right)^k\left(\frac{2np}{|\mathcal{C}|}\right)^k + o(1).$$

By now setting $k = \frac{np}{\ln n}$ and for $|\mathcal{C}| \geq 18\frac{mnp^2}{\ln n}$ we then have that, with probability $1 - o(1)$, there is no label $l \in \mathcal{M}$ such that $Y_l \geq \frac{np}{\ln n}$.

For a label $l \in \mathcal{M}$ now let W_l be the number of vertices $v \in L_l$ such that $shade(v) = c_l(v)$ but they remained uncoloured, hence included in $\mathcal{H}$. In order for a vertex $v \in L_l$ to be counted in W_l, there should exist a label j prior to l (i.e. a label among $1, \ldots l-1$) such that $v \in L_j$ and there is another vertex $u \in L_j$ with $shade(u) = shade(v)$. The probability that this happens is at most $p\left(1-(1-p)^{Z_{shade(v)}}\right)\left(1+(1-p)+(1-p)^2+\cdots\right) = 1-(1-p)^{Z_{shade(v)}}$. The crucial observation now is that, because choices of labels by vertices (of the same shade or not) is done independently and because the vertices counted in W_l have (by definition of the colouring $c_l(\cdot)$ in step 2 of the algorithm) different shades, the inclusion in W_l of any vertex $u \in L_l$ with $shade(u) = c_l(u)$ does not affect the inclusion of another $v \in L_l \backslash \{u\}$ with $shade(v) = c_l(v)$. Hence, taking also into account the concentration bound for $Z_{shade(v)}$ and $|L_l|$, we have that

$$\Pr(\exists l : W_l \geq k') \leq m\binom{(1+\beta_3)np}{k'}\left(1-(1-p)^{(1+\beta_1)\frac{n}{|\mathcal{C}|}}\right)^{k'} + o(1).$$

By now setting $k' = \frac{np}{\ln n}$ and using the relation $(1-x)^y \sim 1-xy$, valid for all x, y such that $xy = o(1)$, we have that when $|\mathcal{C}| \geq 18\frac{mnp^2}{\ln n}$, there is no label l such that $W_l \geq \frac{np}{\ln n}$, with high probability.

We have then proved that the number of vertices in U of the algorithm that have chosen a specific label is with high probability at most $\frac{2np}{\ln n}$. Since, for any vertex v in $G_{n,m,p}$ has $|S_v| \leq (1+\beta_2)mp$, we conclude that the maximum degree in $\mathcal{H}$ satisfies $\max_{v \in \mathcal{H}} degree_{\mathcal{H}}(v) \leq (1+\beta_2)mp\frac{2np}{\ln n}$. It is then evident that we can colour $\mathcal{H}$ greedily, in polynomial time, using $\frac{2.1nmp^2}{\ln n}$ more colours, with high probability. Hence, we can colour $G_{n,m,p}$ in polynomial time, using at most $\frac{20.1nmp^2}{\ln n}$ colours in total. □

It is worth noting here that the number of colours used by the algorithm in the case $mp \geq \ln^2 n, p = O\left(\frac{1}{\sqrt[4]{m}}\right)$ and $m = n^\alpha, \alpha < 1$ is of the correct order of magnitude. Indeed, by the concentration of the values of $|S_v|$ around mp

for any vertex v with high probability, one can use the results of [28] for the uniform random intersection graphs model $G_{n,m,\lambda}$, with $\lambda \sim mp$ to provide a lower bound on the chromatic number. Indeed, it can be easily verified that the independence number of $G_{n,m,\lambda}$, for $\lambda = mp \geq \ln^2 n$ is at most $\Theta\left(\frac{\ln n}{mp^2}\right)$, which implies that the chromatic number of $G_{n,m,\lambda}$ (and hence of the $G_{n,m,p}$ because of the concentration of the values of $|S_v|$) is at least $\Omega\left(\frac{nmp^2}{\ln n}\right)$.

5 Colouring Random Hypergraphs

The model of random intersection graphs $\mathcal{G}_{n,m,p}$ could also be though of as generating random Hypergraphs. The Hypergraphs generated have vertex set V and edge set $\mathcal{M}$. There is a huge amount of literature concerning colouring hypergraphs. However, the question about colouring there seems to be different from the one we answer in this paper. More specifically, a proper colouring of a hypergraph seems to be any assignment of colours to the vertices, so that no monochromatic edge exists. This of course implies that fewer colours than the the chromatic number (studied in this paper) are needed in order to achieve this goal.

We would also like to mention that as far as $\mathcal{G}_{n,m,p}$ is concerned, the problem of finding a colouring such that no label is monochromatic seems to be quite easier when p is not too small.

Theorem 4. *Let $G_{n,m,p}$ be a random instance of the model $\mathcal{G}_{n,m,p}$, for $p = \omega(\frac{\ln m}{n})$ and $m = n^{\alpha}$, for any fixed $\alpha > 0$. Then with high probability, there is a polynomial time algorithm that finds a k-colouring of the vertices such that no label is monochromatic, for any fixed integer $k \geq 2$.*

Proof. Due to lack of space we refer the interested reader to the full version of the paper [25]. □

6 Future Work

We are currently trying to extend the applicability of our colouring methods to other ranges of p, and to the related model $\mathcal{G}_{n,m,\lambda}$, where each vertex selects u.a.r. λ labels to form its corresponding label set.

References

1. Alon, N., Spencer, J.: "The Probabilistic Method". John Wiley & Sons, Inc., Chichester (2000)
2. Alon, N., Kahale, N.: A spectral technique for colouring random 3-colourable graphs. In: The Proceedings of the 26th Annual ACM Symposium on Theory of Computing, pp. 346–355 (1994)
3. Beck, J.: An Algorithmic Approach to the Lovász Local Lemma. Random Structures & Algorithms 2, 343–365 (1991)

4. Behrisch, M., Taraz, A., Ueckerdt, M.: Colouring random intersection graphs and complex networks. SIAM J. Discrete Math. 23(1), 288–299 (2009)
5. Bollobás, B.: "The chromatic number of random graphs". Combinatorica 8(1), 49–55 (1988)
6. Busch, C., Magdon-Ismail, M., Sivrikaya, F., Yener, B.: Contention-free MAC Protocols for Asynchronous Wireless Sensor Networks. Distributed Computing 21(1), 23–42 (2008)
7. Caragiannis, I., Kaklamanis, C., Kranakis, E., Krizanc, D., Wiese, A.: Communication in Wireless Networks with Directional Antennas. In: The Proceedings of the 20th Annual ACM Symposium on Parallelism in Algorithms and Architectures (SPAA 2008), pp. 344–351 (2008)
8. Deijfen, M., Kets, W.: Random Intersection Graphs With Tunable Degree Distribution and Clustering. CentER Discussion Paper Series No. 2007-08, http://ssrn.com/abstract=962359
9. Diaz, J., Lotker, Z., Serna, M.J.: The Distant-2 Chromatic Number of Random Proximity and Random Geometric Graphs. Inf. Process. Lett. 106(4), 144–148 (2008)
10. Dolev, S., Gilbert, S., Guerraoui, R., Newport, C.C.: Secure Communication over Radio Channels. In: The ACM Symposium on Principles of Distributed Computing (PODC), pp. 105–114 (2008)
11. Dolev, S., Lahiani, L., Yung, M.: SECRET SWARM UNIT reactive k –Secret sharing. In: Srinathan, K., Rangan, C.P., Yung, M. (eds.) INDOCRYPT 2007. LNCS, vol. 4859, pp. 123–137. Springer, Heidelberg (2007)
12. Efthymiou, C., Spirakis, P.G.: On the existence of hamiltonian cycles in random intersection graphs. In: Caires, L., Italiano, G.F., Monteiro, L., Palamidessi, C., Yung, M. (eds.) ICALP 2005. LNCS, vol. 3580, pp. 690–701. Springer, Heidelberg (2005)
13. Erdös, P., Selfridge, J.: On a Combinatorial Game. J. Combinatorial Th (A) 14, 298–301 (1973)
14. Fill, J.A., Sheinerman, E.R., Singer-Cohen, K.B.: Random Intersection Graphs when $m = \omega(n)$: An Equivalence Theorem Relating the Evolution of the $G(n,m,p)$ and $G(n,p)$ models, http://citeseer.nj.nec.com/fill98random.html
15. Frieze, A., Mubayi, D.: Colouring Simple Hypergraphs, arXiv:0901.3699v1 (2008)
16. Frieze, A.: On the Independence Number of Random Graphs. Disc. Math. 81, 171–175 (1990)
17. Godehardt, E., Jaworski, J.: Two models of Random Intersection Graphs for Classification. In: Opitz, O., Schwaiger, M. (eds.) Studies in Classification, Data Analysis and Knowledge Organization, pp. 67–82. Springer, Heidelberg (2002)
18. Karoński, M., Scheinerman, E.R., Singer-Cohen, K.B.: On Random Intersection Graphs: The Subgraph Problem. Combinatorics, Probability and Computing Journal 8, 131–159 (1999)
19. Kothapalli, K., Scheideler, C., Schindelhauer, C., Onus, M.: Distributed Colouring in $O(\sqrt{\log n})$ bits. In: The 20th IEEE International Parallel & Distributed Processing Symposium (IPDPS) (2006)
20. Leone, P., Moraru, L., Powell, O., Rolim, J.D.P.: Localization algorithm for wireless ad-hoc sensor networks with traffic overhead minimization by emission inhibition. In: Nikoletseas, S.E., Rolim, J.D.P. (eds.) ALGOSENSORS 2006. LNCS, vol. 4240, pp. 119–129. Springer, Heidelberg (2006)
21. Łuczak, T.: The chromatic number of random graphs. Combinatorica 11(1), 45–54 (2005)

22. Molloy, M., Reed, B.: Graph Colouring and the Probabilistic Method. Springer, Heidelberg (2002)
23. Molloy, M., Reed, B.: Colouring Graphs whose Chromatic Number is almost their Maximum Degree. In: The Proceedings of Latin American Theoretical Informatics, pp. 216–225 (1998)
24. Molloy, M., Reed, B.: Further Algorithmic Aspects of the Local Lemma. In: The Proceedings of the 30th ACM Symposium on Theory of Computing, pp. 524–529 (1998)
25. Nikoletseas, S., Raptopoulos, C., Spirakis, P.: Colouring Non-Sparse Random Intersection Graphs, `http://wwwhni.uni-paderborn.de/alg/mitarbeiter/raptopox/`
26. Nikoletseas, S.E., Raptopoulos, C., Spirakis, P.G.: The existence and efficient construction of large independent sets in general random intersection graphs. In: Díaz, J., Karhumäki, J., Lepistö, A., Sannella, D. (eds.) ICALP 2004. LNCS, vol. 3142, pp. 1029–1040. Springer, Heidelberg (2004)
27. Nikoletseas, S., Raptopoulos, C., Spirakis, P.: Large Independent Sets in General Random Intersection Graphs. Theoretical Computer Science (TCS) Journal, Special Issue on Global Computing (2008)
28. Nikoletseas, S., Raptopoulos, C., Spirakis, P.: On the Independence Number and Hamiltonicity of Uniform Random Intersection Graphs. In: accepted in the 23rd IEEE International Parallel and Distributed Processing Symposium (IPDPS) (2009)
29. Raptopoulos, C., Spirakis, P.G.: Simple and efficient greedy algorithms for hamilton cycles in random intersection graphs. In: Deng, X., Du, D.-Z. (eds.) ISAAC 2005. LNCS, vol. 3827, pp. 493–504. Springer, Heidelberg (2005)
30. Ross, S.M.: "Stochastic Processes", 2nd edn. John Wiley & Sons, Inc., Chichester (2000)
31. Schindelhauer, C., Voss, K.: Oblivious Parallel Probabilistic Channel Utilization without Control Channels. In: The 20th IEEE International Parallel & Distributed Processing Symposium (IPDPS) (2006)
32. Singer-Cohen, K.B.: Random Intersection Graphs, PhD thesis. John Hopkins University (1995)
33. Stark, D.: The Vertex Degree Distribution of Random Intersection Graphs. Random Structures & Algorithms 24(3), 249–258 (2004)

On the Structure of Optimal Greedy Computation (for Job Scheduling)

Periklis A. Papakonstantinou

University of Toronto
papakons@cs.toronto.edu

Abstract. We consider Priority Algorithm [4] as a syntactic model of formulating the concept of greedy algorithm for Job Scheduling, and we study the computation of optimal priority algorithms. A Job Scheduling subproblem $\mathbb{S}$ is determined by a (possibly infinite) set of jobs, every finite subset of which potentially forms an input to a scheduling algorithm. An algorithm is optimal for $\mathbb{S}$, if it gains optimal profit on every input. To the best of our knowledge there is no previous work about such arbitrary subproblems of Job Scheduling. For a finite $\mathbb{S}$, it is coNP-hard to decide whether $\mathbb{S}$ admits an optimal priority algorithm [12]. This indicates that meaningful characterizations of subproblems admitting optimal priority algorithms may not be possible. In this paper we consider those $\mathbb{S}$ that do admit optimal priority algorithms, and we show that the way in which all such algorithms compute has non-trivial and interesting structural features.

1 Introduction

Finding structure in computation of general models is a difficult task, and given the current state of knowledge a far-to-reach goal. In this paper we aim to find non-trivial structure in the computation of models which are syntactically restricted.

Greedy algorithms find many applications due to their conceptual simplicity, computational efficiency, and amenability to analysis. Borodin, Nielsen and Rackoff [4] introduced the "priority algorithm" model, aiming to syntactically formulate the concept of greedy algorithm by generalizing on-line computation. Much previous work on priority algorithms [1,2,4,11], [12,13] and its generalizations [3,9] focuses on priority algorithms for scheduling problems. Unlike most previous research this work is about optimal priority algorithms. Here too we consider priority algorithms for Job Scheduling where the goal is to maximize the total profit by scheduling weighted jobs, with given release, deadline and processing time, on identical multiple machines (see Section 2 for definitions).

A priority algorithm for Job Scheduling proceeds in stages. In the first part of a stage, the algorithm specifies an ordering, or priority, on the set of all possible jobs in the input. In the second part of a stage, the algorithm is given the job from the actual input that has highest priority, and the algorithm must make an irrevocable decision whether or not to schedule the job, and if so, how.

R. Královič and D. Niwiński (Eds.): MFCS 2009, LNCS 5734, pp. 612–623, 2009.

In [4] the general model is restricted in three ways: a *fixed-priority* algorithm computes an ordering only once in the beginning; a *greedy-priority* does not reject a job that can be scheduled, whereas a *memoryless* depends its choices only on the jobs scheduled so far, i.e. it ignores the rejected jobs. These concepts are formalized in Definition 1 (Section 2). In this paper, the term priority algorithm is understood as priority algorithm for JOB SCHEDULING. JOB SCHEDULING is defined in Section 2.

In the context of priority algorithm, subproblems of JOB SCHEDULING are identified by a (typically infinite) set of jobs $\mathbb{S}$. Every finite subset $S \subseteq \mathbb{S}$ is a possible input. The set (subproblem) $\mathbb{S}$ admits an optimal priority algorithm $\mathcal{A}$, if $\mathcal{A}$ gains optimal profit on every input. Common examples of subproblems are Interval Scheduling and Proportional Profit Job Scheduling. In this paper we consider *arbitrary* subproblems. Apart from the theoretical interest, in practice our results apply in a range of areas: from previously studied problems and algorithms, to specialized industrial problems.

The JOB SCHEDULING problem (determined by all possible jobs) does not admit optimal priority algorithms even for one-machine environment and restricted types of jobs [4]. Meaningful characterizations of JOB SCHEDULING subproblems admitting optimal algorithms is beyond our reach, due to their highly complex structure. For example, checking even for a finite subproblem whether it admits an optimal priority algorithm is coNP-hard [12]. Instead, we study properties of optimal priority algorithms in the most general setting. Consider any such subproblem and any priority algorithm optimal on this subproblem, where no explicit description is given either for the subproblem or for the algorithm. Is there anything interesting that can be said on how the algorithm computes? Finding such structure seems to be a non-trivial task. An optimal priority algorithm can be very wild during its computation. Let us start with a very restricted but interesting example.

Example 1. Consider the subproblem specified by the jobs in Figure 1, for one-machine scheduling.

Although in practice, many interesting subproblems are of infinite size for schedules on multiple-machine environments, such a finite subproblem with inputs from these 5 jobs is a good starting point. This example is used in the

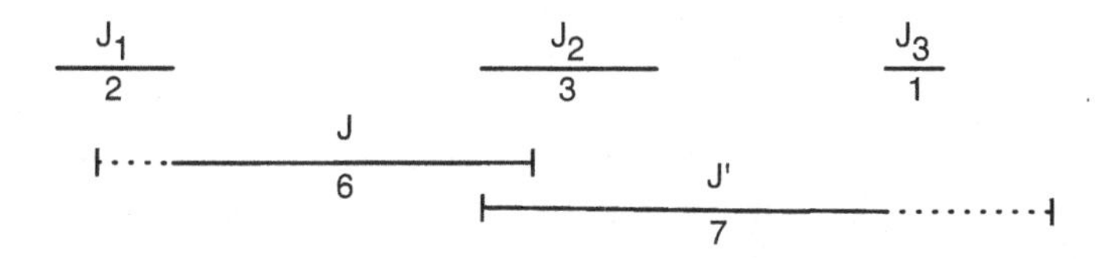

Fig. 1. Jobs J_1, J_2, J_3 are interval jobs (processing time = release - deadline). J and J' are jobs where their processing time is denoted by the solid horizontal line and their release and deadline times are denoted by the small vertical "line-bounds". The job profits are written below each job. Some schedules: $\{J_1, J\}$, $\{J_2, J\}$, $\{J_2, J'\}$, $\{J_3, J'\}$.

proof of Lemma 1. For the moment we just raise the following questions as a challenge. The depicted finite subproblem admits an optimal priority algorithm: the challenge is to find one. Does there exist an optimal *fixed*-priority algorithm? Theorem 2 (below) states that the profit gained from the most profitable parts of the input is insensitive to the presence of jobs of smaller profit. This statement becomes more interesting by observing that it does not imply any order in which an optimal priority algorithm may process the input. What does the example in Figure 1 tell us about this? These questions will be answered below.

Examples even for subproblems of small size reveal a big spectrum of possible optimal priority algorithms, with algorithms exhibiting a conceptually complicated behavior.

Motivation. Previous attempts formulate *optimal* greedy algorithms through connections to matroids [6], greedoids [10] and matroid embeddings [8]. Observe that the greedy algorithm for matroids is a priority algorithm whereas the optimal priority algorithm "Earliest Finishing Time First" does not have natural formulations as a matroid-related subset system (see [5] p.345). In this sense our work generalizes research relating matroids and greedy algorithm.

The choice of JOB SCHEDULING, as a case-study, is not arbitrary. JOB SCHEDULING is a well-studied NP-hard problem. In particular for priority algorithms, JOB SCHEDULING has the advantage of a natural input formulation in terms of jobs where revealing one does not restrict subsequent ones. This is not true if for example we have some other problem where the input consists of a graph, where revealing e.g. a vertex and its neighbors restricts the valid inputs to those containing the neighbors. Finally, an important motivation is the application to real-world problems. JOB SCHEDULING is a well-studied problem and studying efficient algorithms for JOB SCHEDULING subproblems is an area of intensive interest for practitioners. In particular, the general setting (arbitrary subproblem and algorithm) in which we derive our results allow us to apply them to industrial applications, where the scheduling subproblems are tailored to very specialized settings.

Roadmap and Contribution. In Section 2 we give definitions and we show Lemma 1 which rules-out a possible normal form for optimal priority algorithms. According to this not every optimal priority algorithm can be in the restricted class of *fixed-priority* algorithms. Section 3 is devoted to the proofs of Theorem 1, 2 and 3. Theorem 1 complements Lemma 1 by stating that the greedy and memoryless restrictions are not actual restrictions for optimal greedy algorithms. Our main technical contribution is theorems 2 and 3. The first shows that an optimal priority algorithm always gains the more profit from the more profitable parts of the input. The latter says that algorithms optimal on subproblems with jobs of distinct profit, always schedule the same set of jobs. Although the proofs of these two theorems are somehow involved, their details seem to matter to the extent that they reveal aspects of optimal greedy computation.

2 Definitions and Preliminaries

2.1 Definitions and Notation

A *job* J is $(id, r, d, p, w) \in \mathbb{N}^5$, where id is the job descriptor, r, d, p are its release, deadline and processing time respectively and w is its profit (weight). Furthermore, $r < d$ and $p \leq d - r$. We omit id if no confusion arises. When $p = d - r$ then we call this job an *interval.* In case of *proportional profit* $p = w$, a job is denoted by (id, r, d, p) (or (r, d, p)). An instance of the non-preemptive JOB SCHEDULING on identical machines $P|r_j|\sum w_j\bar{U}_j$ (in [7] notation) consists of m machines and n jobs. We want to assign jobs to machines so as to maximize the profit from properly scheduled jobs, rather than the equivalent but more common scheduling-theory formulation where every job must be scheduled according to the release time constraint and we aim to minimize the weight of late scheduled jobs. Also, w_J denotes the profit of job J and $\mathsf{profit}(S) = \sum_{J \in S} w_J$. For a priority algorithm $\mathcal{A}$ and an input S we denote by $\mathcal{A}(S)$ just the set of scheduled jobs (i.e. not where they are scheduled); e.g. $\mathsf{profit}(\mathcal{A}(S))$ denotes the profit of $\mathcal{A}$ on input S. For a given number of machines we denote by $\mathsf{opt}(S)$ the total profit of an optimal schedule on a finite set of jobs S.

Definition 1 (Priority Algorithms for Job Scheduling subproblems) *To simplify notation we assume that the number of machines is given to the algorithm in the beginning of each execution. Let $\mathbb{S}$ be a set of jobs which specifies a* JOB SCHEDULING *subproblem; input instances will be finite subsets of $\mathbb{S}$. A priority algorithm for $\mathbb{S}$ consists of two sets of functions, the ordering and the decision functions. We denote by $\mathcal{O}(\mathbb{S})$ the set of all total orderings on $\mathbb{S}$. Let D be the set of all possible decisions on whether and where to schedule a job. The computation for an input of n jobs proceeds in $n \in \mathbb{N}$ rounds.*

At the beginning of the k-th round the algorithm has already read from the input $k-1$ jobs. For each round k the ordering function *has the following form:*

$$r^{(k)} : (\mathbb{S} \times D)^{k-1} \to \mathcal{O}(\mathbb{S})$$

The decision function *for the k-th round has the following form:*

$$d^{(k)} : (\mathbb{S} \times D)^{k-1} \times \mathbb{S} \to D$$

We place no complexity (or computability) restrictions on the ordering and decision functions. We now describe how this algorithm operates on a finite input instance $S \subseteq \mathbb{S}$. Consider the k-th round, for $k \leq |S|$; $k-1$ jobs from S have already been processed and $\alpha \in (\mathbb{S} \times D)^{k-1}$ consists of the decisions made so far. First, from the jobs remaining in the input, the job J of highest priority according to $r^{(k)}(\alpha)$ is presented to the algorithm. Then, $d^{(k)}(\alpha, J)$ is applied to make an irrevocable *decision to reject or to schedule J (specifying the processor and its starting time). We sometimes refer to this as a* general *or* adaptive *priority algorithm and denote it by* PRIORITY.

Restrictions to the model: *In the* fixed-priority *model the ordering is determined only once in the beginning; this is denoted by the prefix* FIXED *(or F* *for short). According to the* greedy *restriction, when a* greedy-priority *algorithm* GREEDY *(G) reads a job that can be scheduled it necessarily schedules it. The* memoryless *restriction refers to* MEMORYLESS *(M) algorithms where the $r^{(k)}$'s and $d^{(k)}$'s depend only on the scheduled jobs.*

We say that the priority algorithm $\mathcal{A}$ is optimal for the subproblem determined by $\mathbb{S}$ *if for every (finite) input $S \subseteq \mathbb{S}$, $\mathcal{A}$ gains optimal profit on S.*

2.2 Our Results

In general, fixed, greedy and memoryless restrict the power of the model [11]. This is not true for optimal priority algorithms for one-machine scheduling on sets of intervals, where there always exists an optimal MEMORYLESS-FIXED-GREEDY-PRIORITY [12]. For multiple-machines and general types of jobs the following simple lemma states that the situation is different, even for subproblems with jobs of distinct and proportional profits.

Lemma 1. *For any number of machines, there exists an arbitrarily large set of jobs $\mathbb{S}$ and a* MEMORYLESS-GREEDY-PRIORITY *algorithm optimal on $\mathbb{S}$ such that no* FIXED-PRIORITY *algorithm is optimal on $\mathbb{S}$. Furthermore, $\mathbb{S}$ consists of jobs of distinct, proportional profit.*

Proof. We first present a proof for one-machine environments and a finite subproblem and then we show how to generalize it. Say that $\mathbb{S} = \{J_1, J_2, J_3, J, J'\}$ is depicted in Figure 1, where $J_1 = (1,3,2), J_2 = (8,11,3), J_3 = (15,16,1), J = (2,9,6), J' = (8,18,7)$. Here is a M-G-PRIORITY algorithm optimal on $\mathbb{S}$.

- The initial ordering is $J_2 > J_1 > J_3 > J > J'$. Read the first job and schedule it (if the first job is J or J' choose an arbitrary starting point).
- After reading the first job, if J_2 is scheduled then determine the ordering $J > J' > J_1 > J_3$ and schedule greedily. That is, schedule (if present) J starting at 2 (leftmost) and J' starting at 11 (rightmost).
- Else, if J_1 is the first scheduled job then determine the ordering $J > J' > J_3$ and schedule greedily. That is, if J or J' is present then schedule at the earliest available time wrt the so far scheduled jobs.
  Similarly, if J_3 is the first scheduled job (the ordering is $J' > J$).
  If the first scheduled job is among $\{J, J'\}$ then schedule the remaining one - if any.

It is easy to check optimality of the above algorithm. This together with the argument that rules-out FIXED-PRIORITY algorithms, and the generalization to multiple machines and infinite subproblems, are presented in the full version. □

Properties of optimal priority algorithms. For the rest of the paper we implicitly consider a number of machines k, which depends on the inputs we mention.

It is not hard to show the following normal form for optimal priority algorithms. The proof of this and the next two theorems are given in Section 3.

Theorem 1 (Normal form). *Given a set of jobs $\mathbb{S}$ and an algorithm $\mathcal{A} \in$ PRIORITY optimal on $\mathbb{S}$, there exists $\mathcal{A}' \in$ M-G-PRIORITY which is also optimal on $\mathbb{S}$.*

In general it is impossible to give a meaningful characterization of where an optimal priority algorithm schedules the input jobs. It is straightforward to construct subproblems where for inputs of n jobs the number of optimal schedules is e.g. $n!^{\Omega(n!)}$; consider many small jobs with very large release and deadline times. Moreover for each such schedule there is an optimal priority algorithm.

Our main contribution shows that there is structure on the decision of an optimal priority algorithm to gain profit by scheduling a job from particular subsets of the input. This takes us to the first strongly non-trivial fact about greedy computation. Although we do not have explicit knowledge neither of the algorithm nor of the subproblem, we know that the algorithm is going to make more profit from the more profitable parts of the input. Specifically, consider a run of an optimal priority algorithm on some input S. Populate this input set with jobs of smaller profit $\hat{S}$. Theorem 2 says that when an optimal priority algorithm computes on $S \cup \hat{S}$ the total profit gained from the scheduled jobs from S it is the same as if the algorithm were computing on S alone. This statement becomes more interesting by realizing that this "semantically greedy" behavior does not imply something for the order in which the algorithm processes jobs from the input. This is clear when we consider subproblems with non-overlapping jobs, but more importantly when considering subproblems where every optimal algorithm must consider jobs of smaller profit first. Our running example of Figure 1 determines a simple such subproblem: note that we cannot process first neither J nor J'. Still, although smaller jobs are considered first the profit gained from the most profitable parts (which is not revealed yet to the algorithm) is the same no matter what!

Theorem 2 (Main structural theorem). *Let $\mathbb{S}$ be a set of jobs and $\mathcal{A} \in$ PRIORITY be optimal on $\mathbb{S}$. Let $S \subseteq \mathbb{S}$ and $\hat{S} \subseteq \mathbb{S}$ be finite subsets, such that every job in $\hat{S}$ has profit (strictly) smaller than every job in S. If $\mathcal{A}(S \cup \hat{S}) = S' \cup \hat{S}'$, where $S' \subseteq S$ and $\hat{S}' \subseteq \hat{S}$, then* $\mathsf{profit}(S') = \mathsf{profit}(\mathcal{A}(S)) = \mathsf{opt}(S)$.

For subproblems where the jobs are of distinct profit all priority algorithms are bound to schedule the same sets of jobs. Once again, in general we cannot tell where a job is scheduled. Note that there are trivial examples where Theorem 3 fails in case of non-distinct profits; e.g. consider subproblems where any two interval-jobs overlap and they are of the same profit.

Theorem 3 (Distinct profit invariance). *Let $\mathbb{S}$ be a set of* distinct profit *jobs which admits an optimal priority algorithm. Then, every two priority algorithms optimal on $\mathbb{S}$ schedule the same set of jobs on every finite subset of $\mathbb{S}$.*

3 Proofs of Theorems 1, 2 and 3

Theorem 1 directly follows from Lemmas 4 and 5. We show Theorem 2 by first proving a restricted version in Lemma 2. The proof of Theorem 2 is given at the end, and it is a corollary of lemmas 2 and 5. Theorem 3 is a corollary of lemmas 3 and 6, where Lemma 3 is a restricted version of Theorem 3.

Perhaps, it worths noting that the proofs themselves in some sense make transparent issues concerning (optimal) greedy computation.

The following lemma is a restriction of Theorem 2 to memoryless algorithms.

Lemma 2. *Let $\mathbb{S}$ be a set of jobs and $\mathcal{A} \in$ M-PRIORITY be optimal on $\mathbb{S}$. Let $S \subseteq \mathbb{S}$ and $\hat{S} \subseteq \mathbb{S}$ be finite subsets, such that every job in $\hat{S}$ has profit (strictly) smaller than every job in S. If $\mathcal{A}(S \cup \hat{S}) = S' \cup \hat{S}'$, where $S' \subseteq S$ and $\hat{S}' \subseteq \hat{S}$, then* $\mathsf{profit}(\mathcal{A}(S)) = \mathsf{profit}(S')$.

Proof. By the optimality of $\mathcal{A}$, for every finite subset $S \subseteq \mathbb{S}$, $\mathcal{A}(S)$ is the set of jobs corresponding to an optimal schedule. We proceed by induction on the size of S. The induction basis ($S = \emptyset$) trivially holds. Suppose that for every S, where $|S| < k$ the induction claim holds. We wish to show that for every S where $|S| = k$ the induction claim is true. For the inductive step we assume the contrary and we show how to construct an input where $\mathcal{A}$ is non-optimal. That is, assume that there exist S and $\hat{S}$, $|S| = k$ where $\mathsf{profit}(\mathcal{A}(S)) \neq \mathsf{profit}(S')$. Since $\mathcal{A}$ is optimal we have that $\mathsf{opt}(S) = \mathsf{profit}(\mathcal{A}(S)) > \mathsf{profit}(S')$.

Outline. We proceed by first excluding a simple "boundary" case. Then we probe some points of the computation of $\mathcal{A}$ on $S \cup \hat{S}$ which guide us to the construction of two inputs. Finally, by "sieving" the computation of $\mathcal{A}$ when it runs on these two inputs we construct two new inputs where one is a subset of the other but $\mathcal{A}$ gains smaller profit on the bigger input set; i.e. $\mathcal{A}$ is not optimal.

First we observe that there does not exist $J \in S$ s.t. $J \notin \mathcal{A}(S)$ and $J \notin S'$; i.e. the two runs of $\mathcal{A}$ on S and on $S \cup \hat{S}$ cannot agree on rejections regarding jobs from S. If such J exists then since $\mathcal{A}$ is memoryless it schedules the same sets of jobs on $S \cup \hat{S}$ and on $(S \setminus \{J\}) \cup \hat{S}$. Hence by the induction hypothesis, $\mathsf{profit}(\mathcal{A}(S)) = \mathsf{profit}(\mathcal{A}(S \setminus \{J\})) = \mathsf{profit}(S')$, a contradiction.

Now we consider how $\mathcal{A}$ processes jobs from $\hat{S}$ on input $S \cup \hat{S}$. We probe the computation at the first point where the condition in the inductive predicate fails - recall that this is our assumption throughout this argument. Consider the sequence $T_1 = \langle J_1^1, J_2^1, \ldots, J_\alpha^1 \rangle, \alpha \leq |\hat{S}|$ in which $\mathcal{A}$ reads and accepts jobs from $\hat{S}$. Let $P_l = \{J_1^1, \ldots, J_l^1\}$, $1 \leq l \leq |\hat{S}|$, i.e. $\emptyset \equiv P_0 \subsetneq P_1 \subsetneq \ldots \subsetneq P_\alpha \subseteq \hat{S}$. Say that j is minimum s.t. $\mathcal{A}(S \cup P_j) = S_j' \cup P_j$ and $\mathsf{profit}(\mathcal{A}(S)) > \mathsf{profit}(S_j')$, where $S_j' \subseteq S$ is the set of all jobs from S which are scheduled when $\mathcal{A}$ computes on $S \cup P_j$. That is, $\mathcal{A}(S \cup P_{j-1}) = S_{j-1}' \cup P_{j-1}$ and $\mathsf{profit}(\mathcal{A}(S)) = \mathsf{profit}(S_{j-1}')$. Clearly, $j > 0$.

Observe that J_j^1 cannot be the last job scheduled by $\mathcal{A}$ when computing on $S \cup P_j$. If this were the case then $\mathsf{profit}\big(\mathcal{A}(S \cup P_{j-1})\big) > \mathsf{profit}\big(\mathcal{A}(S \cup P_j)\big)$, which contradicts the optimality of $\mathcal{A}$. Therefore, there exists a job from S scheduled by $\mathcal{A}$ after J_j^1.

A run σ of a priority algorithm at some input is the sequence of pairs of the presented jobs and the decisions of the algorithm. Note that a memoryless algorithm does not have knowledge of its decision to reject a job in some previous round. Let σ_j denote the run of $\mathcal{A}$ on $S \cup P_j$, and σ_{j-1} the run of $\mathcal{A}$ on $S \cup P_{j-1}$.

Up to the point that J_j^1 is read in σ_j, the corresponding prefixes of the runs σ_j and σ_{j-1} coincide, and thus the same jobs are scheduled. Record the sequence of jobs read by $\mathcal{A}$ after J_j^1 in σ_j; these jobs are all from S. Let this sequence be $T_2 = \langle J_1^2, J_2^2, \dots, J_m^2, \dots, J_n^2 \rangle$. As before we can easily see (more details are given in the full version) that σ_j and σ_{j-1} may agree only on acceptances. We look for the first job J_m^2 where the two runs of $\mathcal{A}$ disagree on the rejection/acceptance of this job. J_m^2 exists (i.e. $m \geq 1$) since otherwise $\mathsf{profit}(S'_j) = \mathsf{profit}(S'_{j-1}) = \mathsf{profit}(\mathcal{A}(S))$; moreover, all jobs in $\{J_1^2, \dots, J_{m-1}^2\}$ are scheduled in both runs.

We exclude the case that the point J_m^2 of disagreement can be such that $J_m^2 \notin S'_{j-1}$ and $J_m^2 \in S'_j$. Suppose that $J_m^2 \notin S'_{j-1}$ and $J_m^2 \in S'_j$. The two runs σ_j, σ_{j-1} are identical at least up to the point that J_j^1 is read. Since J_m^2 is read after every job in P_j, then every job in P_j is scheduled when $\mathcal{A}$ computes on $(S \setminus \{J_m^2\}) \cup P_j$ as when it computes on $S \cup P_j$. By the induction hypothesis, the fact that $\mathcal{A}$ is memoryless and by construction we have that: $\mathsf{profit}\big(\mathcal{A}((S \setminus \{J_m^2\}) \cup P_j)\big) = \mathsf{profit}\big(\mathcal{A}(S \setminus \{J_m^2\})\big) + \mathsf{profit}\big(P_j\big) = \mathsf{profit}\big(S'_{j-1}\big) + \mathsf{profit}\big(P_j\big) = \mathsf{profit}(\mathcal{A}(S)) + \mathsf{profit}\big(P_j\big) > \mathsf{profit}\big(S'_j\big) + \mathsf{profit}\big(P_j\big) = \mathsf{profit}\big(\mathcal{A}(S \cup P_j)\big)$, which contradicts the optimality of $\mathcal{A}$. Hence, $J_m^2 \in S'_{j-1}$ and $J_m^2 \notin S'_j$.

Now we are ready to show that $\mathcal{A}$ is not optimal by showing that it gains (strictly) smaller profit on a subset of an input than what it gains on this input. Let us denote by P the set of *all* jobs read by $\mathcal{A}$ up to the point that J_j^1 (excluding J_j^1) has been read. We know that when $\mathcal{A}$ runs on $P \cup \{J_1^2, \dots, J_{m-1}^2, J_m^2\}$ all of $\{J_1^2, \dots, J_{m-1}^2, J_m^2\}$ are scheduled. Let $\mathsf{profit}(\mathcal{A}(P)) = s$, $\mathsf{profit}(\{J_1^2, \dots, J_{m-1}^2\}) = s'$ and $\mathsf{profit}(\{J_m^2\}) = \Delta > \delta = \mathsf{profit}(\{J_j^1\})$. Note that since $\mathcal{A}$ considers jobs in the above order, it holds that $\mathcal{A}(P \cup \{J_1^2, \dots, J_m^2\}) = \mathcal{A}(P) \cup \{J_1^2, \dots, J_m^2\}$. Putting everything together we have:

$$\begin{aligned}\mathsf{profit}\big(\mathcal{A}(P \cup \{J_1^2, \dots, J_m^2\} \cup \{J_j^1\})\big) &= s + s' + \delta < s + s' + \Delta \\ &= \mathsf{profit}\big(\mathcal{A}(P \cup \{J_1^2, \dots, J_m^2\})\big)\end{aligned}$$

which contradicts the optimality of $\mathcal{A}$. □

Lemma 3 is the analog of Theorem 3 when we restrict it to memoryless algorithms. As in Lemma 2 here too we do an induction, compare computations (but now of different algorithms), and cut/concatenate them. Though, the argument is quite different. Assuming that two optimal, memoryless algorithms when computing on distinct profit input sets they accept different sets of jobs, we show that the subsets of the output where they disagree are singletons. We conclude using the optimality of the algorithms by showing that the two singleton sets must have the same profit and since each job has distinct profit, these two sets contain the same job.

Lemma 3. *Let $\mathbb{S}$ be a set of* distinct profit *jobs which admits an optimal priority algorithm. Then, every two optimal* MEMORYLESS-PRIORITY *algorithms on $\mathbb{S}$ schedule the same set of jobs on every finite subset of $\mathbb{S}$.*

Proof. By induction on the size of $S \subseteq \mathbb{S}$. The induction basis trivially holds. Assume that for every S, $|S| < k$ the lemma (induction predicate) is true. We wish to show that it holds for every $|S| = k$. Suppose that there exists an S, $|S| = k$ and two memoryless-priority algorithms $\mathcal{A}, \mathcal{A}'$ optimal on $\mathbb{S}$, such that $\mathcal{A}(S) \neq \mathcal{A}'(S)$. Consider the sequence in which $\mathcal{A}$ reads jobs from S: $T_{\mathcal{A}} = \langle J_1, J_2, \ldots, J_k \rangle$. The corresponding sequence for $\mathcal{A}'$ is $T_{\mathcal{A}'} = \langle J_{i_1}, J_{i_2}, \ldots, J_{i_k} \rangle$. Each job in $T_{\mathcal{A}}$ is associated with an acceptance or rejection decision. $\mathcal{A}$ and $\mathcal{A}'$ cannot agree on a rejection decision for a job. If this happens for a job $\hat{J} \in S$ then $\mathcal{A}(S) = \mathcal{A}(S \setminus \{\hat{J}\}) = \mathcal{A}'(S \setminus \{\hat{J}\}) = \mathcal{A}'(S)$, contradiction. Say that $\mathcal{A}(S) = S' \cup S_{\mathcal{A}}$, $\mathcal{A}'(S) = S' \cup S_{\mathcal{A}'}$, where S' contains every job accepted by both algorithms.

Say that $\hat{J}$ is the last job in $T_{\mathcal{A}}$ accepted by $\mathcal{A}$ and rejected by $\mathcal{A}'$. Remove $\hat{J}$ from S. Since $\mathcal{A}'$ is memoryless $\mathcal{A}'(S \setminus \{\hat{J}\}) = \mathcal{A}'(S)$. By the induction hypothesis, on $(S \setminus \{\hat{J}\})$ $\mathcal{A}$ accepts every job from $S_{\mathcal{A}'}$. Furthermore, there are no scheduled jobs from $S_{\mathcal{A}}$. Therefore, when $\mathcal{A}$ runs on S $\hat{J}$ is read before every job from $S_{\mathcal{A}'}$. We argue that $\hat{J}$ is the only job in $S_{\mathcal{A}}$. Suppose that $|S_{\mathcal{A}}| > 1$ (when $\mathcal{A}$ computes on S). Since $\hat{J}$ is the last job accepted by $\mathcal{A}$ and rejected by $\mathcal{A}'$, then there exists a $\hat{J}' \in S_{\mathcal{A}}$ read before $\hat{J}$ (when $\mathcal{A}$ runs on S); o.w. $\mathcal{A}$ gets less profit than $\mathcal{A}'$ on S. If we remove $\hat{J}$ then $\mathcal{A}$ rejects every job from $S_{\mathcal{A}}$. This holds since by the induction hypothesis the two algorithms accept the same set of jobs and since $\mathcal{A}'$ is memoryless, when removing $\hat{J}$, $\mathcal{A}'$ has an identical run as before in which it rejects every job from $S_{\mathcal{A}}$. But, $\mathcal{A}$ cannot reject every job from $S_{\mathcal{A}}$ since $\hat{J}'$ is read and scheduled. Hence, $S_{\mathcal{A}} = \{\hat{J}\}$.

Symmetrically we show that $S_{\mathcal{A}'} = \{\hat{J}'\}$. By the optimality we have $\mathsf{profit}(S' \cup S_{\mathcal{A}}) = \mathsf{profit}(S' \cup S_{\mathcal{A}'})$ which implies that $\mathsf{profit}(\{\hat{J}\}) = \mathsf{profit}(\{\hat{J}'\})$, contradicting the distinct profits assumption. □

It is easy to show Lemma 4 and a simple version of Lemma 5 which together imply the normal form theorem (Theorem 1). However, for the full version of Lemma 5 the proof is more involved. There is a difference in the conclusions of Lemma 4 and 5. In Lemma 4 the optimal greedy-priority algorithm schedules the same jobs as the adaptive priority algorithm. For Lemma 5 a similar thing is not true for the optimal memoryless algorithm and the algorithm with memory. However, with a little more technical effort we can obtain an optimal memoryless algorithm that agrees on the scheduled jobs with the one with memory on at least one input. This additional feature of the full version of Lemma 5 it is not needed for the proof of Theorem 1, but we need it for theorems 2 and 3.

Lemma 4. *Given a set of jobs $\mathbb{S}$ and a priority algorithm $\mathcal{A} \in$ PRIORITY optimal on $\mathbb{S}$, there exists $\mathcal{A}' \in$ G-PRIORITY which is also optimal on $\mathbb{S}$. Furthermore, for every input (finite set) $S \subseteq \mathbb{S}$, $\mathcal{A}(S) = \mathcal{A}'(S)$.*

The argument is based on the fact that an optimal priority algorithm cannot reject preemptively. If a priority algorithm rejects a job that can be scheduled then the algorithm is not optimal on the input that contains every job seen so far. From now on we blur the distinction between optimal algorithms in PRIORITY and G-PRIORITY.

In the following lemma we show that given a priority algorithm $\mathcal{A}$ optimal on a set of jobs, we can construct a family of memoryless priority algorithms optimal on this set. It is quite easy to show that the memoryless algorithm which simulates $\mathcal{A}$ assuming no rejected job to be present in the input, is optimal (that's an easy inductive argument in which we compare with the execution of the optimal $\mathcal{A}$ on the input set that doesn't contain the rejected jobs). However, to conclude the proofs of Lemma 6 and Theorem 2 on a fixed set R we need the memoryless algorithm $\mathcal{A}_M^R$ with the following properties:

i. $\mathcal{A}_M^R$ is optimal on $\mathbb{S}$ and
ii. On every input $S \subseteq \mathbb{S}$ if the set of all jobs rejected by $\mathcal{A}$ is R, then $\mathcal{A}(S) = \mathcal{A}_M^R(S)$.

In light of [11] it is not clear whether a memoryless algorithm with properties (i) and (ii) exists. Moreover, it is easy to show that there is $\mathbb{S}$ and optimal $\mathcal{A} \in$ G-PRIORITY such that for every optimal memoryless $\mathcal{A}_M$, in general it is not the case that $\mathcal{A}(S) = \mathcal{A}_M(S)$, if the set of jobs rejected by $\mathcal{A}$ is $R_1 \subseteq \mathbb{S}$ or $R_2 \subseteq \mathbb{S}$. However, if instead of two we have one set R, then there exists an algorithm $\mathcal{A}_M^R$ with the above two properties. This algorithm simulates $\mathcal{A}$ "assuming that jobs from R that (i) they are consistent with the computation and (ii) they could be rejected, are rejected". Here by "consistent" we mean that if at some round there is a job in R that should be rejected but this conflicts with previous rounds of the computation, then we ignore this job. There are a few more technical details on the implementation of $\mathcal{A}_M^R$. The details (formal description of $\mathcal{A}_M^R$) and the proof of Lemma 5 are given in the full version.

We denote the collection of $\mathcal{A}_M^R$ algorithms (one for each $R \subseteq \mathbb{S}$) by $\mathcal{F}_{\mathcal{A}}$.

Lemma 5. *Let $\mathbb{S}$ be a set of jobs, and $\mathcal{A} \in$ G-PRIORITY optimal on $\mathbb{S}$. Then, the family $\mathcal{F}_{\mathcal{A}}$ of* M-PRIORITY *algorithms is also optimal on $\mathbb{S}$. Furthermore, for every input S, where the set of jobs rejected by $\mathcal{A}$ equals $R \subseteq S$, the algorithm $\mathcal{A}_M^R$ produces an identical schedule to $\mathcal{A}$.*

Lemma 6. *Let $\mathbb{S}$ be a set of* distinct profit *jobs that admits an optimal priority algorithm. Then, for every optimal priority algorithm $\mathcal{A}$ there exists an optimal memoryless-priority algorithm $\mathcal{A}_M$ such that for every finite $S \subseteq \mathbb{S}$, $\mathcal{A}(S) = \mathcal{A}_M(S)$.*

Proof. Corollary of lemmas 3 and 5. By Lemma 3 and since $\mathbb{S}$ is a set of distinct profit jobs every optimal memoryless-priority algorithm on $\mathbb{S}$ schedules the same set of jobs on every subset of $\mathbb{S}$. By Lemma 5 we have a family of optimal memoryless-priority algorithms. Fix a memoryless-priority algorithm $\mathcal{A}_M$ optimal on $\mathbb{S}$. For an arbitrary finite $S \subseteq \mathbb{S}$, let R be the set of jobs rejected by $\mathcal{A}$ when running on S. Say that $\mathcal{A}_M^R$ is the optimal memoryless algorithm associated with the set of rejected jobs R as in the proof of Lemma 5. Then, $\mathcal{A}_M(S) = \mathcal{A}_M^R(S) = \mathcal{A}(S)$. □

Theorem 3 follows by Lemma 3 and 6. The proof of Theorem 2 follows.

Proof (Theorem 2). This is a corollary of lemmas 2 and 5. Fix S and $\hat{S}$ such that $\mathsf{profit}(\mathcal{A}(S \cup \hat{S})) \neq \mathsf{profit}(S')$, where $S, S', \hat{S}$ are defined as in the statement of the theorem. Let R be the set of jobs rejected by $\mathcal{A}$ on $(S \cup \hat{S})$. Consider the optimal memoryless-priority algorithm $\mathcal{A}^R_M$. Say that $\mathcal{A}^R_M(S \cup \hat{S}) = S'_{\mathcal{A}^R_M} \cup \hat{S}_{\mathcal{A}^R_M}$, as in Lemma 2. By Lemma 2 we have that $\mathsf{profit}(\mathcal{A}^R_M(S)) = \mathsf{profit}(S'_{\mathcal{A}^R_M})$. Since $\mathcal{A}^R_M$ and $\mathcal{A}$ are optimal on S we have that $\mathsf{profit}(\mathcal{A}(S)) = \mathsf{profit}(\mathcal{A}^R_M(S))$. Also, $\mathcal{A}^R_M$ and $\mathcal{A}$ have identical schedules on $(S \cup \hat{S})$. Hence, $S'_{\mathcal{A}^R_M} = S'$. Therefore, $\mathsf{profit}(\mathcal{A}(S)) = \mathsf{profit}(S')$. □

4 Conclusions

Previous research deals with the approximation power of priority algorithms for several variations of classic scheduling problems. The *approximation* power of priority algorithms for Job Scheduling is well-understood. Subsets of scheduling problems have received much attention both in practical and theoretical research. In this work we raise the question of what happens when we restrict the model to subsets of Job Scheduling that admit optimal priority algorithms. We set this question in a general framework where we do not need to have an explicit description either of the subset of Job Scheduling or of the algorithm. In this general setting we show that a priority algorithm optimal on a subproblem is bound to make decisions which are also "semantically greedy". We systematically remove possible restrictions from the model: memoryless and greedy, and we show that such optimal algorithms always attempt to maximize what they gain from the most profitable parts of the input. As the contrapositive of Theorem 2 states, an optimal priority algorithm cannot compensate by scheduling jobs of smaller profit for not scheduling more profitable jobs.

It is worth noting that our techniques merely exploit the general primitives of the model. For example, our arguments do not explicitly involve the fact that jobs may overlap. In this sense our proofs can be extended without modification to more general models for maximization packing problems where (i) the profit gained by a specific input element is always the same, i.e. in case of Job Scheduling it applies to identical machine environments, and (ii) input elements do not reveal information about future input elements.

Finally, we remark that our results hold also for the more general case where the inputs of the subproblems have underlying subset systems satisfying the hereditary property.

It seems that the most interesting research direction is to adjust a similar study for problems with more complicated and perhaps less natural descriptions of inputs. We believe that understanding the structure of computation of syntactically defined models is an issue worth pursuing, in the general program of classifying the intuitive concept of algorithmic paradigms.

Acknowledgments

I'd like to thank an anonymous reviewer for corrections and many useful suggestions in the presentation of the proofs. I'm thankful to Allan Borodin and to

Charles Rackoff for the useful remarks and their encouragement. I'd also like to thank Spyros Angelopoulos for his remarks on a previous draft of this work.

References

1. Angelopoulos, S.: Randomized priority algorithms. In: Solis-Oba, R., Jansen, K. (eds.) WAOA 2003. LNCS, vol. 2909, pp. 27–40. Springer, Heidelberg (2004)
2. Angelopoulos, S., Borodin, A.: On the power of priority algorithms for facility location and set cover. In: Jansen, K., Leonardi, S., Vazirani, V.V. (eds.) APPROX 2002. LNCS, vol. 2462, pp. 26–39. Springer, Heidelberg (2002)
3. Borodin, A., Cashman, D., Magen, A.: How well can primal-dual and local-ratio algorithms perform? In: Caires, L., Italiano, G.F., Monteiro, L., Palamidessi, C., Yung, M. (eds.) ICALP 2005. LNCS, vol. 3580, pp. 943–955. Springer, Heidelberg (2005)
4. Borodin, A., Nielsen, M.N., Rackoff, C. (Incremental) priority algorithms. Algorithmica (also SODA 2002) 37(4), 295–326 (2003)
5. Cormen, T.H., Leiserson, C.E., Rivest, R.L.: Introduction to Algorithms. MIT Press, Cambridge (2000)
6. Edmonds, J.: Matroids and the greedy algorithm. Math. Programming 1, 127–136 (1971)
7. Graham, R.L., Lawler, E.L., Lenstra, J.K., Rinnooy Kan, A.H.G.: Optimization and approximation in deterministic sequencing and scheduling: a survey. Annals of Discrete Mathematics 5, 287–326 (1979)
8. Helman, P., Moret, B.M.E., Shapiro, H.D.: An exact characterization of greedy structures. SIAM J. Discrete Math. 6(2), 274–283 (1993)
9. Horn, S.L.: One-pass algorithms with revocable acceptances for job interval selection. Master's thesis, University of Toronto (January 2004)
10. Korte, B., Lovasz, L.: Mathematical structures underlying greedy algorithms. In: Bekic, H. (ed.) Programming Languages and their Definition. LNCS, vol. 177, pp. 205–209. Springer, Heidelberg (1984)
11. Papakonstantinou, P.A.: Hierarchies for priority algorithms for job scheduling. Theoretical Computer Science 352(1-3), 181–189 (2006)
12. Papakonstantinou, P.A., Rackoff, C.W.: Characterizing sets of jobs that admit optimal greedy-like algorithms. Journal of Scheduling (2007) (to appear)
13. Regev, O.: Priority algorithms for makespan minimization in the subset model. Information Processing Letters 84(3), 153–157 (2003)

A Probabilistic PTAS for Shortest Common Superstring

Kai Plociennik

TU Chemnitz, Straße der Nationen 62, 09107 Chemnitz, Germany
kai.plociennik@informatik.tu-chemnitz.de

Abstract. We consider approximation algorithms for the shortest common superstring problem (SCS). It is well-known that there is a constant $f > 1$ such that there is no efficient approximation algorithm for SCS achieving a factor of at most f in the worst case, unless P = NP. We study SCS on random inputs and present an approximation scheme that achieves, for every $\varepsilon > 0$, a $1+\varepsilon$-approximation in expected polynomial time. This result applies not only if the letters are chosen independently at random, but also to the more realistic mixing model, which allows dependencies among the letters of the random strings. Our result is based on a sharp tail bound on the optimal compression, which improves a previous result by Frieze and Szpankowski.

1 Introduction and Results

In this paper, we consider the problem *Shortest Common Superstring* (SCS). For a finite alphabet Σ, a *string* s of *length* l over Σ is a sequence $s = s_1 s_2 \dots s_l$ of *letters* $s_i \in \Sigma$. We write $s \in \Sigma^l$ and denote s's length by $|s|$. For $0 \leq k \leq |s|$, s's *prefix* and *suffix* of length k are $s_1 \dots s_k$ and $s_{|s|-k+1} \dots s_{|s|}$, respectively. For strings s, t with $|s| \leq |t|$, we say that s is a *substring* of t if s is a contiguous subsequence of t, i.e., for some $1 \leq j \leq |t| - |s| + 1$, we have $s_i = t_{j-1+i}$, $i = 1, \dots, |s|$. In this case, the smallest such j is the *leftmost occurrence* of s in t. Given strings s, t, the *overlap* $\mathrm{ov}(s,t)$ of s and t is the longest suffix of s that is also a prefix of t. Now, given a multiset $S = \{s^1, \dots, s^n\}$ (in the following, n always denotes $|S|$) of strings over Σ, the problem SCS is to find an as short as possible *superstring* for S, i.e., a string t over Σ such that all $s^i \in S$ are substrings of t. The length of the shortest superstring is denoted $\mathrm{opt_l}(S)$. We measure the running time of algorithms with respect to the input length $||S|| := \sum_{i=1}^{n} |s^i|$.

Studying SCS is motivated by applications like DNA sequencing (see e.g. Gusfield [6]) and data compression, e.g., sequencing a DNA string is often done by sequencing short random fragments and then heuristically finding a short superstring for them. Due to its applications, fast algorithms for SCS are of interest, but since it is NP-hard (see Middendorf [9]), an efficient (i.e., polynomial worst case running time) algorithm finding optimal solutions unlikely exists. Thus, approximation algorithms have been studied. For a superstring t of an input S, let the (*length*) *approximation ratio* be $\mathrm{ar_l}(t) := |t|/\mathrm{opt_l}(S)$. Let t's *compression* be $c(t) := ||S|| - |t|$ and the *optimal compression* of S be $\mathrm{opt_c}(S) := ||S|| - \mathrm{opt_l}(S)$.

R. Královič and D. Niwiński (Eds.): MFCS 2009, LNCS 5734, pp. 624–635, 2009.

Then, the (compression) approximation ratio of t is $\mathrm{ar_c}(t) := \mathrm{opt_c}(S)/c(t)$. For both length and compression, an algorithm *achieves factor* f, $f \geq 1$, if it always computes a superstring with approximation ratio at most f. Tarhio and Ukkonen [12] and Turner [14] independently considered the now well-known algorithm *Greedy*. It repeatedly removes two strings s, t from S with largest overlap $|\mathrm{ov}(s,t)|$, overlaps them maximally to produce a string $s \oplus t$, and puts $s \oplus t$ back in S. In the end, the only string in S is the superstring. Both [12] and [14] conjectured that *Greedy* has factor 2 in the length measure (Blum, Jiang, Li, Tromp, and Yannakakis [4] showed factor 4), but no proof has been found yet. They *did* prove that it has factor 2 for compression, but it is easy to see that achieving some constant factor in one measure does not imply any constant factor for the other. In a sequence of papers, efficient algorithms with improved constant factor for length were presented, see e.g. [4], Teng and Yao [13], and Armen and Stein [2,3]. Currently, the best algorithm is by Sweedyk [11] and has factor 2.5. Unfortunately, SCS seems to resist arbitrarily good approximation (no PTAS exists): Since it is maxSNP-hard in both measures, for some constant $f > 1$ no efficient algorithm with factor f exists, unless P = NP. Vassilevska [15] showed that assuming P $\neq$ NP, explicit lower bounds for achievable factors f are 1.00082 for length and 1.00093 for compression. On the other hand, many problems with such inapproximability results are solved satisfactorily in practice, i.e., by "fast" algorithms computing "good" solutions. Hence, there is a discrepancy between the theoretical results from worst case analyses and empirical observations. Sometimes, this can be explained by an *average case* analysis. Instead of the worst case behavior over all inputs, the expected behavior for a random input from some distribution over all inputs is analyzed. The problem is that results for one input distribution say little about other distributions or inputs occurring in practice. Hence, Spielman and Teng [10] introduced the so called *smoothed analysis* of algorithms: An algorithm's behavior is analyzed for inputs which are chosen by an adversary and then slightly perturbed by small random modifications. Thus, it is a mixture of worst case and average case analysis. Formally, for an input I, the expected behavior regarding the random perturbation of I is analyzed, and we consider the worst expected behavior over all I. While an average case analysis may yield that we perform good for almost all inputs, a smoothed analysis may show that this even holds in a small neighborhood of every input. Especially for SCS, a smoothed analysis seems very reasonable: Laboratory methods introduce random errors when sequencing DNA, and DNA itself is evolved by random mutations from some ancestor's DNA. In this paper, we perform a probabilistic analysis for SCS "in the spirit of a smoothed analysis". The reason why we say "in the spirit" is that normally, one shows a dependence between the amount of random noise and the analyzed parameter, which here is the expected running time. In our result (see Theorem 1), we neglect the random noise's influence on the running time (we only show it to be polynomial), only the minimum total string length $||S||$ we can handle depends on the amount of noise. Nevertheless, we have a perturbation model of random inputs. In the paper, we also perform a classical average case analysis. Its random input models are introduced next.

Models of Random Inputs. To produce a random input S, we fix the alphabet Σ and the lengths $|s^i|$ of S's strings in some way. The s^i are then generated by a stationary and ergodic stochastic process $\{X_k\}_{k=-\infty}^{\infty}$ with random variables $X_k \in \Sigma$. For $r \leq s \in \mathbb{Z}$, let $X_r^s := \{X_k\}_{k=r}^{s}$ be the subsequence of $\{X_k\}$ between r and s. Then, a string $s^i \in S$ is generated by choosing the letters $X_1^{|s^i|}$ produced by our process, and the s^i are chosen independently. Let the *l-th order probability distribution* of $\{X_k\}$ be $P^{(l)}(x) := \Pr[X_i = x_i,\ 1 \leq i \leq l]$, $x \in \Sigma^l$. For brevity, let $p(x) := P^{(1)}(x)$, $x \in \Sigma$, let $p_{\min} := \min_{x\in\Sigma} p(x)$, and $p_{\max} := \max_{x\in\Sigma} p(x)$. In the *Bernoulli model*, the letters in our process are chosen independently. Since in many applications (e.g., DNA sequencing) this is unrealistic, we also consider the *mixing model* (see e.g. Łuczak and Szpankowski [7]), where our process has the *mixing property*. For $r \leq s$, let $\mathcal{F}_r^s$ be the σ-field generated by X_r^s. The mixing property is that a function $\alpha : \mathbb{N} \longrightarrow [0,1)$ with $\lim_{g\to\infty} \alpha(g) = 0$ exists such that for every $m \in \mathbb{Z}$, $g \in \mathbb{N}$, and events $\mathcal{A} \in \mathcal{F}_{-\infty}^{m}$ and $\mathcal{B} \in \mathcal{F}_{m+g}^{\infty}$,

$$(1-\alpha(g))\Pr[\mathcal{A}]\Pr[\mathcal{B}] \leq \Pr[\mathcal{A}\cap\mathcal{B}] \leq (1+\alpha(g))\Pr[\mathcal{A}]\Pr[\mathcal{B}] \ . \tag{1}$$

Since without dependencies, $\Pr[\mathcal{A}\cap\mathcal{B}] = \Pr[\mathcal{A}]\Pr[\mathcal{B}]$, (1) says that dependencies weaken with increasing distance of letters, and α characterizes their quantity. We note that the mixing model generalizes the Markovian model. In the Bernoulli model, where the *entropy* (the "amount of randomness") is $H = -\sum_{x\in\Sigma} p(x)\ln p(x)$, Alexander [1] analyzed the optimal compression. Essentially, a random input S_n with n strings is divided into a set $S_n^{(\mathrm{l})}$ of long strings and a set $S_n^{(\mathrm{s})}$ of short ones. For $v_n := (|S_n^{(\mathrm{l})}|\ln|S_n^{(\mathrm{l})}|)/H + \sum_{s\in S_n^{(\mathrm{s})}} |s|$, we have $\lim_{n\to\infty} \mathrm{E}[|\mathrm{opt_c}(S_n)/v_n - 1|] = 0$. Thus, if all strings are long, $(1-\varepsilon)(n\ln n)/H \leq \mathrm{E}[\mathrm{opt_c}(S_n)] \leq (1+\varepsilon)(n\ln n)/H$ for all $\varepsilon > 0$ and n large enough (we always assume n large where necessary). Later, Frieze and Szpankowski [5] and Yang and Zhang [17] considered the mixing model and the Bernoulli model, respectively, and analyzed the optimal compression and the one produced by some greedy algorithms. In [5], mixing model inputs S_n with n strings are considered. In this model, the entropy is $H = \lim_{l\to\infty} -\frac{\mathrm{E}[\ln P^{(l)}(X_1^l)]}{l}$. It is shown that for any $\varepsilon > 0$,

$$\Pr[(1-\varepsilon)(n\ln n)/H \leq \mathrm{opt_c}(S_n) \leq (1+\varepsilon)(n\ln n)/H] = 1 - o(1) \ ,$$

and the same holds for *Greedy*'s compression. Thus, *Greedy* compresses optimally in expectation, and $\Pr[\mathrm{opt_c}(S_n) > (1+\varepsilon)(n\ln n)/H] = o(1)$. Our first result is to strengthen the latter statement in Lemma 1 and Corollary 1: We prove an exponentially small upper bound on the probability for a limited class of models. Our second result is given in Theorem 1: We give an approximation scheme that for every $\varepsilon > 0$ has factor $1+\varepsilon$ in the length measure and polynomial expected running time. We note that Ma [8] has performed a smoothed analysis showing that *Greedy* has factor $1+o(1)$ in expectation. However, our approach and the one in [5] and [8] are different: While these papers consider efficient algorithms and *expected* solution quality, we demand *guaranteed* approximation ratio $1+\varepsilon$ and polynomial running time only in expectation. Before stating our results, it remains to introduce our smoothed (perturbation) model. For a random input

S, we assume that for every letter s^i_j in S's strings s^i, the adversary can specify a probability distribution $p^i_j : \Sigma \rightarrow [0,1]$. Then, the s^i_j are drawn independently according to the p^i_j. Notice that this generalizes the Bernoulli model, where all letters share the same distribution. Without restrictions, deterministic worst case inputs can be chosen, prohibiting our results. We hence demand that for a fixed $\hat{\varepsilon} \in (0,1/2]$, $(p^i_j)_{\max} = \max_{x\in\Sigma} p^i_j(x) \leq 1-\hat{\varepsilon}$ for all s^i_j. Thus, $p_{\max} \leq 1-\hat{\varepsilon}$ for $p_{\max} := \max_{s^i_j}(p^i_j)_{\max}$. Observe that we can model fixed strings with random noise: Fix a mutation probability $p_{\mathrm{m}} \in (0,1/2]$. For deterministically chosen strings s^i, we mutate every letter s^i_j independently with probability p_{m}. If a letter s^i_j with original value $x^i_j \in \Sigma$ is mutated, we choose the new value uniformly from $\Sigma \setminus \{x^i_j\}$. For every s^i_j, the resulting distribution p^i_j is $p^i_j(x^i_j) = 1-p_{\mathrm{m}}$ and $p^i_j(x') = p_{\mathrm{m}}/(|\Sigma|-1)$ for $x' \neq x^i_j$. It is easy to see that for $\hat{\varepsilon} := p_{\mathrm{m}}$, $p_{\max} \leq 1-\hat{\varepsilon}$. Thus, we can model this process. Notice that $\hat{\varepsilon}$ limits the amount of random noise and is the equivalent of the standard deviation in [10]. We prove our results for the smoothed and Bernoulli model in Sect. 2.1–2.2 and generalize them to the mixing model in Sect. 2.3. For clarity, we omit any rounding of non-integer values in the paper.

2 Results

For an input S for SCS, we let $\Delta(S) := \max_{1\leq i,j\leq n} |s^i| - |s^j| + 1$. In Sect. 2.2, we give an approximation scheme $ApproxSCS(S,c)$, $c > 0$ rational.

Theorem 1. *Fix a Bernoulli or smoothed model,* $\varepsilon \in (0,1)$, *and* $k \in \mathbb{N}$. *Set* $c := 2(1+k)/|\ln p_{\max}|$. *Then, for a random input* S *with* $||S|| \geq (2c/\varepsilon)(n \ln n)$ *and* $\Delta(S) \leq n^k$, $ApproxSCS(S,c)$ *has factor* $1+\varepsilon$ *for the length measure and polynomial expected running time.*

Notice that individual strings are not required to have a minimum length, the $\Omega(n \ln n)$ lower bound on $||S||$ demands only a $\Omega(\ln n)$ average length. With respect to applications, assuming $\Delta(S) = \mathrm{poly}(n)$ seems not very restrictive. For a Bernoulli model with a uniform distribution on Σ, since $p(x) = |\Sigma|^{-1}$ for all $x \in \Sigma$, we have $\frac{H}{|\ln p_{\max}|} = -\frac{\sum_{x\in\Sigma} p(x)\ln p(x)}{|\ln p_{\max}|} = \frac{-\ln(|\Sigma|^{-1})}{|\ln(|\Sigma|^{-1})|} = 1$. The following lemma applies hence to distributions close to the uniform one. Observe that the upper bound $e^{-\Omega(n \ln n)}$ on $\Pr[\mathrm{opt_c}(S) > (1+\varepsilon)(n \ln n)/H]$ can only be improved by a constant factor in the exponent: The smallest (nonzero) probability that *any* event regarding a random input S can have is the probability that all letters are chosen the value $x \in \Sigma$ with $p(x) = p_{\min}$. For inputs with $||S|| = O(n \ln n)$, the probability of this event is $p_{\min}^{O(n \ln n)} = e^{-O(n \ln n)}$.

Lemma 1. *Fix* $\varepsilon > 0$ *and a Bernoulli model with* $H/|\ln p_{\max}| < 1+\varepsilon$. *For a random input* S *with* $\Delta(S) = \mathrm{polylog}(n)$, *it holds that* $\Pr[\mathrm{opt_c}(S) > (1+\varepsilon)(n \ln n)/H] = e^{-\Omega(n \ln n)}$.

2.1 Upper Bounding the Optimal Compression

The following algorithm isolates, given an input S for SCS, a set S_{rel} of "relevant" (regarding $\mathrm{opt}_{\mathrm{l}}(S)$) strings.

Algorithm Partition(S)

1. Set $S_{\mathrm{rel}} := S$. Then, while there are two strings $s, t \in S_{\mathrm{rel}}$ with s a substring of t, set $S_{\mathrm{rel}} := S_{\mathrm{rel}} \setminus \{s\}$.
2. Set $S_{\mathrm{sub}} := S \setminus S_{\mathrm{rel}}$ and output $(S_{\mathrm{rel}}, S_{\mathrm{sub}})$.

It is easily seen that finally every string in S_{sub} is a substring of one in S_{rel}, and S_{rel} is *substring-free*: for no two strings $s, t \in S_{\mathrm{rel}}$, s is a substring of t. Furthermore, S and S_{rel} have the same possible superstrings. Thus, $\mathrm{opt}_{\mathrm{l}}(S) = \mathrm{opt}_{\mathrm{l}}(S_{\mathrm{rel}})$. Obviously, *Partition* can be implemented with running time $O(\mathrm{poly}(||S||))$.

Definition 1. *For an input $S = \{s^1, \ldots, s^n\}$ for SCS, the* string graph *is the complete, undirected, weighted graph $G_{\mathrm{s}}(S) = (S, E)$ on vertex set S with edge weights as follows: For an edge $e = \{s^i, s^j\} \in E$, if s^i is a substring of s^j or vice versa, $w(e) = \min\{|s^i|, |s^j|\}$. Otherwise, $w(e) = \max\{|\mathrm{ov}(s^i, s^j)|, |\mathrm{ov}(s^j, s^i)|\}$. Furthermore, we let $\mathrm{mxt}(S)$ be the weight of a maximum spanning tree in $G_{\mathrm{s}}(S)$.*

Lemma 2. *For every input S for SCS, $\mathrm{opt}_{\mathrm{c}}(S) \leq \mathrm{mxt}(S)$.*

Proof. We use a known property of substring-free inputs $\widetilde{S}$. Let $\widetilde{t}^*$ be a shortest superstring for $\widetilde{S}$, and $\widetilde{s}^1, \ldots, \widetilde{s}^{|\widetilde{S}|}$ be $\widetilde{S}$'s strings in order of increasing leftmost occurrence in $\widetilde{t}^*$. Then (see Vazirani [16]), the optimal compression for $\widetilde{S}$ fulfills

$$\mathrm{opt}_{\mathrm{c}}(\widetilde{S}) = \textstyle\sum_{i=1}^{|\widetilde{S}|-1} |\mathrm{ov}(\widetilde{s}^i, \widetilde{s}^{i+1})| \ . \tag{2}$$

Set $(S_{\mathrm{rel}}, S_{\mathrm{sub}})$:=*Partition(S)*. Let t^* be a shortest superstring for S_{rel}, and let $s_{\mathrm{r}}^1, \ldots, s_{\mathrm{r}}^{n'}$ be the strings of S_{rel} in order of increasing leftmost occurrence in t^*. Since S_{rel} is substring-free, $\mathrm{opt}_{\mathrm{c}}(S_{\mathrm{rel}}) = ||S_{\mathrm{rel}}|| - |t^*| = \sum_{i=1}^{n'-1} |\mathrm{ov}(s_{\mathrm{r}}^i, s_{\mathrm{r}}^{i+1})| \leq \sum_{i=1}^{n'-1} w(\{s_{\mathrm{r}}^i, s_{\mathrm{r}}^{i+1}\})$ for the weights $w(\{s_{\mathrm{r}}^i, s_{\mathrm{r}}^{i+1}\})$ of the edges $\{s_{\mathrm{r}}^i, s_{\mathrm{r}}^{i+1}\} \in E$ in the string graph $G_{\mathrm{s}}(S) = (S, E)$. Hence, by choosing $P := \{\{s_{\mathrm{r}}^i, s_{\mathrm{r}}^{i+1}\} \in E \,|\, 1 \leq i \leq n' - 1\}$, we get a path P connecting all strings in S_{rel} with weight $w(P) \geq ||S_{\mathrm{rel}}|| - |t^*|$. We set $T := P$ and extend T to a spanning tree of $G_{\mathrm{s}}(S)$. For every string $s^i \in S_{\mathrm{sub}}$, there is a string $s^{r_i} \in S$ such that *Partition* removed s^i from S_{rel} since it found s^i to be a substring of s^{r_i}. Now, for every $s^i \in S_{\mathrm{sub}}$, we add $\{s^i, s^{r_i}\}$ to T. Since *Partition* never considers a string again once it is removed from S_{rel}, it is easy to see that adding the edges $\{s^i, s^{r_i}\}$ to T does not create cycles. Therefore, in the end, T contains $|S_{\mathrm{rel}}| - 1 + |S_{\mathrm{sub}}| = n - 1$ edges and no cycles and hence is a tree. Notice that $w(T) - w(P) = \sum_{s^i \in S_{\mathrm{sub}}} w(\{s^i, s^{r_i}\}) = \sum_{s^i \in S_{\mathrm{sub}}} |s^i| = ||S_{\mathrm{sub}}||$. Using $w(P) \geq ||S_{\mathrm{rel}}|| - |t^*|$, we get $w(T) \geq ||S_{\mathrm{rel}}|| - |t^*| + ||S_{\mathrm{sub}}|| = ||S|| - |t^*|$. Now, since t^* is also a shortest superstring for S, it follows that $\mathrm{opt}_{\mathrm{c}}(S) = ||S|| - |t^*| \leq w(T) \leq \mathrm{mxt}(S)$. □

For a random input S for SCS and a rational number $c > 0$, the following efficient algorithm either certifies by outputting "success" that $\mathrm{opt}_{\mathrm{c}}(S) \leq cn \ln n$, or outputs "fail". The latter happens only with exponentially small probability.

Algorithm BoundComp($S = \{s^1, \ldots, s^n\}, c$)

1. Compute mxt(S). Output "success" if $\text{mxt}(S) \leq cn \ln n$ and "fail" otherwise.

Lemma 3. *Fix a Bernoulli or smoothed model and $\varepsilon > 0$. For a random input S for SCS and $c \geq (1 + (\ln \Delta(S))/\ln n)/|\ln p_{\max}| + \varepsilon$, $c = O(1)$,* BoundComp(S, c) *runs in time $O(\text{poly}(||S||))$. It outputs "fail" with probability $e^{-\Omega(n \ln n)}$. If it outputs "success", $\text{opt}_c(S) \leq cn \ln n$.*

Proof. The algorithm clearly has running time $O(\text{poly}(||S||))$. Lemma 2 yields that if "success" is output, $\text{opt}_c(S) \leq \text{mxt}(S) \leq cn \ln n$. It remains to analyze the failure probability, which is $\Pr[\text{mxt}(S) > cn \ln n]$. If this happens, $\text{mxt}(S) = (cn \ln n) + \delta =: f(\delta)$ for some $\delta \in \{1, \ldots, ||S|| - cn \ln n\}$, since $\text{mxt}(S) \leq ||S||$. Fix such a δ. If $\text{mxt}(S) = f(\delta)$, there is a spanning tree $T \subseteq E$ in the string graph G_s with edge weights $w(e) \geq 0$ for $e \in T$ such that $\sum_{e \in T} w(e) = f(\delta)$. By definition of G_s, this implies the following: For every edge $e = \{s^i, s^j\} \in T$, there is a *direction* $(a, b) \in \{(i,j), (j,i)\}$ such that $|\text{ov}(s^a, s^b)| = w(e)$ (e has *type* overlap) or s^a is a substring of s^b and $|s^a| = w(e)$ (e has type substring). Furthermore, for a "substring" edge, there is a leftmost occurrence $l_{a,b} \in \{1, \ldots, |s^b| - |s^a| + 1\}$ where s^a appears in s^b. A spanning tree T together with its edge weights, directions, types, and leftmost occurrences is called a *spanning tree configuration*. An example of a string graph with a spanning tree configuration is shown in Fig. 1 (only the tree's edges with their weights are shown). An arrow from s^i to s^j at an edge $e = \{s^i, s^j\}$ means that e's direction is (i, j), and the type is indicated by 'o' (overlap) or 's, l' (substring with leftmost occurrence l). We upper bound the number of possible spanning tree configurations: There are $n^{n-2} \leq e^{n \ln n}$ spanning trees in a graph on n labeled vertices. For natural numbers $x \geq y$, the number of ordered partitions of x into y nonnegative summands is $\binom{x+y-1}{y-1}$, and $\binom{x}{y} \leq (ex/y)^y$. Hence, we can partition $f(\delta)$ into $n-1$ edge weights in $\binom{f(\delta)+n-2}{n-1} \leq \binom{2f(\delta)}{n} \leq (2ef(\delta)/n)^n = e^{n \ln(f(\delta)/n) + O(n)}$ ways. The edge directions and types can be chosen in $4^{n-1} = e^{O(n)}$ ways. For the leftmost occurrence of a string s^i in s^j, we have $|s^j| - |s^i| + 1 \leq \Delta(S)$ choices. Thus, for the leftmost occurrences of the at most $n - 1$ "substring" edges, we have at most $(\Delta(S))^{n-1} \leq e^{n \ln \Delta(S)}$ choices. Let $\widehat{\mathcal{C}}$ be the set of all spanning tree configurations. For brevity, let $k := (\ln \Delta(S))/\ln n$. Then, $n \ln \Delta(S) = kn \ln n$. Using this, we conclude

$$|\widehat{\mathcal{C}}| \leq e^{(n \ln n) + n \ln(f(\delta)/n) + (n \ln \Delta(S)) + O(n)} = e^{(1+k)(n \ln n) + n \ln(f(\delta)/n) + O(n)} \, . \quad (3)$$

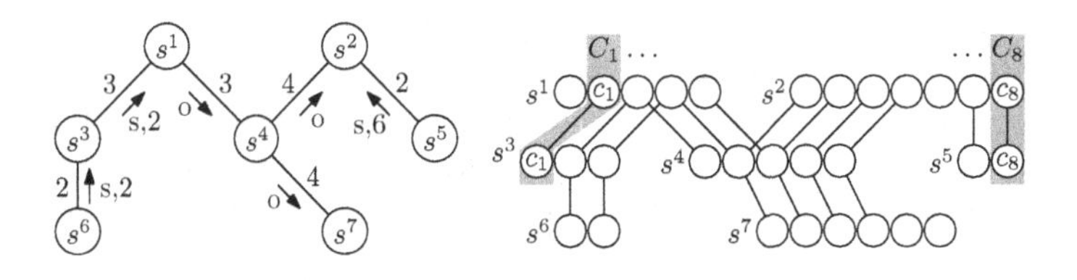

Fig. 1. A Spanning Tree Configuration $\mathcal{C}$ and the Resulting Equality Graph $G_e(\mathcal{C})$

Now, we upper bound the probability that a fixed configuration $\mathcal{C} \in \widehat{\mathcal{C}}$ appears, i.e., the string graph of S contains the maximum spanning tree T from $\mathcal{C}$. We model the properties of S's strings implied by $\mathcal{C}$'s appearance using the *equality graph* $G_e(\mathcal{C}) = (V, E)$, which looks as follows: For every string $s^i \in S$, it contains the vertices s^i_j, $j = 1, \ldots, |s^i|$. In words, for every s^i, it contains $|s^i|$ vertices, regardless of the actual assignment of elements from the alphabet Σ to the letters s^i_j. To construct the edge set E, we consider all edges $e = \{s^i, s^j\} \in T$. Let (a, b) be the direction of e. If e's type is substring with leftmost occurrence $l_{a,b}$, if $\mathcal{C}$ appears, s^a appears as a substring at position $l_{a,b}$ in s^b, i.e., $s^a_1 \ldots s^a_{|s^a|} = s^b_{l_{a,b}} \ldots s^b_{l_{a,b}+|s^a|-1}$. Hence, we add the edges $\{s^a_1, s^b_{l_{a,b}}\}, \ldots, \{s^a_{|s^a|}, s^b_{l_{a,b}+|s^a|-1}\}$ to E. Then, if $\mathcal{C}$ appears, any two letters connected by an edge in G_e are chosen equal elements from Σ. Now, we assume that e has type overlap. If $\mathcal{C}$ appears, the suffix of s^a of length $w(e)$ matches the prefix of s^b of that length, or equivalently $s^a_{|s^a|-w(e)+1} \ldots s^a_{|s^a|} = s^b_1 \ldots s^b_{w(e)}$. Thus, we add the $w(e)$ edges $\{s^a_{|s^a|-w(e)+1}, s^b_1\}, \ldots, \{s^a_{|s^a|}, s^b_{w(e)}\}$ to $G_e(\mathcal{C})$. In Fig. 1, the equality graph for the configuration on the left is shown. Let $C = \{C_1, \ldots, C_{|C|}\}$ contain the connected components of size at least two in G_e, and $L := \bigcup_{i=1}^{|C|} C_i$. Since G_e's edges are inserted according to the tree T containing no cycles, G_e contains no cycles. Hence, all components C_i are trees. The number of vertices in a tree is larger than its edge number by one. Since G_e contains $\sum_{e \in T} w(e) = f(\delta)$ edges, we get $|L| = \sum_{i=1}^{|C|} |C_i| = f(\delta) + |C|$. If $\mathcal{C}$ appears, for every C_i there is a $c_i \in \Sigma$ such that c_i is assigned to all letters in C_i. In both models, the probability that this happens for a single component C_i is at most $p_{\max}^{|C_i|-1}$: Let $l_1, \ldots, l_{|C_i|}$ be the letters in C_i. For a fixed $c_i \in \Sigma$, $\Pr[\bigwedge_{j=1}^{|C_i|} l_j = c_i] = \prod_{j=1}^{|C_i|} \Pr[l_j = c_i] \leq \Pr[l_1 = c_i] \cdot p_{\max}^{|C_i|-1}$, since $\Pr[l_j = c_i] \leq p_{\max}$ for all l_j. Now, the probability that for some $c_i \in \Sigma$, $l_j = c_i$ for all l_j is $\sum_{c_i \in \Sigma} \Pr[\bigwedge_{j=1}^{|C_i|} l_j = c_i] \leq \sum_{c_i \in \Sigma} \Pr[l_1 = c_i] \cdot p_{\max}^{|C_i|-1} = p_{\max}^{|C_i|-1}$. With $|L| = f(\delta) + |C|$, we get

$$\begin{aligned} \Pr[\mathcal{C} \text{ appears}] &\leq \prod_{i=1}^{|C|} p_{\max}^{|C_i|-1} = e^{\sum_{i=1}^{|C|}(|C_i|-1)\ln(p_{\max})} \\ &= e^{\ln(p_{\max})(|L|-|C|)} = e^{\ln(p_{\max}) f(\delta)} \quad . \end{aligned} \tag{4}$$

Clearly, $\Pr[\mathrm{mxt}(S) = f(\delta)] \leq \Pr[\exists \mathcal{C} \in \widehat{\mathcal{C}} : \mathcal{C} \text{ appears}]$. With (3) and (4), we get

$$\Pr[\mathrm{mxt}(S) = f(\delta)] \leq e^{(1+k)(n \ln n) + n \ln(f(\delta)/n) + \ln(p_{\max}) f(\delta) + O(n)} \tag{5}$$

$$= e^{((1+k) - |\ln p_{\max}| c)(n \ln n) + O(n \ln \ln n) + O(n)} \quad . \tag{6}$$

To get (6), consider $n \ln(f(\delta)/n) + \ln(p_{\max}) f(\delta) = n \ln(((cn \ln n) + \delta)/n) + \ln(p_{\max}) \cdot ((cn \ln n) + \delta)$ in (5). Standard analysis shows that it is at most $n \ln(c \ln n) + \ln(p_{\max}) cn \ln n = O(n \ln \ln n) - |\ln p_{\max}| cn \ln n$ (here, we used that $c = O(1)$). Remember that $k = (\ln \Delta(S))/\ln n$. Therefore, by choice of c in the lemma, $c \geq (1 + k)/|\ln p_{\max}| + \varepsilon$, or equivalently $|\ln p_{\max}| c \geq 1 + k + \varepsilon |\ln p_{\max}| = 1 + k + \varepsilon'$, $\varepsilon' > 0$. Thus, $(1 + k) - |\ln p_{\max}| c \leq -\varepsilon'$, and (6) yields $\Pr[\mathrm{mxt}(S) = f(\delta)] \leq e^{-\varepsilon'(n \ln n) + O(n \ln \ln n) + O(n)} = e^{-\Omega(n \ln n)}$. Throughout the paper, we assume that $||S|| \leq 2^n$ (otherwise, we can solve SCS optimally in time $O(\text{poly}(||S||))$, see Sect. 2.2). Then, since $1 \leq \delta \leq ||S|| \leq 2^n$, we get $\Pr[\text{mxt}(S) > cn \ln n] = \sum_{\delta=1}^{||S||} \Pr[\text{mxt}(S) = f(\delta)] \leq 2^n \cdot e^{-\Omega(n \ln n)} = e^{O(n)-\Omega(n \ln n)} = e^{-\Omega(n \ln n)}$. □

2.2 The Approximation Scheme

It is well known that in a complete, directed, weighted graph $G = (V, E)$, one can compute a maximum weight hamiltonian path (a path $(v_1, v_2, \dots, v_{|V|})$ of vertices with $v_i \neq v_j$ for $i \neq j$) in time $O(|V|^2 2^{|V|})$ with dynamic programming. Algorithm *OptSCS* below uses this to compute a shortest superstring in time $O(n^2 2^n)$. We modify the definition of the string graph $G_s(S)$ from Sect. 2.1 to yield the *directed string graph* $\vec{G}_s(S)$. It is a complete (without self-loops), directed, weighted graph on vertex set S, and the weight of an edge (s^i, s^j), $i \neq j$, is $w(s^i, s^j) = |\text{ov}(s^i, s^j)|$.

Algorithm OptSCS($S = \{s^1, \dots, s^n\}$)

1. Set $(S_{\text{rel}}, S_{\text{sub}})$:=*Partition*(S). Let $\{s_r^1, \dots, s_r^{n'}\}$ be the strings in S_{rel}.
2. Compute a maximum weight hamiltonian path $(s_r^{h_1}, \dots, s_r^{h_{n'}})$ in $\vec{G}_s(S_{\text{rel}})$.
3. Output $t := s_r^{h_1} \oplus s_r^{h_2} \oplus \dots \oplus s_r^{h_{n'}}$ (maximally overlap strings along the path).

Lemma 4. *For every input S for SCS,* OptSCS(S) *computes a shortest superstring in time $O(2^n \cdot \text{poly}(||S||))$.*

The proof is simple: Since S and S_{rel} share the same superstrings (see Sect. 2.1), to show that t is a shortest superstring for S, it suffices to show this for S_{rel}. Clearly, t is a superstring for S_{rel}. We show it to be optimal. Let t^* be a shortest superstring for S_{rel} and $s_r^{i_1}, \dots, s_r^{i_{n'}}$ be S_{rel}'s strings in order of increasing leftmost occurrence in t^*. Since $(s_r^{h_1}, \dots, s_r^{h_{n'}})$ is a maximum weight hamiltonian path in $\vec{G}_s(S_{\text{rel}})$, with (2) we get $c(t) = \sum_{j=1}^{n'-1} |\text{ov}(s_r^{h_j}, s_r^{h_{j+1}})| \geq \sum_{j=1}^{n'-1} |\text{ov}(s_r^{i_j}, s_r^{i_{j+1}})| = \text{opt}_c(S_{\text{rel}})$. Thus, $|t| = \text{opt}_l(S_{\text{rel}})$. We turn to the running time. *Partition* runs in time $O(\text{poly}(||S||))$. With Step 2's running time $O(n^2 2^n)$, a running time of $O(\text{poly}(||S||) + n^2 2^n) = O(2^n \cdot \text{poly}(||S||))$ follows for *OptSCS*. We can now prove Theorem 1. Consider the following algorithm.

Algorithm ApproxSCS($S = \{s^1, \dots, s^n\}, c$)

1. Run *BoundComp*(S, c). If it succeeds, output *Greedy*(S).
2. Run *OptSCS*(S) and output the superstring computed.

Proof (of Theorem 1). We start with the running time. *BoundComp*(S, c) runs in time $O(\text{poly}(||S||))$ by Lemma 3. Since the same holds for *Greedy*, Step 1 has polynomial worst case running time. To prove a polynomial expected running time, it thus suffices to upper bound Step 2's expected running time, which is the product of the time spent if it is executed and its execution probability. The former is $O(2^n \cdot \text{poly}(||S||))$ by Lemma 4. We upper bound the latter. Since $\Delta(S) \leq n^k$, we have $(\ln \Delta(S))/\ln n \leq k$. Thus, by choice of c,

$c \geq (1 + (\ln \Delta(S))/\ln n)/|\ln p_{\max}| + \varepsilon'$ for some $\varepsilon' > 0$, and Lemma 3 is applicable. The lemma yields that *BoundComp*(S, c) fails in Step 1 with probability $e^{-\Omega(n \ln n)}$. Since Step 2 is only executed if this happens, its execution probability is $e^{-\Omega(n \ln n)}$, and its expected running time is $O(2^n \cdot \text{poly}(||S||) \cdot e^{-\Omega(n \ln n)}) = O(\text{poly}(||S||)$, since $2^n \cdot e^{-\Omega(n \ln n)} = o(1)$. We turn to the factor for the length measure. A solution output in Step 2 has optimal approximation ratio 1. Now assume that Step 1 outputs $t :=$ *Greedy*(S). This happens only if *BoundComp*(S, c) succeeds, i.e., if $\text{opt}_c(S) \leq cn \ln n$ due to Lemma 3. With $c \leq \varepsilon||S||/(2n \ln n)$ due to $||S|| \geq (2c/\varepsilon)(n \ln n)$ in the lemma, this yields that $\text{opt}_c(S) \leq (\varepsilon/2)||S||$, which in turn yields $\text{opt}_l(S) = ||S|| - \text{opt}_c(S) \geq (1 - \varepsilon/2)||S||$. Clearly, $|t| \leq ||S||$. With $\varepsilon < 1$, we get an approximation ratio of $\frac{|t|}{\text{opt}_l(S)} \leq \frac{||S||}{(1-\varepsilon/2)||S||} = \frac{1}{1-\varepsilon/2} \leq 1+\varepsilon$. □

We prove Lemma 1. Set $c := (1 + \varepsilon)/H$. If $\text{opt}_c(S) > (1 + \varepsilon)(n \ln n)/H]$, *BoundComp*$(S, c)$ fails, which has probability $e^{-\Omega(n \ln n)}$ by Lemma 3. It is usable since $\Delta(S) = \text{polylog}(n)$, yielding $\Delta(S) \leq (\ln n)^k$ for a fixed $k \in \mathbb{N}$. Thus, $(\ln \Delta(S))/\ln n = O(\ln \ln n)/\ln n = o(1)$. Since $H/|\ln p_{\max}| < 1+\varepsilon$, $1/|\ln p_{\max}| = (1+\varepsilon)/H - \varepsilon' = c - \varepsilon'$ for $\varepsilon' > 0$. Since $(1 + (\ln \Delta(S))/\ln n)/|\ln p_{\max}| + \varepsilon'/2 = (1 + o(1))(c - \varepsilon') + \varepsilon'/2 \leq c$, Lemma 3's assumptions are fulfilled.

2.3 Generalization to the Mixing Model

We generalize our results to the mixing model. For a length $l \in \mathbb{N}$, we set $p^l_{\max} := \max_{x \in \Sigma^l} P^{(l)}(x)$ be the maximum probability of a string of length l.

Lemma 5. *Fix a mixing model. Then,* $p^l_{\max} = e^{-\Omega(l)}$.

Proof. For the mixing model, given $l \in \mathbb{N}$, choose $x^* \in \Sigma^l$ with $P^{(l)}(x^*) = p^l_{\max}$. For the function $\alpha\colon \mathbb{N} \to [0, 1)$ with $\lim_{g \to \infty} \alpha(g) = 0$ limiting the dependencies in the model (cf. (1)), we choose $g_0 \in \mathbb{N}$ with $\alpha(g_0) < (1 - p_{\max})/p_{\max}$. Let $r \in \Sigma^l$ be a random string, and set $m := \lfloor (l-1)/g_0 \rfloor = \Omega(l)$. For $k \in \{0, \ldots, m\}$, we let $\mathcal{E}_k$ be the event that $r_{1+k \cdot g_0} = x^*_{1+k \cdot g_0}$, and $\widehat{\mathcal{E}}_k := \bigcap_{k'=0}^{k-1} \mathcal{E}_{k'}$. Then, $p^l_{\max} = \Pr[r = x] \leq \Pr[\bigcap_{k=0}^{m} \mathcal{E}_k]$. Observe that for every k, the letters concerning the events $\mathcal{E}_k$ and $\widehat{\mathcal{E}}_k$ have a distance of at least g_0. By choice of g_0, $(1 + \alpha(g_0))p_{\max} < 1$. With (1), it follows that

$$\Pr[\mathcal{E}_k \mid \widehat{\mathcal{E}}_k] = \frac{\Pr[\mathcal{E}_k \cap \widehat{\mathcal{E}}_k]}{\Pr[\widehat{\mathcal{E}}_k]} \leq (1 + \alpha(g_0)) \Pr[\mathcal{E}_k] \leq (1 + \alpha(g_0))p_{\max} =: c < 1 \ . \quad (7)$$

Now, $p^l_{\max} \leq \Pr[\bigcap_{k=0}^{m} \mathcal{E}_k] = \prod_{k=0}^{m} \Pr[\mathcal{E}_k \mid \widehat{\mathcal{E}}_k] \leq c^{m+1} \leq e^{m \ln c} = e^{-\Omega(l)}$. □

Since $p^l_{\max} = e^{-\Omega(l)}$, a constant $c > 0$ with $p^l_{\max} \leq e^{-cl}$ exists. Thus, $|\ln p^l_{\max}|/l \geq |\ln e^{-cl}|/l = c$. Given a mixing model, we define $c_{\text{mix}} := \sup\{c > 0 \mid \exists l_0 \in \mathbb{N}\colon \forall l \geq l_0\colon |\ln p^l_{\max}|/l \geq c\}$.

Lemma 6. *With* $|\ln p_{\max}|$ *replaced by* c_{mix}*, Lemma 3 holds in the mixing model.*

Proof. We adapt the part of Lemma 3's proof that upper bounds the probability that a configuration $\mathcal{C}$ appears. Consider the equality graph $G_e(\mathcal{C})$. We assign numbers ν to its vertices, i.e., the letters of S's strings, by traversing the spanning tree T in the string graph $G_s(S)$. We start at s^1 and set $\nu(s^1_j) := j$, $j = 1, \ldots, |s^1|$. Then, while unnumbered strings s^i exist, choose one with an edge $e = \{s^i, s^{i'}\} \in T$ such that $s^{i'}$ is numbered (such an s^i always exists, and e is unique, so no numbering conflicts arise). For the letters s^i_j connected to a letter $s^{i'}_{j'}$ by an edge in G_e, we set $\nu(s^i_j) := \nu(s^{i'}_{j'})$. Then, we assign numbers to the not yet numbered letters of s^i such that $\nu(s^i_{j+1}) = \nu(s^i_j) + 1$ for $j = 1, \ldots, |s^i| - 1$. The numbering for the equality graph in Fig. 1 is shown in Fig. 2. For a *block length*

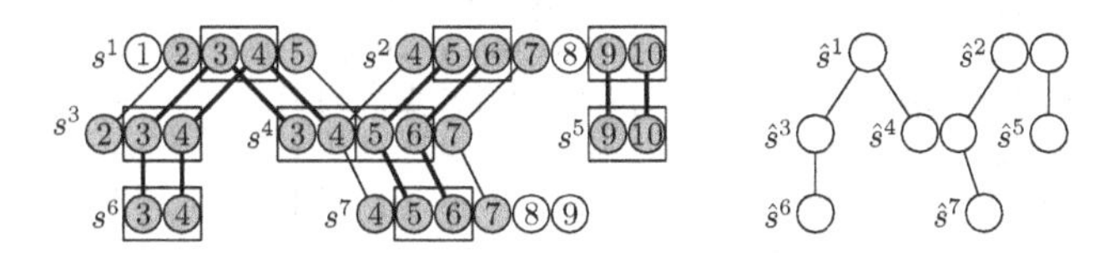

Fig. 2. An Equality Graph G_e with Strings Divided into Blocks

$l \in \mathbb{N}$, we divide the s^i into blocks (substrings) of length at most l. For a string s^i, all letters s^i_j with the same value $\lceil \nu(s^i_j)/l \rceil$ form a block. In Fig. 2, $l = 2$ is used, and s^1 contains three blocks $s^1_1 s^1_2$, $s^1_3 s^1_4$, and s^1_5. Let again C contain the connected components C_i of size at least two in G_e, and $L := \bigcup_{i=1}^{|C|} C_i$ (in Fig. 2, letters in L are grey). For each s^i, let the *block string* $\hat{s}^i = \hat{s}^i_1 \ldots \hat{s}^i_{|\hat{s}^i|}$ be the sequence of the blocks in s^i from left to right which have full length l and contain only letters from L (in Fig. 2, these blocks are marked with boxes, e.g., $\hat{s}^4 = \hat{s}^4_1 \hat{s}^4_2$). Since each block is a length l substring of s^i, $\hat{s}^i$ is a string over the alphabet Σ^l. Notice that the blocks $\hat{s}^i_j$ are chosen by the l-th order probability distribution $P^{(l)} \colon \Sigma^l \to [0, 1]$ of our model. First, we neglect block dependencies and get a Bernoulli model over alphabet Σ^l in which we estimate the probability of the configuration $\mathcal{C}$. Later, we consider the dependencies. Consider blocks $\hat{s}^i_j, \hat{s}^{i'}_{j'}$ in two block strings $\hat{s}^i, \hat{s}^{i'}$ such that for $k = 1, \ldots, l$, the two letters in $\hat{s}^i_j$ and $\hat{s}^{i'}_{j'}$ at position k are connected by an edge in G_e. We call such blocks *matching*. In Fig. 2, $\hat{s}^1_1$ and $\hat{s}^3_1$ have this property. If $\mathcal{C}$ appears, $\hat{s}^i_j = \hat{s}^{i'}_{j'}$. Hence, we create a *block equality graph* $\widehat{G}_e = (\widehat{V}, \widehat{E})$ for the block strings $\hat{s}^i$. We let $\widehat{V}$ be the set of all blocks, and $\{\hat{s}^i_j, \hat{s}^{i'}_{j'}\} \in \widehat{E}$ if $\hat{s}^i_j$ and $\hat{s}^{i'}_{j'}$ are matching. In Fig. 2, the block equality graph for the equality graph on the left is shown. We lower bound the number $|\widehat{E}|$ of edges. Consider an edge $e = \{s^i, s^{i'}\}$ in the spanning tree T of $\mathcal{C}$. There are indices a, a' such that G_e contains the $w(e)$ edges $\{s^i_{a+k}, s^{i'}_{a'+k}\}$, $k = 0, \ldots, w(e) - 1$. Let L' be the set of all letters in these edges. For $l - 1 \le k \le w(e) - l$, regardless of the positions of the blocks in s^i, s^i_{a+k} lies in a block of full length l in $L' \subseteq L$. Thus, there are at least $(w(e) - 2(l - 1))/l$ blocks $\hat{s}^i_j$ with all letters in L. In our numbering ν, two letters connected by an edge in G_e have the same number. Thus, due to the edges $\{s^i_{a+k}, s^{i'}_{a'+k}\}$, for every above block $\hat{s}^i_j$, the letters in $s^{i'}$ connected to $\hat{s}^i_j$ by these edges form a block $\hat{s}^{i'}_{j'}$ in L, and thus $\hat{s}^i_j$ and $\hat{s}^{i'}_{j'}$ are matching.

It follows that for every edge $e \in T$, $\widehat{G}_{\mathrm{e}}$ contains at least $(w(e) - 2(l-1))/l$ edges. In the following, we set $l := \ln\ln n$. With $w(T) = f(\delta) = (cn\ln n) + \delta$ we get $|\widehat{E}| \geq \sum_{e\in T}(w(e) - 2(l-1))/l = w(T)/l - O(n) = (1-o(1))f(\delta)/l$. Since we assume a Bernoulli model, $\widehat{G}_{\mathrm{e}}$ is an equality graph as above. For clarity, we denote probabilities assuming independence of blocks by $\mathrm{Pr_i}$ and ones with dependencies with $\mathrm{Pr_d}$. With $|\widehat{E}| = (1-o(1))f(\delta)/l$, (4) yields

$$\mathrm{Pr_i}[\mathcal{C} \text{ appears}] \leq e^{\ln(p_{\max}^l)(1-o(1))f(\delta)/l} \leq e^{-(1-\varepsilon')c_{\mathrm{mix}}f(\delta)} \tag{8}$$

for every fixed $\varepsilon' > 0$ by definition of c_{mix}. We choose ε' later. We now consider block dependencies and correct (8) accordingly. In G_{e}, for the set L of all letters contained in C's components, $|L| = f(\delta)+|C| \leq 2f(\delta)$ since $|C| \leq f(\delta)$. The total number of blocks $\sum_{i=1}^n |\hat{s}^i|$ is hence at most $2f(\delta)/l = o(1)f(\delta)$ since $l = \omega(1)$. Let an *assignment* be a function $\hat{a}$ from the set of all blocks $\hat{s}^i_j$ in our strings to Σ^l assigning to each block $\hat{s}^i_j$ a value $\hat{a}^i_j$. For a string $\hat{s}^i$, we let $\mathcal{E}^i_j$ be the event that $\hat{s}^i_j = \hat{a}^i_j$ and $\widehat{\mathcal{E}}^i_j := \bigcap_{j'=1}^{j-1} \mathcal{E}^i_{j'}$. Remember that the function α limiting the dependencies in the mixing model has range $[0,1)$ i.e., $\alpha(1) < 1$. Analogously to (7), we get for events $\mathcal{E}^i_j$ and $\widehat{\mathcal{E}}^i_j$ that $\mathrm{Pr_d}[\mathcal{E}^i_j \mid \widehat{\mathcal{E}}^i_j] \leq (1+\alpha(1))\,\mathrm{Pr_d}[\mathcal{E}^i_j] < 2\,\mathrm{Pr_i}[\mathcal{E}^i_j]$ since $\mathrm{Pr_d}[\mathcal{E}^i_j] = \mathrm{Pr_i}[\mathcal{E}^i_j]$. Accordingly,

$$\mathrm{Pr_d}[\widehat{\mathcal{E}}^i_{|\hat{s}^i|+1}] = \prod_{j=1}^{|\hat{s}^i|} \mathrm{Pr_d}[\mathcal{E}^i_j \mid \widehat{\mathcal{E}}^i_j] < \prod_{j=1}^{|\hat{s}^i|} 2\,\mathrm{Pr_i}[\mathcal{E}^i_j] = 2^{|\hat{s}^i|}\,\mathrm{Pr_i}[\widehat{\mathcal{E}}^i_{|\hat{s}^i|+1}] \ , \tag{9}$$

i.e., the dependencies increase the probability that $\hat{s}^i$ is chosen according to $\hat{a}$ by a factor of $2^{|\hat{s}^i|}$. Let A be the subset of all possible assignments such that $\mathcal{C}$ appears iff for some $\hat{a} \in A$, we have $\hat{s}^i_j = \hat{a}^i_j$ for all blocks. Using (9) and $\sum_{i=1}^n |\hat{s}^i| = o(1)f(\delta)$, we get for a fixed assignment $\hat{a}$

$$\begin{aligned}\mathrm{Pr_d}[\widehat{\mathcal{E}}^i_{|\hat{s}^i|+1},\ i=1,\ldots,n] &= \prod_{i=1}^n \mathrm{Pr_d}[\widehat{\mathcal{E}}^i_{|\hat{s}^i|+1}] < \prod_{i=1}^n 2^{|\hat{s}^i|}\,\mathrm{Pr_i}[\widehat{\mathcal{E}}^i_{|\hat{s}^i|+1}] \\ &= 2^{\sum_{i=1}^n |\hat{s}^i|}\prod_{i=1}^n \mathrm{Pr_i}[\widehat{\mathcal{E}}^i_{|\hat{s}^i|+1}] = 2^{o(1)f(\delta)}\,\mathrm{Pr_i}[\widehat{\mathcal{E}}^i_{|\hat{s}^i|+1},\ i=1,\ldots,n] \ . \end{aligned} \tag{10}$$

Since (with or without dependencies) $\Pr[\mathcal{C} \text{ appears}] = \sum_{\hat{a}\in A}\Pr[\widehat{\mathcal{E}}^i_{|\hat{s}^i|+1},\ i = 1,\ldots,n]$, (10) yields $\mathrm{Pr_d}[\mathcal{C} \text{ appears}] = 2^{o(1)f(\delta)}\,\mathrm{Pr_i}[\mathcal{C} \text{ appears}]$. Thus, the additional term $o(1)f(\delta)$ introduced in the exponent of (8) is compensated by the choice of ε'. We finish considering the depedencies by concluding that in the proof of Lemma 3, we can replace (4) by $\mathrm{Pr_d}[\mathcal{C} \text{ appears}] \leq e^{-(1-\varepsilon')c_{\mathrm{mix}}f(\delta)}$. Thus, in the calculations following (4), $|\ln p_{\max}|$ is replaced by $(1-\varepsilon')c_{\mathrm{mix}}$, and (6) becomes

$$\Pr[\mathrm{mxt}(S) = f(\delta)] \leq e^{((1+k)-(1-\varepsilon')c_{\mathrm{mix}}c)(n\ln n)+O(n\ln\ln n)+O(n)} \ . \tag{11}$$

It is easy to see that due to $c \geq (1+k)/c_{\mathrm{mix}} + \varepsilon$ in the lemma, an $\varepsilon' > 0$ exists with $c \geq (1+k)/((1-\varepsilon')c_{\mathrm{mix}}) + \varepsilon'$, yielding $(1-\varepsilon')c_{\mathrm{mix}}c \geq 1+k+\varepsilon''$, $\varepsilon'' > 0$. We fix ε' in that way. Now, in (11), $(1+k) - (1-\varepsilon')c_{\mathrm{mix}}c \leq -\varepsilon''$, yielding $\Pr[\mathrm{mxt}(S) = f(\delta)] = e^{-\Omega(n\ln n)}$ as before. The rest works as before. □

Corollary 1. *With $|\ln p_{\max}|$ replaced by c_{mix}, Theorem 1 and Lemma 1 hold in the mixing model.*

References

1. Alexander, K.S.: Shortest Common Superstrings for Strings of Random Letters. In: Crochemore, M., Gusfield, D. (eds.) CPM 1994. LNCS, vol. 807, pp. 164–172. Springer, Heidelberg (1994)
2. Armen, C., Stein, C.: A 2 3/4 approximation algorithm for the shortest superstring problem. Technical Report PCS-TR94-214, Department of Comp. Sc., Dartmouth College, Hanover, New Hampshire (1994)
3. Armen, C., Stein, C.: A 2 2/3-Approximation Algorithm for the Shortest Superstring Problem. In: Hirschberg, D.S., Meyers, G. (eds.) CPM 1996. LNCS, vol. 1075, pp. 87–101. Springer, Heidelberg (1996)
4. Blum, A., Jiang, T., Li, M., Tromp, J., Yannakakis, M.: Linear Approximation of Shortest Superstrings. J. ACM 41(4), 630–647 (1994)
5. Frieze, A.M., Szpankowski, W.: Greedy Algorithms for the Shortest Common Superstring that are Asymptotically Optimal. In: Díaz, J., Serna, M.J. (eds.) ESA 1996. LNCS, vol. 1136, pp. 194–207. Springer, Heidelberg (1996)
6. Gusfield, D.: Algorithms on Strings, Trees, and Sequences. Cambridge University Press, Cambridge (1997)
7. Łuczak, T., Szpankowski, W.: A Lossy Data Compression Based on String Matching: Preliminary Analysis and Suboptimal Algorithms. In: Crochemore, M., Gusfield, D. (eds.) CPM 1994. LNCS, vol. 807, pp. 102–112. Springer, Heidelberg (1994)
8. Ma, B.: Why Greed Works for Shortest Common Superstring Problem. In: Ferragina, P., Landau, G.M. (eds.) CPM 2008. LNCS, vol. 5029, pp. 244–254. Springer, Heidelberg (2008)
9. Middendorf, M.: More on the Complexity of Common Superstring and Supersequence Problems. Theor. Comp. Sc. 125(2), 205–228 (1994)
10. Spielman, D.A., Teng, S.: Smoothed analysis of algorithms: why the simplex algorithm usually takes polynomial time. In: Proc. 33rd Ann. ACM Symp. on Th. of Comp., pp. 296–305. ACM, New York (2001)
11. Sweedyk, Z.: A 2 1/2 Approximation Algorithm for Shortest Superstring. SIAM J. Comput. 29(3), 954–986 (1999)
12. Tarhio, J., Ukkonen, E.: A Greedy Approximation Algorithm for Constructing Shortest Common Superstrings. Theor. Comp. Sc. 57, 131–145 (1988)
13. Teng, S., Yao, F.: Approximating Shortest Superstring. In: 34th Ann. Symp. on Found. of Comp. Sc., pp. 158–165. IEEE Press, New York (1993)
14. Turner, J.S.: Approximation Algorithms for the Shortest Common Superstring Problem. Inf. and Comput. 83, 1–20 (1989)
15. Vassilevska, V.: Explicit Inapproximability Bounds for the Shortest Superstring Problem. In: Jedrzejowicz, J., Szepietowski, A. (eds.) MFCS 2005. LNCS, vol. 3618, pp. 793–800. Springer, Heidelberg (2005)
16. Vazirani, V.V.: Approximation Algorithms. Springer, Heidelberg (2001)
17. Yang, E.H., Zhang, Z.: The shortest common superstring problem: Average case analysis for both exact and approximate matching. IEEE Transact. on Inf. Th. 45(6), 1867–1886 (1999)

The Cost of Stability in Network Flow Games

Ezra Resnick[1], Yoram Bachrach[2], Reshef Meir[1], and Jeffrey S. Rosenschein[1]

[1] The Hebrew University of Jerusalem, Israel
{ezrar,reshef24,jeff}@cs.huji.ac.il
[2] Microsoft Research Ltd., Cambridge, UK
t-yobach@microsoft.com

Abstract. The core of a cooperative game contains all stable distributions of a coalition's gains among its members. However, some games have an empty core, with every distribution being unstable. We allow an external party to offer a supplemental payment to the grand coalition, which may stabilize the game, if the payment is sufficiently high. We consider the cost of stability (CoS)—the minimal payment that stabilizes the game.

We examine the CoS in threshold network flow games (TNFGs), where each agent controls an edge in a flow network, and a coalition wins if the maximal flow it can achieve exceeds a certain threshold. We show that in such games, it is coNP-complete to determine whether a given distribution (which includes an external payment) is stable. Nevertheless, we show how to bound and approximate the CoS in general TNFGs, and provide efficient algorithms for computing the CoS in several restricted cases.

1 Introduction

Many artificial intelligence settings involve multiple self-interested agents. Although self-interested, the agents may still benefit from cooperation. A natural tool for analyzing such strategic situations is, of course, cooperative game theory. In *cooperative games*, every subset (*coalition*) of agents can achieve a certain utility by cooperating. A natural question which arises is how to divide the gains obtained by a coalition among its members, since the *total* utility generated by the coalition is (by assumption) of little interest to each individual agent. Each possible division of the coalition's gains among its members is called an *imputation*.

Cooperative game theory solution concepts seek to define appropriate ways of distributing a coalition's gains among its members, so as to meet some desirable criteria. A prominent solution concept is the *core* [7], which is the set of all *stable* imputations—those where no subset of agents has a rational incentive to split off from the grand coalition (the set of all agents). Some games have infinitely many imputations in their core, while others have empty cores. In games where the core is empty, any imputation would be unstable. Thus, as opposed to normal-form games where the existence of a stable solution in the form of a (mixed-strategy) Nash equilibrium is guaranteed, some cooperative domains are inherently unstable.

We examine the possibility of stabilizing a cooperative game using external payments, based on a model introduced by Bachrach *et al* [1]. In this model, an external party is

R. Královič and D. Niwiński (Eds.): MFCS 2009, LNCS 5734, pp. 636–650, 2009.

interested in inducing all the agents to cooperate. This is done by offering the grand coalition a supplemental payment, given to the grand coalition as a whole, and provided only if this coalition is formed. The game's *cost of stability* (CoS) is the minimal external payment that allows a stable division of the grand coalition's adjusted gains.

In this work we consider games defined over network flow domains, where agents must cooperate to allow flow through the network. Such games can model situations where some commodity (traffic, liquid, information) flows through a network with various capacity constraints, and different entities own the different links (roads, pipes, cables) along the way. Such games have been studied in several works [8, 9, 2, 4]. We examine *threshold network flow games* (TNFGs), where each agent controls an edge in the network, and a coalition "wins" if the maximal flow it allows from the source vertex to the sink vertex exceeds a certain threshold. Computing the core of such a game enables finding a stable distribution of the rewards obtained from operating the network among the various agents. However, in many such games the core would be empty. In fact, we will see that unless there exists some *veto agent*, without which no coalition can achieve the required flow, the game's core would be empty and no imputation would be stable. There might exist some external party (e.g., a government) that would be willing to pay in order to ensure the cooperation of all agents in allowing flow through the network. Naturally, this external party would want to minimize its costs.

We explore the CoS in TNFGs. We show that it is coNP-complete to determine whether a given profit division, allowed by some external payment, makes a certain TNFG stable. Despite this hardness result, we nevertheless show how to bound and approximate the CoS in general TNFGs, and provide efficient algorithms for computing the CoS and finding optimal super-imputations in several restricted forms of TNFGs. We give an upper bound on the CoS in TNFGs based on the max-flow value of the network, which can also be used to approximate the CoS. We consider the CoS in *connectivity games*, a restricted form of TNFGs, and show that in these games the CoS is equal to the max-flow value of the network. We generalize this result, considering TNFGs with equal edge capacities. We also consider the case of *serial* TNFGs, built by serially connecting several component TNFGs. We show that the CoS of a serial TNFG is equal to the minimal CoS among the component TNFGs, and that this value may be computed efficiently if the number of edges in each component is not too large. Finally, we consider the relationship between the CoS in TNFGs and the CoS in another well-known cooperative domain—weighted voting games.

2 Preliminaries

We now define certain game-theoretic concepts necessary for our analysis of the cost of stability. We also define the domain on which we will be concentrating, the threshold network flow game.

2.1 Cooperative Games

A (transferable utility) *cooperative game* (also called a *coalitional game*) is defined by specifying the collective utility that can be achieved by every coalition of agents. In this work, the term *game* always refers to a cooperative game.

Definition 1. *A* cooperative game *consists of a finite set of agents N and a function $v : 2^N \rightarrow \mathbb{R}$. The function v is called the* characteristic function *of the game.*

The characteristic function maps every coalition of agents to the total utility that can be achieved by those agents together. In many typical cooperative games, adding more agents to a coalition never reduces the achievable utility. Such games are called *increasing*.

Definition 2. *A cooperative game $\langle N, v \rangle$ is* increasing *if $v(C') \leq v(C)$ for any $C' \subseteq C \subseteq N$.*

The TNFG domain considered in this work is one where a coalition can either win or lose. Such domains can be modeled as *simple* cooperative games.

Definition 3. *A cooperative game $\langle N, v \rangle$ is* simple *if v only takes the values* 0 *or* 1, *i.e., $v : 2^N \rightarrow \{0, 1\}$. We say a coalition $C \subseteq N$ is a* winning *coalition if $v(C) = 1$, and it is a* losing *coalition if $v(C) = 0$.*

In such games, it is usually assumed that $v(\emptyset) = 0$ and $v(N) = 1$. An agent without which no coalition can win is called a *veto agent*.

Definition 4. *In a simple cooperative game $\langle N, v \rangle$, an agent $a \in N$ is a* veto agent *if for any coalition $C \subseteq N$ it holds that $v(C \setminus \{a\}) = 0$.*

2.2 Flow Networks

Flow networks are useful for modeling systems where some fluid commodity travels through a network with capacity constraints. A flow network consists of a directed graph $\langle V, E \rangle$, with capacities on the edges $c : E \rightarrow \mathbb{R}_+$, a distinguished source vertex $s \in V$, and a distinguished sink vertex $t \in V$ ($s \neq t$). A *flow* through the network is a function $f : E \rightarrow \mathbb{R}_+$ which obeys the capacity constraints and conserves the flow at each vertex (except for the source and sink), meaning that the total flow entering a vertex must equal the total flow leaving that vertex. The *value* of a flow f (denoted $|f|$) is the net amount flowing out of the source (and into the sink). A *cut* of a flow network is a partition of the vertexes into two subsets S, T (where $S \cup T = V$ and $S \cap T = \emptyset$) such that $s \in S$ and $t \in T$. The capacity of a cut $\langle S, T \rangle$ is defined as the sum of the capacities of the edges crossing the cut (from S to T). We call a minimal capacity cut a *min-cut* and a maximal value flow a *max-flow*. The *max-flow min-cut theorem* states that in any flow network, the max-flow value is equal to the min-cut capacity.

Max-flow min-cut theorem. *The value of a flow f in a flow network is maximal if and only if there exists a cut of the network with capacity equal to $|f|$.*

Many efficient algorithms for finding a maximal value flow for a given network are known. Note that if all the edge capacities in a network are integers, the Ford-Fulkerson algorithm [6] produces an integer max-flow. This implies the following lemma:

Lemma 1. *In a flow network $\langle V, E, c, s, t \rangle$, if $c(e) \in \mathbb{N}$ for all $e \in E$ then there exists a max-flow f such that $f(e) \in \mathbb{N}$ for all $e \in E$.*

A general graph theory problem which may be solved efficiently using flow networks is that of finding the maximal number of edge-disjoint paths between two vertexes in a directed graph. This is done by assigning each edge a capacity of 1, and computing the max-flow value in the resulting flow network.

Lemma 2. *Given a flow network $\langle V, E, c, s, t \rangle$, if $c(e) = 1$ for all $e \in E$ then the maximal number of edge-disjoint paths from s to t in the directed graph $\langle V, E \rangle$ is equal to the max-flow value of the flow network.*

Proof. Since all capacities are 1, Lemma 1 implies that there exists a max-flow f such that $f(e) = 0$ or $f(e) = 1$ for all $e \in E$. This means that f must define $|f|$ edge-disjoint paths from s to t (with a flow of 1 through each). There cannot exist more than $|f|$ edge-disjoint paths, since then we could construct a flow whose value was greater than $|f|$, contradicting the assumption that f is a max-flow. □

2.3 Threshold Network Flow Games

A *threshold network flow game* (TNFG) is a cooperative game defined over a flow network, where each agent controls an edge in the network. Coalitions of agents may cooperate in order to send a certain flow from the source to the sink, and a coalition wins if the max-flow value allowed when using only the edges in the coalition exceeds a certain threshold (this threshold variant of network flow games has been studied by Kalai and Zemel [8] and Bachrach and Rosenschein [2]).

Definition 5. *A* threshold network flow domain *consists of a flow network $\langle V, E, c, s, t \rangle$ and a threshold $k \in \mathbb{R}_+$.*

Definition 6. *Given a threshold network flow domain $\langle V, E, c, s, t, k \rangle$, a* threshold network flow game (TNFG) *is the cooperative game $\langle N, v \rangle$ where $N = E$ and the characteristic function is defined as:*

$$v(C) = \begin{cases} 1 & \textit{if there exists a flow } f \textit{ in the network such that } |f| \geq k \textit{ and} \\ & \forall e \in E \setminus C : f(e) = 0 \\ 0 & \textit{otherwise} \end{cases}$$

By definition, TNFGs are simple games. They are also increasing games, since adding more edges to a coalition can only increase the value of the max-flow. It is easy to check whether a given coalition is a winning coalition by computing the max-flow value of the network which contains only the edges in the coalition and checking whether that value exceeds the threshold.

2.4 Imputations and the Core

The characteristic function of a cooperative game defines only the *total* gains a coalition achieves, but does not offer a way of distributing those gains among the agents in the coalition. Such a division is called an *imputation* (or a *payoff vector*).

Definition 7. *Given a cooperative game $\langle N, v \rangle$, an* imputation *is a vector $p \in \mathbb{R}_+^N$ such that $\sum_{a \in N} p_a = v(N)$. We call p_a the* payoff *of agent a, and denote the payoff of a coalition $C \subseteq N$ as $p(C) = \sum_{a \in C} p_a$.*

Cooperative game theory solution concepts offer ways of choosing an imputation, so as to satisfy some criteria. A basic criterion is *individual rationality*, which requires that $p_a \geq v(\{a\})$ for any agent $a \in N$—otherwise, some agent has an incentive to leave the coalition and work alone. A stronger criterion is that of *coalitional rationality*, based on the notions of *blocking* coalitions and *stable* imputations.

Definition 8. *In a cooperative game $\langle N, v \rangle$, a coalition $C \subseteq N$* blocks *an imputation p if $p(C) < v(C)$.*

Definition 9. *In a cooperative game $\langle N, v \rangle$, an imputation p is* stable *if it is not blocked by any coalition, i.e., for every coalition $C \subseteq N$, $p(C) \geq v(C)$.*

If the coalition C blocks the imputation p, the members of C could leave the grand coalition, derive the gains of $v(C)$, give each member $a \in C$ its previous gains p_a—and still some utility remains, so each agent could get more utility. If an unstable imputation is chosen, we cannot expect all agents to remain in the grand coalition. The *core* is the set of all stable imputations.

Definition 10. *The* core *of a cooperative game is the set of all imputations that are stable.*

Some games have infinitely many imputations in their core, while other games have empty cores. If we divide the gains of the grand coalition using an imputation in the core, then no subset of agents has an incentive to break off and work alone. However, if the core is empty, then any possible division of the grand coalition's gains is unstable: there will always be some coalition with an incentive to break away. In simple games, there is a well-known characterization of the core based on the game's veto agents: the core consists of all imputations which divide the grand coalition's gains only among the veto agents. Consequently, the core of a simple game (such as a TNFG) is nonempty if and only if there exists at least one veto agent. Note that we can compute the core of a TNFG in polynomial time, simply by finding all the veto agents (a given edge is a veto agent if and only if the coalition of all other edges is a losing coalition).

What should we do if we are faced with a game whose core is empty, but we still wish to ensure that no coalition has an incentive to leave the grand coalition? In the next section we suggest a solution, using external payments.

3 The Cost of Stability

We now consider the possibility of stabilizing a cooperative game using external payments, leading to the definition of the *cost of stability*, as introduced by Bachrach *et al* [1]. If a game is increasing, the maximal utility is achieved by the grand coalition. However, if the game's core is empty, it is impossible to distribute the gains of the grand coalition in a stable manner among the agents. This impedes the agents' cooperation, rendering the grand coalition unstable. Consider an external party that would like to induce all the agents to cooperate. One way to do this is by offering the grand coalition a *supplemental payment* if all agents cooperate. This external payment is offered to the grand coalition as a whole, and is provided only if this coalition is formed. The *adjusted game* is defined based on the original game and the supplemental payment.

Definition 11. *Given a cooperative game $G = \langle N, v\rangle$ and a* supplemental payment $\Delta \in \mathbb{R}_+$, *the* adjusted game *is the cooperative game $G(\Delta) = \langle N, v'\rangle$ where the characteristic function is defined as:*

$$v'(C) = \begin{cases} v(C) & \text{if } C \neq N \\ v(C) + \Delta & \text{if } C = N \end{cases}$$

We call $v'(N) = v(N) + \Delta$ the grand coalition's *adjusted gains*. We call a division of the adjusted gains in the adjusted game a *super-imputation*.

Definition 12. *Given an adjusted game $G(\Delta) = \langle N, v'\rangle$, a* super-imputation *is a vector $p \in \mathbb{R}_+^N$ such that $\sum_{a \in N} p_a = v'(N) = v(N) + \Delta$.*

We will sometimes talk about super-imputations without explicitly defining the adjusted game—in such a case the supplemental payment is implied by the sum of the super-imputation's payments.

Even if the core of the original game G was empty, the core of the adjusted game $G(\Delta)$ may not be empty—if the supplemental payment is high enough. Naturally, the external party would prefer to minimize the supplemental payment. The *cost of stability* (CoS) is defined as the *minimal* sum of payments such that a stable super-imputation exists in the adjusted game.

Definition 13. *The* cost of stability *of a cooperative game $G = \langle N, v\rangle$ is defined as follows:*

$$\mathrm{CoS}(G) = \min_{\Delta \in \mathbb{R}_+} \{v(N) + \Delta : \text{the core of } G(\Delta) \text{ is nonempty}\}$$

Note that for any simple game G, $\mathrm{CoS}(G) \geq 1$, and $\mathrm{CoS}(G) = 1$ if and only if the core of G is nonempty. For simple games, we can give additional lower and upper bounds on the CoS.

Theorem 1. *If there exist m pairwise-disjoint winning coalitions in a simple game $G = \langle N, v\rangle$, then $\mathrm{CoS}(G) \geq m$.*

Proof. Let $C_1, \ldots, C_m$ be pairwise-disjoint winning coalitions in G. Let p be a super-imputation such that $p(N) < m$. This means there must exist a winning coalition C_i ($1 \leq i \leq m$) such that $p(C_i) < 1$ (otherwise we would get $p(N) \geq \sum_{j=1}^{m} p(C_j) \geq \sum_{j=1}^{m} 1 = m$). This means that C_i blocks p and p is unstable. Therefore, any stable super-imputation p' must satisfy $p'(N) \geq m$, so $\mathrm{CoS}(G) \geq m$. □

Theorem 2. *Let $G = \langle N, v\rangle$ be a simple game and let $S \subseteq N$ be a subset of agents. If every winning coalition C in G satisfies $C \cap S \neq \emptyset$, then $\mathrm{CoS}(G) \leq |S|$.*

Proof. We define a super-imputation p as follows:

$$\forall a \in N : p_a = \begin{cases} 1 & \text{if } a \in S \\ 0 & \text{otherwise} \end{cases}$$

Any winning coalition includes at least one agent from S, and so is paid at least 1. This means that p is stable, therefore: $\mathrm{CoS}(G) \leq p(N) = |S|$. □

4 Hardness of Determining Stability of Super-Imputations in TNFGs

This work focuses on the CoS in TNFGs. We first consider the problem of testing whether a given super-imputation, allowed by a certain supplemental payment, is actually stable in a TNFG. We show that this problem is in fact coNP-complete.

Definition 14. TNFG-SUPER-IMPUTATION-STABILITY (TNFG-SIS): *Given a TNFG $G = \langle V, E, c, s, t, k\rangle$, a supplemental payment Δ, and a super-imputation p in the adjusted game $G(\Delta)$, decide whether p is stable, i.e., whether there exists some blocking coalition for p in $G(\Delta)$.*

Theorem 3. TNFG-SIS *is* coNP-*complete.*

Proof. TNFG-SIS is in coNP, since we can easily verify instability: given a potentially blocking coalition, we can check whether it is a winning coalition and whether the sum of payments to the coalition members is less than 1, in polynomial time. We show that TNFG-SIS is coNP-hard by polynomially reducing SUBSET-SUM to the complement of TNFG-SIS. SUBSET-SUM is a well-known NP-complete problem, where we are given a set of positive integers $A = \{a_1, \ldots, a_n\}$ and a positive integer b, and are asked to determine whether there exists a subset $A' \subseteq A$ such that the sum of the elements in A' is exactly b. Given a SUBSET-SUM instance, we construct the following TNFG:

$$V = \{s, t\} \cup \{v_1, \ldots, v_n\}$$
$$E = \{(s, v_i) : 1 \leq i \leq n\} \cup \{(v_i, t) : 1 \leq i \leq n\}$$
$$\forall\, 1 \leq i \leq n : \; c(s, v_i) = c(v_i, t) = a_i$$
$$k = b$$

In other words, for each element a_i we add a path from s to t with capacity a_i, and we define the threshold to be the target sum b (see Figure 1). We now define a super-imputation p as follows:

$$\forall\, 1 \leq i \leq n : \; p_{(s,v_i)} = p_{(v_i,t)} = \frac{a_i}{2(b+1)}$$

We show that this super-imputation is unstable if and only if the given SUBSET-SUM instance is a "yes" instance.

First, assume p is unstable. This means there is some winning coalition C such that $p(C) < 1$. We can assume that if $(s, v_i) \in C$ for some $1 \leq i \leq n$ then also $(v_i, t) \in C$

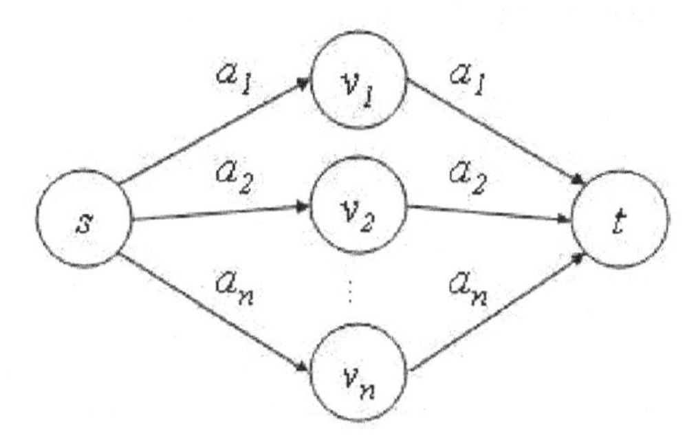

Fig. 1. Reduction of a SUBSET-SUM instance $\langle\{a_1, a_2, \ldots, a_n\}, b\rangle$ to an instance of TNFG-SIS. The game's threshold is b. We consider the stability of a super-imputation giving an edge with capacity a_i a payoff of $\frac{a_i}{2(b+1)}$.

(otherwise, we could remove (s, v_i) from C and C would still block p). Likewise, we can assume that if $(v_i, t) \in C$ for some $1 \leq i \leq n$ then also $(s, v_i) \in C$. Let $I \subseteq \{1, \ldots, n\}$ be the subset of indexes such that $C = \{(s, v_i) : i \in I\} \cup \{(v_i, t) : i \in I\}$. We assumed $p(C) < 1$, so:

$$1 > p(C) = 2\sum_{i\in I} \frac{a_i}{2(b+1)} \Rightarrow b + 1 > \sum_{i\in I} a_i$$

The max-flow value allowed by C is $\sum_{i\in I} a_i$, and we assumed C is a winning coalition, so $\sum_{i\in I} a_i \geq b$. Altogether, we get:

$$b + 1 > \sum_{i\in I} a_i \geq b$$

But since all a_i are integers, we conclude that $\sum_{i\in I} a_i = b$, so the given SUBSET-SUM instance is a "yes" instance.

On the other hand, assume the given SUBSET-SUM instance is a "yes" instance. This means there is some subset of indexes I such that $\sum_{i\in I} a_i = b$. Define the coalition $C = \{(s, v_i) : i \in I\} \cup \{(v_i, t) : i \in I\}$. The max-flow value allowed by C is $\sum_{i\in I} a_i = b$, so C is a winning coalition. However:

$$p(C) = 2\sum_{i\in I} \frac{a_i}{2(b+1)} = \frac{b}{b+1} < 1$$

So the coalition C blocks p, and p is unstable. □

Another interesting question is whether finding the CoS itself is computationally hard in TNFGs. The answer to that question is *yes*, although this result does not follow from Theorem 3. The proof is based on a reduction from the well-known PARTITION problem, and is omitted due to space constraints.

5 The Cost of Stability in TNFGs

We now show how to bound and approximate the CoS in general TNFGs, and provide efficient algorithms for computing the CoS and finding optimal super-imputations[1] in restricted classes of TNFGs.

5.1 Connectivity Games

We first show that the CoS can be computed efficiently in *connectivity games*, where a coalition wins if it contains a path from the network's source to its sink.

Definition 15. *A* connectivity game *is a TNFG where the capacities of all edges are* 1 *and the threshold is also* 1.

Theorem 4. *The CoS of a connectivity game is equal to the max-flow value of the underlying flow network.*

Proof. Let C be the set of edges crossing a min-cut in a connectivity game G. Notice that $|C|$ is equal to the max-flow value of the network due to the max-flow min-cut theorem (since all capacities are 1). Any winning coalition in G (containing a path from s to t) must include some edge in C, so from Theorem 2 we get $\mathrm{CoS}(G) \leq |C|$. On the other hand, Lemma 2 guarantees the existence of $|C|$ edge-disjoint paths from s to t, each of which is a winning coalition, so from Theorem 1 we get $\mathrm{CoS}(G) \geq |C|$. We conclude that $\mathrm{CoS}(G) = |C|$. □

5.2 Bounding the CoS in TNFGs

We now give an upper bound on the CoS in general TNFGs, based on the max-flow value of the underlying flow network.[2]

Theorem 5. *Let G be a TNFG with threshold k, and let F be the max-flow value of the underlying flow network. Then* $\mathrm{CoS}(G) \leq \frac{F}{k}$.

Proof. Let E be the edge set of G, and let S be the set of edges crossing a min-cut of G. We define the super-imputation p as follows:

$$\forall e \in E : p_e = \begin{cases} \frac{c(e)}{k} & \text{if } e \in S \\ 0 & \text{otherwise} \end{cases}$$

Notice that due to the max-flow min-cut theorem:

$$p(E) = \sum_{e \in S} \frac{c(e)}{k} = \frac{F}{k}$$

[1] A super-imputation is optimal if it is stable and the sum of payments is equal to the CoS.

[2] Note that there is a trivial upper bound on the CoS in any simple game—the CoS is never greater than the number of agents in the game (this is implied by Theorem 2).

Let C be a winning coalition in G. This means that:

$$\sum_{e \in C \cap S} c(e) \geq k$$

And so:

$$p(C) = \sum_{e \in C} p_e = \sum_{e \in C \cap S} \frac{c(e)}{k} \geq 1$$

So p is stable and $\mathrm{CoS}(G) \leq p(E) = \frac{F}{k}$. □

A corollary of Theorem 5 is that the ratio between the max-flow value (F) and the threshold (k) of a TNFG (which is easy to compute) can serve as an approximation for the game's CoS (an $\frac{F}{k}$-approximation). Of course, this approximation is tighter the smaller the ratio. Also, the proof of Theorem 5 shows us how to efficiently find a stable super-imputation with adjusted gains equal to this ratio.

5.3 Equal Capacity TNFGs

We now generalize Theorem 4, showing an efficient way to compute the CoS of a TNFG with equal edge capacities.

Theorem 6. *If G is a TNFG where the capacities of all edges are equal to b and the threshold is rb (for some $b \in \mathbb{R}_+$ and $r \in \mathbb{N}$), then $\mathrm{CoS}(G) = \frac{F}{rb}$, where F is the max-flow value of the underlying flow network.*

Proof. We know that $\mathrm{CoS}(G) \leq \frac{F}{rb}$ by Theorem 5, so it suffices to prove that $\mathrm{CoS}(G) \geq \frac{F}{rb}$.[3]

Denote $d = \frac{F}{b}$. Note that $d \in \mathbb{N}$, since a min-cut in G contains d edges (each with capacity b). We claim that there must exist d edge-disjoint paths from s to t in G. This follows from Lemma 2, because if we changed all the capacities in the network to 1, the max-flow value would be d (any min-cut in the original network is still a min-cut after the change).

Let $C_1, \ldots, C_d$ denote edge-disjoint paths from s to t in G. Let p be a stable super-imputation in G. Since the threshold is rb, any coalition containing r of the paths C_i ($1 \leq i \leq d$) is a winning coalition. In other words, for any subset of indexes $I \subseteq \{1, \ldots, d\}$ where $|I| = r$, it must hold that:

$$\sum_{i \in I} p(C_i) \geq 1$$

We can write $\binom{d}{r}$ such inequalities, and each $p(C_i)$ appears in an equal number of them, so summing all the inequalities yields:

$$\frac{r}{d}\binom{d}{r}\sum_{i=1}^{d} p(C_i) \geq \binom{d}{r} \Rightarrow \sum_{i=1}^{d} p(C_i) \geq \frac{F}{rb}$$

[3] The proof of Theorem 5 also provides an efficient method for finding an optimal super-imputation in this case.

Since this is true for any stable super-imputation p, we conclude that: $\text{CoS}(G) \geq \frac{F}{rb}$. □

Note that Theorem 4 is actually a special case of Theorem 6, where $r = b = 1$.

5.4 Serial TNFGs

We now examine the special case of *serial* TNFGs, built by serially connecting a sequence of component TNFGs. Such games can model scenarios where the flow must pass through a series of bottlenecks. We show that in such a case, the CoS of the entire sequence is equal to the minimal CoS among the component TNFGs.

Definition 16. *Given a set of TNFGs $\{G_1, \ldots, G_n\}$ all with the same threshold k, a* serial TNFG *is the TNFG with threshold k over the flow network obtained by merging the sink of G_i with the source of G_{i+1} for every $1 \leq i < n$.*

Theorem 7. *If G is a serial TNFG composed of the TNFGs $\{G_1, \ldots, G_n\}$, then* $\text{CoS}(G) = \min_{1 \leq i \leq n} \text{CoS}(G_i)$.

Proof. We will prove the theorem for the case where $n = 2$, and the general case follows by induction. Assume w.l.o.g. that $\text{CoS}(G_1) \leq \text{CoS}(G_2)$. Denote by E_1 and E_2 the edge sets of G_1 and G_2 respectively, and denote by $E = E_1 \cup E_2$ the edge set of G. Let p' be an optimal super-imputation in G_1 (i.e., p' is stable and $p'(E_1) = \text{CoS}(G_1)$). We define the super-imputation p in G as follows:

$$\forall e \in E : p_e = \begin{cases} p'_e & \text{if } e \in E_1 \\ 0 & \text{if } e \in E_2 \end{cases}$$

Notice that $p(E) = p'(E_1) = \text{CoS}(G_1)$. We will show that p is optimal in G, which implies that $\text{CoS}(G) = \text{CoS}(G_1)$.

First, let $C \subseteq E$ be a winning coalition in G. C must contain a subset $C' \subseteq C \cap E_1$ which is a winning coalition in G_1. p' is stable in G_1, so $p(C) = p'(C') \geq 1$, meaning that p is stable in G.

On the other hand, let $\tilde{p}$ be a super-imputation in G such that $\tilde{p}(E) < p(E) = \text{CoS}(G_1)$. Write $\tilde{p}(E_1) = \alpha\tilde{p}(E)$ and $\tilde{p}(E_2) = (1 - \alpha)\tilde{p}(E)$ for some $0 \leq \alpha \leq 1$. Assume w.l.o.g. $\alpha > 0$. There must exist a winning coalition $C_1 \subseteq E_1$ in G_1 such that $\tilde{p}(C_1) < \alpha$, otherwise the super-imputation $\frac{1}{\alpha}\tilde{p}$ would be stable in G_1 with adjusted gains smaller than $\text{CoS}(G_1)$, which would be a contradiction. Likewise, there must exist a winning coalition $C_2 \subseteq E_2$ in G_2 such that $\tilde{p}(C_2) \leq (1 - \alpha)$.[4] The coalition $C_1 \cup C_2$ is then a winning coalition in G, but $\tilde{p}(C_1 \cup C_2) < \alpha + (1 - \alpha) = 1$. We conclude that $\tilde{p}$ is unstable in G and so $p(E) = \text{CoS}(G)$.

Altogether, this shows that p is optimal in G, which implies that $\text{CoS}(G) = p(E) = \text{CoS}(G_1)$. So the theorem is proved for the case where $n = 2$. The general case follows by induction. □

Using Theorem 7, we now show how the CoS of a serial TNFG can be computed efficiently, as long as the number of edges in each component TNFG is not too large.

[4] Here the inequality is not strict, since if $\alpha = 1$ then $\tilde{p}$ is 0 for any coalition in G_2.

Definition 17. *A* B-bounded serial TNFG *is a serial TNFG with components* $\{G_1, \ldots, G_n\}$ *where the number of edges in each component TNFG* G_i $(1 \leq i \leq n)$ *is bounded by some constant number* B.[5]

Theorem 8. *The CoS of a B-bounded serial TNFG can be computed in polynomial time.*

Proof. Let G be a B-bounded serial TNFG whose component TNFGs are $\{G_1, \ldots, G_n\}$. We present an algorithm for computing $\text{CoS}(G)$ in time linear in n, although the runtime includes a constant factor which is exponential in B. Therefore, this algorithm is only tractable if the bound B is small.

For each TNFG G_i, we can describe $\text{CoS}(G_i)$ as a linear program. Let E_i denote the set of edges in G_i. For every $e \in E_i$ we define a variable p_e. The linear program is:

Minimize: $\sum_{e \in E_i} p_e$

Under the constraints:
$\forall e \in E_i : p_e \geq 0$
$\forall C \subseteq E_i : \sum_{e \in C} p_e \geq v(C)$

Recall that $v(C)$ equals 1 if C is a winning coalition and 0 otherwise. The number of constraints in the linear program is exponential in $|E_i|$, but $|E_i|$ is bounded by the constant B. Linear programs can be solved efficiently, so we can calculate $\text{CoS}(G_i)$ in constant time with respect to n (although exponential with respect to B).

Once we have computed $\text{CoS}(G_i)$ for all n component TNFGs, we can get $\text{CoS}(G)$ by using Theorem 7: $\text{CoS}(G) = \min_{1 \leq i \leq n} \text{CoS}(G_i)$. □

6 Weighted Voting Games and TNFGs

We now examine the relationship between the CoS in TNFGs and in *weighted voting games* (WVGs), a well-known game theoretic model of cooperative decision making.[6]

Definition 18. *Given a set of agents* N, *a weight function* $w : N \to \mathbb{R}_+$ *and a threshold* $q \in \mathbb{R}_+$, *a* weighted voting game *is the simple cooperative game where a coalition* $C \subseteq N$ *is a winning coalition if and only if the sum of the weights of the agents in* C *exceeds the threshold* q, *that is* $w(C) = \sum_{a \in C} w(a) \geq q$.

We can define a WVG based on any subset of agents in a TNFG: given a TNFG $\langle V, E, c, s, t, k \rangle$ and a subset of agents $F \subseteq E$, we define the WVG $W_F = \langle F, w, k \rangle$ where $w(e) = c(e)$ for every agent $e \in F$. We also denote the CoS of the new game W_F as $\text{CoS}(F)$.

We now show that the CoS of a TNFG is bounded by the CoS of any WVG induced by the set of edges crossing a cut of the flow network.

[5] Note that the number of components n is not bounded.

[6] Analysis of the CoS in WVGs is given by Bachrach *et al* [1].

Theorem 9. *Let $G = \langle V, E, c, s, t, k \rangle$ be a TNFG instance, let $F \subseteq E$ be the set of edges crossing a cut of G, and let p_F be a super-imputation in the WVG W_F. If p_F is stable in W_F, then the super-imputation p is stable in G, where $p(e) = p_F(e)$ if $e \in F$ and $p(e) = 0$ otherwise.*

As a direct corollary we get that $\mathrm{CoS}(G) \leq \mathrm{CoS}(F)$.

Proof. Let $C \subseteq E$ be a winning coalition in G, i.e., the agents in C allow a flow with value k from s to t. In particular, it must hold that $w(F \cap C) = c(F \cap C) \geq k$, so $F \cap C$ is a winning coalition in W_F. Since p_F is stable in W_F, we know that $p_F(F \cap C) \geq 1$, and therefore:

$$p(C) \geq p(F \cap C) = p_F(F \cap C) \geq 1 \ .$$

So p is stable in G. □

We can now supply alternative proofs to some of the theorems in this work using WVGs. Theorem 5 is a direct corollary of Theorem 9, if we consider the edges crossing a min-cut of a TNFG. The hardness of TNFG-SIS (Theorem 3) follows from the hardness of the equivalent problem for WVGs, since we can reduce a WVG to a TNFG: given a WVG $W = \langle N, w, q \rangle$, we define the TNFG $G_W = \langle V, E, c, s, t, k \rangle$ by setting $V = \{s, t\}$; $c = w$; $k = q$ and $E = N$, where all edges are from s to t.[7] By similar arguments to those in the proof of Theorem 9, a super-imputation is stable in W if and only if it is stable in G_W. Bachrach *et al.* [1] prove that testing for super-imputation stability in WVGs is coNP-hard, so it follows that TNFG-SIS is coNP-hard as well.

7 Related Work

The concept of the core was introduced by Gillies [7]. Similar concepts are the least-core and the nucleolus [11], which are guaranteed to be nonempty. A different solution concept is the Shapley value [12], which aims for fairness rather than stability.

Elkind *et al.* [3] discuss various solution concepts in WVGs, showing that in this domain computing the core can be done in polynomial time, while many questions relating to other solution concepts are NP-hard. Elkind and Pasechnik [5] show a pseudo-polynomial algorithm for computing the nucleolus of WVGs.

Bachrach and Rosenschein [2] examine calculating power indexes in TNFGs. Power indexes attempt to measure how much "real power" each player has in a given game. It is shown that for TNFGs, computing the Shapley-Shubik index is NP-hard and computing the Banzhaf index is #P-complete. However, an efficient algorithm for the restricted case of connectivity games over bounded layer graphs is provided. Elkind *et al.* [4] show how to compute power indexes in the special case of series-parallel TNFGs.

While our work focuses on TNFGs, much research has considered the *cardinal* network flow game (CNFG), where a coalition's utility equals the max-flow value it can achieve. Computing the core in CNFGs can be done in polynomial time; Kalai and Zemel [8, 9] show that numerous families of CNFGs have nonempty cores.

[7] This requires allowing a multigraph, but we could avoid that by splitting every edge into two equivalent edges.

Yokoo *et al.* [13] demonstrate that various cooperative solution concepts (such as the core, nucleolus and Shapley value) are vulnerable to manipulations in open anonymous environments. They use a more fine-grained model of cooperative games, where each agent has a set of skills, and values are defined for different subsets of skills (rather than subsets of agents). They show that agents may sometimes profit from manipulations such as submitting false names, collusion, and hiding skills.

Monderer and Tennenholtz [10] investigate the case of an interested party who wishes to influence the behavior of agents in a game which is not under its control. The approach taken is close to the one we take here in spirit, although that work deals with *normal-form* games, not cooperative games. In that model, the interested party may commit to making non-negative payments to the agents if certain strategy profiles are selected. Payments are given to agents individually, but they are dependent on the strategies selected by all agents. As in our work, it is assumed that the interested party wishes to minimize its expenses. Determining the optimal monetary offers to be made in order to implement a desired outcome is shown to be NP-hard in general, but becomes tractable under certain modifications.

The CoS concept that we use here was first defined by Bachrach *et al.* [1], who examined the CoS in WVGs. It was shown that it is coNP-complete to test whether a given super-imputation in such a game is stable, but the CoS may be computed efficiently if either the player weights or payments are bounded. An efficient approximation algorithm for the CoS in general WVGs was also given.

8 Conclusion

We examined stabilizing cooperative games using external payments, and considered the CoS—the minimal total payment that allows a stable division of the grand coalition's gains among the agents, in the context of network flow games (TNFGs). We showed that it is coNP-complete to determine whether a given super-imputation in a TNFG is stable. We provided an upper bound on the CoS based on the network's max-flow, which can be used to approximate the CoS. We showed that in connectivity games and in equal capacity TNFGs , both the CoS and an optimal super-imputation may be found efficiently. We also showed how to compute the CoS in serial TNFGs with a small number of edges per component. Finally, we showed that the CoS of any TNFG can be bounded by the CoS of a WVG induced by some cut of the flow network.

In future work, we could examine the CoS in various other cooperative games. Additionally, it might be interesting to define the CoS for any coalition (not only for the grand coalition), and perhaps for various coalitional structures. Finally, we could investigate the relationship between the CoS and other cooperative solution concepts such as the least-core, nucleolus, and Shapley value.

References

[1] Bachrach, Y., Meir, R., Zuckerman, M., Rothe, J., Rosenschein, J.S.: The cost of stability in weighted voting games (extended abstract). In: The International Joint Conference on Autonomous Agents and Multiagent Systems (AAMAS), Budapest, Hungary (May 2009)

[2] Bachrach, Y., Rosenschein, J.S.: Power in threshold network flow games. Journal of Autonomous Agents and Multi-Agent Systems 18(1), 106–132 (2009)
[3] Elkind, E., Goldberg, L.A., Goldberg, P.W., Wooldridge, M.: Computational complexity of weighted threshold games. In: The National Conference on Artificial Intelligence (AAAI), pp. 718–723. AAAI Press, Menlo Park (2007)
[4] Elkind, E., Goldberg, L.A., Goldberg, P.W., Wooldridge, M.: A tractable and expressive class of marginal contribution nets and its applications. In: The International Joint Conference on Autonomous Agents and Multiagent Systems (AAMAS), pp. 1007–1014 (2008)
[5] Elkind, E., Pasechnik, D.V.: Computing the nucleolus of weighted voting games. In: The ACM-SIAM Symposium on Discrete Algorithms, SODA (2009)
[6] Ford, L.R., Fulkerson, D.R.: Maximal flow through a network. Canadian Journal of Mathematics 8, 399–404 (1956)
[7] Gillies, D.B.: Some theorems on n-person games. PhD thesis. Princeton University, Princeton (1953)
[8] Kalai, E., Zemel, E.: On totally balanced games and games of flow. Discussion Papers 413, Northwestern University (January 1980)
[9] Kalai, E., Zemel, E.: Generalized network problems yielding totally balanced games. Operations Research 30, 998–1008 (1982)
[10] Monderer, D., Tennenholtz, M.: K-implementation. In: EC 2003: Proceedings of the 4th ACM conference on electronic commerce, pp. 19–28. ACM, New York (2003)
[11] Schmeidler, D.: The nucleolus of a characteristic function game. SIAM Journal on Applied Mathematics 17(6), 1163–1170 (1969)
[12] Shapley, L.S.: A value for n-person games. Contributions to the Theory of Games, 31–40 (1953)
[13] Yokoo, M., Conitzer, V., Sandholm, T., Ohta, N., Iwasaki, A.: Coalitional games in open anonymous environments. In: The National Conference on Artificial Intelligence (AAAI), pp. 509–515 (2005)

(Un)Decidability of Injectivity and Surjectivity in One-Dimensional Sand Automata

Gaétan Richard

Laboratoire d'informatique fondamentale de Marseille,
Aix-Marseille université, CNRS,
39 rue Joliot-Curie, 13 453 Marseille, France
gaetan.richard@lif.univ-mrs.fr

Abstract. Extension of sand pile models, one-dimensional sand automata are an intermediate discrete dynamical system between one dimensional cellular automata and two-dimensional cellular automata. In this paper, we shall study the decidability problem of global behavior of this system. In particular, we shall focus on the problem of injectivity and surjectivity which have the property of being decidable for one-dimensional cellular automata and undecidable for two-dimensional one. We prove the following quite surprising property that surjectivity is undecidable whereas injectivity is decidable. For completeness, we also study these properties on some classical restrictions of configurations (finite, periodic and bounded ones).

Introduction

Complex systems are systems made of a great number of well known entities interacting locally with each other in a fully determined way. Despite the fact that the local behavior is completely known, the global behavior of the system may be very complex and even unpredictable. One simple formal model of complex systems is cellular automata which consist on entities endowed with a state chosen among a finite set, arranged on a regular grid of fixed dimension. Dynamics of this system is obtained by applying uniformly and synchronously a local transition function. In such systems, it was proved that in for one dimensional grid, injectivity and surjectivity are decidable [1] whereas these properties are undecidable in higher dimensions [2].

Introduced as a generalisation of sand-piles models [3], sand automata [4] are a variant of cellular automata where states are integers and where the local transition function works according to the gap between the neighbour value and the cell's one. To keep some locality, the difference is said to be infinite if it excess the radius, thus the local transition function has only a finite number of cases. In a topological way, those systems can be seen as an intermediate model between one-dimensional and two-dimensional cellular automata (see [5,6]). Therefore, one natural question is the it makes sense to study decidability questions on injectivity and surjectivity in these models.

R. Královič and D. Niwiński (Eds.): MFCS 2009, LNCS 5734, pp. 651–662, 2009.

This paper is divided as follows: in section 1, we give formal definitions needed. Then in section 2, we prove undecidability of surjectivity in general as well as for classical restrictions on configurations. In section 3, we deal with the case of injectivity proving that it is decidable in general but not for all classical restrictions on configurations.

1 Definitions

In the rest of the paper, for any $a, b \in \mathbb{Z}$ with $a \leq b$ and $I \subseteq \mathbb{Z}$, let $[\![a, b]\!]$ be the set $\{a, a+1, \ldots, b-1, b\}$ and $\overline{I}$ the set $I \cup \{-\infty, +\infty\}$.

1.1 Sand Automata

A *sand automaton* is a pair (r, f) where $r \in \mathbb{N}$ is the *radius* and $f : \overline{[\![-r, r]\!]}^{2r+1} \to [\![-r, r]\!]$ is the *local transition rule*. This system acts on elements $c \in \overline{\mathbb{Z}}^{\mathbb{Z}}$ called *configurations*. A configuration c is *bounded* if there exists $b \in \mathbb{N}$ such that $\forall i \in \mathbb{Z}$, $|c(i)| < b$. It is *finite* if c is constant except for a finite numbers of elements (*i.e.*, there exists $l, k \in \mathbb{Z}$, such that for all $z \in \mathbb{Z}$ such that $|z| > l$, $c(z) = k$). It is *weakly periodic* if there exists $p \in \mathbb{Z}^+, d \in \mathbb{Z}$ such that $\forall z \in \mathbb{Z}, c(z+p) = c(z)+d$. In the case where $d = 0$ in the previous definition, the configuration is called *strongly periodic*.

For any $r, l \in \mathbb{N}$, the *l-local view function* $v_l : \overline{\mathbb{Z}}^r \times \mathbb{Z} \times \overline{\mathbb{Z}}^r \to \overline{[\![-l, l]\!]}^{2r+1}$ is defined as:

$$v_l(z_{-r}, \ldots, z_0, \ldots, z_r)(i) = \begin{cases} -\infty & \text{if } (z_i - z_0) < -l \\ (z_i - z_0) & \text{if } |(z_i - z_0)| \leq l \\ +\infty & \text{if } (z_i - z_0) > l \end{cases}$$

This definition is extended to any configuration $c \in \overline{\mathbb{Z}}^{\mathbb{Z}}$ and position $z \in \mathbb{Z}$ as $v_l(c)(z) = v_l(c(z-r), \ldots, c(z+r))$ provided that $c(z) \notin \{-\infty, +\infty\}$. The *global function* $G : \overline{\mathbb{Z}}^{\mathbb{Z}} \to \overline{\mathbb{Z}}^{\mathbb{Z}}$ of a sand automaton (r, f) is defined, for all $c \in \mathbb{Z}^{\mathbb{Z}}$ and $i \in \mathbb{Z}$ by:

$$G(c)(i) = \begin{cases} -\infty & \text{if } c(i) = -\infty \\ c(i) + f(v_r(c)(i)) & \text{if } c(i) \in \mathbb{Z} \\ +\infty & \text{if } c(i) = +\infty \end{cases}$$

An sand automaton is *injective* (resp. *surjective*) if its global function is. It is injective (resp. surjective) on finite configurations if the restriction of its global configuration to finite configurations is injective (resp. surjective). The same holds for bounded, weakly periodic or strongly periodic configurations. The links between those different properties can be found in the article of J. Cervelle, E. Formenti and B. Masson [7]. In this rest of this paper, we shall study whether those properties are decidable.

1.2 Two-Counter Machines

In this paper, we shall obtain undecidability result using reduction from the halting problem of two-counter machines. Let $\Upsilon = \{0, +\}$ and $\Phi = \{-, 0, +\} \times \{0, 1\}$ be respectively the set of *test values* and *counter operations*. For all $(\phi, j) \in \Phi$, *testing* $\tau : \mathbb{N}^2 \to \mathbb{N}^2$ and *modifying* $\theta : \Phi \times \mathbb{N}^2 \to \mathbb{N}^2$ actions are respectively defined for any $i \in \{0, 1\}$ and $v \in \mathbb{N}^2$ by:

$$\tau(v)_i = \begin{cases} 0 & \text{if } v_i = 0 \\ + & \text{if } v_i > 0 \end{cases} \qquad \theta(\phi)(v)_i = \begin{cases} \max(0, v_i - 1) & \text{if } \phi = (i, -) \\ v_i + 1 & \text{if } \phi = (i, +) \\ v_i & \text{otherwise} \end{cases}$$

Introduced by M. Minsky [8], *two-counter machines* (CM-2) are quadruplet (Q, q_0, q_f, t) where Q is a finite set of *states*, $q_0, q_f \in Q$ are respectively the *initial* and *final* state and $t : Q \times \Upsilon^2 \to Q \times \Phi$ the *local transition rule*. Those machines act on *configurations* $c \in Q \times \mathbb{N}^2$ by the *global transition rule* $T : Q \times \mathbb{N}^2 \to Q \times \mathbb{N}^2$ defined as $T(q, v) = (q', \theta(\phi)(v))$ when $(q', \phi) = t(q, \tau(v))$. The configuration $(q_f, (0, 0)) \in Q \times \mathbb{N}^2$ is the *halting* configuration. With these definitions, a two-counter machine is *halting* if starting from the configuration $(q_0, (0, 0))$ it eventually reaches an halting configuration. An *evolution* of a 2-CM is a sequence $(c_0, \ldots c_{n-1}) \subset (Q \times \mathbb{N}^2)^n$ where for all $i \in [\![0, n-2]\!]$, $c_{i+1} = T(c_i)$. Such evolution is *halting* if $c_0 = (q_i, (0, 0))$ and $c_{n-1} = (q_f, (0, 0))$. Thought seeming simple, this system can achieve universal computation and thus the following theorem holds:

Theorem 1 (M. Minsky [8], 1967). *The halting problem for two-counter machines is undecidable.*

2 Surjectivity

In this section, we shall reduce the previous halting problem proving undecidability of surjectivity in sand automaton.

Theorem 2. *Given a sand-automaton $S = (r, f)$, it is undecidable to know whether it is surjective.*

The reduction use the following sketch. We first define an encoding of any evolution of any CM-2 inside a configuration of a sand automaton. Then, for each two-counter machine, we define a sand automaton that is surjective on all configurations except those containing the encoding of an halting evolution. To do this, the constructed sand automaton "checks" locally whether the configuration seems to be a correct evolution of the machine. In this case, the automaton does a XOR on some additional *checking bits*. The main point is that those bits are positioned such that a valid halting evolution creates a finite cycle and thus prevents surjectivity. The idea of this technique is similar to the one used in the proof of undecidability of surjectivity over finite configurations in two-dimensional cellular automata by J. Kari [9] whereas realisation is trickier due to additional restrictions encountered.

2.1 Proof of the Theorem

Encoding two-counter machine evolution. This section is devoted to explain how to encode an evolution of a two-counter machine inside a sand automaton configuration. One trick in the encoding is that all data are encoded by sequence of integers which are all multiple of 10. Intermediate values being only used to achieve unambiguity. Therefore, to ease reading and understanding, values in configurations of sand automaton are given as one digit numbers (ex: 0.5).

For any configuration $c = (q, l, r) \in Q \times \mathbb{N}^2$, any integer $h \in [\![-r-1, l+1]\!]$ and any array of *checking bits* ($x = (x_0, \ldots, x_7) \in \{0,1\}^8$), the $c-h-x$ *snapshot* is the sequence of values (depicted in Fig. 1) obtained by concatenating the following sub-sequences where $l_? = 1$ iff $l = 0$ (resp. $r_? = 1$ iff $r = 0$):

- $(l+1, l+1.3, l+1+x_0, l+1+x_1)$ to encode first counter;
- $(-r-1, -r-1+0.4, -r-1+x_2, -r-1+x_3)$ to encode second counter;
- $(0, .1, x_4, x_5)$ to encode zero;
- $(h, h+0.2, h+q, h+l_?, h+r_?, h+x_6, h+x_7)$ to encode the state.

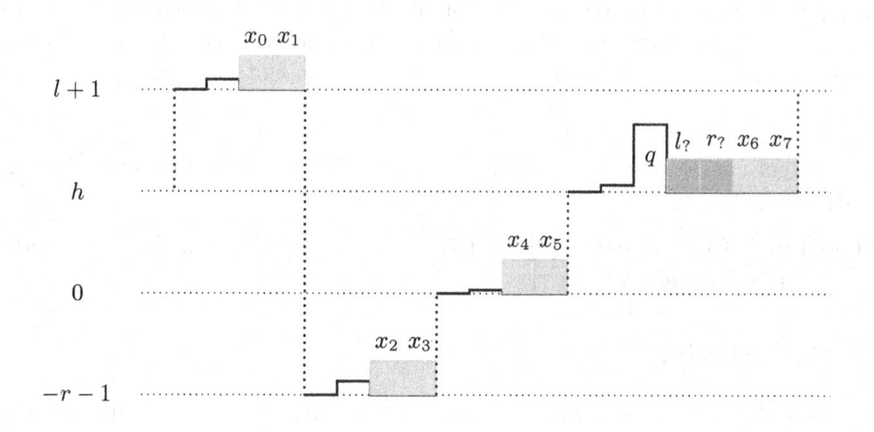

Fig. 1. A $(q, l, r) - h - x$ *snapshot*

A $c-h$ *snapshot* (denoted as $\mathcal{S}_c^h$) a $c-h-x$ snapshot for an arbitrary $x \in \{0,1\}^8$. With this notation, a configuration $c = (q, l, r)$ is encoded as any $c-0$ snapshot and a transition between $c = (q, l, r)$ and $c' = (q', l', r')$ is encoded as following sequence of snapshots:

$$\mathcal{T}_{c,c'} = \mathcal{S}_c^1 \ldots \mathcal{S}_c^l \mathcal{S}_{(q,l',r)}^{l+1} \mathcal{S}_{(q,l',r)}^{l} \mathcal{S}_{(q,l',r)}^{l-1} \cdots \mathcal{S}_{(q,l',r)}^{-r} \mathcal{S}_{(q,l',r')}^{-r-1} \mathcal{S}_{(q,l',r')}^{-r} \cdots \mathcal{S}_{(q,l',r')}^{-1}$$

One way to depict this encoding is to represent each $\mathcal{S}_c^i$ as a four valuated function (one value for $-l-1, 0, i$ and $r+1$) and join those points leading to the figure 2.

With this encoding, to any 2-CM (Q, q_0, q_f, t) and any evolution $(c_0, \ldots c_{n-1}) \subset (Q \times \mathbb{N}^2)^n$, one the set $S \in \mathbb{N}^k$ of partial configuration of sand automata on the form:

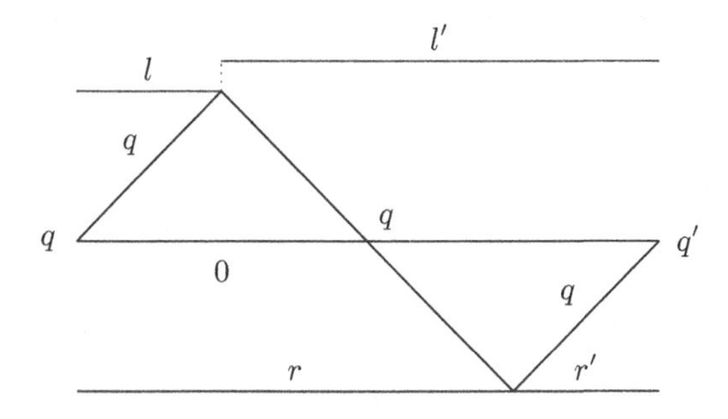

Fig. 2. Encoding a transition $(q, l, r) \vdash (q', l', r')$

$$\mathcal{S}^0_{c_0} \mathcal{T}_{c_0,c_1} \mathcal{S}^0_{c_1} \dots \mathcal{T}_{c_{n-2},c_{n-1}} \mathcal{S}^0_{c_{n-1}}$$

The idea is now to construct, for any 2-CM, a sand automaton which is surjective on all configuration except those containing an encoding of an halting configuration.

Construction of the automaton. The main point of the constructed sand-automaton is to "check" whether the current configuration contains the encoding of an halting evolution. To do this, we use a neigbourhood of large size which is more than twice the size of a snapshot which ensure that our local view contains at least one neighbouring snapshot if it exists. As snapshots can be of arbitrary large height, one cannot see the whole contents of the snapshot though the local view. However, using the small "bumps" (gaps between 0.1 and 0.4) in the encoding, one can determine in which section (head, stacks or zero) is the current position and whether this is compatible with the same section in the left and right neighbours.

The local transition function is chosen to be identity except for position corresponding to one of the checking bit x_i where the local view correspond to a partial correct encoding (see Fig. 3). In this case, we choose either to do a XOR with the value of the corresponding checking bit of either the left or the right neighbour. The former is denoted as x_i^- whereas the latter is denoted as x_i^+. With this definition, the sand automaton behaves as identity except for valid portions of encoding where it behaves as a one dimensional XOR.

The last point of the construction is to define which neighbour is used in each situation and add some additional rules to ensure that the line of xored bits go trough the whole encoding. Thus, we define the order of checking presented in figure 4. For example, when seeing two consecutive snapshots on the form $\mathcal{S}^c_l \mathcal{S}^{(q,l',r)}_{l+1}$ the checking bit x_0 is xored with the checking bit x_6 rather than using the next x_0 bit. Using the depicted order, the resulting line of xored bits is compound either of repetitions of the pattern $x_0^+ x_6^- x_4^- x_7^- x_1^+ x_6^+ x_4^- x_7^+$ or $x_2^- x_7^+ x_5^+ x_6^+ x_3^- x_7^- x_5^+ x_6^-$. One important thing to notice is that all data need for applying the transition of the counter machine is included in the head portion

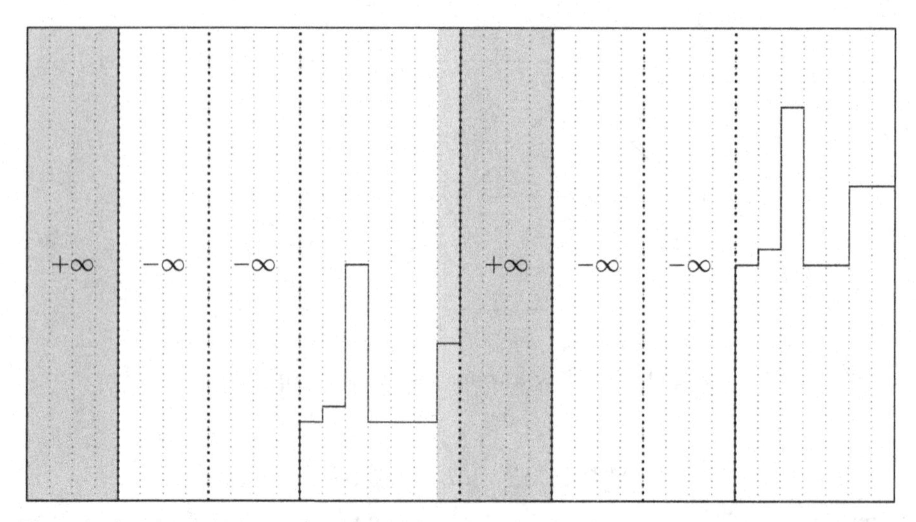

Here the local transition rule enforce the central red column to decrease by one since the local view correspond to a possibly valid encoding and the values of the x_7 in the current and next blocs are both one.

Fig. 3. Example of non-zero transition for x_7^+

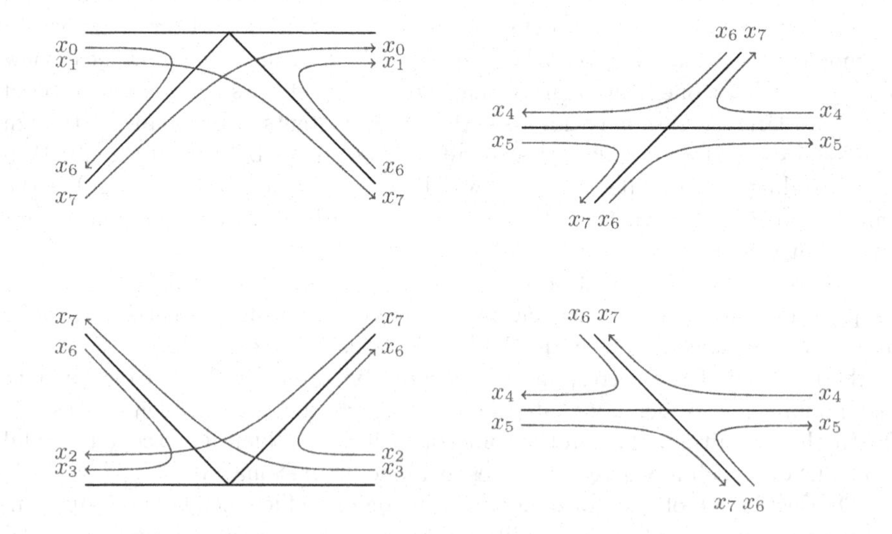

Fig. 4. Order of checks

$(q, l_?, r_?)$ and thus can be read when the sequence of snapshot change the value of l or r (as for example for $\mathcal{S}_l^c\mathcal{S}_{l+1}^{(q,l',r)}$). For the case of initial or halting configuration, we add the following additional order of checking $(x_6^- x_1^+)(x_2^- x_5^+)(x_3^- x_0^+)$ which links the two previously introduced patterns.

Lemma 1. *For the constructed sand-automaton, a configuration contains a cycle of xored bits if and only if it encodes an halting evolution of the associated CM-2.*

Proof. It is clear that the encoding of a correct evolution implies the existence of a cycle of xored bits.

Let us now look at the converse and assume there exists a cycle of xored bits. The first easy remark is that the cycle is restricted by the order of x_i defined previously. Do to this choice, the cycle is made of convex polyominoes. This ensure that the cycle is compound of succession of triangles and trapezes which form a valid upper or lower part.

From this remark, it can be deduced that there exists exactly one starting and one halting configurations and that the two parts (lower and upper) are coherent. Thus, looking around checking bits involved in the cycle, there is the encoding of a valid evolution of the associated CM-2 from the initial configuration to the halting one inducing that the CM-2 is halting. □

This lemma concludes the proof of the theorem: since one dimensional XOR is surjective on finite configuration and infinite one but not for cyclic configurations of fixed size, the constructed automaton is not surjective if and only if the CM-2 is halting.

2.2 Specific Cases

Using some variations of the previous construction, we can achieve to proove undecidability of surjectivity on several restrictions of configurations[1].

Proposition 1. *Given a sand-automaton $S = (r, f)$, it is undecidable to know whether it is surjective on finite configurations.*

Proof. On the one hand, in our construction, for a sufficiently large n the ∞-local view 0^{2r+1} does not encode any valid configuration. This implies that our sand automaton acts as identity on it. Therefore, the only predecessor of pattern 0^{2r+1} is 0 and, if one of the constructed automaton is surjective, it is surjective on finite configuration (note that this implication is not true in general).

On the other hand, if the constructed automaton has a cycle of xored bits, then it exists a finite configuration with this cycle. Hence, it is not surjective on finite configuration. □

Proposition 2. *Given a sand-automaton $S = (r, f)$, it is undecidable to know whether it is surjective on bounded, weakly periodic or strongly periodic configurations.*

Proof. For this result, it is sufficient to remark that the constructed automaton is either surjective for any of these classes of configuration if the 2-CM halts

[1] Note that some of those results can also be achieved using equivalences found in the work of J. Cervelle, E. Formenti and B. Masson [4].

or has a strongly periodic (hence also weakly periodic) and bounded counter-example that can be constructed by repeating the non-finite portion of the finite counter example.

3 Injectivity

Now let us procced with injectivity. In a first part, we use again the previous construction to prove undecidability of injectivity over finite, bounded and strongly periodic configurations. Then, we give a full new proof for decidability of the general and weakly periodic case.

3.1 On Finite, Bounded and Strongly Periodic Configurations

Proposition 3. *Given a sand-automaton, it is undecidable to know whether it is injective on finite configurations.*

Proof. If we take the previous construction, one can see that the automaton is injective unless on configurations containing infinite lines of xored bits or cycles. As previously cycles correspond to halting whereas infinite lines cannot occur in a finite configuration. □

Proposition 4. *Given a sand-automaton, it is undecidable to know whether it is injective on bounded or strongly periodic configurations.*

Proof (sketch). The basic idea of this proof is the same as the previous case, that is, to get rid of the case of infinite xored bit lines which are not cycles. To do this it is sufficient to add in our encoding a constant shift between to consecutive snapshots such that any portion (head, counters or zero) is not horizontal. With this condition, any infinite xored line is necessarily unbounded and thus cannot occur in bounded or strongly periodic configuration. □

3.2 In General and Weakly Periodic Configuration

At this point, one could think that every property is undecidable in one-dimensional sand-automata. In fact, this is not the case and injectivity in decidable in the general case. This result is very interesting since it make the status of those two properties distinct and even make distinction inside injectivity. The rest of this section is thus devoted to prove the following theorem.

Theorem 3. *It is decidable to known whether a sand-automaton is injective.*

Proof of the theorem. The idea of the proof is to show that if a sand-automaton is not injective, then there exist a pair of weakly periodic configurations with the same image and whose perdiod can be bounded. This proof is somewhat similar to the proof in the case of one-dimensional cellular automata (see [1]).

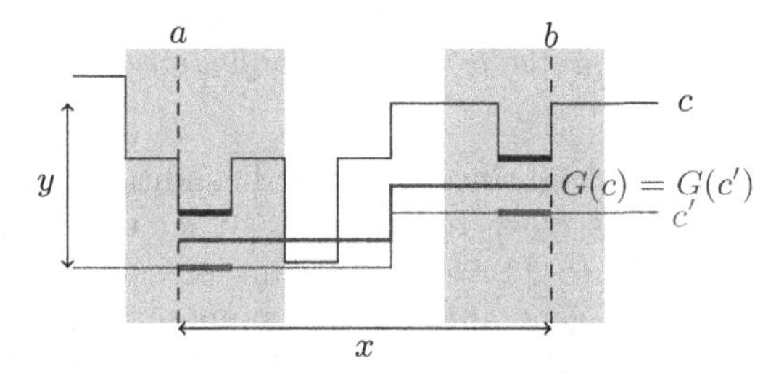

Fig. 5. Example of (x, y) mutually erasable pattern

Let us fix a sand automaton (r, f) with global transition rule G. Let us take $\delta : \mathbb{N} \to \mathbb{N}$ and $\pi : \mathbb{N} \to \mathbb{N}$ defined as $d_0 = 2r + 1$, $\pi_0 = 0$ and for all $n \in \mathbb{N}$, $\delta_{n+1} = 4\delta_n(1 + (2\pi_n + 1)^{2(2r+1)})$ and $\pi_{n+1} = 4\delta_{n+1}(2r + 1)$.

Let I be an interval of $\mathbb{Z}$, two configurations $c, c' \in \overline{\mathbb{Z}}^{\mathbb{Z}}$ are I *mutually erasable* if $G(c)_{|I} = G(c')_{|I}$ and for any sub interval $I' \subseteq I$ such that $|I'| > 2r + 1$, there exists $p \in I'$ such that $c(p) \neq c'(p)$ Inside such a pair, a position $z \in \mathbb{Z}$ is at *level* l if $c(z) \neq c(z')$, for all $i \in [\![0, l - 1]\!]$, $[\![\pi_i, \pi_{i+1} - 1]\!] \cap v_\infty(c)(z) \neq \emptyset$ and $[\![\pi_i, \pi_{i+1} - 1]\!] \cap v_\infty(c')(z) \neq \emptyset$. The set of positions at level l is denoted as $\Delta_l(c, c')$.

A (x, y) *mutually erasable pattern* ((x, y)-mep) is a pair $(c, c') \in [\![a-r, b+r]\!] \to [\![c, d]\!])$ such that $b-a < x$, $d-c < y$, $v_\infty(c)(a) = v_\infty(c')(a)$, $v_\infty(c)(b) = v_\infty(c')(b)$, $G(c)_{|[\![a,b]\!]} = G(c')_{|[\![a,b]\!]}$, $c(a) \neq c'(a)$ and $c(b) \neq c'(b)$ (see Fig. 5). Intuitively, x, y mutually erasable patterns are bounded distinct portions of configuration with the same image and such that local view is the same at extremities of each configuration. The first easy result is that such patterns can be turned into two weakly periodic configurations with the same image.

Lemma 2. *if there exists a (x, y)-mep, then the automaton is not injective.*

Proof. Let us take (c, c') a (x, y)-mep. The basic idea is to construct a configuration by gluing successive repetitions of those patterns. To do this, let us consider the configuration $\tilde{c} : \mathbb{Z} \to \overline{\mathbb{Z}}$ defined as, for any $z \in \mathbb{Z}$,

$$\tilde{c}(z) = c(a + (z \mod (b - a))) + (G(c)(b) - G(c)(a)) \left\lfloor \frac{z}{b - a} \right\rfloor$$

This construction can also be done on c' to obtain the configuration $\tilde{c}'$. One first property is that for all $z \in \mathbb{Z}$, $\tilde{c}(z + (b - a)) = \tilde{c}(z) + (G(c)(b) - G(c)(a))$. As the same can be said on $\tilde{c}'$ and since $G(c')(b) - G(c')(a) = G(c)(b) - G(c)(a)$, it is sufficient to show that $G(\tilde{c})$ and $G(\tilde{c}')$ coincide on $[\![a, b]\!]$. However, as $v_\infty(c)(a) = v_\infty(c)(b)$,we have, for any $z \in [\![a - r, b + r]\!]$, $\tilde{c}(z) = c(z)$. The same applies for c'. Since $G(c)$ and $G(c')$ coincide on $[\![a, b]\!]$ and $c(a) \neq c'(a)$, we constructed two distinct configurations with the same image. □

In the other direction, we shall prove that any non-injective sand-automata do have some mep with a computable bounded size.

Lemma 3 (H_n)**.** *Let take a non-injective sand automaton then either it has a* (d_n, π_n)*-mep or for any* I *mutually erasable configuration* c, c' *where* $|I| \geq 2\delta_n$*,* $\Delta_n(c, c') \neq \emptyset$*.*

Proof. The case $n = 0$ is trivial since the second condition is always true.

Now, assume that H_n is true. To prove H_{n+1}, let us assume that there is no (d_n, π_n)-mep and take (c, c') two I-mutually erasable configurations with $|I| > 2\delta_n$. Without loss of generality, we can suppose that $I = [\![-\delta_n, \delta_n]\!]$ and that we have some position $p \in [\![-r, r]\!]$ such that $0 = c(p) \neq c(p')$.

The first step consist on "clipping" the configuration Let us look at the set $S = \{c(z) \mid z \in I\} \cup \{c'(z) \mid z \in I\}$. This set has at most $4\delta_{n+1}$ values thus there exists $u \in [\![1, 1 + \pi_{n+1}]\!]$ and $l \in [\![-\pi_{n+1} - 1, -1]\!]$ such that $S \cap [\![u - r, u + r]\!] = \emptyset$ and $S \cap [\![l - r, l + r]\!] = \emptyset$. Now let us construct the two elements $d, d' \in \mathbb{Z} \to [\![-\pi_{n+1}, \pi_{n+1}]\!]$ defined, for any $z \in I$, as:

$$d(z) = \begin{cases} -\infty & \text{if } c(z) < l \\ c(z) & \text{if } l \leq c(z) \leq u \\ \infty & \text{if } c(z) > u \end{cases}$$

The same can be done for d'. This operation intuitively consists on "clipping" both configurations between l and u. By construction, it can be easily seen that d and d' are I mutually erasable configurations and that $\Delta_0(d, d') \subset \Delta_0(c, c')$.

Now let us look more in details at $\Delta_0(d, d')$. By construction, we have $p \in \Delta_0(d, d')$. Since $\delta_{n+1} = 4\delta_n(1 + (2\pi_n + 1)^{2(2r+1)})$, we can divide our interval into $1 + 2(2\pi_n + 1)^{2(2r+1)}$ distinct sub-intervals of size $2\delta_n$. Now, let us prove that at least half of them do have a point at level n. The basic idea is to make use of the recurrence hypothesis and the obvious fact that a (δ_n, π_n)-mep is a $(\delta_{n+1}, \pi_{n+1})$-mep. To apply this hypothesis on any sub-interval I', we must ensure that (d, d') is I'-mutually erasable. Since (d, d') is I erasable, the fact that their image by the transition function is the same is trivial. The more difficult point is to show that $\Delta_0(d, d')$ is "dense" on the left or on the right of p. To do this, we shall proof the following lemma:

Lemma 4. *If there exists* $l < p < u$ *such that* $\Delta_0(d, d') \cap [\![l - r, l + r]\!] = \Delta_0(d, d') \cap [\![u - r, u + r]\!] = \emptyset$ *then there exists a* $(u - l + 2r + 1, \pi_{n+1})$*-mep.*

If we are in this conditions, one can easily obtain a mep from this by "gluing" the identical portions as depicted in figure 6. More formally, we consider the configuration e defined as:

$$e(z) = \begin{cases} d(z) & \text{if } z \in [\![l - r, u + r]\!] \\ d(z + u - l + 2r) & \text{if } z \in [\![l - 3r - 1, l - r]\!] \\ d(z - u + l + 2r) & \text{if } z \in [\![u + r, u + 3r + 1]\!] \\ +\infty & \text{otherwise} \end{cases}$$

The same can be done to obtain e' from d'. Since $d(p) \neq d'(p)$, (e, e') is a $(\delta_{n+1}, \pi_{n+1})$-mep, concluding the proof of lemma 4.

With the previous claim, we have found at least $(2\pi_n + 1)^{2(2r+1)}$ positions at level n. Since this number is more that the square the number of possible

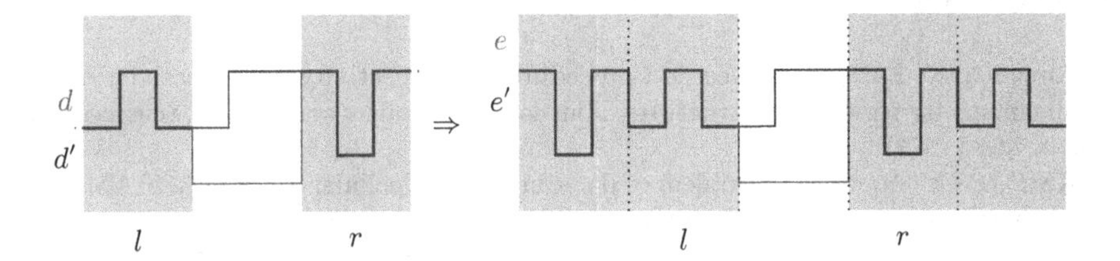

Fig. 6. Gluing by identical portion

elements in v_{π_n}, there are two positions z, z' such that $v_{\pi_n}(d)(z) = v_{\pi_n}(d)(z')$ and $v_{\pi_n}(d')(z) = v_{\pi_n}(d')(z')$. If this condition would also be true for v_∞ then we would have a $(\delta_{n+1}, \pi_{n+1})$-mep. It follows that either $v_\infty(d)(z)$ or $v_\infty(d)(z')$ contains a value larger than π_n which is neither $-\infty$ nor ∞. As d has values into $[\![-\pi_{n+1}, \pi_{n+1}]\!]$, then either z or z' is at level $n+1$.

To sum up, starting for c, c' two I mutually erasable configurations with $|I| \leq 2\delta_{n+1}$ and assuming that there is no $(\delta_{n+1}, \pi_{n+1})$-mep, we have shown that there exists a point at level n (either z or z'). □

To finish the proof, it is sufficient to note that a (f, r) sand automaton cannot have any level $2r+2$ position since v_∞ contains at most $2r+1$ values and that any non injective sand-automaton has either two δ_{2r+1} mutually erasable configuration or a $(\delta_{2r+1}, \pi_{2r+1})$-mep by the same gluing argument as previously. It follows that any non-injective sand-automata have a $(\delta_{2r+1}, \pi_{2r+1})$-mep. As those mep are in finite number (up to some vertical translation), the injectivity problem is decidable for one-dimensional sand-automaton.

4 Conclusion

Those two results of decidability confirm the place of sand automata as an intermediate model between one and two dimensional cellular automata. The fact that status of injectivity and surjectivity differ is very interesting and could perhaps help understanding better these two notions. Even if they use the same global idea as for cellular automata, the two proofs are more subtle. The proof of undecidability of surjectivity is more powerful by working under several additional restrictions as the one for cellular automata whereas the proof of decidability of injectivity is an extension of the "cut and glue" idea used for cellular automata. For the later case, the fact that some restrictions become undecidable is also very interesting as they can all be seen as providing a way to "fix" some origin. In this way, we have somehow the same duality as between the classical halting problem and the immortality problem. It could be interesting to see if sand-automata could help provide a model where the first is undecidable whereas the second is decidable. To conclude, we can note that the bound on size of mep is rough and can probably be improved if trying to consider the complexity of deciding injectivity.

References

1. Amoroso, S., Patt, Y.N.: Decision procedures for surjectivity and injectivity of parallel maps for tessellation structures. Journal of Computer and System Sciences 6(5), 448–464 (1972)
2. Kari, J.: The nilpotency problem of one-dimensional cellular automata. SIAM Journal on Computing 21(3), 571–586 (1992)
3. Goles, E., Kiwi, M.A.: Games on line graphs and sand piles. Theoretical Computer Science 115(2), 321–349 (1993)
4. Cervelle, J., Formenti, E., Masson, B.: From sandpiles to sand automata. Theoretical Computer Science 381(1-3), 1–28 (2007)
5. Dennunzio, A., Guillon, P., Masson, B.: Stable dynamics of sand automata. In: Fifth Ifip International Conference on Theoretical Computer Science, TCS 2008, vol. 273, pp. 157–169 (2008)
6. Dennunzio, A., Guillon, P., Masson, B.: Sand automata as cellular automata. Accepted to TCS (2008) (under revision)
7. Cervelle, J., Formenti, E., Masson, B.: Basic properties for sand automata. In: Jedrzejowicz, J., Szepietowski, A. (eds.) MFCS 2005. LNCS, vol. 3618, pp. 192–211. Springer, Heidelberg (2005)
8. Minsky, M.: Computation: Finite and Infinite Machines. Prentice Hall, Englewoods Cliffs (1967)
9. Kari, J.: Reversibility and surjectivity problems of cellular automata. J. Comput. Syst. Sci. 48(1), 149–182 (1994)

Quantum Algorithms to Solve the Hidden Shift Problem for Quadratics and for Functions of Large Gowers Norm

Martin Rötteler

NEC Laboratories America
4 Independence Way, Suite 200
Princeton, NJ 08540, U.S.A.
mroetteler@nec-labs.com

Abstract. Most quantum algorithms that give an exponential speedup over classical algorithms exploit the Fourier transform in some way. In Shor's algorithm, sampling from the quantum Fourier spectrum is used to discover periodicity of the modular exponentiation function. In a generalization of this idea, quantum Fourier sampling can be used to discover hidden subgroup structures of some functions much more efficiently than it is possible classically. Another problem for which the Fourier transform has been recruited successfully on a quantum computer is the hidden shift problem. Quantum algorithms for hidden shift problems usually have a slightly different flavor from hidden subgroup algorithms, as they use the Fourier transform to perform a *correlation* with a given reference function, instead of sampling from the Fourier spectrum directly. In this paper we show that hidden shifts can be extracted efficiently from Boolean functions that are quadratic forms. We also show how to identify an unknown quadratic form on n variables using a linear number of queries, in contrast to the classical case were this takes $\Theta(n^2)$ many queries to a black box. What is more, we show that our quantum algorithm is robust in the sense that it can also infer the shift if the function is close to a quadratic, where we consider a Boolean function to be close to a quadratic if it has a large Gowers U_3 norm.

1 Introduction

Fourier analysis has a wide range of applications in computer science including signal processing, cryptography, Boolean functions, just to name a few. The fast Fourier transform (FFT) algorithm provides an efficient way to compute the discrete Fourier transform of length N in time $O(N \log N)$. This is a significant improvement over the naive $O(N^2)$ implementation and allows to apply Fourier analysis to correlation problems, to image and audio processing, efficient decoding of error-correcting codes, data compression, etc. In a more theoretical context, the Fourier transform over the Boolean hypercube—also called Walsh-Hadamard transform—is used to study certain classes of Boolean functions, for instance monotone functions, functions with constant depth, and functions with variables of high influence.

In quantum computing, Fourier transforms have turned out to be extremely successful tools and feature prominently in quantum algorithms that achieve exponential speedups. The prime examples are Shor's algorithms for discrete log and factoring

R. Královič and D. Niwiński (Eds.): MFCS 2009, LNCS 5734, pp. 663–674, 2009.

[Sho97]. Indeed, the quantum computer can sample from the Fourier spectrum on N points in quantum time $O(\log^2 N)$, a big advantage over the classical case. Here "quantum time" is measured in terms of elementary quantum gates that are needed to implement the unitary operation corresponding to the Fourier transform. This possibility of performing a quantum Fourier transform more efficiently than in the classical case has a tremendous upside and much of the power of quantum computing stems from there. This fact has been leveraged for instance for the solution of the abelian hidden subgroup problem (HSP) which essentially is solved by sampling from the Fourier spectrum of a given function [ME98, BH97, Kit97]. The hidden subgroup, a secret property of the function, can then be inferred by a subsequent classical post-processing.

However, the high hopes that Fourier sampling might lead to efficient quantum algorithms for HSPs over general non-abelian groups, including cases that would encompass the famous graph isomorphism problem, have been somewhat dampened recently, as [HMR⁺06] showed that new techniques to design highly entangling measurements would be required in order for the standard approach to succeed. Perhaps for this reason, the field of quantum algorithms has seen a shift towards other algebraic problems such as the algorithm for finding hidden nonlinear structures [CSV07]. The techniques to tackle those problems are still based on Fourier analysis but have a different flavor than the HSP.

Classically, besides allowing for *sampling* from the spectrum the importance of the Fourier transform for performing *correlation* tasks cannot be overstated. Therefore, it is very natural to try to leverage the quantum computer's exponential speedup at computing Fourier transforms to compute correlations efficiently. It turns out, however, that this task is an extremely challenging one. First of all, it can be shown that it is impossible to compute correlations between two *unknown* vectors of data due to requirement for the time evolution to be unitary and the fact that the correlation between two inputs is a non-linear map of the inputs. For some special problems, however, in which one of the inputs is a fixed, *known* vector of data, correlations can be computed. This question becomes relevant in particular for hidden shift problems, where correlations can be used in a particularly fruitful way. These problems ask to identify a hidden shift provided that access to a function $f(x)$ and a shifted version $g(x) = f(x + s)$ of the function is given. Formally, the hidden shift problem is defined as follows:

Given: Finite group G, finite set R, maps $f, g : G \to R$.
Promise: There exists $s \in G$ such that $g(x) = f(x + s)$ for all $x \in G$.
Task: Find s.

The first example of a problem of this kind that was solved on a quantum computer was $f(x)$ being the Legendre symbol and s being an unknown element of the cyclic group $\mathbb{Z}_p$ modulo a prime. As shown in [DHI03], for the Legendre symbol the hidden shift s can be found efficiently on a quantum computer. The key observation is that the Legendre function is an eigenfunction of the Fourier transform for the cyclic group $\mathbb{Z}_p$. This fact can be used to compute a correlation of a shifted Legendre symbol with the Legendre symbol itself by using the convolution theorem, involving the application of two discrete Fourier transforms over $\mathbb{Z}_p$.

Our results. We present an efficient quantum algorithm to solve the hidden shift problem for a class of quadratic Boolean functions for which the associated quadratic form is non-degenerate. Those functions are special cases of what is known as *bent* functions [Rot76]. An intriguing property of these functions is that, in absolute values, they have a perfectly flat Fourier spectrum. In general, bent functions are those Boolean functions for which the Hamming distance to the set of all linear Boolean functions is maximum, where distance is measured by Hamming distance between their truth tables. A quantum algorithm to solve the hidden shift problem for bent functions has been studied in [Röt08], where the emphasis is on the richness of different classes of bent functions for which a hidden shift problem can be defined and solved. In this paper, in contrast, we restrict ourselves to just one class of bent functions, namely the quadratic forms, and study a different question: is it possible to solve the hidden shift problem also in cases where a given function f is actually not a quadratic form, but close to a quadratic form? We answer this question in the affirmative, provided that f is not "too far" from a quadratic form, where we measure closeness by the Gowers norm. We give a quantum algorithm that can find a hidden shift for functions that are close to quadratics by using a simple idea: first, we give a quantum algorithm that finds this quadratic form. Then we solve the hidden shift problem for this quadratic form by resorting to the hidden shift algorithm for the bent function case (the case where the corresponding quadratic form is not of full rank can be taken care without major complications), and finally we use a test to determine whether the resulting candidate shift is indeed the correct answer. Overall, we obtain an algorithm that solves the hidden shift for functions of large Gowers norms using $O(n)$ queries to the functions. The classical lower bound for such functions is at least $\Omega(n^2)$ (for the case of perfect quadratics), but we conjecture that for the case of functions that are close to quadratics, actually the classical query complexity scales exponentially.

Related work. We already mentioned [Röt08] which addressed the hidden shift problem for bent functions and which constitutes a building block for our algorithm. The hidden shift problem itself goes back to [DHI03], in which an algorithm similar to our Algorithm 2 was used in order to correlate a shifted function with a given reference function, thereby solving a deconvolution problem. The main difference with the present work is the departure from functions that have perfectly flat Fourier spectrum.

Our algorithm in Section 3 to identify the quadratic function is similar to the methods used in [CSV07, DDW09, BCD05] to extract information about functions that have been encoded into the phases of quantum states. Related to the considered hidden shift problem is also the work by Russell and Shparlinski [RS04] who considered shift problems for the case of $\chi(f(x))$, where f is a polynomial on a finite group G and χ a character of G, a general setup that includes our scenario. The two cases for which algorithms were given in [RS04] are the reconstruction of a monic, square-free polynomial $f \in \mathbb{F}_p[X]$, where χ is the quadratic character (Legendre symbol) over $\mathbb{F}_p$ and the reconstruction of a hidden shift over a finite group $\chi(sx)$, where χ is the character of a known irreducible representation of G. The technique used in [RS04] is a generalization to the technique of [DHI03]. It should be noted that we use a different technique in our algorithm, namely we combine and entangle two states that are obtained from querying the function, whereas [RS04] has more the flavor of a "single register" algorithm. Another difference is that our algorithm is time efficient, i. e., fully polynomial in the input size, whereas [RS04] is query efficient only.

In a broader context, related to the hidden shift problem is the problem of unknown shifts, i. e., problems in which we are given a supply of quantum states of the form $|D + s\rangle$, where s is random, and D has to be identified. Problems of this kind have been studied by Childs, Vazirani, and Schulman [CSV07], where D is a sphere of unknown radius, Decker, Draisma, and Wocjan [DDW09], where D is a graph of a function, and Montanaro [Mon09], where D is the set of points of a fixed Hamming-weight. The latter paper also considers the cases where D hides other Boolean functions such as juntas, a problem that was also studied in [AS07].

2 Fourier Analysis of Boolean Functions

First we briefly recall the Fourier representation of a real valued function $f : \mathbb{Z}_2^n \to \mathbb{R}$ on the n-dimensional Boolean hypercube. For any subset $S \subseteq [n] = \{1, \ldots, n\}$ there is a character of $\mathbb{Z}_2^n$ via $\chi_S : x \mapsto (-1)^{Sx^t}$, where $x \in \mathbb{Z}_2^n$ (the transpose is necessary as we assume that all vectors are row vectors) and $S \in \mathbb{Z}_2^n$ in the natural way. The inner product of two functions on the hypercube is defined as $\langle f, g \rangle = \frac{1}{2^n} \sum_x f(x)g(x) = \mathbb{E}_x(fg)$. The χ_S are inequivalent character of $\mathbb{Z}_2^n$, hence they obey the orthogonality relation $\mathbb{E}_x(\chi_S \chi_T) = \delta_{S,T}$. The Fourier transform of f is a function $\widehat{f} : \mathbb{Z}_2^n \to \mathbb{R}$ defined by

$$\widehat{f}(S) = \mathbb{E}_x(f\chi_S) = \frac{1}{2^n} \sum_{x \in \mathbb{Z}_2^n} \chi_S(x) f(x), \tag{1}$$

$\widehat{f}(S)$ is the Fourier coefficient of f at frequency S, the set of all Fourier coefficients is called the Fourier spectrum of f and we have the representation $f = \sum_S \widehat{f}(S)\chi_S$. The convolution property is useful for our purposes, namely that $\widehat{f * g}(S) = \widehat{f}(S)\widehat{g}(S)$ for all S where the convolution $(f * g)$ of two functions f, g is the function defined as $(f * g)(x) = \frac{1}{2^n} \sum_{y \in \mathbb{Z}_2^n} f(x + y)g(y)$. In quantum notation the Fourier transform on the Boolean hypercube differs slightly in terms of the normalization and is given by the unitary matrix

$$H_{2^n} = \frac{1}{\sqrt{2^n}} \sum_{x,y \in \mathbb{Z}_2^n} (-1)^{xy^t} |x\rangle \langle y|,$$

which is also sometimes called Hadamard transform [NC00]. Note that the Fourier spectrum defined with respect to the Hadamard transform which differs from (1) by a factor of $2^{-n/2}$. It is immediate from the definition of H_{2^n} that it can be written in terms of a tensor (Kronecker) product of the Hadamard matrix of size 2×2, namely $H_{2^n} = (H_2)^{\otimes n}$, a fact which makes this transform appealing to use on a quantum computer since can be computed using $O(n)$ elementary operations.

For Boolean functions $f : \mathbb{Z}_2^n \to \mathbb{Z}_2$ with values in $\mathbb{Z}_2$ we tacitly assume that the real valued function corresponding to f is actually $F : x \mapsto (-1)^{f(x)}$. The Fourier transform is then defined with respect to F, i. .e, we obtain that

$$\widehat{F}(w) = \frac{1}{2^n} \sum_{x \in \mathbb{Z}_2^n} (-1)^{wx^t + f(x)}, \tag{2}$$

where we use $w \in \mathbb{Z}_2^n$ instead of $S \subseteq [n]$ to denote the frequencies. Other than this notational convention, the Fourier transform used in (2) for Boolean valued functions and the Fourier transform used in (1) for real valued functions are the same. In the paper we will sloppily identify $\widehat{f} = \widehat{F}$ and it will be clear from the context which definition has to be used.

We review some basic facts about Boolean quadratic functions. Recall that any quadratic Boolean function f has the form $f(x_1, \ldots, x_n) = \sum_{i<j} q_{i,j} x_i x_j + \sum_i \ell_i x_i$ which can be written as $f(x) = xQx^t + Lx^t$, where $x = (x_1, \ldots, x_n) \in \mathbb{Z}_2^n$. Here, $Q \in \mathbb{F}_2^{n \times n}$ is an upper triangular matrix and $L \in \mathbb{F}_2^n$. Note that since we are working over the Boolean numbers, we can without loss of generality assume that the diagonal of Q is zero (otherwise, we can absorb the terms into L). It is useful to consider the associated symplectic matrix $B = (Q + Q^t)$ with zero diagonal which defines a symplectic form $\mathcal{B}(u, v) = uBv^t$. This form is non-degenerate if and only if $\text{rank}(B) = n$. The coset of $f + R(n, 1)$ of the first order Reed-Muller code is described by the rank of B. This follows from Dickson's theorem [MS77] which gives a complete classification of symplectic forms over $\mathbb{Z}_2$:

Theorem 1 (Dickson [MS77]). *Let $B \in \mathbb{Z}_2^{n \times n}$ be symmetric with zero diagonal (such matrices are also called symplectic matrices). Then there exists $R \in \mathrm{GL}(n, \mathbb{Z}_2)$ and $h \in [n/2]$ such that $RBR^t = D$, where D is the matrix $(\mathbf{1}_h \otimes \sigma_x) \oplus \mathbf{0}_{n-2h}$ considered as a matrix over $\mathbb{Z}_2$ (where σ_x is the permutation matrix corresponding to $(1, 2)$). In particular, the rank of B is always even. Furthermore, under the base change given by R, the function f becomes the quadratic form $ip_h(x_1, \ldots, x_{2h}) + L'(x_1, \ldots, x_n)$ where we used the inner product function ip_h and a linear function L'.*

Let $f(x) = xQx^t + Lx^t$ be a quadratic Boolean function such that the associated symplectic matrix $B = (Q + Q^t)$ satisfies $\text{rank}(B) = 2h = n$. Then the corresponding quadratic form is a so-called *bent function* [Rot76, Dil75, MS77]. In general, bent functions are characterized as the functions f whose Fourier coefficients $\widehat{f}(w) = \frac{1}{2^n} \sum_{x \in \mathbb{Z}_2^n} (-1)^{wx^t + f(x)}$ satisfy $|\widehat{f}(w)| = 2^{-n/2}$ for all $w \in \mathbb{Z}_2^n$, i.e., the spectrum of f is flat. It is easy to see that bent functions can only exist if n is even and that affine transforms of bent functions are again bent functions. Indeed, let f be a bent function, let $A \in \mathrm{GL}(n, \mathbb{Z}_2)$ and $b \in \mathbb{Z}_2^n$, and define $g(x) := f(xA + b)$. Then also $g(x)$ is a bent function and $\widehat{g}(w) = (-1)^{-wb} \widehat{f}(w(A^{-1})^t)$ for all $w \in \mathbb{Z}_2^n$. A very simple, but important observation is that if f is bent, then this implicitly defines another Boolean function via $2^{n/2} \widehat{f}(w) =: (-1)^{\widetilde{f}(w)}$. Then this function $\widetilde{f}$ is again a bent function and called the dual bent function of f. By taking the dual twice we obtain f back: $\widetilde{\widetilde{f}} = f$.

Theorem 1 allows us to define a whole class of bent functions, namely the Boolean quadratics for which $B = (Q + Q^t)$ has maximal rank. It is easy to see that under suitable choice of Q, so instance the inner product function $ip_n(x_1, \ldots, x_n) = \sum_{i=1}^{n/2} x_{2i-1} x_{2i}$ can be written in this way. Using affine transformations we can easily produce other bent functions from the inner product function and Theorem 1 also implies that up to affine transformations the quadratic bent functions are equivalent to the inner product function. From this argument also follows that the dual of a quadratic bent function is again a quadratic bent function, a fact that will be used later on in the algorithm for the hidden shift problem over quadratic bent functions.

3 The Hidden Shift Problem for Quadratics

Let $n \geq 1$ and let $\mathcal{O}$ be an oracle which gives access to two Boolean functions $f, g : \mathbb{Z}_2^n \to \mathbb{Z}_2$ such that there exists $s \in \mathbb{Z}_2^n$ such that $g(x) = f(x+s)$ for all $x \in \mathbb{Z}_2^n$. The hidden shift problem is to find s by making as few queries to $\mathcal{O}$ as possible. If f is a bent function, whence also g since it is an affine transform of f, then the hidden shift can be efficiently extracted using the following *standard algorithm.* Recall that Boolean functions are assumed to be computed into the phase. This is no restriction, as whenever we have a function implemented as $|x\rangle|0\rangle \mapsto |x\rangle|f(x)\rangle$, we can also compute f into the phase as $|x\rangle \mapsto (-1)^{f(x)}$ by applying f to a qubit initialized in $\frac{1}{\sqrt{2}}(|0\rangle - |1\rangle)$.

Algorithm 2 (Standard algorithm for the hidden shift problem [DHI03])
Input: Boolean functions f, g such that $g(x) = f(x+s)$. Output: hidden shift s.

(i) *Prepare the initial state $|0\rangle$.*
(ii) *Apply Fourier transform $H_2^{\otimes n}$ to prepare equal distribution of all inputs:*

$$\frac{1}{\sqrt{2^n}} \sum_{x \in \mathbb{Z}_2^n} |x\rangle .$$

(iii) *Compute the shifted function into the phase to get*

$$\frac{1}{\sqrt{2^n}} \sum_{x \in \mathbb{Z}_2^n} (-1)^{f(x+s)} |x\rangle .$$

(iv) *Apply $H_2^{\otimes n}$ to get*

$$\sum_w (-1)^{sw^t} \hat{f}(w) |w\rangle = \frac{1}{\sqrt{2^n}} \sum_w (-1)^{sw^t} (-1)^{\tilde{f}(w)} |w\rangle .$$

(v) *Compute the function $|w\rangle \mapsto (-1)^{\tilde{f}(w)}$ into the phase resulting in*

$$\frac{1}{\sqrt{2^n}} \sum_w (-1)^{sw^t} |w\rangle .$$

(vi) *Finally, apply another Hadamard transform $H_2^{\otimes n}$ to get $|s\rangle$ and measure s.*

The function $\tilde{f}$ that has been used in Step (iv) can only be applied by means of a unitary operation if the Fourier spectrum of f is flat, in other words if f is a bent function. See also [Röt08] for several classes of bent functions to which this algorithm has been applied. Note that Algorithm 2 requires only one query to g and one query to $\tilde{f}$. Furthermore, the quantum running time is $O(n)$ and the algorithm is exact, i. e., zero error. Note that Step (iii) of Algorithm 2 assumes that the Fourier transform of f is flat.

There is an intriguing connection between the hidden shift problem for injective functions f, g and the hidden subgroup problem over semidirect products of the form $A \rtimes \mathbb{Z}_2$ where the action is given by inversion in A [Kup05, FIM+03]. In our case the functions are not injective, however, it is possible to exploit the property of being bent to derive another injective "quantum" function: $F(x) := \frac{1}{\sqrt{2^n}} \sum_y (-1)^{f(x+y)} |y\rangle$ (similarly a function G can be derived from g). Now, an instance of an abelian hidden subgroup problem in $\mathbb{Z}_2^n \rtimes \mathbb{Z}_2$ can be defined via the hiding function $H(x,b)$ that evaluates to $F(x)$, if $b=0$, and to $G(x)$, if $b=1$. This reduction leads to an algorithm that is different from Algorithm 2, but also can be used to compute the shift.

Now, we consider a different task: we begin with an arbitrary quadratic Boolean function (not necessarily bent) f, which is given by an oracle $\mathcal{O}$. We show that f can be discovered using $O(n)$ quantum queries to $\mathcal{O}$, whereas showing a lower bound of $\Omega(n^2)$ classical queries is straightforward. Recall that Bernstein and Vazirani [BV97] solved the case of linear function f. We use quadratic forms $f(x_1,\ldots,x_n) = \sum_{i<j} q_{i,j} x_i x_j + \sum_i \ell_i x_i$ written as $f(x) = xQx^t + Lx^t$, where $x = (x_1,\ldots,x_n) \in \mathbb{Z}_2^n$. Here, $Q \in \mathbb{Z}_2^{n\times n}$ is an upper triangular matrix and $L \in \mathbb{Z}_2^n$. Using the oracle we can compute the function into the phase and obtain the state

$$|\psi\rangle = \frac{1}{\sqrt{2^n}} \sum_{x\in\mathbb{Z}_2^n} (-1)^{xQx^t + Lx + b} |x\rangle . \tag{3}$$

We will show next, that Q and L can be obtained from a linear number of copies of $|\psi\rangle$. The method uses two such states at a time and combines them using the unitary transform defined by

$$T : |x,y\rangle \mapsto \frac{1}{\sqrt{2^n}} \sum_{z\in\mathbb{Z}_2} (-1)^{zy^t} |x+y, z\rangle .$$

Note that T can be implemented efficiently on a quantum computer as it is just a controlled not between each qubit in the y register as source to the corresponding qubit in the x register as target, followed by a Hadamard transform of each qubit in the y register. The following computation shows that T can be used to extract information about Q from two copies of $|\psi\rangle$.

$$\begin{aligned}
T|\psi\rangle \otimes |\psi\rangle &= T\left(\frac{1}{2^n}\sum_{x,y}(-1)^{xQx^t + yQy^t + L(x+y)^t}|x,y\rangle\right)\\
&= \frac{1}{\sqrt{2^{3n}}}\sum_{x,y,z}(-1)^{xQx^t + yQy^t + L(x+y)^t}(-1)^{zy^t}|x+y,z\rangle\\
&= \frac{1}{\sqrt{2^{3n}}}\sum_{x,u,z}(-1)^{uQu^t + u(Q+Q^t)x^t + Lu^t + z(x+u)^t}|u,z\rangle\\
&= \frac{1}{\sqrt{2^n}}\sum_{u}(-1)^{uQ^tu^t + Lu^t}\left|u, u(Q+Q^t)\right\rangle .
\end{aligned}$$

Hence this state has the form $\frac{1}{\sqrt{2^n}} \sum_u (-1)^{p(u)} |u, u(Q+Q^t)\rangle$, where p is the quadratic Boolean function $p(u) = uQ^tu^t + Lu^t$.

We now describe a direct way to recover f from sampling from these states. Suppose we sample $k = O(n)$ times, obtaining pairs (u_i, v_i) from this process. The goal is to identify the matrix Q. Observe that learning what Q is equivalent to learning what

$M := (Q + Q^t)$ is since Q is an upper triangular matrix with zero diagonal. Now, arrange the sampled vectors u_i into a matrix $U = (u_1|\ldots|u_k)$ and similarly $V = (v_1|\ldots|v_k)$. Then $U^t M = V^t$ is a system of linear equations for each of the n columns of M. Since the matrix U was chosen at random, we obtain that it is invertible with constant probability, i. e., we can find M with constant probability of success.

We shall now improve this in order to obtain a method that is more robust regarding errors in the input state $|\psi\rangle$. Instead of sampling k times from $T|\psi\rangle^{\otimes 2}$, we consider the coherent superposition $|\psi\rangle^{\otimes 2k}$ and apply $T^{\otimes k}$ to it. The resulting state has the form

$$\sum_{u_1,\ldots,u_k} \varphi(u_1,\ldots,u_k)\,|u_1,\ldots,u_k\rangle\,|Mu_1,\ldots,Mu_k\rangle\,, \tag{4}$$

with certain phases, indicated by φ. Next, note that there is an efficient classical algorithm which on input U and V computes the matrix M. We can compute this algorithm in a reversible fashion and apply to the state (4). The resulting state has constant overlap with a state that is the superposition of the cases for which the Gauss algorithm computation was successful (returning M) and those cases for which it was unsuccessful (returning $\perp$): using the shorthand notation $\mathbf{u} = (u_1,\ldots,u_k)$, we obtain the state

$$\left(\sum_{\mathbf{u}\ \mathrm{good}} \varphi(\mathbf{u})\,|\mathbf{u}\rangle\,|\mathbf{Mu}\rangle\right)|M\rangle + \left(\sum_{\mathbf{u}\ \mathrm{bad}} \varphi(\mathbf{u})\,|\mathbf{u}\rangle\,|\mathbf{Mu}\rangle\right)|\perp\rangle\,.$$

Measuring this state will yield M with constant probability. Once M has been found, we can infer Q and uses this information to compute it into the phases in equation (3) in order to cancel the quadratic part out. From the resulting states we can efficiently determine L from a constant number of subsequent Fourier samplings.

Relation to learning parity with errors. We now return to the hidden shift problem. In the following we argue that the quantum algorithm for finding a shift for quadratic functions has an advantage over classical attempts to do so, since it can even handle cases where the function is *close* to a quadratic function. It is easy to see that the shift problem for quadratic functions themselves can be solved classically in $\Theta(n)$ queries: the lower bound is a straightforward information-theoretic argument. For the upper bound we show that from knowledge of the quadratics and the promise that there is a shift s such that $g(x) = f(x+s)$, we can determine s. Indeed, it is sufficient to query at points $(0,\ldots,0)$, and e_i, where e_i denotes the ith vector in the standard basis to get equations of the form $su_i^t = b_i$, where $u_i \in \mathbb{Z}_2^n$ and $b_i \in \mathbb{Z}_2$. With constant probability after n trials the solution is uniquely characterized and can be efficiently found, e. g., by Gaussian elimination. The problem with this approach is that if f and g are not perfect quadratics, the resulting equations will be

$$su_1^t \approx_\varepsilon b_1,\ su_2^t \approx_\varepsilon b_2,\ \ldots$$

where the $\approx_\varepsilon$ symbol means that each equation can be incorrect with probability $1-\varepsilon$. As it turns out from Theorem 4 below, we will be able to tolerate noise of the order $\varepsilon = O(1/n)$. It is perhaps interesting to note that similar equations with errors have been studied in learning. The best known algorithm is the Blum-Kalai-Wasserman sieve [BKW03], running in subexponential time in n, albeit able to tolerate constant error ε.

We show that the following algorithm for computing an approximating quadratic form is robust with respect to errors in the input function:

Algorithm 3 [Find-Close-Quadratic]. *The following algorithm takes as input a black-box for a Boolean function f. The output is a quadratic Boolean function which approximates f.*

- *Prepare $2k$ copies of the state $\frac{1}{\sqrt{2^n}}\sum_x(-1)^{f(x)}|x\rangle$.*
- *Group them into pairs of 2 registers and apply the transformation T to each pair.*
- *Rearrange the register pairs $[1,2]$, $[3,4]$, ..., $[2k-1,2k]$ into a list of the form $[1,3,\ldots,k,2,4,\ldots,2k]$. Next, apply the reversible Gauss algorithm to the sequence of registers.*
- *Measure the register holding the result of the Gauss algorithm computation and obtain $M \in \mathbb{Z}_2^{n\times n}$. Use M to uncompute the quadratic phase and extract the linear term via Fourier sampling.*

Theorem 4. *Let $f,g:\mathbb{Z}_2^n \to \mathbb{Z}_2$ be Boolean functions, let $g=\sum_{i,j} q_{i,j}x_ix_j+\sum_i \ell_i x_i$ be a quadratic polynomial, and assume that $|\langle f,g\rangle| > (1-\varepsilon)$. Then algorithm running* **Find-Close-Quadratic** *on input f finds the quadratic form corresponding to g, and thereby g itself with probability $p_{success} \geq c(1-n\varepsilon)$, where c is a constant independent of n.*

Proof. First note that $|\langle f,g\rangle| > (1-\varepsilon)$ implies that f and g disagree on at most $\varepsilon 2^n$ of the inputs. Hence the two quantum states $|\psi_f\rangle = \frac{1}{\sqrt{2^n}}\sum_x f(x)|x\rangle$ and $|\psi_g\rangle = \frac{1}{\sqrt{2^n}}\sum_x g(x)|x\rangle$ satisfy $|\langle\psi_f|\psi_g\rangle| > (1-\varepsilon)$.

Next, observe that the algorithm can be seen as application of a unitary operation U. We first study the "perfect" case, where we apply U to the state $\left|\psi_g^{\otimes k}\right\rangle$ and then study the effect of replacing this with the input corresponding to f. Notice that the algorithm can also be seen as a POVM $\mathcal{M}$ which consists of rank 1 projectors $\{E_i : i \in I\}$ such that $\sum_{i\in I} E_i = \mathbf{1}$. Since the algorithm identifies M with constant probability, we obtain that the POVM element E_M, which corresponds to the correct answer satisfies $Pr(\text{measure M}) = \text{tr}\left(E_M\left|\psi_g^{\otimes k}\right\rangle\left\langle\psi_g^{\otimes k}\right|\right) = p_0 \geq \Omega(1)$.

For vectors v, w we have that $\|v-w\|_2^2 = 2-2|\langle v,w\rangle|$, we get using $|\langle\psi_f^{\otimes k}|\psi_g^{\otimes k}\rangle| > (1-\varepsilon)^k \sim (1-k\varepsilon)+O(\varepsilon^2)$. For the difference $|\delta\rangle := \left|\psi_f^{\otimes}\right\rangle - \left|\psi_g^{\otimes}\right\rangle$ we therefore get that $\|\delta\|^2 < 2k\varepsilon$. Denoting $E_M = |\varphi\rangle\langle\varphi|$ with normalized vector $|\varphi\rangle$, we obtain for the probability of identifying M on input f:

$$\begin{aligned}\text{tr}\left(E_M\left|\psi_f^{\otimes k}\right\rangle\left\langle\psi_f^{\otimes k}\right|\right) &= \langle\psi_g|E_M|\psi_g^{\otimes k}\rangle + \langle\delta|E_M|\psi_g^{\otimes k}\rangle + \langle\psi_g^{\otimes k}|E_M|\delta\rangle + \langle\delta|E_M|\delta\rangle \\ &\geq p_0 + 2\langle\delta|\varphi\rangle\langle\varphi|\psi_g^{\otimes k}\rangle + |\langle\delta|\varphi\rangle|^2.\end{aligned}$$

By Cauchy-Schwartz, we finally get that $|\langle\delta|\varphi\rangle| \leq \|\delta\|\|\varphi\| \leq \sqrt{2k\varepsilon}$. Hence, we obtain for the overall probability of success $p_{success} \geq p_0 - \sqrt{8k\varepsilon}$. □

We give an application of Theorem 4 to the problem of efficiently finding an approximation of a function of large Gowers U_3 norm in the following section.

4 Polynomials and the Gowers Norm

Recall that the Gowers norms measure the extent to which a function $f : \mathbb{F}^n \to \mathbb{C}$ behaves like a phase polynomial. For $k \geq 1$, the Gowers norm is defined by

$$\|f\|_{U^k(\mathbb{F}^n)} := \left(\mathbb{E}_{x,h_1,\ldots,h_k \in \mathbb{F}^n} \Delta_{h_1} \ldots \Delta_{h_k} f(x)\right)^{1/2^k},$$

where $\Delta_h f(x) = f(x+h) - f(x)$ for all $h \in \mathbb{F}^n$. It is immediate that if $|f(x)| \leq 1$ for all x, then $\|f\|_{U^k(F^n)} \in [0,1]$. Moreover, degree k polynomials are characterized precisely by the vanishing of $\Delta_{h_1} \ldots \Delta_{h_k} f(x)$ for all h_i. It is furthermore easy to see that $\|f\|_{U^k(\mathbb{F}^n)} = 1$ if and only if f is a phase polynomial of degree less than k [GT08].

Theorem 5 (Inverse theorem for the Gowers U_3 norm [GT08]). *Let $f : \mathbb{F}^n \to \mathbb{C}$ be a function that is bounded as $|f(x)| \leq 1$ for all x. Suppose that the kth Gowers norm of f satisfies $\|f\|_{U^k(F^n)} \geq 1 - \varepsilon$. Then there exists a phase polynomial g of degree less than k such that $\|f - g\| = o(1)$. For fixed field $\mathbb{F}$ and degree k, the $o(1)$ term approaches zero as ε goes to zero.*

Before we state the algorithm we recall a useful method to compare two unknown quantum states for equality. This will be useful for a one-sided test that the output of the algorithm indeed is a valid shift.

Lemma 1 (SWAP test [Wat00, Buh01]). *Let $|\psi\rangle$, $|\varphi\rangle$ be quantum states, and denote by SWAP the quantum operation which maps $|\psi\rangle|\varphi\rangle \mapsto |\varphi\rangle|\psi\rangle$, and by $\Lambda(SWAP)$ the same operations but controlled to a classical bit. Apply $(H_2 \otimes \mathbf{1})\Lambda(SWAP)(H_2 \otimes \mathbf{1})$ to the state $|0\rangle|\varphi\rangle|\psi\rangle$, measure the first qubit in the standard basis to obtain a bit b and return the result (where result $b = 1$ indicates that the states are different). Then $Pr(b=1) = \frac{1}{2} - \frac{1}{2}|\langle\varphi|\psi\rangle|^2$.*

Lemma 1 has many uses in quantum computing, see for instance [Wat00, Buh01]. Basically, it is useful whenever given $|\varphi\rangle$ and $|\psi\rangle$ two cases have to be distinguished: (i) are the two states equal, or (ii) do they have inner product at most δ. For this case it provides a one-sided test such that $Pr(b=1) = 0$ if $|\psi\rangle = |\varphi\rangle$ and $Pr(b=1) \geq \frac{1}{2}(1-\delta^2)$ if $|\psi\rangle \neq |\varphi\rangle$ and $|\langle\varphi|\psi\rangle| \leq \delta$.

Algorithm 6 [Shifted-Large-U3]. *The following algorithm solves the hidden shift problem for an oracle $\mathcal{O}$ which hides (f,g), where $g(x) = f(x+s)$ for $s \in \mathbb{Z}_2^n$ and where $\|f\|_{U_3(\mathbb{Z}_2)} \geq (1-\varepsilon)$.*

1. *Solve the hidden quadratic problem for f. This gives a quadratic $g(x) = xQx^t + Lx^t$.*
2. *Compute the dual quadratic function corresponding to the Fourier transform of g.*
3. *Solve the hidden shift problem for $f(x)$, $f(x+s)$, and g. Obtain a candidate $s \in \mathbb{Z}_2^n$.*
4. *Verify s using the SWAP test.*

Theorem 7. *Let f be a Boolean function with $\|f\|_{U_3} \geq 1 - \varepsilon$. Then Algorithm 6 (**Shifted-Large-U3**) solves the hidden shift problem for f with probability $p_{success} > c(1-\varepsilon)$, where c is a universal constant.*

Proof sketch. In general the fact that large Gowers U_3 norm implies large correlation with a quadratic follows from the inverse theorem for Gowers U_3 norm [GT08, Sam07]. For the special case of the field $\mathbb{Z}_2$ and the large Gowers norm $(1-\varepsilon)$ we are interested in, we use [AKK+03] to obtain a stronger bound on the correlation with the quadratics. The claimed result follows from [AKK+03] and the robustness of Algorithm 6 against errors in the input functions. □

Remark 1. It should be noted that in the form stated, Algorithm 6 only applies to the case where the rank $h = \mathrm{rk}(Q + Q^t)/2 = n/2$ is maximum, as only this case corresponds to bent functions. However, it is easy to see that it can be applied in case $h < n/2$ as well. There the matrix $(Q + Q^t)$ has a non-trivial kernel, defining a $n - 2h$ dimensional linear subspace of $\mathbb{Z}_2^n$. In the Fourier transform, the function is supported on an affine shift of dual space, i. e., the function has 2^{2h} non-zero Fourier coefficients, all of which have the same absolute value 2^{-h}. Now, the hidden shift algorithm can be applied in this case too: instead of the dual bent function we compute the Boolean function corresponding to the first $2h$ rows of $(R^{-1})^t$, where R is as in Theorem 1 into the phase. This will have the effect of producing a shift s lying in an affine space $s + V$ of dimension $n - 2h$. For $h < n/2$ the shift is no longer uniquely determined, however, we can describe the set of all shifts efficiently in that case by giving one shift and identifying a basis for V.

5 Conclusions and Open Problems

It is an interesting question is whether the quantum algorithm to find approximations for functions for large Gowers norms U_2 and U_3 can be used to find new linear and quadratic tests for Boolean functions. Furthermore, it would be interesting to study the tradeoff between number of queries and soundness for quantum tests, in analogy to the results that have been shown in the classical case [ST06].

Acknowledgment

I would like to thank the anonymous referees for valuable comments, including the argument presented after Algorithm 2 which shows how to relate the hidden shift problem to an abelian hidden subgroup problem.

References

[AKK+03] Alon, N., Kaufman, T., Krivelevich, M., Litsyn, S.N., Ron, D.: Testing low-degree polynomials over *GF*(2). In: Arora, S., Jansen, K., Rolim, J.D.P., Sahai, A. (eds.) RANDOM 2003 and APPROX 2003. LNCS, vol. 2764, pp. 188–199. Springer, Heidelberg (2003)

[AS07] Atici, A., Servedio, R.: Quantum algorithms for learning and testing juntas. Quantum Information Processing 6(5), 323–348 (2007)

[BCD05] Bacon, D., Childs, A., van Dam, W.: From optimal measurement to efficient quantum algorithms for the hidden subgroup problem over semidirect product groups. In: Proc. FOCS 2005, pp. 469–478 (2005)

[BH97] Brassard, G., Høyer, P.: An exact polynomial–time algorithm for Simon's problem. In: Proceedings of Fifth Israeli Symposium on Theory of Computing and Systems, ISTCS, pp. 12–33. IEEE Computer Society Press, Los Alamitos (1997)

[BKW03] Blum, A., Kalai, A., Wasserman, H.: Noise-tolerant learning, the parity problem, and the statistical query model. Journal of the ACM 50(4), 506–519 (2003)

[Buh01] Buhrman, H., Cleve, R., Watrous, J., de Wolf, R.: Quantum fingerprinting. Phys. Rev. Letters 87, 167902 (4 pages) (2001)

[BV97] Bernstein, E., Vazirani, U.: Quantum complexity theory. SIAM Journal on Computing 26(5), 1411–1473 (1997); Conference version in Proc. STOC 1993, pp. 11–20 (1993)

[CSV07] Childs, A., Schulman, L.J., Vazirani, U.: Quantum algorithms for hidden nonlinear structures. In: Proc. FOCS 2007, pp. 395–404 (2007)

[DDW09] Decker, T., Draisma, J., Wocjan, P.: Efficient quantum algorithm for identifying hidden polynomials. Quantum Information and Computation 9, 215–230 (2009)

[DHI03] van Dam, W., Hallgren, S., Ip, L.: Quantum algorithms for some hidden shift problems. In: Proc. SODA 2003, pp. 489–498 (2003)

[Dil75] Dillon, J.: Elementary Hadamard difference sets. In: Hoffman, F., et al. (eds.) Proc. 6th S-E Conf. on Combinatorics, Graph Theory, and Computing, pp. 237–249. Winnipeg Utilitas Math. (1975)

[FIM+03] Friedl, K., Ivanyos, G., Magniez, F., Santha, M., Sen, P.: Hidden translation and orbit coset in quantum computing. In: Proc. STOC 2003, pp. 1–9 (2003)

[GT08] Green, B., Tao, T.: An inverse theorem for the Gowers $U^3(G)$ norm. Proc. Edin. Math. Soc. (2008); see also arxiv preprint math.NT/0503014 (to appear)

[HMR+06] Hallgren, S., Moore, C., Rötteler, M., Russell, A., Sen, P.: Limitations of quantum coset states for graph isomorphism. In: Proc. STOC 2006, pp. 604–617 (2006)

[Kit97] Yu, A.: Kitaev. Quantum computations: algorithms and error correction. Russian Math. Surveys 52(6), 1191–1249 (1997)

[Kup05] Kuperberg, G.: A subexponential-time quantum algorithm for the dihedral hidden subgroup problem. SIAM Journal on Computing 35(1), 170–188 (2005)

[ME98] Mosca, M., Ekert, A.: The hidden subgroup problem and eigenvalue estimation on a quantum computer. In: Williams, C.P. (ed.) QCQC 1998. LNCS, vol. 1509, pp. 174–188. Springer, Heidelberg (1999)

[Mon09] Montanaro, A.: Quantum algorithms for shifted subset problems. Quantum Information and Computation 9(5&6), 500–512 (2009)

[MS77] MacWilliams, F.J., Sloane, N.J.A.: The Theory of Error–Correcting Codes. North–Holland, Amsterdam (1977)

[NC00] Nielsen, M., Chuang, I.: Quantum Computation and Quantum Information. Cambridge University Press, Cambridge (2000)

[Rot76] Rothaus, O.S.: On "bent" functions. Journal of Combinatorial Theory, Series A 20, 300–305 (1976)

[Röt08] Rötteler, M.: Quantum algorithms for highly non-linear boolean functions. arXiv Preprint 0811.3208 (2008)

[RS04] Russell, A., Shparlinski, I.: Classical and quantum function reconstruction via character evaluation. Journal of Complexity 20(2-3), 404–422 (2004)

[Sam07] Samorodnitsky, A.: Low-degree tests at large distances. In: Proceedings of the 39th Annual ACM Symposium on Theory of Computing (STOC 2007), pp. 506–515 (2007)

[Sho97] Shor, P.: Polynomial-time algorithms for prime factorization and discrete logarithms on a quantum computer. SIAM Journal on Computing 26(5), 1484–1509 (1997)

[ST06] Samorodnitsky, A., Trevisan, T.: Gowers uniformity, influence of variables, and PCPs. In: Proc. STOC 2006, pp. 11–20 (2006)

[Wat00] Watrous, J.: Succinct quantum proofs for properties of finite groups. In: Proc. FOCS 2000, pp. 537–546 (2000)

From Parity and Payoff Games to Linear Programming

Sven Schewe

University of Liverpool
sven.schewe@liverpool.ac.uk

Abstract. This paper establishes a surprising reduction from parity and mean payoff games to linear programming problems. While such a connection is trivial for solitary games, it is surprising for two player games, because the players have opposing objectives, whose natural translations into an optimisation problem are minimisation and maximisation, respectively. Our reduction to linear programming circumvents the need for concurrent minimisation and maximisation by replacing one of them, the maximisation, by approximation. The resulting optimisation problem can be translated to a linear programme by a simple space transformation, which is inexpensive in the unit cost model, but results in an exponential growth of the coefficients. The discovered connection opens up unexpected applications – like μ-calculus model checking – of linear programming in the unit cost model, and thus turns the intriguing academic problem of finding a polynomial time algorithm for linear programming in this model of computation (and subsequently a strongly polynomial algorithm) into a problem of paramount practical importance: All advancements in this area can immediately be applied to accelerate solving parity and payoff games, or to improve their complexity analysis.

1 Introduction

This paper links two intriguing open complexity problems, solving parity and payoff games and solving linear programming problems in the unit cost model.

Linear programming. [1,2,3,4,5,6,7], the problem of maximising $\mathbf{c}^T\mathbf{x}$ under the side conditions $A\mathbf{x} \leq \mathbf{b}$ and $\mathbf{x} \geq \mathbf{0}$, is one of the most researched problems in computer science and discrete mathematics. The interest in linear programming has two sources: It has a significant practical impact, because a wide range of optimisation problems in economy and operations research can be approached with linear programming, and it is the source of a range of challenges that remained unresolved for years.

The most prominent open challenge is Smale's 9th problem [8], which asks if linear programming has a polynomial time solution in the unit cost model. In the unit cost model, we assume that all arithmetic operations have an identical unit cost. This model is inspired by the desire of mathematicians to use real numbers instead of rationals, but it can also be applied to handle large numbers, whose representation in binary led to an exponential (or higher) blow-up in the size of the problem description.

Dantzig's simplex algorithm addresses both complexity models alike. While highly efficient in practice [1], Klee and Minty [2] showed that the worst case running time of

R. Královič and D. Niwiński (Eds.): MFCS 2009, LNCS 5734, pp. 675–686, 2009.

the simplex algorithm is exponential in the size of the linear programme. Their original proof referred to a particular Pivot rule, but it has proven to be very flexible with respect to the chosen Pivot rule, and could be extended to every suggested deterministic Pivot rule that does not depend on the history of previous updates. The proof is also independent of the chosen cost model, because the path constructed by the Pivot rule has exponential length.

While the complexity for the unit cost model is still open, polynomial time algorithms for the Turing model like Kachian's ellipsoid method [3] and interior point methods such as Karmarkar's algorithm [4] are known for decades. Unfortunately, these algorithms depend on the size of the binary representation of the numbers, and do not provide any insight for the unit cost model. Apart from the mathematical interest in determining the unit cost complexity of solving linear programmes, such an algorithm is a prerequisite of finding a *strongly polynomial* time algorithm. The requirement for an algorithm to be strongly polynomial is slightly higher: It also requires that the intermediate arithmetic operations can be performed in polynomial time, which usually depends on the representation of the numbers. Thus, a polynomial time algorithm in the unit cost model will usually provide a polynomial time algorithm for some representation of the numbers different from the usual binary representation.

Current attempts to finding a strongly polynomial time algorithm focus on the Simplex algorithms. Randomised update strategies, for example, have been suggested early on for the simplex technique, but their complexity has not yet been analysed successfully. For the simplest of these techniques, which uses a random edge Pivot rule that chooses a profitable base change uniformly at random, the number of arithmetic steps needed on non-degenerated simplices is merely known to be at least quadratic in the size of the constraint system [5]. Shadow vertex techniques [6,7] for exploring the simplex allowed for randomised polynomial time procedures for the approximation [6] and computation [7] of the solution to linear programmes.

Parity and Payoff Games are finite two player zero sum games of infinite duration. They are played on labelled directed graphs, whose vertices are partitioned into two sets of vertices owned by two players with opposing objectives. Intuitively, they are played by placing a pebble on the game graph. In each step, the owner of a vertex chooses a successor vertex of the digraph. This way, an infinite run of the game is constructed, where the objectives of the players is to minimise and maximise the average payoff of the moves in mean payoff games, or to enforce parity or imparity of the highest colour occurring infinitely many times in parity games, respectively. These games play a central role in model checking [9,10,11,12,13], satisfiability checking [11,9,14,15], and synthesis [16,17], and numerous algorithms for solving them have been studied [9,18,19,20,21,22,23,24,25,26,27,28,29,30,31,32,33,34]. Mean payoff games [34,35,36] have further applications in economic game theory.

The complexity of solving parity games is equivalent to the complexity of μ-calculus model checking [9]. The complexity of solving parity and mean payoff games is known to be in the intersection of UP and coUP [22], but its membership in P is still open. Up to a recent complexity analysis by Friedman [37], strategy improvement algorithms

[25,26,27,28,29] have been considered candidates for a polynomial running time. To the best of my knowledge, all algorithms proposed so far for solving parity or mean payoff games are now known not to be in P.

The Reduction proposed in this paper reduces testing if there are states in a mean payoff game with value ≥ 0 to solving a linear programming problem. It is simple to extend this non-emptiness test to finding the 0 mean partition of a mean payoff game and hence to solving parity and mean payoff games.

Our reduction goes through an intermediate representation of the non-emptiness problem to a non-standard optimisation problem that treats minimisation and maximisation quite differently: While minimisation is represented in a standard way — replacing minimisation $v = \min\{t_1, t_2, \ldots, t_n\}$ over n terms by n inequalities $v \leq t_i$ and maximising over the possible outcome — maximisation is replaced by an approximation within small margins. We use $v = \log_b(b^{t_1} + b^{t_2} + b^{t_n})$ instead of $v = \max\{t_1, t_2, \ldots, t_n\}$, which keeps the error of the operation sufficiently small if the basis b is big enough:

$$\max\{t_1, t_2, \ldots, t_n\} \leq \log_b\left(b^{t_1} + b^{t_2} + b^{t_n}\right) \leq \log_b n + \max\{t_1, t_2, \ldots, t_n\}.$$

From there it is a small step to building a linear programme that computes *the exponent* of the solution to this non-standard optimisation problem.

Starting from parity games or from mean payoff games with a binary representation of the edge weights, this linear programme is cheap to compute in the unit cost model (bi-linear in the number of vertices and edges and linear in the size of the representation), but the constants are exponential in the edge weights of the mean payoff game, and doubly exponential in the number of colours for parity games.

The benefit of the proposed reduction for game solving is therefore not immediate, but depends on future results in linear programming. There has, however, been recent progress in the quest for polynomial time algorithms in the unit cost model, or even strongly polynomial algorithms. Shadow vertex techniques [6,7], for example, seem to be promising candidates; Kelner and Spielman [7] use this quest as the main motivation for their randomised polynomial time algorithm.

But the benefit of the reduction is bi-directional. While solving finite games of infinite duration automatically profits from future progress in the theory of linear programming, linear programming profits from the problem: The proposed reduction contributes an important natural problem class that can be reduced to linear programming, but requires a polynomial time algorithm in the unit cost model, because the constants are too large for efficient binary representation. Opening up a range of model checking problems to linear programming, the proposed reduction lifts the problem of finding such algorithms from a problem of mere academic interest to one of practical importance. Furthermore, it is unintuitive that the complexity of inherently discrete combinatorial problems like μ-calculus model checking or solving parity games should depend on the cost model of arithmetic operations. Provided a polynomial time algorithm for solving linear programming problems in the unit cost model is found, it thus seems likely that a polynomial time algorithm for solving parity games can be inferred.

2 Finite Games of Infinite Duration

Finite games of infinite duration (ω-games) are played by two players, a Maximiser and a Minimiser, with opposing objectives. ω-games are composed of a finite arena and an evaluation function.

Arenas. A finite *arena* is a triple $\mathcal{A} = (V_{\max}, V_{\min}, E)$ consisting of

- a set $V = V_{\max} \cup V_{\min}$ of vertices that is partitioned into two disjoint sets $V_{\max}$ and $V_{\min}$, called the vertices owned by the *Maximiser* and *Minimiser*, respectively, and
- a set $E \subseteq V \times V$ of edges, such that (V,E) is a directed graph without sinks.

Plays. Intuitively, a game is played by placing a token on a vertex of the arena. If the token is on a vertex $v \in V_{\max}$, the Maximiser chooses an edge $e = (v,v') \in E$ originating in v to a vertex $v' \in V$ and moves the token to v'. Symmetrically, the Minimiser chooses a successor in the same manner if the token is on one of her vertices $v \in V_{\min}$. In this way, they successively construct an infinite *play* $\vartheta = v_0v_1v_2v_3\ldots \in V^{\omega}$.

Strategies. For a finite arena $\mathcal{A} = (V_{\max}, V_{\min}, E)$, a (memoryless) *strategy* for the Maximiser is a function $f : V_{\max} \to V$ that maps every vertex $v \in V_{\max}$ of the Maximiser to a vertex $v' \in V$ such that there is an edge $(v,v') \in E$ from v to v'. A play is called *f-conform* if every decision of the Maximiser in the play is in accordance with f. For a strategy f of the Maximiser, we denote with $\mathcal{A}_f = (V_{\max}, V_{\min}, E_f)$ the arena obtained from $\mathcal{A}$ by deleting the transitions from vertices of the Maximiser that are not in accordance with f. The analogous definitions are made for the Minimiser.

Mean Payoff Games. A *mean payoff game* is a game $\mathcal{M} = (V_{\max}, V_{\min}, E, w)$ with arena $\mathcal{A} = (V_{\max}, V_{\min}, E)$ and a weight function $w : E \to \mathbb{Z}$ from the edges of the mean payoff game to the integers. Each play $\vartheta = v_0v_1v_2\ldots$ of a mean payoff game is evaluated to $value(\vartheta) = \liminf\limits_{n\to\infty} \frac{1}{n}\sum_{i=1}^{n} w\big((v_{i-1},v_i)\big)$.

As a variant, we can allow for real valued weight functions $w : E \to \mathbb{R}$. We then refer explicitly to a *real valued mean payoff game*.

The objective of the Maximiser and Minimiser are to maximise and minimise this value, respectively. For single player games where all vertices are owned by one player (or, likewise, all vertices owned by the other player have exactly one successor), the optimal strategy for this player from some vertex v is to proceed to a cycle with maximal or minimal average weight a and henceforth follow it. The outcome $value(v) = a$ of this game when started in v is called the value of v. Mean payoff games are memoryless determined:

Proposition 1. *[34,35,36] For every real valued mean payoff game $\mathcal{M}$, there are Minimiser and Maximiser strategies f and g, respectively, such that the value of every vertex in $\mathcal{M}_f$ equals the value of every vertex in $\mathcal{M}_g$.* □

The 0 *mean partition* $M_{\geq 0} = \{v \in V \mid value(v) \geq 0\}$ of a mean payoff game is thus well-defined. We say that a vertex v is *winning* for the Maximiser if it is in $M_{\geq 0}$, and winning for the Minimiser otherwise.

Corollary 1. *[34,35,36] The* 0 *mean partition is well-defined, and both players have winning strategies for their respective winning region $M_{\geq 0}$ and $V \setminus M_{\geq 0}$.* □

Solving a mean payoff game can be reduced to finding the 0 mean partition of a number of games played on sub-arenas with a slightly adjusted weight function, because the average weight of a cycle is a multiple of $\frac{1}{n}$ for some $n \leq |V|$, and it hence suffices to know that the value is within an interval $[\frac{i}{n^2}, \frac{i+1}{n^2}[$ for some integer $i \in \mathbb{Z}$.

Corollary 2. *A mean payoff game with n vertices and maximal absolute edge weight a can be solved in time $O\big(n\log(a+n)\big)$ when using an oracle for the construction of* 0 *mean partitions.* □

Parity Games. A *parity game* is a game $\mathcal{P} = (V_{\max}, V_{\min}, E, \alpha)$ with arena $\mathcal{A} = (V_{\max}, V_{\min}, E)$ and a colouring function $\alpha : V \to C \subset \mathbb{N}$ that maps each vertex of $\mathcal{P}$ to a natural number. C denotes the finite set of colours.

Each play is evaluated by the highest colour that occurs infinitely often. The Maximiser wins a play $\vartheta = v_0v_1v_2v_3\ldots$ if the highest colour occurring infinitely often in the sequence $\alpha(\vartheta) = \alpha(v_0)\alpha(v_1)\alpha(v_2)\alpha(v_3)\ldots$ is even, while the Minimiser wins if the highest colour occurring infinitely often in $\alpha(\vartheta)$ is odd. Without loss of generality, we assume that the highest occurring colour is bounded by the number of vertices in the arena. It is simple to reduce solving parity games to finding the 0 mean partition of a mean payoff game: One can simply translate a colour c to the weight $|V|^c$ [34].

Corollary 3. *[34] Parity games are memoryless determined, and solving them can be reduced in time $O(mn)$ to solving the* 0 *mean partition problem of a mean payoff game with the same arena, such that the Minimiser and Maximiser have the same winning regions and winning strategies.* □

3 Reduction

In this section, we describe a reduction from finding the 0 mean partition of a mean payoff game, to which solving parity and payoff games can be reduced in polynomial time by Corollaries 2 and 3, to solving a linear programming problem. We first focus on the slightly simpler problem of testing $M_{\geq 0}$ for emptiness, and reduce this question to a linear programming problem.

The first important observation for our reduction is that membership in $M_{\geq 0}$ is invariant under increasing the weight function slightly: If every edge weight in a mean payoff game with n vertices is increased by some value in $[0, \frac{1}{n}[$, then the weight of every cycle is increased by a value in $[0,1[$, and hence non-negative if, and only if, the original integer valued weight of the cycle is non-negative.

Lemma 1. *If we increase the weight function of an (integer valued) mean payoff game $\mathcal{M} = (V_{\max}, V_{\min}, E, w)$ with $n = |V|$ vertices for every edge by some non-negative value $< \frac{1}{n}$, then the same cycles as before have non-negative weight in the resulting real valued mean payoff game, and the* 0 *mean partition does not change.* □

This observation is used to replace maximisation in a natural representation of the objectives of both players in a mean payoff game (Subsection 3.1) by a logarithmic expression in Subsection 3.2, which is subsequently translated into a boundedness test for a linear programming problem in Subsection 3.3. In Subsection 3.4, we show how this boundedness test, which refers to a non-emptiness test of $M_{\geq 0}$, can be adjusted to a bounded optimisation problem that provides $M_{\geq 0}$, and discuss the complexity of the transformations in Subsection 3.5. An example that illustrates these transformations is provided in Section 4.

3.1 Basic Inequations

We now devise a set of inequalities that have a non-trivial solution if, and only if, $M_{\geq 0}$ is non-empty. For our reduction, we extend addition from $\mathbb{R}$ to $\mathbb{R} \cup \{-\infty\}$ in the usual way by choosing $a + (-\infty) = -\infty = -\infty + a$ for all $a \in \mathbb{R} \cup \{-\infty\}$. This motivates the definition of a family of *basic inequalities* for a mean payoff game $\mathcal{M} = (V_{\max}, V_{\min}, E, w)$ that contains one inequality

$$v \leq w\big((v,v')\big) + v'$$

for every edge (v,v') originating from a Minimiser vertex $v \in V_{\min}$, and one inequality

$$v \leq \max\big\{w\big((v,v')\big) \mid v' \in suc(v)\big\} + c_v$$

for every Maximiser vertex $v \in V_{\max}$, where $suc(v)$ denotes the set of successor vertices of v, and each $c_v \in [0, \frac{1}{|V|}[$ can be any sufficiently small slip value (cf. Lemma 1).

Every such system of inequalities has a trivial solution that assigns $-\infty$ to every vertex; but it also has a real valued solution for all vertices in $M_{\geq 0}$.

Lemma 2. *For every such system of inequalities for a mean payoff game $\mathcal{M} = (V_{\max}, V_{\min}, E, w)$, a vertex $v \in V$ has a real valued solution if, and only if, v is in $M_{\geq 0}$.*

Proof. '$\Leftarrow$:' Let us fix an optimal strategy for the Maximiser in the mean payoff game and consider the system of inequalities that contain one equation $v \leq w(e) + v'$ for every edge of the resulting singleton game. (A solution to this set of inequalities is obviously a solution to the original set of inequalities.) This set of inequalities has obviously a solution that is real valued for every vertex $v \in M_{\geq 0}$ (and sets $v = -\infty$ for every vertex v not in $M_{\geq 0}$)[1].

'$\Rightarrow$' A real valued solution for a vertex v defines a strategy for the Maximiser that witnesses $value(v) \geq 0$: If $v \leq \max\big\{w\big((v,v')\big) \mid v' \in suc(v)\big\} + c_v$ holds true, then $v \leq w\big((v,v')\big) + v' + c_v$ holds for some $v' \in suc(v)$ in particular, and we choose a Maximiser strategy that fixes such a successor for every Maximiser vertex. By a simple inductive argument, every vertex u reachable from v in the resulting singleton game is real valued, and, for every cycle reachable from v, the sum of the edge weights and vertex slips is

[1] Starting with the digraph with states $M_{\geq 0}$ and the respective edges defined by the fixed Maximiser strategy, we can apply the following algorithm until values are assigned to all vertices in $M_{\geq 0}$: (1) pick a vertex v in a leaf component of the digraph that is the minimum of the weighted distance to plus the value assigned to any vertex v' that is already removed from the graph (or an arbitrary value if no such vertex exists), and then (2) remove v from the graph.

non-negative. As the sum of the vertex slips is strictly smaller than 1 in every cycle, the sum of the edge weights is strictly greater than -1, and hence non-negative. □

Naturally, having one real valued solution implies having unbounded solutions, because adding the same value $r \in \mathbb{R}$ to every value of a solution provides a new solution.

3.2 Logarithmic Inequations

These observations set the ground for a reduction to linear programming: For a sufficiently large basis $b > 1$, $\log_b \sum_{v' \in suc(v)} b^{w((v,v'))} b^{v'}$ equals $\max\{w((v,v')) + v' \mid v' \in suc(v)\} + c_v$ for some slip value $c_v \in [0, \frac{1}{n}[$, because

$$\max_{v' \in suc(v)} \{w((v,v')) + v'\} \leq \log_b \sum_{v' \in suc(v)} b^{w((v,v'))} b^{v'} \leq \log_b |suc(v)| + \max_{v' \in suc(v)} \{w((v,v')) + v'\}$$

holds true. (For the extension to $\mathbb{R} \cup \{-\infty\}$ we use the usual convention $b^{-\infty} = 0$ and $\log_b 0 = -\infty$.) Choosing a basis $b > {n_{out}}^{|V|}$ that is greater than the $|V|$-th power of the maximal out-degree n_{out} of Maximiser vertices guarantees $\log_b i < \frac{1}{n}$, and hence that the small error caused by moving from minimisation to the logarithm of the sum of the exponents is within the margins allowed for by Lemma 1.

Corollary 4. *The system of inequalities consisting of the Minimiser inequalities and the adjusted Maximiser inequalities have a real valued solution for a vertex $v \in V$ if, and only if, $v \in M_{\geq 0}$.* □

Note that the reduction uses estimation from below (through the inequality) as well as estimations from above (through the slip) at the same time, which is sound only because the slip values are within the small margins allowed for by Lemma 1.

3.3 Linear Inequations

The resulting optimisation problem can be translated into a standard linear programming problem by a simple space transformation: As the exponential function $v \mapsto b^v$ is a strictly monotone ascending mapping from $\mathbb{R} \cup \{-\infty\}$ onto $\mathbb{R}^{\geq 0}$, we can simply replace the Minimiser inequalities by

$$b^v \leq b^{w((v,v'))} \cdot b^{v'},$$

for every edge (v,v') originating from a Minimiser vertex $v \in V_{\min}$, and the adjusted Maximiser inequalities by

$$b^v \leq \sum_{v' \in suc(v)} b^{w((v,v'))} b^{v'}$$

for every vertex $v \in V_{\max}$ owned by the Maximiser, and require $b^v \geq 0$ for all vertices $v \in V$ of the game.

Reading the b^v as variables, this provides us with a linear constraint system

$$A\mathbf{x} \leq \mathbf{0}, \text{ subject to } \mathbf{x} \geq \mathbf{0},$$

and Corollary 4 implies that this constraint system has a solution different from $\mathbf{x}=\mathbf{0}$ if, and only if, $M_{\geq 0}$ is non-empty for the defining mean payoff game. As every positive multiple of a solution to $A\mathbf{x} \leq \mathbf{0}$ and $\mathbf{x} \geq \mathbf{0}$ is again a solution, this implies the following corollary:

Corollary 5. *The resulting linear programme* maximise $\mathbf{1}^T\mathbf{x}$ for $A\mathbf{x} \leq \mathbf{0}$ and $\mathbf{x} \geq \mathbf{0}$ *is unbounded if $M_{\geq 0}$ is non-empty, and the constraint system has $\mathbf{x}=\mathbf{0}$ as the only solution if $M_{\geq 0} = \emptyset$ is empty.* □

3.4 From Qualitative to Quantitative Solutions

While solving the linear programme introduced in the previous subsection answers only the qualitative question of whether the linear programming problem is bounded, and hence if $M_{\geq 0}$ is non-empty, it is simple to extend the approach to a qualitative solution that provides us with $M_{\geq 0}$ and a strategy for the Maximiser that witnesses this. To achieve this, it suffices to bound the value of every vertex from above, for example, by adding a constraint $\mathbf{x} \leq \mathbf{1}$, or any other constraint $\mathbf{x} \leq \mathbf{d}$ for some constant vector $\mathbf{d} > \mathbf{0}$. (Where $>$ for the vector requires $>$ for every row.)

Proposition 2. *For every constant vector $\mathbf{d} > \mathbf{0}$, the solution to the linear programming problem* maximise $\mathbf{c}^T\mathbf{x}$ for $A\mathbf{x} \leq \mathbf{0}$, $\mathbf{x} \leq \mathbf{d}$, and $\mathbf{x} \geq \mathbf{0}$ *assigns a value $\neq 0$ to a variable if, and only if, it is in $M_{\geq 0}$. A witnessing strategy for the Maximiser in the defining mean payoff game can be inferred from the solution.*

Proof. For the solution of the linear programming problem it holds that if a Maximiser vertex v has some successor with non-zero value, or if a Minimiser vertex v has only non-zero successors, than the value b^v assigned to v by the solution is also non-zero. (Otherwise we could increase it, and hence $\mathbf{1}^T\mathbf{x}$, without changing any other value.) Hence, the logarithms of the solution define a solution to the system of logarithmic inequalities from the previous subsection, and we can infer a witnessing strategy for the Maximiser as described in the proof of Lemma 2.

Now consider a solution to the new linear programming problem defined by the sub-game of the mean payoff game that contains only the vertices with 0 values. If it had a solution different to $\mathbf{0}$, we could increase the solution of the linear programming problem we started with by ε times the solution of the new liner programming problem for a sufficiently small $\varepsilon > 0$. Hence $\mathbf{0}$ is the only solution to the new problem, and therefore there is no real valued solution for the basic or logarithmic inequalities defined by this sub-game. By Corollary 1 the Minimiser has thus a witnessing strategy for the 0 mean partition in the sub-game, which is also a witnessing strategy in the full game. □

3.5 Translation Complexity

The proposed translation of a given mean payoff game to a linear programming problem is cheap in the unit cost model:

Proposition 3. *A mean payoff game $\mathcal{M}$ with n vertices and m edges, and edge weights represented in binary can be translated in time $O(|\mathcal{M}|+nm)$ in the unit cost model. (Where $|\mathcal{M}|$ denotes the length of the representation of $\mathcal{M}$.)*

Proof. We have to compute the linear constraint $A\mathbf{x} \leq \mathbf{0}$, which requires the computation of the non-zero constants of A, and filling up A with 0s. As the rows of A refer to Maximiser vertices or edges originating from Minimiser vertices, and the columns refer to vertices, the latter requires $O(nm)$ steps.

Each edge refers to exactly one non-zero constant in A, and we need to translate the edge weight $w(e)$ to $b^{w(e)}$. We compute $b_0 = b$ (computing b is well within $O(nm)$), and then $b_i = b^{(2^i)} = b_{i-1}^2$ for all $i \leq \log_2 \max\{|w(e)| \mid e \in E\}$. $b^{w(e)}$ can then be expressed as a product of the respective b_i if $w(e) > 0$ is positive, as its reciprocal if $w(e) < 0$ is negative, and by 1 if $w(e) = 0$. The required time for the computation of $b^{w(e)}$ is therefore linear in the binary representation of $w(e)$, and computing all constants $b^{w(e)}$ requires $O(|\mathcal{M}|)$ operations. □

The translation of parity games to mean payoff games [34] discussed in Section 2 implies a likewise bound for parity games.

Corollary 6. *A parity game $\mathcal{P}$ with n vertices and m edges can be translated in time $O(nm)$ in the unit cost model.* □

Note that this also implies a polynomial bound on the cost of translating a mean payoff game whose edge weights are represented in unary — and hence of parity games with a bounded number of colours — in the Turing model of computation.

Corollary 7. *For parity games with a bounded number of colours and mean payoff games where the edge weights are represented in unary, the reduction results in a linear programme in binary representation that can be constructed in polynomial time.* □

As a result, the known polynomial bounds [3,4] for solving linear programming problems in the Turing model of computation imply a polynomial bound for these subproblems.

Corollary 8. *Parity games with a bounded number of colours and mean payoff games whose edge weights are represented unary can be solved in polynomial time.* □

Remark 1. If the algorithm requires non-degenerated linear programmes, then we can first apply the strongly polynomial standard ε-perturbation technique [38].

While the bounds provided by Corollary 8 are not new, they can be considered as a sanity check for new techniques: Besides its potential for parity and mean payoff games in general, the reduction is good enough to infer the relevant known polynomial bounds.

4 Example

This section contains an example reduction from solving the small parity game from Figure 1(a) to a linear programming problem.

Finding the winning region for the player that wins when the highest colour occurring infinitely many times is even can be reduced to finding the 0 mean partition of the mean payoff game from Figure 1(b). By Lemma 2, finding this 0 mean partition reduces to determining which variables can have a real value in a solution to any set

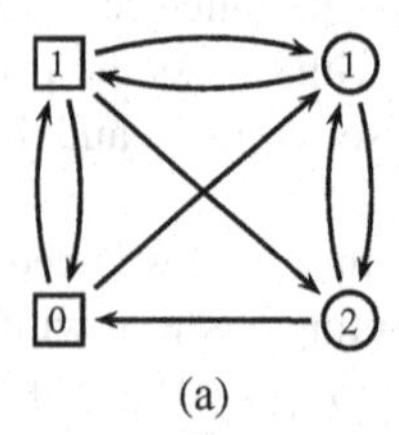

(a)

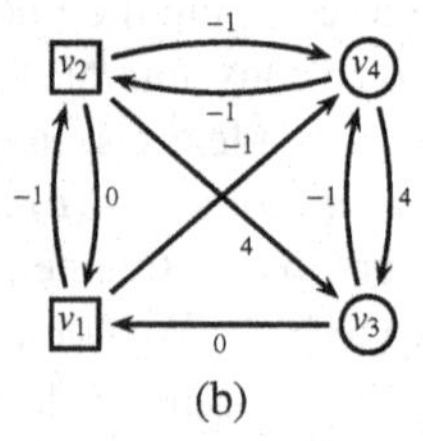

(b)

Fig. 1. Figure 1(a) shows a small example parity game. The vertices of the player with the objective to ensure parity are depicted as squares, while the positions of her opponent are depicted as circles. The vertices of the parity game are decorated with their respective colour. The parity game of Figure 1(a) is translated into the mean payoff game $\mathcal{M}$ of Figure 1(b). The edges of $\mathcal{M}$ are decorated with their weights, and the vertices with their name.

$$\begin{array}{lll} v_1 \leq \max\{v_2-1,\ v_4-1\}+c_{v_1} & v_3 \leq v_1 & v_4 \leq v_2-1 \\ v_2 \leq \max\{v_1,\ v_3+4,\ v_4-1\}+c_{v_2} & v_3 \leq v_4-1 & v_4 \leq v_3+4 \end{array}$$

of inequations, where $c_{v_1}, c_{v_2} < \frac{1}{4}$ can be any non-negative constant smaller than the reciprocal of the size of the game.

The maximal out-degree of a Maximiser vertex is 3, and choosing a basis b big enough to provide $\log_b 3 < \frac{1}{4}$, which holds for all $b > 3^4$, we can seek a solution to the inequations

$$\begin{array}{lll} v_1 \leq \log_b(b^{v_2-1}+b^{v_4-1}) & v_3 \leq v_1 & v_4 \leq v_2-1 \\ v_2 \leq \log_b(b^{v_1}+b^{v_3+4}+b^{v_4-1}) & v_3 \leq v_4-1 & v_4 \leq v_3+4 \end{array}$$

instead by Corollary 4, because $\log_b(b^{v_2}+b^{v_4-1}) = \max\{v_2,\ v_4-1\}+c_{v_1}$ and $\log_b(b^{v_1-1}+b^{v_3+4}+b^{v_4}) = \max\{v_1-1,\ v_3+4,\ v_4\}+c_{v_2}$ holds for some $c_{v_1} \leq \log_b 2 < \frac{1}{4}$ and $c_{v_2} \leq \log_b 3 < \frac{1}{4}$, respectively.

This system of inequations on the domain $[-\infty, \infty[$ can be rewritten as the system

$$\begin{array}{lll} b^{v_1} \leq b^{-1}\cdot b^{v_2}+b^{-1}\cdot b^{v_4} & b^{v_3} \leq b^{v_1} & b^{v_4} \leq b^{-1}\cdot b^{v_2} \\ b^{v_2} \leq b^{v_1}+b^{4}\cdot b^{v_3}+b^{-1}\cdot b^{v_4}) & b^{v_3} \leq b^{-1}\cdot b^{v_4} & b^{v_4} \leq b^{4}\cdot b^{v_3} \end{array}$$

of inequations. Finally, the individual b^{v_i} can be treated as variables after adding the constraints $0 \leq b^{v_1},\ b^{v_2},\ b^{v_3},\ b^{v_4},\ b^{v_5},\ b^{v_6}$.

For finding a witnessing strategy for the Maximiser—and hence a winning strategy for the player that wants to ensure parity in the game from Figure 1(a)—it suffices to add the additional constraint $b^{v_1},\ b^{v_2},\ b^{v_3},\ b^{v_4},\ b^{v_5},\ b^{v_6} \leq 1$ and maximise $b^{v_1}+b^{v_2}+b^{v_3}+b^{v_4}+b^{v_5}+b^{v_6}$.

Note that the constraints in the linear programming problem reach $b^4 = 45212176$ for $b = 82$ even in this tiny example.

5 Discussion

The introduced reduction from solving parity and mean payoff games to linear programming opens up the well developed class of linear programming techniques to the analysis of these classes of ω-games. It also links their complexity to the complexity of linear programming in the unit cost model.

As the unit cost complexity of linear programming is not known, there is no immediate *practical* benefit attached to this reduction, but the drawn connections between linear programmes and finite games of infinite durations link two intriguing open problems. The potential benefit for the two areas are quite different in nature: The linear programming community gains a natural and important class of problems that would benefit from a polynomial time algorithm for linear programming, while the game solving community will automatically profit from future developments of polynomial time algorithms.

References

1. Smale, S.: On the average number of steps of the simplex method of linear programming. Mathematical Programming 27, 241–262 (1983)
2. Klee, F., Minty, G.J.: How good is the simplex algorithm? Inequalities III, 159–175 (1972)
3. Khachian, L.G.: A polynomial algorithm in linear programming. Doklady Akademii Nauk SSSR 244, 1093–1096 (1979)
4. Karmarkar, N.: A new polynomial-time algorithm for linear programming. In: Proceedings of STOC 1984, pp. 302–311. ACM Press, New York (1984)
5. Gärtner, B., Henk, M., Ziegler, G.M.: Randomized simplex algorithms on Klee-Minty cubes. Combinatorica 18, 502–510 (1994)
6. Spielman, D.A., Teng, S.H.: Smoothed analysis of algorithms: Why the simplex algorithm usually takes polynomial time. Journal of the ACM 51, 385–463 (2004)
7. Kelner, J.A., Spielman, D.A.: A randomized polynomial-time simplex algorithm for linear programming. In: Proceedings of STOC 2006, pp. 51–60. ACM Press, New York (2006)
8. Smale, S.: Mathematical problems for the next century. The Mathematical Inteligencer 20, 7–15 (1998)
9. Kozen, D.: Results on the propositional μ-calculus. Theoretical Computer Science 27, 333–354 (1983)
10. Emerson, E.A., Jutla, C.S., Sistla, A.P.: On model-checking for fragments of μ-calculus. In: Courcoubetis, C. (ed.) CAV 1993. LNCS, vol. 697, pp. 385–396. Springer, Heidelberg (1993)
11. Wilke, T.: Alternating tree automata, parity games, and modal μ-calculus. Bulletin of the Belgian Mathematical Society 8 (2001)
12. de Alfaro, L., Henzinger, T.A., Majumdar, R.: From verification to control: Dynamic programs for omega-regular objectives. In: Proceedings of LICS 2001, pp. 279–290. IEEE Computer Society Press, Los Alamitos (2001)
13. Alur, R., Henzinger, T.A., Kupferman, O.: Alternating-time temporal logic. Journal of the ACM 49, 672–713 (2002)
14. Vardi, M.Y.: Reasoning about the past with two-way automata. In: Larsen, K.G., Skyum, S., Winskel, G. (eds.) ICALP 1998. LNCS, vol. 1443, pp. 628–641. Springer, Heidelberg (1998)
15. Schewe, S., Finkbeiner, B.: Satisfiability and finite model property for the alternating-time μ-calculus. In: Ésik, Z. (ed.) CSL 2006. LNCS, vol. 4207, pp. 591–605. Springer, Heidelberg (2006)
16. Piterman, N.: From nondeterministic Büchi and Streett automata to deterministic parity automata. Journal of Logical Methods in Computer Science 3 (2007)
17. Schewe, S., Finkbeiner, B.: Synthesis of asynchronous systems. In: Puebla, G. (ed.) LOPSTR 2006. LNCS, vol. 4407, pp. 127–142. Springer, Heidelberg (2006)

18. Emerson, E.A., Lei, C.: Efficient model checking in fragments of the propositional μ-calculus. In: Proceedings of LICS 1986, pp. 267–278. IEEE Computer Society Press, Los Alamitos (1986)
19. Emerson, E.A., Jutla, C.S.: Tree automata, μ-calculus and determinacy. In: Proceedings of FOCS 1991, pp. 368–377. IEEE Computer Society Press, Los Alamitos (1991)
20. McNaughton, R.: Infinite games played on finite graphs. Annals of Pure and Applied Logic 65, 149–184 (1993)
21. Browne, A., Clarke, E.M., Jha, S., Long, D.E., Marrero, W.: An improved algorithm for the evaluation of fixpoint expressions. Theoretical Computer Science 178, 237–255 (1997)
22. Jurdziński, M.: Deciding the winner in parity games is in UP $\cap$ co-UP. Information Processing Letters 68, 119–124 (1998)
23. Zielonka, W.: Infinite games on finitely coloured graphs with applications to automata on infinite trees. Theoretical Computer Science 200, 135–183 (1998)
24. Jurdziński, M.: Small progress measures for solving parity games. In: Reichel, H., Tison, S. (eds.) STACS 2000. LNCS, vol. 1770, pp. 290–301. Springer, Heidelberg (2000)
25. Ludwig, W.: A subexponential randomized algorithm for the simple stochastic game problem. Information and Computation 117, 151–155 (1995)
26. Puri, A.: Theory of hybrid systems and discrete event systems. PhD thesis, Computer Science Department, University of California, Berkeley (1995)
27. Vöge, J., Jurdziński, M.: A discrete strategy improvement algorithm for solving parity games. In: Emerson, E.A., Sistla, A.P. (eds.) CAV 2000. LNCS, vol. 1855, pp. 202–215. Springer, Heidelberg (2000)
28. Björklund, H., Vorobyov, S.: A combinatorial strongly subexponential strategy improvement algorithm for mean payoff games. Discrete Applied Mathematics 155, 210–229 (2007)
29. Schewe, S.: An optimal strategy improvement algorithm for solving parity and payoff games. In: Kaminski, M., Martini, S. (eds.) CSL 2008. LNCS, vol. 5213, pp. 368–383. Springer, Heidelberg (2008)
30. Obdržálek, J.: Fast μ-calculus model checking when tree-width is bounded. In: Hunt Jr., W.A., Somenzi, F. (eds.) CAV 2003. LNCS, vol. 2725, pp. 80–92. Springer, Heidelberg (2003)
31. Berwanger, D., Dawar, A., Hunter, P., Kreutzer, S.: DAG-width and parity games. In: Durand, B., Thomas, W. (eds.) STACS 2006. LNCS, vol. 3884, pp. 524–536. Springer, Heidelberg (2006)
32. Schewe, S.: Solving parity games in big steps. In: Arvind, V., Prasad, S. (eds.) FSTTCS 2007. LNCS, vol. 4855, pp. 449–460. Springer, Heidelberg (2007)
33. Jurdziński, M., Paterson, M., Zwick, U.: A deterministic subexponential algorithm for solving parity games. SIAM Journal of Computing 38, 1519–1532 (2008)
34. Zwick, U., Paterson, M.S.: The complexity of mean payoff games on graphs. Theoretical Computer Science 158, 343–359 (1996)
35. Ehrenfeucht, A., Mycielski, J.: Positional strategies for mean payoff games. International Journal of Game Theory 2, 109–113 (1979)
36. Gurvich, V.A., Karzanov, A.V., Khachivan, L.G.: Cyclic games and an algorithm to find minimax cycle means in directed graphs. USSR Computational Mathematics and Mathematical Physics 28, 85–91 (1988)
37. Friedmann, O.: A super-polynomial lower bound for the parity game strategy improvement algorithm as we know it. In: Proceedings of LICS (2009)
38. Megiddo, N., Chandrasekaran, R.: On the ε-perturbation method for avoiding degeneracy. Operations Research Letters 8, 305–308 (1989)

Partial Randomness and Dimension of Recursively Enumerable Reals

Kohtaro Tadaki

Research and Development Initiative, Chuo University
JST, CREST
1–13–27 Kasuga, Bunkyo-ku, Tokyo 112-8551, Japan
tadaki@kc.chuo-u.ac.jp

Abstract. A real α is called recursively enumerable ("r.e." for short) if there exists a computable, increasing sequence of rationals which converges to α. It is known that the randomness of an r.e. real α can be characterized in various ways using each of the notions; program-size complexity, Martin-Löf test, Chaitin Ω number, the domination and Ω-likeness of α, the universality of a computable, increasing sequence of rationals which converges to α, and universal probability. In this paper, we generalize these characterizations of randomness over the notion of partial randomness by parameterizing each of the notions above by a real $T \in (0, 1]$, where the notion of partial randomness is a stronger representation of the compression rate by means of program-size complexity. As a result, we present ten equivalent characterizations of the partial randomness of an r.e. real. The resultant characterizations of partial randomness are powerful and have many important applications. One of them is to present equivalent characterizations of the dimension of an individual r.e. real. The equivalence between the notion of Hausdorff dimension and compression rate by program-size complexity (or partial randomness) has been established at present by a series of works of many researchers over the last two decades. We present ten equivalent characterizations of the dimension of an individual r.e. real.

Keywords: algorithmic randomness, recursively enumerable real, partial randomness, dimension, Chaitin Ω number, program-size complexity, universal probability.

1 Introduction

A real α is called *recursively enumerable* ("r.e." for short) if there exists a computable, increasing sequence of rationals which converges to α. The randomness of an r.e. real α can be characterized in various ways using each of the notions; *program-size complexity*, *Martin-Löf test*, *Chaitin Ω number*, the *domination* and *Ω-likeness* of α, the *universality* of a computable, increasing sequence of rationals which converges to α, and *universal probability*. These equivalent characterizations of randomness of an r.e. real are summarized in Theorem 6 (see

R. Královič and D. Niwiński (Eds.): MFCS 2009, LNCS 5734, pp. 687–699, 2009.

Section 3), where the equivalences are established by a series of works of Martin-Löf [9], Schnorr [14], Chaitin [4], Solovay [15], Calude, Hertling, Khoussainov and Wang [1], Kučera and Slaman [7], and Tadaki [20], between 1966 and 2006. In this paper, we generalize these characterizations of randomness over the notion of *partial randomness*, which was introduced by Tadaki [18,19] and is a stronger representation of *the compression rate* by means of program-size complexity. We introduce many characterizations of partial randomness for an r.e. real by parameterizing each of the notions above on randomness by a real $T \in (0, 1]$. In particular, we introduce the notion of T*-convergence* for a computable, increasing sequence of rationals and then introduce the same notion for an r.e. real. The notion of T-convergence plays a crucial role in these our characterizations of partial randomness. We then prove the equivalence of all these characterizations of partial randomness in Theorem 8, our main result, in Section 4.

On the other hand, by a series of works of Ryabko [12,13], Staiger [16,17], Tadaki [18,19], Lutz [8], and Mayordomo [10] over the last two decades, the equivalence between the notion of compression rate by program-size complexity (or partial randomness) and Hausdorff dimension seems to be established at present. The subject of the equivalence seems to be one of the most active areas of the recent research of algorithmic randomness. In the context of the subject, we can consider the notion of the *dimension* of an individual real in particular, and this notion plays a crucial role in the subject. As one of the main applications of our main result on partial randomness, i.e., Theorem 8, we can present many equivalent characterizations of the dimension of an individual r.e. real.

The paper is organized as follows. We begin in Section 2 with some preliminaries to algorithmic information theory and partial randomness. In Section 3, we review the previous results on the equivalent characterizations of randomness of an r.e. real. Our main result on partial randomness of an r.e. real is presented in Section 4. In Section 5 we apply our main result on partial randomness to give many equivalent characterizations of the dimension of an r.e. real. In Section 6, we investigate further properties of the notion of T-convergence, which plays a crucial role in our characterizations of the partial randomness and dimension of r.e. reals. We conclude this paper with a mention of the future direction of this work in Section 7. Due to the 12-page limit, we omit most proofs. A full paper which describes all the proofs and other related results is in preparation.

2 Preliminaries

We start with some notation about numbers and strings which will be used in this paper. $\mathbb{N} = \{0, 1, 2, 3, \dots\}$ is the set of natural numbers, and $\mathbb{N}^+$ is the set of positive integers. $\mathbb{Q}$ is the set of rational numbers, and $\mathbb{R}$ is the set of real numbers. A sequence $\{a_n\}_{n\in\mathbb{N}}$ of numbers (rationals or reals) is called *increasing* if $a_{n+1} > a_n$ for all $n \in \mathbb{N}$.

$\{0, 1\}^* = \{\lambda, 0, 1, 00, 01, 10, 11, 000, \dots\}$ is the set of finite binary strings, where λ denotes the *empty string*. For any $s \in \{0, 1\}^*$, $|s|$ is the *length* of s. A subset S of $\{0, 1\}^*$ is called *prefix-free* if no string in S is a prefix of another string in S. For any partial function f, the domain of definition of f is denoted by $\mathrm{dom}\, f$. We write "r.e." instead of "recursively enumerable."

Normally, $o(n)$ denotes any function $f\colon \mathbb{N}^+ \to \mathbb{R}$ such that $\lim_{n\to\infty} f(n)/n = 0$. On the other hand, $O(1)$ denotes any function $g\colon \mathbb{N}^+ \to \mathbb{R}$ such that there is $C \in \mathbb{R}$ with the property that $|g(n)| \le C$ for all $n \in \mathbb{N}^+$.

Let α be an arbitrary real. For any $n \in \mathbb{N}^+$, we denote by $\alpha\restriction_n \in \{0,1\}^*$ the first n bits of the base-two expansion of $\alpha - \lfloor \alpha \rfloor$ with infinitely many zeros, where $\lfloor \alpha \rfloor$ is the greatest integer less than or equal to α. Thus, in particular, if $\alpha \in [0,1)$, then $\alpha\restriction_n$ denotes the first n bits of the base-two expansion of α with infinitely many zeros. For example, in the case of $\alpha = 5/8$, $\alpha\restriction_6 = 101000$.

A real α is called *r.e.* if there exists a computable, increasing sequence of rationals which converges to α. An r.e. real is also called a *left-computable* real. Let α and β be arbitrary r.e. reals. Then $\alpha+\beta$ is r.e. If α and β are non-negative, then $\alpha\beta$ is r.e. On the other hand, a real α is called *right-computable* if $-\alpha$ is left-computable. We say that a real α is *computable* if there exists a computable sequence $\{a_n\}_{n\in\mathbb{N}}$ of rationals such that $|\alpha - a_n| < 2^{-n}$ for all $n \in \mathbb{N}$. It is then easy to see that, for every $\alpha \in \mathbb{R}$, α is computable if and only if α is both left-computable and right-computable. A sequence $\{a_n\}_{n\in\mathbb{N}}$ of reals is called *computable* if there exists a total recursive function $f\colon \mathbb{N}\times\mathbb{N} \to \mathbb{Q}$ such that $|a_n - f(n,m)| < 2^{-m}$ for all $n, m \in \mathbb{N}$. See e.g. Weihrauch [23] for the detail of the treatment of the computability of reals and sequences of reals.

2.1 Algorithmic Information Theory

In the following we concisely review some definitions and results of algorithmic information theory [4,5]. A *computer* is a partial recursive function $C\colon \{0,1\}^* \to \{0,1\}^*$ such that $\mathrm{dom}\, C$ is a prefix-free set. For each computer C and each $s \in \{0,1\}^*$, $H_C(s)$ is defined as $\min\{\, |p| \mid p \in \{0,1\}^* \;\&\; C(p) = s \,\}$ (may be ∞). A computer U is said to be *optimal* if for each computer C there exists $d \in \mathbb{N}$ with the following property; if $p \in \mathrm{dom}\, C$, then there is $q \in \mathrm{dom}\, U$ for which $U(q) = C(p)$ and $|q| \le |p| + d$. It is easy to see that there exists an optimal computer. We choose a particular optimal computer U as the standard one for use, and define $H(s)$ as $H_U(s)$, which is referred to as the *program-size complexity* of s or the *Kolmogorov complexity* of s. It follows that for every computer C there exists $d \in \mathbb{N}$ such that, for every $s \in \{0,1\}^*$, $H(s) \le H_C(s) + d$.

For any optimal computer V, *Chaitin's halting probability* Ω_V of V is defined as $\sum_{p\in\mathrm{dom}\, V} 2^{-|p|}$. The real Ω_V is also called *Chaitin Ω number.*

Definition 1 (weak Chaitin randomness, Chaitin [4,5]). *For any $\alpha \in \mathbb{R}$, we say that α is weakly Chaitin random if there exists $c \in \mathbb{N}$ such that $n - c \le H(\alpha\restriction_n)$ for all $n \in \mathbb{N}^+$.* □

Chaitin [4] showed that, for every optimal computer V, Ω_V is weakly Chaitin random.

Definition 2 (Martin-Löf randomness, Martin-Löf [9]). *A subset C of $\mathbb{N}^+ \times \{0,1\}^*$ is called a Martin-Löf test if C is an r.e. set and*

$$\forall n \in \mathbb{N}^+ \quad \sum_{s \in C_n} 2^{-|s|} \leq 2^{-n},$$

where $C_n = \{\, s \mid (n,s) \in C \,\}$. For any $\alpha \in \mathbb{R}$, we say that α is Martin-Löf random if for every Martin-Löf test C, there exists $n \in \mathbb{N}^+$ such that, for every $k \in \mathbb{N}^+$, $\alpha\restriction_k \notin C_n$. □

Theorem 1 (Schnorr [14]). *For every $\alpha \in \mathbb{R}$, α is weakly Chaitin random if and only if α is Martin-Löf random.* □

The program-size complexity $H(s)$ is originally defined using the concept of program-size, as stated above. However, it is possible to define $H(s)$ without referring to such a concept, i.e., as in the following, we first introduce a *universal probability* m, and then define $H(s)$ as $-\log_2 m(s)$. A universal probability is defined as follows [24].

Definition 3 (universal probability). *A function $r\colon \{0,1\}^* \to [0,1]$ is called a lower-computable semi-measure if $\sum_{s\in\{0,1\}^*} r(s) \leq 1$ and the set $\{(a,s) \in \mathbb{Q} \times \{0,1\}^* \mid a < r(s)\}$ is r.e. We say that a lower-computable semi-measure m is a universal probability if for every lower-computable semi-measure r, there exists $c \in \mathbb{N}^+$ such that, for all $s \in \{0,1\}^*$, $r(s) \leq cm(s)$.* □

The following theorem can be then shown (see e.g. Theorem 3.4 of Chaitin [4] for its proof).

Theorem 2. *For every optimal computer V, the function $2^{-H_V(s)}$ of s is a universal probability.* □

By Theorem 2, we see that $H(s) = -\log_2 m(s) + O(1)$ for every universal probability m. Thus it is possible to define $H(s)$ as $-\log_2 m(s)$ with a particular universal probability m instead of as $H_U(s)$. Note that the difference up to an additive constant is nonessential to algorithmic information theory.

2.2 Partial Randomness

In the works [18,19], we generalized the notion of the randomness of a real so that *the degree of the randomness*, which is often referred to as *the partial randomness* recently [2,11,3], can be characterized by a real T with $0 < T \leq 1$ as follows.

Definition 4 (weak Chaitin T-randomness). *Let $T \in \mathbb{R}$ with $T \geq 0$. For any $\alpha \in \mathbb{R}$, we say that α is weakly Chaitin T-random if there exists $c \in \mathbb{N}$ such that $Tn - c \leq H(\alpha\restriction_n)$ for all $n \in \mathbb{N}^+$.* □

Definition 5 (Martin-Löf T-randomness). *Let $T \in \mathbb{R}$ with $T \geq 0$. A subset C of $\mathbb{N}^+ \times \{0,1\}^*$ is called a Martin-Löf T-test if C is an r.e. set and*

$$\forall n \in \mathbb{N}^+ \quad \sum_{s \in C_n} 2^{-T|s|} \leq 2^{-n}.$$

For any $\alpha \in \mathbb{R}$, we say that α is Martin-Löf T-random if for every Martin-Löf T-test $\mathcal{C}$, there exists $n \in \mathbb{N}^+$ such that, for every $k \in \mathbb{N}^+$, $\alpha\restriction_k \notin \mathcal{C}_n$. □

In the case where $T = 1$, the weak Chaitin T-randomness and Martin-Löf T-randomness result in weak Chaitin randomness and Martin-Löf randomness, respectively. Tadaki [19] generalized Theorem 1 over the notion of T-randomness as follows.

Theorem 3 (Tadaki [19]). *Let T be a computable real with $T \geq 0$. Then, for every $\alpha \in \mathbb{R}$, α is weakly Chaitin T-random if and only if α is Martin-Löf T-random.* □

Definition 6 (T-compressibility). *Let $T \in \mathbb{R}$ with $T \geq 0$. For any $\alpha \in \mathbb{R}$, we say that α is T-compressible if $H(\alpha\restriction_n) \leq Tn + o(n)$, which is equivalent to* $\limsup_{n\to\infty} H(\alpha\restriction_n)/n \leq T$. □

For every $T \in [0, 1]$ and every $\alpha \in \mathbb{R}$, if α is weakly Chaitin T-random and T-compressible, then

$$\lim_{n\to\infty} \frac{H(\alpha\restriction_n)}{n} = T. \tag{1}$$

The left-hand side of (1) is referred to as the *compression rate* of a real α in general. Note, however, that (1) does not necessarily imply that α is weakly Chaitin T-random. Thus, the notion of partial randomness is a stronger representation of compression rate.

In the works [18,19], we generalized Chaitin Ω number to $\Omega(T)$ as follows. For each optimal computer V and each real $T > 0$, the *generalized halting probability* $\Omega_V(T)$ of V is defined by

$$\Omega_V(T) = \sum_{p\in\operatorname{dom} V} 2^{-\frac{|p|}{T}}.$$

Thus, $\Omega_V(1) = \Omega_V$. If $0 < T \leq 1$, then $\Omega_V(T)$ converges and $0 < \Omega_V(T) < 1$, since $\Omega_V(T) \leq \Omega_V < 1$. The following theorem holds for $\Omega_V(T)$.

Theorem 4 (Tadaki [18,19]). *Let V be an optimal computer and let $T \in \mathbb{R}$.*

(i) If $0 < T \leq 1$ and T is computable, then $\Omega_V(T)$ is weakly Chaitin T-random and T-compressible.
(ii) If $1 < T$, then $\Omega_V(T)$ diverges to ∞. □

3 Previous Results on the Randomness of an r.e. Real

In this section, we review the previous results on the randomness of an r.e. real. First we review some notions on r.e. reals.

Definition 7 (Ω-likeness). *For any r.e. reals α and β, we say that α dominates β if there are computable, increasing sequences $\{a_n\}$ and $\{b_n\}$ of rationals and $c \in \mathbb{N}^+$ such that $\lim_{n\to\infty} a_n = \alpha$, $\lim_{n\to\infty} b_n = \beta$, and $c(\alpha - a_n) \geq \beta - b_n$ for all $n \in \mathbb{N}$. An r.e. real α is called Ω-like if it dominates all r.e. reals.* □

Solovay [15] showed the following theorem. For its proof, see also Theorem 4.9 of [1].

Theorem 5 (Solovay [15]). *For every r.e. reals α and β, if α dominates β then $H(\beta\restriction_n) \leq H(\alpha\restriction_n) + O(1)$ for all $n \in \mathbb{N}^+$.* □

Definition 8 (universality). *A computable, increasing and converging sequence $\{a_n\}$ of rationals is called universal if for every computable, increasing and converging sequence $\{b_n\}$ of rationals there exists $c \in \mathbb{N}^+$ such that $c(\alpha - a_n) \geq \beta - b_n$ for all $n \in \mathbb{N}$, where $\alpha = \lim_{n\to\infty} a_n$ and $\beta = \lim_{n\to\infty} b_n$.* □

The previous results on the equivalent characterizations of randomness for an r.e. real are summarized in the following theorem.

Theorem 6 ([14,4,15,1,7,20]). *Let α be an r.e. real with $0 < \alpha < 1$. Then the following conditions are equivalent:*

(i) The real α is weakly Chaitin random.
(ii) The real α is Martin-Löf random.
(iii) The real α is Ω-like.
(iv) For every r.e. real β, $H(\beta\restriction_n) \leq H(\alpha\restriction_n) + O(1)$ for all $n \in \mathbb{N}^+$.
(v) There exists an optimal computer V such that $\alpha = \Omega_V$.
(vi) There exists a universal probability m such that $\alpha = \sum_{s\in\{0,1\}^*} m(s)$.
(vii) Every computable, increasing sequence of rationals which converges to α is universal.
(viii) There exists a universal computable, increasing sequence of rationals which converges to α. □

The historical remark on the proofs of equivalences in Theorem 6 is as follows. Schnorr [14] showed that (i) and (ii) are equivalent to each other. Chaitin [4] showed that (v) implies (i). Solovay [15] showed that (v) implies (iii), (iii) implies (iv), and (iii) implies (i). Calude, Hertling, Khoussainov, and Wang [1] showed that (iii) implies (v), and (v) implies (vii). Kučera and Slaman [7] showed that (ii) implies (vii). Finally, (vi) was inserted in the course of the derivation from (v) to (viii) by Tadaki [20].

4 New Results on the Partial Randomness of an r.e. Real

In this section, we generalize Theorem 6 above over the notion of partial randomness. For that purpose, we first introduce some new notions. Let T be an arbitrary real with $0 < T \leq 1$ throughout the rest of this paper. These notions are parametrized by the real T.[1]

[1] The parameter T corresponds to the notion of "temperature" in the statistical mechanical interpretation of algorithmic information theory developed by Tadaki [21,22].

Definition 9 (T-convergence). *An increasing sequence $\{a_n\}$ of reals is called T-convergent if $\sum_{n=0}^{\infty}(a_{n+1}-a_n)^T<\infty$. An r.e. real α is called T-convergent if there exists a T-convergent computable, increasing sequence of rationals which converges to α, i.e., if there exists an increasing sequence $\{a_n\}$ of rationals such that (i) $\{a_n\}$ is T-convergent, (ii) $\{a_n\}$ is computable, and (iii) $\lim_{n\to\infty}a_n=\alpha$.* □

Note that every increasing and converging sequence of reals is 1-convergent, and thus every r.e. real is 1-convergent. In general, based on the following lemma, we can freely switch from "T-convergent computable, increasing sequence of reals" to "T-convergent computable, increasing sequence of rationals."

Lemma 1. *For every $\alpha\in\mathbb{R}$, α is an r.e. T-convergent real if and only if there exists a T-convergent computable, increasing sequence of reals which converges to α.* □

The following argument illustrates the way of using Lemma 1: Let V be an optimal computer, and let $p_0,p_1,p_2,\dots$ be a recursive enumeration of the r.e. set dom V. Then $\Omega_V(T)=\sum_{i=0}^{\infty}2^{-|p_i|/T}$, and the increasing sequence $\{\sum_{i=0}^{n}2^{-|p_i|/T}\}_{n\in\mathbb{N}}$ of reals is T-convergent since $\Omega_V=\sum_{i=0}^{\infty}2^{-|p_i|}<1$. If T is computable, then this sequence of reals is computable. Thus, by Lemma 1 we have Theorem 7 below.

Theorem 7. *Let V be an optimal computer. If T is computable, then $\Omega_V(T)$ is an r.e. T-convergent real.* □

Definition 10 ($\Omega(T)$-likeness). *An r.e. real α is called $\Omega(T)$-like if it dominates all r.e. T-convergent reals.* □

Note that an r.e. real α is $\Omega(1)$-like if and only if α is Ω-like.

Definition 11 (T-universality). *A computable, increasing and converging sequence $\{a_n\}$ of rationals is called T-universal if for every T-convergent computable, increasing and converging sequence $\{b_n\}$ of rationals there exists $c\in\mathbb{N}^+$ such that $c(\alpha-a_n)\geq\beta-b_n$ for all $n\in\mathbb{N}$, where $\alpha=\lim_{n\to\infty}a_n$ and $\beta=\lim_{n\to\infty}b_n$.* □

Note that a computable, increasing and converging sequence $\{a_n\}$ of rationals is 1-universal if and only if $\{a_n\}$ is universal.

Using the notions introduced above, Theorem 6 is generalized as follows.

Theorem 8 (main result). *Let α be an r.e. real with $0<\alpha<1$. Suppose that T is computable. Then the following conditions are equivalent:*

(i) The real α is weakly Chaitin T-random.
(ii) The real α is Martin-Löf T-random.
(iii) The real α is $\Omega(T)$-like.
(iv) For every r.e. T-convergent real β, $H(\beta\restriction_n)\leq H(\alpha\restriction_n)+O(1)$ for all $n\in\mathbb{N}^+$.

(v) *For every r.e. T-convergent real* $\gamma > 0$*, there exist an r.e. real* $\beta \geq 0$ *and a rational* $q > 0$ *such that* $\alpha = \beta + q\gamma$*.*

(vi) *For every optimal computer* V*, there exist an r.e. real* $\beta \geq 0$ *and a rational* $q > 0$ *such that* $\alpha = \beta + q\Omega_V(T)$*.*

(vii) *There exist an optimal computer* V *and an r.e. real* $\beta \geq 0$ *such that* $\alpha = \beta + \Omega_V(T)$*.*

(viii) *There exists a universal probability* m *such that* $\alpha = \sum_{s\in\{0,1\}^*} m(s)^{\frac{1}{T}}$*.*

(ix) *Every computable, increasing sequence of rationals which converges to* α *is T-universal.*

(x) *There exists a T-universal computable, increasing sequence of rationals which converges to* α*.* □

We see that Theorem 8 is a massive expansion of Theorem 3 in the case where the real α is r.e. with $0 < \alpha < 1$. The condition (vii) of Theorem 8 corresponds to the condition (v) of Theorem 6. Note, however, that, in the condition (vii) of Theorem 8, a non-negative r.e. real β is needed. The reason is as follows: In the case of $\beta = 0$, the possibility that α is weakly Chaitin T'-random with a real $T' > T$ is excluded by the T-compressibility of $\Omega_V(T)$ imposed by Theorem 4 (i). However, this exclusion is inconsistent with the condition (i) of Theorem 8.

Theorem 8 can be proved by generalizing the proof of Theorem 6 over the notion of partial randomness. For example, using Lemma 2 below, the implication (ii) ⇒ (v) of Theorem 8 is proved as follows, in which the notion of T-convergence plays an important role.

Proof (of (ii) ⇒ (v) of Theorem 8). Suppose that γ is an arbitrary r.e. T-convergent real with $\gamma > 0$. Then there exists a T-convergent computable, increasing sequence $\{c_n\}$ of rationals which converges to γ. Since $\gamma > 0$, without loss of generality we can assume that $c_0 = 0$. We choose any one rational $\varepsilon > 0$ such that $\sum_{n=0}^{\infty}[\varepsilon(c_{n+1} - c_n)]^T \leq 1$. Such ε exists since the sequence $\{c_n\}$ is T-convergent. Note that the sequence $\{\varepsilon(c_{n+1} - c_n)\}$ is a computable sequence of positive rationals. Thus, since α is a positive r.e. real and also Martin-Löf T-random by the assumption, it follows from Lemma 2 below that there exist a computable, increasing sequence $\{a_n\}$ of rationals and a rational $r > 0$ such that $a_{n+1} - a_n > r\varepsilon(c_{n+1} - c_n)$ for every $n \in \mathbb{N}$, $a_0 > 0$, and $\alpha = \lim_{n\to\infty} a_n$. We then define a sequence $\{b_n\}$ of positive rationals by $b_n = a_{n+1} - a_n - r\varepsilon(c_{n+1} - c_n)$. It follows that $\{b_n\}$ is a computable sequence of rationals and $\sum_{n=0}^{\infty} b_n$ converges to $\alpha - a_0 - r\varepsilon(\gamma - c_0)$. Thus we have $\alpha = a_0 + \sum_{n=0}^{\infty} b_n + r\varepsilon\gamma$, where $a_0 + \sum_{n=0}^{\infty} b_n$ is a positive r.e. real. This completes the proof. □

Lemma 2. *Let* α *be an r.e. real, and let* $\{d_n\}$ *be a computable sequence of positive rationals such that* $\sum_{n=0}^{\infty} {d_n}^T \leq 1$*. If* α *is Martin-Löf T-random, then for every* $\varepsilon > 0$ *there exist a computable, increasing sequence* $\{a_n\}$ *of rationals and a rational* $q > 0$ *such that* $a_{n+1} - a_n > qd_n$ *for every* $n \in \mathbb{N}$*,* $a_0 > \alpha - \varepsilon$*, and* $\alpha = \lim_{n\to\infty} a_n$*.* □

Lemma 2 can be proved, based on the generalization of the techniques used in the proof of Theorem 2.1 of Kučera and Slaman [7] over partial randomness.

In addition to the proof of Lemma 2, the complete proof of Theorem 8 will be described in a full version of this paper, which is in preparation.

Theorem 8 has many important applications. One of the main applications is to give many characterizations of the dimension of an individual r.e. real, some of which will be presented in the next section. As another consequence of Theorem 8, we can obtain Corollary 1 below for example, which follows immediately from the implication (vii) $\Rightarrow$ (iv) of Theorem 8 and Theorem 7.

Corollary 1. *Suppose that T is computable. Then, for every two optimal computers V and W, $H(\Omega_V(T)\restriction_n) = H(\Omega_W(T)\restriction_n) + O(1)$ for all $n \in \mathbb{N}^+$.* □

Note that the computability of T is important for Theorem 8 to hold. For example, we cannot allow T to be simply an r.e. real in Theorem 8.

The notion of T-convergence has many interesting properties, in addition to the properties which we saw above. In Section 6, we investigate further properties of the notion of T-convergence.

5 New Characterizations of the Dimension of an r.e. Real

In this section we apply Theorem 8 to give many characterizations of dimension for an individual r.e. real. In the works [18,19], we introduced the notions of six "algorithmic dimensions", 1st, 2nd, 3rd, 4th, upper, and lower algorithmic dimensions as fractal dimensions for a subset F of N-dimensional Euclidean space $\mathbb{R}^N$. These notions are defined based on the notion of partial randomness and compression rate by means of program-size complexity. We then showed that all the six algorithmic dimensions equal to the Hausdorff dimension for any self-similar set which is computable in a certain sense. The class of such self-similar sets includes familiar fractal sets such as the Cantor set, von Koch curve, and Sierpiński gasket. In particular, the notion of lower algorithmic dimension for a subset F of $\mathbb{R}$ is defined as follows.

Definition 12 (lower algorithmic dimension, Tadaki [19]). *Let F be a nonempty subset of $\mathbb{R}$. The lower algorithmic dimension $\underline{\dim}_A F$ of F is defined by $\underline{\dim}_A F = \sup_{x\in F} \liminf_{n\to\infty} H(x\restriction_n)/n$.* □

Thus, for every $\alpha \in \mathbb{R}$,

$$\underline{\dim}_A \{\alpha\} = \liminf_{n\to\infty} \frac{H(\alpha\restriction_n)}{n}. \tag{2}$$

Independently of us, Lutz [8] introduced the notion of constructive dimension of an individual real α using the notion of lower semicomputable s-supergale with $s \in [0,\infty)$, and then Mayordomo [10] showed that, for every real α, the constructive dimension of α equals to the right-hand side of (2). Thus, the constructive dimension of α is precisely the lower algorithmic dimension $\underline{\dim}_A\{\alpha\}$ of α for every real α.

Using Lemma 3 below, we can convert each of all the conditions in Theorem 8 into a characterization of the lower algorithmic dimension $\underline{\dim}_A\{\alpha\}$ for any r.e. real α.

Lemma 3. *Let $\alpha \in \mathbb{R}$. For every $t \in [0, \infty)$, α is weakly Chaitin t-random if $t < \underline{\dim}_A\{\alpha\}$, and α is not weakly Chaitin t-random if $t > \underline{\dim}_A\{\alpha\}$.*

Proof. Let $\alpha \in \mathbb{R}$, and let $t \in [0, \infty)$. Assume first that $t < \underline{\dim}_A\{\alpha\}$. Then, since $\underline{\dim}_A\{\alpha\}n \leq H(\alpha\restriction_n) + o(n)$ for all $n \in \mathbb{N}^+$, we see that

$$tn + \left(\underline{\dim}_A\{\alpha\} - t - \frac{o(n)}{n}\right) n \leq \underline{\dim}_A\{\alpha\}n - o(n) \leq H(\alpha\restriction_n)$$

for all $n \in \mathbb{N}^+$. Thus, since $\underline{\dim}_A\{\alpha\} - t - o(n)/n > 0$ for all sufficiently large n, we see that α is weakly Chaitin t-random.

On the other hand, assume that α is weakly Chaitin t-random. Then we see that $t \leq \liminf_{n\to\infty} H(\alpha\restriction_n)/n = \underline{\dim}_A\{\alpha\}$. Thus, if $t > \underline{\dim}_A\{\alpha\}$ then α is not weakly Chaitin t-random. This completes the proof. □

For example, using Lemma 3, the condition (iii) in Theorem 8 is converted as follows. In this paper, we interpret the supremum $\sup \emptyset$ of the empty set as 0.

Theorem 9. *Let α be an r.e. real. Then, for every $t \in (0, 1]$, α is $\Omega(t)$-like if $t < \underline{\dim}_A\{\alpha\}$, and α is not $\Omega(t)$-like if $t > \underline{\dim}_A\{\alpha\}$. Thus,*

$$\underline{\dim}_A\{\alpha\} = \sup\{\, t \in (0, 1] \mid \alpha \textit{ is } \Omega(t)\textit{-like} \,\}.$$ □

On the other hand, the condition (viii) in Theorem 8 is converted as follows, using Lemma 3. Here $\mathbb{R}_c$ denotes the set of all computable reals.

Theorem 10. *Let α be an r.e. real with $0 < \alpha < 1$, Then, for every $t \in (0, 1] \cap \mathbb{R}_c$, if $t < \underline{\dim}_A\{\alpha\}$ then $\alpha = \sum_{s\in\{0,1\}^*} m(s)^{\frac{1}{t}}$ for some universal probability m, and if $t > \underline{\dim}_A\{\alpha\}$ then $\alpha \neq \sum_{s\in\{0,1\}^*} m(s)^{\frac{1}{t}}$ for any universal probability m. Thus, $\underline{\dim}_A\{\alpha\} = \sup S$, where S is the set of all $t \in (0, 1] \cap \mathbb{R}_c$ such that $\alpha = \sum_{s\in\{0,1\}^*} m(s)^{\frac{1}{t}}$ for some universal probability m.* □

In the same manner, using Lemma 3 we can convert each of the remaining eight conditions in Theorem 8 also into a characterization of the lower algorithmic dimension of an r.e. real. In a full version of this paper, we will describe the complete list of the ten characterizations of the lower algorithmic dimension obtained from Theorem 8.

6 Further Properties of T-Convergence

In this section, we investigate further properties of the notion of T-convergence. First, as one of the applications of Theorem 8, the following theorem can be obtained.

Theorem 11. *Suppose that T is computable. For every r.e. real α, if α is T-convergent, then α is T-compressible.*

Proof. Using (vii) $\Rightarrow$ (iv) of Theorem 8, we see that, for every r.e. T-convergent real α, $H(\alpha\restriction_n) \leq H(\Omega_U(T)\restriction_n) + O(1)$ for all $n \in \mathbb{N}^+$. It follows from Theorem 4 (i) that α is T-compressible for every r.e. T-convergent real α. □

In the case of $T < 1$, the converse of Theorem 11 does not hold, as seen in Theorem 12 below in a sharper form. Theorem 12 can be proved partly using (vii) $\Rightarrow$ (ix) of Theorem 8.

Theorem 12. *Suppose that T is computable and $T < 1$. Then there exists an r.e. real η such that (i) η is weakly Chaitin T-random and T-compressible, and (ii) η is not T-convergent.* □

Let T_1 and T_2 be arbitrary computable reals with $0 < T_1 < T_2 < 1$, and let V be an arbitrary optimal computer. By Theorem 4 (i) and Theorem 11, we see that the r.e. real $\Omega_V(T_2)$ is not T_1-convergent and therefore every computable, increasing sequence $\{a_n\}$ of rationals which converges to $\Omega_V(T_2)$ is not T_1-convergent. At this point, conversely, the following question arises naturally: Is there any computable, increasing sequence of rationals which converges to $\Omega_V(T_1)$ and which is not T_2-convergent ? We can answer this question affirmatively in the form of Theorem 13 below.

Theorem 13. *Let T_1 and T_2 be arbitrary computable reals with $0 < T_1 < T_2 < 1$. Then there exist an optimal computer V and a computable, increasing sequence $\{a_n\}$ of rationals such that (i) $\Omega_V(T_1) = \lim_{n\to\infty} a_n$, (ii) $\{a_n\}$ is T-convergent for every $T \in (T_2, \infty)$, and (iii) $\{a_n\}$ is not T-convergent for every $T \in (0, T_2]$.* □

7 Concluding Remarks

In this paper, we have generalized the equivalent characterizations of randomness of a recursively enumerable real over the notion of partial randomness, so that the generalized characterizations are all equivalent to the weak Chaitin T-randomness. As a stronger notion of partial randomness of a real α, Tadaki [18,19] introduced the notion of the Chaitin T-randomness of α, which is defined as the condition on α that $\lim_{n\to\infty} H(\alpha\restriction_n) - Tn = \infty$.[2] Thus, future work may aim at modifying our equivalent characterizations of partial randomness so that they become equivalent to the Chaitin T-randomness.

Acknowledgments. This work was supported by KAKENHI, Grant-in-Aid for Scientific Research (C) (20540134), by SCOPE from the Ministry of Internal Affairs and Communications of Japan, and by CREST from Japan Science and Technology Agency.

[2] The actual separation of the Chaitin T-randomness from the weak Chaitin T-randomness is done by Reimann and Stephan [11].

References

1. Calude, C.S., Hertling, P.H., Khoussainov, B., Wang, Y.: Recursively enumerable reals and Chaitin Ω numbers. Theoret. Comput. Sci. 255, 125–149 (2001)
2. Calude, C.S., Staiger, L., Terwijn, S.A.: On partial randomness. Annals of Pure and Applied Logic 138, 20–30 (2006)
3. Calude, C.S., Stay, M.A.: Natural halting probabilities, partial randomness, and zeta functions. Inform. and Comput. 204, 1718–1739 (2006)
4. Chaitin, G.J.: A theory of program size formally identical to information theory. J. Assoc. Comput. Mach. 22, 329–340 (1975)
5. Chaitin, G.J.: Algorithmic Information Theory. Cambridge University Press, Cambridge (1987)
6. Downey, R.G., Reimann, J.: Algorithmic randomness. Scholarpedia 2(10), 2574 (2007); revision #37278, `http://www.scholarpedia.org/article/Algorithmic_randomness`
7. Kučera, A., Slaman, T.A.: Randomness and recursive enumerability. SIAM J. Comput. 31(1), 199–211 (2001)
8. Lutz, J.H.: Gales and the constructive dimension of individual sequences. In: Welzl, E., Montanari, U., Rolim, J.D.P. (eds.) ICALP 2000. LNCS, vol. 1853, pp. 902–913. Springer, Heidelberg (2000)
9. Martin-Löf, P.: The definition of random sequences. Information and Control 9, 602–619 (1966)
10. Mayordomo, E.: A Kolmogorov complexity characterization of constructive Hausdorff dimension. Inform. Process. Lett. 84, 1–3 (2002)
11. Reimann, J., Stephan, F.: On hierarchies of randomness tests. In: Proceedings of the 9th Asian Logic Conference, Novosibirsk, Russia, August 16-19. World Scientific Publishing, Singapore (2005)
12. Ya, B., Ryabko: Coding of combinatorial sources and Hausdorff dimension. Soviet Math. Dokl. 30, 219–222 (1984)
13. Ya, B., Ryabko: Noiseless coding of combinatorial sources, Hausdorff dimension, and Kolmogorov complexity. Problems Inform. Transmission 22, 170–179 (1986)
14. Schnorr, C.-P.: Process complexity and effective random tests. J. Comput. System Sci. 7, 376–388 (1973)
15. Solovay, R.M.: Draft of a paper (or series of papers) on Chaitin's work ... done for the most part during the period of September–December 1974. IBM Thomas J. Watson Research Center, Yorktown Heights, New York, 215 p. (May 1975) (unpublished manuscript)
16. Staiger, L.: Kolmogorov complexity and Hausdorff dimension. Inform. and Comput. 103, 159–194 (1993)
17. Staiger, L.: A tight upper bound on Kolmogorov complexity and uniformly optimal prediction. Theory Comput. Systems 31, 215–229 (1998)
18. Tadaki, K.: Algorithmic information theory and fractal sets. In: Proceedings of 1999 Workshop on Information-Based Induction Sciences (IBIS 1999), Syuzenji, Shizuoka, Japan, August 26-27, pp. 105–110 (1999) (in Japanese)
19. Tadaki, K.: A generalization of Chaitin's halting probability Ω and halting self-similar sets. Hokkaido Math. J. 31, 219–253 (2002)
20. Tadaki, K.: An extension of Chaitin's halting probability Ω to a measurement operator in an infinite dimensional quantum system. Math. Log. Quart. 52, 419–438 (2006)

21. Tadaki, K.: A statistical mechanical interpretation of algorithmic information theory. In: Local Proceedings of Computability in Europe 2008 (CiE 2008), June 15-20, pp. 425–434. University of Athens, Greece (2008); Extended and Electronic Version, `http://arxiv.org/abs/0801.4194v1`
22. Tadaki, K.: Fixed point theorems on partial randomness. In: Artemov, S., Nerode, A. (eds.) LFCS 2009. LNCS, vol. 5407, pp. 422–440. Springer, Heidelberg (2008)
23. Weihrauch, K.: Computable Analysis. Springer, Berlin (2000)
24. Zvonkin, A.K., Levin, L.A.: The complexity of finite objects and the development of the concepts of information and randomness by means of the theory of algorithms. Russian Math. Surveys 25(6), 83–124 (1970)

Partial Solution and Entropy

Tadao Takaoka

Department of Computer Science, University of Canterbury
Christchurch, New Zealand
tad@cosc.canterbury.ac.nz

Abstract. If the given problem instance is partially solved, we want to minimize our effort to solve the problem using that information. In this paper we introduce the measure of entropy $H(S)$ for uncertainty in partially solved input data $S(X) = (X_1, ..., X_k)$, where X is the entire data set, and each X_i is already solved. We use the entropy measure to analyze three example problems, sorting, shortest paths and minimum spanning trees. For sorting X_i is an ascending run, and for shortest paths, X_i is an acyclic part in the given graph. For minimum spanning trees, X_i is interpreted as a partially obtained minimum spanning tree for a subgraph. The entropy measure, $H(S)$, is defined by regarding $p_i = |X_i|/|X|$ as a probability measure, that is, $H(S) = -n\Sigma_{i=1}^{k} p_i \log p_i$, where $n = \Sigma_{i=1}^{k}|X_i|$. Then we show that we can sort the input data $S(X)$ in $O(H(S))$ time, and solve the shortest path problem in $O(m + H(S))$ time where m is the number of edges of the graph. Finally we show that the minimum spanning tree is computed in $O(m + H(S))$ time.

Keywords: entropy, complexity, adaptive sort, minimal mergesort, ascending runs, shortest paths, nearly acyclic graphs, minimum spanning trees.

1 Introduction

The concept of entropy is successfully used in information and communication theory. In algorithm research, the idea is used explicitly or implicitly. In [7], entropy is explicitly used to navigate the computation of the knapsack problem. On the other hand, entropy is used implicitly to analyze the computing time of various adaptive sorting algorithms [10]. In this paper, we develop a more unified approach to the analysis of algorithms using the concept of entropy. We regard the entropy measure as the uncertainty of the input data of the given problem instance, that is, the computational difficulty of the given problem instance.

First let us describe the framework of amortized analysis. Let S_0, S_1, ..., S_N be the states of data such that S_0 is the initial state and S_N is the final state. The computation in this paper is to transform S_{i-1} to S_i at the i-th step for $i = 1, , , , , N$. The potential of state S is denoted by $\Phi(S)$, which describes some positive aspect of data. That is, increasing the potential will ease the computation at later steps. The actual time and amortized time for the i-th step are denoted by t_i and a_i. We use the words "time" and "cost" interchangeably. The amortized time is defined by the accounting equation

R. Královič and D. Niwiński (Eds.): MFCS 2009, LNCS 5734, pp. 700–711, 2009.

$$a_i = t_i - \Delta\Phi(S_i), \tag{1}$$

where $\Delta\Phi(S_i) = \Phi(S_i) - \Phi(S_{i-1})$. That is, the amortized time is the actual time minus the increase of potential at the i-th step. By summing up the equation over i, we have

$$\Sigma a_i = \Sigma t_i + \Phi(S_0) - \Phi(S_N), \text{ or } \Sigma t_i = \Sigma a_i - \Phi(S_0) + \Phi(S_N).$$

Let the state of data be given by a decomposition of a set X as $S(X) = (X_1, ..., X_k)$. Let $|X| = n$, $n_i = |X_i|$ and $p_i = n_i/n$. Note that $\sum p_i = 1$. We define the entropy of a decomposition of X, $H(S(X))$, abbreviated as $H(S)$, by

$$H(S) = -n \sum_{i=1}^{k} p_i \log p_i = \sum_{i=1}^{k} |X_i| \log(|X|/|X_i|) \tag{2}$$

Normally entropy is defined without the factor of n, the size of the data set. We include this to deal with a dynamic situation where the size of the data set changes. Logarithm is taken with base 2 unless otherwise specified. Since p_i $(i = 1, \cdots, k)$ can be regarded as a probability measure, we have

$$0 \leq H(S) \leq n \log k$$

and the maximum is obtained when $|X_i| = n/k$ $(i = 1, \cdots, k)$. We capture the computational process as a process of decreasing the entropy in the given data set X. We assume $H(S_0) \geq H(S_1) \geq ... \geq H(S_N)$. We use $-H(S)$ for the potential in equation (1). The accounting equation becomes $a_i = t_i - \Delta H(S_i)$, where $\Delta H(S_i) = H(S_{i-1}) - H(S_i)$. That is, the amortized time is the actual time minus the decrease of entropy at the i-th step. The entropy is regarded as a negative aspect of the data, i.e., the less entropy, the closer to the solution.

Let T and A be the actual total time and the amortized total time. Summing up a_i for $i = 1, ..., N$, we have $A = T + H(S_N) - H(S_0)$, or $T = A + H(S_0) - H(S_N)$. We call this process of summing up amortized times over the computational steps "summing-up". In the following we see three applications, where A can be easily obtained. In many applications, $H(S_N) = 0$, meaning that the total time is A plus the initial entropy. We also have a reasonable assumption that $-\sum_{i=1}^{k} p_i \log p_i > 0$ for the initial state, meaning $H(S_0) = \Omega(n)$.

Let $S'(X) = (X'_1, ..., X'_{k'})$ be a refinement of $S(X) = (X_1, ..., X_k)$, that is, $S'(X)$ is a decomposition of X and for any X'_i there is X_j such that $X'_i \subseteq X_j$. Then we have $H(S) \leq H(S')$. As the entropy is a measure of uncertainty, we can say $S(X)$ is more solved than $S'(X)$.

The concept of amortized cost and actual cost is a relative one. That is, the actual time itself may be formulated as amortized time and actual time at a lower level of computation. Specifically, the actual time t_i in the accounting equation can be like $t_{ij} - (\Psi_{i,j-1} - \Psi_{i,j})$, where Ψ is another potential associated with the lower level computation. In such a case, summing-up takes place over indices i and j. Thanks to the linear property of the accounting equation, we can analyze amortized time on the upper level and lower level computation separately. We will see an example of a two-level amortized analysis in Section 5.

In later sections, we show three interpretations of X_i's. In sorting, X_i's are ascending runs which are regarded as solved. In shortest paths, X_i's are acyclic parts of the given graph, which can be processed without the effort of finding the minimum in the priority queue. In the minimum spanning tree problem, X_i is the set of vertices of a partially solved minimum spanning tree for a subgraph.

The main point of the paper is to offer a new method for algorithm analysis rather than designing new algorithms.

2 Application to Adaptive Sort

Adaptive sorting is to sort the list of n numbers into increasing order as efficiently as possible by utilizing the structure of the list which reflects some presortedness. See Estivill-Castro and Wood [3] for a general survey on adaptive sorting. There are many measures of disorder or presortedness. The simplest one is the number of ascending runs in the list. Let the given list $X = (a_1, a_2, \cdots, a_n)$ be divided into k ascending runs X_i $(i = 1, \cdots, k)$, that is, $S(X) = (X_1, X_2, \cdots, X_k)$ where $X_i = (a_1^{(i)}, \cdots, a_{n_i}^{(i)})$ and $a_1^{(i)}$ is the $|X_1| + \cdots + |X_{i-1}| + 1$-th element in X. We denote the length of list X by $|X|$. $S(X)$ is abbreviated as S. Note that $a_1^{(i)} \leq \cdots \leq a_{n_i}^{(i)}$ for each X_i and $a_{n_i}^{(i)} > a_1^{(i+1)}$ if X_i is not the last list. The sort algorithm called natural merge sort [5] sorts X by merging two adjacent lists for each phase halving the number of ascending runs after each phase so that sorting is completed in $O(n \log k)$ time. Mannila [6] proved that this method is optimal under the measure of the number of ascending runs.

In this paper we generalize the measure $RUNS(S)$ of the number of ascending runs into that of the entropy of ascending runs in X, denoted by $H(S)$. Then we analyze a sorting algorithm, called minimal merge sort, that sorts X by merging two minimal length runs successively until we have the sorted list. We show that the time for this algorithm is $O(H(S))$ and is optimal under the measure of $H(S)$. Hence the measure $H(S)$ derived from runs-entropy is sharper than $RUNS$ measure of $O(n \log k)$.

The idea of merging two shortest runs may be known. The algorithm style based on "meta-sort" in the next section is due to [10].

3 Minimal Mergesort

All lists are maintained in linked list structures in this section. Let $S(X) = (X_1, \cdots, X_k)$ be the given input list such that each X_i is sorted in ascending order. Re-arrange X into $S'(X) = (X_{i_1}, \cdots, X_{i_k})$ in such a way that $|X_{i_j}| \leq |X_{i_{j+1}}|$ $(j = 1, \cdots, k-1)$, that is, $(X_1, \cdots, X_k)$ is sorted with $|X_i|$ as key. We call this "meta-sort." Since each $|X_{i_j}|$ is an integer up to n, we can obtain $S'(X)$ in $O(n)$ time by radix sort. Now we sort $S'(X)$ by merging two shortest lists repeatedly. Formally we have the following. Let M and L be lists of lists, whereas W_i $(i = 1, 2)$ and W are ordinary lists. By the operation $M \Leftarrow L$, the leftmost list in L is moved to the rightmost part of M. By the operation $W_i \Leftarrow M$ $(i = 1, 2)$

the leftmost list of M is moved to W_i. By the operation $M \Leftarrow W$, W is moved to the rightmost part of M. $First(L)$ is the first list in L.

Algorithm 1 (Minimal mergesort)

1 Meta-sort $S(X)$ into $S'(X)$ by length of X_i;
2 Let $L = S'(X)$;
3 $M := \emptyset$;
4 $M \Leftarrow L$;
5 **if** $L \neq \emptyset$ **then** $M \Leftarrow L$;
6 **for** $i := 1$ **to** $k - 1$ **do begin**
7 $W_1 \Leftarrow M$;
8 $W_2 \Leftarrow M$;
9 $W :=$ merge(W_1, W_2);
10 **while** $L \neq \emptyset$ **and** $|W| > |\text{first}(L)|$ **do** $M \Leftarrow L$;
11 $M \Leftarrow W$
12 **end**
$\{W$ is the sorted list$\}$.

Lemma 1. *If W_2 is not an original X_i for any i at line 9, it holds that $|W_2| \leq \frac{2}{3}|W|$.*

Proof. Suppose to the contrary that $|W_2| > 2|W_1|$. Then for the previously merged lists V_1 and V_2, that is, $W_2 =$ merge(V_1, V_2), we have $|V_1| > |W_1|$ or $|V_2| > |W_1|$. Thus V_2 or V_1 must have been merged with W_1 or a shorter list, a contradiction. ■

We measure the computing time by the number of key comparisons in the merge operation at line 9, where the straight-forward merging is done with $|W_1|+|W_2|-1$ key comparisons.

Lemma 2. *We slightly modify the definition of amortized time; it is the actual time minus constant times decrease of entropy. Then the amortized time for the i-th merge is not greater than zero.*

Proof. Let $n_i = |X_i|$ for $i = 1, ..., k$. In particular, $|W_1| = n_1$ and $|W_2| = n_2$. The change of entropy occurs only with n_1 and n_2. Thus, noting $n_1 \leq n_2 \leq 2n_1$, the decrease of entropy is

$$\begin{aligned}\Delta H &= n_1 \log(n/n_1) + n_2 \log(n/n_2) - (n_1 + n_2)\log(n/(n_1 + n_2)) \\ &= n_1 \log(1 + n_2/n_1) + n_2 \log(1 + n_1/n_2) \\ &\geq n_1 \log 2 + n_2 \log(3/2) \geq \log(3/2)(n_1 + n_2)\end{aligned}$$

Thus

$$a_i = n_1 + n_2 - 1 - \Delta H / \log(3/2) \leq 0$$

■

Theorem 1. *The algorithm minimal mergesort sorts $S(X) = (X_1, \cdots, X_k)$ where each X_i is an ascending sequence in $O(H(S))$ time.*

Proof. Theorem follows from Lemma 2 and the initial entropy is given by $H(S)$. ■

Example. Let $|X_1| = 2$, $|X_i| = 2^{i-1}$ $(i = 2, \cdots, k-1)$ and $n = 2^k$. Then minimal mergesort sorts $S(X)$ in $O(n)$ time, since $H(S) = O(n)$, whereas natural mergesort takes $O(n \log \log n)$ time to sort $S(X)$.

Lemma 3. *Any sorting algorithm takes at least $\Omega(H(S))$ time when the entropy of ascending runs in $S(X)$ is $H(S)$ and $|X_i| \geq 2$ for $i = 1, \cdots, k$.*

Proof. Sorting $S(X)$ into $S'(X) = (a'_1, \cdots, a'_n)$ where $a'_1 \leq \cdots \leq a'_n$ means that $S'(X)$ is a permutation of $S(X)$. To establish a lower bound, we can assume that all elements in X are different. Let $X_i = (a_1^{(i)}, \cdots, a_{n_i}^{(i)})$. Let $a_{n_i}^{(i)}$ $(i = 1, \cdots, k)$ be fixed to be the i-th largest element in X. Then there are $\binom{n-k}{n_1-1}$ possibilities of X_1 being scattered in X'. Since the constraint of $a_{n_1}^{(1)} > a_1^{(2)}$ is satisfied by the choice of $a_{n_1}^{(1)}$, we have $\binom{n-k-n_1+1}{n_2-1}$ possibilities of X_2 being scattered in X'. Repeating this calculation yields the number of possibilities N as

$$
\begin{aligned}
N = & \frac{(n-k)!}{(n_1-1)!\,(n-k-n_1+1)!} \times \\
& \frac{(n-k-n_1+1)!}{(n_2-1)!\,(n-k-n_1-n_2+2)!} \times \\
& \cdots \frac{(n_{k-1}-1)!}{(n_k-1)!\,0!} \\
= & \frac{n!}{n_1! \cdots n_k!} \cdot \frac{n_1 \cdots n_k}{n(n-1) \cdots (n-k+1)}.
\end{aligned}
$$

Since the number of possible permutations is not fewer than this, we have the lower bound T on the computing time based on the binary decision tree model approximated by $T = \log N$. In the following we use natural logarithm for notational convenience. The result should be multiplied by $\log_2 e$. We use the following integral approximation.

$$n \log n - n + 1 \leq \textstyle\sum_{j=1}^{n} \log j \leq n \log n - n + \log n$$

T is evaluated by using the first inequality for n and the second for n_i,

$$
\begin{aligned}
T = \log N &\geq \log n! - \textstyle\sum_{i=1}^{k} \log n_i! + \sum_{i=1}^{k} (\log n_i - \log(n-i+1)) \\
&= \textstyle\sum_{i=1}^{k} n_i \log \frac{n}{n_i} - k \log n + 1
\end{aligned}
$$

Since $\sum n_i \log \frac{n}{n_i}$ is minimum when $n_1 = \cdots = n_{k-1} = 2$ and $n_k = n - 2(k-1)$,

$$
\begin{aligned}
& 2T - H(S) \\
& \geq \sum n_i \log \frac{n}{n_i} - 2k \log n + 2 \\
& \geq 2(k-1) \log \frac{n}{2} + (n-2k+2) \log \frac{n}{n-2k+2} - 2k \log n + 2 \\
& = (n-2k) \log \frac{n}{n-2k+2} - 2 \log(n-2k+2) + 4 \\
& \geq -2 \log(n-2k+2),
\end{aligned}
$$

since $1 \leq k \leq n/2$. On the other hand we can show $T \geq \log(n-2k+2)$. Thus we have $T \geq H(S)/4 = \Omega(H(S))$. ∎

If n_i=1 for some i, $a_{n_i}^{(i)}$ and $a_1^{(i+1)}$ form a part of a descending sequence. By reversing the descending sequences, we can guarantee that the sequence is decomposed into ascending runs of length al least 2. Let us extend minimal mergesort with this extra scanning in linear time, and define the entropy on the modified sequence. From this extension and the above lemma we see that minimal mergesort is asymptotically optimal for any sequence under the entropy measure. We can define entropy by decomposing the given sequence in non-consecutive portions. Minimal mergesort is not optimal under the entropy measure defined in this way. There are more entropy measures defined in [10].

4 Application to Shortest Paths for Nearly Acyclic Graphs

Let $G = (V, E)$ be a directed graph where V is the set of vertices with $|V| = n$ and E is the set of edges with $|E| = m$. The non-negative cost of edge (v_i, v_j) is denoted by $c(v_i, v_j)$. Let $OUT(v)$ (also $IN(v)$) be the list of edges from (to) v expressed by the set of the other end points of edges from (to) v. A brief description of Dijkstra's algorithm follows. Let S be the solution set, to which shortest distances have been established by the algorithm. The vertices in $V - S$ have tentative distances that are those of the shortest paths that go through S except for the end points. We take a vertex in $V - S$ that has the minimum distance, finalize it, and update the distances to other vertices in $V - S$ using edge list $OUT(v)$. If we organize Q by a Fibonacci heap or 2-3 heap [9], we can show the single source shortest path problem can be solved in $O(m + n \log n)$ time. We call this algorithm with one of those priority queues the standard single source algorithm. We assume the graph is connected from the source. Note that we use the same symbol S for the state of data and the solution set, hoping this is not a source of confusion.

We give the following well known algorithm [11] and its correctness for acyclic graphs for the sake of completeness. See [11] for the proof. It runs in $O(m)$ time, that is, we do not need an operation of finding the minimum in the priority queue.

ALGORITHM 2 $\{G = (V, E)$ is an acyclic graph.$\}$

1 Topologically sort V and assume without loss of generality); $V = \{v_1, \cdots, v_n\}$
   where $(v_i, v_j) \in E \Leftrightarrow i < j$;
2 $d[v_1] := 0$; $\{v_1$ is the source$\}$
3 **for** $i := 2$ **to** n **do** $d[v_i] := \infty$;
4 **for** $i := 1$ **to** n **do**
5   **for** v_j such that $(v_i, v_j) \in E$ **do**
6         $d[v_j] := \min\{d[v_j], d[v_i] + c(v_i, v_j)\}$.

Lemma 4. *At the beginning of Line 5 in Algorithm 2, the shortest distances from v_1 to v_j $(j < i)$ are computed. Also at the beginning of line 5, distances computed in $d[v_j]$ $(j \geq i)$ are those of shortest paths that lie in $\{v_1, \cdots, v_{i-1}\}$*

except for v_j. Thus at the end shortest distances $d[v_i]$ are computed correctly for all $i (1 \leq i \leq n)$.

Abuaiadh and Kingston [1] gave a result by restricting the given graph to being nearly acyclic. When they solve the single source problem, they distinguish between two kinds of vertices in $V - S$. One is the set of vertices, "easy" ones, to which there are no edges from $V - S$, e.g., only edges from S. The other is the set of vertices, "difficult" ones, to which there are edges from $V - S$. To expand S, if there are easy vertices, those are included in S and distances to other vertices in $V - S$ are updated. If there are no easy vertices, the vertex with minimum tentative distance is chosen to be included in S. If the number of such delete-minimum operations is t, the authors show that the single source problem can be solved in $O(m + n \log t)$ time with use of a Fibonacci heap. That is, the second term of the complexity is improved from $n \log n$ to $n \log t$. If the graph is acyclic, $t = 1$ and we have $O(m+n)$ time. Since we have $O(m + n \log n)$ when $t = n$, the result is an improvement of Fredman and Tarjan with use of the new parameter t. The authors claim that if the given graph is nearly acyclic, t is expected to be small and thus we can have a speed up.

The definition of near acyclicity and the estimate of t under it is not clear, however. We will show that the second term can be bounded by the entropy derived from a structural property of the given graph.

ALGORITHM 3 {*Single source shortest paths with v_0 being the source*} *[1], [8]*

1 **for** $v \in V$ **do if** $v = v_0$ **then** $d[v] := 0$ **else** $d[v] := \infty$;
2 Organize V in a priority queue Q with $d[v]$ as key;
3 $S := \emptyset$;
4 **while** $S \neq V$ **do begin**
5   **if** there is a vertex v in $V - S$ with no incoming edge from $V - S$ **then**
6   Choose v
7   **else**
8   Choose v from $V - S$ such that $d[v]$ is minimum;
9   Delete v from Q;
10    $S := S \cup \{v\}$;
11   **for** $w \in OUT(v) \cap (V - S)$ **do** $d[w] := min\{d[w], d[v] + c(v, w)\}$
12 **end**.

It is shown in [1] that a sequence of n delete, m decrease-key and t find-min operations is processed in $O(m + n \log t)$ time, meaning that the single source shortest path problem can be solved in the same amount of time.

We use the 2-3 heap for priority queue Q with the additional operation of delete. Let $v_1, , , , , v_k$ be deleted between two consecutive find-min operations such that v_1 is found at a find-min operation at line 8, and v_k is found at line 6 immediately before the next find-min. Each induced subgraph from them forms an acyclic graph, and they are topologically sorted in the order in which vertices are chosen at line 6. Thus they can be deleted from the heap without the effort of find-min operations. Let V_1, ..., V_t be the sets of vertices such that V_i is the acyclic set chosen following the i-th find-min operation and just before the next find-min. We call this set the i-th acyclic set. Note that the source is chosen by the first find-min. Then $S(V) = (V_1, ..., V_t)$ forms a decomposition of the set V. We denote the entropy of this decomposition by $H(S)$. Lemma 6 in the next section shows that t find-min operations with $|V_1|+...+|V_t|$ deletes interleaved in Algorithm 3 (each i-th find-min followed by $|V_i|$ deletes) can be done in $O(H(S))$ time. Let m_s be the number of edges examined at line 11 between the s-th find-min operation and the $(s+1)$-th find-min operation. Between these operations, $O(m_s)$ amortized time is spent at line 11. The total time for line 11 becomes $O(m)$.

When $t = 1$, the whole graph is acyclic, and we can solve the single source problem in $O(m)$ time by Lemma 4. The time for building Q at line 2 is absorbed in $O(m)$. Thus we have the following theorem.

Theorem 2. *Algorithm 3 solves the single source shortest path problem in $O(m+H(S))$ time.*

5 Analysis of Delete Operations

We maintain the priority queue for the single source shortest path problem by a 2-3 heap [9]. In traditional priority queues, decrease-key, insert and delete-min operations are defined. We define a delete operation on a 2-3 heap. When we delete node v, we remove the subtree rooted at v similarly to decrease-key on v, entailing a reshape of the work space. After destroying v, we merge the subtrees of v at the root level. The amortized time for a delete is proportional to the number of children, which is $O(\log n_v)$, where n_v is the number of descendants of node v to be deleted. A delete is defined on a Fibonacci heap in [1].

Let us delete nodes $v_j (j = 1, ..., k)$ in the batch mode from a 2-3 heap of size n. In the batch mode, we disconnect all children of all v_j and merge them at the root level. In other words, we do not process v_j one by one. Assume the number of descendants of v_j is n_j. The total amortized time T of deleting $v_1, ..., v_k$ in the batch mode is $T = O(\log n_1 + ... + \log n_k)$. Noting that $n_1 + ... + n_k \leq cn$ for some constant c, T is maximized as $T = O(k \log(n/k))$ when $n_1 = ... = n_k = cn/k$. Thus

Lemma 5. *k consecutive delete operations on a 2-3 heap of size n can be done in $O(k \log(n/k))$ time.*

Now we perform t batches of delete operations. Assume the i-th batch has k_i delete operations. Let the time for the i-th batch of delete operations be denoted by T_i. Since $T_i = O(k_i \log(n/k_i))$ by Lemma 5, we have the total time for all deletes bounded within a constant factor by

$k_1 \log(n/k_1) + ... + k_t \log(n/k_t) = n(\Sigma_{i=1}^{t}(k_i/n)\log(k_i/n)) = n(-\Sigma_{i=1}^{t} p_i \log p_i)$, where $p_i = k_i/n$. We define $H(S) = -n\Sigma_{i=1}^{t} p_i \log p_i$. Let us perform those t batches of delete operations after t find-min operations; each batch after each find-min. One find-min operation can be done in $O(\log n)$ time. Thus the total time becomes $O(t \log n + H(S))$, which is further simplified to $O(H(S))$ by the following lemma.

Lemma 6. *For $t \geq 2$, $t \log n \leq O(H(S))$. Thus the time for heap operations described above is bounded by $O(H(S))$.*

Proof. $H(S)$ is minimum when $k_1 = ... = k_{t-1} = 1$, and $k_t = n - t + 1$. Thus $2H(S) \geq 2(t-1)\log n + 2(n-t+1)\log(n/(n-t+1)) \geq t \log n$ ■

Remark. In terms of amortized analysis in Section 1, we can define a_i and t_i in the following way. Let $V^{(s)} = V_s \cup ... \cup V_t$. We define the state of data set, S_s, to be the data set $V^{(s)}$ decomposed as above. The initial state $S = S_1$ is given by $V_1 \cup ... \cup V_t$. The entropy of the state of data is defined for the beginning of the s-th find-min at line 8 using $V^{(s)}$ by

$$H(S_s) = \Sigma_{i=s}^{t} |V_i| \log(|V^{(s)}|/|V_i|)$$

Noting that $|V^{(s)}| \geq |V^{(s+1)}|$, the decrease of entropy at the beginning of the next find-min operation is

$$\begin{aligned} \Delta H(S_{s+1}) &= \Sigma_{i=s}^{t} |V_i| \log(|V^{(s)}|/|V_i|) - \Sigma_{i=s+1}^{t} |V_i| \log(|V^{(s+1)}|/|V_i|) \\ &\geq |V_s| \log(|V^{(s)}|/|V_s|) \end{aligned}$$

We define the actual time and amortized time to be those for the stage from the s-th find-min to the $(s+1)$-th. The actual time t_s for stage s is given by

$$t_s = O(m_s + |V_s| \log(|V^{(s)}|/|V_s|) + \log n).$$

This is because we inspect $O(m_s)$ edges, perform $O(V_s)$ deletes in the heap of size $|V^{(s)}|$, and spend $O(\log n)$ time for a find-min. We interpret the above formula as $t_s \leq c_s(m_s + |V_s| \log(|V^{(s)}|/|V_s|) + \log n)$ with some constant $c_s > 0$.

The amortized time for the s-th stage is slightly modified with constant c_s, and given as

$$\begin{aligned} a_s &= t_s - c_s \Delta H(S_{s+1}) \\ &\leq c_s(m_s + |V_s| \log(|V^{(s)}|/|V_s|) + \log n) - c_s |V_s| \log(|V^{(s)}|/|V_s|) \\ &\leq c_s(m_s + \log n) \end{aligned}$$

Using constant $c = max\{c_s\}$ and noting $H(S_t) = 0$, we have

$$T \leq A + c(H(S_1) - H(S_t)) \leq O(m + t \log n + H(S_1)) = O(m + H(S))$$

Note that when we perform summing-up over a_s, we also perform summing-up over actual and amortized times for operations on the 2-3 heap. In this sense, the above is a "two-level" amortized analysis.

Remark. In [1], $O(t \log n + H(S))$ is bounded by $O(n \log t)$. Thus our analysis of $O(t \log n + H(S)) \leq O(H(S))$ is sharper.

6 Relationship with 1-Dominator

As a definition of near-acyclicity, the definition and algorithm for a 1-dominator decomposition is given in [8]. The decomposition is given by the set of disjoint sets, called 1-dominator sets, whose union is V. A 1-dominator set dominated by a trigger v, A_v, is the maximal set of vertices w such that any path from outside A_v to w must go through v, and the subgraph induced by A_v is an acyclic graph. Let $IN(v) = \{u|(u,v) \in E\}$. A_v is formally defined by the maximal set satisfying the following formula for any w.

The induced graph from A_v is acyclic, $v \in A_v$ and
$(w \in A_v)\&(w \neq v) \rightarrow (IN(w) \neq \phi)\&(IN(w) \subseteq A_v)$
In [8] it is shown V is uniquely decomposed into several A_v's, and the time for this decomposition is $O(m)$. The 1-dominator decomposition is used for identifying the set of the triggers, R. Only triggers are maintained in the heap. Once the distance to a trigger is finalized, the distances to members of the 1-dominator set are finalized through Algorithm 2 in time proportional to the number of edges in the set. At the border of the set. the distances to other triggers are updated. The time for the single source problem becomes $O(m + r \log r)$, where r is the number of triggers, that is, $r = |R|$.

We show that the entropy $H(S)$ in Section 4 is bounded by the entropy defined by the 1-dominator decomposition.

Theorem 3. *The decomposition by 1-dominator sets is a refinement of the decomposition defined by Algorithm 3.*

Proof. Suppose a vertex v is obtained by find-min at line 8 and v is not a trigger. Then v is inside some 1-dominator set. Since the distance to the corresponding trigger is smaller, the trigger must be included in the solution set earlier, and v must have subsequently been deleted from the heap, a contradiction. Thus v is a trigger. Then the 1-dominator set is subsequently deleted from the heap, and possibly more 1-dominator sets. Thus the 1-dominator decomposition is a refinemment of the decomposition $S(V)$. ∎

The decomposition by Algorithm 3 is dynamically defined, i.e., it cannot be defined statically before the algorithm starts. On the other hand, the algorithm based on the 1-dominator decomposition is more predictable as the preprocessing can reveal the 1-dominator decomposition and its entropy, which bounds the entropy defined by Algorithm 3.

In [8], the single source algorithm is given in a slightly different way. It maintains only triggers in the heap, and the distances between triggers are given by those of pseudo edges, which are defined between triggers through the intervening acyclic part and obtained in $O(m)$ time. In other words, the standard single source algorithm runs on this reduced graph. Thus the time becomes $O(m+r \log r)$, which may be better than $O(m+H(S))$ of Algorithm 3. However Algorithm 3 and the single source algorithm in [8] are not incompatible. We can run Algorithm 3 on the reduced graph obtained through the 1-dominator decomposition. Then the time will become $O(m + H(S'))$, where $H(S')$ is the entropy defined by the algorithm run on the reduced graph.

7 Remaining Work for the MST Problem

Let $G = (V, E)$ be an undirected graph with edge cost function $c(u, v)$ for the edge (u, v). Let Kruskal's algorithm continue to work for the minimum (cost) spanning tree (MST) problem after the problem has been solved partially by the same algorithm. We estimate how much more time is needed to complete the work by using the concept of entropy. Let $G_1 = (V_1, E_1)$, ..., $G_k = (V_k, E_k)$ be subgraphs of G such that V_1, ..., V_k form a decomposition of V and G_i is the induced sub-graph from V_i. We assume the MST problem has been solved for G_i with spanning trees T_i for $i = 1, ..., k$. The state of data $S(V)$ is defined by $S(V) = (V_1, ..., V_k)$, and the entropy of the state is defined by (2) where X_i is interpreted as V_i.

The remaining work is to keep merging two trees by connecting them by the best possible edge. We use array *name* to keep track of names of trees to which vertices belong. If the two end points of an edge have different names, it connects distinct trees successfully. Otherwise it would form a cycle, not desirable situation, resulting in skipping the edge. The following algorithm completes the work from line 4.

ALGORITHM 4 {*To complete the MST problem*}

1 Let the sorted edge list L has been partially scanned
2 Minimum spanning trees for G_1, ..., G_k have been obtained
3 Let $name[v] = i$ for $v \in V_i$ have been set for $i = 1, ..., k$
4 **while** $k > 1$ **do begin**
5   Remove the first edge (u, v) from L
6   **if** u and v belong to different subtrees T_1 and T_2 (without loss of generality)
7   **then begin**
8   Connect T_1 and T_2 by (u, v);
9   Change the names of the nodes in the smaller tree to that of the larger tree;
10       $k := k - 1$;
11   **end**
12 **end.**

The analysis is similar to the proof of Lemma 2. We first analyze the time for name changes at line 9. Let t_i and a_i be the actual time and amortized time for the i-th operation that merges T_1 and T_2, where $|T_1| = n_1$ and $|T_2| = n_2$ and V_1 and V_2 are the sets of vertices corresponding to those spanning sub-trees. We measure the time by the number of name changes. Let the state of data after the i-th merge be S_i. The change of entropy occurs only with n_1 and n_2. Thus the decrease of entropy is

$$\Delta H(S_i) = n_1 \log(n/n_1) + n_2 \log(n/n_2) - (n_1 + n_2) \log(n/(n_1 + n_2))$$
$$= n_1 \log(1 + n_2/n_1) + n_2 \log(1 + n_1/n_2) \geq min\{n_1, n_2\}$$

Noting that $t_i = min\{n_1, n_2\}$, amortized time becomes

$$a_i = t_i - \Delta H(S_i) \leq 0$$

The rest of work is bounded by $O(m)$. Thus the total time for the remaining work becomes $O(m + H(S))$, where $H(S)$ is the initial entropy at the beginning of line 4. Note that the condition for $a_i \leq 0$ is crucial. In the analysis of minimal merge sort, it is satisfied by the fact that the two shortest ascending runs are merged, whereas in the MST problem in this section, it is satisfied by merging the smaller tree to the larger tree.

8 Concluding Remarks

We captured computation as a process of reducing entropy, starting from some positive value and ending in zero. The amortized time for a step of the computation is the sum of the actual time minus the reduction of entropy. If the analysis of a single amortized time is easier than the analysis of the total actual time, this method by entropy will be useful for analysis. We showed that three specific problems of sorting, shortest paths and minimum spanning trees can be analyzed by this unified entropy analysis.

If the computation process is a merging process of two sets in the decomposition, our method may be used. The definition of entropy and actual time needs care depending on the specifics of each problem. It remains to be seen if more difficult problems can be analyzed by this method.

Acknowledgment. The author gratefully acknowledges many constructive comments given by the reviewers. This work was partially done at Kansai University.

References

1. Abuaiadh, D., Kingston, J.H.: Are Fibonacci heaps optimal? In: Du, D.-Z., Zhang, X.-S. (eds.) ISAAC 1994. LNCS, vol. 834, pp. 442–450. Springer, Heidelberg (1994)
2. Dijkstra, E.W.: A note on two problems in connection with graphs. Numer. Math. 1, 269–271 (1959)
3. Estivill-Castro, V., Wood, D.: A survey of adaptive sorting algorithms. ACM Computing Surveys 24, 441–476 (1992)
4. Fredman, M.L., Tarjan, R.E.: Fibonacci heaps and their use in improved network optimization problems. J. ACM 34(3), 596–615 (1987)
5. Knuth, D.E.: The Art of Computer Programming. Sorting and Searching, vol. 3. Addison-Wesley, Reading (1974)
6. Mannila, H.: Measures of presortedness and optimal sorting algorithms. IEEE Trans. Comput. C-34, 318–325 (1985)
7. Nakagawa, Y.: A Difficulty Estimation Method for Multidimensional Nonlinear 0-1 Knapsack Problem Using Entropy. Transactions of the Institute of Electronics, Communication and Information J87-A(3), 406–408 (2004)
8. Saunders, S., Takaoka, T.: Solving shortest paths efficiently on nearly acyclic directed graphs. Theoretical Computer Science 370(1-3), 94–109 (2007)
9. Takaoka, T.: Theory of 2-3 Heaps. Discrete Applied Math. 126, 115–128 (2003)
10. Takaoka, T.: Entropy – Measure of Disorder. In: Proc. CATS (Computation: Australasian Theory Symposium), pp. 77–85 (1998)
11. Tarjan, R.E.: Data Structures and Network Algorithms, Regional Conference Series in Applied math. 44 (1983)

On Pebble Automata for Data Languages with Decidable Emptiness Problem

Tony Tan

Department of Computer Science, Technion – Israel Institute of Technology
School of Informatics, University of Edinburgh
tantony@cs.technion.ac.il

Abstract. In this paper we study a subclass of pebble automata (PA) for data languages for which the emptiness problem is decidable. Namely, we show that the emptiness problem for weak 2-pebble automata is decidable, while the same problem for weak 3-pebble automata is undecidable. We also introduce the so-called *top view* weak PA. Roughly speaking, top view weak PA are weak PA where the equality test is performed only between the data values seen by the two most recently placed pebbles. The emptiness problem for this model is still decidable.

1 Introduction

Logic and automata for words over finite alphabets are relatively well understood and recently there is broad research activity on logic and automata for words and trees over infinite alphabets. Partly, the study of infinite alphabets is motivated by the need for formal verification and synthesis of infinite-state systems and partly, by the search for automated reasoning techniques for XML. Recently, there has been a significant progress in this field, see [1,2,4,5,8,10].

Roughly speaking, there are two approaches to studying data languages: logic and automata. Below is a brief survey on both approaches. For a more comprehensive survey, we refer the reader to [10]. The study of data languages, which can also be viewed as languages over infinite alphabets, starts with the introduction of finite-memory automata (FMA) in [5], which are also known as *register automata* (RA). The study of RA was continued and extended in [8], in which *pebble automata* (PA) were also introduced. Each of both models has its own advantages and disadvantages. Languages accepted by FMA are closed under standard language operations: intersection, union, concatenation, and Kleene star. In addition, from the computational point of view, FMA are a much easier model to handle. Their emptiness problem is decidable, whereas the same problem for PA is not. However, the PA languages possess a very nice logical property: closure under *all* boolean operations, whereas FMA languages are not closed under complementation.

Later in [2] first-order logic for data languages was considered, and, in particular, the so-called *data automata* was introduced. It was shown that data automata define the fragment of existential monadic second order logic for data

R. Královič and D. Niwiński (Eds.): MFCS 2009, LNCS 5734, pp. 712–723, 2009.

languages in which the first order part is restricted to two variables only. An important feature of data automata is that their emptiness problem is decidable, even for the infinite words, but is at least as hard as reachability for Petri nets. The automata themselves always work nondeterministically and seemingly cannot be determinized, see [1]. It was also shown that the satisfiability problem for the three-variable first order logic is undecidable.

Another logical approach is via the so called *linear temporal logic with n register freeze quantifier over the labels Σ*, denoted $\mathrm{LTL}_n^{\downarrow}(\Sigma, \mathtt{X}, \mathtt{U})$, see [4]. It was shown that one way alternating n register automata accept all $\mathrm{LTL}_n^{\downarrow}(\Sigma, \mathtt{X}, \mathtt{U})$ languages and the emptiness problem for one way alternating one register automata is decidable. Hence, the satisfiability problem for $\mathrm{LTL}_1^{\downarrow}(\Sigma, \mathtt{X}, \mathtt{U})$ is decidable as well. Adding one more register or past time operators to $\mathrm{LTL}_1^{\downarrow}(\Sigma, \mathtt{X}, \mathtt{U})$ makes the satisfiability problem undecidable.

In this paper we continue the study of PA, which are finite state automata with a finite number of pebbles. The pebbles are placed on/lifted from the input word in the stack discipline – first in last out – and are intended to mark positions in the input word. One pebble can only mark one position and the most recently placed pebble serves as the head of the automaton. The automaton moves from one state to another depending on the current label and the equality tests among data values in the positions currently marked by the pebbles, as well as, the equality tests among the positions of the pebbles.

Furthermore, as defined in [8], there are two types of PA, according to the position of the new pebble placed. In the first type, the ordinary PA, also called *strong* PA, the new pebbles are placed at the beginning of the string. In the second type, called *weak* PA, the new pebbles are placed at the position of the most recent pebble. Obviously, two-way weak PA is just as expressive as two-way ordinary PA. However, it is known that one-way nondeterministic weak PA are weaker than one-way ordinary PA, see [8, Theorem 4.5.].

In this paper we show that the emptiness problem for one-way weak 2-pebble automata is decidable, while the same problem for one-way weak 3-pebble automata is undecidable. We also introduce the so-called *top view* weak PA. Roughly speaking, top view weak PA are one-way weak PA where the equality test is performed only between the data values seen by the two most recently placed pebbles. Top view weak PA are quite robust: alternating, nondeterministic and deterministic top view weak PA have the same recognition power. To the best of our knowledge, this is the first model of computation for data language with such robustness. It is also shown that top view weak PA can be simulated by one-way alternating one-register RA. Therefore, their emptiness problem is decidable. Another interesting feature is top view weak PA can simulate all $\mathrm{LTL}_1^{\downarrow}(\Sigma, \mathtt{X}, \mathtt{U})$ languages.

This paper is organized as follows. In Section 2 we review the models of computations for data languages considered in this paper. Section 3 and Section 4 deals with the decidability and the complexity issues of weak PA, respectively. In Section 5 we introduce top view weak PA. Finally, we end our paper with a brief observation in Section 6.

2 Model of Computations

In Subsection 2.1 we recall the definition of weak PA from [8]. We will use the following notation. We always denote by Σ a finite alphabet of *labels* and by $\mathfrak{D}$ an infinite set of *data values.* A *Σ-data word* $w = \binom{\sigma_1}{a_1}\binom{\sigma_2}{a_2}\cdots\binom{\sigma_n}{a_n}$ is a finite sequence over $\Sigma \times \mathfrak{D}$, where $\sigma_i \in \Sigma$ and $a_i \in \mathfrak{D}$. A *Σ-data language* is a set of Σ-data words.

We assume that neither of Σ and $\mathfrak{D}$ contain the left-end marker $\triangleleft$ or the right-end marker $\triangleright$. The input word to the automaton is of the form $\triangleleft w \triangleright$, where $\triangleleft$ and $\triangleright$ mark the left-end and the right-end of the input word. Finally, the symbols $\nu, \vartheta, \sigma, \ldots$, possibly indexed, denote labels in Σ and the symbols $a, b, c, d, \ldots$, possibly indexed, denote data values in $\mathfrak{D}$.

2.1 Pebble Automata

Definition 1. *(See [8, Definition 2.3]) A* one-way alternating weak k-pebble automaton *(k-PA) over Σ is a system $\mathcal{A} = \langle \Sigma, Q, q_0, F, \mu, U\rangle$ whose components are defined as follows.*

- *Q, $q_0 \in Q$ and $F \subseteq Q$ are a finite set of* states, *the* initial state, *and the set of* final states, *respectively;*
- *$U \subseteq Q - F$ is the set of* universal *states; and*
- *$\mu \subseteq \mathcal{C} \times \mathcal{D}$ is the transition relation, where*
  - *$\mathcal{C}$ is a set whose elements are of the form (i, σ, V, q) where $1 \le i \le k$, $\sigma \in \Sigma$, $V \subseteq \{i+1, \ldots, k\}$ and $q \in Q$; and*
  - *$\mathcal{D}$ is a set whose elements are of the form $(q, \mathtt{act})$, where $q \in Q$ and $\mathtt{act}$ is either $\mathtt{stay}$, $\mathtt{right}$, $\mathtt{place\text{-}pebble}$ or $\mathtt{lift\text{-}pebble}$.*

  Elements of μ will be written as $(i, \sigma, V, q) \to (p, \mathtt{act})$.

Given a word $w = \binom{\sigma_1}{a_1}\cdots\binom{\sigma_n}{a_n} \in (\Sigma \times \mathfrak{D})^*$, a *configuration of $\mathcal{A}$ on $\triangleleft w \triangleright$* is a triple $[i, q, \theta]$, where $i \in \{1, \ldots, k\}$, $q \in Q$, and $\theta : \{i, i+1, \ldots, k\} \to \{0, 1, \ldots, n, n+1\}$, where 0 and $n+1$ are positions of the end markers $\triangleleft$ and $\triangleright$, respectively. The function θ defines the position of the pebbles and is called the *pebble assignment.* The *initial* configuration is $\gamma_0 = [k, q_0, \theta_0]$, where $\theta_0(k) = 0$ is the *initial* pebble assignment. A configuration $[i, q, \theta]$ with $q \in F$ is called an *accepting* configuration.

A transition $(i, \sigma, V, p) \to \beta$ *applies to a configuration* $[j, q, \theta]$, if

(1) $i = j$ and $p = q$,
(2) $V = \{l > i : a_{\theta(l)} = a_{\theta(i)}\}$, and
(3) $\sigma_{\theta(i)} = \sigma$.

Note that in a configuration $[i, q, \theta]$, pebble i is in control, serving as the head pebble.

Next, we define the transition relation $\vdash_{\mathcal{A}}$ as follows: $[i, q, \theta] \vdash_{\mathcal{A}} [i', q', \theta']$, if there is a transition $\alpha \to (p, \mathtt{act}) \in \mu$ that applies to $[i, q, \theta]$ such that $q' = p$, $\theta'(j) = \theta(j)$, for all $j > i$, and

- if $\texttt{act} = \texttt{stay}$, then $i' = i$ and $\theta'(i) = \theta(i)$;
- if $\texttt{act} = \texttt{right}$, then $i' = i$ and $\theta'(i) = \theta(i) + 1$;
- if $\texttt{act} = \texttt{lift-pebble}$, then $i' = i + 1$;
- if $\texttt{act} = \texttt{place-pebble}$, then $i' = i - 1$, $\theta'(i-1) = \theta'(i) = \theta(i)$.

As usual, we denote the reflexive transitive closure of $\vdash_{\mathcal{A}}$ by $\vdash^*_{\mathcal{A}}$. When the automaton $\mathcal{A}$ is clear from the context, we will omit the subscript $\mathcal{A}$.

Remark 1. Note the pebble numbering that differs from that in [8]. In the above definition we adopt the pebble numbering from [3] in which the pebbles placed on the input word are numbered from k to i and not from 1 to i as in [8]. The reason for this reverse numbering is that it allows us to view the computation between placing and lifting pebble i as a computation of an $(i-1)$-pebble automaton.

Furthermore, the automaton is no longer equipped with the ability to compare positional equality, in contrast with the ordinary PA introduced in [8]. Such ability no longer makes any difference because of the "weak" manner in which the new pebbles are placed.

The acceptance criteria is based on the notion of *leads to acceptance* below. For every configuration $\gamma = [i, q, \theta]$,

– if $q \in F$, then γ leads to acceptance;
– if $q \in U$, then γ leads to acceptance if and only if for all configurations γ' such that $\gamma \vdash \gamma'$, γ' leads to acceptance;
– if $q \notin F \cup U$, then γ leads to acceptance if and only if there is at least one configuration γ' such that $\gamma \vdash \gamma'$ and γ' leads to acceptance.

A Σ-data word $w \in (\Sigma \times \mathfrak{D})^*$ is accepted by $\mathcal{A}$, if γ_0 leads to acceptance. The language $L(\mathcal{A})$ consists of all data words accepted by $\mathcal{A}$.

The automaton $\mathcal{A}$ is *nondeterministic*, if the set $U = \emptyset$, and it is *deterministic*, if there is exactly one transition that applies for each configuration. It turns out that weak PA languages are quite robust.

Theorem 1. *For all $k \geq 1$, alternating, non-deterministic and deterministic weak k-PA have the same recognition power.*

The proof is quite standard and the details will appear in the journal version of this paper. In view of this, we will always assume that our weak k-PA is deterministic.

We end this subsection with an example of language accepted by weak 2-PA. This example will be useful in the subsequent section.

Example 1. Consider a Σ-data language $L_\sim$ defined as follows. A Σ-data word $w = \binom{\sigma_1}{a_1} \cdots \binom{\sigma_n}{a_n} \in L_\sim$ if and only if for all $i, j = 1, \ldots, n$, if $a_i = a_j$, then $\sigma_i = \sigma_j$. That is, $w \in L_\sim$ if and only if whenever two positions in w carry the same data value, their labels are the same.

The language $L_\sim$ is accepted by weak 2-PA which works in the following manner. Pebbles 2 iterates through all possible positions in w. At each iteration, the automaton stores the label seen by pebble 2 in the state and places pebble 1. Then, pebble 1 scans through all the positions to the right of pebble 2, checking whether there is a position with the same data value of pebble 2. If there is such a position, then the labels seen by pebble 1 must be the same as the label seen by pebble 2, which has been previously stored in the state.

3 Decidability and Undecidability of Weak PA

In this section we will discuss the decidability issue of weak PA. We show that the emptiness problem for weak 3-PA is undecidable, while the same problem for weak 2-PA is decidable. The proof of the decidability of the emptiness problem for weak 2-PA will be the basis of the proof of the decidability of the same problem for top view weak PA.

Theorem 2. *The emptiness problem for weak 3-PA is undecidable.*

The proof of Theorem 2 is similar to the proof of the undecidability of the emptiness problem for weak 5-PA in [8]. The main technical step in the proof is to show that the following Σ-data language

$$L_{\mathrm{ord}} = \left\{ \binom{\sigma}{a_1} \cdots \binom{\sigma}{a_n} \binom{\$}{d} \binom{\sigma}{a_1} \cdots \binom{\sigma}{a_n} : a_1, \ldots, a_n \text{ are pairwise different} \right\}$$

is accepted by weak 5-PA, where $\Sigma = \{\sigma, \$\}$. We observe that weak 3-PA is sufficient. From this step, the undecidability can be easily obtained via a reduction from the Post Correspondence Problem (PCP).

Now we are going to show that the emptiness problem for weak 2-PA is decidable. The proof is by simulating weak 2-PA by one-way alternating one register automata (1-RA). See [4] for the definition of alternating RA.

In fact, the simulation can be easily generalized to arbitrary number of pebbles. That is, weak k-PA can be simulated by one-way alternating $(k-1)$-RA. (Here $(k-1)$-RA stands for $(k-1)$-register automata.) This result settles a question left open in [8]: Can weak PA be simulated by alternating RA?

Theorem 3. *For every weak k-PA $\mathcal{A}$, there exists a one-way alternating $(k-1)$-RA $\mathcal{A}'$ such that $L(\mathcal{A}) = L(\mathcal{A}')$. Moreover, the construction of $\mathcal{A}'$ from $\mathcal{A}$ is effective.*

Now, by Theorem 3, we immediately obtain the decidability of the emptiness problem for weak 2-PA because the same problem for one-way alternating 1-RA is decidable [4, Theorem 4.4].

Corollary 1. *The emptiness problem for weak 2-PA is decidable.*

We devote the rest of this section to the proof of Theorem 3 for $k = 2$. Its generalization to arbitrary $k \geq 3$ is pretty straightforward.

Recall that we always assume that weak PA is deterministic. Let $\mathcal{A} = \langle \Sigma, Q, q_0, \mu, F \rangle$ be a weak 2-PA. We normalize the behavior of $\mathcal{A}$ as follows.

- Pebble 1 is lifted only after it reads the right-end marker $\triangleright$.
- Only pebble 2 can enter a final state and it does so after it reads the right-end marker $\triangleright$.
- Immediately after pebble 2 moves right, pebble 1 is placed.
- Immediately after pebble 1 is lifted, pebble 2 moves right.

Such normalization can be obtained in a pretty standard manner. The details will appear in the journal version of this paper.

On input word $w = \binom{\sigma_1}{d_1} \cdots \binom{\sigma_n}{d_n}$, the run of $\mathcal{A}$ on $\triangleleft w \triangleright$ can be depicted as a tree shown in Figure 1.

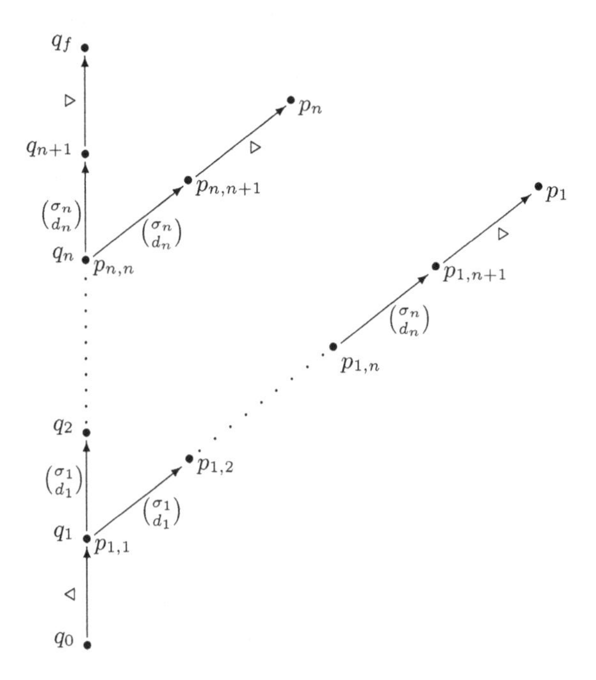

Fig. 1. The tree representation of a run of $\mathcal{A}$ on the data word $w = \binom{\sigma_1}{d_1} \cdots \binom{\sigma_n}{d_n}$

The meaning of the tree is as follows.

- $q_0, q_1, \ldots, q_n, q_{n+1}$ are the states of $\mathcal{A}$ when pebble 2 is the head pebble reading the positions $0, 1, \ldots, n, n+1$, respectively, that is, the symbols $\triangleleft, \binom{\sigma_1}{d_1}, \ldots, \binom{\sigma_n}{d_n}, \triangleright$, respectively.
- q_f is the state of $\mathcal{A}$ after pebble 2 reads the symbol $\triangleright$.
- For $1 \leq i \leq j \leq n$, $p_{i,j}$ is the state of $\mathcal{A}$ when pebble 1 is the head pebble reading the position j, while pebble 2 is above the position i.
- For $1 \leq i \leq n$, the state p_i is the state of $\mathcal{A}$ immediately after pebble 1 is lifted and pebble 2 is above the position i.
  It must be noted that there is a transition $(2, \sigma_i, \emptyset, p_i) \rightarrow (q_{i+1}, \texttt{right})$ applied by $\mathcal{A}$ that is not depicted in the figure.

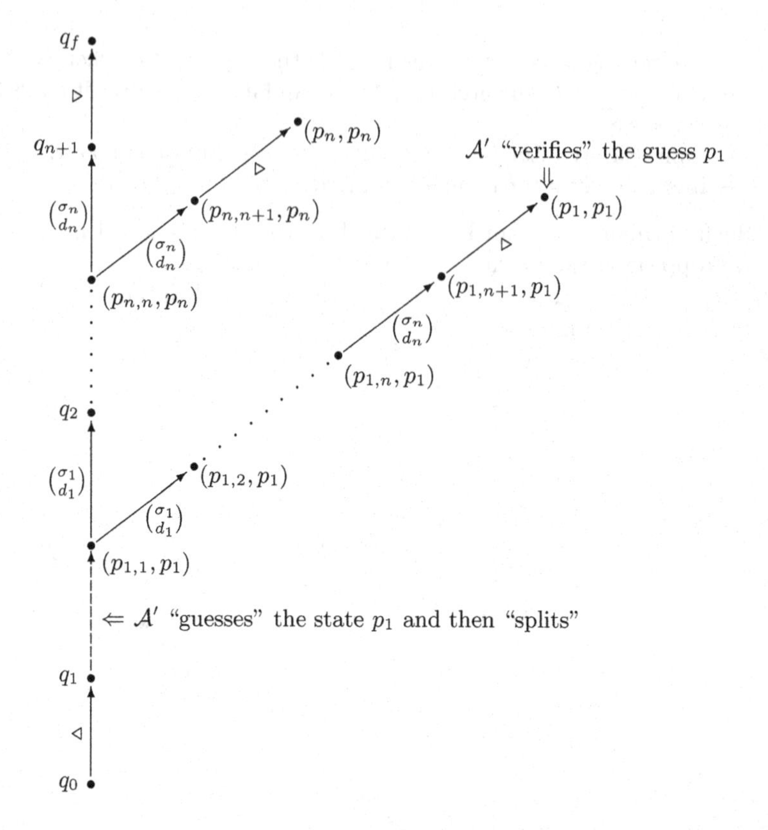

Fig. 2. The corresponding run of $\mathcal{A}'$ the data word $w = \binom{\sigma_1}{d_1} \cdots \binom{\sigma_n}{d_n}$ to the one in Figure 1

Now the simulation of $\mathcal{A}$ by a one-way alternating 1-RA $\mathcal{A}'$ becomes straightforward if we view the tree in Figure 1 as a tree depicting the computation of $\mathcal{A}'$ on the same word w. Figure 2 shows the corresponding run of $\mathcal{A}'$ on the same word. Roughly, the automaton $\mathcal{A}'$ is defined as follows.

- The states of $\mathcal{A}'$ are elements of $Q \cup (Q \times Q)$;
- the initial state is q_0; and
- the set of final states is $F \cup \{(p,p) : p \in Q\}$.

For each placement of pebble 1 on position i, the automaton performs the following "Guess–Split–Verify" procedure which consists of the following steps.

1. From the state q_i, $\mathcal{A}'$ "guesses" (by disjunctive branching) the state in which pebble 1 is eventually lifted, i.e. the state p_i. It stores p_i in its internal state and simulates the transition $(2, \sigma_i, \emptyset, \emptyset, q_i) \to (p_{i,i}, \texttt{place-pebble}) \in \mu$ to enter into the state $(p_{i,i}, p_i)$.
2. $\mathcal{A}'$ "splits" its computation (by conjunctive branching) into two branches.
   - In one branch, assuming that the guess p_i is correct, $\mathcal{A}'$ moves right and enters into the state q_{i+1}, simulating the transition $(2, \emptyset, p_i) \to$

$(q_{i+1}, \mathtt{right})$. After this, it recursively performs the Guess–Split–Verify procedure for the next placement of pebble 1 on position $(i+1)$.
- In the other branch $\mathcal{A}'$ stores the data value d_i in its register and simulates the run of pebble 1 on $\binom{\sigma_i}{d_i} \cdots \binom{\sigma_n}{d_n}$, starting from the state $p_{i,i}$, to "verify" that the guess p_i is correct.
  During the simulation, to remember the guess p_i, the states of $\mathcal{A}'$ are $(p_{i,i}, p_i), \ldots, (p_{i,n+1}, p_i)$. $\mathcal{A}'$ accepts when the simulation ends in the state (p_i, p_i), that is, when the guess p_i is "correct."

4 Complexity of Weak 2-PA

In this subsection we are going to study the time complexity of three specific problems related to weak 2-PA.

Emptiness problem. The emptiness problem for weak 2-PA. That is, given a weak 2-PA $\mathcal{A}$, is $L(\mathcal{A}) = \emptyset$?
Labelling problem. Given a deterministic weak 2-PA $\mathcal{A}$ over the labels Σ and a sequence of data values $d_1 \cdots d_n \in \mathfrak{D}^n$, is there a sequence of labels $\sigma_1 \cdots \sigma_n \in \Sigma^n$ such that $\binom{\sigma_1}{d_1} \cdots \binom{\sigma_n}{d_n} \in L(\mathcal{A})$?
Data value membership problem. Given a deterministic weak 2-PA $\mathcal{A}$ over the labels Σ and a sequence of finite labels $\sigma_1 \cdots \sigma_n \in \Sigma^n$, is there a sequence of data values $d_1 \cdots d_n \in \mathfrak{D}^n$ such that $\binom{\sigma_1}{d_1} \cdots \binom{\sigma_n}{d_n} \in L(\mathcal{A})$?

The emptiness problem, as we have seen in the previous section, is decidable. The labelling and data value membership problem are in NP. To solve the labelling problem, one simply guesses a sequence $\sigma_1 \cdots \sigma_n \in \Sigma^n$ and runs $\mathcal{A}$ to check whether $\binom{\sigma_1}{d_1} \cdots \binom{\sigma_n}{d_n} \in L(\mathcal{A})$. Similarly, to solve the data value membership problem, one can guess a sequence of data values $d_1 \cdots d_n$ and run $\mathcal{A}$ to check whether $\binom{\sigma_1}{d_1} \cdots \binom{\sigma_n}{d_n} \in L(\mathcal{A})$.

We will show that the emptiness problem is not primitive recursive, while both the labelling and data value membership problems are NP-complete.

Theorem 4. *The emptiness problem for weak 2-PA is not primitive recursive.*

The proof of theorem 4 is by simulation of incrementing counter automata and follows closely the proof of similar lower bound for one-way alternating 1-RA [4, Theorem 2.9]. In short, the main technical step is to show that the following Σ-data language L_{inc} which consists of the data words of the form: $\binom{\alpha}{a_1} \cdots \binom{\alpha}{a_m} \binom{\beta}{b_1} \cdots \binom{\beta}{b_n}$; where

- $\Sigma = \{\alpha, \beta\}$;
- the data values $a_1, \ldots, a_m$ are pairwise different;
- the data values $b_1, \ldots, b_n$ are pairwise different;
- each a_i appears among $b_1, \ldots, b_n$;

is accepted by weak 2-PA. The intuition of this language is to represent the inequality $m \leq n$, which is important in the simulation of incrementing counter automata, of which the emptiness problem is known to be decidable [9], but not primitive recursice [7].

Now we are going to show the NP-hardness of the labelling problem. It is by a reduction from graph 3-colorability problem which states as follows. Given an undirected graph $G = (V, E)$, is G is 3-colorable?

Let $V = \{1, \ldots, n\}$ and $E = \{(i_1, j_1), \ldots, (i_m, j_m)\}$. Assuming that $\mathfrak{D}$ contains the natural numbers, we take $i_1 j_1 \cdots i_m j_m$ as the sequence of data values. Then, we construct a weak 2-PA $\mathcal{A}$ over the alphabet $\Sigma = \{\vartheta_R, \vartheta_G, \vartheta_B\}$ that accepts data words of even length in which the following hold.

- For all odd position x, the label on position x is different from the label on position $x + 1$.
- For every two positions x and y, if they have the same data value, then they have the same label. (See Example 1.)

Thus, the graph G is 3-colorable if and only if there exists $\sigma_1 \cdots \sigma_{2m} \in \{\vartheta_R, \vartheta_G, \vartheta_B\}^*$ such that $\binom{\sigma_1}{i_1}\binom{\sigma_2}{j_1} \cdots \binom{\sigma_{2m-1}}{i_m}\binom{\sigma_{2m}}{j_m} \in L(\mathcal{A})$, and the NP-hardness, hence the NP-completeness, of the labelling problem follows.

The NP-hardness of data value membership problem can established in a similar spirit. The reduction is from the following variant of graph 3-colorability, called 3-colorability with constraint. Given a graph $G = (V, E)$ and three integers n_r, n_g, n_b, is the graph G 3-colorable with the colors R, G and B such that the numbers of vertices colored with R, G and B are n_r, n_g and n_b, respectively?

The polynomial time reduction to data value membership problem is as follows. Let $V = \{1, \ldots, n\}$ and $E = \{(i_1, j_1), \ldots, (i_m, j_m)\}$.

We define $\Sigma = \{\vartheta_R, \vartheta_G, \vartheta_B, \nu_1, \ldots, \nu_n\}$ and take

$$\nu_{i_1}\nu_{j_1} \cdots \nu_{i_m}\nu_{j_m} \underbrace{\vartheta_R \cdots \vartheta_R}_{n_r \text{ times}} \underbrace{\vartheta_G \cdots \vartheta_G}_{n_g \text{ times}} \underbrace{\vartheta_B \cdots \vartheta_B}_{n_b \text{ times}}$$

as the sequence of finite labels. Then, we construct a weak 2-PA over Σ that accepts data words of the form

$$\binom{\nu_{i_1}}{c_1}\binom{\nu_{j_1}}{d_1} \cdots \binom{\nu_{i_m}}{c_m}\binom{\nu_{j_m}}{d_m}\binom{\vartheta_R}{a_1} \cdots \binom{\vartheta_R}{a_{n_r}}\binom{\vartheta_G}{a'_1} \cdots \binom{\vartheta_G}{a'_{n_g}}\binom{\vartheta_B}{a''_1} \cdots \binom{\vartheta_B}{a''_{n_b}},$$

where

- $\nu_{i_1}, \nu_{j_1}, \ldots, \nu_{i_m}, \nu_{j_m} \in \{\nu_1, \ldots, \nu_n\}$;
- in the sub-word $\binom{\nu_{i_1}}{c_1}\binom{\nu_{j_1}}{d_1} \cdots \binom{\nu_{i_m}}{c_m}\binom{\nu_{j_m}}{d_m}$, every two positions with the same labels have the same data value (see Example 1);
- the data values $a_1, \ldots, a_{n_r}, a'_1, \ldots, a'_{n_g}, a''_1, \ldots, a''_{n_b}$ are pairwise different;
- For each $i = 1, \ldots, m$, the data values c_i, d_i appear among $a_1, \ldots, a_{n_r}$, $a'_1, \ldots, a'_{n_g}$, $a''_1, \ldots, a''_{n_b}$ such that the following holds:
  - if c_i appears among $a_1, \ldots, a_{n_r}$, then d_i appears among $a'_1, \ldots, a'_{n_g}$ or $a''_1, \ldots, a''_{n_b}$;
  - if c_i appears among $a'_1, \ldots, a'_{n_g}$, then d_i appears either among $a_1, \ldots, a_{n_r}$ or $a''_1, \ldots, a''_{n_b}$; and

- if c_i appears among $a''_1, \ldots, a''_{n_b}$, then d_i appears among $a_1, \ldots, a_{n_r}$ or $a'_1, \ldots, a'_{n_g}$.

Note that we can store the integers r, g, b and m in the internal states of $\mathcal{A}$, thus, enable $\mathcal{A}$ to "count" up to n_r, n_g, n_b and m. We have each state for the numbers $1, \ldots, n_r$, $1, \ldots, n_g$, $1, \ldots, n_b$ and $1, \ldots, m$. The number of states in $\mathcal{A}$ is still polynomial on n.

Now the graph G is 3-colorable with the constraints n_r, n_g, n_b if and only if there exits $c_1 d_1 \cdots c_m d_m a_1 \cdots a_{n_r} a'_1 \cdots a'_{n_g} a''_1 \cdots a''_{n_b}$ such that

$$\binom{\nu_{i_1}}{c_1}\binom{\nu_{j_1}}{d_1}\cdots\binom{\nu_{i_m}}{c_m}\binom{\nu_{j_m}}{d_m}\binom{\vartheta_R}{a_1}\cdots\binom{\vartheta_R}{a_{n_r}}\binom{\vartheta_G}{a'_1}\cdots\binom{\vartheta_G}{a'_{n_g}}\binom{\vartheta_B}{a''_1}\cdots\binom{\vartheta_B}{a''_{n_b}}$$

is accepted by $\mathcal{A}$. The NP-hardness, hence the NP-completeness, of the data value membership problem then follows.

5 Top View Weak k-PA

In this section we are going to define *top view* weak PA. Roughly speaking, top view weak PA are weak PA where the equality test is performed only between the data values seen by the last and the second last placed pebbles. That is, if pebble i is the head pebble, then it can only compare the data value it reads with the data value read by pebble $(i+1)$. It is not allowed to compare its data value with those read by pebble $(i+2), (i+3), \ldots, k$.

Formally, top view weak k-PA is a tuple $\mathcal{A} = \langle \Sigma, Q, q_0, \mu, F \rangle$ where Q, q_0, F are as usual and μ consists of transitions of the form: $(i, \sigma, V, q) \rightarrow (q', \mathtt{act})$, where V is either $\emptyset$ or $\{i+1\}$.

The criteria for the application of transitions of top view weak k-PA is defined by setting

$$V = \begin{cases} \emptyset, & \text{if } a_{\theta(i+1)} \neq a_{\theta(i)} \\ \{i+1\}, & \text{if } a_{\theta(i+1)} = a_{\theta(i)} \end{cases}$$

in the definition of transition relation in Subsection 2.1. Note that top view weak 2-PA and weak 2-PA are the same.

Remark 2. We can also define the alternating version of top view weak k-PA. However, just like in the case of weak k-PA, alternating, nondeterministic and deterministic top view weak k-PA have the same recognition power. Furthermore, by using the same proof presented in Section 4, it is straightforward to show that the emptiness problem, the labelling problem, and the data value membership problem have the same complexity lower bound for top view weak k-PA, for each $k = 2, 3, \ldots$.

The following theorem is a stronger version of Theorem 3.

Theorem 5. *For every top view weak k-PA $\mathcal{A}$, there is a one-way alternating 1-RA $\mathcal{A}'$ such that $L(\mathcal{A}') = L(\mathcal{A})$. Moreover, the construction of $\mathcal{A}'$ is effective.*

Proof. The proof is a straightforward generalization of the proof of Theorem 3. Each placement of a pebble is simulated by "Guess–Split–Verify" procedure. Since each pebble i can only compare its data value with the one seen by pebble $(i+1)$, $\mathcal{A}'$ does not need to store the data values seen by pebbles $(i+2), \ldots, k$. It only needs to store the data value seen by pebble $(i+1)$, thus, one register is sufficient for the simulation.

Following Theorem 5, we immediately obtain the decidability of the emptiness problem for top view weak k-PA.

Corollary 2. *The emptiness problem for top view weak k-PA is decidable.*

Since the emptiness problem for ordinary 2-PA (See [6, Theorem 4]) and for weak 3-PA is already undecidable, it seems that top view weak PA is a tight boundary of a subclass of PA languages for which the emptiness problem is decidable.

Remark 3. In [11] it is shown that for every sentence $\psi \in \mathrm{LTL}_1^{\downarrow}(\Sigma, \mathtt{X}, \mathtt{U})$, there exists a weak k-PA $\mathcal{A}_\psi$, where $k = \mathsf{fqr}(\psi) + 1$, such that $L(\mathcal{A}_\psi) = L(\psi)$. We remark that the proof actually shows that the automaton $\mathcal{A}_\psi$ is top view weak k-PA. Thus, it shows that the class of top view weak k-PA languages contains the languages definable by $\mathrm{LTL}_1^{\downarrow}(\Sigma, \mathtt{X}, \mathtt{U})$.

6 Concluding Remark

We end this paper with a quick observation on top view weak PA. We note that the finiteness of the number of pebbles for top view weak PA is not necessary. In fact, we can just define top view weak PA with unbounded number of pebbles, which we call top view weak unbounded PA.

We elaborate on it in the following paragraphs. Let $\mathcal{A} = \langle \Sigma, Q, q_0, \mu, F \rangle$ be top view weak unbounded PA. The pebbles are numbered with the numbers $1, 2, 3, \ldots$. The automaton $\mathcal{A}$ starts the computation with only pebble 1 on the input word. The transitions are of the form: $(\sigma, \chi, q) \rightarrow (p, \mathtt{act})$, where $\chi \in \{0, 1\}$ and $\sigma, q, p, \mathtt{act}$ are as in the ordinary weak PA.

Let $w = \binom{\sigma_1}{a_1} \cdots \binom{\sigma_n}{a_n}$ be an input word. A *configuration of $\mathcal{A}$ on $\triangleleft w \triangleright$* is a triple $[i, q, \theta]$, where $i \in \mathbb{N}$, $q \in Q$, and $\theta : \mathbb{N} \rightarrow \{0, 1, \ldots, n, n+1\}$. The *initial* configuration is $[1, q_0, \theta_0]$, where $\theta_0(1) = 0$. The accepting configurations are defined similarly as in ordinary weak PA.

A transition $(\sigma, \chi, p) \rightarrow \beta$ *applies to a configuration* $[i, q, \theta]$, if

(1) $p = q$, and $\sigma_{\theta(i)} = \sigma$,
(2) $\chi = 1$ if $a_{\theta(i-1)} = a_{\theta(i)}$, and $\chi = 0$ if $a_{\theta(i-1)} \neq a_{\theta(i)}$,

The transition relation $\vdash$ and the acceptance criteria can be defined in a similar manner as in Section 2.1.

It is straightforward to show that 1-way deterministic 1-RA can be simulated by top view weak unbounded PA. Each time the register automaton change the content of the register, the top view weak unbounded PA places a new pebble.

Furthermore, top view weak unbounded PA can be simulated by 1-way alternating 1-RA. Each time a pebble is placed, the register automaton performs "Guess–Split–Verify" procedure described in Section 3. Thus, the emptiness problem for top view unbounded weak PA is still decidable.

Acknowledgment. The author would like to thank Michael Kaminski for his invaluable directions and guidance related to this paper and for pointing out the notion of unbounded pebble automata.

References

1. Björklund, H., Schwentick, T.: On Notions of Regularity for Data Languages. In: Csuhaj-Varjú, E., Ésik, Z. (eds.) FCT 2007. LNCS, vol. 4639, pp. 88–99. Springer, Heidelberg (2007)
2. Bojańczyk, M., Muscholl, A., Schwentick, T., Segoufin, L., David, C.: Two-Variable Logic on Words with Data. In: Proceedings of the 21th IEEE Symposium on Logic in Computer Science (LICS 2006), pp. 7–16 (2006)
3. Bojańczyk, M., Samuelides, M., Schwentick, T., Segoufin, L.: Expressive Power of Pebble Automata. In: Bugliesi, M., Preneel, B., Sassone, V., Wegener, I. (eds.) ICALP 2006. LNCS, vol. 4051, pp. 157–168. Springer, Heidelberg (2006)
4. Demri, S., Lazic, R.: LTL with the Freeze Quantifier and Register Automata. In: Proceedings of the 21th IEEE Symposium on Logic in Computer Science (LICS 2006), pp. 17–26 (2006)
5. Kaminski, M., Francez, N.: Finite-memory automata. Theoretical Computer Science 134, 329–363 (1994)
6. Kaminski, M., Tan, T.: A Note on Two-Pebble Automata over Infinite Alphabets. Technical Report CS-2009-02, Department of Computer Science, Technion – Israel Institute of Technology (2009)
7. Mayr, R.: Undecidable problems in unreliable computations. Theoretical Computer Science 297(1-3), 337–354 (2003)
8. Neven, F., Schwentick, T., Vianu, V.: Finite state machines for strings over infinite alphabets. ACM Transactions on Computational Logic 5(3), 403–435 (2004)
9. Schnoebelen, P.: Verifying lossy channel systems has nonprimitive recursive complexity. Information Processing Letters 83(5), 251–261 (2002)
10. Segoufin, L.: Automata and Logics for Words and Trees over an Infinite Alphabet. In: Ésik, Z. (ed.) CSL 2006. LNCS, vol. 4207, pp. 41–57. Springer, Heidelberg (2006)
11. Tan, T.: Graph Reachability and Pebble Automata over Infinite Alphabets. To appear in the Proceedings of the 24th IEEE Symposium on Logic in Computer Science (LICS 2009)

Size and Energy of Threshold Circuits Computing Mod Functions

Kei Uchizawa[1,*], Takao Nishizeki[1], and Eiji Takimoto[2,**]

[1] Graduate School of Information Sciences, Tohoku University, Aramaki Aoba-aza 6-6-05, Aoba-ku, Sendai, 980-8579, Japan

[2] Department of Informatics, Graduate School of Information Science and Electrical Engineering, Kyushu University, 744 Motooka, Nishi-ku, Fukuoka 819-0395, Japan

{uchizawa,nishi}@ecei.tohoku.ac.jp, eiji@i.kyushu-u.ac.jp

Abstract. Let C be a threshold logic circuit computing a Boolean function $\mathrm{MOD}_m : \{0,1\}^n \to \{0,1\}$, where $n \geq 1$ and $m \geq 2$. Then C outputs "0" if the number of "1"s in an input $\boldsymbol{x} \in \{0,1\}^n$ to C is a multiple of m and, otherwise, C outputs "1." The function MOD_2 is the so-called PARITY function, and MOD_{n+1} is the OR function. Let s be the size of the circuit C, that is, C consists of s threshold gates, and let e be the energy complexity of C, that is, at most e gates in C output "1" for any input $\boldsymbol{x} \in \{0,1\}^n$. In the paper, we prove that a very simple inequality $n/(m-1) \leq s^e$ holds for every circuit C computing MOD_m. The inequality implies that there is a tradeoff between the size s and energy complexity e of threshold circuits computing MOD_m, and yields a lower bound $e = \Omega((\log n - \log m)/\log\log n)$ on e if $s = O(\mathrm{polylog}(n))$. We actually obtain a general result on the so-called generalized mod function, from which the result on the ordinary mod function MOD_m immediately follows. Our results on threshold circuits can be extended to a more general class of circuits, called unate circuits.

1 Introduction

A circuit of threshold gates is a theoretical model of a neural circuit in the brain, and is well studied through decades [10,11,13,14]. An input-output characteristic of a biological neuron is roughly represented by a threshold gate, but the mechanism of energy consumption of a neuron is quite different from an electrical circuit: a neural "firing" consumes substantially more energy than a "non-firing" [8,9], while a gate in an electrical circuit consumes almost the same amount of energy in either case of outputting "1" and outputting "0" [1,7]. A biological study reports that, due to the asymmetricity of the energy consumption, the fraction of neurons firing concurrently is possibly fewer than 1% [8]. Based on the biological fact above, the energy complexity e of a threshold circuit C is defined as the maximum number of threshold gates outputting "1" over all inputs to C [16]. We then confront the following natural question from the

* Supported by MEXT Grant-in-Aid for Young Scientists (B) No.21700003.
** Supported by MEXT Grant-in-Aid for Scientific Research (C) No.20500001.

R. Královič and D. Niwiński (Eds.): MFCS 2009, LNCS 5734, pp. 724–735, 2009.

point of computational complexity: what Boolean functions can or cannot be computed by reasonably small threshold circuits with small energy complexity? It has been shown that the energy complexity strongly influences the computational power of threshold circuits [16,18]. In particular, if a Boolean function f has high communication complexity, there exists a tradeoff among the following three complexities: size (that is, the number of gates) s, depth d, and energy complexity e of threshold circuits computing f [18]. However, the mod function $\mathrm{MOD}_m : \{0,1\}^n \to \{0,1\}$ has low communication complexity, and hence the result in [18] does not yield any interesting tradeoff for MOD_m, where n and m are positive integers, and $\mathrm{MOD}_m(\boldsymbol{x})$ is 0 if the number of "1"s in an input $\boldsymbol{x} \in \{0,1\}^n$ is a multiple of m and, otherwise, $\mathrm{MOD}_m(\boldsymbol{x})$ is 1. MOD_m is the PARITY function if $m = 2$, and is the OR function if $m = n+1$.

In the paper, we deal with a fairly large class of Boolean functions, called the generalized mod function [2,3,5], and show that there exists a tradeoff between the size s and energy complexity e of threshold circuits C computing the generalized mod function. The result immediately yields a very simple tradeoff for the ordinary mod function MOD_m. More precisely, we prove that $n/(m-1) \le s^e$, that is, $\log(n/(m-1)) \le e \log s$, for every circuit C computing MOD_m. Both n and m, and hence $n/(m-1)$, do not depend on the design of C, while s^e is monotonically increasing with respect to s and e. Therefore, s and e cannot be simultaneously small. That is, if s is small, then e must be large, and if e is small, then s must be large. The tradeoff $n/(m-1) \le s^e$ immediately implies a lower bound on the size s expressed by n, m and e: $(n/(m-1))^{1/e} \le s$. If $s = O(\mathrm{polylog}(n))$, then the tradeoff also implies a lower bound on e: $e = \Omega((\log n - \log m)/\log\log n)$. The lower bound on e is tight up to a constant factor. Our results on threshold circuits can be extended to a more general class of circuits, called "unate circuits," as stated in Section 4.

It is well known that there exists a tradeoff between the size s and depth d of a threshold circuit computing the PARITY function. Siu *et al.* proved that $n \le (s/d)^{d+\epsilon}$ for any fixed $\epsilon > 0$ if the weights of the threshold gates are integers and their absolute values are sufficiently small [15]. Impagliazzo *et al.* proved that $n/2 \le s^{2(d-1)}$ even if the absolute values of weights are arbitrarily large [6]. Our tradeoff between s and e holds even if the absolute values of weights are arbitrarily large. It should be noted that the inequality $d \le e$ does not necessarily holds, and that if a Boolean function f can be computed by a polynomial-size threshold circuit C of energy complexity e then the function f can be computed by a polynomial-size threshold circuit C' of depth $d' = 2e + 1$ [17].

2 Preliminaries

For $\boldsymbol{x} = (x_1, x_2, \cdots, x_n) \in \{0,1\}^n$, we denote by $[\boldsymbol{x}]_m$ the Hamming weight of $\boldsymbol{x}$ modulo m and hence $[\boldsymbol{x}]_m = \sum_{i=1}^n x_i \pmod m$. Let $M = \{0, 1, \cdots, m-1\}$, then $m = |M|$. For a set $A \subseteq M$, the *generalized mod function* $\mathrm{MOD}_m^A : \{0,1\}^n \to \{0,1\}$ is defined as follows [2,3,5]:

$$\mathrm{MOD}_m^A(\boldsymbol{x}) = \begin{cases} 0 \text{ if } [\boldsymbol{x}]_m \in A; \\ 1 \text{ otherwise.} \end{cases} \tag{1}$$

Let $a = \min(|A|, |M - A|)$. We may assume that the generalized mod function MOD_m^A is not trivial, and hence

$$1 \le a \le \left\lfloor \frac{m}{2} \right\rfloor \tag{2}$$

and

$$2 \le m \le n + 1. \tag{3}$$

If $A = \{0\}$, then MOD_m^A is the ordinary mod function MOD_m, and

$$\mathrm{MOD}_m(\boldsymbol{x}) = \begin{cases} 0 \text{ if } [\boldsymbol{x}]_m = 0; \\ 1 \text{ otherwise.} \end{cases}$$

If $m = 2$ and $A = \{0\}$, then MOD_m^A is the so-called PARITY function. If $m = n + 1$ and $A = \{0\}$, then MOD_m^A is the OR function. If $m = n + 1$ and $A = \{0, 1, \cdots, \lfloor n/2 \rfloor\}$, then MOD_m^A is the MAJORITY function. Thus, the class of generalized mod functions MOD_m^A is fairly large.

In the paper, a *threshold gate* is the so-called linear threshold logic gate, and can have an arbitrary number k of inputs. For every input $\boldsymbol{z} = (z_1, z_2, \cdots, z_k) \in \{0,1\}^k$ to a threshold gate g with weights $w_1, w_2, \cdots, w_k$ and a threshold t, the output $g(\boldsymbol{z})$ of the gate g for $\boldsymbol{z}$ is defined as follows:

$$g(\boldsymbol{z}) = \begin{cases} 1 \text{ if } \sum_{i=1}^{k} w_i z_i \ge t; \\ 0 \text{ otherwise,} \end{cases} \tag{4}$$

where $w_1, w_2, \cdots, w_k$ and t are arbitrary real numbers.

A *threshold (logic) circuit* C is a combinatorial circuit of threshold gates, and is expressed by a directed acyclic graph as illustrated in Fig. 1. Let n be the

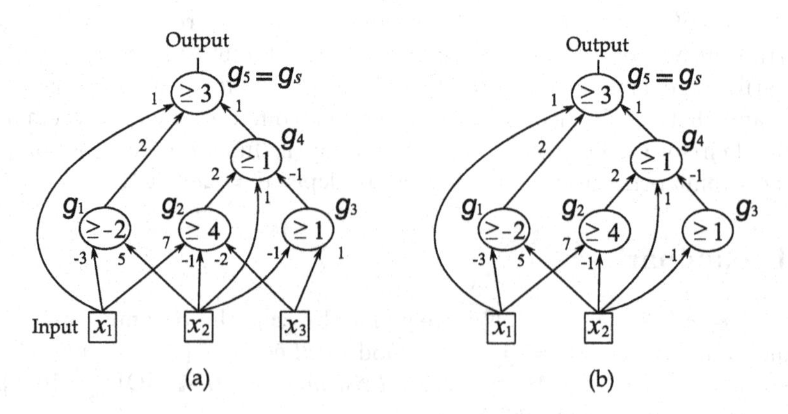

Fig. 1. (a) A threshold circuit C with $n = 3$ and $s = 5$; and (b) the 0-fixed circuit C_0 of C

Table 1. Various designs of circuits computing the PARITY function of n variables

Designs \ Complexities	s	e	d	Notes
Small d	$n+1$	$n+1$	2	Fig. 2(a)
Small e	$n+1$	2	$n+1$	Fig. 2(b)
Moderate s, e and d	$\log n$	$\log n$	$\log n$	Ref. [14]
Moderate s, e and fairly small d	polylog(n)	polylog(n)	$\log n/\log\log n$	Ref. [12]
Moderate s, d and fairly small e	polylog(n)	$\log n/\log\log n$	polylog(n)	Sect. 3.1
Small e and d	$2^{n-1}+1$	1	2	Truth table

number of input variables to C, then C has n input nodes of in-degree 0, each of which corresponds to one of the n input variables $x_1, x_2, \cdots, x_n$.

The *size* s of a threshold circuit C is the number of threshold gates in C. Figure 1(a) depicts a threshold circuit with $n = 3$ and $s = 5$, while Fig. 1(b) depicts a circuit with $n = 2$ and $s = 5$. (Impagliazzo *et al.* define the "size" of C to be the number of wires in C, and obtained a tradeoff between the "size" and the depth [6].)

Let C be a threshold circuit of size s, let $g_1, g_2, \cdots, g_s$ be the gates in C, and let $\boldsymbol{x} = (x_1, x_2, \cdots, x_n) \in \{0,1\}^n$ be an input to C. Then the input $\boldsymbol{z}_i$ to a gate g_i, $1 \le i \le s$, either consists of the inputs $x_1, x_2, \cdots, x_n$ to C and the outputs of the gates other than g_i or consists of some of them. However, we denote the output $g_i(\boldsymbol{z}_i)$ of g_i for $\boldsymbol{z}_i$ by $g_i[\boldsymbol{x}]$, because $\boldsymbol{x}$ decides $g_i(\boldsymbol{z}_i)$. Thus $g_i[\boldsymbol{x}] = g_i(\boldsymbol{z}_i)$. Let g_s be one of the gates of out-degree 0, and we regard the output $g_s[\boldsymbol{x}]$ of g_s as the *output* $C(\boldsymbol{x})$ *of* C. Thus, $C(\boldsymbol{x}) = g_s[\boldsymbol{x}]$ for every input $\boldsymbol{x} \in \{0,1\}^n$. The gate g_s is called the *output gate* of C.

A threshold circuit C *computes* a Boolean function $f : \{0,1\}^n \to \{0,1\}$ if $C(\boldsymbol{x}) = f(\boldsymbol{x})$ for every input $\boldsymbol{x} \in \{0,1\}^n$.

The *depth* d of a circuit C is the number of gates in the longest path from an input node to the output gate g_s, and corresponds to the parallel computation time.

We define the energy complexity e of a threshold circuit C as

$$e = \max_{\boldsymbol{x} \in \{0,1\}^n} \sum_{i=1}^{s} g_i[\boldsymbol{x}].$$

Thus, the energy complexity e is the maximum number of gates outputting "1" over all inputs $\boldsymbol{x} \in \{0,1\}^n$. Clearly $0 \le e \le s$. We may assume without loss of generality that $e \ge 1$.

As summarized in the Table 1, there are various designs of threshold circuits computing the PARITY function MOD_2. Figure 2 illustrates two of them, for which $n = 4$ and $s = n + 1 = 5$. For the circuit in Fig. 2(a) $d = 2$ and $e = n = 4$. On the other hand, for the circuit in Fig 2(b) $d = n + 1 = 5$ and $e = 2$; if the number i of "1"s in an input is odd, then only the two gates g_i and g_s output "1"; and otherwise only g_i outputs "1."

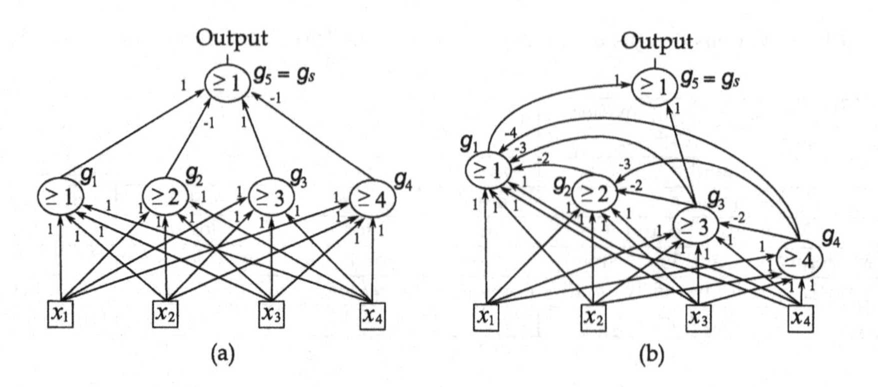

Fig. 2. Threshold circuits computing the PARITY function of $n = 4$ variables; (a) $s = 5$, $d = 2$, $e = 4$; and (b) $s = 5$, $d = 5$, $e = 2$

3 Size-Energy Tradeoff

In Section 3.1, we present, as Theorem 1, our main result on the size-energy tradeoff for circuits computing the generalized mod function MOD_m^A. The theorem immediately yields a tradeoff for circuits computing the ordinary mod function MOD_m. In Section 3.2, we present four lemmas, and using them we prove Theorem 1. In Section 3.3, we present a tradeoff better than that in Theorem 1 if $e \geq 5$.

3.1 Main Theorem and Corollaries

Our main result is the following theorem:

Theorem 1. *Let C be a threshold circuit computing the generalized mod function MOD_m^A of n variables, and let $a = \min\{|A|, |M - A|\}$. Then the size s and energy complexity e of C satisfy*

$$\frac{n+1-a}{m-a} \leq s^e. \tag{5}$$

The ordinary mod function MOD_m is MOD_m^A for the case where $A = \{0\}$ and hence $a = 1$. The PARITY function is MOD_m for the case $m = 2$. We thus have the following corollary.

Corollary 1
(a) If a threshold circuit C computes the ordinary mod function MOD_m, then $n/(m-1) \leq s^e$.
(b) If a threshold circuit C computes the PARITY function, then $n \leq s^e$ and hence $\log n \leq e \log s$.

If n, m and a are fixed, then the left side $(n+1-a)/(m-a)$ of Eq. (5) is a constant and does not depend on the design of C. On the other hand, s and e

depend on the design of C, and the right side s^e is monotonically increasing with regards to s and e. Thus Eq. (5) implies that there exists a tradeoff between e and s.

One can know that the lower bound $(n+1-a)/(m-a)$ on s^e in Eq. (5) cannot be improved much, as follows. For the case where $m = n+1$ and $A = \{0\}$, MOD_m^A is the OR function, and can be computed by a circuit C with $s = e = 1$, and hence Eq. (5) holds in equality for the circuit C. Thus, for any $\epsilon > 0$, the inequality

$$(1+\epsilon)\left(\frac{n+1-a}{m-a}\right) \leq s^e$$

does not hold in general. For the case where $m = 2$ and $A = \{0\}$, MOD_m^A is the PARITY function MOD_2, which can be computed by a circuit C such that $s = n+1$ and $e = 2$ as illustrated in Fig. 2(b). In this case, the right side s^e of Eq. (5) is $(n+1)^2$ for the circuit C, while the left side is n. Therefore, for any $\epsilon > 0$, the inequality

$$\left(\frac{n+1-a}{m-a}\right)^{2+\epsilon} \leq s^e$$

does not hold if n is sufficiently large.

Equation (5) immediately implies

$$\left(\frac{n+1-a}{m-a}\right)^{1/e} \leq s,$$

which is a lower bound on s expressed in terms of n, m, a and e. One can easily know from the bound that $s = \Omega(\sqrt{n})$ if $e \leq 2$ and $m = O(1)$.

From Theorem 1, one can immediately obtain a lower bound on e expressed in terms of n and m as follows.

Corollary 2. *Let C be a threshold circuit computing MOD_m. If $s = O(\mathrm{polylog}(n))$, then*

$$e = \Omega\left(\frac{\log n - \log m}{\log\log n}\right).$$

Corollary 2 implies that if $m = o(n)$ then MOD_m cannot be computed by any threshold circuit C such that $s = O(\mathrm{polylog}(n))$ and $e = o(\log n/\log\log n)$. Similarly to the corollary above, Sung and Nishino [12] prove that $d = \Theta(\log n/\log\log n)$ if a threshold circuit C with depth d computes the PARITY function and $s = O(\mathrm{polylog}(n))$. Slightly modifying a circuit given in [12], one can construct a threshold circuit of size $s = O(\mathrm{polylog}(n))$ and energy $e = O(\log n/\log\log n)$ that computes the PARITY function of n variables. (See Table 1.) Thus, the lower bound on e in Corollary 2 is best possible within a constant factor for the case where $m = 2$.

3.2 Proof of Theorem 1

Let a threshold circuit C consist of gates $g_1, g_2, \cdots, g_s$, and let g_s be the output gate of C: $g_s[\boldsymbol{x}] = C(\boldsymbol{x})$ for every $\boldsymbol{x} \in \{0,1\}^n$. For an input $\boldsymbol{x} \in \{0,1\}^n$, we define a *pattern* $\boldsymbol{p}_C(\boldsymbol{x}) \in \{0,1\}^s$ of C for $\boldsymbol{x}$ as $\boldsymbol{p}_C(\boldsymbol{x}) = (g_1[\boldsymbol{x}], g_2[\boldsymbol{x}], \cdots, g_s[\boldsymbol{x}])$. We often denote $\boldsymbol{p}_C(\boldsymbol{x})$ simply by $\boldsymbol{p}(\boldsymbol{x})$. We denote by $P(C)$ the set of all patterns that arise in C: $P(C) = \{\boldsymbol{p}_C(\boldsymbol{x}) \mid \boldsymbol{x} \in \{0,1\}^n\}$.

One can easily prove the following lemma.

Lemma 1. *For an arbitrary threshold circuit C, $|P(C)| \leq s^e + 1$.*

Proof. If $s = 1$, then $|P(C)| \leq 2$, $s^e + 1 = 2$ and hence Lemma 1 holds. We may thus assume that $s \geq 2$. Since the energy complexity of C is e, at most e of the s gates output "1" for any input $\boldsymbol{x}$. Therefore, we have

$$|P(C)| \leq \sum_{i=0}^{e} \binom{s}{i} \tag{6}$$

$$\leq 1 + s + \frac{1}{2}\left(s^2 + s^3 + \cdots + s^e\right)$$

$$\leq 1 + s + \frac{s^2(s^{e-1} - 1)}{2(s-1)}. \tag{7}$$

From Eq. (7) and $s \leq 2(s-1)$, we obtain

$$|P(C)| \leq 1 + s + s(s^{e-1} - 1) = 1 + s^e.$$

□

For every input $\boldsymbol{x} \in \{0,1\}^n$, we define an *extended pattern* $\boldsymbol{q}_C(\boldsymbol{x}) \in \{0,1\}^s \times M$ of a threshold circuit C for $\boldsymbol{x}$ as follows: $\boldsymbol{q}_C(\boldsymbol{x}) = (\boldsymbol{p}_C(\boldsymbol{x}), [\boldsymbol{x}]_m)$, where $M = \{0, 1, \cdots, m-1\}$. We often denote $\boldsymbol{q}_C(\boldsymbol{x})$ simply by $\boldsymbol{q}(\boldsymbol{x})$. We denote by $Q(C)$ the set of all extended patterns that arise in C: $Q(C) = \{\boldsymbol{q}_C(\boldsymbol{x}) \mid \boldsymbol{x} \in \{0,1\}^n\}$. Since $|M| = m$, we have $|Q(C)| \leq |P(C)| \cdot m$. A better upper bound on $|Q(C)|$ can be obtained for a circuit C computing MOD_m^A, as follows.

Lemma 2. *Let C be a threshold circuit computing MOD_m^A, and let $a = |A|$. Then*

$$|Q(C)| \leq (|P(C)| - 1)(m - a) + a. \tag{8}$$

Proof. We give a proof only for the case where $|A| \leq |M - A|$ and hence $a = |A|$, because the proof for the other case where $|A| > |M - A|$ is similar.

The set $P(C)$ can be partitioned into the following two subsets $P_1(C)$ and $P_0(C)$:

$$P_1(C) = \{\boldsymbol{p}(\boldsymbol{x}) \mid \boldsymbol{x} \in \{0,1\}^n, C(\boldsymbol{x}) = 1\}$$

and

$$P_0(C) = \{\boldsymbol{p}(\boldsymbol{x}) \mid \boldsymbol{x} \in \{0,1\}^n, C(\boldsymbol{x}) = 0\}.$$

Since g_s is the output gate of C, we have $g_s[\boldsymbol{x}] = 1$ if $C(\boldsymbol{x}) = 1$, and $g_s[\boldsymbol{x}] = 0$ if $C(\boldsymbol{x}) = 0$. Thus $P_1(C) \cap P_0(C) = \emptyset$. Similarly, the set $Q(C)$ can be partitioned into the following two subsets $Q_1(C)$ and $Q_0(C)$:

$$Q_1(C) = \{\boldsymbol{q}(\boldsymbol{x}) \mid \boldsymbol{x} \in \{0,1\}^n, C(\boldsymbol{x}) = 1\}$$

and

$$Q_0(C) = \{\boldsymbol{q}(\boldsymbol{x}) \mid \boldsymbol{x} \in \{0,1\}^n, C(\boldsymbol{x}) = 0\}.$$

Clearly

$$|P(C)| = |P_1(C)| + |P_0(C)| \tag{9}$$

and

$$|Q(C)| = |Q_1(C)| + |Q_0(C)|. \tag{10}$$

If $C(\boldsymbol{x}) = \mathrm{MOD}_m^A(\boldsymbol{x}) = 1$, then $[\boldsymbol{x}]_m \in M - A$ by Eq. (1). We thus have

$$|Q_1(C)| \leq |P_1(C)| \cdot (m - a). \tag{11}$$

On the other hand, if $C(\boldsymbol{x}) = 0$ then $[\boldsymbol{x}]_m \in A$. We thus have

$$|Q_0(C)| \leq |P_0(C)| \cdot a \tag{12}$$

Substituting Eqs. (11) and (12) to Eq. (10), we have

$$|Q(C)| \leq |P_1(C)| \cdot (m - a) + |P_0(C)| \cdot a. \tag{13}$$

Equations (9) and (13) imply that

$$\begin{aligned} |Q(C)| &\leq (|P(C)| - |P_0(C)|) \cdot (m - a) + |P_0(C)| \cdot a \\ &= |P(C)| \cdot (m - a) - |P_0(C)| \cdot (m - 2a). \end{aligned} \tag{14}$$

By Eq. (2) we have $m - 2a \geq 0$. Therefore, the right side of Eq. (14) is non-increasing with respect to $|P_0(C)|$. Since $a \geq 1$ by Eq. (2), we have $A \neq \emptyset$. Since $A \subseteq M = \{0, 1, \cdots, m-1\}$ and $m - 1 \leq n$ by Eq. (3), there is an input $\boldsymbol{x} \in \{0,1\}^n$ such that $[\boldsymbol{x}]_m \in A$ and hence $C(\boldsymbol{x}) = \mathrm{MOD}_m^A(\boldsymbol{x}) = 0$. Therefore, $\boldsymbol{p}(\boldsymbol{x}) \in P_0(C)$ and hence $|P_0(C)| \geq 1$. Thus, the right side of Eq. (14) takes the maximum value when $|P_0(C)| = 1$, and hence Eq. (8) holds. □

For a threshold circuit C with $n(\geq 2)$ inputs, we denote by C_0 a circuit obtained from C by fixing the n-th variable x_n of input $\boldsymbol{x} = (x_1, x_2, \cdots, x_n)$ to the constant 0. As illustrated in Fig. 1, one can obtain C_0 from C by deleting the n-th input node for x_n and all the wires linked from the node. We call C_0 the 0-*fixed* circuit of C. The 0-fixed circuit C_0 has $n - 1$ inputs, but the size of C_0 is the same as that of C.

Define $X_0 \subseteq \{0,1\}^n$ as follows: $X_0 = \{(x_1, x_2, \cdots, x_n) \in \{0,1\}^n \mid x_n = 0\}$. For each input $\boldsymbol{x}' = (x_1, x_2, \cdots, x_{n-1}) \in \{0,1\}^{n-1}$ to the 0-fixed circuit C_0 of C, let $\boldsymbol{x} \in X_0$ be the input to C such that $\boldsymbol{x} = (x_1, x_2, \cdots, x_{n-1}, 0)$. Then clearly $[\boldsymbol{x}']_m = [\boldsymbol{x}]_m$ and $\boldsymbol{p}_{C_0}(\boldsymbol{x}') = \boldsymbol{p}_C(\boldsymbol{x})$. We thus have

$$\begin{aligned} P(C_0) &= \{\boldsymbol{p}_{C_0}(\boldsymbol{x}') \mid \boldsymbol{x}' \in \{0,1\}^{n-1}\} \\ &= \{\boldsymbol{p}_C(\boldsymbol{x}) \mid \boldsymbol{x} \in X_0\} \subseteq P(C) \end{aligned} \tag{15}$$

and

$$Q(C_0) = \{(\boldsymbol{p}_{C_0}(\boldsymbol{x}'), [\boldsymbol{x}']_m) \mid \boldsymbol{x}' \in \{0,1\}^{n-1}\}$$
$$= \{(\boldsymbol{p}_C(\boldsymbol{x}), [\boldsymbol{x}]_m) \mid \boldsymbol{x} \in X_0\} \subseteq Q(C). \quad (16)$$

If a threshold circuit C computes the function MOD_m^A of n variables, then clearly the 0-fixed circuit C_0 computes the function MOD_m^A of $n-1$ variables. We now have the following key lemma on $Q(C)$ and $Q(C_0)$.

Lemma 3. *If a threshold circuit C computes the function MOD_m^A of $n(\geq 1)$ variables, then $|Q(C_0)| + 1 \leq |Q(C)|$ where $|Q(C_0)|$ is assumed to be 1 if $n = 1$.*

Sketchy proof. Clearly Lemma 3 holds for the case where $n = 1$, and hence one may assume that $n \geq 2$. Suppose that a threshold circuit C computes the function MOD_m^A of $n(\geq 2)$ variables, and that C consists of s threshold gates $g_1, g_2, \cdots, g_s$. One may assume that $g_1, g_2, \cdots, g_s$ are topologically ordered with respect to the underlying directed acyclic graph of C, and that g_s is the output gate of C. Assume for a contradiction that Lemma 3 does not hold. Then, $Q(C_0) = Q(C)$ because $Q(C_0) \subseteq Q(C)$ by Eq. (16). Let X_1 be the subset of $\{0,1\}^n$ such that $X_1 = \{(x_1, x_2, \cdots, x_n) \in \{0,1\}^n \mid x_n = 1\}$. Let Q_1 be the subset of $Q(C)$ such that $Q_1 = \{(\boldsymbol{p}(\boldsymbol{x}), [\boldsymbol{x}]_m) \mid \boldsymbol{x} \in X_1\}$. Then we have $Q_1 \subseteq Q(C) = Q(C_0)$.

Let $h = |P(C)|$. To derive a contradiction, we construct the following sequence of $2h+1$ inputs to C:

$$\boldsymbol{x}_0 \to \boldsymbol{y}_0 \to \boldsymbol{x}_1 \to \boldsymbol{y}_1 \to \cdots \to \boldsymbol{x}_{h-1} \to \boldsymbol{y}_{h-1} \to \boldsymbol{x}_h, \quad (17)$$

where $\boldsymbol{x}_j \in X_1$ and $\boldsymbol{y}_j \in X_0$ for every index j. We arbitrarily choose $\boldsymbol{x}_0$ from the set $X_1(\neq \emptyset)$, and choose $\boldsymbol{y}_0, \boldsymbol{x}_1, \cdots, \boldsymbol{x}_h$ as in the following (a) and (b):

(a) From $\boldsymbol{x}_j$, $0 \leq j \leq h-1$, we obtain $\boldsymbol{y}_j \in X_0$ such that $\boldsymbol{p}(\boldsymbol{y}_j) = \boldsymbol{p}(\boldsymbol{x}_j)$ and $(\boldsymbol{p}(\boldsymbol{y}_j), y_j') \notin Q(C_0)$ for $y_j' = [\boldsymbol{y}_j]_m + 1 \pmod{m}$.

(b) From $\boldsymbol{y}_j \in X_0$, $0 \leq j \leq h-1$, we obtain $\boldsymbol{x}_{j+1} \in X_1$ simply by flipping the n-th input of $\boldsymbol{y}_j \in X_0$.

The sequence (17) corresponds to the following sequence of patterns:

$$\boldsymbol{p}(\boldsymbol{x}_0) \to \boldsymbol{p}(\boldsymbol{y}_0) \to \boldsymbol{p}(\boldsymbol{x}_1) \to \boldsymbol{p}(\boldsymbol{y}_1) \to \cdots \to \boldsymbol{p}(\boldsymbol{x}_{h-1}) \to \boldsymbol{p}(\boldsymbol{y}_{h-1}) \to \boldsymbol{p}(\boldsymbol{x}_h). \quad (18)$$

One can prove that $\boldsymbol{p}(\boldsymbol{x}_j) = \boldsymbol{p}(\boldsymbol{y}_j) \neq \boldsymbol{p}(\boldsymbol{x}_{j+1})$ for each j, $0 \leq j \leq h-1$. Therefore, the sequence (18) contains $h+1$ patterns $\boldsymbol{p}(\boldsymbol{x}_0), \boldsymbol{p}(\boldsymbol{x}_1), \cdots, \boldsymbol{p}(\boldsymbol{x}_h)$, but $h = |P(C)|$. Thus, there is a pair of indices l and r, $0 \leq l < r \leq h$, such that $\boldsymbol{p}(\boldsymbol{x}_l) = \boldsymbol{p}(\boldsymbol{x}_r)$. We now consider the following subsequence of (18)

$$\boldsymbol{p}(\boldsymbol{x}_l) \to \boldsymbol{p}(\boldsymbol{y}_l) \to \boldsymbol{p}(\boldsymbol{x}_{l+1}) \to \boldsymbol{p}(\boldsymbol{y}_{l+1}) \to \cdots \to \boldsymbol{p}(\boldsymbol{x}_{r-1}) \to \boldsymbol{p}(\boldsymbol{y}_{r-1}) \to \boldsymbol{p}(\boldsymbol{x}_r),$$

and find a sequence of gates $g_{i_l}, g_{i_{l+1}}, \cdots, g_{i_{r-1}}$, as follows. Since $\boldsymbol{p}(\boldsymbol{y}_j) \neq \boldsymbol{p}(\boldsymbol{x}_{j+1})$ for each j, $l \leq j \leq r-1$, there are one or more gates that output $b \in \{0,1\}$ for $\boldsymbol{y}_j$ and output the complement $\bar{b}$ of b for $\boldsymbol{x}_{j+1}$. Let g_{i_j} be the gate with the smallest index among all these gates. Let i_t, $l \le t \le r-1$, be the smallest index among $i_l, i_{l+1}, \cdots, i_{r-1}$. Then one can prove that the n-th input node x_n of C is directly connected to the gate g_{i_t}, and the weight w_n is not zero. From the fact one can derive $g_{i_t}[\boldsymbol{x}_l] \ne g_{i_t}[\boldsymbol{x}_r]$, which contradicts to $\boldsymbol{p}(\boldsymbol{x}_l) = \boldsymbol{p}(\boldsymbol{x}_r)$. The details are omitted, due to the page limitation. □

From Lemma 3 one can easily prove the following lower bound on $|Q(C)|$.

Lemma 4. *If a threshold circuit C computes the function MOD_m^A of $n(\ge 1)$ variables, then*

$$n + 1 \le |Q(C)|. \tag{19}$$

Proof. By Eq. (3) we have $n \ge m-1$, and hence we prove by induction on n that Eq. (19) holds for every integer n such that $n \ge m-1$.

For the inductive basis, we assume that $n = m-1$. Clearly, for every integer $i \in M$, there exists an input $\boldsymbol{x} \in \{0,1\}^n$ such that $[\boldsymbol{x}]_m = i$. Thus $|Q(C)| \ge |M| = m = n+1$, and hence Eq. (19) holds.

For the inductive hypothesis, we assume that $n \ge m (\ge 2)$ and that Eq. (19) holds for every threshold circuit computing the function MOD_m^A of $(n-1)$ variables. Let C be a threshold circuit computing MOD_m^A of n variables. Since the 0-fixed circuit C_0 of C computes the function MOD_m^A of $n-1$ variables, the induction hypothesis implies that $|Q(C_0)| \ge (n-1)+1 = n$. Therefore, by Lemma 3 we have $|Q(C)| \ge |Q(C_0)| + 1 \ge n+1$. □

There exists a threshold circuit C computing the function MOD_m^A of n variables such that $|Q(C)| = n+1$, as illustrated in Fig. 2(a) for $m = 2$ and $A = \{0\}$. Therefore, the lower bound on $|Q(C)|$ in Lemma 4 is best possible.

Using Lemmas 1, 2 and 4, one can easily prove Theorem 1, as follows.

Proof of Theorem 1. By Lemma 2 and Lemma 4 we have

$$n + 1 \le (|P(C)| - 1)(m - a) + a. \tag{20}$$

Slightly modifying Eq. (20) and using Lemma 1, we have

$$\frac{n+1-a}{m-a} \le |P(C)| - 1 \le s^e.$$ □

3.3 Theorem 2

In the section, we present a tradeoff which is better than that in Theorem 1 if $e \ge 5$.

Applying a counting argument ([4, p.102, p.122]) and the Stirling's formula to Eq. (6), one can easily prove the following upper bound on $|P(C)|$, which is better than the bound in Lemma 1 if $e \ge 5$:

$$|P(C)| \le \frac{1}{\sqrt{2\pi e}} \cdot \left(\frac{2c_{npr} \cdot s}{e} \right)^e \tag{21}$$

where $c_{npr} \cong 2.718$ is the Napier's (or mathematical) constant. Similarly to the proof of Theorem 1, we can prove the following theorem from Eqs. (20) and (21).

Theorem 2. *Let C be a threshold circuit computing the function MOD_m^A of n variables. Then the size s and energy complexity e of C satisfy*

$$\frac{n+1-a}{m-a} + 1 \leq \frac{1}{\sqrt{2\pi e}} \cdot \left(\frac{2c_{npr} \cdot s}{e}\right)^e . \tag{22}$$

4 Conclusions

In the paper, we show that there exists a very simple tradeoff $(n+1-a)/(m-a) \leq s^e$ between the size s and the energy complexity e of a threshold circuit computing MOD_m^A, where n is the number of input variables, $2 \leq m \leq n+1$, and $a = \min\{|A|, |M-A|\}$. The main idea of the proof of our result is to show that the number of patterns of a circuit is at most $s^e + 1$ and the number of extended patterns is at least $n+1$.

We have so far considered circuits of threshold logic gates, but our result can be extended to a more general class of circuits, called "unate circuits." A function $g(z_1, z_2, \cdots, z_k) : \{0,1\}^k \to \{0,1\}$ is said to be *unate in variable* z_i if

$$g(z_1, \cdots, z_{i-1}, 0, z_{i+1}, \cdots, z_k) \leq g(z_1, \cdots, z_{i-1}, 1, z_{i+1}, \cdots, z_k)$$

or

$$g(z_1, \cdots, z_{i-1}, 1, z_{i+1}, \cdots, z_k) \leq g(z_1, \cdots, z_{i-1}, 0, z_{i+1}, \cdots, z_k)$$

holds for all $z_1, \cdots, z_{i-1}, z_{i+1}, \cdots, z_k \in \{0,1\}$. A function g is said to be *unate* if g is unate in every variable z_i, $1 \leq i \leq k$. A *unate gate* is a logical gate computing a unate function. Clearly, a threshold gate, OR gate, AND gate, *etc.* are unate gates, while there is a unate function which cannot be computed by any single threshold gate. A *unate circuit* is a combinatorial circuit C consisting of unate gates, let the size s of C be the number of gates in C, and let the energy complexity e of C be the maximum number of gates outputting "1" over all inputs. Then one can observe that our proof scheme for threshold circuits can be applied to unate circuits and yields the same tradeoffs as in Theorem 1 and Theorem 2.

References

1. Aggarwal, A., Chandra, A., Raghavan, P.: Energy consumption in VLSI circuits. In: Proceedings of the 20th Annual ACM Symposium on Theory of Computing, pp. 205–216 (1988)
2. Beigel, R., Maciel, A.: Upper and lower bounds for some depth-3 circuit classes. Computational Complexity 6(3), 235–255 (1997)
3. Chattopadhyay, A., Goyal, N., Pudlak, P., Therien, D.: Lower bounds for circuits with MOD_m gates. In: Proceedings of the 47th Annual IEEE Symposium on Foundations of Computer Science, pp. 709–718 (2006)

4. Cormen, T.H., Leiserson, C.E., Rivest, R.L.: Introduction to Algorithms. MIT Press, Cambridge (1989)
5. Grolmusz, V., Tardos, G.: Lower bounds for (MOD_p - MOD_m) circuits. SIAM Journal on Computing 29(4), 1209–1222 (2000)
6. Impagliazzo, R., Paturi, R., Saks, M.E.: Size-depth trade-offs for threshold circuits. SIAM Journal on Computing 26(3), 693–707 (1997)
7. Kissin, G.: Upper and lower bounds on switching energy in VLSI. Journal of the Association for Computing Machinery 38, 222–254 (1991)
8. Lennie, P.: The cost of cortical computation. Current Biology 13, 493–497 (2003)
9. Margrie, T.W., Brecht, M., Sakmann, B.: In vivo, low-resistance, whole-cell recordings from neurons in the anaesthetized and awake mammalian brain. Pflugers Arch. 444(4), 491–498 (2002)
10. Minsky, M., Papert, S.: Perceptrons: An Introduction to Computational Geometry. MIT Press, Cambridge (1988)
11. Parberry, I.: Circuit Complexity and Neural Networks. MIT Press, Cambridge (1994)
12. Shao-Chin, S., Nishino, T.: The complexity of threshold circuits for parity functions. IEICE Transactions on Information and Systems 80(1), 91–93 (1997)
13. Sima, J., Orponen, P.: General-purpose computation with neural networks: A survey of complexity theoretic results. Neural Computation 15, 2727–2778 (2003)
14. Siu, K.Y., Roychowdhury, V., Kailath, T.: Discrete Neural Computation; A Theoretical Foundation. Prentice-Hall, Inc., Upper Saddle River (1995)
15. Siu, K.Y., Roychowdhury, V.P., Kailath, T.: Rational approximation techniques for analysis of neural networks. IEEE Transactions on Information Theory 40(2), 455–466 (1994)
16. Uchizawa, K., Douglas, R., Maass, W.: On the computational power of threshold circuits with sparse activity. Neural Computation 18(12), 2994–3008 (2006)
17. Uchizawa, K., Nishizeki, T., Takimoto, E.: Energy complexity and depth of threshold circuits. In: Proceedings of the 17th International Symposium on Fundamentals of Computation Theory. Springer, Heidelberg (to appear)
18. Uchizawa, K., Takimoto, E.: Exponential lower bounds on the size of threshold circuits with small energy complexity. Theoretical Computer Science 407(1-3), 474–487 (2008)

Points on Computable Curves of Computable Lengths*

Robert Rettinger[1] and Xizhong Zheng[2,3,**]

[1] Lehrgebiet Algorithmen und Komplexität
FernUniversität Hagen, 58084 Hagen, Germany
[2] Department of Computer Science and Mathematics
Arcadia University, Glenside, PA 19038, USA
[3] Department of Mathematics, Jiangsu University
Zhenjiang 212013, China
ZhengX@Arcadia.edu

Abstract. A computable plane curve is defined as the image of a computable real function from a closed interval to the real plane. As it is showed by Ko [7] that the length of a computable curve is not necessarily computable, even if the length is finite. Therefore, the set of the computable curves of computable lengths is different from the set of the computable curves of finite lengths. In this paper we show further that the points covered by these two sets of curves are different as well. More precisely, we construct a computable curve K of a finite length and a point z on the curve K such that the point z does not belong to any computable curve of computable length. This gives also a positive answer to an open question of Gu, Lutz and Mayordomo in [4].

1 Introduction

The computability and complexity over real numbers are of fundamental importance both for the practical applications (e.g in scientific computations) and for the theoretical interests. One of very realistic approachs to computability and complexity over real numbers is the Turing-machine-based bit model (see [10,2,6]). In this model, every real number x is represented by a sequence (x_n) of rational numbers which converges to x effectively in the sense that $|x-x_n| \leq 2^{-n}$ for all n. This sequence serves as a name of the real number x. A real number x is called computable if it has a computable name. Similarly a real function f is computable if there is a Turing machine which transfers each name of a real number x in the domain of f into a name of $f(x)$. By the same principle, we can define the computability of other mathematical objects by introducing proper "naming systems", for example, the computability of subsets of the Euclidean space [1], of semi-continuous functions [11], of functional spaces [12], etc.

A prominent and important example of such mathematical objects is the curve. In mathematical analysis, a curve can be defined as the range of a continuous real function. Thus, computability of curves can be introduced by means

* This work is supported by DFG (446 CHV 113/266/0-1) and NSFC (10420130638).
** Corresponding author.

R. Královič and D. Niwiński (Eds.): MFCS 2009, LNCS 5734, pp. 736–743, 2009.

of computable real functions (see [10,3]). Physically, a curve can be regarded as the trace of a particle motion. If the particle moves according to some algorithmically definable laws, its trace should be naturally regarded as computable. This leads to the definition of computable curves as the ranges of computable functions $f : [0;1] \to \mathbb{R}^n$. The function f is then called a *parametrization* of the curve C if $C = \text{range}(f)$. Notice that this is not the only possible definition of computable curves (see Lemma 3 below). For example, one could restrict the parameterizations to one-to-one functions. From the non-computable point of view these gives the same notion of curves.

Recently, Gu, Lutz and Mayordomo [3,4] investigated in details the differences of these two notions of computable curves and moreover the points covered by computable curves. Among others, they showed in [4] that there is a computable curve Γ of finite length such that any of its computable parameterizations f must retrace. That is, f traces some segments of the curve Γ more than once. This means that two versions of computable curves mentioned above are different. This does not happen if the curve have a computable length because then there exists always a one-to-one computable parametrization (except possibly one point). This shows a significant difference between computable curves of finite and computable lengths. Therefore they raise the question whether *there exists a point which lies on a computable curve of finite length but not on any computable curve of computable length?* in [4]. The main theorem of this paper will give a positive answer to this question. We actually show that there is a computable curve K of a finite length and an one-to-one parameterization such that K contains a point z which does not belong to any computable curve of a computable length.

Our paper is organized as follows. In Section 2 we will shortly recall some basic definitions and show a technical lemma which will be used in the proof of the main theorem. Section 3 shows some basic facts related to computable curves of computable lengths. As a simple example we will show that the set of points covered by computable curves of computable lengths is not exhausted by computable polygons. In the last Section 4 we prove our main theorem.

2 A Technical Lemma

We consider only plane curves in this paper. A plane curve C is a subset of $\mathbb{R}^2$ so that there exists a continuous function $f : [0;1] \to \mathbb{R}^2$ with $f([0;1]) = C$. Any such a function f is then called a *parameterization* of C. The curve C is called *simple* if there exists a parameterization which is either an injection on the closed interval $[0;1]$ or an injection on $(0;1]$ with the additional condition $f(0) = f(1)$. In both cases we call the function f an *one-to-one* parameterization of C (even if f may be not one-to-one on $[0,1]$) and the curve of the latter case is called *closed*. Thus, a simple curve never intersects itself.

If $f : [0;1] \to \mathbb{R}^2$ is an one-to-one parameterization of a simple curve C, then, according to Jordan [5], the *length* $l(C)$ of the curve C is defined by

$$l(C) := \sup \sum_{i=0}^{n-1} |f(t_i) - f(t_{i+1})| \tag{1}$$

where the supremum is taken over all partitions $0 = t_0 < t_1 < t_2 < \cdots < t_n = 1$, and $|x|$ is the Euclidean norm of the point $x \in \mathbb{R}^2$, i.e., $|f(t_i) - f(t_{i+1})|$ is the length of the straight line connecting the points $f(t_i)$ and $f(t_{i+1})$. A curve of a finite length is also called *rectifiable*. It is well known that the length of a simple curve defined by (1) is independent of the choice of the one-to-one parameterization f. A curve is called *differentiable* if it has a differentiable parameterization.

The computability of real numbers and real functions are defined based on the (extended) Turing machine model (see [10] for details). We say that a sequence $(x_n)_{n\in\mathbb{N}}$ of real numbers converges effectively if $|x_{n+1} - x_n| \leq 2^{-n}$ for all n. A real number x is computable if there is a computable sequence $(x_n)_{n\in\mathbb{N}}$ of rational numbers which converges to x effectively. A real function $f : [0; 1] \to \mathbb{R}$ is computable if there is a Turing machine M such that, for any input of sequence $(x_n)_{n\in\mathbb{N}}$ of rational numbers converging effectively to an $x \in [0; 1]$, $M((x_n)_{n\in\mathbb{N}})$ outputs a sequence $(y_n)_{n\in\mathbb{N}}$ of rational numbers which converges to $f(x)$ effectively. Equivalently, f is computable iff there is a computable sequence $(p_n)_{n\in\mathbb{N}}$ of computable rational polynomials which converges uniformly and effectively to f. Naturally, a function $f : [0; 1] \to \mathbb{R}^n$ is computable if all of its component functions are computable. Finally, we call a curve C *computable* if it has a computable parameterization, i.e., $C = \text{range} f$ for a computable function $f : [0; 1] \to \mathbb{R}^n$.

Notice that we can also define curves by images of open intervals and even $\mathbb{R}$. This does not change anything presented below, because any point on such a curve belongs trivially to the image of the corresponding parameterization restricted to some subinterval $[a; b]$ where a and b is rational.

The next lemma shows a simple fact related to two curves which will allow to separat curves by points later on unless one of the curve is a subcurve of the other.

Lemma 1. *Let C and C' be two rectifiable, non-closed simple curves and let $g : [0; 1] \to \mathbb{R}^2$ be a parameterization of C'. If we have $C' \cap U_z \neq \emptyset$ for any point $z \in C$ and any open neighborhood U_z of z, then there exists an interval $[a; b] \subseteq [0; 1]$ such that $g([a; b]) = C$.*

Proof. Suppose that C, C' are rectifiable, non-closed simple curves. If $C' \cap U_z \neq \emptyset$ for any point $z \in C$ and any open neighborhood U_z of z, then C must be a part of C', i.e., $C \subseteq C'$. Otherwise, by the compactness of C', we can find a point z in $C \backslash C'$ which has a positive distance from C' and hence some open neighborhood of z is disjointed from C which contradicts the hypothesis.

Because C' is a rectifiable simple curve, there exists an one-to-one parameterization $f : [0; 1] \to C'$. This parameterization f must be injective since C' is non-closed. Therefore the inverse function f^{-1} exists which is also continuous and maps particularly two end points of C to $u, v \in [0; 1]$. Suppose w.l.o.g. that $u < v$. Then we have $f([u; v]) = C$.

Let $h : [0; 1] \to [0; 1]$ be a continuous function defined by $h := f^{-1} \circ g$. Since $f([0; 1]) = C \subseteq C' = g([0; 1])$, we have $[u; v] \subseteq h([0; 1])$. By the continuity of h, there exist $a \in h^{-1}(u)$ and $b \in h^{-1}(v)$ such that $h([a; b]) = [u; v]$ (we suppose w.l.o.g that $a < b$). This implies immediately that $g([a; b]) = C$.

By Lemma 1, if a curve C is not contained completely in another curve C', then there exist a point z in C and a small neighborhood U_z around z such that U is totally disjoint from the curve C'. Particularly, if C is longer than C', then C cannot be completely contained in C'. If in addition C is a rational polygon and C' is a computable curve, then such a point z and the corresponding neighborhood U_z can be effectively found. That is, we have the following lemma.

Lemma 2. *Let C be a rational polygon and let C' be a computable curve. If the length of C is larger than the length of C', then we can effectively find a rational point z on C and a rational neighborhood U_z of z such that $C' \cap U_z = \emptyset$.*

3 Computable Curves of Computable Length

In this section we investigate the properties of computable curves of computable lengths. We will see that computable curves can be equivalently defined in several different ways. Finally, by a simple example, we will show that there is a point which lies on a computable curve of computable length but not on any computable polygon.

It is well known that a curve of infinite length can be significantly different from that of finite length. While the later has always a Lebesgue measure zero, a curve of infinite length can almost fill the complete space. For computable curves, it makes a big difference whether a computable curve has a computable or non-computable length. For example, Gu, Lutz and Mayordomo showed in [4] that the arc-length parameterization of a curve is relatively computable in its length. Therefore, any computable curve of computable length has always a computable one-to-one parameterization. But [4] shows also that there exists a computable curve of finite length which must retrace no matter what computable parameterization we choose, and hence it does not have a one-to-one computable parameterization at all.

By a similar technique used in [4] we can show that computable curves of a computable length can be equivalently characterized in several ways.

Lemma 3. *Let C be a curve of a computable length l. Then the following conditions are equivalent.*

1. *C is a computable curve, i.e., C has a computable parameterization;*
2. *C has a computable one-to-one parameterization;*
3. *There is a computable sequence of rational polygon functions $(p_n)_{n\in\mathbb{N}}$ which converges uniformly effectively to C and the sequence $(l(p_n))_n$ of the polygon-lengths converges effectively as well*
4. *There is a computable sequence $(U_s)_{s\in\mathbb{N}}$ of finite sets of rational open balls such that the Hausdorff distance*

$$d_H\left(\bigcup U_s, C\right) \leq 2^{-s}$$

for all s. Here a rational open ball is a set $B(a,\delta) := \{x \in \mathbb{R}^2 : |x-a| < \delta\}$ for some rational point a and positive rational number δ.

5. *C has a computable arc length parameterization $f : [0; l] \to \mathbb{R}^2$ where l is the length of C. Here f is an arc length parameterization means that the segment $f([0; t])$ has the length t for any $t \in [0; l]$.*

However, in another paper [9] we have shown that all of the above conditions are different if the curve does not have a computable length.

Many curves we are familiar with do have computable length. The following lemma gives a simple sufficient condition that a curve has computable length.

Lemma 4. *If an one-to-one parameterization of a simple curve C has a computable derivative, then C has computable length.*

Proof. Let $f(t) := \langle x(t), y(t) \rangle$ be a one-to-one parameterization of C such that the derivative $f'(t) = \langle x'(t), y'(t) \rangle$ is computable as well. Then the arc length of C can be calculated by $l(C) = \int_0^1 \sqrt{(x'(t))^2 + (y'(t))^2} dt$ which is computable (see for example [8]).

Thus, by Lemma 4, line segments connecting two computable points, computable polygons (connecting finitely many computable points by straight lines), computable circles, etc, have computable length.

The polygons are probably among the simplest curves. On the other hand, any computable curve can be approximated effectively by a computable sequence of computable polygons. It would be quite natural to guess that the set of points on computable curves of computable length should be covered by all computable polygons. However, the following simple example shows that this is not the case. Even computable circles can contain points which do not lie on any computable polygon.

Example 1. Let $a \in \mathbb{R}^2$ be any computable point and let $r > 0$ be any computable real number. Then the circle centered at a of the radius r is computable curve C of a computable length.

Notice that any computable line segment intersects the circle C in at most two points. The number of computable line segments is countable. Since the circle C contains uncountably many points, there must be points on C which do not lie on any computable line segment. This implies immediately that the sets of points covered by computable curves of computable length and by computable polygons are different.

The same argument as in the example shows that there are points on computable line segments which are not on any computable circle. Therefore, this does not mean that circles are more complicated than the polygons. It might be interesting to determine which hierarchy of curves can be separated properly by points.

4 Computable Curves of Non-computable Length

In this section we will have a closer look at computable curves of non-computable length and prove our main theorem.

From Lemma 3 we know that any computable curve of a computable length has one-to-one parameterizations. However, Gu, Lutz and Mayordomo showed in [4] that this is not generally true for computable curves. They constructed a computable curve Γ of left computable length such that any computable parameterization $f : [0;1] \to \Gamma$ of Γ must retrace some portion arbitrarily many times. More precisely, for any natural number n, there are disjoint closed subintervals $I_0, I_1, \cdots, I_n \subseteq [0;1]$ such that $f(I_i) = f(I_0)$ for all $i \leq n$. This shows a significant difference between computable curves of finite and computable length. Then they asked, if the points covered by these two types of curves are also different. The following theorem gives a positive answer to this question.

Theorem 1. *There is a point which lies on a simple computable curve of finite length but not on any simple computable curve of computable length.*

Proof. We will construct a computable curve K of finite length and a point z on the curve K such that z does not lie on any computable curve of computable length. The curve K is constructed in stages by a finite injury priority method.

By Lemma 3, if C is a computable curve of computable length, then there is a computable sequence $(p_n)_{n\in\mathbb{N}}$ of rational polygons which converges to C effectively. In addition, there is also a computable sequence $(c_n)_{n\in\mathbb{N}}$ of rational numbers which converges effectively to the length l of C. Let $(M_i)_{i\in\mathbb{N}}$ be an effective enumeration of all Turing machines whose possible outputs are pairs of computable sequences of rational polygons and sequences of rational numbers. If Turing machine M_i outputs a pair $\langle (p_s^i)_{s\in\mathbb{N}}, (c_s^i)_{s\in\mathbb{N}} \rangle$ of sequences of rational polygons and rational numbers which satisfy the following conditions

$$|p_s^i - p_{s+1}^i| \leq 2^{-s} \qquad \text{and} \qquad |c_s^i - c_{s+1}^i| \leq 2^{-s} \tag{2}$$

for all s, then $C_i := \lim_{s\to\infty} p_s^i$ is a computable curve of the computable length $l_i := \lim_{s\to\infty} c_s^i$. For technical simplicity, let $C_i = \emptyset$ and $l_i = 0$, if M_i does not output a pair satisfying (2). Then we have an effective enumeration $\langle C_i, l_i \rangle$ which includes all computable curves C_i of computable lengths l_i. Now it suffices to construct a computable curve K of finite length and a point z on K which satisfies for all i the following requirements

R_i: the point z does not lie on the curve C_i.

To satisfy the single requirement R_i, fix a neighborhood, say, a ball $B(a, \delta)$ with rational center a and rational radius δ where a is a point of the already constructed polygon K. By means of the sequences (p_s^i) and (c_s^i), we first try to shrink the ball $B(a, \delta)$ further to some open subset U of $B(a, \delta)$ with $U \cap K \neq \emptyset$ so that only one connected part of C_i could possibly be in $U \cap K$. Notice that we can always find such U if C_i is indeed simple. For convenience we assume that $B(a, \delta)$ is small enough to fulfill this condition. Otherwise replace $B(a, \delta)$ by U. Next we estimate the length of the segment of C_i in the neighborhood $B(a, \delta)$ to the sufficient precision. If it is very close to the length of K in this ball, then replace the straight line of K in this neighborhood by sufficiently small "zig-zag". In this way we can increase the length of K in the neighborhood so that it is longer than the length of C_i in the neighborhood by at least, say, 2^{-i}. If we do that at the step s, we should restrict the "heights" of all "zig-zag" less than 2^{-s} to guarantee that the constructed polygon sequence converges effectively. Then, by Lemma 2, we can effectively find a point z on K and a new (sub)neighborhood $B' \subseteq B(a, \delta)$ of z such that the curve C_i does not intersect the neighborhood B'. This new neighborhood can be used for the actions of another computable curve of computable length.

To satisfy all requirements R_i simultaneously we arrange that R_i has a higher priority than R_j if $i < j$. We will construct a computable sequence $(K_s)_{s\in\mathbb{N}}$ of rational polygons which converges effectively to a curve K, where K_s is defined at the stage s. In order to make sure that K contains a point z which does not lie on any curve C_i, we will use the technique described above to find a point z_i on K and a rational ball $B_i(z_i, \delta_i)$ such that C_i is disjoint from B_i. Since K_s changes during the construction, we can only find possible candidates $z_{i,s}$ on K_s and and a ball $B_{i,s}$ centered at z_i at each stage s such that $B_{i,s}$ is disjointed from C_i. Possibly, they have to be changed later. More precisely, we will actually construct K_s, $z_{i,s}$ and $B_{i,s}$ at the stage s which satisfy all the following conditions:

(a) $|K_t - K_s| \leq 2^{-s}$ for all $t \geq s$ and hence $K := \lim_{s\to\infty} K_s$ is a computable curve;
(c) For each s, $z_{i,s}$ is on the curve K_s and $B_{i,s} = B(z_{i,s}, \delta_{i,s})$ (for some $\delta_{i,s}$) is disjointed from C_i;
(d) For each i, $B_{i+1,s}$ is a subset of $B_{i,s}$ and $z_{i,s}$ and $B_{i,s}$ can be changed only finitely many times hence they converge to z_i and B_i, respectively.
(e) The radiuses of B_i converges to zero if i increases to infinite.

The sequences mentioned above can be constructed by a standard finite injury priority method. Therefore the limit $z := \lim_{i\to\infty} z_i$ is a point on the computable curve K and it does not lie on any computable curve of computable length.

It remains to show that the curve K has finite length. Notice that the actions for C_i could be destroyed at most $2^{(i-1)}$ times due to the actions for C_j of higher priority. If, at the stage s, we construct a new polygon K_s according to the strategy for C_i, then we should make sure that the difference of the lengths between K_s and the previous K_{s-1} does not exceed 2^{-2i}. Therefore the actions for C_i in the construction can contribute a length-change of K at most 2^{-i} totally and the final length of the curve K must be finite.

Notice that the computable curve K constructed in the proof of the Theorem 1 is an effective limit of a computable sequence of polygon functions. So it has actually a computable parameterization without retrace. Thus Theorem 1 can actually be strengthen to the following.

Theorem 2. *There is simple computable curve C and a point z on C which satisfy the following conditions:*

1. *C has a finite (non-computable) length;*
2. *C has a computable one-to-one parameterization;*
3. *z does not lie on any simple computable curve of computable length.*

References

1. Brattka, V., Weihrauch, K.: Computability on subsets of Euclidean space I: Closed and compact subsets. Theoretical Computer Science 219, 65–93 (1999)
2. Braverman, M., Cook, S.: Computing over real numbers: Foundation for scientific computing. Notics of AMS 53(3), 318–329 (2006)
3. Gu, X., Lutz, J.H., Mayordomo, E.: Points on computable curves. In: Proceedings of FOCS 2006, pp. 469–474. IEEE Computer Society Press, Los Alamitos (2006)
4. Gu, X., Lutz, J.H., Mayordomo, E.: Curves that must be retraced (preprint, 2008)
5. Jordan, C.: Cours d'analyse de l'Ecole Polytechnique. Publications mathématiques d'Orsay (1882)
6. Ko, K.-I.: Complexity Theory of Real Functions. Progress in Theoretical Computer Science. Birkhäuser, Boston (1991)
7. Ko, K.-I.: A polynomial-time computable curve whose interior has a nonrecursive measure. Theoretical Computer Science 145, 241–270 (1995)
8. Pour-El, M.B., Richards, J.I.: Computability in Analysis and Physics. Perspectives in Mathematical Logic. Springer, Berlin (1989)
9. Rettinger, R., Zheng, X.: On the computability of rectifiable simple curve (in preparation, 2009)
10. Weihrauch, K.: Computable Analysis, An Introduction. Springer, Heidelberg (2000)
11. Weihrauch, K., Zheng, X.: Computability on continuous, lower semi-continuous and upper semi-continuous real functions. In: Jiang, T., Lee, D.T. (eds.) COCOON 1997. LNCS, vol. 1276, pp. 166–175. Springer, Heidelberg (1997)
12. Zhong, N., Weihrauch, K.: Computability theory of generalized functions. Informatik Berichte 276, FernUniversität Hagen, Hagen (September 2000)

The Expressive Power of Binary Submodular Functions*

Stanislav Živný[1], David A. Cohen[2], and Peter G. Jeavons[1]

[1] Computing Laboratory, University of Oxford, UK
{stanislav.zivny,peter.jeavons}@comlab.ox.ac.uk
[2] Department of Computer Science, Royal Holloway, University of London, UK
d.cohen@rhul.ac.uk

Abstract. We investigate whether all Boolean submodular functions can be decomposed into a sum of binary submodular functions over a possibly larger set of variables. This question has been considered within several different contexts in computer science, including computer vision, artificial intelligence, and pseudo-Boolean optimisation. Using a connection between the expressive power of valued constraints and certain algebraic properties of functions, we answer this question negatively.

Our results have several corollaries. First, we characterise precisely which submodular polynomials of arity 4 can be expressed by binary submodular polynomials. Next, we identify a novel class of submodular functions of arbitrary arities which can be expressed by binary submodular functions, and therefore minimised efficiently using a so-called expressibility reduction to the MIN-CUT problem. More importantly, our results imply limitations on this kind of reduction and establish for the first time that it cannot be used in general to minimise arbitrary submodular functions. Finally, we refute a conjecture of Promislow and Young on the structure of the extreme rays of the cone of Boolean submodular functions.

Keywords: Decomposition of submodular functions, Min-Cut, Pseudo-Boolean optimisation, Submodular function minimisation.

1 Introduction

A function $f : 2^V \to \mathbb{R}$ is called *submodular* if for all $S, T \subseteq V$,

$$f(S \cap T) + f(S \cup T) \leq f(S) + f(T).$$

Submodular functions are a key concept in operational research and combinatorial optimisation [27,26,33,32,14,21,17]. Examples include cut capacity functions, matroid rank functions, and entropy functions. Submodular functions are often considered to be a discrete analogue of convex functions [24].

* The authors would like to thank Martin Cooper for fruitful discussions on submodular functions and in particular for help with the proof of Theorem 1. The first author gratefully acknowledges the support of EPSRC grant EP/F01161X/1.

R. Královič and D. Niwiński (Eds.): MFCS 2009, LNCS 5734, pp. 744–757, 2009.

Both minimising and maximising submodular functions, possibly under some additional conditions, have been considered extensively in the literature. Submodular function maximisation is easily shown to be NP-hard [32] since it generalises many standard NP-hard problems such as the maximum cut problem. In contrast, the problem of *minimising* a submodular function (SFM) can be solved efficiently with only polynomially many oracle calls, see [17]. The time complexity of the fastest known general algorithm for SFM is $O(n^6 + n^5 L)$, where n is the number of variables and L is the time required to evaluate the function [28].

The minimisation of submodular functions on sets is equivalent to the minimisation of submodular functions on distributive lattices [32]. Krokhin and Larose have also studied the more general problem of minimising submodular functions on non-distributive lattices [22].

An important and well-studied sub-problem of SFM is the minimisation of submodular functions of bounded arity (SFM_b), also known as *locally defined* submodular functions [8], or submodular functions with *succinct representation* [12]. In this scenario the submodular function to be minimised is defined as the sum of a collection of functions which each depend only on a bounded number of variables. Locally defined optimisation problems of this kind occur in a wide variety of contexts:

- In the context of pseudo-Boolean optimisation, such problems involve the minimisation of Boolean polynomials of bounded *degree* [2].
- In the context of artificial intelligence, they have been studied as *valued constraint satisfaction problems* (VCSP) [31], also known as *soft* or *weighted* constraint satisfaction problems.
- In the context of computer vision, such problems are often formulated as *Gibbs energy minimisation* problems or *Markov Random Fields* (also known as *Conditional Random Fields*) [23].

We will present our results primarily in the language of pseudo-Boolean optimisation. Hence an instance of SFM_b with n variables will be represented as a polynomial in n Boolean variables, of some fixed bounded degree.

However, the concept of submodularity is important in a wide variety of fields within computer science, and our results have direct consequences for Constraint Satisfaction Problems [10,7,19,11] and Computer Vision [20]. Due to space restrictions we will not elaborate on these connections.

A general algorithm for SFM can always be used for the more restricted SFM_b, but the special features of this more restricted problem sometimes allow more efficient special-purpose algorithms to be used. (Note that we are focusing on *exact* algorithms which find an optimal solution.) In particular, it has been shown that certain cases can be solved much more efficiently by reducing to the MIN-CUT problem; that is, the problem of finding a minimum cut in a directed graph which includes a given source vertex and excludes a given target vertex. For example, it has been known since 1965 that the minimisation of *quadratic* submodular polynomials is equivalent to finding a minimum cut in a corresponding directed graph [16,2]. Hence quadratic submodular polynomials can be minimised in $O(n^3)$ time, where n is the number of variables.

A Boolean polynomial in at most 2 variables has degree at most 2, so any *sum* of binary Boolean polynomials has degree at most 2; in other words, it is quadratic. It follows that an efficient algorithm, based on reduction to MIN-CUT, can be used to minimise any class of functions that can be written as a sum of binary submodular polynomials. We will say that a polynomial that can be written in this way, perhaps with additional variables to be minimised over, is *expressible* by binary submodular polynomials (see Section 2.1). The following classes of functions have all been shown to be expressible by binary submodular polynomials in this way[1], over the past four decades:

- polynomials where all terms of degree 2 or more have negative coefficients (also known as *negative-positive* polynomials) [30];
- cubic submodular polynomials [1];
- $\{0,1\}$-valued submodular functions (also known as 2-monotone functions) [10,6];
- a class recently found by Živný and Jeavons [35] and independently in [34].

All these classes of functions have been shown to be expressible by binary submodular polynomials and hence minimisable in cubic time (in the total number of variables). Moreover, several broad classes of submodular functions over non-Boolean domains have also been shown to be expressible by binary submodular functions and hence minimisable in cubic time [3,5,6]. This series of positive expressibility results naturally raises the following question:

Question 1. Are *all* submodular polynomials expressible by binary submodular polynomials, over a possibly larger set of variables?

Each of the above expressibility results was obtained by an ad-hoc construction, and no general technique[2] has previously been proposed which is sufficiently powerful to address Question 1.

1.1 Contributions

Cohen et al. recently developed a novel algebraic approach to characterising the expressive power of valued constraints in terms of certain algebraic properties of those constraints [4].

Using this systematic algebraic approach we are able to give a negative answer to Question 1: we show that there exist submodular polynomials of degree 4 that cannot be expressed by binary submodular polynomials. More precisely, we

[1] In fact, it is known that *all* Boolean polynomials (of arbitrary degree) are expressible by binary polynomials [2], but the general construction does not preserve submodularity; that is, the resulting binary polynomials are not necessarily submodular.

[2] For example, standard combinatorial counting techniques cannot resolve this question because we allow arbitrary real-valued coefficients in submodular polynomials. We also allow an arbitrary number of additional variables.

characterise exactly which submodular polynomials of arity 4 are expressible by binary submodular polynomials and which are not.

On the way to establishing these results we show that two broad families of submodular functions, known as *upper fans* and *lower fans*, are all expressible by binary submodular functions. This provides a new class of submodular polynomials of all arities which are expressible by binary submodular polynomials and hence solvable efficiently by reduction to MIN-CUT. We use the expressibility of this family, and the existence of non-expressible functions, to refute a conjecture from [29] on the structure of the extreme rays of the cone of Boolean submodular functions, and suggest a more refined conjecture of our own.

2 Preliminaries

In this section, we introduce the basic definitions and the main tools used throughout the paper.

2.1 Cost Functions and Expressibility

We denote by $\overline{\mathbb{R}}$ the set of all real numbers together with (positive) infinity. For any fixed set D, a function ϕ from D^n to $\overline{\mathbb{R}}$ will be called a *cost function* on D of arity n. If the range of ϕ lies entirely within $\mathbb{R}$, then ϕ is called a *finite-valued* cost function. If the range of ϕ is $\{0, \infty\}$, then ϕ can be viewed as a predicate, or *relation*, allowing just those tuples $t \in D^n$ for which $\phi(t) = 0$.

Cost functions can be added and multiplied by arbitrary real values, hence for any given set of cost functions, Γ, we define the convex cone generated by Γ, as follows.

Definition 1. *For any set of cost functions Γ, the* cone generated by Γ, *denoted* Cone(Γ), *is defined by:*

$$\text{Cone}(\Gamma) = \{\alpha_1\phi_1 + \cdots + \alpha_r\phi_r \mid r \geq 1;\ \phi_1, \ldots, \phi_r \in \Gamma;\ \alpha_1, \ldots, \alpha_r \geq 0\}.$$

Definition 2. *A cost function ϕ of arity n is said to be* expressible *by a set of cost functions Γ if $\phi = \min_{y_1,\ldots,y_j} \phi'(x_1, \ldots, x_n, y_1, \ldots, y_j) + \kappa$, for some $\phi' \in \text{Cone}(\Gamma)$ and some constant κ.*

The variables $y_1, \ldots, y_j$ are called extra *(or* hidden*) variables, and ϕ' is called a* gadget *for ϕ over Γ.*

We denote by $\langle\Gamma\rangle$ the *expressive power* of Γ, which is the set of all cost functions expressible by Γ.

It was shown in [4] that the expressive power of a set of cost functions is characterised by certain algebraic properties of those cost functions called fractional polymorphisms. For the results of this paper, we will only need a certain subset of these algebraic properties, called *multimorphisms* [7]. These are defined in Definition 3 below (see also Figure 1).

The i-th component of a tuple t will be denoted by $t[i]$. Note that any operation on a set D can be extended to tuples over the set D in a

$$\begin{array}{lcc}
t_1 & t_1[1]\ t_1[2]\ \ldots\ t_1[n] & \phi(t_1) \\
t_2 & t_2[1]\ t_2[2]\ \ldots\ t_2[n] \xrightarrow{\phi} & \phi(t_2) \\
\vdots & \vdots & \vdots \\
t_k & t_k[1]\ t_k[2]\ \ldots\ t_k[n] & \phi(t_k)
\end{array} \Bigg\} \sum_{i=1}^{k} \phi(t_i)$$

$$\text{IV}$$

$$\begin{array}{lcc}
t'_1 = f_1(t_1,\ldots,t_k) & t'_1[1]\ t'_1[2]\ \ldots\ t'_1[n] & \phi(t'_1) \\
t'_2 = f_2(t_1,\ldots,t_k) & t'_2[1]\ t'_2[2]\ \ldots\ t'_2[n] \xrightarrow{\phi} & \phi(t'_2) \\
\vdots & \vdots & \vdots \\
t'_k = f_k(t_1,\ldots,t_k) & t'_k[1]\ t'_k[2]\ \ldots\ t'_k[n] & \phi(t'_k)
\end{array} \Bigg\} \sum_{i=1}^{k} \phi(t'_i)$$

Fig. 1. Inequality establishing $\mathcal{F} = \langle f_1, \ldots, f_k \rangle$ as a multimorphism of cost function ϕ (see Definition 3)

standard way, as follows. For any function $f : D^k \rightarrow D$, and any collection of tuples $t_1, \ldots, t_k \in D^n$, define $f(t_1, \ldots, t_k) \in D^n$ to be the tuple $\langle f(t_1[1], \ldots, t_k[1]), \ldots, f(t_1[n], \ldots, t_k[n]) \rangle$.

Definition 3 ([7]). *Let $\mathcal{F} : D^k \rightarrow D^k$ be the function whose k-tuple of output values is given by the tuple of functions $\mathcal{F} = \langle f_1, \ldots, f_k \rangle$, where each $f_i : D^k \rightarrow D$.*

For any n-ary cost function ϕ, we say that $\mathcal{F}$ is a k-ary multimorphism *of ϕ if, for all $t_1, \ldots, t_k \in D^n$,*

$$\sum_{i=1}^{k} \phi(t_i) \geq \sum_{i=1}^{k} \phi(f_i(t_1, \ldots, t_k)).$$

For any set of cost functions, Γ, we will say that $\mathcal{F}$ is a multimorphism of Γ if $\mathcal{F}$ is a multimorphism of every cost function in Γ. The set of all multimorphisms of Γ will be denoted $\mathsf{Mul}(\Gamma)$.

Note that multimorphisms are preserved under expressibility. In other words, if $\mathcal{F} \in \mathsf{Mul}(\Gamma)$, and $\phi \in \langle \Gamma \rangle$, then $\mathcal{F} \in \mathsf{Mul}(\{\phi\})$ [7,4]. This has two important corollaries. First, if $\langle \Gamma_1 \rangle = \langle \Gamma_2 \rangle$, then $\mathsf{Mul}(\Gamma_1) = \mathsf{Mul}(\Gamma_2)$. Second, if there exists $\mathcal{F} \in \mathsf{Mul}(\Gamma)$ such that $\mathcal{F} \notin \mathsf{Mul}(\{\phi\})$, then ϕ is not expressible by Γ, that is, $\phi \notin \langle \Gamma \rangle$.

2.2 Lattices and Submodularity

Recall that L is a *lattice* if L is a partially ordered set in which every pair of elements (a, b) has a unique supremum and a unique infimum. For a finite lattice L and a pair of elements (a, b), we will denote the unique supremum of a and b by $a \vee b$, and the unique infimum of a and b by $a \wedge b$.

For any finite lattice-ordered set D, a cost function $\phi : D^n \rightarrow \overline{\mathbb{R}}$ is called *submodular* if for every $u, v \in D^n$, $\phi(u \wedge v) + \phi(u \vee v) \leq \phi(u) + \phi(v)$ where both $\wedge$ and $\vee$ are applied coordinate-wise on tuples u and v [27]. This standard definition can be reformulated very simply in terms of multimorphisms: ϕ is submodular if $\langle \wedge, \vee \rangle \in \mathsf{Mul}(\{\phi\})$.

Using results from [7] and [32], it can be shown that any submodular cost function ϕ can be expressed as the sum of a finite-valued submodular cost function ϕ_{fin}, and a submodular relation ϕ_{rel}, that is, $\phi = \phi_{\mathsf{fin}} + \phi_{\mathsf{rel}}$.

Moreover, it is known that all submodular *relations* are binary decomposable (that is, equal to the sum of their binary projections) [18], and hence expressible by binary submodular relations. Therefore, when considering which cost functions are expressible by binary submodular cost functions, we can restrict our attention to *finite-valued* cost functions without any loss of generality.

Next we define some particular families of submodular cost functions, first described in [29], which will turn out to play a central role in our analysis.

Definition 4. *Let L be a lattice. We define the following cost functions on L:*

- *For any set A of pairwise incomparable elements $\{a_1, \dots, a_m\} \subseteq L$, such that each pair of distinct elements (a_i, a_j) has the same least upper bound, $\bigvee A$, the following cost function is called an* upper fan*:*

$$\phi_A(x) = \begin{cases} -2 & \text{if } x \geq \bigvee A, \\ -1 & \text{if } x \not\geq \bigvee A, \text{ but } x \geq a_i \text{ for some } i, \\ 0 & \text{otherwise.} \end{cases}$$

- *For any set B of pairwise incomparable elements $\{a_1, \dots, a_m\} \subseteq L$, such that each pair of distinct elements (a_i, a_j) has the same greatest lower bound, $\bigwedge B$, the following cost function is called a* lower fan*:*

$$\phi_B(x) = \begin{cases} -2 & \text{if } x \leq \bigwedge B, \\ -1 & \text{if } x \not\leq \bigwedge B, \text{ but } x \leq a_i \text{ for some } i, \\ 0 & \text{otherwise.} \end{cases}$$

We call a cost function a *fan* if it is either an upper fan or a lower fan. Note that our definition of fans is slightly more general than the definition in [29]. In particular, we allow the set A to be empty, in which case the corresponding upper fan ϕ_A is a constant function. It is not hard to show that all fans are submodular [29].

2.3 Boolean Cost Functions and Polynomials

In this paper we will focus on problems over Boolean domains, that is, where $D = \{0, 1\}$.

Any cost function of arity n can be represented as a table of values of size D^n. Moreover, a finite-valued cost function $\phi : D^n \to \mathbb{R}$ on a Boolean domain $D = \{0, 1\}$ can also be represented as a unique *polynomial* in n (Boolean) variables with coefficients from $\mathbb{R}$ (such functions are sometimes called *pseudo-Boolean functions* [2]). Hence, in what follows, we will often refer to a finite-valued cost function on a Boolean domain and its corresponding polynomial interchangeably.

For polynomials over Boolean variables there is a standard way to define *derivatives* of each order (see [2]). For example, the second-order derivative of a polynomial p, with respect to the first two indices, denoted $\delta_{1,2}(\mathbf{x})$, is defined as $p(1,1,\mathbf{x})-p(1,0,\mathbf{x})-p(0,1,\mathbf{x})+p(0,0,\mathbf{x})$. Derivatives for other pairs of indices are defined analogously. It was shown in [13] that a polynomial $p(x_1,\ldots,x_n)$ over Boolean variables $x_1,\ldots,x_n$ represents a submodular cost function if, and only if, its second-order derivatives $\delta_{i,j}(\mathbf{x})$ are non-positive for all $1 \leq i < j \leq n$ and all $\mathbf{x} \in D^{n-2}$. An immediate corollary is that a quadratic polynomial represents a submodular cost function if, and only if, the coefficients of all quadratic terms are non-positive.

Note that a cost function is called *supermodular* if all its second-order derivatives are non-negative. Clearly, f is submodular if, and only if, $-f$ is supermodular, so it is straightforward to translate results about supermodular functions, such as those given in [6] and [29], into similar results for submodular functions, and we will use this observation several times below. Cost functions which are both submodular and supermodular (in other words, all second-order derivatives are equal to zero) are called *modular*, and polynomials corresponding to modular cost functions are linear [2].

Example 1. For any set of indices $I = \{i_1,\ldots,i_m\} \subseteq \{1,\ldots,n\}$ we can define a cost function ϕ_I in n variables as follows:

$$\phi_I(x_1,\ldots,x_n) = \begin{cases} -1 & \text{if } (\forall i \in I)(x_i = 1), \\ 0 & \text{otherwise.} \end{cases}$$

The polynomial representation of ϕ_I is $p(x_1,\ldots,x_n) = -x_{i_1}\ldots x_{i_m}$, which is a polynomial of degree m. Note that it is straightforward to verify that ϕ_I is submodular by checking the second-order derivatives of p.

However, the function ϕ_I is also expressible by *binary* submodular polynomials, using a single extra variable, y, as follows:

$$\phi_I(x_1,\ldots,x_n) = \min_{y\in\{0,1\}}\{-y + y\sum_{i\in I}(1-x_i)\}.$$

We remark that this is a special case of the expressibility result for negative-positive polynomials first obtained in [30].

Note that when $D = \{0,1\}$, the set D^n with the product ordering is isomorphic to the lattice of all subsets of an n-element set ordered by inclusion. Hence, a cost function on a Boolean domain can be viewed as a cost function defined on a lattice of subsets, and we can apply Definition 4 to identify certain Boolean functions as upper fans or lower fans, as the following example indicates.

Example 2. Let $A = \{I_1,\ldots,I_r\}$ be a set of subsets of $\{1,2,\ldots,n\}$ such that for all $i \neq j$ we have $I_i \not\subseteq I_j$ and $I_i \cup I_j = \bigcup A$.

By Definition 4, the corresponding upper fan function ϕ_A has the following polynomial representation:

$$p(x_1,\ldots,x_n) = (r-2)\prod_{i\in\bigcup A} x_i - \prod_{i\in I_1} x_i - \cdots - \prod_{i\in I_r} x_i.$$

We remark that any permutation of a set D gives rise to an automorphism of cost functions over D. In particular, for any cost function f on a Boolean domain D, the *dual* of f is the corresponding cost function which results from exchanging the values 0 and 1 for all variables. In other words, if p is the polynomial representation of f, then the dual of f is the cost function whose polynomial representation is obtained from p by replacing all variables x with $1-x$. Observe that, due to symmetry, taking the dual preserves submodularity and expressibility by binary submodular cost functions.

It is not hard to see that upper fans are duals of lower fans and vice versa.

3 Results

In this section, we present our main results. First, we show that fans of all arities are expressible by binary submodular cost functions. Next, we characterise the multimorphisms of binary submodular cost functions. Combining these results, we then characterise precisely which 4-ary submodular cost functions are expressible by binary submodular cost functions. More importantly, we show that some submodular cost functions are *not* expressible by binary submodular cost functions, and therefore cannot be minimised using the MIN-CUT problem via an expressibility reduction. Finally, we consider the complexity of recognizing which cost functions are expressible by binary submodular cost functions.

3.1 Expressibility of Upper Fans and Lower Fans

We denote by $\Gamma_{\mathsf{sub},n}$ the set of all finite-valued submodular cost functions of arity at most n on a Boolean domain D, and we set $\Gamma_{\mathsf{sub}} = \bigcup_n \Gamma_{\mathsf{sub},n}$.

We denote by $\Gamma_{\mathsf{fans},n}$ the set of all fans of arity at most n on a Boolean domain D, and we set $\Gamma_{\mathsf{fans}} = \bigcup_n \Gamma_{\mathsf{fans},n}$.

Our next result shows that $\Gamma_{\mathsf{fans}} \subseteq \langle \Gamma_{\mathsf{sub},2} \rangle$. The proof is omitted due to space restrictions.

Theorem 1. *Any fan on a Boolean domain D is expressible by binary submodular functions on D using at most* $1 + \lfloor m/2 \rfloor$ *extra variables, where m is the degree of its polynomial representation.*

Many of the earlier expressibility results mentioned in Section 1 can be obtained as simple corollaries of Theorem 1, as the following examples indicate.

Example 3. Any negative monomial $-x_1 x_2 \cdots x_m$ is a positive multiple of an upper fan, and the positive linear monomial x_1 is equal to $-(1-x_1)+1$, so it is a positive multiple of a lower fan, plus a constant. Hence all negative-positive submodular polynomials are contained in Cone(Γ_{fans}), and by Theorem 1, they are expressible by binary submodular polynomials, as originally shown in [30].

Example 4. A polynomial is called *homogeneous* [1] or *polar* [9] if it can be expressed as a sum of terms of the form $a x_1 x_2 \ldots x_k$ or $a(1-x_1)(1-x_2)\ldots(1-x_k)$ with positive coefficients a, together with a constant term. It was observed in [1] that all polar polynomials are supermodular, so all negated polar polynomials are submodular. As every negated term $-ax_1x_2\ldots x_k$, is a positive multiple of an upper fan, and every negated term $-a(1-x_1)(1-x_2)\ldots(1-x_k)$, is a positive multiple of a lower fan, by Theorem 1, all cost functions which are the negations of polar polynomials are expressible by binary submodular polynomials, and hence solvable by reduction to Min-Cut, as originally shown in [1].

Example 5. Any cubic submodular polynomial can be expressed as a positive sum of upper fans [29]. Hence, by Theorem 1, all cubic submodular polynomials are expressible by binary submodular polynomials, as originally shown in [1].

Example 6. A Boolean cost function ϕ is called *2-monotone* [10] if there exist two sets $R, S \subseteq \{1,\ldots,n\}$ such that $\phi(\mathbf{x}) = 0$ if $R \subseteq \mathbf{x}$ or $\mathbf{x} \subseteq S$ and $\phi(\mathbf{x}) = 1$ otherwise (where $R \subseteq \mathbf{x}$ means $\forall i \in R, x[i] = 1$ and $\mathbf{x} \subseteq S$ means $\forall i \notin S, x[i] = 0$). It was shown in [6, Proposition 2.9] that a 2-valued Boolean cost function is 2-monotone if, and only if, it is submodular.

For any 2-monotone cost function defined by the sets of indices R and S, it is straightforward to check that $\phi = \min_{y\in\{0,1\}} y(1+\phi_A/2) + (1-y)(1+\phi_B/2)$ where ϕ_A is the upper fan defined by $A = \{R\}$ and ϕ_B is the lower fan defined by $B = \{\overline{S}\}$. Note that the function $y\phi_A$ is an upper fan, and the function $(1-y)\phi_B$ is a lower fan. Hence, by Theorem 1, all 2-monotone polynomials are expressible by binary submodular polynomials, and solvable by reduction to Min-Cut, as originally shown in [10].

However, Theorem 1 also provides many new functions of all arities which have not previously been shown to be expressible by binary submodular functions, as the following example indicates.

Example 7. The function $2x_1x_2x_3x_4 - x_1x_2x_3 - x_1x_2x_4 - x_1x_3x_4 - x_2x_3x_4$ belongs to $\Gamma_{\mathsf{fans},4}$, but does not belong to any class of submodular functions which has previously been shown to be expressible by binary submodular functions. In particular, it does not belong to the class Γ_{new} identified in [34,35].

3.2 Characterisation of $\mathsf{Mul}(\Gamma_{\mathsf{sub},2})$

Since we have seen that a cost function can only be expressed by a given set of cost functions if it has the same multimorphisms, we now investigate the multimorphisms of $\Gamma_{\mathsf{sub},2}$.

A function $\mathcal{F} : D^k \to D^k$ is called *conservative* if, for each possible choice of $x_1,\ldots,x_k$, the tuple $\mathcal{F}(x_1,\ldots,x_k)$ is a permutation of $x_1,\ldots,x_k$ (though different inputs may be permuted in different ways).

For any two tuples $\mathbf{x} = \langle x_1,\ldots,x_k\rangle$ and $\mathbf{y} = \langle y_1,\ldots,y_k\rangle$ over D, we denote by $H(\mathbf{x},\mathbf{y})$ the *Hamming distance* between $\mathbf{x}$ and $\mathbf{y}$, which is the number of positions at which the corresponding values are different.

Theorem 2. *For any Boolean domain D, and any $\mathcal{F} : D^k \to D^k$, the following are equivalent:*

1. $\mathcal{F} \in \mathsf{Mul}(\Gamma_{\mathsf{sub},2})$.
2. $\mathcal{F} \in \mathsf{Mul}(\Gamma^{\infty}_{\mathsf{sub},2})$, *where* $\Gamma^{\infty}_{\mathsf{sub},2}$ *denotes the set of binary submodular cost functions taking finite or infinite values.*
3. $\mathcal{F}$ *is conservative and Hamming distance non-increasing.*

The proof is omitted due to space restrictions.

3.3 Non-expressibility of Γ_{sub} over $\Gamma_{\mathsf{sub},2}$

Theorem 2 characterises the multimorphisms of $\Gamma_{\mathsf{sub},2}$, and hence enables us to systematically search (for example, using MATHEMATICA) for multimorphisms of $\Gamma_{\mathsf{sub},2}$ which are not multimorphisms of Γ_{sub}. In this way, we have identified the function $\mathcal{F}_{sep} : \{0,1\}^5 \to \{0,1\}^5$ defined in Figure 2. We will show in this section that this function can be used to characterise all the submodular functions of arity 4 which are expressible by binary submodular functions on a Boolean domain. Using this result, we show that some submodular functions are *not* expressible in this way.

	0 0 0 0 0 0 0 0 0 0 0 0 0 0 0 0 1 1 1 1 1 1 1 1 1 1 1 1 1 1 1 1
	0 0 0 0 0 0 0 0 1 1 1 1 1 1 1 1 0 0 0 0 0 0 0 0 1 1 1 1 1 1 1 1
$\mathbf{x}$	0 0 0 0 1 1 1 1 0 0 0 0 1 1 1 1 0 0 0 0 1 1 1 1 0 0 0 0 1 1 1 1
	0 0 1 1 0 0 1 1 0 0 1 1 0 0 1 1 0 0 1 1 0 0 1 1 0 0 1 1 0 0 1 1
	0 1 0 1 0 1 0 1 0 1 0 1 0 1 0 1 0 1 0 1 0 1 0 1 0 1 0 1 0 1 0 1
	0 0 0 0 0 0 0 0 0 0 0 0 0 0 0 0 0 0 0 1 0 0 0 1 0 0 0 1 0 0 0 1
	0 0 0 0 0 0 0 0 0 0 0 0 0 1 0 1 0 0 0 0 0 0 0 0 0 0 0 0 0 1 1 1
$\mathcal{F}_{sep}(\mathbf{x})$	0 0 0 0 0 0 1 1 0 0 0 1 0 0 1 1 0 0 0 0 0 1 1 1 1 1 1 1 1 1 1 1
	0 0 0 1 0 1 0 1 1 1 1 1 1 1 1 1 1 1 1 1 1 1 1 1 1 1 1 1 1 1 1 1
	0 1 1 1 1 1 1 1 0 1 1 1 1 1 1 1 0 1 1 1 1 1 1 1 0 1 1 1 1 1 1 1

Fig. 2. Definition of $\mathcal{F}_{sep}$

Proposition 1. $\mathcal{F}_{sep}$ *is conservative and Hamming distance non-increasing.*

Proof. Straightforward exhaustive verification. □

Theorem 3. *For any function* $f \in \Gamma_{\mathsf{sub},4}$ *the following are equivalent:*

1. $f \in \langle \Gamma_{\mathsf{sub},2} \rangle$.
2. $\mathcal{F}_{sep} \in \mathsf{Mul}(\{f\})$.
3. $f \in \mathrm{Cone}(\Gamma_{\mathsf{fans},4})$.

Proof. First, we show (1) ⇒ (2). Proposition 1 and Theorem 2 imply that $\mathcal{F}_{sep}$ is a multimorphism of any binary submodular function on a Boolean domain. Hence having $\mathcal{F}_{sep}$ as a multimorphism is a necessary condition for any submodular cost function on a Boolean domain to be expressible by binary submodular cost functions.

Next, we show (2) $\Rightarrow$ (3). Consider the complete set of inequalities on the values of a 4-ary cost function resulting from having the multimorphism $\mathcal{F}_{sep}$, as specified in Definition 3. A routine calculation in MATHEMATICA shows that, out of 16^5 such inequalities, there are 4635 which are distinct. After removing from these all those which are equal to the sum of two others, we obtain a system of just 30 inequalities which must be satisfied by any 4-ary submodular cost function which has the multimorphism $\mathcal{F}_{sep}$. Using the double description method [25], we obtain from these 30 inequalities an equivalent set of 31 extreme rays which generate the same polyhedral cone of cost functions. These extreme rays all correspond to fans or sums of fans.

Finally, we show (3) $\Rightarrow$ (1). By Theorem 1, all fans are expressible over $\Gamma_{\mathsf{sub},2}$. It follows that any cost function in this cone of functions is also expressible over $\Gamma_{\mathsf{sub},2}$. □

Next we show that there are indeed 4-ary submodular cost functions which do not have $\mathcal{F}_{sep}$ as a multimorphism and therefore are not expressible by binary submodular cost functions.

Definition 5. *For any Boolean tuple t of arity 4 containing exactly 2 ones and 2 zeros, we define the 4-ary cost function θ_t as follows:*

$$\theta_t(x_1,x_2,x_3,x_4) \;=\; \begin{cases} -1 & \textit{if } (x_1,x_2,x_3,x_4)=(1,1,1,1) \textit{ or } (0,0,0,0),\\ 1 & \textit{if } (x_1,x_2,x_3,x_4)=t,\\ 0 & \textit{otherwise.}\end{cases}$$

Cost functions of the form θ_t were introduced in [29], where they are called *quasi-indecomposable* functions. We denote by Γ_{qin} the set of all (six) quasi-indecomposable cost functions of arity 4. It is straightforward to check that they are submodular, but the next result shows that they are *not* expressible by binary submodular functions.

Proposition 2. *For all $\theta \in \Gamma_{\mathsf{qin}}$, $\mathcal{F}_{sep} \notin \mathsf{Mul}(\{\theta\})$.*

Proof. The following table shows that $\mathcal{F}_{sep} \notin \mathsf{Mul}(\{\theta_{(1,1,0,0)}\})$.

$$\begin{array}{r c c c l}
 & 1\,0\,1\,0 & & 0 & \\
 & 1\,0\,0\,1 & & 0 & \\
 & 0\,1\,0\,1 & \overset{\theta_{(1,1,0,0)}}{\longrightarrow} & 0 & \left.\right\} \sum = 0\\
 & 0\,1\,1\,0 & & 0 & \\
 & 0\,0\,1\,1 & & 0 & \\
\hline
 & 0\,0\,1\,0 & & 0 & \\
 & 0\,0\,0\,1 & & 0 & \\
\mathcal{F}_{sep} & 1\,1\,0\,0 & \overset{\theta_{(1,1,0,0)}}{\longrightarrow} & 1 & \left.\right\} \sum = 1\\
 & 1\,0\,1\,1 & & 0 & \\
 & 0\,1\,1\,1 & & 0 &
\end{array}$$

Permuting the columns appropriately establishes the result for all other $\theta \in \Gamma_{\mathsf{qin}}$. □

Corollary 1. *For all $\theta \in \Gamma_{\mathsf{qin}}$, $\theta \notin \langle \Gamma_{\mathsf{sub},2} \rangle$.*

Proof. By Theorem 3 and Proposition 2. □

Are there any other 4-ary submodular cost functions which are not expressible over $\Gamma_{\mathsf{sub},2}$? Promislow and Young characterised the extreme rays of the cone of all 4-ary submodular cost functions and established that $\Gamma_{\mathsf{sub},4} = \mathrm{Cone}(\Gamma_{\mathsf{fans},4} \cup \Gamma_{\mathsf{qin}})$ – see Theorem 5.2 of [29]. Hence the results in this section characterise the expressibility of all 4-ary submodular functions.

Promislow and Young conjectured that for $k \neq 4$, all extreme rays of $\Gamma_{\mathsf{sub},k}$ are fans [29]; that is, they conjectured that for all $k \neq 4$, $\Gamma_{\mathsf{sub},k} = \mathrm{Cone}(\Gamma_{\mathsf{fans},k})$. However, if this conjecture were true it would imply that all submodular functions of arity 5 and above were expressible by binary submodular functions, by Theorem 1. This is clearly not the case, because inexpressible cost functions such as those identified in Corollary 1 can be extended to larger arities (for example, by adding dummy arguments) and remain inexpressible. Hence our results refute this conjecture for all $k \geq 5$. However, we suggest that this conjecture can be refined to a similar statement concerning just those submodular functions which are expressible by binary submodular functions, as follows:

Conjecture 1. For all k, $\Gamma_{\mathsf{sub},k} \cap \langle \Gamma_{\mathsf{sub},2} \rangle = \mathrm{Cone}(\Gamma_{\mathsf{fans},k})$.

This conjecture was previously known to be true for $k \leq 3$ [29]; Theorem 1 shows that $\mathrm{Cone}(\Gamma_{\mathsf{fans},k}) \subseteq \Gamma_{\mathsf{sub},k} \cap \langle \Gamma_{\mathsf{sub},2} \rangle$ for all k, and Theorem 3 confirms that equality holds for $k = 4$.

3.4 The Complexity of Recognising Expressible Functions

Finally, we show that we can test efficiently whether a submodular polynomial of arity 4 is expressible by binary submodular polynomials.

Definition 6. *Let $p(x_1, x_2, x_3, x_4)$ be the polynomial representation of a 4-ary submodular cost function f. We denote by a_I the coefficient of the term $\prod_{i \in I} x_i$. We say that f satisfies condition* **Sep** *if for each $\{i,j\}, \{k,\ell\} \subset \{1,2,3,4\}$, with i,j,k,ℓ distinct, we have $a_{\{i,j\}} + a_{\{k,\ell\}} + a_{\{i,j,k\}} + a_{\{i,j,\ell\}} \leq 0$.*

Theorem 4. *For any $f \in \Gamma_{\mathsf{sub},4}$, the following are equivalent:*

1. *$f \in \langle \Gamma_{\mathsf{sub},2} \rangle$.*
2. *f satisfies condition* **Sep**.

Proof. As in the proof of Theorem 3, we construct a set of 30 inequalities corresponding to the multimorphism $\mathcal{F}_{sep}$. Each of these inequalities on the values of a cost function can be translated into inequalities on the coefficients of the corresponding polynomial representation by a straightforward linear transformation. This calculation shows that 24 of the resulting inequalities impose the condition of submodularity, and the remaining 6 impose condition **Sep**. Hence a submodular cost function of arity 4 has the multimorphism $\mathcal{F}_{sep}$ if, and only if, its polynomial representation satisfies condition **Sep**. The result then follows from Theorem 3. □

Using Theorem 4, we can test whether optimisation problems given as a sum of submodular functions of arity 4 can be reduced to the Min-Cut problem via the expressibility reduction. These problems arise in Computer Vision and in Valued Constraint Satisfaction Problems.

Furthermore, by Theorem 1, the number of extra variables needed in this reduction is rather small compared to the theoretical upper bound given in [4].

It is known that the problem of recognising whether an arbitrary degree-4 polynomial is submodular is co-NP-complete [9,15]. At the moment, the complexity of the recognition problem for submodular polynomials of degree 4 that are expressible by binary submodular polynomials is open.

References

1. Billionet, A., Minoux, M.: Maximizing a supermodular pseudo-Boolean function: a polynomial algorithm for cubic functions. Discrete Applied Math. 12, 1–11 (1985)
2. Boros, E., Hammer, P.L.: Pseudo-Boolean optimization. Discrete Applied Mathematics 123(1-3), 155–225 (2002)
3. Burkard, R., Klinz, B., Rudolf, R.: Perspectives of Monge properties in optimization. Discrete Applied Mathematics 70, 95–161 (1996)
4. Cohen, D., Cooper, M., Jeavons, P.: An algebraic characterisation of complexity for valued constraints. In: Benhamou, F. (ed.) CP 2006. LNCS, vol. 4204, pp. 107–121. Springer, Heidelberg (2006)
5. Cohen, D., Cooper, M., Jeavons, P., Krokhin, A.: A maximal tractable class of soft constraints. Journal of Artificial Intelligence Research 22, 1–22 (2004)
6. Cohen, D., Cooper, M., Jeavons, P., Krokhin, A.: Supermodular functions and the complexity of Max-CSP. Discrete Applied Mathematics 149, 53–72 (2005)
7. Cohen, D., Cooper, M., Jeavons, P., Krokhin, A.: The complexity of soft constraint satisfaction. Artificial Intelligence 170, 983–1016 (2006)
8. Cooper, M.: Minimization of Locally Defined Submodular Functions by Optimal Soft Arc Consistency. Constraints 13(4), 437–458 (2008)
9. Crama, Y.: Recognition problems for special classes of polynomials in 0-1 variables. Mathematical Programming 44, 139–155 (1989)
10. Creignou, N., Khanna, S., Sudan, M.: Complexity Classification of Boolean Constraint Satisfaction Problems. SIAM, Philadelphia (2001)
11. Deineko, V., Jonsson, P., Klasson, M., Krokhin, A.: The approximability of Max CSP with fixed-value constraints. Journal of the ACM 55(4) (2008)
12. Feige, U., Mirrokni, V.S., Vondrák, J.: Maximizing non-monotone submodular functions. In: Proceedings of FOCS 2007 (2007)
13. Fisher, M., Nemhauser, G., Wolsey, L.: An analysis of approximations for maximizing submodular set functions-I. Mathematical Programming 14, 265–294 (1978)
14. Fujishige, S.: Submodular Functions and Optimization, 2nd edn. Annals of Discrete Mathematics, vol. 58. North-Holland, Amsterdam (2005)
15. Gallo, G., Simeone, B.: On the supermodular knapsack problem. Mathematical Programming 45, 295–309 (1988)
16. Hammer, P.L.: Some network flow problems solved with pseudo-Boolean programming. Operations Research 13, 388–399 (1965)
17. Iwata, S.: Submodular function minimization. Mathematical Programming 112, 45–64 (2008)

18. Jeavons, P., Cohen, D., Cooper, M.: Constraints, consistency and closure. Artificial Intelligence 101(1-2), 251–265 (1998)
19. Jonsson, P., Klasson, M., Krokhin, A.: The approximability of three-valued MAX CSP. SIAM Journal on Computing 35(6), 1329–1349 (2006)
20. Kolmogorov, V., Zabih, R.: What energy functions can be minimized via graph cuts? IEEE Transactions on PAMI 26(2), 147–159 (2004)
21. Korte, B., Vygen, J.: Combinatorial Optimization, 4th edn. Algorithms and Combinatorics, vol. 21. Springer, Heidelberg (2007)
22. Krokhin, A., Larose, B.: Maximizing supermodular functions on product lattices, with application to maximum constraint satisfaction. SIAM Journal on Discrete Mathematics 22(1), 312–328 (2008)
23. Lauritzen, S.L.: Graphical Models. Oxford University Press, Oxford (1996)
24. Lovász, L.: Submodular functions and convexity. In: Bachem, A., Grötschel, M., Korte, B. (eds.) Math. Programming, pp. 235–257. Springer, Berlin (1983)
25. Motzkin, T., Raiffa, H., Thompson, G., Thrall, R.: The double description method. Contributions to the Theory of Games, vol. 2, pp. 51–73. Princeton Univ. Press, Princeton (1953)
26. Narayanan, H.: Submodular Functions and Electrical Networks (1997)
27. Nemhauser, G., Wolsey, L.: Integer and Combinatorial Optimization (1988)
28. Orlin, J.B.: A faster strongly polynomial time algorithm for submodular function minimization. Mathematical Programming 118, 237–251 (2009)
29. Promislow, S., Young, V.: Supermodular functions on finite lattices. Order 22(4), 389–413 (2005)
30. Rhys, J.: A selection problem of shared fixed costs and network flows. Management Science 17(3), 200–207 (1970)
31. Rossi, F., van Beek, P., Walsh, T.: The Handbook of Constraint Programming (2006)
32. Schrijver, A.: Combinatorial Optimization: Polyhedra and Efficiency. Algorithms and Combinatorics, vol. 24. Springer, Heidelberg (2003)
33. Topkis, D.: Supermodularity and Complementarity (1998)
34. Zalesky, B.: Efficient determination of Gibbs estimators with submodular energy functions. arXiv:math/0304041v1 (February 2008)
35. Živný, S., Jeavons, P.G.: Classes of submodular constraints expressible by graph cuts. In: Stuckey, P.J. (ed.) CP 2008. LNCS, vol. 5202, pp. 112–127. Springer, Heidelberg (2008)

Author Index